39b $\displaystyle\int \sin^n u \cos^m u \, du = \frac{\sin^{n+1} u \cos^{m-1} u}{n+m} + \frac{m-1}{n+m} \int \sin^n u \cos^{m-2} u \, du$  if $m \neq -n$

40 $\displaystyle\int u \sin u \, du = \sin u - u \cos u + C$

41 $\displaystyle\int u \cos u \, du = \cos u + u \sin u + C$

42 $\displaystyle\int u^n \sin u \, du = -u^n \cos u + n \int u^{n-1} \cos u \, du$

43 $\displaystyle\int u^n \cos u \, du = u^n \sin u - n \int u^{n-1} \sin u \, du$

## FORMS INVOLVING $\sqrt{u^2 \pm a^2}$

44 $\displaystyle\int \sqrt{u^2 \pm a^2} \, du = \frac{u}{2} \sqrt{u^2 \pm a^2} \pm \frac{a^2}{2} \ln|u + \sqrt{u^2 \pm a^2}| + C$

45 $\displaystyle\int \frac{du}{\sqrt{u^2 \pm a^2}} = \ln|u + \sqrt{u^2 \pm a^2}| + C$

46 $\displaystyle\int \frac{\sqrt{u^2 + a^2}}{u} \, du = \sqrt{u^2 + a^2} - a \ln\left(\frac{a + \sqrt{u^2 + a^2}}{u}\right) + C$

47 $\displaystyle\int \frac{\sqrt{u^2 - a^2}}{u} \, du = \sqrt{u^2 - a^2} - a \sec^{-1} \frac{u}{a} + C$

48 $\displaystyle\int u^2 \sqrt{u^2 \pm a^2} \, du = \frac{u}{8} (2u^2 \pm a^2) \sqrt{u^2 \pm a^2} - \frac{a^4}{8} \ln|u + \sqrt{u^2 \pm a^2}| + C$

49 $\displaystyle\int \frac{u^2 \, du}{\sqrt{u^2 \pm a^2}} = \frac{u}{2} \sqrt{u^2 \pm a^2} \mp \frac{a^2}{2} \ln|u + \sqrt{u^2 \pm a^2}| + C$

50 $\displaystyle\int \frac{du}{u^2 \sqrt{u^2 \pm a^2}} = \mp \frac{\sqrt{u^2 \pm a^2}}{a^2 u} + C$

51 $\displaystyle\int \frac{\sqrt{u^2 \pm a^2}}{u^2} \, du = -\frac{\sqrt{u^2 \pm a^2}}{u} + \ln|u + \sqrt{u^2 \pm a^2}| + C$

52 $\displaystyle\int \frac{du}{(u^2 \pm a^2)^{3/2}} = \frac{\pm u}{a^2 \sqrt{u^2 \pm a^2}} + C$

53 $\displaystyle\int (u^2 \pm a^2)^{3/2} \, du = \frac{u}{8} (2u^2 \pm 5a^2) \sqrt{u^2 \pm a^2} + \frac{3a^4}{8} \ln|u + \sqrt{u^2 \pm a^2}| + C$

## FORMS INVOLVING $\sqrt{a^2 - u^2}$

54 $\displaystyle\int \sqrt{a^2 - u^2} \, du = \frac{u}{2} \sqrt{a^2 - u^2} + \frac{a^2}{2} \sin^{-1} \frac{u}{a} + C$

55 $\displaystyle\int \frac{\sqrt{a^2 - u^2}}{u} \, du = \sqrt{a^2 - u^2} - a \ln\left|\frac{a + \sqrt{a^2 - u^2}}{u}\right| + C$

56 $\displaystyle\int \frac{u^2 \, du}{\sqrt{a^2 - u^2}} = -\frac{u}{2} \sqrt{a^2 - u^2} + \frac{a^2}{2} \sin^{-1} \frac{u}{a} + C$

57 $\displaystyle\int u^2 \sqrt{a^2 - u^2} \, du = \frac{u}{8} (2u^2 - a^2) \sqrt{a^2 - u^2} + \frac{a^4}{8} \sin^{-1} \frac{u}{a} + C$

58 $\displaystyle\int \frac{du}{u^2 \sqrt{a^2 - u^2}} = -\frac{\sqrt{a^2 - u^2}}{a^2 u} + C$

59 $\displaystyle\int \frac{\sqrt{a^2 - u^2}}{u^2} \, du = -\frac{\sqrt{a^2 - u^2}}{u} - \sin^{-1} \frac{u}{a} + C$

60 $\displaystyle\int \frac{du}{u \sqrt{a^2 - u^2}} = -\frac{1}{a} \ln\left|\frac{a + \sqrt{a^2 - u^2}}{u}\right| + C$

61 $\displaystyle\int \frac{du}{(a^2 - u^2)^{3/2}} = \frac{u}{a^2 \sqrt{a^2 - u^2}} + C$

62 $\displaystyle\int (a^2 - u^2)^{3/2} \, du = \frac{u}{8} (5a^2 - 2u^2) \sqrt{a^2 - u^2} + \frac{3a^4}{8} \sin^{-1} \frac{u}{a} + C$

## EXPONENTIAL AND LOGARITHMIC FORMS

63 $\displaystyle\int u e^u \, du = (u - 1) e^u + C$

64 $\displaystyle\int u^n e^u \, du = u^n e^u - n \int u^{n-1} e^u \, du$

65 $\displaystyle\int \ln u \, du = u \ln u - u + C$

66 $\displaystyle\int u^n \ln u \, du = \frac{u^{n+1}}{n+1} \ln u - \frac{u^{n+1}}{(n+1)^2} + C$

67 $\displaystyle\int e^{au} \sin bu \, du = \frac{e^{au}}{a^2 + b^2} (a \sin bu - b \cos bu) + C$

68 $\displaystyle\int e^{au} \cos bu \, du = \frac{e^{au}}{a^2 + b^2} (a \cos bu + b \sin bu) + C$

*(Continued inside back cover)*

# CALCULUS PROGRAMS DISK

Designed to accompany CALCULUS AND ANALYTIC GEOMETRY by C. H. Edwards, Jr. and David E. Penney and CALCULUS AND THE PERSONAL COMPUTER by C. H. Edwards, Jr.

This IBM-PC compatible software contains the optional BASIC programs included in CALCULUS AND ANALYTIC GEOMETRY and CALCULUS AND THE PERSONAL COMPUTER. It allows the student or instructor to run any of 67 programs.

The programs can be used either in a computer laboratory setting or for self-study. Topics include:

1. *Numerical and graphical solution of equations and of maximum-minimum problems*
2. *Limits and numerical differentiation*
3. *Numerical integration and summation of infinite series*
4. *Numerical Solution of velocity and acceleration problems (elementary differential equations)*

# CALCULUS AND THE PERSONAL COMPUTER

*by C. H. Edwards, Jr.*

This text introduces and illustrates the central concepts of calculus with concrete examples and applications on a personal computer. No prior knowledge of calculus and only a modest acquaintance with BASIC is necessary. The text emphasizes the interaction between calculus and computing, and includes the following:

1. *The historical background and development of algorithms in calculus*
2. *Mathematical modeling and effective programming techniques*
3. *Programs designed to highlight and clarify mathematical ideas*

The text includes an Appendix on computerized practice testing of differentiation and integration skills, and includes program listings, data printouts with screen displays, illustrations, and suggested readings.

## ORDER YOUR COPY TODAY!

---

Please send me:

☐ **CALCULUS PROGRAMS DISK**
(Title Code #11152–6) . . . . . . . . . . . . . . . $ 6.33

☐ **CALCULUS AND THE PERSONAL COMPUTER**
(Title Code #11231–8) . . . . . . . . . . . . . . . $20.89

| QUANTITY | TITLE/AUTHOR | TITLE CODE | PRICE | TOTAL |
|---|---|---|---|---|
| (Sample Entry) 2 | Edwards/CALCULUS PROGRAMS DISK | 11152–6 | $6.33 | $12.66 |
| | | | | |
| | | | | |
| | | | | |

**TOTAL PRICE:** _____

**SAVE!**
If payment accompanies order, plus your state's sales tax where applicable, Prentice Hall pays postage and handling charges. Same return privilege refund guaranteed.

☐ PAYMENT ENCLOSED—shipping and handling to be paid by publisher (please include your state's tax where applicable).

**MAIL TO:**
PRENTICE HALL
Book Distribution Center
Route 59 at Brook Hill Drive
West Nyack, NY 10995

NAME _____

ORGANIZATION _____

STREET ADDRESS _____

CITY _____ STATE _____ ZIP _____

I Prefer to charge my ☐ VISA ☐ MasterCard
Card Number _____ Expiration Date _____

Signature _____

All prices in this catalog are subject to change without notice. Prices not valid outside the U.S.

Dept. 1                                                    D–CPDP–FU(9)

# EPIC: EXPLORATION PROGRAMS IN CALCULUS, 2/E

*by James W. Burgmeier and Larry L. Kost*

This interactive software package for the IBM Personal Computer is a collection of programs that allows users to explore a variety of calculus concepts, including: equations of lines, intersections of curves, maxima and minima, differentiation, and anti-derivatives. The hardware requirements include a 256K memory, one disk drive, and a graphics card. The software also offers the following features:

1. *Users may superimpose graphs of functions with parameters.*
2. *Users have complete control over the range used for the x and y variables on graphs.*
3. *Functions may include parameters that users can alter (e.g., users may enter asin bx and specify a,b later).*
4. *Users can invoke options or gain information by a single keystroke (e.g., they can change the range along the x axis, get the max and min of the function, or trace polar curves).*

A departmental version of this software package is available and offers the same features listed above, as well as:

1. *A built-in math calculator with 25 memory addresses and 10 formula saving registers*
2. *Help screens and on-screen reminders*
3. *Extended options for: function plotting, integration, and parametric/polar equations*

## ORDER YOUR COPY TODAY!

Please send me:

# Calculus and Analytic Geometry

# ALTERNATE SECOND EDITION

**Prentice Hall**
**Englewood Cliffs, New Jersey 07632**

# Calculus and Analytic Geometry

**C. H. Edwards, Jr.**

The University of Georgia, Athens

**David E. Penney**

The University of Georgia, Athens

**Library of Congress Cataloging-in-Publication Data**

Edwards, C. H. (Charles Henry), (date)
    Calculus and analytic geometry.

    Bibliography: p.
    Includes index.
    1. Calculus.   2. Geometry, Analytic.    I. Penney,
David E.   II. Title.
QA303.E223   1987          515'.16          87-13933
ISBN 0-13-111469-7

Editorial/production supervision: *Maria McColligan*
Interior design: *Walter A. Behnke*
Chapter opening and cover design: *Anne T. Bonanno*
Illustrator: *Eric G. Hieber*
Manufacturing buyer: *Paula Benevento*

**Calculus and Analytic Geometry,** Alternate Second Edition
C. H. Edwards, Jr. and David E. Penney

 © 1988, 1986, 1982 by Prentice-Hall, Inc.
A Division of Simon & Schuster
Englewood Cliffs, New Jersey 07632

Printed in the United States of America
10   9   8   7   6   5   4   3   2   1

ISBN 0-13-111469-7   01

Figures 18.1 to 18.6 and 18.12 to 18.14 are taken from
*Elementary Differential Equations with Applications*
by C. H. Edwards, Jr. and David E. Penney, © 1985
by Prentice-Hall, Inc., pages 4, 19, 43, 49, 53, 137, 138.
Reprinted by permission of Prentice-Hall, Inc., Englewood
Cliffs, N.J.

*Cover photograph: Centroid by Peter Aldridge, Steuben Glass. This geometric crystal sculpture, based on a cube, is deeply cut to form eight sections radiating from the center. Emerging from within each face of the cube, multiple planes reflect together as starlike bursts of light.*

Prentice-Hall International (UK) Limited, *London*
Prentice-Hall of Australia Pty. Limited, *Sydney*
Prentice-Hall Canada Inc., *Toronto*
Prentice-Hall Hispanoamericana, S.A., *Mexico*
Prentice-Hall of India Private Limited, *New Delhi*
Prentice-Hall of Japan, Inc., *Tokyo*
Simon & Schuster Asia Pte. Ltd., *Singapore*
Editora Prentice-Hall do Brasil, Ltda., *Rio de Janeiro*

# Contents

# Preface

Calculus is one of the supreme accomplishments of the human intellect. Many of the scientific discoveries that have shaped our civilization during the past three centuries would have been impossible without the use of calculus. Today this body of computational technique continues to serve as the principal quantitative language of science and technology.

We prepared this revision with the goal of making the riches of calculus more attractive and understandable to the increasing number of men and women who take the standard calculus course for science, mathematics, and engineering students. This edition (like its predecessors) was written with five related objectives in constant view: *concreteness*, *readability*, *motivation*, *applicability*, and *accuracy*.

## CONCRETENESS

The power of calculus is impressive in its precise answers to realistic questions and problems. In the necessary development of the theory, we keep in mind the central question: How does one actually *compute* it? We place special emphasis on concrete examples, applications, and problems that serve both to highlight the development of the theory and to demonstrate the remarkable versatility of calculus in the investigation of important scientific questions.

## READABILITY

Difficulties in learning mathematics often are complicated by language difficulties. Our writing style stems from the belief that crisp exposition, both intuitive and precise, makes mathematics more accessible—and hence more readily learned—with no loss of rigor. We hope our language is clear and attractive to students and that they can and actually will read it, thereby enabling the instructor to concentrate class time on the less routine aspects of teaching calculus.

## MOTIVATION

Our exposition is centered around examples of the use of calculus to solve real problems of interest to real people. In selecting such problems for our examples and exercises, we took the view that stimulating interest and motivating effective study go hand in hand. We attempt to make it clear to students how the knowledge gained with each new concept or technique will be worth the effort expended. In theoretical discussions, especially, we try to provide an intuitive picture of the goal before we set off in pursuit of it.

## APPLICATIONS

Its diverse applications are what attract many students to calculus, and realistic applications provide valuable motivation and reinforcement for all students. Section 1-1 contains a list of twenty *sample* applications that the student can anticipate for later study. This list illustrates the unusually broad range of applications that we include, but it is neither necessary nor desirable that the course cover all the applications in the book. Each section or subsection that may be omitted without loss of continuity is marked with an asterisk. This provides flexibility for each instructor to steer his or her own path between theory and applications.

## ACCURACY

To help ensure authoritative and complete coverage of calculus, both this edition and its predecessors were subjected to a comprehensive reviewing process. With regard to the selection, sequence, and treatment of mathematical topics, our approach is traditional. With regard to the level of rigor, we favor an intuitive and conceptual treatment that is careful and precise in the formulation of definitions and the statements of theorems. Some proofs that may be omitted at the discretion of the instructor are placed at the ends of sections. Others (such as the proofs of the intermediate value theorem and of the integrability of continuous functions) are deferred to the book's appendices. In this way we leave ample room for variation in seeking the proper balance between rigor and intuition.

## SPECIAL FEATURES OF THIS EDITION

Substantial portions of this alternate edition are based on our *Calculus and Analytic Geometry*, second edition (1986). The present edition is intended for use in those courses where a later introduction of trigonometric functions is desired. Here the calculus of trigonometric functions is delayed until Chapter 8. Other changes for this revision include an earlier introduction to the chain rule (in Section 3-3) for use in differentiating algebraic functions, and a more intuitive introduction to exponential and logarithmic functions (in Section 7-1).

***Trigonometric Functions*** A review of the elementary trigonometry needed for calculus has been inserted in Section 8-1, preceding the first appearance of trigonometric limits in Section 8-2. Derivatives and integrals of trigonometric functions first appear in Sections 8-3 and 8-4.

The paragraphs that follow describe features that this alternate edition shares with the second edition.

*Additional Problems*   This edition contains over 6000 problems. Most of the new problems are drill or practice exercises. They have been inserted mainly at the beginnings of problem sets to insure that students gain sufficient confidence and computational skill before moving on to less routine problems.

*New Examples and Computational Details*   In many sections throughout the book, we have inserted a simpler first example as an initial illustration of the main ideas of the section. Moreover, we have inserted an additional line or two of computational detail in many of the worked-out examples to make them easier for student readers to follow.

*Split Sections*   We divided a number of the longer sections in the first edition into two sections for this revision. For instance, each of the following *pairs* of sections corresponds to a single original section: Sections 1-2 and 1-3 (real numbers and functions), Sections 2-1 and 2-5 (limits), Sections 3-5 and 3-6 (maxima and minima), Sections 4-4 and 4-5 (the first derivative test and graphs of polynomials), Sections 5-4 and 5-5 (evaluation of integrals and the fundamental theorem of calculus), Sections 11-1 and 11-2 (indeterminate forms and l'Hopital's rule), Sections 14-4 and 14-5 (space curves and curvature), Sections 15-2 and 15-3 (functions of several variables and limits), Sections 16-1 and 16-2 (double integrals), and Sections 17-2 and 17-3 (line integrals). In each case the separation of sections enabled us to add more explanatory discussions for the benefit of the student.

*Optional Computer Applications*   We have included twenty-one optional programming notes for supplementary reading by those students who might be motivated by computer applications. Each of these notes appears at the end of a section (following the problems) and applies very simple BASIC programming to illustrate the ideas of the section. These programming notes are completely optional—we never assume that any have been included in the calculus course, and we never refer to them in the text proper. Their purpose is to stimulate interest in calculus in the rapidly increasing population of students who are already interested in computers. Those who would like to explore this topic further may consult Edwards: *Calculus and the Personal Computer* (Englewood Cliffs, N.J.: Prentice-Hall, 1986) for self-study or as a computer calculus laboratory text.

*Introductory Chapters*   The initial chapter of the first edition has been divided into two shorter chapters for this edition. This permits the inclusion of more review material and a slightly slower pace at the beginning of the course. But we still retain the objective of a quick start on calculus itself. Section 1-6 gives a first look at the derivative, and it serves to motivate the formal treatment of limits in Chapter 2.

*Differentiation Chapters*   We have substantially reordered the sequence of topics on differentiation in Chapters 3 and 4. Our objective is to build student confidence by introducing topics more nearly in order of increasing difficulty. We cover the basic techniques for differentiating algebraic functions in Sections 3-2 through 3-4 before discussing maxima and minima in Sections 3-5 and 3-6. Section 3-9 on Newton's method has been simplified. The mean value theorem and its applications are deferred to Chapter 4. All curve-sketching techniques now appear consecutively in Sections 4-5 through 4-7. In Section 4-8 we have all but eliminated the alternative $D^{-1}$ notation for antiderivatives that appeared in the first edition.

*Integration Chapters*   The proof of the fundamental theorem of calculus in Section 5-5 is preceded by an intuitive treatment in Section 5-4. We have also inserted a number of additional and simpler examples in Chapters 5 and 6, as well as in Chapter 9 (techniques of integration).

*Infinite Series and Taylor's Formula*   Taylor's formula and polynomial approximations appear in Section 11-3. The extension to Taylor series is now delayed until Section 12-7 in the chapter on infinite series.

*Analytic Geometry and Vectors*   Vectors in the plane and vectors in space now appear in the consecutive Chapters 13 and 14; these are now easy to combine as a single unit if the instructor so wishes. We have augmented substantially the discussion of vector fields in Section 17-1.

*Differential Equations*   Sections 7-6 and 7-8 introduce the very simplest separable differential equations and their impressive applications. Nevertheless, these are optional sections, and the instructor may delay them until Chapter 18 (on differential equations) is covered. Chapter 18 has been revised substantially—it now ends with Section 18-7 on elementary power series methods and Section 18-8 on elementary numerical methods. Some of the lengthier applications have been deleted but are now included in Edwards and Penney, *Elementary Differential Equations with Applications* (Englewood Cliffs, N.J.; Prentice-Hall, 1985).

*Computer Graphics*   The ability of students to visualize surfaces and graphs of functions of two variables should be enhanced by the IBM-PC graphics that appear in Chapters 14 and 15. For these excellent computer graphics, fully integrated with text discussions, we are indebted to John K. Edwards. He developed and programmed them using the APL*PLUS/PC System from STSC, Inc. (with the exception of Figures 15.22–15.33, for which he used the PLOT-CALL System from Golden Software, Inc.).

## ANSWERS AND MANUALS

Answers to most of the odd-numbered problems appear in the back of the book. Solutions to most problems (other than those odd-numbered ones for which an answer alone is sufficient) are available in an Instructor's Manual. A subset of this manual, containing solutions to problems numbered 1, 4, 7, 10, . . . is available as a Student Manual. The statements of Problems 1, 4, 7, 10, . . . , in the first twelve chapters, followed by their solutions and renumbered conventionally, are available as *Worked Problems in Calculus*. A collection of some 1400 additional problems suitable for use as test questions, *Calculus Test Item File*, is available for use by instructors.

## ACKNOWLEDGMENTS

All experienced textbook authors know the value of critical reviewing during the preparation and revision of a manuscript. In carrying out the revision on which this edition is based we benefited greatly from the advice of the following exceptionally able reviewers:

Leon E. Arnold, Delaware County Community College

H.L. Bentley, University of Toledo

Michael L. Berry, West Virginia Wesleyan College

William Blair, Northern Illinois University

Wil Clarke, Atlantic Union College

Peter Colwell, Iowa State University

Robert Devaney, Boston University

Dianne H. Haber, Westfield State College

W. Cary Huffman, Loyola University of Chicago

Lois E. Knouse, Le Tourneau College

Morris Kalka, Tulane University

Catherine Lilly, Westfield State College

Barbara Moses, Bowling Green University

James P. Qualey, Jr., University of Colorado

Lawrence Runyan, Shoreline Community College

John Spellman, Southwest Texas State University

Virginia Taylor, University of Lowell

Samuel A. Truitt, Jr., Middle Tennessee State University

Robert Urbanski, Middlesex County College

Robert Whiting, Villanova University

John C. Higgins, Brigham Young University

Barbara L. Osofsky, Rutgers University—New Brunswick

James W. Daniel, University of Texas at Austin

We would like to offer special thanks to Janet Jenness, Jim Sosebee, Randall Turner, Chris Vickery, David Hammett, Robert Martin, Anthony Barcellos, Marshall Saade, Ted Shifrin, Carl Pomerance, Peter Rice, Robert Varley, Harvey Blau, Robin Giles, Frank Huggins, Gene Ortner, Pamela Hoveland, Woody Lichtenstein, Pete Pedersen, Jay Sultan, Trinette Draffin, B.J. Ball, J.G. Horne, J.F. Epperson, Stephen McCleary, George Wilson, Jon Carlson, John Hollingsworth, and Roy Smith.

Many of our best improvements must be credited to colleagues and users of the first edition throughout the country (and abroad). We are grateful to all those, especially students, who have written to us, and hope that they will continue to do so. We also believe that the quality of the finished book itself is adequate testimony to the skill, diligence, and talent of an exceptional staff at Prentice Hall; our special thanks go to David Ostrow, mathematics editor; Maria McColligan production editor; Walter A. Behnke and Anne T. Bonanno, designers; and Eric G. Hieber, illustrator. Finally, we cannot adequately thank Alice Fitzgerald Edwards and Carol Wilson Penney for their continued assistance, encouragement, support, and patience.

*Athens, Georgia*                                         C. H. E., Jr.
                                                          D. E. P.

# 1

## Prelude to Calculus

## Introduction

We live in a world of ceaseless change, filled with bodies in motion and with phenomena of ebb and flow. The principal object of the body of computational methods known as **calculus** is the analysis of problems of change and motion. This mathematical discipline stems from the seventeenth-century investigations of Isaac Newton (1642–1727) and Gottfried Wilhelm Leibniz (1646–1716). Many (if not most) of the scientific discoveries that have shaped our civilization during the past three centuries would have been impossible without the use of calculus, and today it continues to serve as the principal quantitative language of science and technology. We list below some of the problems that you will learn to solve as you study this book. The first two problems will be discussed in this chapter, and the others will be covered in later chapters.

1. *The Fence Problem.* What is the maximum rectangular area that can be enclosed with a fence of perimeter 140 m?

2. *The Refrigerator Problem.* The manager of an appliance store buys refrigerators at a wholesale price of $250 each. On the basis of past experience, the manager knows she can sell 20 refrigerators each month at $400 each and an additional refrigerator each month for each $3 reduction in selling price. What selling price will maximize the monthly profit of the store?

3. A cork ball of specific gravity $\frac{1}{4}$ is thrown into water. How deep will it sink? (Section 3-9)

4. What is the maximum possible radius of a "black hole" with the same mass as the sun? (Section 4-9)

5. If you have enough spring in your legs to jump straight up 4 ft on the earth, could you blast off under your own power from an asteroid of diameter 3 mi? (Section 4-9)

6. How much power must a rocket engine produce in order to put a satellite into orbit around the earth? (Section 6-5)

7. If the population of the earth continues to grow at its present rate, when will there be standing room only? (Section 7-5)

8. Suppose that you deposit $100 each month in a savings account that pays 7.5% interest compounded continuously. How much will you have in the bank after 10 years? (Section 7-5)

9. The factories polluting a certain lake are ordered to cease immediately. How long will it take for natural processes to restore the lake to an acceptable level of purity? (Section 7-6)

10. According to newspaper accounts, it is possible to survive a free fall (without parachute) from a height of 20,000 ft. Can this be true? (Section 7-6)

11. How can a pendulum clock be used to determine the altitude of a mountain peak? (Section 8-6)

12. What is the best shape for the reflector in a solar heater? (Section 10-4)

13. How often can a fixed dose of a drug be administered without producing a dangerous level of the drug in the patient's bloodstream? (Section 12-3)

**14** How do we know that $\pi = 3.14159265\ldots$? (Section 12-7)

**15** At what angle should the curve on a race track be banked to best accommodate cars traveling at 150 mi/h? (Section 13-5)

**16** How does a satellite of the earth use its thrusters to transfer from one circular orbit to another? (Section 13-7)

**17** Does a baseball pitch actually curve, or is it some sort of optical illusion? (Section 14-4)

**18** What temperature can a mercury thermometer withstand before its bulb bursts? (Section 15-7)

**19** How can two companies that make the same product conspire to maximize their total profits? (Section 15-11)

**20** A coin, a hoop, and a baseball roll down a hill. Which will reach the bottom first? (Section 16-5)

This first chapter contains some review material and material preliminary to your study of calculus: real numbers and inequalities, functions and their graphs, straight lines and slopes, and equations of circles and of parabolas. In Section 1-6 we introduce (quite informally) *limits* of functions and the problem of finding tangent lines to curves. This leads to the key concept of the *derivative* of a function, which can be used to solve problems such as the first two just given. This preliminary discussion is intended to motivate the more detailed and formal treatment of limits in Chapter 2 and of derivatives in Chapter 3.

---

**1-2**

**Real Numbers**

The **real numbers** are already familiar to you; they are just those numbers ordinarily used in most measurements. The mass, speed, temperature, and charge of a body are measured by real numbers. Real numbers can be represented by **terminating** or **nonterminating** decimal expansions. Any terminating decimal can be written in nonterminating form by adding zeros:

$$\tfrac{3}{8} = 0.375 = 0.375000000\ldots.$$

Any **repeating** nonterminating decimal, such as

$$\tfrac{7}{22} = 0.31818181818\ldots,$$

represents a **rational** number, one that is the quotient of two integers. Conversely, every rational number is represented by a repeating decimal expansion (as displayed above). The decimal expansion of an **irrational** number (one that is not rational), such as

$$\sqrt{2} = 1.414213562\ldots$$

or

$$\pi = 3.141592653589793\ldots,$$

is both nonterminating and nonrepeating.

The geometric representation of real numbers as points on the **real line** $\mathscr{R}$ should also be familiar to you. Each real number is represented by precisely one point of $\mathscr{R}$, and each point of $\mathscr{R}$ represents precisely one real

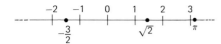

**1.1** The real line $\mathscr{R}$

number. By convention, the positive numbers lie to the right of zero and the negative numbers to its left, as indicated in Fig. 1.1.

If the real number $a$ lies (on $\mathscr{R}$) to the left of the number $b$, then we write $a < b$ (or $b > a$) and say that $a$ is *less than* b and that $b$ is *greater than* a. The number $a$ is *positive* if $a > 0$; if $a < 0$, then $a$ is said to be *negative*. The following properties of inequalities of real numbers are fundamental and are often used.

$$\text{If } a < b \text{ and } b < c, \text{ then } a < c.$$
$$\text{If } a < b \text{ then } a + c < b + c.$$
$$\text{If } a < b \text{ and } c > 0, \text{ then } ac < bc.$$
$$\text{If } a < b \text{ and } c < 0, \text{ then } ac > bc.$$

(1)

The last two statements mean that an inequality is preserved when its members are multiplied by a *positive* number but *reversed* when they are multiplied by a *negative* number.

In the problems at the end of this section, we ask you to deduce from the properties in (1) that the sum of two positive numbers is positive, while the sum of two negative numbers is negative; the product of two positive numbers *or* of two negative numbers is positive, while the product of a positive number and a negative number is negative. Moreover,

$$a > b \quad \text{if and only if} \quad a - b > 0. \tag{2}$$

### ABSOLUTE VALUE

The (nonnegative) distance along the real line between zero and the real number $a$ is the **absolute value** of $a$, written $|a|$. Equivalently,

$$|a| = \begin{cases} a & \text{if } a \geqq 0; \\ -a & \text{if } a < 0. \end{cases} \tag{3}$$

The notation $a \geqq 0$ means that $a$ is *either* greater than zero *or* equal to zero. Note that (3) implies that $|a| \geqq 0$ for every real number $a$, while $|a| = 0$ if and only if $a = 0$. For example,

$$|4| = 4, \qquad |-3| = 3, \qquad |0| = 0, \quad \text{and} \quad |\sqrt{2} - 2| = 2 - \sqrt{2},$$

the latter being true because $2 > \sqrt{2}$. Thus $\sqrt{2} - 2 < 0$, and hence

$$|\sqrt{2} - 2| = -(\sqrt{2} - 2) = 2 - \sqrt{2}.$$

The following properties of absolute values are frequently used.

$$|a| = |-a| = \sqrt{a^2} \geqq 0,$$
$$|ab| = |a|\,|b|,$$
$$-|a| \leqq a \leqq |a|,$$
$$|a| < b \quad \text{if and only if} \quad -b < a < b,$$
$$|a| = b \geqq 0 \quad \text{if and only if} \quad a = b \text{ or } a = -b, \qquad \text{and}$$
$$|a| > b \geqq 0 \quad \text{if and only if} \quad a < -b \text{ or } a > b.$$

(4)

The **distance** between the real numbers $a$ and $b$ is defined to be $|a - b|$ — or, if you prefer, $|b - a|$, which is equal to $|a - b|$. This distance is simply the length of the line segment of the real line $\mathcal{R}$ with endpoints $a$ and $b$, as indicated in Fig. 1.2.

The properties of inequalities and of absolute values listed above imply the following important fact.

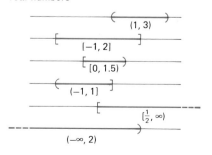

1.2 The distance between $a$ and $b$

---

> ***Triangle Inequality*** $\quad |a + b| \leq |a| + |b|.$ $\hspace{2cm}$ (5)

---

***Proof*** We add the inequalities $-|a| \leq a \leq |a|$ and $-|b| \leq b \leq |b|$. This gives us

$$-(|a| + |b|) \leq a + b \leq |a| + |b|. \hspace{2cm} (6)$$

But the properties in (4) imply that

$$|c| \leq d \quad \text{if and only if} \quad -d \leq c \leq d. \hspace{2cm} (7)$$

We set $c = a + b$ and $d = |a| + |b|$, and it follows that (5) and (6) are equivalent. $\hspace{1cm}$ ∎

### INTERVALS

Suppose that $S$ is a set (collection) of real numbers. Then we write $x \in S$ if and only if $x$ is an element of $S$. The set $S$ may be described by means of the notation

$$S = \{x : \cdots\},$$

where the ellipsis represents a condition that the real number $x$ satisfies exactly when $x$ belongs to $S$. The most important sets of real numbers in calculus are **intervals.** If $a < b$, then the **open interval** $(a, b)$ is defined to be the set

$$(a, b) = \{x : a < x < b\}$$

of real numbers, and the **closed interval** $[a, b]$ is

$$[a, b] = \{x : a \leq x \leq b\}.$$

Thus a closed interval contains its endpoints, while an open interval does not. We also use the **half-open intervals**

$$[a, b) = \{x : a \leq x < b\}$$

and

$$(a, b] = \{x : a < x \leq b\}.$$

Thus the open interval $(1, 3)$ is the set of those real numbers $x$ such that $1 < x < 3$; the closed interval $[-1, 2]$ is the set of those real numbers $x$ such that $-1 \leq x \leq 2$; and the half-open interval $(-1, 2]$ is the set of those real numbers $x$ such that $-1 < x \leq 2$. In Fig. 1.3 we show examples of such intervals, as well as some **unbounded intervals,** which have forms such as

$$[a, \infty) = \{x : x \geq a\},$$

$$(-\infty, a] = \{x : x \leq a\},$$

$$(a, \infty) = \{x : x > a\}, \quad \text{and}$$

$$(-\infty, a) = \{x : x < a\}.$$

1.3 Some examples of intervals of real numbers

(1, 3)

[−1, 2]

[0, 1.5)

(−1, 1]

$[\frac{1}{2}, \infty)$

(−∞, 2)

We emphasize that the symbol $\infty$, denoting infinity, is merely a notational device and does *not* represent a real number—the real line $\mathscr{R}$ does *not* have "endpoints at infinity." The use of this symbol is justified by the brief and natural descriptions of the sets $[\frac{1}{2}, \infty)$ and $(-\infty, 2)$ for the sets of all real numbers $x$ such that $x \geq \frac{1}{2}$ and $x < 2$, respectively.

The set of solutions of an inequality involving a variable $x$ is often an interval or a union of intervals, as in the next examples. The **solution set** of such an inequality is simply the set of all those real numbers $x$ that satisfy the inequality.

**EXAMPLE 1**   Solve the inequality $2x - 1 < 4x + 5$.

*Solution*   Using the properties of inequalities listed in (4), we proceed much as if we were solving an equation for $x$: We isolate $x$ on one or the other side of an inequality. Here we begin with

$$2x - 1 < 4x + 5$$

and it follows that

$$-1 < 2x + 5;$$
$$-6 < 2x;$$
$$-3 < x.$$

Hence the solution set is the unbounded interval $(-3, \infty)$.

**EXAMPLE 2**   Solve the inequality $-13 < 1 - 4x \leq 7$.

*Solution*   We simplify the given inequality as follows.

$$-13 < \quad 1 - 4x \leq 7;$$
$$-7 \leq 4x - 1 \quad < 13;$$
$$-6 \leq 4x \qquad < 14;$$
$$-\tfrac{3}{2} \leq \quad x \qquad < \tfrac{7}{2}.$$

Thus the solution set of the given inequality is the half-open interval $[-\frac{3}{2}, \frac{7}{2})$.

**EXAMPLE 3**   Solve the inequality $|3 - 5x| < 2$.

*Solution*   From the fourth property of absolute values in (4), we see that the given inequality is equivalent to

$$-2 < 3 - 5x < 2.$$

We now simplify as in the previous two examples:

$$-5 < -5x < -1;$$
$$\tfrac{1}{5} < \quad x < 1.$$

Thus the solution set is the open interval $(\frac{1}{5}, 1)$.

**EXAMPLE 4**  Solve the inequality

$$\frac{5}{|2x - 3|} < 1.$$

*Solution*  It is usually best to begin by eliminating a denominator containing the unknown. Here we multiply each term by the *positive* quantity $|2x - 3|$ to obtain the equivalent inequality

$$|2x - 3| > 5.$$

It follows from the last property in (4) that this is so if and only if either

$$2x - 3 < -5 \quad \text{or} \quad 2x - 3 > 5.$$

The solutions of these *two* inequalities are the open intervals $(-\infty, -1)$ and $(4, \infty)$, respectively. Hence the solution set of the original inequality consists of all those numbers $x$ that lie in *either* of these two open intervals.

---

The **union** of the two sets $S$ and $T$ is the set $S \cup T$ given by

$$S \cup T = \{x : \text{either} \quad x \in S \quad \text{or} \quad x \in T \quad \text{or both}\}.$$

Thus the solution set in Example 4 is the union $(-\infty, -1) \cup (4, \infty)$.

**EXAMPLE 5**  In accord with Boyle's law, the pressure $p$ (pounds per square inch) and volume $v$ (cubic inches) of a certain gas satisfy the condition $pv = 100$. Suppose that $50 \leq v \leq 150$. What is the range of possible values of the pressure?

*Solution*  If we substitute $v = 100/p$ in the given inequality $50 \leq v \leq 150$, we get

$$50 \leq \frac{100}{p} \leq 150.$$

It follows that *both*

$$50 \leq \frac{100}{p} \quad \textit{and} \quad \frac{100}{p} \leq 150;$$

that is, that both

$$p \leq 2 \quad \text{and} \quad p \geq \frac{2}{3}.$$

Thus the pressure $p$ must lie in the closed interval $[\frac{2}{3}, 2]$.

---

The **intersection** of the two sets $S$ and $T$ is the set $S \cap T$ defined as follows:

$$S \cap T = \{x : \text{both} \quad x \in S \quad \text{and} \quad x \in T\}.$$

Thus the solution set in Example 5 is the intersection $(-\infty, 2] \cap [\frac{2}{3}, \infty) = [\frac{2}{3}, 2]$.

Simplify each expression in Problems 1–12 by writing it without using absolute value symbols.

**1** $|3 - 17|$

**2** $|-3| - |17|$

**3** $\left|-0.25 - \frac{1}{4}\right|$

**4** $|5| - |-7|$

**5** $|(-5)(4 - 9)|$

**6** $\dfrac{|-6|}{|4| + |-2|}$

**7** $|(-3)^3|$

**8** $|3 - \sqrt{3}|$

**9** $\left|\pi - \frac{22}{7}\right|$

**10** $-|7 - 4|$

**11** $|x - 3|$, given $x < 3$

**12** $|x - 5| + |x - 10|$, given $|x - 7| < 1$

Solve each of the inequalities in Problems 13–31. Write each solution set in terms of intervals.

**13** $2x - 7 < -3$

**14** $1 - 4x > 2$

**15** $3x - 4 \geqq 17$

**16** $2x + 5 \leqq 9$

**17** $2 - 3x < 7$

**18** $6 - 5x > -9$

**19** $-3 < 2x + 5 < 7$

**20** $4 \leqq 3x - 5 \leqq 10$

**21** $-6 \leqq 5 - 2x < 2$

**22** $3 < 1 - 5x < 7$

**23** $|3 - 2x| < 5$

**24** $|5x + 3| \leqq 4$

**25** $|1 - 3x| > 2$

**26** $1 < |7x - 1| < 3$

**27** $2 \leqq |4 - 5x| \leqq 4$

**28** $\dfrac{1}{2x + 1} > 3$

**29** $\dfrac{2}{7 - 3x} \leqq -5$

**30** $\dfrac{2}{|3x - 4|} < 1$

**31** $\dfrac{1}{|1 - 5x|} \geqq -\dfrac{1}{3}$

**32** Solve the inequality $x^2 - x - 6 > 0$. (*Suggestion:* Conclude from the factorization $x^2 - x - 6 = (x - 3)(x + 2)$ that the quantities $x - 3$ and $x + 2$ are either both positive or both negative. Consider the two cases separately to deduce that the solution set is $(-\infty, -2) \cup (3, \infty)$.)

Use the method of Problem 32 to solve the inequalities in Problems 33–36.

**33** $x^2 - 2x - 8 > 0$

**34** $x^2 - 3x + 2 < 0$

**35** $4x^2 - 8x + 3 \geqq 0$

**36** $2x \geqq 15 - x^2$

**37** In accord with Boyle's law, the pressure $p$ (in pounds per square inch) and volume $v$ (in cubic inches) of a certain gas satisfy the condition $pv = 800$. What is the range of possible values of the pressure, given $100 \leqq v \leqq 200$?

**38** The relationship between the Fahrenheit temperature $F$ and the Celsius temperature $C$ is given by $F = 32 + \frac{9}{5}C$. If the temperature on a certain day ranged from a low of $70°F$ to a high of $90°F$, what was the range of the temperature in degrees Celsius?

**39** An electrical circuit contains a battery supplying $E$ volts in series with a resistance of $R$ ohms, as shown in Fig. 1.4. Then the current of $I$ amperes that flows in the circuit satisfies Ohm's law, $E = IR$. If $E = 100$ and $25 < R < 50$, what is the range of possible values of $I$?

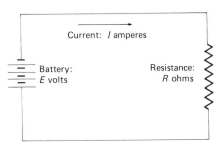

**1.4** A simple electric circuit

**40** The period $T$ (in seconds) of a simple pendulum of length $L$ (in feet) is given by $T = 2\pi\sqrt{L/32}$. If $3 < L < 4$, what is the range of possible values of $T$?

**41** Use the second property in (1) to show that the sum of two positive numbers is positive.

**42** Use the third property in (1) to show that the product of two positive numbers is positive.

**43** Prove that the product of two negative numbers is positive, while the product of a positive number and a negative number is negative.

**44** Suppose that $a < b$ and that $a$ and $b$ are either both positive or both negative. Prove that $1/a > 1/b$.

**45** Apply the triangle inequality twice to show that

$$|a + b + c| \leqq |a| + |b| + |c|$$

for arbitrary real numbers $a$, $b$, and $c$.

**46** Write $a = (a - b) + b$ to deduce from the triangle inequality that

$$|a| - |b| \leqq |a - b|$$

for arbitrary real numbers $a$ and $b$.

**47** Deduce from the definition in (3) that $|a| < b$ if and only if $-b < a < b$.

The key to the mathematical analysis of a geometric or scientific situation is often the recognition of relationships between the variables that describe the situation. Such a relationship may be a formula that expresses one variable as a function of another. For example, the area $A$ of a circle of radius $r$ is given by $A = \pi r^2$. The volume $V$ and surface area $S$ of a sphere of radius $r$ are given by $V = \frac{4}{3}\pi r^3$ and $S = 4\pi r^2$, respectively. After $t$ seconds a body that has been dropped from rest has fallen a distance $s = \frac{1}{2}gt^2$ feet and has speed $v = gt$ feet per second, where $g \approx 32$ ft/s$^2$ is gravitational acceleration. The volume $V$ (in liters) of 3 g of carbon dioxide ($CO_2$) at 27°C is given in terms of its pressure $p$ (in atmospheres) by $V = 1.68/p$. These are all examples of real-valued functions of a real variable.

> *Definition of a Function*
>
> A real-valued **function** $f$ defined on a set $D$ of real numbers is a rule that assigns to each number $x$ in $D$ exactly one real number $f(x)$.

The set $D$ of all those numbers $x$ for which $f(x)$ is defined is called the **domain** (or **domain of definition**) of the function $f$. The number $f(x)$, read "$f$ of $x$," is called the **value** of the function $f$ at the number or point $x$. The set of all values $y = f(x)$ is called the **range** of $f$. That is, the range of $f$ is the set

$$\{y : y = f(x) \quad \text{for some } x \text{ in } D\}.$$

In this section we will be more concerned with the domain of a function than with its range.

A function $f$ with domain $D$ and range $R$ may be thought of as a machine that accepts a number $x$ from $D$ and then puts out (or displays or prints) the number $f(x)$, as indicated in Fig. 1.5. Such a machine is the familiar pocket calculator with a $\sqrt{\ }$ button. When a nonnegative number $x$ is entered and this button is pressed, the calculator displays (an approximation to) the number $\sqrt{x}$. If such a calculator were not limited to eight- or ten-digit accuracy, it would be a perfect model for the square root function. In this case it is easy to see that the domain $D$ is the set of all nonnegative real numbers and the range is the same (because $\sqrt{x}$ always denotes the *nonnegative* square root of $x$).

Often a function is described by means of a formula that specifies how to compute the number $f(x)$ in terms of the number $x$. For example, the formula

$$f(x) = x^2 + x - 3 \tag{1}$$

describes the rule of a function $f$ having as its domain the entire real line $\mathscr{R}$. Some typical values of $f$ are $f(-2) = -1$, $f(0) = -3$, and $f(3) = 9$.

When we describe the function $f$ by writing a formula $y = f(x)$, we call $x$ the **independent variable** and $y$ the **dependent variable** because the value of

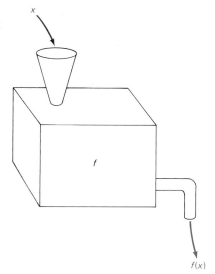

1.5 A "function machine"

$x$

$f$

$f(x)$

$y$ depends—through $f$—upon the choice of $x$. As $x$ changes, or varies, then so does $y$, and the way that $y$ varies with $x$ is determined by the function. For instance, if $f$ is the function of Equation (1), then $y = -1$ when $x = -2$, $y = -3$ when $x = 0$, and $y = 9$ when $x = 3$.

The symbol $f(\ )$ may be regarded as an operation that is to be performed whenever a value from the domain of $f$ is inserted between the parentheses. For example, with the function $f(x) = x^2 + x - 3$ of (1) we get

$$f(4) = (4)^2 + (4) - 3 = 17,$$

$$f(c) = c^2 + c - 3,$$

$$f(2 + h) = (2 + h)^2 + (2 + h) - 3$$
$$= (4 + 4h + h^2) + (2 + h) - 3 = 3 + 5h + h^2,$$

and

$$f(a + h) - f(a) = (a + h)^2 + (a + h) - 3 - (a^2 + a - 3)$$
$$= (a^2 + 2ah + h^2 + a + h - 3) - (a^2 + a - 3)$$
$$= 2ah + h^2 + h.$$

Not every function is defined by means of a one-part formula. For instance, if we write

$$f(x) = \begin{cases} x^2 & \text{if } x \geq 0, \\ -x & \text{if } x < 0, \end{cases}$$

we have defined a perfectly good function. Some of its values are $f(-3) = +3$, $f(0) = 0$, and $f(2) = 4$. The function in Example 1 is defined initially by means of a verbal description rather than by means of formulas.

**EXAMPLE 1**  For each real number $x$, let $f(x)$ denote the greatest integer that is less than or equal to $x$. For instance $f(-\frac{1}{2}) = -1$, $f(0) = 0$, $f(\sqrt{2}) = 1$, and $f(3) = 3$. If $n$ is an integer, then $f(x) = n$ for every $x$ in the half-open interval $[n, n + 1)$. This function $f$ is called the **greatest integer function** and is often denoted by

$$f(x) = [\![x]\!]. \tag{2}$$

Thus $[\![-\frac{1}{2}]\!] = -1$, $[\![0]\!] = 0$, $[\![\sqrt{2}]\!] = 1$, and $[\![3]\!] = 3$. Note that while $[\![x]\!]$ is defined for all real $x$, the range of the greatest integer function consists only of the set of all integers.

---

What is the domain of a function supposed to be when it is not specified? This is a common situation, and it occurs when we give a function $f$ by writing its formula $y = f(x)$. If no domain is given, we agree for convenience that the domain $D$ is the set of all real numbers $x$ for which the expression $f(x)$ makes sense. For example, the domain of $f(x) = 1/x$ is the set of all nonzero real numbers (because $1/x$ is defined only for $x \neq 0$).

**EXAMPLE 2**  Find the domain of the function $g$ with formula

$$g(x) = \frac{1}{\sqrt{2x + 4}}.$$

*Solution*   Note that the value of $\sqrt{x}$ is defined only if $x$ itself is nonnegative. So $\sqrt{2x+4}$ is defined only if $2x+4 \geqq 0$—that is, for $x \geqq -2$. In order that the reciprocal $1/\sqrt{2x+4}$ be defined, we also require that $\sqrt{2x+4} \neq 0$ and thus that $x \neq -2$. Hence the domain of $g$ is the interval $D = (-2, +\infty)$.

**EXAMPLE 3**   Express the volume $V$ of a cube as a function of its total surface area $S$.

*Solution*   Suppose that $e$ is the length of each edge of the cube. Then $V = e^3$ and $S = 6e^2$ (there are six square faces). Hence $e = \sqrt{S/6}$, and so

$$V(S) = e^3 = \left(\frac{S}{6}\right)^{3/2}.$$

The domain of $V$ is the (unbounded) open interval $(0, +\infty)$.

---

In the two examples that follow, we begin the solutions of the fence problem and the refrigerator problem, the first two problems listed in Section 1-1.

**EXAMPLE 4**   In connection with the fence problem, write the area $A$ of a rectangle of perimeter 140 as a function of the length $x$ of its base.

*Solution*   As indicated in Fig. 1.6, let $y$ denote the height of the rectangle. Then its area is given by

$$A = xy. \tag{3}$$

1.6   The rectangle of Example 4

Note, however, that this is a formula for $A$ in terms of the *two* variables $x$ and $y$ rather than a function of the single variable $x$. To eliminate $y$ and thereby to get $A$ as a function of $x$ alone, we use the fact that the perimeter of the rectangle is $2x + 2y = 140$, so that $y = 70 - x$. Therefore, Equation (3) yields $A = x(70 - x)$.

In addition to this latter formula, we should also specify the domain of the function $A$. Only values $x > 0$ will produce actual rectangles, but we will find it convenient to include the value $x = 0$ as well. This value of $x$ corresponds to a "degenerate rectangle" with base zero and height 70. For similar reasons, we have the restriction $y \geqq 0$. Because $y = 70 - x$, it follows that $x \leqq 70$. Thus the complete definition of our area function is

$$A(x) = x(70 - x), \qquad 0 \leqq x \leqq 70. \tag{4}$$

---

Example 4 illustrates an important part of the solution of a typical problem involving applications of mathematics. The domain of a function is a necessary part of its definition, and for each function we must specify the domain of values of the independent variable. In applications, we use the values of the independent variable that are relevant to the problem at hand.

For convenience, we restate here the second problem in Section 1-1.

THE REFRIGERATOR PROBLEM  The manager of an appliance store buys refrigerators at a wholesale price of $250 each. On the basis of past experience, the manager knows she can sell 20 refrigerators each month at $400 each and an additional refrigerator each month for each $3 reduction in selling price. What selling price will maximize the monthly profit of the store?

EXAMPLE 5  In connection with the refrigerator problem, express the monthly profit $P$ as a function of the number $x$ of refrigerators sold monthly.

*Solution*  We interpret the statement of the problem as meaning that the selling price $p$ of each refrigerator is set at the beginning of the month, and that all refrigerators are then sold at this same price $p$. Because the wholesale price is $250 each, the profit on each refrigerator sold is $p - 250$, and hence the total monthly profit $P$ on the sale of $x$ refrigerators is given by

$$P = x(p - 250). \tag{5}$$

To express $P$ as a function of $x$ alone, we must eliminate the variable $p$.

Let $n$ denote the number of $3 reductions below the original selling price, so that

$$p = 400 - 3n. \tag{6}$$

Then $n$ more refrigerators (more, that is, than the original 20) can be sold, and thus

$$x = n + 20; \quad \text{that is,} \quad n = x - 20.$$

We therefore see from (6) that

$$p = 400 - 3(x - 20) = 460 - 3x. \tag{7}$$

Upon substitution of this value of $p$ in (5), we get the formula

$$P = x(210 - 3x) = 3x(70 - x)$$

for the total monthly profit $P$ as a function of the number $x$ of refrigerators sold each month.

Finally, we must find the relevant domain of values of $x$. Certainly $x \geq 0$. On the other hand, a negative profit would be unacceptable, so $x \leq 70$. Therefore the complete description of our profit function is

$$P(x) = 3x(70 - x), \quad 0 \leq x \leq 70. \tag{8}$$

Once the value of $x$ that maximizes $P(x)$ has been found, the optimal selling price $p$ is then given by Equation (7), $p = 460 - 3x$.

---

Examples 4 and 5 reveal a surprising connection between the fence problem and the refrigerator problem: The only difference between $A(x)$ in (4) and $P(x)$ in (8) is the constant 3. So the same value of $x$, whatever it may be, will maximize both $A(x)$ and $P(x)$. As difficult as the fence problem and the refrigerator problem seem at first glance, each—when reduced to essentials—amounts to the problem of finding the maximum value of $f(x) = x(70 - x)$

for $0 \le x \le 70$. Much of the marvelous power and efficiency of calculus stems from the use of functional language to recognize such similarities between diverse types of applied problems and from the exploitation of these similarities by the application of general mathematical methods for solving such problems.

When confronted with a verbally stated applied problem such as the refrigerator problem, anyone's first question is, How on earth do we get started on it? The function concept is the key to getting a handle on such a situation. If we can express the quantity to be maximized—the dependent variable—as a function of some independent variable, then we have something tangible to do: to find the maximum value attained by this function.

For instance, we might attack the refrigerator problem by calculating a table of values of the profit function $P(x)$ in (8). Such a table is shown in Fig. 1.7. The data in this table suggest strongly that the maximum profit is $P = 3675$, attained with $x = 35$. To check this, we compute the values of $P(x)$ from $x = 30$ to $x = 40$, as shown in the table of Fig. 1.8. These data make it quite clear that if only *integral* values of $x$ are considered, the maximum value of $P(x)$ for such $x$ is $P(35) = 3675$.

As a practical matter, this finishes the refrigerator problem, because only an *integral* number of refrigerators can be sold. From equation (7), we see that the manager should set the selling price at

$$p = 460 - (3)(35) = 355$$

dollars per refrigerator and thereby sell 35 refrigerators each month for a maximal profit of $3675.

In the fence problem, however, there remains the possibility that the base $x$ of the rectangle of perimeter 140 and maximal area is *not* an integer, in which case tables like those in Figs. 1.7 and 1.8 do not settle the matter. A new mathematical idea is needed in order to find the maximum value of $f(x) = x(70 - x)$ for *all* $x$ in the interval $[0, 70]$. We shall attack this problem in Section 1-5 after a review of rectangular coordinates in Section 1-4.

| $x$ | $P(x)$ | |
|---|---|---|
| 0 | 0 | |
| 5 | 975 | |
| 10 | 1800 | |
| 15 | 2475 | |
| 20 | 3000 | |
| 25 | 3375 | |
| 30 | 3600 | |
| 35 | 3675 | ← |
| 40 | 3600 | |
| 45 | 3375 | |
| 50 | 3000 | |
| 55 | 2475 | |
| 60 | 1800 | |
| 65 | 975 | |
| 70 | 0 | |

**1.7** Profit $P(x)$ at selling price $x$

| $x$ | $P(x)$ | |
|---|---|---|
| 30 | 3600 | |
| 31 | 3627 | |
| 32 | 3648 | |
| 33 | 3663 | |
| 34 | 3672 | |
| 35 | 3675 | ← |
| 36 | 3672 | |
| 37 | 3663 | |
| 38 | 3648 | |
| 39 | 3627 | |
| 40 | 3600 | |

**1.8** Profit $P(x)$ at selling price $x$

## 1-3 PROBLEMS

In each of Problems 1–4, find each of the following values:
(a) $f(-a)$;   (b) $f(\frac{1}{a})$;   (c) $f(\sqrt{a})$;   (d) $f(a^2)$

**1** $f(x) = \dfrac{1}{x}$

**2** $f(x) = x^2 + 5$

**3** $f(x) = \dfrac{1}{x^2 + 5}$

**4** $f(x) = \sqrt{1 + x^2 + x^4}$

In each of Problems 5–10, find all values of $a$ such that $g(a) = 5$.

**5** $g(x) = 3x + 4$

**6** $g(x) = \dfrac{1}{2x - 1}$   ?

**7** $g(x) = \sqrt{x^2 + 16}$

**8** $g(x) = x^3 - 3$

**9** $g(x) = \sqrt[3]{x + 25}$

**10** $g(x) = 2x^2 - x + 4$

In each of Problems 11–16, compute the quantity $f(a + h) - f(a)$.

**11** $f(x) = 3x - 2$

**12** $f(x) = 1 - 2x$

**13** $f(x) = x^2$

**14** $f(x) = x^2 + 2x$

**15** $f(x) = \dfrac{1}{x}$

**16** $f(x) = \dfrac{2}{x + 1}$

In each of Problems 17–20, find the range of values of the given function.

**17** $f(x) = \dfrac{x}{|x|}$ if $x \ne 0$, while $f(0) = 0$.

**18** $f(x) = [\![3x]\!]$, where $[\![x]\!]$ is the largest integer not exceeding $x$.

**19** $f(x) = (-1)^{[\![x]\!]}$

**20** $f(x)$ is the first-class postage (in cents) for a letter mailed in the United States and weighing $x$ ounces, $0 < x < 12$. In 1985 the postage rate for such a letter was 22¢ for the first ounce plus 17¢ for each additional ounce or fraction thereof.

In each of Problems 21–35, find the largest domain (of real numbers) on which the given formula determines a function.

**21** $f(x) = 10 - x^2$  **22** $f(x) = x^3 + 5$

**23** $f(t) = \sqrt{t^2}$  **24** $g(t) = (\sqrt{t})^2$

**25** $f(x) = \sqrt{3x - 5}$  **26** $g(t) = \sqrt[3]{t + 4}$

**27** $f(t) = \sqrt{1 - 2t}$  **28** $g(x) = \dfrac{1}{(x + 2)^2}$

**29** $f(x) = \dfrac{2}{3 - x}$  **30** $g(t) = \left(\dfrac{2}{3 - t}\right)^{1/2}$

**31** $f(x) = \sqrt{x^2 + 9}$  **32** $h(z) = \dfrac{1}{\sqrt{4 - z^2}}$

**33** $f(x) = (4 - \sqrt{x})^{1/2}$  **34** $f(x) = \left(\dfrac{x + 1}{x - 1}\right)^{1/2}$

**35** $g(t) = \dfrac{t}{|t|}$

**36** Express the area $A$ of a square as a function of its perimeter $P$.

**37** Express the circumference $C$ of a circle as a function of its area $A$.

**38** Express the volume $V$ of a sphere as a function of its surface area $S$.

**39** Given: $0°C$ is the same as $32°F$ and a change of $1°C$ is the same as a change of $1.8°F$. Express the Celsius temperature $C$ as a function of the Fahrenheit temperature $F$.

**40** A rectangular box has volume 125 and square base of edge length $x$. Write its total surface area $A$ as a function of $x$.

**41** A rectangle with base of length $x$ is inscribed in a circle of radius 2. Express the area $A$ of the rectangle as a function of $x$.

**42** An oil field containing 20 wells has been producing 4000 barrels of oil daily. For each new well that is drilled, suppose that the daily production of each well decreases by 5 barrels. Write the total daily production of the oil field as a function of the number $x$ of new wells drilled.

**43** A rectangle has area 100. Express its perimeter $P$ as a function of the length $x$ of its base.

**44** A rectangle of fixed perimeter 36 is rotated about one of its sides $S$ to generate a right circular cylinder. Express the volume $V$ of this cylinder as a function of the length $x$ of the side $S$.

**45** A right circular cylinder has a volume of 1000 in.$^3$ and the radius of its base is $x$. Express the total surface area $A$ of the cylinder as a function of $x$.

**46** A rectangular box has a total surface area of 600 in.$^2$ and a square base with edge length $x$. Express the volume $V$ of the box as a function of $x$.

**47** An open-topped box is to be made from a square piece of cardboard with edge length 50 in. First, four small squares each having edge length $x$ inches are cut from the four corners of the cardboard, as indicated in Fig. 1.9. Then the four resulting flaps are turned up to form the four sides of the box, which will have a square base and a depth of $x$ inches. Express its volume $V$ as a function of $x$.

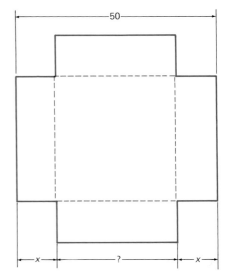

**1.9** Fold the edges up to make a box.

## *1-3 Optional Computer Application

Repetitive computations of the sort required to produce tables of values of functions (as in Figs. 1.7 and 1.8) can be tedious and boring. This kind of labor can be lightened by the use of the pocket computers that are now widely available and relatively inexpensive (to say nothing of the microcomputers and terminals of mainframe computers that are available to students on many campuses). A surprising amount can be accomplished with a very modest knowledge of BASIC programming, such as a student might acquire from an hour's study of a typical manual for a pocket computer.

The following program can be used to compute values of the profit function
$$P(x) = 3x(70 - x), \quad 0 \le x \le 70 \quad \text{of the refrigerator problem.}$$

```
10    INPUT  X
20    P = 3*X*(70 −X)
30    PRINT  X,  P
```

When line 10 is executed, the computer displays a question mark. After we enter the desired value of  $x$,  the computer calculates the value of  $P(x)$  in line 20 and then prints (displays) both  $x$  and the result at line 30. But with this simple program, we must run (execute) the program anew for each new value of  $x$.

The following extension of the simple program just given was used to calculate the data shown in Fig. 1.7, which lists the values of  $x$  and  $P(x)$  from  $x = 0$  to  $x = 70$  in increments of  5.

```
10    X = 0
20    P = 3*X*(70 − X)
30    PRINT  "  X  = ";  X;  "  P  = ";  P;
40    X = X + 5
50    IF  X < 75  THEN GOTO  20
60    END
```

When this program is executed, the computer begins with  $x = 0$  (line 10), computes the value of  $P(x)$  (line 20), and prints the current values of  $x$  and  $P$  (line 30). At a typical step in the computation we see the display

```
X = 35    P = 3675
```

telling us that  $P(35) = 3675$. In line 40 the value of  $x$  is increased by  5.  If  $x < 75$,  then line 50 transfers execution back to line 20 to calculate the next value of  $P(x)$.

The BASIC programming language varies slightly from one computer to another. With some computers the semicolon at the end of line 30 and the word GOTO in line 50 are unnecessary and should be deleted. With others, the semicolons should be replaced by commas.

**Exercise 1** Key in and run the second program above. Make whatever corrections are necessary for its proper execution on the computer you are using. Compare the results with the data shown in Fig. 1.7.

**Exercise 2** Modify the second program so that, when executed, it will produce the data shown in Fig. 1.8—the values of  $P(x)$  from  $x = 30$  to  $x = 40$  by increments of  1  (in  $x$).

**Exercise 3** If you are a moderately experienced programmer, write a program that first asks you to enter the end points  $a$  and  $b$  of an interval  $[a, b]$  and a positive integer  $n$  and then prints the values of  $P(x)$  from  $x = a$  to  $x = b$  by increments in  $x$  in the amount  $h = (b − a)/n$. Use this program to check the data in Fig. 1.8.

**Exercise 4** Modify this program so that improper input data will be rejected: if  $a < 0$, if  $b > 70$, if  $a \geq b$, or if  $n$  is not a positive integer. The program should print explanatory messages to the user ("Sorry—  N  is not an integer") and then request the needed value anew.

## The Coordinate Plane and Straight Lines

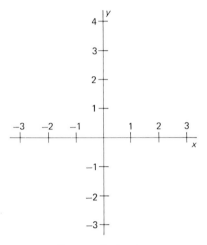

**1.10** The coordinate plane

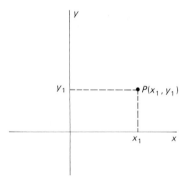

**1.11** The point $P$ has rectangular coordinates $(x_1, y_1)$.

Imagine the flat, featureless two-dimensional plane of Euclid's **geometry.** Install a copy of the real number line $\mathscr{R}$, with the line horizontal and the positive numbers to the right, as in Fig. 1.10. Add another copy of $\mathscr{R}$ perpendicular to the first, with the two lines crossing where zero is located on each. The second line should have the positive numbers above and the negative numbers below the first line. The first line is called the **x-axis** and the second is the **y-axis.**

With these added features, we call the plane the **coordinate plane,** because it is now possible to locate any point there by a pair of numbers called the *coordinates of the point.* If $P$ is a point in the plane, draw perpendiculars from $P$ to the coordinate axes, as shown in Fig. 1.11. One perpendicular meets the $x$-axis in the **x-coordinate** of $P$, called $x_1$. The other meets the $y$-axis in the **y-coordinate** $y_1$ of $P$. The pair of numbers $(x_1, y_1)$ is called the **coordinate pair** for $P$, or simply the **coordinates** of $P$. In compact notation, we speak of "the point $P(x_1, y_1)$." The numbers $x_1$ and $y_1$ also are called the **abscissa** and **ordinate,** respectively, of the point $P$.

This coordinate system is called the **rectangular coordinate system,** or sometimes the **Cartesian coordinate system** (because its use in geometry was popularized, beginning in the 1630s, by the French mathematician and philosopher René Descartes (1596–1650)). The plane, thus coordinatized, is sometimes denoted by $\mathscr{R}^2$ because of the two copies of $\mathscr{R}$ that are used and is sometimes called the **Cartesian plane.**

Rectangular coordinates are easy to use because $P(x_1, y_1)$ and $Q(x_2, y_2)$ denote the same point when and only when $x_1 = x_2$ and $y_1 = y_2$. Thus when you know that $P$ and $Q$ are different points, you may conclude that $P$ and $Q$ have different abscissas or different ordinates (or both).

The point of symmetry $(0, 0)$ where the coordinate axes cross is called the **origin.** The points on the $x$-axis all have coordinates of the form $(x, 0)$, and, while the *real number* $x$ is not the same as the *geometric point* $(x, 0)$, there are situations in which it is useful to think of the two as the same. Similar remarks can be made about the points $(0, y)$ that constitute the $y$-axis.

The concept of distance in the plane is based on the **Pythagorean theorem:** If $ABC$ is a right triangle with right angle $C$ and hypotenuse $c$ (see Fig. 1.12), then

$$c^2 = a^2 + b^2. \tag{1}$$

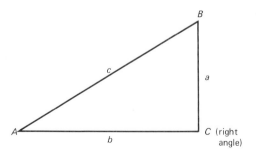

**1.12** The Pythagorean theorem

The converse of the Pythagorean theorem is also true—that is, if the three sides of a given triangle satisfy the Pythagorean relation in (1), then the angle opposite side $c$ is a right angle.

The *distance* $d(P_1, P_2)$ between the points $P_1$ and $P_2$ is (by definition) the length of the straight line segment joining $P_1$ and $P_2$. The following formula gives $d(P_1, P_2)$ in terms of the coordinates of the two points.

---

*Distance Formula*

The **distance** between $P_1(x_1, y_1)$ and $P_2(x_2, y_2)$ is

$$d(P_1, P_2) = \sqrt{(x_2 - x_1)^2 + (y_2 - y_1)^2}. \qquad (2)$$

---

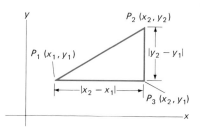

1.13 Use this triangle to deduce the distance formula.

**Proof** If $x_1 \neq x_2$ and $y_1 \neq y_2$, then the formula in (2) follows from an application of the Pythagorean theorem. Use the right triangle with vertices $P_1$, $P_2$, and $P_3(x_2, y_1)$ shown in Fig. 1.13.

If $x_1 = x_2$, then $P_1$ and $P_2$ lie in a vertical line. In this case

$$d(P_1, P_2) = |y_2 - y_1| = \sqrt{(y_2 - y_1)^2}.$$

This agrees with the formula in (2) because $x_1 = x_2$. The remaining case, in which $y_1 = y_2$, is similar. ∎

**EXAMPLE 1** Show that the triangle $PQR$ with vertices $P(1, 0)$, $Q(5, 4)$, and $R(-2, 3)$ is a right triangle. (This triangle is shown in Fig. 1.14.)

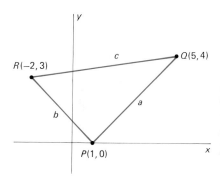

**Solution** The distance formula gives

$$a^2 = [d(P, Q)]^2 = (5 - 1)^2 + (4 - 0)^2 = 32,$$
$$b^2 = [d(P, R)]^2 = (-2 - 1)^2 + (3 - 0)^2 = 18, \quad \text{and}$$
$$c^2 = [d(Q, R)]^2 = (-2 - 5)^2 + (3 - 4)^2 = 50.$$

Because $a^2 + b^2 = c^2$, the *converse* of the Pythagorean theorem implies that $RPQ$ is a right angle. The right angle is at $P$, since it is the vertex opposite the longest side $QR$.

1.14 Is this a right triangle? See Example 1.

1.15 The midpoint $M$

Another application of the distance formula is an expression for the co-ordinates of the midpoint $M$ of the line segment $P_1P_2$ with endpoints $P_1$ and $P_2$, shown in Fig. 1.15. Recall from geometry that $M$ is the (one and only) point of the line segment $P_1P_2$ equally distant from $P_1$ and $P_2$. The following formula tells us that the coordinates of $M$ are the *averages* of the corresponding coordinates of $P_1$ and $P_2$.

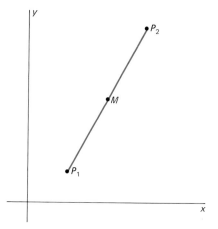

---

*Midpoint Formula*

The **midpoint** of the line segment with endpoints $P_1(x_1, y_1)$ and $P_2(x_2, y_2)$ is the point $M(\bar{x}, \bar{y})$ with coordinates

$$\bar{x} = \tfrac{1}{2}(x_1 + x_2), \qquad \bar{y} = \tfrac{1}{2}(y_1 + y_2). \qquad (3)$$

---

***Proof*** If you substitute the coordinates of $P_1$, $P_2$, and $M$ in the distance formula, you will find that $d(P_1, M) = d(P_2, M)$. All that remains is to show that $M$ is on the line segment $P_1P_2$. We ask you to do this in Problem 31. ∎

## STRAIGHT LINES AND SLOPE

We need to define the *slope* of a straight line, a measure of its rate of rise or fall from left to right. Given a line $L$ in the plane, with $L$ not vertical, choose two points $P_1(x_1, y_1)$ and $P_2(x_2, y_2)$ on $L$. Consider the **increments** $\Delta x$ and $\Delta y$ (read "delta $x$" and "delta $y$") in the $x$- and $y$-coordinates from $P_1$ to $P_2$. These are defined to be

$$\Delta x = x_2 - x_1 \quad \text{and} \quad \Delta y = y_2 - y_1. \tag{4}$$

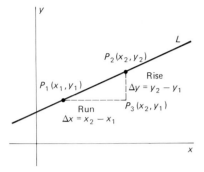

**1.16** The slope of a straight line

Then $\Delta y$ is the **rise** from $P_1$ to $P_2$, and $\Delta x$ is the **run** from $P_1$ to $P_2$, as illustrated in Fig. 1.16. The **slope** $m$ of the nonvertical line $L$ is then given by the ratio of rise to run; that is,

$$m = \frac{\Delta y}{\Delta x} = \frac{y_2 - y_1}{x_2 - x_1}. \tag{5}$$

Recall that corresponding sides of similar (that is, equal-angled) triangles have equal ratios. Now note that if $P_3(x_3, y_3)$ and $P_4(x_4, y_4)$ are two other points on $L$, then the similarity of the triangles of Fig. 1.17 implies that

$$\frac{y_4 - y_3}{x_4 - x_3} = \frac{y_2 - y_1}{x_2 - x_1}.$$

Hence the slope $m$ as defined in Equation (5) does *not* depend upon the particular choice of $P_1$ and $P_2$.

If the line $L$ is horizontal, then $\Delta y = 0$; thus Equation (5) gives $m = 0$. If $L$ is vertical, then $\Delta x = 0$; the slope of $L$ is *not defined* in this case. Thus we have the following statements:

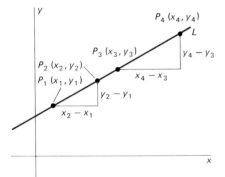

**1.17** The result of the slope computation does not depend upon which two points of $L$ are used.

> Horizontal lines have slope zero.
> Vertical lines have no slope at all.

That is, the slope of a horizontal line is 0, whereas the slope of a vertical line is undefined.

**EXAMPLE 2** (a) The slope of the line through the points $(3, -2)$ and $(-1, 4)$ is

$$m = \frac{4 - (-2)}{(-1) - 3} = \frac{6}{-4} = -\frac{3}{2}.$$

(b) The points $(3, -2)$ and $(7, -2)$ have the same $y$-coordinate. Therefore, the line through them is horizontal and hence has slope $m = 0$.
(c) The points $(3, -2)$ and $(3, 4)$ have the same $x$-coordinate. Therefore, the line through them is vertical and so its slope is undefined.

## EQUATIONS OF STRAIGHT LINES

Our immediate goal is to be able to write equations of given straight lines. That is, if $L$ is a straight line, we wish to construct a sentence about points $(x, y)$ in the plane, a sentence that is true when $(x, y)$ is on $L$ and false otherwise. Of course, the sentence is merely an algebraic equation involving the variables $x$ and $y$, together with some constants having to do with the particular line $L$. To this end, the concept of the slope of $L$ is almost essential.

Suppose, then, that $P_0(x_0, y_0)$ is a fixed point on the nonvertical line $L$ of slope $m$. Let $P(x, y)$ be any *other* point on $L$. We apply Equation (5) with $P$ and $P_0$ in place of $P_2$ and $P_1$ to find that

$$m = \frac{y - y_0}{x - x_0};$$

that is,

$$y - y_0 = m(x - x_0). \tag{6}$$

Because the point $(x_0, y_0)$ satisfies Equation (6), as does any other point $(x, y)$ of $L$, and because no other point of the plane can do so, Equation (6) is indeed an equation for the given line $L$.

---

### The Point-Slope Equation

The point $P(x, y)$ lies on the line with slope $m$ through the fixed point $(x_0, y_0)$ if and only if its coordinates satisfy the equation

$$y - y_0 = m(x - x_0). \tag{6}$$

---

**EXAMPLE 3**   Write an equation for the straight line $L$ through the points $P_1(1, -1)$ and $P_2(3, 5)$.

*Solution*   The slope of $L$ is

$$m = \frac{5 - (-1)}{3 - 1} = 3.$$

We take $P_1(1, -1)$ as the fixed point. Then, with the aid of (6), the point-slope equation of $L$ is

$$y + 1 = 3(x - 1);$$

simplification gives $y - 3x + 4 = 0$.

---

Equation (6) can be written in the form

$$y = mx + b, \tag{7}$$

where $b = y_0 - mx_0$ is a constant. Because $y = b$ when $x = 0$, the **y-intercept** of $L$ is the point $(0, b)$ shown in Fig. 1.18. Equations (6) and (7) are two different forms of the equation of a straight line.

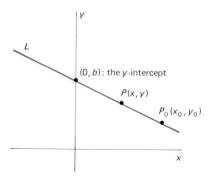

**1.18** Finding the slope-intercept equation of a line

---

### The Slope-Intercept Equation

The point $P(x, y)$ lies on the line with slope $m$ and $y$-intercept $b$ if and only if its coordinates satisfy the equation

$$y = mx + b. \tag{7}$$

---

Note that Equations (6) and (7) each can be written in the form of the general linear equation

$$Ax + By = C \qquad (8)$$

where $A$, $B$, and $C$ are constants. Conversely, if $B \neq 0$, then Equation (8) can be written in the form of Equation (7) by division of each term by $B$. Therefore, Equation (8) represents a straight line with its slope being the coefficient of $x$ *after solution of the equation for* $y$. If $B = 0$, then (8) reduces to the equation of a vertical line: $x = K$ ($K$ a constant). If $A = 0$, then (8) reduces to the equation of a horizontal line: $y = H$ ($H$ a constant). Thus we see that Equation (8) is always an equation of a straight line unless $A = B = 0$. Conversely, every straight line in the plane—even a vertical line—has an equation of the form in (8).

### PARALLEL AND PERPENDICULAR LINES

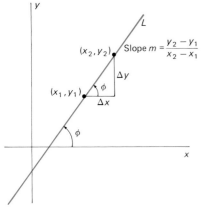

**1.19** How is the angle of inclination $\phi$ related to the slope $m$?

If the line $L$ is not horizontal, it must cross the $x$-axis. Then its **angle of inclination** is the angle $\phi$ measured counterclockwise from the positive $x$-axis to $L$. It follows that $0° < \phi < 180°$ if $\phi$ is measured in degrees. The angle of inclination of a horizontal line is $\phi = 0°$. Figure 1.19 makes it clear that this angle $\phi$ and the slope $m$ of a nonvertical line are related by the equation

$$m = \frac{\Delta y}{\Delta x} = \tan \phi. \qquad (9)$$

This is true because, if $\phi$ is an acute angle in a right triangle, then $\tan \phi$ is the ratio of the leg opposite $\phi$ to the leg adjacent to $\phi$ (a *leg* of a right triangle is either of its two *shorter* sides).

It is equally clear that two lines are parallel if and only if they have the same angle of inclination. So it follows from (9) that two parallel and nonvertical lines have the same slope and that two lines with the same slope must be parallel. This completes the proof of Theorem 1.

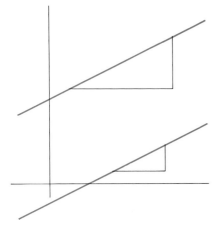

**1.20** Two parallel lines

**1.21** The demand curve for the refrigerator problem. See Example 5.

---

*Theorem 1   Slopes of Parallel Lines*

Two nonvertical lines are parallel if and only if they have the same slope.

---

Theorem 1 can also be proved without use of the tangent function. The two lines shown in Fig. 1.20 are parallel if and only if the two right triangles are similar, in which case the two slopes are equal.

**EXAMPLE 4**   Write an equation of the line $L$ that passes through the point $P(3, -2)$ and is parallel to the line $L'$ with equation $x + 2y = 6$.

***Solution***   When we solve the equation of $L'$ for $y$, we get $y = -\frac{1}{2}x + 3$. So $L'$ has slope $m = -\frac{1}{2}$. Because $L$ has the same slope, its point-slope equation is then

$$y + 2 = -\tfrac{1}{2}(x - 3);$$

that is, $x + 2y + 1 = 0$.

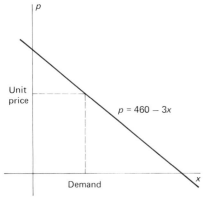

**EXAMPLE 5** In our discussion of the refrigerator problem in Example 5 of Section 1-3, we found that the selling price $p$ and the *demand* $x$ (the number of refrigerators $x$ that could be sold monthly at unit price $p$) were related by the equation

This is the equation of a straight line in the $xp$-plane, and it is shown in Fig. 1.21. This line is the **demand curve** (or **line**) for the store's refrigerators. (Demand curves are usually *not* straight lines.)

---

> **Theorem 2** *Slopes of Perpendicular Lines*
>
> Two lines $L_1$ and $L_2$ with slopes $m_1$ and $m_2$ are perpendicular if and only if
>
> $$m_1 m_2 = -1. \tag{10}$$
>
> That is, the slope of each is the *negative reciprocal* of the slope of the other.

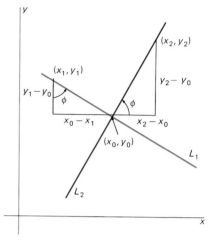

**1.22** Illustration of the proof of Theorem 2

**1.23** Positive and negative slope; effect on $\phi$

**Proof** If the two lines are perpendicular and the slope of each exists, then neither is horizontal or vertical. Thus the situation is like the one illustrated in Fig. 1.22, with the two lines intersecting at the point $(x_0, y_0)$. It is easy to see that the two right triangles are similar, so equality of ratios of corresponding sides yields

$$m_2 = \frac{y_2 - y_0}{x_2 - x_0} = \frac{x_0 - x_1}{y_1 - y_0}$$

$$= -\frac{x_1 - x_0}{y_1 - y_0} = -\frac{1}{m_1}.$$

Thus (10) holds if the two lines are perpendicular. This argument can be reversed to prove the converse—that the lines are perpendicular if $m_1 m_2 = -1$. ∎

**EXAMPLE 6** Write an equation of the line $L$ through the point $P(3, -2)$ that is perpendicular to the line $L'$ with equation $x + 2y = 6$.

**Solution** As we saw in Example 4, the slope of $L'$ is $m' = -\frac{1}{2}$. By Theorem 2, the slope of $L$ is $m = -1/m' = 2$, so $L$ has the point-slope equation

$$y + 2 = 2(x - 3);$$

equivalently, $2x - y - 8 = 0$.

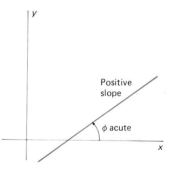

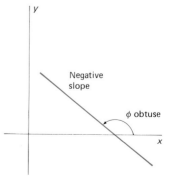

---

Finally, note that the *sign* of the slope $m$ of the line $L$ indicates whether $L$ runs upward or downward from left to right. If $m > 0$, then (because $m = \tan \phi$) $\phi$ must be an acute angle. In this case, $L$ "runs upward." If $m < 0$, then $\phi$ is obtuse, so that $L$ "runs downward." Figure 1.23 shows the geometry behind these observations.

Three points, $A$, $B$, and $C$, lie on a single straight line if and only if the slope of $AB$ is equal to that of $BC$. In each of Problems 1–4, plot the three given points, and then determine whether or not they lie on a single line.

1  $A(-1, -2)$, $B(2, 1)$, $C(4, 3)$
2  $A(-2, 5)$, $B(2, 3)$, $C(8, 0)$
3  $A(-1, 6)$, $B(1, 2)$, $C(4, -2)$
4  $A(-3, 2)$, $B(1, 6)$, $C(8, 14)$

In Problems 5 and 6, use the concept of slope to show that the points $A$, $B$, $C$, and $D$ are the vertices of a parallelogram.

5  $A(-1, 3)$, $B(5, 0)$, $C(7, 4)$, $D(1, 7)$
6  $A(7, -1)$, $B(-2, 2)$, $C(1, 4)$, $D(10, 1)$

In Problems 7 and 8, show that the given points $A$, $B$, and $C$ are the vertices of a right triangle.

7  $A(-2, -1)$, $B(2, 7)$, $C(4, -4)$
8  $A(6, -1)$, $B(2, 3)$, $C(-3, -2)$

In each of Problems 9–13, find the slope $m$ and $y$-intercept $b$ of the line with the given equation; then sketch the line.

9  $2x = 3y$
10  $x + y = 1$
11  $2x - y + 3 = 0$
12  $3x + 4y = 6$
13  $2x = 3 - 5y$

In Problems 14–23, write an equation of the straight line $L$ that is described.

14  $L$ is vertical and has $x$-intercept 7.
15  $L$ is horizontal and passes through $(3, -5)$.
16  $L$ has $x$-intercept 2 and $y$-intercept $-3$.
17  $L$ passes through $(2, -3)$ and $(5, 3)$.
18  $L$ passes through $(-1, -4)$ and has slope $\frac{1}{2}$.
19  $L$ passes through $(4, 2)$ and has angle of inclination $135°$.
20  $L$ has slope 6 and $y$-intercept 7.
21  $L$ passes through $(1, 5)$ and is parallel to the line with equation $2x + y = 10$.
22  $L$ passes through $(-2, 4)$ and is perpendicular to the line with equation $x + 2y = 17$.
23  $L$ is the perpendicular bisector of the line segment that has endpoints $(-1, 2)$ and $(3, 10)$.

24  Find the perpendicular distance from the **point** $(2, 1)$ to the line with equation $y = x + 1$.
25  Find the perpendicular distance between the parallel lines $y = 5x + 1$ and $y = 5x + 9$.
26  The points $A(-1, 6)$, $B(0, 0)$, and $C(3, 1)$ are three consecutive vertices of a parallelogram. Find the fourth vertex.
27  Prove that the two diagonals of the parallelogram of Problem 26 bisect each other.
28  Show that the points $A(-1, 2)$, $B(3, -1)$, $C(6, 3)$, and $D(2, 6)$ are the vertices of a *rhombus*—a parallelogram with all sides of equal length. Then prove that the diagonals of this rhombus are perpendicular to each other.
29  The points $A(2, 1)$, $B(3, 5)$, and $C(7, 3)$ are the vertices of a triangle. Show that the line joining the midpoints of $AB$ and $BC$ is parallel to $AC$.
30  A median of a triangle is a line joining a vertex to the midpoint of the opposite side. Show that the three medians of the triangle of Problem 29 intersect in a single point.
31  Complete the proof of the midpoint formula in Equation (3). It is necessary to show that the point $M$ actually lies on the segment $P_1P_2$. Do so by showing that the slope of $P_1M$ is equal to the slope of $MP_2$.
32  Let $P(x_0, y_0)$ be a point of the circle with center $C(0, 0)$ and radius $r$. Then the tangent line to the circle at $P$ is perpendicular to the line segment $CP$. Show that the equation of this tangent line is $x_0x + y_0y = r^2$.
33  The Fahrenheit temperature $F$ and the absolute temperature $K$ satisfy a linear equation. Given that $K = 273$ when $F = 32$ and that $K = 373$ when $F = 212$, express $K$ in terms of $F$. What is the value of $F$ when $K = 0$?
34  The length $L$ (in centimeters) of a copper rod is a linear function of its Celsius temperature $C$. If $L = 124.942$ when $C = 20$ and $L = 125.134$ when $C = 110$, express $L$ in terms of $C$.
35  The owner of a grocery store finds that he can sell 980 gal. of milk each week at \$1.69/gal. and 1220 gal. of milk each week at \$1.49. Assume a linear relationship between selling price and demand. How many gallons could he sell weekly at \$1.56/gal.?

---

**1-5**

## Graphs of Equations and Functions

We saw in Section 1-4 that the points $(x, y)$ satisfying the linear equation $Ax + By = C$ form a very simple set: a straight line. By contrast, the set of points $(x, y)$ that satisfy the equation

$$x^4 - 4x^3 + 3x^2 + 2x^2y^2 = y^2 + 4xy^2 - y^4$$

is the exotic curve of Fig. 1.24 (after you have studied Chapter 10, you will be able to demonstrate this with a brief computation; it is certainly not obvious now). But both the straight line and the complicated curve are examples of graphs.

---

**Definition** *Graph of an Equation*

The **graph** of an equation in two variables $x$ and $y$ is the set of all points $(x, y)$ in the plane that satisfy the equation.

---

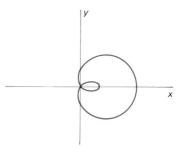

**1.24** The graph of the equation

$$x^4 - 4x^3 + 3x^2 + 2x^2y^2 = y^2 + 4xy^2 - y^4$$

For instance, the distance formula tells us that the graph of the equation

$$x^2 + y^2 = r^2 \tag{1}$$

is the circle of radius $r$ centered at the origin $(0, 0)$. More generally, the graph of the equation

$$(x - h)^2 + (y - k)^2 = r^2 \tag{2}$$

is the circle of radius $r$ with center $(h, k)$. This follows immediately from the distance formula because the distance between the points $(x, y)$ and $(h, k)$ in Fig. 1.25 is $r$.

We may regard the general circle with equation in (2) as a *translate* of the origin-centered circle with equation in (1). Suppose that every point of the $xy$-plane is translated $h$ units to the right and $k$ units upward. (If $h < 0$, the translation would be $|h|$ units to the left; if $k < 0$, the translation would be $|k|$ units downward.) Then the origin-centered circle $C$ of Equation (1) would be moved to the position of the circle $C_T$ of Equation (2), a circle of the same radius but with new center at $(h, k)$. Note in Fig. 1.26 that a point $(x, y)$ lies on the translated circle $C_T$ if and only if the point $(x - h, y - k)$ lies on the original circle $C$. This corresponds to the fact that Equation (2) can be obtained from Equation (1) by replacing $x$ by $x - h$ and $y$ by $y - k$. This discussion can be generalized to graphs of arbitrary equations, thereby establishing the following useful principle.

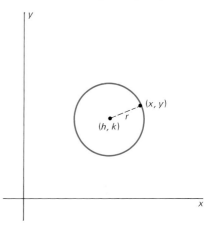

**1.25** A translated circle

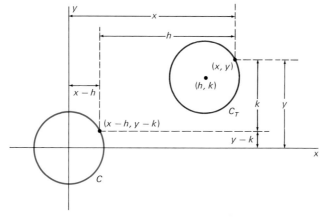

**1.26** Illustration of the translation principle

> **Translation Principle**
>
> Suppose that the curve $C$ is the graph of the equation $F(x, y) = 0$ (here, $F(x, y)$ simply denotes an expression, or formula, involving the two variables $x$ and $y$ together with some numerical constants). Let $C_T$ be the curve obtained by translating each point $(x, y)$ of $C$ to the point $(x + h, y + k)$. Then $C_T$ is the graph of the equation
>
> $$F(x - h, y - k) = 0. \tag{3}$$

That is, if the translation $(x, y) \to (x + h, y + k)$ moves the curve $C$ onto the curve $C_T$, then the replacement of $x$ by $x - h$ and of $y$ by $y - k$ converts the equation of $C$ into the equation of $C_T$.

Observe that the equation of a translated circle in (2) can be written in the general form

$$x^2 + y^2 + ax + by = c. \tag{4}$$

Conversely, when we encounter an equation of this form, we can determine the center and radius of the circle it represents by using the elementary technique of *completing the square*. To do so, note that

$$x^2 + ax = \left( x + \frac{a}{2} \right)^2 - \frac{a^2}{4},$$

which shows that $x^2 + ax$ can be made into a perfect square by adding to it the square of *half* the coefficient of $x$.

**EXAMPLE 1**   Find the center and radius of the circle having the equation

$$x^2 + y^2 - 4x + 6y = 12.$$

*Solution*   We complete the square separately in each of the two variables. This gives

$$(x^2 - 4x + 4) + (y^2 + 6y + 9) = 12 + 4 + 9,$$

which is equivalent to

$$(x - 2)^2 + (y + 3)^2 = 25.$$

Hence the circle has center $(2, -3)$ and radius 5.

The graph of a function is a special case of the graph of an equation.

> **Definition**   *Graph of a Function*
>
> The **graph** of the function $f$ is the graph of the equation $y = f(x)$.

Thus the graph of the function $f$ is the set of all points in the plane of the form $(x, f(x))$, where $x$ is in the domain of $f$. Because the second coordinate of such a point is uniquely determined by its first coordinate, it follows that no *vertical* line can intersect the graph of a function in more than one point. Alternatively, suppose that $f$ is a function with domain $D$; think of $D$ as a set of points on the $x$-axis. Then each vertical line through a point of $D$ intersects the graph of $f$ in exactly one point.

If you examine Fig. 1.24 (at the beginning of this section), you will see from these remarks that the graph shown there cannot be the graph of a *function*, although it *is* the graph of an equation.

**EXAMPLE 2**   Construct the graph of the absolute value function $f(x) = |x|$.

*Solution*   Recall that $|x| = -x$ if $x \leq 0$ and that $|x| = x$ for $x \geq 0$. So the graph of $y = |x|$ consists of the left half of the line $y = -x$, together with the right half of the line $y = x$, as shown in Fig. 1.27.

**EXAMPLE 3**   Construct the graph of the parabola $y = x^2$.

*Solution*   We plot some points in a short table of values.

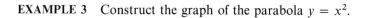

| $x$ | $-3$ | $-2$ | $-1$ | $0$ | $1$ | $2$ | $3$ |
|-----|------|------|------|-----|-----|-----|-----|
| $y = x^2$ | $9$ | $4$ | $1$ | $0$ | $1$ | $4$ | $9$ |

When we draw a smooth curve through these points, we obtain the curve of Fig. 1.28. The parabola $y = -x^2$ would look similar except that it would open downward instead of upward. More generally, the graph of the equation

$$y = ax^2 \tag{5}$$

is a parabola with its *vertex* at the origin, provided $a \neq 0$. This parabola opens upward if $a > 0$ and downward if $a < 0$. (For the time being, we may regard the vertex of a parabola as the point at which it "changes direction." The vertex of a parabola of the form $y = ax^2$ $(a \neq 0)$ will always be at the origin. A precise definition of the vertex of a parabola appears in Chapter 10.)

**EXAMPLE 4**   Construct the graphs of the functions $y = \sqrt{x}$ and $y = -\sqrt{x}$.

*Solution*   After plotting and connecting points as in Example 3, we obtain the parabola $y^2 = x$ shown in Fig. 1.29. Note that the parabola opens to the right. The upper half is the graph of $y = \sqrt{x}$, while the lower half is the graph of $y = -\sqrt{x}$. Thus the union of the graphs of these *two* functions is the graph of the *single* equation $y^2 = x$. More generally, the graph of the equation

$$x = by^2 \tag{6}$$

is a parabola with its vertex at the origin, provided $b \neq 0$. This parabola opens to the right (as in Fig. 1.29) if $b > 0$, but it opens to the left if $b < 0$.

**EXAMPLE 5**   Show that the graph of $y = 70x - x^2$ is a translated parabola. Locate its vertex.

*Solution*   We complete the square in the variable $x$:

$$y = 70x - x^2 = -(x^2 - 70x)$$
$$= 1225 - (x^2 - 70x + 1225),$$

so that

$$y - 1225 = -(x - 35)^2.$$

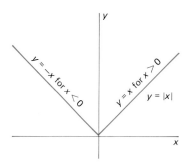

**1.27**   The graph of the absolute value function $y = |x|$

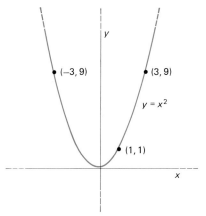

**1.28**   The graph of the parabola $y = x^2$

**1.29**   The graph of the parabola $x = y^2$

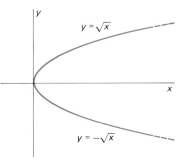

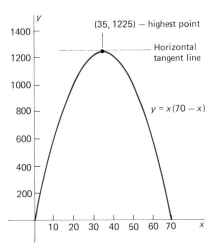

**1.30** The graph of $f(x) = x(70 - x)$
$0 \leq x \leq 70$

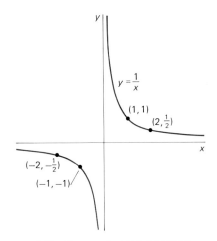

**1.31** The graph of the reciprocal function $y = 1/x$

**1.32** The greatest integer ("staircase") function $f(x) = [\![x]\!]$

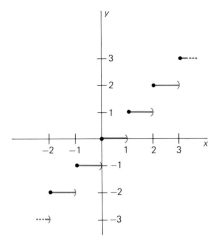

The translation principle therefore implies that the graph of the equation $y = 70x - x^2$ is shaped exactly like that of the parabola $y = -x^2$ but with its vertex at $(35, 1225)$ rather than at $(0, 0)$. Because the latter is the highest point of the graph of $y = -x^2$, the former must be the highest point of $y = 70x - x^2$; see Fig. 1.30.

NOTE  In this figure we have used different scales along the two axes: a 10-unit interval on the $x$-axis has the same length as a 150-unit interval on the $y$-axis. In an applied problem in which there is no physical relation between the units used to measure the independent and dependent variables, there is no necessity to use the same scale on the two axes. As in this example, you may choose the scale for convenience in sketching the graph. Only in a geometric problem involving angles or distances will you find it important to use the same scale on the two axes.

More generally, the graph of any equation of the form

$$y = ax^2 + bx + c \qquad (a \neq 0) \tag{7}$$

can be recognized as a translated parabola by first completing the square in $x$ and then applying the translation principle. The size of $|a|$ determines the width of the parabola, while the sign of $a$ determines whether the parabola opens upward or downward.

**EXAMPLE 6**  Sketch the graph of the reciprocal function $f(x) = 1/x$.

***Solution***  We note that $|f(x)| = 1/|x|$ is a very large positive number when $|x|$ is near zero, and that when $|x|$ is a large positive number, then $|f(x)|$ is near zero. Also, $f(x) = 1/x$ has the same sign as $x$. To get started with the graph, we can plot a few points, such as $(1, 1)$, $(10, 0.1)$, $(-10, -0.1)$, $(-1, -1)$, $(0.1, 10)$, and $(-0.1, -10)$. The other information above strongly suggests that the actual graph is much like the one of Fig. 1.31.

Figure 1.31 exhibits a "gap" or "discontinuity" in the graph of $y = 1/x$ at $x = 0$. Indeed, it might be called an "infinite discontinuity," because $y$ *decreases* without bound as $x$ approaches zero from the left, while $y$ increases without bound as $x$ approaches zero from the right. This phenomenon is often signaled by the presence of denominators that are zero at certain points, as in the case of the functions

$$f(x) = \frac{1}{1 - x} \quad \text{and} \quad f(x) = \frac{1}{x^2}$$

that we ask you to graph in the problems.

**EXAMPLE 7**  Figure 1.32 shows the graph of the greatest integer function $f(x) = [\![x]\!]$ of Example 1 in Section 1-3. Note the "jumps" that occur at the integral values of $x$.

**EXAMPLE 8**  Graph the function $f$ with formula

$$f(x) = x - [\![x]\!] - \tfrac{1}{2}.$$

***Solution*** Think of $x$ as temporarily fixed. If $n$ is an integer such that $n \leq x < n + 1$, then $[\![x]\!] = n$. If so, then

$$f(x) = x - n - \tfrac{1}{2}.$$

Because $y = x - n - \tfrac{1}{2}$ has as its graph a straight line of slope 1 and $f(n) = n - n - \tfrac{1}{2}$, it follows that the graph of $f$ takes the form shown in Fig. 1.33. This *sawtooth function* is another example of a discontinuous function. The values of $x$ where the value of $f(x)$ makes a jump are called **points of discontinuity** of the function $f$. Thus the points of discontinuity of the sawtooth function are the integers—as $x$ approaches the integer $n$ from the left, the value of $f(x)$ approaches $+\tfrac{1}{2}$, but it abruptly jumps to $-\tfrac{1}{2}$ when $x = n$. A precise definition of continuity and discontinuity for functions is given in Section 2-4.

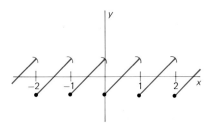

**1.33** The sawtooth function
$f(x) = x - [\![x]\!] - \tfrac{1}{2}$

In Section 1-3 we saw that the solutions of both the fence problem and the refrigerator problem reduce to the same mathematical problem—that of finding the maximum value of the function

$$f(x) = x(70 - x), \qquad 0 \leq x \leq 70.$$

Now we can resolve the matter, because we saw in Example 5 that the graph of $y = x(70 - x)$ is the parabola shown in Fig. 1.30, with its vertex $(35, 1225)$ being its highest point. Indeed, the fact that

$$f(x) = x(70 - x) = 1225 - (x - 35)^2$$

*proves* that the maximum value of $f$ is $f(35) = 1225$ because $(x - 35)^2 > 0$ except when $x = 35$.

In Example 4 of Section 1-3, we saw that the area of a rectangle with perimeter 140 and base $x$ is

$$A(x) = x(70 - x) = f(x), \qquad 0 \leq x \leq 70.$$

Now that we know that $f(x)$ is maximal when $x = 35$, it follows that the rectangle with perimeter 140 and *maximal* area is a square with base 35.

Next, in Example 5 of Section 1-3 (the refrigerator problem), we saw that if $x$ refrigerators are sold monthly at a price of $p = 460 - 3x$ dollars each, then the monthly profit is

$$P(x) = 3x(70 - x) = 3f(x), \qquad 0 \leq x \leq 70.$$

The fact that the maximum value of $f$ is $f(35) = 1225$ now can be seen to imply that the maximum profit of $3675 monthly is made by selling 35 refrigerators each month at a price of $355 each.

So, if we need to find the maximum (or minimum) value of a function $f$ and the formula for $f$ is that of a parabola or the upper or lower half of a semicircle, we can use the technique of completing the square, together with the translation principle, to find this maximum. This is a very limited technique, and one of our primary goals in differential calculus is to develop a more efficient technique that also can be applied to a far wider variety of functions.

The basis of this technique is the following observation: Visual inspection of the graph of $f(x) = x(70 - x)$ in Fig. 1.30 suggests that the tangent line to the curve at its highest point is horizontal. If we *knew* that the tangent line to a graph at its highest point must be horizontal, then our problem would reduce to showing that (35, 1225) is the only point of the graph of $f$ where the tangent line is horizontal.

But what is meant by the *tangent line* to an arbitrary curve? We pursue this question in Section 1-6. The answer will open the door to the possibility of finding maximum and minimum values of virtually arbitrary functions.

## 1-5 PROBLEMS

Sketch the graphs of the functions in Problems 1–25.

1 $f(x) = 2 - 5x, -1 \leqq x \leqq 1$

2 $f(x) = 2 - 5x, 0 \leqq x < 2$

3 $f(x) = 10 - x^2$      4 $f(x) = 1 + 2x^2$

5 $f(x) = x^3$      6 $f(x) = x^4$

7 $f(x) = \sqrt{4 - x^2}$      8 $f(x) = -\sqrt{9 - x^2}$

9 $f(x) = \sqrt{x^2 - 9}$      10 $f(x) = \dfrac{1}{1 - x}$

11 $f(x) = \dfrac{1}{x + 2}$      12 $f(x) = \dfrac{1}{x^2}$

13 $f(x) = \dfrac{1}{(x - 1)^2}$      14 $f(x) = \dfrac{|x|}{x}$

15 $f(x) = \dfrac{1}{2x + 3}$      16 $f(x) = \dfrac{1}{(2x + 3)^2}$

17 $f(x) = \sqrt{1 - x}$      18 $f(x) = \dfrac{1}{\sqrt{1 - x}}$

19 $f(x) = \dfrac{1}{\sqrt{2x + 3}}$      20 $f(x) = |2x - 2|$

21 $f(x) = |x| + x$      22 $f(x) = |x - 3|$

23 $f(x) = |2x + 5|$

24 $f(x) = |x|$ if $x < 0$, while $f(x) = x^2$ if $x \geqq 0$

25 $f(x) = 2 - x$ if $x < 2$, while $f(x) = (x - 2)^2$ if $x \geqq 2$

The graph of the equation $(x - h)^2 + (y - k)^2 = C$ is a circle if $C > 0$, is the single point $(h, k)$ if $C = 0$, and contains *no* points if $C < 0$. (Why?) Find the graphs of the equations in Problems 26–30. In case the graph is a circle, give its center and radius.

26 $x^2 + y^2 - 2x + 4y + 1 = 0$

27 $x^2 + y^2 - 6x + 8y = 0$

28 $x^2 + y^2 - 2x + 2y + 2 = 0$

29 $x^2 + y^2 + 2x + 6y + 20 = 0$

30 $2x^2 + 2y^2 - 2x + 6y + 5 = 0$

Sketch the translated parabolas in Problems 31–34 and indicate the vertex of each.

31 $y = x^2 + 2x + 4$      32 $2y = x^2 - 4x + 8$

33 $y = 5x^2 + 20x + 23$      34 $y = x - x^2$

Graph the functions given in Problems 35–40, indicating any points of discontinuity.

35 $f(x) = 0$ if $x < 0$, while $f(x) = 1$ if $x \geqq 0$

36 $f(x) = 1$ if $x$ is an integer; otherwise $f(x) = 0$

37 $f(x) = [\![2x]\!]$

38 $f(x) = \dfrac{x - 1}{|x - 1|}$

39 $f(x) = [\![x]\!] - x$

40 $f(x) = [\![x]\!] + [\![-x]\!] + 1$

41 If a ball is thrown straight upward with an initial velocity of 96 ft/s, then its height $t$ seconds later is $y = 96t - 16t^2$ ft. Determine the maximum height the ball attains by constructing the graph of $y$ as a function of $t$.

42 In Problem 42 of Section 1-3, you were asked to express the daily production of a certain oil field as a function $P = f(x)$ of the number $x$ of new oil wells that were drilled. Now construct the graph of $f$ and use it to find the value of $x$ that maximizes $P$.

---

## 1-6

## Tangent Lines and the Derivative — A First Look

At the end of the last section we saw that certain applied problems raise the question of what is meant by the *tangent line* at a point $P$ of a general curve $y = f(x)$. In this section we will see that this question leads to the introduction of a new function called the *derivative* of $f$. In turn, the derivative involves the new concept of *limits*. Our use of limits here will be quite intuitive and informal. Indeed, the main purpose of this section is to provide background and motivation for the formal treatment of limits in Chapter 2, after which we will initiate a detailed study of derivatives in Chapter 3.

To begin our investigation of tangent lines, we first need to decide how to *define* the tangent line $L$ at an arbitrary point $P$ of a curve $y = f(x)$. Our intuitive idea is this: The tangent line $L$ should be the straight line through $P$ that has the same direction as does the curve there. Because the direction of a line is determined by its slope, our plan for defining the tangent line amounts to finding an appropriate "slope-prediction formula"—one that will give the proper slope of the desired tangent line.

Our first example illustrates this approach in the case of one of the simplest of all nonstraight curves, the parabola with equation $y = x^2$.

**EXAMPLE 1** Determine the slope of the tangent line to the parabola $y = x^2$ at the point $P(a, a^2)$.

*Solution* In Fig. 1.34, we show the parabola $y = x^2$ with a typical point $P(a, a^2)$ marked on it. The figure also shows the result of a visual guess about the position of the desired tangent line $L$. Our problem, then, is this: Find the slope of $L$.

To begin with, we *can* compute the slope of a straight line passing through two points. Such a line is the **secant line** of Fig. 1.35; it passes through $P(a, a^2)$ and the nearby point $Q$ of the parabola. We take care to choose $Q \neq P$, though we do take $Q$ quite close to $P$. Thus the abscissa of $Q$ is not $a$; it is some nearby number, so it can be written in the form $a + h$, where $h$ is some small nonzero number.

In effect, think of starting with the point $P(a, a^2)$, then changing the abscissa $a$ of $P$ by a small amount $h \neq 0$ to obtain a different number $a + h$, and finally using $x = a + h$ in the formula $y = x^2$ to get the ordinate of $Q$. Thus we can describe the point $Q$ in terms of the single variable $h$, for we can write its coordinates in those terms: $Q = Q(a + h, (a + h)^2)$.

**1.34** The tangent line at $P$ should have the same direction as the curve does there.

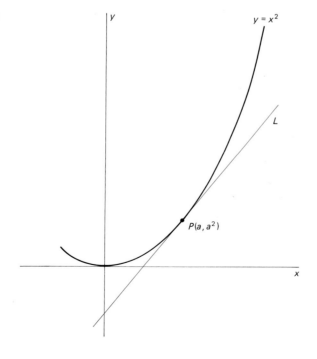

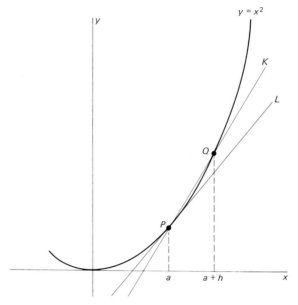

**1.35** The secant line *K* passes through two points —*P* and *Q*—which we can use to compute its slope.

Now $Q \neq P$, so we can use the definition of slope to compute the slope of the secant line *K* through *P* and *Q*. We denote this slope with the function notation $m(h)$ because the slope *m* is actually a function of *h*; if you change the value of *h*, you change the line *K*, and thereby change its slope. Therefore,

$$m(h) = \frac{(a + h)^2 - a^2}{(a + h) - a} = \frac{2ah + h^2}{h}. \tag{1}$$

Because *h* is nonzero, we may cancel, and we find that

$$m(h) = 2a + h. \tag{2}$$

Now imagine what happens as the point *Q* draws nearer and nearer to the point *P*. This corresponds to *h* approaching zero. The line *K* still passes through *P* and *Q*, but it pivots around the fixed point *P*. As *h* approaches zero, the secant line comes more and more nearly into coincidence with the tangent line *L*. This phenomenon is suggested in Fig. 1.36, which shows the secant line *K* almost coinciding with the tangent line *L*.

Our idea is that the tangent line *L* must, by definition, lie in the limiting position of the secant line *K*. To see precisely what this means, we can examine what happens to the slope of *K* as *K* pivots into coincidence with *L*:

As *h* approaches zero,
*Q* approaches *P*, and so
*K* approaches *L*; meanwhile,
*the slope of K approaches the slope of L.*

But we found above that the slope of the secant line *K* is

$$m(h) = 2a + h.$$

If we think of numerically smaller and smaller values of *h*—such as $h = 0.1$, $h = 0.01$, and $h = 0.001$—then it is quite clear that $m(h) = 2a + h$ approaches $2a$ as *h* approaches zero. Consequently we *must* define the tangent

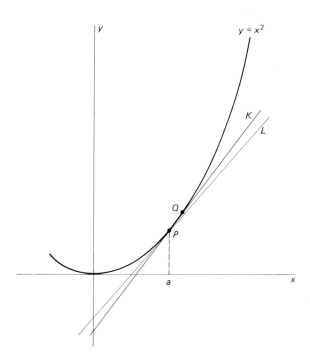

**1.36** As $h \to 0$, $Q$ approaches $P$, and $K$ moves into coincidence with the tangent line $L$.

line $L$ to be that straight line through $P$ with slope

$$m = 2a.$$

We summarize this discussion by saying that $m$ is the **limit of $m(h)$** as $h$ approaches zero, and we write

$$m = \lim_{h \to 0} m(h) = \lim_{h \to 0} (2a + h) = 2a. \tag{3}$$

Now that we have the slope of the tangent line, we can write its equation. For instance, to find the equation of the line tangent to the parabola $y = x^2$ at the point $(2, 4)$, we take $a = 2$ in (3) and get $m = 4$. The point-slope equation of the tangent line is then

$$y - 4 = 4(x - 2);$$

that is, $y = 4x - 4$.

---

The general case $y = f(x)$ is scarcely more complicated than the special case $y = x^2$. Suppose that $y = f(x)$ is given and that we want to find the slope of the tangent line $L$ to its graph at the point $(a, f(a))$. As shown in Fig. 1.37, let $K$ be the secant line passing through the point $P(a, f(a))$ and a nearby point $Q(a + h, f(a + h))$ on the graph. The slope of this secant line $K$ is

$$m_{\text{sec}} = m(h) = \frac{f(a + h) - f(a)}{h} \qquad \text{(for } h \neq 0\text{)}. \tag{4}$$

We now force $Q$ to approach $P$ along the curve by making $h$ approach zero. Suppose that $m_{\text{sec}}$ approaches the number $m$ as $h$ gets numerically smaller and smaller, approaching zero. Then the *tangent line $L$* to the curve $y = f(x)$

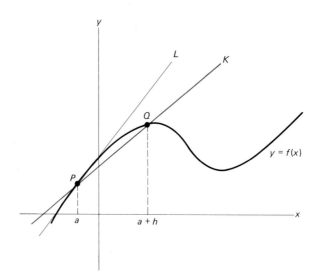

**1.37** As $h \to 0$, $Q \to P$, and the slope of $K$ approaches the slope of the tangent line $L$.

at the point $P(a, f(a))$ is, *by definition*, the line through $P$ whose slope is this number $m$.

To describe the fact that $m_{\text{sec}}$ approaches $m$ as $h$ approaches zero, we call $m$ the **limit of** $\mathbf{m_{sec}}$ as $h$ approaches zero, and we write

$$m = \lim_{h \to 0} m_{\text{sec}} = \lim_{h \to 0} \frac{f(a + h) - f(a)}{h}. \tag{5}$$

The slope $m$ in Equation (5) depends both on the function $f$ and on the number $a$. We indicate this by writing

$$m = f'(a) = \lim_{h \to 0} \frac{f(a + h) - f(a)}{h}. \tag{6}$$

We now replace $a$ by the more generic $x$ (because nothing was said about the particular value of $a$, it changes nothing to write $x$ in place of $a$). Thus we see that in actuality we are giving the definition of a new function:

$$f'(x) = \lim_{h \to 0} \frac{f(x + h) - f(x)}{h}. \tag{7}$$

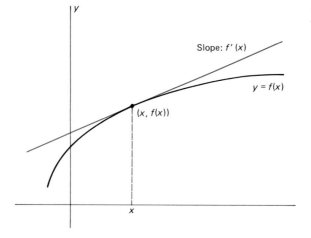

**1.38** The slope of the tangent at $(x, f(x))$ is $f'(x)$.

Slope: $f'(x)$

$y = f(x)$

$(x, f(x))$

This new function $f'$ is defined for all values of $x$ such that the limit above exists. It is called the **derivative** of the original function $f$, and the process of computing the derivative $f'$ (read "$f$ prime") is called **differentiation** of $f$.

Our discussion leading up to Equation (7) shows that the derivative has the following important geometric interpretation:

$f'(x)$ is the slope of the line tangent to
the curve $y = f(x)$ at the point $(x, f(x))$.

This interpretation can be illustrated as shown in Fig. 1.38.

There is an alternative notation for the derivative, and it originates from the early custom of writing $\Delta x$ in place of $h$ (because $h = \Delta x$ is an increment in $x$) and $\Delta y = f(x + \Delta x) - f(x)$ for the resulting change (or increment) in $y$. The slope of the secant line $K$ of Fig 1.39 is then

$$m_{\text{sec}} = \frac{\Delta y}{\Delta x}, \tag{8}$$

and the slope of the tangent line is

$$m = \frac{dy}{dx} = \lim_{\Delta x \to 0} \frac{\Delta y}{\Delta x}. \tag{9}$$

Hence, if $y = f(x)$, we often write

$$\frac{dy}{dx} = f'(x). \tag{10}$$

For example, the derivative of the function $y = x^2$ of Example 1 may be written as

$$\frac{dy}{dx} = 2x.$$

The two symbols $f'(x)$ and $dy/dx$ for the derivative of the function $y = f(x)$ are used interchangeably in mathematics and its applications, so familiarity with both is necessary. It is also important to understand that $dy/dx$ is a single symbol representing the derivative and is *not* the quotient of two separate quantities $dy$ and $dx$.

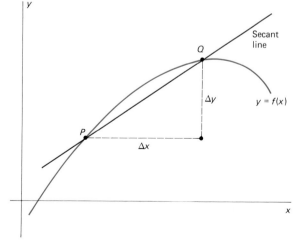

**1.39** Genesis of the $dy/dx$ notation

Now that the definition in (7),

$$f'(x) = \lim_{h \to 0} \frac{f(x + h) - f(x)}{h},$$

has been formulated, the discussion in Example 1 could be simplified greatly. But it is just as easy to discuss a more general function of the form $f(x) = ax^2 + bx + c$ (where $a$, $b$, and $c$ are constants).

---

**Theorem**   *Differentiation of Quadratic Functions*
If $f(x) = ax^2 + bx + c$, then

$$f'(x) = 2ax + b. \tag{11}$$

---

***Proof***   The definition of the derivative in (7) calls for us to

**1** Recall the definition of $f'(x)$.
**2** Substitute the given function $f$.
**3** Make algebraic simplifications, until we can complete Step 4.
**4** Recognize the value of the limit as $h \to 0$.

We should remember that $x$ may be thought of as *constant* throughout this computation—it is $h$ that is the variable in this four-step process. With $f(x) = ax^2 + bx + c$, these four steps yield

$$f'(x) = \lim_{h \to 0} \frac{f(x + h) - f(x)}{h}$$

$$= \lim_{h \to 0} \frac{[a(x + h)^2 + b(x + h) + c] - [ax^2 + bx + c]}{h}$$

$$= \lim_{h \to 0} \frac{2axh + bh + ah^2}{h}$$

$$= \lim_{h \to 0} (2ax + b + ah).$$

Thus we find that

$$f'(x) = 2ax + b$$

because the value of $ah$ approaches zero as $h \to 0$.   ∎

Note that the symbol $\lim_{h \to 0}$ signifies an operation to be performed, and so we continue to write it until the final step—when the operation of taking the limit *is* performed.

In geometric terms, the theorem above tells us this: The slope-prediction formula for curves of the form $y = ax^2 + bx + c$ is

$$m = 2ax + b. \tag{12}$$

The line through the point $P$ on the curve $y = f(x)$ that is perpendicular to the tangent line there is called the **normal line** at $P$. By Theorem 2 in Section 1-4, its slope is $-1/m$ (provided $m \neq 0$). A normal line is shown in Fig. 1.40.

Now that the derivative formula in (11) is available, we can simply apply it to differentiate specific quadratic functions instead of retracing the steps

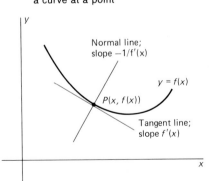

**1.40**   The tangent and normal lines to a curve at a point

in the proof of the theorem. For instance, if

$$f(x) = 2x^2 - 3x + 5,$$

then we can immediately write

$$f'(x) = (2)(2)x - 3 = 4x - 3.$$

**EXAMPLE 2**   Write equations for both the tangent line and the normal line to the parabola $y = 2x^2 - 3x + 5$ at the point $P(-1, 10)$.

*Solution*   We use the derivative that we just computed. The slope of the tangent line at $(-1, 10)$ is

$$f'(-1) = (4)(-1) - 3 = -7.$$

Hence the point-slope equation of the desired tangent line is

$$y - 10 = -7(x + 1).$$

The normal line has slope $m' = -1/(-7) = \frac{1}{7}$, so its point-slope equation is

$$y - 10 = \tfrac{1}{7}(x + 1).$$

**EXAMPLE 3**   Differentiate the function

$$f(x) = \frac{1}{x + 1}.$$

*Solution*   No formula for this derivative has yet been established, so we have no recourse but to carry out the four steps that ensue from the definition of the derivative. We get

$$f'(x) = \lim_{h \to 0} \frac{f(x + h) - f(x)}{h}$$

$$= \lim_{h \to 0} \frac{1}{h} \left( \frac{1}{x + h + 1} - \frac{1}{x + 1} \right)$$

$$= \lim_{h \to 0} \frac{1}{h} \left[ \frac{(x + 1) - (x + h + 1)}{(x + h + 1)(x + 1)} \right]$$

$$= \lim_{h \to 0} \frac{1}{h} \frac{-h}{(x + h + 1)(x + 1)}$$

$$= \lim_{h \to 0} \frac{-1}{(x + h + 1)(x + 1)}$$

It is apparent that $\lim_{h \to 0} (x + h + 1) = x + 1$. Hence we finally get

$$f'(x) = -\frac{1}{(x + 1)^2}.$$

It is possible for a function to be defined everywhere but to fail to have a derivative—that is, not be differentiable—at some points in its domain. Example 4 shows this.

TERMINOLOGY   The function $f$ is said to be **differentiable** at the point $x$ provided that the limit in Equation (7) (the definition of the derivative $f'(x)$) exists.

**EXAMPLE 4**   Show that the function $f(x) = |x|$ is *not* differentiable at $x = 0$.

*Solution*   When $x = 0$, we have

$$\frac{f(x + h) - f(x)}{h} = \frac{|h|}{h} = \begin{cases} -1 & \text{if } h < 0; \\ +1 & \text{if } h > 0. \end{cases}$$

Thus there is no *single* value that the quotient

$$\frac{f(0 + h) - f(0)}{h}$$

approaches as $h \to 0$. This means that the limit in question does not exist, and so $f(x) = |x|$ is not differentiable at $x = 0$.

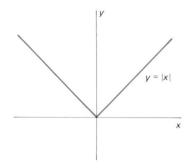

**1.41**   The graph of $f(x) = |x|$ has a corner point at (0, 0).

We have shown the graph of the function $f$ of Example 4 in Fig. 1.41. The graph has a sharp corner at the point (0, 0), and this explains why there can be no tangent line there—no single straight line is a good approximation to the shape of the graph at (0, 0). On the other hand, the figure makes it evident that $f'(x)$ exists if $x \ne 0$. Indeed,

$$f'(x) = \begin{cases} -1 & \text{if } x < 0; \\ +1 & \text{if } x > 0. \end{cases}$$

In conclusion, we apply the derivative to wrap up our continuing discussion of the refrigerator problem. Recall that the store manager has been purchasing refrigerators at a wholesale price of \$250 each and has been selling 20 refrigerators per month at \$400 each, for a monthly profit of $(20)(400 - 250) = 3000$ (dollars). Also, past experience indicates that an additional refrigerator can be sold each month for each \$3 reduction in selling price. In Example 5 of Section 1-3, we found that if $x$ refrigerators per month are sold for $p = 460 - 3x$ dollars each, then the monthly profit will be

$$P(x) = 3x(70 - x) = -3x^2 + 210x, \qquad 0 \le x \le 70.$$

This is the profit function that we seek to maximize.

Let us accept the intuitive fact—we shall give a rigorous proof in Chapter 3—that the maximum value of $P(x)$ occurs at a point where the tangent line to the graph of $y = P(x)$ is horizontal. Then we can apply the theorem on differentiation of quadratic functions (see Equation (11)) to find this maximum point. The slope at an arbitrary point $(x, P(x))$ is

$$m = P'(x) = -6x + 210.$$

We ask when $m = 0$, and we find that this happens when

$$-6x + 210 = 0,$$

and thus when $x = 35$. In agreement with the result we obtained by algebraic methods in Section 1-5, we find that the maximum profit is

$$P(35) = (3)(35)(70 - 35) = 3675$$

dollars. So just a bit of mathematics enables the store manager to add \$675 to her monthly profit.

In each of Problems 1–14, apply the theorem on differentiation of quadratic functions (see Equation (11)) to write the derivative $f'(x)$ of the given function $f(x)$.

**1** $f(x) = 5$

**2** $f(x) = x$

**3** $f(x) = x^2$

**4** $f(x) = 1 - 2x^2$

**5** $f(x) = 4x - 5$

**6** $f(x) = 7 - 3x$

**7** $f(x) = 2x^2 - 3x + 4$

**8** $f(x) = 5 - 3x - x^2$

**9** $f(x) = 2x(x + 3)$

**10** $f(x) = 3x(5 - x)$

**11** $f(x) = 2x - \left(\dfrac{x}{10}\right)^2$

**12** $f(x) = 4 - (3x + 2)^2$

**13** $f(x) = (2x + 1)^2 - 4x$

**14** $f(x) = (2x + 3)^2 - (2x - 3)^2$

In Problems 15–20, find the point or points of the curve $y = f(x)$ at which the tangent line is horizontal.

**15** $y = 10 - x^2$

**16** $y = 10x - x^2$

**17** $y = x^2 - 2x + 1$

**18** $y = x^2 + x - 2$

**19** $y = x - \left(\dfrac{x}{10}\right)^2$

**20** $y = x(100 - x)$

Apply the definition of the derivative—that is, the four-step process of this section—to find $f'(x)$ for each of the functions in Problems 21–32. Then write an equation for the line tangent to the curve $y = f(x)$ at the point $(2, f(2))$.

**21** $f(x) = 3x - 1$

**22** $f(x) = x^2 - x - 2$

**23** $f(x) = 2x^2 - 3x + 5$

**24** $f(x) = 70x - x^2$

**25** $f(x) = (x - 1)^2$

**26** $f(x) = x^3$

**27** $f(x) = \dfrac{1}{x}$

**28** $f(x) = x^4$

**29** $f(x) = \dfrac{1}{x^2}$

**30** $f(x) = x^2 + \dfrac{3}{x}$

**31** $f(x) = \dfrac{2}{x - 1}$

**32** $f(x) = \dfrac{x}{x - 1}$

In the remaining problems, use Equation (11) to find the derivative of a quadratic function where necessary. In Problems 33–35, write equations for the tangent line and the normal line to the curve $y = f(x)$ at the point $P$ given in the problem.

**33** $y = x^2$;  $P(-2, 4)$

**34** $y = 5 - x - 2x^2$;  $P(-1, 4)$

**35** $y = 2x^2 + 3x - 5$;  $P(2, 9)$

**36** Prove that the line tangent to the parabola $y = x^2$ at the point $(x_0, y_0)$ intersects the $x$-axis at the point $(x_0/2, 0)$.

**37** In Problem 42 of Section 1-3, you were asked to express the daily production of an oil field as a function $P = f(x)$ of the number $x$ of new oil wells drilled. Determine the maximum possible production by finding the point where $f'(x) = 0$ and the value of $P$ there.

**38** If a projectile is fired at an angle of $45°$ from the horizontal, with initial position the origin in the $xy$-plane and with an initial velocity of $100\sqrt{2}$ ft/s, then its trajectory is the parabola $y = x - (x/25)^2$.
(a) How far does the projectile travel (horizontally) before it hits the ground?
(b) What is the maximum height above the ground that the projectile attains?

**39** One of the two lines that pass through the point $(3, 0)$ and are tangent to the parabola $y = x^2$ is the $x$-axis itself. Find an equation for the *other* line. (*Suggestion:* Sketch the parabola and the other line. Let $(a, a^2)$ be the point at which this other line is tangent to the parabola; first find $a$.)

**40** Write equations for the two straight lines through the point $(2, 5)$ that are tangent to the parabola $y = 4x - x^2$. (See the suggestion for the previous problem.)

## CHAPTER 1 REVIEW: Definitions, Concepts, Results

Use the list below as a guide to ideas that you may need to review.

**1** Rational and irrational numbers
**2** The real line
**3** Properties of inequalities
**4** Absolute value of a real number
**5** Properties of absolute values
**6** Triangle inequality
**7** Intervals (open and closed)
**8** Solving inequalities
**9** Unions and intersections of sets
**10** Definition of a function
**11** Domain of a function
**12** Independent and dependent variables
**13** Coordinate plane
**14** Pythagorean theorem
**15** Distance formula
**16** Midpoint formula
**17** Slope of a straight line
**18** Point-slope equation of a line
**19** Slope-intercept equation of a line
**20** Slope relationship of parallel lines
**21** Slope relationship of perpendicular lines
**22** Graph of an equation
**23** Graph of a function
**24** Translation principle

**25** Equations of circles  
**26** Parabolas and the graph of $y = ax^2 + bx + c$  
**27** Tangent and normal lines to the graph of a function  

**28** Definition of the derivative of a function  
**29** Four-step process for finding the derivative  
**30** The derivative of $f(x) = ax^2 + bx + c$  

## MISCELLANEOUS PROBLEMS

Express the solutions of the inequalities in Problems 1–9 in terms of intervals.

**1** $3x + 5 < -10$      **2** $2x + 17 \geq 33$  
**3** $2 - 5x < 17 + 2x$      **4** $0 < 3x + 4 < 20$  
**5** $-3 < 1 - 2x < -1$      **6** $x^2 + 1 > 2x$  
**7** $-7 \leq 1 - 4x < 3$      **8** $3 < |2x - 5| \leq 6$  

**9** $-2 \leq \dfrac{3}{4x - 1} \leq 4$

In each of Problems 10–17, find the domain of definition of the function with the given formula.

**10** $f(x) = \dfrac{1}{2 - x}$      **11** $f(x) = \dfrac{1}{x^2 - 9}$

**12** $f(x) = \dfrac{x}{x^2 + 1}$      **13** $f(x) = (1 + \sqrt{x})^3$

**14** $f(x) = \dfrac{x + 1}{x^2 - 2x}$      **15** $f(x) = \sqrt{2 - 3x}$

**16** $f(x) = \dfrac{1}{\sqrt{9 - x^2}}$      **17** $f(x) = (x - 2)(4 - x)$

**18** The height of a circular cylinder is equal to its radius. Express its total surface area $A$ (including both ends) as a function of its volume.  
**19** Express the area $A$ of an equilateral triangle as a function of its perimeter $P$.  
**20** A piece of wire 100 in. long is cut into two pieces of lengths $x$ and $100 - x$. The first piece is bent into the shape of a square and the second is bent into the shape of a circle. Express as a function of $x$ the sum $A$ of the areas of the square and the circle.

In each of Problems 21–26, write an equation of the straight line $L$ described.

**21** $L$ passes through $(-3, 5)$ and $(1, 13)$.  
**22** $L$ passes through $(4, -1)$ and has slope $-3$.  
**23** $L$ has slope $\frac{1}{2}$ and $y$-intercept $-5$.  
**24** $L$ passes through $(2, -3)$ and is parallel to $3x - 2y = 4$.  
**25** $L$ passes through $(-3, 7)$ and also is perpendicular to $y - 2x = 10$.

**26** $L$ is the perpendicular bisector of the segment joining $(1, -5)$ and $(3, -1)$.

Sketch the graphs of the equations and functions given in Problems 27–35.

**27** $2x - 5y = 7$      **28** $|x - y| = 1$  
**29** $x^2 + y^2 = 2x$      **30** $x^2 + y^2 = 4y - 6x + 3$  
**31** $y = 2x^2 - 4x - 1$      **32** $y = 4x - x^2$

**33** $f(x) = \dfrac{1}{x + 5}$      **34** $f(x) = \dfrac{1}{4 - x^2}$

**35** $f(x) = |x - 3|$  
**36** Write an equation of the circle with center $(2, 3)$ that is tangent to the line $x + y + 3 = 0$.

Apply the formula for the derivative of $f(x) = ax^2 + bx + c$ to differentiate the functions in Problems 37–42. Then write an equation for the tangent line to the curve $y = f(x)$ at the point $(1, f(1))$.

**37** $f(x) = 3 + 2x^2$      **38** $f(x) = x - 5x^2$  
**39** $f(x) = 3x^2 + 4x - 5$      **40** $f(x) = 1 - 2x - 3x^2$

**41** $f(x) = (x - 1)(2x + 1)$      **42** $f(x) = \dfrac{x}{3} - \left(\dfrac{x}{4}\right)^2$

In each of Problems 43–49, apply the definition of the derivative directly to find $f'(x)$.

**43** $f(x) = 2x^2 + 3x$      **44** $f(x) = x - x^3$

**45** $f(x) = \dfrac{1}{3 - x}$      **46** $f(x) = \dfrac{1}{2x + 1}$

**47** $f(x) = x - \dfrac{1}{x}$      **48** $f(x) = \dfrac{x}{x + 1}$

**49** $f(x) = \dfrac{x + 1}{x - 1}$

**50** Find the derivative of $f(x) = 3x - x^2 + |2x + 3|$ at the points where it is differentiable. Find the point where it is *not* differentiable. Sketch the graph of $f$.  
**51** Write equations of the two lines through $(3, 4)$ that are tangent to the parabola $y = x^2$. (*Suggestion:* Let $(a, a^2)$ denote either point of tangency; first solve for $a$.)

# 2

# Limits and Continuity

## Limits and the Limit Laws

In Section 1-6 we defined the derivative $f'(a)$ of the function $f$ at the number $a$ to be

$$f'(a) = \lim_{h \to 0} \frac{f(a + h) - f(a)}{h}. \tag{1}$$

The picture that motivated this definition appears in Fig. 2.1, with $a + h$ relabeled as $x$ (so that $h = x - a$). We see that $x$ approaches $a$ as $h$ approaches zero, so (1) can be rewritten in the form

$$f'(a) = \lim_{x \to a} \frac{f(x) - f(a)}{x - a}. \tag{2}$$

Thus the computation of $f'(a)$ amounts to the determination of the limit, as $x$ approaches $a$, of the function

$$F(x) = \frac{f(x) - f(a)}{x - a}. \tag{3}$$

To prepare for the differentiation of more complicated functions than we could manage in Section 1-6, we now turn our attention to the meaning of the statement

$$\lim_{x \to a} F(x) = L. \tag{4}$$

This is read, "The limit of $F(x)$, as $x$ approaches $a$, is $L$." We shall sometimes write (4) in the concise form $F(x) \to L$ as $x \to a$.

To discuss the limit of $F$ at $a$ does not require that $F$ be defined at $a$. Neither its value there, nor the possibility that it has no value at $a$, is of importance. All we require is that $F$ be defined in some **deleted neighborhood**

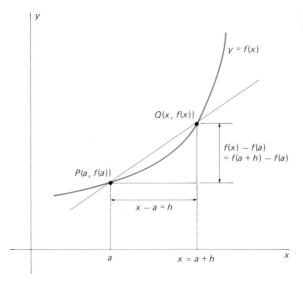

**2.1** The derivative can be defined in this way:

$$f'(a) = \lim_{x \to a} \frac{f(x) - f(a)}{x - a}.$$

of $a$—that is, in a set obtained by deleting the single point $a$ from some open interval containing $a$. For instance, it would suffice for $F$ to be defined on the deleted neighborhood $(a - \frac{1}{2}, a) \cup (a, a + \frac{1}{2})$, though not defined at the point $a$ itself. Note that this is exactly the situation in the case of a function like the one in (3) above, which is defined *except at a*.

The following statement gives the meaning of Equation (4) in intuitive language.

---

*Idea of the Limit*

We say that the number $L$ is the *limit* of $F(x)$ as $x$ approaches $a$ provided that the number $F(x)$ can be made as close to $L$ as one pleases merely by choosing $x$ sufficiently near, though not equal to, the number $a$.

---

What this means, very roughly, is that $F(x)$ tends to get closer and closer to $L$ as $x$ gets closer and closer to $a$. Once we decide how close to $L$ we want $F(x)$ to be, it is necessary that $F(x)$ be that close to $L$ for *all* $x$ sufficiently close to (but not equal to) $a$. The actual definition of the limit makes this requirement precise.

---

*Definition of the Limit*

We say that the number $L$ is the **limit** of $F(x)$ as $x$ approaches $a$ provided that, given any number $\varepsilon > 0$, there exists a number $\delta > 0$ such that

$$|F(x) - L| < \varepsilon$$

for all $x$ such that

$$0 < |x - a| < \delta.$$

---

Figure 2.2 gives a geometric illustration of the meaning of this definition. The points on the graph of $y = F(x)$ that satisfy the inequality $|F(x) - L| < \varepsilon$

**2.2** Geometric illustration of the limit definition.

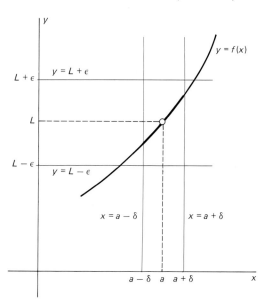

are those that lie between the two (horizontal) lines $y = L - \varepsilon$ and $y = L + \varepsilon$. The points on this graph that satisfy the inequality $|x - a| < \delta$ are those that lie between the two (vertical) lines $x = a - \delta$ and $x = a + \delta$. Consequently the definition implies that $\lim_{x \to a} F(x) = L$ if and only if the following is true. Suppose that the two horizontal lines $y = L - \varepsilon$ and $y = L + \varepsilon$ (with $\varepsilon > 0$) are given in advance. Then it is possible to choose two vertical lines $x = a - \delta$ and $x = a + \delta$ (with $\delta > 0$) with the following property: Every point (with $x \neq a$) on the graph of $y = F(x)$ that lies between the two vertical lines also lies between the two horizontal lines.

An inspection of Fig. 2.2 suggests the likelihood that the closer together are the two horizontal lines, the closer together the two vertical lines will have to be. This is the meaning of "making $F(x)$ closer to $L$ by making $x$ closer to $a$." In some cases in which $\lim_{x \to a} F(x) = L$, we will find that $F(x)$ does not get *steadily* closer to $L$ as $x$ moves steadily toward $a$; this is perfectly permissible, as you can see either by a careful examination of the definition of limit or by study of Fig. 2.2.

In this section we explore the "idea of the limit," mainly through the investigation of specific examples. We defer further discussion of the precise definition of the limit to Section 2-6.

**EXAMPLE 1**   Evaluate $\lim_{x \to 2} x^2$.

*Investigation*   Note that we call this an "investigation" rather than a solution. Our method here is useful for gleaning preliminary information about what the value of a limit might be. This method reinforces our intuitive guess as to what the value of a limit should be, but it does not constitute a proof.

We form a table of values of $x^2$, using values of $x$ approaching $a = 2$. Such a table is shown in Fig. 2.3. We use regularly changing values of $x$ because that makes the behavior exhibited in the table easier to understand. We use values of $x$ that are "nice" in the decimal system for the same reason.

In any case, examine the table—read down the column for $x$, because down is the table's direction for "approaches"—to see what happens to the values of $x^2$. In spite of the fact that we are using only special values of $x$, the table provides convincing evidence that

$$\lim_{x \to 2} x^2 = 4.$$

Surely it is almost certain, and perhaps clear to the intuition, that the limit of $x^2$ is 4 as $x$ approaches 2.

| $x$ | $x^2$ | (Values rounded) |
|---|---|---|
| 2.1 | 4.410 | |
| 1.9 | 3.610 | |
| 2.01 | 4.040 | |
| 1.99 | 3.960 | |
| 2.001 | 4.004 | |
| 1.999 | 3.996 | |
| 2.0001 | 4.0004 | |
| 1.9999 | 3.9996 | |
| 2.00001 | 4.0000 | |
| 1.99999 | 4.0000 | |

**2.3**   What happens to $x^2$ as $x$ approaches 2?

IMPORTANT   *Note that we did not substitute the value $x = 2$ into the function $F(x) = x^2$ to obtain the value 4 of the limit.* While such substitution produces the correct answer in this particular case, in many important limits it produces either no answer at all or else an incorrect answer.

**EXAMPLE 2**   Evaluate $\lim_{x \to 3} \dfrac{x - 1}{x + 2}$.

*Investigation*   Your natural guess is, of course, that the limit is $\frac{2}{5} = 0.4$. The data shown in Fig. 2.4 reinforce this guess. If you experiment with other

**2.4**   Investigating the limit in Example 2

| $x$ | $\dfrac{x - 1}{x + 2}$ | (Values rounded) |
|---|---|---|
| 3.1 | 0.41176 | |
| 3.01 | 0.40120 | |
| 3.001 | 0.40012 | |
| 3.0001 | 0.40001 | |
| 3.00001 | 0.40000 | |
| 3.000001 | 0.40000 | |

Please keep in mind, however, that the computer printed LIMIT REACHED only because we told it to do so, not because it or we know—on the basis of these computations alone—that the limit actually is 2.71828 to five decimal places or even that this limit exists. The numerical results above are only suggestive, not conclusive.

Another question the computer ignores is this: What is meant by the expression $(1 + x)^{1/x}$ if $x$ is *irrational?* This question will be discussed in Chapter 7. There we will see that the limit in (17) does exist and that its value is

$$e \approx 2.718281828459045.$$

This number $e$ plays a very special role in calculus.

> **Exercise 1** The program above uses values of $x$ only on the positive side of zero. Alter it in order to calculate the values of $f(x)$ for $x = -1$, $-0.01$, $-0.001$, and so on.
>
> **Exercise 2** Alter lines 10–40 of the program above so you can use it to carry out the numerical investigation requested in Problem 33 of Section 2-1.
>
> **Exercise 3** Alter lines 10–40 of the program above so you can use it to carry out the numerical investigation requested in Problem 36 of Section 2-1.
>
> **Exercise 4** Alter the program above so you can examine values of $(1 + x)^{1/x}$ as $x \to 0$ through random values. (On most computers, the command X = RND(0) or some variation of this assigns a random number in $[0, 1]$ to the variable X.)

## 2-2

## One-Sided Limits

In Example 5 of Section 2-1 we examined the function

$$f(x) = \frac{x}{|x|} = \begin{cases} 1 & \text{if } x > 0; \\ -1 & \text{if } x < 0. \end{cases}$$

The graph of $y = f(x)$ is shown in Fig. 2.10. We argued that the limit of $f(x) = x/|x|$ as $x \to 0$ does not exist because $f(x)$ approaches $+1$ as $x$ approaches zero from the right, while $f(x) \to -1$ as $x$ approaches zero from the left. A natural way of describing this situation is to say that at $x = 0$ the *right-hand limit* of $f(x)$ is $+1$, while the *left-hand limit* of $f(x)$ is $-1$.

In this section we define and investigate these one-sided limits. Their definitions will be stated initially in the informal language used to describe the "idea of the limit" in Section 2-1. To define the right-hand limit of $f(x)$ at $x = a$, we must assume that $f$ is defined on an open interval immediately to the right of $a$; to define the left-hand limit, we must assume that $f$ is defined on an open interval just to the left of $a$.

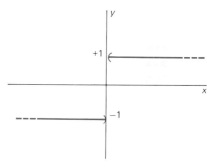

**2.10** The graph of $f(x) = x/|x|$ again

> ### The Right-Hand Limit of a Function
>
> Suppose that $f$ is defined on an open interval $(a, c)$. Then we say that the number $L$ is the **right-hand limit** of $f(x)$ as $x$ approaches $a$, and we write
>
> $$\lim_{x \to a^+} f(x) = L, \tag{1}$$
>
> provided that the number $f(x)$ can be made as close to $L$ as one pleases merely by choosing the point $x$ in $(a, c)$ sufficiently near the number $a$.

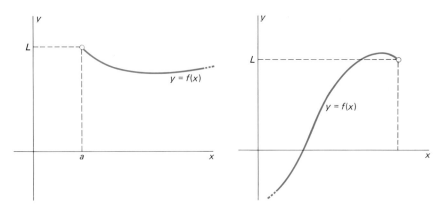

**2.11**   (a) The right-hand limit of $f(x)$
is $L$.
(b) The left-hand limit of $f(x)$ is $L$.

We may describe the right-hand limit in (1) by saying that $f(x) \to L$ as $x \to a^+$, or as $x$ approaches $a$ from the right; the symbol $a^+$ connotes the right-hand, or positive, side of $a$. More precisely, (1) means that, given $\varepsilon > 0$, there exists $\delta > 0$ such that

$$\left| f(x) - L \right| < \varepsilon \tag{2}$$

for all $x$ such that

$$a < x < a + \delta. \tag{3}$$

See Fig. 2.11 for a geometric interpretation of these one-sided limits.

---

### The Left-Hand Limit of a Function

Suppose that $f$ is defined on an open interval $(c, a)$. Then we say that the number $L$ is the **left-hand limit** of $f(x)$ as $x$ approaches $a$, and we write

$$\lim_{x \to a^-} f(x) = L, \tag{4}$$

provided that the number $f(x)$ can be made as close to $L$ as one pleases merely by choosing the point $x$ in $(c, a)$ sufficiently near the number $a$.

---

Note that a consequence of each definition is that the value of $f(a)$ itself is not relevant to the existence or value of the one-sided limits, just as it is not relevant to the existence or value of the two-sided limit.

We may describe the left-hand limit in (4) by saying that $f(x) \to L$ as $x \to a^-$, or as $x$ approaches $a$ from the left; the symbol $a^-$ connotes the left, or negative, side of $a$. We get a precise definition of the left-hand limit in (4) merely by changing the open interval in (3) to $a - \delta < x < a$.

Our preliminary discussion of the function $f(x) = x/|x|$ amounts to saying that its one-sided limits at $x = 0$ are

$$\lim_{x \to 0^+} \frac{x}{|x|} = 1 \quad \text{and} \quad \lim_{x \to 0^-} \frac{x}{|x|} = -1.$$

In Example 5 of Section 2-1 we argued further that, because these two limits are not equal, *the* (two-sided) limit of $x/|x|$ as $x \to 0$ does not exist. The following theorem is readily proved on the basis of precise definitions of all the limits involved.

> **Theorem    One-Sided and Two-Sided Limits**
>
> Suppose that the function $f$ is defined on a deleted neighborhood of the point $a$. Then the limit $\lim\limits_{x \to a} f(x)$ exists and is equal to the number $L$ if and only if the one-sided limits $\lim\limits_{x \to a^+} f(x)$ and $\lim\limits_{x \to a^-} f(x)$ both exist and both equal the number $L$.

Note that this theorem will be particularly useful in proving that certain (two-sided) limits do *not* exist.

**EXAMPLE 1**    The graph of the greatest integer function $f(x) = [\![x]\!]$ is shown in Fig. 2.12. It should be apparent that if $a$ is not an integer, then

$$\lim_{x \to a^+} [\![x]\!] = \lim_{x \to a^-} [\![x]\!] = \lim_{x \to a} [\![x]\!] = [\![a]\!].$$

On the other hand, if $a = n$, an integer, then

$$\lim_{x \to n^-} [\![x]\!] = n - 1 \quad \text{and} \quad \lim_{x \to n^+} [\![x]\!] = n.$$

Because these left-hand and right-hand limits are not equal, it follows from the theorem that the limit of $f(x) = [\![x]\!]$ does not exist as $x$ approaches an integer $n$.

**EXAMPLE 2**    According to the root law in Section 2-1,

$$\lim_{x \to a} \sqrt{x} = \sqrt{a} \qquad \text{if } a > 0.$$

But the limit of $f(x) = \sqrt{x}$ as $x \to 0^-$ is not defined because the square root of a negative number is undefined, and hence $f$ is undefined on a deleted neighborhood of zero. What we *can* say at $a = 0$ is that

$$\lim_{x \to 0^+} \sqrt{x} = 0,$$

although the left-hand limit $\lim\limits_{x \to 0^-} \sqrt{x}$ is not defined.

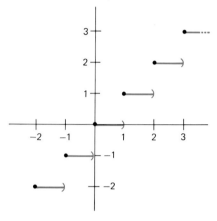

**2.12**  The graph of the greatest integer function $f(x) = [\![x]\!]$

To each of the limit laws stated in Section 2-1, there correspond two *one-sided limit laws*—a right-hand version and a left-hand version. We apply these one-sided limit laws wherever possible to evaluate one-sided limits.

**EXAMPLE 3**    Upon application of the appropriate one-sided limit theorems, we find that

$$\lim_{x \to 3^-} \left( \frac{x^2}{x^2 + 1} + \sqrt{9 - x^2} \right) = \frac{\lim\limits_{x \to 3^-} x^2}{\lim\limits_{x \to 3^-} (x^2 + 1)} + \sqrt{\lim\limits_{x \to 3^-} (9 - x^2)}$$

$$= \frac{9}{9 + 1} + \sqrt{0} = \frac{9}{10}.$$

Note that the (two-sided) limit as $x \to 3$ is not defined here because $\sqrt{9 - x^2}$ is not defined when $x > 3$.

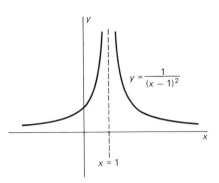

**2.13** The graph of the function $f(x) = 1/(x - 1)^2$

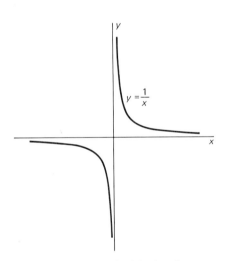

**2.14** The graph of the function $f(x) = 1/x$

## INFINITE LIMITS

In Example 9 of Section 2-1, we investigated the function $f(x) = 1/(x - 1)^2$; the graph of $f$ is shown in Fig. 2.13. The value of $f(x)$ *increases without bound* (eventually exceeds any preassigned number) as $x$ approaches 1 either from the right or from the left. This situation is sometimes described by writing

$$\lim_{x \to 1^-} \frac{1}{(x - 1)^2} = +\infty = \lim_{x \to 1^+} \frac{1}{(x - 1)^2}, \tag{5}$$

and we may say that each of these one-sided limits equals "plus infinity."

CAUTION    Merely to write the expression

$$\lim_{x \to 1^+} \frac{1}{(x - 1)^2} = +\infty \tag{6}$$

does not mean that there exists an "infinite real number" denoted by $+\infty$—there does not! Neither does it mean that the limit on the left-hand side in (6) exists—it does not! To the contrary, the expression in (6) is simply a convenient way of saying *why* the right-hand limit of $1/(x - 1)^2$ as $x \to 1^+$ does not exist—it is because the quantity $1/(x - 1)^2$ increases without bound as $x \to 1^+$.

With similar provisos we may write

$$\lim_{x \to 1} \frac{1}{(x - 1)^2} = +\infty \tag{7}$$

despite the fact that the (two-sided) limit of $1/(x - 1)^2$ as $x \to 1$ does not exist. Our understanding is that the expression in (7) is only a convenient way of saying that the limit in (7) fails to exist because $1/(x - 1)^2$ increases without bound as $x \to 1$ from either side.

Now consider the function $f(x) = 1/x$; its graph is shown in Fig. 2.14. Note that $1/x$ increases without bound as $x$ approaches zero from the right but decreases without bound—eventually becomes less than any preassigned negative number—as $x$ approaches zero from the left. We therefore write

$$\lim_{x \to 0^-} \frac{1}{x} = -\infty \quad \text{and} \quad \lim_{x \to 0^+} \frac{1}{x} = +\infty. \tag{8}$$

**EXAMPLE 4**    Investigate the behavior of the function

$$f(x) = \frac{2x + 1}{x - 1}$$

near the point $x = 1$, where its limit does not exist.

***Solution***    First we look at the behavior of $f(x)$ just to the right of the number 1. If $x$ is greater than 1 but still close to 1, then $2x + 1$ is close to 3, while $x - 1$ is a small *positive* number. In this case the quotient $(2x + 1)/(x - 1)$ will be a large positive number, and the closer $x$ is to 1, the larger (without bound) this positive quotient will be. This reasoning makes it apparent that

$$\lim_{x \to 1^+} \frac{2x + 1}{x - 1} = +\infty, \qquad (9)$$

| $x$ | $\dfrac{2x + 1}{x - 1}$ | $x$ | $\dfrac{2x + 1}{x - 1}$ |
|---|---|---|---|
| 1.1 | 32 | 0.9 | −28 |
| 1.01 | 302 | 0.99 | −298 |
| 1.001 | 3002 | 0.999 | −2998 |
| 1.0001 | 30002 | 0.9999 | −29998 |

as the data in Fig. 2.15 also indicate.

**2.15** The behavior of $f(x) = (2x + 1)/(x - 1)$ for $x$ near 1.

If $x$ is less than 1 but close to 1; then $2x + 1$ is still close to 3, but now $x - 1$ is a small *negative* number. In this case the quotient $(2x + 1)/(x - 1)$ will be a (numerically) large negative number that decreases without bound as $x \to 1^-$. Hence we conclude that

$$\lim_{x \to 1^-} \frac{2x + 1}{x - 1} = -\infty. \qquad (10)$$

The two results in (9) and (10) provide a concise description of the behavior of $f(x) = (2x + 1)/(x - 1)$ near the point $x = 1$. Finally, to remain consistent with the theorem on one-sided and two-sided limits, we say in this case that

$$\lim_{x \to 1} \frac{2x + 1}{x - 1} \text{ does not exist.}$$

## 2-2 PROBLEMS

Use one-sided limit laws to find the limits in Problems 1–20 or to determine that they do not exist.

1. $\lim\limits_{x \to 0^+} (3 - \sqrt{x})$

2. $\lim\limits_{x \to 0^+} (4 + 3x^{3/2})$

3. $\lim\limits_{x \to 1^-} \sqrt{x - 1}$

4. $\lim\limits_{x \to 4^-} \sqrt{4 - x}$

5. $\lim\limits_{x \to 2^+} \sqrt{x^2 - 4}$

6. $\lim\limits_{x \to 3^+} \sqrt{9 - x^2}$

7. $\lim\limits_{x \to 5^-} \sqrt{x(5 - x)}$

8. $\lim\limits_{x \to 2^-} x\sqrt{4 - x^2}$

9. $\lim\limits_{x \to 4^+} \sqrt{\dfrac{4x}{x - 4}}$

10. $\lim\limits_{x \to -3^+} \sqrt{6 - x - x^2}$

11. $\lim\limits_{x \to 5^-} \dfrac{x - 5}{|x - 5|}$

12. $\lim\limits_{x \to -4^+} \dfrac{16 - x^2}{\sqrt{16 - x^2}}$

13. $\lim\limits_{x \to 3^+} \dfrac{\sqrt{x^2 - 6x + 9}}{x - 3}$

14. $\lim\limits_{x \to 2^+} \dfrac{x - 2}{x^2 - 5x + 6}$

15. $\lim\limits_{x \to 2^+} \dfrac{2 - x}{|x - 2|}$

16. $\lim\limits_{x \to 7^-} \dfrac{7 - x}{|x - 7|}$

17. $\lim\limits_{x \to 1^+} \dfrac{1 - x^2}{1 - x}$

18. $\lim\limits_{x \to 0^-} \dfrac{x}{x - |x|}$

19. $\lim\limits_{x \to 5^+} \dfrac{\sqrt{(5 - x)^2}}{5 - x}$

20. $\lim\limits_{x \to -4^-} \dfrac{4 + x}{\sqrt{(4 + x)^2}}$

For each of the functions in Problems 21–30, there is exactly one point $a$ where the right-hand and left-hand limits of $f(x)$ both fail to exist. Describe (as in Example 4) the behavior of $f(x)$ for $x$ near $a$.

21. $f(x) = \dfrac{1}{x - 1}$

22. $f(x) = \dfrac{2}{3 - x}$

23. $f(x) = \dfrac{x - 1}{x + 1}$

24. $f(x) = \dfrac{2x - 5}{5 - x}$

25. $f(x) = \dfrac{1 - x^2}{x + 2}$

26. $f(x) = \dfrac{1}{(x - 5)^2}$

27. $f(x) = \dfrac{|1 - x|}{(1 - x)^2}$

28. $f(x) = \dfrac{x + 1}{x^2 + 6x + 9}$

29. $f(x) = \dfrac{x - 2}{4 - x^2}$

30. $f(x) = \dfrac{x - 1}{x^2 - 3x + 2}$

In Problems 31–35, $[\![x]\!]$ denotes the greatest integer function. First find the limits $\lim\limits_{x \to n^-} f(x)$ and $\lim\limits_{x \to n^+} f(x)$ for each integer $n$ (the answer will normally be in terms of $n$); then sketch the graph of $f$.

31. $f(x) = 2$ if $x$ is not an integer; $f(x) = 2 + (-1)^x$ if $x$ is an integer.

32. $f(x) = x$ if $x$ is not an integer; $f(x) = 0$ otherwise.

33. $f(x) = (-1)^{[\![x]\!]}$

34. $f(x) = \left[\!\left[ \dfrac{x}{2} \right]\!\right]$

35. $f(x) = 1 + [\![x]\!] - x$

## Combinations of Functions and Inverse Functions

In order to find a limit of a function, we must analyze the definition or construction of the function—the way it is "put together." Many important functions are built of simpler ones, and the results can be quite complex. Here we discuss some of the ways of combining functions to obtain new ones.

Suppose that $f$ and $g$ are functions, and that $c$ is a fixed number. The **(scalar) multiple** $cf$, the **sum** $f + g$, the **difference** $f - g$, the **product** $f \cdot g$, and the **quotient** $f/g$ are the new functions determined by the formulas

$$(cf)(x) = cf(x), \tag{1}$$

$$(f + g)(x) = f(x) + g(x), \tag{2}$$

$$(f - g)(x) = f(x) - g(x), \tag{3}$$

$$(f \cdot g)(x) = f(x) \cdot g(x), \qquad \text{and} \tag{4}$$

$$\left(\frac{f}{g}\right)(x) = \frac{f(x)}{g(x)}. \tag{5}$$

The combinations in (2)–(4) are defined for every number $x$ in both the domain of $f$ and the domain of $g$. In (5), we must require that $g(x) \neq 0$.

For example, let $f(x) = x^2 + 1$ and $g(x) = x - 1$. Then

$$(3f)(x) = 3(x^2 + 1),$$

$$(f + g)(x) = (x^2 + 1) + (x - 1) = x^2 + x,$$

$$(f - g)(x) = (x^2 + 1) - (x - 1) = x^2 - x + 2,$$

$$(f \cdot g)(x) = (x^2 + 1)(x - 1) = x^3 - x^2 + x - 1, \qquad \text{and}$$

$$(f/g)(x) = \frac{x^2 + 1}{x - 1} \qquad (x \neq 1).$$

If $f(x) = \sqrt{1 - x}$ for $x \leq 1$ and $g(x) = \sqrt{1 + x}$ for $x \geq -1$, then the sum or product of $f$ and $g$ is defined where *both* $f$ and $g$ are defined. In this case, the domain would be the closed interval $[-1, 1]$.

You can build very complicated functions from simple functions using only the operations of Equations (1)–(5). For example, from the *identity* function $I(x) = x$ and the constant function $1(x) = 1$ (a useful abuse of notation), you can construct the function $f(x) = x^2 - 2x - 3$ by forming

$$f = I \cdot I - 2 \cdot I - 3.$$

In fact, you can "build" *any* polynomial in similar fashion, for

$$p(x) = a_n x^n + a_{n-1} x^{n-1} + \cdots + a_1 x + a_0$$

is the sum and product of constant functions and the identity function. By dividing the polynomial $p(x)$ by $q(x)$, you obtain also the **general rational function**

$$r(x) = \frac{p(x)}{q(x)},$$

a quotient of two polynomials.

## COMPOSITION OF FUNCTIONS

Some important functions cannot be constructed in any of the ways suggested above; for example, the function

$$k(x) = \sqrt{1 - x^2}, \qquad -1 \leqq x \leqq 1.$$

The graph of $k$ is the upper half of the circle with equation $x^2 + y^2 = 1$, as shown in Fig. 2.16.

Certainly $f(x) = \sqrt{x}$ and $g(x) = 1 - x^2$ are simple functions. But they cannot be added, subtracted, multiplied, or divided to produce $k(x)$. On the other hand, if you start with $x$, let $g$ act upon it,

$$x \xrightarrow[g]{} 1 - x^2$$

and next use the *output* of $g$ as the *input* for $f$,

$$x \xrightarrow[g]{} 1 - x^2 \xrightarrow[f]{} \sqrt{1 - x^2},$$

then the *net* effect is that $x$ is transformed into $\sqrt{1 - x^2}$. That is exactly how $k$ is supposed to act on the input $x$.

When you think of a function as a sort of calculator, one that accepts values of $x$ and produces values of the function, then the way we have combined $f(x) = x^{1/2}$ and $g(x) = 1 - x^2$ to produce $k(x)$ could be represented in the way illustrated in Fig. 2.17. Moreover, a natural way to write $k$ in terms of $f$ and $g$ would be this:

$$k(x) = f(g(x)).$$

The expression $g(x)$ appears as the input or argument of $f$ to signify that the *value* $g(x)$ is to be acted on by the function $f$. This is exactly what we need to produce the effect of the function $k$.

WARNING    The order in which the two functions $g$ and $f$ are to be applied is this: *first $g$, then $f$*. Though the notation $k(x) = f(g(x))$ shows $f$ appearing first, you should note upon close examination that *$f$ is to be applied last.*

The way in which $f$ and $g$ are combined to form $k$ is called the *composition* of $f$ and $g$ and is perhaps the most important way in which two functions can be combined. The reason is that many human activities are performed in the same way. Certain operations are to be performed in a certain order and each step acts on the result of the previous step: It's much like assembling an automobile, making a pizza, dancing, or learning calculus.

What about the domain of $f(g(x))$? Suppose that $g$ has domain $D$ and range $V$ and that $f$ has domain $C$ that at least partly overlaps $V$, as we show in Fig. 2.18. Then the **composition** of $f$ and $g$ is denoted by $f(g)$, and its value at $x$ is given by

$$(f(g))(x) = f(g(x)). \tag{6}$$

The domain $E$ of $f(g)$ is the set of all numbers in $D$ that give values in $C$.

Another common notation for the composition is $f \circ g$, with a small circle between $f$ and $g$.

The composition notation is very different from the product notation. The product of the two functions $f$ and $g$ is written $f \cdot g$, or simply $fg$. But

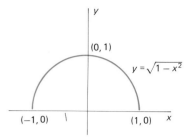

2.16    The graph of $k$

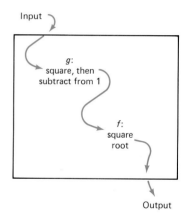

2.17    The action of $k(x) = \sqrt{1 - x^2}$

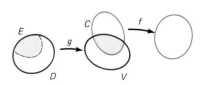

2.18    The domain $E$ of $f(g)$

the composition of $f$ and $g$ may involve no multiplication whatsoever and is written $f(g)$. This is read "$f$ of $g$," and its formula $f(g(x))$ is read "$f$ of $g$ of $x$."

A sure way to see that products differ from compositions is this: The product of $f$ and $g$ is the same whether $f$ or $g$ is written first. But when we reverse the order of composition, a different function is usually the result. With $f(x) = x^{1/2}$ and $g(x) = 1 - x^2$ as above, we find that

$$g(f(x)) = 1 - (\sqrt{x})^2 = 1 - x, \qquad x \geqq 0.$$

This is not at all the same as

$$f(g(x)) = \sqrt{1 - x^2}, \qquad -1 \leqq x \leqq 1.$$

**EXAMPLE 1**  Find $f(g)$ and $g(f)$, given $f(x) = x^3$ and $g(x) = x - 2$.

*Solution*  First, $f(g)$ has the rule

$$f(g(x)) = f(x - 2) = (x - 2)^3.$$

The domain of $g$ is all of $\mathscr{R}$. Since every possible value of $g$ lies in the domain (also $\mathscr{R}$) of $f$, the domain of $f(g)$ is just $\mathscr{R}$.

Next, $g(f)$ has the rule

$$g(f(x)) = g(x^3) = x^3 - 2,$$

and its domain is also all of $\mathscr{R}$.

Note that the two possible results of composition in Example 1 are *not* the same function. In addition, neither is equal to the product of $f$ and $g$:

$$(f \cdot g)(x) = f(x)g(x) = x^3(x - 2).$$

Sometimes we are given a function $h$ and want to find two simpler functions $f$ and $g$ whose composition is $h$; that is, $h(x) = f(g(x))$. We can think of $f$ and $g$ as "machines," as illustrated in Fig. 2.19. The "inner" function $g$

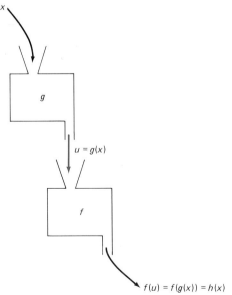

**2.19**  The composition of $f$ and $g$

$x$

$g$

$u = g(x)$

$f$

$f(u) = f(g(x)) = h(x)$

accepts $x$ as input and produces the output $u = g(x)$. The "outer" function $f$ represents the final step in manufacturing the result $h(x)$; it accepts the output $u = g(x)$ as its input and produces the ultimate output $f(u) = f(g(x)) = h(x)$. Thus the sequence of operations is

$$x \xrightarrow{\ g\ } g(x) = u \xrightarrow{\ f\ } f(u) = f(g(x)).$$

**EXAMPLE 2**  Given the function $h(x) = (x^2 + 4)^{3/2}$, find two functions $f$ and $g$ such that $f(g(x)) = h(x)$.

*Solution*  It is technically correct, but useless, simply to let $g(x) = x$ and $f(u) = (u^2 + 4)^{3/2}$. We seek a nontrivial solution here. So we observe that to calculate $(x^2 + 4)^{3/2}$, we first must calculate $x^2 + 4$, so we choose $g(x) = x^2 + 4$ as our inner function. The last step is to raise $g(x)$ to the power $3/2$, so we take $f(u) = u^{3/2}$ as our outer function. Thus if

$$f(x) = x^{3/2} \quad \text{and} \quad g(x) = x^2 + 4,$$

then $f(g(x)) = f(x^2 + 4) = (x^2 + 4)^{3/2} = h(x)$.

Alternatively, we could have decided first to take $f(x) = x^{3/2}$, because the last operation in calculating a value of $(x^2 + 4)^{3/2}$ is to raise a number to the power $3/2$. In this case we ask what $u = g(x)$ should be in order that

$$f(u) = u^{3/2} = (x^2 + 4)^{3/2} = h(x).$$

If we raise each of the terms above to the power $2/3$, we get $u = x^2 + 4$, so we take $g(x) = x^2 + 4$.

## INVERSE FUNCTIONS

We often encounter pairs of functions that are natural opposites of one another, in the sense that each undoes the result of applying the other. We show a schematic of this phenomenon in Fig. 2.20, but actual examples may be more informative: In Example 3 we give some pairs of functions that are inverses of each other. The domain of every function is, unless otherwise specified, the set of all real numbers for which its formula is meaningful.

**EXAMPLE 3**  The following are some pairs of inverse functions.

$$f(x) = x + 1 \quad \text{and} \quad g(x) = x - 1$$

(Adding 1 and subtracting 1 are inverse operations; doing one undoes the other.)

$$f(x) = 2x \quad \text{and} \quad g(x) = \frac{x}{2}$$

(Doubling and halving are inverse operations.)

$$f(x) = \frac{1}{x} \quad \text{and} \quad g(x) = \frac{1}{x}$$

(A function can be its own inverse.)

$$f\colon (0, \infty) \to \mathscr{R} \quad \text{and} \quad g\colon (0, \infty) \to \mathscr{R}$$
$$\text{by } f(x) = x^2 \qquad \text{by } g(x) = \sqrt{x}$$

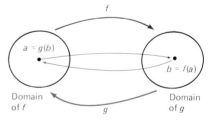

**2.20**  The action of inverse functions $f$ and $g$

(Squaring and taking the square root are inverse operations when only positive numbers are involved.)

Each pair $f$ and $g$ of functions given above has the property that

$$f(g(x)) = x \quad and \quad g(f(x)) = x. \tag{7}$$

When this pair of equations holds for all values of $x$ in the appropriate domains, then we say that $f$ and $g$ are *inverses* of each other.

---

**Definition** *Inverse Functions*

Two functions $f : A \to B$ and $g : D \to E$ are said to be **inverse functions,** or **inverses** of each other, provided that:

(i) The domain of each is the range of the other — $A = E$ and $D = B$; and
(ii) $f(g(x)) = x$ for all $x$ in $D$ and $g(f(x)) = x$ for all $x$ in $A$.

---

Condition (i) of this definition implies that the range of values of $f$ is the domain of definition of $g$ and that the range of $g$ is the domain of $f$. Care frequently is required in specifying the domains of $f$ and $g$ to insure that this condition is satisfied. For example, the rule $f(x) = x^2$ makes sense for all real $x$, but it is not the same as the function

$$f(x) = x^2, \qquad x \geq 0$$

that is the inverse of $g(x) = \sqrt{x}, x \geq 0$. On the other hand, the functions

$$f(x) = x^3 \quad and \quad g(x) = x^{1/3}$$

are inverses of each other with no restriction on their domains.

Condition (ii) of the definition is the same as the statement that

$$f(x) = y \quad if \text{ and only if} \quad g(y) = x$$

for all $x$ in the domain of $f$ and all $y$ in the domain of $g$. That is, the graphs of the two equations

$$y = f(x) \quad and \quad x = g(y)$$

are *identical*. On the other hand, the graphs of the two equations

$$x = g(y) \quad and \quad y = g(x)$$

are *reflections* of one another across the line $y = x$. This is shown in Fig. 2.21. The reason for this reflection phenomenon is that either of the last two equations may be obtained from the other simply by interchanging $x$ and $y$; geometrically, this corresponds to interchanging the $x$-axis with the $y$-axis (provided that the same units of measurement are used on the two axes). This is the reason for the "one-dimensional mirror" along the line $y = x$. Theorem 1 follows.

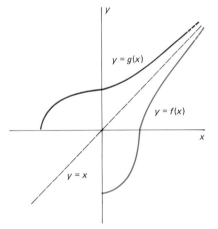

**2.21** The graphs of two inverse functions are reflections of one another through the line $y = x$.

---

**Theorem 1** *Reflection Property of Inverse Functions*

Two functions $f$ and $g$ are inverse functions if and only if their graphs are reflections of each other across the line $y = x$.

---

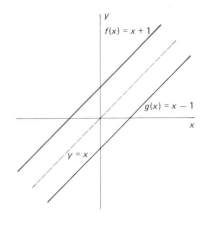

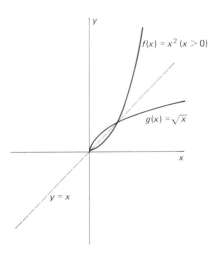

**2.22** The inverse function pair
$f(x) = x + 1$ and $g(x) = x - 1$

**2.23** The inverse function pair
$f(x) = 2x$ and $g(x) = x/2$

**2.24** The inverse function pair $f(x) = x^2$
$(x > 0)$ and $g(x) = \sqrt{x}$ $(x > 0)$

We illustrate this property with the graphs of the pairs of inverse functions of Example 2, shown in Figs. 2.22–2.25.

The reflection property gives us a criterion for determining whether a given function has an inverse function. Suppose that $f$ and $g$ are inverses of each other. Then each vertical line intersects the graph of $g$ in at most one point. It then follows from the reflection property that each horizontal line intersects the graph of $f$ in at most one point. Consequently, $f$ is a **one-to-one** function; that is, given $y_1$ in the range of $f$, there is only one number $x_1$ in the domain of $f$ such that $f(x_1) = y_1$. By reasoning along such lines, one can construct a proof of the following theorem.

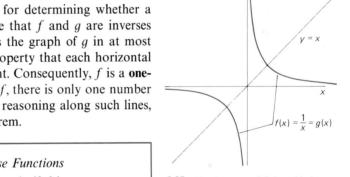

**2.25** The inverse of $f(x) = 1/x$ is $g(x) = 1/x = f(x)$.

---

**Theorem 2** *One-to-One Property of Inverse Functions*

The function $f$ has an inverse function if and only if $f$ is a one-to-one function.

---

If the function $f$ defined by the formula $y = f(x)$ has an inverse function $g$, then we can attempt to find a formula for $g$ by solving the equation $y = f(x)$ for $x = g(y)$.

**EXAMPLE 4** The function

$$f(x) = \frac{1}{x^2 + 1}, \qquad x \geqq 0,$$

is one-to-one (because $0 \leqq a < b$ implies $0 \leqq a^2 < b^2$). Find its inverse function $g$.

**Solution** We solve the equation $y = 1/(x^2 + 1)$ and obtain

$$x^2 + 1 = \frac{1}{y}$$

so that

$$x = \sqrt{\frac{1}{y} - 1} = \sqrt{\frac{1 - y}{y}} = g(y).$$

We take the positive root because $x \geq 0$. The domain of $g$ is the **range** (0, 1] of $f$. Hence, taking $x$ as the independent variable, we find that $g$ is the function

$$g(x) = \sqrt{\frac{1 - x}{x}}, \qquad 0 < x \leq 1.$$

## 2-3  PROBLEMS

In Problems 1–5, find $f + g$, $f \cdot g$, and $f/g$, and give the domain of each of these new functions.

**1** $f(x) = x + 1$, $g(x) = x^2 + 2x - 3$

**2** $f(x) = \dfrac{1}{x - 1}$, $g(x) = \dfrac{1}{2x + 1}$

**3** $f(x) = \sqrt{x}$, $g(x) = \sqrt{x - 2}$

**4** $f(x) = \sqrt{x + 1}$, $g(x) = \sqrt{5 - x}$

**5** $f(x) = \sqrt{x^2 - 1}$, $g(x) = 1/\sqrt{4 - x^2}$

In Problems 6–10, find $f(g(x))$ and $g(f(x))$.

**6** $f(x) = 1 - x^2$, $g(x) = 2x + 3$

**7** $f(x) = -17$, $g(x) = |x|$

**8** $f(x) = \sqrt{x^2 - 3}$, $g(x) = x^2 + 3$

**9** $f(x) = x^2 + 1$, $g(x) = \dfrac{1}{x^2 + 1}$

**10** $f(x) = x^3 - 4$, $g(x) = \sqrt[3]{x + 4}$

In each of Problems 11–20, find a function of the form $f(x) = x^k$ (you must specify $k$) and a function $g(x)$ such that $f(g(x)) = h(x)$.

**11** $h(x) = (2 + 3x)^2$      **12** $h(x) = (4 - x)^3$

**13** $h(x) = \sqrt{2x - x^3}$      **14** $h(x) = (1 + x^4)^{17}$

**15** $h(x) = (5 - x^2)^{3/2}$      **16** $h(x) = \sqrt[3]{(4x - 6)^4}$

**17** $h(x) = \dfrac{1}{x + 1}$      **18** $h(x) = \dfrac{1}{1 + x^2}$

**19** $h(x) = \dfrac{1}{\sqrt{x + 10}}$      **20** $h(x) = \dfrac{1}{(1 + x + x^2)^3}$

In Problems 21–25, find $g$ such that $f(g(x)) = h(x)$.

**21** $f(x) = 2x + 3$, $h(x) = 2x + 5$

**22** $f(x) = x + 1$, $h(x) = x^3$

**23** $f(x) = x^2$, $h(x) = x^4 + 1$

**24** $f(x) = \dfrac{1}{x}$, $h(x) = x^5$

**25** $f(x) = \sqrt{x - 2}$, $h(x) = \sqrt{x + 2}$

In Problems 26–30, determine whether $f$ and $g$ are inverse functions.

**26** $f(x) = 4x - 5$, $g(x) = \dfrac{x + 5}{4}$

**27** $f(x) = 2x + 3$, $g(x) = \dfrac{x + 3}{2}$

**28** $f(x) = x^2 + 3$, $g(x) = \sqrt{x + 3}$

**29** $f(x) = \dfrac{1}{\sqrt{x + 1}}$, $g(x) = \dfrac{1 - x^2}{x^2}$

**30** $f(x) = x^3 + 1$, $g(x) = \sqrt[3]{x - 1}$

In Problems 31–36, find the inverse function of $f$.

**31** $f(x) = 13x - 5$

**32** $f(x) = \sqrt{3x + 7}$

**33** $f(x) = 3x^3 + 1$

**34** $f(x) = x^2 - 4$, $0 \leq x \leq 2$

**35** $f(x) = (x^3 - 1)^{17}$, $0 \leq x \leq 1$

**36** $f(x) = \dfrac{x + 1}{x - 1}$

**37** Suppose that $f(x) = ax + b$ ($a \neq 0$). Prove that $f$ has an inverse function defined for all $x$.

**38** Suppose that $f(x) = x^2 + bx + c$ with domain the set of all real numbers. Prove that $f$ does not have an inverse function.

**39** Suppose that $g_1$ and $g_2$ are each inverse functions of $f$. Prove that $g_1 = g_2$. That is, prove that $g_1$ and $g_2$ have the same domain and that $g_1(x) = g_2(x)$ for each $x$ in that domain.

**40** The function $f$ is called an **increasing function** provided that $x_1 < x_2$ implies $f(x_1) < f(x_2)$ for all $x_1$ and $x_2$ in the domain of $f$. Prove that every increasing function has an inverse function.

**41** Apply the result of Problem 40 to show that $f(x) = x^3 + 2x$ has an inverse function.

**42** Example 1 and the discussion preceding it give examples of pairs of functions $f$ and $g$ such that $f(g(x)) \neq g(f(x))$. Give two more such examples.

**43** Find two different pairs of functions $f$ and $g$ such that $f(g(x)) = g(f(x))$ for all $x$.

**44** Find two different pairs of functions $f$ and $g$ such that $f$ and $g$ are inverses of each other.

**45** It is clear how the composition $k(x) = f(g(h(x)))$ of three functions can be defined. Express $k(x) = (1 + x^3)^{1/2}$ as the composition of three functions none of which is the identity function $I(x) = x$.

## Continuous Functions

When we stated the substitution law for limits in Section 2-1, we needed for its validity the condition

$$\lim_{x \to a} f(x) = f(a) \tag{1}$$

on the outer function in the composition $f(g(x))$. (The law as originally stated has $L$ in place of $a$, but of course this is unimportant.) A function that satisfies Equation (1) is said to be *continuous* at the number $a$.

---

*Definition of Continuity at a Point*

Suppose that the function $f$ is defined in a neighborhood of $a$. We say that $f$ is **continuous at $a$** provided that $\lim_{x \to a} f(x)$ exists and, moreover, that the value of this limit is $f(a)$; that is, that Equation (1) holds.

---

Briefly, continuity of $f$ at $a$ means that the limit of $f$ at $a$ is equal to its value there. Another way to put it is this: The limit of $f$ at $a$ is its "expected" value—the value that you would assign if you knew the values of $f$ in a deleted neighborhood of $a$ and you knew $f$ to be "predictable." Alternatively, continuity of $f$ at $a$ means this: When $x$ is close to $a$, then $f(x)$ is close to $f(a)$.

Analyzing this definition, we see that in order to be continuous at the point $a$, the function $f$ must satisfy the following three conditions:

**1** $f$ must be defined at $a$ (so that $f(a)$ exists);
**2** the limit of $f(x)$ as $x$ approaches $a$ must exist; and
**3** $\lim_{x \to a} f(x)$ must equal $f(a)$.

If any one of these conditions is not satisfied, then $f$ is not continuous at $a$.

Suppose that $f$ is defined on an open interval or on a union of open intervals. Then we say simply that $f$ is **continuous** if it is continuous at each point of its domain of definition. The definition of continuity in terms of limits makes precise the intuitive idea of a function with a graph that is a connected curve—connected in the sense that it can be traced with a "continuous" motion of the pencil without lifting the pencil from the paper.

If the function $f$ is *not* continuous at $a$, we say that it is **discontinuous** there, or that $a$ is a **discontinuity** of $f$. Intuitively, a discontinuity of $f$ is a point where its graph has a "gap" or "jump" of some sort. For instance, the function

$$f(x) = \begin{cases} 1 & \text{if } x \neq 0; \\ 0 & \text{if } x = 0, \end{cases}$$

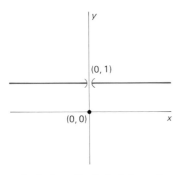

**2.26** A discontinuity in $f$ at $x = 0$

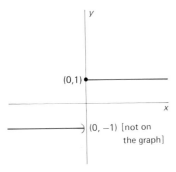

**2.27** A discontinuity in $g$ at $x = 0$

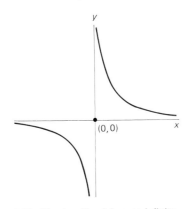

**2.28** The function $h$ has an infinite discontinuity at $x = 0$.

(its graph appears in Fig. 2.26) jumps in value from 1 to 0 and back as $x$ increases through the value zero. And $f$ is not continuous at $x = 0$ because, as noted in Example 6 of Section 2-1,

$$\lim_{x \to 0} f(x) = 1 \neq 0 = f(0);$$

the limit of $f$ at 0 is not equal to its value there.

Also, the function

$$g(x) = \begin{cases} 1 & \text{if } x \geq 0; \\ -1 & \text{if } x < 0, \end{cases}$$

is discontinuous at $x = 0$. As its graph (in Fig. 2.27) shows, there is a jump from $-1$ to 1 at 0. We saw in Example 5 of Section 2-1 that the limit of $g(x)$ as $x \to 0$ does not exist.

The graph of

$$h(x) = \begin{cases} \dfrac{1}{x} & \text{if } x \neq 0; \\ 0 & \text{if } x = 0, \end{cases}$$

appears in Fig. 2.28. This function has what might be called an "infinite discontinuity" at zero. Finally, the greatest integer function $f(x) = [\![x]\!]$, which has the graph shown in Fig. 2.12, is continuous at every *noninteger* point but discontinuous at every integer point.

We are more interested in functions that *are* continuous; most (though not all) of the functions we study in calculus are continuous on their natural domains of definition. For example, it follows readily from the sum and product laws for limits that *every polynomial function*

$$p(x) = b_n x^n + b_{n-1} x^{n-1} + \cdots + b_1 x_1 + b_0$$

*is continuous at every point of the real line;* in short, every polynomial is continuous everywhere.

If $p(x)$ and $q(x)$ are polynomials, then the quotient law for limits and the continuity of polynomials imply that

$$\lim_{x \to a} \frac{p(x)}{q(x)} = \frac{\lim_{x \to a} p(x)}{\lim_{x \to a} q(x)} = \frac{p(a)}{q(a)}$$

provided that $q(a) \neq 0$. Thus every rational function $f(x) = p(x)/q(x)$ is continuous wherever it is defined—that is, wherever the denominator polynomial $q(x)$ is nonzero.

**EXAMPLE 1** Suppose that

$$f(x) = \frac{x - 2}{x^2 - 3x + 2}. \tag{2}$$

We factor the denominator: $x^2 - 3x + 2 = (x - 1)(x - 2)$; this shows that $f$ is not defined either at $x = 1$ or at $x = 2$. Thus the rational function $f$ defined in (2) is continuous except at these two points. In the case $x \neq 2$, we may cancel $x - 2$ to obtain

$$f(x) = \frac{x - 2}{(x - 1)(x - 2)} = \frac{1}{x - 1}. \tag{3}$$

But $1/(x - 1)$ *is* defined at $x = 2$; its value there is 1. This tells us that although the function $f$ as originally defined in (2) is discontinuous at both $x = 1$ and $x = 2$, it can be given the value $f(2) = 1$ in order to "remove" the discontinuity at $x = 2$. Thus $x = 1$ is the only point where $f$ cannot be defined in order to be continuous—it is obvious that the one-sided limits of $1/(x - 1)$ as $x \to 1$ do not exist.

The following theorem has a valuable consequence: Those functions that are built by forming compositions of continuous functions are themselves continuous.

---

**Theorem 1   Continuity of Compositions**

The composition of two continuous functions is itself continuous. More precisely, if $g$ is continuous at $a$ and $f$ is continuous at $g(a)$, then $f(g(x))$ is continuous at $a$.

---

**Proof**   The continuity of $g$ at $a$ means that $g(x) \to g(a)$ as $x \to a$, and the continuity of $f$ at $g(a)$ implies that $f(g(x)) \to f(g(a))$ as $g(x) \to g(a)$. Hence the substitution law for limits (Section 2-1) yields

$$\lim_{x \to a} f(g(x)) = f\left( \lim_{x \to a} g(x) \right) = f(g(a)),$$

as desired.   ∎

The following theorem is normally proved in a course in advanced calculus; it tells us that the inverse of a continuous function is continuous.

---

**Theorem 2   Continuity of Inverse Functions**

Let $f$ be a continuous function with domain the open interval $I$. Suppose that $f$ has an inverse function $g$ with domain the open interval $J$. Then $g$ is continuous (at each point of $J$).

---

For example, consider the $n$th power function $f(x) = x^n$ (here, $n$ is a positive integer). If $n$ is odd, we take $I$ as the whole real line $(-\infty, +\infty)$. If $n$ is even, we take $I$ to be the positive half-line $(0, +\infty)$. Then $f$ has as its inverse function the $n$th-root function $g(x) = \sqrt[n]{x}$. The domain of $g$ is the interval $J$, where $J = (-\infty, +\infty)$ if $n$ is odd and $J = (0, +\infty)$ if $n$ is even. Since $f(x) = x^n$ is continuous on $I$, Theorem 2 implies that $g(x) = \sqrt[n]{x}$ is continuous on $J$. Thus

$$\lim_{x \to a} \sqrt[n]{x} = \sqrt[n]{a},$$

with the stipulation that $a > 0$ if $n$ is even. This proves the root law of Section 2-1.

We may combine this result with Theorem 1. Then we see that a root of a continuous function is continuous where defined. That is, the composition

$$g(x) = [f(x)]^{1/n}$$

is continuous at $a$ if $f$ is, assuming that $f(a) > 0$ if $n$ is even (so that $\sqrt[n]{f(a)}$ will be defined).

**EXAMPLE 2** Show that the function

$$f(x) = \left(\frac{x - 7}{x^2 + 2x + 2}\right)^{2/3}$$

is continuous on the whole real line.

*Solution* Note first that the denominator

$$x^2 + 2x + 2 = (x + 1)^2 + 1$$

is never zero. Hence the rational function $r(x) = (x - 7)/(x^2 + 2x + 2)$ is defined and continuous everywhere. It then follows from Theorem 1 and the continuity of the cube root function that

$$f(x) = ([r(x)]^2)^{1/3}$$

is continuous everywhere. Hence (for example)

$$\lim_{x \to -1} \left(\frac{x - 7}{x^2 + 2x + 2}\right)^{2/3} = f(-1) = (-8)^{2/3} = 4.$$

## CONTINUOUS FUNCTIONS ON CLOSED INTERVALS

An applied problem often involves a function with a domain that is a *closed interval*. For example, in the fence problem we found that the area of a rectangle with perimeter 140 and base $x$ is $f(x) = x(70 - x)$. Though the formula for $f$ is meaningful for all $x$, only values of $x$ in the closed interval $[0, 70]$ correspond to rectangles and are thereby pertinent to the stated problem.

The function $f$ defined on the closed interval $[a, b]$ is said to be **continuous on** $[a, b]$ provided that it is continuous at each point of the open interval $(a, b)$ *and* that

$$\lim_{x \to a^+} f(x) = f(a) \quad \text{and} \quad \lim_{x \to b^-} f(x) = f(b).$$

The last two conditions mean that at each endpoint, the value of the function is equal to its limit from within the interval. For instance, every polynomial is continuous on any closed interval.

Continuous functions defined on closed intervals are very special: Every such function has the *maximum value property* and the *intermediate value property* that are described in Theorems 3 and 4 of this section. Proofs of these theorems are given in Appendix B.

---

*Definition*    *Maximum and Minimum Value*

If $c$ is in $[a, b]$, then $f(c)$ is called the **maximum value** of $f(x)$ on $[a, b]$ if $f(c) \geq f(x)$ for all $x$ in $[a, b]$. Similarly, if $d$ is in $[a, b]$, then the value $f(d)$ is called the **minimum value** of $f(x)$ on $[a, b]$ if $f(d) \leq f(x)$ for all $x$ in $[a, b]$.

---

Thus the maximum value of $f(x)$ is simply a value no less than any other value of $f(x)$. The point of the next theorem is not merely that there is a number $M$ such that $M \geq f(x)$ for all $x$, but that if $f$ is continuous and its domain is a closed interval, then $M$ is actually a *value* of $f(x)$.

---

**Theorem 3    Maximum Value Property**

If $f$ is continuous on the closed interval $[a, b]$, then there exists a number $c$ in $[a, b]$ such that $f(c)$ is the maximum value of $f$ on $[a, b]$.

---

By applying this result to the function $g(x) = -f(x)$, we can see that $f(x)$ must also attain a minimum value at some point of the interval. In short, a continuous function on a closed interval attains both a maximum value and a minimum value at points of the interval.

In Chapter 1 we saw that both the fence problem and the refrigerator problem amount to finding the maximum value of $f(x) = x(70 - x)$ for $x$ in $[0, 70]$. Now we see that it is the continuity of $f$ on the closed interval $[0, 70]$ that insures that the maximum value of $f(x)$ exists and is attained at some point of the interval $[0, 70]$.

Suppose that the function $f$ is defined on the interval $I$. The following examples show that if *either* $f$ is not continuous *or* $I$ is not closed, then $f$ may fail to attain maximum and minimum values at points of $I$. Thus both hypotheses in Theorem 3 are necessary.

**EXAMPLE 3**    Let the continuous function $f(x) = 2x$ be defined only for $0 \leq x < 1$, so that its domain of definition is a half-open interval rather than a closed interval. From the graph shown in Fig. 2.29, it is clear that $f$ attains its minimum value 0 at $x = 0$. But $f(x) = 2x$ attains *no* maximum value at any point of $[0, 1)$—the only candidate for a maximum value is 2 at $x = 1$, but $f(1)$ is not defined.

**EXAMPLE 4**    The function $f$ defined on the closed interval $[0, 1]$ with formula

$$f(x) = \begin{cases} \dfrac{1}{x} & \text{if } 0 < x \leq 1; \\ 1 & \text{if } x = 0, \end{cases}$$

is not continuous on $[0, 1]$ because $\lim_{x \to 0^+} (1/x)$ does not exist (see Fig. 2.30). This function does attain its minimum value of 1 at $x = 0$ and also at $x = 1$, but attains no maximum value on $[0, 1]$ because $1/x$ can be made arbitrarily large by choosing $x$ positive but very close to zero.

---

For a variation on Example 4, the function $g(x) = 1/x$ with domain the *open* interval $(0, 1)$ attains neither a maximum nor a minimum value there.

We suggested earlier that continuity of a function is related to the possibility of tracing its graph without lifting the pencil from the paper. The following theorem expresses this fact more precisely.

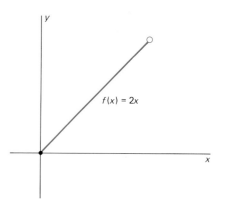

**2.29**    The graph of the function of Example 3

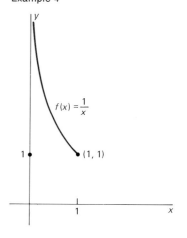

**2.30**    The graph of the function of Example 4

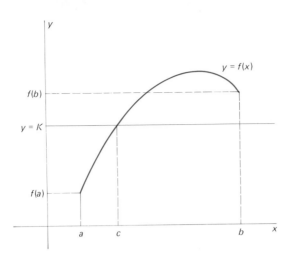

2.31 The continuous function f attains the intermediate value K at x = c.

---

**Theorem 4**    *Intermediate Value Property*

Suppose that the function $f$ is continuous on the closed interval $[a, b]$. Then $f(x)$ assumes every intermediate value between $f(a)$ and $f(b)$. Thus if $K$ is any number between $f(a)$ and $f(b)$, then there exists at least one point $c$ in $(a, b)$ such that $f(c) = K$.

---

Figure 2.31 shows the graph of a fairly typical continuous function $f(x)$ with domain the closed interval $[a, b]$. The number $K$ is located on the $y$-axis, somewhere between $f(a)$ and $f(b)$. In the figure, we have $f(a) < f(b)$, but that is not essential. The horizontal line through $K$ must cut the graph of $f$ somewhere, and the $x$-coordinate of the point where graph and line meet is the value of $c$, the existence of which is assured by the intermediate value property of $f$.

Thus the intermediate value property implies that each horizontal line meeting the $y$-axis somewhere between $f(a)$ and $f(b)$ must cut the graph of this continuous function somewhere. This is a way of saying that the graph has no gaps and suggests that the idea of being able to trace such a graph without lifting the pencil from the paper is accurate.

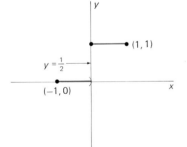

2.32 This discontinuous function does not have the intermediate value property.

**EXAMPLE 5**    The discontinuous function defined on $[-1, 1]$ as

$$f(x) = \begin{cases} 0 & \text{if} \quad x < 0; \\ 1 & \text{if} \quad x \geqq 0, \end{cases}$$

does *not* attain the intermediate value $\frac{1}{2}$. See Fig. 2.32.

**EXAMPLE 6**    Show that the equation

$$x^3 + x - 1 = 0 \tag{4}$$

has a solution between 0 and 1.

***Solution***    The function $f(x) = x^3 + x - 1$ is continuous on $[0, 1]$, and $f(0) = -1$ while $f(1) = 1$. Thus $f(0) < 0 < f(1)$, so the intermediate value property of $f$ implies that it attains the value 0 at some point between $x = 0$

and $x = 1$. That is, $f(c) = 0$ for some number $c$ in $(0, 1)$; in other words, $c^3 + c - 1 = 0$.

---

The intermediate value property provides the basis for the *method of bisection*, a natural technique for obtaining numerical approximations to solutions of equations of the form

$$f(x) = 0. \tag{5}$$

Of course, any equation in one variable can easily be put into the form in (5), so the method is quite general.

As indicated in Fig. 2.33, the solutions of Equation (5) are simply the points where the graph $y = f(x)$ crosses the $x$-axis. Suppose that $f$ is continuous and that we can find a closed interval $[a, b]$ (like the interval $[0, 1]$ in Example 6) such that the value of $f(x)$ is positive at one endpoint of $[a, b]$ and negative at the other. That is, $f(x)$ *changes sign* on the closed interval $[a, b]$. Then the intermediate value property insures that $f(x) = 0$ at some point of $[a, b]$.

Now let $m$ be the midpoint of $[a, b]$. If $f(m) = 0$, we are done. Otherwise $f(x)$ must change sign on either $[a, m]$ or $[m, b]$. If $f(x)$ changes sign on $[a, m]$, then $[a, m]$ contains a solution of $f(x) = 0$; we can bisect $[a, m]$ and continue this process for as many steps as may be needed to find the solution to the desired accuracy. We proceed in the same way if $f(x)$ changes sign on $[m, b]$. Theoretically there is no limit to the accuracy we can attain with sufficient patience, because the length of the interval containing the desired solution halves at each stage.

The data shown in Fig. 2.34 result from carrying the method of bisection through eight steps for the equation $x^3 + x - 1 = 0$. At the last step we see that the root lies between $\frac{87}{128} \approx 0.6797$ and $\frac{175}{256} \approx 0.6836$. Thus the root is approximately 0.68, accurate to two decimal places. By continuing this work through sufficiently many steps, you can verify that the first five decimal places of the root are, in fact, 0.68233.

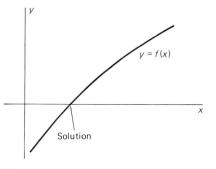

**2.33** A root of the equation $f(x) = 0$

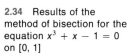

**2.34** Results of the method of bisection for the equation $x^3 + x - 1 = 0$ on $[0, 1]$

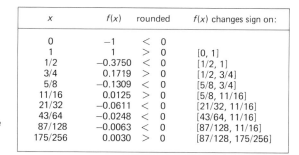

| $x$ | $f(x)$ | rounded | $f(x)$ changes sign on: |
|---|---|---|---|
| 0 | $-1$ | $< 0$ | |
| 1 | 1 | $> 0$ | [0, 1] |
| 1/2 | $-0.3750$ | $< 0$ | [1/2, 1] |
| 3/4 | 0.1719 | $> 0$ | [1/2, 3/4] |
| 5/8 | $-0.1309$ | $< 0$ | [5/8, 3/4] |
| 11/16 | 0.0125 | $> 0$ | [5/8, 11/16] |
| 21/32 | $-0.0611$ | $< 0$ | [21/32, 11/16] |
| 43/64 | $-0.0248$ | $< 0$ | [43/64, 11/16] |
| 87/128 | $-0.0063$ | $< 0$ | [87/128, 11/16] |
| 175/256 | 0.0030 | $> 0$ | [87/128, 175/256] |

## CONTINUITY AND DIFFERENTIABILITY

We have seen that a wide variety of functions are continuous. Though there exist continuous functions so exotic that they are nowhere differentiable, the following theorem tells us that every differentiable function is continuous.

**Theorem 5  Differentiability Implies Continuity**

Suppose that $f$ is defined in a neighborhood of $a$. If $f$ is differentiable at $a$, then $f$ is continuous at $a$.

***Proof***  Because $f'(a)$ exists, the product law for limits yields this:

$$\lim_{x \to a} \left[ f(x) - f(a) \right] = \lim_{x \to a} (x - a) \cdot \frac{f(x) - f(a)}{x - a}$$

$$= \left( \lim_{x \to a} (x - a) \right) \left( \lim_{x \to a} \frac{f(x) - f(a)}{x - a} \right)$$

$$= 0 \cdot f'(a) = 0.$$

Thus $\lim_{x \to a} f(x) = f(a)$, and so $f$ is continuous at $a$.  ∎

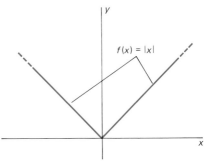

2.35  The graph of $f(x) = |x|$

Although it is difficult to describe a nowhere differentiable but everywhere continuous function, the absolute value function $f(x) = |x|$ provides us with a simple example of an everywhere continuous function that is not differentiable at the one point $x = 0$ (as we saw in Example 4 of Section 1-6). The "corner point" on the graph (see Fig. 2.35) indicates that $y = |x|$ has no tangent line at $(0, 0)$ and thus that $f(x) = |x|$ cannot be differentiable there.

## 2-4  PROBLEMS

In each of Problems 1–16, tell where the function with the given formula is continuous. Recall that when the domain of a function is not specified, it is the set of all real numbers for which the formula of the function is meaningful.

1  $f(x) = 2x + \sqrt[3]{x}$

2  $f(x) = x^2 + \dfrac{1}{x}$

3  $f(x) = \dfrac{1}{x + 3}$

4  $f(x) = \dfrac{5}{5 - x}$

5  $f(x) = \dfrac{1}{x^2 + 1}$

6  $f(x) = \dfrac{1}{x^2 - 1}$

7  $f(x) = \dfrac{x - 5}{|x - 5|}$

8  $f(x) = \dfrac{x^2 + x + 1}{x^2 + 1}$

9  $f(x) = \dfrac{x^2 + 4}{x - 2}$

10  $f(x) = \sqrt[4]{4 + x^4}$

11  $f(x) = \sqrt[3]{\dfrac{x + 1}{x - 1}}$

12  $f(x) = \sqrt[3]{3 - x^3}$

13  $f(x) = \dfrac{3}{x^2 - x}$

14  $f(x) = \sqrt{9 - x^2}$

15  $f(x) = \dfrac{x}{\sqrt{4 - x^2}}$

16  $f(x) = \sqrt{\dfrac{1 - x^2}{4 - x^2}}$

In Problems 17–26, find the points of discontinuity of each given function. For each such point $a$, tell whether the value $f(a)$ can be defined in order to make $f$ continuous at $a$.

17  $f(x) = \dfrac{x}{(x + 3)^3}$

18  $f(x) = \dfrac{x}{x^2 - 1}$

19  $f(x) = \dfrac{x - 2}{x^2 - 4}$

20  $f(x) = \dfrac{x + 1}{x^2 - x - 6}$

21  $f(x) = \dfrac{1}{1 - |x|}$

22  $f(x) = \dfrac{|x - 1|}{(x - 1)^3}$

23  $f(x) = \dfrac{x - 17}{|x - 17|}$

24  $f(x) = \dfrac{x^2 + 5x + 6}{x + 2}$

25  $f(x) = \begin{cases} -x & \text{if } x < 0; \\ x^2 & \text{if } x > 0 \end{cases}$

26  $f(x) = \begin{cases} x + 1 & \text{if } x < 1; \\ 3 - x & \text{if } x > 1 \end{cases}$

In each of Problems 27–34, tell whether the given function attains a maximum value or a minimum value (or both) on the given interval. (*Suggestion:* Begin by sketching the graph of the function.)

27  $f(x) = 1 - x, \quad -1 \le x < 1$

28  $f(x) = 2x + 1, \quad -1 \le x < 1$

**29** $f(x) = |x|, \quad -1 < x < 1$

**30** $f(x) = \dfrac{1}{\sqrt{x}}, \quad 0 < x \le 1$

**31** $f(x) = |x - 2|, \quad 1 < x \le 4$
**32** $f(x) = 5 - x^2, \quad -1 \le x < 2$
**33** $f(x) = x^3 + 1, \quad -1 \le x \le 1$

**34** $f(x) = \dfrac{1}{1 + x^2}, \quad -\infty < x < \infty$

In each of Problems 35–40, apply the intermediate value property of continuous functions to show that the given equation has a solution in the given interval.

**35** $x^2 - 5 = 0$ on $[2, 3]$
**36** $x^3 + x + 1 = 0$ on $[-1, 0]$
**37** $x^3 - 3x^2 + 1 = 0$ on $[0, 1]$
**38** $x^3 = 5$ on $[1, 2]$
**39** $x^4 + 2x - 1 = 0$ on $[0, 1]$
**40** $x^5 - 5x^3 + 3 = 0$ on $[-3, -2]$

In each of Problems 41–42, show that the given equation has three distinct roots by calculating the values of $f(x)$ at $x = -3$, $-2$, $-1$, $0$, $1$, $2$, and $3$ and then applying the intermediate value property of continuous functions on appropriate closed intervals.

**41** $x^3 - 4x + 1 = 0$      **42** $x^3 - 3x^2 + 1 = 0$

**43** Apply the intermediate value property of continuous functions to show that every positive number $a$ has a square root. That is, given $a > 0$, there is a number $r$ such that $r^2 = a$.

**44** Apply the intermediate value property to show that every real number has a cube root.

**45** Determine where the function $f(x) = [\![x/3]\!]$ is continuous.

**46** Determine where the function $f(x) = x - [\![x]\!]$ is continuous.

**47** Suppose that $f(x) = 0$ if $x$ is rational, while $f(x) = 1$ if $x$ is irrational. Show that $f$ is discontinuous everywhere.

**48** Suppose that $f(x) = 0$ if $x$ is rational, while $f(x) = x^2$ if $x$ is irrational. Show that $f$ is continuous only at the single point $x = 0$.

**49** Use the method of bisection to approximate the solution of $x^4 + 2x - 1 = 0$ in $[0, 1]$ to two-place accuracy.

**50** Use the method of bisection to approximate the solution of $x^5 + x = 1$ in $[0, 1]$ to three-place accuracy.

## *2-4   Optional Computer Application

In principle, the method of bisection can be used to approximate a root of an equation $f(x) = 0$ to any desired degree of accuracy. In practice, however, too many bisections are needed for this to be an efficient method for use in manual computations. This is another type of problem that calls for the use of a computer that very quickly can bisect a given interval, determine on which of the resulting two subintervals the function $f$ changes sign, and automatically iterate this process as many times as necessary. The strategy behind a BASIC program to accomplish these goals is this:

**1** Obtain the endpoints A and B of the interval.
**2** Let M be the midpoint of the current interval.
**3** Evaluate the function at A (call the value E) and at M (call that value F).
**4** If F = 0 (unlikely), print M and stop (do this by jumping to Step 8).
**5** If E and F have opposite signs, replace B by M (so that the new interval is [A, M]).
**6** If E and F have the same sign, replace A by M (so that the new interval is [M, B]).
**7** In either case, if B − A (using the new values) exceeds 0.000001, go back to Step 2. Otherwise, go on to Step 8.
**8** Let M be the average of A and B; print M and stop.

The following BASIC program follows the strategy above for solving the equation

$$f(x) = x^3 + x - 1 = 0 \tag{4}$$

of Example 6 with five-place accuracy.

```
10    INPUT  "A";  A
20    INPUT  "B";  B
30    M = (A + B)/2
40    E = A*A*A + A − 1
50    F = M*M*M + M − 1
60    IF  F = 0  THEN GOTO  110
70    IF  E*F < 0  THEN LET  B = M
80    IF  E*F > 0  THEN LET  A = M
90    IF  B − A > 0.000001  THEN GOTO  30
100   M = (A + B)/2
110   PRINT  "  SOLUTION:  ";  M
120   END
```

At lines 10 and 20 the computer prompts us to enter the endpoints of an interval $[A, B]$ on which we have found that $f(x) = x^3 + x − 1$ changes sign. Lines 40 and 50 calculate the values $E = f(A)$ and $F = f(M)$, where $M$ is the midpoint of $[A, B]$. Lines 70 and 80 choose the subinterval $[A, M]$ for retention if $EF < 0$, and choose the subinterval $[M, B]$ if $EF > 0$. In either case, this subinterval on which $f(x)$ changes sign is renamed $[A, B]$, so the same notation can be used if iteration of this process is required. Line 90 directs that the bisection process be repeated until the length of the last subinterval chosen does not exceed $0.000001$, at which point its midpoint is displayed as the solution sought.

When this program is executed and the values $A = 0$ and $B = 1$ are used, the result

SOLUTION:  0.6823277473

is displayed. What this actually means is that the desired solution lies in an interval of length no greater than $0.000001$ and having midpoint $M = 0.6823277473$. Hence the value of this root, rounded to five decimal places, must be $0.68233$.

> **Exercise 1** Alter the program above so you can use it to find the three solutions of the equation $x^3 − 4x + 1 = 0$ of Problem 41 of this section.

> **Exercise 2** Alter the program above so you can use it to find the three solutions of the equation $x^3 − 3x^2 + 1 = 0$.

> **Exercise 3** The program above is designed with the assumption that you have verified in advance the condition $f(A)f(B) < 0$; that is, that $f$ changes sign on $[A, B]$. Alter this program so that, if $f(A)f(B) > 0$, it immediately displays the phrase "no solution found" instead of proceeding with the bisection process.

## *2-5

## An Appendix: Proofs of the Limit Laws

Recall the definition of the limit:

$$\lim_{x \to a} F(x) = L$$

provided that, given $\varepsilon > 0$, there exists a $\delta > 0$ such that

$$0 < |x − a| < \delta \quad \text{implies} \quad |F(x) − L| < \varepsilon. \tag{1}$$

Note that the number $\varepsilon > 0$ comes *first. Then* a value of $\delta > 0$ must be found so that (1) holds. In order to prove that $F(x) \to L$ as $x \to a$, you must, in effect, be able to stop the first person you see on the street and ask him or her to pick a positive number $\varepsilon$ at random. Then *you* must be able to respond with a positive number $\delta$, which is chosen so that you can prove that (1) holds with your $\delta$ and his or her $\varepsilon$.

To do this you will ordinarily have to give an explicit method—a recipe—for producing the value $\delta$ that works for a given value of $\varepsilon$. As the next few examples show, this method or recipe will depend on the particular function $F$ involved and on the numbers $a$ and $L$.

**EXAMPLE 1**   Prove that $\lim\limits_{x \to 3} (2x - 1) = 5$.

**Solution**   Given $\varepsilon > 0$, we must find a $\delta > 0$ such that
$$|(2x - 1) - 5| < \varepsilon \quad \text{if} \quad 0 < |x - 3| < \delta.$$
Now
$$|(2x - 1) - 5| = |2x - 6| = 2|x - 3|.$$
So
$$0 < |x - 3| < \frac{\varepsilon}{2} \quad \text{implies} \quad |(2x - 1) - 5| < 2 \cdot \frac{\varepsilon}{2} = \varepsilon.$$

Hence, given $\varepsilon > 0$, it is sufficient to choose $\delta = \varepsilon/2$. This illustrates the observation that the required $\delta$ is generally a function of the given $\varepsilon$.

**EXAMPLE 2**   Prove that $\lim\limits_{x \to 2} (3x^2 + 5) = 17$.

**Solution**   Given $\varepsilon > 0$, we must find $\delta > 0$ such that
$$0 < |x - 2| < \delta \quad \text{implies} \quad |(3x^2 + 5) - 17| < \varepsilon.$$
Now
$$|(3x^2 + 5) - 17| = |3x^2 - 12| = 3 \cdot |x + 2| \cdot |x - 2|.$$
Our problem, therefore, involves showing that $|x + 2| \cdot |x - 2|$ can be made as small as we please by choosing $|x - 2|$ sufficiently small. The idea is that $|x + 2|$ cannot be too large if $|x - 2|$ is fairly small. For instance, if $|x - 2| < 1$, then
$$|x + 2| = |(x - 2) + 4| \leq |x - 2| + 4 < 5.$$
Therefore,
$$0 < |x - 2| < 1 \quad \text{implies} \quad |(3x^2 + 5) - 17| < 15 \cdot |x - 2|.$$

Consequently, let us choose $\delta$ to be the minimum of the two numbers 1 and $\varepsilon/15$. Then
$$0 < |x - 2| < \delta \quad \text{implies} \quad |(3x^2 + 5) - 17| < 15 \cdot \frac{\varepsilon}{15} = \varepsilon,$$
as desired.

**EXAMPLE 3**   Prove that
$$\lim_{x \to a} \frac{1}{x} = \frac{1}{a} \quad \text{if} \quad a \neq 0.$$

**Solution** We take the liberty of considering only the case $a > 0$, since that will simplify the notation; the method is quite similar in the case $a < 0$. Now suppose that $\varepsilon > 0$ is given. We must find a number $\delta > 0$ such that

$$0 < |x - a| < \delta \quad \text{implies} \quad \left|\frac{1}{x} - \frac{1}{a}\right| < \varepsilon.$$

Now

$$\left|\frac{1}{x} - \frac{1}{a}\right| = \left|\frac{a - x}{ax}\right| = \frac{|x - a|}{a|x|}.$$

The idea is that $1/|x|$ cannot be too large if $|x - a|$ is fairly small. For example, if $|x - a| < a/2$, then $a/2 < x < 3a/2$. Therefore,

$$|x| > \frac{a}{2}, \quad \text{so that} \quad \frac{1}{|x|} < \frac{2}{a}.$$

It follows that

$$\left|\frac{1}{x} - \frac{1}{a}\right| < \frac{2}{a^2} \cdot |x - a|$$

if $|x - a| < a/2$. Thus if we choose $\delta$ to be the minimum of the two numbers $a/2$ and $a^2 \cdot \varepsilon/2$, then

$$0 < |x - a| < \delta \quad \text{implies} \quad \left|\frac{1}{x} - \frac{1}{a}\right| < \frac{2}{a^2} \cdot \frac{a^2 \varepsilon}{2} = \varepsilon,$$

as desired.

---

We are now ready to give proofs of the limit laws that we stated in Section 2-1.

---

> **Constant Law**
> If $C$ is a constant, then
> $$\lim_{x \to a} C = C.$$

---

**Proof** Since $|C - C| = 0$, we merely choose $\delta = 1$, regardless of the previously given value of $\varepsilon > 0$. Then, if $0 < |x - a| < \delta$, it is automatic that $|C - C| < \varepsilon$. ∎

---

> **Addition Law**
> If
> $$\lim_{x \to a} F(x) = L \quad \text{and} \quad \lim_{x \to a} G(x) = M,$$
> then
> $$\lim_{x \to a} [F(x) + G(x)] = L + M.$$

***Proof*** Let $\varepsilon > 0$ be given. We must find $\delta > 0$ such that

$$0 < |x - a| < \delta \quad \text{implies} \quad |(F(x) + G(x)) - (L + M)| < \varepsilon.$$

Since $L$ is the limit of $F(x)$ as $x \to a$, there is a number $\delta_1 > 0$ such that

$$0 < |x - a| < \delta_1 \quad \text{implies} \quad |F(x) - L| < \frac{\varepsilon}{2}.$$

Because $M$ is the limit of $G(x)$ as $x \to a$, there is a number $\delta_2 > 0$ such that

$$0 < |x - a| < \delta_2 \quad \text{implies} \quad |G(x) - M| < \frac{\varepsilon}{2}.$$

So we let $\delta$ be the minimum of the two numbers $\delta_1$ and $\delta_2$. Then $0 < |x - a| < \delta$ implies that

$$|(F(x) + G(x)) - (L + M)| \leqq |F(x) - L| + |G(x) - M|$$

$$< \frac{\varepsilon}{2} + \frac{\varepsilon}{2} = \varepsilon,$$

as desired. ∎

---

### Product Law

If

$$\lim_{x \to a} F(x) = L \quad \text{and} \quad \lim_{x \to a} G(x) = M,$$

then

$$\lim_{x \to a} [F(x) \cdot G(x)] = L \cdot M.$$

---

***Proof*** Given $\varepsilon > 0$, we must find a number $\delta > 0$ such that

$$0 < |x - a| < \delta \quad \text{implies} \quad |F(x) \cdot G(x) - L \cdot M| < \varepsilon.$$

But first, note that the triangle inequality gives the result

$$|F(x) \cdot G(x) - L \cdot M| = |F(x)G(x) - L \cdot G(x) + L \cdot G(x) - LM|$$

$$\leqq |G(x)| \cdot |F(x) - L| + |L| \cdot |G(x) - M|. \qquad (2)$$

Since $\lim_{x \to a} F(x) = L$, there exists $\delta_1 > 0$ such that

$$0 < |x - a| < \delta_1 \quad \text{implies} \quad |F(x) - L| < \frac{\varepsilon}{2(|M| + 1)}. \qquad (3)$$

And since $\lim_{x \to a} G(x) = M$, there is a number $\delta_2 > 0$ such that

$$0 < |x - a| < \delta_2 \quad \text{implies} \quad |G(x) - M| < \frac{\varepsilon}{2(|L| + 1)}. \qquad (4)$$

Moreover, there is a *third* number $\delta_3 > 0$ such that

$$0 < |x - a| < \delta_3 \quad \text{implies} \quad |G(x) - M| < 1,$$

which in turn implies that

$$|G(x)| < |M| + 1. \qquad (5)$$

We now choose $\delta$ to be the least of the three positive numbers $\delta_1$, $\delta_2$, and $\delta_3$. Then we substitute (3), (4), and (5) into (2), and finally see that $0 < |x - a| < \delta$ implies

$$|F(x)G(x) - LM| < (|M| + 1) \cdot \frac{\varepsilon}{2(|M| + 1)} + |L| \cdot \frac{\varepsilon}{2(|L| + 1)}$$

$$< \frac{\varepsilon}{2} + \frac{\varepsilon}{2} = \varepsilon.$$

This establishes the product law. The use of $|M| + 1$ and $|L| + 1$ in the above denominators takes care of the possibility that $L$ or $M$ might be zero. ∎

---

### Substitution Law

If

$$\lim_{x \to a} g(x) = L \quad \text{and} \quad \lim_{x \to L} f(x) = f(L),$$

then

$$\lim_{x \to a} f(g(x)) = f(L).$$

---

**Proof**   Let $\varepsilon > 0$ be given. We must find a number $\delta > 0$ such that

$$0 < |x - a| < \delta \quad \text{implies} \quad |f(g(x)) - f(L)| < \varepsilon.$$

Since $\lim_{y \to L} f(y) = f(L)$, there exists $\delta_1 > 0$ such that

$$0 < |y - L| < \delta_1 \quad \text{implies} \quad |f(y) - f(L)| < \varepsilon. \tag{6}$$

And since $\lim_{x \to a} g(x) = L$, we can choose $\delta > 0$ such that

$$0 < |x - a| < \delta \quad \text{implies} \quad |g(x) - L| < \delta_1,$$

or

$$|y - L| < \delta_1,$$

where $y = g(x)$. From (6) we see that

$$0 < |x - a| < \delta \quad \text{implies} \quad |f(g(x)) - f(L)| = |f(y) - f(L)| < \varepsilon,$$

exactly as desired. ∎

---

### Reciprocal Law

If

$$\lim_{x \to a} g(x) = L \quad \text{and} \quad L \neq 0,$$

then

$$\lim_{x \to a} \frac{1}{g(x)} = \frac{1}{L}.$$

---

**Proof**   Let $f(x) = 1/x$. Then, as we saw in Example 3,

$$\lim_{x \to L} f(x) = \lim_{x \to L} \frac{1}{x} = \frac{1}{L} = f(L).$$

Hence the substitution law gives us the result

$$\lim_{x \to a} \frac{1}{g(x)} = \lim_{x \to a} f(g(x)) = f(L) = \frac{1}{L},$$

and this completes the proof. ∎

---

*Quotient Law*

Suppose that

$$\lim_{x \to a} F(x) = L \quad \text{and} \quad \lim_{x \to a} G(x) = M \neq 0.$$

Then

$$\lim_{x \to a} \frac{F(x)}{G(x)} = \frac{L}{M}.$$

---

*Proof*  It follows immediately from the product and reciprocal laws that

$$\lim_{x \to a} \frac{F(x)}{G(x)} = \lim_{x \to a} F(x) \cdot \frac{1}{G(x)}$$

$$= \left( \lim_{x \to a} F(x) \right) \cdot \left( \lim_{x \to a} \frac{1}{G(x)} \right) = L \cdot \frac{1}{M} = \frac{L}{M}. \quad \blacksquare$$

---

*Squeeze Law*

Suppose that $f(x) \leq g(x) \leq h(x)$ in some deleted neighborhood of $a$ and also that

$$\lim_{x \to a} f(x) = L = \lim_{x \to a} h(x).$$

Then

$$\lim_{x \to a} g(x) = L$$

as well.

---

*Proof*  Given $\varepsilon > 0$, we choose $\delta_1 > 0$ and $\delta_2 > 0$ such that

$$0 < |x - a| < \delta_1 \quad \text{implies} \quad |f(x) - L| < \varepsilon$$

and

$$0 < |x - a| < \delta_2 \quad \text{implies} \quad |h(x) - L| < \varepsilon.$$

If we now choose $\delta > 0$ as the smaller of the two numbers $\delta_1$ and $\delta_2$, then $0 < |x - a| < \delta$ implies that $f(x)$ and $h(x)$ are both points of the open interval $(L - \varepsilon, L + \varepsilon)$. So

$$L - \varepsilon < f(x) \leq g(x) \leq h(x) < L + \varepsilon.$$

Thus

$$0 < |x - a| < \delta \quad \text{implies} \quad |g(x) - L| < \varepsilon,$$

as desired. This completes the proof of the squeeze law. ∎

In Problems 1–10, apply the definition of the limit to establish the given equality.

**1** $\lim\limits_{x \to a} x = a$

**2** $\lim\limits_{x \to 2} 3x = 6$

**3** $\lim\limits_{x \to 2} (x + 3) = 5$

**4** $\lim\limits_{x \to -3} (2x + 1) = -5$

**5** $\lim\limits_{x \to 1} x^2 = 1$

**6** $\lim\limits_{x \to a} x^2 = a^2$

**7** $\lim\limits_{x \to -1} (2x^2 - 1) = 1$

**8** $\lim\limits_{x \to a} \dfrac{1}{x^2} = \dfrac{1}{a^2}$ if $a \neq 0$

**9** $\lim\limits_{x \to a} \dfrac{1}{x^2 + 1} = \dfrac{1}{a^2 + 1}$

**10** $\lim\limits_{x \to a} \dfrac{1}{\sqrt{x}} = \dfrac{1}{\sqrt{a}}$ if $a > 0$

**11** Suppose that $\lim\limits_{x \to a} f(x) = L$ and that $\lim\limits_{x \to a} f(x) = M$. Prove that $L = M$. Thus a limit of a function is unique if it exists.

**12** Suppose that $C$ is a constant and that $\lim\limits_{x \to a} f(x) = L$.

Apply the definition of the limit to prove that

$$\lim\limits_{x \to a} C \cdot f(x) = C \cdot L.$$

**13** Suppose that $L \neq 0$ and that $\lim\limits_{x \to a} f(x) = L$. Use the method of Example 3 and the definition of the limit to show directly that

$$\lim\limits_{x \to a} \frac{1}{f(x)} = \frac{1}{L}.$$

**14** In this problem use the algebraic identity

$$x^n - a^n = (x - a)(x^{n-1} + x^{n-2}a + \cdots$$
$$+ xa^{n-2} + a^{n-1}).$$

Show directly from the definition of the limit that $\lim\limits_{x \to a} x^n = a^n$ if $n$ is a positive integer.

**15** Apply the identity

$$\left| \sqrt{x} - \sqrt{a} \right| = \frac{|x - a|}{\sqrt{x} + \sqrt{a}}$$

to show directly from the definition of the limit that $\lim\limits_{x \to a} \sqrt{x} = \sqrt{a}$ if $a > 0$.

**16** Suppose that $f$ is continuous at $a$ and that $f(a) > 0$. Prove that $f(x)$ is positive on some neighborhood of $a$; that is, that there exists $\delta > 0$ such that

$$|x - a| < \delta \quad \text{implies} \quad f(x) > 0.$$

**17** (*Challenge*) Let $f$ be defined as follows:

$$f(x) = \begin{cases} 0 & \text{if } x \text{ is irrational;} \\ 1 & \text{if } x = 0; \\ \dfrac{1}{q} & \text{if } x = \dfrac{p}{q} \neq 0, \end{cases}$$

where the integers $p$ and $q > 0$ have no common factors larger than 1.

(a) Show that $f$ is discontinuous at $a$ if $a$ is rational.
(b) Show that $f$ is continuous at $a$ if $a$ is irrational.
(*Suggestion:* Note first that, given $\varepsilon > 0$, there are only finitely many positive integers $q$ such that $1/q \geq \varepsilon$. Then show that, for each fixed positive integer $q_0$, the neighborhood $(a - 1, a + 1)$ of $a$ contains only finitely many rational numbers of the form $p/q_0$.)

## CHAPTER 2 REVIEW: Definitions, Concepts, Results

Use the list below as a guide to ideas that you may need to review.

**1** Limit of $f(x)$ as $x$ approaches $a$
**2** Limit laws: constant, addition, product, quotient, root, substitution, squeeze.
**3** Right-hand and left-hand limits
**4** Relation between one-sided and two-sided limits
**5** Infinite limits
**6** Composition of functions
**7** Inverse functions; one-to-one functions

**8** Continuity of a function at a point
**9** Continuity of polynomials and rational functions
**10** Continuity of compositions of functions
**11** Continuity of inverse functions
**12** Continuity of a function on a closed interval
**13** Maximum value property
**14** Intermediate value property
**15** Method of bisection
**16** Differentiability implies continuity

Apply the limit laws to evaluate the limits in Problems 1–40 or to show that the indicated limit does not exist, as appropriate.

**1** $\lim\limits_{x \to 0} (x^2 - 3x + 4)$

**2** $\lim\limits_{x \to -1} (3 - x + x^3)$

**3** $\lim\limits_{x \to 2} (4 - x^2)^{10}$

**4** $\lim\limits_{x \to 1} (x^2 + x - 1)^{17}$

**5** $\lim\limits_{x \to 2} \dfrac{1 + x^2}{1 - x^2}$

**6** $\lim\limits_{x \to 3} \dfrac{2x}{x^2 - x - 3}$

**7** $\lim\limits_{x \to 1} \dfrac{x^2 - 1}{1 - x}$

**8** $\lim\limits_{x \to -2} \dfrac{x + 2}{x^2 + x - 2}$

**9** $\lim\limits_{t \to -3} \dfrac{t^2 + 6t + 9}{9 - t^2}$

**10** $\lim\limits_{x \to 0} \dfrac{4x - x^3}{3x + x^2}$

**11** $\lim\limits_{x \to 3} (x^2 - 1)^{2/3}$

**12** $\lim\limits_{x \to 2} \sqrt{\dfrac{2x^2 + 1}{2x}}$

**13** $\lim\limits_{x \to 3} \left(\dfrac{5x + 1}{x^2 - 8}\right)^{3/4}$

**14** $\lim\limits_{x \to 1} \dfrac{x^4 - 1}{x^2 + 2x - 3}$

**15** $\lim\limits_{x \to 7} \dfrac{\sqrt{x + 2} - 3}{x - 7}$

**16** $\lim\limits_{x \to 1^+} (x - \sqrt{x^2 - 1})$

**17** $\lim\limits_{x \to -4} \dfrac{(1/\sqrt{13 + x}) - \frac{1}{3}}{x + 4}$

**18** $\lim\limits_{x \to 1^+} \dfrac{1 - x}{|1 - x|}$

**19** $\lim\limits_{x \to 2^+} \dfrac{2 - x}{\sqrt{4 - 4x + x^2}}$

**20** $\lim\limits_{x \to -2^-} \dfrac{x + 2}{|x + 2|}$

**21** $\lim\limits_{x \to 4^+} \dfrac{x - 4}{|x - 4|}$

**22** $\lim\limits_{x \to 3^-} \sqrt{x^2 - 9}$

**23** $\lim\limits_{x \to 2^+} \sqrt{4 - x^2}$

**24** $\lim\limits_{x \to -3} \dfrac{x}{(x + 3)^2}$

**25** $\lim\limits_{x \to 2} \dfrac{x + 2}{(x - 2)^2}$

**26** $\lim\limits_{x \to 1^-} \dfrac{x}{x - 1}$

**27** $\lim\limits_{x \to 3^+} \dfrac{x}{x - 3}$

**28** $\lim\limits_{x \to 1^-} \dfrac{x - 2}{x^2 - 3x + 2}$

**29** $\lim\limits_{x \to 1^-} \dfrac{x + 1}{(x - 1)^3}$

**30** $\lim\limits_{x \to 5^+} \dfrac{25 - x^2}{x^2 - 10x + 25}$

**31** $\lim\limits_{x \to 0} \dfrac{3x - x^3}{x}$

**32** $\lim\limits_{x \to 0} \dfrac{5x + x^2}{x}$

**33** $\lim\limits_{x \to 0} \dfrac{3x - x^2}{2x - x^2}$

**34** $\lim\limits_{x \to 0} \dfrac{4x - x^2}{x^3}$

**35** $\lim\limits_{x \to 0^+} \dfrac{x}{\sqrt{x} - x\sqrt{x}}$

**36** $\lim\limits_{x \to 0^+} \dfrac{1 - 3x}{\sqrt{x}}$

**37** $\lim\limits_{x \to 0} \dfrac{9x^2 - x^4}{4x^2}$

**38** $\lim\limits_{x \to 0} \dfrac{x^3(2 - x^2)}{(3x - x^3)^2}$

**39** $\lim\limits_{x \to 0} \dfrac{6x - 2x^3}{3x(1 - 2x^2)^2}$

**40** $\lim\limits_{x \to 0} \dfrac{x^2(1 - 3x^2)}{3x - 4x^3}$

In each of Problems 41–44, find the inverse function of $f$.

**41** $f(x) = \sqrt{5 - 2x}$

**42** $f(x) = 1 - 3x$

**43** $f(x) = \dfrac{1}{(x^3 - 1)^5}$

**44** $f(x) = \dfrac{1}{\sqrt{2x + 1}}$

**45** Suppose that $f(x) = 1 + x^2$. Find $g$ so that $f(g(x)) = 1 + x^2 - 2x^3 + x^4$.

**46** Suppose that $g(x) = 1 + \sqrt{x}$. Find $f$ so that $f(g(x)) = 3 + 2\sqrt{x} + x$.

Explain why each of the functions in Problems 47–50 is continuous wherever it is defined by the given formula. For each point $a$ where $f$ is not defined by the formula, tell whether $f(a)$ can be defined in order to make $f$ continuous at $a$.

**47** $f(x) = \dfrac{1 - x}{1 - x^2}$

**48** $f(x) = \dfrac{1 - x}{(2 - x)^2}$

**49** $f(x) = \dfrac{x^2 + x - 2}{x^2 + 2x - 3}$

**50** $f(x) = \dfrac{|x^2 - 1|}{x^2 - 1}$

**51** Apply the intermediate value property of continuous functions to prove that the equation $x^5 + x = 1$ has a solution.

**52** Apply the intermediate value property of continuous functions to prove that the equation $x^3 - 3x^2 + 1 = 0$ has three different solutions.

**53** Show that there is a number $x$ between 1 and 2 such that $x^3 - 1 = x$.

**54** Show that there is a positive number $x$ such that $x^2 = 4 - x$. (*Suggestion:* First sketch the graphs of $y = x^2$ and $y = 4 - x$.)

**55** (*Challenge*) Suppose that $f'$ is continuous on the set of all real numbers, that $f'(0) \geqq 0$, and that $f = f^{-1}$. Prove that $f(x) = x$ for all $x$, or show that this assertion is false by finding a different function with the given properties.

# 3

# Differentiation

## The Derivative and Rates of Change

Our preliminary investigation of tangent lines and derivatives in Chapter 1 led to the study of limits in Chapter 2. Armed with this knowledge of limits, we are now prepared to study derivatives more fully.

In Section 1-6 we introduced the derivative $f'(x)$ as the slope of the tangent line to the graph of the function $f$ at the point $(x, f(x))$. More precisely, we were motivated by the geometry to *define* the tangent line to the graph at the point $P(a, f(a))$ to be the straight line through $P$ with slope

$$m = \lim_{h \to 0} \frac{f(a + h) - f(a)}{h}, \tag{1}$$

as indicated in Fig. 3.1. If we replace the arbitrary number $a$ in (1) with the independent variable $x$, we get a new function $f'$, the derivative of the original function $f$.

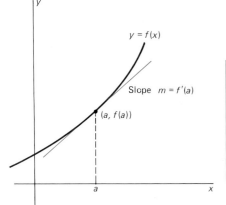

**3.1** The geometric motivation for the definition of the derivative

---

*Definition of the Derivative*

The **derivative** of the function $f$ is the function $f'$ defined by

$$f'(x) = \lim_{h \to 0} \frac{f(x + h) - f(x)}{h} \tag{2}$$

for all $x$ such that this limit exists.

---

The last phrase implies, in particular, that $f$ must be defined in a neighborhood of $x$.

We emphasized in Section 1-6 that we hold $x$ *fixed* in (2) while $h$ approaches zero. When we are specifically interested in the value of the derivative $f'$ at $x = a$, we sometimes rewrite (2) in the form

$$f'(a) = \lim_{h \to 0} \frac{f(a + h) - f(a)}{h}$$

$$= \lim_{x \to a} \frac{f(x) - f(a)}{x - a}. \tag{3}$$

The second limit here is obtained from the first by writing $x = a + h$, $h = x - a$, and noting that $x \to a$ as $h \to 0$. The statement that these equivalent limits exist is sometimes abbreviated as "$f'(a)$ exists." In this case we say that the function $f$ is **differentiable** at $x = a$. Finally, the process of finding the derivative $f'$ is called **differentiation** of $f$.

In Sections 1-6 and 2-1, we used several examples to illustrate the process of differentiating a given function $f$ by direct evaluation of the limit in (2). This involves carrying out these four steps:

**1** Write the definition in (2) of the derivative.

**2** Substitute the expressions $f(x + h)$ and $f(x)$ as determined by the particular given function $f$.

**3** Simplify the result by algebraic methods to make Step 4 possible.

**4** Evaluate the limit—typically, by application of the limit laws.

**EXAMPLE 1**  Differentiate $f(x) = \dfrac{x}{x+3}$.

**Solution**

$$f'(x) = \lim_{h \to 0} \frac{1}{h} [f(x+h) - f(x)]$$

$$= \lim_{h \to 0} \frac{1}{h} \left( \frac{x+h}{x+h+3} - \frac{x}{x+3} \right)$$

$$= \lim_{h \to 0} \frac{(x+h)(x+3) - x(x+h+3)}{h(x+h+3)(x+3)}$$

$$= \lim_{h \to 0} \frac{3h}{h(x+h+3)(x+3)}$$

$$= \frac{3}{\left( \lim\limits_{h \to 0} (x+h+3) \right)\left( \lim\limits_{h \to 0} (x+3) \right)}.$$

Therefore,

$$f'(x) = \frac{3}{(x+3)^2}.$$

Even when the function $f$ is rather simple, this process for computing $f'$ directly from the definition of the derivative can be tedious. Also, Step 3 may require considerable ingenuity. Moreover, it would be very repetitious to continue to rely on the four-step process above. To avoid tedium, we want a fast, easier, and shorter method.

That method is the main subject of this chapter: the development of systematic methods, or "rules," for differentiating those functions that occur most frequently. These functions include polynomials, rational functions, and compositions of such functions. Once these general differentiation rules have been established, they can be applied formally, almost mechanically, to compute derivatives. Only rarely will recourse to the definition of the derivative be necessary.

An example of a "differentiation rule" is the theorem of Section 1-6 on differentiation of quadratic functions:

$$\text{If} \quad f(x) = ax^2 + bx + c, \quad \text{then} \quad f'(x) = 2ax + b. \tag{4}$$

Once we know this rule, we need never again apply the definition of the derivative to differentiate a quadratic function. For instance, if $f(x) = 3x^2 - 4x + 5$, we can apply (4) immediately to write

$$f'(x) = (2)(3)x + (-4) = 6x - 4.$$

Similarly, if $g(t) = 2t - 5t^2$, then

$$g'(t) = (2) + (2)(-5)t = 2 - 10t,$$

because it makes no difference whether we write $t$ or $x$ for the independent variable in (4). Such flexibility is valuable, and so you should learn every differentiation rule in a form independent of the notation used to state it.

We will develop additional differentiation rules in Sections 3-2 and 3-3. First, however, we introduce a new interpretation of the derivative.

## RATE OF CHANGE

In Section 1-6 we introduced the derivative of a function as the slope of the tangent line to its graph. Here we introduce the equally important interpretation of the derivative of a function as the function's rate of change with respect to the independent variable.

We begin with the instantaneous rate of change of a function having time $t$ as its independent variable. Suppose that $Q$ is a quantity that varies with time, and write $Q = f(t)$ for its value at time $t$. For instance, $Q$ might be

 1 The size of a population (such as rabbits, people, or bacteria);
 2 The number of dollars in a bank account;
 3 The volume of a balloon that is being inflated;
 4 The amount of water in a reservoir with variable inflow and outflow;
 5 The amount of a certain chemical product produced in a reaction up to time $t$; or
 6 The distance traveled in time $t$ since the beginning of a journey.

The change in $Q$ from time $t$ to time $t + \Delta t$ is the **increment**

$$\Delta Q = f(t + \Delta t) - f(t).$$

The **average rate of change** of $Q$ (per unit of time) is, by definition, the quotient

$$\frac{\Delta Q}{\Delta t} = \frac{f(t + \Delta t) - f(t)}{\Delta t}. \tag{5}$$

We define the **instantaneous rate of change** of $Q$ (per unit of time) to be the limit of this average rate as $\Delta t \to 0$. That is, the instantaneous rate of change is

$$\lim_{\Delta t \to 0} \frac{\Delta Q}{\Delta t} = \lim_{\Delta t \to 0} \frac{f(t + \Delta t) - f(t)}{\Delta t}. \tag{6}$$

But this latter limit is simply the derivative $f'(t)$. So we see that the instantaneous rate of change of $Q = f(t)$ is

$$\frac{dQ}{dt} = f'(t). \tag{7}$$

On the left-hand side in (7) we use the differential notation

$$\frac{dQ}{dt} = \lim_{\Delta t \to 0} \frac{\Delta Q}{\Delta t}$$

for the derivative that was mentioned in Section 1-6.

This concept of the instantaneous rate of change agrees with our intuitive idea of what it should be. If we think of $Q$ as changing with time, but then suddenly, at time $t$, continuing in the direction of its graph at time $t$ without subsequent curvature, the graph of $Q$ would appear as shown in Fig. 3.2. The broken line in that figure is meant to indicate how $Q = f(t)$ might normally behave. But if $Q$ were to follow the straight line, that would be "change at a constant rate." Because the straight line is tangent to the graph of $Q$, we can interpret $dQ/dt$ as the instantaneous rate of change of $Q$ at time $t$. In brief,

> The instantaneous rate of change of $Q = f(t)$ at time $t$ is equal to the slope of the tangent line to the curve $Q = f(t)$ at the point $(t, f(t))$.

There are important additional conclusions that we can draw. Because a positive slope corresponds to a rising tangent line and a negative slope to a falling tangent line, we say that

$$Q \text{ is increasing at time } t \quad \text{if} \quad \frac{dQ}{dt} > 0;$$

$$Q \text{ is decreasing at time } t \quad \text{if} \quad \frac{dQ}{dt} < 0.$$

(8)

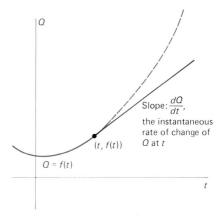

**3.2** The relation between the tangent line at $(t, f(t))$ and the instantaneous rate of change at $t$

Note this: The meaning of the phrase "$Q = f(t)$ is increasing *over* (or during) *the time interval* $a \leq t \leq b$" should be intuitively clear. What we have in (8) is a way to make precise what is meant by "$Q = f(t)$ is increasing *at time* $t$"—that is, at the instant $t$.

**EXAMPLE 2**    A cylindrical tank with vertical axis is initially filled with 200,000 gal. of water. This tank takes 50 min. to empty after a drain in its bottom is opened. Suppose that the drain is opened at time $t = 0$. A consequence of Torricelli's law is that the volume $V$ of water remaining in the tank after $t$ minutes is

$$V(t) = (200{,}000)\left(1 - \frac{t}{50}\right)^2;$$

that is,

$$V(t) = 200{,}000 - 8000t + 80t^2$$

(gallons; $t$ is in minutes). Find the instantaneous rate at which the water is flowing out of the tank when $t = 30$.

**Solution**    We simply need to find the value of $dV/dt$ when $t = 30$. But

$$\frac{dV}{dt} = (-8000) + (2)(80)t = -8000 + 160t$$

as long as the water is flowing, and the value of $dV/dt$ at time $t = 30$ is simply

$$V'(30) = -8000 + (160)(30) = -3200.$$

Because the units for $V$ are gallons and those for $t$ are minutes, the units for "rate of change of $V$ with respect to $t$" are gallons per minute. Thirty minutes after the drain is opened, the water is flowing out at 3200 gal./min.

That $V'(30)$ is negative means that $V$ is *decreasing* at time $t = 30$. Had $dV/dt$ been a positive number, we would be forced to conclude that the tank

| $t$ | Year | U.S. Pop. (millions) |
|---|---|---|
| 0 | 1800 | 5.3 |
| 10 | 1810 | 7.2 |
| 20 | 1820 | 9.6 |
| 30 | 1830 | 12.9 |
| 40 | 1840 | 17.1 |
| 50 | 1850 | 23.2 |
| 60 | 1860 | 31.4 |
| 70 | 1870 | 38.6 |
| 80 | 1880 | 50.2 |
| 90 | 1890 | 62.9 |
| 100 | 1900 | 76.0 |

**3.3** Data for Example 3

was being filled rather than emptied (or else that we had made an error in our computations!).

Our examples of functions up to this point have been restricted to those having either formulas or verbal descriptions. In science we often work with tables of values obtained from observations or from empirical measurements. Our next example shows how the instantaneous rate of change of such a tabulated function can be estimated.

**EXAMPLE 3** The table in Fig. 3.3 gives the U.S. population $P$ (in millions) at 10-year intervals in the nineteenth century. Estimate the instantaneous rate of population growth in 1850.

*Solution* We take $t = 0$ (years) in 1800, so that $t = 50$ in 1850. In Fig. 3.4 we have plotted the data of our example and added a freehand sketch of a smooth curve that fits these data.

However it may be obtained, a curve that fits the data ought to be a good approximation to the true graph of the unknown function $P = f(t)$. The instantaneous rate of change $dP/dt$ is the slope of the tangent line at the point $(50, 23.2)$. We draw the tangent as accurately as we can by visual inspection and then measure the base and height of the triangle of Fig. 3.4. Thus we approximate the slope of the tangent as

$$\frac{dP}{dt} \approx \frac{36}{51} \approx 0.71$$

millions of people per year (in 1850). Although there was no national census in 1851, we would expect the population of the United States then to have been approximately $23.2 + 0.7 = 23.9$ million people.

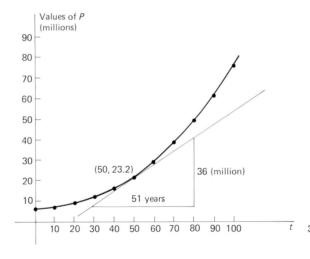

**3.4** A smooth curve that fits the data well

## VELOCITY AND ACCELERATION

A particle moves along a straight line with **position function** $s = f(t)$. Thus we make the line of motion a coordinate axis: To so specify $f$ is to have chosen an origin and a positive direction on the line of motion. Also, $f(t)$ is just the coordinate of the particle at time $t$ (see Fig. 3.5).

**3.5** The particle in motion is at the point $s = f(t)$ at time $t$

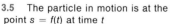
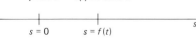

CHAP. 3: Differentiation

Think of the time interval from $t$ to $t + \Delta t$. The particle moves from $f(t)$ to $f(t + \Delta t)$. Its displacement is then the increment

$$\Delta s = f(t + \Delta t) - f(t).$$

The **average velocity** of the particle during this time interval is

$$\bar{v} = \frac{\text{displacement}}{\text{elapsed time}} = \frac{\Delta s}{\Delta t} = \frac{f(t + \Delta t) - f(t)}{\Delta t}. \tag{9}$$

We define the **instantaneous velocity** $v$ of the particle at the time $t$ to be the limit of this average velocity $\bar{v}$ as $\Delta t \to 0$. That is,

$$v = \lim_{\Delta t \to 0} \frac{\Delta s}{\Delta t} = \lim_{\Delta t \to 0} \frac{f(t + \Delta t) - f(t)}{\Delta t}. \tag{10}$$

We recognize the right-hand limit—it is the definition of the derivative of $f$ at time $t$. So the velocity of the moving particle at time $t$ is

$$v = \frac{ds}{dt} = f'(t). \tag{11}$$

Thus *velocity is instantaneous rate of change of position.* The velocity of a moving particle is positive or negative, depending on whether it is moving in the positive or negative direction along the line of motion. The *speed* is the absolute value $|v|$ of the particle's velocity.

**EXAMPLE 4** If the position function of the particle is $s(t) = 5t^2 + 100$, then its velocity at time $t$ is $v(t) = s'(t) = 10t$. For instance, at time $t = 10$ its position is $s(10) = (5)(10^2) + 100 = 600$ and its velocity at that time is $v(10) = (10)(10) = 100$.

---

The case of vertical motion under the influence of gravity is of special interest. If a particle is projected straight upward from a height $s_0$ above the ground, at time $t = 0$ (seconds), and with initial velocity $v_0$, then its height $s$ above the ground at time $t$ is

$$s = -\tfrac{1}{2}gt^2 + v_0 t + s_0. \tag{12}$$

Here $g$ denotes the gravitational acceleration. Near the surface of the earth, $g \approx 32$ ft/s$^2$, or $g \approx 980$ cm/s$^2$. If we differentiate $s$, we obtain the velocity of the particle at time $t$:

$$v = \frac{ds}{dt} = -gt + v_0. \tag{13}$$

The **acceleration** of the particle is the instantaneous rate of change, or derivative, of its velocity:

$$a = \frac{dv}{dt} = -g. \tag{14}$$

Your intuition and probably your experience tell you that a body so projected upward will reach its maximum height at the instant that its velocity vanishes—when $v = 0$. (We shall see why this is true in Section 3-5.)

**EXAMPLE 5** Find the maximum height attained by a ball that is thrown straight upward from the ground with initial velocity $v_0 = +96$ ft/s.

*Solution*  We use Equation (13), with $v_0 = 96$ and $g = 32$. We find the velocity of the ball to be

$$v = -32t + 96.$$

It attains its maximum height at the instant when $v = 0$; that is, when

$$-32t + 96 = 0.$$

This occurs when $t = 3$ (seconds). Upon substitution of this value of $t$ in (12), we find that the maximum height is

$$s_{max} = s(3) = (-16)(3^2) + (96)(3) = 144 \text{ ft}$$

because $s_0 = 0$.

---

The derivative of any function—not merely a function of time—may be interpreted as its instantaneous rate of change with respect to the independent variable. If $y = f(x)$, then the **average rate of change** of $y$ (per unit change in $x$) on the interval $[x, x + \Delta x]$ is the quotient

$$\frac{\Delta y}{\Delta x} = \frac{f(x + \Delta x) - f(x)}{\Delta x}.$$

The **instantaneous rate of change of $y$ with respect to $x$** is the limit as $\Delta x \to 0$ of this average rate of change. That is,

$$\lim_{\Delta x \to 0} \frac{\Delta y}{\Delta x} = \frac{dy}{dx} = f'(x). \tag{15}$$

The following example illustrates the fact that a dependent variable may sometimes be expressed as two different functions of two different independent variables. The derivatives of those two functions are then rates of change of the dependent variable with respect to the two different independent variables.

**EXAMPLE 6**  A square with edge length $x$ (centimeters) has area $A = x^2$, so the derivative of $A$ with respect to $x$,

$$\frac{dA}{dx} = 2x, \tag{16}$$

is the rate of change of its area $A$ (in square centimeters *per centimeter*) with respect to $x$. Now suppose that the edge length of the square is increasing with time: $x = 5t$, with time $t$ in seconds. Then the area of the square at time $t$ is

$$A = (5t)^2 = 25t^2.$$

The derivative of $A$ with respect to $t$ is

$$\frac{dA}{dt} = (2)(25)t = 50t; \tag{17}$$

this is the rate of change of $A$ (in square centimeters *per second*) with respect to time $t$. For instance, when $t = 10$—so that $x = 50$—the values of the two derivatives of $A$ in (16) and (17) are

$$\left.\frac{dA}{dx}\right|_{x=50} = (2)(50) = 100 \quad \text{cm}^2/\text{cm}$$

and

$$\left.\frac{dA}{dt}\right|_{t=10} = (50)(10) = 500 \quad \text{cm}^2/\text{s}.$$

In each of Problems 1–10, find the indicated derivative by using the differentiation rule in (4):

$$\text{If} \quad f(x) = ax^2 + bx + c, \quad \text{then} \quad f'(x) = 2ax + b.$$

1  $f(x) = 4x - 5$;   find $f'(x)$
2  $g(t) = 100 - 16t^2$;   find $g'(t)$
3  $h(z) = z(25 - z)$;   find $h'(z)$
4  $f(x) = 16 - 49x$;   find $f'(x)$
5  $y = 2x^2 + 3x - 17$;   find $dy/dx$
6  $x = 16t - 100t^2$;   find $dx/dt$
7  $z = 5u^2 + 3u$;   find $dz/du$
8  $v = 5y(100 - y)$;   find $dv/dy$
9  $x = -5y^2 + 17y + 300$;   find $dx/dy$
10  $u = 7t^2 + 13t$;   find $du/dt$

In each of Problems 11–20, apply the definition of the derivative (as in Example 1) to find $f'(x)$.

11  $f(x) = 2x - 1$
12  $f(x) = 2 - 3x$
13  $f(x) = x^2 + 5$
14  $f(x) = 3 - 2x^2$

15  $f(x) = \dfrac{1}{2x + 1}$
16  $f(x) = \dfrac{1}{3 - x}$

17  $f(x) = \sqrt{2x + 1}$
18  $f(x) = \dfrac{1}{\sqrt{x + 1}}$

19  $f(x) = \dfrac{x}{1 - 2x}$
20  $f(x) = \dfrac{x + 1}{x - 1}$

In each of Problems 21–25, the position function $s = f(t)$ of a body moving in a straight line is given. Find its location $s$ when its velocity $v$ is zero.

21  $s = 100 - 16t^2$
22  $s = -16t^2 + 160t + 25$
23  $s = -16t^2 + 80t - 1$
24  $s = 100t^2 + 50$
25  $s = 100 - 20t - 5t^2$

In each of Problems 26–29, the height $s(t)$ of a ball thrown vertically upward (in feet after $t$ seconds) is given. Find the maximum height that the ball attains.

26  $s = -16t^2 + 160t$
27  $s = -16t^2 + 64t$
28  $s = -16t^2 + 128t + 25$
29  $s = -16t^2 + 96t + 50$
30  The Celsius temperature $C$ is given in terms of Fahrenheit temperature $F$ by $C = \frac{5}{9}(F - 32)$. Find the rate of change of $C$ with respect to $F$ and also the rate of change of $F$ with respect to $C$.

31  Find the rate of change of the area $A$ of a circle with respect to its circumference $C$.

32  A stone dropped into a pond (at time $t = 0$ s) causes a circular ripple that travels out from the point of impact at 5 ft/s. At what rate (in square feet per second) is the area within this circle increasing when $t = 10$ s?

33  A car is traveling at 100 ft/s when the driver suddenly applies the brakes ($s = 0$, $t = 0$). The position function of the skidding car is $s = 100t - 5t^2$. How long and how far does the car skid before coming to a stop?

34  A water bucket containing 10 gal. of water develops a leak at time $t = 0$, and the volume of water in the bucket $t$ seconds later is

$$V = 10\left(1 - \frac{t}{100}\right)^2.$$

(a) At what rate is water leaking from the bucket after 1 minute has passed?
(b) When is the instantaneous rate of change of $V$ equal to the average rate of change of $V$ from $t = 0$ to $t = 100$ s?

35  A certain population of rodents numbers

$$P = 100[1 + (0.3)t + (0.04)t^2]$$

after $t$ months.
(a) How long does it take for this population to double its initial ($t = 0$) size?
(b) What is the rate of growth of the population when $P = 200$?

36  The following data describe the growth of the population $P$ (in thousands) of a certain city during the 1970s. Use the graphical method of Example 3 to estimate its rate of growth in 1975.

| Year | 1970 | 1972 | 1974 | 1976 | 1978 | 1980 |
|------|------|------|------|------|------|------|
| $P$  | 265  | 293  | 324  | 358  | 395  | 437  |

37  The following data give the distance $s$ in feet traveled by an accelerating car (starting from rest) in the first $t$ seconds. Use the graphical method of Example 3 to estimate

its speed (in miles per hour) when $t = 20$ s and again when $t = 40$ s.

| $t$ | 0 | 10 | 20 | 30 | 40 | 50 | 60 |
|-----|---|-----|-----|------|------|------|------|
| $s$ | 0 | 224 | 810 | 1655 | 2686 | 3850 | 5109 |

In Problems 38–43, use the fact (proved in Section 3-2) that the derivative of $y = ax^3 + bx^2 + cx + d$ is $dy/dx = 3ax^2 + 2bx + c$.

**38** Show that the rate of change of the volume of a cube with respect to its edge length is equal to half the surface area of the cube.

**39** Show that the rate of change of the volume of a sphere with respect to its radius is equal to its surface area.

**40** The height of a changing cylinder is always twice its radius. Show that the rate of change of its volume with respect to its radius is equal to its total surface area.

**41** A spherical balloon with an initial radius of 5 in. begins to leak at $t = 0$, and its radius $t$ seconds later is $r = (60 - t)/12$ in. At what rate (in cubic inches per second) is air leaking from the balloon when $t = 30$ s?

**42** The volume $V$ (in liters) of 3 g of $CO_2$ at $27°C$ is given in terms of its pressure $p$ (in atmospheres) by the formula $V = 1.68/p$. What is the rate of change of $V$ with respect to $p$ when $p = 2$ atm? (Use the fact that the derivative of $f(x) = c/x$ is $f'(x) = -c/x^2$ if $c$ is a constant; you can establish this by using the definition of the derivative if you wish.)

**43** As a snowball with an initial radius of 12 cm melts, its radius decreases at a constant rate. It begins to melt when $t = 0$ (hours) and takes 12 h to disappear.
(a) What is its rate of change of volume when $t = 6$?
(b) What is its average rate of change of volume from $t = 3$ to $t = 9$?

**44** A ball thrown vertically upward at 96 ft/s from a height of 112 ft (when $t = 0$ s) has height function $y = -16t^2 + 96t + 112$.
(a) What is the maximum height attained by the ball?
(b) When and with what impact speed does it hit the ground?

**45** A spaceship approaching touchdown on a distant planet has height $y$ (meters) at time $t$ (seconds) given by $y = 100 - 100t + 25t^2$. When and with what speed does it hit the ground?

**46** In 1970 a certain city had population (in thousands) given by $P = 100[1 + (0.04)t + (0.003)t^2]$, with $t$ in years and with $t = 0$ corresponding to 1970.
(a) What was the rate of change of $P$ in 1975?
(b) What was the average rate of change of $P$ from 1973 to 1978?

---

## 3-2

## Basic Differentiation Rules

In this section we begin our development of formal rules for finding the derivative $f'$ of the function $f$:

$$f'(x) = \lim_{h \to 0} \frac{f(x + h) - f(x)}{h}. \tag{1}$$

Some alternative notation for derivatives will aid us.

When we interpreted the derivative in Section 3-1 as a rate of change, we found it useful to employ the dependent-independent variable notation

$$y = f(x), \qquad \Delta x = h, \qquad \Delta y = f(x + \Delta x) - f(x). \tag{2}$$

This led to the "differential notation"

$$\frac{dy}{dx} = \lim_{\Delta x \to 0} \frac{\Delta y}{\Delta x} = \lim_{\Delta x \to 0} \frac{f(x + \Delta x) - f(x)}{\Delta x} \tag{3}$$

for the derivative. When you use this notation, it is important to remember that the symbol $dy/dx$ is simply an alternative notation for the derivative $f'(x)$; it is *not* the quotient of two separate entities $dy$ and $dx$.

A third notation is commonly used for the derivative $f'(x)$; it is $Df(x)$. Here, think of $D$ as a "machine" that "operates" on the function $f$ to produce the derivative function $Df$, as indicated in Fig. 3.6. Thus we can write the derivative of $y = f(x) = x^2$ in any of three ways:

$$f'(x) = \frac{dy}{dx} = Dx^2 = 2x.$$

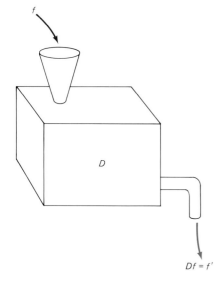
**3.6** The "differentiation machine" $D$

These three notations for the derivative—the functional notation $f'(x)$, the differential notation $dy/dx$, and the operator notation $Df(x)$—are used interchangeably in mathematical and scientific writing, so familiarity with each is necessary.

## DERIVATIVES OF POLYNOMIALS

Our first differentiation rule says that *the derivative of a constant function is identically zero*. This is geometrically obvious, because the graph of a constant function is a horizontal straight line and therefore has slope zero at each point.

---

**Theorem 1**   *Derivative of a Constant*

If $f(x) = c$ (a constant) for all $x$, then $f'(x) = 0$ for all $x$. That is,

$$\frac{dc}{dx} = Dc = 0. \tag{4}$$

---

**Proof**   Because $f(x + h) = f(x) = c$, we see that

$$f'(x) = \lim_{h \to 0} \frac{f(x + h) - f(x)}{h}$$

$$= \lim_{h \to 0} \frac{c - c}{h} = \lim_{h \to 0} \frac{0}{h} = 0. \qquad \blacksquare$$

As motivation for the next rule, consider the following list of derivatives, all computed in Chapters 1 and 2.

| | |
|---|---|
| $Dx = 1$ | Special cases of the |
| $Dx^2 = 2x$ | theorem in Section 1-6 |
| $Dx^3 = 3x^2$ | (Section 1-6, Problem 26) |
| $Dx^{-1} = -x^{-2}$ | (Section 1-6, Problem 27) |
| $Dx^{-2} = -2x^{-3}$ | (Section 1-6, Problem 29) |
| $Dx^{1/2} = \frac{1}{2}x^{-1/2}$ | (Section 2-1, Example 12) |
| $Dx^{-1/2} = -\frac{1}{2}x^{-3/2}$ | (Section 2-1, Problem 37) |

Each of these formulas fits the simple pattern

$$Dx^n = nx^{n-1}. \tag{5}$$

As an inference from the instances listed above, (5) is only a conjecture. But many discoveries in mathematics are made by detecting such apparent patterns, and later proving that they hold universally.

Eventually we shall see that Formula (5), called the **power rule**, is valid for all real numbers $n$. At this time we give a proof only for the case when the exponent $n$ is a *positive integer*. We need the *binomial formula* from high-school algebra, which says that

$$(a + b)^n = a^n + na^{n-1}b + \frac{n(n-1)}{1 \cdot 2} a^{n-2}b^2$$

$$+ \cdots + \frac{n(n-1)\cdots(n-k+1)}{1 \cdot 2 \cdot 3 \cdots k} a^{n-k}b^k$$

$$+ \cdots + nab^{n-1} + b^n \tag{6}$$

if $n$ is a positive integer. The cases $n = 2$ and $n = 3$ are the familiar formulas

$$(a + b)^2 = a^2 + 2ab + b^2$$

and

$$(a + b)^3 = a^3 + 3a^2b + 3ab^2 + b^3.$$

There are $n + 1$ terms on the right-hand side of (6), but all we need to know here is the exact form of the first two terms, and the fact that all the others include $b^2$ as a factor.

---

**Theorem 2  Power Rule for n a Positive Integer**

If $n$ is a positive integer and $f(x) = x^n$, then $f'(x) = nx^{n-1}$.

---

***Proof***   The binomial formula gives

$$(x + h)^n = x^n + nx^{n-1}h + \{(n - 1) \text{ terms each involving } h^2\}.$$

Therefore,

$$f'(x) = \lim_{h \to 0} \frac{(x + h)^n - x^n}{h}$$

$$= \lim_{h \to 0} \frac{1}{h} \left[ (x^n + nx^{n-1}h + \cdots) - x^n \right]$$

$$= \lim_{h \to 0} \left[ nx^{n-1} + \{(n - 1) \text{ terms each involving } h\} \right]$$

$$= nx^{n-1},$$

because each of the $n - 1$ terms having $h$ as a factor has limit zero.   ∎

For example, $Dx^{17} = 17x^{16}$ and $Dx^{1000} = 1000x^{999}$. Of course, we need not always use $x$ as the independent variable. We may write the power rule with other symbols for the independent variable as

$$Dt^n = nt^{n-1}, \quad Du^n = nu^{n-1}, \quad \text{and} \quad Dy^n = ny^{n-1}.$$

To use the power rule to differentiate polynomials, we need to know how to differentiate linear combinations. A **linear combination** of the functions $f$ and $g$ is a function of the form $af + bg$, where $a$ and $b$ are constants. It follows from the sum and product laws for limits that

$$\lim_{x \to c} \left[ af(x) + bg(x) \right] = a \lim_{x \to c} f(x) + b \lim_{x \to c} g(x) \tag{7}$$

provided that the two right-hand limits exist. Formula (7) is called the **linearity property** of the limit operation. It implies an analogous linearity property of differentiation.

---

**Theorem 3**   *Derivative of a Linear Combination*

If $f$ and $g$ are differentiable functions, then

$$D[af(x) + bg(x)] = aDf(x) + bDg(x). \tag{8}$$

With $u = f(x)$ and $v = g(x)$, this takes the form

$$\frac{d(au + bv)}{dx} = a\frac{du}{dx} + b\frac{dv}{dx}. \tag{8'}$$

---

***Proof***   The linearity property of limits immediately gives

$$D[af(x) + bg(x)] = \lim_{h \to 0} \frac{[af(x + h) + bg(x + h)] - [af(x) + bg(x)]}{h}$$

$$= a\left(\lim_{h \to 0} \frac{f(x + h) - f(x)}{h}\right) + b\left(\lim_{h \to 0} \frac{g(x + h) - g(x)}{h}\right)$$

$$= aDf(x) + bDg(x). \qquad \blacksquare$$

Now take $a = c$ and $b = 0$ in Equation (8). The result is

$$D[cf(x)] = cDf(x) \quad \text{or} \quad \frac{d(cu)}{dx} = c\frac{du}{dx}. \tag{9}$$

That is, *the derivative of a constant multiple of a function is the same constant multiple of its derivative.*

Next, take $a = b = 1$ in Equation (8). We find that

$$D[f(x) + g(x)] = Df(x) + Dg(x) \quad \text{or} \quad \frac{d(u + v)}{dx} = \frac{du}{dx} + \frac{dv}{dx}. \tag{10}$$

Thus *the derivative of the sum of two functions is the sum of their derivatives.* Similarly, for differences we have

$$\frac{d(u - v)}{dx} = \frac{du}{dx} - \frac{dv}{dx}. \tag{11}$$

Repeated application of (10) to a sum of a finite number of differentiable functions gives

$$\frac{d(u_1 + u_2 + \cdots + u_n)}{dx} = \frac{du_1}{dx} + \frac{du_2}{dx} + \cdots + \frac{du_n}{dx}. \tag{12}$$

When we apply (9) and (12) and the power rule to the polynomial

$$p(x) = a_n x^n + a_{n-1} x^{n-1} + \cdots + a_2 x^2 + a_1 x + a_0,$$

we find the derivative *as fast as we can write it;* it is

$$p'(x) = na_n x^{n-1} + (n - 1)a_{n-1} x^{n-2} + \cdots + 2a_2 x + a_1. \tag{13}$$

With this result, it becomes a routine matter to write an equation for a tangent line to the graph of a polynomial.

**EXAMPLE 1**    Find the tangent line to the curve $y = 2x^3 - 7x^2 + 3x + 4$ at the point $(1, 2)$.

*Solution*    We compute the derivative as in (13) and find it to be

$$\frac{dy}{dx} = (2)(3x^2) - (7)(2x) + 3 = 6x^2 - 14x + 3.$$

We substitute $x = 1$ in the derivative and find that the slope of the tangent line at $(1, 2)$ is $m = -5$. So the point-slope equation of the tangent line is

$$y - 2 = -5(x - 1).$$

**EXAMPLE 2**    The volume in cubic centimeters of a quantity of water varies with changing temperature $T$. For $T$ between $0°C$ and $30°C$, the relationship is almost exactly

$$V = V_0(1 + \alpha T + \beta T^2 + \gamma T^3),$$

where $V_0$ is the volume at $0°C$, and the three constants have the values

$$\alpha = -0.06427 \times 10^{-3}, \qquad \beta = 8.5053 \times 10^{-6}, \qquad \gamma = -6.7900 \times 10^{-8}.$$

The rate of change of volume with respect to temperature is

$$\frac{dV}{dT} = V_0(\alpha + 2\beta T + 3\gamma T^2).$$

Suppose that $V_0 = 10^5$ cm$^3$ and that $T = 20°C$. Then $V \approx 100{,}157$ cm$^3$ and $dV/dT \approx 19.4$ cm$^3/°C$. We conclude that, at $T = 20°C$, this amount of water should increase in volume by approximately 19.4 cc for each rise of $1°C$ in temperature. By comparison, direct substitution into the above formula for $V$ gives

$$V(21) - V(20) \approx 19.9 \quad \text{cm}^3.$$

## THE PRODUCT AND QUOTIENT RULES

It might be natural to conjecture that the derivative of a product $f(x)g(x)$ is the product of the derivatives. This is *false*. For example, if $f(x) = g(x) = x$, then

$$D[f(x)g(x)] = Dx^2 = 2x$$

while

$$(Df(x))(Dg(x)) = (Dx)(Dx) = 1 \cdot 1 = 1.$$

In general, the derivative of a product is *not* merely the product of the derivatives. The following theorem tells what it *is*.

> **Theorem 4    The Product Rule**
>
> If $f$ and $g$ are differentiable at $x$, then $fg$ is differentiable at $x$, and
>
> $$D[f(x)g(x)] = f'(x)g(x) + f(x)g'(x). \tag{14}$$
>
> With $u = f(x)$ and $v = g(x)$, this takes the form
>
> $$\frac{d(uv)}{dx} = v\frac{du}{dx} + u\frac{dv}{dx}. \tag{14'}$$
>
> When it is unmistakably clear what the independent variable is, we can write the product rule even more briefly:
>
> $$(uv)' = u'v + uv'. \tag{14''}$$

**Proof**    We use an "add and subtract" device.

$$D[f(x)g(x)] = \lim_{h \to 0} \frac{f(x+h)g(x+h) - f(x)g(x)}{h}$$

$$= \lim_{h \to 0} \frac{f(x+h)g(x+h) - f(x)g(x+h) + f(x)g(x+h) - f(x)g(x)}{h}$$

$$= \lim_{h \to 0} \frac{f(x+h)g(x+h) - f(x)g(x+h)}{h}$$

$$\quad + \lim_{h \to 0} \frac{f(x)g(x+h) - f(x)g(x)}{h}$$

$$= \left( \lim_{h \to 0} \frac{f(x+h) - f(x)}{h} \right) \left( \lim_{h \to 0} g(x+h) \right)$$

$$\quad + f(x)\left( \lim_{h \to 0} \frac{g(x+h) - g(x)}{h} \right)$$

$$= f'(x)g(x) + f(x)g'(x).$$

In this proof we used the sum and product laws for limits, the definitions of $f'(x)$ and $g'(x)$, and the fact that

$$\lim_{h \to 0} g(x+h) = g(x).$$

The last equation holds because $g$ is differentiable and therefore continuous at $x$ (by Theorem 5 in Section 2-4). ∎

In words, the product rule says that *the derivative of a product is equal to the second factor times the derivative of the first factor, plus the first factor times the derivative of the second.*

**EXAMPLE 3**    Find the derivative of

$$f(x) = (1 - 4x^3)(3x^2 - 5x + 2)$$

without first multiplying the two factors.

*Solution*

$$D[(1 - 4x^3)(3x^2 - 5x + 2)]$$
$$= [D(1 - 4x^3)](3x^2 - 5x + 2) + (1 - 4x^3)[D(3x^2 - 5x + 2)]$$
$$= (-12x^2)(3x^2 - 5x + 2) + (1 - 4x^3)(6x - 5)$$
$$= -60x^4 + 80x^3 - 24x^2 + 6x - 5.$$

We can apply the product rule repeatedly to find the derivative of a product of three or more functions $u_1, u_2, \ldots, u_n$ of $x$. For example,

$$D[u_1 u_2 u_3] = D[(u_1 u_2) \cdot u_3]$$
$$= [D(u_1 u_2)] \cdot u_3 + (u_1 u_2) \cdot Du_3$$
$$= [(Du_1)u_2 + u_1(Du_2)]u_3 + (u_1 u_2)Du_3$$
$$= (Du_1)u_2 u_3 + u_1(Du_2)u_3 + u_1 u_2(Du_3).$$

The general result is

$$D(u_1 u_2 \cdots u_n) = (Du_1)u_2 \cdots u_n + u_1(Du_2)u_3 \cdots u_n$$
$$+ \cdots + u_1 u_2 \cdots u_{n-1}(Du_n), \tag{15}$$

where the last sum has one term for each of the $n$ factors. It is easy to establish this result (Problem 62) if we use the following strategy for a *proof by induction on n.*

---

*Principle of Mathematical Induction*

Suppose it is required to prove true for all positive integers $n$ a statement or formula involving the variable $n$. It is sufficient to show the following.

(i) The statement holds when $n = 1$.
(ii) The assumption that the statement holds for $n = k$ (a fixed but arbitrary positive integer) implies that it also holds for $n = k + 1$.

---

**EXAMPLE 4**   Use the principle of mathematical induction to establish the power rule for $n$ a positive integer.

*Solution*   We are to show that $Dx^n = nx^{n-1}$ for all positive integers $n$. This statement is clearly true when $n = 1$; that is, $Dx^1 = 1x^0$. Assume that it holds when $n = k$; that is, assume that $Dx^k = kx^{k-1}$ for some integer $k \geq 1$. The product rule then gives

$$Dx^{k+1} = D(x \cdot x^k) = (Dx) \cdot x^k + x \cdot (Dx^k).$$

Our assumption about the truth of the formula for $n = k$ now lets us continue these computations, and we find that

$$Dx^{k+1} = 1 \cdot x^k + x \cdot kx^{k-1} = x^k + kx^k = (k + 1)x^k.$$

This is the correct form of the power rule in the case $n = k + 1$. Thus we satisfy both conditions of the induction principle, and we conclude that $Dx^n = nx^{n-1}$ for all integral $n \geq 1$.

Our next theorem tells us how to find the derivative of the reciprocal of a function if we know the function's derivative.

---

*Theorem 5    The Reciprocal Rule*

If $f$ is differentiable at $x$ and $f(x) \neq 0$, then

$$D\frac{1}{f(x)} = -\frac{f'(x)}{[f(x)]^2}. \qquad (16)$$

With $u = f(x)$ the reciprocal rule takes the form

$$D\frac{1}{u} = -\frac{1}{u^2}\frac{du}{dx} \quad \text{or} \quad \left(\frac{1}{u}\right)' = -\frac{u'}{u^2}. \qquad (16')$$

---

**Proof**    As in the proof of Theorem 4, we use the limit laws, the definition of the derivative, and the fact that a function is continuous wherever it is differentiable (by Theorem 5 in Section 2-4). Here $f(x + h) \neq 0$ for $h$ near zero because $f(x) \neq 0$ and $f$ is continuous at $x$. Therefore,

$$D\frac{1}{f(x)} = \lim_{h \to 0}\frac{1}{h}\left(\frac{1}{f(x + h)} - \frac{1}{f(x)}\right)$$

$$= \lim_{h \to 0}\frac{f(x) - f(x + h)}{hf(x + h)f(x)}$$

$$= -\left(\lim_{h \to 0}\frac{1}{f(x + h)f(x)}\right)\left(\lim_{h \to 0}\frac{f(x + h) - f(x)}{h}\right)$$

$$= -\frac{f'(x)}{[f(x)]^2}. \qquad \blacksquare$$

For instance,

$$D\frac{1}{x^2 + 1} = -\frac{D(x^2 + 1)}{(x^2 + 1)^2} = -\frac{2x}{(x^2 + 1)^2}.$$

We now combine the reciprocal rule with the power rule for positive integral exponents to establish the power rule for negative integral exponents.

---

*Theorem 6    Power Rule for n a Negative Integer*

If $n$ is a negative integer, then $Dx^n = nx^{n-1}$.

---

**Proof**    Let $m = -n$, so that $m$ is a positive integer. Then

$$Dx^n = D\frac{1}{x^m} = -\frac{D(x^m)}{(x^m)^2}$$

$$= -\frac{mx^{m-1}}{x^{2m}} = (-m)x^{(-m)-1} = nx^{n-1}. \qquad \blacksquare$$

This proof also shows that the rule of Theorem 6 holds exactly when the function being differentiated is defined: when $x \neq 0$. As an illustration, $Dx^{-7} = -7x^{-8}$ when $x \neq 0$.

Now we apply the product and reciprocal rules to get a rule for differentiation of the quotient of two functions.

---

**Theorem 7**    *The Quotient Rule*

If $f$ and $g$ are differentiable at $x$ and $g(x) \neq 0$, then $f/g$ is differentiable at $x$ and

$$D\left(\frac{f(x)}{g(x)}\right) = \frac{f'(x)g(x) - f(x)g'(x)}{[g(x)]^2}. \tag{17}$$

With $u = f(x)$ and $v = g(x)$, this rule takes the forms

$$D\left(\frac{u}{v}\right) = \frac{v\dfrac{du}{dx} - u\dfrac{dv}{dx}}{v^2} \quad \text{and} \quad \left(\frac{u}{v}\right)' = \frac{u'v - uv'}{v^2}. \tag{17'}$$

---

***Proof***   We apply the product rule to the factorization

$$\frac{f(x)}{g(x)} = f(x) \cdot \frac{1}{g(x)}.$$

This gives

$$D\left(\frac{f(x)}{g(x)}\right) = (Df(x)) \cdot \frac{1}{g(x)} + f(x) \cdot D\frac{1}{g(x)}$$

$$= \frac{f'(x)}{g(x)} + f(x)\left(-\frac{g'(x)}{[g(x)]^2}\right)$$

$$= \frac{f'(x)g(x) - f(x)g'(x)}{[g(x)]^2}. \qquad \blacksquare$$

Note that the numerator in (17) is *not* the derivative of the product of $f$ and $g$. And the minus sign means that the order of the terms in the numerator is important.

**EXAMPLE 5**   Find $dz/dt$ if

$$z = \frac{1 - t^3}{1 + t^4}.$$

***Solution***   With $t$ rather than $x$ as the independent variable, the quotient rule gives

$$\frac{dz}{dt} = \frac{[D(1 - t^3)](1 + t^4) - (1 - t^3)[D(1 + t^4)]}{(1 + t^4)^2}$$

$$= \frac{(-3t^2)(1 + t^4) - (1 - t^3)(4t^3)}{(1 + t^4)^2} = \frac{t^6 - 4t^3 - 3t^2}{(1 + t^4)^2}.$$

**98**

Apply the differentiation rules of this section to find the derivatives of the functions given in Problems 1–40.

**1** $f(x) = 3x^2 - x + 5$      **2** $g(t) = 1 - 3t^2 - 2t^4$

**3** $f(x) = (2x + 3)(3x - 2)$

**4** $g(x) = (2x^2 - 1)(x^3 + 2)$

**5** $h(x) = (x + 1)^3$      **6** $g(t) = (4t - 7)^2$

**7** $f(y) = y(2y - 1)(2y + 1)$

**8** $f(x) = 4x^4 - \dfrac{1}{x^2}$

**9** $g(x) = \dfrac{1}{x + 1} - \dfrac{1}{x - 1}$

**10** $f(t) = \dfrac{1}{4 - t^2}$      **11** $h(x) = \dfrac{3}{x^2 + x + 1}$

**12** $f(x) = \dfrac{1}{1 - (2/x)}$

**13** $g(t) = (t^2 + 1)(t^3 + t^2 + 1)$

**14** $f(x) = (2x^3 - 3)(17x^4 - 6x + 2)$

**15** $g(z) = \dfrac{1}{2z} - \dfrac{1}{3z^2}$

**16** $f(x) = \dfrac{2x^3 - 3x^2 + 4x - 5}{x^2}$

**17** $g(y) = 2y(3y^2 - 1)(y^2 + 2y + 3)$

**18** $f(x) = \dfrac{x^2 - 4}{x^2 + 4}$      **19** $g(t) = \dfrac{t - 1}{t^2 + 2t + 1}$

**20** $u(x) = \dfrac{1}{(x + 2)^2}$      **21** $v(t) = \dfrac{1}{(t - 1)^3}$

**22** $h(x) = \dfrac{2x^3 + x^2 - 3x + 17}{2x - 5}$

**23** $g(x) = \dfrac{3x}{x^3 + 7x - 5}$      **24** $f(t) = \dfrac{1}{[t + (1/t)]^2}$

**25** $g(x) = \dfrac{(1/x) - (2/x^2)}{(2/x^3) - (3/x^4)}$

**26** $f(x) = \dfrac{x^3 - [1/(x^2 + 1)]}{x^4 + [1/(x^2 + 1)]}$

**27** $y = x^3 - 6x^5 + \frac{3}{2}x^{-4} + 12$

**28** $y = \dfrac{3}{x} - \dfrac{4}{x^2} - 5$      **29** $y = \dfrac{5 - 4x^2 + x^5}{x^3}$

**30** $y = \dfrac{2x - 3x^2 + 2x^4}{5x^2}$      **31** $y = 3x - \dfrac{1}{4x^2}$

**32** $y = \dfrac{1}{x(x^2 + 2x + 2)}$      **33** $y = \dfrac{x}{x - 1} + \dfrac{x + 1}{3x}$

**34** $y = \dfrac{1}{1 - 4x^{-2}}$      **35** $y = \dfrac{x^3 - 4x + 5}{x^2 + 9}$

**36** $y = x^2\left(2x^3 - \dfrac{3}{4x^4}\right)$      **37** $y = \dfrac{2x^2}{3x - 4/(5x^4)}$

**38** $y = \dfrac{4}{(x^2 - 3)^2}$

**39** $y = \dfrac{x^2}{x + 1}$      **40** $y = \dfrac{x + 10}{x^2}$

In each of Problems 41–50, write an equation of the tangent line to the curve $y = f(x)$ at the given point $P$ on the curve. Write the answer in the form $ax + by = c$.

**41** $y = x^3$;   $P(2, 8)$

**42** $y = 3x^2 - 4$;   $P(1, -1)$

**43** $y = \dfrac{1}{x - 1}$;   $P(2, 1)$

**44** $y = 2x - \dfrac{1}{x}$;   $P(0.5, -1)$

**45** $y = x^3 + 3x^2 - 4x - 5$;   $P(1, -5)$

**46** $y = \left(\dfrac{1}{x} - \dfrac{1}{x^2}\right)^{-1}$;   $P(2, 4)$

**47** $y = \dfrac{3}{x^2} - \dfrac{4}{x^3}$;   $P(-1, 7)$

**48** $y = \dfrac{3x - 2}{3x + 2}$;   $P(2, 0.5)$

**49** $y = \dfrac{3x^2}{x^2 + x + 1}$;   $P(-1, 3)$

**50** $y = \dfrac{6}{1 - x^2}$;   $P(2, -2)$

**51** Apply the formula of Example 2 to answer the following two questions.

(a) If 1000 cm³ of water at 0°C is heated, does it initially expand or contract?

(b) What is the rate (in cubic centimeters per degree Celsius) at which it contracts or expands?

**52** Susan's weight in pounds is given by the formula $W = 2 \times 10^9/R^2$, where $R$ is her distance in miles from the center of the earth. What is the rate of change of $W$ with respect to $R$ when $R = 3960$ mi? If Susan climbs a mountain, at what rate in ounces per (vertical) mile does her weight decrease?

**53** The conical tank shown in Fig. 3.7 has radius 160 cm and height 800 cm. Water is running out a small hole in the bottom of the tank. When the height $h$ of water in the tank is 600 cm, what is the rate of change of its volume $V$ with respect to $h$?

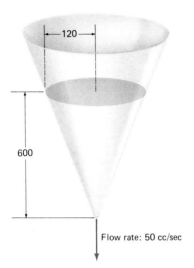

**3.7** The leaky tank of Problem 53

**54** Find the intercepts of the straight line that is tangent to the curve $y = x^3 + x^2 + x$ at the point $(1, 3)$.

**55** Find the line through the point $(1, 5)$ that is tangent to the curve $y = x^3$.

**56** Find *two* lines through the point $(2, 8)$ that are tangent to the curve $y = x^3$.

**57** Show that no straight line can be tangent to the curve $y = x^2$ at two different points.

**58** Find the two straight lines of slope $-2$ that are tangent to the curve $y = 1/x$.

**59** Find the $x$-intercept of the line that is tangent to the curve $y = x^n$ at the point $P(x_0, y_0)$.

**60** Show that the curve $y = x^5 + 2x$ has no horizontal tangent line. What is the smallest slope that a tangent to this curve can have?

**61** Apply Equation (15) with $n = 3$ and $u_1 = u_2 = u_3 = f(x)$ to show that

$$D([f(x)]^3) = 3[f(x)]^2 f'(x).$$

**62** Use the principle of mathematical induction to prove Equation (15).

**63** Apply (15) to show that

$$D([f(x)]^n) = n[f(x)]^{n-1} f'(x)$$

if $n$ is a positive integer and $f'(x)$ exists.

**64** Use the result of Problem 63 to find $D(x^2 + x + 1)^{100}$.

**65** Find $g'(x)$ if $g(x) = (x^3 - 17x + 35)^{17}$.

**66** Find the constants $a$, $b$, $c$, and $d$ if the curve $y = ax^3 + bx^2 + cx + d$ has horizontal tangent lines at the points $(0, 1)$ and $(1, 0)$.

---

## 3-3

## The Chain Rule

In Section 3-2 we saw how to differentiate polynomials and rational functions. But one often needs to differentiate *powers* of such functions. For example, suppose that

$$y = (x^2 - 1)^3. \qquad (1)$$

In order to find the derivative $dy/dx$ using the techniques of Section 3-2, we could first apply the binomial formula to write

$$y = x^6 - 3x^4 + 3x^2 - 1.$$

Then

$$\frac{dy}{dx} = 6x^5 - 12x^3 + 6x$$

$$= 6x(x^4 - 2x^2 + 1) = 6x(x^2 - 1)^2.$$

Thus

$$D(x^2 - 1)^3 = 6x(x^2 - 1)^2. \qquad (2)$$

Observe that the derivative of $(x^2 - 1)^3$ is **not** simply the quantity $3(x^2 - 1)^2$ that one might naively guess in analogy with the correct formula $Dx^3 = 3x^2$. There is an additional factor of $2x$ that we may spotlight by writing

$$D(x^2 - 1)^3 = 3(x^2 - 1)^2(2x). \tag{3}$$

We can explain the origin of this extra factor by writing $y = (x^2 - 1)^3$ in the form

$$y = u^3 \quad \text{with} \quad u = x^2 - 1.$$

Then

$$\frac{dy}{dx} = D(x^2 - 1)^3,$$

$$\frac{dy}{du} = 3u^2 = 3(x^2 - 1)^2, \quad \text{and}$$

$$\frac{du}{dx} = 2x,$$

so the derivative formula in (3) takes the form

$$\frac{dy}{dx} = \frac{dy}{du}\frac{du}{dx}. \tag{4}$$

The **chain rule** tells us that this last formula holds for *any* two differentiable functions $y = g(u)$ and $u = f(x)$. The formula in (3) is simply the special case of Equation (4) with $g(u) = u^3$ and $f(x) = x^2 - 1$.

**EXAMPLE 1** If

$$y = (3x + 5)^{17},$$

it would be impractical to write the binomial expansion of the seventeenth power of $3x + 5$: The result would be a polynomial with 18 terms, and some of the coefficients would have 14 digits. But if we write

$$y = u^{17} \quad \text{with} \quad u = 3x + 5,$$

then

$$\frac{dy}{du} = 17u^{16} \quad \text{and} \quad \frac{du}{dx} = 3.$$

Hence the chain rule in (4) yields

$$D(3x + 5)^{17} = \frac{dy}{dx} = \frac{dy}{du}\frac{du}{dx}$$

$$= (17u^{16})(3)$$

$$= 17(3x + 5)^{16}(3) = 51(3x + 5)^{16}.$$

---

The formula in (4) is one that, once learned, is truly unlikely to be forgotten. Although $dy/du$ and $du/dx$ are *not* fractions—they are merely symbols representing the derivatives $g'(u)$ and $f'(x)$—it is just as though they *were* fractions, with the $du$ in the first "canceling" the $du$ in the second. Of course, such "cancellation" no more proves the chain rule than canceling $d$'s proves that

$$\frac{dy}{dx} = \frac{y}{x} \qquad \text{(an absurdity)}.$$

It is nevertheless an excellent way to *remember* the chain rule. Such manipulations with differentials are so suggestive (even when invalid) that they played a substantial role in the early development of calculus in the seventeenth and eighteenth centuries. For thus were produced formulas that were later proved valid (as well as some formulas that were incorrect).

Though the formula in (4) is a memorable statement of the chain rule in differential notation, it suffers from the defect of not specifying where each of the derivatives is evaluated. This is better said in the notation in which derivatives are expressed as functions. Let us write

$$y = g(u), \qquad u = f(x), \qquad y = h(x) = g(f(x)). \qquad (5)$$

Then

$$\frac{du}{dx} = f'(x), \qquad \frac{dy}{dx} = h'(x),$$

and

$$\frac{dy}{du} = g'(u) = g'(f(x)). \qquad (6)$$

Substitution of these derivatives in (4) now recasts the chain rule in the form

$$h'(x) = g'(f(x)) \cdot f'(x). \qquad (7)$$

In this form the chain rule gives the derivative of the *composition* of two functions in terms of *their* derivatives.

---

**Theorem** *The Chain Rule*

Suppose that $f$ is differentiable at $x$ and that $g$ is differentiable at $f(x)$. Then the composition $h = g(f)$ is differentiable at $x$, and its derivative there is

$$D[g(f(x))] = g'(f(x)) \cdot f'(x). \qquad (8)$$

---

This is important: While the derivative of $h = g(f)$ is a product of the derivatives of $f$ and $g$, these latter derivatives are evaluated at *different* points. For $g'$ is evaluated at $f(x)$, while $f'$ is evaluated at $x$. For a particular number $x = x_0$, (3) tells us that

$$h'(x_0) = g'(u_0)f'(x_0) \quad \text{where} \quad u_0 = f(x_0). \qquad (9)$$

For example, if $h(x) = g(f(x))$, where $f$ and $g$ are differentiable functions, with

$$f(2) = 17, \qquad f'(2) = -3, \quad \text{and} \quad g'(17) = 5,$$

then the chain rule gives

$$h'(2) = g'(f(2)) \cdot f(2) = g'(17) \cdot f'(2) = (5)(-3) = -15.$$

To *outline* a proof of the chain rule, suppose that we are given $y = g(u)$ and $u = f(x)$ and want to compute the derivative

$$\frac{dy}{dx} = \lim_{\Delta x \to 0} \frac{\Delta y}{\Delta x} = \lim_{\Delta x \to 0} \frac{g(f(x + \Delta x)) - g(f(x))}{\Delta x}. \tag{10}$$

The differential form of the chain rule suggests the factorization

$$\frac{\Delta y}{\Delta x} = \frac{\Delta y}{\Delta u} \frac{\Delta u}{\Delta x}. \tag{11}$$

The product law of limits then gives

$$\frac{dy}{dx} = \lim_{\Delta x \to 0} \frac{\Delta y}{\Delta u} \frac{\Delta u}{\Delta x} \tag{12}$$

$$= \left( \lim_{\Delta u \to 0} \frac{\Delta y}{\Delta u} \right) \left( \lim_{\Delta x \to 0} \frac{\Delta u}{\Delta x} \right) = \frac{dy}{du} \frac{du}{dx}.$$

This will suffice to prove the chain rule *provided that*

$$\Delta u = f(x + \Delta x) - f(x) \tag{13}$$

is a *nonzero* quantity that approaches zero as $\Delta x \to 0$. Certainly $\Delta u \to 0$ as $\Delta x \to 0$ because $f$ is differentiable and thus continuous. But it is quite possible that $\Delta u$ *is* zero for some (even all) nonzero values of $\Delta x$. In such a case, the proposed factorization

$$\frac{\Delta y}{\Delta x} = \frac{\Delta y}{\Delta u} \frac{\Delta u}{\Delta x}$$

would involve the *invalid* step of division by zero. Thus our proof is incomplete; a complete proof of the chain rule is given at the end of this section.

If we substitute $f(x) = u$ and $f'(x) = du/dx$ in (8), we get the hybrid form

$$D_x g(u) = g'(u) \frac{du}{dx} \tag{14}$$

of the chain rule that often is the most useful for purely computational purposes. The subscript $x$ in $D_x$ specifies that $g(u)$ is being differentiated with respect to $x$, rather than with respect to $u$ (as would be understood if we wrote $Dg(u)$ without any subscript).

Let us set $g(u) = u^n$ in (14), where $n$ is an integer. Because $g'(u) = nu^{n-1}$, we thereby obtain the *chain rule version*

$$D_x u^n = nu^{n-1} \frac{du}{dx} \tag{15}$$

of the power rule. If $u = f(x)$ is a differentiable function, then Equation (15) implies that

$$D[f(x)]^n = n[f(x)]^{n-1} f'(x). \tag{16}$$

(If $n - 1 < 0$, we must add the proviso that $f(x) \neq 0$ in order that the right-hand side in (16) be meaningful.) We refer to this chain rule version of the power rule as the **generalized power rule.**

**EXAMPLE 2**  To differentiate

$$y = \frac{1}{(2x^3 - x + 7)^2},$$

we first write

$$y = (2x^3 - x + 7)^{-2}$$

in order to apply the generalized power rule in (16) with $n = -2$. This gives

$$\frac{dy}{dx} = (-2)(2x^3 - x + 7)^{-3}D(2x^3 - x + 7)$$

$$= (-2)(2x^3 - x + 7)^{-3}(6x^2 - 1)$$

$$= \frac{2(1 - 6x^2)}{(2x^3 - x + 7)^3}.$$

**EXAMPLE 3**  Find the derivative $h'(z)$ of the function

$$h(z) = \left(\frac{z - 1}{z + 1}\right)^5.$$

*Solution*  The key to applying the generalized power rule is observing *what* the given function is a power *of*. Here,

$$h(z) = u^5 \quad \text{where} \quad u = \frac{z - 1}{z + 1},$$

and $z$ is the independent variable instead of $x$. Hence we apply first (15), then the quotient rule, to get

$$h'(z) = 5u^4 \frac{du}{dx}$$

$$= 5\left(\frac{z - 1}{z + 1}\right)^4 D_z\left(\frac{z - 1}{z + 1}\right)$$

$$= 5\left(\frac{z - 1}{z + 1}\right)^4 \cdot \frac{(1)(z + 1) - (z - 1)(1)}{(z + 1)^2}$$

$$= 5\left(\frac{z - 1}{z + 1}\right)^4 \cdot \frac{2}{(z + 1)^2} = \frac{10(z - 1)^4}{(z + 1)^6}.$$

The importance of the chain rule goes far beyond the power function differentiations illustrated in Examples 1–3. In Chapters 7 and 8 we will learn how to differentiate exponential, logarithmic, and trigonometric functions. Each time we learn a new differentiation formula—for the derivative $g'(x)$ of a new function $g(x)$—the formula in (14) immediately provides us with the chain rule version

$$D_x g(u) = g'(u)D_x u$$

of that formula. The step from the power rule $Dx^n = nx^{n-1}$ to the generalized power rule $D_x u^n = nu^{n-1}D_x u$ is our first instance of this general phenomenon.

## RATE-OF-CHANGE APPLICATIONS

Suppose that the physical or geometric quantity $p$ depends on the quantity $q$, which in turn depends on time $t$. Then the *dependent* variable $p$ is a function both of the *intermediate* variable $q$ and of the *independent* variable $t$, and the derivatives that appear in the chain rule formula

$$\frac{dp}{dt} = \frac{dp}{dq}\frac{dq}{dt}$$

are rates of change (as in Section 3-1) of these variables with respect to one another. For example, suppose that a spherical balloon is being inflated or deflated. Then its volume $V$ and its radius $r$ are changing with time $t$, and

$$\frac{dV}{dt} = \frac{dV}{dr}\frac{dr}{dt}.$$

Remember that a positive derivative signals an increasing quantity and that a negative derivative signals a decreasing quantity.

**EXAMPLE 4**  A spherical balloon is being inflated, and its radius $r$ is increasing at the rate of 0.2 in./s when $r = 5$ in. At what rate is the volume $V$ of the balloon increasing at that instant?

**Solution**  Given $dr/dt = 0.2$ in./s when $r = 5$ in., we want to find $dV/dt$. Since the volume of the balloon is

$$V = \tfrac{4}{3}\pi r^3,$$

we see that $dV/dr = 4\pi r^2$. So the chain rule gives

$$\frac{dV}{dt} = \frac{dV}{dr}\frac{dr}{dt} = 4\pi r^2\frac{dr}{dt}$$

$$= 4\pi(5)^2(0.2) \approx 62.83$$

in.$^3$/s at the instant mentioned.

---

Observe that in Example 4 we did not need to know $r$ explicitly as a function of $t$. On the other hand, suppose we are told that after $t$ seconds the radius (in inches) of an inflating balloon is $r = 3 + (0.2)t$ (at least until the balloon bursts). Then the volume of the balloon is

$$V = \frac{4}{3}\pi r^3 = \frac{4}{3}\pi\left(3 + \frac{t}{5}\right)^3,$$

so $dV/dt$ is given explicitly as a function of $t$ by

$$\frac{dV}{dt} = \frac{4}{3}\pi(3)\left(3 + \frac{t}{5}\right)^2\left(\frac{1}{5}\right)$$

$$= \frac{4}{5}\pi\left(3 + \frac{t}{5}\right)^2.$$

**EXAMPLE 5**  Imagine a spherical waterdrop falling through water vapor in the air. Suppose that the vapor adheres to its surface in such a way that the time rate of increase of the droplet's mass $M$ is proportional to the droplet's surface area $S$. If the drop starts its fall with a radius that is effectively zero, and $r = 1$ mm after 20 s, when is the radius 3 mm?

***Solution*** We are given that

$$\frac{dM}{dt} = kS$$

where $k$ is some constant that depends upon atmospheric conditions. Now

$$M = \tfrac{4}{3}\rho\pi r^3 \quad \text{and} \quad S = 4\pi r^2$$

where $\rho$ is the density of water. Hence the chain rule gives

$$4\pi k r^2 = kS = \frac{dM}{dt} = \frac{dM}{dr}\frac{dr}{dt},$$

or

$$4\pi k r^2 = 4\pi\rho r^2 \frac{dr}{dt}.$$

This implies that

$$\frac{dr}{dt} = \frac{k}{\rho}, \qquad \text{a constant.}$$

So the radius of the droplet grows at a *constant* rate. Thus if it takes 20 s for $r$ to grow to 1 mm, it will take 1 min for it to grow to 3 mm.

---

## PROOF OF THE CHAIN RULE

To prove the chain rule, we need to show that if $f$ is differentiable at $a$ and $g$ is differentiable at $f(a)$, then

$$\lim_{h \to 0} \frac{g(f(a + h)) - g(f(a))}{h} = g'(f(a))f'(a). \tag{17}$$

If the quantities $h$ and

$$k(h) = f(a + h) - f(a) \tag{18}$$

are nonzero, then we can write the difference quotient on the left-hand side in (17) as

$$\frac{g(f(a + h)) - g(f(a))}{h} = \frac{g(f(a) + k(h)) - g(f(a))}{k(h)} \cdot \frac{k(h)}{h}. \tag{19}$$

To investigate the first factor on the right-hand side in (19), we define the function $\phi$ as follows:

$$\phi(k) = \begin{cases} \dfrac{g(f(a) + k) - g(f(a))}{k} & \text{if } k \neq 0; \\[2ex] g'(f(a)) & \text{if } k = 0. \end{cases} \tag{20}$$

By the definition of the derivative of $g$, we see from (20) that $\phi$ is continuous at $k = 0$; that is,

$$\lim_{k \to 0} \phi(k) = g'(f(a)). \tag{21}$$

Next,

$$\lim_{h \to 0} k(h) = \lim_{h \to 0} \left[ f(a + h) - f(a) \right] = 0 \tag{22}$$

because $f$ is continuous at $x = a$, and $\phi(0) = g'(f(a))$. It therefore follows from (21) that

$$\lim_{h \to 0} \phi(k(h)) = g'(f(a)). \tag{23}$$

We are now ready to assemble all this information. Note from (19) that if $h \neq 0$, then

$$\frac{g(f(a + h)) - g(f(a))}{h} = \phi(k(h)) \cdot \frac{f(a + h) - f(a)}{h} \tag{24}$$

even if $k(h) = 0$, in which case both sides in (24) are zero. Hence the product rule for limits yields

$$\lim_{h \to 0} \frac{g(f(a + h)) - g(f(a))}{h} = \lim_{h \to 0} \phi(k(h)) \frac{f(a + h) - f(a)}{h}$$
$$= g'(f(a)) \cdot f'(a),$$

a consequence of Equation (23) and the definition of the derivative of the function $f$. We have therefore established the chain rule in the form in Equation (17). ∎

## 3-3 PROBLEMS

Find $dy/dx$ in Problems 1–12.

**1** $y = (3x + 4)^5$

**2** $y = (2 - 5x)^3$

**3** $y = \dfrac{1}{3x - 2}$

**4** $y = \dfrac{1}{(2x + 1)^3}$

**5** $y = (x^2 + 3x + 4)^3$

**6** $y = (7 - 2x^3)^{-4}$

**7** $y = (2 - x)^4(3 + x)^7$
**8** $y = (x + x^2)^5(1 + x^3)^2$

**9** $y = \dfrac{x + 2}{(3x - 4)^3}$

**10** $y = \dfrac{(1 - x^2)^3}{(4 + 5x + 6x^2)^2}$

**11** $y = [1 + (1 + x)^3]^4$
**12** $y = [x + (x + x^2)^{-3}]^{-5}$

In each of Problems 13–20, express the derivative $dy/dx$ in terms of $x$.

**13** $y = (u + 1)^3$ and $u = \dfrac{1}{x^2}$

**14** $y = \dfrac{1}{2u} - \dfrac{1}{3u^2}$ and $u = 2x + 1$

**15** $y = (1 + u^2)^3$ and $u = (4x - 1)^2$

**16** $y = u^5$ and $u = \dfrac{1}{3x - 2}$

**17** $y = u(1 - u)^3$ and $u = \dfrac{1}{x^4}$

**18** $y = \dfrac{u}{u + 1}$ and $u = \dfrac{x}{x + 1}$

**19** $y = u^2(u - u^4)^3$ and $u = \dfrac{1}{x^2}$

**20** $y = \dfrac{u}{(2u + 1)^4}$ and $u = x - \dfrac{2}{x}$

In each of Problems 21–26, identify a function $u$ of $x$ and an integer $n \neq 1$ such that $f(x) = u^n$. Then compute $f'(x)$.

**21** $f(x) = (2x - x^2)^3$

**22** $f(x) = \dfrac{1}{2 + 5x^3}$

**23** $f(x) = \dfrac{1}{(1 - x^2)^4}$

**24** $f(x) = (x^2 - 4x + 1)^3$

**25** $f(x) = \left(\dfrac{x + 1}{x - 1}\right)^7$

**26** $f(x) = \dfrac{(x^2 + x + 1)^7}{(x + 1)^4}$

Differentiate each of the functions given in Problems 27–36.

**27** $g(y) = y + (2y - 3)^5$

**28** $h(z) = z^2(z^2 + 4)^3$

29 $F(s) = \left(s - \dfrac{1}{s^2}\right)^3$

30 $G(t) = \left[t^2 + \left(1 + \dfrac{1}{t}\right)\right]^2$

31 $f(u) = (1 + u)^3(1 + u^2)^4$

32 $g(w) = (w^2 - 3w + 4)(w + 4)^5$

33 $h(v) = \left[v - \left(1 - \dfrac{1}{v}\right)^{-1}\right]^{-2}$

34 $p(t) = \left(\dfrac{1}{t} + \dfrac{1}{t^2} + \dfrac{1}{t^3}\right)^{-4}$

35 $F(z) = \dfrac{1}{(3 - 4z + 5z^5)^{10}}$

36 $G(x) = \{1 + [x + (x^2 + x^3)^4]^5\}^6$

In each of Problems 37–44, $dy/dx$ can be found in two ways—one way using the chain rule, the other way without using it. Use both techniques to find $dy/dx$, then compare the answers (they should agree!).

37 $y = (x^3)^4 = x^{12}$

38 $y = x = \left(\dfrac{1}{x}\right)^{-1}$

39 $y = (x^2 - 1)^2 = x^4 - 2x^2 + 1$

40 $y = (1 - x)^3 = 1 - 3x + 3x^2 - x^3$

41 $y = (x + 1)^4 = x^4 + 4x^3 + 6x^2 + 4x + 1$

42 $y = (x + 1)^{-2} = \dfrac{1}{x^2 + 2x + 1}$

43 $y = (x^2 + 1)^{-1} = \dfrac{1}{x^2 + 1}$

44 $y = (x^2 + 1)^2 = (x^2 + 1)(x^2 + 1)$

In Chapter 8 we will establish that $D \sin x = \cos x$ (provided that $x$ is in radian measure). Use this fact and the chain rule to find the derivatives of the functions given in Problems 45–48.

45 $f(x) = \sin(x^3)$      46 $g(t) = (\sin t)^3$

47 $g(z) = (\sin 2z)^3$      48 $k(u) = \sin(1 + \sin u)$

49 At what rate is the area of a circle increasing when its radius is 10 in. and is increasing at the rate of 2 in./s?

50 At what rate is the radius of a circle decreasing when its area is $75\pi$ in.$^2$ and is decreasing at the rate of $2\pi$ in.$^2$/s?

51 At what rate is the area of a square increasing when its edge is 10 in. and is increasing at 2 in./s?

52 At what rate is the area of an equilateral triangle increasing when its edge is 10 in. and is increasing at 2 in./s?

53 A cubical block of ice is melting in such a way that its edge decreases steadily by 2 in. every hour. At what rate is its volume decreasing when its edge is 10 in.?

54 Find $f'(-1)$ given $f(y) = h(g(y))$, $g(-1) = 2$, $h'(2) = -1$, and $g'(-1) = 7$.

55 Given: $G(t) = f(h(t))$, $h(1) = 4$, $f'(4) = 3$, and $h'(1) = -6$. Find $G'(1)$.

56 Suppose that $f(0) = 0$ and that $f'(0) = 1$. Calculate the value of the derivative of $f(f(f(x)))$ at $x = 0$.

57 Air is being pumped into a spherical balloon in such a way that its radius is increasing at the rate of $dr/dt = 1$ cm/s. What is the time rate of increase, in cubic centimeters per second, of the balloon's volume when its radius $r$ is 10 cm?

58 Suppose that the air is being pumped into the balloon of Problem 57 at the constant rate of $200\pi$ cm$^3$/s. What is the time rate of increase of the radius when $r = 5$ cm?

59 Air is escaping from a spherical balloon at the constant rate of $300\pi$ cm$^3$/s. What is the radius of the balloon when its radius is decreasing at 3 cm/s?

60 A spherical hailstone is losing mass by melting uniformly over its surface as it falls. At a certain time, its radius is 2 cm and its volume is decreasing at the rate of 0.1 cm$^3$/s. How fast is its radius decreasing at that time?

61 A spherical snowball is melting in such a way that the rate of decrease of its volume is proportional to its surface area. At 10 A.M. its volume was 500 in.$^3$, and at 11 A.M. its volume was 250 in.$^3$ When did the snowball finish melting? (See Example 5.)

62 A cubical block of ice with edge 20 in. begins to melt at 8 A.M. Its edge decreases steadily thereafter, and at 4 P.M. is 8 in. What was the rate of change of the volume of the block of ice at 12 noon?

63 Suppose that $u$ is a function of $v$, that $v$ is a function of $w$, that $w$ is a function of $x$, and that all these functions are differentiable. Explain why it follows from the chain rule that

$$\frac{du}{dx} = \frac{du}{dv}\frac{dv}{dw}\frac{dw}{dx}.$$

---

## 3-4

## Derivatives of Algebraic Functions

In Section 3-3 we saw that the chain rule yields the differentiation formula

$$D[f(x)]^n = n[f(x)]^{n-1}f'(x)$$

if $f(x)$ is a differentiable function and the exponent $n$ is an integer. According to Theorem 1 of this section, this **generalized power rule** holds not only when the exponent is an integer, but also when it is a rational number $r = p/q$ (where $p$ and $q$ are integers). Recall that rational powers are defined in terms of integral roots and powers as follows:

$$u^{p/q} = \sqrt[q]{u^p} = (\sqrt[q]{u})^p.$$

---

**Theorem 1   Generalized Power Rule**

If $f$ is differentiable at $x$ and $r$ is a rational number, then

$$D[f(x)]^r = r[f(x)]^{r-1}f'(x) \tag{1}$$

at all points where the right-hand side is meaningful: the points where $f(x) \neq 0$ if $r - 1 < 0$ and $f(x) > 0$ if an even root is involved.

---

In terms of the variable $u = f(x)$, we can write the generalized power rule in the form

$$D_x u^r = \frac{d(u^r)}{dx} = ru^{r-1}\frac{du}{dx}. \tag{1'}$$

Recall that the subscript $x$ in $D_x$ specifies that $u^r$ is being differentiated with respect to $x$, rather than with respect to $u$ (as would be understood if we wrote $Du^r$). When $u = f(x) = x$, we see that the rule in (1') takes the simpler form

$$Dx^r = rx^{r-1}, \tag{2}$$

the power rule for rational exponents. For example,

$$Dx^{1/2} = \tfrac{1}{2}x^{-1/2},$$

$$Dx^{3/2} = \tfrac{3}{2}x^{1/2},$$

$$Dx^{2/3} = \tfrac{2}{3}x^{-1/3},$$

and so forth.

To differentiate a function involving roots (or radicals) we first "prepare" it for application of the generalized power rule: We rewrite it as a power function with fractional exponents. The first three examples illustrate this technique.

**EXAMPLE 1**   If $y = 5\sqrt{x^3} - \dfrac{2}{\sqrt[3]{x}}$

then

$$y = 5x^{3/2} - 2x^{-1/3},$$

so

$$\frac{dy}{dx} = 5\left(\frac{3}{2}\right)x^{1/2} - 2\left(-\frac{1}{3}\right)x^{-4/3}$$

$$= \frac{15}{2}x^{1/2} + \frac{2}{3}x^{-4/3} = \frac{15}{2}\sqrt{x} + \frac{2}{3\sqrt[3]{x^4}}$$

provided that $x > 0$.

**EXAMPLE 2** With $f(x) = 2x^2 - 3x + 5$ and $r = \frac{1}{2}$, the generalized power rule gives

$$D\sqrt{2x^2 - 3x + 5} = D(2x^2 - 3x + 5)^{1/2}$$

$$= \frac{1}{2}(2x^2 - 3x + 5)^{-1/2}D(2x^2 - 3x + 5)$$

$$= \frac{4x - 3}{2\sqrt{2x^2 - 3x + 5}}.$$

**EXAMPLE 3** If $y = [5x + \sqrt[3]{(3x - 1)^4}]^{10}$, then (1') with $u = 5x + (3x - 1)^{4/3}$ gives

$$\frac{dy}{dx} = 10u^9 \frac{du}{dx}$$

$$= 10[5x + (3x - 1)^{4/3}]^9 D[5x + (3x - 1)^{4/3}]$$

$$= 10[5x + (3x - 1)^{4/3}]^9 [D(5x) + D(3x - 1)^{4/3}]$$

$$= 10[5x + (3x - 1)^{4/3}]^9 \left[5 + \left(\frac{4}{3}\right)(3x - 1)^{1/3}(3)\right]$$

$$= 10[5x + (3x - 1)^{4/3}]^9 [5 + 4\sqrt[3]{3x - 1}].$$

Note that in Example 3 we applied the generalized power rule twice in succession—first in the form $D_x u^{10} = 10u^9 D_x u$, a second time in computing $D_x u$.

### PROOF OF THE GENERALIZED POWER RULE

In Section 3-3 we verified that the formula

$$D[f(x)]^r = r[f(x)]^{r-1}f'(x)$$

of Equation (1) holds when the exponent $r$ is an *integer*. In extending the generalized power rule from integral to rational exponents, the most difficult step is the case in which $f(x) = x$ (so that $f'(x) \equiv 1$) and $r = 1/q$ where $q$ is a *positive* integer:

$$Dx^{1/q} = \frac{1}{q}x^{(1/q)-1}. \tag{3}$$

Because $x^{1/q} = \sqrt[q]{x}$, it is natural to call the formula in (3) the **root rule** of differentiation. The chain rule version of the root rule is

$$D_x u^{1/q} = \frac{1}{q}u^{(1/q)-1}D_x u \tag{3'}$$

with $u$ a differentiable function of $x$.

The root rule is established (independently) in Theorem 3 of this section. Assuming it for now, we can complete the proof of the generalized power rule

$$D[f(x)]^{p/q} = \frac{p}{q}[f(x)]^{(p/q)-1}f'(x)$$

for the rational exponent $r = p/q$ in the following way. First we write

$$[f(x)]^{p/q} = \{[f(x)]^p\}^{1/q} = u^{1/q}$$

with $q$ a positive integer and

$$u = [f(x)]^p, \quad \text{so that} \quad D_x u = p[f(x)]^{p-1}f'(x) \tag{4}$$

by the integral case of the generalized power rule ($p$ being an integer). Then

$$D_x[f(x)]^{p/q} = D_x u^{1/q}$$

$$= \frac{1}{q}u^{(1/q)-1}D_x u \qquad \text{(by Eq. (3'))}$$

$$= \frac{1}{q}\{[f(x)]^p\}^{(1/q)-1}p[f(x)]^{p-1}f'(x) \qquad \text{(using (4))}$$

$$= \frac{p}{q}[f(x)]^{p[(1/q)-1]+p-1}f'(x)$$

$$= \frac{p}{q}[f(x)]^{(p/q)-1}f'(x),$$

as desired (subject to the restrictions mentioned in the statement of Theorem 1). In Chapter 7 we will see that the generalized power rule continues to hold even when the exponent $r$ in (1) is an *irrational* number (such as $\sqrt{2}$ or $\pi$).

## DERIVATIVES OF INVERSE FUNCTIONS

Recall from Section 2-3 that $f$ and $g$ are called inverse functions provided that $f(g(x)) = x$ and $g(f(x)) = x$ whenever these compositions are defined. For example, $f(x) = x^2$ and $g(x) = x^{1/2}$, each with domain $x > 0$, are inverse functions. Recall also that a given function $f$ has an inverse function $g$ if and only if $f$ is one-to-one. The following theorem tells us how to differentiate $g$, provided that we already know how to differentiate $f$.

---

**Theorem 2** *Differentiation of an Inverse Function*

Suppose that the one-to-one function $f$ is differentiable with $f'(x) > 0$ for all $x$ in its domain. Then its inverse function $g$ is also differentiable, and

$$g'(x) = \frac{1}{f'(g(x))}. \tag{5}$$

---

COMMENT   This theorem is also true when the condition $f'(x) > 0$ is replaced by the condition $f'(x) < 0$. In Section 4-3 we will see that the truth of either of these inequalities on an interval $I$ implies that $f$ must be one-to-one on $I$, and thus that the inverse $f^{-1}$ of $f$ exists.

The formula in (5) is easy to remember in differential notation. Let us write $x = f(y)$ and $y = g(x)$. Then $dy/dx = g'(x)$ and $dx/dy = f'(y)$. So (5) becomes the seemingly inevitable formula

$$\frac{dy}{dx} = \frac{1}{\dfrac{dx}{dy}} \tag{5'}$$

It is important to remember that $dy/dx$ above is to be evaluated at $x$, while $dx/dy$ is to be evaluated at the corresponding value of $y$; namely, $y = g(x)$.

***Proof of Theorem 2***  Figure 3.8 helps make it clear that

$$\text{if} \quad k = g(x + h) - g(x), \quad \text{then} \quad h = f(y + k) - f(y).$$

Hence

$$g'(x) = \lim_{h \to 0} \frac{g(x + h) - g(x)}{h}$$

$$= \lim_{k \to 0} \frac{k}{f(y + k) - f(y)} \qquad \text{(Because } g \text{ is continuous by Theorem 2}$$
$$\text{of Section 2-4, } k \to 0 \text{ as } h \to 0.)$$

$$= \frac{1}{\displaystyle\lim_{k \to 0} \frac{f(y + k) - f(y)}{k}} = \frac{1}{f'(y)} = \frac{1}{f'(g(x))} \qquad \blacksquare$$

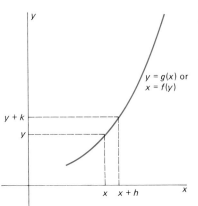

**3.8**  The relation between $h$ and $k$ in the proof of Theorem 2.

We can now use the fact that $f(x) = x^q$ and $g(x) = x^{1/q}$ are mutual inverses to help us compute the derivative of $x^{1/q}$.

---

***Theorem 3***  *The Root Rule*

$$Dx^{1/q} = \frac{1}{q} x^{(1/q) - 1}$$

if $x \neq 0$, and with the proviso that $x > 0$ if the positive integer $q$ is even.

---

***Proof***  If $y = x^{1/q}$ and $x = y^q$, then (5') gives

$$\frac{dy}{dx} = \frac{1}{\dfrac{dx}{dy}} = \frac{1}{qy^{q-1}} = \frac{1}{q(x^{1/q})^{q-1}},$$

so that

$$\frac{dy}{dx} = \frac{1}{q} x^{(1/q) - 1},$$

as desired.  $\blacksquare$

**3.9**  The graph of the cube root function

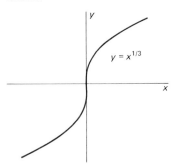

$y = x^{1/3}$

**EXAMPLE 4**  The root rule gives $Dx^{1/3} = \frac{1}{3}x^{-2/3}$, which increases without bound as $x$ approaches zero. How is this information related to the graph of $y = x^{1/3}$? The graph appears in Fig. 3.9. The derivative $dy/dx$ does not exist at $x = 0$. So the curve does not have a tangent line (as defined in Section 1-6) at the origin. Nevertheless, in this case we may regard the vertical line $x = 0$ as the tangent line to $y = x^{1/3}$ at $(0, 0)$. Examples such as this motivate the following definition.

Note that $f$ must be defined at $x = a$ if it is to have a tangent line at $(a, f(a))$.

If $f$ is defined (and differentiable) on only one side of $x = a$, we mean in this definition that $|f'(x)| \to \infty$ as $x$ approaches $a$ from that side.

**EXAMPLE 5**  Find the points on the curve

$$y = f(x) = x\sqrt{1 - x^2}, \qquad -1 \le x \le 1$$

at which the tangent line is either horizontal or vertical.

*Solution*  We differentiate using the product rule:

$$f'(x) = \sqrt{1 - x^2} + x\left(\frac{1}{2}\right)(1 - x^2)^{-1/2}(-2x)$$

$$= (1 - x^2)^{-1/2}[(1 - x^2) - x^2]$$

$$= \frac{1 - 2x^2}{\sqrt{1 - x^2}}.$$

Then $f'(x) = 0$ only when the numerator $1 - 2x^2$ is zero—that is, when $x = \pm 1/\sqrt{2}$. Because $f(\pm 1/\sqrt{2}) = \pm\frac{1}{2}$, the curve has a horizontal tangent line at each of the two points $(1/\sqrt{2}, \frac{1}{2})$ and $(-1/\sqrt{2}, -\frac{1}{2})$.

We observe also that the denominator $\sqrt{1 - x^2}$ approaches zero as $x \to -1^+$ and as $x \to 1^-$. Because $f(\pm 1) = 0$, we see that the curve has a vertical tangent line at each of the two points $(1, 0)$ and $(-1, 0)$.

---

## 3-4  PROBLEMS

Differentiate the functions given in Problems 1–34.

**1** $f(x) = \sqrt{x^3 + 1}$

**2** $f(x) = \dfrac{1}{(x^4 + 3)^2}$

**3** $f(x) = \sqrt{2x^2 + 1}$

**4** $f(x) = \dfrac{x}{\sqrt{1 + x^4}}$

**5** $f(t) = \sqrt{2t^3}$

**6** $g(t) = \sqrt{1/3t^5}$

**7** $f(x) = (2x^2 - x + 7)^{3/2}$

**8** $g(z) = (3z^2 - 4)^{97}$

**9** $g(x) = \dfrac{1}{(x - 2x^3)^{4/3}}$

**10** $f(t) = (t^2 + [1 + t]^4)^5$

**11** $f(x) = x\sqrt{1 - x^2}$

**12** $g(x) = \sqrt{\dfrac{2x + 1}{x - 1}}$

**13** $f(t) = \sqrt{\dfrac{t^2 + 1}{t^2 - 1}}$

**14** $h(y) = \left(\dfrac{y + 1}{y - 1}\right)^{17}$

**15** $f(x) = \left(x - \dfrac{1}{x}\right)^3$

**16** $g(z) = \dfrac{z^2}{\sqrt{1 + z^2}}$

**17** $f(v) = \dfrac{\sqrt{v + 1}}{v}$

**18** $h(x) = \left(\dfrac{x}{1 + x^2}\right)^{5/3}$

**19** $f(x) = \sqrt[3]{1 - x^2}$

**20** $g(x) = \sqrt{x + \sqrt{x}}$

**21** $f(x) = x(3 - 4x)^{1/2}$

**22** $g(t) = \dfrac{1}{t^2}\left[t - (1 + t^2)^{1/2}\right]$

**23** $f(x) = (1 - x^2)(2x + 4)^{4/3}$

**24** $f(x) = (1 - x)^{1/2}(2 - x)^{1/3}$

**25** $g(t) = \left(1 + \dfrac{1}{t}\right)^2 (3t^2 + 1)^{1/2}$

**26** $f(x) = x(1 + 2x + 3x^2)^{10}$

**27** $f(x) = \dfrac{2x - 1}{(3x + 4)^5}$

**28** $h(z) = (z - 1)^4(z + 1)^6$

**29** $f(x) = \dfrac{(2x + 1)^{1/2}}{(3x + 4)^{1/3}}$

**30** $f(x) = (1 - 3x^4)^5(4 - x)^{1/3}$

**31** $h(y) = \dfrac{(1 + y)^{1/2} + (1 - y)^{1/2}}{y^{5/3}}$

**32** $f(x) = (1 - x^{1/3})^{1/2}$

**33** $g(t) = [t + (t + t^{1/2})^{1/2}]^{1/2}$

**34** $f(x) = x^3\left(1 - \dfrac{1}{x^2 + 1}\right)^{1/2}$

For each curve given in Problems 35–40, find all points on the curve where the tangent line is either horizontal or vertical.

**35** $y = x^{2/3}$

**36** $y = x\sqrt{4 - x^2}$

**37** $y = x^{1/2} - x^{3/2}$

**38** $y = \dfrac{1}{\sqrt{9 - x^2}}$

**39** $y = \dfrac{x}{\sqrt{1 - x^2}}$

**40** $y = \sqrt{(1 - x^2)(4 - x^2)}$

**41** The period of oscillation, $P$ seconds, of a simple pendulum of length $L$ feet is given by $P = 2\pi\sqrt{L/g}$, where $g = 32$ ft/s². Find the rate of change of $P$ with respect to $L$ when $P = 2$.

**42** Find the rate of change of the volume $V$ of a sphere with respect to its surface area $S$ when its radius is 10.

**43** Find the two points on the circle $x^2 + y^2 = 1$ at which the slope of the tangent line is $-2$.

**44** Find the two points on the circle $x^2 + y^2 = 1$ at which the slope of the tangent line is 3.

**45** Find a line through the point $P(3, 0)$ that is normal to the parabola $y = x^2$ at some point $Q(a, a^2)$. (*Suggestion:* You will obtain a cubic equation in the unknown $a$. Find by inspection a small integral root $r$. The cubic polynomial is then the product of $a - r$ and a quadratic polynomial; you can find the quadratic by division of $a - r$ into the cubic.)

**46** Find three distinct lines through the point $(3, 10)$ that are normal to the parabola $y = x^2$. See the suggestion for Problem 45.

**47** Find two lines through the point $(0, \frac{5}{2})$ that are normal to the curve $y = x^{2/3}$.

---

**3-5**

## Maxima and Minima of Functions on Closed Intervals

In applications we often need to find the maximum or minimum value that a specified quantity can attain. The fence problem and the refrigerator problem stated in Section 1-1 are simple but typical examples of applied maximum-minimum problems. In Section 1-3 we saw that each of these two problems is equivalent to the purely mathematical problem of finding the maximum value attained by the function $f(x) = x(70 - x)$ on the closed interval $[0, 70]$.

In this section we discuss the general problem of finding the maximum and minimum values attained by a *continuous* function $f$ on a *closed* interval $[a, b]$. By the maximum value property of continuous functions (stated in Section 2-4), we know that $f$ *does* have maximum and minimum values on $[a, b]$. The question, then, is this: Exactly where *are* these values located?

In Section 1-6 we solved the fence and refrigerator problems on the basis of the geometrically motivated assumption that the function $f(x) = x(70 - x)$ attains its maximum value on $[0, 70]$ at an interior point of that interval, one at which the tangent line is horizontal. Theorems 1 and 2 provide a rigorous basis for the method we used there.

We say that the value $f(c)$ is a **local maximum value** of the function $f$ if $f(x) \leq f(c)$ for all $x$ sufficiently near $c$. More precisely, if this inequality holds for all $x$ that are simultaneously in the domain of $f$ and in some open interval containing $c$, then $f(c)$ is a local maximum of $f$. Similarly, we say that the value $f(c)$ is a **local minimum value** of $f$ if $f(x) \geq f(c)$ for all $x$ sufficiently near $c$. Thus a local maximum occurs where the value of $f$ is at least as large as it is anywhere nearby; a local minimum occurs where the function is at least as small as at nearby points. As the graph of Fig. 3.10 shows, a local maximum is a point such that no nearby points on the graph are higher, and a local minimum is one such that no nearby points on the curve are lower. A **local extremum** is a value of $f$ that is either a local maximum or a local minimum.

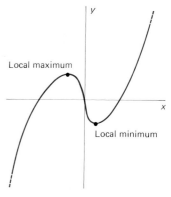

Local maximum

Local minimum

**3.10** Local extrema

---

> **Theorem 1**  *Local Maxima and Minima*
>
> If $f$ is differentiable at $c$ and is defined on an open interval containing $c$, and if $f(c)$ is either a local maximum value or a local minimum value of $f$, then $f'(c) = 0$.

---

Thus a local extremum of a *differentiable* function on an *open* interval can occur only at a point where the derivative is zero and, therefore, where the tangent line to the graph is horizontal.

**Proof**  Suppose, for instance, that $f(c)$ is a local maximum value. The fact that $f'(c)$ exists means that the right-hand and left-hand limits

$$\lim_{h \to 0^+} \frac{f(c + h) - f(c)}{h} \quad \text{and} \quad \lim_{h \to 0^-} \frac{f(c + h) - f(c)}{h}$$

both exist and are equal to $f'(c)$.

If $h > 0$, then

$$\frac{f(c + h) - f(c)}{h} \leqq 0$$

because $f(c) \geqq f(c + h)$ for all small positive values of $h$. Hence, by a one-sided version of the squeeze property for limits (Section 2-1), the above inequality will be preserved when we take the limit as $h \to 0$. We thus find that

$$f'(c) = \lim_{h \to 0^+} \frac{f(c + h) - f(c)}{h} \leqq \lim_{h \to 0^+} 0 = 0.$$

Similarly, in the case $h < 0$, we find that

$$\frac{f(c + h) - f(c)}{h} \geqq 0.$$

So

$$f'(c) = \lim_{h \to 0^-} \frac{f(c + h) - f(c)}{h} \geqq \lim_{h \to 0^-} 0 = 0.$$

Since both $f'(c) \leq 0$ and $f'(c) \geqq 0$, we may conclude that $f'(c) = 0$. This establishes Theorem 1. ∎

BEWARE   The converse of Theorem 1 is false. That is, the fact that $f'(c) = 0$ is *not enough* to imply that $f(c)$ is a local extremum. For example, consider the function $f(x) = x^3$. Its derivative $f'(x) = 3x^2$ vanishes at $x = 0$. But a glance at its graph, shown in Fig. 3.11, shows us that $f(0) = 0$ is *not* a local extremum of $x^3$.

**3.11**   No extremum at $x = 0$ even though the derivative is zero there

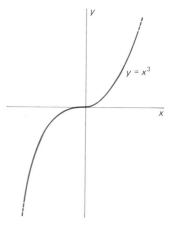

$y = x^3$

Thus the equation $f'(c) = 0$ is a *necessary* condition for $f(c)$ to be a local maximum or minimum value of $f$ (for a function $f$ differentiable on an open interval). It is *not* a *sufficient* condition. For $f'(x)$ can well be zero at points other than local maxima and minima. We shall give sufficient conditions for local maxima and minima in Chapter 4.

Now in most sorts of optimization problems, our interest is not in local maxima and minima as such, but rather in the global or *absolute* maximum and minimum values attained by a given continuous function. If $f$ is a function with domain $D$, we call $f(c)$ the **absolute maximum value** of $f$ on $D$ provided that $f(c) \geqq f(x)$ for *all* $x$ in the domain $D$. Briefly, $f(c)$ is the largest value of $f$ on $D$. It should be clear how to define the global or absolute minimum. The graph in Fig. 3.12 illustrates some local and global extrema. Note that every global extremum is, of course, local as well; on the other hand, the graph shows some local extrema that are not global.

Theorem 2 tells us that the absolute maximum and absolute minimum values of the continuous function $f$ on the closed interval $[a, b]$ both occur either at one of the end points $a$ and $b$ or at a critical point of $f$. The number $c$ in the domain of the function $f$ is called a **critical point** of $f$ if either

$$\text{(i) } f'(c) = 0, \qquad \text{or} \qquad \text{(ii) } f'(c) \text{ does not exist.}$$

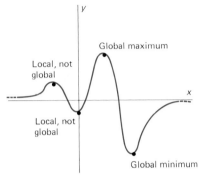

**3.12** Some extrema are global; others are merely local.

---

*Theorem 2    Absolute Maxima and Minima*

Suppose that $f(c)$ is the absolute maximum (or minimum) value of the continuous function $f$ on the closed interval $[a, b]$. Then $c$ is either a critical point of $f$ or one of the end points $a$ and $b$.

---

***Proof***    This follows almost immediately from Theorem 1. If $c$ is not an end point of $[a, b]$, then $f(c)$ is a local extremum of $f$ on the open interval $(a, b)$. In this case Theorem 1 implies that $f'(c) = 0$, provided that $f$ is differentiable at $c$. ∎

As a consequence of Theorem 2, we can find the (absolute) maximum and minimum values of $f$ on $[a, b]$ by

**1** First locating the critical points of $f$ in $[a, b]$; and
**2** Then finding the value of $f$ at each of these critical points *and* at the two end points.

The largest of these values must then be the absolute maximum value of $f$, and the smallest will be its absolute minimum value. We call this procedure the **closed interval maximum-minimum method.**

**EXAMPLE 1**    Find the maximum and minimum values of

$$f(x) = 2x^3 - 3x^2 - 12x + 15$$

on the closed interval $[0, 3]$.

***Solution***    The derivative of $f$ is

$$f'(x) = 6x^2 - 6x - 12 = 6(x - 2)(x + 1).$$

CHAP. 3:  Differentiation

Hence the critical points of $f$ are $-1$ and $+2$, but only the latter lies in $[0, 3]$. Thus the only critical point of $f$ in $[0, 3]$ is $x = 2$. Including the two end points, our list of possibilities for a maximum or minimum consists of $x = 0, 2,$ and $3$. We evaluate the function $f$ at each of these points:

$$f(0) = \phantom{-}15, \quad \longleftarrow \text{Maximum}$$

$$f(2) = -5, \quad \longleftarrow \text{Minimum}$$

$$f(3) = \phantom{-}6.$$

Therefore, the maximum value of $f(x)$ on $[0, 3]$ is $f(0) = 15$; the minimum value is $f(2) = -5$.

In Example 1 the function $f$ was differentiable everywhere. Examples 2 and 3 illustrate the case of an extremum at a critical point at which the function is not differentiable.

**EXAMPLE 2**  Find the maximum and minimum values of the function $f(x) = 3 - |x - 2|$ on the interval $[1, 4]$.

*Solution*  If $x \leq 2$, then $x - 2 \leq 0$; in this case

$$f(x) = 3 + (x - 2) = x + 1.$$

If $x \geq 2$ then $x - 2 \geq 0$, and so

$$f(x) = 3 - (x - 2) = 5 - x.$$

Consequently the graph of $f$ looks like the one shown in Fig. 3.13. The only critical point of $f$ in $[1, 4]$ is the point $x = 2$, because $f$ is not differentiable there. (Why?) Evaluation of $f$ at this critical point and at the two end points yields

$$f(1) = 2,$$

$$f(2) = 3, \quad \longleftarrow \text{Maximum}$$

$$f(4) = 1. \quad \longleftarrow \text{Minimum}$$

**3.13**  Graph of the function of Example 2

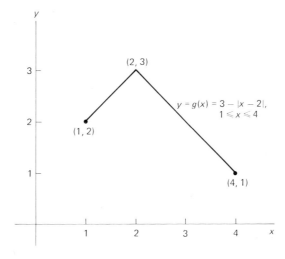

**EXAMPLE 3** Find the maximum and minimum values of

$$f(x) = 5x^{2/3} - x^{5/3}$$

on the closed interval $[-1, 4]$.

*Solution* Differentiation of $f$ yields

$$f'(x) = \frac{10}{3} x^{-1/3} - \frac{5}{3} x^{2/3}$$

$$= \frac{5}{3} x^{-1/3}(2 - x) = \frac{5(2 - x)}{3x^{1/3}}.$$

Hence $f$ has two critical points in the interval: $x = 2$, where the derivative is zero, and $x = 0$, where $f'$ does not exist—the graph has a vertical tangent line at $(0, 0)$. When we evaluate $f(x)$ at these two critical points and at the two end points, we get

$$f(-1) = 6, \qquad \longleftarrow \text{Maximum}$$

$$f(0) = 0, \qquad \longleftarrow \text{Minimum}$$

$$f(2) = (5)(2^{2/3}) - 2^{5/3} \approx 4.76,$$

$$f(4) = (5)(4^{2/3}) - 4^{5/3} \approx 2.52.$$

Thus the maximum value $f(-1) = 6$ occurs at an end point, and the minimum value $f(0) = 0$ occurs at the point where $f$ is not differentiable.

By plotting sufficiently many points, you could verify that the graph of $f$ looks as it is shown in Fig. 3.14. (In Chapter 4 we will study more efficient ways of sketching graphs.) But in the usual case of a continuous function having only finitely many critical points in a given closed interval, the closed interval maximum-minimum method suffices to determine its maximum and minimum values without our having any detailed knowledge of its graph.

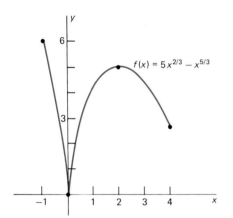
$f(x) = 5x^{2/3} - x^{5/3}$

**3.14** Graph of the function of Example 3

## 3-5 PROBLEMS

In each of Problems 1–30, find the maximum and minimum values attained by the given function on the indicated closed interval.

**1** $f(x) = 3x + 2$ on $[-2, 3]$

**2** $g(x) = 4 - 3x$ on $[-1, 5]$

**3** $h(x) = 4 - x^2$ on $[1, 3]$

**4** $f(x) = x^2 + 3$ on $[0, 5]$

**5** $g(x) = (x - 1)^2$ on $[-1, 4]$

**6** $h(x) = x^2 + 4x + 7$ on $[-3, 0]$

**7** $f(x) = x^3 - 3x$ on $[-2, 4]$

**8** $g(x) = 2x^3 - 9x^2 + 12x$ on $[0, 4]$

**9** $h(x) = x + \dfrac{4}{x}$ on $[1, 4]$

**10** $f(x) = x^2 + \dfrac{16}{x}$ on $[1, 3]$

**11** $f(x) = 3 - 2x$ on $[-1, 1]$

**12** $f(x) = x^2 - 4x + 3$ on $[0, 3]$

**13** $f(x) = 5 - 12x - 9x^2$ on $[-1, 1]$

**14** $f(x) = 2x^2 - 4x + 7$ on $[0, 2]$

**15** $f(x) = x^3 - 3x^2 - 9x + 5$ on $[-2, 4]$

**16** $f(x) = x^3 + x$ on $[-1, 2]$

**17** $f(x) = 3x^5 - 5x^3$ on $[-2, 2]$

**18** $f(x) = |2x - 3|$ on $[1, 2]$

**19** $f(x) = 5 + |7 - 3x|$ on $[1, 5]$

**20** $f(x) = |x + 1| + |x - 1|$ on $[-2, 2]$

**21** $f(x) = 50x^3 - 105x^2 + 72x$ on $[0, 1]$

**22** $f(x) = 2x + \dfrac{1}{2x}$ on $[1, 4]$

**23** $f(x) = \dfrac{x}{x + 1}$ on $[0, 3]$

**24** $f(x) = \dfrac{x}{1 + x^2}$ on $[0, 3]$

**25** $f(x) = \dfrac{1 - x}{x^2 + 3}$ on $[-2, 5]$

for its base and any appropriate rectangular piece to make into its curved side so long as the given conditions are met. What is the greatest possible volume of such a can?

**19** Three large squares of tin, each of edge length 1 m, have four equal small squares cut from their corners. All twelve resulting small squares are to be the same size. The three large cross-shaped pieces are then folded to form boxes without tops, and the twelve small squares are used to make two small cubes. How should this be done to maximize the total volume of all five boxes?

**20** A wire of length 100 cm is to be cut into two pieces. One piece is bent into a circle, the other into a square. How should the cut be made to maximize the sum of the areas of the square and the circle. How should it be done to minimize that sum?

**21** A farmer has 600 m of fencing with which she plans to enclose a rectangular divided pasture adjacent to a long existing wall. She plans to build one fence parallel to the wall, two to form the ends of the enclosure, and a fourth (parallel to the two ends of the enclosure) to divide it. What is the maximum area that she can enclose in this way?

**22** A rectangular outdoor pen is to be added to an animal house with a corner notch, as shown in Fig. 3.21. If 85 m of new fence are available, what should be the dimensions of the pen in order to maximize its area? Of course, no fence need be used along the walls of the animal house.

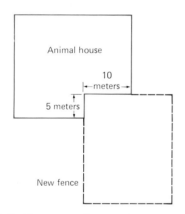

**3.21** The rectangular pen of Problem 22

**23** Suppose that a post office can accept a package for mailing by parcel post only if the sum of its length and its girth (the circumference of its cross section) is at most 100 in. What is the maximum volume of a rectangular box with square cross section that can be mailed?

**24** Repeat Problem 23, except use a package that is cylindrical; its cross section is circular.

**25** A printing company has eight presses, each of which can print 3600 copies per hour. It costs $5.00 to set up each press for a run, and $10 + 6n$ dollars to run $n$ presses for 1 hour. How many presses should be used in order to print 50,000 copies of a poster most profitably?

**26** A farmer wants to hire workers to pick 900 bushels of beans. Each worker can pick 5 bu/h and is paid $1.00/bu. The farmer must also pay a supervisor $10 per hour while the picking is in progress, and he has miscellaneous additional expenses of $8 for each worker hired. How many workers should he hire in order to minimize the total cost? What will then be the cost per bushel picked?

**27** The heating and cooling costs for a certain uninsulated house are $500 per year, but with $x \leq 10$ in. of insulation the costs would be $1000/(2 + x)$ dollars per year. It will cost $150 for each inch (thickness) of insulation installed. How many inches of insulation should be installed in order to minimize the *total* (initial plus annual) costs over a 10-year period? What will then be the annual savings resulting from this optimal insulation?

**28** A concessionaire had been selling 5000 wieners each game night at 50¢ each. When she raised the price to 70¢ each, sales dropped to 4000 per night. If she has fixed costs of $1000 per night and each wiener costs her 25¢, what price will maximize her nightly profit?

**29** A commuter train carries 600 passengers each day from a suburb to a city. It costs $1.50 per person to ride the train. It is found that 40 fewer people will ride the train for each 5¢ increase in the fare. What fare should be charged to make the largest possible revenue?

**30** Find the shape of the cylinder of maximum volume that can be inscribed in a sphere of radius $R$. Show that the ratio of the cylinder's height to its radius is $\sqrt{2}$, and that the ratio of the sphere's volume to that of the cylinder is $\sqrt{3}$.

**31** Find the dimensions of the right circular cylinder of greatest volume that can be inscribed in a right circular cone of radius $R$ and height $H$.

**32** In a circle of radius 1 is inscribed a trapezoid with the longer of its parallel sides coincident with a diameter of the circle. What is the maximum possible area of such a trapezoid? (*Suggestion:* A positive quantity is maximized when its square is maximized.)

**33** Show that the rectangle with maximum perimeter that can be inscribed in a circle is a square.

**34** Find the dimensions of the rectangle (with vertical and horizontal sides) of maximum area that can be inscribed in the ellipse with equation $x^2/25 + y^2/9 = 1$ (shown in Fig. 3.22).

**3.22** The ellipse of Problem 34

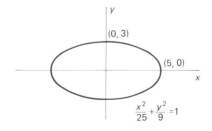

**35** A right circular cone of radius $r$ and height $h$ has slant height $L = (r^2 + h^2)^{1/2}$. What is the maximum possible volume of a cone with slant height 10?

**36** Two vertical poles are each 10 ft tall and stand 10 ft apart. Find the length of the shortest rope that can reach from the top of one pole to a point between them on the ground and then to the top of the other pole.

**37** The sum of two nonnegative numbers is 16. Find the maximum and the minimum possible value of the sum of their cube roots.

**38** A straight wire 60 cm long is bent into the shape of an $L$. What is the shortest possible distance between the two ends of the wire?

**39** What is the shortest possible distance from a point of the parabola $y = x^2$ to the point $(0, 1)$?

**40** There is exactly one point on the graph of $y = (3x - 4)^{1/3}$ closest to the origin. Find it. (*Suggestion:* Solve the equation that you obtain by inspection.)

**41** Find the dimensions that maximize the cross-sectional area of the four planks that can be cut from the four pieces of the circular log of Example 5, the pieces that remain after cutting a square beam (see Fig. 3.18).

**42** Assume that the strength of a rectangular beam is proportional to the product of its width and the *cube* of its depth. Find the dimensions of the strongest rectangular beam that can be cut from a cylindrical log of radius $R$.

**43** A small island is 2 km offshore in a large lake. A woman on the island can row her boat 10 km/h and can run 20 km/h. Where should she land her boat in order to most quickly reach a village on the straight shoreline of the lake and 6 km down the shoreline from the point nearest the island?

**44** A factory is located on one bank of a straight river that is 2000 m wide. On the opposite bank but 4500 m downstream is a power station from which the factory must draw its electricity. Assume that it costs three times as much per meter to lay an underwater cable as to lay a cable over the ground. What path should a cable connecting the power station to the factory take to minimize the cost of laying the cable?

**45** A company has plants that are located (in an appropriate coordinate system) at the points $A(0, 1)$, $B(0, -1)$, and $C(3, 0)$. The company plans to construct a distribution center at the point $P(x, 0)$. What should be the value of $x$ to minimize the sum of the distances of $P$ from $A$, $B$, and $C$? (The object is to minimize transportation costs from the plants to the center.)

**46** Light travels with speed $c$ in air and with a somewhat slower speed $v$ in water. (The constant $c$ is approximately $3 \times 10^{10}$ cm/s; the ratio $n = c/v$, known as the **index of refraction** of water, depends upon the color of the light, but is approximately 1.33.) In Fig. 3.23 we show the path of a light ray traveling from point $A$ in air to point $B$ in water, with what appears to be a sudden change in direction as the ray moves through the air-water interface.

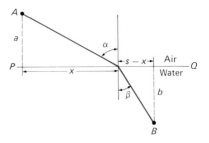

**3.23** Snell's law gives the path of refracted light.

(a) Write the time $T$ required for the ray to travel from $A$ to $B$ in terms of the variable $x$ and the constants $a, b, c, s,$ and $v$, all of which have been defined or are shown in the figure.

(b) Show that the equation $T'(x) = 0$ for minimizing $T$ is equivalent to the condition

$$\frac{\sin \alpha}{\sin \beta} = \frac{c}{v} = n.$$

This is **Snell's law:** The ratio of the sines of the angles of incidence and refraction is equal to the index of refraction.

**47** The mathematics of Snell's law is applicable to situations other than the refraction of light. Figure 3.24 shows a geological fault running from west to east, separating two towns at $A$ and $B$. Assume that $A$ is $a$ miles north of the fault, that $B$ is $b$ miles south of the fault, and that $B$ is $L$ miles east of $A$. We want to build a road from $A$ to $B$. Because of differences in terrain, the cost of construction is $C_1$ (in millions of dollars per mile) north of the fault and $C_2$ south of it. Where should the point $P$ be placed to minimize the total cost of road construction?

(a) Using the notation in the figure, show that the cost is minimal when $C_1 \sin \theta_1 = C_2 \sin \theta_2$.

(b) Take the case $a = b = C_1 = 1$, $C_2 = 2$, and $L = 4$. Show that the equation in part (a) is equivalent to

$$f(x) = 3x^4 - 24x^3 + 51x^2 - 32x + 64 = 0.$$

**3.24** Illustration for Problem 47

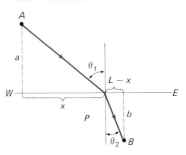

To approximate the desired solution of this equation, calculate $f(0)$, $f(1)$, $f(2)$, $f(3)$, and $f(4)$. You should find that $f(3)$ is positive, while $f(4)$ is negative. Interpolate between $x = 3$ and $x = 4$ to approximate the root of the above equation.

**48** The sum of the surface areas of a cube and a sphere is 1000 in.² What should be their dimensions in order to minimize the sum of their volumes? In order to maximize it?

**49** A kite frame is to be made of six pieces of wood, as shown in Fig. 3.25. The four outer pieces with the indicated lengths have already been cut. What should be the lengths of the two diagonal pieces in order to maximize the area of the kite?

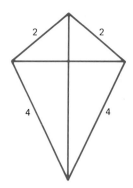

**3.25** The kite frame

<hr>

## 3-7
## Implicit Differentiation

An equation in two variables $x$ and $y$ may have one or more solutions for $y$ in terms of $x$ or for $x$ in terms of $y$. These solutions are functions that we say are **implicitly defined** by the equation. In this section we discuss the differentiation of such functions and the use of their derivatives in solving certain optimization problems.

For example, the equation $y^2 - x = 0$ of a parabola implicitly defines two continuous functions of $x$:

$$y = \sqrt{x} \quad \text{and} \quad y = -\sqrt{x}.$$

Each has domain the half line $x \geq 0$. The graphs of these two functions are the upper and lower branches of the parabola of Fig. 3.26. The whole parabola cannot be the graph of a function of $x$ because no vertical line can meet the graph of a function in more than one point.

The equation of the unit circle, $x^2 + y^2 = 1$, implicitly defines four functions (among others):

$$y = +\sqrt{1 - x^2} \quad \text{for} \quad x \text{ in } [-1, 1],$$

$$y = -\sqrt{1 - x^2} \quad \text{for} \quad x \text{ in } [-1, 1],$$

$$x = +\sqrt{1 - y^2} \quad \text{for} \quad y \text{ in } [-1, 1], \quad \text{and}$$

$$x = -\sqrt{1 - y^2} \quad \text{for} \quad y \text{ in } [-1, 1].$$

The four graphs are highlighted against the four unit circles in Fig. 3.27.

The equation $2x^2 + 2y^2 + 3 = 0$ implicitly defines *no* function, because this equation has no real solution $(x, y)$.

In advanced calculus one studies conditions that will guarantee when an implicitly defined function is actually differentiable. Here we will proceed on the assumption that our implicitly defined functions are differentiable at most points in their domains. (The functions that have the graphs shown in Fig. 3.27 are *not* differentiable at the endpoints of their domains.) This is a reasonable assumption whenever the equation we are working with is itself "nice," such as a polynomial in two variables.

When we assume differentiability, we can use the chain rule to differentiate the given equation, thinking of $x$ as the independent variable. We can

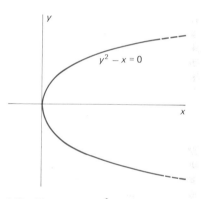

**3.26** The parabola $y^2 - x = 0$

**3.27** Continuous functions defined implicitly by $x^2 + y^2 = 1$

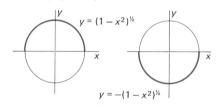

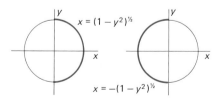

then solve the resulting equation for the derivative $dy/dx = f'(x)$ of the implicitly defined function. This process is called **implicit differentiation.**

**EXAMPLE 1**   Use implicit differentiation to find the derivative of a differentiable function $y = f(x)$ implicitly defined by the equation $x^2 + y^2 = 1$.

*Solution*   The equation $x^2 + y^2 = 1$ is to be thought of as an *identity* that implicitly defines $y$ as a function of $x$. Since $x^2 + y^2$ is then a function of $x$, it has the same derivative as the constant function 1 on the other side of the identity. Thus we may differentiate both sides of $x^2 + y^2 = 1$ with respect to $x$ and equate the results. We obtain

$$2x + 2y\frac{dy}{dx} = 0.$$

In this step, we must remember that $y$ is a function of $x$, so that $D_x(y^2) = 2yD_xy$.

Then we solve for $dy/dx$:

$$\frac{dy}{dx} = -\frac{x}{y}. \tag{1}$$

It may be surprising to see a formula for $dy/dx$ containing both $x$ *and* $y$, but such a formula is frequently just as useful as one containing only $x$. For example, the above formula tells us that the slope of the tangent line to the circle $x^2 + y^2 = 1$ at the point $(1/\sqrt{5}, 2/\sqrt{5})$ is

$$\frac{dy}{dx}\bigg|_{(1/\sqrt{5},\, 2/\sqrt{5})} = -\frac{1/\sqrt{5}}{2/\sqrt{5}} = -\frac{1}{2}.$$

We introduce here a self-explanatory notational device for specifying where $dy/dx$ is evaluated.

Note that if $y = \pm\sqrt{1 - x^2}$, then

$$\frac{dy}{dx} = \frac{-x}{\pm\sqrt{1 - x^2}} = -\frac{x}{y},$$

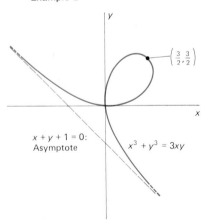

**3.28**  The folium of Descartes; see Example 2

$x + y + 1 = 0$: Asymptote

$x^3 + y^3 = 3xy$

$\left(\frac{3}{2}, \frac{3}{2}\right)$

in agreement with Equation (1). Thus (1) simultaneously gives us the derivatives of both the functions $y = +\sqrt{1 - x^2}$ and $y = -\sqrt{1 - x^2}$ defined implicitly by the equation $x^2 + y^2 = 1$.

**EXAMPLE 2**   The *folium of Descartes* is the graph of the equation

$$x^3 + y^3 = 3xy. \tag{2}$$

This curve was originally proposed by René Descartes as a challenge to Pierre Fermat (1601–1665) to find its tangent line at an arbitrary point. The graph of the folium of Descartes appears in Fig. 3.28; it has the line $x + y + 1 = 0$ as an asymptote. We indicate in Problem 24 of Section 13-1 how this graph can be constructed. Here we want to find the slope of its tangent line.

CHAP. 3:   Differentiation

***Solution*** We use implicit differentiation (rather than the ad hoc methods with which Fermat met Descartes' challenge). We differentiate both sides of (2) and find that

$$3x^2 + 3y^2 \frac{dy}{dx} = 3y + 3x \frac{dy}{dx}.$$

We solve for the derivative:

$$\frac{dy}{dx} = \frac{y - x^2}{y^2 - x}. \qquad (3)$$

For instance, at the point $(\frac{3}{2}, \frac{3}{2})$ on the folium, the slope of the tangent line is

$$\left. \frac{dy}{dx} \right|_{(3/2,\, 3/2)} = \frac{(\frac{3}{2}) - (\frac{3}{2})^2}{(\frac{3}{2})^2 - (\frac{3}{2})} = -1,$$

and this result agrees with our intuition about the figure. At the point $(\frac{2}{3}, \frac{4}{3})$, Equation (3) gives

$$\left. \frac{dy}{dx} \right|_{(2/3,\, 4/3)} = \frac{(\frac{4}{3}) - (\frac{2}{3})^2}{(\frac{4}{3})^2 - (\frac{2}{3})} = \frac{4}{5}.$$

Thus the equation of the tangent line at this point is

$$y - \tfrac{4}{3} = \tfrac{4}{5}(x - \tfrac{2}{3}), \quad \text{or} \quad 4x - 5y + 4 = 0.$$

## MAXIMUM-MINIMUM PROBLEMS WITH CONSTRAINTS

In applied maximum-minimum problems, the quantity $Q$ to be maximized or minimized often appears initially as a function of two variables, $Q = Q(x, y)$, with $x$ and $y$ related by an equation of the form

$$C(x, y) = K, \qquad (4)$$

where $K$ is some constant. That is, we are to find the maximum or minimum value of $Q(x, y)$ subject to the side condition or *constraint* that $x$ and $y$ satisfy Equation (4). For example, the problem of minimizing the total surface area of a box with a square base (as shown in Fig. 3.29) required to have volume $V = 1000$ is that of minimizing $A = 2x^2 + 4xy$ subject to the constraint $x^2 y = 1000$.

In Section 3-6 our method of solving such a problem was to solve the constraint equation (4) for $y$ as a function of $x$ and then substitute $y = y(x)$ into the expression $Q(x, y)$, the quantity to be maximized or minimized. This produced a function $Q = Q(x, y(x)) = f(x)$ of the single variable $x$, to which we then applied methods of calculus. The constraint equation was thus used to eliminate the "auxiliary" variable $y$.

Sometimes, however, it is difficult or even impossible to solve the constraint equation (4) for $y$ as a function of $x$. If so, an alternative method is available. It involves implicit differentiation and is called the **method of auxiliary variables.** Here, one works with both $x$ as the independent variable *and* the auxiliary variable $y$, now thought of as a function of $x$. Under the assumption that $dy/dx$ exists, we begin by differentiating $Q(x, y)$ and $C(x, y)$

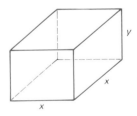

**3.29** A box with square base of edge $x$ and with height $y$

implicitly. Then

$$\frac{dQ}{dx} = 0 \tag{5}$$

at the maximum-minimum point being sought, while

$$\frac{dC}{dx} = 0 \tag{6}$$

also holds, the result of implicit differentiation of the constraint equation $C(x, y) = K$ (a constant).

It is always easy to eliminate $dy/dx$ from Equations (5) and (6); we use this technique in the examples below. The result is a single equation in $x$ and $y$. The final step is to solve this equation and the constraint equation (4) simultaneously. The resulting solutions $(x, y)$ are the points corresponding to horizontal tangents in our earlier method.

**EXAMPLE 3** Use the method of auxiliary variables to solve the sawmill problem (Example 5 in Section 3-6), the problem of finding the rectangle of largest area inscribed in the unit circle $x^2 + y^2 = 1$.

*Solution* We inscribe the rectangle with its sides parallel to the coordinate axes to keep matters simple. Let $(x, y)$ denote the first-quadrant vertex of the rectangle, as in Fig. 3.30. We want to maximize the area

$$A = 4xy$$

subject to the constraint

$$C = x^2 + y^2 = 1.$$

Equations (5) and (6) above now take the particular forms

$$\frac{dA}{dx} = 4y + 4x\frac{dy}{dx} = 0$$

and

$$\frac{dC}{dx} = 2x + 2y\frac{dy}{dx} = 0.$$

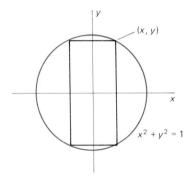

**3.30** The sawmill problem

The cases $x = 0$ and $y = 0$ do not produce maxima, so we may assume throughout that $x > 0$ and $y > 0$. With this assumption, the two equations above become

$$\frac{dy}{dx} = -\frac{y}{x} \quad \text{and} \quad \frac{dy}{dx} = -\frac{x}{y},$$

respectively. Thus

$$-\frac{y}{x} = -\frac{x}{y}, \quad \text{or} \quad x^2 = y^2.$$

The only first-quadrant solution of the equations

$$x^2 + y^2 = 1 \quad \text{and} \quad x^2 = y^2$$

is the point $(1/\sqrt{2}, 1/\sqrt{2})$, which determines a square with edge length $\sqrt{2}$. This is, of course, the same solution as the one we found in Example 5 of Section 3-6.

**EXAMPLE 4**  One common nuclear reactor design involves a cylindrical core with radius $r$ and height $h$. For a certain mixture of moderator and fissionable material, the reactor will reach critical mass and thus produce a self-sustaining nuclear reaction if $r$ and $h$ (in centimeters) satisfy the equation

$$\frac{\lambda^2}{r^2} + \frac{\pi^2}{h^2} = B^2 \qquad (7)$$

where $\lambda = 2.405$ and $B = 0.04$. Find the dimensions for which the critical mass volume $V = \pi r^2 h$ is minimal; that is, the least expensive reactor.

*Solution*  Let us regard $h$ as a function of $r$. Then

$$\frac{dV}{dr} = 2\pi r h + \pi r^2 \frac{dh}{dr}.$$

We also differentiate the constraint condition (7) and obtain

$$-\frac{2\lambda^2}{r^3} - \frac{2\pi^2}{h^3}\frac{dh}{dr} = 0.$$

We write the equation $dV/dr = 0$ and find that

$$2\pi r h + \pi r^2 \frac{dh}{dr} = 0.$$

We solve the last two equations for $dh/dr$:

$$\frac{dh}{dr} = -\frac{2h}{r} = -\frac{\lambda^2 h^3}{\pi^2 r^3},$$

and it follows that

$$h^2 = \frac{2\pi^2 r^2}{\lambda^2}.$$

We substitute this value of $h^2$ in Equation (7), and it is now easy to solve for

$$r = \frac{\lambda\sqrt{3}}{B\sqrt{2}} = \frac{(2.405)\sqrt{3}}{(0.04)\sqrt{2}} \approx 73.638$$

cm, or about 2.42 ft. This is the radius of the minimal volume critical mass core. The corresponding height is

$$h = \frac{\pi\sqrt{2r}}{\lambda} \approx \frac{\pi\sqrt{2(73.638)}}{2.405} \approx 136.04$$

cm, or about 4.46 ft.

In Problems 1–10, find $dy/dx$ by implicit differentiation.

1  $x^2 - y^2 = 1$      2  $xy = 1$
3  $16x^2 + 25y^2 = 400$      4  $x^3 + y^3 = 1$
5  $x^{1/2} + y^{1/2} = 1$      6  $x^2 + xy + y^2 = 9$
7  $x^{2/3} + y^{2/3} = 1$      8  $(x - 1)y^2 = x + 1$
9  $x^2(x - y) = y^2(x + y)$      10  $(x^2 + y^2)^2 = 4xy$

In each of Problems 11–20, first find $dy/dx$ by implicit differentiation; then write an equation of the tangent line to the graph of the equation at the given point.

11  $x^2 + y^2 = 25$;  $(3, -4)$      12  $xy = -8$;  $(4, -2)$
13  $x^2y = x + 2$;  $(2, 1)$
14  $x^{1/4} + y^{1/4} = 4$;  $(16, 16)$
15  $xy^2 + x^2y = 2$;  $(1, -2)$

16  $\dfrac{1}{x + 1} + \dfrac{1}{y + 1} = 1$;  $(1, 1)$

17  $12(x^2 + y^2) = 25xy$;  $(3, 4)$
18  $x^2 + xy + y^2 = 7$;  $(3, -2)$

19  $\dfrac{1}{x^3} + \dfrac{1}{y^3} = 2$;  $(1, 1)$

20  $(x^2 + y^2)^3 = 8x^2y^2$;  $(1, -1)$
21  Find $dy/dx$ given $xy^3 - x^5y^2 = 4$. Then find the slope of the tangent line to the graph of the given equation at the point (1, 2).
22  Show that the graph of $xy^5 + x^5y = 1$ has no horizontal tangents.
23  Find all points on the graph of the equation $x^3 + y^3 = 3xy - 1$ at which the tangent line is horizontal.
24  Find all points on the graph of the equation $x^4 + y^4 + 2 = 4xy^3$ at which the tangent line is horizontal.

25  Find all points on the graph of $x^2 + y^2 = 4x + 4y$ at which the tangent line is horizontal.
26  Find the first quadrant points of the folium of Example 2 at which the tangent line is either horizontal ($dy/dx = 0$) or vertical (that is, where $dx/dy = 1/(dy/dx) = 0$).
27  The graph of the equation $x^2 - xy + y^2 = 9$ is the "rotated ellipse" shown in Fig. 3.31. Find the tangent lines to this curve at the two points where it intersects the $y$-axis, and show that these two lines are parallel.
28  Find the points on the curve $x^2 - xy + y^2 = 9$ where the tangent line is horizontal ($dy/dx = 0$); then find the points where it is vertical ($dx/dy = 0$).
29  Find the points of the curve $x^2 - xy + y^2 = 9$ that are closest to and farthest from the origin. (*Suggestion:* Maximize the quantity $d^2 = x^2 + y^2$ subject to the constraint $x^2 - xy + y^2 = 9$.)
30  The lemniscate with equation $(x^2 + y^2)^2 = x^2 - y^2$ is shown in Fig. 3.32. Find by implicit differentiation the four points on the lemniscate where the tangent line is horizontal. Also find the two points where the tangent line is vertical—that is, where $dx/dy = 1/(dy/dx) = 0$.

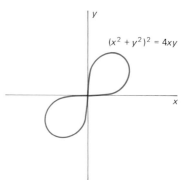

**3.32**   The lemniscate of Problem 30

**3.33**   Another lemniscate (see Problem 31)

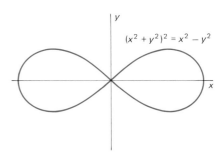

**3.31**   The rotated ellipse of Problem 27

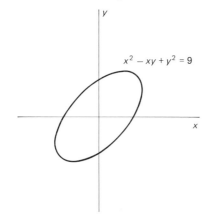

**31** The lemniscate $(x^2 + y^2)^2 = 4xy$ resembles the figure-8 curve shown in Fig. 3.33. Use the method of auxiliary variables to find the points on it that are farthest from the origin. (*Suggestion:* Maximize the square of the distance.)

Use the method of auxiliary variables to solve Problems 32–37.

**32** A poster is to contain 384 in.$^2$ of print, and each copy must have 6-in. margins at top and bottom and 4-in. margins on each side. What are the dimensions of such a poster having the least possible total area?

**33** Find the maximum and minimum values of the expression $x^2y^2$ on the ellipse with equation $x^2 - xy + y^2 = 3$.

**34** Find the maximum and minimum values of the expression $x - y$ on the ellipse with equation $x^2 + xy + y^2 = 16$.

**35** A rectangle has each diagonal of length 5. What is the maximum possible perimeter that such a rectangle can have?

**36** A box with no top has a square base and total surface area 24 ft$^2$. What is the maximum possible volume of such a box?

**37** Find the points on the folium with equation $x^3 + y^3 = 6xy$ that are closest to (2, 2). (See Fig. 3.34.)

Use the method of auxiliary variables to solve the following applied maximum-minimum problems in Section 3-6.

**38** Problem 4          **39** Problem 5
**40** Problem 8          **41** Problem 11

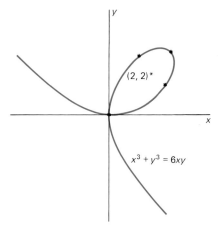

**3.34**   The folium of Problem 37

**42** Problem 18          **43** Problem 21
**44** Problem 22          **45** Problem 23
**46** Problem 30          **47** Problem 31
**48** Problem 34          **49** Problem 35
**50** Problem 48
**51** (*Challenge*) As a follow-up to Problem 23, show that the graph of the equation $x^3 + y^3 = 3xy - 1$ consists of the straight line with equation $x + y + 1 = 0$, together with the isolated point (1, 1).

---

### 3-8

### Related Rates

A **related-rates problem** involves two or more quantities that vary with time and an equation that expresses some relationship between them. Typically, the values of these quantities at some instant are given, together with all their time rates of change but one. The problem is usually to find the time rate of change that is *not* given, at some instant specified in the problem. One common method for solving such a problem is to begin by implicit differentiation of the equation that relates the given quantities.

For example, suppose that $x(t)$ and $y(t)$ are the $x$- and $y$-coordinates at time $t$ of a point moving around the circle with equation

$$x^2 + y^2 = 25. \tag{1}$$

Let us use the chain rule to differentiate both sides of this equation *with respect to time t*. This produces the equation

$$2x\frac{dx}{dt} + 2y\frac{dy}{dt} = 0. \tag{2}$$

If the values of $x$, $y$, and $dx/dt$ at a certain instant $t$ are known, then this last equation can be solved for the value of $dy/dt$ at time $t$. Note that it is *not* necessary to know $x$ and $y$ as functions of $t$. Indeed, it is common for a related-rates problem to contain insufficient information to express $x$ and $y$ as functions of $t$.

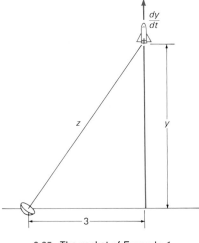

**3.35** The rocket of Example 1

For instance, suppose that we are given $x = 3$, $y = 4$, and $dx/dt = 12$ at a certain instant. Substitution of these values in Equation (2) yields

$$(2)(3)(12) + (2)(4)\frac{dy}{dt} = 0,$$

so we find that $dy/dt = -9$ at the same instant.

**EXAMPLE 1**   A rocket is launched vertically and is tracked by a radar station, which is located on the ground 3 mi from the launch site. What is the vertical speed of the rocket at the instant when its distance from the radar station is 5 mi and this distance is increasing at the rate of 5000 mi/h?

*Solution*   Figure 3.35 illustrates this situation. We denote the altitude of the rocket by $y$ and its distance from the radar station by $z$. We are given that

$$\frac{dz}{dt} = 5000 \quad \text{(mi/h)} \quad \text{when} \quad z = 5 \quad \text{(mi)}.$$

We want to find the value of $dy/dt$ at this instant.

We apply the Pythagorean theorem to the right triangle in the figure and obtain

$$y^2 + 9 = z^2$$

as a relation between $y$ and $z$. From this we see that $y = 4$ when $z = 5$. Implicit differentiation now gives

$$2y\frac{dy}{dt} = 2z\frac{dz}{dt}.$$

We substitute $y = 4$, $z = 5$, and $dz/dt = 5000$. We thus find that

$$\frac{dy}{dt} = 6250 \quad \text{(mi/h)}$$

at the instant in question.

---

Example 1 illustrates the following steps in the solution of a typical related rates problem of the sort that involves a geometric situation:

1 Draw a diagram and label as variables the various quantities involved in the problem.
2 Record the values of the variables and their rates of change, as given in the problem.
3 Read from the diagram an equation that relates the important variables of the problem.
4 Differentiate this equation implicitly with respect to time $t$.
5 Substitute the given numerical data into the resulting equation, and then solve for the unknown.

WARNING   The most common error to be avoided is the premature substitution of the given data, before rather than after implicit differentiation. If, in Example 1, we had substituted $z = 5$ to begin with, our equation would have been $y^2 + 9 = 25$, and implicit differentiation would have given the absurd result that $dy/dt = 0$.

In the following example, we use similar triangles (rather than the Pythagorean theorem) to discover the needed relation between the variables.

**EXAMPLE 2**  A man 6 ft tall walks with a speed of 8 ft/s away from a street light atop an 18-ft pole. How fast is the tip of his shadow moving along the ground when he is 100 ft from the light pole?

**Solution**  Let $x$ be the man's distance from the pole and $z$ the distance of the tip of his shadow from the base of the pole. Note that, though $x$ and $z$ are functions of $t$, we do *not* attempt to obtain explicit formulas for either.

We are given that $dx/dt = 8$ (ft/s), and we want to find $dz/dt$ when $x = 100$ (feet). We equate ratios of corresponding sides of the two similar triangles of Fig. 3.36 and find that

$$\frac{z}{18} = \frac{z - x}{6}.$$

Thus

$$2z = 3x.$$

Implicit differentiation now gives

$$2\frac{dz}{dt} = 3\frac{dx}{dt}.$$

We substitute $dx/dt = 8$, and find that

$$\frac{dz}{dt} = \frac{3}{2} \cdot \frac{dx}{dt} = \frac{3}{2} \cdot (8) = 12.$$

So the tip of the man's shadow is moving at 12 ft/s.

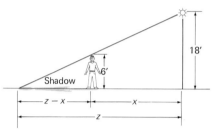

**3.36**  The moving shadow

---

Example 2 is somewhat unusual in that the answer is independent of the man's distance from the light pole—the given value $x = 100$ is superfluous. The next example is a related rates problem with two relationships between the variables, not quite so unusual.

**EXAMPLE 3**  Two radar stations at $A$ and $B$, with $B$ 6 mi east of $A$, are tracking a ship. At a certain instant, the ship is 5 mi from $A$, and this distance is increasing at the rate of 28 mi/h. At the same instant the ship is also 5 mi from $B$, while this distance is increasing at only 4 mi/h. Where is the ship, how fast is it moving, and in what direction is it moving?

**3.37**  Radar stations tracking a ship

**Solution**  With the distances indicated in Figure 3.37, we find—again with the aid of the Pythagorean theorem—that

$$x^2 + y^2 = u^2 \quad \text{and} \quad (6 - x)^2 + y^2 = v^2.$$

We are given that $u = v = 5$, that $du/dt = 28$, and that $dv/dt = 4$ at the instant in question. Since the ship is equally distant from $A$ and $B$, it is clear that $x = 3$. (Why?) So $y = 4$. Thus the ship is 3 mi east and 4 mi north of $A$.

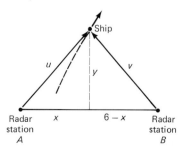

We differentiate implicitly the two equations above, and we obtain

$$2x \frac{dx}{dt} + 2y \frac{dy}{dt} = 2u \frac{du}{dt}$$

and

$$-2(6 - x) \frac{dx}{dt} + 2y \frac{dy}{dt} = 2v \frac{dv}{dt}.$$

When we substitute the numerical data given or deduced, we find that

$$3 \frac{dx}{dt} + 4 \frac{dy}{dt} = 140$$

and

$$-3 \frac{dx}{dt} + 4 \frac{dy}{dt} = 20.$$

These equations are easy to solve; $dx/dt = dy/dt = 20$. Therefore the ship is sailing northeast at a speed of $\sqrt{(20)^2 + (20)^2} = 20\sqrt{2}$ mi/h ... *if* the figure is correct! For a mirror along the line $AB$ will reflect *another* ship, 3 mi east and 4 mi *south* of $A$, sailing *southeast* at a speed of $20\sqrt{2}$ mi/h.

The lesson? Figures are important, helpful, often essential—and potentially misleading. Try to avoid taking anything for granted when you draw a figure. In this example there would be no real problem, for both radar stations would be able to tell whether the ship was generally to the north or to the south.

## 3-8  PROBLEMS

In each of Problems 1–5, a relation between $x$ and $y$ is given, together with the values at a particular instant of three of the four quantities $x$, $y$, $dx/dt$, and $dy/dt$. Find the value of the one not given.

**1** $x^2 + y^2 = 13$;  $x = 3$,  $\dfrac{dx}{dt} = -4$,  $\dfrac{dy}{dt} = 6$

**2** $y^2 = x^3 + 17$;  $x = 2$,  $y = -5$,  $\dfrac{dx}{dt} = 10$

**3** $x^3 + 3xy + y^3 = 15$;  $x = 1$,  $y = 2$,  $\dfrac{dy}{dt} = -3$

**4** $x^4 + x^2y^2 + y^4 = 21$;  $x = 2$,  $\dfrac{dx}{dt} = 2$,  $y = 1$

**5** $x^{2/3} + y^{2/3} = 13$;  $x = -8$,  $y = 27$,  $\dfrac{dy}{dt} = 6$

**6** Water is being collected from a block of ice with a square base. The water is produced because the ice is melting in such a way that the edge of the base of the block is decreasing at 2 in./h, while the block's height is decreasing at 3 in./h. What is the rate of flow of water into the collecting pan when the base has edge length 20 in. and the block's height is 15 in? Make the simplifying assumption that the water and ice have the same density.

**7** Sand being emptied from a hopper at the rate of 10 ft$^3$/s forms a conical pile whose height is always twice its radius. At what rate is the radius of the pile increasing when its height is 5 ft?

**8** Suppose that water is being emptied from a spherical tank of radius 10 ft. If the depth of water in the tank is 5 ft and is decreasing at the rate of 3 ft/s, at what rate is the radius of the top surface of the water decreasing?

**9** A circular oil slick of uniform thickness is caused by a spill of 1 m$^3$ of oil. The thickness of the oil slick is decreasing at the rate of 0.1 cm/h. At what rate is the radius of the slick increasing when it is 8 m?

**10** Suppose that an ostrich 5 ft tall is walking at a speed of 4 ft/s directly toward a street light 10 ft high. How fast is the tip of the ostrich's shadow moving along the ground? At what rate is the ostrich's shadow decreasing in length?

**11** The width of a rectangle is half its length. At what rate is its area increasing if its width is 10 cm and is increasing at 0.5 cm/s?

**12** At what rate is the area of an equilateral triangle increasing if its base is 10 cm and is increasing at 0.5 cm/s?

**13** A gas balloon is being filled at the rate of $100\pi$ cm$^3$ of gas per second. At what rate is its radius increasing when the radius is 10 cm?

**14** The volume $V$ (in cubic inches) and pressure $P$ (in pounds per square inch) of a certain gas sample satisfy the equation $PV = 1000$. At what rate is the volume of the sample changing if the pressure is 100 lb/in.$^2$ and is increasing at the rate 2 lb/in.$^2$ per second?

**15** A kite in the air at an altitude of 400 ft is being blown horizontally at the rate of 10 ft/s away from the person holding the kite string at ground level. At what rate is the string being payed out when 500 ft of string are already out?

**16** A weather balloon is rising vertically and is being observed from a point on the ground 2500 ft from the spot directly beneath the balloon. At what speed is the balloon rising when radar indicates that the distance from the observer to the balloon is 6500 ft and this distance is increasing at 120 ft/min?

**17** An airplane flying horizontally at an altitude of 3 mi and a speed of 480 mi/h passes directly above an observer on the ground. How fast is the distance from the observer to the airplane increasing 30 s later?

**18** The volume $V$ of water in a partially filled spherical tank of radius $a$ is $V = \frac{1}{3}\pi y^2(3a - y)$, where $y$ is the maximum depth of the water. Suppose that water is being drained from a spherical tank of radius 5 ft at the rate of 100 gal./min. Find the rate at which the depth $y$ of water is decreasing when

(a) $y = 7$ ft;

(b) $y = 3$ ft.

(*Note:* One gallon of water occupies a volume of approximately 0.1337 ft$^3$.)

**19** Repeat Problem 18, except use a tank that is hemispherical (flat side upward) with radius 10 ft.

**20** A swimming pool is 50 ft long and 20 ft wide. Its depth varies uniformly from 2 ft at the shallow end to 12 ft at the deep end. Suppose that the pool is being filled at the rate of 1000 gal./min. At what rate is the depth of water at the deep end increasing when it is 6 ft? (*Note:* One gallon of water occupies a volume of approximately 0.1337 ft$^3$.)

**21** A ladder 41 ft long has been leaning against a vertical wall. It begins to slip, so that its top slides down the wall while its bottom moves along the ground; the bottom moves at a constant speed of 10 ft/s. How fast is the top of the ladder moving when it is 9 ft above the ground?

**22** The base of a rectangle is increasing at 4 cm/s, while its height is decreasing at 3 cm/s. At what rate is its area changing when its base is 20 cm and its height is 12 cm?

**23** The height of a cone is decreasing at 3 cm/s, while its radius is increasing at 2 cm/s. When the radius is 4 cm and the height is 6 cm, is the volume of the cone increasing or is it decreasing? At what rate is the volume changing?

**24** A square is expanding. When its edge is 10 in., its area

is increasing at 120 in.$^2$/s. At what rate is the length of its edge increasing then?

**25** A rocket is launched vertically and is tracked by a radar station located on the ground 4 mi from the launch site. What is the vertical speed of the rocket at the instant when its distance from the radar station is 5 mi and this distance is increasing at the rate of 3600 mi/h?

**26** Two straight roads intersect at right angles. At 10 A.M. a car passes through the intersection headed due east at 30 mi/h. At 11 A.M. a truck passes through the intersection heading due north at 40 mi/h. Assume that the two vehicles maintain the given speeds and directions. At what rate are they separating at 1 P.M.?

**27** A 10-ft ladder is leaning against a wall. The bottom of the ladder begins to slide away from the wall at a speed of 1 mi/h. How fast is the top of the ladder moving when it is 4 ft above the ground?

**28** Two ships are sailing toward the same small island. One ship, the *Pinta*, is east of the island and is sailing due west at 15 mi/h. The other ship, the *Niña*, is north of the island and is sailing due south at 20 mi/h. At a certain time the *Pinta* is 30 mi from the island and the *Niña* is 40 mi from it. Are the two ships drawing closer together or farther apart at that time? At what rate?

**29** At time $t = 0$, a single-engine military jet is flying due east at 12 mi/min. At the same altitude and 208 mi directly ahead of it, still at time $t = 0$, is a commercial jet, flying due north at 8 mi/min. When are the two planes closest to each other? What is the minimum distance between them?

**30** A ship with a long anchor chain is anchored in 11 fathoms of water. The anchor chain is being wound in at the rate of 10 fathoms/min, causing the ship to move toward the spot directly over where the anchor is resting on the ocean's bottom. The hawsehole—the point of contact between ship and chain—is located 1 fathom above the waterline. At what speed is the ship moving when there are exactly 13 fathoms of chain still out?

**31** A water tank is in the shape of a cone with vertical axis and vertex downward. The tank's radius is 3 ft and the tank is 5 ft high. At first the tank is full of water, but at time $t = 0$ (in seconds), a small hole at the vertex is opened, and the water begins to drain. When the height of water in the tank has dropped to 3 ft, the water is flowing out at 0.02 ft$^3$/s. At what rate, in feet per second, is the water level dropping then?

**32** When a spherical tank with a radius of 10 ft contains water with a maximum depth of $y$ feet, the volume of water in the tank is $V = (\pi/3)(30y^2 - y^3)$. If the tank is being filled at the rate of 200 gal./min, how fast is the water level rising when $y = 5$ (feet)? (*Note:* One gallon is approximately 0.1337 ft$^3$.)

**33** A water bucket is shaped like the frustum of a cone with height 2 ft and lower and upper base radii 6 in. and 12 in., respectively. Water is leaking from the bucket at 10 in.$^3$/min. At what rate is the water level falling when the depth of water in the bucket is 1 ft? (*Note:* The volume of a conical frustum with height $h$ and base radii $a$ and $b$ is $V = \pi h(a^2 + ab + b^2)/3$.)

**34** Suppose that the radar stations at $A$ and $B$ of Example 3 are now 12.6 mi apart. At a certain instant, the ship is 10.4 mi from $A$, and its distance from $A$ is increasing at 19.2 mi/h. At the same instant, its distance from $B$ is 5 mi, and this distance is decreasing at 0.6 mi/h. Find the location, speed, and direction of motion of the ship.

**35** An airplane flying horizontally at an altitude of 3 mi passes directly over a radar station. A later reading shows the plane to be 5 mi from the station and that this distance is increasing at the rate of 8 mi/min. Find the speed of the plane in mi/h.

**36** The water tank of Problem 32 is completely full when a plug at its bottom is pulled. According to Torricelli's law, the water drains in such a way that $dV/dt = -k\sqrt{y}$, where $k$ is a positive empirical constant.
(a) Find $dy/dt$ as a function of $y$.
(b) Find the depth of water when the water level is falling the *least* rapidly. (You will need to compute the derivative of $dy/dt$ with respect to $y$.)

**37** Sand is pouring from a pipe at the rate of $120\pi$ ft³/s. The falling sand forms a conical pile on the ground; the altitude of the cone is always $\frac{1}{3}$ the radius of its base. How fast is the altitude increasing when the pile is 20 ft high?

**38** A man 6 ft tall walks at 5 ft/s along one edge of a road that is 30 ft wide. On the other edge of the road is a light atop a pole 18 ft high. How fast is the length of the man's shadow (on the horizontal ground) increasing when he is 40 ft beyond the point directly across the road from the pole?

---

### 3-9

## Successive Approximations and Newton's Method

The quadratic formula is one tool for obtaining an exact solution of any second degree equation $ax^2 + bx + c = 0$. Formulas are known for the exact solution of third and fourth degree equations, but they are rarely used because they are complicated. And it has been proved that there is *no* general formula possible for giving the roots of an arbitrary fifth degree (or higher) polynomial equation in terms of algebraic operations. Thus the exact solution (for all its roots) of an equation such as

$$x^5 - 3x^3 + x^2 - 23x + 19 = 0$$

may be quite difficult, or even—as a practical matter—impossible.

Actually, there is a question as to what it means to solve even so simple an equation as

$$x^2 - 2 = 0. \tag{1}$$

The positive exact solution is $x = \sqrt{2}$. But the number $\sqrt{2}$ is irrational and hence cannot be expressed as a terminating or repeating decimal. Thus if we mean by a *solution* an exact decimal value for $x$, even Equation (1) can be solved only approximately.

Over two millennia ago, ancient Babylonian mathematicians devised an effective way of generating a sequence of better and better approximations to $\sqrt{A}$, the square root of a given positive number $A$. Here is the *Babylonian square root method:* We begin with a first guess $x_0$ for the value of $\sqrt{A}$. For $\sqrt{2}$, we might guess $x_0 = 1.5$. If $x_0$ is too large—that is, if $x_0 > \sqrt{A}$—then $A/x_0 < A/\sqrt{A} = \sqrt{A}$, so $A/x_0$ is too small an estimate of $\sqrt{A}$. Similarly, if $x_0$ is too small (if $x_0 < \sqrt{A}$), then $A/x_0$ is too large an estimate of $\sqrt{A}$; that is, $A/x_0 > \sqrt{A}$.

Thus in each case one of the two numbers $x_0$ and $A/x_0$ is an underestimate of $\sqrt{A}$ and the other is an overestimate. The Babylonian idea was that we might get a better estimate of $\sqrt{A}$ by *averaging* $x_0$ and $A/x_0$. This yields a first approximation

$$x_1 = \frac{1}{2}\left(x_0 + \frac{A}{x_0}\right) \tag{2}$$

to $\sqrt{A}$. But why not repeat this process? We can average $x_1$ and $A/x_1$ to get a second approximation $x_2$, average $x_2$ and $A/x_2$ to get $x_3$, and so on. By repeating this process, we can generate a sequence

$$x_1, x_2, x_3, x_4, \ldots$$

which we have every right to hope will consist of better and better approximations to $\sqrt{A}$.

Specifically, having calculated the $n$th approximation $x_n$, we calculate the next one by means of the *iterative formula*

$$x_{n+1} = \frac{1}{2}\left(x_n + \frac{A}{x_n}\right). \tag{3}$$

In other words, we plow each approximation to $\sqrt{A}$ back into the right-hand side in (3) to calculate the next approximation. This is an *iterative* process—the words *iteration* and *iterative* stem from the Latin verb *iterare*, to plow again.

Suppose we find that, after sufficiently many steps in this iteration, $x_{n+1} \approx x_n$ to the number of decimal places we are retaining in our computations. Then Equation (3) yields

$$x_n \approx x_{n+1} = \frac{1}{2}\left(x_n + \frac{A}{x_n}\right) = \frac{1}{2x_n}(x_n^2 + A),$$

so $2x_n^2 \approx x_n^2 + A$, and hence $x_n^2 \approx A$ to some degree of accuracy.

**EXAMPLE 1**  With $A = 2$ we begin with the crude first guess $x_0 = 1$ to $\sqrt{2}$. Then successive application of the formula in (3) yields

$$x_1 = \frac{1}{2}\left(1 + \frac{2}{1}\right) = \frac{3}{2},$$

$$x_2 = \frac{1}{2}\left(\frac{3}{2} + \frac{2}{3/2}\right) = \frac{17}{12} \approx 1.416666667,$$

$$x_3 = \frac{1}{2}\left(\frac{17}{12} + \frac{2}{17/12}\right) = \frac{577}{408} \approx 1.414215686,$$

$$x_4 = \frac{1}{2}\left(\frac{577}{408} + \frac{2}{577/408}\right) = \frac{665857}{470832} \approx 1.414213562,$$

rounding our decimal results to nine places. It happens that $x_4$ gives $\sqrt{2}$ accurate to all nine places!

---

The Babylonian iteration defined in (3) is a method for generating a sequence of approximations to a root $r = \sqrt{A}$ of the particular equation $x^2 - A = 0$. There are numerous iterative methods currently in use for approximating a root $r$ of an equation $f(x) = 0$, and one of the most effective is Newton's method, which we describe after Example 2. First we need a brief discussion of convergence and interpolation.

We say that the sequence of approximations $x_1, x_2, x_3, \ldots$ *converges* to the number $r$ provided that we can make $x_n$ as close to $r$ as we please merely by choosing $n$ sufficiently large. More precisely, for any given $\varepsilon > 0$, there exists a positive integer $N$ such that $|x_n - r| < \varepsilon$ for all $n \geq N$. As a practical matter this means that—as illustrated in Example 1—for any positive integer $k$, $x_n$ and $r$ agree to $k$ or more decimal places once $n$ becomes sufficiently large.

## LINEAR INTERPOLATION

Any approximation method requires an initial approximation $x_0$ to the solution of the equation $f(x) = 0$ that we wish to solve. Suppose that $a$ and $b$ are two points such that $f(a)$ and $f(b)$ differ in sign and $f$ is continuous on $[a, b]$. Then we know from the intermediate value property of continuous functions (Section 2-4) that the equation $f(x) = 0$ has at least one solution in $(a, b)$. If, in addition, the curve $y = f(x)$ either is steadily rising or is steadily falling (from left to right) on $[a, b]$, then it follows that the equation $f(x) = 0$ has *exactly one* solution $x_*$ in $(a, b)$.

A common way to obtain an initial approximation $x_0$ to $x_*$ is by **linear interpolation.** In this method we choose $x_0$ as the point where the line segment joining $(a, f(a))$ and $(b, f(b))$ meets the $x$-axis. From the similar triangles in Figure 3.38, where $f(a)$ is negative and $f(b)$ is positive, we obtain the proportion

$$\frac{x_0 - a}{-f(a)} = \frac{b - x_0}{f(b)}. \tag{4}$$

It is then easy to solve this equation for

$$x_0 = \frac{af(b) - bf(a)}{f(b) - f(a)}. \tag{4'}$$

The case $f(a) > 0 > f(b)$ leads to the same formula for the interpolated value $x_0$. In practice it will probably be more convenient to set up the proportion than to remember the formula in (4').

**EXAMPLE 2**    A large cork ball has radius 1 ft, and its density is $\frac{1}{4}$ that of water. Archimedes' law of buoyancy implies that, when the ball floats in water, $\frac{1}{4}$ of its volume ($\frac{1}{4}$ of $4\pi/3$, which is $\pi/3$) is submerged. Find an equa-

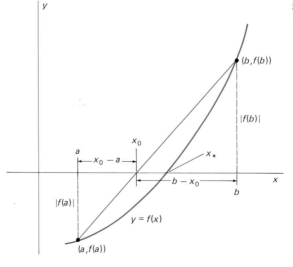

**3.38** Obtaining $x_0$ by linear interpolation

tion determining the depth $x$ to which the cork ball sinks, and use linear interpolation to make an initial approximation to its solution.

**Solution**  The volume of a spherical segment of height $x$ and radius $r$ (the part beneath the water in Fig. 3.39) is given by the formula

$$V = \frac{\pi x}{6}(3r^2 + x^2).$$

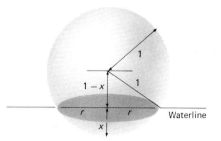

**3.39**  The floating cork ball

From the right triangle of Fig. 3.39, we see that

$$r^2 + (1 - x)^2 = 1^2,$$

so that $r^2 = 2x - x^2$. We combine Archimedes' law and the volume formula above to obtain

$$\frac{\pi x^2}{3}(3 - x) = \frac{\pi}{3}; \quad \text{that is,} \quad 3x^2 - x^3 = 1.$$

So we must solve the equation

$$f(x) = x^3 - 3x^2 + 1 = 0. \tag{5}$$

This last equation has no rational solutions; we cannot expect to solve it by entirely elementary methods. We note, however, that $f(0) = 1$ and $f(1) = -1$, so there is a root between 0 and 1. This is the solution we seek, since it is physically evident that the desired depth is between 0 and 1. With $a = 0$, $b = 1$, $f(a) = 1$, and $f(b) = -1$, the proportion in (4) is

$$\frac{x_0 - 0}{1} = \frac{1 - x_0}{1},$$

with solution $x_0 = \frac{1}{2}$. This will be our initial approximation to the root of Equation (5) that lies in $(0, 1)$.

## NEWTON'S METHOD

In Fig. 3.40 we illustrate Newton's method of constructing a rapidly convergent sequence of successive approximations to a root $x_*$ of the equation $f(x) = 0$. The tangent line at $(x_n, f(x_n))$ is used to construct a better

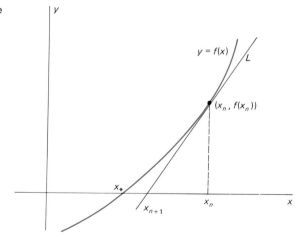

**3.40**  Geometry of the iteration of Newton's method

approximation $x_{n+1}$ to $x_*$ as follows. Begin at the point $x_n$ on the $x$-axis. Go vertically up (or down) to the point $(x_n, f(x_n))$ on the curve $y = f(x)$. Then follow the tangent line $L$ there to the point where it intersects the $x$-axis. That point will be $x_{n+1}$.

Here is a formula for $x_{n+1}$. We obtain it by computing the slope of the line $L$ in two ways: from the derivative and from the two-point definition of slope. Thus

$$f'(x_n) = \frac{f(x_n) - 0}{x_n - x_{n+1}},$$

and we solve easily for $x_{n+1}$:

$$x_{n+1} = x_n - \frac{f(x_n)}{f'(x_n)}. \tag{6}$$

This equation is the **iterative formula** for Newton's method, so called because in about 1669, Newton introduced an algebraic procedure (rather than the geometric construction above) that is equivalent to the iterative use of Equation (6). Newton's first example was the cubic equation $x^3 - 2x - 5 = 0$, for which he found the root $x_* \approx 2.0946$ (as we ask you to do in Problem 18).

Suppose now that we want to apply Newton's method to solve the equation

$$f(x) = 0 \tag{7}$$

with an accuracy of $k$ decimal places. Remember that an equation must be written precisely in the form in (7) to use the formula in (6). If we reach the point in our iteration at which $x_n$ and $x_{n+1}$ agree to $k$ decimal places, it then follows that

$$x_n \approx x_{n+1} = x_n - \frac{f(x_n)}{f'(x_n)};$$

$$0 \approx -\frac{f(x_n)}{f'(x_n)};$$

$$f(x_n) \approx 0.$$

Thus we have found an approximate root $x_n \approx x_{n+1}$ of our equation. In practice, then, we retain $k$ decimal places in our computations and persist until $x_n = x_{n+1}$ to this degree of accuracy. (We do not consider here the possibility of roundoff error, an important subject in numerical analysis.)

**EXAMPLE 3**   Use Newton's method to find $\sqrt{2}$ accurate to nine decimal places.

*Solution*   More generally, consider the square root of the positive number $A$ as the positive root of the equation

$$f(x) = x^2 - A = 0.$$

Because $f'(x) = 2x$, Equation (6) gives the iterative formula

$$x_{n+1} = x_n - \frac{x_n^2 - A}{2x_n} = \frac{1}{2}\left(x_n + \frac{A}{x_n}\right). \tag{8}$$

Thus we have derived the Babylonian iteration formula as a special case of Newton's method. The use of (8) with $A = 2$ therefore yields the values $x_1$, $x_2$, $x_3$, and $x_4$ we calculated in Example 1, and upon another iteration we find that

$$x_5 = \frac{1}{2}\left(x_4 + \frac{2}{x_4}\right) = 1.414213562,$$

in agreement with $x_4$ to nine decimal places. The very rapid convergence here is an important characteristic of Newton's method. As a general rule (with some exceptions), each iteration doubles the number of decimal places of accuracy.

**EXAMPLE 4** Use Newton's method to solve the cork ball equation

$$f(x) = x^3 - 3x^2 + 1 = 0.$$

**Solution** Here $f'(x) = 3x^2 - 6x$, so the iterative Formula (6) becomes

$$x_{n+1} = x_n - \frac{x_n^3 - 3x_n^2 + 1}{3x_n^2 - 6x_n}. \tag{9}$$

With $x_0 = 0.5$, this gives the values

$$x_0 = 0.5,$$

$$x_1 = 0.5 - \frac{(0.5)^3 - (3)(0.5)^2 + 1}{(3)(0.5)^2 - (6)(0.5)} = 0.6667,$$

$$x_2 = 0.6528,$$

$$x_3 = 0.6527,$$

$$x_4 = 0.6527.$$

Thus we obtain the root $x_* \approx 0.6527$, retaining only four decimal places.

---

The equation $x^3 - 3x^2 + 1 = 0$ has two additional solutions. This is because the continuous function $f(x) = x^3 - 3x^2 + 1$ has the intermediate value property on each of the intervals $[-1, 0]$, $[0, 1]$, and $[2, 3]$. The starting values of $x_0$ shown in the table below are estimated by linear interpolation.

| $x$ | $f(x)$ | |
|---|---|---|
| $-1$ | $-3$ | |
| $0$ | $1$ | $x_0 = -0.25$ |
| $1$ | $-1$ | $x_0 = +0.50$ |
| $2$ | $-3$ | |
| $3$ | $1$ | $x_0 = +2.75$ |

Beginning with $x_0 = -0.25$ and, subsequently, with $x_0 = 2.75$, the iteration in (9) produces the two sequences

$$
\begin{aligned}
x_0 &= -0.25, & x_0 &= 2.75, \\
x_1 &= -0.7222, & x_1 &= 2.8939, \\
x_2 &= -0.5626, & x_2 &= 2.8795, \\
x_3 &= -0.5331, & x_3 &= 2.8794, \\
x_4 &= -0.5321, & x_4 &= 2.8794. \\
x_5 &= -0.5321;
\end{aligned}
$$

Thus the other two roots are $-0.5321$ and $2.8794$ (to four decimal places).

Iterative methods are particularly well adapted to use with programmable calculators. With a calculator that is not programmable, the computations can be simplified a bit by replacing $f'(x_n)$ by the *constant* $f'(x_0)$ in the denominator in (6). This gives the *modified Newton's method*

$$
x_{n+1} = x_n - \frac{f(x_n)}{f'(x_0)}. \tag{10}
$$

For example, with $f(x) = x^3 - 3x^2 + 1$ and $x_0 = 0.5$, as in Example 4, (10) simplifies to

$$
x_{n+1} = x_n + \tfrac{4}{9}(x_n^3 - 3x_n^2 + 1).
$$

This iteration gives the values

$$
\begin{aligned}
x_0 &= 0.5, \\
x_1 &= 0.6667, & x_4 &= 0.6526, \\
x_2 &= 0.6502, & x_5 &= 0.6527, \\
x_3 &= 0.6531, & x_6 &= 0.6527.
\end{aligned}
$$

We need a few more iterations, but each step is a little simpler.

Newton's method is one for which "the proof is in the pudding." If it works, it's obvious that it does, and everything's fine! When Newton's method fails, it may do so spectacularly. For instance, suppose that we want to solve the equation

$$
f(x) = x^{1/3} = 0.
$$

Note that $x_* = 0$ is the only solution. The iterative formula in (6) becomes

$$
x_{n+1} = x_n - \frac{(x_n)^{1/3}}{(\tfrac{1}{3})(x_n)^{-2/3}} = x_n - 3x_n = -2x_n.
$$

If we begin with $x_0 = 1$, Newton's method yields $x_1 = -2$, $x_2 = +4$, $x_3 = -8$, and so on. Figure 3.41 indicates why our "approximations" are not converging.

When Newton's method fails to provide a sequence of approximations that converge to a solution, the implication is simply that some other method should be used. For instance, the method of bisection discussed in Section 2-4 can always be used as a last resort. But Newton's method is much faster (fewer iterations are required), easier to program, and therefore preferable when it succeeds.

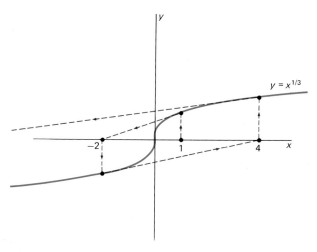

**3.41** A failure of Newton's method

## *THE METHOD OF REPEATED SUBSTITUTION

Occasionally it is advantageous to rewrite a given equation in the form

$$x = G(x). \tag{11}$$

For an equation in this form, the iterative formula

$$x_{n+1} = G(x_n) \tag{12}$$

sometimes produces a sequence $x_1, x_2, x_3, \ldots$ of approximations that converge to a solution of (11).

For instance, the cork ball equation

$$x^3 - 3x^2 + 1 = 0$$

of Example 4 may be transformed as follows:

$$3x^2 - x^3 = 1;$$

$$x^2(3 - x) = 1;$$

$$x = \frac{1}{\sqrt{3 - x}}.$$

The final equation here is of the form in (11) with $G(x) = 1/\sqrt{3 - x}$, so the *repeated substitution* formula corresponding to (12) is

$$x_{n+1} = \frac{1}{\sqrt{3 - x_n}}.$$

If we begin with $x_0 = 0.5$ and retain four decimal places, we get

$$x_1 = 0.6325$$

$$x_2 = 0.6499$$

$$x_3 = 0.6523$$

$$x_4 = 0.6526$$

$$x_5 = 0.6527 = x_6.$$

SEC. 3-9: Successive Approximations and Newton's Method

SEC. 3-9:  Successive Approximations and Newton's Method

**147**

As suggested by this example, the method of repeated substitution usually requires more steps than Newton's method, although it's normal for each step to be simpler.

While the method of repeated substitution is simple to apply in the case of an equation written in the form in (11), the method also fails frequently. For example, another two equivalent forms of the cork ball equation $x^3 - 3x^2 + 1 = 0$ are

$$x = 3 - \frac{1}{x^2} \quad \text{and} \quad x = (3x^2 - 1)^{1/3}.$$

If you apply the method of repeated substitution to either of these equations, beginning with $x_0 = 0.5$ in the hope of finding the nearby solution $x_* = 0.6527$, you will witness some of the drawbacks of the method. When repeated substitution fails, the next natural recourse is Newton's method.

## 3-9 PROBLEMS

In each of Problems 1–20, use Newton's method to find the solution of the given equation $f(x) = 0$ in the indicated interval $[a, b]$ accurate to four decimal places. Choose the initial estimate of the solution by using the interpolation formula in (4′).

1 $x^2 - 5 = 0$; $[2, 3]$    (to find the positive square root of 5)

2 $x^3 - 2 = 0$; $[1, 2]$    (to find the cube root of 2)

3 $x^5 - 100 = 0$; $[2, 3]$    (to find the fifth root of 100)

4 $x^{3/2} - 10 = 0$; $[4, 5]$    (to find $10^{2/3}$)

5 $x^2 + 3x - 1 = 0$; $[0, 1]$

6 $x^3 + 4x - 1 = 0$; $[0, 1]$

7 $x^6 + 7x^2 - 4 = 0$; $[-1, 0]$

8 $x^3 + 3x^2 + 2x = 10$; $[1, 2]$

9 $x^6 - 60 = 0$; $[1, 2]$

10 $2x^3 = 50x - 27$; $[0, 1]$

11 $2x^3 = 50x - 27$; $[4, 5]$

12 $2x^3 = 50x - 27$; $[-6, -5]$

13 $x^5 + x^4 = 100$; $[2, 3]$

14 $x^5 + 2x^4 + 4x = 5$; $[0, 1]$

15 $x^{10} - 1000 = 0$; $[1, 2]$

16 $x^{10} - 10x - 1000 = 0$; $[1, 2]$

17 $x^3 - 10 = 0$; $[2, 3]$

18 $x^3 - 2x - 5 = 0$; $[2, 3]$    (Newton's own example)

19 $x^5 - 5x - 10 = 0$; $[1, 2]$

20 $x^5 = 32$; $[0, 5]$

21 (a) Show that Newton's method applied to the equation $x^3 - a = 0$ yields the iteration

$$x_{n+1} = \frac{1}{3}\left(2x_n + \frac{a}{x_n^2}\right)$$

for approximating the cube root of $a$.
(b) Use this iteration to find $\sqrt[3]{2}$ accurate to five decimal places.

22 (a) Show that Newton's method yields the iteration

$$x_{n+1} = \frac{1}{k}\left[(k - 1)x_n + \frac{a}{(x_n)^{k-1}}\right]$$

for approximating the $k$th root of the positive number $a$.
(b) Use this iteration to find $\sqrt[10]{100}$ accurate to five decimal places.

23 With many inexpensive hand-held calculators, you can calculate square roots directly, but not higher roots because there are no keys for exponentiation. Such a calculator can nevertheless be used to compute roots. For example, if $x = \sqrt[17]{a}$ for some positive number $a$, then $x^{17} = a$. So $x^{16} = a/x$, and thus

$$x = \sqrt{\sqrt{\sqrt{\sqrt{a/x}}}}.$$

Use this formula and repeated substitution to find $\sqrt[17]{10}$ accurate to four decimal places.

24 Modify the method of Problem 23 in order to approximate $\sqrt[3]{5}$ accurate to four decimal places using only the square root key (and the four arithmetic keys).

25 Repeat Problem 24 for $\sqrt[7]{10}$.

26 Show that Newton's method applied to the equation $(1/x) - a = 0$ yields the iteration $x_{n+1} = 2x_n - a(x_n)^2$, and thus provides a method for approximating the reciprocal $1/a$ without performing any divisions. The reason for using such a method is that in most high-speed computers, the operation of division is *much* more time-consuming than even several additions and multiplications.

In each of Problems 27–36, use Newton's method to find all real roots of the given equation, accurate to two places to the right of the decimal.

27 $x^4 = 20$

28 $x^3 - 5x + 2 = 0$

29 $x^5 - 5x + 2 = 0$

30 $x^4 + x^3 - 5 = 0$

**31** $x^3 - 5x^2 + 1 = 0$

**32** $x^3 + 2x = 2$

**33** $x^5 + x = 1$

**34** $1 + \dfrac{1}{x} = \dfrac{3}{x^2}$

**35** $\dfrac{1}{x} + \dfrac{1}{x + 1} = 1$

**36** $\dfrac{1}{x^2} + \dfrac{1}{x^2 + 4} = 1$

**37** Explain why the equation $x^7 - 3x^3 + 1 = 0$ must have at least one solution. Then use Newton's method to find one solution.

**38** Use Newton's method to approximate $\sqrt[3]{5}$ to three-place accuracy.

**39** To find a solution of the quadratic equation

$$x^2 + x - 1 = 0,$$

first rewrite it as follows:

$$x(x + 1) - 1 = 0;$$

$$x = \frac{1}{x + 1}.$$

The last equation has the form in Equation (11) with $G(x) = 1/(x + 1)$. Now apply the method of repeated substitution beginning with $x_0 = 1$.

**40** Apply the method of Problem 39 to find a solution of the quadratic equation $x^2 + 3x - 2 = 0$.

**41** In Problem 47 of Section 3-6, we dealt with the problem of minimizing the cost of joining by a road two points lying on opposite sides of a geological fault. This problem led to the equation

$$f(x) = 3x^4 - 24x^3 + 51x^2 - 32x + 64 = 0.$$

Use Newton's method to find to four-place accuracy the root of this equation lying in (3, 4).

**42** The equation $x^3 - 9x = 1$ has exactly three real roots. Use Newton's method to find them all.

**43** A great problem of Archimedes was that of using a plane to cut a sphere into two segments with volumes having a given (preassigned) ratio. Archimedes showed that the volume of a segment of height $h$ of a sphere of radius $a$ is

$$V = \tfrac{1}{3}\pi h^2 (3a - h).$$

If a plane at distance $x$ from the center of the unit sphere cuts it into two segments, one with twice the volume of the other, show that

$$3x^3 - 9x + 2 = 0.$$

Then use Newton's method to find $x$ accurate to four decimal places.

**44** The equation

$$f(x) = x^3 - 4x + 1 = 0$$

has three distinct real roots. Locate them by calculating the value of $f$ for $x = -3, -2, -1, 0, 1, 2,$ and 3. Then use Newton's method to approximate each of the three roots to four-place accuracy.

---

## *3-9  Optional Computer Application

Iterative processes virtually cry out for the facility of the computer to perform repetitive computations. The following program was written to implement the Babylonian square root method on a pocket computer.

```
10    INPUT  "A",  A
20    X = A/2
30    PRINT  X
40    Y = 0.5*(X + A/X)
50    IF  ABS(Y − X) < 1.0E−06  THEN GOTO  80
60    X = Y
70    GOTO  30
80    PRINT  "SQRT";  A;  " = ";  Y
90    END
```

The computer first prompts us to enter the *positive* number  A  whose square root we want to compute. It then sets  $x_0 = A/2$,  so our initial guess at  $\sqrt{A}$  is the number  A/2.  At each stage in the computation  X  denotes  $x_n$  and  Y  denotes  $x_{n+1}$.  The Babylonian iterative formula appears in line 40. The program prints the successive approximations  $x_1, x_2, x_3, \ldots$  until  $|x_{n+1} - x_n| <$  0.000001,  at which point it prints the result  SQRT A  $= x_{n+1}$  (accurate to five decimal places).

The next program implements Newton's method to solve the cubic equation $x^3 - 3x^2 + 1 = 0$ of Examples 2 and 4 of this section. At the $n$th step in the iteration we let $F = f(x_n) = (x_n)^3 - 3(x_n)^2 + 1$ and $D = f'(x_n) = 3(x_n)^2 - 6x_n$, so the iterative formula of Newton's method is $x_{n+1} = x_n - F/D$ (see line 50).

```
10    INPUT "X0"; X
20    PRINT X
30    F = X*X*X − 3*X*X + 1
40    D = 3*X*X − 6*X
50    Y = X − F/D
60    IF ABS(Y − X) < 1. 0E−06 THEN GOTO 90
70    X = Y
80    GOTO 20
90    PRINT "SOLUTION: "; Y
100   END
```

Observe that the structure of this program is the same as that of the previous one, with $X$ denoting $x_n$ and $Y$ denoting $x_{n+1}$ during the $n$th iteration, and with the iteration continuing until $|x_{n+1} - x_n| < 0.000001$, at which point the result is printed.

**Exercise 1** Enter the Babylonian method program on a pocket computer or microcomputer; use it to compute several square roots.

**Exercise 2** Use the iterative formula of Problem 21 to alter the Babylonian method program to obtain a cube root program.

**Exercise 3** Alter the Newton's method program above to find all solutions of the equation $x^3 - 9x = 1$ of Problem 42.

**Exercise 4** The *secant method* for solving the equation $f(x) = 0$, sometimes called the "method of false position," may be described as follows. We begin with an interval $[a, b]$ on which $f(x)$ changes sign. Then let $x_0$ be the value obtained by interpolation,

$$x_0 = \frac{af(b) - bf(a)}{f(b) - f(a)}. \tag{4'}$$

Thus (see Fig. 3.38) $x_0$ is the point at which the secant line through $(a, f(a))$ and $(b, f(b))$ intersects the $x$-axis. Having calculated $x_0, x_1, x_2, x_3, \ldots, x_n$, we let the new interval $[a, b]$ be that one of the intervals $[a, x_n]$ and $[x_n, b]$ on which $f(x)$ changes sign. Then the next approximation $x_{n+1}$ is given by the right-hand side in Equation (4'). Write a program to implement the secant method. You may find it convenient to use part of the bisection method program in the Computer Application in Section 2-4.

**Exercise 5** Because

$$f(x) = x^3 - 117x^2 - 53x - 77$$

is a polynomial of odd degree, the equation $f(x) = 0$ has at least one solution. Use Newton's method to find one.

*Differentiation Formulas*

$$D_x(cu) = c\frac{du}{dx}$$

$$D_x(u + v) = \frac{du}{dx} + \frac{dv}{dx}$$

$$D_x(uv) = u\frac{dv}{dx} + v\frac{du}{dx}$$

$$D_x\frac{u}{v} = \frac{vD_xu - uD_xv}{v^2}$$

$$D_x(u^r) = ru^{r-1}\frac{du}{dx}$$

$$D_xg(u) = g'(u)\frac{du}{dx}$$

Use the list below as a guide to ideas that you may need to review.

1 Definition of the derivative
2 Average rate of change of a function
3 Instantaneous rate of change of a function
4 Position function; velocity and acceleration
5 Differential, function, and operator notation for derivatives

6 The binomial formula
7 The power rule
8 Linearity of differentiation
9 The product rule
10 The quotient rule
11 Principle of mathematical induction
12 Local minimum and maximum
13 $f'(c) = 0$ as a necessary condition for local extrema
14 Critical point
15 Closed interval maximum-minimum method
16 Generalized power rule
17 Differentiation of inverse functions
18 Vertical tangent lines
19 Steps in the solution of an applied maximum-minimum problem
20 Chain rule
21 Implicitly defined functions
22 Implicit differentiation
23 Maximum-minimum problems with constraints
24 Method of auxiliary variables
25 Solution of related rates problems
26 Linear interpolation
27 Newton's method
28 Method of repeated substitution

## MISCELLANEOUS PROBLEMS

Find $dy/dx$ in Problems 1–35.

1 $y = x^2 + \dfrac{3}{x^2}$

2 $y^2 = x^2$

3 $y = \sqrt{x} + \dfrac{1}{\sqrt[3]{x}}$

4 $y = (x^2 + 4x)^{5/2}$

5 $y = (x - 1)^7(3x + 2)^9$

6 $y = \dfrac{x^4 + x^2}{x^2 + x + 1}$

7 $y = \left(3x - \dfrac{1}{2x^2}\right)^4$

8 $y = x^{10}\sqrt{1 - x^2}$

9 $xy = 9$

10 $y = \sqrt{1/5x^6}$

11 $y = \dfrac{1}{\sqrt{(x^3 - x)^3}}$

12 $y = \sqrt[3]{2x + 1}\sqrt[5]{3x - 2}$

13 $y = \dfrac{1}{1 + u^2}$, where $u = \dfrac{1}{1 + x^2}$

14 $x^3 = y^2 + \sqrt{1 + y^2}$

15 $y = (\sqrt{x} + \sqrt[3]{2x})^{7/3}$

16 $y = \sqrt{3x^5 - 4x^2}$

17 $y = \dfrac{u + 1}{u - 1}$, $u = \sqrt{x + 1}$

18 $y = (1 + 2\sqrt{2x})^3$

19 $x^2y^2 = x + y$

20 $y = \sqrt{1 - \sqrt{3x}}$

21 $y = \sqrt{x + \sqrt{2x + \sqrt{3x}}}$

22 $y = \dfrac{x + \sqrt{x}}{x^2 + 1}$

23 $x^{1/3} + y^{1/3} = 4$

24 $x^3 + y^3 = xy$

25 $y = (1 + 2u)^3$, where $u = \dfrac{1}{(1 + x)^3}$

26 $y = \sqrt{1 + (1 + \sqrt{x})^2}$

27 $y = \sqrt{x(1 + \sqrt{x})}$

28 $y = (1 + \sqrt{x})^3(1 - 2\sqrt[3]{x})^4$

**29** $y = \dfrac{x}{\sqrt{x+1}}$

**30** $x^3 - x^2y + xy^2 - y^3 = 4$

**31** $y = x^2\sqrt{1-x^3}$

**32** $y = [1 + (2 + 3x)^{-3/2}]^{2/3}$

**33** $y = \left(x + \dfrac{1}{x}\right)^{10}$

**34** $\sqrt{x+y} = \sqrt[3]{x-y}$

**35** $y = \sqrt[3]{1+x^4}$

In each of Problems 36–39, find the tangent line to the given curve at the indicated point.

**36** $y = \dfrac{x+1}{x-1}; \quad (0, -1)$

**37** $x = \sqrt{25 - y^2}; \quad (3, 4)$

**38** $x^2 - 3xy + 2y^2 = 0; \quad (2, 1)$

**39** $y^3 = x^2 + x; \quad (0, 0)$

**40** If a hemispherical bowl with radius 1 ft is filled with water to a depth of $x$ inches, the volume of liquid in the bowl turns out to be $V = (\pi/3)(36x^2 - x^3)$ in.³ If the water flows out a hole at the bottom of the bowl at the rate of $36\pi$ in.³/s, how fast is $x$ decreasing when $x = 6$ in.?

**41** Falling sand forms a conical sandpile with height which always remains twice its radius $r$ while both are increasing. If sand is falling onto the pile at the rate of $25\pi$ ft³/min, how fast is $r$ increasing when $r = 5$ ft?

In each of Problems 42–47, identify two functions $f$ and $g$ such that $h(x) = g(f(x))$, and then apply the chain rule to find $h'(x)$.

**42** $h(x) = \sqrt[3]{x + x^4}$

**43** $h(x) = \dfrac{1}{\sqrt{x^2 + 25}}$

**44** $h(x) = \sqrt{\dfrac{x}{x^2 + 1}}$

**45** $h(x) = \sqrt[3]{(x - 1)^5}$

**46** $h(x) = \dfrac{(x+1)^{10}}{(x-1)^{10}}$

**47** $h(x) = \dfrac{1}{x^2 + 1}$

**48** The period $T$ of oscillation (in seconds) of a pendulum of length $L$ (in feet) is given by $T = 2\pi(L/32)^{1/2}$. What is the rate of change of $T$ with respect to $L$ when $L = 4$ (ft)?

**49** What is the rate of change of the volume $V = \frac{4}{3}\pi r^3$ of a sphere with respect to its surface area $S = 4\pi r^2$?

**50** What is an equation for the straight line through $(1, 0)$ that is tangent to the graph of $h(x) = x + 1/x$ at a point in the first quadrant?

**51** An isosceles triangle has its vertex at the origin and its horizontal base below the origin—the end points of the base lie on the graph of $y = x^2 - 27$. Find the maximum possible area that such a triangle can have.

**52** An oil field containing 20 wells has been producing 4000 barrels of oil daily. Suppose that, for each new well that is drilled, the daily production of each well decreases by 5 barrels. How many new wells should be drilled in order to maximize the total daily production of the oil field?

**53** A triangle is inscribed in a circle of radius $R$ with one side of the triangle coincident with a diameter of the circle. What, in terms of $R$, is the maximum possible area of such a triangle?

**54** Five rectangular pieces of sheet metal measure 210 by 336 cm each. Equal squares are to be cut from all their corners, and the resulting five cross-shaped pieces of metal are to be folded and welded to form five boxes without tops. The 20 little squares left over are to be assembled in groups of four into five larger squares, and these five larger squares are to be assembled into a cubical box with no top. What is the maximum possible total volume of the six boxes that can be so obtained?

**55** A mass of clay of volume $V$ is formed into two spheres. For what distribution of clay is the total surface area of the two spheres a maximum? A minimum?

**56** A right triangle has legs of lengths 3 and 4 m. What is the maximum possible area of a rectangle inscribed in the triangle in the "obvious way"—that is, with one corner at the triangle's right angle, the adjacent sides of the rectangle lying on the triangle's legs, and the opposite corner on the triangle's hypotenuse?

**57** What is the maximum possible volume of a right circular cone inscribed in a sphere of radius $R$?

**58** A farmer has 400 ft of fencing with which to build a rectangular corral. He will use part or even all of an existing straight wall 100 ft long as part of the perimeter of the corral. What is the maximum area that he can enclose?

**59** In one simple model of the spread of a contagious disease among members of a population of $M$ people, the incidence of the disease, measured as the number of new cases per day, is given in terms of the number $x$ of individuals already infected by

$$R(x) = kx(M - x) = kMx - kx^2$$

where $k$ is some positive constant. How many individuals in the population are infected when the incidence $R$ is the highest?

**60** Three sides of a trapezoid have side length $L$, a constant. What should be the length of the fourth side if the trapezoid is to have maximum area?

**61** A box with no top must have a base twice as long as it is wide, and the total surface area of the box is to be 54 ft². What is the maximum possible volume of such a box?

**62** A small right circular cone is inscribed in a larger one, as shown in Fig. 3.42. The larger cone has fixed radius $R$ and fixed altitude $H$. What is the largest fraction of the volume of the larger cone that the smaller one can occupy?

**63** Two vertices of a trapezoid are at $(-2, 0)$ and $(2, 0)$, and the other two lie on the semicircle $x^2 + y^2 = 4, y \geq 0$. What is the maximum possible area of the trapezoid? (*Note:*

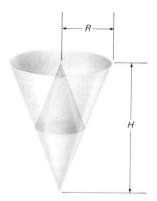

**3.42** A small cone inscribed in a larger one

The area of a trapezoid of height $h$ and with bases $b_1$ and $b_2$ is $A = h(b_1 + b_2)/2$.)

**64** Show by the method of auxiliary variables that, if $Q(x, y)$ is the point of the curve $y = f(x)$ that is closest to the point $P(x_0, y_0)$ not on the curve, then

$$f'(x) = -\frac{x - x_0}{y - y_0}$$

at $Q$. Conclude that the segment $PQ$ is perpendicular to the tangent line to the curve at $Q$. (*Suggestion:* Minimize the square of the distance $PQ$.)

**65** Use the result of Problem 64 to show that the minimum distance from the point $(x_0, y_0)$ to a point of the straight line $Ax + By + C = 0$ is

$$\frac{|Ax_0 + By_0 + C|}{\sqrt{A^2 + B^2}}.$$

**66** A race track is to be built in the shape of two parallel and equal straightaways connected by semicircles on each end. The length of the track, one lap, is to be exactly 5 km. What should be its design to maximize the rectangular area within it, indicated by the shading in Fig. 3.43?

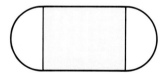

**3.43** Design the race track to maximize the shaded area

**67** Two towns are located near the straight shore of a lake. Their nearest distances to points on the shore are 1 mi and 2 mi, respectively, and these points on the shore are 6 mi apart. Where should a fishing pier be located to minimize the total amount of paving necessary to build a straight road from each town to the pier?

**68** A rectangular garden is to be constructed with one side on your next-door neighbor's property line. Your neighbor agrees to pay half the cost of the fence along his property line; you pay the other half as well as the total cost of the fence along the other three sides of the garden. If the garden is to enclose exactly 1728 ft$^2$, how should you design the garden so that the amount *you* pay for the fence is minimal?

**69** When an arrow is shot from the origin with initial velocity $v_0$ and initial angle of inclination $\alpha$ (from the horizontal $x$-axis, which represents the ground), then its trajectory is the curve

$$y = mx - \frac{16}{v_0^2}(1 + m^2)x^2$$

where $m = \tan \alpha$.
(a) Find the maximum height reached by the arrow (in terms of $m$ and $v_0$).
(b) For what $m$ (and hence, for what $\alpha$) does the arrow travel the greatest horizontal distance?

**70** A man has 1020 ft of fencing for enclosing two separate lots, one of which is to be square and the other twice as long as it is wide. Find the dimensions of each lot so that the total area enclosed will be a minimum.

In each of Problems 71–82, use Newton's method to find the solution of the given equation $f(x) = 0$ in the indicated interval $[a, b]$ accurate to four decimal places.

**71** $x^2 - 7 = 0$;  $[2, 3]$  (to find the positive square root of 7)
**72** $x^3 - 3 = 0$;  $[1, 2]$  (to find the cube root of 3)
**73** $x^5 - 75 = 0$;  $[2, 3]$  (to find the fifth root of 75)
**74** $x^{4/3} - 10 = 0$;  $[5, 6]$  (to approximate $10^{3/4}$)
**75** $x^3 - 3x - 1 = 0$;  $[-1, 0]$
**76** $x^3 - 4x - 1 = 0$;  $[-1, 0]$
**77** $x^6 + 7x^2 - 4 = 0$;  $[0, 1]$
**78** $x^3 - 3x^2 + 2x + 10 = 0$;  $[-2, -1]$
**79** $x^4 + 8x - 12 = 0$;  $[1, 2]$
**80** $2x^3 - 14x^2 + 2x + 5 = 0$;  $[6, 7]$
**81** $x^3 - 3x^2 + 3 = 0$;  $[-1, 0]$
**82** $x^4 + 4x^3 - 6x^2 - 20x = 23$;  $[2, 3]$
**83** Find the depth to which a wooden ball with radius 2 ft sinks in water if its density is $\frac{1}{3}$ that of water. A useful formula appears in Problem 43, Section 3-9.
**84** The equation $x^2 + 1 = 0$ has no real solutions. Try finding a solution by using Newton's method and report what happens. Use the initial estimate $x_0 = 2$.
**85** At the beginning of Section 3-9, we mentioned the fifth degree equation

$$x^5 - 3x^3 + x^2 - 23x + 19 = 0,$$

which has at least one and no more than five distinct real solutions. Find *all* real solutions by Newton's method.

**86** Find all four solutions of the equation

$$x^4 - 4x^3 - 6x^2 + 20x + 9 = 0$$

by Newton's method.

**87** Criticize the following "proof" that $3 = 2$. Begin by writing

$$x^3 = x \cdot (x^2)$$
$$= x^2 + x^2 + x^2 + \cdots + x^2 \qquad (x \text{ summands})$$

Differentiate to obtain $3x^2 = 2x + 2x + 2x + \cdots + 2x$ ($x$ summands). Thus $3x^2 = 2x^2$, and so $3 = 2$.

If we substitute $z = x + h$ in the definition of the derivative, the result is

$$f'(x) = \lim_{z \to x} \frac{f(z) - f(x)}{z - x}.$$

Use this form in Problems 88–89, together with the formula

$$a^3 - b^3 = (a - b)(a^2 + ab + b^2)$$

for factoring the difference of two cubes.

**88** Show that

$$Dx^{3/2} = \lim_{z \to x} \frac{z^{3/2} - x^{3/2}}{z - x} = \frac{3}{2} x^{1/2}.$$

(*Suggestion:* Factor the numerator as a difference of cubes and the denominator as a difference of squares:

$$z - x = (z^{1/2})^2 - (x^{1/2})^2.)$$

**89** Show that

$$Dx^{2/3} = \lim_{z \to x} \frac{z^{2/3} - x^{2/3}}{z - x} = \frac{2}{3} x^{-1/3}.$$

(*Suggestion:* Factor the numerator as a difference of squares and the denominator as a difference of cubes.)

**90** A rectangular block with square base is being squeezed in such a way that its height $y$ is decreasing at the rate of 2 in./min while its volume remains constant. At what rate is the edge $x$ of its base increasing when $x = 30$ inches and $y = 20$ in.?

**91** Air is being pumped into a spherical balloon at the constant rate of 10 in.$^3$/s. At what rate is the surface area of the balloon increasing when its radius is 5 in.?

**92** A ladder 10 ft long is leaning against a wall. If the bottom of the ladder slides away from the wall at the constant rate of 1 mi/h, how fast (in miles per hour) is the top of the ladder moving when it is 1/100 ft above the ground?

**93** A water tank has the shape of an inverted cone (axis vertical and vertex downward) with a top radius of 5 ft and a height of 10 ft. The water is flowing out of the tank, through a hole at the vertex, at the rate of 50 ft$^3$/min. What is the time rate of change of the depth of the water in the tank at the instant when it is 6 ft deep?

**94** Plane $A$ is flying west toward an airport at an altitude of 2 mi; plane $B$ is flying south toward the same airport at an altitude of 3 mi. When both planes are 2 mi (ground distance) from the airport, the speed of plane $A$ is 500 mi/h and the distance between the two planes is decreasing at 600 mi/h. What is the speed of plane $B$?

**95** A water tank is shaped so that the volume of water in the tank is $V = 2y^{3/2}$ in.$^3$ when its depth is $y$ inches. If water flows out a hole at the bottom at the rate of $3\sqrt{y}$ in.$^3$/min, at what rate does the water level in the tank fall? Can you think of a practical application of such a water tank?

**96** Water is being poured into the conical tank of Problem 93 at the rate of 50 ft$^3$/min and is draining out the hole at the bottom at the rate of $10\sqrt{y}$ ft$^3$/min, where $y$ is the depth of water in the tank.
(a) At what rate is the water level rising when $y = 5$ ft?
(b) Suppose that the tank is initially empty, water is poured in at 25 ft$^3$/min, and water continues to drain at $10\sqrt{y}$ ft$^3$/min. What is the maximum depth attained by the water?

**97** Let $L$ be a straight line passing through the fixed point $P(x_0, y_0)$ and also tangent to the parabola $y = x^2$ at the point $Q(a, a^2)$.
(a) Show that $a^2 - 2ax_0 + y_0 = 0$.
(b) Apply the quadratic formula to show that if $y_0 < x_0^2$ (that is, if $P$ lies below the parabola), then there are two possible values for $a$ and thus two lines through $P$ that are tangent to the parabola.
(c) Similarly, show that if $y_0 > x_0^2$ ($P$ lies above the parabola), then no line through $P$ is also tangent to the parabola.

# 4

## Applications of Derivatives and Antiderivatives

## Introduction

In Chapter 3 you learned to differentiate a wide variety of algebraic functions, and you saw that derivatives have such diverse applications as maximum-minimum problems, related-rates problems, and the solution of equations by Newton's method. The further applications of differentiation that we discuss in this chapter all depend ultimately upon a single fundamental question. Suppose that $y = f(x)$ is a differentiable function defined on the closed interval $[a, b]$ of length $\Delta x = b - a$. Then the *increment* $\Delta y$ in the value of $f(x)$ as $x$ changes from $x = a$ to $x = b$ is

$$\Delta y = f(b) - f(a). \tag{1}$$

The question is this: How is the increment $\Delta y$ related to the derivative—the rate of change—of the function $f$ at the points of the interval $[a, b]$?

An *approximate* answer is given in Section 4-2. If the function continued throughout the interval with the same rate of change $f'(a)$ that it had at $x = a$, then the change in its value would be $f'(a)(b - a) = f'(a)\,\Delta x$. This observation motivates the tentative approximation

$$\Delta y \approx f'(a)\,\Delta x. \tag{2}$$

A precise answer to the question above is provided by the mean value theorem of Section 4-3. This theorem implies that the exact increment is given by

$$\Delta y = f'(c)\,\Delta x \tag{3}$$

for some point $c$ in $(a, b)$. The mean value theorem is the central theoretical result of differential calculus, and it also is the key to many of the more advanced applications of derivatives.

In Section 4-8 we introduce the operation of **antidifferentiation**—the inverse of differentiation. If $g'(x) = f(x)$, then the function $g$ is called an **antiderivative** of $f$. The importance of antiderivatives results partly from the fact that scientific laws often specify the *rates of change* of quantities. The quantities themselves are then found by antidifferentiation. For example, in Section 4-9 we begin with the given acceleration of a moving particle. From this we find its velocity and then its position function, using successive antidifferentiation.

The methods of calculus are important not only in engineering and the physical sciences but in the biological, behavioral, and economic sciences as well. In Section 4-10 we discuss the use of elementary calculus in the analysis of business management decisions.

## Increments, Differentials, and Linear Approximation

Sometimes we need a quick and simple estimate of the change in $f(x)$ that results from a given change in $x$. Write $y$ for $f(x)$, and suppose that the change in the independent variable is the *increment* $\Delta x$, so that $x$ changes from its original value to the new value $x + \Delta x$. The actual change in the value of $y$ is the **increment** $\Delta y$, computed by subtracting the old value of $y$

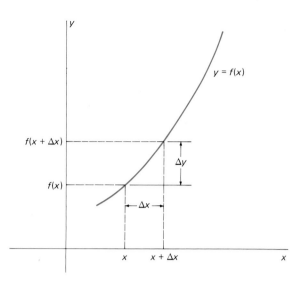

**4.1** The increments $\Delta x$ and $\Delta y$

from its new value:

$$\Delta y = f(x + \Delta x) - f(x). \tag{1}$$

The increments $\Delta x$ and $\Delta y$ are represented geometrically in Fig. 4.1.

Now we compare the actual increment $\Delta y$ with the change that *would* occur in $y$ *if* it continued to change at the *fixed* rate $f'(x)$ as $x$ changes to $x + \Delta x$. This hypothetical change in $y$ is the **differential**

$$dy = f'(x)\,\Delta x. \tag{2}$$

As Fig. 4.2 shows, $dy$ is the change in height of a point that moves along the tangent line at the point $(x, f(x))$, rather than along the curve $y = f(x)$.

Think of $x$ as fixed. Then (2) shows that the differential $dy$ is a *linear* function of the increment $\Delta x$. For this reason, $dy$ is called the **linear approximation** to the actual increment $\Delta y$. We can approximate $f(x + \Delta x)$ by writing $dy$ in place of $\Delta y$:

$$f(x + \Delta x) = y + \Delta y \approx y + dy.$$

**4.2** The estimate $dy$ of the actual increment $\Delta y$

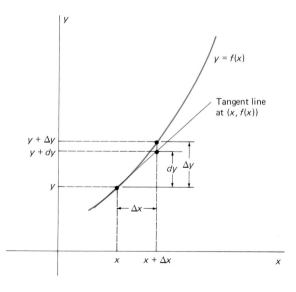

Since $y = f(x)$ and $dy = f'(x) \, \Delta x$, this gives the **linear approximation formula**

$$f(x + \Delta x) \approx f(x) + f'(x) \, \Delta x. \tag{3}$$

The idea is that this last approximation is a "good" one, at least when $\Delta x$ is relatively small. If we combine (1), (2), and (3), we see that

$$\Delta y \approx f'(x) \, \Delta x = dy. \tag{3'}$$

Thus the differential $dy = f'(x) \, \Delta x$ is an approximation to the actual increment $\Delta y = f(x + \Delta x) - f(x)$.

**EXAMPLE 1** Use the linear approximation formula to estimate $(122)^{2/3}$. Note that $(125)^{2/3} = [(125)^{1/3}]^2 = (5)^2 = 25$.

*Solution* Since we plan to estimate a particular value of $x^{2/3}$, we take $f(x) = x^{2/3}$. Thus, given the fact that $f(125) = 25$, we seek an estimate of $f(122)$. Now $f'(x) = \frac{2}{3} x^{-1/3}$, so we also have the exact value

$$f'(125) = \frac{2}{3}(125)^{-1/3} = \frac{2}{15}.$$

We take $x = 125$ and $\Delta x = -3$ in the linear approximation formula (3) and find that

$$(122)^{2/3} = f(122) \approx f(125) + (-3)f'(125)$$
$$= 25 + (-3)(\tfrac{2}{15}) = 24\tfrac{3}{5};$$

that is, $(122)^{2/3} \approx 24.6$. The actual value of $(122)^{2/3}$ is about 24.59838 (the digits given are correct), so Formula (3) gives us a rather good estimate.

**EXAMPLE 2** A hemispherical bowl of radius 10 in. is filled with water to a depth of $x$ inches. The volume of water in the bowl is given by the known formula

$$V = \frac{\pi}{3}(30x^2 - x^3)$$

($V$ is in cubic inches). Suppose that you *measure* the depth of water in the bowl, and you find it to be 5 in. with a maximum possible measurement error of $\frac{1}{16}$ in. Estimate the maximum error in the calculated volume of water in the bowl.

*Solution* The error $\Delta V$ in the calculated volume $V$ caused by an error $\Delta x$ in the measured depth $x$ is approximately equal to the differential

$$dV = \frac{dV}{dx} \, \Delta x = \frac{\pi}{3}(60x - 3x^2) \, \Delta x = \pi(20x - x^2) \, \Delta x.$$

We take $x = 5$ and $\Delta x = \pm\frac{1}{16}$, and obtain

$$dV = \pi[(20)(5) - 5^2](\pm\tfrac{1}{16}) \approx \pm 14.73$$

in.$^3$ With $x = 5$, the formula for $V$ gives $V(5) \approx 654.50$ in.$^3$, but we see now that this may be in error by as much as 15 in.$^3$ either way.

The **relative error** in an estimated or measured value is defined by

$$\text{Relative error} = \frac{\text{error}}{\text{value}}. \tag{4}$$

Thus, in Example 2, a relative error in the measured depth $x$ of

$$\frac{\Delta x}{x} = \frac{\frac{1}{16}}{5} = 0.0125 = 1.25\%$$

leads to a relative error in the estimated volume of

$$\frac{dV}{V} = \frac{14.73}{654.50} = 0.0225 = 2.25\%$$

The relationship between these two relative errors is of some interest. The formulas for $dV$ and $V$ above give

$$\frac{dV}{V} = \frac{\pi(20x - x^2)\,\Delta x}{\frac{1}{3}\pi(30x^2 - x^3)} = \frac{3(20 - x)}{30 - x} \cdot \frac{\Delta x}{x}.$$

In particular, when $x = 5$, this gives

$$\frac{dV}{V} = (1.80)\frac{\Delta x}{x}.$$

Hence, in order to estimate the volume of water in the bowl with a relative error of at most 0.5%, we would need to measure the depth with a relative error of at most $(0.5\%)/(1.8)$, or about 0.3%.

Now we take up the question of how closely the differential $dy$ approximates the actual increment $\Delta y$. It is apparent in Fig. 4.2 that the smaller is $\Delta x$, the closer are the corresponding points on the curve $y = f(x)$ and its tangent line. Since the difference in the heights of two such points is the value of $\Delta y - dy$ determined by a particular choice of $\Delta x$, we conclude that $\Delta y - dy \to 0$ as $\Delta x \to 0$.

But even more is true: As $\Delta x \to 0$, the difference $\Delta y - dy$ is small *even in comparison with* $\Delta x$. For

$$\frac{\Delta y - dy}{\Delta x} = \frac{f(x + \Delta x) - f(x) - f'(x)\,\Delta x}{\Delta x}$$

$$= \frac{f(x + \Delta x) - f(x)}{\Delta x} - f'(x);$$

that is,

$$\frac{\Delta y - dy}{\Delta x} = \varepsilon(\Delta x) \tag{5}$$

where, by the definition of the derivative $f'(x)$, we see that $\varepsilon(\Delta x)$ is a function of $\Delta x$ that approaches zero as $\Delta x \to 0$. We multiply both sides of (5) by $\Delta x$, write $\varepsilon$ for $\varepsilon(\Delta x)$, and obtain

$$\Delta y = dy + \varepsilon\,\Delta x \tag{6}$$

where $\varepsilon \to 0$ as $\Delta x \to 0$. If $\Delta x$ is "very small," so that $\varepsilon$ is also "very small," we might well describe the product $\varepsilon\,\Delta x = \Delta y - dy$ as "very *very* small."

**EXAMPLE 3** Suppose that $y = x^3$. Verify that $\varepsilon = (\Delta y - dy)/\Delta x$ approaches zero as $\Delta x \to 0$.

*Solution*   Simple computations give

$$\Delta y = (x + \Delta x)^3 - x^3 = 3x^2 \, \Delta x + 3x(\Delta x)^2 + (\Delta x)^3$$

and $dy = 3x^2 \, \Delta x$. Hence

$$\varepsilon = \frac{\Delta y - dy}{\Delta x} = \frac{3x(\Delta x)^2 + (\Delta x)^3}{\Delta x} = 3x \, \Delta x + (\Delta x)^2,$$

which obviously approaches zero as $\Delta x \to 0$.

### DIFFERENTIALS

The linear approximation formula (3) is often written with $dx$ in place of $\Delta x$, in the form

$$f(x + dx) \approx f(x) + f'(x) \, dx. \tag{7}$$

In this case $dx$ is an independent variable, called the **differential** of $x$, while $x$ is fixed. Thus the differentials of $x$ and $y$ are defined by

$$dx = \Delta x \quad \text{and} \quad dy = f'(x) \, \Delta x = f'(x) \, dx. \tag{8}$$

With this definition it follows immediately that

$$\frac{dy}{dx} = \frac{f'(x) \, dx}{dx} = f'(x),$$

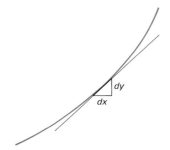

**4.3**  The slope of the tangent line as the ratio of the infinitesimals $dy$ and $dx$

in agreement with the notation we have been using. Indeed, Leibniz originated differential notation by visualizing infinitesimal increments $dx$ and $dy$ as in Fig. 4.3, with their ratio $dy/dx$ being the slope of the tangent line.

Differential notation also provides us with a convenient way to write derivative formulas. For suppose that $z = g(u)$, so that $dz = g'(u) \, du$. With the particular choice $g(u) = u^n$ we get the formula

$$d(u^n) = nu^{n-1} \, du. \tag{9}$$

In this way we can write differentiation rules in "differential form," without the need to identify the independent variable. The sum, product, and quotient rules take the forms

$$d(u + v) = du + dv, \tag{10}$$

$$d(uv) = u \, dv + v \, du, \quad \text{and} \tag{11}$$

$$d\left(\frac{u}{v}\right) = \frac{v \, du - u \, dv}{v^2}. \tag{12}$$

If $u = f(x)$ and $z = g(u)$, we may substitute $du = f'(x) \, dx$ in the formula $dz = g'(u) \, du$. This gives

$$dz = g'(f(x)) \cdot f'(x) \, dx. \tag{13}$$

This is the differential form of the chain rule

$$Dg(f(x)) = g'(f(x))f'(x).$$

Thus the chain rule appears here as though it were the result of mechanical manipulations of the differential notation. This compatibility with the chain rule is one reason for the extraordinary usefulness of differential notation in calculus.

In Problems 1–16, write $dy$ in terms of $x$ and $dx$.

**1** $y = 3x^2 - \dfrac{4}{x^2}$

**2** $y = 2\sqrt{x} - \dfrac{3}{\sqrt[3]{x}}$

**3** $y = x - \sqrt{4 - x^3}$

**4** $y = \dfrac{1}{x - \sqrt{x}}$

**5** $y = 3x^2(x - 3)^{3/2}$

**6** $y = \dfrac{x}{x^2 - 4}$

**7** $y = x(x^2 + 25)^{1/4}$

**8** $y = \dfrac{1}{(x^2 - 1)^{4/3}}$

**9** $y = (1 + \sqrt{x})^3$

**10** $y = x^2(1 + x)^5$

**11** $y = 2x\sqrt{1 + x^4}$

**12** $y = (1 + 3x)^{4/3}$

**13** $y = \dfrac{\sqrt{1 + 2x}}{3x}$

**14** $y = \dfrac{(1 - 2x)^3}{\sqrt{x}}$

**15** $y = \dfrac{1}{1 - \sqrt{2x + 1}}$

**16** $y = \left(1 + \dfrac{2}{x}\right)^{3/2}$

In Problems 17–21, calculate $\Delta y$ and $dy$ with the indicated values of $x$ and $\Delta x$.

**17** $f(x) = 3x + 4$, $\quad x = 2$, $\Delta x = 1$
**18** $f(x) = x^2$, $\quad x = 10$, $\Delta x = 1$
**19** $f(x) = \sqrt{x}$, $\quad x = 10$, $\Delta x = 1$
**20** $f(x) = x^{3/2}$, $\quad x = 4$, $\Delta x = 0.1$

**21** $f(x) = \dfrac{1}{x^2 + 1}$, $\quad x = 0$, $\Delta x = -0.2$

In each of Problems 22–32, estimate the indicated number by linear approximation, as in Example 1.

**22** $\sqrt{102}$

**23** $\sqrt[3]{25}$

**24** $\sqrt{80}$

**25** $(15)^{1/4}$

**26** $(80)^{3/4}$

**27** $(65)^{-2/3}$

**28** $\sqrt[3]{130}$

**29** $\sqrt[3]{997}$

**30** $\sqrt[4]{250}$

**31** $(0.999)^{100}$

**32** $(1.01)^{10}$

In Problems 33–37, compute the differential of each side of the given equation, regarding $x$ and $y$ as *dependent* variables (as if both were functions of some third unspecified variable). Then solve for $dy/dx$.

**33** $x^2 + y^2 = 1$

**34** $x^{2/3} + y^{2/3} = 4$

**35** $x^3 + y^3 = 3xy$

**36** $(x^2 + y^2)^2 = x^2 - y^2$

**37** $x\sqrt{1 + y} = 1$

In each of Problems 38–45, use linear approximation to estimate the change in the indicated quantity.

**38** The circumference of a circle, if its radius is increased from 10 to 10.5 in.

**39** The area of a square, if its edge is decreased from 10 to 9.8 in.

**40** The surface area of a sphere, if its radius is increased from 5 to 5.2 in.

**41** The volume of a cylinder having height equal to its radius, if both are decreased from 15 to 14.7 cm.

**42** The volume of a conical sandpile of height 7 in. and radius 14 in., if its height is increased to 7.1 in.

**43** The range $R = \frac{1}{32}v_0^2$ (in feet) of a projectile fired at a 45° angle from the horizontal over level ground, if its initial velocity $v_0$ is increased from 80 to 81 ft/sec.

**44** The range $R = 6400/g$ (in feet) of a projectile fired at a 45° angle from the horizontal over level ground from a gun with a muzzle velocity of 80 ft/sec, if the gun is moved from a location where $g = 32.00$ ft/sec$^2$ to one where $g = 32.16$ ft/sec$^2$.

**45** The wattage $W = RI^2$ of a floodlight with resistance $R = 10\ \Omega$ (ohms), if the current $I$ is increased from 3 to 3.1 A (amperes).

**46** The equatorial radius of the earth is approximately 3960 mi. Suppose that a wire is wrapped tightly around the earth at the equator. How much must this wire be lengthened if it is to be strung on poles 10 ft above the ground, all the way around the earth?

**47** The radius of a spherical ball is measured as 10 in., with a maximum error of $\frac{1}{16}$ in. What is the maximum resulting error in its calculated volume?

**48** With what accuracy must the radius of the ball of Problem 47 be measured to insure an error of at most 1 in.$^3$ in its calculated volume?

**49** The radius of a hemispherical dome is measured as 100 m with a maximum error of 1 cm. What is the maximum resulting error in its calculated surface area?

**50** With what accuracy must the radius of a hemispherical dome be measured in order to have an error of at most 0.01% in its calculated surface area?

## The Mean Value Theorem and Applications

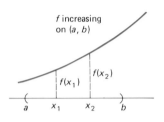

f increasing on (a, b)

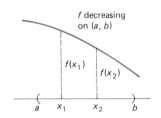

**4.4** An increasing function; a decreasing function

**4.5** A graph rising at *x*; a graph falling at *x*

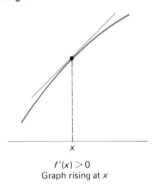

$f'(x) > 0$
Graph rising at *x*

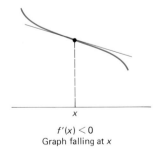

$f'(x) < 0$
Graph falling at *x*

The significance of the *sign* of the first derivative is simple but crucial:

$f$ is increasing on an interval where $f'(x) > 0$;

$f$ is decreasing on an interval where $f'(x) < 0$.

Geometrically, this means that where $f'(x) > 0$, the graph of $f$ is rising, as you scan it from left to right. Where $f'(x) < 0$, the graph is falling. We can make the terms *increasing* and *decreasing* more precise as follows.

---

*Definition*   *Increasing and Decreasing Functions*

The function $f$ is **increasing** on the interval $I$ if, for each two numbers $x_1$ and $x_2$ in $I$ with $x_1 < x_2$,

$$f(x_1) < f(x_2).$$

That is,

$$x_1 < x_2 \quad \text{implies} \quad f(x_1) < f(x_2).$$

The function $f$ is **decreasing** on $I$ provided that

$$x_1 < x_2 \quad \text{implies} \quad f(x_1) > f(x_2)$$

for any two points $x_1$ and $x_2$ in $I$.

---

Figure 4.4 illustrates this definition.

Note that we speak of a function as increasing or decreasing *on an interval*, not at a single point. Nevertheless, if we consider the sign of $f'$ at a single point, we get a useful intuitive picture of the significance of the sign of the derivative. This is because the derivative $f'(x)$ is the slope of the tangent line at the point $(x, f(x))$ on the graph of $f$. If $f'(x) > 0$, the tangent line has positive slope, and therefore rises as you scan from left to right. Intuitively, it seems likely that a rising tangent corresponds to a rising graph, and thus to an increasing function. Similarly, we expect to see a falling graph where $f'(x)$ is negative. Figure 4.5 shows a pair of graphs illustrating this intuitive idea. One caution: In order to determine whether a function $f$ is increasing or decreasing, we must examine the sign of $f'$ on a whole interval, not merely at a single point.

### THE MEAN VALUE THEOREM

Though pictures of rising and falling graphs are quite suggestive, they provide no actual proof of the significance of the sign of the derivative. To establish the connection between the graph's rising or falling and the sign of the derivative, we need the **mean value theorem,** stated below. This theorem is the principal theoretical tool of differential calculus, and we shall see that it has many important applications.

As an introduction to the mean value theorem, we pose the following question. Suppose that $P$ and $Q$ are two points in the plane, with $Q$ lying generally to the east of $P$, as shown in Fig. 4.6. Is it possible to sail a boat from $P$ to $Q$, sailing always roughly east, without *ever* (even for an instant) sailing in the exact direction from $P$ to $Q$? That is, can we sail from $P$ to $Q$ without our instantaneous line of motion ever being parallel to the line $PQ$?

The mean value theorem says that the answer to this question is no; there will always be at least one instant when we are sailing parallel to the line $PQ$, no matter which path we choose.

Here is a mathematical paraphrase: Let the path of the sailboat be the graph of a differentiable function, $y = f(x)$, with end points $P(a, f(a))$ and $Q(b, f(b))$. Then we are saying that there must be some point on this graph where the tangent line to the curve is parallel to the line $PQ$ joining its end points. This is a *geometric interpretation* of the mean value theorem.

But the slope of the tangent line at the point $(c, f(c))$, shown in Fig. 4.7, is $f'(c)$, while the slope of the line $PQ$ is

$$\frac{f(b) - f(a)}{b - a}.$$

We may think of this last quotient as the average (or *mean*) value of the slope of $f$. The mean value theorem guarantees that there is a point $c$ in $(a, b)$ where the tangent line at $(c, f(c))$ is indeed parallel to the line $PQ$. In the language of algebra, there's a number $c$ in $(a, b)$ such that

$$f'(c) = \frac{f(b) - f(a)}{b - a}. \tag{1}$$

We first give a preliminary result, a lemma to expedite the proof of the mean value theorem. This lemma is called **Rolle's theorem,** after the Frenchman Michel Rolle (1652–1719), who discovered it in 1690. Rolle studied the emerging subject of calculus in his youth but later renounced it; he argued that it was based on logical fallacies, and today he is remembered only for a single theorem that bears his name. It is ironic that his theorem plays an important role in the rigorous proofs of several calculus theorems.

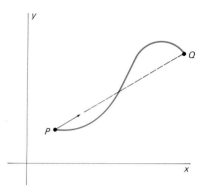

**4.6** Can you sail from $P$ to $Q$ without ever sailing—at least for an instant— in the direction $PQ$ (the direction of the arrow)?

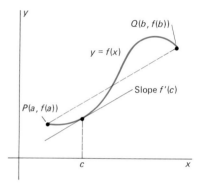

**4.7** The sailboat problem in mathematical terminology

---

> *Rolle's Theorem*
>
> Suppose that the function $f$ is continuous on the closed interval $[a, b]$ and is differentiable on its interior $(a, b)$. If $f(a) = 0 = f(b)$, then $f'(c) = 0$ for some number $c$ in $(a, b)$.

Thus between each pair of zeros of a differentiable function there is *at least one* point where the tangent line is horizontal. Some possible pictures of the situation are indicated in Fig. 4.8.

**4.8** The existence of the horizontal tangent is a consequence of Rolle's theorem.

*Proof*  Since $f$ is continuous on $[a, b]$, it must attain both a maximum and a minimum value on $[a, b]$ (by the maximum value property of Section 2-4). If $f$ has any positive values at all, consider its maximum value $f(c)$. Since $f(a) = f(b) = 0$, $c$ must be a point of $(a, b)$. Since $f$ is then differentiable at $c$, it follows from Theorem 1 of Section 3-5 that $f'(c) = 0$.

Similarly, if $f$ has any negative values, we may consider its minimum value $f(c)$ and conclude that $f'(c) = 0$. If $f$ has neither positive nor negative

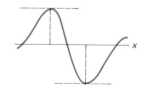

values, then $f$ is identically zero on $[a, b]$, and it follows that $f'(c) = 0$ for *every* $c$ in $(a, b)$. Thus we see that the conclusion of Rolle's theorem is justified in every case. ∎

**EXAMPLE 1**   Suppose that $f(x) = x^{1/2} - x^{3/2}$ on $[0, 1]$. Find a number $c$ satisfying the conclusion of Rolle's theorem.

***Solution***   Note that $f$ is continuous on $[0, 1]$ and differentiable on $(0, 1)$; because of the presence of the term $x^{1/2}$, $f$ is *not* differentiable at $x = 0$. Also, $f(0) = f(1) = 0$, so all hypotheses of Rolle's theorem are satisfied. Since

$$f'(x) = \tfrac{1}{2}x^{-1/2} - \tfrac{3}{2}x^{1/2} = \tfrac{1}{2}x^{-1/2}(1 - 3x),$$

we see that $f'(c) = 0$ for $c = \tfrac{1}{3}$.

**EXAMPLE 2**   Suppose that $f(x) = 1 - x^{2/3}$ on $[-1, 1]$. Then $f$ satisfies the hypotheses of Rolle's theorem *except* for the fact that $f'(0)$ does not exist. It is clear from the graph of $f$ (Fig. 4.9) that there is *no* point where the tangent line is horizontal. Indeed,

$$f'(x) = -\frac{2}{3}x^{-1/3} = -\frac{2}{3x^{1/3}} \neq 0$$

for $x \neq 0$, and we see that $|f'(x)| \to \infty$ as $x \to 0$, so the graph of $f$ has a vertical tangent line—rather than a horizontal one—at the point $(0, 1)$.

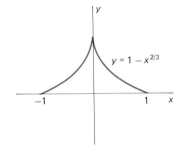

$y = 1 - x^{2/3}$

**4.9**   The function of Example 2; Rolle's hypotheses are not satisfied

Now we are ready to state formally and prove the mean value theorem.

---

> ***The Mean Value Theorem***
> Suppose that the function $f$ is continuous on $[a, b]$ and differentiable on $(a, b)$. Then
> $$f(b) - f(a) = f'(c)(b - a) \qquad (2)$$
> for some number $c$ in $(a, b)$.

---

COMMENT   Because Equation (2) is equivalent to Equation (1), the conclusion of the mean value theorem is that there must be at least one point on the curve $y = f(x)$ at which the tangent line is parallel to the line joining its endpoints $P(a, f(a))$ and $Q(b, f(b))$.

***Motivation for the Proof***   We consider the "auxiliary function" $\phi$ suggested by Fig. 4.10. The value $\phi(x)$ is by definition the vertical height difference over $x$ of the point $(x, f(x))$ on the curve and the corresponding point on the line $PQ$. It appears that a point on the curve $y = f(x)$, where the tangent line is parallel to $PQ$, corresponds to a maximum or minimum of $\phi$. It's clear also that $\phi(a) = \phi(b) = 0$, so Rolle's theorem can be applied to the function $\phi$. So our plan for proving the mean value theorem is this: First, we obtain a formula for the function $\phi$. Second, we locate the point $c$ such that $\phi'(c) = 0$. Finally, we show that this number $c$ is exactly the number needed to satisfy Equation (2).

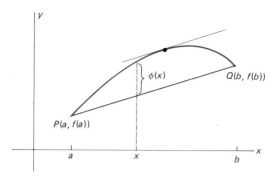

**4.10** The construction of the auxiliary function $\phi$

**Proof of the Mean Value Theorem**  Since the line $PQ$ passes through $(a, f(a))$ with slope

$$m = \frac{f(b) - f(a)}{b - a},$$

the point-slope formula for the equation of a straight line gives us the following equation for $PQ$:

$$y = y_{\text{line}} = f(a) + m(x - a).$$

Thus

$$\phi(x) = y_{\text{curve}} - y_{\text{line}} = f(x) - f(a) - m(x - a).$$

You may verify by direct substitution that $\phi(a) = \phi(b) = 0$. And, since $\phi$ is certainly continuous on $[a, b]$ and differentiable on $(a, b)$, we may apply Rolle's theorem to it. Thus there is a point $c$ somewhere in the open interval $(a, b)$ at which $\phi'(c) = 0$. But

$$\phi'(x) = f'(x) - m = f'(x) - \frac{f(b) - f(a)}{b - a}.$$

So, since $\phi'(c) = 0$, we find that

$$0 = f'(c) - \frac{f(b) - f(a)}{b - a},$$

and so

$$f(b) - f(a) = f'(c)(b - a). \qquad \blacksquare$$

Observe that the proof of the mean value theorem is an application of Rolle's theorem, while Rolle's theorem is the special case $f(a) = 0 = f(b)$ of the mean value theorem.

**EXAMPLE 3**  Suppose that we travel from Eugene, Oregon to Corvallis—a road distance of exactly 50 mi (let us say)—in a time of precisely 1 hr, from time $t = 0$ to time $t = 1$. Let $f(t)$ denote the distance we have traveled at time $t$, and assume that $f$ is a differentiable function. Then the mean value theorem implies that

$$50 = f(1) - f(0) = f'(c)(1 - 0) = f'(c)$$

at some instant $c \in (0, 1)$. But $f'(c)$ is our instantaneous velocity at time

$t = c$. Thus if our *average* velocity for the trip is 50 mi/hr, then we must have an *instantaneous* velocity of exactly 50 mi/hr at least once during the trip.

## CONSEQUENCES OF THE MEAN VALUE THEOREM

The first of these consequences is the *non*trivial converse of the trivial fact that the derivative of a constant function is identically zero. That is, we prove that there is *no* unknown exotic function that is nonconstant but which has derivative identically zero.

---

**Corollary 1** *Functions with Zero Derivatives*

If $f'(x) = 0$ for all $x$ in $[a, b]$, then $f$ is a constant on $[a, b]$. That is, there is a constant $C$ such that $f(x) = C$ for all $x$ in $[a, b]$.

---

***Proof*** Apply the mean value theorem to the function $f$ on the interval $[a, x]$. The right-hand end point of the interval is indeed $x$, rather than $b$. We think of $x$ as a "temporary constant," chosen in the range $a < x \leq b$. We find that

$$f(x) - f(a) = f'(c) \cdot (x - a)$$

for some number $c$ between $a$ and $x$.

But $f'$ is always zero on the interval $[a, b]$, and so $f'(c) = 0$. Thus $f(x) - f(a) = 0$, and so $f(x) = f(a)$.

Now we release $x$ from the condition of being "temporarily constant" and observe that the result $f(x) = f(a)$ holds for *all* $x$ in the range $a < x \leq b$. That is, $f(x)$ has the fixed value $C = f(a)$ regardless of the choice of $x$ in $[a, b]$. This establishes Corollary 1. ∎

The ease and simplicity of the above proof suggests that the mean value theorem is a powerful tool.

Corollary 1 is usually applied in a different but equivalent form, which we prove next.

---

**Corollary 2** *Functions with Equal Derivatives*

Suppose that $f$ and $g$ are two differentiable functions such that $f'(x) = g'(x)$ for all $x$ in the closed interval $[a, b]$. Then $f$ and $g$ differ by a constant on $[a, b]$. That is, there is a constant $K$ such that

$$f(x) = g(x) + K$$

for all $x$ in $[a, b]$.

---

***Proof*** Given the hypotheses, let $h(x) = f(x) - g(x)$. Then

$$h'(x) = f'(x) - g'(x) = 0$$

because $f' = g'$ on $[a, b]$. So, by Corollary 1, $h$ is a constant $K$ on $[a, b]$. That is, $f(x) - g(x) = K$ for all $x$ in $[a, b]$, and so

$$f(x) = g(x) + K$$

for all $x$ in $[a, b]$. This establishes Corollary 2. ∎

The following consequence of the mean value theorem verifies the remarks about increasing and decreasing functions with which we opened this section.

---

**Corollary 3**  *Increasing and Decreasing Functions*

Let $f$ be a function that is continuous on $[a, b]$ and differentiable on $(a, b)$. If $f'(x) > 0$ for all $x$ in $(a, b)$, then $f$ is an increasing function on $[a, b]$. If $f'(x) < 0$ for all $x$ in $(a, b)$, then $f$ is a decreasing function on $[a, b]$.

---

***Proof***  Suppose, for example, that $f'(x) > 0$ for all $x$ in $(a, b)$. We need to show this: If $x_1$ and $x_2$ are points of $[a, b]$ with $x_1 < x_2$, then $f(x_1) < f(x_2)$. We apply the mean value theorem to $f$, but on the interval $[x_1, x_2]$. This gives

$$f(x_2) - f(x_1) = f'(c)(x_2 - x_1)$$

for some $c$ in $(x_1, x_2)$. Since $x_2 > x_1$ and since, by hypothesis, $f'(c) > 0$, it follows that

$$f(x_2) - f(x_1) > 0, \quad \text{or} \quad f(x_2) > f(x_1),$$

as we wanted to show. The proof is similar in the case that $f'$ is negative on $(a, b)$. ∎

**EXAMPLE 4**  Where is the function $f(x) = x^2 - 4x + 5$ increasing and where is it decreasing?

***Solution***  The derivative of $f$ is $f'(x) = 2x - 4$. It is clear that $f'(x) > 0$ if $x > 2$, while $f'(x) < 0$ if $x < 2$. Hence $f$ is decreasing on $(-\infty, 2)$ and increasing on $(2, +\infty)$. Together with the facts that $f(2) = 1$ and $f'(2) = 0$, this information is enough to obtain a sketch of the graph of $f$. It appears in Fig. 4.11.

**EXAMPLE 5**  Show that the equation $x^3 + x - 1 = 0$ has exactly one (real) root.

***Solution***  Let $f(x) = x^3 + x - 1$. Since $f(0) = -1 < 0$ and $f(1) = +1 > 0$, the intermediate value property guarantees that $f(x)$ has *at least* one zero in $[0, 1]$. But

$$f'(x) = 3x^2 + 1 > 0$$

for all $x$, so Corollary 3 shows that $f$ is an increasing function on the whole real line. Thus $f$ cannot have more than one zero. (Why?)

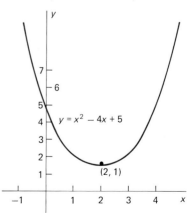

4.11  The parabola of Example 4

**EXAMPLE 6** Determine the open intervals of the $x$-axis on which the function

$$f(x) = 3x^4 - 4x^3 - 12x^2 + 5$$

is increasing and those on which it is decreasing.

*Solution* The derivative of $f$ is

$$f'(x) = 12x^3 - 12x^2 - 24x$$
$$= 12x(x^2 - x - 2) = 12x(x + 1)(x - 2).$$

The critical points $x = -1, 0, 2$ separate the $x$-axis into the open intervals $(-\infty, -1)$, $(-1, 0)$, $(0, 2)$, and $(2, \infty)$. In the given table we have recorded the sign of each of the three factors of $f'(x)$ on each of these four open intervals. We note the resulting sign of $f'(x)$ on each interval and then apply Corollary 3; it follows that $f$ is decreasing on $(-\infty, -1)$ and on $(0, 2)$, while $f$ is increasing on $(-1, 0)$ and on $(2, \infty)$.

| Interval | $x + 1$ | $12x$ | $x - 2$ | $f'(x)$ | $f$ |
|----------|---------|-------|---------|---------|-----|
| $(-\infty, -1)$ | Neg. | Neg. | Neg. | Neg. | Decreasing |
| $(-1, 0)$ | Pos. | Neg. | Neg. | Pos. | Increasing |
| $(0, 2)$ | Pos. | Pos. | Neg. | Neg. | Decreasing |
| $(2, \infty)$ | Pos. | Pos. | Pos. | Pos. | Increasing |

## 4-3 PROBLEMS

For each of the functions in Problems 1–14, determine (as in Example 6) the open intervals on the $x$-axis on which it is increasing as well as those on which it is decreasing.

**1** $f(x) = 3x + 2$      **2** $f(x) = 4 - 5x$
**3** $f(x) = 4 - x^2$      **4** $f(x) = 4x^2 + 8x + 13$
**5** $f(x) = 6x - 2x^2$      **6** $f(x) = x^3 - 12x + 17$

**7** $f(x) = x^4 - 2x^2 + 1$      **8** $f(x) = \dfrac{x}{x + 1}$

**9** $f(x) = 3x^4 + 4x^3 - 12x^2$
**10** $f(x) = x\sqrt{x^2 + 1}$      **11** $f(x) = 8x^{1/3} - x^{4/3}$
**12** $f(x) = 2x^3 + 3x^2 - 12x + 5$

**13** $f(x) = \dfrac{(x - 1)^2}{x^2 - 3}$      **14** $f(x) = x^2 + \dfrac{16}{x^2}$

In Problems 15–18, show that the given function satisfies the hypotheses of Rolle's theorem on the indicated interval $[a, b]$, and find all numbers $c$ in $[a, b]$ that satisfy the conclusion of the theorem.

**15** $f(x) = x^2 - 2x$   on $[0, 2]$
**16** $f(x) = 9x^2 - x^4$   on $[-3, 3]$

**17** $f(x) = \dfrac{1 - x^2}{1 + x^2}$   on $[-1, 1]$

**18** $f(x) = 5x^{2/3} - x^{5/3}$   on $[0, 5]$

In Problems 19–21, show that the given function $f$ does not satisfy the conclusion of Rolle's theorem on the indicated interval. Which of the hypotheses does it fail to satisfy?

**19** $f(x) = 1 - |x|$   on $[-1, 1]$
**20** $f(x) = 1 - (2 - x)^{2/3}$   on $[1, 3]$
**21** $f(x) = x^4 + x^2$   on $[0, 1]$

In Problems 22–26, show that the given function $f$ satisfies the hypotheses of the mean value theorem on the indicated interval, and find all numbers $c$ in $(a, b)$ that satisfy the conclusion of that theorem.

**22** $f(x) = x^3$   on $[-1, 1]$
**23** $f(x) = 3x^2 + 6x - 5$   on $[-2, 1]$
**24** $f(x) = \sqrt{x - 1}$   on $[2, 5]$
**25** $f(x) = (x - 1)^{2/3}$   on $[1, 2]$

**26** $f(x) = x + \dfrac{1}{x}$   on $[1, 2]$

In Problems 27–30, show that the given function satisfies neither the hypotheses nor the conclusion of the mean value theorem on the indicated interval.

**27** $f(x) = |x - 2|$   on $[1, 4]$
**28** $f(x) = 1 + |x - 1|$   on $[0, 3]$
**29** $f(x) = [\![x]\!]$ (the greatest integer function) on $[-1, 1]$
**30** $f(x) = 3x^{2/3}$   on $[-1, 1]$

In each of Problems 31–33, show that the given equation has exactly one root in the indicated interval.

**31** $x^5 + 2x - 3 = 0$ in $[0, 1]$
**32** $x^{10} = 1000$ in $[1, 2]$
**33** $x^4 - 3x = 20$ in $[2, 3]$
**34** Show that the function $f(x) = x^{2/3}$ does not satisfy the hypotheses of the mean value theorem on $[-1, 27]$, but nevertheless there is a number $c$ in $(-1, 27)$ such that

$$f'(c) = \frac{f(27) - f(-1)}{27 - (-1)}.$$

**35** Show that the function $f(x) = (1 + x)^{3/2} - \frac{3}{2}x - 1$ is increasing on $(0, +\infty)$. Conclude that $(1 + x)^{3/2} > 1 + \frac{3}{2}x$ for all $x > 0$.
**36** Suppose that $f'$ is constant on the interval $[a, b]$. Prove that $f$ must be a linear function (a function which has a graph that is a straight line).
**37** Suppose that $f'(x)$ is a polynomial of degree $n - 1$ on the interval $[a, b]$. Prove that $f(x)$ must be a polynomial of degree $n$ on $[a, b]$.
**38** Suppose that there are $k$ different points of $[a, b]$ at which the differentiable function $f$ vanishes (is zero). Show that $f'$ must vanish at at least $k - 1$ points of $[a, b]$.
**39** (a) Apply the mean value theorem to $f(x) = \sqrt{x}$ on $[100, 101]$ to show that $\sqrt{101} = 10 + 1/(2\sqrt{c})$ for some number $c$ in $(100, 101)$.
(b) Show that if $100 < c < 101$, then $10 < \sqrt{c} < 10.5$,

and use this fact to conclude from part (a) that $10.0475 < \sqrt{101} < 10.0500$.
**40** Show that the equation $x^7 + x^5 + x^3 + 1 = 0$ has exactly one real root.
**41** (a) Suppose that $f(x) = 1/(x - 1)$ and that $g(x) = x/(x - 1)$. Show that $f'(x) = g'(x)$ if $x > 1$.
(b) Conclude that for all $x > 1$,

$$\frac{1}{x - 1} = \frac{x}{x - 1} + C$$

for some constant $C$. Then find the value of $C$.
**42** Explain why the mean value theorem does not apply to the function $f(x) = |x|$ on the interval $[-1, 2]$.
**43** Suppose that the function $f$ is differentiable on the interval $[-1, 2]$, with $f(-1) = -1$ and $f(2) = 5$. Show that there is a point on the graph of $f$ at which the tangent line is parallel to the line $y = 2x$.
**44** Let $f(x) = x^4 - x^3 + 7x^2 + 3x - 11$. Show that the graph of $f$ has at least one horizontal tangent line. (*Suggestion:* Do not try to find the coordinates of the point of tangency; instead, apply the intermediate value property of continuous functions.)
**45** Suppose that $y$ is implicitly defined as a function of $x$ through the equation $(y + 1)^3 = x^2$. Then $y = 0$ when $x = -1$ and also when $x = 1$. May you apply Rolle's theorem to conclude that $dy/dx = 0$ for some value of $x$ between $-1$ and 1? Explain your answer.

---

**4-4**

Suppose that $f$ is a function that is differentiable on the open interval $I$ and that $f$ has a local extremum in $I$. Theorem 1 of Section 3-5 tells us this: That extremum must occur at the sort of critical point where $f'(x) = 0$. But if $f'(c) = 0$, it does not follow that there is a local extremum at $c$. What we need is a way of finding, if $f'(c) = 0$, whether $f(c)$ is a local maximum value, a local minimum value, or neither.

Figure 4.12 suggests how such distinctions might be developed. If $f$ is decreasing on the left of $c$ and increasing on the right of $c$, then $f(c)$ should be a local minimum value. On the other hand, if $f$ is increasing on the left of

## The First Derivative Test

**4.12** The first derivative test

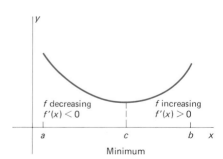

Minimum

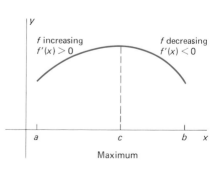

Maximum

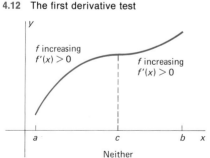

Neither

$c$ and decreasing on its right, then $f(c)$ should be a local maximum. And we can tell, with the aid of Corollary 3 to the mean value theorem, where the function $f$ is increasing and where it is decreasing; such behavior is determined by the sign of $f'(x)$. Thus we obtain the following test for local maxima and minima.

---

**Theorem** *The First Derivative Test*

Suppose that the function $f$ is continuous on the open interval $(a, b)$ and is differentiable there except possibly at $c$.

(i) If $f'(x) < 0$ on $(a, c)$ and $f'(x) > 0$ on $(c, b)$, then $f(c)$ is the minimum value of $f$ on $(a, b)$.
(ii) If $f'(x) > 0$ on $(a, c)$ and $f'(x) < 0$ on $(c, b)$, then $f(c)$ is the maximum value of $f$ on $(a, b)$.
(iii) If $f'(x) > 0$ or if $f'(x) < 0$ for all $x$ in $(a, b)$ except for $x = c$, then $f(c)$ is neither a maximum nor a minimum value for $f$.

---

Thus $f(c)$ is a local extremum if the first derivative $f'(x)$ changes sign as $x$ increases through $c$, and the direction of this sign change determines whether $f(c)$ is a local maximum or a local minimum. A good way to remember parts (i) and (ii) of the first derivative test is simply to visualize Fig. 4.12.

***Proof*** We shall prove part (i); the proofs of the other two parts are similar. Let $z$ be a point of $(a, b)$ other than $c$. If $z < c$ then the fact that $f'(x) < 0$ on $(z, c)$ implies that $f$ is decreasing on $[z, c]$. So it follows that $f(z) > f(c)$. If $z > c$, the fact that $f'(x) > 0$ on $(c, z)$ implies that $f$ is increasing on $[c, z]$, so it follows that $f(c) < f(z)$. Thus $f(c) < f(z)$ for all $z$ in $(a, b)$ other than $c$, and this means that $f(c)$ is the minimum value of $f$ on $(a, b)$. ∎

**EXAMPLE 1** Find and classify the critical points of the function

$$f(x) = 2x^3 - 3x^2 - 36x + 7.$$

***Solution*** The derivative is

$$f'(x) = 6x^2 - 6x - 36 = 6(x + 2)(x - 3),$$

so the critical points (where $f'(x) = 0$) are $x = -2$ and $x = 3$. They separate the $x$-axis into the open intervals $(-\infty, -2)$, $(-2, 3)$, and $(3, \infty)$. The given table exhibits the sign of $f'(x)$ in each of these intervals.

| Interval | $x + 2$ | $x - 3$ | $f'(x)$ | $f$ |
|---|---|---|---|---|
| $(-\infty, -2)$ | Neg. | Neg. | Pos. | Increasing |
| $(-2, 3)$ | Pos. | Neg. | Neg. | Decreasing |
| $(3, \infty)$ | Pos. | Pos. | Pos. | Increasing |

Thus $f'(x)$ is positive to the left and negative to the right of the critical point $x = -2$, so by part (ii) of the theorem $f(-2)$ is a local maximum value of $f$. At the critical point $x = 3$, $f(x)$ is negative to the left and positive to the right, so $f(3)$ is a local minimum value by part (i).

A sphere with given radius $a$ is inscribed in a pyramid with a square base so that the sphere touches the base of the pyramid and also each of its four sides. Show that the minimum possible volume of the pyramid is $8/\pi$ times the volume of the sphere. (*Suggestion:* Use the two right triangles in Fig. 4.20 to show that the volume of the pyramid is

$$V = V(y) = \frac{4a^2y^2}{3(y - 2a)}.$$

This can be done fairly easily with the aid of the angle $\theta$ and *without* the formula for $\tan(\theta/2)$.)

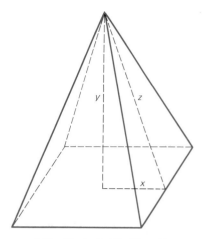

**4.21** The tent of Problem 39

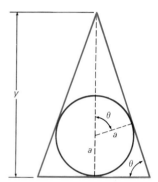

**4.20** Cross section through the centers of the sphere and pyramid of Problem 37

**38** Two noisy discothèques, one of them four times as noisy as the other, are located on opposite ends of a block 1000 ft long. What is the quietest point on the block between the two discos? Assume that the intensity of noise at a point removed from its source is proportional to the noisiness and inversely proportional to the square of the distance from the source.

**39** A floored tent with given volume $V$ is to be shaped like a pyramid with a square base and congruent sides, as in Fig. 4.21. What should be its height $y$ and base edge $2x$ in order to minimize its total surface area (including its floor)?

**40** Given a rectangle of area $A$ in the plane, suppose that an ellipse is circumscribed about it in the natural way—so that each vertex of the rectangle lies on the ellipse. One such ellipse has minimal area $E$. Find the ratio of $E$ to $A$. (*Suggestion:* If the rectangle has its center at the origin and its sides parallel to the coordinate axes, then the ellipse has an equation of the form

$$\frac{x^2}{a^2} + \frac{y^2}{b^2} = 1,$$

and the area of an ellipse with such an equation is $\pi ab$. The numbers $a$ and $b$ have the following interpretation: The longest chord of the ellipse passing through its center is called its *major axis* and the shortest chord of the ellipse passing through its center is called its *minor axis*. The major axis has length the larger of $2a$ and $2b$ and the minor axis has length the smaller of $2a$ and $2b$. Thus $a$ and $b$ play the roles of the longest and shortest "radii" of the ellipse. This makes the formula for its area easy to remember.)

---

## 4-5

We can construct a reasonably accurate graph of the polynomial function

$$f(x) = a_nx^n + a_{n-1}x^{n-1} + \cdots + a_2x^2 + a_1x + a_0 \qquad (1)$$

by assembling the following information.

## Graphs of Polynomials

**1** The critical points of $f$. These include points on the graph where the tangent line is horizontal.

**2** The intervals on which $f$ is increasing and those on which it is decreasing.

**3** The behavior of $f(x)$ as $x \to +\infty$ and as $x \to -\infty$.

To examine the behavior of $f$ "at infinity," we write $f(x)$ in the form

$$f(x) = x^n \left( a_n + \frac{a_{n-1}}{x} + \cdots + \frac{a_1}{x^{n-1}} + \frac{a_0}{x^n} \right).$$

Thus we conclude that the behavior of $f(x)$ as $x \to \pm\infty$ is much the same as that of its *leading term* $a_n x^n$, because the terms with powers of $x$ in the denominator all approach zero as $x \to \pm\infty$. In particular, if $a_n > 0$, then

$$\lim_{x \to \infty} f(x) = +\infty, \tag{2}$$

meaning that $f(x)$ increases without bound as $x \to +\infty$. Also,

$$\lim_{x \to -\infty} f(x) = \begin{cases} +\infty & \text{if } n \text{ is even;} \\ -\infty & \text{if } n \text{ is odd.} \end{cases} \tag{3}$$

If $a_n < 0$, reverse the signs on the right-hand sides of (2) and (3).

Every polynomial, including $f(x)$ above, is differentiable everywhere. So the critical points of $f(x)$ are the roots of the polynomial equation $f'(x) = 0$, or

$$na_n x^{n-1} + (n-1)a_{n-1}x^{n-2} + \cdots + 2a_2 x + a_1 = 0. \tag{4}$$

Sometimes one can find all (real) solutions of such an equation by factoring, and sometimes one must resort to numerical methods aided by calculator or computer. But suppose that we have somehow found *all* the solutions $c_1, c_2, \ldots, c_k$ of Equation (4). Then these solutions are the critical points of $f(x)$. If they are arranged in increasing order, as in Fig. 4.22, they separate the $x$-axis into the finitely many open intervals

$$(-\infty, c_1), (c_1, c_2), (c_2, c_3), \ldots, (c_{k-1}, c_k), (c_k, +\infty)$$

**4.22** The zeros of $f'$ divide the $x$-axis into intervals on each of which $f'$ does not change sign.

that also appear in the figure. The intermediate value property applied to $f'(x)$ tells us that $f'$ can change sign only at the critical points of $f$, so that $f'$ has only one sign on each of these open intervals. It is typical for $f'(x)$ to be positive on some and negative on the others. And it is easy to find the sign of $f'$; we need only substitute *any* convenient point of the interval into $f'(x)$.

Once we know the sign of $f'$ on each of these intervals, we know where $f$ is increasing and where it is decreasing. We then apply the first derivative test to find which of the critical values are local maxima, which are local minima, and which are neither—merely places where the tangent line is horizontal. With this information, the knowledge of the behavior of $f$ as $x \to \pm\infty$, and the fact that $f$ is continuous, we can sketch its graph. We plot the critical points $(c_i, f(c_i))$, and connect them with a smooth curve that is consistent with our other data.

It may also be helpful to plot the $y$-intercept $(0, f(0))$, and also any $x$-intercepts that are easy to find. But we recommend that (until inflection points are introduced in Section 4-6) you plot *only* these points—critical points and intercepts—and rely otherwise on the increasing-decreasing behavior of $f$.

**EXAMPLE 1**  Sketch the graph of $f(x) = x^3 - 27x$.

*Solution*  Since the leading term is $x^3$, we see that

$$\lim_{x \to \infty} f(x) = +\infty \quad \text{and} \quad \lim_{x \to -\infty} f(x) = -\infty.$$

Because

$$f'(x) = 3x^2 - 27 = 3(x + 3)(x - 3),$$

the critical points are $x = -3$ and $x = 3$. The corresponding points on the graph of $f$ are $(-3, 54)$ and $(3, -54)$. The critical points separate the $x$-axis into the three open intervals

$$(-\infty, -3), \quad (-3, 3) \quad \text{and} \quad (3, \infty).$$

The given table shows the behavior of $f$ on each of these intervals.

| Interval | $x + 3$ | $x - 3$ | $f'(x)$ | $f$ |
|----------|---------|---------|---------|-----|
| $(-\infty, -3)$ | Neg. | Neg. | Pos. | Increasing |
| $(-3, 3)$ | Pos. | Neg. | Neg. | Decreasing |
| $(3, +\infty)$ | Pos. | Pos. | Pos. | Increasing |

We use this information to plot the critical points and the intercepts $(0, 0)$, $(3\sqrt{3}, 0)$, and $(-3\sqrt{3}, 0)$, and thus obtain the graph sketched in Fig. 4.23.

**4.23**  Graph of the function of Example 1

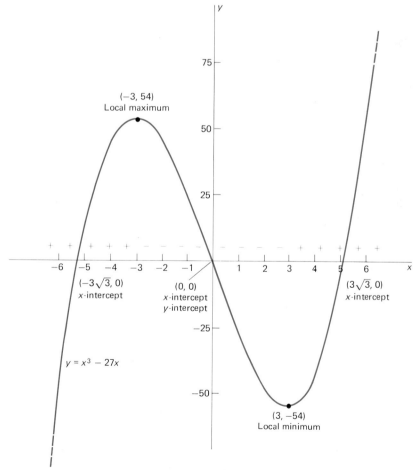

In the figure we used plus and minus signs to mark the sign of $f'(x)$ in each interval. This makes it clear that $(-3, 54)$ is a local maximum and that $(3, -54)$ is a local minimum.

**EXAMPLE 2**   Sketch the graph of $f(x) = 8x^5 - 5x^4 - 20x^3$.

*Solution*   Because

$$f'(x) = 40x^4 - 20x^3 - 60x^2 = 20x^2(x + 1)(2x - 3),$$

the critical points are $-1$, $0$, and $\frac{3}{2}$. The behavior of $f$ on the four open intervals determined by these critical points is indicated in the table.

| Interval | $x + 1$ | $x^2$ | $2x - 3$ | $f'(x)$ | $f$ |
|---|---|---|---|---|---|
| $(-\infty, -1)$ | Neg. | Pos. | Neg. | Pos. | Increasing |
| $(-1, 0)$ | Pos. | Pos. | Neg. | Neg. | Decreasing |
| $(0, \frac{3}{2})$ | Pos. | Pos. | Neg. | Neg. | Decreasing |
| $(\frac{3}{2}, \infty)$ | Pos. | Pos. | Pos. | Pos. | Increasing |

**4.24** The graph of the function of Example 2

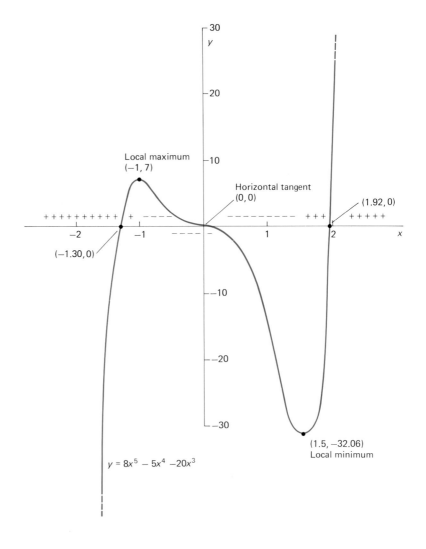

The points on the graph that correspond to the critical points are $(-1, 7)$, $(0, 0)$, and $(1.5, -32.06)$ (the last ordinate is an approximation).

We write $f(x)$ in the form

$$f(x) = x^3(8x^2 - 5x - 20);$$

we can use the quadratic formula to find the $x$-intercepts: $(-1.30, 0)$ and $(1.92, 0)$ (approximations again), and the origin $(0, 0)$. The latter is also the $y$-intercept. We apply the first derivative test to the increasing-decreasing behavior shown in the table, and thus see that $(-1, 7)$ is a local maximum, $(1.5, -32.06)$ is a local minimum, and $(0, 0)$ is neither. The graph looks like the one shown in Fig. 4.24.

---

In our next example, the function is not a polynomial. Nevertheless, the methods of this section suffice for sketching its graph.

**EXAMPLE 3** Sketch the graph of

$$f(x) = x^{2/3}(x^2 - 2x - 6) = x^{8/3} - 2x^{5/3} - 6x^{2/3}.$$

*Solution* The derivative of $f$ is

$$f'(x) = \frac{8}{3}x^{5/3} - \frac{10}{3}x^{2/3} - \frac{12}{3}x^{-1/3}$$

$$= \frac{2}{3}x^{-1/3}(4x^2 - 5x - 6) = \frac{2(4x + 3)(x - 2)}{3x^{1/3}}.$$

The tangent line is horizontal at the two critical points $x = -\frac{3}{4}$ and $x = 2$. In addition, $|f'(x)| \to \infty$ as $x \to 0$, so $x = 0$ (a critical point because $f$ is not differentiable there) is a point where the tangent line is vertical. These three critical points give the points $(-0.75, -3.25)$, $(0, 0)$, and $(2, -9.52)$ on the graph. (We use approximations where appropriate.) The following table summarizes our analysis of the sign of $f'(x)$ on each of the open intervals into which these critical points divide the $x$-axis.

| Interval | $4x + 3$ | $x^{-1/3}$ | $x - 2$ | $f'(x)$ | $f$ |
|----------|----------|-----------|---------|---------|-----|
| $(-\infty, -\frac{3}{4})$ | Neg. | Neg. | Neg. | Neg. | Decreasing |
| $(-\frac{3}{4}, 0)$ | Pos. | Neg. | Neg. | Pos. | Increasing |
| $(0, 2)$ | Pos. | Pos. | Neg. | Neg. | Decreasing |
| $(2, \infty)$ | Pos. | Pos. | Pos. | Pos. | Increasing |

The first derivative test now shows local minima at $(-0.75, -3.25)$ and $(2, -9.52)$, and a local maximum at $(0, 0)$. We note that the function is continuous everywhere: the only possible discontinuity is at $x = 0$, but $f(x)$ is close to zero when $x$ is, so $\lim_{x \to 0} f(x) = 0 = f(0)$.

We use the quadratic formula to find the $x$-intercepts in addition to the origin; they are $(1 \pm \sqrt{7}, 0)$. We then plot the approximations $(-1.65, 0)$

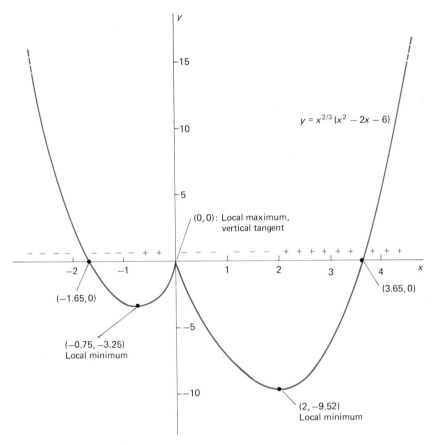

$y = x^{2/3}(x^2 - 2x - 6)$

(0,0): Local maximum, vertical tangent

(−1.65, 0)

(−0.75, −3.25) Local minimum

(3.65, 0)

(2, −9.52) Local minimum

**4.25** The technique is valid for nonpolynomial functions, as in Example 3.

and (3.65, 0). Finally, we note that $f(x) \to \infty$ as $x \to \pm\infty$. So the graph has the shape shown in Fig. 4.25.

An important application of curve-sketching techniques is to the solution of an equation of the form

$$f(x) = 0. \tag{5}$$

The *real* solutions of this equation are simply the x-intercepts of the graph of $y = f(x)$, so by sketching this graph with reasonable accuracy we can glean information about the number of real solutions of Equation (5) and their approximate locations.

To illustrate this approach, let us consider a cubic equation of the special form

$$f(x) = x^3 + px + q = 0, \tag{6}$$

containing no second degree term (see Problem 38). The derivative of $f$ is

$$f'(x) = 3x^2 + p. \tag{7}$$

If $p \geq 0$, then $f'(x) > 0$ everywhere, with the single exception that $f'(0) = 0$ if $p = 0$. It follows that in this case the function $f$ is an increasing function on the whole real line. Because $f$ takes on arbitrarily large positive and negative values and has the intermediate value property, we may conclude that Equation (6) has precisely one real solution.

The more interesting case is $p < 0$. In this case we see from (7) that $f$ has two distinct critical points,

$$c_1 = -\sqrt{\frac{-p}{3}} \quad \text{and} \quad c_2 = +\sqrt{\frac{-p}{3}}, \tag{8}$$

with $c_1 < 0 < c_2$. In Problem 35 we ask you to show that $f(c_1) > f(c_2)$. It follows that the curve $y = f(x)$ is as shown in Fig. 4.26. The five horizontal lines represent all the possibilities for the location of the x-axis relative to the curve. We may conclude from this figure that Equation (6) has

- One real solution if $f(c_1) < 0$ or $f(c_2) > 0$;
- Two real solutions if $f(c_1) = 0$ or $f(c_2) = 0$;
- Three real solutions if $f(c_1) > 0 > f(c_2)$.

The assumption that the cubic equation in (6) contained no term of the form $ax^2$ simplified this discussion, but a similar analysis can be carried out with an arbitrary cubic equation

$$f(x) = x^3 + ax^2 + bx + c = 0 \tag{9}$$

to determine whether it has one, two, or three real solutions.

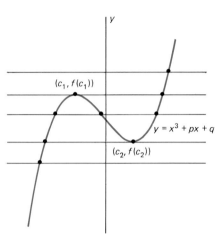

**4.26** Real solutions of the cubic equation in (6)

## 4-5 PROBLEMS

For each of the functions in Problems 1-34, find the intervals on which it is increasing and those on which it is decreasing. Sketch the graph of the function and label the local maxima and minima.

1 $f(x) = 3x^2 - 6x + 5$        2 $f(x) = 5 - 8x - 2x^2$
3 $f(x) = x^3 - 12x$
4 $f(x) = 2x^3 + 3x^2 - 12x$
5 $f(x) = x^3 - 6x^2 + 9x$      6 $f(x) = x^3 + 6x^2 + 9x$
7 $f(x) = x^3 + 3x^2 + 9x$      8 $f(x) = x^3 - 27x$
9 $f(x) = (x - 1)^2(x + 2)^2$
10 $f(x) = (x - 2)^2(2x + 3)^2$
11 $f(x) = 3\sqrt{x} - x\sqrt{x}$    12 $f(x) = x^{2/3}(5 - x)$
13 $f(x) = 3x^5 - 5x^3$           14 $f(x) = x^4 + 4x^3$

15 $f(x) = x^4 - 8x^2 + 7$      16 $f(x) = \dfrac{1}{x}$

17 $f(x) = 2x^2 - 3x - 9$      18 $f(x) = 6 - 5x - 6x^2$
19 $f(x) = 2x^3 + 3x^2 - 12x$
20 $f(x) = x^3 + 4x$
21 $f(x) = 50x^3 - 105x^2 + 72x$
22 $f(x) = x^3 - 3x^2 + 3x - 1$
23 $f(x) = 3x^4 - 4x^3 - 12x^2 + 8$
24 $f(x) = x^4 - 2x^2 + 1$
25 $f(x) = 3x^5 - 20x^3$

26 $f(x) = 3x^5 - 25x^3 + 60x$
27 $f(x) = 2x^3 + 3x^2 + 6x$
28 $f(x) = x^4 - 4x^3$
29 $f(x) = 8x^4 - x^8$          30 $f(x) = 1 - x^{1/3}$
31 $f(x) = x^{1/3}(4 - x)$       32 $f(x) = x^{2/3}(x^2 - 16)$
33 $f(x) = x(x - 1)^{2/3}$       34 $f(x) = x^{1/3}(2 - x)^{2/3}$
35 Verify that if $c_1$ and $c_2$ are the critical points in (8) of the cubic equation in (6), then $f(c_1) > f(c_2)$.

The *discriminant D* of the special cubic equation

$$f(x) = x^3 + px + q = 0$$

with $p \neq 0$ is defined to be

$$D = 4p^3 + 27q^2.$$

The following two problems, together with the discussion at the end of this section, show that the *sign* of $D$ determines the number of distinct real solutions of the special cubic equation.

36 Let $c_1$ and $c_2$ be the critical points defined in (8). Show that

$$f(c_1)f(c_2) = \tfrac{1}{27}(4p^3 + 27q^2).$$

37 Deduce from the result in Problem 36 and from the discussion at the end of this section that the special cubic

equation $x^3 + px + q = 0$ has

- One real solution if $D > 0$;
- Two real solutions if $D = 0$;
- Three real solutions if $D < 0$.

**38** Show that the substitution $x = z - k$, where $k = a/3$, transforms the general cubic equation

$$x^3 + ax^2 + bx + c = 0$$

into the special cubic equation $z^3 + pz + q = 0$, with $p = b - 2ak + 3k^2$ and $q = c - bk + ak^2 - k^3$. Consequently our results on the number of roots of a spe-

cial cubic equation can be applied to the general cubic equation.

**39** Use the methods of Problems 37 and 38 to determine the number of distinct real solutions of each of these cubic equations.
(a) $x^3 + 6x^2 + 9x + 3 = 0$
(b) $x^3 + 6x^2 + 9x + 4 = 0$
(c) $x^3 + 6x^2 + 9x + 5 = 0$

**40** Show by constructing a graph that the equation

$$x^5 - 5x^3 - 20x + 17 = 0$$

has precisely three real solutions.

---

**4-6**

## Higher Derivatives and Concavity

We saw in Section 4-3 that the sign of the first derivative $f'$ tells whether the graph of the function $f$ is rising or falling. In this section we shall see that the sign of the *second* derivative of $f$, the derivative of $f'$, tells which way the curve $y = f(x)$ is *bending*, upward or downward.

### HIGHER DERIVATIVES

The **second derivative** of $f$ is denoted by $f''$, and its value at $x$ is

$$f''(x) = D(f'(x)) = D(Df(x)) = D^2 f(x).$$

The derivative of $f''$ is the **third derivative** $f'''$ of $f$, with

$$f'''(x) = D(f''(x)) = D(D^2 f(x)) = D^3 f(x).$$

The third derivative is also denoted by $f^{(3)}$. More generally, the result of starting with the function $f$ and differentiating $n$ times in succession is the **$n$th derivative** $f^{(n)}$ of $f$, with $f^{(n)}(x) = D^n f(x)$.

If $y = f(x)$, then the first $n$ derivatives may be written as

$$D_x y, \quad D_x^2 y, \quad D_x^3 y, \quad \dots, \quad D_x^n y$$

or

$$y', \quad y'', \quad y''', \dots, \quad y^{(n)},$$

or finally as

$$\frac{dy}{dx}, \quad \frac{d^2 y}{dx^2}, \quad \frac{d^3 y}{dx^3}, \dots, \quad \frac{d^n y}{dx^n}$$

in differential notation. The history of the curious placement of superscripts in the differential notation for higher derivatives involves the metamorphosis

$$\frac{d}{dx}\left(\frac{dy}{dx}\right) \longrightarrow \frac{d}{dx}\frac{dy}{dx} \longrightarrow \frac{(d)^2 y}{(dx)^2} \longrightarrow \frac{d^2 y}{dx^2}.$$

CHAP. 4: Applications of Derivatives and Antiderivatives

**EXAMPLE 1** Find the first four derivatives of

$$f(x) = 2x^3 + \frac{1}{x^2} + 16x^{7/2}.$$

*Solution*

$$f'(x) = 6x^2 - \frac{2}{x^3} + 56x^{5/2},$$

$$f''(x) = 12x + \frac{6}{x^4} + 140x^{3/2},$$

$$f'''(x) = 12 - \frac{24}{x^5} + 210x^{1/2}, \quad \text{and}$$

$$f^{(4)}(x) = \frac{120}{x^6} + \frac{105}{x^{1/2}}.$$

The following example shows how higher derivatives of implicitly defined functions may be found.

**EXAMPLE 2** Find the second derivative $y''$ of the function $y = y(x)$ if

$$x^2 - xy + y^2 = 9.$$

*Solution* Implicit differentiation of the given equation with respect to $x$ gives us

$$2x - y - xy' + 2yy' = 0,$$

so

$$y' = \frac{y - 2x}{2y - x}.$$

We obtain $y''$ by differentiating implicitly again, using the quotient rule. After that, we make substitutions for $y'$ by using the result just found.

$$y'' = D_x\left(\frac{y - 2x}{2y - x}\right)$$

$$= \frac{(y' - 2)(2y - x) - (y - 2x)(2y' - 1)}{(2y - x)^2}$$

$$= \frac{3xy' - 3y}{(2x - y)^2}$$

$$= \frac{3x[(y - 2x)/(2y - x)] - 3y}{(2y - x)^2}.$$

Thus

$$y'' = -\frac{6(x^2 - xy + y^2)}{(2y - x)^3}.$$

## CONCAVITY AND THE SECOND DERIVATIVE TEST

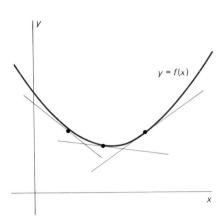

Now we investigate the significance of the *sign* of the second derivative. Suppose first that $f''(x) > 0$ on the interval $I$. Then $f'(x)$ is an increasing function on $I$ because *its* derivative $f''(x)$ is positive. Thus, as we scan the graph $y = f(x)$ from left to right, we see the tangent line turning in a counterclockwise direction, as illustrated in Fig. 4.27. This situation may be described by saying that the curve $y = f(x)$ is **bending upward.** Note that a curve can be bending upward without rising, as illustrated in Fig. 4.28.

On the other hand, if $f''(x) < 0$ on the interval $I$, then $f'(x)$ is decreasing on $I$, so the tangent line turns clockwise as $x$ increases. In this case we say that the curve $y = f(x)$ is **bending downward.** Figures 4.29 and 4.30 illustrate two ways this can happen.

A comparison of Figs. 4.27 and 4.28 with Figs. 4.29 and 4.30 suggests that the question of whether the curve $y = f(x)$ is bending upward or downward is closely related to the question of whether it lies above or below its tangent lines. The latter question refers to the important property of *concavity.*

**4.27**  The graph is bending upward ('concave upward').

---

*Definition of Concavity*

Suppose that the function $f$ is differentiable at the point $a$ and that $L$ is the tangent line to $y = f(x)$ at $(a, f(a))$. Then the function $f$ (or its graph) is

(i) **Concave upward** at $a$ if on some open interval containing $a$, the graph of $f$ lies *above* $L$;

(ii) **Concave downward** at $a$ if on some open interval containing $a$, the graph of $f$ lies *below* $L$.

---

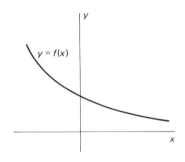

**4.28**  Another graph bending upward (concave upward)

The following theorem establishes the connection between concavity and the sign of the second derivative. That connection is the one suggested by our discussion of bending.

**4.29**  A graph bending downward ('concave downward')

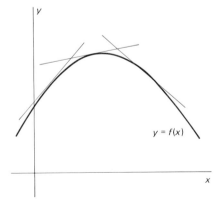

**4.30**  Another graph bending downward (concave downward)

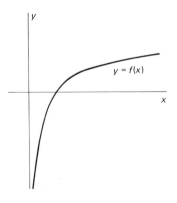

CHAP. 4:  Applications of Derivatives and Antiderivatives

> **Theorem 1   Test for Concavity**
>
> Suppose that $f'$ is differentiable on an open interval containing $a$. Then $f$ is
>
> (i) Concave upward at $a$ if $f''(a) > 0$;
> (ii) Concave downward at $a$ if $f''(a) < 0$.

**Proof**   We give the proof for Case (i), illustrated by Fig. 4.31. Write $y_{\text{curve}} = f(x)$ for the curve and

$$y_{\text{line}} = f(a) + f'(a)(x - a)$$

for the tangent line at $(a, f(a))$. Then it is enough to show that

$$y_{\text{curve}} - y_{\text{line}} = f(x) - f(a) - f'(a)(x - a)$$

is positive for all $x \neq a$ sufficiently near $a$. We apply the mean value theorem to $f$ on the interval $[a, x]$. Thus we find that

$$f(x) - f(a) = f'(c)(x - a)$$

where $c$ is some number between $a$ and $x$. If we substitute this information into the previous equation, we see that

$$
\begin{aligned}
y_{\text{curve}} - y_{\text{line}} &= f'(c)(x - a) - f'(a)(x - a) \\
&= [f'(c) - f'(a)](x - a) \\
&= \frac{f'(c) - f'(a)}{c - a}(c - a)(x - a).
\end{aligned}
$$

Now $c - a$ and $x - a$ have the same sign because $c$ is between $a$ and $x$. So all we must do is show that

$$\frac{f'(c) - f'(a)}{c - a} > 0$$

for $x$—and, therefore, for $c$—sufficiently near $a$.

But this last inequality follows from the hypothesis that $f''(a) > 0$. For, because $f''(a)$ exists, it is the limit of the above fraction; moreover, since that limit is positive, the fraction itself must be positive for all values of $c$ sufficiently near $a$.

If we need to deal with values of $x < a$, the same proof works, except that we apply the mean value theorem to $f$ on the interval $[x, a]$ instead. Hence we have established Theorem 1.   ■

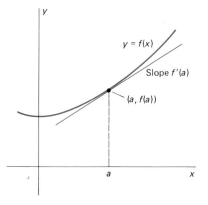

**4.31**   Case (i) of Theorem 1

NOTE   Theorem 1 deals with concavity of a function at a single point. But to determine whether a curve $y = f(x)$ is bending upward or bending downward, we must examine the sign of $f''$ on a whole interval, not merely at a single point.

Now suppose that, in addition to the differentiability hypothesis of Theorem 1, $f'(a) = 0$, so that $a$ is a critical point of the sort where the tangent line is *horizontal*. Then $f$ has a local minimum at $a$ if, near $a$, its graph lies above this horizontal tangent line; by contrast, $f$ has a local maximum at $a$

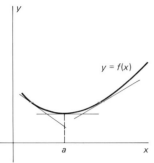

$f''(a) > 0$;
tangent turning counterclockwise;
graph concave up;
local minimum at $a$.

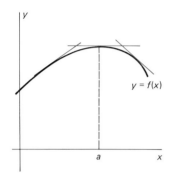

$f''(a) < 0$;
tangent turning clockwise;
graph concave down;
local maximum at $a$.

**4.32** The second derivative test
(Theorem 2)

**4.33** No conclusion possible if
$f'(a) = 0 = f''(a)$

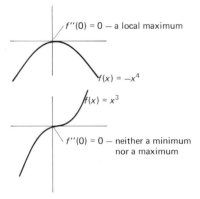

$f(x) = x^4$

$f''(0) = 0$ — a local minimum

$f''(0) = 0$ — a local maximum

$f(x) = -x^4$

$f(x) = x^3$

$f''(0) = 0$ — neither a minimum
nor a maximum

if, near $a$, its graph lies below this horizontal tangent line. These observations together with Theorem 1 yield the following *sufficient* condition for a local extremum.

---

**Theorem 2  Second Derivative Test**

Suppose that the function $f$ is differentiable on an open interval containing the critical point $a$, where $f'(a) = 0$. Then

(i) $f(a)$ is a local minimum value if $f''(a) > 0$;

(ii) $f(a)$ is a local maximum value if $f''(a) < 0$.

---

Rather than memorizing Conditions (i) and (ii) verbatim, it is easier and more reliable to remember the second derivative test by visualizing the graphs shown in Fig. 4.32.

Note that the second derivative test says nothing about what happens if $f''(a) = 0$. Consider the three functions $f(x) = x^4$, $f(x) = -x^4$, and $f(x) = x^3$. For each of these, $f'(0) = 0$ and $f''(0) = 0$. Their graphs, shown in Fig. 4.33, demonstrate that *anything* can happen at such a point.

Similarly, the test for concavity (Theorem 1) says nothing about the case $f''(a) = 0$. A point where the second derivative vanishes *may or may not* be a point where the function changes from concave upward on one side to concave downward on the other. A point where the concavity of a function does change in this manner is called an inflection point. More precisely, the point $x = a$ is an **inflection point** of the function $f$ provided that $f$ is concave upward on one side of $a$, concave downward on the other side, and continuous at $x = a$. We may also refer to $(a, f(a))$ as an inflection point on the graph of $f$.

---

**Theorem 3  Inflection Point Test**

The point $a$ is an inflection point of the continuous function $f$ provided there is an open interval $I$ containing $a$ such that, for points $x$ in $I$, either

(i) $f''(x) > 0$ if $x < a$ and $f''(x) < 0$ if $x > a$, or

(ii) $f''(x) < 0$ if $x < a$ and $f''(x) > 0$ if $x > a$.

---

The fact that a point where the second derivative changes sign is an inflection point follows immediately from Theorem 1 and the definition of inflection point. *Note:* At the inflection point $a$ itself, either $f''(a) = 0$ or $f''(a)$ does not exist. Some of the various possibilities are indicated in Fig. 4.34. Note how we have marked the intervals of upward concavity and downward concavity with small cups opening upward and downward, respectively.

The tests for concavity, local extrema, and inflection points in Theorems 1–3 enable us to flesh out the curve-sketching techniques of Section 4-5.

**EXAMPLE 3**  Sketch the graph of $f(x) = 8x^5 - 5x^4 - 20x^3$, indicating local extrema, inflection points, and concave structure.

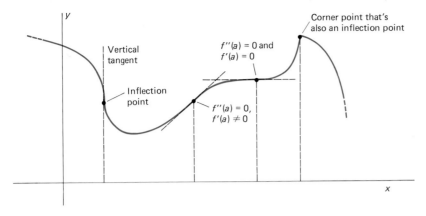

4.34  Some inflection points

**Solution**   We sketched this curve in Example 2 of Section 4-5; see Fig. 4.24 for the graph. In that example we found the first derivative to be

$$f'(x) = 40x^4 - 20x^3 - 60x^2 = 20x^2(x + 1)(2x - 3),$$

so the critical points are $x = -1, 0,$ and $\frac{3}{2}$. The second derivative is

$$f''(x) = 160x^3 - 60x^2 - 120x = 160x(x^2 - \frac{3}{8}x - \frac{3}{4}).$$

When we compute $f''(x)$ at each critical point, we find that $f''(-1) = -100 < 0, f''(0) = 0,$ and $f''(\frac{3}{2}) = 225 > 0$. Hence the second derivative test tells us that $f$ has a local maximum at $x = -1$ and a local minimum at $x = \frac{3}{2}$. The second derivative test is not enough to determine the behavior of $f$ at $x = 0$.

Because $f''$ exists everywhere, the possible inflection points are the solutions of the equation

$$f''(x) = 160x(x^2 - \frac{3}{8}x - \frac{3}{4}) = 0.$$

Clearly one solution is $x = 0$. To find the other two, we use the quadratic formula to solve the equation

$$x^2 - \frac{3}{8}x - \frac{3}{4} = 0.$$

This gives

$$x = \frac{1}{2}[\frac{3}{8} \pm \sqrt{(\frac{3}{8})^2 + 3}] \approx -0.70, 1.07.$$

Thus the three possible inflection points of $f$ are $x = 0, x \approx -0.70,$ and $x \approx 1.07$. We take these two approximations as sufficiently exact and write

$$f''(x) = 160x(x + 0.70)(x - 1.07).$$

This lets us analyze the concave structure of $f$ in a manner like our analysis of increasing-decreasing behavior in Section 4-5. To do this we construct a table, using the open intervals into which the zeros of $f''$ separate the $x$-axis.

| Interval | $x + 0.70$ | $160x$ | $x - 1.07$ | $f''(x)$ | $f$ |
|---|---|---|---|---|---|
| $(-\infty, -0.70)$ | Neg. | Neg. | Neg. | Neg. | Concave Down |
| $(-0.70, 0)$ | Pos. | Neg. | Neg. | Pos. | Concave Up |
| $(0, 1.07)$ | Pos. | Pos. | Neg. | Neg. | Concave Down |
| $(1.07, +\infty)$ | Pos. | Pos. | Pos. | Pos. | Concave Up |

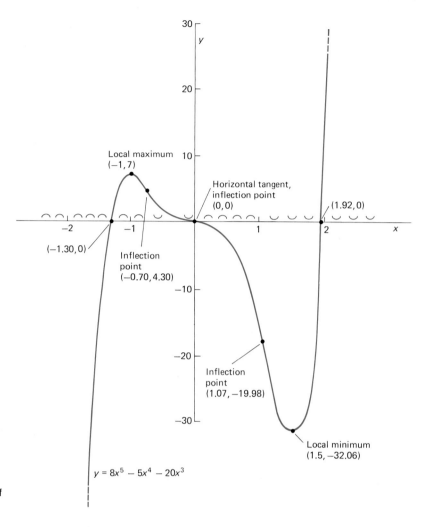

Local maximum
$(-1,7)$

Horizontal tangent,
inflection point
$(0,0)$

$(1.92,0)$

$(-1.30,0)$

Inflection
point
$(-0.70,4.30)$

Inflection
point
$(1.07,-19.98)$

Local minimum
$(1.5,-32.06)$

$y = 8x^5 - 5x^4 - 20x^3$

**4.35** The graph of the function of
Example 3

From the table we see that the direction of concavity of $f$ changes at each of the points $x = -0.70$, $x = 0$, and $x = 1.07$. So these three points are indeed inflection points. This information is shown in the graph sketched in Fig. 4.35.

**EXAMPLE 4** Sketch the graph of $f(x) = 4x^{1/3} + x^{4/3}$, indicating local extrema, inflection points, and concave structure.

*Solution* First,

$$f'(x) = \frac{4}{3} x^{-2/3} + \frac{4}{3} x^{1/3} = \frac{4(x + 1)}{3x^{2/3}},$$

so the critical points are $x = -1$ (where the tangent line is horizontal) and $x = 0$ (where the tangent line is vertical). Next,

$$f''(x) = -\frac{8}{9} x^{-5/3} + \frac{4}{9} x^{-2/3} = \frac{4(x - 2)}{9x^{5/3}},$$

so the possible inflection points are $x = 2$ (where $f''(x) = 0$) and $x = 0$ (where $f''(x)$ does not exist).

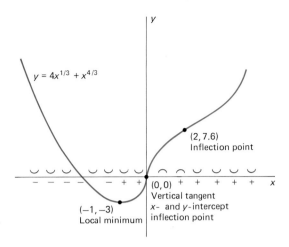

**4.36** The graph of the function of Example 4

$y = 4x^{1/3} + x^{4/3}$

(2, 7.6)
Inflection point

(0, 0)
Vertical tangent
$x$- and $y$-intercept
inflection point

(−1, −3)
Local minimum

To determine where $f$ is increasing and decreasing, we construct the given table.

| Interval | $x + 1$ | $x^{2/3}$ | $f'(x)$ | $f$ |
|----------|---------|-----------|---------|-----|
| $(-\infty, -1)$ | Neg. | Pos. | Neg. | Decreasing |
| $(-1, 0)$ | Pos. | Pos. | Pos. | Increasing |
| $(0, +\infty)$ | Pos. | Pos. | Pos. | Increasing |

Thus $f$ is decreasing when $x < -1$ and increasing when $x > -1$.

To determine the concavity of $f$, we construct a table to find the sign of $f''(x)$ on each of the intervals separated by its zeros. It shows that $f$ is concave downward on $(0, 2)$ and concave upward for $x < 0$ and for $x > 2$.

| Interval | $x^{5/3}$ | $x - 2$ | $f''(x)$ | $f$ |
|----------|-----------|---------|----------|-----|
| $(-\infty, 0)$ | Neg. | Neg. | Pos. | Concave upward |
| $(0, 2)$ | Pos. | Neg. | Neg. | Concave downward |
| $(2, +\infty)$ | Pos. | Pos. | Pos. | Concave upward |

We note in addition that $f(x) \to \infty$ as $x \to \pm\infty$, and we mark the intervals on the $x$-axis with plus signs where $f$ is increasing, minus signs where it is decreasing, cups opening upward where $f$ is concave upward, and cups opening downward where $f$ is concave downward. We plot (approximately) the points on the graph of $f$ corresponding to the zeros and discontinuities of $f'$ and $f''$; these are $(-1, -3)$, $(0, 0)$, and $(2, 6\sqrt[3]{2})$. Finally, we use all this information to draw the smooth curve of Fig. 4.36.

---

## 4-6  PROBLEMS

Calculate the first three derivatives of the functions given in Problems 1–15.

**1** $f(x) = 2x^4 - 3x^3 + 6x - 17$

**2** $f(x) = 2x^5 + x^{3/2} - \dfrac{1}{2x}$

**3** $f(x) = \dfrac{2}{(2x - 1)^2}$

**4** $g(t) = t^2 + \sqrt{t + 1}$

**5** $g(t) = (3t - 2)^{4/3}$

**6** $f(x) = x\sqrt{x + 1}$

**7** $h(y) = \dfrac{y}{y + 1}$

**8** $f(x) = (1 + \sqrt{x})^3$

$9\ g(t) = \dfrac{1}{2t^{1/2}} - \dfrac{3}{(1-t)^{1/3}}$

$10\ h(z) = \dfrac{z^2}{z^2 + 4}$

$11\ f(x) = (2x + 3)^{7/2}$

$12\ f(x) = \left(1 + \dfrac{1}{x}\right)^3$

$13\ g(t) - \dfrac{t-1}{t+1}$

$14\ h(y) = \dfrac{1}{1 - \sqrt{y}}$

$15\ f(x) = \dfrac{1}{1 + x^2}$

In each of Problems 16–22, calculate $y'$ and $y''$, assuming that $y$ is defined implicitly as a function of $x$ by the given equation. Primes denote derivatives with respect to $x$.

$16\ x^2 + y^2 = 4$

$17\ x^2 + xy + y^2 = 3$

$18\ x^{1/3} + y^{1/3} = 1$

$19\ y^3 + x^2 + x = 5$

$20\ \dfrac{1}{x} + \dfrac{1}{y} = 2$

$21\ (x + y)^2 = xy$

$22\ \sqrt{x} + \sqrt{y} = 4$

Apply the second derivative test to find the local maxima and local minima of the functions given in Problems 23–33, and apply the inflection point test to find all inflection points.

$23\ f(x) = x^2 - 4x + 3$

$24\ f(x) = 5 - 6x - x^2$

$25\ f(x) = x^3 - 3x + 1$

$26\ f(x) = x^3 - 3x^2$

$27\ f(x) = x^3$

$28\ f(x) = x^4$

$29\ f(x) = x^5 + 2x$

$30\ f(x) = x^4 - 8x^2$

$31\ f(x) = x^2(x - 1)^2$

$32\ f(x) = x^3(x + 2)^2$

$33\ f(x) = (x - 1)^2(x - 2)^3$

Sketch the graphs of the functions in Problems 34–48, indicating all critical points and inflection points. Apply the second derivative test at each critical point. Show the correct concave structure in your sketches, and indicate the behavior of $f(x)$ as $x \to \pm\infty$.

$34\ f(x) = 12x - x^3$

$35\ f(x) = 2x^3 - 3x^2 - 12x + 3$

$36\ f(x) = 3x^4 - 4x^3 - 5$

$37\ f(x) = 6 + 8x^2 - x^4$

$38\ f(x) = 3x^5 - 5x^3$

$39\ f(x) = 3x^4 - 4x^3 - 12x^2 - 1$

$40\ f(x) = 3x^5 - 25x^3 + 60x$

$41\ f(x) = x^3(x - 1)^4$

$42\ f(x) = (x - 1)^2(x + 2)^3$

$43\ f(x) = 1 + x^{1/3}$

$44\ f(x) = 2 - (x - 3)^{1/3}$

$45\ f(x) = (x + 3)\sqrt{x}$

$46\ f(x) = x^{2/3}(5 - 2x)$

$47\ f(x) = (4 - x)\sqrt[3]{x}$

$48\ f(x) = x^{1/3}(6 - x)^{2/3}$

$49$ Find all the nonzero derivatives of $f(x) = (x + 1)^5$.

$50$ Suppose that $f(x) = x^n$, where $n$ is a positive integer. Show by induction that $f^{(n)}(x) = n! = n(n - 1) \cdots 3 \cdot 2 \cdot 1$.

$51$ Conclude from the result of Problem 50 that if $f(x)$ is a polynomial of degree $n$, then $f^{(k)}(x) \equiv 0$ (is identically zero; $f^{(k)}(x) = 0$ for all $x$) if $k > n$.

$52$ (a) Suppose that $f''(x) \equiv 0$ on the interval $I$. Apply Corollary 1 in Section 4-3 twice in succession to show that there exist constants $A$ and $B$ such that $f(x) = Ax + B$ on $I$.

(b) Suppose that $f''(x) \equiv g''(x)$ on the interval $I$. Conclude from part (a) that there exist constants $A$ and $B$ such that $f(x) = g(x) + Ax + B$ on $I$.

$53$ Suppose that $z = g(y)$ and that $y = f(x)$. Show that

$$\frac{d^2z}{dx^2} = \frac{d^2z}{dy^2}\left(\frac{dy}{dx}\right)^2 + \frac{dz}{dy}\frac{d^2y}{dx^2}.$$

$54$ Prove that the graph of a quadratic polynomial has no inflection points.

$55$ Prove that the graph of a cubic polynomial (one of degree 3) has exactly one inflection point.

$56$ Prove that the graph of a polynomial function of degree 4 has either no inflection points or exactly two of them.

$57$ Suppose that the pressure $P$ (in atmospheres), volume $V$ (in cubic centimeters), and temperature $T$ (in degrees Kelvin) of $n$ moles of carbon dioxide satisfies van der Waals' equation

$$\left(P + \frac{n^2a}{V^2}\right)\left(V - nb\right) = nRT$$

where $a$, $b$, and $R$ are empirical constants. The following experiment was carried out in order to find the values of these constants.

One mole of $CO_2$ was compressed at the constant temperature $T = 304°$ K. The measured pressure-volume ($PV$) data were then plotted as in Fig. 4.37, with the $PV$ curve showing a *horizontal* inflection point at $V = 128.1$, $P = 72.8$. Use this information to calculate $a$, $b$, and $R$. (*Suggestion:* Solve van der Waals' equation for $P$ and then calculate $dP/dV$ and $d^2P/dV^2$.)

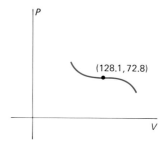

**4.37** A problem involving van der Waals' equation

$58$ Suppose that $f''$ is defined at each point of the open interval $I$ of real numbers and that for each two points $a$ and $b$ of $I$, the chord joining $(a, f(a))$ and $(b, f(b))$ lies entirely above the graph of $f$ (except at $x = a$ and $x = b$). Prove that the graph of $f$ is concave upward on $I$. (*Suggestion:* Suppose that $f''(c) < 0$ at some point $c$ of $I$. Choose *h* sufficiently small that $c - h$ and $c + h$ are points of *I*. Use the fact—which you can establish rigorously after you study Section 11-1—that

$$f''(x) = \lim_{h \to 0} \frac{f(x + h) - 2f(x) + f(x - h)}{h^2}.$$

By choosing *h* sufficiently small and positive, the quotient shown above will have the same sign as $f''(x)$. Use this to obtain a contradiction. *Comment:* It is also true that if *f* is concave upward on the open interval *I*, then any chord like the one described above lies above the graph of *f*.)

4-7

# Curve Sketching and Asymptotes

Before we turn to scientific applications of differentiation, we want to extend the limit concept to include infinite limits and limits at infinity. This extension of the limit concept adds a new weapon to our arsenal of curve-sketching techniques, the notion of an *asymptote* to a curve—a straight line that the curve approaches arbitrarily closely (in a sense soon to be specified).

Recall from Section 2-2 that $f(x)$ is said to **increase without bound,** or **become infinite,** as *x* approaches *a*, and we write

$$\lim_{x \to a} f(x) = +\infty, \tag{1}$$

provided that $f(x)$ can be made arbitrarily large by choosing *x* sufficiently close (though not equal) to *a*. More precisely, (1) means that, given $M > 0$, there exists $\delta > 0$ such that

$$0 < |x - a| < \delta \quad \text{implies} \quad f(x) > M.$$

For example, it is apparent that

$$\lim_{x \to 2} \frac{1}{(x - 2)^2} = +\infty$$

because $(x - 2)^2$ is positive and is approaching zero as $x \to 2$.

The statement that $f(x)$ **decreases without bound,** or **becomes negatively infinite,** as $x \to a$, written

$$\lim_{x \to a} f(x) = -\infty, \tag{2}$$

has an analogous definition.

One-sided versions of (1) and (2) also make sense. For instance, if *n* is an *odd* integer, then it is apparent that

$$\lim_{x \to 2^-} \frac{1}{(x - 2)^n} = -\infty \quad \text{while} \quad \lim_{x \to 2^+} \frac{1}{(x - 2)^n} = +\infty$$

because $(x - 2)^n$ is negative when *x* is to the left of 2 and positive when *x* is to the right of 2.

We say that the line $x = a$ is a **vertical asymptote** for the curve $y = f(x)$ provided that

$$\lim_{x \to a} |f(x)| = \infty. \tag{3}$$

The geometric significance of a vertical asymptote is illustrated by the graphs of $y = 1/(x - 1)$ and $y = 1/(x - 1)^2$ in Figs 4.38 and 4.39. In each case, as $x \to 1$ and $f(x) \to \pm\infty$, the point $(x, f(x))$ on the curve approaches the vertical asymptote $x = 1$.

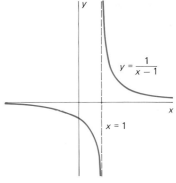

**4.38** The graph of $y = 1/(x - 1)$

**4.39** The graph of $y = 1/(x - 1)^2$

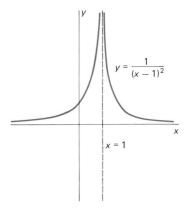

The most typical occurrence of a vertical asymptote is for a rational function $f(x) = p(x)/q(x)$ at a point $x = a$ where $q(a) = 0$ but $p(a) \neq 0$. Several examples appear shortly.

In Section 4-5 we mentioned infinite limits at infinity in connection with the behavior of a polynomial as $x \to \pm\infty$. There is also such a thing as a finite limit at infinity. We say that $f(x)$ **approaches the number $L$ as $x$ increases without bound,** and write

$$\lim_{x \to \infty} f(x) = L, \tag{4}$$

provided that $|f(x) - L|$ can be made arbitrarily small by choosing $x$ sufficiently large. That is, given $\varepsilon > 0$, there exists $M > 0$ such that

$$x > M \quad \text{implies} \quad |f(x) - L| < \varepsilon. \tag{5}$$

The statement that

$$\lim_{x \to -\infty} f(x) = L$$

has a definition of similar form: merely replace the condition $x > M$ by the condition $x < -M$.

The analogues for limits at infinity of the limit laws of Section 2-1 all hold, including in particular the sum, product, and quotient laws. In addition, it is not difficult to show that, if

$$\lim_{x \to \infty} f(x) = L \quad \text{and} \quad \lim_{x \to \infty} g(x) = \pm\infty,$$

then

$$\lim_{x \to \infty} \frac{f(x)}{g(x)} = 0.$$

It follows from this result that

$$\lim_{x \to \infty} \frac{1}{x^k} = 0 \tag{6}$$

for any choice of the positive rational number $k$.

Using (6) and the limit laws, limits at infinity of rational functions are easy to evaluate. The general method is this: First divide each term in both numerator and denominator by the highest power of $x$ that appears in any of the terms. Then apply the limit laws.

**EXAMPLE 1** Find

$$\lim_{x \to \infty} f(x) \quad \text{if} \quad f(x) = \frac{3x^3 - x}{2x^3 + 7x^2 - 4}.$$

***Solution*** We begin by dividing both the numerator and the denominator by $x^3$. Thus

$$\lim_{x \to \infty} \frac{3x^3 - x}{2x^3 + 7x^2 - 4} = \lim_{x \to \infty} \frac{3 - 1/x^2}{2 + 7/x - 4/x^3}$$

$$= \frac{\lim_{x \to \infty} (3 - 1/x^2)}{\lim_{x \to \infty} (2 + 7/x - 4/x^3)}$$

$$= \frac{3 - 0}{2 + 0 - 0} = \frac{3}{2}.$$

Note that the same computation, but with $x \to -\infty$, also gives the result $\lim\limits_{x \to -\infty} f(x) = \frac{3}{2}$.

**EXAMPLE 2**  Find $\lim\limits_{x \to \infty} (\sqrt{x + a} - \sqrt{x})$.

*Solution*  We use the familiar divide and multiply technique.

$$\lim_{x \to \infty} (\sqrt{x + a} - \sqrt{x}) = \lim_{x \to \infty} (\sqrt{x + a} - \sqrt{x}) \cdot \frac{\sqrt{x + a} + \sqrt{x}}{\sqrt{x + a} + \sqrt{x}}$$

$$= \lim_{x \to \infty} \frac{a}{\sqrt{x + a} + \sqrt{x}} = 0.$$

---

The geometric meaning of the statement $\lim\limits_{x \to \infty} f(x) = L$ is that the point $(x, f(x))$ on the curve $y = f(x)$ approaches the horizontal line $y = L$ as $x \to \infty$. In particular, with the numbers $M$ and $\varepsilon$ of Condition (5), the part of the curve with $x > M$ lies between the horizontal lines $y = L - \varepsilon$ and $y = L + \varepsilon$ (see Fig. 4.40). We therefore say that the line $y = L$ is a **horizontal asymptote** for the curve $y = f(x)$ if either

$$\lim_{x \to \infty} f(x) = L \quad \text{or} \quad \lim_{x \to -\infty} f(x) = L.$$

**EXAMPLE 3**  Sketch the graph of $f(x) = x/(x - 2)$. Indicate any horizontal or vertical asymptotes.

*Solution*  First we note that $x = 2$ is a vertical asymptote because $|f(x)| \to \infty$ as $x \to 2$. Also,

$$\lim_{x \to \pm\infty} \frac{x}{x - 2} = \lim_{x \to \pm\infty} \frac{1}{1 - (2/x)} = 1.$$

So $y = 1$ is a horizontal asymptote. The first two derivatives of $f$ are readily

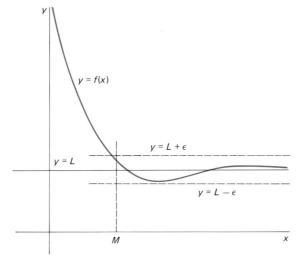

**4.40**  Geometry of the definition of horizontal asymptote

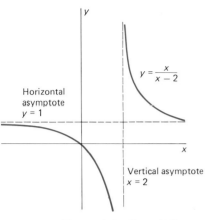

Horizontal
asymptote
$y = 1$

$y = \dfrac{x}{x - 2}$

Vertical asymptote
$x = 2$

**4.41** The graph for Example 3

found to be

$$f'(x) = -\frac{2}{(x - 2)^2} \quad \text{and} \quad f''(x) = \frac{4}{(x - 2)^3}.$$

Note that neither $f'(x)$ nor $f''(x)$ vanishes anywhere, so the function $f$ has no critical points and no inflection points. Because $f'(x) < 0$ for $x \neq 2$, we see that $f(x)$ is decreasing on the open intervals $(-\infty, 2)$ and $(2, +\infty)$. Since $f''(x) < 0$ for $x < 2$ while $f''(x) > 0$ for $x > 2$, we see also that $f$ is concave downward on $(-\infty, 2)$ and concave upward on $(2, \infty)$. We show the graph of $f$ in Fig. 4.41.

---

The curve-sketching techniques of Section 4-5 and 4-6, amalgamated with those of the present section, can be summarized as a list of steps. If you follow these steps, loosely rather than rigidly, you should obtain a rather accurate sketch of the graph of a given function $y = f(x)$.

1 Solve the equation $f'(x) = 0$ and also find where $f'(x)$ does not exist. This gives the critical points of $f$. Note whether the tangent is horizontal, vertical, or nonexistent.

2 Determine the intervals on which $f$ is increasing and those on which it is decreasing.

3 Solve the equation $f''(x) = 0$ and also find where $f''(x)$ does not exist. These will be the *possible* inflection points of the graph.

4 Determine the intervals on which $f$ is concave upward and those on which it is concave downward.

5 Find the $y$-intercept and the $x$-intercepts (if any) of the graph.

6 Plot and label the critical points, possible inflection points, and intercepts.

7 Determine the asymptotes (if any), discontinuities and behavior of $f$ nearby, and the behavior of $f$ as $x \to \pm\infty$.

8 Finally, join the plotted points with a curve that is consistent with the information you have accumulated. Remember that corner points are rare and that straight sections of graph are even rarer.

Of course, you may follow these steps in any convenient order and omit any that present formidable computational difficulties. Many examples will require fewer than the eight steps; see Example 3. But our next example requires them all.

**EXAMPLE 4**  Sketch the graph of

$$f(x) = \frac{2 + x - x^2}{(x - 1)^2}.$$

*Solution*  We notice immediately that

$$\lim_{x \to 1} f(x) = \infty,$$

because the numerator approaches 2 as $x \to 1$ while the denominator approaches zero through *positive* values. So the line $x = 1$ is a vertical asymptote. Also

$$\lim_{x \to \pm\infty} \frac{2 + x - x^2}{(x - 1)^2} = \lim_{x \to \pm\infty} \frac{(2/x^2) + (1/x) - 1}{(1 - (1/x))^2} = -1,$$

so the line $y = -1$ is a horizontal asymptote.

CHAP. 4:  Applications of Derivatives and Antiderivatives

Next, we apply the quotient rule and simplify to find that

$$f'(x) = \frac{x - 5}{(x - 1)^3}.$$

So the only critical point is $x = 5$, and we plot the point $(5, f(5)) = (5, -\frac{9}{8})$ on a convenient coordinate plane. We note first that there is a horizontal tangent at $(5, f(5))$. To determine the increasing-decreasing behavior of $f$, we use both the critical point $x = 5$ and the point $x = 1$ (where $f'$ is not defined) to separate the $x$-axis into open intervals. Here are the results:

| Interval | $(x - 1)^3$ | $x - 5$ | $f'(x)$ | $f$ |
|---|---|---|---|---|
| $(-\infty, 1)$ | Neg. | Neg. | Pos. | Increasing |
| $(1, 5)$ | Pos. | Neg. | Neg. | Decreasing |
| $(5, +\infty)$ | Pos. | Pos. | Pos. | Increasing |

After some simplification, we find the second derivative to be

$$f''(x) = \frac{2(7 - x)}{(x - 1)^4}.$$

The only possible inflection point is at $x = 7$, corresponding to the point $(7, -\frac{10}{9})$ on the graph. We use both $x = 7$ and the point $x = 1$ where $f''$ is not defined to separate the $x$-axis into open intervals. The concave structure of the curve can be deduced with the aid of the next table.

| Interval | $(x - 1)^4$ | $7 - x$ | $f''(x)$ | $f$ |
|---|---|---|---|---|
| $(-\infty, 1)$ | Pos. | Pos. | Pos. | Concave upward |
| $(1, 7)$ | Pos. | Pos. | Pos. | Concave upward |
| $(7, +\infty)$ | Pos. | Neg. | Neg. | Concave downward |

The $y$-intercept of $f$ is $(0, 2)$, and the equation $2 + x - x^2 = 0$ readily yields the $x$-intercepts $(-1, 0)$ and $(2, 0)$. We plot these intercepts, sketch the asymptotes, and finally sketch the graph with the aid of the two tables above—their information now appears along the $x$-axis with plus signs above the intervals where $f$ is increasing, minus signs where $f$ is decreasing, cups opening upward where $f$ is concave upward, and cups opening downward where $f$ is concave downward. The result appears in Fig. 4.42.

Asymptotes may be inclined, as well as being either horizontal or vertical. The nonvertical line $y = mx + b$ is an **asymptote** for the curve $y = f(x)$ provided that

$$\lim_{x \to \pm\infty} [f(x) - (mx + b)] = 0. \tag{7}$$

This condition means that, as $x \to \pm\infty$, the vertical distance between the point $(x, f(x))$ on the curve and the point $(x, mx + b)$ on the line approaches zero.

If $f(x) = p(x)/q(x)$ is a rational function with the degree of $p$ greater by 1 than that of $q$, then by long division of $q(x)$ into $p(x)$ we find that $f(x)$ has the form

$$f(x) = mx + b + g(x),$$

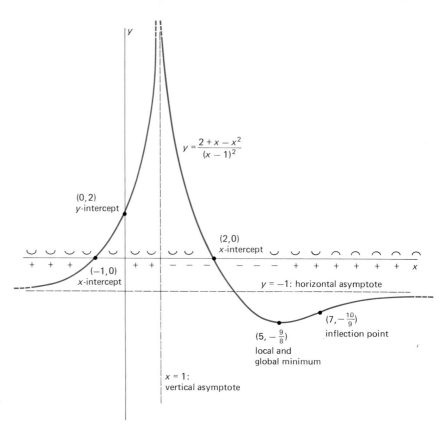

$$y = \frac{2 + x - x^2}{(x - 1)^2}$$

(0, 2)
y-intercept

(2, 0)
x-intercept

(-1, 0)
x-intercept

$y = -1$: horizontal asymptote

$(7, -\frac{10}{9})$
inflection point

$(5, -\frac{9}{8})$
local and
global minimum

$x = 1$:
vertical asymptote

**4.42** Graphing the function of
Example 4

where

$$\lim_{x \to \pm\infty} g(x) = 0.$$

Thus $y = mx + b$ is an asymptote for $y = f(x)$.

**EXAMPLE 5**   Sketch the graph of

$$f(x) = \frac{x^2 + x - 1}{x - 1}.$$

**Solution**   The long division suggested above takes this form:

$$
\begin{array}{r}
x + 2 \\
x - 1 \overline{\smash{)}\ x^2 + x - 1} \\
\underline{x^2 - x} \\
2x - 1 \\
\underline{2x - 2} \\
1
\end{array}
$$

Thus

$$f(x) = x + 2 + \frac{1}{x - 1}.$$

So $y = x + 2$ is an asymptote for the curve. Also,

$$\lim_{x \to 1} |f(x)| = \infty,$$

CHAP. 4:   Applications of Derivatives and Antiderivatives

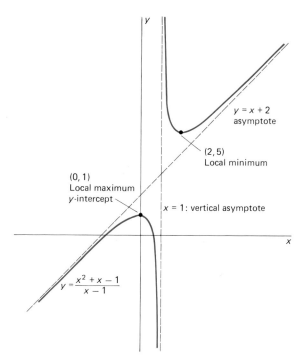

$y = x + 2$
asymptote

$(2, 5)$
Local minimum

$(0, 1)$
Local maximum
$y$-intercept

$x = 1$: vertical asymptote

$y = \dfrac{x^2 + x - 1}{x - 1}$

**4.43**  A function with asymptote the line $y = x + 2$

so $x = 1$ is a vertical asymptote. The first two derivatives of $f$ are

$$f'(x) = 1 - \frac{1}{(x-1)^2} = \frac{x(x-2)}{(x-1)^2}$$

and

$$f''(x) = \frac{2}{(x-1)^3}.$$

It follows that $f$ has critical points at $x = 0$ and at $x = 2$ but no inflection points. The sign of $f'(x)$ tells us that $f$ is increasing on $(-\infty, 0)$ and on $(2, \infty)$ and decreasing on $(0, 1)$ and on $(1, 2)$. And examination of $f''(x)$ reveals that $f$ is concave downward on $(-\infty, 1)$ and concave upward on $(1, \infty)$. In particular, $f(0) = 1$ is a local maximum value and $f(2) = 5$ is a local minimum value. So the graph of $f$ looks much like the one in Fig. 4.43.

## 4-7  PROBLEMS

Investigate the limits in Problems 1–16.

**1** $\displaystyle\lim_{x \to \infty} \frac{x}{x + 1}$

**2** $\displaystyle\lim_{x \to -\infty} \frac{x^2 + 1}{x^2 - 1}$

**3** $\displaystyle\lim_{x \to 1} \frac{x^2 + x - 2}{x - 1}$

**4** $\displaystyle\lim_{x \to 1} \frac{x^2 - x - 2}{x - 1}$

**5** $\displaystyle\lim_{x \to \infty} \frac{2x^2 - 1}{x^2 - 3x}$

**6** $\displaystyle\lim_{x \to -\infty} \frac{x^2 + 3x}{x^3 - 5}$

**7** $\displaystyle\lim_{x \to -1} \frac{x^2 + 2x + 1}{(x + 1)^2}$

**8** $\displaystyle\lim_{x \to \infty} \frac{5x^3 - 2x + 1}{7x^3 + 4x^2 - 2}$

**9** $\displaystyle\lim_{x \to 4} \frac{x - 4}{\sqrt{x} - 2}$

**10** $\displaystyle\lim_{x \to \infty} \frac{2x + 1}{x - x^{3/2}}$

**11** $\displaystyle\lim_{x \to -\infty} \frac{8 - \sqrt[3]{x}}{2 + x}$

**12** $\displaystyle\lim_{x \to \infty} \frac{2x^2 - 17}{x^3 - 2x + 27}$

**13** $\lim\limits_{x \to \infty} \sqrt{\dfrac{4x^2 - x}{x^2 + 9}}$

**14** $\lim\limits_{x \to -\infty} \dfrac{\sqrt[3]{x^3 - 8x + 1}}{3x - 4}$

**15** $\lim\limits_{x \to -\infty} (\sqrt{x^2 + 2x} - x)$

**16** $\lim\limits_{x \to -\infty} (2x - \sqrt{4x^2 - 5x})$

Sketch the graphs of the functions in Problems 17–42. Identify and label all extrema, inflection points, intercepts, and asymptotes. Show the concave structure clearly, as well as the behavior of the graph for $|x|$ large and for $x$ near any discontinuities of the function.

**17** $f(x) = \dfrac{2}{x - 3}$

**18** $f(x) = \dfrac{4}{5 - x}$

**19** $f(x) = \dfrac{3}{(x + 2)^2}$

**20** $f(x) = \dfrac{-4}{(3 - x)^2}$

**21** $f(x) = \dfrac{1}{(2x - 3)^3}$

**22** $f(x) = \dfrac{x + 1}{x - 1}$

**23** $f(x) = \dfrac{x^2}{x^2 + 1}$

**24** $f(x) = \dfrac{2x}{x^2 + 1}$

**25** $f(x) = \dfrac{1}{x^2 - 9}$

**26** $f(x) = \dfrac{x}{4 - x^2}$

**27** $f(x) = \dfrac{1}{x^2 + x - 6}$

**28** $f(x) = \dfrac{2x^2 + 1}{x^2 - 2x}$

**29** $f(x) = x + \dfrac{1}{x}$

**30** $f(x) = 2x + \dfrac{1}{x^2}$

**31** $f(x) = \dfrac{x^2}{x - 1}$

**32** $f(x) = \dfrac{2x^3 - 5x^2 + 4x}{x^2 - 2x + 1}$

**33** $f(x) = \dfrac{1}{(x - 1)^2}$

**34** $f(x) = \dfrac{1}{x^2 - 4}$

**35** $f(x) = \dfrac{x}{x + 1}$

**36** $f(x) = \dfrac{1}{(x + 1)^3}$

**37** $f(x) = \dfrac{1}{x^2 - x - 2}$

**38** $f(x) = \dfrac{1}{(x - 1)(x + 1)^2}$

**39** $f(x) = \dfrac{x^2 - 4}{x}$

**40** $f(x) = \dfrac{x}{x^2 - 1}$

**41** $f(x) = \dfrac{x^3 - 4}{x^2}$

**42** $f(x) = \dfrac{x^2 + 1}{x - 2}$

**43** Suppose that $f(x) = x^2 + (2/x)$. Note that

$$\lim\limits_{x \to \pm\infty} (f(x) - x^2) = 0,$$

so the curve $y = f(x)$ approaches the parabola $y = x^2$ as $x \to \pm\infty$. Use this observation to make an accurate sketch of the graph of $f$.

**44** Use the method of Problem 43 to make an accurate sketch of the graph of $f(x) = x^3 - [12/(x - 1)]$.

---

**4-8**

**Antiderivatives**    The language of change is the natural language for the statement of most scientific laws and principles. For instance, Newton's law of cooling says that the *rate of change* of the temperature $T$ of a body is proportional to the difference between $T$ and the temperature of the surrounding medium. That is,

$$\frac{dT}{dt} = k(A - T) \tag{1}$$

where $k$ is a positive constant and $A$, normally assumed to be constant, is the surrounding temperature. Similarly, the *rate of change* of a population $P$ with constant birth and death rates is proportional to the size of the population:

$$\frac{dP}{dt} = kP \qquad (k \text{ constant}). \tag{2}$$

Torricelli's law implies that the *rate of change* of the volume $V$ of water in a draining tank is proportional to the square root of the depth $y$ of the water; that is,

$$\frac{dV}{dt} = -k\sqrt{y} \qquad (k \text{ constant}). \tag{3}$$

Thus mathematical models of real-world situations frequently involve equations containing *derivatives* of unknown functions. Equations such as (1)–(3) are called **differential equations.**

The simplest kind of differential equation has the form

$$\frac{dy}{dx} = f(x),$$

where $f$ is a given (known) function and the function $y$ of $x$ is unknown. The process of finding a function from its derivative is the opposite of differentiation and is therefore called **antidifferentiation.** If we can find a function $F$ having $f$ as its derivative, we call $F$ an *antiderivative* of $f$.

---

*Definition* *Antiderivative*

An **antiderivative** of the function $f$ is a function $F$ such that

$$F'(x) = f(x)$$

wherever $f$ is defined.

---

The table exhibits some examples of functions and antiderivatives.

| Function $f(x)$ | Antiderivative $F(x)$ |
|:---:|:---:|
| $1$ | $x$ |
| $2x$ | $x^2$ |
| $x^3$ | $\frac{1}{4}x^4$ |
| $\sqrt{x}$ | $\frac{2}{3}x^{3/2}$ |
| $\dfrac{1}{x^2}$ | $-\dfrac{1}{x}$ |

Observe that if a function has one antiderivative, then it has many antiderivatives. By contrast, a function can have only one derivative. While $F(x) = x^3$ is an antiderivative of $f(x) = 3x^2$, so are the functions $G(x) = x^3 + 17$, $H(x) = x^3 + \pi$, and $K(x) = x^3 - \sqrt{2}$. Indeed, $x^3 + C$ is an antiderivative of $3x^2$ for *any* choice of the constant $C$.

More generally, if $F(x)$ is an antiderivative of $f(x)$, then so is $F(x) + C$ for any constant $C$. The converse of this statement is more subtle: If $F(x)$ is one antiderivative of $f(x)$ *on the interval I*, then *every* antiderivative of $f(x)$ on $I$ is of the form $F(x) + C$. This follows immediately from Corollary 2 of the mean value theorem in Section 4-3, according to which two functions with the same derivative on an interval differ only by a constant. Hence the following theorem describes the set of *all* antiderivatives of $f$.

---

*Theorem 1* *The Most General Antiderivative*

If $F'(x) = f(x)$ at each point of the open interval $I$, then every antiderivative $G$ of $f$ on $I$ has the form

$$G(x) = F(x) + C, \tag{4}$$

where $C$ is a constant.

---

Thus if $F$ is one antiderivative of $f$ on $I$, then the *most general antiderivative* of $f$ has the form $F(x) + C$ given in (4). We have noted the different antiderivatives $x^3$, $x^3 + 17$, $x^3 + \pi$, and $x^3 - \sqrt{2}$ of the function $f(x) = 3x^2$. Each of these is of the form in (4) with $F(x) = x^3$ and is said to be a **particular** antiderivative of $f(x) = 3x^2$.

On certain occasions, when we want to emphasize the fact that antidifferentiation is the opposite (or inverse) of differentiation, we use the notation

$$D^{-1}f(x) = F(x) + C \qquad (5a)$$

or

$$D_x^{-1}f(x) = F(x) + C. \qquad (5b)$$

for the most general antiderivative of $f$. Thus

$$D^{-1}f(x) = F(x) + C \quad \text{if and only if} \quad DF(x) = f(x),$$

where $D$ denotes the operation of differentiation. For instance, we may write $D^{-1}3x^2 = x^3 + C$ to state concisely the form of the most general antiderivative of $3x^2$.

Every differentiation formula yields an immediate antidifferentiation formula. We ordinarily write antidifferentiation formulas in terms of most general antiderivatives, as in the following theorem.

---

**Theorem 2** *Power Rule Antidifferentiation Formula*
If $r \neq -1$, then the most general antiderivative of

$$f(x) = x^r \quad \text{is} \quad F(x) = \frac{x^{r+1}}{r+1} + C. \qquad (6)$$

---

Each such formula may be checked by differentiation of the right-hand side. This is the sure-fire way to check any antidifferentiation: To verify that $F$ is an antiderivative of $f$, compute $F'$ and see whether it is equal to $f$.

The linearity of the operation of differentiation implies immediately that antidifferentiation is linear in the following sense. If $F$ and $G$ are antiderivatives of $f$ and $g$, respectively, and $a$ is a constant, then the most general antiderivative of

$$af(x) \quad \text{is} \quad aF(x) + C; \qquad (7)$$

$$f(x) + g(x) \quad \text{is} \quad F(x) + G(x) + C. \qquad (8)$$

In essence, we may antidifferentiate a sum of terms by antidifferentiating each of them individually.

**EXAMPLE 1** Find the most general antiderivative of

$$f(x) = x^3 + 3\sqrt{x} - \frac{4}{x^2}.$$

*Solution* To prepare the function $f$ for antidifferentiation, we write it in the form

$$f(x) = x^3 + 3x^{1/2} - 4x^{-2}.$$

Then Equations (6), (7), and (8) yield

$$F(x) = \frac{x^4}{4} + 3\left(\frac{x^{3/2}}{3/2}\right) - 4\left(\frac{x^{-1}}{-1}\right) + C$$

$$= \frac{1}{4}x^4 + 2x^{3/2} + \frac{4}{x} + C$$

for the most general antiderivative $F$ of $f$.

---

A common technique of antidifferentiation is the application of the chain rule in reverse. The chain rule in the form

$$D_x g(u) = g'(u)\frac{du}{dx}$$

yields the following result.

---

**Theorem 3   Chain Rule Antidifferentiation**

If $u$ is a differentiable function of $x$ and $g(u)$ is a differentiable function of $u$, then the most general antiderivative of

$$f(x) = g'(u)\frac{du}{dx} \quad \text{is} \quad F(x) = g(u) + C, \tag{9}$$

where $u$ is expressed in terms of $x$ on the right-hand side.

---

When we combine the formula in (9) with the antidifferentiation formula of Theorem 2, we find that if $r \neq -1$, then the most general antiderivative of

$$f(x) = u^r\frac{du}{dx} \quad \text{is} \quad F(x) = \frac{u^{r+1}}{r+1} + C, \tag{10}$$

where $u$ is expressed in terms of $x$ on the right-hand side.

NOTE   Just as in differentiation using the chain rule, the factor $du/dx$ in Equation (10) is vital. It results from the fact that we are antidifferentiating with respect to $x$ rather than with respect to $u$.

The key to applying the formula in (10) is to spot the appropriate function $u(x)$ such that $f(x)$ is the product of a function of $u$ and the derivative $du/dx$. We then need only antidifferentiate that function of $u$.

**EXAMPLE 2**   Find the most general antiderivative of

$$f(x) = (x^2 + 1)^{10}(2x).$$

**Solution**   If we choose $u = x^2 + 1$, then $du/dx = 2x$, so

$$f(x) = u^{10}\frac{du}{dx}.$$

Hence the formula in (10) yields

$$F(x) = \frac{u^{11}}{11} + C = \frac{1}{11}(x^2 + 1)^{11} + C$$

for the most general antiderivative $F$ of $f$. The resubstitution of $x^2 + 1$ for $u$ is essential because the symbol $u$ does not appear in the original problem.

**EXAMPLE 3**  Find the most general antiderivative of

$$f(x) = x\sqrt{1 + x^2}.$$

**Solution**  Be willing to experiment. The required factor $du/dx$ will be $2x$ if we choose $u = 1 + x^2$. The term $x$ is present in $f(x)$, but $2x$ is not. As a general rule, if the error is only a constant *multiplicative* factor (here, we have $x$ rather than $2x$), the process still succeeds. So we cheerfully continue by letting $u = x^2 + 1$ and write

$$f(x) = \frac{1}{2}(1 + x^2)^{1/2}(2x) = \frac{1}{2}u^{1/2}\frac{du}{dx}.$$

Now the formula in (10) yields the antiderivative

$$F(x) = \frac{1}{2}\frac{u^{3/2}}{3/2} + C$$

$$= \frac{1}{3}u^{3/2} + C = \frac{1}{3}(x^2 + 1)^{3/2} + C.$$

At this point in our studies, antidifferentiation is largely educated guessing, backed up by our experience in finding derivatives. The following example illustrates how we may make more effective guesses: Once the proper function $u(x)$ for use in Equation (10) has been identified, the remainder of the computation often can be carried out quite systematically.

**EXAMPLE 4**  Find the most general antiderivative of

$$f(x) = \frac{2 - x^2}{\sqrt[3]{6x - x^3}}.$$

**Solution**  We note that if $u = 6x - x^3$, then $du/dx = 6 - 3x^2$, so $2 - x^2 = (\frac{1}{3})(du/dx)$. Hence

$$f(x) = \frac{(\frac{1}{3})(du/dx)}{\sqrt[3]{u}} = \frac{1}{3}u^{-1/3}\frac{du}{dx}.$$

Therefore the formula in (10) yields the antiderivative

$$F(x) = \frac{1}{3}\frac{u^{2/3}}{2/3} + C = \frac{1}{2}(6x - x^3)^{2/3} + C.$$

**EXAMPLE 5**  Find the most general antiderivative of

$$g(t) = \frac{(1 + \sqrt{t})^4}{\sqrt{t}}.$$

**Solution** We begin with this observation:

$$\text{If } u = 1 + \sqrt{t} \quad \text{then} \quad \frac{du}{dt} = \frac{1}{2\sqrt{t}}.$$

Thus

$$g(t) = 2 \cdot (1 + \sqrt{t})^4 \cdot \frac{1}{2\sqrt{t}} = 2u^4 \frac{du}{dt},$$

so the formula in (10) yields the antiderivative

$$G(t) = 2 \cdot \frac{u^5}{5} + C = \frac{2}{5}(1 + \sqrt{t})^5 + C.$$

---

## *THE SUSPENSION BRIDGE

This is a simple example of the use of antiderivatives to explain a non-trivial physical phenomenon—the hanging suspension bridge. We assume that the bridge supports a horizontally uniform load. This is close to the true situation in most suspension bridges, such as New York's Verrazano Narrows Bridge, shown in Fig. 4.44. The weight of the cables will be neglected in comparison with the weight of the roadway itself, and the latter may be assumed horizontally uniform.

Now suppose that the shape of the cable from which the bridge is suspended is the graph of the function $y = f(x)$. Our task is to find the function $f$: We really want an explicit formula for it, as if we planned on designing

**4.44** The Verrazano Narrows Bridge. (Photograph courtesy of H. Armstrong Roberts, Philadelphia; and DePascale & Associates, Athens, Georgia.)

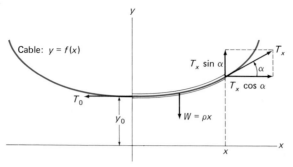

**4.45** Suspending cable for a suspension bridge

and building our own suspension bridge. If we can find $f$, we have determined the shape of the supporting cable.

As suggested in Fig. 4.45, we assume that the solution is symmetric about the $y$-axis if that axis is arranged to pass through the lowest point of the graph. We concentrate on finding the formula for $y = f(x)$ for $x \geq 0$, and in fact consider the part of the cable (or graph) that lies above the interval $[0, x]$ on the $x$-axis.

A consequence of elementary physics is this: The vertical and horizontal forces exerted on the segment of cable above $[0, x]$ must balance, or else the cable would not remain stationary. Let $T_0$ (an unknown constant) denote the tension force exerted on the left end of this portion of the cable, and let $T_x$ be the tension force exerted on the right end. We are correct in using the last subscript $x$, since the value of $T_x$ *does* depend on $x$. What's pulling on the segment to produce $T_0$ is the part of the cable to the left of the $y$-axis and other structures, such as the towers from which the cable is hung. The force $T_x$ is produced by the cable to the right of $x$.

Now $T_0$ is a horizontal force, since we are assuming that it is measured at the lowest point of the cable, at which $x = 0$ and the tangent to the cable is horizontal. The force $T_x$ is inclined at the angle $\alpha$, where $\tan \alpha = dy/dx = f'(x)$. The reason for this is that the direction of $T_x$ must be the same as the cable's direction where it's measured.

Finally, the assumption that the load on the cable is horizontally uniform means that the weight supported by our portion of cable over $[0, x]$ is given by $W = \rho x$, where $\rho$ is a constant. This weight is a *vertical* force; the horizontal component is zero.

We know that the forces on the segment of cable over $[0, x]$ balance, so we equate the vertical forces to get the first equation below and the horizontal forces to get the second.

$$T_x \sin \alpha = \rho x, \tag{11}$$

$$T_x \cos \alpha = T_0. \tag{12}$$

Now $dy/dx = \tan \alpha$, so we divide the first equation by the second and find that

$$\frac{dy}{dx} = \tan \alpha = \frac{\rho}{T_0} x.$$

Since $\rho/T_0$ is a constant, a single antidifferentiation gives us the equation

$$y = \frac{\rho}{2T_0} x^2 + C$$

CHAP. 4: Applications of Derivatives and Antiderivatives

for some constant $C$. We denote by $y_0$ the minimum height of the cable (above the $x$-axis in Fig. 4.45), and it follows that $C = y_0$. Thus

$$y = f(x) = \frac{\rho}{2T_0} x^2 + y_0 \tag{13}$$

is the equation of the cable. We have found that the hanging cable with uniform horizontal load takes the shape of a *parabola*.

## 4-8 PROBLEMS

Find the most general antiderivatives of the functions in Problems 1–65.

1 $f(x) = 3x^2 + 2x + 1$

2 $g(t) = 3t^4 + 5t - 6$

3 $h(x) = 1 - 2x^2 + 3x^3$

4 $j(t) = -\dfrac{1}{t^2}$

5 $f(x) = \dfrac{3}{x^3} + 2x^{3/2} - 1$

6 $h(x) = x^{5/2} - \dfrac{5}{x^4} - \sqrt{x}$

7 $f(t) = \frac{3}{2}t^{1/2} + 7$

8 $h(x) = \dfrac{2}{x^{3/4}} - \dfrac{3}{x^{2/3}}$

9 $f(x) = \sqrt[3]{x^2} + \dfrac{4}{\sqrt[4]{x^5}}$

10 $g(x) = (x + 1)^4$

11 $f(x) = 4x^3 - 4x + 6$

12 $g(t) = \frac{1}{4}t^5 - 6t^{-2}$

13 $h(x) = 7$

14 $f(x) = 4x^{2/3} - \dfrac{5}{x^{1/3}}$

15 $g(x) = 2x^{3/2} - \dfrac{1}{x^{1/2}}$

16 $h(t) = (t + 1)^{10}$

17 $f(x) = \dfrac{1}{(x - 10)^7}$

18 $g(z) = \sqrt{z + 1}$

19 $f(x) = x^2(x^3 + 2)^{1/3}$

20 $f(t) = 6t(3t^2 - 1)^7$

21 $f(x) = \dfrac{1}{x^2}\left(1 + \dfrac{1}{x}\right)^{10}$

22 $h(t) = \dfrac{3t^2}{(2t^3 + 1)^{1/2}}$

23 $f(x) = x^7(x^8 + 9)^{10}$

24 $f(x) = x^2\sqrt{3x^3 + 4}$

25 $f(x) = \sqrt{x}(1 + x\sqrt{x})^{1/2}$

26 $g(t) = \dfrac{t^3}{(t^4 + 1)^5}$

27 $f(x) = (x^4 + x^2)^7(2x^3 + x)$

28 $h(x) = \sqrt{24x + 1}$

29 $f(x) = (x^2 + 1)^3$

30 $g(t) = (t - 3)^{3/2}$

31 $h(x) = x\sqrt{1 - x^2}$

32 $f(x) = \dfrac{3x}{(x^2 + 16)^{1/2}}$

33 $g(x) = \dfrac{3x - 12}{x^3}$

34 $h(x) = \dfrac{7}{(x + 77)^2}$

35 $f(x) = 5 - 2x + x^3$

36 $g(t) = t(1 - 2t^2)^{1/3}$

37 $f(x) = x(x^{1/2} - x^{-1/2})^2$

38 $g(x) = \dfrac{3}{\sqrt{(x - 1)^3}}$

39 $f(x) = \dfrac{1}{x^2}\left(1 - \dfrac{1}{x}\right)^4$

40 $g(x) = \dfrac{x - \sqrt{x}}{x^3}$

41 $f(x) = \left(x + \dfrac{1}{x}\right)^2$

42 $f(t) = \dfrac{1}{t^{5/3}}$

43 $g(x) = x^2(x^7 - x^5)$

44 $h(x) = \dfrac{x^3 + 4}{x^2}$

45 $f(x) = \dfrac{6 - 5x^4}{x^2}$

46 $g(x) = \dfrac{2x^3}{(1 + x^4)^2}$

47 $f(x) = \sqrt{a + bx}$ ($a$ and $b$ are constants)
48 $g(x) = (a + bx)^2$ ($a$ and $b$ are constants)

49 $f(x) = (x + 1)^{17}$

50 $f(x) = \dfrac{x}{\sqrt[3]{1 + 2x^2}}$

51 $h(t) = (2t - 1)^5$
53 $g(x) = 3x^2(x^3 + 7)^{5/3}$

52 $f(x) = 2x\sqrt{x^2 + 1}$
54 $h(t) = \sqrt{t + 1}$

55 $f(x) = \dfrac{x}{\sqrt{x^2 + 1}}$

56 $g(t) = \dfrac{t^3 + 3t^2 - 1}{\sqrt{t}}$

57 $f(x) = \left(\sqrt[3]{x} + \dfrac{1}{\sqrt[3]{x}}\right)^2$

58 $h(t) = t(1 - t^2)^{10}$

59 $f(x) = \dfrac{(1 + \sqrt{x})^{3/2}}{\sqrt{x}}$

60 $f(x) = \dfrac{2x^3 - x}{(x^2 - x^4)^2}$

61 $f(x) = \dfrac{x + 2}{\sqrt{x^2 + 4x + 5}}$

62 $j(x) = \dfrac{(1 + \sqrt{x})^{10}}{\sqrt{x}}$

63 $f(t) = \dfrac{t}{(a + bt^2)^3}$   (a and b are constants)

64 $g(x) = \dfrac{x^2}{(a + bx^3)^2}$   (a and b are constants)

65 $f(x) = \dfrac{2x + 3}{\sqrt{x^2 + 3x}}$

Problems 66–68 deal with a suspension cable that supports a bridge with a uniform horizontal load of $\rho$ tons per linear foot. We choose the origin $(0, 0)$ as the lowest point of the cable, so its shape is that given in Equation (13) with $y_0 = 0$ and with $T_0$ in tons.

66 (a) Deduce from Equations (11) and (12) that the tension $T_x$ at the point $(x, \rho x^2/2T_0)$ is given by

$$(T_x)^2 = (T_0)^2 + \rho^2 x^2.$$

(b) Eliminate $T_0$ to show that if $x \neq 0$, then

$$T_x = \frac{\rho x}{2y}(x^2 + 4y^2)^{1/2}.$$

67 Suppose that the bridge of Problem 66 is 200 ft long and weighs 200 tons, and that the two suspension towers each reach 20 ft above the lowest point of the bridge. Use the results of Problem 66 to calculate:
(a) the tension $T_0$ (in tons) at the lowest point;
(b) the tension $T_{100}$ at the points where the cable is attached to the towers; and
(c) the angle between the cable and the towers at the point of attachment.

68 Repeat the computations of Problem 67 in the case that the two suspension towers each reach 40 ft above the lowest point of the bridge. Compare your answers with those of the previous problem. You should observe that increasing the height of the towers decreases the tension in the supporting cable.

---

### 4-9

## Velocity and Acceleration

Antidifferentiation is the tool that enables us, in many important cases, to analyze the motion of a particle (or "mass point") in terms of the forces acting on it. We consider first the motion of a particle moving along a straight line under the influence of a *constant* accelerating force. If we regard the line of motion as the $x$-axis, then—as in Section 3-1—the motion of the particle is described by its **position function**

$$x = f(t) \tag{1}$$

giving its $x$-coordinate at time $t$. Generally the function $f(t)$ will be unknown to begin with, and our problem will be to find a formula for it, using such data as the initial position, the initial velocity, and the constant acceleration of the particle.

Recall from Section 3-1 that the *velocity* $v(t)$ of the moving particle is the derivative of its position function,

$$v = f'(t); \quad \text{that is,} \quad v = \frac{dx}{dt}, \tag{2}$$

and that its *acceleration* $a(t)$ is the derivative of its velocity,

$$a = v'(t) = f''(t); \quad \text{that is,} \quad a = \frac{dv}{dt} = \frac{d^2x}{dt^2}. \tag{3}$$

So if the acceleration is *constant*, we begin with the equation

$$\frac{dv}{dt} = a \quad \text{(a constant).} \tag{4}$$

The antiderivatives $v$ and $at$ of the expressions on each side of this equation can differ only by a constant, so we conclude that

$$v = at + C. \tag{5}$$

The constant $C$ is ordinarily evaluated by substitution of $t = 0$ in *both* sides in (5); this gives

$$v_0 = v(0) = a \cdot 0 + C = C,$$

so $C$ turns out to be the *initial velocity* $v_0$. Hence the velocity $v = dx/dt$ of the particle at time $t$ is

$$\frac{dx}{dt} = v = at + v_0. \tag{6}$$

To find the position function $x(t)$, we antidifferentiate each side in Equation (6): The two antiderivatives $x$ of $dx/dt$ and $\frac{1}{2}at^2 + v_0t$ of $at + v_0$ again can differ only by a constant, so it follows that

$$x = \tfrac{1}{2}at^2 + v_0t + C \tag{7}$$

is the position of the particle at time $t$.

Here we should mention that the constants of antidifferentiation in Equations (5) and (7) have nothing to do with one another, despite the usual practice of denoting both by the same letter $C$. So we must, as before, evaluate the constant $C$ in Equation (7), and—also as before—we do so by substituting $t = 0$ into both sides of (7). This gives us

$$x_0 = x(0) = \tfrac{1}{2}a(0)^2 + v_0 \cdot (0) + C,$$

so that $C = x_0$, the *initial position* of the particle. And thus the position function of our particle is

$$x = \tfrac{1}{2}at^2 + v_0t + x_0. \tag{8}$$

WARNING  Formulas (6) and (8) are valid only in the case of *constant* acceleration $a$. They do *not* apply to problems in which the acceleration varies.

**EXAMPLE 1**  The skid marks made by an automobile indicate that its brakes were fully applied for a distance of 225 ft before it came to a stop. Suppose it is known that the car in question has a constant deceleration of 50 ft/s$^2$ under the conditions we describe. How fast was the car going when its brakes were applied?

*Solution*  The introduction of a convenient coordinate system is often crucial to the successful solution of a physical problem. Here we take the $x$-axis as positively oriented in the direction of motion of the car. We choose the origin so that $x_0 = 0$ when $t = 0$, as indicated in Fig. 4.46. In this coordinate system, the car's velocity $v$ will be a decreasing function of time $t$, so that $a = -50$ (ft/s$^2$) rather than $a = +50$. Hence Equations (6) and (8) take the forms

$$v = -50t + v_0,$$

$$x = -25t^2 + v_0t.$$

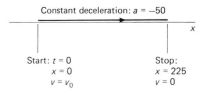

4.46  Skid marks 225 ft long

The fact that the skid marks are 225 ft long tells us that $x = 225$ when the car came to a stop; that is, when $v = 0$. From the first of the equations above, we find that $t = v_0/50$ when $v = 0$, and substitution of this value of

$t$ together with the corresponding value $x = 225$ into the second equation gives

$$225 = -25\left(\frac{v_0}{50}\right)^2 + v_0\left(\frac{v_0}{50}\right).$$

When we solve this equation for $v_0$, we find that

$$v_0 = (100 \cdot 225)^{1/2} = 150$$

ft/s, or about 102 mi/h, was the velocity of the car when the brakes were first applied.

## VERTICAL MOTION WITH CONSTANT GRAVITATIONAL ACCELERATION

One common application of Equations (4), (6), and (8) involves vertical motion near the earth's surface. A particle in such motion is subject to a downward acceleration denoted by $g$, which is equal to about 32 ft/s². If we neglect air resistance, we may assume that this acceleration of gravity is the only outside influence on the moving particle; moreover, if the motion involved is fairly close to the earth's surface, we may also assume that $g$ remains constant. (If you need more accurate values for $g$, you may use $g = 32.16$ ft/s² in the fps system, $g = 980$ cm/s² in the cgs system, or $g = 9.80$ m/s² in the mks system.)

Since we deal with vertical motion here, it is natural to choose the $y$-axis as the coordinate system for position. If we choose the upward direction as the positive direction, then the effect of gravity on the particle is to *decrease* its height and also to *decrease* its velocity $v = dy/dt$, so we see that the particle's acceleration is

$$a = \frac{dv}{dt} = -g = -32 \text{ (ft/s}^2).$$

Equations (6) and (8) then become

$$v = -32t + v_0 \tag{6'}$$

and

$$y = -16t^2 + v_0 t + y_0. \tag{8'}$$

Here $y_0$ is the initial height of the particle in feet and $v_0$ its initial velocity in feet per second.

**EXAMPLE 2**  A ball is thrown straight downward from the top of a tall building, with an initial speed of 30 ft/s. Suppose that the ball strikes the ground with a speed of 190 ft/s. How tall is the building?

*Solution*  We set up the coordinate system illustrated in Fig. 4.47, with ground level corresponding to $y = 0$, with the ball thrown at time $t = 0$, and with the positive direction being the upward direction.

Since the height $y$ of the ball is a decreasing function of time, the velocity $v = dy/dt$ is negative. So the data of the problem amount to this: $v_0 = -30$ and $v = -190$ when $y = 0$. We want to find $y_0$. Equations (6')

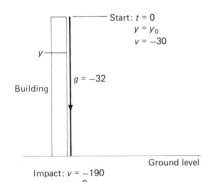

Start: $t = 0$
$y = y_0$
$v = -30$

$y$

$g = -32$

Building

Ground level

Impact: $v = -190$
$y = 0$

**4.47**  A ball is thrown down from the top of a building.

CHAP. 4:  Applications of Derivatives and Antiderivatives

and (8') give

$$y = -16t^2 - 30t + y_0,$$
$$v = -32t - 30.$$

If we use the fact that $v = -190$ when $y = 0$, the second equation will tell us when the ball hits the ground:

$$-190 = -32t - 30,$$

so that the solution $t = 5$ means that the ball strikes the ground 5 s after it's thrown. The other of the two equations may then be used, for we know that $y = 0$ when $t = 5$, and we find that

$$0 = (-16)(25) - (30)(5) + y_0,$$

so that $y_0 = 550$. Thus the building is 550 ft high.

## SIMPLE DIFFERENTIAL EQUATIONS

The antidifferentiation technique by which we derived the velocity and acceleration equations in (6) and (8) can be used to solve any differential equation of the form

$$\frac{dy}{dx} = f(x),$$

*provided* we can find an antiderivative of the given function $f(x)$.

**EXAMPLE 3**  Find a solution $y(x)$ of the differential equation

$$\frac{dy}{dx} = 2x + 3$$

that satisfies the condition $y(1) = 2$.

*Solution*  Antidifferentiating each side of the given equation, we get

$$y = D^{-1}(2x + 3) = x^2 + 3x + C.$$

Because we want $y = 2$ when $x = 1$, we substitute these values:

$$2 = (1)^2 + (3)(1) + C.$$

Hence $C = -2$, and so the desired solution is

$$y = x^2 + 3x - 2.$$

**EXAMPLE 4**  Find a solution of the differential equation

$$\frac{dy}{dx} = 2xy^2$$

such that $y(0) = 1$.

*Solution*  Because the right-hand side is not a function of $x$ alone, we first prepare the given differential equation for antidifferentiation by dividing both sides by $y^2$. This yields

$$\frac{1}{y^2} \frac{dy}{dx} = 2x.$$

Now we can antidifferentiate each side with respect to $x$. The left-hand side is the derivative of $-1/y$ with respect to $x$, and the right-hand side is the derivative of $x^2$. It follows that

$$-\frac{1}{y} = x^2 + C.$$

We substitute $x = 0$ and $y = 1$ ($y(0) = 1$ was given) and thereby find that $C = -1$. Consequently

$$-\frac{1}{y} = x^2 - 1,$$

and therefore the desired solution is

$$y = \frac{1}{1 - x^2}.$$

## *NEWTON'S INVERSE-SQUARE LAW OF GRAVITATION

Since we have already used whatever calculus is necessary in deriving the two key formulas in (6) and (8), such problems as that in Example 2—though interesting and even important—challenge only one's algebraic skills. Let us turn to some more exciting problems, involving motion of objects (meteors? spacecraft?) at considerable distances from the earth (and from other astronomical bodies). To do so, we need first to mention Newton's second law of motion and his inverse-square law of gravitation. According to the law of motion, the force $F$ acting on a particle of mass $m$ is proportional to the acceleration $a$ that it produces:

$$F = ma. \tag{9}$$

According to the law of gravitation, the gravitational force of attraction between two point masses $m$ and $M$ located at a distance $r$ apart is given by

$$F = \frac{GMm}{r^2} \tag{10}$$

where $G$ is a certain empirical constant. The formula also applies if either of the two masses is not merely a point mass, but a homogeneous sphere; in this case, the distance $r$ is measured between the centers of the spheres. (We shall derive this result in Chapter 17.)

Let $M$ denote the mass of the earth and $R$ its radius. We obtain the gravitational acceleration $a = g$ of a particle of mass $m$ at the earth's surface by using Equations (9) and (10) above simultaneously:

$$ma = mg = \frac{GMm}{r^2} = \frac{GMm}{R^2},$$

so that

$$g = \frac{GM}{R^2}. \tag{11}$$

We shall use Equation (11) from time to time to remove the necessity for the actual determination of $G$ in our problems and examples.

We mentioned that Equation (10) holds for point masses, and that we shall later show that it also holds for spheres. It appears that Newton was aware of the inverse-square law of gravitation for point masses before 1670, but was unable to show that it applies to massive objects (such as spherical planets) until some years later. This may have been the reason for the delay until 1687 of the publication of his *Philosophiae Naturalis Principia Mathematica* (*Mathematical Principles of Natural Philosophy*), the founding document of modern exact science.

**EXAMPLE 5**  If a woman has enough "spring" in her legs to jump vertically to a height of 4 ft on the earth, how high could she jump on the moon? Use the fact that the mass $\bar{M}$ and the radius $\bar{R}$ of the moon are given in terms of the mass $M$ and radius $R$ of the earth by $\bar{M} \approx (0.0123)M$ and $\bar{R} \approx (0.2725)R$. Because her mass is presumably the same anywhere in the universe, it is also reasonable to assume that she attains the same initial velocity before liftoff on the moon as on the earth.

*Solution*  First we need to find the initial velocity $v_0$ required to jump 4 ft high on the earth. Then we can calculate how high she can jump on the moon with the same initial velocity.

On the earth, we use the standard equations (8′)

$$y = -16t^2 + v_0 t + y_0$$

and (6′)

$$v = -32t + v_0.$$

In the woman's jump on the earth, we have $v = v_0$ when $t = 0$, $y_0 = 0$ (with $y = 0$ at ground level), and $v = 0$ when $y = 4$. So we substitute $v = 0$ in (6′), and find that the time to reach her maximum height of 4 ft is $t = v_0/32$ s.

Hence Equation (8′) gives

$$4 = -16\left(\frac{v_0}{32}\right)^2 + v_0\left(\frac{v_0}{32}\right) = \frac{v_0^2}{64}.$$

So $v_0 = +16$ ft/s.

Now we shift our attention to the moon. In order to find the correct versions of Equations (6′) and (8′) for the moon, we need to find the gravitational acceleration on its surface. But Equation (11) tells us that, if $\bar{g}$ denotes the gravitational acceleration for the moon,

$$\bar{g} = \frac{G\bar{M}}{(\bar{R})^2} = G\frac{(0.0123)M}{(0.2725R)^2} = (0.1656)\frac{GM}{R^2}$$

$$= (0.1656)g = (0.1656)(32) = 5.3$$

ft/s² (approximately).

On the moon, then,

$$\frac{dv}{dt} = -5.3,$$

$$v = \frac{dy}{dt} = (-5.3)t + v_0, \tag{6″}$$

and

$$y = (-2.65)t^2 + v_0t + y_0. \tag{8''}$$

In the moon jump, we have $y_0 = 0$ and $v_0 = 16$, from our previous work with the earth jump. In addition, $v = 0$ at maximum height $y$. So (6″) gives

$$0 = (-5.3)t + 16$$

then, and thus $t \approx 3.02$ s at her maximum height $y_{\text{max}}$. We substitute this value of $t$ into (8″), and obtain

$$y_{\text{max}} \approx (-2.65)(3.02)^2 + (16)(3.02) \approx 24.15$$

ft as the height she can jump on the moon. Note that she is off the ground over 6 s on the moon but only 1 s on earth.

## *ESCAPE VELOCITY

As an example of linear motion with *variable* acceleration, we now calculate the "escape velocity" from the earth—the minimum initial velocity with which an object must be launched straight upward from the earth's surface so that it will continue forever to move away from the earth.

The condition that the upward motion will continue forever becomes the relation

$$v = \frac{dy}{dt} > 0 \quad \text{for all} \quad t \geqq 0.$$

We will use $y = y(t)$ to denote the distance from the object to the earth's *center* because of the way that Newton's law of gravitation is phrased. Our launch site, though, will be the earth's *surface*, where we take $y_0 = y(0) = 6370$ km, or 6,370,000 m. The solution to the escape velocity problem will be the initial velocity $v_0 = v(0)$ that is just sufficient to insure that $v = v(t)$ is never zero or negative.

Now imagine the launched object as shown in Fig. 4.48, with mass $m$, at distance $y = y(t)$ from the earth's center at some time $t > 0$, with velocity $v = v(t)$ then. The only force acting on the mass $m$ is the pull of the earth's gravity, given by Equation (10), Newton's law of gravitation, as

$$F = -\frac{GMm}{y^2}.$$

The resulting acceleration of $m$ is then given, with the *same* values of $F$ and $m$, by Newton's second law of motion:

$$F = ma. \tag{9}$$

We eliminate $F$ by equating the last two right-hand sides. The minus sign in the gravitation equation belongs there because the force is acting opposite to the direction of increasing values of $y$. No minus sign is needed in the law of motion, because the sign of the acceleration $a$ is built in automatically. And so

$$ma = m\frac{dv}{dt} = -\frac{GMm}{y^2},$$

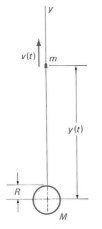

**4.48** Escape velocity

CHAP. 4:  Applications of Derivatives and Antiderivatives

which we simplify to

$$\frac{dv}{dt} = -\frac{GM}{y^2}.$$

To solve this differential equation, we use the chain rule:

$$\frac{dv}{dt} = \frac{dv}{dy}\frac{dy}{dt} = v\frac{dv}{dy}.$$

Thus

$$v\frac{dv}{dy} = -\frac{GM}{y^2}.$$

That is,

$$D_y\left(\frac{1}{2}v^2\right) = D_y\left(\frac{GM}{y}\right).$$

Note that we think of $y$ as the independent variable, rather than the more natural $t$. It follows immediately that

$$\frac{1}{2}v^2 = \frac{GM}{y} + C.$$

To evaluate the constant $C$ we use the fact that when $t = 0$, $v = v_0$ and $y = R$ (the earth's radius). Thus

$$\frac{1}{2}v_0^2 = \frac{GM}{R} + C,$$

and substitution of

$$C = \frac{1}{2}v_0^2 - \frac{GM}{R}$$

into the equation

$$\frac{1}{2}v^2 = \frac{GM}{y} + C$$

tells us that

$$v^2 = v_0^2 + 2GM\left(\frac{1}{y} - \frac{1}{R}\right).$$

In particular,

$$v^2 > v_0^2 - \frac{2GM}{R}.$$

Therefore $v$ will remain positive provided that

$$v_0^2 \geqq \frac{2GM}{R}.$$

So the escape velocity for the earth is given by

$$v_0 = \sqrt{\frac{2GM}{R}} = \left(2\frac{GM}{R^2}R\right)^{1/2} = \sqrt{2gR}. \tag{12}$$

**4.49** Escape velocity is *definitely* finite! (NASA photograph)

With $g = 9.8$ m/s$^2$ and $R = 6.37 \times 10^6$ m, this gives $v_0 \approx 11{,}174$ m/s, about 36,660 ft/s, about 24,995 mi/h, or about 6.94 mi/s.

The point is that escape velocity is *finite* and relatively easy to attain, as you might deduce from the existence of the photograph of Fig. 4.49.

**EXAMPLE 6** Suppose that you are stranded—your rocket engine has failed—on an asteroid of diameter 3 mi and of the same density as the earth. If you have enough spring in your legs to jump 4 ft straight up on earth while wearing your present life-support equipment, can you blast off from this asteroid using leg power alone?

***Solution*** By the computation shown in the first steps of the solution to Example 5, we find that an initial velocity of $v_0 = 16$ ft/s is required to jump 4 ft high on earth. We must determine if this initial velocity is adequate for escape from the asteroid.

Let $\rho$ denote the ratio of the radius of the asteroid to that of the earth, so that

$$\rho = \frac{1.5}{3960} = \frac{1}{2640}.$$

CHAP. 4: Applications of Derivatives and Antiderivatives

# 5

## The Integral

## Introduction

The earlier chapters have dealt with **differential calculus,** which is one of two closely related parts of *the* calculus. Differential calculus is centered on the concept of the *derivative*. Recall that the original motivation for the derivative was the problem of defining tangent lines to graphs of functions and calculating the slopes of such lines. By contrast, the importance of the derivative stems from its applications to diverse problems that may seem, upon initial inspection, to have little connection with tangent lines to graphs.

**Integral calculus** centers around the concept of the *integral*. The definition of the integral is motivated by the problem of defining and calculating the area of the region lying between the graph of a positive-valued function $f$ and the $x$-axis over a closed interval $[a, b]$. The area of the region $R$ of Fig. 5.1 is given by the **integral** of $f$ from $a$ to $b$, denoted by the symbol $\int_a^b f(x)dx$. But the importance of the integral, like that of the derivative, is due to its applications in many problems that may appear unrelated to its original motivation.

The principal theorem of this chapter is the *fundamental theorem of calculus* in Section 5-5. It provides a vital connection between the operations of differentiation and integration and also a method of computing values of integrals. In this chapter we concentrate on the definition and the basic computational properties of the integral, and reserve applications—other than those closely associated with area computations—to Chapter 6.

## Elementary Area Computations

Perhaps everyone's first contact with the concept of area is the formula $A = bh$, which gives the area $A$ of a rectangle as the product of its base $b$ and its height $h$. We next learn that the area of a triangle is half the product of its base and height. This follows because any triangle can be split into

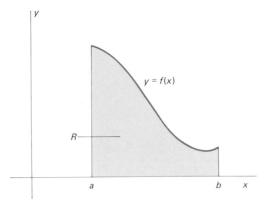

5.1 The problem of computing the area of $R$ motivates integral calculus.

two *right* triangles, and every right triangle is exactly half a rectangle (see Fig. 5.2).

Given the formula $A = \frac{1}{2}bh$ for the area of a triangle, we can find the area of an arbitrary polygonal figure (a plane set bounded by a closed "curve" that consists of straight line segments). The reason is that any polygonal figure can be divided into nonoverlapping triangles, as shown in Fig. 5.3, and the area of the polygonal figure is then the sum of the areas of these triangles. This approach to area dates back several thousand years to the ancient civilizations of Egypt and Babylonia.

The ancient Greeks began the investigation of areas of *curvilinear* figures in the fourth and fifth centuries B.C.; the approach they devised ultimately led—much later—to the integral as described in the introduction. We take it as obvious on intuitive grounds that every reasonably nice plane set $S$ bounded by closed curves has an area $a(S)$, a nonnegative real number with the following properties:

**1** If the set $S$ is contained in the set $T$, then $a(S) \leq a(T)$.

**2** If $S$ and $T$ are nonoverlapping sets—they intersect only in points of their boundary curves, if at all—then the area of their union $S \cup T$ is $a(S \cup T) = a(S) + a(T)$.

**3** If the sets $S$ and $T$ are congruent (have the same size and shape), then $a(S) = a(T)$.

To investigate the area of a curvilinear set $S$, we might inscribe in it a sequence of polygons $P_1, P_2, P_3, \ldots$ with increasing areas that gradually fill or "exhaust" the set $S$. We can then attempt to compute the limit $\lim\limits_{n \to \infty} a(P_n)$ of the areas of these inscribed polygons. Recall from the definition of the limit of a sequence (Section 3-9) that

$$A = \lim_{n \to \infty} a(P_n)$$

means that $a(P_n)$ can be made as close to $A$ as we please by choosing $n$ sufficiently large.

Suppose that we also circumscribe about $S$ a sequence of polygons $Q_1, Q_2, Q_3, \ldots$ with decreasing areas that gradually "close down" on the set $S$, and attempt to compute the limit $\lim\limits_{n \to \infty} a(Q_n)$ of the areas of these circumscribed polygons. Our earlier inscribed polygons and these circumscribed polygons should enable us, as Fig. 5.4 suggests, to obtain arbitrarily good estimates of the actual area $a(S)$. For by Property 1 of area above, we see that

$$a(P_n) \leq a(S) \leq a(Q_n).$$

By the squeeze law for limits of sequences (analogous to the squeeze law for limits of functions, stated in Section 2.1), it then follows that

$$\lim_{n \to \infty} a(P_n) \leq a(S) \leq \lim_{n \to \infty} a(Q_n),$$

provided that both these limits exist. If, moreover, the values of these two limits turn out to be equal, then the area of the set $S$ we started with must be given by

$$a(S) = \lim_{n \to \infty} a(P_n) = \lim_{n \to \infty} a(Q_n). \tag{1}$$

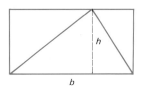

**5.2** The formula for the area of a triangle follows with the aid of this figure.

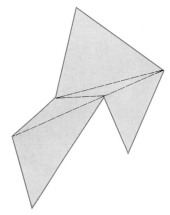

**5.3** Every polygon can be represented as the union of nonoverlapping triangles.

**5.4** Estimating the area $a(S)$ of the set $S$ by using areas of inscribed and circumscribed polygons.

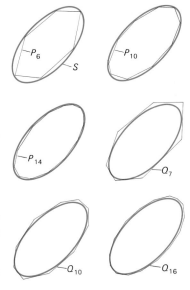

## THE NUMBER $\pi$

In the third century B.C., Archimedes, the greatest mathematician of antiquity, used an approach similar to that outlined above to derive the famous estimate

$$\tfrac{223}{71} = 3\tfrac{10}{71} < \pi < 3\tfrac{1}{7} = \tfrac{22}{7}.$$

The number $\pi$ is sometimes defined as the ratio of the circumference of a circle to its diameter, and sometimes as the ratio of the area of a circle to the square of its radius. It may also be defined to be the area of the unit circle $x^2 + y^2 \leq 1$. With this last definition, the problem of computing $\pi$ is that of finding the area of the unit circle.

Let $P_n$ and $Q_n$ be $n$-sided regular polygons, with $P_n$ inscribed in the unit circle and $Q_n$ circumscribed about it, as shown in Fig. 5.5. Since the polygons are regular, all their sides and angles are equal, so all we need to find is the area of *one* of the triangles that we've shown making up $P_n$ and one of those making up $Q_n$.

Let $\alpha_n$ be the central angle subtended by *half* of one of the sides. The angle $\alpha_n$ is the same whether we work with $P_n$ or with $Q_n$. In degrees,

$$\alpha_n = \frac{360°}{2n} = \frac{180°}{n}.$$

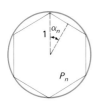

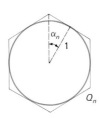

**5.5** Estimating $\pi$ by using inscribed and circumscribed regular polygons and the unit circle.

We can read various dimensions and proportions from Fig. 5.5; we see that the area of $P_n$ is given by

$$a(P_n) = n \cdot 2 \cdot \frac{1}{2} \sin \alpha_n \cos \alpha_n = \frac{n}{2} \sin\left(\frac{360°}{n}\right) \qquad (2)$$

and that

$$a(Q_n) = n \cdot 2 \cdot \frac{1}{2} \tan \alpha_n = n \tan\left(\frac{180°}{n}\right). \qquad (3)$$

| $n$ | $a(P_n)$ | $a(Q_n)$ |
|---|---|---|
| 6 | 2.598076 | 3.464102 |
| 12 | 3.000000 | 3.215390 |
| 24 | 3.105829 | 3.159660 |
| 48 | 3.132629 | 3.146086 |
| 96 | 3.139350 | 3.142715 |
| 180 | 3.140955 | 3.141912 |
| 360 | 3.141433 | 3.141672 |
| 720 | 3.141553 | 3.141613 |
| 1440 | 3.141583 | 3.141598 |
| 2880 | 3.141590 | 3.141594 |
| 5760 | 3.141592 | 3.141593 |

**5.6** Data in estimating $\pi$ (rounded to six-place accuracy)

We substitute selected values of $n$ in the formulas in (2) and (3), and thereby obtain the entries of the table shown in Fig. 5.6. Because $a(P_n) \leq \pi \leq a(Q_n)$ for all $n$, we see that $\pi \approx 3.14159$ to five decimal places. We should remark that Archimedes' reasoning was *not* "circular"—he used a direct method for computing the sines and cosines in (2) and (3) that does not depend upon a prior knowledge of the value of $\pi$ (see Problem 36).

## SUMMATION NOTATION

For convenient application of the method of inscribed and circumscribed polygons to compute areas of rather general figures, we need a compact notation for sums of many numbers. We use the symbol $\sum_{i=1}^{n} a_i$ to abbreviate the sum of the $n$ numbers $a_1, a_2, \ldots, a_n$, so that

$$\sum_{i=1}^{n} a_i = a_1 + a_2 + \cdots + a_n.$$

The $\Sigma$ (Greek capital sigma) indicates the sum of the terms $a_i$ as the **summation index** $i$ runs through integer values from 1 to $n$. For example, the sum of the squares of the first ten positive integers is

$$\sum_{i=1}^{10} i^2 = 1^2 + 2^2 + \cdots + 10^2$$

$$= 1 + 4 + \cdots + 100 = 385.$$

The particular letter used for the summation index is immaterial; we may write

$$\sum_{i=1}^{10} i^2 = \sum_{k=1}^{10} k^2 = \sum_{r=1}^{10} r^2 = 385.$$

The simple rules of summation

$$\sum_{i=1}^{n} ca_i = c\left(\sum_{i=1}^{n} a_i\right) \tag{4}$$

and

$$\sum_{i=1}^{n} (a_i + b_i) = \sum_{i=1}^{n} a_i + \sum_{i=1}^{n} b_i \tag{5}$$

are easy to verify by writing out each sum in full.

Note that if $a_i = a$ (constant) for each $i = 1, 2, 3, \ldots, n$, then (5) yields

$$\sum_{i=1}^{n} (a + b_i) = \sum_{i=1}^{n} a + \sum_{i=1}^{n} b_i$$

$$= \underbrace{(a + a + \cdots + a)}_{(n \text{ times})} + \sum_{i=1}^{n} b_i,$$

so that

$$\sum_{i=1}^{n} (a + b_i) = na + \sum_{i=1}^{n} b_i.$$

In particular,

$$\sum_{i=1}^{n} 1 = n.$$

The sum of the $k$th powers of the first $n$ positive integers,

$$\sum_{i=1}^{n} i^k = 1^k + 2^k + 3^k + \cdots + n^k,$$

often occurs in area computations. The values of this sum for $k = 1, 2, 3$, and 4 are given by the following formulas. They may be established by mathematical induction (see Problems 27–30).

$$\sum_{i=1}^{n} i = \frac{n}{2}(n + 1) = \frac{n^2}{2} + \frac{n}{2}. \tag{6}$$

$$\sum_{i=1}^{n} i^2 = \frac{n}{6}(n + 1)(2n + 1) = \frac{n^3}{3} + \frac{n^2}{2} + \frac{n}{6}. \tag{7}$$

$$\sum_{i=1}^{n} i^3 = \left[\frac{n}{2}(n + 1)\right]^2 = \frac{n^4}{4} + \frac{n^3}{2} + \frac{n^2}{4}. \tag{8}$$

$$\sum_{i=1}^{n} i^4 = \frac{n^5}{5} + \frac{n^4}{2} + \frac{n^3}{3} - \frac{n}{30}. \tag{9}$$

In fact, in the general case, it is known that

$$\sum_{i=1}^{n} i^k = \frac{n^{k+1}}{k+1} + \frac{n^k}{2} + \text{(lower powers of } n). \tag{10}$$

**EXAMPLE 1**   Use the summation formulas above to evaluate the sum

$$\sum_{i=1}^{10} (2i^3 - 3i^2 + 7) = 6 + 11 + 34 + \cdots + 1707.$$

*Solution*

$$\sum_{i=1}^{10} (2i^3 - 3i^2 + 7) = 2\left(\sum_{i=1}^{10} i^3\right) - 3\left(\sum_{i=1}^{10} i^2\right) + 7\left(\sum_{i=1}^{10} 1\right)$$
$$= 2[\tfrac{10}{2}(10 + 1)]^2 - 3 \cdot \tfrac{10}{6}(10 + 1)(20 + 1) + 7 \cdot 10$$
$$= 2(3025) - 3(385) + 70 = 4965.$$

**EXAMPLE 2**   Evaluate

$$\lim_{n \to \infty} \frac{1^2 + 2^2 + \cdots + n^2}{n^3}.$$

*Solution*   Using Formula (7) we obtain

$$\lim_{n \to \infty} \frac{1^2 + 2^2 + \cdots + n^2}{n^3} = \lim_{n \to \infty} \frac{1}{n^3}\left(\frac{n^3}{3} + \frac{n^2}{2} + \frac{n}{6}\right)$$
$$= \lim_{n \to \infty} \left(\frac{1}{3} + \frac{1}{2n} + \frac{1}{6n^2}\right) = \frac{1}{3}.$$

We use the fact that the terms $1/2n$ and $1/6n^2$ approach zero as $n \to \infty$.

## AREAS UNDER GRAPHS

Let $f$ be a continuous, positive-valued function on $[a, b]$, and suppose that we want to calculate the **area $A$ under the graph of $f$ from $a$ to $b$.** That is, $A$ is the area of the region $R$ that is bounded by the curve $y = f(x)$, the $x$-axis, and the vertical lines $x = a$ and $x = b$. This is the region shown earlier in Fig. 5.1.

Our plan is to get high and low (but close) estimates of $A = a(R)$ by using certain special inscribed and circumscribed polygons. These polygons will be unions of rectangles with vertical sides, each having its base on the $x$-axis. We use such polygons because they are particularly easy to work with.

We start with a fixed positive integer $n$. Given $n$, we **partition,** or subdivide, the interval $[a, b]$ into $n$ equal subintervals by choosing points $x_0$, $x_1, x_2, \ldots, x_n$ such that

$$a = x_0 < x_1 < x_2 < \cdots < x_{n-1} < x_n = b,$$

with the length of the $i$th subinterval $[x_{i-1}, x_i]$ given by

$$\Delta x = x_i - x_{i-1} = \frac{b - a}{n}.$$

for each $i$ ($1 \leq i \leq n$). The point $x_i$ will be

$$x_i = x_0 + i\,\Delta x = a + \frac{i}{n}(b - a).$$

Now let $x_i^\flat$ be a point of $[x_{i-1}, x_i]$ at which $f$ attains its minimum value, so that $f(x_i^\flat)$ is the minimum value of $f(x)$ on the $i$th subinterval $[x_{i-1}, x_i]$. The rectangle with base $[x_{i-1}, x_i]$ and height $f(x_i^\flat)$ then lies *within* the region $R$. (See Fig. 5.7.) The union of these rectangles for $i = 1, 2, \ldots, n$ is the **inscribed rectangular polygon** $P_n$ associated with our partition of $[a, b]$ into $n$ equal subintervals. Its area is the sum of the areas of the $n$ rectangles with base length $\Delta x$ and heights $f(x_1^\flat), f(x_2^\flat), \ldots, f(x_n^\flat)$. So

$$a(P_n) = \sum_{i=1}^{n} f(x_i^\flat)\,\Delta x. \tag{11}$$

| Symbol | Pronounced |
| --- | --- |
| $x_i^\flat$ | $x_i$-flat |
| $x_i^\#$ | $x_i$-sharp |

Next, let $x_i^\#$ be a point of $[x_{i-1}, x_i]$ such that $f(x_i^\#)$ is the maximum value of $f(x)$ on the $i$th subinterval $[x_{i-1}, x_i]$. For each $i$ ($1 \leq i \leq n$), consider the rectangle wih base $[x_{i-1}, x_i]$, height $f(x_i^\#)$, and area $f(x_i^\#)\,\Delta x$. (See Fig. 5.8.) The union of these $n$ rectangles contains the region $R$ under the graph; it is the circumscribed rectangular polygon $Q_n$ associated with the partition of $[a, b]$ into $n$ equal subintervals. Its area is

$$a(Q_n) = \sum_{i=1}^{n} f(x_i^\#)\,\Delta x. \tag{12}$$

Since the region $R$ contains $P_n$ but lies within $Q_n$, we conclude from (11) and (12) that its area $A$ satisfies the inequalities

$$\sum_{i=1}^{n} f(x_i^\flat)\,\Delta x \leq A \leq \sum_{i=1}^{n} f(x_i^\#)\,\Delta x. \tag{13}$$

We could therefore attempt to estimate the value of $A$ by computing (for a fixed value of $n$) the values of the two sums (11) and (12).

**5.7** The area of the inscribed polygon $P_n$ is an underestimate of the area $A = a(R)$.

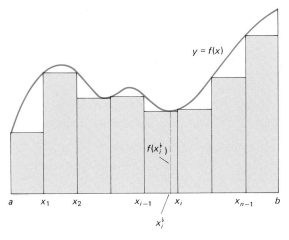

**5.8** The rectangular circumscribed polygon gives an overestimate of the area $A = a(R)$.

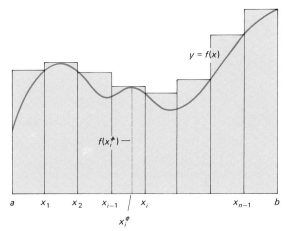

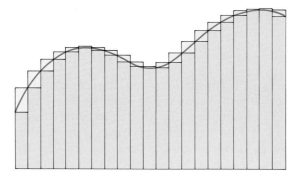

**5.9** The areas of the inscribed and circumscribed polygons should be nearly equal when $\Delta x$ is small.

Figure 5.9 suggests that if $n$ is very large, and hence $\Delta x$ is very small, then the areas $a(P_n)$ and $a(Q_n)$ of the inscribed and circumscribed rectangular polygons will be very close to $A = a(R)$. For the difference in the heights of the inscribed and circumscribed rectangles over $[x_{i-1}, x_i]$ is $f(x_i^\#) - f(x_i^\flat)$, which should be small when $\Delta x$ is small; so, intuitively, the difference between the two sums in (13) will be small when $n$ is very large. Indeed, the assumption that $f$ is continuous is enough to insure that this is so; in Section 5-3 we shall state a theorem which implies that

$$A = \lim_{n \to \infty} \sum_{i=1}^{n} f(x_i^\flat)\, \Delta x = \lim_{n \to \infty} \sum_{i=1}^{n} f(x_i^\#)\, \Delta x. \tag{14}$$

In this section we are proceeding on the basis of an intuitive concept of area. When the necessary groundwork has been laid, we shall see that the area under the graph of $f$ from $a$ to $b$ must be *defined* by means of (14). The following two examples illustrate the use of (14) for the explicit computation of areas.

**EXAMPLE 3**   Find the area $A$ under the graph of $f(x) = x^4$ from $x = 0$ to $x = b > 0$.

*Solution*   Since the two limits in (14) are equal, it is unnecessary to work with both inscribed and circumscribed rectangular polygons. We shall use only circumscribed figures, as shown in Fig. 5.10. First we subdivide the interval $[0, b]$ into $n$ equal subintervals, each with length $\Delta x = b/n$. To do so we must have $x_i = i\, \Delta x = ib/n$. Since $f(x) = x^4$ is an increasing function on $[0, b]$, $f$ attains its maximum value on the $i$th subinterval $[x_{i-1}, x_i]$ at the right-hand end point $x_i^\# = x_i = ib/n$. Hence

$$\sum_{i=1}^{n} f(x_i^\#)\, \Delta x = \sum_{i=1}^{n} \left(\frac{ib}{n}\right)^4 \left(\frac{b}{n}\right) = \frac{b^5}{n^5} \sum_{i=1}^{n} i^4.$$

Now we use Formula (9) for $\displaystyle\sum_{i=1}^{n} i^4$. This gives a formula for the area of $Q_n$, the circumscribed polygon made up of $n$ rectangles:

$$a(Q_n) = \frac{b^5}{n^5} \left(\frac{n^5}{5} + \frac{n^4}{2} + \frac{n^3}{3} - \frac{n}{30}\right)$$

$$= b^5 \left(\frac{1}{5} + \frac{1}{2n} + \frac{1}{3n^2} - \frac{1}{30n^4}\right).$$

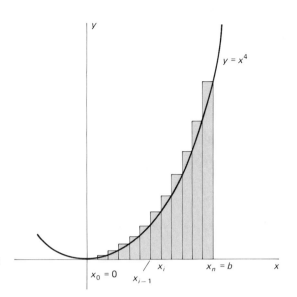

**5.10**  An area computation using circumscribed rectangles

Finally, we take the limit as $n \to \infty$. Since the terms $1/2n$, $1/3n^2$, and $1/30n^4$ approach zero as $n \to \infty$, we find that

$$A = \lim_{n \to \infty} \sum_{i=1}^{n} f(x_i^{\#}) \, \Delta x$$

$$= \lim_{n \to \infty} b^5 \left( \frac{1}{5} + \frac{1}{2n} + \frac{1}{3n^2} - \frac{1}{30n^4} \right) = \frac{b^5}{5}.$$

Note that the answer is *not* an approximation; it is *exact*.

**EXAMPLE 4**  Find the area $A$ under the graph of $f(x) = 100 - 3x^2$ from $x = 1$ to $x = 5$.

**Solution**  The function $f(x) = 100 - 3x^2$ is decreasing on $[1, 5]$ and therefore attains its minimum value on $[x_{i-1}, x_i]$ at the right-hand end point $x_i$. Hence $x_i^{\flat} = x_i$, and it is a bit more convenient to use inscribed rectangular polygons as indicated in Fig. 5.11. We subdivide $[1, 5]$ into $n$

**5.11**  An area computation using inscribed rectangles

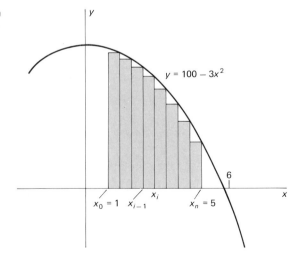

subintervals each having length $\Delta x = 4/n$. Then

$$x_i^\flat = x_i = 1 + i\,\Delta x = 1 + \frac{4i}{n}.$$

And so

$$\sum_{i=1}^{n} f(x_i^\flat)\,\Delta x = \sum_{i=1}^{n} \left[ 100 - 3\left(1 + \frac{4i}{n}\right)^2 \right]\left(\frac{4}{n}\right)$$

$$= \sum_{i=1}^{n} \left[ 97 - \frac{24i}{n} - \frac{48i^2}{n^2} \right]\left(\frac{4}{n}\right)$$

$$= \frac{388}{n} \sum_{i=1}^{n} 1 - \frac{96}{n^2} \sum_{i=1}^{n} i - \frac{192}{n^3} \sum_{i=1}^{n} i^2$$

$$= \frac{388}{n}(n) - \frac{96}{n^2}\left(\frac{n^2}{2} + \frac{n}{2}\right) - \frac{192}{n^3}\left(\frac{n^3}{3} + \frac{n^2}{2} + \frac{n}{6}\right)$$

$$= 276 - \frac{144}{n} - \frac{32}{n^2}.$$

We applied Formulas (6) and (7) in the next-to-last step. Consequently, the desired area is

$$A = \lim_{n \to \infty} \left( 276 - \frac{144}{n} - \frac{32}{n^2} \right) = 276$$

since the terms $144/n$ and $32/n^2$ approach zero as $n \to \infty$.

Examples 3 and 4 illustrate the fact that if $f$ is a continuous function, then we can take our choice between the sums

$$\sum_{i=1}^{n} f(x_i^\flat)\,\Delta x \quad \text{and} \quad \sum_{i=1}^{n} f(x_i^\#)\,\Delta x,$$

using whichever of the two seems the more convenient.

## 5-2 PROBLEMS

Use Formulas (6) through (9) to find the sums in Problems 1–10.

**1** $\displaystyle\sum_{i=1}^{10} (4i - 3)$

**2** $\displaystyle\sum_{j=1}^{8} (5 - 2j)$

**3** $\displaystyle\sum_{i=1}^{10} (3i^2 + 1)$

**4** $\displaystyle\sum_{k=1}^{6} (2k - 3k^2)$

**5** $\displaystyle\sum_{r=1}^{8} (r - 1)(r + 2)$

**6** $\displaystyle\sum_{i=1}^{5} (i^3 - 3i + 2)$

**7** $\displaystyle\sum_{i=1}^{6} (i^4 - i^3)$

**8** $\displaystyle\sum_{k=1}^{10} (2k - 1)^2$

**9** $\displaystyle\sum_{i=1}^{1000} i^2$

**10** $\displaystyle\sum_{i=1}^{100} i^3$

Use the method of Example 2 to evaluate the limits in Problems 11–13.

**11** $\displaystyle\lim_{n \to \infty} \frac{1 + 2 + \cdots + n}{n^2}$

**12** $\displaystyle\lim_{n \to \infty} \frac{1^3 + 2^3 + \cdots + n^3}{n^4}$

**13** $\displaystyle\lim_{n \to \infty} \frac{1^4 + 2^4 + \cdots + n^4}{n^5}$

Use Formulas (6) through (9) to derive compact formulas in terms of $n$ for the sums in Problems 14–16.

**14** $\displaystyle\sum_{i=1}^{n} (2i - 1)$

**15** $\displaystyle\sum_{i=1}^{n} (2i - 1)^2$

**16** $\displaystyle\sum_{i=1}^{n} (n^2 - i^2)^2$

In each of Problems 17–26, use the method of Examples 3 and 4 to find the area under the graph of $f$ from $a$ to $b$.

**17** $f(x) = 2x + 5; \quad a = 0, b = 3$
**18** $f(x) = 13 - 3x; \quad a = 0, b = 4$

**19** $f(x) = x^2$;    $a = 0, b = 1$

**20** $f(x) = x^3$;    $a = 0, b = 4$

**21** $f(x) = 3x^2 + 2$;    $a = 1, b = 5$

**22** $f(x) = 10 - x^2$;    $a = 1, b = 3$

**23** $f(x) = 3x^2 + 5x + 2$;    $a = 2, b = 4$

**24** $f(x) = 4x^3 + 2x$;    $a = 0, b = 3$

**25** $f(x) = x^2$;    $a = 0, b = b$

**26** $f(x) = x^3$;    $a = 0, b = b$

**27** Establish Formula (6) by mathematical induction.

**28** Establish Formula (7) by mathematical induction.

**29** Derive Formula (6) by adding the equations

$$\sum_{i=1}^{n} i = 1 + 2 + \cdots + n$$

and

$$\sum_{i=1}^{n} i = n + (n - 1) + \cdots + 1.$$

**30** Write the $n$ equations obtained by substituting the values $k = 1, 2, 3, \ldots, n$ into the identity

$$(k + 1)^3 - k^3 = 3k^2 + 3k + 1.$$

Add these $n$ equations, and use their sum to deduce Formula (7) from Formula (6).

**31** Derive the formula $A = \frac{1}{2}bh$ for the area of a right triangle by using circumscribed rectangular polygons to find the area under the graph of $f(x) = hx/b$ from 0 to $b$.

In Problems 32 and 33, let $A$ denote the area and $C$ the circumference of a circle of radius $r$ and let $A_n$ and $C_n$ denote the area and perimeter, respectively, of a regular $n$-sided polygon inscribed in this circle.

**32** Show that

$$A_n = nr^2 \sin\left(\frac{180°}{n}\right) \cos\left(\frac{180°}{n}\right)$$

and that

$$C_n = 2nr \sin\left(\frac{180°}{n}\right).$$

**33** Deduce that $A = \frac{1}{2}rC$ by taking the limit of $A_n/C_n$ as $n \to \infty$, using the fact that

$$\lim_{n \to \infty} \cos\left(\frac{180°}{n}\right) = \cos 0° = 1.$$

Then, given that $A = \pi r^2$, conclude from this result that $C = 2\pi r$.

**34** Use the method of Example 2 to deduce from Formula (10) that

$$\lim_{n \to \infty} \frac{1^k + 2^k + \cdots + n^k}{n^{k+1}} = \frac{1}{k + 1}.$$

**35** Use circumscribed rectangular polygons and the result of Problem 34 to show that the area under the graph of

$f(x) = x^k$ ($k$ is a positive integer) from $x = 0$ to $x = b$ is $A = b^{k+1}/(k + 1)$.

**36** The formulas in Equations (2) and (3) imply that the areas of the inscribed and circumscribed regular $n$-sided polygons are

$$a(P_n) = n \sin \alpha_n \cos \alpha_n$$

and

$$a(Q_n) = \frac{n \sin \alpha_n}{\cos \alpha_n},$$

where $\alpha = 180°/n$. Thus $\alpha_6 = 30°$, so $\sin \alpha_6 = \frac{1}{2}$ and $\cos \alpha_6 = \sqrt{3}/2$. Now $\alpha_{12} = \frac{1}{2}\alpha_6$, so the half-angle identities yield

$$\sin \alpha_{12} = \sqrt{\frac{1 - \cos \alpha_6}{2}} = \frac{1}{2}\sqrt{2 - \sqrt{3}},$$

$$\cos \alpha_{12} = \sqrt{\frac{1 + \cos \alpha_6}{2}} = \frac{1}{2}\sqrt{2 + \sqrt{3}}.$$

Once $\sin \alpha_{12}$ and $\cos \alpha_{12}$ have been calculated, the half-angle identities can be applied again to calculate $\sin \alpha_{24}$ and $\cos \alpha_{24}$, and so on.

(a) Use this approach—not touching the sine or cosine keys on your calculator—to verify the entries in Fig. 5.6 for $n = 6, 12, 24, 48, 96$. This is how far Archimedes got.

(b) Verify that your results for $n = 96$ imply Archimedes' inequality

$$3\tfrac{10}{71} < \pi < 3\tfrac{1}{7}.$$

**37** Surely you noticed the remarkable coincidence—a consequence of the formulas in (6) and (8)—that

$$(1 + 2 + 3 + \cdots + n)^2 = 1^3 + 2^3 + 3^3 + \cdots + n^3$$

for each positive integer $n$. If we write

$$f_k(n) = 1^k + 2^k + 3^k + \cdots + n^k$$

for each positive integer $n$, we can abbreviate this fact in the form

$$[f_1(n)]^2 = f_3(n) \qquad (15)$$

for all $n = 1, 2, 3, \ldots$. Use the fact that $f_k(n)$ is a polynomial in the variable $n$ for each positive integer $k$ together with the formula in Equation (10) to show that the only solution of the "functional equation"

$$[f_p(n)][f_q(n)] = f_r(n)$$

(for all $n$) for positive integers $p$, $q$, and $r$ is that given in (15).

**38** In continuation of Problem 37, it is known that the term of lowest degree in the polynomial $f_k(n)$ is, for each $k$, either $\alpha n$ or $\alpha n^2$, where $\alpha$ is a nonzero rational number. Use this information to show that the only way in which one of the polynomials $f_k(n)$ is the product of two or more others is that given in (15).

## 5-3

## Riemann Sums and the Integral

In the previous section we saw that the area $A$ under the graph from $x = a$ to $x = b$ of the *continuous, positive-valued* function $f$ satisfies the inequalities

$$\sum_{i=1}^{n} f(x_i^{\flat}) \, \Delta x \leq A \leq \sum_{i=1}^{n} f(x_i^{\#}) \, \Delta x, \qquad (1)$$

where $f(x_i^{\flat})$ and $f(x_i^{\#})$ are the minimum and maximum values of $f$ on the $i$th subinterval $[x_{i-1}, x_i]$ of a partition of $[a, b]$ into $n$ equal subintervals of length $\Delta x$. The two approximating sums in (1) are both of the form

$$\sum_{i=1}^{n} f(x_i^*) \, \Delta x, \qquad (2)$$

where $x_i^*$ denotes an arbitrary point of the $i$th subinterval $[x_{i-1}, x_i]$. Sums of the form of (2) appear as approximations in a wide range of applications and also constitute the basis for the definition of the integral. Motivated by our discussion of area in Section 5-2, we want to define the integral of $f$ from $a$ to $b$ as some sort of limit as $\Delta x \to 0$ of sums such as the one in (2). Our goal is to begin with a fairly general function $f$ and define a computable real number $I$ (its integral), which—in the special case when $f$ is continuous and positive-valued on $[a, b]$—will equal the area under the graph of $y = f(x)$.

We start with a function $f$ on $[a, b]$ *not* necessarily either continuous or positive-valued. A **partition** $P$ of $[a, b]$ is a collection of subintervals

$$[x_0, x_1], \quad [x_1, x_2], \quad [x_2, x_3], \quad \ldots, \quad [x_{n-1}, x_n]$$

of $[a, b]$ such that

$$a = x_0 < x_1 < x_2 < \cdots < x_{n-1} < x_n = b.$$

The **mesh** of the partition $P$ is the largest of the lengths

$$\Delta x_i = x_i - x_{i-1}$$

of the subintervals in $P$ and is denoted by $|P|$. To get a sum such as (2) we need a point $x_i^*$ of the $i$th subinterval for each $i$, $1 \leq i \leq n$. A collection of points

$$S = \{x_1^*, x_2^*, x_3^*, \ldots, x_n^*\}$$

with $x_i^*$ in $[x_{i-1}, x_i]$ (for each $i$) is called a **selection** for the partition $P$.

---

*Definition*    *Riemann Sum*

Let $f$ be a function defined on the interval $[a, b]$. If $P$ is a partition of $[a, b]$ and $S$ is a selection for $P$, then the **Riemann sum** for $f$ determined by $P$ and $S$ is

$$R = \sum_{i=1}^{n} f(x_i^*) \, \Delta x_i. \qquad (3)$$

We also say that this Riemann sum is **associated with** the partition $P$.

---

The German mathematician G. F. B. Riemann (1826–1866) provided a rigorous definition of the integral. Various special types of "Riemann sums" had appeared in area and volume computations since the time of Archimedes, but it was Riemann who framed the definition above in its full generality.

The point $x_i^*$ in (3) is an arbitrary point of the $i$th subinterval $[x_{i-1}, x_i]$—that is, it can be *any* point of this subinterval. But when we actually compute Riemann sums, we usually choose the points of the selection $S$ in some systematic manner, as in the following example.

**EXAMPLE 1**  Let $f(x) = x^2$ on $[0, 2]$. We are going to calculate three different Riemann sums associated with the partition of $[0, 2]$ into $n = 10$ equal subintervals, each having length

$$\Delta x_i = \Delta x = \tfrac{2}{10} = \tfrac{1}{5}.$$

The endpoints of these subintervals are the integral multiples of $\tfrac{1}{5}$ that appear in the inequalities

$$0 < \tfrac{1}{5} < \tfrac{2}{5} < \tfrac{3}{5} < \cdots < \tfrac{8}{5} < \tfrac{9}{5} < \tfrac{10}{5} = 2.$$

The $i$th subinterval is $[x_{i-1}, x_i]$, where $x_i = i/5$ for $i = 1, 2, 3, \ldots, 10$. We shall select $x_i^*$ in $[x_{i-1}, x_i]$ in the same way for each of these ten subintervals.

(a) If we choose $x_i^* = x_{i-1} = (i-1)/5$, the *left-hand end point* of $[x_{i-1}, x_i]$, then (3) is the Riemann sum

$$R_{\text{left}} = \sum_{i=1}^{n} f(x_{i-1})\, \Delta x_i = \sum_{i=1}^{n} \left(\frac{i-1}{5}\right)^2 \left(\frac{1}{5}\right)$$

$$= \frac{1}{125}(0^2 + 1^2 + 2^2 + \cdots + 9^2) = \frac{285}{125} = \frac{57}{25} = 2.28.$$

(b) If we choose $x_i^* = x_i = i/5$, the *right-hand end point* of $[x_{i-1}, x_i]$, then (3) is the Riemann sum

$$R_{\text{right}} = \sum_{i=1}^{n} f(x_i)\, \Delta x_i = \sum_{i=1}^{n} \left(\frac{i}{5}\right)^2 \left(\frac{1}{5}\right)$$

$$= \frac{1}{125}(1^2 + 2^2 + 3^3 + \cdots + 10^2)$$

$$= \frac{385}{125} = \frac{77}{25} = 3.08.$$

(c) If we choose $x_i^*$ to be the *midpoint*

$$m_i = \frac{1}{2}(x_{i-1} + x_i)$$

$$= \frac{1}{2}\left(\frac{i-1}{5} + \frac{i}{5}\right) = \frac{1}{10}(2i - 1),$$

then (3) is the Riemann sum

$$R_{\text{mid}} = \sum_{i=1}^{n} f(m_i)\, \Delta x_i = \sum_{i=1}^{n} \left(\frac{2i-1}{10}\right)^2 \left(\frac{1}{5}\right)$$

$$= \frac{1}{500}(1^2 + 3^2 + 5^2 + \cdots + 19^2) = \frac{1330}{500} = 2.66.$$

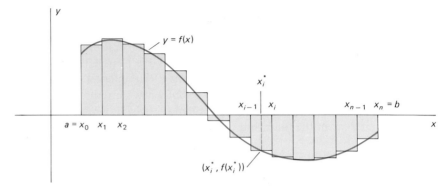

**5.12** A geometric interpretation of the Riemann sum in (3)

The Riemann sum (3) can be interpreted geometrically, as in Fig. 5.12. On each subinterval $[x_{i-1}, x_i]$, we construct a rectangle with width $\Delta x_i$ and "height" $f(x_i^*)$; if $f(x_i^*) > 0$, then this rectangle stands above the x-axis, while if $f(x_i^*) < 0$, it lies below the x-axis. The Riemann sum $R$ is then the sum of the **signed** areas of these rectangles—that is, the sum of the areas of those rectangles that lie above the x-axis *minus* the sum of the areas of those that lie below the x-axis.

If the widths $\Delta x_i$ of these rectangles are all very small, that is, the mesh $|P|$ is very small, then it appears that the Riemann sum $R$ ought to closely approximate the area from $a$ to $b$ under $y = f(x)$ above the x-axis minus the area over $y = f(x)$ below the x-axis. This suggests that the integral of $f$ from $a$ to $b$ be defined by taking the limit of the Riemann sum as the mesh $|P|$ approaches zero:

$$I = \lim_{|P| \to 0} \sum_{i=1}^{n} f(x_i^*) \, \Delta x_i. \tag{4}$$

The actual definition of the integral is obtained by saying precisely what this limit means. It means that the Riemann sum can be made as close to the number $I$ as we like by taking the mesh of the partition $P$ sufficiently small.

---

*Definition*    *The (Definite) Integral*

The **(definite) integral of the function $f$ from $a$ to $b$** is the number

$$I = \lim_{|P| \to 0} \sum_{i=1}^{n} f(x_i^*) \, \Delta x_i \tag{4}$$

provided that this limit exists, in which case we say that $f$ is **integrable** on $[a, b]$. The meaning of (4) is that, for each number $\varepsilon > 0$, there exists a number $\delta > 0$ such that

$$\left| I - \sum_{i=1}^{n} f(x_i^*) \, \Delta x_i \right| < \varepsilon$$

for every Riemann sum associated with any partition $P$ of $[a, b]$ with $|P| < \delta$.

---

The customary notation for the integral of $f$ from $a$ to $b$, due to the German mathematician and philosopher G. W. Leibniz (1646–1716), is

$$I = \int_a^b f(x) \, dx = \lim_{|P| \to 0} \sum_{i=1}^{n} f(x_i^*) \, \Delta x_i. \tag{5}$$

Considering $I$ as the area under $y = f(x)$ from $a$ to $b$, Leibniz first thought of a narrow strip with height $f(x)$ and "infinitesimally" small width $dx$ (see Fig. 5.13), so that its area would be the product $f(x)\, dx$. He regarded the integral as a sum of areas of such strips, denoting this sum by the elongated capital $S$ (for *sum*) that appears as the integral symbol in (5).

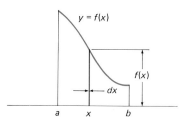

5.13 Origin of Leibniz's notation for the integral

We shall see that this integral notation is not only highly suggestive, but also exceedingly useful in manipulations with integrals. The numbers $a$ and $b$ are called the **lower limit** and **upper limit,** respectively, of the integral; they are merely the end points of the interval of integration. The function $f(x)$ that appears between the integral sign and $dx$ is sometimes called the **integrand.** The symbol $dx$ following $f(x)$ in (5) should be thought of as simply an indication of what the independent variable is. Like the index in a summation, $x$ is merely a dummy variable, and we may write

$$\int_a^b f(x)\, dx = \int_a^b f(t)\, dt = \int_a^b f(u)\, du.$$

The definition above applies only if $a < b$, but it is convenient to include the cases $a = b$ and $a > b$ as well. The integral is *defined* in these cases by

$$\int_a^a f(x)\, dx = 0 \tag{6}$$

and

$$\int_a^b f(x)\, dx = -\int_b^a f(x)\, dx \tag{7}$$

provided the right-hand integral exists. Thus *interchanging the limits of integration reverses the sign of the integral.*

Not every function is integrable. Suppose that $c$ is a point of $[a, b]$ such that $f(x) \to \infty$ as $x \to c$. If $[x_{k-1}, x_k]$ is the subinterval of the partition $P$ that contains $c$, then the Riemann sum in (3) can be made arbitrarily large by choosing $x_k^*$ sufficiently close to $c$. For our purposes, however, it is sufficient to know that every continuous function is integrable. The following theorem is proved in Appendix A.

---

**Theorem    Existence of the Integral**

If the function $f$ is continuous on $[a, b]$, then $f$ is integrable on $[a, b]$.

---

Although we omit the details, it is not difficult to show that the definition of the integral can be reformulated in terms of sequences of Riemann sums, as follows.

---

**Theorem    The Integral as a Limit of a Sequence**

The function $f$ is integrable on $[a, b]$ with integral $I$ if and only if

$$\lim_{n \to \infty} R_n = I \tag{8}$$

for every sequence $\{R_n\}_1^\infty$ of Riemann sums associated with a sequence of partitions $\{P_n\}_1^\infty$ of $[a, b]$ such that $|P_n| \to 0$ as $n \to \infty$.

---

This reformulation is advantageous because it is easier to visualize a specific sequence of Riemann sums than the vast totality of all possible Riemann sums. In the case of a continuous function $f$ that is known to be integrable (by the existence theorem above), the situation can be simplified even more by using only Riemann sums associated with partitions consisting of subintervals all having the same length,

$$\Delta x_1 = \Delta x_2 = \cdots = \Delta x_n = \frac{b - a}{n} = \Delta x.$$

Such a partition of $[a, b]$ into equal subintervals is called a **regular** partition of $[a, b]$.

Any Riemann sum associated with a regular partition can be written in the form

$$\sum_{i=1}^{n} f(x_i^*)\, \Delta x, \tag{9}$$

where the absence of a subscript in $\Delta x$ signifies that the sum is associated with a regular partition. In such a case, the conditions $|P| \to 0$, $\Delta x \to 0$, and $n \to \infty$ are equivalent, so the integral of a *continuous* function can be defined quite simply by

$$\int_a^b f(x)\, dx = \lim_{n \to \infty} \sum_{i=1}^{n} f(x_i^*)\, \Delta x = \lim_{\Delta x \to 0} \sum_{i=1}^{n} f(x_i^*)\, \Delta x. \tag{10}$$

Therefore, because we are concerned for the most part only with integrals of continuous functions, in our subsequent discussions we will employ only regular partitions.

For continuous functions, two special sums are of particular importance. As in Section 5-2, let $f(x_i^\flat)$ and $f(x_i^\#)$ denote the minimum and maximum value, respectively, of $f$ on the $i$th subinterval $[x_{i-1}, x_i]$ of a regular partition $P$ of $[a, b]$. If $S = \{x_1^*, x_2^*, \ldots, x_n^*\}$ is any selection for $P$, then it follows that

$$\sum_{i=1}^{n} f(x_i^\flat)\, \Delta x \le \sum_{i=1}^{n} f(x_i^*)\, \Delta x \le \sum_{i=1}^{n} f(x_i^\#)\, \Delta x. \tag{11}$$

The sums on the right and left in (11) are called the **upper Riemann sum** and **lower Riemann sum**, respectively, for $f$ associated with the partition $P$. The following example illustrates how integrals can sometimes be computed by "trapping" arbitrary Riemann sums between upper and lower Riemann sums.

**EXAMPLE 2**   Use Riemann sums to compute $\int_0^b x^2\, dx$, $b > 0$.

*Solution*   We start with a regular partition $P_n$ of $[0, b]$ into $n$ equal subintervals by means of the points

$$0 = x_0 < x_1 < x_2 < \cdots < x_{n-1} < x_n = b$$

where $x_i = bi/n$. If $x_i^* \in [x_{i-1}, x_i]$, then

$$(x_{i-1})^2 \le (x_i^*)^2 \le (x_i)^2,$$

so

$$\frac{b^2(i-1)^2}{n^2} \leq (x_i^*)^2 \leq \frac{b^2 i^2}{n^2}$$

for $i = 1, 2, \ldots, n$. We multiply each term by $\Delta x = b/n$, then sum from $i = 1$ to $i = n$. This yields

$$\frac{b^3}{n^3} \sum_{i=1}^{n} (i-1)^2 \leq \sum_{i=1}^{n} (x_i^*)^2 \, \Delta x \leq \frac{b^3}{n^3} \sum_{i=1}^{n} i^2.$$

But the formula in Equation (7) of Section 5-2 gives

$$\sum_{i=1}^{n} i^2 = \frac{n^3}{3} + \frac{n^2}{2} + \frac{n}{6}$$

and so

$$\sum_{i=1}^{n} (i-1)^2 = 1^2 + 2^2 + \cdots + (n-1)^2$$

$$= \left( \sum_{i=1}^{n} i^2 \right) - n^2$$

$$= \left( \frac{n^3}{3} + \frac{n^2}{2} + \frac{n}{6} \right) - n^2$$

$$= \frac{n^3}{3} - \frac{n^2}{2} + \frac{n}{6}.$$

It follows that

$$b^3 \left( \frac{1}{3} - \frac{1}{2n} + \frac{1}{6n^2} \right) \leq \sum_{i=1}^{n} (x_i^*)^2 \, \Delta x \leq b^3 \left( \frac{1}{3} + \frac{1}{2n} + \frac{1}{6n^2} \right).$$

We take limits as $n \to \infty$, and the squeeze law for limits of sequences gives

$$\int_0^b x^2 \, dx = \lim_{n \to \infty} \sum_{i=1}^{n} (x_i^*)^2 \, \Delta x = \frac{b^3}{3}. \tag{12}$$

For example, with $b = 2$ the result of Example 2 yields

$$\int_0^2 x^2 \, dx = \frac{8}{3} \approx 2.67.$$

Thus our third Riemann sum in Example 1, $R_{\text{mid}} = 2.66$, was quite close to the true value of the integral. Note also that (7) yields

$$\int_2^0 x^2 \, dx = -\int_0^2 x^2 \, dx = -\frac{8}{3}.$$

A final remark: In Example 2 we worked simultaneously with *two* sequences of Riemann sums. Only half this work is needed; as a consequence of the existence theorem and the discussion preceding Equation (10), it suffices—when a continuous function is involved—to work with any *single* sequence of Riemann sums. For instance, we could have derived the integral in

(12) simply by writing

$$\int_0^b x^2 \, dx = \lim_{n \to \infty} \sum_{i=1}^n \left(\frac{bi}{n}\right)^2 \left(\frac{b}{n}\right)$$

$$= \lim_{n \to \infty} \frac{b^3}{n^3} \sum_{i=1}^n i^2$$

$$= \lim_{n \to \infty} \frac{b^3}{n^3} \left(\frac{n^3}{3} + \frac{n^2}{2} + \frac{n}{6}\right)$$

$$= \lim_{n \to \infty} b^3 \left(\frac{1}{3} + \frac{1}{2n} + \frac{1}{6n^2}\right);$$

therefore

$$\int_0^b x^2 \, dx = \frac{b^3}{3}.$$

## 5-3  PROBLEMS

In each of Problems 1–8, compute the Riemann sum

$$\sum_{i=1}^n f(x_i^*) \, \Delta x$$

for the indicated function and a regular partition of the given interval into $n$ equal subintervals. Use $x_i^* = x_i$, the right-hand endpoint of the $i$th subinterval $[x_{i-1}, x_i]$.

1  $f(x) = x^2$  on  $[0, 1]$;  $n = 5$
2  $f(x) = x^3$  on  $[0, 1]$;  $n = 5$

3  $f(x) = \dfrac{1}{x}$  on  $[1, 6]$;  $n = 5$

4  $f(x) = \sqrt{x}$  on  $[0, 5]$;  $n = 5$
5  $f(x) = 2x + 1$  on  $[1, 4]$;  $n = 6$
6  $f(x) = x^2 + 2x$  on  $[1, 4]$;  $n = 6$
7  $f(x) = x^3 - 3x$  on  $[1, 4]$;  $n = 5$
8  $f(x) = 1 + 2\sqrt{x}$  on  $[2, 3]$;  $n = 5$

9–16  Repeat each of Problems 1–8, except with $x_i^* = x_{i-1}$, the left-hand endpoint.
17–24  Repeat each of Problems 1–8, except with $x_i^* = m_i = (x_{i-1} + x_i)/2$, the midpoint of the $i$th subinterval.
25  Work Problem 3 with $x_i^* = (3x_{i-1} + 2x_i)/5$.
26  Work Problem 4 with $x_i^* = (x_{i-1} + 2x_i)/3$.

Use the method of Example 2 to evaluate the integrals in Problems 27 and 28.

27  $\displaystyle\int_0^2 x \, dx$ 　　　　 28  $\displaystyle\int_0^4 x^3 \, dx$

In Problems 29–32, evaluate the given integral by computing

$$\lim_{n \to \infty} \sum_{i=1}^n f(x_i^*) \, \Delta x$$

for a regular partition of the interval of integration. Take $x_i^* = x_i$ in each case.

29  $\displaystyle\int_0^3 (2x + 1) \, dx$ 　　　 30  $\displaystyle\int_1^5 (4 - 3x) \, dx$

31  $\displaystyle\int_0^3 (3x^2 + 1) \, dx$ 　　　 32  $\displaystyle\int_0^4 (x^3 - x) \, dx$

33  Show by the method of Problems 29–32 that

$$\int_0^b x \, dx = \tfrac{1}{2} b^2$$

if $b > 0$.

34  Show by the method of Problems 29–32 that

$$\int_0^b x^3 \, dx = \tfrac{1}{4} b^4$$

if $b > 0$.

35  Let $f(x) = x$, and let $\{x_0, x_1, x_2, \ldots, x_n\}$ be an arbitrary partition of the closed interval $[a, b]$. For each $i$ $(1 \le i \le n)$, let $x_i^* = (x_{i-1} + x_i)/2$. Then show that

$$\sum_{i=1}^n x_i^* \, \Delta x_i = \tfrac{1}{2} b^2 - \tfrac{1}{2} a^2.$$

Explain why this computation proves that

$$\int_a^b x \, dx = \frac{b^2 - a^2}{2}.$$

36  Suppose that $f$ is a continuous function on $[a, b]$ and that $k$ is a constant. Use Riemann sums to prove that

$$\int_a^b kf(x) \, dx = k \int_a^b f(x) \, dx.$$

37  Suppose that $f(x) = c$, a constant. Use Riemann sums to prove that

$$\int_a^b c \, dx = c(b - a).$$

The program below can be used to compute Riemann sums of the form

$$\sum_{i=1}^{n} f(x_i)\, \Delta x \tag{13}$$

for the function $f(x) = x^2$, corresponding to a regular partition of $[a, b]$ into $n$ equal subintervals of length $\Delta x = (b - a)/n$.

```
10    INPUT  "A",  A
20    INPUT  "B",  B
30    INPUT  "N",  N
40    H = (B − A)/N
50    X = A + H
60    S = 0
70    FOR  I = 1  TO  N
80      F = X*X
90      S = S + F*H
100     X = X + H
110   NEXT  I
120   PRINT  "RIEMANN SUM:  ";  S
```

Lines 10–30 call for you to enter the end points of the interval and the desired number of subintervals. Line 40 computes the value of $\Delta x$, denoted by H. In line 50 the program computes the first right-hand end point, $x_1 = a + \Delta x = $ A + H. After the Ith pass through the FOR-NEXT loop in lines 70-110, S denotes the sum of the first I terms of the sum in (13). In each pass through this loop, another term is calculated and added to the running sum S. When all N terms have been calculated and added, the result is printed (or displayed) upon execution of line 120.

In order to compute Riemann sums for a different function $f(x)$, it is necessary only to alter line 80 accordingly. For example, we would use

```
80    F = 1/X
```

if the function in question were $f(x) = 1/x$.

In order to alter the choice of $x_i^*$, we need only change line 50:

```
50    X = A
```

chooses $x_i^* = x_{i-1}$, the left-hand end point of the $i$th subinterval, while

```
50    X = A + H/2
```

chooses $x_i^*$ as the midpoint.

**Exercise 1** Use the program above (and the indicated alterations) to check our numerical results in Example 1.

**Exercise 2** Use the program above (and appropriate alterations) to compute the three Riemann sums as in Example 1, except for the function $f(x) = 1/x$ on the interval $[1, 2]$ with $n = 10$ equal subintervals.

**Exercise 3** Write a program to compute and display the values of *both* the Riemann sums

$$L = \sum_{i=1}^{n} f(x_{i-1}) \, \Delta x \quad \text{and} \quad R = \sum_{i=1}^{n} f(x_i) \, \Delta x,$$

as well as their average $(L + R)/2$.

## 5-4

### Evaluation of Integrals

The evaluation of integrals using Riemann sums as in Section 5-3 is tedious and time-consuming. Fortunately, we will seldom find it necessary to evaluate an integral in this way. In 1666 Isaac Newton, while still a student at Cambridge University, discovered a much more efficient way of doing it. A few years later Gottfried Wilhelm Leibniz, working with a different approach, discovered this method independently.

Newton's key idea was that in order to evaluate the *number* $\int_a^b f(x) \, dx$, we should first introduce the *function* $A(x)$ defined as follows:

$$A(x) = \int_a^x f(t) \, dt. \tag{1}$$

Note that the independent variable $x$ appears in the *upper limit* of the integral in (1); the dummy variable $t$ is used in the integrand to avoid duplication of notation. If $f$ is positive-valued and $x > a$, then $A(x)$ is the area under the curve $y = f(x)$ over the interval $[a, x]$. (See Fig. 5.14.)

It is apparent from Fig. 5.14 that $A(x)$ increases as $x$ increases. When $x$ increases by $\Delta x$, $A$ increases by the area $\Delta A$ of the narrow strip in Fig. 5.14 with base $[x, x + \Delta x]$. If $\Delta x$ is very small, it appears that the area of this strip is very close to the area $f(x) \, \Delta x$ of the rectangle with base $[x, x + \Delta x]$ and height $f(x)$. Thus

$$\Delta A \approx f(x) \, \Delta x,$$

$$\frac{\Delta A}{\Delta x} \approx f(x). \tag{2}$$

Moreover, the figure makes it plausible that we get equality in the limit as $\Delta x \to 0$:

$$\frac{dA}{dx} = \lim_{\Delta x \to 0} \frac{\Delta A}{\Delta x} = f(x).$$

5.14 The area function $A(x)$

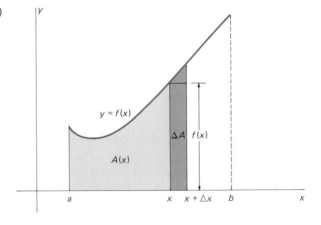

That is,
$$A'(x) = f(x), \tag{3}$$
so *the derivative of the area function $A(x)$ is the curve's height function $f(x)$.* The other way around, (3) implies that $A(x)$ is an *antiderivative* of $f(x)$.

Now suppose that $G(x)$ is any other antiderivative of $f(x)$—perhaps one found by the methods of Section 4-8. Then
$$A(x) = G(x) + C \tag{4}$$
because two antiderivatives of the same function can differ only by a constant. Because
$$A(a) = \int_a^a f(t)\, dt = 0 \tag{5}$$
and
$$A(b) = \int_a^b f(t)\, dt = \int_a^b f(x)\, dx, \tag{6}$$
it finally follows that
$$\int_a^b f(x)\, dx = A(b) - A(a)$$
$$= [G(b) + C] - [G(a) + C],$$
and thus
$$\int_a^b f(x)\, dx = G(b) - G(a).$$

Our intuitive discussion has led us to the statement of the following theorem.

---

**Theorem** *Evaluation of Integrals*

If $G$ is an antiderivative of the continuous function $f$ on the interval $[a, b]$, then
$$\int_a^b f(x)\, dx = G(b) - G(a). \tag{7}$$

---

In Section 5-5 we will fill in the details in the discussion above, thus giving a rigorous proof of this theorem (which is part of the fundamental theorem of calculus).

Here we concentrate on the computational applications of this theorem. The difference $G(b) - G(a)$ is customarily abbreviated to $[G(x)]_a^b$, so the theorem implies that
$$\int_a^b f(x)\, dx = \Big[ G(x) \Big]_a^b \tag{8}$$
if $G$ is any antiderivative of the continuous function $f$. Thus if we can find the antiderivative $G$, we can instantly evaluate the integral *without* recourse to the paraphernalia of limits of Riemann sums.

For instance, if $f(x) = x^n$ with $n \neq -1$, then an antiderivative of $f$ is $G(x) = x^{n+1}/(n + 1)$, so (8) yields
$$\int_a^b x^n\, dx = \left[ \frac{x^{n+1}}{n + 1} \right]_a^b = \frac{b^{n+1} - a^{n+1}}{n + 1}. \tag{9}$$

Contrast the immediacy of this result with the complexity of the computations in Example 2 of Section 5-3.

The theorem on evaluation of integrals says that if $G'(x) = f(x)$, then

$$\int_a^b f(x)\, dx = \Big[ G(x) \Big]_a^b = G(b) - G(a) \tag{10}$$

(under the assumption that $f$ is continuous on $[a, b]$). Thus the fact that $Dx^3 = 3x^2$ implies immediately, with $G(x) = x^3$ and $f(x) = 3x^2$ in (10), that

$$\int_1^2 3x^2\, dx = \Big[ x^3 \Big]_1^2 = 8 - 1 = 7.$$

**EXAMPLE 1**  Some immediate applications of the evaluation theorem are shown below.

$$\int_0^2 x^5\, dx = \left[ \frac{x^6}{6} \right]_0^2 = \frac{64}{6} - 0 = \frac{32}{3}.$$

$$\int_1^9 (2x - x^{-1/2} - 3)\, dx = \Big[ x^2 - 2\sqrt{x} - 3x \Big]_1^9 = 52.$$

$$\int_0^1 x^2(x^3 + 1)^{1/3}\, dx = \left[ \frac{1}{4}(x^3 + 1)^{4/3} \right]_0^1 = \frac{1}{4}(2^{4/3} - 1).$$

$$\int_1^2 \frac{1}{x^5}\, dx = \left[ -\frac{1}{4x^4} \right]_1^2 = -\frac{1}{64} + \frac{1}{4} = \frac{15}{64}.$$

We have not shown the details of finding the antiderivatives, but you can (and should) check each of these results by showing that the derivative of the function within the brackets on the right is equal to the integrand function in the integral on the left. In Example 2 we show the details.

**EXAMPLE 2**  Evaluate $\int_3^4 x\sqrt{25 - x^2}\, dx$.

*Solution*  If $u = 25 - x^2$, then $du/dx = -2x$, so

$$x\sqrt{25 - x^2} = -\frac{1}{2} u^{1/2} \frac{du}{dx}.$$

Hence the most general antiderivative of $x\sqrt{25 - x^2}$ is

$$-\frac{1}{2} \cdot \frac{u^{3/2}}{3/2} + C = -\frac{1}{3} u^{3/2} + C$$

$$= -\tfrac{1}{3}(25 - x^2)^{3/2} + C.$$

Therefore

$$\int_3^4 x\sqrt{25 - x^2}\, dx = \Big[ -\tfrac{1}{3}(25 - x^2)^{3/2} \Big]_3^4$$

$$= -\tfrac{1}{3}[(9)^{3/2} - (16)^{3/2}]$$

$$= -\tfrac{1}{3}(27 - 64) = \tfrac{37}{3}.$$

---

If the derivative $F'(x)$ of the function $F(x)$ is continuous, then the evaluation theorem, with $F'(x)$ and $F(x)$ in place of $f(x)$ and $G(x)$, respectively, yields

$$\int_a^b F'(x)\, dx = \Big[ F(x) \Big]_a^b = F(b) - F(a). \tag{11}$$

**EXAMPLE 3** Suppose that an animal population $P(t)$ initially numbers $P(0) = 100$, and its rate of growth after $t$ months is observed to be given by the formula

$$P'(t) = 10 + t + (0.06)t^2.$$

What is the population after 10 months?

*Solution* By the formula in (11), we know that

$$P(10) - P(0) = \int_0^{10} P'(t)\, dt$$

$$= \int_0^{10} (10 + t + 0.06t^2)\, dt$$

$$= \left[ 10t + \tfrac{1}{2}t^2 + 0.02t^3 \right]_0^{10} = 170.$$

Thus $P(10) = 100 + 170 = 270$ individuals.

**EXAMPLE 4** Evaluate

$$\lim_{n \to \infty} \sum_{i=1}^{n} \frac{2i}{n^2}$$

by recognizing this limit as the value of an integral.

*Solution* If we write

$$\sum_{i=1}^{n} \frac{2i}{n^2} = \sum_{i=1}^{n} \left( \frac{2i}{n} \right) \left( \frac{1}{n} \right),$$

we recognize that we have a Riemann sum for the function $f(x) = 2x$, associated with a partition of the interval $[0, 1]$ into $n$ equal subintervals. The $i$th point of subdivision is $x_i = i/n$, and $\Delta x = 1/n$. Hence it follows from the definition of the integral and from the evaluation theorem that

$$\lim_{n \to \infty} \sum_{i=1}^{n} \frac{2i}{n^2} = \lim_{n \to \infty} \sum_{i=1}^{n} 2x_i\, \Delta x$$

$$= \lim_{n \to \infty} \sum_{i=1}^{n} f(x_i)\, \Delta x$$

$$= \int_0^1 f(x)\, dx = \int_0^1 2x\, dx;$$

therefore

$$\lim_{n \to \infty} \sum_{i=1}^{n} \frac{2i}{n^2} = \left[ x^2 \right]_0^1 = 1.$$

**BASIC PROPERTIES OF INTEGRALS**

Elementary proofs of the properties stated below are outlined in Problems 38–40 at the end of this section. We assume that $f$ is integrable on $[a, b]$.

> **Integral of a Constant**
>
> $$\int_a^b c \, dx = c(b - a).$$

This is intuitively obvious because the area represented by the **integral is** simply a rectangle with base $b - a$ and height $c$ (see Fig. 5.15).

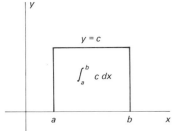

**5.15** The integral of a constant is the area of a rectangle.

> **Constant Multiple Property**
>
> $$\int_a^b cf(x) \, dx = c \int_a^b f(x) \, dx.$$

Thus a constant can be "moved out from under the integral sign." For instance,

$$\int_0^1 7x^2 \, dx = 7 \int_0^1 x^2 \, dx = 7 \left[ \tfrac{1}{3}x^3 \right]_0^1 = (7)(\tfrac{1}{3}) = \tfrac{7}{3}.$$

> **Interval Union Property**
>
> If $a < c < b$, then
>
> $$\int_a^b f(x) \, dx = \int_a^c f(x) \, dx + \int_c^b f(x) \, dx.$$

Figure 5.16 indicates the plausibility of this property, which allows manipulations such as this:

$$\int_0^3 \sqrt{x} \, dx = \int_0^1 \sqrt{x} \, dx + \int_1^3 \sqrt{x} \, dx.$$

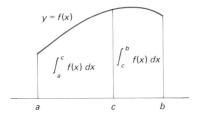

**5.16** The way the interval union property works

> **Comparison Property**
>
> If $m \leq f(x) \leq M$ for all $x$ in $[a, b]$, then
>
> $$m(b - a) \leq \int_a^b f(x) \, dx \leq M(b - a).$$

**5.17** Plausibility of the comparison property

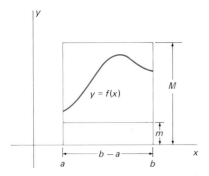

The plausibility of this property is indicated in Fig. 5.17. Note that $m$ and $M$ are not necessarily the minimum and maximum values of $f(x)$ on $[a, b]$. For instance, suppose that $2 \leq x \leq 3$. Then

$$5 \leq x^2 + 1 \leq 10,$$

and so

$$m = \frac{1}{10} \leq \frac{1}{x^2 + 1} \leq \frac{1}{5} = M.$$

Hence the comparison property yields

$$\frac{1}{10} \leq \int_2^3 \frac{1}{x^2 + 1} \, dx \leq \frac{1}{5}.$$

The properties of integrals stated above are frequently used in computations and will be applied in the proof of the fundamental theorem of calculus in Section 5-5.

Apply the evaluation theorem in this section to evaluate the integrals in Problems 1–30.

1　$\int_{-1}^{1} x(x^2 + 1)^4 \, dx$

2　$\int_{1}^{3} \frac{6}{x^2} \, dx$

3　$\int_{-1}^{2} x^2(x^3 + 1)^{1/2} \, dx$

4　$\int_{-2}^{-1} \frac{1}{x^4} \, dx$

5　$\int_{0}^{1} (x^4 - x^3) \, dx$

6　$\int_{1}^{2} (x^4 - x^3) \, dx$

7　$\int_{-1}^{0} (x + 1)^3 \, dx$

8　$\int_{1}^{3} \frac{x^4 + 1}{x^2} \, dx$

9　$\int_{0}^{4} x^{1/2} \, dx$

10　$\int_{1}^{4} x^{-1/2} \, dx$

11　$\int_{-1}^{2} (3x^2 + 2x + 4) \, dx$

12　$\int_{0}^{1} x^{99} \, dx$

13　$\int_{-1}^{1} x^{99} \, dx$

14　$\int_{0}^{4} (7x^{5/2} - 5x^{3/2}) \, dx$

15　$\int_{1}^{2} 2x(x^2 + 1)^3 \, dx$

16　$\int_{1}^{2} (x^2 + 1)^3 \, dx$

17　$\int_{-1}^{0} (2x + 1)^3 \, dx$

18　$\int_{0}^{1} \frac{3x^2}{(x^3 + 1)^3} \, dx$

19　$\int_{1}^{8} x^{2/3} \, dx$

20　$\int_{1}^{9} (1 + \sqrt{x})^2 \, dx$

21　$\int_{0}^{1} (x^2 - 3x + 4) \, dx$

22　$\int_{1}^{4} \sqrt{3t} \, dt$

23　$\int_{1}^{9} \left( \sqrt{x} - \frac{2}{\sqrt{x}} \right) dx$

24　$\int_{2}^{3} \frac{du}{u^2}$　$\left( \text{Note the abbreviation for } \int_{2}^{3} \frac{1}{u^2} \, du. \right)$

25　$\int_{1}^{4} \frac{x^2 - 1}{\sqrt{x}} \, dx$

26　$\int_{1}^{4} (t^2 - 2)\sqrt{t} \, dt$

27　$\int_{0}^{3} x\sqrt{x^2 + 4} \, dx$

28　$\int_{4}^{9} x^{5/2} \, dx$

29　$\int_{-1}^{1} 51x^2 \, dx$

30　$\int_{0}^{\pi} 64x^3 \, dx$

In each of Problems 31–36, evaluate the given limit by first recognizing the indicated sum as a Riemann sum associated with a regular partition of $[0, 1]$ and then evaluating the corresponding integral.

31　$\lim_{n \to \infty} \sum_{i=1}^{n} \left( \frac{2i}{n} - 1 \right) \frac{1}{n}$

32　$\lim_{n \to \infty} \sum_{i=1}^{n} \frac{i^2}{n^3}$

33　$\lim_{n \to \infty} \frac{1 + 2 + 3 + \cdots + n}{n^2}$

34　$\lim_{n \to \infty} \frac{1^3 + 2^3 + 3^3 + \cdots + n^3}{n^4}$

35　$\lim_{n \to \infty} \frac{\sqrt{1} + \sqrt{2} + \sqrt{3} + \cdots + \sqrt{n}}{n\sqrt{n}}$

36　$\lim_{n \to \infty} \sum_{i=1}^{n} \frac{1}{n} \sqrt{1 + \frac{i}{n}}$

37 Evaluate the integral $\int_{0}^{5} \sqrt{25 - x^2} \, dx$ by interpreting it as the area under the graph of an appropriate function.

38 Use sequences of Riemann sums to establish the constant multiple property.

39 Use sequences of Riemann sums to establish the interval union property of the integral. Note that if $R'_n$ and $R''_n$ are Riemann sums for $f$ on the intervals $[a, c]$ and $[c, b]$ respectively, then $R_n = R'_n + R''_n$ is a Riemann sum for $f$ on $[a, b]$.

40 Use Riemann sums to establish the comparison property for integrals. Show first that if $m \leq f(x) \leq M$ for all $x$ in $[a, b]$, then

$$m(b - a) \leq R \leq M(b - a)$$

for every Riemann sum $R$ for $f$ on $[a, b]$.

41 Suppose that a tank initially contains 1000 gal. of water and that the rate of change of its volume after draining $t$ minutes is $V'(t) = (0.8)t - 40$ gal./min. How much water does the tank contain after it has been draining a half hour?

42 Suppose that the population of a city in 1960 was 125 (thousand) and that its rate of growth $t$ years later was $P'(t) = 8 + (0.5)t + (0.03)t^2$ (thousand) per year. What was its population in 1980?

43 Find a lower and an upper bound for

$$\int_{1}^{2} \frac{1}{x} \, dx$$

using the comparison property and a subdivision of $[1, 2]$ into five equal subintervals.

44 Find a lower and an upper bound for

$$\int_{0}^{1} \frac{1}{x^2 + 1} \, dx$$

using the comparison property and a subdivision of $[0, 1]$ into four equal subintervals.

## The Fundamental Theorem of Calculus

Newton and Leibniz are generally credited with the invention of calculus in the latter part of the seventeenth century. Actually, others had earlier calculated areas essentially equivalent to integrals, and tangent line slopes essentially equivalent to derivatives. The great accomplishment of Newton and Leibniz was the discovery and computational exploitation of the inverse relationship between differentiation and integration. This relationship is embodied in the **fundamental theorem of calculus.** One part of this theorem is the evaluation theorem in Section 5-4: In order to evaluate $\int_a^b f(x)\, dx$, it suffices to find an antiderivative of $f$ on $[a, b]$. The other part of the fundamental theorem tells us that this is usually possible, at least in theory: Every continuous function has an antiderivative.

### THE AVERAGE VALUE OF A FUNCTION

As a preliminary to the proof of the fundamental theorem of calculus, and also as a matter of independent interest, we investigate the notion of *average value*. For example, let the temperature $T$ during a particular 24-hour day be described by $T = f(t)$, $t \in [0, 24]$. We might define the average temperature $\bar{T}$ for the day as the (ordinary arithmetical) average of the hourly temperatures:

$$\bar{T} = \frac{1}{24} \sum_{i=1}^{24} f(i) = \frac{1}{24} \sum_{i=1}^{24} f(t_i)$$

where $t_i = i$. If we subdivided the day into $n$ equal subintervals $[t_{i-1}, t_i]$ rather than into twenty-four 1-hour intervals, we would obtain the more general average

$$\bar{T} = \frac{1}{n} \sum_{i=1}^{n} f(t_i).$$

The larger $n$ is, the closer would we expect $\bar{T}$ to be to the "true" average temperature for the entire day. It is therefore plausible to define the true average temperature by letting $n$ increase without bound. This gives

$$\bar{T} = \lim_{n \to \infty} \frac{1}{n} \sum_{i=1}^{n} f(t_i).$$

The right-hand side resembles a Riemann sum, and we can make it into a Riemann sum by introducing the factor $\Delta t = (b - a)/n$ where $a = 0$ and $b = 24$. Then

$$\bar{T} = \lim_{n \to \infty} \frac{1}{b - a} \sum_{i=1}^{n} f(t_i) \frac{b - a}{n}$$

$$= \frac{1}{b - a} \lim_{n \to \infty} \sum_{i=1}^{n} f(t_i)\, \Delta t$$

$$= \frac{1}{b - a} \int_a^b f(t)\, dt = \frac{1}{24} \int_0^{24} f(t)\, dt$$

under the assumption that $f$ is continuous, so that the Riemann sums converge to the integral as $n \to \infty$. This example motivates the following definition.

---

**Definition** *Average Value of a Function*

Suppose that the function $f$ is integrable on $[a, b]$. Then the **average value** $\bar{y}$ of $y = f(x)$ on $[a, b]$ is

$$\bar{y} = \frac{1}{b - a} \int_a^b f(x)\, dx. \qquad (1)$$

---

For example, the average value of $f(x) = x^2$ on $[0, 2]$ is

$$\bar{y} = \frac{1}{2} \int_0^2 x^2\, dx = \frac{1}{2}\left[\frac{x^3}{3}\right]_0^2 = \frac{4}{3}.$$

**EXAMPLE 1**   A thin rod 10 m long occupies the interval $0 \leq x \leq 10$ on the $x$-axis. Its temperature $T = f(x)$ in degrees Celsius at the point $x$ is given by

$$f(x) = 10 + \tfrac{3}{5}x^2,$$

so that $T = 10$ at $x = 0$ and $T = 70$ at $x = 10$. Find the average temperature in this rod.

**Solution**   With $a = 0$ and $b = 10$, Equation (1) gives

$$T = \frac{1}{10 - 0} \int_0^{10} \left(10 + \frac{3}{5}x^2\right) dx$$

$$= \frac{1}{10}\left[10x + \frac{1}{5}x^3\right]_0^{10} = 30 \quad \text{(degrees Celsius)}.$$

---

The following theorem tells us that every continuous function on a closed interval *attains* its average value at some point of the interval.

---

*Average Value Theorem*

If $f$ is continuous on $[a, b]$, then

$$f(\bar{x}) = \frac{1}{b - a} \int_a^b f(x)\, dx \qquad (2)$$

for some point $\bar{x}$ of $[a, b]$.

---

**Proof**   Let $m = f(c)$ be the minimum value of $f$ on $[a, b]$, and let $M = f(d)$ be its maximum value there. Then, by the comparison property of Section 5-4,

$$m = f(c) \leq \bar{y} = \frac{1}{b - a} \int_a^b f(x)\, dx \leq f(d) = M.$$

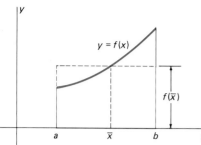
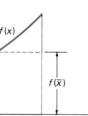

**5.18** Rectangle illustrating the average value theorem

Since $f$ is continuous, we can now apply the intermediate value property. Because the number $\bar{y}$ is between the two values $m$ and $M$ of $f$, $\bar{y}$ itself must be a value of $f$; specifically, $\bar{y} = f(\bar{x})$ for some number $\bar{x}$ between $a$ and $b$. This gives (2). ■

Note that (2) can be rewritten as

$$\int_a^b f(x)\, dx = f(\bar{x})(b - a).$$

If $f$ is positive-valued on $[a, b]$, this equation implies that the area under $y = f(x)$ over $[a, b]$ is equal to the area of a rectangle with base $b - a$ and height $f(\bar{x})$. (See Fig. 5.18.)

### THE FUNDAMENTAL THEOREM

We state the fundamental theorem of calculus in two parts. The first part is the fact that every function $f$ continuous on an interval $I$ has an antiderivative on $I$. In particular, an antiderivative of $f$ can be obtained by integrating $f$ in a certain way. Intuitively, in the case $f(x) > 0$, we let $F(x)$ denote the area under the graph of $f$ from a fixed point $a$ of $I$ to $x$, a point of $I$ with $x > a$. We shall prove that $F'(x) = f(x)$. We show the function $F(x)$ geometrically in Fig. 5.19. More precisely, we define the function $F$ by

$$F(x) = \int_a^x f(t)\, dt,$$

where we use the dummy variable $t$ in the integrand to avoid confusion with the upper limit $x$. The proof that $F'(x) = f(x)$ will be independent of the fact that $x > a$.

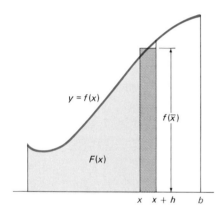

**5.19** The area function $F$ is an antiderivative of $f$.

---

**The Fundamental Theorem of Calculus**

Let $f$ be a continuous function defined on $[a, b]$.
*Part 1:* If the function $F$ is defined on $[a, b]$ by

$$F(x) = \int_a^x f(t)\, dt$$

then $F$ is an antiderivative of $f$. That is, $F'(x) = f(x)$ for $x$ in $(a, b)$.
*Part 2:* If $G$ is any antiderivative of $f$ on $[a, b]$, then

$$\int_a^b f(x)\, dx = \Big[ G(x) \Big]_a^b = G(b) - G(a).$$

---

***Proof of Part 1*** By the definition of the derivative,

$$F'(x) = \lim_{h \to 0} \frac{F(x + h) - F(x)}{h}$$

$$= \lim_{h \to 0} \frac{1}{h} \left[ \int_a^{x+h} f(t)\, dt - \int_a^x f(t)\, dt \right].$$

But

$$\int_a^{x+h} f(t)\, dt = \int_a^x f(t)\, dt + \int_x^{x+h} f(t)\, dt$$

by the interval union property (Section 5-4). Thus

$$F'(x) = \lim_{h \to 0} \frac{1}{h} \int_x^{x+h} f(t)\, dt.$$

The average value theorem above tells us that

$$\frac{1}{h} \int_x^{x+h} f(t)\, dt = f(\bar{t})$$

for some number $\bar{t}$ in $[x, x + h]$. Finally, we note that $\bar{t} \to x$ as $h \to 0$. Thus, since $f$ is continuous, we see that

$$F'(x) = \lim_{h \to 0} \frac{1}{h} \int_x^{x+h} f(t)\, dt$$

$$= \lim_{h \to 0} f(\bar{t}) = \lim_{\bar{t} \to x} f(\bar{t}) = f(x).$$

Hence the function $F$ defined by (3) is, indeed, an antiderivative of $f$. ∎

**Proof of Part 2** Here we apply Part 1 to give a proof of the evaluation theorem in Section 5-4. If $G$ is *any* antiderivative of $f$, then—because it and the function $F$ of Part 1 are both antiderivatives of $f$ on the interval $[a, b]$—we know that

$$G(x) = F(x) + C$$

on $[a, b]$ for some constant $C$. To evaluate $C$, we substitute $x = a$, and obtain

$$C = G(a) - F(a) = G(a)$$

because

$$F(a) = \int_a^a f(t)\, dt = 0.$$

Hence $G(x) = F(x) + G(a)$, or

$$F(x) = G(x) - G(a)$$

for all $x$ in $[a, b]$. With $x = b$ this gives

$$G(b) - G(a) = F(b) = \int_a^b f(x)\, dx,$$

which establishes Equation (4). ∎

Sometimes the fundamental theorem of calculus is interpreted as a statement to the effect that differentiation and integration are *inverse processes*. Part 1 may be written in the form

$$\frac{d}{dx} \left( \int_a^x f(t)\, dt \right) = f(x)$$

if $f$ is continuous. That is, if we first integrate the function $f$ (with *variable* upper limit of integration $x$) and then differentiate with respect to $x$, the result is the function $f$ again. So differentiation "cancels" the effect of integration.

Moreover, Part 2 of the fundamental theorem may be written in the form

$$\int_a^x G'(t)\, dt = G(x) - G(a),$$

if we assume that $G'$ is continuous. If so, this equation means that if we first differentiate the function $G$ and then integrate the result from $a$ to $x$, the result can differ from the original function $G(x)$ by, at worst, the constant $G(a)$. If $a$ is chosen so that $G(a) = 0$, this means that integration "cancels" the effect of differentiation.

For our first application of the fundamental theorem of calculus, we use it to establish the **linearity property** of the integral:

$$\int_a^b [Af(x) + Bg(x)]\, dx = A \int_a^b f(x)\, dx + B \int_a^b g(x)\, dx \tag{5}$$

if $A$ and $B$ are constants and the functions $f$ and $g$ are continuous on $[a, b]$. For example,

$$\int_0^1 (3x^2 - 2x\sqrt{x^2 + 1})\, dx = 3 \int_0^1 x^2\, dx - 2 \int_0^1 x\sqrt{x^2 + 1}\, dx.$$

Thus linearity enables us to "divide and conquer," to split a relatively complicated integral into simpler ones.

The linearity of integration follows—via the fundamental theorem of calculus—from the linearity of differentiation. The linearity formula (5) is a unified statement of the following two properties of integration.

$$\int_a^b Cf(x)\, dx = C \int_a^b f(x)\, dx, \tag{6}$$

$$\int_a^b [f(x) + g(x)]\, dx = \int_a^b f(x)\, dx + \int_a^b g(x)\, dx. \tag{7}$$

Formula (6) is merely the constant multiple property of Section 5-4. To prove (7), let $F(x)$ be an antiderivative of $f(x)$ and let $G(x)$ be an antiderivative of $g(x)$. Then

$$D[F(x) + G(x)] = f(x) + g(x),$$

because differentiation is linear. This means that $F + G$ is an antiderivative of $f + g$, which implies that

$$\int_a^b [f(x) + g(x)]\, dx = \Big[F(x) + G(x)\Big]_a^b$$
$$= [F(b) + G(b)] - [F(a) + G(a)]$$
$$= [F(b) - F(a)] + [G(b) - G(a)]$$
$$= \Big[F(x)\Big]_a^b + \Big[G(x)\Big]_a^b$$
$$= \int_a^b f(x)\, dx + \int_a^b g(x)\, dx.$$

Note that we have invoked the fundamental theorem of calculus twice in this derivation.

One consequence of linearity is the fact that integration preserves inequalities between functions. That is, if $f$ and $g$ are continuous functions with $f(x) \leq g(x)$ for all $x$ in $[a, b]$, then

$$\int_a^b f(x)\, dx \leq \int_a^b g(x)\, dx. \tag{8}$$

To prove this, let $H(x) = g(x) - f(x)$. Then $H(x) \geq 0$ on $[a, b]$, so it follows from the comparison property of Section 5-4 that

$$\int_a^b H(x)\, dx \geq 0.$$

So

$$\int_a^b g(x)\, dx - \int_a^b f(x)\, dx = \int_a^b H(x)\, dx \geq 0,$$

which is equivalent to (8).

The verification of the following consequence of (8) is left for the problems:

$$\left| \int_a^b f(x)\, dx \right| \leq \int_a^b |f(x)|\, dx. \tag{9}$$

Examples 1 and 2 of Section 5-4 illustrate the use of Part 2 of the fundamental theorem in the evaluation of integrals. Additional examples appear in the problems and in the following section. The example below illustrates the necessity of splitting an integral into a sum of integrals when its integrand has different antiderivative formulas on different intervals.

**EXAMPLE 2**   Evaluate $\int_{-1}^2 |x^3 - x|\, dx$.

**Solution**   We note that $x^3 - x \geq 0$ on $[-1, 0]$, that $x^3 - x \leq 0$ on $[0, 1]$, and that $x^3 - x \geq 0$ on $[1, 2]$. So we write

$$\int_{-1}^2 |x^3 - x|\, dx = \int_{-1}^0 (x^3 - x)\, dx + \int_0^1 (x - x^3)\, dx + \int_1^2 (x^3 - x)\, dx$$

$$= \left[ \tfrac{1}{4}x^4 - \tfrac{1}{2}x^2 \right]_{-1}^0 + \left[ \tfrac{1}{2}x^2 - \tfrac{1}{4}x^4 \right]_0^1 + \left[ \tfrac{1}{4}x^4 - \tfrac{1}{2}x^2 \right]_1^2$$

$$= \tfrac{1}{4} + \tfrac{1}{4} + [2 - (-\tfrac{1}{4})] = \tfrac{11}{4} = 2.75.$$

The first part of the fundamental theorem of calculus says that the derivative of an integral with respect to its upper limit is equal to the value of the integrand *at* the upper limit. For instance, if

$$y = \int_0^x t^3 \sqrt{1 + t}\, dt$$

then

$$\frac{dy}{dx} = x^3 \sqrt{1 + x}.$$

The following example is a bit more complicated in that the upper limit of the integral is a function of the independent variable.

**EXAMPLE 3**   Find $h'(x)$ given

$$h(x) = \int_0^{x^2} t^3 \sqrt{1 + t}\, dt.$$

**Solution**   Let $y = h(x)$ and $u = x^2$. Then

$$y = \int_0^u t^3 \sqrt{1 + t}\, dt,$$

so

$$\frac{dy}{du} = u^3 \sqrt{1 + u}$$

by the fundamental theorem of calculus. Then the chain rule yields

$$h'(x) = \frac{dy}{dx} = \frac{dy}{du}\frac{du}{dx}$$

$$= (u^3 \sqrt{1 + u})(2x) = 2x^7 \sqrt{1 + x^2}.$$

## 5-5   PROBLEMS

In Problems 1–11, find the average value of the given function on the specified interval.

**1** $f(x) = x^3;$   $[0, 2]$   **2** $g(x) = \sqrt{x};$   $[1, 4]$
**3** $h(x) = 3x^2\sqrt{x^3 + 1};$   $[0, 2]$
**4** $f(x) = 8x;$   $[0, 4]$   **5** $g(x) = 8x;$   $[-4, 4]$
**6** $h(x) = x^2;$   $[-4, 4]$   **7** $f(x) = x^3;$   $[0, 5]$
**8** $g(x) = 1/\sqrt{x};$   $[1, 4]$
**9** $f(x) = \sqrt{x + 1};$   $[0, 3]$

**10** $g(x) = \dfrac{1}{x^2};$   $[1, 2]$   **11** $f(x) = |x|;$   $[-1, 1]$

Evaluate the integrals in Problems 12–30.

**12** $\displaystyle\int_{-1}^{2} (4 - 3x + 2x^2)\, dx$

**13** $\displaystyle\int_{-1}^{3} dx$   (Here $dx$ stands for $1\, dx$.)

**14** $\displaystyle\int_{1}^{2} (y^5 - 1)\, dy$   **15** $\displaystyle\int_{1}^{4} \frac{dx}{\sqrt{9x^3}}$

**16** $\displaystyle\int_{-1}^{1} (x^3 + 2)^2\, dx$   **17** $\displaystyle\int_{1}^{3} \frac{3t - 5}{t^4}\, dt$

**18** $\displaystyle\int_{-2}^{-1} \frac{x^2 - x + 3}{\sqrt[3]{x}}\, dx$   **19** $\displaystyle\int_{0}^{1} x\sqrt{1 - x^2}\, dx$

**20** $\displaystyle\int_{-1}^{2} |x|\, dx$   **21** $\displaystyle\int_{1}^{2} \left(t - \frac{1}{2t}\right)^2 dt$

**22** $\displaystyle\int_{-1}^{1} \frac{x^2 - 4}{x + 2}\, dx$   **23** $\displaystyle\int_{-2}^{-1} x^{-3}\, dx$

**24** $\displaystyle\int_{0}^{2} |x - \sqrt{x}|\, dx$   **25** $\displaystyle\int_{-2}^{2} |x^2 - 1|\, dx$

**26** $\displaystyle\int_{1}^{4} (x - 1)\sqrt{x}\, dx$   **27** $\displaystyle\int_{2}^{7} \sqrt{x + 2}\, dx$

**28** $\displaystyle\int_{5}^{10} \frac{dx}{\sqrt{x - 1}}$   **29** $\displaystyle\int_{0}^{3} x\sqrt{x^2 + 16}\, dx$

**30** $\displaystyle\int_{0}^{4} \frac{x\, dx}{\sqrt{9 + x^2}}$

In each of Problems 31–35, apply the first part of the fundamental theorem of calculus to find the derivative of the given function.

**31** $f(x) = \displaystyle\int_{-1}^{x} (t^2 + 1)^{17}\, dt$

**32** $g(t) = \displaystyle\int_{0}^{t} (x^2 + 25)^{1/2}\, dx$

**33** $h(z) = \displaystyle\int_{2}^{z} (u - 1)^{1/3}\, du$   **34** $A(x) = \displaystyle\int_{1}^{x} \frac{dt}{t}$

**35** $f(x) = \displaystyle\int_{x}^{10} \left(t + \frac{1}{t}\right) dt$   $\left(\text{Suggestion: Use the fact that}\right.$

$\displaystyle\int_{a}^{b} g(t)\, dt = -\int_{b}^{a} g(t)\, dt.\Bigg)$

In Problems 36–39, $G(x)$ is the integral of the given function $f(t)$ over the specified interval of the form $[a, x]$, $x > a$. Apply the first part of the fundamental theorem of calculus to find $G'(x)$.

**36** $f(t) = \dfrac{t}{t^2 + 1};$   $[2, x]$

**37** $f(t) = \sqrt{t + 4};$   $[0, x]$
**38** $f(t) = t(t + 1)^{99};$   $[0, x]$
**39** $f(t) = \sqrt{t^3 + 1};$   $[1, x]$

In Problems 40–46, differentiate the function $f$ by first writing $f(x)$ in the form $g(u)$, where $u$ is the upper limit of the integral.

**40** $f(x) = \displaystyle\int_{0}^{x^2} \sqrt{1 + t^3}\, dt$

**41** $f(x) = \displaystyle\int_{1}^{1/x} \frac{dt}{t}$   **42** $f(x) = \displaystyle\int_{0}^{\sqrt{x}} \sqrt{1 - t^2}\, dt$

**43** $f(x) = \displaystyle\int_{1}^{\sqrt{x}} \frac{dt}{1 + t^2}$   **44** $f(x) = \displaystyle\int_{1}^{x^{10}} (t^2 + 1)^3\, dt$

**45** $f(x) = \displaystyle\int_{1}^{x^2 + 1} \frac{dt}{t}$   **46** $f(x) = \displaystyle\int_{1}^{x^5} \sqrt{1 + t^2}\, dt$

**47** The fundamental theorem of calculus *seems* to say that

$$\int_{-1}^{1} \frac{dx}{x^2} = \left[-\frac{1}{x}\right]_{-1}^{1} = -2,$$

in apparent contradiction to the fact that $1/x^2$ is always positive. What's wrong here?

**48** Note that $\pm f(x) \le |f(x)|$, and thereby deduce Inequality (9) from Inequality (8).

**49** Apply Part 2 of the fundamental theorem of calculus to establish the interval union property (Section 5-4) of the integral.

**50** Show that the average rate of change

$$\frac{f(b) - f(a)}{b - a}$$

of the function $f$ on $[a, b]$ is equal to the average value of its derivative on $[a, b]$.

**51** If a ball is dropped from a height of 400 ft, find its average height and its average velocity between the time it is dropped and the time it strikes the ground.

**52** Find the average value on $[0, 10]$ of the animal population $P(t) = 100 + 10t + \frac{1}{2}t^2 + (0.02)t^3$ of Example 3 in Section 5-4.

**53** Suppose that a 5000-gal. water tank takes 10 min to drain and that after $t$ minutes, the amount of water remaining in the tank is $V(t) = 50(10 - t)^2$ gallons. What is the average amount of water in the tank while it is draining?

**54** A thin rod occupies the interval $a \le x \le b$ on the $x$-axis, and its temperature $T$ at the point $x$ is a linear function of $x$:

$$T(x) = A + Bx$$

where $A$ and $B$ are constants. Show that its average temperature $\bar{T}$ is equal to the temperature at its midpoint $\bar{x} = \frac{1}{2}(a + b)$.

**55** A straight line segment of length $a > 0$ is divided into two pieces by choosing a point at random on the segment. What is the average area of the rectangle with adjacent sides the two pieces?

---

**5-6**

**Integration by Substitution**

We can write the computational part of the fundamental theorem of calculus in the form

$$\int_a^b f(x)\, dx = \left[ D^{-1}f(x) \right]_a^b. \tag{1}$$

Because of this relationship between integration and antidifferentiation, it is customary to write

$$\int f(x)\, dx = D^{-1}f(x) = F(x) + C \tag{2}$$

if $F'(x) = f(x)$. The expression $\int f(x)\, dx$, with no limits on the integral symbol, is called the **indefinite integral** of the function $f$, in contrast with the **definite integral,** which has upper and lower limits. Thus the indefinite integral of $f$ is simply the most general antiderivative of $f$, and indefinite integration is simply antidifferentiation. For example,

$$\int (3x^2 - 4)\, dx = x^3 - 4x + C, \qquad \text{and}$$

$$\int (1 + t)^{10}\, dt = \tfrac{1}{11}(1 + t)^{11} + C.$$

Note that the indefinite integral is a *function* (actually, a collection of functions), while the definite integral is a *number*. The relationship between definite and indefinite integration is obtained by rewriting (1) as

$$\int_a^b f(x)\, dx = \left[ \int f(x)\, dx \right]_a^b. \tag{3}$$

In the notation of indefinite integrals, the antidifferentiation formulas in (6)–(8) of Section 4-8 take the forms

$$\int af(x)\, dx = a \int f(x)\, dx \qquad (a \text{ is a constant}), \tag{4}$$

$$\int [f(x) + g(x)]\, dx = \int f(x)\, dx + \int g(x)\, dx, \tag{5}$$

$$\int x^r\, dx = \frac{x^{r+1}}{r+1} + C \qquad (\text{if } r \ne -1). \tag{6}$$

A common sort of indefinite integral takes the form

$$\int f(g(x))g'(x)\,dx.$$

If we write $u = g(x)$, then the differential of $u$ is $du = g'(x)\,dx$, so a purely mechanical substitution gives the *tentative* formula

$$\int f(g(x))g'(x)\,dx = \int f(u)\,du. \tag{7}$$

One of the beauties of differential notation is that Formula (7) is not only plausible, but in fact true—with the understanding that $u$ is to be replaced by $g(x)$ after the indefinite integration on the right-hand side has been performed. Indeed, (7) is merely an indefinite integral version of the chain rule. For if $F'(x) = f(x)$, then

$$D_x F(g(x)) = F'(g(x))g'(x) = f(g(x))g'(x)$$

by the chain rule, so that

$$\int f(g(x))g'(x)\,dx = \int F'(g(x))g'(x)\,dx$$

$$= \int D[F(g(x))]\,dx = F(g(x)) + C$$

$$= F(u) + C \qquad (u = g(x))$$

$$= \int f(u)\,du.$$

Formula (7) is the basis for the powerful technique of indefinite **integration by substitution.** It may be used whenever the integrand function is recognizable in the form $f(g(x))g'(x)$.

**EXAMPLE 1** Find $\int x^2 \sqrt{x^3 + 9}\,dx$.

***Solution*** Note that $x^2$ is, to within a constant factor, the derivative of $x^3 + 9$. We can therefore substitute

$$u = x^3 + 9, \qquad du = 3x^2\,dx.$$

The constant factor 3 can be supplied if we compensate by multiplying the integral by $\frac{1}{3}$. This gives

$$\int x^2 \sqrt{x^3 + 9}\,dx = \frac{1}{3} \int (x^3 + 9)^{1/2} 3x^2\,dx$$

$$= \frac{1}{3} \int u^{1/2}\,du = \frac{1}{3} \frac{u^{3/2}}{\frac{3}{2}} + C$$

$$= \frac{2}{9} u^{3/2} + C = \frac{2}{9}(x^3 + 9)^{3/2} + C.$$

Another valid way to carry out the same substitution is to solve $du = 3x^2\,dx$ for $x^2\,dx = \frac{1}{3}\,du$, and then write

$$\int (x^3 + 9)^{1/2} x^2\,dx = \int u^{1/2} \cdot \tfrac{1}{3}\,du = \tfrac{1}{3} \int u^{1/2}\,du.$$

Three items worth noting appear upon examination of the solution to Example 1:

- The differential $dx$ along with the rest of the integrand is "transformed," or replaced, in terms of $u$ and $du$.
- Once the actual integration has been performed, the constant $C$ of integration is added.
- A final resubstitution is necessary to write the answer in terms of the original variable $x$.

The method of integration by substitution can also be used with definite integrals. Only one additional step is required—evaluation of the final antiderivative at the original limits of integration. For example, the substitution of Example 1 gives

$$\int_0^3 x^2 \sqrt{x^3 + 9}\,dx = \tfrac{1}{3} \int_*^{**} u^{1/2}\,du$$

$$= \tfrac{1}{3} \left[ \tfrac{2}{3} u^{3/2} \right]_*^{**}$$

$$= \tfrac{2}{9} \left[ (x^3 + 9)^{3/2} \right]_0^3$$

$$= \tfrac{2}{9}[216 - 27] = 42.$$

The limits * and ** on $u$ are so indicated because they were not calculated—there was no need to know them—and because it would be only a coincidence if they were the same as the limits on $x$.

But sometimes it is more convenient to determine the limits of integration with respect to the new variable $u$. Under the above substitution $u = x^3 + 9$, the lower limit $x = 0$ corresponds to $u = 9$, and the upper limit $x = 3$ corresponds to $u = 36$. Hence we may alternatively write

$$\int_0^3 x^2 \sqrt{x^3 + 9}\,dx = \tfrac{1}{3} \int_9^{36} u^{1/2}\,du$$

$$= \tfrac{1}{3} \left[ \tfrac{2}{3} u^{3/2} \right]_9^{36} = 42.$$

The following theorem tells how to transform the limits $x = a$ and $x = b$ under the substitution $u = g(x)$. The new lower limit is $u = g(a)$ and the new upper limit is $u = g(b)$, *whether or not $g(b)$ is greater than $g(a)$.*

---

**Theorem**   *Definite Integration by Substitution*

Suppose that the function $g$ has a continuous derivative on $[a, b]$, and that $f$ is continuous on $g([a, b])$. Let $u = g(x)$. Then

$$\int_a^b f(g(x))g'(x)\,dx = \int_{g(a)}^{g(b)} f(u)\,du. \qquad (8)$$

---

***Proof***   Choose an antiderivative $F$ of $f$, so that $F' = f$. Then, by the chain rule,

$$D[F(g(x))] = F'(g(x))g'(x) = f(g(x))g'(x).$$

So

$$\int_a^b f(g(x))g'(x)\,dx = \Big[F(g(x))\Big]_a^b = F(g(b)) - F(g(a))$$

$$= \Big[F(u)\Big]_{u=g(a)}^{g(b)} = \int_{g(a)}^{g(b)} f(u)\,du.$$

We used the fundamental theorem to obtain the first and last equalities in this argument.   ∎

**EXAMPLE 2**   Evaluate $\displaystyle\int_4^9 \frac{\sqrt{x}\,dx}{(30 - x^{3/2})^2}$.

***Solution***   Note that $30 - x^{3/2}$ is nonzero on $[4, 9]$, so the integrand is continuous. We substitute

$$u = 30 - x^{3/2}, \quad \text{so that} \quad du = -\tfrac{3}{2}x^{1/2}\,dx$$

or $\sqrt{x}\,dx = -\tfrac{2}{3}\,du$. To change the limits, note that

$$\text{if } x = 4, \quad \text{then } u = 22, \qquad \text{and}$$
$$\text{if } x = 9, \quad \text{then } u = 3.$$

Hence our substitution gives

$$\int_4^9 \frac{\sqrt{x}\,dx}{(30 - x^{3/2})^2} = \int_{22}^3 \frac{-\tfrac{2}{3}\,du}{u^2}$$

$$= \frac{2}{3}\int_3^{22} \frac{du}{u^2} = \frac{2}{3}\left[-\frac{1}{u}\right]_3^{22}$$

$$= \frac{2}{3}\left(-\frac{1}{22} + \frac{1}{3}\right) = \frac{19}{99}.$$

**EXAMPLE 3**   Evaluate $\displaystyle\int_0^3 x^2\sqrt{9 + x^3}\,dx$.

***Solution***   We substitute

$$u = 9 + x^3, \quad \text{so that} \quad du = 3x^2\,dx.$$

Then

$$u = 9 \quad \text{when } x = 0, \quad \text{and} \quad u = 36 \quad \text{when } x = 3.$$

Hence

$$\int_0^3 x^2\sqrt{9 + x^3}\,dx = \int_9^{36} \tfrac{1}{3}u^{1/2}\,du$$

$$= \Big[\tfrac{2}{9}u^{3/2}\Big]_9^{36} = \tfrac{2}{9}(216 - 27) = 42.$$

**260**

Use the given substitution to compute the most general antiderivative of $f(x)$ in Problems 1–12.

1 $f(x) = x^3\sqrt{1 + x^4}$; $u = x^4$

2 $f(x) = \sqrt{x}(1 + x\sqrt{x})$; $u = 1 + x\sqrt{x}$

3 $f(x) = \dfrac{1}{\sqrt{x}(1 + \sqrt{x})^2}$; $u = 1 + \sqrt{x}$

4 $f(x) = \dfrac{1}{\sqrt{x}(1 + \sqrt{x})^2}$; $u = \sqrt{x}$

5 $f(x) = \dfrac{x^2}{\sqrt{1 + 4x^3}}$; $u = 1 + 4x^3$

6 $f(x) = x(x + 1)^{14}$; $u = x + 1$

7 $f(x) = x(x^2 + 1)^{14}$; $u = x^2 + 1$

8 $f(x) = x\sqrt{2x^2 + 3}$; $u = 2x^2 + 3$

9 $f(x) = x\sqrt{4 - x}$; $u = 4 - x$

10 $f(x) = \dfrac{x + 2x^3}{(x^4 + x^2)^3}$; $u = x^4 + x^2$

11 $f(x) = \dfrac{2x^3}{\sqrt{1 + x^4}}$; $u = x^4$

12 $f(x) = (2x + 1)(x^2 + x)^{-1/2}$; $u = x^2 + x$

Evaluate the integrals in Problems 13–36.

13 $\displaystyle\int (4x - 3)^5 \, dx$

14 $\displaystyle\int x\sqrt{x^2 - 1} \, dx$

15 $\displaystyle\int x\sqrt{2 - 3x^2} \, dx$

16 $\displaystyle\int 3t(1 - 2t^2)^{10} \, dt$

17 $\displaystyle\int \dfrac{x^2 \, dx}{(x^3 + 5)^4}$

18 $\displaystyle\int \dfrac{t \, dt}{\sqrt{2t^2 + 1}}$

19 $\displaystyle\int x^2 \sqrt[3]{2 - 4x^3} \, dx$

20 $\displaystyle\int \dfrac{(x + 1) \, dx}{(x^2 + 2x + 5)^2}$

21 $\displaystyle\int \dfrac{t^2 + 1}{(t^3 + 3t)^2} \, dt$

22 $\displaystyle\int \dfrac{1}{(a + bx)^3} \, dx$  ($a$ and $b$ are constants)

23 $\displaystyle\int_{-1}^{-2} \dfrac{dt}{(t + 3)^3}$

24 $\displaystyle\int_0^4 x\sqrt{x^2 + 9} \, dx$

25 $\displaystyle\int_0^4 \dfrac{dx}{\sqrt{2x + 1}}$

26 $\displaystyle\int_{-1}^1 \dfrac{(x + 1) \, dx}{\sqrt{x^2 + 2x + 2}}$

27 $\displaystyle\int_0^8 t(t + 1)^{1/2} \, dt$  (*Suggestion:* Try $u = t + 1$.)

28 $\displaystyle\int_0^\pi (\pi^2 x - x^3) \, dx$

29 $\displaystyle\int_0^5 \sqrt{4 + x} \, dx$

30 $\displaystyle\int_0^a (\sqrt{a} - \sqrt{x})^2 \, dx$  ($a$ is a constant)

31 $\displaystyle\int_0^a x\sqrt{a^2 - x^2} \, dx$  ($a$ is a constant)

32 $\displaystyle\int_1^4 \dfrac{(1 + \sqrt{x})^4}{\sqrt{x}} \, dx$

33 $\displaystyle\int \dfrac{x^3 - 1}{(x^4 - 4x)^{2/3}} \, dx$

34 $\displaystyle\int \dfrac{1}{x^3}\left(1 + \dfrac{1}{x^2}\right)^{5/3} dx$

35 $\displaystyle\int (2 - t^2)\sqrt[4]{6t - t^3} \, dt$

36 $\displaystyle\int \dfrac{2 - x^2}{(x^3 - 6x + 1)^5} \, dx$

In each of Problems 37–40, verify the given formula by differentiating the right-hand side.

37 $\displaystyle\int \dfrac{dx}{x^2\sqrt{x^2 + a^2}} = -\dfrac{\sqrt{x^2 + a^2}}{a^2 x} + C$

38 $\displaystyle\int (x^2 - a^2)^{3/2} \, dx = \dfrac{x}{4}(x^2 - a^2)^{3/2}$

$$- \dfrac{3}{4}a^2 \int \sqrt{x^2 - a^2} \, dx$$

39 $\displaystyle\int x\sqrt{x + 1} \, dx = \tfrac{2}{15}(3x - 2)(x + 1)^{3/2} + C$

40 $\displaystyle\int \dfrac{x \, dx}{(1 - x^2)^2} = \dfrac{x^2}{2(1 - x^2)} + C$

41 Substitute $u = 1 - x^2$ to show that

$$\int \dfrac{x \, dx}{(1 - x^2)^2} = \dfrac{1}{2(1 - x^2)} + C.$$

Is this answer consistent with the formula of Problem 40? Explain.

42 Substitute $u = x + 1$ to derive the formula of Problem 39.

43 Suppose that $f$ is an **odd** function, meaning that $f(-x) = -f(x)$ for all $x$. Substitute $u = -x$ in the integral $\int_{-a}^0 f(x) \, dx$ to show that

$$\int_{-a}^a f(x) \, dx = 0$$

if $f$ is also continuous on $[-a, a]$.

44 If $f$ is an **even** function, meaning that $f(-x) = f(x)$ for all $x$, use the method of Problem 43 to show that

$$\int_{-a}^a f(x) \, dx = 2\int_0^a f(x) \, dx$$

if $f$ is also continuous on $[-a, a]$.

## Computing Areas by Integration

In Section 5-2 we discussed the area $A$ under the graph of a positive-valued continuous function $f$ on the interval $[a, b]$. This discussion motivated our definition in Section 5-3 of the integral of $f$ from $a$ to $b$, with the result that

$$A = \int_a^b f(x)\, dx \qquad (1)$$

by definition.

Here we consider the more general problem of finding the area of a region that is bounded by the graphs of *two* functions. Suppose that the functions $f$ and $g$ are continuous on $[a, b]$, and that $f(x) \geq g(x)$ for all $x$ in $[a, b]$. We are interested in the area $A$ of the region $R$ shown in Fig. 5.20, in which $R$ is bounded by the graphs of $f$ and $g$ and the vertical lines $x = a$ and $x = b$.

In order to approximate $A$, we consider a regular partition of $[a, b]$ into $n$ equal subintervals, each with length $\Delta x$. If $\Delta A_i$ denotes the area of the region between the graphs of $f$ and $g$ and lying above the $i$th subinterval $[x_{i-1}, x_i]$, and $x_i^*$ is an arbitrary number chosen in that subinterval (all this for $i = 1, 2, \ldots, n$), then $\Delta A_i$ is approximately equal to the area of a rectangle with height $f(x_i^*) - g(x_i^*)$ and width $\Delta x$ (see Fig. 5.21). Hence

$$\Delta A_i \approx [f(x_i^*) - g(x_i^*)]\, \Delta x,$$

and so

$$A = \sum_{i=1}^n \Delta A_i \approx \sum_{i=1}^n [f(x_i^*) - g(x_i^*)]\, \Delta x.$$

We introduce the function $h(x) = f(x) - g(x)$, and observe that $A$ is given approximately by a Riemann sum for $h$ associated with our regular

**5.20** A region between two graphs

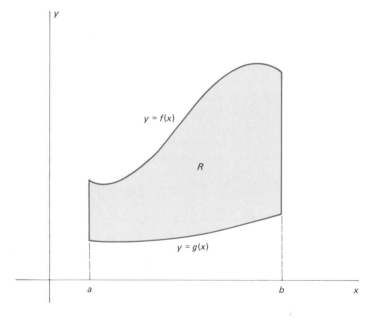

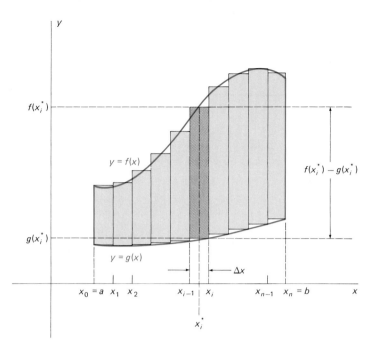

**5.21** A partition of $[a, b]$ divides $R$ into vertical strips that we approximate with rectangular strips.

partition of $[a, b]$:

$$A \approx \sum_{i=1}^{n} h(x_i^*)\, \Delta x.$$

Intuition and reason both suggest that this approximation can be made arbitrarily accurate by choosing $n$ sufficiently large (and hence $\Delta x = (b - a)/n$ sufficiently small). We would therefore conclude that

$$A = \lim_{\Delta x \to 0} \sum_{i=1}^{n} h(x_i^*)\, \Delta x$$

$$= \int_{a}^{b} h(x)\, dx = \int_{a}^{b} \left[ f(x) - g(x) \right] dx.$$

Since our discussion is based on an intuitive concept rather than on a logical definition of area, it does *not* constitute a proof of the formula above. It does, however, provide justification for the following *definition* of the area in question.

---

*Definition*   *The Area Between Two Curves*

Let $f$ and $g$ be continuous with $f(x) \geqq g(x)$ for $x$ in $[a, b]$. Then the **area** $A$ of the region bounded by the curves $y = f(x)$ and $y = g(x)$ and the vertical lines $x = a$ and $x = b$ is

$$A = \int_{a}^{b} \left[ f(x) - g(x) \right] dx. \qquad (2)$$

---

Note that our earlier Formula (1) is the case $g(x) \equiv 0$ of (2). On the other hand, if $f(x) \equiv 0$ and $g(x) \leqq 0$ on $[a, b]$, then (2) reduces to

$$A = -\int_{a}^{b} g(x)\, dx \quad \text{or} \quad \int_{a}^{b} g(x)\, dx = -A.$$

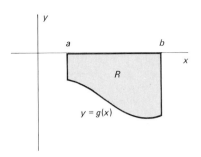

5.22 The integral gives the negative of the geometric area for a region lying below the x-axis.

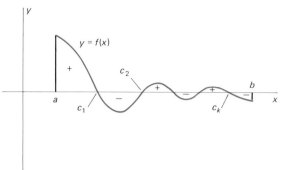

5.23 The integral computes the area above the x-axis *minus* the area below the x-axis.

In this case, the area lies below the x-axis, as in Fig. 5.22. Thus the integral from a to b of a negative-valued function is the *negative* of the area of the region bounded by its graph, the x-axis, and the vertical lines $x = a$ and $x = b$.

More generally, consider a continuous function $f$ with a graph that crosses the x-axis at finitely many points $c_1, c_2, \ldots, c_k$ between $a$ and $b$, as shown in Fig. 5.23. We write

$$\int_a^b f(x) \, dx = \int_a^{c_1} f(x) \, dx + \int_{c_1}^{c_2} f(x) \, dx + \cdots + \int_{c_k}^b f(x) \, dx.$$

Thus we see that $\int_a^b f(x) \, dx$ is equal to the area under $y = f(x)$ *above* the x-axis *minus* the area over $y = f(x)$ *below* the x-axis.

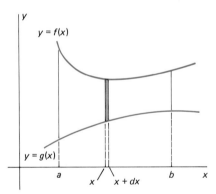

5.24 Heuristic approach to setting up area integrals

The following *heuristic* (suggestive, though not rigorous) way of setting up integral formulas like that in (2) is sometimes useful. Consider the vertical strip of area lying above the interval $[x, x + dx]$, which is shaded in Fig. 5.24. We think of the length $dx$ of the interval $[x, x + dx]$ as being so small that this strip may be regarded as a rectangle with width $dx$ and height $f(x) - g(x)$, so that its area is

$$dA = [f(x) - g(x)] \, dx.$$

Think now of the region over $[a, b]$ between $y = f(x)$ and $y = g(x)$ as made up of many such strips. Its area may be thought of as a sum of areas of such rectangles. If we write $\int$ for *sum*, this gives the formula

$$A = \int dA = \int_a^b [f(x) - g(x)] \, dx.$$

This heuristic approach bypasses the subscript notation associated with Riemann sums. Nevertheless, it *is not and should not be regarded as a complete derivation of the formula*. It is best used only as a convenient memory device. For instance, in the figures for illustrative examples, we shall often show a strip of width $dx$ as a visual aid in properly setting up the correct integral.

**EXAMPLE 1** Find the area $A$ of the region $R$ that is bounded by the line $y = x$ and the parabola $y = 6 - x^2$.

*Solution* The region $R$ is shown in Fig. 5.25. It is clear that we should use the formula in (2), taking $f(x) = 6 - x^2$ and $g(x) = x$. The limits $a$ and $b$

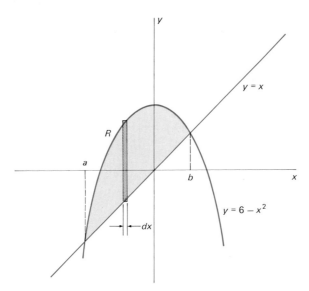

5.25 The region $R$ of Example 1

will be the $x$-coordinates of the two points of intersection of the line and the parabola. To find $a$ and $b$, we therefore equate $f(x)$ and $g(x)$ and solve the resulting equation for $x$:

$$x = 6 - x^2;$$

$$x^2 + x - 6 = 0;$$

$$(x - 2)(x + 3) = 0;$$

$$x = -3, 2.$$

Thus $a = -3$ and $b = 2$, so the formula in (2) gives

$$A = \int_{-3}^{2} (6 - x^2 - x)\, dx$$

$$= \left[ 6x - \frac{x^3}{3} - \frac{x^2}{2} \right]_{-3}^{2}$$

$$= \left[ 6(2) - \frac{1}{3}(2)^3 - \frac{1}{2}(2)^2 \right] - \left[ 6(-3) - \frac{1}{3}(-3)^2 - \frac{1}{2}(-3)^2 \right]$$

$$= \frac{125}{6}.$$

As in the following example, it is sometimes necessary to subdivide a region before applying Formula (2).

**EXAMPLE 2**  Find the area $A$ of the region $R$ bounded by the line $y = x$ and the parabola $y^2 = 6 - x$.

*Solution*  The region $R$ is shown in Fig. 5.26. The points of intersection $(-3, -3)$ and $(2, 2)$ are found by equating $y = x$ and $y = \pm\sqrt{6 - x}$ and then solving for $x$. The lower boundary of $R$ is given by $y = -\sqrt{6 - x}$ on

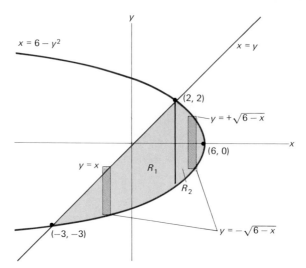

$x = 6 - y^2$

$x = y$

$(2, 2)$

$y = +\sqrt{6 - x}$

$(6, 0)$

$y = x$

$R_1$

$R_2$

$(-3, -3)$

$y = -\sqrt{6 - x}$

**5.26** In Example 2, we split the region $R$ into $R_1$ and $R_2$.

$[-3, 6]$. But the upper boundary of $R$ is given by

$$y = x \quad \text{on} \quad [-3, 2],$$
$$y = +\sqrt{6 - x} \quad \text{on} \quad [2, 6].$$

We must therefore subdivide $R$ into the two regions $R_1$ and $R_2$ as indicated in Fig. 5.26. Then Formula (2) gives

$$A = \int_{-3}^{2} [x - (-\sqrt{6 - x})] \, dx + \int_{2}^{6} [\sqrt{6 - x} - (-\sqrt{6 - x})] \, dx$$

$$= \int_{-3}^{2} (x + \sqrt{6 - x}) \, dx + 2 \int_{2}^{6} \sqrt{6 - x} \, dx$$

$$= \left[ \tfrac{1}{2}x^2 - \tfrac{2}{3}(6 - x)^{3/2} \right]_{-3}^{2} + 2 \left[ -\tfrac{2}{3}(6 - x)^{3/2} \right]_{2}^{6}$$

$$= (2 - \tfrac{16}{3}) - (\tfrac{9}{2} - 18) + 2(0) - 2(-\tfrac{16}{3}) = \tfrac{125}{6}.$$

The region of Example 2 appears simpler if it is considered to be bounded by graphs of functions of $y$ rather than by functions of $x$. Figure 5.27 shows a region $R$ bounded by the curves $x = f(y)$ and $x = g(y)$, with $f(y) \geq g(y)$ for $y$ in $[c, d]$, and by the horizontal lines $y = c$ and $y = d$. To approximate the area $A$ of $R$, we begin with a regular partition of $[c, d]$ into $n$ subintervals, each with length $\Delta y$. We choose a point $y_i^*$ in the $i$th subinterval $[y_{i-1}, y_i]$ for each $i = 1, 2, \ldots, n$. The strip of $R$ lying opposite $[y_{i-1}, y_i]$ is approximated by a rectangle with width $\Delta y$ and length $f(y_i^*) - g(y_i^*)$. Hence

$$A \approx \sum_{i=1}^{n} [f(y_i^*) - g(y_i^*)] \, \Delta y.$$

Recognition of this sum as a Riemann sum for the integral

$$\int_{c}^{d} [f(y) - g(y)] \, dy$$

motivates the following definition.

**266**

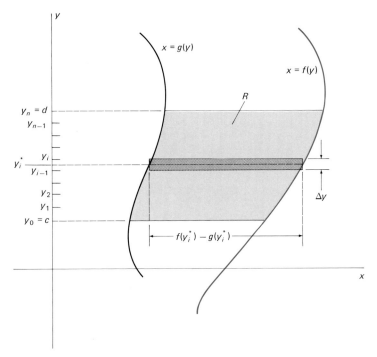

5.27   Finding area using an integral with respect to $y$

---

**Definition**   *The Area Between Two Curves*

Let $f$ and $g$ be continuous with $f(y) \geq g(y)$ for $y$ in $[c, d]$. Then the **area** $A$ of the region bounded by the curves $x = f(y)$ and $x = g(y)$ and the horizontal lines $y = c$ and $y = d$ is

$$A = \int_c^d [f(y) - g(y)]\, dy. \tag{3}$$

---

In a more advanced course, one would prove that the formulas in (2) and (3) yield the same area $A$ for a region that can be described both in the manner shown in Fig. 5.20 and in the manner shown in Fig. 5.27.

Comparison of Example 2 with the one following illustrates the advantage of choosing the "right" variable of integration—the one that makes the resulting computation simpler.

**EXAMPLE 3**   Integrate with respect to $y$ to find the area of the region $R$ of Example 2.

*Solution*   We see from Fig. 5.28 that the formula in (3) applies with $f(y) = 6 - y^2$ and $g(y) = y$ for $y$ in $[-3, 2]$. This gives

$$A = \int_{-3}^{2} [6 - y^2 - y]\, dy = \left[ 6y - \frac{y^3}{3} - \frac{y^2}{2} \right]_{-3}^{2} = \frac{125}{6}.$$

**EXAMPLE 4**   Use calculus to derive the formula $A = \pi r^2$ for the area of a circle of radius $r$.

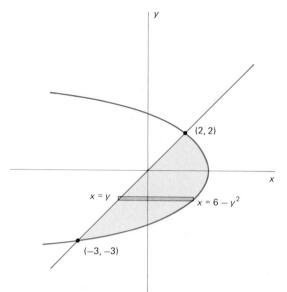

**5.28** Recomputation of the area of Example 2

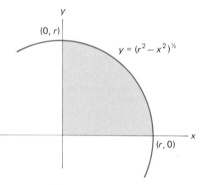

**5.29** The number $\pi$ is four times the shaded area.

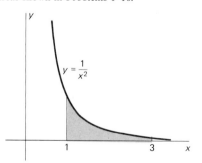

**5.30** The shaded area can be written as an integral.

**Solution** We begin with the *definition* (in Section 5-2) of the number $\pi$ as the area of the *unit* circle $x^2 + y^2 \leq 1$. Then, with the aid of Fig. 5.29, we may write

$$\pi = 4 \int_0^1 \sqrt{1 - x^2} \, dx, \tag{4}$$

because the integral in (4) is, by Formula (1), the area of the first quadrant of the unit circle. We apply Formula (1) to the first quadrant of the circle of radius $r$ of Fig. 5.30, and we find the total area $A$ of that circle to be

$$A = 4 \int_0^r \sqrt{r^2 - x^2} \, dx = 4r \int_0^r \sqrt{1 - (x/r)^2} \, dx$$

$$= 4r \int_0^1 \sqrt{1 - u^2} \, r \, du \quad \left( \text{Substitution: } u = \frac{x}{r}, \, dx = r \, du. \right)$$

$$= 4r^2 \int_0^1 \sqrt{1 - u^2} \, du = 4r^2 \int_0^1 \sqrt{1 - x^2} \, dx.$$

So, by Equation (4) above, $A = \pi r^2$.

## 5-7 PROBLEMS

Find the areas shown in Problems 1–10.

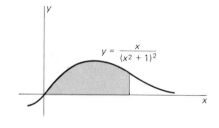

1

2

**3**

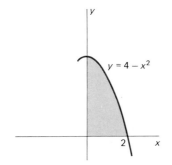

$y = 4 - x^2$

2

**4**

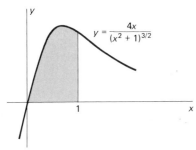

$y = \dfrac{4x}{(x^2 + 1)^{3/2}}$

1

**5**

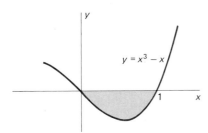

$y = x^3 - x$

1

**6**

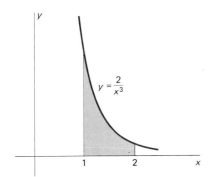

$y = \dfrac{2}{x^3}$

1   2

**7**

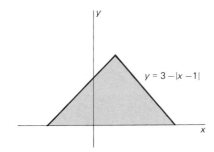

$y = 3 - |x - 1|$

**8**

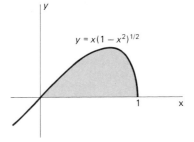

$y = x(1 - x^2)^{1/2}$

1

**9**

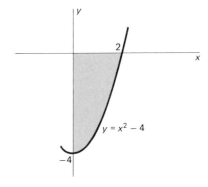

2

$y = x^2 - 4$

−4

**10**

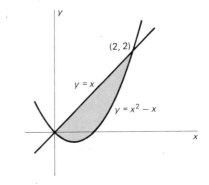

(2, 2)

$y = x$

$y = x^2 - x$

Find the areas of the regions described in Problems 11–20.

**11** The region bounded below by the graph of $y = x^3$ and above by the graph of $y = x$ over the interval $[0, 1]$.

**12** The region between the graph of $y = 1/(x + 1)^2$ and the x-axis over the interval $[1, 3]$.

**13** The region bounded above by the graph of $y = x^3$ and below by the graph of $y = x^4$ over the interval $[0, 1]$.

**14** The region bounded above by the graph of $y = x^2$ and below by the horizontal line $y = -1$ over the interval $[-1, 2]$.

**15** The region bounded above by the graph of $y = 1/(x + 1)^3$ and below by the x-axis over the interval $[0, 2]$.

**16** The region bounded above by the graph of $y = 4x - x^2$ and below by the x-axis.

**17** The region bounded on the left by the graph of $x = y^2$ and on the right by the vertical line $x = 4$.

**18** The region between the graphs of $y = x^4 - 4$ and $y = 3x^2$.

**19** The region between the graphs of $x = 8 - y^2$ and $x = y^2 - 8$.

**20** The region between the graphs of $y = x^{1/3}$ and $y = x^3$.

In each of Problems 21–42, sketch the region bounded by the given curves and then find its area.

**21** $y = 0$, $\quad y = 25 - x^2$
**22** $y = x^2$, $\quad y = 4$
**23** $y = x^2$, $\quad y = 8 - x^2$
**24** $x = 0$, $\quad x = 16 - y^2$
**25** $x = y^2$, $\quad x = 25$
**26** $x = y^2$, $\quad x = 32 - y^2$
**27** $y = x^2$, $\quad y = 2x$
**28** $y = x^2$, $\quad x = y^2$
**29** $y = x^2$, $\quad y = x^3$
**30** $y = 2x^2$, $\quad y = 5x - 3$
**31** $x = 4y^2$, $\quad x + 12y + 5 = 0$
**32** $y = x^2$, $\quad y = 3 + 5x - x^2$
**33** $x = 3y^2$, $\quad x = -y^2 + 12y - 5$
**34** $y = x^2$, $\quad y = 4(x - 1)^2$
**35** $x = y^2 - 2y - 2$, $\quad x = -2y^2 + y + 4$
**36** $y = x^4$, $\quad y = 32 - x^4$
**37** $y = x^3$, $\quad y = 32\sqrt{x}$
**38** $y = x^3$, $\quad y = 2x - x^2$
**39** $y = x^2$, $\quad y = x^{2/3}$
**40** $y^2 = x$, $\quad y^2 = 2(x - 3)$
**41** $y = x^3$, $\quad y = 2x^3 + x^2 - 2x$
**42** $y = x^3$, $\quad x + y = 0$, $\quad y = x + 6$

**43** The *ellipse* $x^2/a^2 + y^2/b^2 = 1$ is shown in Fig. 5.31. Use the method of Example 4 to show that the area of the region it bounds is $A = \pi ab$, a pleasing generalization of the area formula for the circle.

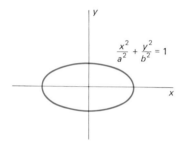

**5.31** The ellipse of Problem 43

**44** Let $A$ and $B$ be the points of intersection of the parabola $y = x^2$ and the line $y = x + 2$, and let $C$ be the point on the parabola where the tangent line is parallel to the graph of $y = x + 2$. Show that the area of the parabolic segment cut off from the parabola by the line (see Fig. 5.32) is four-thirds the area of triangle $ABC$.

**45** Find the area of the unbounded region shaded in Fig. 5.33, regarding it as the limit as $b \to \infty$ of the region bounded by $y = 1/x^2$, $y = 0$, $x = 1$, and $x = b > 1$.

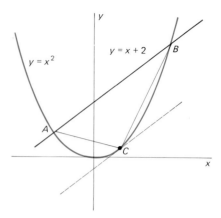

**5.32** The parabolic segment of Problem 44

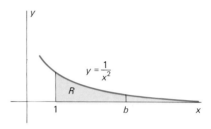

**5.33** The unbounded region of Problem 45

**46** Find the total area of the bounded regions that are bounded by the x-axis and the curve $y = 2x^3 - 2x^2 - 12x$.

**47** Suppose that the quadratic function $f(x) = px^2 + qx + r$ is never negative on $[a, b]$. Show that the area under the graph of $f$ from $a$ to $b$ is

$$A = \frac{h}{3}[f(a) + 4f(m) + f(b)]$$

where $h = \frac{1}{2}(b - a)$ and $m = \frac{1}{2}(a + b)$. (*Suggestion:* By horizontal translation of this region, it may be assumed that $a = -h$, $m = 0$, and $b = h$.)

In Problems 48–50, the area $A(u)$ between the graph of the positive and continuous function $y = f(x)$ and the x-axis over the interval $0 \leq x \leq u$ is given. Find $f(x)$.

**48** $A(u) = u^5$
**49** $A(u) = \frac{2}{3}u^{3/2}$
**50** $A(u) = [f(u)]^2$

---

## 5-8

## Numerical Integration

The fundamental theorem of calculus, $\int_a^b f(x)\, dx = [G(x)]_a^b$, can be used to evaluate the integral only if a convenient formula for the antiderivative $G(x)$ of $f(x)$ can be found. But there are rather simple functions with antiderivatives that are not elementary functions. An **elementary function** is one that can be expressed in terms of polynomial, trigonometric, exponential, and logarithmic functions by means of finite combinations of sums, differences, products, quotients, roots, and function composition.

The problem is that elementary functions can have nonelementary antiderivatives. For example, it is known that the elementary function $2^{-x^2}$ has no elementary antiderivative. Consequently, we cannot use the fundamental theorem of calculus to evaluate an integral such as

$$\int_0^1 2^{-x^2}\,dx.$$

In this section we discuss the use of Riemann sums to numerically approximate integrals that cannot conveniently be evaluated exactly, whether or not nonelementary functions are involved. Given a continuous function $f$ on $[a, b]$ with an integral that is to be approximated, consider a regular partition $P$ of $[a, b]$ into $n$ subintervals, each with length $\Delta x = (b - a)/n$. Recall from Section 5-3 that the value of the integral lies between the upper and lower Riemann sums for $f$ associated with the partition $P$:

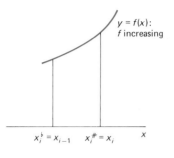

$$\sum_{i=1}^n f(x_i^{\flat})\,\Delta x \leqq \int_a^b f(x)\,dx \leqq \sum_{i=1}^n f(x_i^{\#})\,\Delta x, \tag{1}$$

where $f(x_i^{\flat})$ and $f(x_i^{\#})$ are the minimum and maximum values, respectively, of $f$ on the $i$th subinterval $[x_{i-1}, x_i]$ of the partition.

The relation in (1) is particularly easy to apply when $f$ is either an *increasing function* or a *decreasing function* on $[a, b]$. For if $f$ is increasing on $[a, b]$, then the minimum value of $f$ on the $i$th subinterval $[x_{i-1}, x_i]$ occurs at $x_{i-1}$, so that $x_i^{\flat} = x_{i-1}$. It is equally clear that $x_i^{\#} = x_i$. If, instead, $f$ is a decreasing function on $[a, b]$, then $x_i^{\flat} = x_i$ and $x_i^{\#} = x_{i-1}$. The two cases are illustrated in Fig. 5.34.

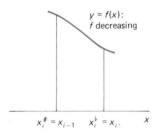

**5.34** If $f$ is increasing or decreasing, its extrema occur at the endpoints of the interval $[x_{i-1}, x_i]$.

The *left-end-point approximation* $L_n$ and the *right-end-point approximation* $R_n$ to the definite integral $\int_a^b f(x)\,dx$, associated with the regular partition of $[a, b]$ into $n$ equal subintervals, are the Riemann sums

$$L_n = \sum_{i=1}^n f(x_{i-1})\,\Delta x \tag{2}$$

and

$$R_n = \sum_{i=1}^n f(x_i)\,\Delta x. \tag{3}$$

The notation for $L_n$ and $R_n$ may be simplified a bit by writing $y_i$ for $f(x_i)$.

---

*Definition* *End-point Approximations*

The **left-end-point approximation** $L_n$ and the **right-end-point approximation** $R_n$ to $\int_a^b f(x)\,dx$ with $\Delta x = (b - a)/n$ are

$$L_n = (\Delta x)(y_0 + y_1 + y_2 + \cdots + y_{n-1}) \tag{2}$$

and

$$R_n = (\Delta x)(y_1 + y_2 + y_3 + \cdots + y_n). \tag{3}$$

---

So Inequality (1) above amounts to this:

$$L_n \leqq \int_a^b f(x)\,dx \leqq R_n \tag{4}$$

for an increasing function $f$, and

$$R_n \leqq \int_a^b f(x)\,dx \leqq L_n \tag{5}$$

for a decreasing function $f$. Thus in the case of a function $f$ that either is increasing on $[a, b]$ or is decreasing on $[a, b]$, one of the two approximations $R_n$ and $L_n$ is an overestimate of $\int_a^b f(x)\,dx$ and the other is an underestimate.

We also have an error estimate available. Note that

$$\left| R_n - L_n \right| = \left| \sum_{i=1}^n \left[ f(x_i) - f(x_{i-1}) \right] \Delta x \right|$$

$$= (\Delta x) \left| f(x_1) - f(x_0) + f(x_2) - f(x_1) + \cdots + f(x_n) - f(x_{n-1}) \right|$$

$$= (\Delta x) \left| f(x_n) - f(x_0) \right|.$$

Thus

$$\left| R_n - L_n \right| = (\Delta x)\left| f(b) - f(a) \right| = \frac{b-a}{n}\left| f(b) - f(a) \right|. \tag{6}$$

Equation (6) enables us to estimate the error when we approximate the integral of an increasing (or a decreasing) function by either $R_n$ or $L_n$. From a geometric point of view Equation (6) is related to the observation that $\left| R_n - L_n \right|$ is the sum of the areas of the shaded small rectangles shown in Fig. 5.35 and that these small rectangles can be moved horizontally to form a vertical stack. The stack, also shown in Fig. 5.35, is a rectangle with base $\Delta x = (b - a)/n$ and height $\left| f(b) - f(a) \right|$.

Note also that Equation (6) can be used to determine how large $n$ must be in order for $L_n$ and $R_n$ to approximate $\int_a^b f(x)\,dx$ with a given degree of accuracy. For example, to obtain three-place accuracy, it suffices to choose $n$ sufficiently large that

$$\frac{b-a}{n}\left| f(b) - f(a) \right| < 0.0005.$$

5.35 Error in the endpoint approximation

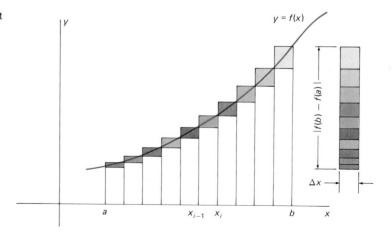

We illustrate these numerical techniques by approximating the integral

$$\int_0^1 \frac{4}{1 + x^2} \, dx. \tag{7}$$

In Section 8-5 we will use the inverse tangent function $y = \arctan x$ to show that

$$\int_0^1 \frac{1}{1 + x^2} \, dx = \frac{\pi}{4} \quad \text{(exactly!)}.$$

Hence the true value of the integral in (7) is

$$\pi = 3.14159\ 265358\ 97932 \ldots .$$

This last fact will provide us with a check on the accuracy of our approximations.

**EXAMPLE 1**   Calculate the left-end-point and right-end-point approximations to the integral (7) with $n = 10$ and $\Delta x = 0.1$.

*Solution*   Note first that $f(x) = 4/(1 + x^2)$ is a decreasing function on $[0, 1]$, so $L_{10}$ will be a high estimate of the integral and $R_{10}$ a low estimate. We use a hand-held calculator and note in writing only the final value of each approximation, because writing the individual terms and (subsequently) their sum would vitiate the eight- or nine-place internal accuracy of such a calculator. Since

$$x_0 = 0, x_1 = 0.1, x_2 = 0.2, \ldots, x_9 = 0.9, x_{10} = 1,$$

we obtain

$$L_{10} = (0.1) \left[ \frac{4}{1.00} + \frac{4}{1.01} + \frac{4}{1.04} + \cdots + \frac{4}{1.81} \right]$$
$$\approx 3.239925990 \approx 3.24$$

and

$$R_{10} = (0.1) \left[ \frac{4}{1.01} + \frac{4}{1.04} + \frac{4}{1.09} + \cdots + \frac{4}{2.00} \right]$$
$$\approx 3.039925990 \approx 3.04.$$

Note that the average

$$\tfrac{1}{2}(L_{10} + R_{10}) \approx \tfrac{1}{2}(3.239925990 + 3.039925990)$$
$$= 3.139925990 \approx 3.14$$

is much closer to the true value of $\pi$ than is either $L_{10}$ or $R_{10}$. The technique of averaging upper and lower estimates plays an important role in numerical integration.

Now $\pi$ lies between $R_{10}$ and $L_{10}$, so these computations show that $\pi = 3.14 \pm 0.10$ if rounded to two decimal places (3.14 is the two-place average of $R_{10}$ and $L_{10}$). Since $f(0) = 4$ and $f(1) = 2$, Equation (6) gives

$$|R_n - L_n| = \frac{2}{n}.$$

In order to achieve the accuracy $|R_n - L_n| \leq 0.0005$, we would therefore need a partition with $n \geq 4000$ subintervals.

The average $T_n = \frac{1}{2}(L_n + R_n)$ of the left-end-point and right-end-point approximations is called the *trapezoidal approximation* to $\int_a^b f(x)\, dx$ associated with the partition of $[a, b]$ into $n$ equal subintervals. Written in full, we see that

$$T_n = \frac{1}{2}(L_n + R_n) = \frac{\Delta x}{2} \sum_{i=1}^{n} [f(x_{i-1}) + f(x_i)];$$

that is,

$$T_n = \frac{\Delta x}{2} [f(x_0) + 2f(x_1) + 2f(x_2) + \cdots + 2f(x_{n-2}) + 2f(x_{n-1}) + f(x_n)].$$

Note the 1–2–2–$\cdots$–2–2–1 pattern of the coefficients.

---

**Definition**  *Trapezoidal Approximation*

The **trapezoidal approximation** to $\int_a^b f(x)\, dx$ with $\Delta x = (b - a)/n$ is

$$T_n = \frac{\Delta x}{2}(y_0 + 2y_1 + 2y_2 + \cdots + 2y_{n-1} + y_n). \tag{8}$$

---

Figure 5.36 shows where $T_n$ gets its name. The partition points $x_0, x_1, \ldots, x_n$ are used to build trapezoids from the $x$-axis to the function's graph. The trapezoid over the $i$th subinterval $[x_{i-1}, x_i]$ has **altitude** $\Delta x$ and its parallel **bases** have lengths $f(x_{i-1})$ and $f(x_i)$. So its area is

$$\frac{\Delta x}{2}[f(x_{i-1}) + f(x_i)].$$

A comparison of this with (8) shows that $T_n$ is merely the sum of the areas of the $n$ trapezoids shown in Fig. 5.36.

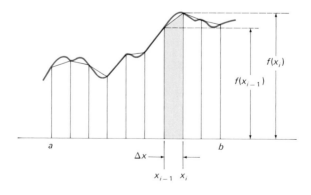

**5.36**  Geometry of the trapezoidal approximation

**EXAMPLE 2**  Calculate the trapezoidal approximation to the integral in (7) with $n = 10$ and $\Delta x = 0.1$.

**Solution**  Since we have already calculated $L_{10}$ and $R_{10}$ in Example 1, we can simply write

$$T_{10} = \tfrac{1}{2}(L_{10} + R_{10})$$
$$\approx \tfrac{1}{2}(3.239925990 + 3.039925990)$$

so

$$T_{10} \approx 3.139925990.$$

Had $L_{10}$ and $R_{10}$ not been available, we would have written

$$T_{10} = \frac{0.1}{2}\left[\frac{(1)4}{1.00} + \frac{(2)4}{1.01} + \cdots + \frac{(2)4}{1.81} + \frac{(1)4}{2.00}\right]$$

$$\approx 3.139925990.$$

Another useful approximation to $\int_a^b f(x)\,dx$ is the *midpoint approximation* $M_n$. It is the Riemann sum obtained by choosing the point $x_i^*$ in $[x_{i-1}, x_i]$ to be its midpoint $m_i = \tfrac{1}{2}(x_{i-1} + x_i)$. Thus

$$M_n = \sum_{i=1}^{n} f(m_i)\,\Delta x = (\Delta x)[f(m_1) + f(m_2) + \cdots + f(m_n)]. \qquad (9)$$

It is sometimes convenient to write $m_i = x_{i-1/2}$ and $f(m_i) = y_{i-1/2}$. With this convention we frame our next definition.

---

**Definition**  *Midpoint Approximation*

The **midpoint approximation** to $\int_a^b f(x)\,dx$ with $\Delta x = (b - a)/n$ is

$$M_n = (\Delta x)(y_{1/2} + y_{3/2} + y_{5/2} + \cdots + y_{n-1/2}). \qquad (9)$$

---

**EXAMPLE 3**  Calculate the midpoint approximation to the integral in (7) with $n = 10$ and $\Delta x = 0.1$.

**Solution**  We still have the integral

$$\int_0^1 \frac{4}{1 + x^2}\,dx \qquad (7)$$

to be approximated. But now $m_1 = 0.05$, $m_2 = 0.15$, $m_3 = 0.25, \ldots,$ and $m_{10} = 0.95$. So

$$M_{10} = (0.1)\left[\frac{4}{1.0025} + \frac{4}{1.0225} + \frac{4}{1.0625} + \cdots + \frac{4}{1.9025}\right]$$

$$\approx 3.142425986 \approx 3.14.$$

Thus $M_{10}$ gives $\pi$ rounded to two decimal places.

---

The midpoint approximation (9) is sometimes called the **tangent line** approximation, because the area of the rectangle with base $[x_{i-1}, x_i]$ and

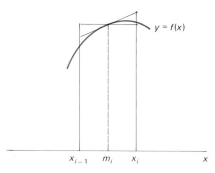

**5.37** The midpoint or tangent rule

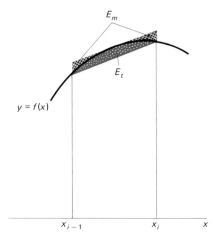

**5.38** Comparison of the midpoint rule error $E_m$ with the trapezoidal rule error $E_t$

height $f(m_i)$ is also the area of another approximating figure. As shown in Fig. 5.37, we draw a tangent segment to the graph of $f$, tangent at the point $(m_i, f(m_i))$ of its graph, and use that segment for one side of a trapezoid, somewhat like the method of the trapezoidal rule. This trapezoid and the rectangle mentioned above have the same area, and so the value of $M_n$ is the sum of the areas of trapezoids like the one of Fig. 5.37.

When we compare the results of Examples 2 and 3, we see that the midpoint approximation $M_{10} \approx 3.1424$ is somewhat closer to $\pi \approx 3.1416$ than is the trapezoidal approximation $T_{10} \approx 3.1399$. The trapezoids shown in Fig. 5.38 suggest why this should happen. Moreover, the area of the trapezoid associated with the midpoint approximation is generally closer to the true value of

$$\int_{x_{i-1}}^{x_i} f(x)\, dx$$

than is the area of the trapezoid associated with the trapezoidal approximation. Figure 5.38 also shows this, in that the midpoint error $E_m$, which is crosshatched in the figure, is generally smaller than the trapezoidal error $E_t$, which is dotted in the figure. Figure 5.38 also indicates that if $y = f(x)$ is concave downward, then $M_n$ will be an overestimate and $T_n$ will be an underestimate of $\int_a^b f(x)\, dx$. If the graph is concave upward, then the situation will be reversed.

Such observations motivate the consideration of a *weighted* average of $M_n$ and $T_n$, with $M_n$ weighted more heavily than $T_n$, for further improving our numerical estimates of the definite integral. The particular weighted average

$$S_{2n} = \tfrac{1}{3}(2M_n + T_n) = \tfrac{2}{3}M_n + \tfrac{1}{3}T_n, \tag{10}$$

is called *Simpson's approximation* to $\int_a^b f(x)\, dx$. The reason for the subscript $2n$ is that it's most natural to regard $S_{2n}$ as being associated with a partition of $[a, b]$ into an *even* number $2n$ of equal subintervals with the endpoints

$$a = x_0 < x_1 < x_2 < \cdots < x_{2n-2} < x_{2n-1} < x_{2n} = b.$$

The midpoint and trapezoidal approximations associated with the $n$ subintervals

$$[x_0, x_2], [x_2, x_4], [x_4, x_6], \ldots, [x_{2n-4}, x_{2n-2}], [x_{2n-2}, x_{2n}]$$

that all have the same length $2\,\Delta x$ can then be written in the form

$$M_n = (2\,\Delta x)[y_1 + y_3 + y_5 + \cdots + y_{2n-1}]$$

and

$$T_n = \frac{2\,\Delta x}{2}[y_0 + 2y_2 + 2y_4 + \cdots + 2y_{2n-2} + y_{2n}].$$

We substitute these formulas for $M_n$ and $T_n$ into (10), and find—after a bit of algebra—that

$$S_{2n} = \frac{\Delta x}{3}[y_0 + 4y_1 + 2y_2 + 4y_3 + 2y_4 + \cdots$$
$$+ 2y_{2n-2} + 4y_{2n-1} + y_{2n}]. \tag{11}$$

> **Definition** *Simpson's Approximation*
>
> **Simpson's approximation** to $\int_a^b f(x)\, dx$ with $\Delta x = (b-a)/2n$, associated with a partition of $[a, b]$ into an even number $2n$ of equal subintervals, is the sum $S_{2n}$ given by Equation (11).

Note that $1-4-2-4-2-\cdots-4-2-4-1$ pattern of coefficients in Simpson's approximation.

**EXAMPLE 4**  Calculate Simpson's approximation $S_{10}$ to the integral

$$\int_0^1 \frac{4}{1+x^2}\, dx. \tag{7}$$

**Solution**  Working with 10 subintervals of $[0, 1]$ and $\Delta x = 0.1$, we obtain

$$S_{10} = \frac{0.1}{3}\left[\frac{(1)4}{1.00} + \frac{(4)4}{1.01} + \frac{(2)4}{1.04} + \frac{(4)4}{1.09} + \cdots + \frac{(2)4}{1.64} + \frac{(4)4}{1.81} + \frac{(1)4}{2.00}\right],$$

or

$$S_{10} \approx 3.141592616.$$

Simpson's approximation correctly gives the first *seven* decimal places of $\pi$!

**EXAMPLE 5**  Calculate Simpson's approximation $S_{20}$ to the above integral (7).

**Solution**  Since we have already calculated $T_{10}$ and $M_{10}$ in Examples 2 and 3, we can use (10):

$$S_{20} = \tfrac{2}{3}M_{10} + \tfrac{1}{3}T_{10}$$
$$\approx \tfrac{2}{3}(3.142425986) + \tfrac{1}{3}(3.139925990)$$

or

$$S_{20} \approx 3.141592654,$$

which is the correct value of $\pi$ rounded to *nine* decimal places!

Although we have defined Simpson's approximation $S_{2n}$ as a weighted average of the midpoint and trapezoidal approximations, it has an important interpretation in terms of **parabolic approximations** to the curve $y = f(x)$. Starting with the partition of $[a, b]$ into $2n$ equal subintervals, we define the parabolic function

$$p_i(x) = A_i + B_i x + C_i x^2$$

on $[x_{2i-2}, x_{2i}]$. We choose the coefficients $A_i$, $B_i$, and $C_i$ so that $p_i(x)$ agrees with $f(x)$ at the three points $x_{2i-2}$, $x_{2i-1}$, and $x_{2i}$ (see Fig. 5.39). This can be done by solving the three equations

$$A_i + B_i x_{2i-2} + C_i(x_{2i-2})^2 = f(x_{2i-2}),$$
$$A_i + B_i x_{2i-1} + C_i(x_{2i-1})^2 = f(x_{2i-1}),$$
$$A_i + B_i x_{2i} \quad + C_i(x_{2i})^2 \quad = f(x_{2i})$$

**5.39**  The parabolic approximation $y = p_i(x)$ to $y = f(x)$ on $[x_{2i-2}, x_{2i}]$

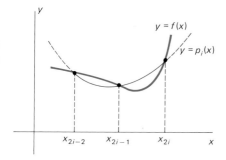

in the three unknowns $A_i$, $B_i$, and $C_i$. A routine, albeit tedious, algebraic computation (see Problem 47 in Section 5-7) then shows that

$$\int_{x_{2i-2}}^{x_{2i}} p_i(x)\,dx = \frac{\Delta x}{3}(y_{2i-2} + 4y_{2i-1} + y_{2i}).$$

We now approximate $\int_a^b f(x)\,dx$ by replacing $f(x)$ by $p_i(x)$ on the interval $[x_{2i-2}, x_{2i}]$ for $i = 1, 2, 3, \ldots, n$. This gives

$$\int_a^b f(x)\,dx = \sum_{i=1}^{n} \int_{x_{2i-2}}^{x_{2i}} f(x)\,dx$$

$$\approx \sum_{i=1}^{n} \int_{x_{2i-2}}^{x_{2i}} p_i(x)\,dx$$

$$= \sum_{i=1}^{n} \frac{\Delta x}{3}(y_{2i-2} + 4y_{2i-1} + y_{2i})$$

$$= \frac{\Delta x}{3}(y_0 + 4y_1 + 2y_2 + 4y_3 + \cdots + 4y_{2n-3}$$

$$+ 2y_{2n-2} + 4y_{2n-1} + y_{2n}).$$

Thus the parabolic approximation described above results in Simpson's approximation $S_{2n}$ to $\int_a^b f(x)\,dx$.

The numerical methods of this section are especially useful for approximating integrals of functions that are available only in graphical or in tabular form. This is often the case with functions derived from empirical data or from experimental measurements.

**EXAMPLE 6** Suppose that the graph of Fig. 5.40 shows the velocity recorded by instruments on board a nuclear submarine traveling under the polar ice cap directly toward the North Pole. Use the trapezoidal approximation and Simpson's approximation to estimate the distance $s = \int_a^b v(t)\,dt$ traveled by the submarine during the 10-hour period from $t = 0$ to $t = 10$.

*Solution* We read the following data from the graph.

| $t$ | 0 | 1 | 2 | 3 | 4 | 5 | 6 | 7 | 8 | 9 | 10 | h |
|---|---|---|---|---|---|---|---|---|---|---|---|---|
| $v$ | 13 | 14 | 17 | 21 | 23 | 21 | 15 | 11 | 11 | 14 | 17 | mi/h |

Using the trapezoidal approximation with $n = 10$ and $\Delta x = 1$, we obtain

$$s = \int_0^{10} v(t)\,dt$$

$$\approx \tfrac{1}{2}[13 + 2(14 + 17 + 21 + 23 + 21 + 15 + 11 + 11 + 14) + 17]$$

$$= 162 \text{ mi}.$$

Using Simpson's approximation with $2n = 10$ and $\Delta x = 1$, we obtain

$$s = \int_0^{10} v(t)\,dt$$

$$\approx \tfrac{1}{3}[13 + 4(14) + 2(17) + 4(21) + 2(23) + 4(21)$$

$$+ 2(15) + 4(11) + 2(11) + 4(14) + 17]$$

$$= 162 \text{ mi}.$$

The submarine has traveled about 162 mi during this 10-hour time period.

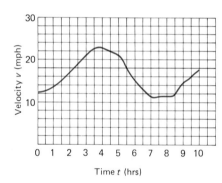

**5.40** Velocity graph for the submarine of Example 6

## *ERROR ESTIMATES

If $f$ is either increasing ($f'(x) > 0$) on $[a, b]$ or decreasing ($f'(x) < 0$) on $[a, b]$, then $\int_a^b f(x)\,dx$ lies between the end-point approximations $L_n$ and $R_n$. If $f$ is either concave upward ($f''(x) > 0$) on $[a, b]$ or concave downward ($f''(x) < 0$) on $[a, b]$, then this integral lies between the midpoint and trapezoidal approximations $M_n$ and $T_n$. So in these four cases the value of the integral can—in principle—be approximated with any desired degree of accuracy by choosing $n$ sufficiently large.

In addition, there are *error estimates* that can be used to predict in advance the maximum possible error in a particular approximation. Here we will simply state and briefly illustrate two such error estimates and defer further discussion until Section 11-5.

---

*Trapezoidal Error Estimate*

Suppose that $f''$ is continuous on $[a, b]$ and that $|f''(x)| \leq M$ for all $x$ in $[a, b]$. Then the difference between $T_n$ and $\int_a^b f(x)\,dx$ is at most $M(b - a)^3/12n^2$.

---

*Simpson's Error Estimate*

Suppose that $f^{(4)}$ is continuous on $[a, b]$ and that $|f^{(4)}(x)| \leq M$ for all $x$ in $[a, b]$. Then the difference between $S_n$ [$n$ even] and $\int_a^b f(x)\,dx$ is at most $M(b - a)^5/180n^4$.

---

For example, beginning with the integrand $f(x) = 4/(1 + x^2)$ of the integral in (7), repeated differentiation gives the second and fourth derivatives as

$$f''(x) = \frac{4(6x^2 - 2)}{(1 + x^2)^3} \quad \text{and} \quad f^{(4)}(x) = \frac{96(5x^4 - 10x^2 + 1)}{(1 + x^2)^5}.$$

By application of the closed-interval maximum-minimum method of Section 3-4, it can be verified (see Example 2 in Section 11-5) that

$$|f''(x)| \leq 8 \quad \text{and} \quad |f^{(4)}(x)| \leq 96$$

for $x$ in $[0, 1]$. Hence with $n = 10$ as in Example 2, the maximum error in the trapezoidal approximation is

$$\frac{(8)(1)^3}{12(10)^2} \approx 0.006666667,$$

so the actual value of $\pi$, according to the integral (7), is

$$3.139925990 \pm 0.006666667;$$

that is, $\pi$ lies between

$$3.133259323 \approx 3.133 \quad \text{and} \quad 3.146592657 \approx 3.147.$$

With $n = 10$ as in Example 4, the maximum error in Simpson's approximation is

$$\frac{(96)(1)^5}{(180)(10)^4} \approx 0.000053333$$

so the actual value of $\pi$, according to the integral (7), is

$$3.141592616 \pm 0.000053333;$$

thus $\pi$ lies between

$$3.141539283 \approx 3.1415 \quad \text{and} \quad 3.141645949 \approx 3.1417.$$

It is fairly typical of numerical integration that the approximations obtained, as in Examples 2 and 4, are actually much more accurate than the above error estimates suggest.

## 5-8  PROBLEMS

In Problems 1–5, calculate the right-end-point and left-end-point approximations to the given integral; use the indicated number of subintervals and round answers to two decimal places.

**1** $\int_1^2 x^2 \, dx, \quad n = 5$     **2** $\int_0^1 \sqrt{x} \, dx, \quad n = 5$

**3** $\int_0^1 \sqrt{1 + x^3} \, dx, \quad n = 4$     **4** $\int_0^1 \frac{dx}{1 + x}, \quad n = 10$

**5** $\int_0^1 \frac{\sqrt{1 + x} - 1}{x} \, dx, \quad n = 8$   (Make the integrand continuous by assuming that its value at $x = 0$ is $\frac{1}{2}$, its limit there.)

**6** Calculate the midpoint approximation to the integral of Problem 4 with $n = 10$, rounding the answer to four decimal places.

**7** Repeat Problem 6, but with the integral of Problem 5.

In Problems 8–12, calculate
(a) The trapezoidal approximation $T_n$, and
(b) Simpson's approximation $S_n$
to the given integral, rounding answers to four decimal places.

**8** $\int_1^3 x^3 \, dx, \quad n = 8$     **9** $\int_0^2 \frac{dx}{1 + x^3}, \quad n = 8$

**10** $\int_1^4 \sqrt{1 + x^4} \, dx, \quad n = 6$   **11** $\int_0^1 \sqrt{2 - x^2} \, dx, n = 10$

**12** $\int_0^1 \frac{\sqrt{1 + x} - 1}{x} \, dx, \quad n = 10$   (See the note to Problem 5.)

In Problems 13 and 14, calculate
(a) The trapezoidal approximation and
(b) Simpson's approximation
to $\int_a^b f(x) \, dx$, where $f$ is the given tabulated function.

**13**

| $x$ | $a = 1.00$ | 1.25 | 1.50 | 1.75 | 2.00 | 2.25 | $2.50 = b$ |
|---|---|---|---|---|---|---|---|
| $f(x)$ | 3.43 | 2.17 | 0.38 | 1.87 | 2.65 | 2.31 | 1.97 |

**14**

| $x$ | $a = 0$ | 1 | 2 | 3 | 4 | 5 | 6 | 7 | 8 | 9 | $10 = b$ |
|---|---|---|---|---|---|---|---|---|---|---|---|
| $f(x)$ | 23 | 8 | $-4$ | 12 | 35 | 47 | 53 | 50 | 39 | 29 | 5 |

**15** The graph in Fig. 5.41 shows the measured rate of water flow (gallons per minute) into a tank during a 10-min period. Using ten subintervals in each case, estimate the total amount of water flowing into the tank during this period using
(a) The trapezoidal approximation, and
(b) Simpson's approximation.

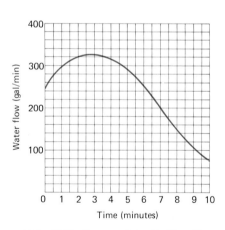

**5.41**  Water flow graph for Problem 15

**16** The graph in Fig. 5.42 shows the daily mean temperatures recorded during a winter month at a certain location. Using ten subintervals in each case, estimate the average temperature during the month using
(a) The trapezoidal approximation, and
(b) Simpson's approximation.

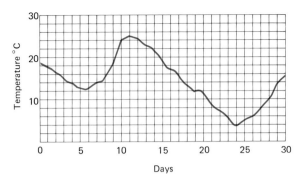

**5.42** Temperature graph for Problem 16

Problems 17–21 deal with the natural logarithm of 2, denoted by ln 2. In Chapter 7 we will see that the value of ln 2 is given by the integral

$$\ln 2 = \int_1^2 \frac{1}{x}\,dx. \tag{12}$$

The correct seven-place value is ln 2 ≈ 0.6931472.

**17** Calculate the end-point approximations $L_n$ and $R_n$ to the integral in (12); take $n = 10$.

**18** Use Equation (6) to determine how large $n$ must be in order to guarantee that $L_n$ and $R_n$ differ by no more than 0.0005.

**19** Explain why the integral in (12) lies between the midpoint approximation $M_n$ and the trapezoidal approximation $T_n$. Then calculate these approximations with $n = 10$ subintervals, rounding the answers to six decimal places.

**20** Use the trapezoidal error estimate to determine how large $n$ must be in order to guarantee that $T_n$ differs from ln 2 by at most 0.0005.

**21** Use the Simpson's error estimate to determine how large $n$ must be in order to guarantee that $S_n$ differs from ln 2 by at most 0.000005.

**22** The base for natural logarithms is a certain number $e$. In Chapter 7 we will see that

$$\int_1^e \frac{1}{x}\,dx = 1.$$

Approximate the integrals

$$\int_1^{2.7} \frac{1}{x}\,dx \quad \text{and} \quad \int_1^{2.8} \frac{1}{x}\,dx$$

with sufficient accuracy to show that $2.7 < e < 2.8$.

## *5-8   Optional Computer Application

It is difficult to think of an area where computer methods are more obviously applicable than in numerical integration. Let us write the trapezoidal approximation in the form

$$T_n = \frac{h}{2} \sum_{i=0}^{n} k_i f(x_i) \tag{13}$$

where $h = \Delta x = (b - a)/n$, $x_i = a + ih$, and

$$k_i = \begin{cases} 1 & \text{if } i = 0 \text{ or if } i = n, \\ 2 & \text{if } 0 < i < n. \end{cases} \tag{14}$$

The following program was written to compute $T_n$ in the case of Example 2, where $f(x) = 4/(1 + x^2)$.

```
10     INPUT  "A",  A
20     INPUT  "B",  B
30     INPUT  "N",  N
40     H = (B − A)/N
50     X = A  :  S = 0
60     FOR  I = 0  TO  N
70        IF  I = 0  THEN LET  K = 1:  GOTO 100
80        IF  I = N  THEN LET  K = 1:  GOTO 100
90        K = 2
100       LET  S = S + K*4/(1 + X*X)
110       X = X + H
120    NEXT  I
130    T = (H/2)*S
140    PRINT  "TRAPEZOIDAL APPROXIMATION:     T
150    END
```

Lines 10–30 call for input of the end points of the interval $[a, b]$ and of the desired number $n$ of subintervals. The running sum of functional values in (13) is denoted by the variable S, which is initialized to zero in line 50. The FOR-NEXT loop in lines 60–120 computes the value of S, which is then multiplied by $h/2$ in line 130 to get the trapezoidal approximation T. Note how the proper value of the multiplier $k$ is obtained in lines 70–90.

Before using this program with a different integrand function $f(x)$, the formula for $f(x)$ must be edited in line 100.

To write a program to compute Simpson's approximation, we write

$$S_n = \frac{h}{3} \sum_{i=0}^{n} k_i f(x_i), \tag{15}$$

where the number $n$ of subintervals is even, $h = \Delta x = (b - a)n$, $x_i = a + ih$, and

$$k_i = \begin{cases} 1 & \text{if } i = 0 \text{ or if } i = n, \\ 4 & \text{if } i \text{ is odd,} \\ 2 & \text{otherwise.} \end{cases} \tag{16}$$

The following program to compute $S_n$ is only a slight alteration of the trapezoidal approximation program above. The new line 90 assigns the multiplier $k$ in accord with (16), and $h/2$ is replaced by $h/3$ in line 130. In addition, the input statement in line 30 reminds us that $n$ must be even for use in Simpson's approximation.

```
10    INPUT  "A",  A
20    INPUT  "B",  B
30    INPUT  "N  EVEN",  N
40    H = (B − A)/N
50    X = A  :  S = 0
60    FOR  I = 0  TO  N
70      IF  I = 0  THEN LET  K = 1:  GOTO  100
80      IF  I = N  THEN LET  K = 1:  GOTO  100
90      K = 3 + (−1)↑(I + 1)
100     LET  S = S + K*4/(1 + X*X)
110     X = X + H
120   NEXT  I
130   S = (H/3)*S
140   PRINT  "SIMPSON'S APPROXIMATION:  ";  S
150   END
```

**Exercise 1** Enter and run the first program to check the result of Example 2.

**Exercise 2** Enter and run the second program to check the result of Example 4.

**Exercise 3** (a) Show first that $S_{2n} = \frac{1}{3}(4T_{2n} - T_n)$. (b) Then write a program that computes $S_{2n}$ by using the trapezoidal approximation program above as a subroutine to compute $T_n$ and $T_{2n}$.

**Exercise 4** Write a program that uses the trapezoidal approximation program as a subroutine to compute the sums $T_{10}$, $T_{20}$, $T_{40}$, $T_{80}, \ldots$, continuing until two consecutive results agree to seven decimal places. Test the program on three different functions (including $f(x) = 4/(x^2 + 1)$) for which you can obtain an answer of that accuracy by other methods.

**Exercise 5** Write a program that uses the Simpson's approximation program as a subroutine to compute the sums $S_{10}, S_{20}, S_{40}, S_{80}, \dots,$ continuing until two consecutive results agree to nine decimal places. Test the program on three different functions (including $f(x) = 4/(x^2 + 1)$) for which you can obtain an answer of that accuracy by other methods.

---

# CHAPTER 5 REVIEW:   Definitions, Concepts, Results

Use the list below as a guide to ideas that you may need to review.

1 Properties of area
2 Summation notation
3 The area under the graph of $f$ from $a$ to $b$
4 Inscribed and circumscribed rectangular polygons
5 A partition of $[a, b]$
6 The mesh of a partition
7 A Riemann sum associated with a partition
8 The (definite) integral of $f$ from $a$ to $b$
9 Existence of the integral of a continuous function
10 The integral as the limit of a sequence of Riemann sums
11 Regular partition; $\displaystyle\lim_{\Delta x \to 0} \sum_{i=1}^{n} f(x_i^*) \, \Delta x$
12 Upper and lower Riemann sums
13 The constant multiple, interval union, and comparison properties of integrals
14 Evaluation of definite integrals using antiderivatives
15 The average value of $f(x)$ on the interval $[a, b]$

16 The average value theorem
17 The fundamental theorem of calculus
18 Linearity of integration
19 Indefinite integral
20 The method of integration by substitution
21 Transforming the limits in definite integration by substitution
22 The area between $y = f(x)$ and $y = g(x)$ by integration with respect to $x$
23 The area between $x = f(y)$ and $x = g(y)$ by integration with respect to $y$
24 The right-end-point and left-end-point approximations
25 The trapezoidal approximation
26 The midpoint approximation
27 Simpson's approximation
28 Error estimates for the trapezoidal and Simpson's approximations

---

# MISCELLANEOUS PROBLEMS

Find the sums in Problems 1–4.

**1** $\displaystyle\sum_{i=1}^{100} 17$

**2** $\displaystyle\sum_{k=1}^{100} \left( \frac{1}{k} - \frac{1}{k+1} \right)$

**3** $\displaystyle\sum_{n=1}^{10} (3n - 2)^2$

**4** $\displaystyle\sum_{n=1}^{16} \sin \frac{n\pi}{2}$

In each of Problems 5–7, find the limit of the given Riemann sum associated with a regular partition of the indicated interval $[a, b]$. First express it as an integral from $a$ to $b$, and then evaluate that integral.

**5** $\displaystyle\lim_{n \to \infty} \sum_{i=1}^{n} \frac{\Delta x}{\sqrt{x_i^*}};$ $[1, 2]$

**6** $\displaystyle\lim_{n \to \infty} \sum_{i=1}^{n} [(x_i^*)^2 - 3x_i^*] \, \Delta x;$ $[0, 3]$

**7** $\displaystyle\lim_{n \to \infty} \sum_{i=1}^{n} 2\pi x_i^* \sqrt{1 + (x_i^*)^2} \, \Delta x;$ $[0, 1]$

**8** Evaluate

$$\lim_{n \to \infty} \frac{1}{n^{11}} (1^{10} + 2^{10} + 3^{10} + \cdots + n^{10})$$

by expressing this limit as an integral from 0 to 1.

**9** Use Riemann sums to prove that, if $f(x) \equiv c$ (a constant), then $\int_a^b f(x)\, dx = c(b - a)$.
**10** Use Riemann sums to prove that, if $f$ is continuous on $[a, b]$ and $f(x) \geq 0$ for all $x$ in $[a, b]$, then $\int_a^b f(x)\, dx \geq 0$.
**11** Use the comparison property of integrals (Section 5-4) to prove that $\int_a^b f(x)\, dx > 0$ if $f$ is a continuous function with $f(x) > 0$ on $[a, b]$.

Evaluate the integrals in Problems 12–25.

**12** $\displaystyle\int_0^1 (1 - x^2)^3 \, dx$

**13** $\displaystyle\int \left( \sqrt{2x} - \frac{1}{\sqrt{3x^3}} \right) dx$

**14** $\displaystyle\int \frac{(1 - \sqrt[3]{x})^2}{\sqrt{x}} \, dx$

**15** $\displaystyle\int \frac{4 - x^3}{2x^2} \, dx$

**16** $\displaystyle\int_0^1 \frac{dt}{(3 - 2t)^2}$

**17** $\displaystyle\int (x + 1)\sqrt{x^2 + 2x + 5} \, dx$

**18** $\displaystyle\int_0^2 x^2 \sqrt{9 - x^3} \, dx$

**19** $\displaystyle\int \frac{(1 + \sqrt{x})^7}{\sqrt{x}} \, dx$

$$20 \quad \int_1^2 \frac{2t + 1}{\sqrt{t^2 + t}}\, dt$$

$$21 \quad \int \frac{\sqrt[3]{u}}{(1 + u^{4/3})^3}\, du$$

$$22 \quad \int_{-1}^1 \frac{t}{\sqrt{t^2 + 1}}\, dt$$

$$23 \quad \int_1^4 \frac{(1 + \sqrt{t})^2}{\sqrt{t}}\, dt$$

$$24 \quad \int \frac{\sqrt[3]{1 - (1/u)}}{u^2}\, du$$

$$25 \quad \int \frac{\sqrt{4x^2 - 1}}{x^4}\, dx$$

Find the areas of the regions bounded by the curves given in Problems 26–32.

26 $y = x^3$ and the lines $x = -1$ and $y = 1$
27 $y = x^4$ and $y = x^5$
28 $y^2 = x$ and $3y^2 = x + 6$
29 $y = x^4$ and $y = 2 - x^2$
30 $y = x^4$ and $y = 2x^2 - 1$
31 $y = (x - 2)^2$ and $y = 10 - 5x$
32 $y = x^{2/3}$ and $y = 2 - x^2$

33 Evaluate the integral $\int_0^2 \sqrt{2x - x^2}\, dx$ by interpreting it as the area of a region.

34 Repeat Problem 33 for the integral

$$\int_1^5 \sqrt{6x - 5 - x^2}\, dx.$$

35 Find a function $f(x)$ such that

$$x^2 = 1 + \int_1^x \sqrt{1 + [f(t)]^2}\, dt$$

for all $x > 1$. (*Suggestion:* Differentiate both sides of the equation with the aid of the fundamental theorem of calculus.)

36 Show that $G'(x) = \phi(h(x))h'(x)$ if

$$G(x) = \int_a^{h(x)} \phi(t)\, dt.$$

37 Use right- and left-end-point approximations to estimate $\int_0^1 \sqrt{1 + x^2}\, dx$ with error not exceeding 0.05.

38 Calculate the trapezoidal and Simpson's approximations to

$$\int_0^1 x\sqrt{1 + x}\, dx$$

with $n = 6$ subintervals. For comparison, calculate the exact value of this integral by using the formula given in Problem 39 of Section 5-6.

39 Calculate the midpoint and trapezoidal approximations to $\int_1^2 \frac{dx}{x + x^2}$ with $n = 5$ subintervals. Explain why the exact value of the integral lies between these two approximations.

In Problems 40–42, let $\{x_0, x_1, x_2, \ldots, x_n\}$ be a partition of $[a, b]$ where $0 < a < b$.

40 For $i = 1, 2, 3, \ldots, n$, let $x_i^*$ be determined by

$$[x_i^*]^2 = \tfrac{1}{3}[(x_{i-1})^2 + x_{i-1}x_i + (x_i)^2].$$

Show first that $x_{i-1} < x_i^* < x_i$, and then use the algebraic identity

$$(c - d)(c^2 + cd + d^2) = c^3 - d^3$$

to show that

$$\sum_{i=1}^n (x_i^*)^2\, \Delta x_i = \tfrac{1}{3}(b^3 - a^3).$$

Explain why this computation proves that

$$\int_a^b x^2\, dx = \tfrac{1}{3}(b^3 - a^3).$$

41 Let $x_i^* = \sqrt{x_{i-1}x_i}$ for $i = 1, 2, 3, \ldots, n$. Show that

$$\sum_{i=1}^n \frac{\Delta x_i}{(x_i^*)^2} = \frac{1}{a} - \frac{1}{b}.$$

Then explain why this computation proves that

$$\int_a^b \frac{dx}{x^2} = \frac{1}{a} - \frac{1}{b}.$$

42 Define $x_i^*$ by means of the equation

$$(x_i^*)^{1/2}(x_i - x_{i-1}) = \tfrac{2}{3}[(x_i)^{3/2} - (x_{i-1})^{3/2}].$$

Show that $x_{i-1} < x_i^* < x_i$, and then use this selection for the given partition to prove that

$$\int_a^b \sqrt{x}\, dx = \tfrac{2}{3}(b^{3/2} - a^{3/2}).$$

43 In Section 5-8 we derived Equation (10) for Simpson's approximation by giving the mid-point approximation a weight of two-thirds and the trapezoidal approximation a weight of one-third. Why not some *other* weights? Read the discussion of parabolic approximations following Example 5 of Section 5-8, and then work Problem 47 of Section 5-7. This should provide most of the answer.

# 6

## Applications of the Integral

## Setting Up Integral Formulas

In Section 5-3 we defined the integral of the function $f$ from $a$ to $b$ as a limit of Riemann sums. Specifically, one begins with a partition $P$ of the interval $[a, b]$. A selection of numbers $x_i^*$ in each subinterval of $P$ produces a Riemann sum associated with the partition $P$; this is a number of the form

$$\sum_{i=1}^{n} f(x_i^*) \, \Delta x_i.$$

Finally, the integral of $f$ on $[a, b]$ is defined to be the limit of such sums as the mesh $|P|$ approaches zero. That is,

$$\int_a^b f(x) \, dx = \lim_{|P| \to 0} \sum_{i=1}^{n} f(x_i^*) \, \Delta x_i. \tag{1}$$

The wide applicability of the definite integral arises from the fact that many geometric and physical quantities can be approximated arbitrarily closely by Riemann sums of the sort that appear in the above definition. Such approximations lead to integral formulas for the computation of these quantities.

For example, our discussion in Sections 5-2 and 5-7 of the area from $a$ to $b$ under the graph of the positive-valued continuous function $f$ can be summarized as follows. Begin with a regular partition of $[a, b]$ into $n$ subintervals, each with length $\Delta x = (b - a)/n$. For each $i$ $(i = 1, 2, 3, \ldots, n)$, select a point $x_i^*$ in the $i$th subinterval $[x_{i-1}, x_i]$. Let $\Delta A_i$ denote the area below the graph of $f$ over this subinterval. Then this area is given approximately by

$$\Delta A_i \approx f(x_i^*) \, \Delta x.$$

Figure 6.1 shows why this approximation should be good and why it should improve as the mesh of the partition approaches zero.

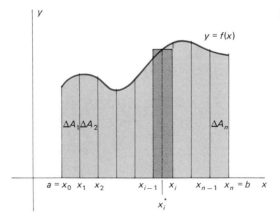

**6.1** Approximating an area by means of a Riemann sum

It follows that the total area $A$ under the graph of $f$ is given approximately by

$$A = \sum_{i=1}^{n} \Delta A_i \approx \sum_{i=1}^{n} f(x_i^*) \, \Delta x. \tag{2}$$

The point is that the approximating sum on the right is a Riemann sum for $f$ on $[a, b]$. Moreover:

1 It is intuitively evident that the Riemann sum in (2) approaches the actual area $A$ as $n \to \infty$.

2 By the definition of the integral, this Riemann sum approaches $\int_a^b f(x) \, dx$ as $n \to \infty$.

These observations justify the *definition* of the area $A$ by means of the formula

$$A = \int_a^b f(x) \, dx. \tag{3}$$

This justification of the area formula in (3) illustrates the following general method of setting up integral formulas. Suppose that we want to compute a certain quantity $Q$, where $Q$ is associated with the interval $[a, b]$ in such a way that subintervals of $[a, b]$ correspond to specific portions of $Q$. Also assume that if $\Delta Q_1, \Delta Q_2, \Delta Q_3, \ldots, \Delta Q_n$ are the portions of $Q$ corresponding to the subintervals of a partition of $[a, b]$ into $n$ subintervals, then

$$Q = \sum_{i=1}^{n} \Delta Q_i. \tag{4}$$

Such assumptions hold, for instance, if $Q$ is:

- An *area* lying over the interval $[a, b]$;
- The *distance* traveled by a moving particle during the time interval $a \leq t \leq b$;
- The *number of births* occurring in a certain population during the time interval $a \leq t \leq b$;
- The *volume* of water flowing into a tank during the time interval $a \leq t \leq b$; or
- The *work* done by a force in moving a particle from the point $x = a$ to the point $x = b$.

For example, if $Q$ is the volume of water flowing into a tank during the time interval $[a, b]$, then the portion $\Delta Q_i$ of $Q$ corresponding to the subinterval $[t_{i-1}, t_i]$ is the volume of water that flows into the tank during this subinterval of time. In any such situation it will generally turn out that the quantity $Q$ can be computed by evaluating a certain integral,

$$Q = \int_a^b f(x) \, dx \quad \text{or} \quad Q = \int_a^b f(t) \, dt, \tag{5}$$

depending on whether $x$ or $t$ is the independent variable. The key step in "setting up" such an integral formula is finding the particular function $f$ to be integrated.

The usual way of finding $f$ is to approximate the portion $\Delta Q_i$ of the quantity $Q$ that corresponds to the subinterval $[x_{i-1}, x_i]$. We are still thinking of a regular partition of $[a, b]$ into $n$ equal subintervals of length $\Delta x$, or into subintervals $[t_{i-1}, t_i]$ of length $\Delta t$ in case $t$ is the independent variable.

Suppose we find that

$$\Delta Q_i \approx f(x_i^*) \, \Delta x \tag{6}$$

for some point $x_i^*$ in $[x_{i-1}, x_i]$. What follows now is the approximation

$$Q = \sum_{i=1}^{n} \Delta Q_i \approx \sum_{i=1}^{n} f(x_i^*) \, \Delta x. \tag{7}$$

The right-hand sum in (7) is a Riemann sum that approaches the integral $\int_a^b f(x) \, dx$ as $n \to \infty$. If it is also evident—for geometric or physical reasons, for example—that this Riemann sum must approach the quantity $Q$ as $n \to \infty$, then (7) justifies our setting up the integral formula

$$Q = \int_a^b f(x) \, dx. \tag{5}$$

The crucial step above is the determination of the particular function $f$ in the approximation in (6): $\Delta Q_i \approx f(x_i^*) \, \Delta x$. If $Q$ is the area under a curve, then (as we have seen) $f(x)$ should be the height of the curve. If $Q$ is the distance traveled by a moving particle, then (as we shall see below) $f(t)$ should be its velocity function. If $Q$ is the volume of water flowing into a tank during a specified time interval, then $f(t)$ should be the rate of water flow (for instance, in gallons per minute). This chapter is devoted largely to ways of setting up integral formulas like (5) in a variety of specific situations. The applications that we consider include computations of distances, volumes, work, fluid pressure, lengths of curves, areas of surfaces of revolution, and centers of mass.

Observe that the integral in (5) results from the summation in (7) when we make the following replacements:

$$\sum_{i=1}^{n} \quad \text{becomes} \quad \int_a^b,$$

$$x_i^* \quad \text{becomes} \quad x,$$

$$\Delta x \quad \text{becomes} \quad dx.$$

**EXAMPLE 1** Suppose that water is pumped into an initially empty tank. It is known that the rate of flow of water into the tank after $t$ minutes is $50 - t$ gallons per minute. How much water flows into the tank during the first half hour?

*Solution*  We want to compute the amount $Q$ of water that flows into the tank during the time interval $[0, 30]$. Think of a regular partition of $[0, 30]$ into $n$ equal subintervals of length $\Delta t = 30/n$ each.

Next choose a point $t_i^*$ in the $i$th subinterval $[t_{i-1}, t_i]$. The rate of water flow between time $t_{i-1}$ and time $t_i$ is approximately $50 - t_i^*$ gallons per minute, so the amount $\Delta Q_i$ of water that flows into the tank during this subinterval of time is approximately given by

$$\Delta Q_i \approx (50 - t_i^*) \, \Delta t$$

(in gallons). Therefore the total amount $Q$ we seek is given approximately by

$$Q = \sum_{i=1}^{n} \Delta Q_i \approx \sum_{i=1}^{n} (50 - t_i^*) \, \Delta t$$

(in gallons, still). We recognize the Riemann sum on the right, and—most important—we see that the function involved is $f(t) = 50 - t$. Hence we may conclude that

$$Q = \lim_{n \to \infty} \sum_{i=1}^{n} (50 - t_i^*) \, \Delta t$$

$$= \int_0^{30} (50 - t) \, dt = \left[ 50t - \tfrac{1}{2}t^2 \right]_0^{30} = 1050 \text{ gal.}$$

**EXAMPLE 2**  Calculate $Q$ if

$$Q = \lim_{n \to \infty} \sum_{i=1}^{n} m_i \sqrt{25 - (m_i)^2} \, \Delta x,$$

where (for each $i = 1, 2, \ldots, n$) $m_i$ denotes the midpoint of the $i$th subinterval $[x_{i-1}, x_i]$ of a partition of the interval $[3, 4]$ into $n$ equal subintervals each of length $\Delta x = 1/n$.

**Solution**  We recognize the given sum as a Riemann sum (with $x_i^* = m_i$) for the function $f(x) = x\sqrt{25 - x^2}$. Hence

$$Q = \int_3^4 x\sqrt{25 - x^2} \, dx$$

$$= \left[ -\tfrac{1}{3}(25 - x^2)^{3/2} \right]_3^4$$

$$= \tfrac{1}{3}(16^{3/2} - 9^{3/2}) = \tfrac{37}{3}.$$

## DISTANCE AND VELOCITY

Consider a particle traveling along a (directed) straight line with velocity $v = f(t)$ at time $t$. We want to compute the **net distance** $s$ it travels between time $t = a$ and time $t = b$; that is, the distance between its initial position at time $t = a$ and its final position at time $t = b$.

We start with a regular partition of $[a, b]$ into $n$ equal subintervals, each of length $\Delta t = (b - a)/n$. If $t_i^*$ is an arbitrary point of the $i$th subinterval $[t_{i-1}, t_i]$, then during this subinterval the velocity of the particle is approximately $f(t_i^*)$. Hence the distance $\Delta s_i$ traveled by the particle between time $t = t_{i-1}$ and $t = t_i$ is given approximately by

$$\Delta s_i \approx f(t_i^*) \, \Delta t.$$

Consequently the net distance $s$ is given approximately by

$$s = \sum_{i=1}^{n} \Delta s_i \approx \sum_{i=1}^{n} f(t_i^*) \, \Delta t. \tag{8}$$

We recognize the right-hand approximation as a Riemann sum for the integral $\int_a^b f(t) \, dt$, and we reason that this approximation should approach the actual net distance $s$ as $n \to \infty$. Thus we conclude that

$$s = \int_a^b f(t) \, dt = \int_a^b v \, dt. \tag{9}$$

Note that the *net distance is the definite integral of the velocity.*

If the velocity function $v = f(t)$ has both positive and negative values for $t$ in $[a, b]$, then forward and backward distances cancel when we compute the net distance $s$ given by the integral formula in (9). The **total distance** traveled, irrespective of direction, may be found by evaluating the integral of the *absolute value* of the velocity:

$$\text{Total distance} = \int_a^b |f(t)| \, dt = \int_a^b |v| \, dt. \tag{10}$$

This integral may be computed by integrating separately over the subintervals where $v$ is positive and those where $v$ is negative and then adding the absolute values of the results.

**EXAMPLE 3**  Suppose that the velocity of a moving particle is $v = 30 - 2t$ feet per second. Find both the net distance and the total distance it travels between the times $t = 0$ and $t = 20$ seconds.

*Solution*  For net distance, we use Formula (9) and find that

$$s = \int_0^{20} (30 - 2t) \, dt = \left[ 30t - t^2 \right]_0^{20} = 200$$

feet. To find the total distance traveled, we note that $v$ is positive if $t < 15$, while $v$ is negative if $t > 15$. Since

$$\int_0^{15} (30 - 2t) \, dt = \left[ 30t - t^2 \right]_0^{15} = 225 \text{ ft}$$

and

$$\int_{15}^{20} (2t - 30) \, dt = \left[ t^2 - 30t \right]_{15}^{20} = 25 \text{ ft},$$

we see that the particle travels 225 ft forward and then 25 ft backward, for a total distance of 250 ft.

---

The derivation of Formula (9) above, based on the intuitive reasoning that the approximation in (8) *ought* to approach $s$ as $n \to \infty$, is a "plausible justification" of the distance formula rather than a rigorous proof of it. Plausible justifications of this sort will suffice for most of our purposes, but on occasion we shall need a rigorous proof of an integral formula. A rigorous proof of distance formula (9) can be given as follows.

Begin as before with a regular partition of $[a, b]$ into $n$ equal subintervals of length $\Delta t$. Let $f(t_i^\flat)$ and $f(t_i^\#)$ be the minimum and maximum values, respectively, of the velocity function for $t$ in the $i$th subinterval $[t_{i-1}, t_i]$. Then the distance $\Delta s_i$ traveled by the particle during this subinterval satisfies the inequalities

$$f(t_i^\flat) \, \Delta t \leq \Delta s_i \leq f(t_i^\#) \, \Delta t.$$

Since

$$s = \sum_{i=1}^n \Delta s_i,$$

addition of these inequalities for $i = 1, 2, 3, \ldots, n$ gives

$$\sum_{i=1}^n f(t_i^\flat) \, \Delta t \leq s \leq \sum_{i=1}^n f(t_i^\#) \, \Delta t.$$

Now we note that each of the last two sums is a Riemann sum for $f$ on $[a, b]$. So, assuming that $f$ is continuous, both sums approach $\int_a^b f(t) \, dt$ as $n \to \infty$.

The squeeze law for limits therefore gives

$$\int_a^b f(t)\, dt \leqq s \leqq \int_a^b f(t)\, dt,$$

which establishes Formula (9).

In each of Problems 1–10, compute both the net distance and the total distance traveled between times $t = a$ and $t = b$ by a particle moving along a line with the given velocity function $v = f(t)$.

1  $v = -32$;   $a = 0, b = 10$
2  $v = 2t + 10$;   $a = 1, b = 5$
3  $v = 4t - 25$;   $a = 0, b = 10$
4  $v = |2t - 5|$;   $a = 0, b = 5$
5  $v = 4t^3$;   $a = -2, b = 3$

6  $v = t - \dfrac{1}{t^2}$;   $a = \dfrac{1}{10}, b = 1$

7  $v = t^2$;   $a = -1, b = 2$
8  $v = t^3$;   $a = -1, b = 2$
9  $v = t^4$;   $a = -2, b = 2$
10  $v = 9t - t^3$;   $a = -1, b = 3$

In each of Problems 11–19, $x_i^*$ denotes an arbitrary point and $m_i$ denotes the midpoint of the $i$th subinterval $[x_{i-1}, x_i]$ of a regular partition of the indicated interval $[a, b]$ into $n$ equal subintervals of length $\Delta x$. Evaluate the given limit by computing the value of the appropriate related integral.

11  $\displaystyle\lim_{n \to \infty} \sum_{i=1}^{n} 2x_i^* \, \Delta x$;   $a = 0, b = 1$

12  $\displaystyle\lim_{n \to \infty} \sum_{i=1}^{n} \frac{\Delta x}{(x_i^*)^2}$;   $a = 1, b = 2$

13  $\displaystyle\lim_{n \to \infty} \sum_{i=1}^{n} \sqrt{x_i^*} \, \Delta x$;   $a = 0, b = 4$

14  $\displaystyle\lim_{n \to \infty} \sum_{i=1}^{n} [3(x_i^*)^2 - 1] \, \Delta x$;   $a = -1, b = 3$

15  $\displaystyle\lim_{n \to \infty} \sum_{i=1}^{n} x_i^* \sqrt{(x_i^*)^2 + 9} \, \Delta x$;   $a = 0, b = 4$

16  $\displaystyle\lim_{n \to \infty} \sum_{i=1}^{n} (x_i)^2 \, \Delta x$;   $a = 2, b = 4$

17  $\displaystyle\lim_{n \to \infty} \sum_{i=1}^{n} (2m_i - 1) \, \Delta x$;   $a = -1, b = 3$

18  $\displaystyle\lim_{n \to \infty} \sum_{i=1}^{n} \sqrt{2m_i + 1} \, \Delta x$;   $a = 0, b = 4$

19  $\displaystyle\lim_{n \to \infty} \sum_{i=1}^{n} \frac{m_i}{\sqrt{(m_i)^2 + 16}} \, \Delta x$;   $a = -3, b = 0$

In Problems 20–23, the notation is the same as in Problems 11–19. Express the given limit as an integral involving the function $f(x)$.

20  $\displaystyle\lim_{n \to \infty} \sum_{i=1}^{n} 2\pi x_i^* f(x_i^*) \, \Delta x$;   $a = 1, b = 4$

21  $\displaystyle\lim_{n \to \infty} \sum_{i=1}^{n} [f(x_i^*)]^2 \, \Delta x$;   $a = -1, b = 1$

22  $\displaystyle\lim_{n \to \infty} \sum_{i=1}^{n} \sqrt{1 + [f(x_i^*)]^2} \, \Delta x$;   $a = 0, b = 10$

23  $\displaystyle\lim_{n \to \infty} \sum_{i=1}^{n} 2\pi m_i \sqrt{1 + [f(m_i)]^2} \, \Delta x$;   $a = -2, b = 3$

24  If a particle is thrown straight upward from the ground with an initial velocity of 160 ft/s, then its velocity after $t$ seconds is $v = -32t + 160$ ft/s, and it attains its maximum height when $t = 5$ s (and $v = 0$). Use the formula in (9) to compute this maximum height. Check your answer by the method of Section 4-9.

25  Suppose that the rate of flow of water into an initially empty tank is $100 - 3t$ gallons per minute after $t$ minutes have elapsed. How much water flows into the tank over the interval from $t = 10$ to $t = 20$ min?

26  Suppose that the birth rate in a certain city $t$ years after 1960 was $13 + t$ thousands of births per year. Set up and evaluate an appropriate integral to compute the total number of births that occurred between 1960 and 1980.

27  Assume that the city of Problem 26 had a death rate of $5 + t/2$ thousands per year $t$ years after 1960. If the population of the city was 125,000 in 1960, what was its population in 1980? Consider both births and deaths in this problem.

28  The average daily rainfall in a certain locale is $r(t)$ inches per day at time $t$ (days), $0 \leq t \leq 365$. Begin with a regular partition of the interval $[0, 365]$ and derive the formula $R = \int_0^{365} r(t)\, dt$ for the average total annual rainfall.

29  Consider a thin straight rod of length $L$ and total mass $m$ that coincides with the interval $[0, L]$ on the $x$-axis. Let the continuous function $\rho(x)$ give the *density* of the rod at each point $x$. This means that, given a segment $\Delta x$ of the rod, the mass $\Delta m$ of this segment satisfies

$$\rho(x^\flat) \, \Delta x \leq \Delta m \leq \rho(x^\#) \, \Delta x,$$

where $\rho(x^\flat)$ and $\rho(x^\#)$ are the minimum and maximum values, respectively, of the density at points of the segment. Begin with a regular partition of $[0, L]$ and use the method by which we established the formula in Equation (9) at the end of this section to derive rigorously the formula

$$m = \int_a^b \rho(x)\, dx$$

for the mass of the rod.

**30** Suppose that the rate of flow of water into a tank is $r(t)$ gallons per minute at time $t$. Use the method by which we established Formula (9) at the end of this section to derive rigorously the formula $Q = \int_a^b r(t)\, dt$ for the amount of water that flows into the tank between times $t = a$ and $t = b$.

**31** Evaluate

$$\lim_{n \to \infty} \frac{\sqrt[3]{1} + \sqrt[3]{2} + \sqrt[3]{3} + \cdots + \sqrt[3]{n}}{n^{4/3}}$$

by first finding a function $f$ such that the limit is equal to $\int_0^1 f(x)\, dx$.

**32** A spherical ball has radius 1 ft and, at distance $r$ from its center, its density is $100(1 + r)$ pounds per cubic foot. Use Riemann sums to find a function $f(r)$ such that the weight of the ball is

$$W = \int_0^1 f(r)\, dr$$

(in pounds). Then compute $W$ by evaluating this integral. (*Suggestion:* The surface area of a sphere of radius $r$ is given by the formula $A = 4\pi r^2$. Given a partition $0 = r_0 < r_1 < \cdots < r_n = 1$ of $[0, 1]$, estimate the weight $\Delta W_i$ of the spherical shell $r_{i-1} \leq r \leq r_i$ of the ball.)

---

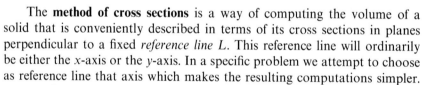

## 6-2

## Volumes by the Method of Cross Sections

In this section we show how to use integrals to calculate the volumes of certain solids or regions in space. We begin with the intuitive idea that volume has properties analogous to the properties of area listed in Section 5-2. That is, we assume that every reasonably nice solid region $R$ has a volume $v(R)$, a nonnegative real number with the following properties:

**1** If the solid region $R$ is contained in the solid region $S$, then $v(R) \leq v(S)$.

**2** If $R$ and $S$ are nonoverlapping solid regions, then the volume of their union is $v(R \cup S) = v(R) + v(S)$.

**3** If the solid regions $R$ and $S$ are congruent (have the same size and shape), then $v(R) = v(S)$.

The **method of cross sections** is a way of computing the volume of a solid that is conveniently described in terms of its cross sections in planes perpendicular to a fixed *reference line L*. This reference line will ordinarily be either the $x$-axis or the $y$-axis. In a specific problem we attempt to choose as reference line that axis which makes the resulting computations simpler.

We first consider the case in which $L$ is the $x$-axis. Suppose that the solid $R$ with volume $V = v(R)$ that we want to calculate lies opposite the interval $[a, b]$ on the $x$-axis. That is, a plane perpendicular to the $x$-axis intersects the solid if and only if this plane meets the $x$-axis in a point of $[a, b]$, as indicated in Fig. 6.2. Let $R_x$ denote the intersection of $R$ with the perpendicular plane that meets the $x$-axis at the point $x$ of $[a, b]$. We call $R_x$ the (plane) **cross section** of the solid $R$ at $x$.

This situation is especially nice if all the cross sections of $R$ are congruent to one another and, in addition, are parallel translates of each other. In this case the solid $R$ is called a **cylinder** with **bases** $R_a$ and $R_b$ and height $h = b - a$. If $R_a$ and $R_b$ are circles, then $R$ is the familiar **circular cylinder.**

In order to calculate volumes by integration, we need a fourth property of volume adjoined to the three listed above:

**4** The volume of any cylinder, circular or not, is the product of its height and the area of its base.

This is a generalization of the familiar formula $V = \pi r^2 h$ for the volume of a circular cylinder with radius $r$ and height $h$.

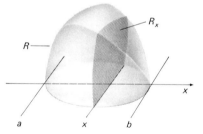

**6.2** $R_x$ is the cross section of $R$ in the plane perpendicular to the $x$-axis at $x$.

Now for each $x$ in $[a, b]$, let $A(x)$ denote the area of the cross section $R_x$ of our original solid $R$:

$$A(x) = a(R_x).\tag{1}$$

We shall assume that the shape of $R$ is nice enough so that this **cross-sectional area function** $A$ is continuous and, therefore, integrable.

To set up an integral formula for $V = v(R)$, we begin with a regular partition of $[a, b]$ into $n$ equal subintervals each with length $\Delta x = (b - a)/n$. Let $R_i$ denote the slab or slice of the solid $R$ that lies opposite the $i$th subinterval $[x_{i-1}, x_i]$, as shown in Fig. 6.3. We denote the volume of this $i$th slice of $R$ by $\Delta V_i = v(R_i)$, so that

$$V = \sum_{i=1}^{n} \Delta V_i.$$

To approximate $\Delta V_i$, we select an arbitrary point $x_i^*$ in $[x_{i-1}, x_i]$, and consider the *cylinder* $C_i$ with height $\Delta x$ and whose base is the cross section $R_{x_i^*}$ of $R$ at $x_i^*$. Figure 6.4 suggests that if $\Delta x$ is small then $v(C_i)$ is a good approximation to $\Delta V_i = v(R_i)$,

$$\Delta V_i \approx v(C_i) = a(R_{x_i^*})\Delta x = A(x_i^*)\Delta x,$$

using Property (4) of volume.

Then we add the volumes of these approximating cylinders for $i = 1, 2, 3, \ldots, n$. We find that

$$V = \sum_{i=1}^{n} \Delta V_i \approx \sum_{i=1}^{n} A(x_i^*)\,\Delta x.$$

We recognize the approximating sum on the right as a Riemann sum that approaches $\int_a^b A(x)\, dx$ as $n \to \infty$. But this sum should also approach the actual

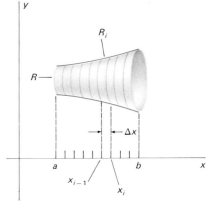

**6.3** Planes through the partition points $x_0, x_1, x_2, \ldots, x_n$ partition the solid $R$ into slabs $R_1, R_2, \ldots, R_n$.

**6.4** The slab $R_i$ is approximated by the cylinder $C_i$ of volume $a(R_{x_i^*})\,\Delta x$.

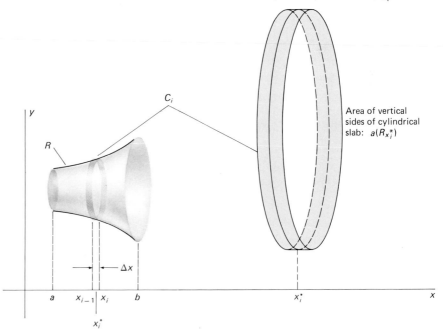

Area of vertical sides of cylindrical slab: $a(R_{x_i^*})$

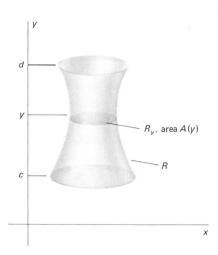

**6.5** $A(y)$ is the area of the cross section $R_y$ in the plane perpendicular to the y-axis at the point y.

volume $V$ as $n \to \infty$. This justifies the following *definition* of the volume of a solid $R$ in terms of its cross-sectional area function $A(x)$.

---

**Definition**   *Volume by Cross Sections*

If the solid $R$ lies opposite the interval $[a, b]$ on the x-axis and has continuous cross-sectional area function $A(x)$, then its volume $V = v(R)$ is

$$V = \int_a^b A(x)\, dx. \tag{2}$$

---

Formula (2) is sometimes called **Cavalieri's principle,** after the Italian mathematician Bonaventura Cavalieri (1598–1647), who systematically exploited the fact that the volume of a solid is determined by the areas of its cross sections perpendicular to a given reference line.

In the case of a solid $R$ that lies opposite the interval $[c, d]$ on the y-axis, we denote by $A(y)$ the area of its cross section $R_y$ in the plane that is perpendicular to the y-axis at the point $y$ of $[c, d]$ (see Fig. 6.5). A similar discussion, starting with a regular partition of $[c, d]$, leads to the volume formula

$$V = \int_c^d A(y)\, dy. \tag{3}$$

### SOLIDS OF REVOLUTION

An important special case of Formula (2) gives the volume of a **solid of revolution.** For example, let the solid $R$ be obtained by revolving around the x-axis the region under the graph of $y = f(x)$ over the interval $[a, b]$, where $f(x) \geqq 0$. The region and the resulting solid are shown in Fig. 6.6.

Since the solid $R$ is obtained by revolution, each cross section of $R$ at $x$ is a circular disk of radius $f(x)$. The cross-sectional area function is then $A(x) = \pi[f(x)]^2$, so Formula (2) gives us

$$V = \int_a^b \pi y^2\, dx = \int_a^b \pi[f(x)]^2\, dx \tag{4}$$

for the **volume of a solid of revolution around the x-axis.**

NOTE:   In the expression $\int \pi y^2\, dx$, the differential $dx$ tells us what the independent variable is. You *must* express $y$ (and any other dependent variables) in terms of $x$ in order to perform the indicated integration.

**6.6**   Volume of revolution about the x-axis

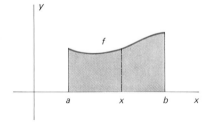

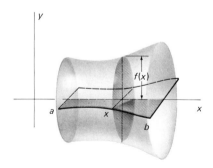

By a similar argument, if the region bounded by the curve $x = g(y)$, the $y$-axis, and the horizontal lines $y = c$ and $y = d$ is rotated about the $y$-axis, then the volume of the resulting solid of revolution (Fig. 6.7) is

$$V = \int_c^d \pi x^2 \, dy = \int_c^d \pi [g(y)]^2 dy. \qquad (5)$$

Again, it is essential to express $x$ (and any other dependent variables) in terms of the independent variable $y$ before attempting to antidifferentiate in (5).

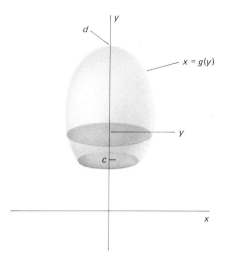

**6.7**  Volume of revolution about the $y$-axis

**EXAMPLE 1**    Use the method of cross sections to verify the familiar formula $V = \frac{4}{3}\pi R^3$ for the volume of a sphere of radius $R$.

**Solution**    We think of this sphere as the solid of revolution obtained by revolving the semicircular plane region of Fig. 6.8 around the $x$-axis. This is the region bounded above by the semicircle $y = \sqrt{R^2 - x^2}$ ($-R \le x \le R$) and below by the interval $[-R, R]$ on the $x$-axis. To use Formula (4), we take $f(x) = \sqrt{R^2 - x^2}$, $a = -R$, and $b = R$. This gives

$$V = \int_{-R}^R \pi(\sqrt{R^2 - x^2})^2 \, dx = \pi \int_{-R}^R (R^2 - x^2) \, dx$$

$$= \pi \left[ R^2 x - \tfrac{1}{3}x^3 \right]_{-R}^R = \tfrac{4}{3}\pi R^3.$$

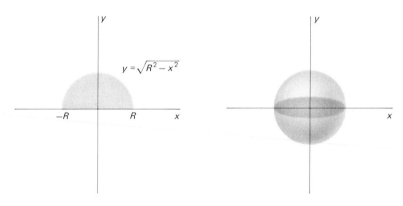

$$y = \sqrt{R^2 - x^2}$$

**6.8**  A sphere generated by rotation of a semicircular region

**6.9**  Generating a cone by rotation (see Example 2)

**EXAMPLE 2**    Use the method of cross sections to verify the familiar formula $V = \frac{1}{3}\pi r^2 h$ for the volume of a right circular cone with base radius $r$ and height $h$.

**Solution**    We may think of the cone as the solid of revolution obtained by revolving the plane region bounded by the $y$-axis and the lines $y = h$ and $x = ry/h$ around the $y$-axis (as in Fig. 6.9). Then Formula (5) with $g(y) = ry/h$ gives

$$V = \int_0^h \pi \left( \frac{ry}{h} \right)^2 dy = \frac{\pi r^2}{h^2} \int_0^h y^2 \, dy$$

$$= \frac{\pi r^2}{h^2} \left[ \frac{1}{3} y^3 \right]_0^h = \frac{1}{3} \pi r^2 h.$$

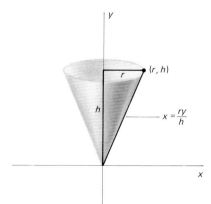

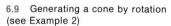

**6.10** The wedge of Example 3

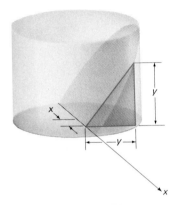

**6.11** A cross section of the wedge—an isosceles triangle

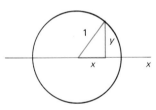

**6.12** The base of the cylinder

**EXAMPLE 3** Find the volume of the "wedge" that is cut from a circular cylinder with unit radius *and* height by a plane that passes through a diameter of the bottom base of the cylinder and through a point on the circumference of its top.

*Solution* The cylinder and wedge are shown in Fig. 6.10. To form such a wedge, fill a cylindrical glass with cider and then drink slowly until half the bottom of the glass is exposed; the remaining cider forms the wedge of Example 3.

We choose as reference line and *x*-axis the line through the "edge of the wedge"—the original diameter of the base of the cylinder. It's easy to verify with similar triangles that each cross section of the wedge perpendicular to the *x*-axis is an isosceles right triangle. One of these triangles is shown in Fig. 6.11. We denote by *y* the (equal) base and height of this triangle.

In order to determine the cross-sectional area function $A(x)$, we must express *y* in terms of *x*. Figure 6.12 shows the unit circular base of the original cylinder. We apply the Pythagorean theorem to the right triangle in this figure and thus find that $y = \sqrt{1 - x^2}$. Hence

$$A(x) = \tfrac{1}{2}y^2 = \tfrac{1}{2}(1 - x^2),$$

so Formula (2) gives

$$V = \int_{-1}^{1} A(x)\,dx = 2 \int_{0}^{1} A(x)\,dx \qquad \text{(by symmetry)}$$

$$= 2 \int_{0}^{1} \tfrac{1}{2}(1 - x^2)\,dx = \left[x - \tfrac{1}{3}x^3\right]_{0}^{1} = \tfrac{2}{3}$$

for the volume of the wedge.

---

A useful habit is the checking of answers for plausibility whenever convenient. For example, we may compare a given solid with one whose volume is already known. Since the volume of the original cylinder in Example 3 is $\pi$, we have found that the wedge occupies the fraction

$$\frac{V_{wedge}}{V_{cyl}} = \frac{\tfrac{2}{3}}{\pi} \approx 21\%$$

of the volume of the cylinder. A glance at Fig. 6.10 indicates that this is plausible. An error in our computation would likely have given an implausible answer, thereby revealing the existence of the error.

The wedge of Example 3 has an ancient history. Its volume was first calculated in the third century B.C. by Archimedes, to whom also is originally due the formula $V = \tfrac{4}{3}\pi r^3$ for the volume of a sphere. His work on the wedge is found in a manuscript that was rediscovered in 1906 after having been lost for many centuries. Archimedes used a method of exhaustion for volumes similar to that discussed for areas in Section 5-2.

Sometimes we need to calculate the volume of a solid generated by revolution of a plane region lying between two given curves. Suppose that $f(x) > g(x) > 0$ for *x* in the interval $[a, b]$, and that the solid *R* is generated by revolving the region between $y = f(x)$ and $y = g(x)$ about the $x$-axis. Then the cross section at $x$ is an **annular ring** bounded by two circles, as shown in Fig. 6.13. The ring has inner radius $r_i = g(x)$ and outer radius $r_0 = f(x)$, so our formula for the cross-sectional area of $R$ at $x$ is

$$A(x) = \pi(r_0)^2 - \pi(r_i)^2 = \pi([f(x)]^2 - [g(x)]^2).$$

Thus, by the formula in (2), the volume $V$ of $R$ is given by

$$V = \int_a^b \pi\left[(r_0)^2 - (r_i)^2\right] dx$$

$$= \int_a^b \pi([f(x)]^2 - [g(x)]^2)\, dx. \qquad (6)$$

Similarly, if $f(y) > g(y) > 0$ for $y$ in $[c, d]$, then the volume of the solid obtained by revolving the region between $x = f(y)$ and $x = g(y)$ about the $y$-axis is

$$V = \int_c^d \pi([f(y)]^2 - [g(y)]^2)\, dy. \qquad (7)$$

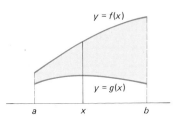

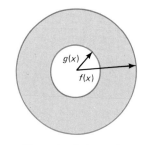

**6.13** The region between two positive graphs is rotated about the $x$-axis; cross sections are annular rings.

**EXAMPLE 4** Consider the region shown in Fig. 6.14, bounded by the curves $y = x^2$ and $y = x^3$ over the interval $0 \le x \le 1$. If this region is rotated about the $x$-axis, then the formula in (6) with $r_0 = x^2$ and $r_i = x^3$ gives the volume swept out as

$$V = \int_0^1 \pi[(x^2)^2 - (x^3)^2]\, dx$$

$$= \pi \int_0^1 (x^4 - x^6)\, dx$$

$$= \pi \left[\frac{x^5}{5} - \frac{x^7}{7}\right]_0^1 = \frac{2\pi}{35}.$$

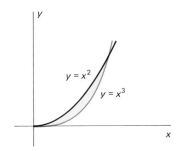

**6.14** The plane region bounded by the graphs of $y = x^2$ and $y = x^3$

If the same region is rotated about the $y$-axis, then each cross section perpendicular to the $y$-axis is an annular ring with outer radius $x = y^{1/3}$ and inner radius $x = y^{1/2}$. Hence the formula in (7) gives the volume of the resulting solid as

$$V = \int_0^1 \pi[(y^{1/3})^2 - (y^{1/2})^2]\, dy$$

$$= \pi \int_0^1 (y^{2/3} - y)\, dy$$

$$= \pi \left[\frac{3}{5} y^{5/3} - \frac{1}{2} y^2\right]_0^1 = \frac{\pi}{10}.$$

**EXAMPLE 5** Suppose instead that the region of Example 4 is rotated about the vertical line $x = -1$. Then each horizontal cross section of the resulting solid is an annular ring with inner radius $r_i = 1 + y^{1/2}$ and outer radius $r_0 = 1 + y^{1/3}$. The area of such a cross section is therefore

$$A(y) = \pi(1 + y^{1/3})^2 - \pi(1 + y^{1/2})^2$$
$$= \pi(y^{2/3} + 2y^{1/3} - y - 2y^{1/2}),$$

and so the volume of the solid is

$$V = \int_0^1 \pi(y^{2/3} + 2y^{1/3} - y - 2y^{1/2})\, dy$$

$$= \pi\left[\frac{3}{5} y^{5/3} + \frac{3}{2} y^{4/3} - \frac{1}{2} y^2 - \frac{4}{3} y^{3/2}\right]_0^1 = \frac{4\pi}{15}.$$

## 6-2 PROBLEMS

In each of Problems 1–20, find the volume of the solid that is generated by rotating about the indicated axis the plane region bounded by the given curves.

1  $y = x^2$, $y = 0$, $x = 1$;   about the x-axis
2  $y = \sqrt{x}$, $y = 0$, $x = 4$;   about the x-axis
3  $y = x^2$, $y = 4$, $x = 0$ (first quadrant only);   about the y-axis
4  $y = 1/x$, $y = 0$, $x = \frac{1}{10}$, $x = 1$;   about the x-axis
5  $y = x - x^3$ (on $[0, 1]$), $y = 0$;   about the x-axis
6  $y = 9 - x^2$, $y = 0$;   about the x-axis
7  $y = x^2$, $x = y^2$;   about the x-axis
8  $y = x^2$, $y = 4x$;   about the line $x = 5$
9  $y = x^2$, $y = 8 - x^2$;   about the x-axis
10  $x = y^2$, $x = y + 6$;   about the y-axis
11  $y = 1 - x^2$, $y = 0$;   about the x-axis
12  $y = x - x^3$, $y = 0$ $(0 \le x \le 1)$;   about the x-axis
13  $y = 1 - x^2$, $y = 0$;   about the y-axis
14  $y = 6 - x^2$, $y = 2$;   about the x-axis
15  $y = 6 - x^2$, $y = 2$;   about the y-axis
16  $y = 1 - x^2$, $y = 0$;   about the vertical line $x = 2$
17  $y = x - x^3$, $y = 0$ $(0 \le x \le 1)$;   about the horizontal line $y = -1$
18  $y = 4$, $x = 0$, $y = x^2$;   about the x-axis
19  $y = 4$, $x = 0$, $y = x^2$;   about the y-axis
20  $x = 16 - y^2$, $x = 0$, $y = 0$;   about the x-axis
21  Find the volume of the solid generated by rotating the region bounded by the parabolas $y^2 = x$ and $y^2 = 2(x - 3)$ about the x-axis.

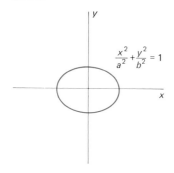

**6.15**   The ellipse of Problems 22 and 23

22  Find the volume of the ellipsoid that is generated by rotating the region bounded by the ellipse with equation $x^2/a^2 + y^2/b^2 = 1$ about the x-axis (see Fig. 6.15).
23  Repeat Problem 22, except rotate the elliptical region about the y-axis.
24  Find the volume of the unbounded solid that is generated by rotating the unbounded region of Fig. 6.16 about the x-axis. This is the region between the graph of $y = 1/x^2$ and the x-axis for $x \ge 1$. (*Method:* Compute the volume from $x = 1$ to $x = b$ (where $b > 1$); then find the limit of this volume as $b \to \infty$.)

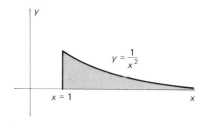

**6.16**   The unbounded plane region of Problem 24

25  The base of a certain solid is a circle with diameter $AB$ of length $2a$. Find the volume of the solid if each cross section perpendicular to $AB$ is a square.
26  Repeat Problem 25 if each cross section of the solid perpendicular to $AB$ is a semicircle rather than a square.
27  Repeat Problem 25 if each cross section of the solid perpendicular to $AB$ is an equilateral triangle.
28  The base of a certain solid is the region in the $xy$-plane bounded by the parabolas $y = x^2$ and $x = y^2$. Find the volume of this solid if every cross section perpendicular to the x-axis is a square with base in the $xy$-plane.
29  The paraboloid generated by rotating about the x-axis the region under the parabola $y^2 = 2px$, $0 \le x \le h$, is shown in Fig. 6.17. Show that the volume of the paraboloid is one-half that of the indicated circumscribed cylinder.
30  A pyramid has height $h$ and square base with area $A$. Show that its volume is $V = \frac{1}{3}Ah$. Note that each cross section parallel to the base is a square.
31  Repeat Problem 30, except make the base a triangle with area $A$.

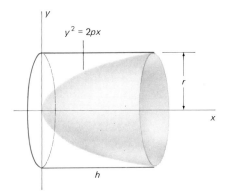

**6.17** The paraboloid and cylinder of Problem 29

**32** Find the volume that remains after a hole of radius 3 is bored through the center of a solid sphere of radius 5.

**33** Two horizontal circular cylinders each have radius $a$, and their axes intersect at right angles. Find the volume of their solid of intersection. (*Suggestion:* Draw the usual $xy$-coordinate axes and add a $z$-axis "coming out of the paper" toward you. Imagine the first cylinder with its axis coinciding with the $x$-axis and the second with its axis coinciding with the $z$-axis. But draw only the quarter of the first cylinder that lies in front of the $xy$-plane and above the $xz$-plane. Draw only the quarter of the second that lies to the right of the $yz$-plane and above the $xz$-plane. Their intersection is one-eighth the total volume and is now relatively easy to sketch.)

**34** Figure 6.18 shows a "spherical segment" of height $h$ cut off from a sphere of radius $r$ by a horizontal plane. Show that its volume is $V = \frac{1}{3}\pi h^2(3r - h)$.

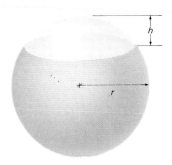

**6.18** A spherical segment

**35** A doughnut-shaped solid called a *torus* is generated by revolving around the $y$-axis the circular disk

$$(x - b)^2 + y^2 \leq a^2$$

centered at the point $(b, 0)$ where $0 < a < b$. Show that the volume of this torus is $V = 2\pi^2 a^2 b$. (*Suggestion:* Note that each cross section perpendicular to the $y$-axis is an annular ring, and recall that

$$\int_0^a \sqrt{a^2 - y^2} \, dy = \frac{1}{4}\pi a^2$$

because the integral represents the area of a quarter-circle of radius $a$.)

**36** The summit of a hill is 100 ft higher than the surrounding level terrain, and each horizontal cross section of the hill is circular. The following table gives the radius $r$ for selected values of the height $h$ above the surrounding terrain. Use Simpson's approximation to estimate the volume of the hill.

| $h$ (ft) | 0 | 25 | 50 | 75 | 100 |
|---|---|---|---|---|---|
| $r$ (ft) | 60 | 55 | 50 | 35 | 0 |

**37** *Newton's Wine Barrel.* Consider a barrel with the shape of the solid generated by revolving around the $x$-axis the region under the parabola $y = R - kx^2$, $-h/2 \leq x \leq h/2$.
(a) Show that the radius of each end of the barrel is $r = R - \delta$, where $4\delta = kh^2$.
(b) Then show that the volume of the barrel is

$$V = \frac{1}{3}\pi h(2R^2 + r^2 - \frac{6}{15}\delta^2).$$

**38** *The Clepsydra, or Water Clock.* Consider a water tank whose lateral surface is generated by rotating the curve $y = kx^4$ about the $y$-axis ($k$ is a positive constant).
(a) Compute $V(y)$, the volume of water in the tank as a function of its depth $y$.
(b) Suppose that water drains from the tank through a small hole at the bottom. Use the chain rule and Torricelli's law (Equation (3) in Section 4-8) to show that the water level in this tank falls at a *constant* rate.

**39** Let $V$ be the volume generated by revolving around the $x$-axis the area under the graph of the continuous and nonnegative function $f$ from $x = a$ to $x = b$. Let $f(x_i^\flat)$ and $f(x_i^\#)$ be the minimum and maximum values of $f$ on the $i$th subinterval of a regular partition of $[a, b]$. Deduce from Properties 1–4 of volume that

$$\sum_{i=1}^{n} \pi [f(x_i^\flat)]^2 \, \Delta x \leq V \leq \sum_{i=1}^{n} \pi [f(x_i^\#)]^2 \, \Delta x.$$

Explain why this implies the validity of Formula (4).

**40** Water evaporates from an open bowl at a rate proportional to the area of the surface of the water. Show that whatever the shape of the bowl, the water level will drop at a constant rate.

**41** A frustum of a right circular cone has height $h$ and volume $V$. Its base is a circle with radius $R$ and its top is a circle with radius $r$. Apply the method of cross sections to show that

$$V = \frac{\pi}{3}(r^2 + rR + R^2)h.$$

The program that follows uses the method of cross sections to approximate the volume of the hemisphere of radius 1 obtained by rotating around the $x$-axis the area under the quarter-circle $y = \sqrt{1 - x^2}$, $0 \le x \le 1$. The interval is subdivided into $n$ equal subintervals, each of length $h = \Delta x = (b - a)/n$, and the FOR-NEXT loop in lines 150–190 computes the sum S of the volumes of the corresponding slices. In line 170 the volume

$$\pi(y_{k-1})^2 \, \Delta x = \pi[1 - (x_{k-1})^2]h$$

is added to the previous value of S.

```
100    INPUT  "A";  A
110    INPUT  "B";  B
120    PI = 3.14159
130    I = 0  :  N = 10
140    X = A  :  S = 0
150    FOR  K = 1  TO  N
160          H = (B − A)/N
170          S = S + PI*(1 − X*X)*H
180          X = X + H
190    NEXT  K
200    J = S
210    PRINT  N,  J
220    IF  ABS(I − J) < 0.001  THEN GOTO  250
230    I = J  :  N  = 2*N
240    GOTO  140
250    PRINT  "ANSWER:  ";  J
260    END
```

The approximating sum  S  is computed with the successive numbers  N = 10, 20, 40, ... of subintervals. At each state,  I  denotes the previous approximation and  J  denotes the new approximation. These two approximations are compared in line 220. Unless $|I - J| < 0.001$, the number  N  of subintervals is doubled (line 230) and the process is repeated.

**Exercise 1** Enter and run this program. The answer should be approximately $2\pi/3$.

**Exercise 2** Alter this program to approximate the volume of the paraboloid obtained by rotating about the $x$-axis the region under the parabola $y^2 = x$, $0 \le x \le 1$. As a consequence of Problem 29, the answer should be approximately $\pi/2$.

## 6-3

## Volume by the Method of Cylindrical Shells

The method of cross sections of Section 6-2 is a technique of approximating a solid with a stack of thin slabs or slices. In the case of a solid of revolution, these slices are circular disks or annular rings. The **method of cylindrical shells** is a second way of computing volumes of solids of revolution. It is a technique of approximating a solid of revolution with a collection of thin cylindrical shells, and it sometimes leads to simpler computations.

A **cylindrical shell** is a region bounded by two concentric circular cylinders of the same height $h$. If, as in Fig. 6.19, the inner cylinder has radius $r_1$ and the outer one has radius $r_2$, then we can write $\bar{r} = (r_1 + r_2)/2$ for the **average radius** of the cylindrical shell and $t = r_2 - r_1$ for its **thickness.** We then get the volume of the cylindrical shell by subtracting the volume of the inner cylinder from that of the outer one. So the shell has volume

$$V = \pi r_2^2 h - \pi r_1^2 h = 2\pi \frac{r_1 + r_2}{2}(r_2 - r_1)h = 2\pi\bar{r}th. \tag{1}$$

**6.19**  A cylindrical shell

In words, the volume of the shell is the product of $2\pi$, its average radius, its thickness, and its height.

Now suppose that we want to find the volume $V$ of revolution generated by revolving around the $y$-axis the region under $y = f(x)$ from $x = a$ to $x = b$. We assume, as in Fig. 6.20, that $0 \leq a < b$ and that $f(x)$ is nonnegative on $[a, b]$. The solid will resemble the one also shown in Fig. 6.20.

To find $V$, we begin with a regular partition of $[a, b]$ into $n$ equal subintervals each of length $\Delta x = (b - a)/n$. Let $x_i^*$ denote the midpoint of the $i$th subinterval $[x_{i-1}, x_i]$, and consider the rectangle with base $[x_{i-1}, x_i]$ and height $f(x_i^*)$. When this rectangle is revolved about the $y$-axis, it sweeps out a cylindrical shell like the one in Fig. 6.20, with average radius $x_i^*$, thickness $\Delta x$, and height $f(x_i^*)$. This cylindrical shell approximates the solid with volume $\Delta V_i$ that is obtained by revolving the region under $y = f(x)$ and over $[x_{i-1}, x_i]$, and thus Formula (1) gives

$$\Delta V_i \approx 2\pi x_i^* f(x_i^*)\,\Delta x.$$

We add the volumes of the $n$ cylindrical shells determined by the partition. This sum should approximate $V$, since—as Fig. 6.20 suggests—the union of

**6.20**  A solid of revolution—note the hole through its center—and a way to approximate it with nested cylindrical shells.

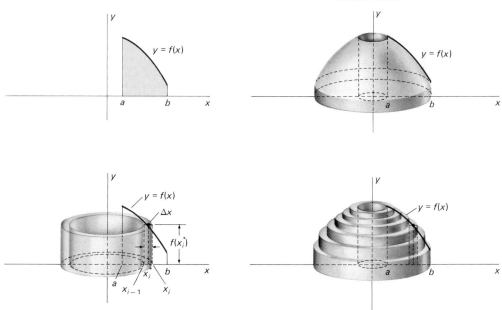

the shells should approximate the solid of revolution. Thus we obtain the approximation

$$V = \sum_{i=1}^{n} \Delta V_i \approx \sum_{i=1}^{n} 2\pi x_i^* f(x_i^*) \, \Delta x.$$

This approximation to the volume $V$ is a Riemann sum **that approaches** $\int_a^b 2\pi x f(x) \, dx$ as $\Delta x \to 0$, so it appears that the volume of our solid of revolution is given by

$$V = \int_a^b 2\pi x f(x) \, dx. \tag{2}$$

A complete discussion would require a proof that this formula gives the same volume as that *defined* by the method of cross sections in Section 6-2 (see the appendix to this chapter).

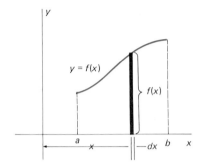

**6.21** Heuristic device for setting up the formula in Equation (2)

It is more reliable to learn how to set up integral formulas than merely to memorize such formulas. A useful heuristic device for setting up the formula in (2) is to picture the very narrow rectangular strip of area shown in Fig. 6.21. When this strip is revolved about the $y$-axis, it produces a thin cylindrical shell of radius $x$, height $y = f(x)$, and thickness $dx$. So, if its volume is denoted by $dV$, we may write $dV = 2\pi x f(x) \, dx$. We think of $V$ as a sum of very many such volumes, nested concentrically to form the solid of revolution. This makes it natural to write

$$V = \int dV = \int_a^b 2\pi x y \, dx = \int_a^b 2\pi x f(x) \, dx.$$

Do not forget to express $y$ (and any other dependent variables) in terms of the independent variable $x$ (identified by the differential $dx$) before integrating.

**EXAMPLE 1**  Find the volume of the solid generated by revolving around the $y$-axis the region under $y = 3x^2 - x^3$ from $x = 0$ to $x = 3$ (this region is shown in Fig. 6.22).

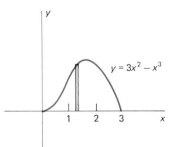

**6.22** The region of Example 1—rotate it about the $y$-axis.

*Solution*  Here it would not be practical to use the method of cross sections, because a cross section perpendicular to the $y$-axis is an annular ring and finding its inner and outer radii would require solution of the equation $y = 3x^2 - x^3$ for $x$ in terms of $y$. We would prefer to avoid this troublesome task, and Formula (2) provides us with an alternative: We take $f(x) = 3x^2 - x^3$, $a = 0$, and $b = 3$. It immediately follows that

$$V = \int_0^3 2\pi x (3x^2 - x^3) \, dx = 2\pi \int_0^3 (3x^3 - x^4) \, dx$$

$$= 2\pi \left[ \frac{3}{4} x^4 - \frac{1}{5} x^5 \right]_0^3 = \frac{243\pi}{10}.$$

**EXAMPLE 2**  Find the volume of the solid that remains after boring a hole of radius $a$ through the center of a solid sphere of radius $r > a$.

*Solution*  We think of the sphere of radius $r$ as being generated by revolving the right half of the circular disk $x^2 + y^2 \leq r^2$ around the $y$-axis, and we think of the hole as vertical and with its centerline coincident with part of the y-axis. Then the upper *half* of our solid is generated by revolving the region shaded in Fig. 6.23 around the y-axis—this is the region under the graph of $y = \sqrt{r^2 - x^2}$ (and over the x-axis) from $x = a$ to $x = r$. The volume of the entire sphere-with-hole is then double that of the upper half, and Formula (2) gives

$$V = 2 \int_a^r 2\pi x \sqrt{r^2 - x^2}\, dx = 4\pi \left[ -\tfrac{1}{3}(r^2 - x^2)^{3/2} \right]_a^r,$$

so that

$$V = \frac{4\pi}{3}(r^2 - a^2)^{3/2}.$$

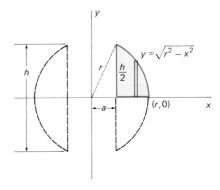

**6.23** Middle cross section of the sphere-with-hole

A way to check this answer is to test it in some extreme cases. If $a = 0$, which corresponds to boring no hole at all, our result reduces to the volume $V = \tfrac{4}{3}\pi r^3$ of the whole sphere. If $a = r$, which corresponds to using a drill bit as large as the sphere, then $V = 0$; this, too, is correct.

Now let $A$ denote the region between the curves $y = f(x)$ and $y = g(x)$ over the interval $[a, b]$, where $0 \leq a < b$ and $g(x) \leq f(x)$ for $x$ in $[a, b]$. Such a region is shown in Fig. 6.24. When $A$ is rotated about the y-axis, it generates a solid of revolution. Suppose that we want to find the volume $V$ of this solid. A development similar to that of Formula (2) leads to the approximation

$$V \approx \sum_{i=1}^n 2\pi x_i^* [f(x_i^*) - g(x_i^*)]\, \Delta x,$$

from which we may conclude that

$$V = \int_a^b 2\pi x [f(x) - g(x)]\, dx. \tag{3}$$

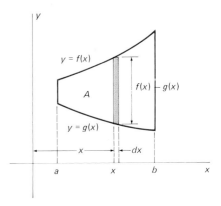

**6.24** The region $A$ between the graphs of $f$ and $g$ over $[a, b]$ is to be rotated about the y-axis.

The method of cylindrical shells is also an effective way to compute volumes of solids of revolution about the x-axis. Figure 6.25 shows the region $A$ bounded by the curves $x = f(y)$ and $x = g(y)$ for $c \leq y \leq d$ and by the horizontal lines $y = c$ and $y = d$. Let $V$ be the volume obtained by revolving the region $A$ about the x-axis. To compute $V$, we begin with a regular partition of $[c, d]$ into $n$ equal subintervals of length $\Delta y$ each. Let $y_i^*$ denote the midpoint of the $i$th subinterval $[y_{i-1}, y_i]$. Then the volume of the cylindrical shell with average radius $y_i^*$, height $f(y_i^*) - g(y_i^*)$, and thickness $\Delta y$ is

$$V_i = 2\pi y_i^* [f(y_i^*) - g(y_i^*)]\, \Delta y.$$

We add the volumes of these cylindrical shells and thus obtain the approximation

$$V \approx \sum_{i=1}^n 2\pi y_i^* [f(y_i^*) - g(y_i^*)]\, \Delta y.$$

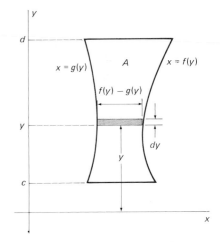

**6.25** The region $A$ is to be rotated about the x-axis.

We recognize the right-hand side as a Riemann sum for an integral with respect to $y$ from $c$ to $d$, and so conclude that the volume of the solid of revolution is given by

$$V = \int_c^d 2\pi y [f(y) - g(y)]\, dy. \tag{4}$$

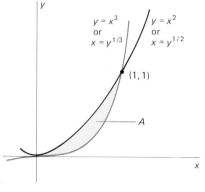

**6.26** The region of Example 3

**EXAMPLE 3** Consider the region in the first quadrant that is bounded by the curves $y = x^2$ and $y = x^3$, shown in Fig. 6.26. Use the method of cylindrical shells to compute the volumes of the solids obtained by revolving this region first about the $y$-axis, then instead about the $x$-axis.

*Solution* For the volume of revolution around the $y$-axis, Formula (3) can be applied with $f(x) = x^2$ and $g(x) = x^3$. This gives

$$V = \int_0^1 2\pi x(x^2 - x^3)\, dx = 2\pi\left[\frac{x^4}{4} - \frac{x^5}{5}\right]_0^1 = \frac{\pi}{10}.$$

For the volume of revolution around the $x$-axis, we use Formula (4) with $f(y) = y^{1/3}$ and $g(y) = y^{1/2}$:

$$V = \int_0^1 2\pi y[y^{1/3} - y^{1/2}]\, dy$$

$$= 2\pi\left[\frac{3}{7}y^{7/3} - \frac{2}{5}y^{5/2}\right]_0^1 = \frac{2\pi}{35}.$$

These answers are the same, of course, as those we obtained using the method of cross sections in Example 4 of Section 6-2.

**EXAMPLE 4** Suppose instead that the region of Example 3 is rotated about the vertical line $x = -1$. Then the area element $dA = (x^2 - x^3)\, dx$ is revolved through a circle of radius $r = x + 1$, so the volume of the resulting cylindrical shell is

$$dV = 2\pi r\, dA$$

$$= 2\pi(x + 1)(x^2 - x^3)\, dx = 2\pi(x^2 - x^4)\, dx.$$

Hence the volume of the solid is

$$V = \int_0^1 2\pi(x^2 - x^4)\, dx$$

$$= 2\pi\left[\frac{1}{3}x^3 - \frac{1}{5}x^5\right]_0^1 = \frac{4\pi}{15}.$$

This is the same answer we obtained in Example 5 of Section 6-2, where the method of cross sections led to a slightly more complicated integral.

## 6-3 PROBLEMS

In each of Problems 1–28, use the method of cylindrical shells to find the volume of the solid generated by revolving about the indicated axis the region bounded by the given curves.

1. $y = x^2$, $y = 0$, $x = 2$; about the $y$-axis
2. $x = y^2$, $x = 4$; about the $y$-axis
3. $y = 25 - x^2$, $y = 0$; about the $y$-axis
4. $y = 2x^2$, $y = 8$; about the $y$-axis
5. $y = x^2$, $y = 8 - x^2$; about the $y$-axis

6. $x = 9 - y^2$, $x = 0$; about the $x$-axis
7. $x = y$, $x + 2y = 3$, $y = 0$; about the $x$-axis
8. $y = x^2$, $y = 2x$; about the line $y = 5$
9. $y = 2x^2$, $y^2 = 4x$; about the $x$-axis
10. $y = 3x - x^2$, $y = 0$; about the $y$-axis
11. $y = 4x - x^3$, $y = 0$; about the $y$-axis
12. $x = y^3 - y^4$, $x = 0$; about the line $y = -2$
13. $y = x - x^3$, $y = 0$ $(0 \le x \le 1)$; about the $y$-axis

**14** $x = 16 - y^2$, $x = 0$, $y = 0$ $(0 \le y \le 4)$; about the $x$-axis

**15** $y = x - x^3$, $y = 0$ $(0 \le x \le 1)$; about the vertical line $x = 2$

**16** $y = x^3$, $y = 0$, $x = 2$; about the $y$-axis

**17** $y = x^3$, $y = 0$, $x = 2$; about the vertical line $x = 3$

**18** $y = x^3$, $y = 0$, $x = 2$; about the $x$-axis

**19** $y = x^2$, $y = 0$, $x = -1$, $x = 1$; about the vertical line $x = 2$

**20** $y = x^2$, $y = x$ $(0 \le x \le 1)$; about the $y$-axis

**21** $y = x^2$, $y = x$ $(0 \le x \le 1)$; about the $x$-axis

**22** $y = x^2$, $y = x$ $(0 \le x \le 1)$; about the horizontal line $y = 2$

**23** $y = x^2$, $y = x$ $(0 \le x \le 1)$; about the vertical line $x = -1$

**24** $x = y^2$, $x = 2 - y^2$; about the $x$-axis

**25** $x = y^2$, $x = 2 - y^2$; about the horizontal line $y = 1$

**26** $y = 4x - x^2$, $y = 0$; about the $y$-axis

**27** $y = 4x - x^2$, $y = 0$; about the vertical line $x = -1$

**28** $y = x^2$, $x = y^2$; about the horizontal line $y = -1$

**29** Verify the formula for the volume of a cone by using the method of cylindrical shells. Apply the method to the figure generated by revolving the triangle with vertices $(0, 0)$, $(r, 0)$, and $(0, h)$ about the $y$-axis.

**30** Use the method of cylindrical shells to compute the volume of the paraboloid of Problem 29 in Section 6-2.

**31** Use the method of cylindrical shells to find the volume of the ellipsoid obtained by revolving the elliptical region bounded by $x^2/a^2 + y^2/b^2 = 1$ about the $y$-axis.

**32** Use the method of cylindrical shells to derive the formula given in Problem 34 of Section 6-2 for the volume of a spherical segment.

**33** Use the method of cylindrical shells to compute the volume of the torus of Problem 35 in Section 6-2. (*Suggestion:* Substitute $u$ for $x - b$ in the integral given by the formula in Equation (2).)

**34** (a) Find the volume of the solid generated by revolving the region bounded by the curves $y = x^2$ and $y = x + 2$ about the line $x = -2$.
(b) Repeat (a), but revolve the region about the line $x = 3$.

**35** Find the volume of the solid generated by revolving the circular disk $x^2 + y^2 \le a^2$ about the line $x = -a$.

**36** The region that lies under the parabola $y = \sqrt{x + 1}$ over the interval $3 \le x \le 8$ is revolved around the $y$-axis. Use the integral formula given in Problem 39 of Section 5-6 to find the volume of the resulting solid of revolution.

**37** In Example 2 of this section, we found that the volume remaining when a hole of radius $a$ is bored through the center of a sphere of radius $r > a$ is

$$V = \frac{4\pi}{3}(r^2 - a^2)^{3/2}.$$

(a) Express the volume $V$ above *without* use of the hole radius $a$; use instead the hole height $h$. (*Suggestion:* Use the right triangle shown in Fig. 6.23.)
(b) What is remarkable about the answer to part (a)?

---

## 6-4

# Arc Length and Surface Area of Revolution

A **smooth arc** is the graph of a smooth function defined on a closed interval; a **smooth function** $f$ on $[a, b]$ is one which has derivative $f'$ that is continuous on $[a, b]$. The continuity of $f'$ rules out the possibility of corner points on the graph of $f$, points where the direction of the tangent line changes abruptly. The graphs of the functions $f(x) = |x|$ and $g(x) = x^{2/3}$ are shown in Fig. 6.27; neither is smooth because each has a corner point at the origin.

6.27 Graphs having corner points

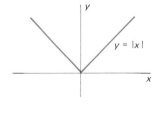

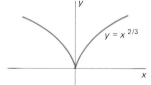

To investigate the length of a smooth arc, we begin with the length of a straight line segment, which is simply the distance between its endpoints. Then, given a smooth arc $C$, we pose the following question: If $C$ were a thin wire and we straightened it without stretching it, how long would the resulting straight wire be? The answer is what we would call the *length* of $C$.

To approximate the length $s$ of the smooth arc $C$, we can inscribe in $C$ a polygonal arc—one made up of straight line segments—and then calculate the length of this polygonal arc. We proceed in the following way, under the assumption that $C$ is the graph of a smooth function $f$ on the closed interval $[a, b]$. Consider a regular partition of $[a, b]$ into $n$ subintervals each having length $\Delta x$. Let $P_i$ denote the point $(x_i, f(x_i))$ on the arc $C$ corresponding to the *i*th subdivision point $x_i$. Our polygonal arc "inscribed in" $C$ is then the

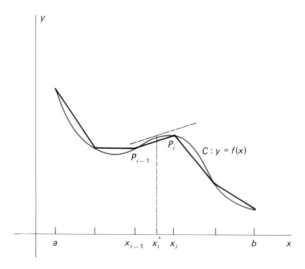

**6.28** A polygonal arc inscribed in the smooth curve C

union of the line segments $P_0P_1, P_1P_2, P_2P_3, \ldots, P_{n-1}P_n$. So our approximation to the length $s$ of $C$ is

$$s \approx \sum_{i=1}^{n} |P_{i-1}P_i|, \tag{1}$$

the sum of the lengths of these line segments (see Fig. 6.28).

The length of $P_{i-1}P_i$ is

$$|P_{i-1}P_i| = [(x_i - x_{i-1})^2 + (f(x_i) - f(x_{i-1}))^2]^{1/2}.$$

We apply the mean value theorem to the function $f$ on the interval $[x_{i-1}, x_i]$, and thereby conclude the existence of a point $x_i^*$ in this interval such that

$$f(x_i) - f(x_{i-1}) = f'(x_i^*)(x_i - x_{i-1}).$$

Hence

$$|P_{i-1}P_i| = \left[1 + \left(\frac{f(x_i) - f(x_{i-1})}{x_i - x_{i-1}}\right)^2\right]^{1/2} (x_i - x_{i-1})$$

$$= [1 + [f'(x_i^*)]^2]^{1/2}\, \Delta x$$

where $\Delta x = x_i - x_{i-1}$.

We next substitute this expression for $|P_{i-1}P_i|$ into (1), which gives the approximation

$$s \approx \sum_{i=1}^{n} \sqrt{1 + [f'(x_i^*)]^2}\, \Delta x.$$

This sum is a Riemann sum for the function $\sqrt{1 + [f'(x)]^2}$ on $[a, b]$, and therefore (because $f'$ is continuous) approaches the integral

$$\int_a^b \sqrt{1 + [f'(x)]^2}\, dx$$

as $\Delta x \to 0$. But our approximation ought also to approach the actual length $s$ as $\Delta x \to 0$. On this basis we *define* the **length** $s$ of the smooth arc $C$ to be

$$s = \int_a^b \sqrt{1 + [f'(x)]^2}\, dx = \int_a^b \sqrt{1 + \left(\frac{dy}{dx}\right)^2}\, dx. \tag{2}$$

**306**

In the case of a smooth arc given as a graph $x = g(y)$ for $y$ in $[c, d]$, a similar discussion starting with a regular partition of $[c, d]$ leads to the formula

$$s = \int_c^d \sqrt{1 + [g'(y)]^2}\, dy = \int_c^d \sqrt{1 + \left(\frac{dx}{dy}\right)^2}\, dy \qquad (3)$$

for its length. The length of a more general curve, such as a circle, can be computed by subdividing it into finitely many smooth arcs, and then applying to each of these arcs whichever of Formulas (2) or (3) is required.

There is a convenient symbolic device that we can employ to remember Formulas (2) and (3) simultaneously. We think of two nearby points $P(x, y)$ and $Q(x + dx, y + dy)$ on the smooth arc $C$, and denote by $ds$ the length of the arc joining $P$ and $Q$. Imagine that $P$ and $Q$ are so close together that $ds$ is, for all practical purposes, equal to the length of the straight line segment $PQ$. Then the Pythagorean theorem, applied to the small right triangle in Fig. 6.29, gives

$$ds = \sqrt{(dx)^2 + (dy)^2} \qquad (4)$$

$$= \sqrt{1 + \left(\frac{dy}{dx}\right)^2}\, dx \qquad (4')$$

$$= \sqrt{1 + \left(\frac{dx}{dy}\right)^2}\, dy. \qquad (4'')$$

Thinking of the whole length $s$ of $C$ as the sum of small pieces like $ds$, we write

$$s = \int ds. \qquad (5)$$

Then formal (symbolic) substitution of Expressions (4′) and (4″) for $ds$ in Equation (5) yields Formulas (2) and (3); only the limits of integration remain to be determined.

**EXAMPLE 1** Find the length of the "semicubical parabola" $y = x^{3/2}$ on $[0, 5]$, shown in Fig. 6.30.

**Solution**   We first compute the integrand of Formula (2):

$$\left[1 + \left(\frac{dy}{dx}\right)^2\right]^{1/2} = \left[1 + \left(\frac{3}{2}x^{1/2}\right)^2\right]^{1/2}$$

$$= \left(1 + \frac{9}{4}x\right)^{1/2} = \frac{1}{2}(4 + 9x)^{1/2}.$$

Hence the length of the arc $y = x^{3/2}$ over the interval $[0, 5]$ is

$$s = \int_0^5 \tfrac{1}{2}(4 + 9x)^{1/2}\, dx = \left[\tfrac{1}{27}(4 + 9x)^{3/2}\right]_0^5 = \tfrac{335}{27}.$$

**EXAMPLE 2**   Find the length of the curve

$$x = \frac{y^3}{6} + \frac{1}{2y}; \qquad 1 \leqq y \leqq 2.$$

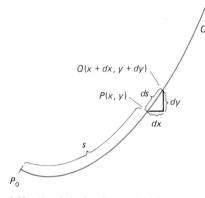

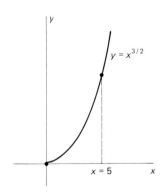

**6.29**   Heuristic development of the arc length formula

**6.30**   The semicubical parabola of Example 1

**Solution** Here $y$ is the natural independent variable. We calculate

$$1 + \left(\frac{dx}{dy}\right)^2 = 1 + \left(\frac{y^2}{2} - \frac{1}{2y^2}\right)^2$$

$$= 1 + \left(\frac{y^4}{4} - \frac{1}{2} + \frac{1}{4y^4}\right)$$

$$= \frac{y^4}{4} + \frac{1}{2} + \frac{1}{4y^4}$$

$$= \left(\frac{y^2}{2} + \frac{1}{2y^2}\right)^2.$$

Thus we can "get out from under the radical" in Formula (3):

$$s = \int_c^d \sqrt{1 + \left(\frac{dx}{dy}\right)^2}\, dy$$

$$= \int_1^2 \left(\frac{y^2}{2} + \frac{1}{2y^2}\right) dy$$

$$= \left[\frac{y^3}{6} - \frac{1}{2y}\right]_1^2 = \frac{17}{12}.$$

**EXAMPLE 3** Figure 6.31 shows a parabolic suspension cable (as in Section 4-8) for a 100-ft bridge. Its shape is described by

$$y = \frac{x^2}{100}, \qquad -50 \leqq x \leqq 50.$$

To find the length of this suspension cable, we first calculate

$$ds = \sqrt{1 + \left(\frac{dy}{dx}\right)^2}\, dx$$

$$= \sqrt{1 + \left(\frac{x}{50}\right)^2}\, dx = \frac{1}{50}\sqrt{x^2 + 2500}\, dx.$$

Hence

$$s = \int_{-50}^{50} \frac{1}{50}\sqrt{x^2 + 2500}\, dx = \int_0^{50} \frac{1}{25}\sqrt{x^2 + 2500}\, dx.$$

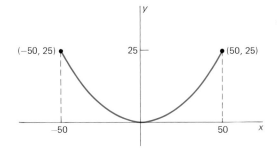

**6.31** The suspension cable of Example 3.

The last integral cannot be evaluated exactly with methods presently available to us, so we approximate it using Simpson's approximation (Section 5-8).

If $x_i$ denotes the $i$th subdivision point of a partition of $[0, 50]$ into $n = 10$ equal subintervals each of length $\Delta x = 5$, and

$$y_i = \tfrac{1}{25}\sqrt{x_i^2 + 2500}$$

for $i = 0, 1, 2, \ldots, 10$, then Simpson's approximation yields

$$\begin{aligned}
s &\approx \tfrac{5}{3}(y_0 + 4y_1 + 2y_2 + 4y_3 + \cdots + 2y_8 + 4y_9 + y_{10}) \\
&= \tfrac{1}{15}(\sqrt{2500} + 4\sqrt{2525} + 2\sqrt{2600} \\
&\quad + \cdots + 4\sqrt{4525} + \sqrt{5000}); \\
s &\approx 114.78 \quad \text{(ft)}.
\end{aligned}$$

Using techniques of integration that we will study in Chapter 9, one can show that this result is accurate to two decimal places.

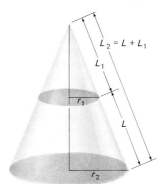

6.32   A frustum of a cone; the slant height is $L$.

### AREA OF SURFACES OF REVOLUTION

A **surface of revolution** is one obtained by revolving an arc or curve about an axis lying in its plane. The surface of a cylinder or of a sphere and the curved surface of a cone are important examples of surfaces of revolution.

Our basic approach to the area of such a surface is this: First we inscribe a polygonal arc in the curve to be revolved. We then regard the area of the surface generated by revolving the polygonal arc as an approximation to the surface generated by revolving the original curve. Since a surface generated by revolving a polygonal arc about an axis consists of frusta (sections) of cones, we can calculate its area in a reasonably simple way.

This approach to surface area originated with Archimedes. For example, it's the method he used to establish the formula $A = 4\pi r^2$ for the surface area of a sphere of radius $r$.

We will need the formula

$$A = 2\pi \bar{r} L \tag{6}$$

for the area of a frustum of a cone with average radius $\bar{r} = \tfrac{1}{2}(r_1 + r_2)$ and *slant height* $L$, as shown in Fig. 6.32. The formula in (6) follows from the formula

$$A = \pi r L \tag{7}$$

for the area of a conical surface with base radius $r$ and slant height $L$, as shown in Fig. 6.33. It is easy to derive the formula in (7) by "unrolling" the conical surface onto a sector of a circle of radius $L$ (also in Fig. 6.33) because the area of this sector is

$$A = \frac{2\pi r}{2\pi L} \cdot \pi L^2 = \pi r L.$$

To derive the formula in (6) from the formula in (7), we think of the frustum as the lower section of a cone with slant height $L_2 = L + L_1$, as indicated in Fig. 6.34. Then subtraction of the area of the upper conical

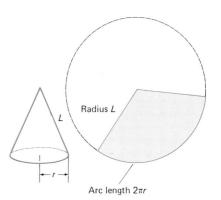

6.33   Surface area of cone: Cut along $L$, and then unroll the cone onto the circular sector.

6.34   Derivation of Formula (6)

section from that of the entire cone gives

$$A = \pi r_2 L_2 - \pi r_1 L_1 = \pi r_2 (L + L_1) - \pi r_1 L_1$$
$$= \pi (r_2 - r_1) L_1 + \pi r_2 L$$

for the area of the frustum. But the similar right triangles of Fig. 6.34 yield the proportion

$$\frac{r_1}{L_1} = \frac{r_2}{L_2} = \frac{r_2}{L + L_1},$$

from which we find that $(r_2 - r_1)L_1 = r_1 L$. Hence the area of our frustum is

$$A = \pi r_1 L + \pi r_2 L = 2\pi \bar{r} L$$

where $\bar{r} = \frac{1}{2}(r_1 + r_2)$. So we have verified Formula (6).

Now suppose that the surface $S$ has area $A$ and is generated by revolving around the $x$-axis the smooth arc $y = f(x)$, $a \le x \le b$; suppose also that $f(x)$ is never negative on $[a, b]$. To approximate $A$ we begin with a regular partition of $[a, b]$ into $n$ equal subintervals of length $\Delta x$. As in our discussion of arc length leading to Formula (2), let $P_i$ denote the point $(x_i, f(x_i))$ on the arc. Then, as before, the line segment $P_{i-1}P_i$ has length

$$L_i = |P_{i-1}P_i| = [1 + (f'(x_i^*))^2]^{1/2} \, \Delta x$$

for some point $x_i^*$ in the $i$th subinterval $[x_{i-1}, x_i]$.

The conical frustum obtained by revolving the segment $P_{i-1}P_i$ around the $x$-axis has slant height $L_i$ and, as shown in Fig. 6.35, average radius

$$\bar{r}_i = \frac{1}{2}[f(x_{i-1}) + f(x_i)].$$

Since $\bar{r}_i$ lies between the values $f(x_{i-1})$ and $f(x_i)$, the intermediate value property (Section 2-4) yields a point $x_i^{**}$ in $[x_{i-1}, x_i]$ such that $\bar{r}_i = f(x_i^{**})$. By Formula (6), the area of this conical frustum is, therefore,

$$2\pi \bar{r}_i L_i = 2\pi f(x_i^{**})[1 + (f'(x_i^*))^2]^{1/2} \, \Delta x.$$

We add the areas of these conical frusta for $i = 1, 2, 3, \ldots, n$; this gives the approximation

$$A \approx \sum_{i=1}^{n} 2\pi f(x_i^{**})[1 + (f'(x_i^*))^2]^{1/2} \, \Delta x.$$

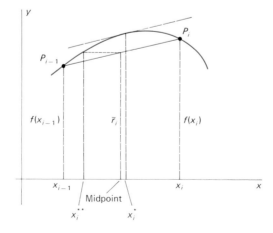

**6.35** Approximation of a surface area of revolution by the surface of a frustum of a cone

If $x_i^*$ and $x_i^{**}$ were the *same* point of the $i$th subinterval $[x_{i-1}, x_i]$, then this approximation would be a Riemann sum for the integral

$$\int_a^b 2\pi f(x)\sqrt{1 + [f'(x)]^2}\, dx.$$

Even though the numbers $x_i^*$ and $x_i^{**}$ are generally unequal, it still follows (from a result stated in the appendix to this chapter) that our approximation approaches the above integral as $\Delta x \to 0$.

We therefore *define* the **area** $A$ of the surface generated by revolving the smooth arc $y = f(x)$, $a \leq x \leq b$, around the $x$-axis by the integral formula

$$A = \int_a^b 2\pi f(x)\sqrt{1 + [f'(x)]^2}\, dx. \tag{8}$$

If we write $y = f(x)$ and $ds = \sqrt{1 + (dy/dx)^2}\, dx$, as in Formula (4'), then Formula (8) can be abbreviated to

$$A = \int 2\pi y\, ds \qquad (x\text{-axis}). \tag{9}$$

This abbreviated formula is conveniently remembered by thinking of $dA = 2\pi y\, ds$ as the area of the narrow frustum obtained by revolving the tiny arc $ds$ around the $x$-axis in a circle of radius $y$, as in Fig. 6.36.

If our smooth arc being revolved around the $x$-axis is given by $x = g(y)$, $c \leq y \leq d$, then an approximation based on a regular partition of $[c, d]$ leads to the area formula

$$A = \int_c^d 2\pi y\sqrt{1 + [g'(y)]^2}\, dy. \tag{10}$$

Note that Formula (10) can be obtained by making the formal substitution $ds = \sqrt{1 + (dx/dy)^2}\, dy$ of Formula (4'') in the abbreviated Formula (9) for surface area of revolution around the $x$-axis and then inserting the limits $c$ and $d$.

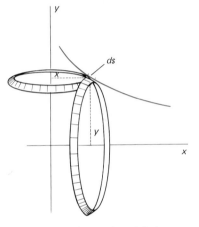

Now let us consider the surface generated by revolving our smooth arc around the $y$-axis rather than the $x$-axis. In Fig. 6.36 we see that the average radius of the narrow frustum obtained by revolving the tiny arc $ds$ is now $x$ instead of $y$. This suggests the abbreviated formula

$$A = \int 2\pi x\, ds \qquad (y\text{-axis}) \tag{11}$$

**6.36** The arc $ds$ may be rotated about either the $x$-axis or the $y$-axis.

for a surface area of revolution about the $y$-axis. If the smooth arc is given by $y = f(x)$, $a \leq x \leq b$, then the symbol  substitution $ds = \sqrt{1 + (dy/dx)^2}\, dx$ gives

$$A = \int_a^b 2\pi x\sqrt{1 + [f'(x)]^2}\, dx. \tag{12}$$

But if our smooth arc is given by $x = g(y)$, $c \leq y \leq d$, then the symbolic substitution $ds = \sqrt{1 + (dx/dy)^2}\, dy$ in (11) gives

$$A = \int_c^d 2\pi g(y)\sqrt{1 + [g'(y)]^2}\, dy. \tag{13}$$

It is possible to verify Formulas (12) and (13) by using approximations similar to the one leading to Formula (8).

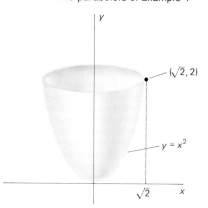

| | Axis of revolution | |
| | x-axis | y-axis |
| Description of curve C | $y = f(x),$ $a \leqslant x \leqslant b$ | $\displaystyle\int_a^b 2\pi f(x)\sqrt{1 + [f'(x)]^2}\ dx$ (8) | $\displaystyle\int_a^b 2\pi x \sqrt{1 + [f'(x)]^2}\ dx$ (10) |
| | $x = g(y),$ $c \leqslant y \leqslant d$ | $\displaystyle\int_c^d 2\pi y \sqrt{1 + [g'(y)]^2}\ dy$ (12) | $\displaystyle\int_c^d 2\pi g(y)\sqrt{1 + [g'(y)]^2}\ dy$ (13) |

**6.37** Area formulas for surfaces of revolution.

Thus we have *four* formulas for areas of surfaces of revolution, summarized in the table in Fig. 6.37. Which of these formulas is appropriate for computing the area of a given surface depends on whether the smooth arc that generates it is described as a function of $x$ or as a function of $y$, and on whether this arc is to be revolved around the $x$-axis or around the $y$-axis. But memorizing the formulas in the table is not necessary. We suggest that you instead remember the abbreviated formulas in (9) and (11) in conjunction with Fig. 6.36 and make either the substitution

$$y = f(x), \qquad ds = \sqrt{1 + \left(\frac{dy}{dx}\right)^2}\ dx$$

or the substitution

$$x = g(y), \qquad ds = \sqrt{1 + \left(\frac{dx}{dy}\right)^2}\ dy,$$

depending on the way in which the smooth arc is described. It may also be helpful to note that each of these four surface area formulas is of the form

$$A = \int 2\pi r\ ds \tag{14}$$

where $r$ denotes the radius of the circle around which the arc length element $ds$ is revolved.

In earlier sections we cautioned you to identify the independent variable through examining the differential, and to express every dependent variable in terms of the independent variable before antidifferentiating. While it is sometimes possible to express $r$ in terms of arc length $s$ in (14), it is almost always much simpler—and quite correct—to express everything (including $ds$) in terms of $x$ or everything in terms of $y$.

**6.38** The paraboloid of Example 4

**EXAMPLE 4** Find the area of the paraboloid shown in Fig. 6.38, obtained by revolving the parabolic arc $y = x^2, 0 \leq x \leq \sqrt{2}$, around the $y$-axis.

**Solution** Following the suggestion that precedes the example, we get

$$A = \int 2\pi x\ ds = \int_a^b 2\pi x \sqrt{1 + \left(\frac{dy}{dx}\right)^2}\ dx$$

$$= \int_0^{\sqrt{2}} 2\pi x \sqrt{1 + (2x)^2}\ dx = \int_0^{\sqrt{2}} \frac{\pi}{4}(1 + 4x^2)^{1/2}(8x)\ dx$$

$$= \left[\frac{\pi}{6}(1 + 4x^2)^{3/2}\right]_0^{\sqrt{2}} = \frac{13\pi}{3}.$$

Alternatively, we could describe our parabolic arc as $x = y^{1/2}$, $0 \le y \le 2$, and get

$$A = \int 2\pi x \, ds = \int_c^d 2\pi x \sqrt{1 + \left(\frac{dx}{dy}\right)^2} \, dy$$

$$= \int_0^2 2\pi y^{1/2} \sqrt{1 + (\tfrac{1}{2}y^{-1/2})^2} \, dy$$

$$= \int_0^2 \pi \sqrt{1 + 4y} \, dy = \left[\frac{\pi}{6}(1 + 4y)^{3/2}\right]_0^2 = \frac{13\pi}{3}.$$

Note that the decision of which abbreviated formula—(9) or (11)—should be used is determined by the axis of revolution. By contrast, the decision whether the variable of integration should be $x$ or $y$ is determined by the way in which the smooth arc is given: as a function of $x$ or as a function of $y$. In some problems either $x$ or $y$ may be used as the variable of integration, but the integral is usually much simpler to evaluate if you make the correct choice (experience is very helpful here).

## 6-4  PROBLEMS

In Problems 1–10, set up and simplify the integral that gives the length of the given smooth arc; do not evaluate the integral.

1  $y = x^2$, $0 \le x \le 1$
2  $y = x^{5/2}$, $1 \le x \le 3$
3  $y = 2x^3 - 3x^2$, $0 \le x \le 2$
4  $y = x^{4/3}$, $-1 \le x \le 1$
5  $y = 1 - x^2$, $0 \le x \le 100$
6  $x = 4y - y^2$, $0 \le y \le 1$
7  $x = y^4$, $-1 \le y \le 2$
8  $x^2 = y$, $1 \le y \le 4$
9  $xy = 1$, $1 \le x \le 2$
10  $x^2 + y^2 = 4$, $0 \le x \le 2$

In Problems 11–20, set up and simplify the integral that gives the surface area of revolution generated by rotation of the given smooth arc about the given axis; do not evaluate the integral.

11  $y = x^2$, $0 \le x \le 4$;  the x-axis
12  $y = x^2$, $0 \le x \le 4$;  the y-axis
13  $y = x - x^2$, $0 \le x \le 1$;  the x-axis
14  $y = x^2$, $0 \le x \le 1$;  the line $y = 4$
15  $y = x^2$, $0 \le x \le 1$;  the line $x = 2$
16  $y = x - x^3$, $0 \le x \le 1$;  the x-axis
17  $y = \sqrt{x}$, $1 \le x \le 4$;  the x-axis
18  $y = \sqrt{x}$, $1 \le x \le 4$;  the y-axis
19  $y = x^{3/2}$, $1 \le x \le 4$;  the line $x = -1$
20  $y = x^{5/2}$, $1 \le x \le 4$;  the line $y = -2$

Find the lengths of the smooth arcs in Problems 21–28.

21  $y = \frac{2}{3}(x^2 + 1)^{3/2}$  from $x = 0$ to $x = 2$
22  $x = \frac{2}{3}(y - 1)^{3/2}$  from $y = 1$ to $y = 5$

23  $y = \dfrac{x^3}{6} + \dfrac{1}{2x}$  from $x = 1$ to $x = 3$

24  $x = \dfrac{y^4}{8} + \dfrac{1}{4y^2}$  from $y = 1$ to $y = 2$

25  $8x^2y - 2x^6 = 1$  from $(1, \frac{3}{8})$ to $(2, \frac{129}{32})$
26  $12xy - 4y^4 = 3$  from $(\frac{7}{12}, 1)$ to $(\frac{67}{24}, 2)$
27  $y^3 = 8x^2$  from $(1, 2)$ to $(8, 8)$
28  $(y - 3)^2 = 4(x + 2)^3$  from $(-1, 5)$ to $(2, 19)$

In each of Problems 29–35, find the area of the surface of revolution generated by revolving the given curve about the indicated axis.

29  $y = \sqrt{x}$, $0 \le x \le 1$;  the x-axis
30  $y = x^3$, $1 \le x \le 2$;  the x-axis

31  $y = \dfrac{x^5}{5} + \dfrac{1}{12x^3}$, $1 \le x \le 2$;  the y-axis

32  $x = \dfrac{y^4}{8} + \dfrac{1}{4y^2}$, $1 \le y \le 2$;  the x-axis

33  $y^3 = 3x$, $0 \le x \le 9$;  the y-axis
34  $y = \frac{2}{3}x^{3/2}$, $1 \le x \le 2$;  the y-axis (*Suggestion:* Make the substitution $u = 1 + x$.)
35  $y = (2x - x^2)^{1/2}$, $0 \le x \le 2$;  the x-axis
36  Use the integral formula given in Problem 39 of Section 5-6 to find the area of the surface obtained by revolving the curve

$$y = \tfrac{2}{3}x^{3/2}, \qquad 3 \le x \le 8,$$

about the y-axis.

**37** Use Simpson's rule with $n = 10$ subintervals to approximate the area of the surface obtained by revolving the curve of Problem 36 about the $x$-axis.

**38** Use Simpson's approximation with $n = 10$ subintervals to estimate the length of the parabola $y = x^2$ from $x = 0$ to $x = 1$.

**39** Verify the formula in (6) for the area of a conical frustum. Think of the frustum as being generated by revolving the line segment from $(r_1, 0)$ to $(r_2, h)$ around the $y$-axis.

**40** By considering a sphere of radius $r$ as a surface of revolution, derive the formula $A = 4\pi r^2$ for its surface area.

**41** Find the total length of the *astroid* shown in Fig. 6.39. The equation of its graph is $x^{2/3} + y^{2/3} = 1$.

**42** Find the area of the surface generated by revolving the astroid of Problem 41 about the $y$-axis.

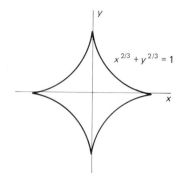

**6.39** The astroid—see Problems 41 and 42

---

### 6-5

### Force and Work

The concept of *work* is introduced to measure the cumulative effect of a force in moving a body from one position to another. In the simplest case, a particle is moved along a straight line by the action of a *constant* force. In this case, the work done by the force is defined to be the product of the force and the distance through which it acts:

$$\text{Work} = (\text{constant force}) \times (\text{distance}). \tag{1}$$

For instance, if a constant horizontal force of 50 lb is applied to push a heavy box a distance of 10 ft along a rough floor, the work done by the force is

$$W = (50 \text{ lb})(10 \text{ ft}) = 500 \text{ ft-lb}.$$

Similarly, in order to lift a weight of 75 lb a vertical distance of 5 ft, a constant force of 75 lb must be applied, and the work done by this force is

$$W = (75 \text{ lb})(5 \text{ ft}) = 375 \text{ ft-lb}.$$

In this section we use the integral to generalize the above definition to the case in which a particle is moved along a straight line by a *variable* force. Given a **force function** $F(x)$ defined at each point $x$ of the straight line segment $[a, b]$, we must first say what properties the work $W$ done by the force $F$ ought to have.

Work should be additive. That is, if we do work $W_1$ in moving a particle from $a$ to $c$ and do work $W_2$ in moving it from $c$ to $b$, then the total work done in moving the particle from $a$ to $b$ should be

$$W = W_1 + W_2. \tag{2}$$

Work should also have the following monotonicity property: If $F_1$ and $F_2$ are constants such that

$$F_1 \leqq F(x) \leqq F_2$$

for all $x$ in $[a, b]$, then the work $W$ done by the force $F$ in moving a particle the distance $\Delta x$ should satisfy the inequalities

$$F_1 \, \Delta x \leqq W \leqq F_2 \, \Delta x. \tag{3}$$

**314**

We shall see that these two properties are enough to *determine* the way in which the work done by a variable force must be defined.

We begin with the usual regular partition of $[a, b]$ into $n$ equal subintervals, each with length $\Delta x$. Let $F(x_i^b)$ and $F(x_i^\#)$ be the minimum and maximum values, respectively, of $F(x)$ on the $i$th subinterval $[x_{i-1}, x_i]$. By Inequalities (3), the work $\Delta W_i$ done by the force in moving the particle from $x_{i-1}$ to $x_i$ satisfies

$$F(x_i^b)\,\Delta x \leqq \Delta W_i \leqq F(x_i^\#)\,\Delta x.$$

If we add these inequalities for $i = 1, 2, 3, \ldots, n$, we find that

$$\sum_{i=1}^{n} F(x_i^b)\,\Delta x \leqq W \leqq \sum_{i=1}^{n} F(x_i^\#)\,\Delta x.$$

These two sums are the lower and upper Riemann sums for the function $F$ on the interval $[a, b]$. If we know that $F$ is continuous on $[a, b]$, it follows that (as $\Delta x \to 0$) both these sums approach the value of the integral $\int_a^b F(x)\,dx$. Consequently the squeeze law of limits leaves us no alternative but to *define* the **work** $W$ done by the force $F(x)$ in moving a particle from $x = a$ to $x = b$ by

$$W = \int_a^b F(x)\,dx. \tag{4}$$

The following heuristic way of setting up Formula (4) is useful in setting up integrals for work problems. Imagine that $dx$ is so small a number that the value of $F$ does not change appreciably on the tiny interval from $x$ to $x + dx$. Then the work done by the force in moving a particle from $x$ to $x + dx$ should be very close to

$$dW = F(x)\,dx.$$

Property (2) then suggests that we could obtain the total work $W$ by adding these tiny elements of work, so that

$$W = \int dW = \int_a^b F(x)\,dx.$$

## ELASTIC SPRINGS

Consider a spring with left end held fixed and right end free to move along the $x$-axis. We assume that the right end is at the origin $x = 0$ when the spring has its **natural length;** that is, when the spring is in its rest position, neither compressed nor stretched by outside forces.

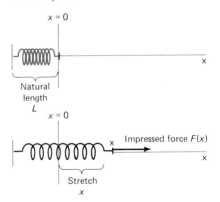

**6.40** The stretch $x$ is proportional to the impressed force $F$.

According to **Hooke's law** for elastic springs, the force $F(x)$ that must be exerted on the spring to hold its right end at the point $x$ is proportional to the displacement $x$ of the right end from its rest position. That is,

$$F(x) = kx \tag{5}$$

for some positive constant $k$. The constant $k$ is a characteristic of the particular spring under study and is called the **spring constant.**

Figure 6.40 shows the arrangement of such a spring along the $x$-axis. The spring is being held with its right end at position $x$ on the $x$-axis by a force $F(x)$. The figure shows the situation for $x > 0$, so that the spring is stretched. The force that the spring exerts on its right-hand end is directed to the left, so—as the figure shows—the external force $F(x)$ must act to the right. The right is the positive direction here, so $F(x)$ must be a positive number. Since $x$ and $F(x)$ have the same sign, $k$ must be positive too. You can mentally check that $k$ is also positive in the case $x < 0$.

**EXAMPLE 1** Suppose that a spring has a natural length of 1 foot and that a force of 10 pounds is required to hold it compressed to a length of 6 inches. How much work is done in stretching the spring from its natural length to a total length of 2 feet?

*Solution* To move the free end from $x = 0$ (the natural length position) to $x = 1$ (stretched by 1 foot), we must exert a variable force $F(x)$ determined by Hooke's law. We are given that $F = -10$ pounds when $x = -\frac{1}{2}$ ft, so Equation (5), $F = kx$, implies that the spring constant for this spring is $k = 20$ lb/ft. Thus $F(x) = 20x$, and so—using Formula (4)—we find that the work done in stretching this spring in the manner given is

$$W = \int_0^1 20x \, dx = \left[10x^2\right]_0^1 = 10 \text{ ft-lb.}$$

**\*WORK AGAINST GRAVITY**

According to Newton's law of gravitation, the force that must be exerted on a body to hold it at a distance $r$ from the earth's center is inversely proportional to $r^2$. That is, if $F(r)$ denotes the holding force, then

$$F(r) = \frac{k}{r^2} \tag{6}$$

for some positive constant $k$. The value of this force at the earth's surface, where $r = R \approx 4000$ miles, is called the **weight** of the body.

Given the weight $F(R)$ of a particular body, we can find the corresponding value of $k$ by using Formula (6):

$$k = R^2 F(R).$$

The work that must be done to lift the body vertically from the surface (where $r = R$) to a distance $R_1$ from the center of the earth is then

$$W = \int_R^{R_1} \frac{k}{r^2} \, dr. \tag{7}$$

If distance is measured in miles and force in pounds, then this integral gives the work in mile-pounds. This is a rather unconventional unit of work; we shall multiply by 5280 (ft/mi) to convert any such result into foot-pounds.

**EXAMPLE 2** (Satellite launch) How much work must be done in order to lift a 1000-lb satellite vertically from the surface of the earth to an orbit 1000 mi above the earth's surface? (Take $R = 4000$ mi as the radius of the earth.)

*Solution* Since $F = 1000$ (lb) when $r = R = 4000$ (mi), we find from Equation (6) that

$$k = (4000)^2(1000) = 16 \times 10^9 \text{ (mi}^2\text{-lb).}$$

Then by Formula (7), the work done is

$$W = \int_{4000}^{5000} \frac{k}{r^2}\, dr = \left[ -\frac{k}{r} \right]_{4000}^{5000}$$

$$= (16 \times 10^9)\left( \frac{1}{4000} - \frac{1}{5000} \right) = 8 \times 10^5 \text{ mi-lb.}$$

We multiply by 5280 and write the answer as $4.224 \times 10^9$, or 4,224,000,000 ft-lb.

We can also express the answer in terms of the power that the launch rocket must provide. **Power** is the rate at which work is performed. For instance, 1 **horsepower** (hp) is defined to be 33,000 ft-lb/min. If it is known that the ascent to orbit takes 15 min and that only 2% of the power generated by the rocket is effective in lifting the satellite (the rest is used in lifting the rocket and its fuel), we can convert the above answer to horsepower. The *average* power that the rocket engine must produce during the 15-min ascent is

$$P = \frac{(50)(4.224 \times 10^9)}{(15)(33,000)} \approx 426,667 \text{ hp.}$$

The term 50 in the numerator comes from the fact that, because of the 2% "efficiency" of the rocket, the total power must be multiplied by $1/(0.02) = 50$.

### WORK TO FILL A TANK

Examples 1 and 2 are applications of Formula (4) for calculating the work done by a variable force in moving a particle a certain distance. Another common type of force-work problem involves the summation of work done by constant forces that act through different distances. For example, consider the problem of pumping a fluid from ground level up into an above-ground tank, like the one shown in Fig. 6.41.

6.41   An above-ground tank

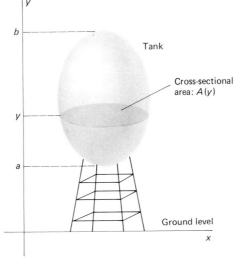

Tank

Cross-sectional area: $A(y)$

Ground level

A convenient way to think of how the tank is filled is this: Think of thin horizontal layers of fluid, each lifted from the ground to its final position in the tank. No matter how the fluid actually behaves as the tank is filled, this simple way of thinking about the process gives a way to compute the work done in the filling process. When we think of filling the tank in this way, different layers of fluid must be lifted different distances to reach their eventual positions in the tank.

Suppose that the bottom of the tank is at height $y = a$ and its top at height $y = b$. Let $A(y)$ be the cross-sectional area of the tank at height $y$. Consider a regular partition of $[a, b]$ into $n$ equal subintervals of length $\Delta y$. Then the volume of the horizontal slice of the tank that corresponds to the $i$th subinterval $[y_{i-1}, y_i]$ is

$$\Delta V_i = \int_{y_{i-1}}^{y_i} A(y)\, dy = A(y_i^*)\, \Delta y$$

for some number $y_i^*$ in $[y_{i-1}, y_i]$; this is a consequence of the average value theorem for integrals (Section 5-5). If $\rho$ is the density of the fluid (for instance, in lb/ft$^3$) then the force required to lift this slice is

$$F_i = \rho\, \Delta V_i = \rho A(y_i^*)\, \Delta y.$$

What about the distance through which this force must act? The fluid in question must be lifted from ground level to the level of the subinterval $[y_{i-1}, y_i]$, so every particle of the fluid is lifted at least the distance $y_{i-1}$ and at most the distance $y_i$. Hence the work $\Delta W_i$ needed to lift this $i$th slice of fluid satisfies the inequalities $F_i y_{i-1} \leqq \Delta W_i \leqq F_i y_i$, or

$$\rho y_{i-1} A(y_i^*)\, \Delta y \leqq \Delta W_i \leqq \rho y_i A(y_i^*)\, \Delta y.$$

Now we add these inequalities for $i = 1, 2, 3, \ldots, n$, and find that the total work $W = \sum \Delta W_i$ satisfies the inequalities

$$\sum_{i=1}^{n} \rho y_{i-1} A(y_i^*)\, \Delta y \leqq W \leqq \sum_{i=1}^{n} \rho y_i A(y_i^*)\, \Delta y.$$

If the three points $y_{i-1}$, $y_i^*$, and $y_i$ of $[y_{i-1}, y_i]$ were the same, then both the above sums would be Riemann sums for the function $f(y) = \rho y A(y)$ on $[a, b]$. Although the three points are not equal, it still follows—from a result stated in the appendix to this chapter—that both sums approach $\int_a^b \rho y A(y)\, dy$ as $\Delta y \to 0$. The squeeze law of limits therefore gives the formula

$$W = \int_a^b \rho y A(y)\, dy. \tag{8}$$

*This is the work $W$ done in pumping fluid of density $\rho$ from the ground into a tank with horizontal cross-sectional area $A(y)$, located between heights $y = a$ and $y = b$ above the ground.*

A quick heuristic way to set up Formula (8) is to think of a thin horizontal slice of fluid with volume $dV = A(y)\, dy$ and weight $\rho\, dV = \rho A(y)\, dy$. The work required to lift this slice a distance $y$ is

$$dW = \rho y A(y)\, dy,$$

so the total work required to fill the tank is

$$W = \int dW = \int_a^b \rho y A(y)\, dy,$$

because the horizontal slices lie between $y = a$ and $y = b$.

**EXAMPLE 3**   Suppose that it took 20 years to construct the great pyramid of Khufu at Gizeh. This pyramid is 500 ft high and has a square base with edge length 750 ft. Suppose also that the pyramid is made of rock with density $\rho = 120$ lb/ft$^3$. Finally, suppose that each laborer did 160 ft-lb/h of effective work in lifting rocks from ground level to position in the pyramid and worked 12 h daily for 330 days per year. How many laborers would have been required to construct this pyramid?

*Solution*   We assume a constant labor force throughout the 20-year period of construction. We think of the pyramid as being made up of thin horizontal slabs of rock, each slab lifted ( just like a slice of water) from ground level to its ultimate height. Hence we can use the formula in (8) to compute the work $W$ required.

Figure 6.42 shows a vertical cross section of the pyramid. The horizontal cross section at height $y$ is a square with edge $s$. From the similar triangles in Figure 6.42 we see that

$$\frac{s}{750} = \frac{500 - y}{500}, \quad \text{so that} \quad s = \frac{3}{2}(500 - y).$$

Hence the cross-sectional area at height $y$ is

$$A(y) = \tfrac{9}{4}(500 - y)^2.$$

Formula (8) therefore gives

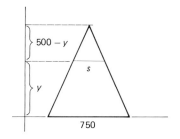

**6.42**   Vertical cross section of Khufu's pyramid

$$W = \int_0^{500} 120 \cdot y \cdot \tfrac{9}{4}(500 - y)^2 \, dy$$

$$= 270 \int_0^{500} (250{,}000y - 1000y^2 + y^3) \, dy$$

$$= 270 \left[ 125{,}000y^2 - \frac{1000}{3}y^3 + \frac{y^4}{4} \right]_0^{500},$$

so that $W \approx 1.406 \times 10^{12}$ ft-lb.

Since each laborer does

$$(160)(12)(330)(20) \approx 1.267 \times 10^7 \text{ ft-lb}$$

of work, the construction of the pyramid would—under our assumptions—have required

$$\frac{1.406 \times 10^{12}}{1.267 \times 10^7},$$

or about 111,000, laborers.

---

Suppose now that the tank shown in Fig. 6.43 is already filled with a liquid of density $\rho$ pounds per cubic foot, and we want to pump all this liquid from the tank up to the level $y = h$ above the top of the tank. We imagine a thin horizontal slice of liquid at height $y$. If its thickness is $dy$, then its volume is $dV = A(y) \, dy$, and so its weight is $\rho \, dV = \rho A(y) \, dy$. This slice must be lifted the distance $h - y$, so the work done to lift the slice is

$$dW = (h - y)\rho \, dV = \rho(h - y)A(y) \, dy.$$

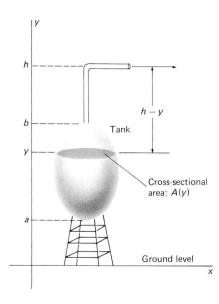

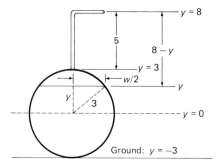

**6.44** End view of the cylindrical tank of Example 4

**6.43** Pumping liquid from a tank to a higher level

Hence the total amount of work done on all the liquid originally in the tank is

$$W = \int_a^b \rho(h - y)A(y)\,dy. \tag{9}$$

In Problem 14 we ask you to use Riemann sums to set up this integral.

**EXAMPLE 4**   A cylindrical tank of radius 3 ft and length 10 ft is lying on its side on horizontal ground. If this tank initially is full of gasoline weighing 40 lb/ft³, how much work is done in pumping this gasoline to a point 5 ft above the top of the tank?

*Solution*   Figure 6.44 shows an end view of the tank. In order to exploit circular symmetry, we choose $y = 0$ at the *center* of the circular vertical section, so the tank lies between $y = -3$ and $y = 3$. A horizontal cross section of the tank is a rectangle of length 10 ft and width $w$. From the right triangle in Fig. 6.44, we see that

$$\tfrac{1}{2}w = \sqrt{9 - y^2},$$

so the area of this cross section is

$$A(y) = 10w = 20\sqrt{9 - y^2}.$$

This cross section must be lifted the distance $8 - y$, so the formula in (9) with $\rho = 40$, $a = -3$, and $b = 3$ yields

$$W = \int_{-3}^{3} (40)(8 - y)(20\sqrt{9 - y^2})\,dy$$

$$= 6400 \int_{-3}^{3} \sqrt{9 - y^2}\,dy - 800 \int_{-3}^{3} y\sqrt{9 - y^2}\,dy.$$

We attack the two integrals separately. First,

$$\int_{-3}^{3} y\sqrt{9 - y^2}\,dy = \left[ -\tfrac{1}{3}(9 - y^2)^{3/2} \right]_{-3}^{3} = 0.$$

**320**

Second,

$$\int_{-3}^{3} \sqrt{9 - y^2} \, dy = \tfrac{1}{2}\pi(3)^2 = \tfrac{9}{2}\pi$$

because the integral is simply the area of a semicircle of radius 3. Hence

$$W = (6400)(\tfrac{9}{2}\pi) = 28{,}800\pi;$$

that is, approximately 90,478 ft-lb.

As in Example 4, you may use as needed in the problems the integral

$$\int_{0}^{a} \sqrt{a^2 - x^2} \, dx = \tfrac{1}{4}\pi a^2 \tag{10}$$

that corresponds to the area of a quarter-circle of radius $a$.

## 6-5   PROBLEMS

In each of Problems 1–5, find the work done by the given force $F(x)$ in moving a particle along the $x$-axis from $x = a$ to $x = b$.

**1** $F(x) = 10;\quad a = -2, b = 1$

**2** $F(x) = 3x - 1;\quad a = 1, b = 5$

**3** $F(x) = \dfrac{10}{x^2};\quad a = 1, b = 10$

**4** $F(x) = -3\sqrt{x};\quad a = 0, b = 4$

**5** $F(x) = x^{2/3};\quad a = -1, b = 1$

**6** A spring has a natural length of 1 ft, and a force of 10 lb is required to hold it stretched to a total length of 2 ft. How much work is done in compressing this spring from its natural length to a length of 6 in.?

**7** A spring has a natural length of 2 ft, and a force of 15 lb is required to hold it compressed to a length of 18 in. How much work is done in stretching this spring from its natural length to a length of 3 ft?

**8** Apply Formula (4) to compute the amount of work done in lifting a 100-lb weight a height of 10 ft, assuming this work is done against the constant force of gravity.

**9** Compute the amount of work (in foot-pounds) done in lifting a 1000-lb weight from an orbit 1000 mi above the earth's surface to one 2000 mi above the earth's surface. Use the value of $k$ given in Example 2.

**10** A cylindrical tank is resting on the ground with its axis vertical; it has radius 5 ft and height 10 ft. Use Formula (8) to compute the amount of work done in filling this tank with water pumped in from ground level. (For the density of water, use $\rho = 62.4 \text{ lb/ft}^3$.)

**11** A conical tank is resting on its base—which is at ground level—and with its axis vertical. The tank has radius 5 ft and height 10 ft. Compute the work done in filling this tank with water ($\rho = 62.4 \text{ lb/ft}^3$) pumped in from ground level.

**12** Repeat Problem 11, except that now the tank is up-ended: its vertex is at ground level, its base is 10 ft above the ground (but its axis is still vertical).

**13** A tank with its lowest point 10 ft above ground has the shape of a cup obtained by rotating the parabola $y = \tfrac{1}{5}x^2$, $-5 \le x \le 5$, around the $y$-axis. The units on the coordinate axes are also in feet. How much work is done in filling this tank with oil weighing 50 lb/ft$^3$ if the oil is pumped in from ground level?

**14** Suppose that the tank of Fig. 6.43 is filled with fluid of density $\rho$ and that all this fluid is to be pumped from the tank to the level $y = h$ above the top of the tank. Use Riemann sums, as in the derivation of the formula in Equation (8), to obtain the formula

$$W = \int_{a}^{b} \rho(h - y)A(y) \, dy$$

for the work required to do this.

**15** Use the formula of Problem 14 to find the amount of work done in pumping the water in the tank of Problem 10 to a height of 5 ft above the top of the tank.

**16** Gasoline at a service station is stored in a cylindrical tank buried on its side, with the highest part of the tank 5 ft below the surface. The tank is 6 ft in diameter and 10 ft long. The density of gasoline is 45 lb/ft$^3$. Assume that the filler cap of each automobile gas tank is 2 ft above the ground. How much work is done in emptying all the gasoline from this tank, which is initially full, into automobiles?

**17** Consider a spherical water tank of radius 10 ft, with center 50 ft above the ground. How much work is required to fill this tank by pumping water up from ground level? (*Suggestion:* It may simplify your computations to take $y = 0$ at the center of the tank and think of the distance each horizontal slice of water must be lifted.)

**18** A hemispherical tank of radius 10 ft is located with its flat side down atop a tower 60 ft high. How much work is required to fill it with oil weighing 50 lb/ft$^3$ if the oil is to be pumped into the tank from ground level?

**19** Water is being drawn from a well 100 ft deep, using a bucket that scoops up 100 lb of water. The bucket is being pulled up at the rate of 2 ft/s, but it has a hole in its bottom through which water leaks out at the rate of $\frac{1}{2}$ lb/s. How much work is done in pulling the bucket to the top? Neglect the weight of the bucket, the weight of the rope and the work done in overcoming friction. (*Suggestion:* Take $y = 0$ at the level of water in the well, so that $y = 100$ at ground level. Let $\{y_0, y_1, y_2, \ldots, y_n\}$ be a subdivision of $[0, 100]$ into $n$ equal subintervals. Estimate the amount of work $W_i$ required to raise the bucket from $y_{i-1}$ to $y_i$. Then set up the sum $W = \sum W_i$ and proceed to the appropriate integral by letting $n \to \infty$.)

**20** A rope 100 ft long and weighing $\frac{1}{4}$ lb per linear foot hangs from the edge of a very tall building. How much work is required to pull this rope to the top of the building?

**21** Suppose we plug the hole in the leaking bucket of Problem 19. How much work is done in lifting the mended bucket, full of water, to the surface, using the rope of Problem 20?

**22** Consider a volume $V$ of gas in a cylinder fitted with a piston at one end, where the pressure $p$ of the gas is a function $p(V)$ of its volume. Such a cylinder is shown in Fig. 6.45. Let $A$ be the area of the face of the piston. Then the force exerted on the piston by gas in the cylinder is $F = pA$. Assume that the gas expands from volume $V_1$ to volume $V_2$. Show that the work done by the force $F$ is then given by

$$W = \int_{V_1}^{V_2} p(V)\, dV.$$

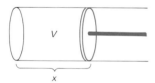

**6.45** A cylinder fitted with a piston

[*Suggestion:* If $x$ is the length of the cylinder (from its fixed end to the face of the piston), then $F = Ap(Ax)$. Apply Formula (4) of this section, and make the substitution $V = Ax$ in the resulting integral.]

**23** The pressure and volume of the steam in a small **steam** engine satisfy the condition $pV^{1.4} = c$ (constant). In one cycle, the steam expands from a volume $V_1 = 50$ in$^3$ to $V_2 = 500$ in$^3$ with an initial pressure of 200 lb/in$^2$. Use the formula of Problem 22 to compute the work (in foot-pounds) done by this engine in each such cycle.

**24** A tank is in the shape of a hemisphere of radius 60 ft, resting on its flat base with the curved surface on top. It is filled with alcohol weighing 40 lb/ft$^3$. How much work is done in pumping all the alcohol to the level of the top of the tank?

**25** A tank is in the shape of the surface generated by rotating the graph of $y = x^4$, $0 \leq x \leq 1$, about the $y$-axis. The tank is initially full of oil weighing 60 lb/ft$^3$. The units on the coordinate axes are in feet. Find how much work is done in pumping all the oil to the level of the top of the tank.

**26** A cylindrical tank of radius 3 ft and length 20 ft is lying on its side on horizontal ground. Gasoline weighing 40 lb/ft$^3$ is available at ground level and is to be pumped into the tank. Find the work required to fill the tank.

**27** A spherical storage tank has radius 12 ft. The base of the tank is at ground level. Find the amount of work done in filling the tank with oil weighing 50 lb/ft$^3$ if all the oil is initially at ground level.

**28** A 20-lb monkey is attached to a 50-ft chain weighing 0.5 lb per (linear) foot. The other end of the chain is attached to the 40-ft-high ceiling of the monkey's cage. Find the amount of work done by the monkey in climbing up her chain to the ceiling.

**29** A kite is flying at a height of 500 ft above the ground. Suppose that the kite string weighs 1/16 oz per (linear) foot, and (for simplicity) that its string is stretched in a straight line that makes an angle of 45° with the ground. How much work was done by the wind in lifting the string from ground level up to its flying position?

## *6-6

**Centroids of Plane Regions and Curves**

According to the **law of the lever,** two masses $m_1$ and $m_2$ on opposite sides and at respective distances $d_1$ and $d_2$ from the fulcrum of a lever will balance provided that $m_1 d_1 = m_2 d_2$ (see Fig. 6.46). Think of the $x$-axis as the location of a (weightless) lever arm supporting various point masses and of the origin as a fulcrum. Then a more general form of the law of the lever states that masses $m_0, m_1, m_2, \ldots, m_n$ with coordinates $x_0, x_1, x_2, \ldots, x_n$ will balance provided that

$$\sum_{i=0}^{n} m_i x_i = m_0 x_0 + m_1 x_1 + \cdots + m_n x_n = 0. \tag{1}$$

Now consider arbitrary masses $m_1, m_2, \ldots, m_n$ at the points $x_1, x_2, \ldots,$ $x_n$. Then a single particle with mass

$$m = m_1 + m_2 + \cdots + m_n = \sum_{i=1}^{n} m_i$$

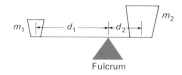

**6.46** Law of the lever: The weights balance when $m_1 d_1 = m_2 d_2$.

and position $-\bar{x}$ will balance these $n$ masses provided that

$$-m\bar{x} + \sum_{i=1}^{n} m_i x_i = 0;$$

that is, provided that

$$\bar{x} = \frac{1}{m} \sum_{i=1}^{n} m_i x_i. \tag{2}$$

Thus the original $n$ masses act on the lever like a single particle of mass $m$ located at the point $\bar{x}$. The point $\bar{x}$ is called the *center of mass*, and the sum $\sum_{i=1}^{n} m_i x_i$ that appears in (2) is called the *moment* about the origin.

Now consider a system of $n$ particles with masses $m_1, m_2, \ldots, m_n$ located in the plane (see Fig. 6.47) at the coordinates $(x_1, y_1), (x_2, y_2), \ldots, (x_n, y_n)$. In analogy with the one-dimensional case above, we define the **moment** $M_y$ of this system **about the y-axis** and its **moment** $M_x$ **about the x-axis** by means of the equations

$$M_y = \sum_{i=1}^{n} m_i x_i \quad \text{and} \quad M_x = \sum_{i=1}^{n} m_i y_i. \tag{3}$$

The **center of mass** of this system of $n$ particles is the point $(\bar{x}, \bar{y})$ with co-ordinates defined to be

$$\bar{x} = \frac{M_y}{m} \quad \text{and} \quad \bar{y} = \frac{M_x}{m} \tag{4}$$

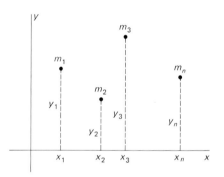

**6.47** Point masses in the plane and their moment arms about the x-axis

where $m = m_1 + m_2 + \cdots + m_n$ is the sum of the masses. Thus $(\bar{x}, \bar{y})$ is the point where a single particle of mass $m$ would have the same moments as the system about the two coordinate axes. In elementary physics it is shown that, if the $xy$-plane were a rigid and weightless plastic sheet with our $n$ particles imbedded in it, then it would balance (horizontally) on the point of an icepick at the point $(\bar{x}, \bar{y})$.

As the number of particles we consider increases while their masses decrease in proportion, their aggregate more and more closely resembles a plane region of varying density. Let us first define the center of mass $(\bar{x}, \bar{y})$ and the moments about the coordinate axes of a thin plate or **lamina** of *constant* density $\rho$, one that occupies a bounded plane region $R$. Since $\rho$ is constant, the numbers $\bar{x}$ and $\bar{y}$ should be independent of its actual value, so we will take $\rho = 1$ (for convenience) in our definition and computations. In this case $(\bar{x}, \bar{y})$ is called the **centroid of the plane region** $R$. We shall also define $M_y(R)$ and $M_x(R)$, the moments of the plane region $R$ about the axes, with $\rho = 1$. The corresponding moments of a lamina of constant density $\rho \neq 1$ would then be $\rho M_y(R)$ and $\rho M_x(R)$.

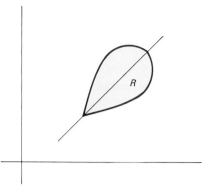

**6.48** A line of symmetry

Our definitions will be based on the following two physical principles. The first one is quite natural: If a region has a line of symmetry, as in Fig. 6.48, then its centroid lies on this line.

---

*Symmetry Principle*

If the plane region $R$ is symmetric with respect to the line $L$—that is, if $R$ is carried onto itself when the plane is rotated through an angle of $180°$ about the line $L$—then the centroid of $R$ (considered as a lamina of constant density) lies on $L$.

---

We will find the second principle very useful in locating centroids of regions that are unions of simple regions.

---

*Additivity of Moments*

If $R$ is the union of the two nonoverlapping regions $S$ and $T$, then

$$M_y(R) = M_y(S) + M_y(T)$$

and                                                                                             (5)

$$M_x(R) = M_x(S) + M_x(T).$$

---

For example, the symmetry principle implies that the centroid of a rectangle is its geometric center: the intersection of the perpendicular bisectors of its sides. We also assume that the moments of a rectangle $R$ with area $A$ and centroid $(\bar{x}, \bar{y})$ are $M_y(R) - A\bar{x}$ and $M_x(R) - A\bar{y}$. Knowing the centroid of a rectangle, our strategy will be to calculate the moments of a more general region by using additivity of moments and the integral, and then finally to define the centroid of this more general region by analogy with Equations (4).

Now suppose that the function $f$ is continuous and nonnegative on $[a, b]$, and suppose also that the region $R$ is the region between the graph of $f$ and the $x$-axis for $a \leq x \leq b$. We begin with a regular partition of $[a, b]$ into $n$ equal subintervals of length $\Delta x$ and denote by $x_i^*$ the midpoint of the $i$th subinterval $[x_{i-1}, x_i]$. As shown in Fig. 6.49, the rectangle with base $[x_{i-1}, x_i]$

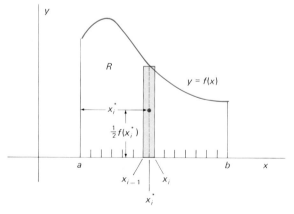

**6.49** Locating the centroid of $R$ by approximating $R$ with rectangles

and height $f(x_i^*)$ has area $f(x_i^*) \, \Delta x$ and centroid $(x_i^*, \frac{1}{2}f(x_i^*))$. If $P_n$ denotes the union of these rectangles for $i = 1, 2, 3, \ldots, n$, then—because moments are additive—the moments of the rectangular polygon $P_n$ about the $y$-axis and the $x$-axis are

$$M_y(P_n) = \sum_{i=1}^{n} x_i^* f(x_i^*) \, \Delta x$$

and

$$M_x(P_n) = \sum_{i=1}^{n} \tfrac{1}{2}f(x_i^*) \cdot f(x_i^*) \, \Delta x,$$

respectively.

We define the moments $M_y(R)$ and $M_x(R)$ of the region $R$ itself by taking the limits of $M_y(P_n)$ and $M_x(P_n)$ as $\Delta x \to 0$. Since the two sums above are Riemann sums, their limits are the definite integrals

$$M_y(R) = \int_a^b xf(x) \, dx \tag{6}$$

and

$$M_x(R) = \int_a^b \tfrac{1}{2} \, [f(x)]^2 \, dx. \tag{7}$$

Finally, we define the centroid $(\bar{x}, \bar{y})$ of $R$ by

$$\bar{x} = \frac{M_y(R)}{A} \quad \text{and} \quad \bar{y} = \frac{M_x(R)}{A}, \tag{8}$$

where $A = \int_a^b f(x) \, dx$ is the area of $R$. Thus

$$\bar{x} = \frac{1}{A} \int_a^b xf(x) \, dx \tag{9}$$

and

$$\bar{y} = \frac{1}{A} \int_a^b \tfrac{1}{2}[f(x)]^2 \, dx \tag{10}$$

are the coordinates of the centroid of the region under $y = f(x)$ from $x = a$ to $x = b$.

By the symmetry principle, the centroid of a circular disk is its center. But the centroid of a semicircle is more interesting.

**EXAMPLE 1**  Find the centroid of the upper half $D$ of the circular disk with center $(0, 0)$ and radius $r$.

**6.50** The half-disk $D$ of Example 1

**Solution**  By symmetry, the centroid of $D$ lies on the $y$-axis, so $\bar{x} = 0$. Our semicircular disk lies under $y = \sqrt{r^2 - x^2}$ from $x = -r$ to $x = r$, as shown in Fig. 6.50. So the formula in (7) gives

$$M_x(D) = \int_{-r}^{r} \tfrac{1}{2}(\sqrt{r^2 - x^2})^2 \, dx$$

$$= \tfrac{1}{2} \int_{-r}^{r} (r^2 - x^2) \, dx = \left[ r^2 x - \frac{x^3}{3} \right]_0^r = \tfrac{2}{3}r^3.$$

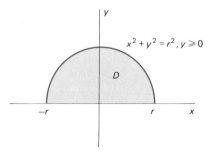

Because $A = \frac{1}{2}\pi r^2$, the formula in (8) gives

$$\bar{y} = \frac{M_x(D)}{A} = \frac{\frac{2}{3}r^3}{\pi r^2/2} = \frac{4r}{3\pi}.$$

Thus the centroid of $D$ is the point $(0, 4r/3\pi)$. Note that our computed value for $\bar{y}$ has the dimensions of a length (because $r$ is a length), as it should. Any answer having other dimensions would be suspect.

**EXAMPLE 2** Find the centroid of the triangle $T$ with vertices $(0, 0)$, $(1, 0)$, and $(0, 1)$.

**Solution** Figure 6.51 shows the triangle $T$. Note first that $\bar{x} = \bar{y}$ by symmetry. The formula in (7) with $y = f(x) = 1 - x$ gives

$$M_y(T) = \int_0^1 x(1 - x)\, dx = \left[\tfrac{1}{2}x^2 - \tfrac{1}{3}x^3\right]_0^1 = \tfrac{1}{6}.$$

So

$$\bar{x} = \frac{M_y(T)}{A} = \frac{\frac{1}{6}}{\frac{1}{2}} = \frac{1}{3}.$$

Thus the centroid of $T$ is $(\tfrac{1}{3}, \tfrac{1}{3})$.

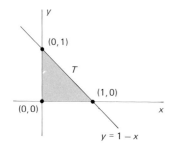

**6.51**  The triangle $T$ of Example 2

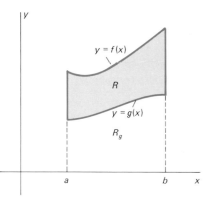

**6.52**  The region $R$ between $y = f(x)$ and $y = g(x)$

Additivity of moments can be used to define the moments and centroid of any plane region that is the union of a finite number of nonoverlapping regions shaped like the one shown in Fig. 6.52. For example, suppose that $f(x) \geq g(x) \geq 0$ on $[a, b]$, and that $R$ is the region between the graphs $y = f(x)$ and $y = g(x)$ from $a$ to $b$. If $R_f$ and $R_g$ denote the regions under the graphs of $f$ and $g$, respectively, then $M_y(R) + M_y(R_g) = M_y(R_f)$ by additivity of moments. Therefore

$$M_y(R) = M_y(R_f) - M_y(R_g)$$
$$= \int_a^b xf(x)\, dx - \int_a^b xg(x)\, dx \qquad \text{(by (6))},$$

so that

$$M_y(R) = \int_a^b x[f(x) - g(x)]\, dx. \qquad (11)$$

Similarly,

$$M_x(R) = M_x(R_f) - M_x(R_g)$$
$$= \int_a^b \tfrac{1}{2}[f(x)]^2\, dx - \int_a^b \tfrac{1}{2}[g(x)]^2\, dx \qquad \text{(by (7))},$$

and thus

$$M_x(R) = \int_a^b \tfrac{1}{2}([f(x)]^2 - [g(x)]^2)\, dx. \qquad (12)$$

We then define the centroid of $R$ by means of Equations (8), with

$$A = \int_a^b [f(x) - g(x)]\, dx. \qquad (13)$$

**326**

**EXAMPLE 3**  Let $R$ be the rectangle of Fig. 6.53. Following two applications of the symmetry principle, we conclude that its centroid is at the point $(4, 2)$. Let us test Equations (11) and (12) to see if they yield the same result. With $a = 2$, $b = 6$, $f(x) \equiv 3$, and $g(x) \equiv 1$, we get

$$M_y(R) = \int_2^6 x(3 - 1)\, dx = \left[x^2\right]_2^6 = 32,$$

$$M_x(R) = \int_2^6 \tfrac{1}{2}\left[(3)^2 - (1)^2\right] dx = \left[4x\right]_2^6 = 16.$$

Because the region has area $A = (4)(2) = 8$, we find using (8) that

$$\bar{x} = \frac{M_y}{A} = \frac{32}{8} = 4, \qquad \bar{y} = \frac{M_x}{A} = \frac{16}{8} = 2,$$

as expected.

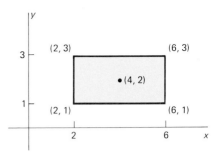

**6.53**  The rectangle of Example 3

---

An important theorem relating centroids and volumes of revolution is named for the Greek mathematician who stated it during the third century A.D.

---

> *First Theorem of Pappus*   *Volume of Revolution*
>
> Suppose that a plane region $R$ is revolved about an axis in its plane, generating a solid of revolution with volume $V$. Assume that the axis does not intersect the interior of $R$. Then $V$ is the product of the area $A$ of $R$ and the distance traveled by the centroid of $R$.

---

***Proof***   (For the special case of a region like the one shown in Fig. 6.54.) This is the region between the two graphs $y = f(x)$ and $y = g(x)$ from $a$ to $b$, and we shall take the axis of revolution to be the $y$-axis. Then, in a revolution about the $y$-axis, the distance traveled by the centroid of $R$ is $d = 2\pi\bar{x}$. By the method of cylindrical shells [see Formula (3) in Section 6-3] the volume of the solid generated is

$$V = \int_a^b 2\pi x[f(x) - g(x)]\, dx$$

$$= 2\pi M_y(R) \qquad \text{(by Formula (11))}$$

$$= 2\pi\bar{x} \cdot A \qquad \text{(by Formula (8))},$$

and thus $V = d \cdot A$, as desired.  ∎

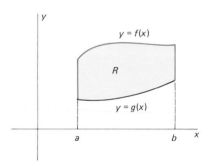

**6.54**  A region $R$ between the graphs of two functions

**EXAMPLE 4**  Find the volume $V$ of the sphere of radius $r$ generated by revolving around the $x$-axis the semicircle $D$ of Example 1.

***Solution***   The area of $D$ is $A = \tfrac{1}{2}\pi r^2$, and we found in Example 1 that $\bar{y} = 4r/3\pi$. Hence Pappus's theorem gives

$$V = 2\pi\bar{y}A = 2\pi \cdot \frac{4r}{3\pi} \cdot \frac{\pi r^2}{2} = \frac{4}{3}\pi r^3.$$

**6.55**  Rotate the circular disk about the $y$-axis to produce a torus (Example 5).

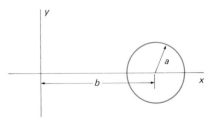

**EXAMPLE 5**  Consider the circular disk of Fig. 6.55, with radius $a$ and center at the point $(b, 0)$; also, $0 < a < b$. Find the volume $V$ of the solid torus generated by revolving this disk around the $y$-axis.

***Solution*** The centroid of the circle is its center $(b, 0)$, so $\bar{x} = b$. Hence this centroid is revolved through a distance $d = 2\pi b$. Consequently

$$V = d \cdot A = (2\pi b)(\pi a^2) = 2\pi^2 a^2 b.$$

Note that this result is dimensionally correct.

## MOMENTS AND CENTROIDS OF CURVES

Moments and centroids of plane *curves* are defined in a way that is quite analogous to the method for plane regions, so we shall present this topic in less detail. The moments $M_y(C)$ and $M_x(C)$ of the curve $C$ about the $y$- and $x$-axes, respectively, are defined to be

$$M_y(C) = \int x \, ds \quad \text{and} \quad M_x(C) = \int y \, ds. \tag{14}$$

The **centroid** $(\bar{x}, \bar{y})$ of $C$ is then defined to be the point with coordinates

$$\bar{x} = \frac{1}{s} M_y(C) \quad \text{and} \quad \bar{y} = \frac{1}{s} M_x(C), \tag{15}$$

where $s$ is the arc length of $C$.

The meaning of the integrals in (14) is that of the notation of Section 6-4. That is, $ds$ is a symbol to be replaced (before evaluation of the integral) by either

$$ds = \sqrt{1 + \left(\frac{dy}{dx}\right)^2} \, dx \quad \text{or} \quad ds = \sqrt{1 + \left(\frac{dx}{dy}\right)^2} \, dy,$$

depending upon whether $C$ is a smooth arc of the form $y = f(x)$ or one of the form $x = g(y)$. For example, if $C$ is described by $y = f(x)$, $a \leq x \leq b$, then

$$M_y(C) = \int_a^b x\sqrt{1 + [f'(x)]^2} \, dx \tag{16}$$

and

$$M_x(C) = \int_a^b f(x)\sqrt{1 + [f'(x)]^2} \, dx. \tag{17}$$

**EXAMPLE 6** Let $J$ denote the upper half of the circle (not the *disk*) of radius $r$, defined by

$$y = \sqrt{r^2 - x^2}, \quad -r \leq x \leq r.$$

Find the centroid of $J$.

***Solution*** Note first that $M_y(J) = 0$ by symmetry. Now

$$\frac{dy}{dx} = -\frac{x}{\sqrt{r^2 - x^2}},$$

so

$$ds = \sqrt{1 + \frac{x^2}{r^2 - x^2}} \, dx = \frac{r \, dx}{\sqrt{r^2 - x^2}}.$$

Hence the second of Formulas (14) gives

$$M_x(J) = \int_{-r}^{r} \sqrt{r^2 - x^2} \, \frac{r \, dx}{\sqrt{r^2 - x^2}}$$

$$= \int_{-r}^{r} r \, dx = 2r^2.$$

Because $s = \pi r$, the coordinates of the centroid of $J$ are

$$\bar{x} = \frac{M_y(J)}{s} = 0 \quad \text{and} \quad \bar{y} = \frac{M_x(J)}{s} = \frac{2r^2}{\pi r} = \frac{2r}{\pi}.$$

The first theorem of Pappus has an analogue for surface area of revolution.

> **Second Theorem of Pappus**  *Area of Revolution*
>
> Let the plane curve $C$ be revolved about an axis in its plane that does not intersect the curve. Then the area $A$ of the surface of revolution generated is equal to the product of the length of $C$ and the distance traveled by the centroid of $C$.

**Proof**  (For the special case in which $C$ is a smooth arc described by $y = f(x)$, $a \leq x \leq b$, and the axis of revolution is the $y$-axis.) The distance traveled by the centroid of $C$ is $d = 2\pi\bar{x}$. By Formula (12) of Section 6-4, the area of the surface of revolution is

$$A = \int_a^b 2\pi x \sqrt{1 + [f'(x)]^2} \, dx$$

$$= 2\pi M_y(C) \qquad \text{(Formula (16))}$$

$$= 2\pi\bar{x}s = d \cdot s \qquad \text{(Formula (15))},$$

as desired.  ∎

**EXAMPLE 7**  Find the surface area $A$ of the sphere of radius $r$ generated by revolving around the $x$-axis the semicircular arc of Example 6.

**Solution**  Since we found that $\bar{y} = 2r/\pi$ and we know that $s = \pi r$, the second theorem of Pappus gives

$$A = 2\pi\bar{y}s = 2\pi \left(\frac{2r}{\pi}\right)(\pi r) = 4\pi r^2.$$

**EXAMPLE 8**  Find the surface area $A$ of the torus of Example 5.

**Solution**  Now we think of revolving the circle (*not* the disk) of radius $a$ centered at the point $(b, 0)$. Of course, the centroid of the circle is located at its center $(b, 0)$; this follows from the symmetry principle or can be verified using computations like those in Example 6. Hence the distance traveled by the centroid is $d = 2\pi b$. Because the circumference of the circle is $s = 2\pi a$, the second theorem of Pappus gives

$$A = (2\pi b)(2\pi a) = 4\pi^2 ab.$$

## 6-6 PROBLEMS

In Problems 1–18, find the centroid of the plane region bounded by the given curves.

1  $x = 0,\ x = 4,\ y = 0,\ y = 6$
2  $x = 1,\ x = 3,\ y = 2,\ y = 4$
3  $x = -1,\ x = 3,\ y = -2,\ y = 4$
4  $x = 0,\ y = 0,\ x + y = 3$
5  $x = 0,\ y = 0,\ x + 2y = 4$
6  $y = 0,\ y = x,\ x + y = 2$
7  $y = 0,\ y = x^2,\ x = 2$
8  $y = x^2,\ y = 9$          9  $y = 0,\ y = x^2 - 4$
10  $x = -2,\ x = 2,\ y = 0,\ y = x^2 + 1$
11  $y = 4 - x^2,\ y = 0$     12  $y = x^2,\ y = 18 - x^2$
13  $y = 3x^2,\ y = 0,\ x = 1$
14  $x = y^2,\ y = 0,\ x = 4$    (*above* the x-axis)
15  $y = x,\ y = 6 - x^2$      16  $y = x^2,\ y^2 = x$
17  $y = x^2,\ y = x^3$
18  $2y^2 = x^3,\ x = 2$

19  Find the centroid of the first quadrant of the circular disk $x^2 + y^2 \leq r^2$ by direct computation, as in Example 1.
20  Apply the first theorem of Pappus to find the centroid of the first quadrant of the circular disk $x^2 + y^2 \leq r^2$. Use the facts that $\bar{x} = \bar{y}$ by symmetry and that revolution of this quarter-disk about either coordinate axis gives a solid hemisphere with volume $V = \frac{2}{3}\pi r^3$.
21  Find the centroid of the arc consisting of the first quadrant portion of the circle $x^2 + y^2 = r^2$ by direct computation, as in Example 6.
22  Apply the second theorem of Pappus to find the centroid of the quarter-circular arc of Problem 21. Note that $\bar{x} = \bar{y}$ by symmetry, and that revolution of this arc about either coordinate axis gives a hemisphere with surface area $A = 2\pi r^2$.
23  Show by direct computation that the centroid of the triangle with vertices $(0, 0)$, $(r, 0)$, and $(0, h)$ is the point $(r/3, h/3)$. Verify that this point lies on the line from the vertex $(0, 0)$ to the midpoint of the opposite side of the triangle and two-thirds of the way from the vertex to the midpoint.
24  Apply the first theorem of Pappus and the result of Problem 23 to verify the formula $V = \frac{1}{3}\pi r^2 h$ for the volume of the cone obtained by revolving the triangle about the y-axis.
25  Apply the second theorem of Pappus to show that the lateral surface area of the cone of Problem 24 is $A = \pi r L$, where $L = \sqrt{r^2 + h^2}$ is the slant height of the cone.
26  (a) Use additivity of moments to find the centroid of the trapezoid shown in Fig. 6.56.

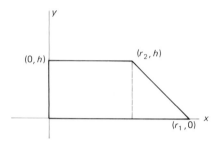

**6.56**  The trapezoid of Problem 26

(b) Apply the first theorem of Pappus and the result of part (a) to show that the volume of the conical frustum generated by revolving the trapezoid around the y-axis is

$$V = \frac{\pi}{3}(r_1^2 + r_1 r_2 + r_2^2)h.$$

27  Apply the second theorem of Pappus to show that the lateral surface area of the conical frustum in Problem 26 is $A = \pi(r_1 + r_2)L$, where

$$L = \sqrt{(r_1 - r_2)^2 + h^2}$$

is its slant height.
28  (a) Apply the second theorem of Pappus to verify that the curved surface area of a cylinder with height $h$ and base radius $r$ is $A = 2\pi rh$.
(b) Explain how this also follows from the result of Problem 27.
29  (a) Use additivity of moments to find the centroid of the plane region shown in Fig. 6.57, which consists of a semicircular region of radius $a$ sitting atop a rectangular region of width $2a$ and height $b$.

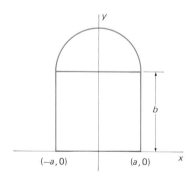

**6.57**  The plane region of Problem 29

(b) Then apply the first theorem of Pappus to find the volume generated by rotating this region about the $x$-axis.

**30** (a) Consider the plane region of Fig. 6.58, bounded by $x^2 = 2py$, $x = 0$, and $y = h = r^2/2p$ $(p > 0)$. Show that its area is $A = 2rh/3$ and that the $x$-coordinate of its centroid is $\bar{x} = 3r/8$.

(b) Use Pappus's theorem and the result of part (a) to show that the volume of a paraboloid of revolution with radius $r$ and height $h$ is $V = \frac{1}{2}\pi r^2 h$.

CHALLENGE   The region in the first quadrant bounded by the graphs of $y = x$ and $y = x^2$ is rotated about the line $y = x$. Find the volume of the solid thus generated. (*Suggestion:* Use the first theorem of Pappus.)

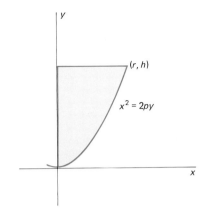

**6.58**   The region of Problem 30

---

**Additional Applications**

We take up moments and centroids of solids in Chapter 16. But here we can quite simply compute moments for a special sort of solid, a solid of revolution. Since the centroid of such a solid will lie on its axis of symmetry, we shall need to compute only one coordinate of the centroid.

The moment of a mass particle in space about a *plane* is the product of its mass and its (signed) perpendicular distance from the plane. For solids (as for plane regions) we assume additivity of moments and symmetry; if a solid is symmetric with respect to a plane, then its centroid lies in that plane. For example, the centroid $C$ of a solid cylinder lies on the axis of the cylinder and halfway between its two bases; thus the moment of the cylinder about a plane parallel to its bases is the product of its volume (we still assume, for simplicity, that density $\rho = 1$) and the distance from $C$ to the plane.

Now suppose that $f$ is a continuous and nonnegative function defined on the interval $[a, b]$ of the $x$-axis. Let $S$ be the solid of revolution generated by revolving about the $x$-axis the region $R$ that lies under the graph $y = f(x)$ for $a \leq x \leq b$. We think of $S$ as a solid with constant density $\rho = 1$. We wish to define the moment of $S$ about the plane that is perpendicular to the $x$-axis at the origin. Since it is natural to introduce a $z$-axis in space perpendicular to both the $x$-axis and the $y$-axis at the origin, we shall refer to this plane as the $yz$-plane.

To define the moment $M_{yz}$ of $S$ about the $yz$-plane, we begin with a regular partition of $[a, b]$ into $n$ equal subintervals each of length $\Delta x$. Suppose that $x_i^*$ is the midpoint of the $i$th subinterval $[x_{i-1}, x_i]$. In forming $S$ by rotation, one "slice" of $S$ is well approximated by the cylinder with axis $[x_{i-1}, x_i]$ and radius $f(x_i^*)$ shown in Fig. 6.59. This cylinder has volume $\pi[f(x_i^*)]^2 \Delta x$ and centroid $(x_i^*, 0)$. Hence its moment about the $yz$-plane is $\pi x_i^*[f(x_i^*)]^2 \Delta x$.

By additivity of moments, the moment about the $yz$-plane of the union of these $n$ cylinders (for $i = 1, 2, 3, \ldots, n$) is

$$\sum_{i=1}^{n} \pi x_i^*[f(x_i^*)]^2 \Delta x.$$

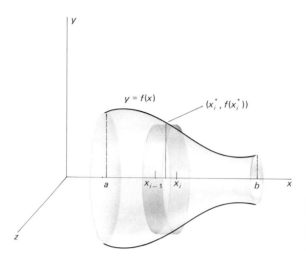

6.59   A solid of revolution is the union of almost-cylindrical slabs whose centroids are easy to locate.

This sum is a Riemann sum for the function $\pi x[f(x)]^2$ on $[a, b]$. Therefore we define the **moment** $M_{yz}$ of the solid of revolution $S$ about the $yz$-plane by means of the equation

$$M_{yz} = \int_a^b \pi x[f(x)]^2 \, dx. \tag{1}$$

We then define the $x$-coordinate of the **centroid** of $S$ to be

$$x = \frac{M_{yz}}{V} = \frac{1}{V} \int_a^b \pi x[f(x)]^2 \, dx, \tag{2}$$

where $V$ is the volume of $S$.

In the case of a solid generated by revolving a region $0 \leqq x \leqq g(y)$, $c \leqq y \leqq d$, around the $y$-axis, the roles of $x$ and $y$ are reversed, and the centroid lies on the $y$-axis. The **moment** $M_{xz}$ of this solid about the $xz$-plane is defined to be

$$M_{xz} = \int_c^d \pi y[g(y)]^2 \, dy, \tag{3}$$

and the $y$-coordinate of its centroid is

$$\bar{y} = \frac{M_{xz}}{V} = \frac{1}{V} \int_c^d \pi y[g(y)]^2 \, dy. \tag{4}$$

6.60   The parabolic solid of Example 1

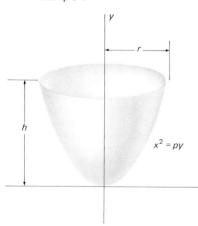

**EXAMPLE 1**   Determine the centroid of the solid paraboloid shown in Fig. 6.60, with "height" $h$, "base" a circle of radius $r$, and "vertex" at the origin, obtained by revolving the parabola $x^2 = py$ (with $p = r^2/h$) around the $y$-axis.

*Solution*   By the method of cross sections, the volume of this solid paraboloid is

$$V = \int \pi x^2 \, dy = \int_0^h \pi p y \, dy$$

$$= \left[ \tfrac{1}{2} \pi p y^2 \right]_0^h = \tfrac{1}{2} \pi p h^2 = \tfrac{1}{2} \pi r^2 h.$$

**332**

Formula (3) with $g(y) = \sqrt{py}$ gives

$$M_{xz} = \int_0^h \pi p y^2 \, dy = \left[\tfrac{1}{3}\pi p y^3\right]_0^h = \tfrac{1}{3}\pi p h^3 = \tfrac{1}{3}\pi r^2 h^2.$$

Then

$$\bar{y} = \frac{M_{xz}}{V} = \frac{\pi r^2 h^2/3}{\pi r^2 h/2} = \frac{2}{3} h.$$

Thus the centroid of a solid paraboloid lies on its axis two-thirds of the way from its vertex to its base.

## FORCE EXERTED BY A LIQUID

The **pressure** $p$ at depth $h$ in a liquid is the force per unit area exerted by the liquid at that depth, and is given by

$$p = \rho h \tag{5}$$

where $\rho$ is the (weight) density of the liquid. For example, at a depth of 10 ft in water, for which $\rho = 62.4$ lb/ft$^3$, the pressure is $(62.4)(10) = 624$ lb/ft$^2$. Hence if a thin flat plate with area 5 ft$^2$ is suspended in a horizontal position at a depth of 10 ft in water, then the water exerts a downward force of $(624)(5) = 3120$ lb on the top face of the plate and an equal upward force on its bottom face.

It is an important fact that at a given depth in a liquid the pressure is the same in all directions. But if a flat plate is submerged in a vertical position in a liquid, then the pressure on the face of the plate is not constant, because by (5) it increases with increasing depth. Consequently the total force exerted on the vertical plate must be computed by integration.

Consider a vertical flat plate submerged in a liquid of density $\rho$ as indicated in Fig. 6.61. The surface of the liquid is the line $y = c$, and the plate lies opposite the interval $a \leq y \leq b$. The width of the plate at depth $c - y$ is some function of $y$, which we denote by $w(y)$.

To compute the total force $F$ exerted by the liquid on a face of this plate, we begin with a regular partition of $[a, b]$ into $n$ equal subintervals of length $\Delta y$, and denote by $y_i^*$ the midpoint of the subinterval $[y_{i-1}, y_i]$. The horizontal strip of the plate opposite this $i$th subinterval is approximated by a rectangle with width $w(y_i^*)$ and height $\Delta y$, and its average depth in the liquid is $c - y_i^*$. Hence the force $\Delta F_i$ exerted by the liquid on this strip of the plate is given approximately by

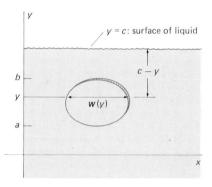

**6.61** A thin plate suspended vertically in a liquid

$$\Delta F_i \approx \rho(c - y_i^*)w(y_i^*) \, \Delta y, \tag{6}$$

and the total force on the entire plate is given approximately by

$$F = \sum_{i=1}^n \Delta F_i \approx \sum_{i=1}^n \rho(c - y_i^*)w(y_i^*) \, \Delta y.$$

We obtain the exact value of $F$ by taking the limit of this Riemann sum as $\Delta y \to 0$:

$$F = \int_a^b \rho(c - y)w(y) \, dy. \tag{7}$$

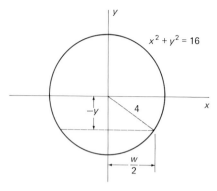

$x^2 + y^2 = 16$

**6.62**  View of one end of the oil tank

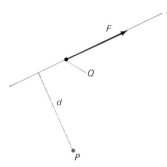

**6.63**  The moment of the force $F$ about the point $P$

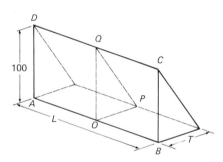

**6.64**  View of a model of a dam

**EXAMPLE 2**  A cylindrical tank 8 ft in diameter is lying on its side and is half full of oil with density $\rho = 75$ lb/ft$^3$. Find the force exerted by the oil on one end of the tank.

***Solution***  We locate the $y$-axis as indicated in Fig. 6.62, so that the surface of the oil is at the level $y = 0$. The oil corresponds to the interval $-4 \leq y \leq 0$, and from the right triangle in the figure we see that the width of the oil at depth $-y$ is $w(y) = 2\sqrt{16 - y^2}$. Hence Formula (7) gives

$$F = \int_{-4}^{0} 75(-y)(2\sqrt{16 - y^2})\, dy$$

$$= 75\left[\tfrac{2}{3}(16 - y^2)^{3/2}\right]_{-4}^{0} = 3200 \quad \text{lb.}$$

### MOMENTS OF FORCES ON A DAM

This subsection is intended mainly for those who have already studied moments of forces in a physics course. Consider a force of magnitude $F$ that acts at the point $Q$ in the direction of the line $L$ through $Q$ shown in Fig. 6.63. Recall that in physics the **moment** of the force $F$ about the point $P$ is defined to be

$$M = Fd$$

where $d$ is the perpendicular distance from the point $P$ to the line $L$.

If a force is distributed over an area, rather than being applied at a single point, then in general its moment about a given point must be computed by integration. We illustrate such a computation by analyzing a dam with the shape indicated in Fig. 6.64. The vertical side $ABCD$ faces the water and has height 100 feet and length $L$ feet. The dam is made of concrete of density $3.6\rho$, where $\rho$ denotes the density of water. We want to determine what the thickness $T$ of the dam at its base should be in order that the weight $W$ of the dam will balance the force $F$ exerted by water 100 feet deep on the vertical face of the dam.

Figure 6.65 shows the vertical midsection $OPQ$ of the dam. The point $Z$ is the centroid of the dam, which by Problem 23 in Section 6-6 is the point $(T/3, 100/3)$ in the indicated $xy$-coordinate system. We assume that the weight $W$ of the dam acts as a force directed downward through $Z$, and that the force

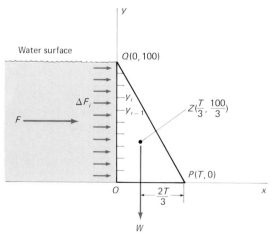

**6.65**  Vertical cross section at the middle of the dam, as seen from the side

$F$ exerted by the water acts as though it were concentrated along the segment $OQ$. We want to determine $T$ so that the moments of $F$ and $W$ about the point $P$ will be equal in magnitude; note that $F$ acts clockwise around $P$ while $W$ acts in a counterclockwise direction.

The moment of $W$ about $P$ is

$$(3.6\rho)\left(\frac{100T}{2}\right)(L)\left(\frac{2T}{3}\right) = 120\rho LT^2.$$

To compute the moment of $F$ about $P$, we begin with a regular partition of $[0, 100]$ into $n$ equal subintervals of length $\Delta y$, and denote by $y_i^*$ the midpoint of $[y_{i-1}, y_i]$. The force $\Delta F_i$ on the horizontal strip of the dam's face corresponding to $[y_{i-1}, y_i]$ is given by (6) with $c = 100$ and $w(y) = L$, so

$$\Delta F_i \approx \rho(100 - y_i^*)L\,\Delta y.$$

The moment $\Delta M_i$ of $\Delta F_i$ about $P$ is therefore given approximately by

$$\Delta M_i \approx \rho y_i^*(100 - y_i^*)L\,\Delta y,$$

and the moment of $F$ about $P$ is obtained—again approximately—by adding these moments for $i = 1, 2, 3, \ldots, n$:

$$M \approx \sum_{i=1}^{n} \rho L(100y_i^* - [y_i^*]^2)\,\Delta y.$$

We take the limit as $\Delta y \to 0$, and this gives

$$M = \int_0^{100} \rho L(100y - y^2)\,dy$$

$$= \rho L\left[50y^2 - \tfrac{1}{3}y^3\right]_0^{100} = \tfrac{5}{3}\rho L \times 10^5.$$

The condition that the moments of $F$ and $W$ balance about $P$ can therefore be expressed in the form of the equation

$$120\rho LT^2 = \tfrac{5}{3}\rho L \times 10^5,$$

and so

$$T = \sqrt{\frac{500{,}000}{360}} \approx 37.27 \text{ ft.}$$

## 6-7 PROBLEMS

In each of Problems 1–8, find the centroid of the solid generated by revolving the region bounded by the given curves around the indicated axis.

1 $y^2 = 2x$, $x = 2$; the $x$-axis
2 $y^2 = 3x$, $x = 0$, $y = 3$; the $y$-axis
3 $y = 2x - x^2$, $y = 0$; the $x$-axis
4 $x = 3y - y^2$, $x = 0$; the $y$-axis
5 $y = x^2$, $y = 3x$; the $x$-axis
6 $y = x^2$, $x = y^2$; the $y$-axis
7 $x^2 - y^2 = 9$, $x = 5$; the $x$-axis

8 $x = \dfrac{1}{y^2}$, $x = 0$, $y = 1$, $y = 2$; the $y$-axis

9 Show that the centroid of a right circular cone lies on the axis of the cone and three-fourths of the way from the vertex to the base.

10 Show that the centroid of a solid hemisphere lies on its axis of symmetry, with its distance from the base equal to three-eighths the radius of the hemisphere.

11 Generalize the result of Problem 10 by finding the centroid of the upper half of the solid ellipsoid obtained by revolving the ellipse $(x/a)^2 + (y/b)^2 = 1$ around the $y$-axis.

12 Consider the first-quadrant region bounded by the coordinate axes and the parabola $y = h^2 - x^2$. If this region is revolved around the $x$-axis, show that the volume of the

resulting solid is $V = \frac{8}{15}\pi h^5$ and that the $x$-coordinate of its centroid is $\bar{x} = \frac{5}{16}h$.

**13** Suppose that $n$ is a positive even integer; consider the solid that is generated by revolving the region bounded by the curve $y = x^n$ and the line $y = h$ around the $y$-axis. Show that $\bar{y} = (n + 2)h/(2n + 2)$ is the $y$-coordinate of its centroid.

**14** A water tank is in the shape of a solid of revolution, so that its horizontal cross sections are circular. The work $W$ required to fill this tank with water pumped up from the ground is given by Formula (8) in Section 6-5. Show that $W = w\bar{y}$, where $w$ is the weight of the water in the full tank and $\bar{y}$ is the $y$-coordinate of the centroid of that mass of water.

**15** A water trough 10 ft long has a square cross section that is 2 ft wide. If the trough is full of water ($\rho = 62.4$ lb/ft$^3$), find the force exerted by the water on one end of the trough.

**16** Repeat Problem 15, where the cross section of the trough is an equilateral triangle with edge 3 ft.

**17** Repeat Problem 15, where the cross section of the trough is a trapezoid 3 ft high, 2 ft wide at the bottom, and 4 ft wide at the top.

**18** Find the force on one end of the cylindrical tank of Example 2 if it is filled with oil weighing 50 lb/ft$^3$. Remember that

$$\int_0^a \sqrt{a^2 - y^2}\, dy = \frac{\pi a^2}{4}$$

because the integral represents the area of a quarter-circle of radius $a$.

In each of Problems 19–22, a gate in the vertical face of a dam is described. Find the total force of water on this gate if its top is 10 ft beneath the surface of the water.

**19** A square of edge 5 ft with its top parallel to the water surface.

**20** A circle with radius 3 ft.

**21** An isosceles triangle 5 ft high and 8 ft wide at the top.

**22** A semicircle with radius 4 ft with its diameter also its top edge and parallel to the water surface.

**23** Suppose that the dam of Fig. 6.64 is $L = 200$ ft long and $T = 30$ ft thick at its base. Find the force of water on the dam if the water is 100 ft deep and the *slanted* side of the dam faces the water.

**24** Deduce from the formula in Equation (7) that the force exerted by fluid on a face of a vertically submerged flat plate with area $A$ is $F = \rho h A$, where $h$ is the depth (beneath the surface of the liquid) of the centroid of the plate.

**25** Apply the result of Problem 24 to find the force on the gate of:
(a) Problem 19;
(b) Problem 20;
(c) Problem 22 (see Example 1 in Section 6-6).

---

**\*6-8**

**Appendix on Approximations and Riemann Sums**

Several times in this chapter, our attempt to compute some quantity $Q$ has led to the following situation. Beginning with a regular partition of an appropriate interval $[a, b]$ into $n$ equal subintervals of length $\Delta x$, we have found an approximation $A_n$ to $Q$ of the form

$$A_n = \sum_{i=1}^{n} g(u_i)h(v_i)\, \Delta x \tag{1}$$

where $u_i$ and $v_i$ are two (generally different) points of the $i$th subinterval $[x_{i-1}, x_i]$. For example, in our discussion of surface area of revolution that precedes the formula in Equation (8) of Section 6-4, we found the approximation

$$\sum_{i=1}^{n} 2\pi f(u_i)[1 + (f'(v_i))^2]^{1/2}\, \Delta x \tag{2}$$

to the area of the surface generated by revolving the curve $y = f(x)$, $a \leqq x \leqq b$, around the $x$-axis. (In Section 6-4, we wrote $x_i^{**}$ for $u_i$ and $x_i^*$ for $v_i$.) Note that the expression in (2) is the same as the right-hand side of Equation (1); take $g(x) = 2\pi f(x)$ and $h(x) = [1 + (f'(x))^2]^{1/2}$.

In such a situation we observe that, if $u_i$ and $v_i$ were the *same* point $x_i^*$ of $[x_{i-1}, x_i]$ for each $i$ ($i = 1, 2, 3, \ldots, n$), then Approximation (1) would be a Riemann sum for the function $g(x)h(x)$ on $[a, b]$. This leads us to suspect that

$$\lim_{\Delta x \to 0} \sum_{i=1}^{n} g(u_i)h(v_i)\, \Delta x = \int_a^b g(x)h(x)\, dx. \qquad (3)$$

In Section 6-4, we assumed the truth of Equation (3) and concluded from the approximation in (2) that surface area of revolution ought to be defined to be

$$A = \lim_{\Delta x \to 0} \sum_{i=1}^{n} 2\pi f(u_i)[1 + (f'(v_i))^2]^{1/2}\, \Delta x$$

$$= \int_a^b 2\pi f(x)[1 + (f'(x))^2]^{1/2}\, dx.$$

The following theorem guarantees that Equation (3) holds under only mild restrictions on the functions $g$ and $h$.

---

**Theorem 1**  *A Generalization of Riemann Sums*

Suppose that $h$ and the derivative $g'$ are continuous on $[a, b]$. Then

$$\lim_{\Delta x \to 0} \sum_{i=1}^{n} g(u_i)h(v_i)\, \Delta x = \int_a^b g(x)h(x)\, dx \qquad (3)$$

where $u_i$ and $v_i$ are arbitrary points of the $i$th subinterval of a regular partition of $[a, b]$ into $n$ equal subintervals each of length $\Delta x$.

---

**Proof**  Let $M_1$ and $M_2$ denote the maximum values on $[a, b]$ of $|g'(x)|$ and $|h(x)|$, respectively. Note that

$$\sum_{i=1}^{n} g(u_i)h(v_i)\, \Delta x = R_n + S_n$$

where

$$R_n = \sum_{i=1}^{n} g(v_i)h(v_i)\, \Delta x$$

is a Riemann sum approaching $\int_a^b g(x)h(x)\, dx$ as $\Delta x \to 0$, and

$$S_n = \sum_{i=1}^{n} [g(u_i) - g(v_i)]h(v_i)\, \Delta x.$$

To prove (3) it is sufficient to show that $S_n \to 0$ as $\Delta x \to 0$. The mean value theorem gives

$$|g(u_i) - g(v_i)| = |g'(\bar{x}_i)| \cdot |u_i - v_i| \qquad (\bar{x}_i \text{ in } (u_i, v_i))$$
$$\leqq M_1\, \Delta x$$

because $u_i$ and $v_i$ are both points of the interval $[x_{i-1}, x_i]$ of length $\Delta x$. Then

$$|S_n| \leqq \sum_{i=1}^{n} |g(u_i) - g(v_i)| \cdot |h(v_i)|\, \Delta x$$

$$\leqq \sum_{i=1}^{n} M_1\, \Delta x \cdot M_2\, \Delta x$$

$$= (M_1 M_2\, \Delta x) \sum_{i=1}^{n} \Delta x = M_1 M_2 (b - a)\, \Delta x,$$

from which it follows that $S_n \to 0$ as $\Delta x \to 0$, as desired.  ∎

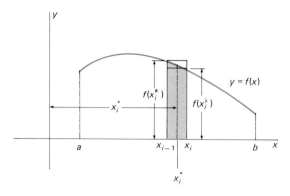

**6.66** A careful estimate of the volume of a solid of revolution around the $y$-axis.

As an application of Theorem 1, let us give a rigorous derivation of Formula (2) in Section 6-3,

$$V = \int_a^b 2\pi x f(x)\, dx, \tag{4}$$

for the volume of the solid generated by revolving the region lying under $y = f(x)$, $a \leq x \leq b$, around the $y$-axis. Starting with the usual regular partition of $[a, b]$, let $f(x_i^\flat)$ and $f(x_i^\#)$ denote the minimum and maximum values of $f$ on the $i$th subinterval $[x_{i-1}, x_i]$. Denote by $x_i^*$ the midpoint of this subinterval. From Fig. 6.66, we see that the part of our solid generated by revolving the region under $y = f(x)$, $x_{i-1} \leq x \leq x_i$, contains a cylindrical shell with average radius $x_i^*$, thickness $\Delta x$, and height $f(x_i^\flat)$, and is contained in another cylindrical shell with the same average radius and thickness, but with height $f(x_i^\#)$. Hence the volume $\Delta V_i$ of this part of the solid satisfies the inequalities

$$2\pi x_i^* f(x_i^\flat)\, \Delta x \leq \Delta V_i \leq 2\pi x_i^* f(x_i^\#)\, \Delta x.$$

We add these inequalities for $i = 1, 2, 3, \ldots, n$, and find that

$$\sum_{i=1}^n 2\pi x_i^* f(x_i^\flat)\, \Delta x \leq V \leq \sum_{i=1}^n 2\pi x_i^* f(x_i^\#)\, \Delta x.$$

Since Theorem 1 implies that each of the last two sums approaches $\int_a^b 2\pi x\, f(x)\, dx$, the squeeze law of limits now implies Formula (4).

We will occasionally need a generalization of Theorem 1 that involves the notion of a continuous function $F(x, y)$ of the two variables $x$ and $y$. We say that $F$ is *continuous* at the point $(x_0, y_0)$ provided that the value $F(x, y)$ can be made arbitrarily close to $F(x_0, y_0)$ merely by choosing the point $(x, y)$ sufficiently close to $(x_0, y_0)$. We shall discuss continuity of functions of two variables in Chapter 15. Here it will suffice to accept the following facts: If $g(x)$ and $h(y)$ are continuous functions of the single variables $x$ and $y$, respectively, then simple combinations such as

$$g(x) \pm h(y), \qquad g(x)h(y), \quad \text{and} \quad [(g(x))^2 + (h(y))^2]^{1/2}$$

are continuous functions of the two variables $x$ and $y$.

Now consider a regular partition of $[a, b]$ into $n$ equal subintervals of length $\Delta x$, and let $u_i$ and $v_i$ denote arbitrary points of the $i$th subinterval $[x_{i-1}, x_i]$. The following theorem—we omit the proof—tells us how to find the limit as $\Delta x \to 0$ of a sum of the form $\sum_{i=1}^n F(u_i, v_i)\, \Delta x$.

Note that Theorem 1 is the special case $F(x, y) = g(x)h(y)$ of Theorem 2. Observe also that the integrand $F(x, x)$ on the right in Equation (5) is merely an ordinary function of the (single) variable $x$. As a formal matter, the integral corresponding to the sum in (5) is obtained by replacing the summation symbol with an integral symbol, changing both $u_i$ and $v_i$ to $x$, and replacing $\Delta x$ by $dx$.

For instance, if the interval $[a, b]$ is $[0, 4]$, then

$$\lim_{\Delta x \to 0} \sum_{i=1}^{n} \sqrt{9u_i^2 + v_i^4}\, \Delta x = \int_0^4 \sqrt{9x^2 + x^4}\, dx$$

$$= \int_0^4 x\sqrt{9 + x^2}\, dx = \left[ \tfrac{1}{3}(9 + x^2)^{3/2} \right]_0^4$$

$$= \tfrac{1}{3}\left[ (25)^{3/2} - (9)^{3/2} \right] = \tfrac{98}{3}.$$

## 6-8   PROBLEMS

In each of Problems 1–7, $u_i$ and $v_i$ are arbitrary points of the $i$th subinterval of a regular partition of $[a, b]$ into $n$ equal subintervals of length $\Delta x$ each. Express the given limit as an integral from $a$ to $b$, and then compute the value of this integral.

**1** $\displaystyle\lim_{\Delta x \to 0} \sum_{i=1}^{n} u_i v_i\, \Delta x; \quad a = 0, b = 1$

**2** $\displaystyle\lim_{\Delta x \to 0} \sum_{j=1}^{n} (3u_j + 5v_j)\, \Delta x; \quad a = -1, b = 3$

**3** $\displaystyle\lim_{\Delta x \to 0} \sum_{i=1}^{n} u_i\sqrt{4 - v_i^2}\, \Delta x; \quad a = 0, b = 2$

**4** $\displaystyle\lim_{\Delta x \to 0} \sum_{i=1}^{n} \frac{u_i\, \Delta x}{\sqrt{16 + v_i^2}}; \quad a = 0, b = 3$

**5** $\displaystyle\lim_{\Delta x \to 0} \sum_{i=1}^{n} u_i(1 + v_i^2)^5\, \Delta x; \quad a = 0, b = 1$

**6** $\displaystyle\lim_{\Delta x \to 0} \sum_{i=1}^{n} (1 + u_i + v_i)^2\, \Delta x; \quad a = 0, b = 3$

**7** $\displaystyle\lim_{\Delta x \to 0} \sum_{k=1}^{n} \sqrt{u_k^4 + v_k^7}\, \Delta x; \quad a = 0, b = 2$

**8** Explain how Theorem 1 applies to show that Formula (8) in Section 6-5 follows from the discussion preceding it in that section.

**9** Use Theorem 1 to derive Formula (10) in Section 6-4.

## CHAPTER 6 REVIEW:   Definitions, Concepts, Results

Use the list below as a guide to ideas that you may need to review.

**1** The general method of setting up an integral formula for a quantity by approximating it and then recognizing the approximation as a Riemann sum corresponding to the desired integral

**2** Net distance traveled as the definite integral of velocity

**3** Total distance as the integral of the absolute value of the velocity

**4** The method of cross sections for computing volumes

**5** The volume of a solid of revolution (around either the $x$-axis or the $y$-axis)

**6** The method of cylindrical shells for computing volumes

**7** The arc length of a smooth arc described either in the form $y = f(x), a \leq x \leq b$, or in the form $x = g(y), c \leq y \leq d$

**8** The area of a frustum of a cone

**9** The area of the surface of revolution generated by revolving a smooth arc, given in either of the forms $y = f(x)$ or $x = g(y)$, about either the $x$-axis or the $y$-axis

**10** The work done by a force function in moving a particle along a straight line segment

**11** Hooke's law and the work done in stretching or compressing an elastic spring

**12** Work done against the force of gravity

**13** Work done in filling a tank, or in pumping the liquid in a tank to another level

**14** The center of mass of a finite system of particles

**15** The symmetry principle (for centroids) and additivity of moments

**16** The centroid of a plane region: definition and formulas

**17** The moments of a plane region about the two coordinate axes

**18** The centroid and moments of a smooth arc or curve

**19** The first theorem of Pappus: volume of revolution

**20** The second theorem of Pappus: surface area of revolution

**21** The centroid of a solid of revolution

**22** The force exerted by liquid on a face of a submerged plate

**23** Recognition of integrals corresponding to limits of Riemann sums (Section 6-8)

## MISCELLANEOUS PROBLEMS

In Problems 1–3, find both the net distance and the total distance traveled between times $t = a$ and $t = b$ by a particle moving along a line with the given velocity function $v = f(t)$.

**1** $v = t^2 - t - 2$;  $a = 0, b = 3$

**2** $v = |t^2 - 4|$;  $a = 1, b = 4$

**3** $v = \dfrac{t}{\sqrt{t^2 + 16}}$;  $a = -3, b = 0$

In Problems 4–8, $u_i$ and $v_i$ denote points of the $i$th subinterval of a regular partition of the indicated interval $[a, b]$ into $n$ equal subintervals of length $\Delta x$. Evaluate the given limit by computing the value of an appropriate related integral.

**4** $\displaystyle \lim_{\Delta x \to 0} \sum_{i=1}^{n} (3u_i^2 + 2)\, \Delta x$;  $a = -1, b = 2$

**5** $\displaystyle \lim_{\Delta x \to 0} \sum_{j=1}^{n} \sqrt{2v_j + 1}\, \Delta x$;  $a = 0, b = 12$

**6** $\displaystyle \lim_{\Delta x \to 0} \sum_{i=1}^{n} u_i \sqrt{v_i}\, \Delta x$;  $a = 0, b = 5$

**7** $\displaystyle \lim_{\Delta x \to 0} \sum_{i=1}^{n} \frac{u_i^3 - 1}{v_i^2}\, \Delta x$;  $a = 1, b = 2$

**8** $\displaystyle \lim_{\Delta x \to 0} \sum_{i=1}^{n} \sqrt{u_i + v_i^{5/2}}\, \Delta x$;  $a = 0, b = 1$

**9** Suppose that rainfall begins at time $t = 0$ and that the rate after $t$ hours is $(t + 6)/12$ inches per hour. How many inches of rain fall during the first 12 h?

**10** The base of a certain solid is the region in the first quadrant that is bounded by the curves $y = x^3$ and $y = 2x - x^2$. Find its volume, if each cross section perpendicular to the $x$-axis is a square with one edge in the base of the solid.

**11** Find the volume of the solid generated by revolving the first-quadrant region of Problem 10 about the $x$-axis.

**12** Find the volume of the solid generated by revolving the region bounded by $y = 2x^4$ and $y = x^2 + 1$ around

(a) the $x$-axis;
(b) the $y$-axis.

**13** A wire is made of copper (density 8.5 gm/cm$^3$) and is shaped like a helix that spirals around the $x$-axis from $x = 0$ to $x = 20$ cm. Each cross section of this wire perpendicular to the $x$-axis is a circular disk of radius 0.25 cm. What is the total mass of the wire?

**14** Derive the formula $V = \frac{1}{3}\pi h(r_1^2 + r_1 r_2 + r_2^2)$ for the volume of a frustum of a cone with height $h$ and base radii $r_1$ and $r_2$.

**15** Suppose that the point $P$ lies on a line perpendicular to the $xy$-plane at the origin $O$, with $|OP| = h$. Consider the "elliptical cone" consisting of all points on line segments from $P$ to points on and within the ellipse $(x/a)^2 + (y/b)^2 = 1$. Show that the volume of this elliptical cone is $V = \frac{1}{3}\pi abh$.

**16** Figure 6.67 shows the region $R$ bounded by the ellipse $(x/a)^2 + (y/b)^2 = 1$ and the line $x = a - h$, where $0 < h < a$. Revolution of $R$ around the $x$-axis generates a "segment of an ellipsoid" with radius $r$, height $h$, and volume $V$. Show that

$$r^2 = \frac{b^2(2ah - h^2)}{a^2} \quad \text{and that} \quad V = \frac{1}{3}\pi r^2 h\, \frac{3a - h}{2a - h}.$$

**6.67**  A segment of an ellipsoid

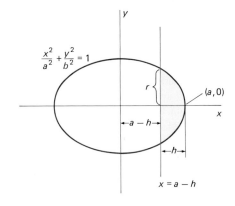

$$\frac{x^2}{a^2} + \frac{y^2}{b^2} = 1$$

$(a, 0)$

$\leftarrow a - h \rightarrow$

$\leftarrow h \rightarrow$

$x = a - h$

**17** Figure 6.68 shows the region $R$ bounded by the hyperbola $(x/a)^2 - (y/b)^2 = 1$ and the line $x = a + h$, where $h > 0$. Revolution of $R$ around the $x$-axis generates a "segment of a hyperboloid" with radius $r$, height $h$, and volume $V$. Show that

$$r^2 = \frac{b^2(2ah + h^2)}{a^2} \quad \text{and that} \quad V = \frac{1}{3}\pi r^2 h \frac{3a + h}{2a + h}.$$

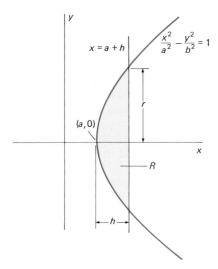

**6.68**   The region $R$ of Problem 17

In Problems 18–20, the function $f(x)$ is nonnegative and continuous for $x \geq 1$. When the region lying under $y = f(x)$ from $x = 1$ to $x = t$ is revolved around the indicated axis, the volume of the resulting solid is $V(t)$. Find the function $f(x)$.

**18** $V(t) = \pi\left(1 - \dfrac{1}{t}\right)$;   the $x$-axis

**19** $V(t) = \frac{1}{6}\pi[(1 + 3t)^2 - 16]$;   the $x$-axis

**20** $V(t) = \frac{2}{9}\pi[(1 + 3t^2)^{3/2} - 8]$;   the $y$-axis

**21** Let $R$ denote the region that lies under the curve

$$y = x\sqrt{1 + x^3}$$

over the interval $0 \leq x \leq 2$. Find the volume of the solid obtained by revolving the region $R$ around
(a) the $x$-axis;
(b) the $y$-axis.

**22** Use the method of cylindrical shells to find the volume of the solid generated by revolving the region bounded by $y = x^2$ and $y = x + 2$ around the line $x = -2$.

**23** Find the length of the curve $y = \frac{1}{3}x^{3/2} - x^{1/2}$ from $x = 1$ to $x = 4$.

**24** Find the area of the surface generated by revolving the curve of Problem 23 around
(a) the $x$-axis;
(b) the $y$-axis.

**25** Find the length of the curve $x = \frac{3}{8}(y^{4/3} - 2y^{2/3})$ from $y = 1$ to $y = 8$.

**26** Find the area of the surface generated by revolving the curve of Problem 25 around
(a) the $x$-axis;
(b) the $y$-axis.

**27** Find the area of the surface generated by revolving the curve of Problem 23 around the line $x = 1$.

**28** If $-r < a < b < r$, then a "spherical zone" of "height" $h = b - a$ is generated by revolving the circular arc

$$y = \sqrt{r^2 - x^2}, \qquad a \leq x \leq b$$

around the $x$-axis. Show that the area of this spherical zone is $A = 2\pi rh$, the same as that of a cylinder with radius $r$ and height $h$.

**29** Apply the result of Problem 28 to show that the surface area of a sphere of radius $r$ is $A = 4\pi r^2$.

Find the centroids of the curves in Problems 30–33.

**30** $y = \dfrac{x^5}{5} + \dfrac{1}{12x^3}, \quad 1 \leq x \leq 2$

**31** $x = \dfrac{y^4}{8} + \dfrac{1}{4y^2}, \quad 1 \leq y \leq 2$

**32** The curve of Problem 23

**33** The curve of Problem 25

**34** Find the centroid of the plane region in the first quadrant that is bounded by the curves $y = x^3$ and $y = 2x - x^2$.

**35** Find the centroid of the plane region bounded by the curves $x = 2y^4$ and $x = y^2 + 1$.

**36** Let $T$ be the triangle with vertices $(0, 0), (a, b)$, and $(c, 0)$. Show that the centroid of the triangle $T$ is the point of intersection of its medians.

**37** Use the first theorem of Pappus to find the $y$-coordinate of the centroid of the upper half of the ellipse $(x/a)^2 + (y/b)^2 = 1$. Employ the facts that the area of this semiellipse is $A = \pi ab/2$, while the volume of the ellipsoid it generates when rotated about the $x$-axis is $V = \frac{4}{3}\pi ab^2$.

**38** (a) Use the first theorem of Pappus to find the centroid of the first-quadrant part of the annular ring with boundary circles $x^2 + y^2 = a^2$ and $x^2 + y^2 = b^2$, with $0 < a < b$.
(b) Show that the limiting position of this centroid as $b \to a$ is the centroid of a quarter-circular arc, as found in Problem 22 of Section 6-6.

**39** Let $T$ be the triangle in the first quadrant with vertex $(0, 0)$ and opposite side $L$ of length $w$ joining the other two vertices $(a, b)$ and $(c, d)$, where $a > c > 0$ and $d > b > 0$. Let $A$ denote the area of $T$, $\bar{y}$ the $y$-coordinate of the centroid of $T$, $p$ the perpendicular distance from $(0, 0)$ to $L$, $V$ the volume generated by revolving $T$ around the $x$-axis, and $S$ the surface area generated by revolving $L$ around the $x$-axis. Then derive the formulas that follow in the order listed.
(a) $A = \frac{1}{2}(ad - bc)$
(b) $\bar{y} = \frac{1}{3}(b + d)$
(c) $V = \frac{1}{3}\pi(b + d)(ad - bc)$

(d) $p = \dfrac{ad - bc}{w}$

(e) $S = \pi(b + d)w$

(f) $V = \frac{1}{3}pS$

**40** Show that the $x$-coordinate of the centroid of the segment of an ellipsoid described in Problem 16 is

$$\bar{x} = \frac{3(2a - h)^2}{4(3a - h)}.$$

**41** Show that the $x$-coordinate of the centroid of the segment of a hyperboloid described in Problem 17 is

$$\bar{x} = \frac{3(2a + h)^2}{4(3a + h)}.$$

**42** Find the $y$-coordinate of the centroid of the solid generated by revolving around the $y$-axis the region bounded by $y = x^2$ and $y = x$ in two ways:
(a) by integration;
(b) by additivity of moments, taking as known the centroids of a cone (Problem 9 in Section 6-7) and a solid paraboloid (Example 1 in Section 6-7).

**43** Find the natural length $L$ of a spring if five times as much work is required to stretch it from a length of 2 ft to a length of 5 ft, as is required to stretch it from a length of 2 ft to a length of 3 ft.

**44** A steel beam weighing 1000 lb hangs from a 50-ft cable; the cable weighs 5 lb per linear foot. How much work is done in winding in 25 ft of the cable with a windlass?

**45** A spherical tank of radius $R$ (in feet) is initially full of oil weighing $\rho$ pounds per cubic foot. Find the total work

done in pumping all the oil from the sphere to a height of $2R$ above the top of the tank.

**46** How much work is done by a colony of ants in building a conical anthill with height and diameter both 1 ft, using sand initially at ground level and with a density of 150 lb/ft$^3$?

**47** Below the surface of the earth the gravitational attraction is directly proportional to the distance from the earth's center. Suppose that a straight cylindrical hole with a radius of 1 ft is dug from the surface of the earth (radius 3960 mi) to its center. Assume that the earth has a uniform density of 350 lb/ft$^3$. How much work, in foot-pounds, is done in lifting a 1-lb weight from the bottom of this hole to its top?

**48** How much work is done in digging the hole of Problem 47; that is, in lifting the material it at first contained to the earth's surface?

**49** Suppose that a dam is shaped like a trapezoid with height 100 ft, 300 ft long at the top and 200 ft long at the bottom. When the water level behind the dam is level with its top, what is the total force that the water exerts on the dam?

**50** In Problem 49, find the moment about the bottom of the dam of the force of water pressure.

**51** Suppose that a dam has the same top and bottom lengths as in Problem 49 and the same vertical height of 100 ft, but that its face toward the water is slanted at an angle of 30° from the vertical. Now what is the total force of water pressure on the dam?

**52** For $c > 0$, the graphs of $y = c^2x^2$ and $y = c$ bound a plane region. Revolve this region about the horizontal line $y = -1/c$ to form a solid. For what value of $c$ is the volume of this solid maximal? Minimal?

# 7

# Exponential and Logarithmic Functions

**Introduction**  Up to this point our study of calculus has been concentrated on power functions and on the algebraic functions that are obtained from power functions by forming finite combinations and compositions. This chapter is devoted to the calculus of exponential and logarithmic functions. This introductory section gives a brief and somewhat informal overview of these non-algebraic functions, partly from the perspective of precalculus mathematics. In addition to reviewing the laws of exponents and the laws of logarithms, our purpose here is to provide some intuitive background for the systematic treatment of exponential and logarithmic functions that begins in Section 7-2.

An *exponential function* is one of the form $f(x) = a^x$. Note that $x$ is the variable; the number $a$ is a constant. Thus an exponential function is "a constant raised to a variable power," whereas the power function $p(x) = x^k$ is "a variable raised to a constant power."

In precalculus courses, the *logarithmic function* with base $a$ is introduced as the inverse of the exponential function $f(x) = a^x$. That is, $\log_a x$ is "the power to which $a$ must be raised to get $x$," so that

$$y = \log_a x \quad \text{means that} \quad a^y = x.$$

But there are some fundamental questions that underlie these seemingly straightforward definitions of exponential and logarithmic functions.

In elementary algebra a *rational* power of the positive real number $a$ is defined in terms of integral roots and powers. There we learn that if $p$ and $q$ are integers (with $q > 0$), then

$$a^{p/q} = \sqrt[q]{a^p} = (\sqrt[q]{a})^p.$$

The following **laws of exponents** are then established for all *rational* exponents $r$ and $s$:

$$a^{r+s} = a^r a^s, \qquad (a^r)^s = a^{rs},$$

$$a^{-r} = \frac{1}{a^r}, \qquad (ab)^r = a^r b^r. \tag{1}$$

Moreover, recall that

$$a^0 = 1$$

for every positive real number $a$. A typical calculation using the laws of exponents is

$$\frac{(2^2)^3(3)^{-4}}{(2)^{-2}} = \frac{2^6 \cdot 2^2}{3^4} = \frac{2^8}{3^4} = \frac{256}{81}.$$

In applications we often need powers with irrational exponents as well as with rational ones. For example, consider a bacteria population $P(t)$ that increases by the same factor in any two time intervals of the same length and, in particular, doubles every hour. Suppose that the initial population is $P(0) = 1$ million. In 3 h, $P$ will increase by a factor of $2 \cdot 2 \cdot 2 = 2^3$, so that

$P(3) = 2^3$ (million). If $k$ is the factor by which $P$ increases in $\frac{1}{3}$ h, then in 1 h $P$ will increase by a factor of $k \cdot k \cdot k = k^3 = 2$, so $k = 2^{1/3}$ and $P(\frac{1}{3}) = 2^{1/3}$ (million). More generally, if $p$ and $q$ are positive integers, then $P(1/q) = 2^{1/q}$; in $p/q$ hours, $P$ will increase $p$ times by a factor of $2^{1/q}$ each time, so $P(p/q) = (2^{1/q})^p = 2^{p/q}$. Thus $P(t) = 2^t$ if $t$ is rational. But since time is not restricted to rational values alone, we surely ought to conclude that $P(t) = 2^t$ for *all* $t > 0$.

But what is meant by an expression such as $2^{\sqrt{2}}$ or $2^\pi$ involving an irrational exponent? To find the value of $2^\pi$ we might work with (rational) finite decimal approximations to the irrational number $\pi = 3.1415926 \ldots$. A calculator gives the (rounded) values

$$2^{3.1} \approx 8.5742$$

$$2^{3.14} \approx 8.8152$$

$$2^{3.141} \approx 8.8214$$

$$2^{3.1415} \approx 8.8244$$

$$2^{3.14159} \approx 8.8250$$

$$2^{3.141592} \approx 8.8250$$

$$2^{3.1415926} \approx 8.8250.$$

Thus it appears that $2^\pi \approx 8.8250$, rounded to four decimal places. (A ten-place calculator actually approximates $2^\pi$ by computing instead $2^{3.141592654} \approx 8.8224977830$ with a rational exponent.)

But the real question is this: What is the precise *definition* of the number $a^x$ if the exponent $x$ is irrational? One answer is to define the number $a^x$ using limits:

$$a^x = \lim_{n \to \infty} a^{r_n}, \tag{2}$$

where $r_1, r_2, r_3, \ldots$ is a sequence of *rational* numbers (like the finite decimal approximations to $\pi$) having $x$ as its limit:

$$\lim_{n \to \infty} r_n = x.$$

In order to carry out this approach we would need to show all the following:

1 The limit in (2) always exists, and its value is independent of the particular sequence of rational numbers used to approximate $x$. (Because a limit is unique when it exists, this implies that $f(x) = a^x$ is a *function*.)

2 The laws of exponents in (1) hold for irrational exponents as well as for rational exponents.

3 The function $f(x) = a^x$ is a continuous, indeed a differentiable, function.

All this can be done, but the complete program requires much in the way of detailed analysis of limit processes.

Fortunately, there is an alternative approach that employs concepts of calculus to circumvent this painstaking work with limits. In this section we will assume the laws of exponents and of logarithms, and proceed to investigate informally the derivatives of the exponential and logarithmic functions. What we learn will tell us how (in subsequent sections) to formulate precise definitions of these functions. These "official" definitions can finally be used

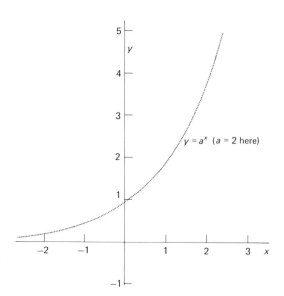

**7.1** Graph of $y = a^x$ has "holes" if only rational values of $x$ are used.

to establish rather simply the laws of exponents and of logarithms, as well as other useful properties of the exponential and logarithmic functions.

Let us investigate the function $f(x) = a^x$ in the case $a > 1$. We first note that $a^r > 1$ if $r$ is a positive rational number. If $r < s$, another rational number, then the laws of exponents give

$$a^r < a^r a^{s-r} = a^s,$$

so $f(r) < f(s)$ whenever $r < s$. That is, $f(x) = a^x$ is an *increasing* function of the rational exponent $r$. If we plot points on the graph of $y = a^x$ for rational values $x$ using a typical fixed value of $a$ such as $a = 2$, we obtain a graph like the one in Fig. 7.1. This graph is shown with a dotted curve to suggest that it is densely filled with "holes" corresponding to irrational values of $x$. In Section 7-4 we shall show that the "holes" in this graph can be filled to obtain the graph of an increasing and continuous function $f$ that is defined for all real $x$ and such that $f(r) = a^r$ for every rational number $r$. We therefore write $f(x) = a^x$ for all $x$ and call $f$ the **exponential function with base $a$.**

In elementary algebra the common logarithm function is introduced as the inverse function of the exponential function $10^x$ with base 10. That is, $y = \log_{10}x$ is the power to which 10 must be raised to obtain $x$, so that

$$y = \log_{10}x \quad \text{if and only if} \quad 10^y = x.$$

Thus $\log_{10}1000 = 3$ because $1000 = 10^3$, and $\log_{10}(0.1) = -1$ because $0.1 = 10^{-1}$. Similarly, the **base $a$ logarithm function $\log_a x$** is the inverse function of the exponential function $f(x) = a^x$ with base $a > 1$. That is,

$$y = \log_a x \quad \text{if and only if} \quad a^y = x. \tag{3}$$

Thus $\log_2 16 = 4$ because $2^4 = 16$, and $\log_3 9 = 2$ because $3^2 = 9$.

It follows from the reflection property of inverse functions (Section 2-3) that the graph of $y = \log_a x$ is the reflection in the line $y = x$ of the graph of $y = a^x$ and therefore has the shape shown in Fig. 7.2. Because $a^0 = 1$, it also follows that

$$\log_a 1 = 0,$$

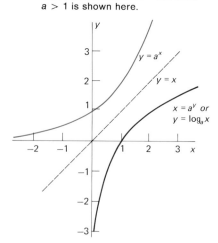

**7.2** The graph of $x = a^y$ is the graph of the inverse function $\log_a x$ of the exponential function $a^x$—the case $a > 1$ is shown here.

so the intercepts in the figure are independent of the choice of $a > 1$. Note also that $\log_a x$ is defined *only* for $x > 0$.

The inverse function relationship between $\log_a x$ and $a^x$ can be used to deduce, from the laws of exponents (1), the following **laws of logarithms:**

$$\log_a xy = \log_a x + \log_a y,$$

$$\log_a \frac{1}{x} = -\log_a x,$$

$$\log_a \frac{x}{y} = \log_a x - \log_a y, \tag{4}$$

$$\log_a x^y = y \log_a x.$$

We will verify these laws of logarithms in Section 7-2. A typical calculation using the laws of logarithms is

$$\log_{10}\left(\frac{3000}{17}\right) = \log_{10} 3000 - \log_{10} 17$$

$$= \log_{10}(3)(10^3) - \log_{10} 17$$
$$= \log_{10} 3 + 3 \log_{10} 10 - \log_{10} 17$$
$$\approx 0.47712 + (3)(1) - 1.23045;$$

$$\log_{10}\left(\frac{3000}{17}\right) \approx 2.24667.$$

We used a calculator to find the values $\log_{10} 3 \approx 0.47712$ and $\log_{10} 17 \approx 1.23045$.

## THE DERIVATIVES

To compute the derivative of the exponential function $f(x) = a^x$, we begin with the definition of the derivative and then use the first law of exponents in (1) to simplify. This gives

$$f'(x) = D_x a^x = \lim_{h \to 0} \frac{f(x + h) - f(x)}{h}$$

$$= \lim_{h \to 0} \frac{a^{x+h} - a^x}{h}$$

$$= \lim_{h \to 0} \frac{a^x a^h - a^x}{h} \qquad \text{(by the law of exponents)}$$

$$= a^x \left( \lim_{h \to 0} \frac{a^h - 1}{h} \right) \qquad \begin{array}{l} \text{(because } a^x \text{ is ``constant''} \\ \text{with respect to } h\text{).} \end{array}$$

Under the assumption that $f(x) = a^x$ is indeed differentiable, it follows that the limit

$$m(a) = \lim_{h \to 0} \frac{a^h - 1}{h} \tag{5}$$

exists. Although its value $m(a)$ depends on $a$, it is a constant as far as $x$ is concerned. Thus we find that the derivative of $a^x$ is a *constant multiple* of $a^x$ itself,

$$D_x a^x = m(a) \cdot a^x. \tag{6}$$

| $h$ | $\dfrac{2^h - 1}{h}$ | $\dfrac{3^h - 1}{h}$ |
|---|---|---|
| 0.1 | 0.718 | 1.161 |
| 0.01 | 0.696 | 1.105 |
| 0.001 | 0.693 | 1.099 |
| 0.0001 | 0.693 | 1.099 |

**7.3** Investigating the values of $m(2)$ and $m(3)$

Because $a^0 = 1$, we see from (6) that the constant $m(a)$ is the slope of the tangent line to the curve $y = a^x$ at the point where $x = 0$.

The numerical data shown in Fig. 7.3 suggest that $m(2) \approx 0.693$ and $m(3) \approx 1.099$ (to three places). The tangent lines with these slopes are shown in Fig. 7.4. Thus it appears that

$$D_x 2^x \approx (0.693)2^x \quad \text{and} \quad D_x 3^x \approx (1.099)3^x. \tag{7}$$

We would like somehow to avoid awkward numerical factors like those in Equation (7). It seems plausible that the value $m(a)$ defined in (5) is a continuous function of $a$. If so, then because $m(2) < 1$ and $m(3) > 1$, the intermediate value property implies that $m(e) = 1$ (exactly) for some number $e$ between 2 and 3. If we use this particular number $e$ as base, it then follows from Equation (6) that the derivative of the resulting exponential function $f(x) = e^x$ is given by

$$D_x e^x = e^x. \tag{8}$$

Thus the function $e^x$ is its own derivative! For this reason we call $f(x) = e^x$ the **natural exponential function.** Its graph is shown in Fig. 7.5.

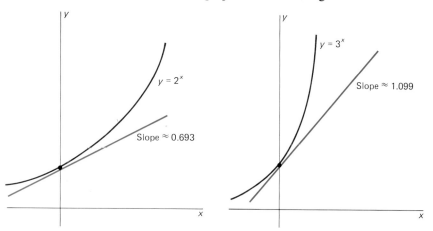

**7.4** The graphs $y = 2^x$ and $y = 3^x$

**7.5** The graph $y = e^x$

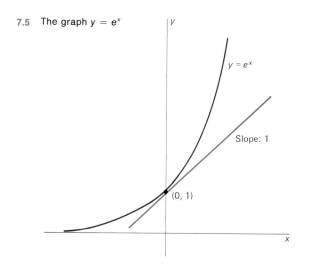

In Section 7-3 we will see that the number $e$ is given by the limit

$$e = \lim_{n \to \infty} \left(1 + \frac{1}{n}\right)^n. \tag{9}$$

| $n$ | $\left(1 + \dfrac{1}{n}\right)^n$ |
|---|---|
| 10 | 2.594 |
| 100 | 2.705 |
| 1,000 | 2.717 |
| 10,000 | 2.718 |
| 100,000 | 2.718 |

Let us investigate this limit numerically. With a calculator we obtain the values in the table of Fig. 7.6. The evidence suggests (although it does not prove) that $e \approx 2.718$ to three decimal places. This number $e$ is one of the most important special numbers in mathematics, and its value to 20 places is

**7.6** Numerical estimate of the number $e$

$$e \approx 2.71828182845904523536.$$

**EXAMPLE 1**  Find $dy/dx$ if $y = x^2 e^x$.

*Solution*  The formula in (8) and the product rule yield

$$\frac{dy}{dx} = (Dx^2)e^x + x^2(De^x)$$

$$= (2x)e^x + x^2(e^x) = x(2 + x)e^x.$$

The logarithm function $g(x) = \log_e x$ with base $e$ is called the **natural logarithm function.** It is commonly denoted by the special logarithm symbol ln, so that

$$\ln x = \log_e x. \tag{10}$$

The natural logarithm function is the inverse of the natural exponential function, so

$$e^{\ln x} = x \qquad \text{for all } x > 0 \tag{11a}$$

and

$$\ln(e^x) = x \qquad \text{for all } x. \tag{11b}$$

In order to find the derivative of the function

$$u = \ln x,$$

we begin with Equation (11a) in the form

$$e^u = x.$$

Because this last equation is actually an identity (for $x > 0$), the derivative of the left-hand side with respect to $x$ is also identically equal (for $x > 0$) to the derivative of the right-hand side with respect to $x$:

$$D_x e^u = D_x x.$$

With the aid of the chain rule we differentiate, and find that

$$D_u(e^u)\frac{du}{dx} = 1.$$

But $D_u(e^u) = e^u$ by (8), so

$$e^u \frac{du}{dx} = 1,$$

and finally

$$\frac{du}{dx} = \frac{1}{e^u} = \frac{1}{e^{\ln x}} = \frac{1}{x}.$$

Hence the derivative $du/dx$ of the natural logarithm function $u = \ln x$ is given by

$$D_x \ln x = \frac{1}{x}. \tag{12}$$

Thus $\ln x$ is the hitherto missing function with derivative $1/x$.

Just as with exponentials, the derivative of a logarithm function with base other than $e$ involves an inconvenient numerical factor. For example, we will see in Section 7-4 that

$$D_x \log_{10} x \approx \frac{0.4343}{x}. \tag{13}$$

The contrast between (12) and (13) illustrates one way in which base $e$ logarithms are "natural."

**EXAMPLE 2**  Find the derivative of

$$y = \frac{\ln x}{x}.$$

**Solution**  The formula in (12) and the quotient rule yield

$$\frac{dy}{dx} = \frac{(D \ln x)(x) - (\ln x)(Dx)}{x^2}$$

$$= \frac{(1/x)(x) - (\ln x)(1)}{x^2} = \frac{1 - \ln x}{x^2}.$$

**EXAMPLE 3**  To find the derivative of $\ln x^2$ we can apply a law of logarithms to simplify before differentiation. Thus

$$D_x(\ln x^2) = D_x(2 \ln x) = 2(D_x \ln x) = \frac{2}{x}.$$

---

In the preliminary discussion of this section, we have introduced in an informal manner

- The number $e \approx 2.71828$,
- The natural exponential function $e^x$, and
- The natural logarithm function $\ln x$.

Our investigation here of the derivatives of $e^x$ and $\ln x$—given by the formulas in (8) and (12)—should be regarded as provisional, pending a more complete discussion of these new functions in Sections 7-2 and 7-3. In any case, this brief preview of how the theory of exponential and logarithmic functions fits together should help you better understand the systematic development in the following two sections.

In subsequent sections of this chapter we shall see that the functions $e^x$ and $\ln x$ play a vital role in the quantitative analysis of a wide range of natural phenomena—including population growth, radioactive decay, spread of epidemics, growth of investments, diffusion of pollutants, and motion with the effect of air resistance taken into account.

Use the laws of exponents to simplify the expressions in Problems 1–10; then write each answer as an integer.

**1** $(2^3)(2^4)$      **2** $(3^2)(3^3)$
**3** $(2^2)^3$      **4** $2^{(2^3)}$
**5** $(3^5)(3^{-5})$      **6** $(10^{10})(10^{-10})$
**7** $(2^{12})^{1/3}$      **8** $(3^6)^{1/2}$
**9** $(4^5)(2^{-6})$      **10** $(6^5)(3^{-5})$

Note that $\log_{10}100 = \log_{10}10^2 = 2\log_{10}10 = 2$. Use a similar technique to evaluate (without use of a calculator) each of the expressions in Problems 11–16.

**11** $\log_2 16$      **12** $\log_3 27$
**13** $\log_5 125$      **14** $\log_7 49$
**15** $\log_{10}1000$      **16** $\log_{12}144$

Use the laws of logarithms to express each of the natural logarithms in Problems 17–26 in terms of the three numbers $\ln 2$, $\ln 3$, and $\ln 5$.

**17** $\ln 8$      **18** $\ln 9$
**19** $\ln 6$      **20** $\ln 15$
**21** $\ln 72$      **22** $\ln 200$

**23** $\ln \dfrac{8}{27}$      **24** $\ln \dfrac{12}{25}$

**25** $\ln \dfrac{27}{40}$      **26** $\ln \dfrac{1}{90}$

**27** Which is larger, $2^{(3^4)}$ or $(2^3)^4$?
**28** Evaluate $\log_{0.5}16$.
**29** By inspection, find two values of $x$ such that $x^2 = 2^x$.
**30** Show that the number $\log_2 3$ is irrational. (*Suggestion:* Assume to the contrary that $\log_2 3 = p/q$ where $p$ and $q$ are positive integers, and then express the consequence of this assumption in exponential form. Can an integral power of 2 equal an integral power of 3?)

In Problems 31–40, solve for $x$ *without* using a calculator.

**31** $2^x = 64$      **32** $10^{-x} = 0.001$
**33** $10^{-x} = 100$      **34** $(3^x)^2 = 81$
**35** $x^x = x^2$ (Find *all* solutions!)
**36** $\log_x 16 = 2$      **37** $\log_3 x = 4$
**38** $e^{5x} = 7$      **39** $3e^x = 3$
**40** $2e^{-7x} = 5$

Find $dy/dx$ in each of Problems 41–54.

**41** $y = xe^x$      **42** $y = x^3 e^x$

**43** $y = \sqrt{x}e^x$      **44** $y = \dfrac{1}{x}e^x$

**45** $y = \dfrac{e^x}{x^2}$      **46** $y = \dfrac{e^x}{\sqrt{x}}$

**47** $y = x \ln x$      **48** $y = x^2 \ln x$

**49** $y = \sqrt{x} \ln x$      **50** $y = \dfrac{\ln x}{\sqrt{x}}$

**51** $y = \dfrac{x}{e^x}$      **52** $y = e^x \ln x$

**53** $y = \ln x^3$ (This means $\ln(x^3)$.)
**54** $y = \ln \sqrt[3]{x}$
**55** Use the chain rule to deduce from $De^x = e^x$ that

$$D_x(e^{kx}) = ke^{kx}$$

if $k$ is a constant.

Apply the formula in Problem 55 to find the derivatives of the functions in Problems 56–58.

**56** $f(x) = e^{3x}$      **57** $f(x) = e^{x/10}$
**58** $f(x) = e^{-10x}$
**59** Substitute $2 = e^{\ln 2}$ in $P(t) = 2^t$, then apply the chain rule to show that $P'(t) = 2^t \ln 2$.

Use the method of Problem 59 to find the derivatives of the functions in Problems 60–63.

**60** $P(t) = 10^t$      **61** $P(t) = 3^t$
**62** $P(t) = 2^{3t}$      **63** $P(t) = 2^{-t}$
**64** If a population $P(t)$ of bacteria (in millions) at time $t$ (in hours) doubles every hour and initially (at time $t = 0$) the population is 1 (million), then its population at time $t \geq 0$ is given by $P(t) = 2^t$. Use the result of Problem 59 to find the instantaneous rate of growth (in millions of bacteria per hour) of this population
(a) at time $t = 0$;
(b) after 4 hours; that is, at time $t = 4$.
(*Note:* The natural logarithm of 2 is $\ln 2 \approx 0.69315$.)

## The Natural Logarithm

We now begin our systematic development of the properties of exponential and logarithmic functions. Though we gave an informal overview of this material in Section 7-1, we now start anew; in this section, we shall include the important details.

It is simplest to make the definition of the natural logarithm our initial step. Guided by the results in Section 7-1, we want to define $\ln x$ for $x > 0$ so that

$$\ln 1 = 0 \quad \text{and} \quad D \ln x = \frac{1}{x}. \tag{1}$$

To do so, we recall part 1 of the fundamental theorem of calculus (Section 5-5), according to which

$$D_x\left(\int_a^x f(t)\,dt\right) = f(x)$$

if $f$ is a continuous function. In order that $\ln x$ satisfy Equations (1), we take $a = 1$ and $f(t) = 1/t$.

---

**Definition**  *The Natural Logarithm*

The **natural logarithm** $\ln x$ of the positive number $x$ is defined to be

$$\ln x = \int_1^x \frac{1}{t}\,dt. \tag{2}$$

---

Note that $\ln x$ is *not* defined for $x \leq 0$. Geometrically, $\ln x$ is the area under the graph of $y = 1/t$ from $t = 1$ to $t = x$ if $x > 1$ (this is the situation shown in Fig. 7.7), the negative of this area if $0 < x < 1$, and 0 if $x = 1$. The fact that $D \ln x = 1/x$ follows immediately from the fundamental theorem of calculus, as indicated above. By Theorem 5 in Section 2-4, the fact that the function $\ln x$ is differentiable implies that it is continuous for $x > 0$.

Because $D(\ln x) = 1/x > 0$ for $x > 0$, we see that $\ln x$ must be an increasing function. Because its second derivative

$$D^2(\ln x) = D\frac{1}{x} = -\frac{1}{x^2}$$

is negative for $x > 0$, it follows from Theorem 1 in Section 4-6 that the graph of $y = \ln x$ is everywhere concave downward. Below we shall use the laws of logarithms to show that

$$\lim_{x \to 0^+} \ln x = -\infty \tag{3}$$

and

$$\lim_{x \to \infty} \ln x = +\infty. \tag{4}$$

When we assemble all these facts, we see that the graph $y = \ln x$ has the shape shown in Fig. 7.8.

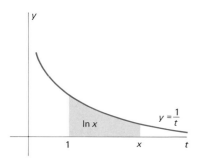

**7.7** The natural logarithm function defined by means of an integral

**7.8** The graph of the natural logarithm function

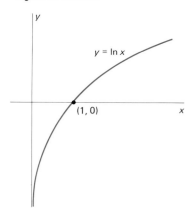

When we combine the formula $D \ln x = 1/x$ with the chain rule, we obtain the differentiation formula

$$D_x \ln u = \frac{D_x u}{u} = \frac{1}{u}\frac{du}{dx}. \tag{5}$$

Here, $u = u(x)$ denotes a positive-valued differentiable function of $x$.

**EXAMPLE 1**   If $y = \ln(x^2 + 1)$, then Formula (5) with $u = x^2 + 1$ gives

$$\frac{dy}{dx} = \frac{D_x(x^2 + 1)}{x^2 + 1} = \frac{2x}{x^2 + 1}.$$

If $y = \ln\sqrt{x^2 + 1}$, then with $u = \sqrt{x^2 + 1}$ we obtain

$$\frac{dy}{dx} = \frac{D_x\sqrt{x^2 + 1}}{\sqrt{x^2 + 1}} = \frac{1}{\sqrt{x^2 + 1}} \cdot \frac{x}{\sqrt{x^2 + 1}}$$

$$= \frac{x}{x^2 + 1}.$$

Note that these two derivatives are consistent with the fact that

$$\ln\sqrt{x^2 + 1} = \tfrac{1}{2}\ln(x^2 + 1)$$

(by a law of logarithms that we have not yet rigorously proved).

**EXAMPLE 2**   To differentiate $f(x) = (\ln x)^2$, we use the chain rule with $u = \ln x$. This gives

$$D(\ln x)^2 = 2u\frac{du}{dx} = 2(\ln x)D \ln x = \frac{2 \ln x}{x}.$$

---

The function $f(x) = \ln|x|$ is defined for all $x \neq 0$. If $x > 0$, then $|x| = x$ and, in this case,

$$f'(x) = D \ln x = \frac{1}{x}.$$

If $x < 0$, then $|x| = -x$ and so

$$f'(x) = D \ln(-x) = \frac{(-1)}{(-x)} = \frac{1}{x}.$$

In this computation, we used (5) with $u = -x$. Thus we have shown that

$$D \ln|x| = \frac{1}{x} \qquad (x \neq 0) \tag{6}$$

whether $x$ is positive or negative.

When we combine (6) with the chain rule, we obtain the formula

$$D_x \ln|u| = \frac{D_x u}{u} = \frac{1}{u}\frac{du}{dx}, \tag{7}$$

valid whenever the differentiable function $u = g(x)$ is nonzero.

The formula in (7) is equivalent to the integral formula

$$\int \frac{g'(x)}{g(x)}\,dx = \ln|g(x)| + C,$$

or simply

$$\int \frac{du}{u} = \ln|u| + C. \tag{8}$$

**EXAMPLE 3**  If $x > 0$, then the formula in (8) with

$$u = 4x + 3, \qquad du = 4\,dx$$

yields

$$\begin{aligned}
\int \frac{dx}{4x + 3} &= \frac{1}{4} \int \frac{4\,dx}{4x + 3} \\
&= \frac{1}{4} \int \frac{du}{u} \\
&= \tfrac{1}{4} \ln|u| + C \\
&= \tfrac{1}{4} \ln|4x + 3| + C \\
&= \tfrac{1}{4} \ln(4x + 3) + C.
\end{aligned}$$

**EXAMPLE 4**  With $u = x^2 - 1$, (8) gives

$$\int \frac{2x}{x^2 - 1}\,dx = \ln|x^2 - 1| + C = \begin{cases} \ln(x^2 - 1) + C & \text{if } |x| > 1; \\ \ln(1 - x^2) + C & \text{if } |x| < 1. \end{cases}$$

Note the dependence on whether $|x| > 1$ or $|x| < 1$. For instance,

$$\int_{\sqrt{2}}^{\sqrt{3}} \frac{2x\,dx}{x^2 - 1} = \left[\ln(x^2 - 1)\right]_{\sqrt{2}}^{\sqrt{3}} = \ln 2 - \ln 1 = \ln 2,$$

while

$$\int_{1/3}^{3/4} \frac{2x\,dx}{x^2 - 1} = \left[\ln(1 - x^2)\right]_{1/3}^{3/4}$$

$$= \ln \frac{7}{16} - \ln \frac{8}{9} = \ln \frac{63}{128}.$$

We now use our ability to differentiate logarithms to establish the laws of logarithms.

---

**Theorem**  *Laws of Logarithms*

If $x$ and $y$ are positive numbers and $r$ is a rational number, then:

$$\ln xy = \ln x + \ln y; \tag{9}$$

$$\ln\left(\frac{1}{x}\right) = -\ln x; \tag{10}$$

$$\ln\left(\frac{x}{y}\right) = \ln x - \ln y; \tag{11}$$

$$\ln(x^r) = r \ln x. \tag{12}$$

---

**Proof of (9)**   We temporarily fix $y$; thus we may regard $x$ as the independent variable and $y$ as a constant. Then

$$D_x \ln xy = \frac{D_x(xy)}{xy} = \frac{y}{xy} = \frac{1}{x} = D_x \ln x.$$

Thus $\ln xy$ and $\ln x$ have the same derivative with respect to $x$. We antidifferentiate each and conclude that

$$\ln xy = \ln x + C$$

for some constant $C$. To evaluate $C$, we substitute $x = 1$ into both sides of the last equation. The fact that $\ln 1 = 0$ then implies that $C = \ln y$, and this establishes Equation (9).

**Proof of (10)**   We differentiate $\ln(1/x)$:

$$D \ln\left(\frac{1}{x}\right) = \frac{(-1/x^2)}{(1/x)} = -\frac{1}{x} = D(-\ln x),$$

so $\ln(1/x)$ and $-\ln x$ have the same derivative. Hence antidifferentiation gives

$$\ln\left(\frac{1}{x}\right) = -\ln x + C$$

where $C$ is a constant. We substitute $x = 1$ into this last equation. Since $\ln 1 = 0$, it follows that $C = 0$, and this proves (10).

**Proof of (11)**   Since $x/y = x \cdot (1/y)$, Equation (11) follows immediately from Equations (9) and (10).

**Proof of (12)**   We know that $Dx^r = rx^{r-1}$ if $r$ is rational. So

$$D \ln x^r = \frac{rx^{r-1}}{x^r} = \frac{r}{x} = D(r \ln x).$$

Antidifferentiation then gives

$$\ln(x^r) = r \ln x + C$$

for some constant $C$. As before, substitution of $x = 1$ gives $C = 0$, which proves (12). In Section 7-4 we will show that (12) holds whether or not $r$ is rational.  ∎

Note that the proofs of (9), (10), and (12) are all quite similar—we differentiate the left-hand side, apply the fact that two functions with the same derivative (on an interval) differ by a constant $C$ (on that interval), and evaluate $C$ using the fact that $\ln 1 = 0$.

The laws of logarithms can often be used to simplify an expression prior to differentiating it, as in the following two examples.

**EXAMPLE 5**   If $f(x) = \ln(3x + 1)^{17}$ then

$$\begin{aligned}
f'(x) &= D_x \ln(3x + 1)^{17} \\
&= D_x[17 \ln(3x + 1)] \\
&= 17\,\frac{3}{3x + 1} = \frac{51}{3x + 1}.
\end{aligned}$$

**EXAMPLE 6**   Find $dy/dx$ if

$$y = \ln \frac{\sqrt{x^2 + 1}}{\sqrt[3]{x^3 + 1}}.$$

**Solution**   Immediate differentiation would require using the quotient rule and the chain rule (*several* times), all the while working with a complicated fraction. Our work is made much easier if we first use the laws of logarithms to simplify the formula for $y$:

$$y = \ln[(x^2 + 1)^{1/2}] - \ln[(x^3 + 1)^{1/3}]$$
$$= \tfrac{1}{2} \ln(x^2 + 1) - \tfrac{1}{3} \ln(x^3 + 1).$$

Finding $dy/dx$ is now no trouble:

$$\frac{dy}{dx} = \frac{1}{2}\left(\frac{2x}{x^2 + 1}\right) - \frac{1}{3}\left(\frac{3x^2}{x^3 + 1}\right)$$

$$= \frac{x}{x^2 + 1} - \frac{x^2}{x^3 + 1}.$$

---

Now we establish the limits of $\ln x$ as $x \to 0^+$ and $x \to \infty$, as stated in (3) and (4). Since $\ln 2$ is the area under $y = 1/x$ from $x = 1$ to $x = 2$, we can inscribe and circumscribe a pair of rectangles, as shown in Fig. 7.9, and conclude that

$$\tfrac{1}{2} < \ln 2 < 1.$$

Law (12) of logarithms then gives

$$\ln(2^n) = n \ln 2 > \frac{n}{2}.$$

Next suppose that $M > 0$. Let $n$ be an integer greater than $2M$. If $x > 2^n$, then—because $\ln x$ is an increasing function—it follows that

$$\ln x > \ln(2^n) > \frac{n}{2} > \frac{2M}{2} = M.$$

Thus we can make $\ln x$ as large as we please by choosing $x$ sufficiently large. This proves (4): $\lim\limits_{x \to \infty} \ln x = \infty$. To prove (3), we use Law (10) of logarithms

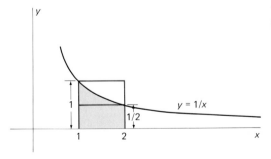

**7.9**  Using rectangles to estimate ln 2

to write

$$\lim_{x \to 0^+} \ln x = -\lim_{x \to 0^+} \ln\left(\frac{1}{x}\right)$$

$$= -\lim_{y \to \infty} \ln y = -\infty,$$

by taking $y = 1/x$ and applying (4).

The graph of the natural logarithm function (Fig. 7.8) suggests that, though $\ln x \to \infty$ as $x \to \infty$, it does so rather slowly. Indeed, the function $\ln x$ increases more slowly than any positive integral power of $x$. By this statement, we mean that

$$\lim_{x \to \infty} \frac{\ln x}{x^n} = 0 \qquad (13)$$

if $n$ is a fixed positive integer. To prove this, note first that

$$\ln x = \int_1^x \frac{dt}{t} \leq \int_1^x \frac{1}{\sqrt{t}} \, dt = 2(\sqrt{x} - 1),$$

because $1/t \leq 1/\sqrt{t}$ if $t \geq 1$. Hence if $x > 1$ then

$$0 < \frac{\ln x}{x^n} \leq \frac{\ln x}{x} \leq \frac{2}{\sqrt{x}} - \frac{2}{x}.$$

The last expression on the right approaches zero as $x \to \infty$. Equation (13) follows by the squeeze law of limits. Moreover, this argument shows that the positive integer $n$ in (13) may be replaced by any rational number $k \geq 1$.

Because $\ln x$ is an increasing function, the intermediate value property implies that the curve $y = \ln x$ crosses the horizontal line $y = 1$ precisely once. This point of intersection has as its abscissa the important number $e \approx 2.71828$ mentioned in Section 7-1 (see Fig. 7.10).

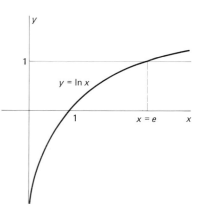

**7.10** The fact that $\ln e = 1$ is expressed graphically here.

---

*Definition of e*

The number $e$ is the unique real number such that

$$\ln e = 1. \qquad (14)$$

---

The letter $e$ has been used to denote the number with natural logarithm 1 ever since this number was introduced by the Swiss mathematician Leonhard Euler (1707–1783); he used $e$ for "exponential."

**EXAMPLE 7**   Sketch the graph of $f(x) = (\ln x)/x$, $x > 0$.

**Solution**   First we compute the derivative

$$f'(x) = \frac{(1/x)(x) - (\ln x)(1)}{x^2} = \frac{1 - \ln x}{x^2}.$$

Thus the only critical point is where $\ln x = 1$; that is, where $x = e$. Since $f'(x) > 0$ if $x < e$ (the same as $\ln x < 1$) and $f'(x) < 0$ if $x > e$ (the same as $\ln x > 1$), we see that $f$ is increasing if $x < e$ and decreasing if $x > e$. Hence $f$ has a local maximum at $x = e$.

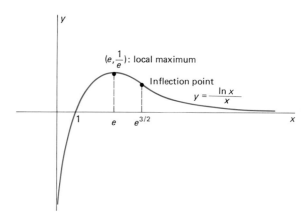

$(e, \frac{1}{e})$: local maximum

Inflection point

$y = \dfrac{\ln x}{x}$

**7.11** The graph for Example 7

The second derivative of $f$ is

$$f''(x) = \frac{(-1/x)(x^2) - (1 - \ln x)(2x)}{x^4} = \frac{2 \ln x - 3}{x^3},$$

so the only inflection point of the graph is where $\ln x = \frac{3}{2}$; that is, at $x = e^{3/2} \approx 4.48$.

Because

$$\lim_{x \to 0^+} \frac{\ln x}{x} = -\infty \quad \text{and} \quad \lim_{x \to \infty} \frac{\ln x}{x} = 0$$

—consequences of (3) and (13), respectively—we conclude that the graph of $f$ looks like the one shown in Fig. 7.11.

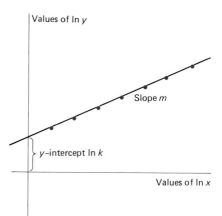

Values of ln $y$

Slope $m$

$y$-intercept ln $k$

Values of ln $x$

**7.12** Plotting the logarithms of data may reveal a hidden relationship.

### LOGARITHMS AND EXPERIMENTAL DATA

Certain empirical data can be explained by assuming that the observed dependent variable $y$ is a **power** function of the independent variable $x$; that is, $y$ is described by a mathematical model of the form

$$y = kx^m$$

where $k$ and $m$ are constants. If so, the laws of logarithms tell us that

$$\ln y = \ln k + m \ln x.$$

The experimenter then plots values of ln $y$ versus values of ln $x$. If the model assumed above is valid, the resulting data points will lie on a straight line with slope $m$ and $y$-intercept ln $k$, as shown in Fig. 7.12. The usefulness of this technique is that it is easy to see whether or not the data lie on a straight line, and—if they do—it is easy to measure the slope and $y$-intercept of the line and thereby find the values $k$ and $m$.

**EXAMPLE 8** (Planetary Motion) The table of Fig. 7.13 gives the period of revolution $T$ and the major semiaxis $a$ of the elliptical orbit of each of the first six planets about the sun, together with the logarithms of these numbers. If we plot ln $T$ against ln $a$, it is immediately apparent that the resulting points lie on a straight line with slope $m = \frac{3}{2}$. Hence $T$ and $a$

**358**

| Planet | T (in days) | a (in 10⁶ km) | ln T | ln a |
|---|---|---|---|---|
| Mercury | 87.97 | 58 | 4.48 | 4.06 |
| Venus | 224.70 | 108 | 5.41 | 4.68 |
| Earth | 365.26 | 149 | 5.90 | 5.00 |
| Mars | 686.98 | 228 | 6.53 | 5.43 |
| Jupiter | 4332.59 | 778 | 8.37 | 6.66 |
| Saturn | 10759.20 | 1426 | 9.28 | 7.26 |

**7.13  Data for Example 8**

satisfy an equation of the form $T = ka^{3/2}$, and so

$$T^2 = Ca^3.$$

This means that the square of the period $T$ is proportional to the cube of the major semiaxis $a$. This is Kepler's third law of planetary motion, which he discovered empirically in 1619.

## 7-2  PROBLEMS

In each of Problems 1–15, find the derivative of the given function $f(x)$.

**1** $\ln(3x - 1)$  **2** $\ln(4 - x^2)$

**3** $\ln\sqrt{1 + 2x}$  **4** $\ln[(1 + x)^3]$

**5** $\ln\sqrt[3]{x^3 - x}$  **6** $\ln x\sqrt{1 - x}$

**7** $(x \ln x) - x$  **8** $(\ln x)^3$

**9** $\dfrac{1}{\ln x}$  **10** $\ln(\ln x)$

**11** $\ln[(2x + 1)^3(x^2 - 4)^4]$

**12** $\ln\sqrt{\dfrac{1 - x}{1 + x}}$  **13** $\ln\sqrt{\dfrac{4 - x^2}{9 + x^2}}$

**14** $\ln\dfrac{\sqrt{4x - 7}}{(3x - 2)^3}$  **15** $\ln\dfrac{\sqrt[4]{(1 - 5x)^3}}{\sqrt[3]{(1 + x^4)^5}}$

Differentiate the functions given in Problems 16–33.

**16** $f(x) = \ln(3x^2 + 2x)$  **17** $f(x) = \ln(\sqrt{3x - 1})$
**18** $f(x) = \ln([x^2 + 4]^{3/2})$  **19** $g(t) = t \ln(t^2)$
**20** $g(t) = t^{3/2}\ln(t + 1)$  **21** $f(x) = \ln(x\sqrt{x^2 + 1})$
**22** $f(x) = \ln(x \ln x)$
**23** $f(x) = \ln[(x^2 + 1)^{1/3}(x^3 - 1)^{1/2}]$

**24** $f(x) = x^2\ln(\ln x)$  **25** $f(t) = \ln\dfrac{t^3}{t^2 + 1}$

**26** $g(t) = t(\ln t)^2$  **27** $g(t) = \sqrt{t}(1 + \ln t)^2$

**28** $f(x) = \ln\dfrac{x + 1}{x - 1}$  **29** $f(t) = \ln\left(\dfrac{t}{t + 1}\right)^{1/2}$

**30** $f(x) = x^2\ln\dfrac{1}{2x + 1}$  **31** $g(t) = \ln\dfrac{t^2}{t^2 + 1}$

**32** $f(x) = \ln\dfrac{\sqrt{x + 1}}{(x - 1)^3}$  **33** $f(x) = \ln\dfrac{1 - x}{x}$

Evaluate the indefinite integrals in Problems 34–50.

**34** $\displaystyle\int \dfrac{dx}{3x + 5}$  **35** $\displaystyle\int \dfrac{x\, dx}{1 + 3x^2}$

**36** $\displaystyle\int \dfrac{x^2\, dx}{4 - x^3}$  **37** $\displaystyle\int \dfrac{(x + 1)\, dx}{2x^2 + 4x + 1}$

**38** $\displaystyle\int \dfrac{dx}{(1 + x)^3}$  **39** $\displaystyle\int \dfrac{(\ln x)^2}{x}\, dx$

**40** $\displaystyle\int \dfrac{dx}{x \ln x}$  **41** $\displaystyle\int \dfrac{1}{x + 1}\, dx$

**42** $\displaystyle\int \dfrac{x\, dx}{1 - x^2}$  **43** $\displaystyle\int \dfrac{2x + 1}{x^2 + x + 1}\, dx$

**44** $\displaystyle\int \dfrac{x + 1}{x^2 + 2x + 3}\, dx$  **45** $\displaystyle\int \dfrac{\ln x}{x}\, dx$

**46** $\displaystyle\int \dfrac{\ln(x^3)}{x}\, dx$  **47** $\displaystyle\int \dfrac{2x^3 + x}{x^4 + x^2}\, dx$

**48** $\displaystyle\int \dfrac{dx}{x(\ln x)^2}$  **49** $\displaystyle\int \dfrac{x^2 - 2x}{x^3 - 3x^2 + 1}\, dx$

**50** $\displaystyle\int \dfrac{dx}{x^{1/2}(1 + x^{1/2})}$  (*Suggestion:* Let $u = 1 + x^{1/2}$.)

Apply (13) to evaluate the limits in Problems 51–56.

**51** $\displaystyle\lim_{x \to \infty} \dfrac{\ln(x^{1/2})}{x}$  **52** $\displaystyle\lim_{x \to \infty} \dfrac{\ln x^3}{x^2}$

**53** $\displaystyle\lim_{x \to \infty} \dfrac{\ln x}{x^{1/2}}$  (*Suggestion:* Substitute $x = u^2$.)

**54** $\displaystyle\lim_{x \to 0^+} x \ln x$  (*Suggestion:* Substitute $x = 1/u$.)

**55** $\lim\limits_{x\to 0^+} x^{1/2}\ln x$  **56** $\lim\limits_{x\to\infty} \dfrac{(\ln x)^2}{x}$

**57** Show that if $x \geq 1$, then
$$\ln(x + \sqrt{x^2 - 1}) = -\ln(x - \sqrt{x^2 - 1}).$$

**58** Find a formula for $f^{(n)}(x)$ given $f(x) = \ln x$.

**59** The heart rate $R$ (in beats per minute) and weight $W$ (in pounds) of various mammals were measured, with the results shown in Fig. 7.14. Use the method of Example 8 to find a relation between the two of the form $R = kW^m$.

| W | R |
|---|---|
| 25 | 131 |
| 67 | 103 |
| 127 | 88 |
| 175 | 81 |
| 240 | 75 |
| 975 | 53 |

**7.14**　Data for Problem 59

**60** During the adiabatic expansion of a certain diatomic gas, its volume $V$ (in liters) and its pressure $P$ (in atmospheres) were measured, with the results shown in Fig. 7.15.

| V | P |
|---|---|
| 1.46 | 28.3 |
| 2.50 | 13.3 |
| 3.51 | 8.3 |
| 5.73 | 4.2 |
| 7.26 | 3.0 |

**7.15**　Data for Problem 60

Use the method of Example 8 to find a relation between $V$ and $P$ of the form $P = kV^m$.

**61** Substitute $y = x^p$ and then apply (13) to show that
$$\lim_{x\to\infty} \frac{\ln x}{x^p} = 0$$
if $0 < p < 1$.

**62** Deduce from Problem 61 that
$$\lim_{x\to\infty} \frac{(\ln x)^k}{x} = 0 \quad \text{if} \quad k > 0.$$

**63** Substitute $y = 1/x$ and then apply (13) to show that
$$\lim_{x\to 0^+} x^k \ln x = 0 \quad \text{if} \quad k > 0.$$

Use the limits in the preceding problems to sketch the graphs, for $x > 0$, of the functions given in Problems 64–67.

**64** $y = x \ln x$  **65** $y = x^2 \ln x$

**66** $y = \sqrt{x} \ln x$  **67** $y = \dfrac{\ln x}{\sqrt{x}}$

**68** Problem 22 of Section 5-8 calls for showing, by numerical integration, that
$$\int_1^{2.7} \frac{dx}{x} < 1 < \int_1^{2.8} \frac{dx}{x}.$$

Explain carefully why this result shows that $2.7 < e < 2.8$.

**69** If $n$ moles of an ideal gas expand at *constant* temperature $T$, then its pressure and volume satisfy the equation $pV = nRT$ ($R$ is a constant). With the aid of Problem 22 in Section 6-5, show that the work $W$ done by the gas in expanding from volume $V_1$ to volume $V_2$ is
$$W = nRT \ln \frac{V_2}{V_1}.$$

**70** "Gabriel's trumpet" is obtained by revolving the curve $y = 1/x, x \geq 1$, around the $x$-axis. Let $A_b$ denote its surface area from $x = 1$ to $x = b$. Show that $A_b \geq 2\pi \ln b$, so— as a consequence—$A_b \to \infty$ as $b \to \infty$. Thus the surface area of Gabriel's trumpet is infinite. Is its volume finite or infinite?

**71** According to the prime number theorem, which was conjectured by the great German mathematician C. F. Gauss in 1792 (when he was 15 years old) but not proved until over a century later, the number of primes between the large positive numbers $a$ and $b$ ($a < b$) is given to a close approximation by the integral
$$\int_a^b \frac{dx}{\ln x}.$$

The midpoint and trapezoidal approximations with $n = 1$ subinterval provide an underestimate and an overestimate of the value of this integral. (Why?) Calculate them with $a = 90{,}000$ and $b = 100{,}000$. The actual number of primes in this range is 879.

## *7-2　Optional Computer Application

In (14) we defined the number $e$ as the unique solution of the equation
$$f(x) = \ln x - 1 = 0.$$

Because $f'(x) = 1/x$, the iterative formula of Newton's method (Section 3-9) is

$$x_{n+1} = x_n - \frac{f(x_n)}{f'(x_n)}$$

$$= x_n - \frac{(\ln x_n) - 1}{1/x_n};$$

thus

$$x_{n+1} = 2x_n - x_n \ln x_n.$$

The following program implements this iteration, starting with $x_0 = 3$.

```
10   X = 3  :  N = 0
20   PRINT  N;  X
30   X = 2*X − X*LOG(X)
40   N = N + 1
50   GOTO  20
```

Note that LOG( ) denotes the natural logarithm function in BASIC. When we run this program, we get the data below, giving $e$ accurate to nine decimal places.

| $n$ | $x_n$ |
| --- | --- |
| 0 | 3 |
| 1 | 2.70416 3134 |
| 2 | 2.71824 5099 |
| 3 | 2.71828 1828 |
| 4 | 2.71828 1828 |

The "infinite loop" in this program is deliberate. With a personal computer you can simply press Control/Break to halt execution when you've seen enough.

## The Exponential Function

We saw in Section 7-2 that the natural logarithm function $\ln x$ is continuous and increasing for $x > 0$, and that it attains arbitrarily large positive and negative values (because of the limits in (3) and (4) of Section 7-2). It follows that $\ln x$ has an inverse function that is defined for all $x$. To see this, let $y$ be any (fixed) real number whatsoever. If $a$ and $b$ are positive numbers such that $\ln a < y < \ln b$, then the intermediate value property gives a number $x > 0$ (between $a$ and $b$) such that $\ln x = y$. Because $\ln$ is an increasing function, there is only *one* such number $x$ with $\ln x = y$. This inverse function to $\ln$ is called the *natural exponential function*, denoted by exp.

> *Definition*    *The Natural Exponential Function*
>
> The (natural) **exponential function** exp is defined for all $x$ by
>
> $$\exp x = y \quad \text{if and only if} \quad \ln y = x. \tag{1}$$

Thus exp $x$ is simply that (positive) number $y$ whose natural logarithm is $x$. It is an immediate consequence of (1) that

$$\ln(\exp x) = x \quad \text{for all} \quad x, \tag{2}$$

and that

$$\exp(\ln y) = y \quad \text{for all} \quad y > 0. \tag{3}$$

It follows from the reflection property of inverse functions (Section 2-3) that the graphs $y = \exp x$ and $y = \ln x$ are reflections of each other in the line $y = x$, like the graphs in Fig. 7.2. Therefore, the graph of the exponential function looks like the one shown in Fig. 7.16. In particular, exp $x$ is positive-valued for all $x$, and

$$\exp 0 = 1, \tag{4}$$

$$\lim_{x \to \infty} \exp x = \infty, \quad \text{and} \tag{5}$$

$$\lim_{x \to -\infty} \exp x = 0. \tag{6}$$

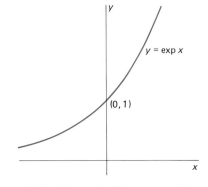

**7.16** The graph of the exponential function, exp

These facts follow from the equation $\ln 1 = 0$ and from the limits in (3) and (4) in Section 7-2.

Recall that the number $e \approx 2.71828$ was defined in Section 7-2 as the number that has natural logarithm 1. If $r$ is any rational number, it follows that

$$\ln(e^r) = r \ln e = r.$$

But Equation (1) tells us that $\ln(e^r) = r$ if and only if

$$\exp r = e^r.$$

Thus exp $x$ is equal to $e^x$ ($e$ to the power $x$) if $x$ is a rational number. We therefore *define* $e^x$ for $x$ irrational as well as rational by

$$e^x = \exp x. \tag{7}$$

This is our first instance of irrational powers.

Equation (7) is the reason for calling exp the natural exponential function. With this notation, Equations (1) through (3) become

$$e^x = y \quad \text{if and only if} \quad \ln y = x, \tag{8}$$

$$\ln(e^x) = x \quad \text{for all } x, \quad \text{and} \tag{9}$$

$$e^{\ln x} = x \quad \text{for all } x > 0. \tag{10}$$

To justify Equation (7), we should show that powers of $e$ satisfy the laws of exponents. We can do this immediately.

---

*Theorem   Laws of Exponents*

If $x$, $x_1$, and $x_2$ are real numbers and $r$ is rational, then

$$e^{x_1}e^{x_2} = e^{x_1 + x_2}, \tag{11}$$

$$e^{-x} = \frac{1}{e^x}, \tag{12}$$

$$(e^x)^r = e^{rx}. \tag{13}$$

---

**Proof**    The laws of logarithms and (9) give

$$\ln(e^{x_1}e^{x_2}) = \ln e^{x_1} + \ln e^{x_2}$$
$$= x_1 + x_2 = \ln(e^{x_1 + x_2}),$$

so (11) follows from the fact that ln is an increasing, and hence a one-to-one, function. Similarly,

$$\ln([e^x]^r) = r \ln[e^x] = rx = \ln(e^{rx}).$$

So (13) follows in the same way. The proof of (12) is almost identical. In Section 7-4, we will see that the restriction that $r$ be rational in (13) is actually unnecessary; that is,

$$(e^x)^y = e^{xy}$$

for *all* real numbers $x$ and $y$.    ■

Because $e^x$ is the inverse of the differentiable and increasing function ln $x$, it follows from Theorem 2 in Section 3-4 that $e^x$ is differentiable and therefore also continuous. We may thus differentiate both sides of the equation (actually, the *identity*)

$$\ln(e^x) = x$$

with respect to $x$. Let $u = e^x$; the above equation becomes

$$\ln u = x,$$

and the derivatives must also be equal:

$$\frac{1}{u}\frac{du}{dx} = 1 \qquad (\text{since } u > 0).$$

So

$$\frac{du}{dx} = e^x;$$

that is,

$$De^x = e^x, \tag{14}$$

as we indicated in Section 7-1.

If $u$ denotes a differentiable function of $x$, then (14) combined with the chain rule gives

$$D_x e^u = e^u \frac{du}{dx}. \tag{15}$$

The corresponding integration formula is

$$\int e^u \, du = e^u + C. \tag{16}$$

The special case of (15) with $u = kx$ ($k$ constant) is worth noting:

$$D_x e^{kx} = ke^{kx}.$$

For example, $D_x e^{5x} = 5e^{5x}$.

**EXAMPLE 1** Find $dy/dx$ if $y = e^{\sqrt{x}}$.

**Solution**  With $u = \sqrt{x}$, the formula in (15) gives

$$\frac{dy}{dx} = e^{\sqrt{x}}D_x(\sqrt{x}) = e^{\sqrt{x}}\left(\frac{1}{2}x^{-1/2}\right) = \frac{e^{\sqrt{x}}}{2\sqrt{x}}.$$

**EXAMPLE 2**  If $y = x^2 e^{-2x^3}$, then the formula in (15) and the product rule yield

$$\frac{dy}{dx} = (Dx^2)e^{-2x^3} + x^2 D(e^{-2x^3})$$

$$= 2xe^{-2x^3} + x^2 e^{-2x^3}D(-2x^3)$$

$$= 2xe^{-2x^3} + x^2 e^{-2x^3}(-6x^2);$$

therefore,

$$\frac{dy}{dx} = (2x - 6x^4)e^{-2x^3}.$$

**EXAMPLE 3**  Find $\int xe^{-3x^2}\,dx$.

**Solution**  We substitute $u = -3x^2$. Since $du = -6x\,dx$, we have $x\,dx = -\frac{1}{6}du$, and we obtain

$$\int xe^{-3x^2}\,dx = -\frac{1}{6}\int e^u\,du$$

$$= -\frac{1}{6}e^u + C = -\frac{1}{6}e^{-3x^2} + C.$$

## ORDER OF MAGNITUDE

The exponential function is remarkable for its high rate of increase with increasing $x$. In fact, $e^x$ increases more rapidly as $x \to \infty$ than *any* fixed power of $x$. In the language of limits,

$$\lim_{x\to\infty}\frac{x^k}{e^x} = 0, \quad \text{or} \quad \lim_{x\to\infty}\frac{e^x}{x^k} = \infty \tag{17}$$

for any fixed $k > 0$. Since we have not yet defined $x^k$ for $k$ irrational, we prove (17) for the case $k$ rational; once we know that (for $x > 1$) the power function $x^k$ is an increasing function of $k$ for all $k$, the general case will follow.

We begin by taking logarithms; we find that

$$\ln\left(\frac{e^x}{x^k}\right) = x - k \ln x = \left(\frac{x}{\ln x} - k\right)\ln x.$$

Because we know (from the formula in (13) of Section 7-2) that $x/(\ln x) \to \infty$ as $x \to \infty$, this makes it clear that

$$\lim_{x\to\infty}\ln\left(\frac{e^x}{x^k}\right) = \infty.$$

Hence $e^x/x^k \to \infty$ as $x \to \infty$, so we have proved (17).

**364**

The formula in (17) may be used to evaluate certain limits using the laws of limits. For example,

$$\lim_{x \to \infty} \frac{2e^x - 3x}{3e^x + x^3} = \lim_{x \to \infty} \frac{2 - 3xe^{-x}}{3 + x^3 e^{-x}}$$

$$= \frac{2 - 0}{3 + 0} = \frac{2}{3}.$$

**EXAMPLE 4**  Sketch the graph of $f(x) = xe^{-x}$.

**Solution**  From (17) we see that $f(x) \to 0$ as $x \to \infty$, while $f(x) \to -\infty$ as $x \to -\infty$. Since

$$f'(x) = e^{-x} - xe^{-x} = e^{-x}(1 - x),$$

the only critical point of $f$ is $x = 1$, at which $y = e^{-1} \approx 0.37$. And since

$$f''(x) = -e^{-x}(1 - x) + e^{-x}(-1) = e^{-x}(x - 2),$$

the only inflection point is $x = 2$, where $y = 2e^{-2} \approx 0.27$. Hence the graph of $f$ looks like the one in Fig. 7.17.

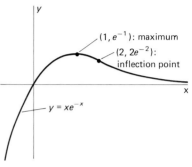

**7.17**  The graph of the function of Example 4

## THE NUMBER *e* AS A LIMIT

We now establish the following limit expression for the exponential function:

$$e^x = \lim_{n \to \infty} \left( 1 + \frac{x}{n} \right)^n. \tag{18}$$

We begin by differentiating $\ln t$ using the definition of the derivative in combination with the fact that we already know that the derivative is $1/t$. Thus

$$\frac{1}{t} = D \ln t = \lim_{h \to 0} \frac{\ln(t + h) - \ln(t)}{h}$$

$$= \lim_{h \to 0} \frac{1}{h} \ln\left( \frac{t + h}{t} \right)$$

$$= \lim_{h \to 0} \ln\left[ \left( 1 + \frac{h}{t} \right)^{1/h} \right] \qquad \text{(by laws of logarithms)}$$

$$= \ln\left[ \lim_{h \to 0} \left( 1 + \frac{h}{t} \right)^{1/h} \right] \qquad \begin{array}{l}\text{(by continuity of the}\\ \text{logarithm function).}\end{array}$$

The substitution $n = 1/h$ allows us to write

$$\frac{1}{t} = \ln\left[ \lim_{n \to \infty} \left( 1 + \frac{1}{nt} \right)^n \right];$$

then the substitution $x = 1/t$ gives

$$x = \ln\left[ \lim_{n \to \infty} \left( 1 + \frac{x}{n} \right)^n \right],$$

from which (18) follows, because $x = \ln y$ implies $e^x = y$. With $x = 1$, we obtain the following important expression of $e$ as a limit:

$$e = \lim_{n \to \infty} \left(1 + \frac{1}{n}\right)^n. \tag{19}$$

## *EXPONENTIALS AND THE HYDROGEN ATOM

In the modern quantum theory of atomic structure, the electron in a hydrogen atom does not travel in a well-defined orbit about the nucleus (the proton) of the atom. Instead it occupies a state known as an **orbital,** which may be visualized as a cloud of negative electricity surrounding the nucleus. The density of this "electron cloud" at a particular point is a measure of the probability that, at a given instant, the electron is near that point.

More precisely, a **spherically symmetric** orbital is described in terms of its **radial probability density function** $f(r)$, a function of the radius $r$, the distance from the nucleus. This function has the property that the probability $P(r, \Delta r)$ of finding the electron at a distance between $r$ and $r + \Delta r$ from the nucleus is approximately

$$P(r, \Delta r) \approx f(r)\, \Delta r,$$

"approximately" in the sense that the error is small when compared with $\Delta r$. Here we are using an intuitive and natural concept of probability; you may think of $P(r, \Delta r)$ as the fraction of the time the electron spends within the spherical shell with radii $r$ and $r + \Delta r$. Though the electron may be zipping about the nucleus, it is reasonable to regard the value of $r$ that maximizes $f(r)$ as the most probable distance of the electron from the nucleus.

For the normal ("unexcited") state of the hydrogen atom, called the $1s$-orbital, the density function is

$$f(r) = \frac{4r^2}{a_0^3}\, e^{-2r/a_0}.$$

The number $a_0$ is the **Bohr radius,** about 0.529 angstroms (1 angstrom is $10^{-10}$ m). Because $f(r) > 0$ for all $r > 0$, it is theoretically possible for the electron to be anywhere in the universe! But the probability of finding the electron more than a few angstroms from the nucleus is *very* small.

To find the most probable distance of the electron from the nucleus, we differentiate $f(r)$ with the goal of eventually maximizing it:

$$f'(r) = \frac{4}{a_0^3}\left(2re^{-2r/a_0} - \frac{2r^2}{a_0}e^{-2r/a_0}\right)$$

$$= \frac{8}{a_0^4}r(a_0 - r)e^{-2r/a_0}.$$

**7.18** Graph of the radial probability density function

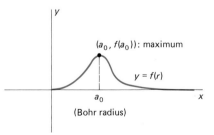

Thus there are two critical points, $r = 0$ and $r = a_0$. Because $f(0) = 0$ and $f(r) \to 0$ as $r \to \infty$ (why?), we see that the graph of $f$ looks like the one in Fig. 7.18. In particular, $f(a_0)$ is the maximum value of $f(r)$, so the significance of the Bohr radius $a_0$ is this: It is the distance of the electron from the nucleus in the $1s$-orbital of the hydrogen atom where the value of the probability density function $f(r)$ is maximal.

Differentiate the functions in Problems 1–35.

**1** $f(x) = e^{2x}$        **2** $f(x) = e^{3x-1}$
**3** $f(x) = e^{x^2}$       **4** $f(x) = e^{4-x^3}$
**5** $f(x) = e^{1/x^2}$     **6** $f(x) = x^2 e^{x^3}$
**7** $g(t) = te^{\sqrt{t}}$        **8** $g(t) = (e^{2t} + e^{3t})^7$
**9** $g(t) = (t^2 - 1)e^{-t}$      **10** $g(t) = \sqrt{e^t - e^{-t}}$
**11** $g(t) = e^{1+\ln t}$        **12** $f(x) = x \ln(1 + e^x)$
**13** $f(x) = \ln(1 - e^{-x})$      **14** $f(x) = [\ln(1 + e^x)]^2$
**15** $f(x) = \ln(x + e^{-x})$      **16** $f(x) = e^x \ln x$
**17** $f(x) = e^{-2x} \ln \sqrt{x}$      **18** $g(t) = \ln(te^{t^2})$
**19** $g(t) = 3(e^t - \ln t)^5$
**20** $g(t) = \ln[(1 + e^{2t})(1 - e^{-t})]$

**21** $f(x) = \dfrac{2 + 3x}{e^{4x}}$        **22** $g(t) = \dfrac{1 + e^t}{1 - e^t}$

**23** $g(t) = \dfrac{1 - e^{-t}}{t}$        **24** $f(x) = e^{-1/x}$

**25** $f(x) = xe^{10 \ln x}$        **26** $f(x) = x^3 e^{4-x^2}$

**27** $f(x) = \dfrac{1 - x}{e^x}$        **28** $f(x) = e^{\sqrt{x}} + e^{-\sqrt{x}}$

**29** $f(x) = e^{(e^x)}$        **30** $f(x) = \sqrt{e^{2x} + e^{-2x}}$
**31** $f(x) = \ln(2e^x)$        **32** $f(x) = \ln(x^2 + e^{-x})$

**33** $f(x) = \dfrac{e^{-x}}{\ln(x + 1)}$        **34** $f(x) = (e^x + e^{-x})^2$

**35** $f(x) = \dfrac{e^{\sqrt{x}}}{\sqrt{x}}$

Find the antiderivatives indicated in Problems 36–53.

**36** $\displaystyle\int e^{3x}\, dx$        **37** $\displaystyle\int e^{1-2x}\, dx$

**38** $\displaystyle\int xe^{x^2}\, dx$        **39** $\displaystyle\int x^2 e^{3x^3+1}\, dx$

**40** $\displaystyle\int \sqrt{x} e^{2x\sqrt{x}}\, dx$        **41** $\displaystyle\int \dfrac{e^{2x}}{1 + e^{2x}}\, dx$

**42** $\displaystyle\int \dfrac{\ln(x + 1)}{x + 1}\, dx$        **43** $\displaystyle\int \dfrac{1}{x} e^{1+\ln x}\, dx$

**44** $\displaystyle\int (e^x + e^{-x})^2\, dx$        **45** $\displaystyle\int \dfrac{x + e^{2x}}{x^2 + e^{2x}}\, dx$

**46** $\displaystyle\int e^{2x+3}\, dx$        **47** $\displaystyle\int te^{-t^2/2}\, dt$

**48** $\displaystyle\int x^2 e^{1-x^3}\, dx$        **49** $\displaystyle\int \dfrac{e^{\sqrt{x}}}{\sqrt{x}}\, dx$

**50** $\displaystyle\int \dfrac{e^{1/t}}{t^2}\, dt$        **51** $\displaystyle\int \dfrac{e^x}{1 + e^x}\, dx$

**52** $\displaystyle\int \exp(x + e^x)\, dx$        **53** $\displaystyle\int \sqrt{x} e^{-\sqrt{x^3}}\, dx$

Apply the formula in (18) to evaluate (in terms of the exponential function) the limits in Problems 54–58.

**54** $\displaystyle\lim_{n\to\infty} \left(1 - \dfrac{1}{n}\right)^n$        **55** $\displaystyle\lim_{n\to\infty} \left(1 + \dfrac{2}{n}\right)^n$

**56** $\displaystyle\lim_{n\to\infty} \left(1 + \dfrac{1}{3n}\right)^n$        **57** $\displaystyle\lim_{h\to 0} (1 + h)^{1/h}$

**58** $\displaystyle\lim_{h\to 0} (1 + 2h)^{1/h}$  (*Suggestion:* Substitute $k = 2h$.)

Evaluate the limits in Problems 59–62 by applying the fact that

$$\lim_{x\to\infty} x^k e^{-x} = 0.$$

**59** $\displaystyle\lim_{x\to\infty} \dfrac{e^x}{x}$        **60** $\displaystyle\lim_{x\to\infty} \dfrac{e^x}{x^{1/2}}$

**61** $\displaystyle\lim_{x\to\infty} \dfrac{e^{(x^{1/2})}}{x}$        **62** $\displaystyle\lim_{x\to\infty} x^2 e^{-x}$

In Problems 63–65, sketch the graph of the given equation. Show and label all extrema, inflection points, and asymptotes; show the concave structure clearly.

**63** $y = x^2 e^{-x}$        **64** $y = x^3 e^{-x}$
**65** $y = e^{-x^2}$

**66** Find the area under the graph of $y = e^x$ from $x = 0$ to $x = 1$.

**67** Find the volume generated by revolving the region of Problem 66 around the $x$-axis.

**68** Let $R$ be the plane figure bounded below by the $x$-axis, above by the graph of $y = \exp(-x^2)$, and on the sides by the vertical lines at $x = 0$ and $x = 1$. Find the volume generated by rotating $R$ about the $y$-axis.

**69** Find the length of the curve $y = (e^x + e^{-x})/2$ from $x = 0$ to $x = 1$.

**70** Find the area of the surface generated by revolving the curve of Problem 69 around the $x$-axis.

**71** Show that the equation $e^{-x} = x - 1$ has a single solution, and use Newton's method to find it to three-place accuracy.

**72** If a plant releases an amount $A$ of pollutant into a canal at time $t = 0$, then the resulting concentration of pollutant at time $t$ in the water at a town on the canal at distance $x_0$ from the plant is

$$C(t) = \dfrac{A}{\sqrt{k\pi t}} \exp\left(-\dfrac{x_0^2}{4kt}\right)$$

where $k$ is a certain constant. Show that the maximum concentration at the town is

$$C_{max} = \dfrac{A}{x_0}\sqrt{\dfrac{2}{\pi e}}.$$

**73** Sketch the graph of $f(x) = x^n e^{-x}$ for $x \geq 0$ ($n$ is a fixed but arbitrary positive integer). In particular, show that the maximum value of $f$ is $f(n) = n^n e^{-n}$.

**74** Approximate the number $e$ as follows. First apply Simpson's approximation with $n = 2$ subintervals to the integral

$$\int_0^1 e^x \, dx = e - 1$$

to obtain the approximation $5e - 4\sqrt{e} - 7 \approx 0$. Then solve for $e$.

**75** Suppose that $f(x) = x^n e^{-x}$, where $n$ is a fixed but arbitrary positive integer. Conclude from Problem 73 that the numbers $f(n - 1)$ and $f(n + 1)$ are each less than $f(n) = n^n e^{-n}$. Deduce from this that

$$\left(1 + \frac{1}{n}\right)^n < e < \left(1 - \frac{1}{n}\right)^{-n}.$$

Substitute $n = 1024$ to show that $2.716 < e < 2.720$. Note that $1024 = 2^{10}$, so that $a^{1024}$ can be computed easily with almost any calculator by entering $a$ and then squaring $a$ ten times in succession.

**76** Suppose that the quadratic equation $am^2 + bm + c = 0$ has the two real roots $m_1$ and $m_2$, and suppose that $C_1$ and $C_2$ are two arbitrary constants. Show that the function

$$y = y(x) = C_1 e^{m_1 x} + C_2 e^{m_2 x}$$

satisfies the differential equation $ay'' + by' + cy = 0$.

**77** Use the result of Problem 76 to find a solution $y(x)$ of the differential equation $y'' + y' - 2y = 0$ such that $y(0) = 5$ and $y'(0) = 2$.

## *7-3  Optional Computer Application

The following program may be used to approximate the value of $e^x$ by means of the limit

$$e^x = \lim_{n \to \infty} \left(1 + \frac{x}{n}\right)^n.$$

After the value of $x$ is entered at line 10, the quantity $(1 + x/n)^n$ is calculated for $n = 10, 100, 1000, \ldots$.

```
10    INPUT "X"; X
20    N = 10
30    E = (1 + X/N)↑N
40    PRINT N; E
50    N = 10*N
60    GOTO 30
70    END
```

When line 40 is executed, the current values of $N$ and $E = e^x$ are printed, and then $N$ is increased by a factor of $10$ in preparation for the next computation.

**Exercise** Run this program with $x = 1$ to check that it gives $e \approx 2.718281828$. Then run it with $x = 0.5$ to check that it gives $\sqrt{e}$ accurately. (Whether or not you get all ten significant figures depends upon the precision of your computer. Any computer retains only a finite number of decimal places and therefore rounds $1 + x/n$ to $1$ when $n$ is sufficiently large; at this point, we erroneously get $(1 + x/n)^n \approx 1$. Moreover, the program "cheats" in that many computers will use the exponential function itself to calculate $(1 + x/n)^n$ by means of the formula $a^n = \exp(n \ln a)$.)

## 7-4

### General Exponential and Logarithmic Functions

The natural exponential $e^x$ and the natural logarithm $\ln x$ are often called the exponential and logarithm with *base* $e$. We now define general exponential and logarithm functions, having the forms $a^x$ and $\log_a x$, with base $a$ a positive number $a \neq 1$. But it is now convenient to reverse the order of treatment in Sections 7-2 and 7-3, so we first consider the general exponential function.

If $r$ is a rational number, then one of the laws of exponents (Equation (13) in Section 7-3) gives

$$a^r = (e^{\ln a})^r = e^{r \ln a}.$$

We therefore *define* arbitrary powers (rational *and* irrational) of the positive number $a$ this way:

$$a^x = e^{x \ln a} \tag{1}$$

for all $x$. Then $f(x) = a^x$ is called the **exponential function with base** $a$. Note that $a^x > 0$ for all $x$ and that $a^0 = e^0 = 1$ for all $a > 0$.

The *laws of exponents* for general exponentials follow almost immediately from Definition (1) and the laws of exponents for the natural exponential.

$$a^x a^y = a^{x+y}, \tag{2}$$

$$a^{-x} = \frac{1}{a^x}, \quad \text{and} \tag{3}$$

$$(a^x)^y = a^{xy} \tag{4}$$

for all $x$ and $y$. To show the first formula, we write

$$a^x a^y = e^{x \ln a} e^{y \ln a} = e^{(x \ln a) + (y \ln a)}$$
$$= e^{(x+y) \ln a} = a^{x+y}.$$

To derive (4), note first from (1) that $\ln a^x = x \ln a$. Then

$$(a^x)^y = e^{y \ln(a^x)} = e^{xy \ln a} = a^{xy}.$$

Observe that this follows for all real numbers $x$ and $y$, so the restriction that $r$ be rational in the formula $(e^x)^r = e^{rx}$ (see Equation (13) in Section 7-3) has now been removed.

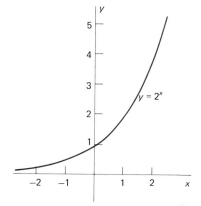

**7.19** The graph of $y = 2^x$

If $a > 1$ so that $\ln a > 0$, then Formulas (5) and (6) in Section 7-3 immediately give us the results

$$\lim_{a \to \infty} a^x = \infty \quad \text{and} \quad \lim_{x \to -\infty} a^x = 0. \tag{5}$$

The values of these two limits are interchanged if $0 < a < 1$, for then $\ln a < 0$.

Because

$$D_x a^x = D_x e^{x \ln a} = a^x \ln a \tag{6}$$

is positive for all $x$ if $a > 1$, we see that—in this case—$a^x$ is an increasing function of $x$. If $a > 1$, the graph of $y = a^x$ resembles that of $f(x) = e^x$, and in fact the graph of $y = 2^x$ appears in Fig. 7.19. But if $0 < a < 1$, then $\ln a < 0$, and it then follows from (6) that $a^x$ is a decreasing function. In this case, the graph of $y = a^x$ will look like the one shown in Fig. 7.20.

If $u = u(x)$ is a differentiable function of $x$, then (6) combined with the chain rule gives

$$D_x a^u = a^u (\ln a) \frac{du}{dx}. \tag{7}$$

**7.20** The graph of $y = a^x$ is decreasing and concave upward if $0 < a < 1$.

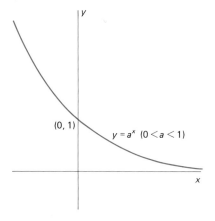

The corresponding integral formula is

$$\int a^u \, du = \frac{a^u}{\ln a} + C. \tag{8}$$

But rather than using these general formulas, it is generally simpler to rely upon Definition (1) alone, as in the following examples.

**EXAMPLE 1**  To differentiate $f(x) = 3^{x^2}$, we first write

$$3^{x^2} = (e^{\ln 3})^{x^2} = e^{x^2 \ln 3}.$$

Then

$$D_x 3^{x^2} = D_x \, e^{x^2 \ln 3}$$
$$= e^{x^2 \ln 3} D_x \, (x^2 \ln 3) = 3^{x^2}(\ln 3)(2x).$$

**EXAMPLE 2**  Find $\int \dfrac{10^{\sqrt{x}}}{\sqrt{x}} \, dx.$

***Solution***  We first write

$$10^{\sqrt{x}} = (e^{\ln 10})^{\sqrt{x}} = e^{\sqrt{x} \ln 10}.$$

Then

$$\int \frac{10^{\sqrt{x}}}{\sqrt{x}} \, dx = \int \frac{e^{\sqrt{x} \ln 10}}{\sqrt{x}} \, dx$$

$$= \int \frac{2e^u}{\ln 10} \, du \qquad \left( u = \sqrt{x} \ln 10, \quad du = \frac{\ln 10}{2\sqrt{x}} \, dx \right)$$

$$= \frac{2e^u}{\ln 10} + C = \frac{2(10^{\sqrt{x}})}{\ln 10} + C.$$

---

Whether or not the exponent $r$ is rational, the **general power function** $f(x) = x^r$ is now defined for $x > 0$ by

$$x^r = e^{r \ln x}.$$

We may now prove the power rule for differentiation for an *arbitrary* (constant) exponent, as follows:

$$D_x x^r = D_x(e^{r \ln x}) = e^{r \ln x} D(r \ln x)$$

$$= x^r \cdot \frac{r}{x} = rx^{r-1}.$$

For example, we now know that

$$Dx^\pi = \pi x^{\pi - 1} \approx (3.14159)x^{2.14159}.$$

If $a > 1$, then the general exponential function $a^x$ is continuous and increasing for all $x$, and attains all positive values (by an argument similar to that in the first paragraph of Section 7-3, using (6) and the limits in (5) above). It therefore has an inverse function that is defined for all $x > 0$. This inverse function to $a^x$ is called the **logarithm function with base $a$** and is denoted by

$\log_a x$. Thus

$$y = \log_a x \quad \text{if and only if} \quad x = a^y. \tag{9}$$

Note that the logarithm function with base $e$ is the natural logarithm function $\log_e x = \ln x$.

The following *laws of logarithms* are easy to derive from the laws of exponents above.

$$\log_a xy = \log_a x + \log_a y, \tag{10}$$

$$\log_a\left(\frac{1}{x}\right) = -\log_a x, \tag{11}$$

$$\log_a x^y = y \log_a x. \tag{12}$$

These formulas hold for any positive base $a \neq 1$ and for all positive values of $x$ and $y$; in the last formula, $y$ may be negative or zero as well.

Logarithms with one base are related to logarithms with another base, and the relationship is most easily expressed by the formula

$$(\log_a b)(\log_b c) = \log_a c. \tag{13}$$

This formula holds for all values of $a$, $b$, and $c$ that make sense; that is, the bases $a$ and $b$ are positive numbers other than 1 and $c$ is positive. The proof of this formula is outlined in Problem 53. Note also how easy it is to remember the formula in (13)—it is as if some arcane cancellation law held.

If we take $c = a$ in the above formula, this gives

$$(\log_a b)(\log_b a) = 1, \tag{14}$$

which in turn, with $b = e$, gives

$$\ln a = \frac{1}{\log_a e}. \tag{15}$$

If in (13) we replace $a$ with $e$, $b$ with $a$, and $c$ with $x$, we get

$$(\log_e a)(\log_a x) = \log_e x,$$

so

$$\log_a x = \frac{\log_e x}{\log_e a} = \frac{\ln x}{\ln a}. \tag{16}$$

On most calculators, the "log" button denotes common (base 10) logarithms: $\log x = \log_{10} x$. In BASIC, however, only the natural logarithm LOG(X) appears explicitly; to get $\log_{10} x$ we write LOG(X)/LOG(10).

Differentiation in (16) yields

$$D_x \log_a x = \frac{1}{x \ln a} = \frac{\log_a e}{x}. \tag{17}$$

For example,

$$D \log_{10} x = \frac{\log_{10} e}{x} \approx \frac{0.4343}{x}.$$

If we now reason as in Equation (6) of Section 7-2, the chain rule yields the general formula

$$D_x \log_a|u| = \frac{1}{u \ln a}\frac{du}{dx} = \frac{\log_a e}{u}\frac{du}{dx} \qquad (u \neq 0) \qquad (18)$$

if $u$ is a differentiable function of $x$. For instance,

$$D_x \log_2 \sqrt{x^2 + 1} = \frac{1}{2} D_x \log_2(x^2 + 1)$$

$$= \frac{1}{2} \cdot \frac{\log_2 e}{x^2 + 1}(2x) \approx \frac{(1.4427)x}{x^2 + 1}.$$

We used here the fact that $\log_2 e = 1/(\ln 2)$ by (15).

## LOGARITHMIC DIFFERENTIATION

The derivatives of certain functions are most conveniently found by first differentiating their logarithms. This process—called **logarithmic differentiation**—involves the following steps for finding $f'(x)$.

**1** Given: $\qquad\qquad\qquad\qquad\qquad\qquad y = f(x).$

**2** Take *natural* logarithms, then simplify
using laws of logarithms: $\qquad\qquad \ln y = \ln f(x).$

**3** Now differentiate with respect to $x$: $\qquad \dfrac{1}{y}\dfrac{dy}{dx} = D_x[\ln f(x)].$

**4** Finally, multiply both sides by $y = f(x)$: $\qquad \dfrac{dy}{dx} = f(x)D_x[\ln f(x)].$

REMARK   If $f(x)$ is not positive-valued everywhere, Steps 1 and 2 should be replaced with $|y| = |f(x)|$ and $\ln|y| = \ln|f(x)|$, respectively. The differentiation in Step 3 nevertheless leads to the same result in Step 4. In effect then, we need not be overly concerned with the sign of $f(x)$ in using this technique.

**EXAMPLE 3**   Find $dy/dx$ if

$$y = \frac{\sqrt{(x^2 + 1)^3}}{\sqrt[3]{(x^3 + 1)^4}}.$$

*Solution*   The laws of logarithms give

$$\ln y = \ln \frac{(x^2 + 1)^{3/2}}{(x^3 + 1)^{4/3}} = \frac{3}{2}\ln(x^2 + 1) - \frac{4}{3}\ln(x^3 + 1).$$

So differentiation with respect to $x$ gives

$$\frac{1}{y}\frac{dy}{dx} = \left(\frac{3}{2}\right)\frac{2x}{x^2 + 1} - \left(\frac{4}{3}\right)\frac{3x^2}{x^3 + 1}$$

$$= \frac{3x}{x^2 + 1} - \frac{4x^2}{x^3 + 1}.$$

Finally, to solve for $dy/dx$, we multiply both sides by

$$y = \sqrt{(x^2 + 1)^3}/\sqrt[3]{(x^3 + 1)^4},$$

and we obtain

$$\frac{dy}{dx} = \left(\frac{3x}{x^2 + 1} - \frac{4x^2}{x^3 + 1}\right)\frac{\sqrt{(x^2 + 1)^3}}{\sqrt[3]{(x^3 + 1)^4}}.$$

**EXAMPLE 4** Find $dy/dx$ if $y = x^{x+1}$.

**Solution** If $y = x^{x+1}$, then

$$\ln y = \ln(x^{x+1}) = (x + 1) \ln x,$$

$$\frac{1}{y}\frac{dy}{dx} = (1) \ln x + (x + 1)\left(\frac{1}{x}\right)$$

$$= 1 + \frac{1}{x} + \ln x.$$

And now multiplication by $y = x^{x+1}$ gives

$$\frac{dy}{dx} = \left(1 + \frac{1}{x} + \ln x\right)x^{x+1}.$$

## 7-4 PROBLEMS

In Problems 1–24, find the derivative of the given function $f(x)$.

1 $10^x$

2 $2^{(1/x^2)}$

3 $\dfrac{3^x}{4^x}$

4 $\log_{10}(\ln x)$

5 $7^{\exp(x^2)}$

6 $2^x 3^{(x^2)}$

7 $2^{x\sqrt{x}}$

8 $\log_{100}(10^x)$

9 $2^{\ln x}$

10 $7^{(8^x)}$

11 $17^x$

12 $2^{\sqrt{x}}$

13 $10^{1/x}$

14 $3^{\sqrt{1-x^2}}$

15 $2^{(2^x)}$

16 $\log_2 x$

17 $\log_3\sqrt{x^2 + 4}$

18 $\log_{10}(e^x)$

19 $\log_3(2^x)$

20 $\log_{10}(\log_{10} x)$

21 $\log_2(\log_3 x)$

22 $\pi^x + x^\pi + \pi^\pi$

23 $\exp(\log_{10} x)$

24 $\pi^{(x^3)}$

Evaluate the integrals given in Problems 25–32.

25 $\displaystyle\int 3^{2x}\,dx$

26 $\displaystyle\int x(10^{-x^2})\,dx$

27 $\displaystyle\int \frac{2^{\sqrt{x}}}{\sqrt{x}}\,dx$

28 $\displaystyle\int \frac{10^{1/x}}{x^2}\,dx$

29 $\displaystyle\int x^2 7^{x^3 + 1}\,dx$

30 $\displaystyle\int \frac{dx}{x \log_{10} x}$

31 $\displaystyle\int \frac{\log_2 x}{x}\,dx$

32 $\displaystyle\int (2^x)3^{(2^x)}\,dx$

In Problems 33–52, find $dy/dx$ by logarithmic differentiation.

33 $y = \sqrt{(x^2 - 4)\sqrt{2x + 1}}$

34 $y = \dfrac{\sqrt[3]{3 - x^2}}{\sqrt[4]{x^4 + 1}}$

35 $y = 2^x$

36 $y = x^x$

37 $y = x^{\ln x}$

38 $y = (1 + x)^{1/x}$

39 $y = \sqrt[3]{\dfrac{(x + 1)(x + 2)}{(x^2 + 1)(x^2 + 2)}}$

40 $y = \sqrt{x + 1}\sqrt[3]{x + 2}\sqrt[4]{x + 3}$

41 $y = (\ln x)^{\sqrt{x}}$

42 $y = (3 + 2^x)^x$

43 $y = \dfrac{(1 + x^2)^{3/2}}{(1 + x^3)^{4/3}}$

44 $y = (x + 1)^x$

45 $y = (x^2 + 1)^{(x^2)}$

46 $y = \left(1 + \dfrac{1}{x}\right)^x$

47 $y = (\sqrt{x})^{\sqrt{x}}$

48 $y = (\ln x)^{\sqrt{x}}$

49 $y = e^x$

50 $y = (\ln x)^{\ln x}$

51 $y = x^{(e^x)}$

52 $y = x^{x^x}$

53 Prove Formula (13). (*Suggestion:* Let $x = \log_a b$, $y = \log_b c$, and $z = \log_a c$. Then show that $a^z = a^{xy}$, and conclude that $z = xy$.)

54 Suppose that $u$ and $v$ are differentiable functions of $x$. Show by logarithmic differentiation that

$$D_x(u^v) = v(u^{v-1})\frac{du}{dx} + u^v(\ln u)\frac{dv}{dx}.$$

Interpret the two terms on the right in relation to each of the two special cases: (i) $u$ is a constant; (ii) $v$ is a constant.

**55** Suppose that $a > 0$. Show that

$$\lim_{x \to \infty} a^{1/x} = 1$$

by examining $\ln(a^{1/x})$. It follows that

$$\lim_{n \to \infty} \sqrt[n]{a} = 1.$$

Test this conclusion by entering some positive number into your calculator and then pressing the square root key repeatedly. Tabulate the results of two such experiments.

**56** Show that

$$\lim_{n \to \infty} \sqrt[n]{n} = 1$$

by showing that

$$\lim_{x \to \infty} x^{1/x} = 1.$$

Use the method of Problem 55.

**57** Show that

$$\lim_{x \to \infty} \frac{x^x}{e^x} = \infty$$

by examining $\ln(x^x/e^x)$. Thus $x^x$ increases faster than the exponential function $e^x$ as $x \to \infty$.

**58** (a) Show that the equation $2^x = x^{10}$ has the same positive solutions as the equation $(\ln x)/x = \frac{1}{10} \ln 2$.
(b) Conclude from the graph of $y = (\ln x)/x$ (Example 7 in Section 7-2) that the equation $2^x = x^{10}$ has exactly two positive solutions.
(c) Show that one of these two solutions is between 1 and 2, while the other is between 50 and 60. Then use Newton's method to approximate each solution with at least two-place accuracy.

**59** Consider the function $f$ given by

$$f(x) = \frac{1}{1 + 2^{1/x}} \qquad \text{for } x \neq 0.$$

Show that the left-hand and right-hand limits of $f(x)$ at $x = 0$ both exist but are unequal.

**60** Find $dy/dx$ if $y = \log_x 2$. (*Suggestion:* Note that $x^y = 2$.)

Problems 61–63 are motivated by the expression

$$y = x^{x^{x^{\cdot^{\cdot^{\cdot)}}}}} = x^{(x^{(x^{\cdot^{\cdot)}}})}, \tag{19}$$

which *apparently* is equivalent to $y = x^y$, and which we define to be the limit of the sequence $y_1, y_2, y_3, \ldots$, where

$$y_1 = x \quad \text{and} \quad y_{n+1} = x^{y_n} \qquad \text{for } n \geq 1.$$

Here we note only that, if the formula in (19) actually is a valid way of giving a formula for a function $y$ of $x$, then it is an implicit solution of the equation $y = x^y$ with $y > 0$.

**61** (a) Assume that the differentiable function $y = y(x)$ satisfies the equation $y = x^y$. Find its derivative $dy/dx$ by logarithmic differentiation.
(b) Assume that the graph of $y = x^y$ is a curve having a tangent line at each point. Show that the tangent line at the point $(e^{1/e}, e)$ is vertical.

**62** (a) Solve the equation $y = x^y$ for $x = y^{1/y}$, and conclude that the graphs $y = x^y$ and $y = x^{1/x}$ are reflections of each other through the line $y = x$.
(b) Use the result of Problem 56 to conclude that the graph of $y = x^y$ looks as shown in Fig. 7.21.

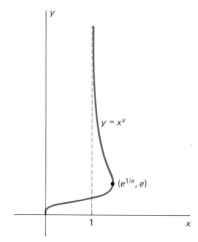

$y = x^y$

$(e^{1/e}, e)$

**7.21** The graph of the function defined implicitly by the equation $y = x^y$

**63** Explain why

$$y = x^{x^{x^{\cdot^{\cdot^{\cdot)}}}}} = x^{(x^{(x^{\cdot)}})}$$

is *not* equivalent to $y = x^y$ on the interval $1 < x \leq e^{1/e}$.

---

## 7-5

### Natural Growth and Decay

Consider a population that numbers $P(t)$ persons—or animals, bacteria, or any sort of entity—at time $t$. We assume that this population has a constant birth rate $\beta$ and a constant death rate $\delta$. Roughly speaking, this means that, during any 1-year period, $\beta P$ births and $\delta P$ deaths occur.

But since $P$ changes during the course of that year, some allowance must be made for changes in the number of births and the number of deaths. To be more precise, we think of a very brief time interval from $t$ to $t + \Delta t$. For very small values of $\Delta t$, the value of $P = P(t)$ will change by such a small amount during the time interval $[t, t + \Delta t]$ that we can regard $P$ as "almost" constant. We require that the numbers of births and deaths during this time interval be given with sufficient accuracy by the approximations

$$\frac{\text{number}}{\text{of births}} \approx \beta P(t) \cdot \Delta t \quad \text{and} \quad \frac{\text{number}}{\text{of deaths}} \approx \delta P(t) \cdot \Delta t. \tag{1}$$

What we mean when we say that the birth rate is $\beta$ and the death rate is $\delta$ is this: The ratios to $\Delta t$ of the errors in the above approximations both approach zero as $\Delta t \to 0$.

We would like to use the information in (1) to deduce, if possible, the form of the function $P(t)$ that describes our population. Our strategy begins with finding the **time rate of change** of $P$. Hence we consider the increment $\Delta P = P(t + \Delta t) - P(t)$ of $P$ during the time interval $[t, t + \Delta t]$. Since $\Delta P$ is simply the number of births minus the number of deaths, we find from (1) that

$$\Delta P = P(t + \Delta t) - P(t) \approx \beta P(t)\, \Delta t - \delta P(t)\, \Delta t.$$

Therefore,

$$\frac{\Delta P}{\Delta t} = \frac{P(t + \Delta t) - P(t)}{\Delta t} \approx (\beta - \delta)P(t).$$

The quotient on the left-hand side approaches the derivative $P'(t)$ as $\Delta t \to 0$ and, by the assumption following (1) above, it also approaches the right-hand side $(\beta - \delta)P(t)$. Hence, when we take the limit as $\Delta t \to 0$, we get the differential equation

$$P'(t) = (\beta - \delta)P(t);$$

that is,

$$\frac{dP}{dt} = kP \qquad \text{where } k = \beta - \delta. \tag{2}$$

This differential equation may be regarded as a *mathematical model* of the changing population.

The differential equation

$$\frac{dx}{dt} = kx \qquad (k \text{ a constant}) \tag{3}$$

serves as a mathematical model for an extraordinarily wide range of natural phenomena. It is easy to solve; we first write it in the form

$$\frac{1}{x}\frac{dx}{dt} = k;$$

that is,

$$D_t(\ln x) = D_t(kt).$$

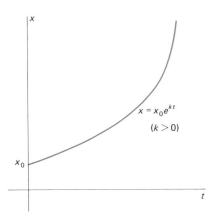

**7.22** Solution of the exponential growth equation for $k > 0$

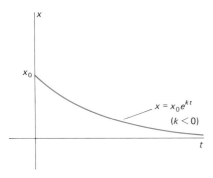

**7.23** Solution of the exponential growth equation—now actually a *decay* equation—for the case $k < 0$

We then antidifferentiate and obtain

$$\ln x = kt + C \qquad \text{(assuming } x > 0\text{)}.$$

Note that the same result is obtained formally if we first "separate the variables" in (3) and then integrate:

$$\frac{dx}{x} = k\,dt;$$

$$\int \frac{dx}{x} = \int k\,dt;$$

$$\ln x = kt + C.$$

In either case, we next apply the exponential function to both sides of this equation to solve for $x$. This gives

$$x = e^{\ln x} = e^{kt+C} = Ae^{kt}.$$

Here, $A = e^C$ is a constant that remains to be determined. But we see that $A$ will simply be the value of $x$ when $t = 0$; that is, $A = x(0) = x_0$. Thus the solution of the differential equation in (3) with the *initial value* $x(0) = x_0$ is

$$x(t) = x_0 e^{kt}. \tag{4}$$

As a result, Equation (3) is often called the **exponential growth equation,** or the **natural growth equation.** We see from (4) that with $x_0 > 0$, the solution $x(t)$ is an increasing function if $k > 0$ and a decreasing function if $k < 0$. These cases are shown in Figs. 7.22 and 7.23, respectively. The remainder of this section is devoted to examples of natural phenomena for which this differential equation serves as a mathematical model.

## POPULATION GROWTH

When we compare Equations (2), (3), and (4), we see that a population $P(t)$ with constant birth rate $\beta$ and constant death rate $\delta$ is given by

$$P(t) = P_0 e^{kt} \tag{5}$$

where $P_0 = P(0)$ and $k = \beta - \delta$. If $t$ is measured in years, then $k$ is called the **annual growth rate,** whether it be positive, negative, or zero. Its value is often given as a percentage (which is its actual decimal value multiplied by 100). If $k$ is not too large, this value is fairly close to the actual percentage increase (or decrease) of the population each year.

**EXAMPLE 1** In mid-1984, the world population was 4.76 billion persons, and it was increasing then at the rate of 220 thousand persons each day. Let us assume constant birth and death rates. We would like to find the answers to these questions:

  **a** What is the annual growth rate $k$?
  **b** How long will it take the world's population to double?
  **c** What will the world's population be in the year 2000?
  **d** When will the world's population be 50 billion (some demographers believe that this is the maximum for which the planet can supply food)?

*Solution*  We measure the world's population $P(t)$ in billions and measure time $t$ in years. We will take $t = 0$ to correspond to 1984, so that $P_0 = 4.76$. The fact that $P$ is increasing by 220 thousand, or 0.00022 billion, persons per day at time $t = 0$ means that

$$P'(0) = (0.00022)(365.25) \approx 0.0804$$

billion per year. From Equation (2) we now obtain

$$k = \left| \frac{1}{P} \frac{dP}{dt} \right|_{t=0} = \frac{P'(0)}{P(0)} = \frac{0.0804}{4.76} \approx 0.0169.$$

Thus the population is growing at the rate of about 1.69% per year.

We use this value of $k$ to conclude that the world population at time $t$ should be

$$P(t) = (4.76)e^{(0.0169)t}.$$

For instance, $t = 16$ yields

$$P(16) = (4.76)e^{(0.0169)(16)} \approx 6.24 \quad \text{(billion)}$$

for the population in the year 2000.

To find when the population will be 9.52 billion, we solve the equation

$$9.52 = (4.76)e^{(0.0169)t}$$

(by taking the natural logarithm of each side) for

$$t = \frac{\ln 2}{0.0169} \approx 41 \quad \text{(years)},$$

which corresponds to the year 2025. The world population will be 50 billion when

$$50 = (4.76)e^{(0.0169)t},$$

$$t = \frac{\ln(50/4.76)}{0.0169} \approx 139;$$

that is, in the year 2123.

## RADIOACTIVE DECAY

Consider a sample of material that contains $N(t)$ atoms of a certain radioactive isotope at time $t$. It has been observed that a constant fraction of these radioactive atoms will spontaneously decay (into atoms of another element or another isotope of the same element) during each unit of time. Consequently the sample behaves exactly like a population with a constant death rate but with no births occurring. To write a model for $N(t)$, we use Equation (2) with $N$ in place of $P$, with $k > 0$ in place of $\delta$, and with $\beta = 0$. We thus get the differential equation

$$\frac{dN}{dt} = -kN. \tag{6}$$

From the solution in (4) of Equation (3), with $k$ replaced by $-k$, we conclude that

$$N(t) = N_0 e^{-kt} \tag{7}$$

where $N_0 = N(0)$, the number of radioactive atoms present in the sample at time $t = 0$.

The value of the *decay constant* $k$ depends upon the particular isotope with which we are dealing. If $k$ is large, then the isotope decays rapidly, while if $k$ is near zero, the isotope decays quite slowly and thus may be a relatively persistent factor in its environment. The decay constant is often specified in terms of another empirical parameter, the **half-life** of the isotope, because this parameter is more convenient. The half-life $\tau$ of a radioactive isotope is the time required for *half* of it to decay. To find the relationship between $k$ and $\tau$, we set

$$t = \tau \quad \text{and} \quad N = \tfrac{1}{2}N_0$$

in Equation (7), so that

$$\tfrac{1}{2}N_0 = N_0 e^{-k\tau}.$$

When we solve for $\tau$, we find that

$$\tau = \frac{\ln 2}{k}. \tag{8}$$

Note that the concept of half-life is meaningful; the value of $\tau$ depends *only* upon $k$ and thus depends only on the radioactive isotope involved. It does *not* depend upon the amount of that isotope present.

**EXAMPLE 2** The half-life of radioactive carbon $C^{14}$ is about 5700 years. A specimen of charcoal found at Stonehenge turns out to contain 63% as much $C^{14}$ as a sample of present-day charcoal. What is the age of the sample?

*Solution* The key to the method of *radiocarbon dating* is that living organic matter maintains a constant level of $C^{14}$ by "breathing" air (or by consuming organic matter that does). Since air contains $C^{14}$ along with the much more common stable isotope $C^{12}$ of carbon, mostly in the gas $CO_2$, the same percentage permeates all life, because organic processes seem to make no distinction between the two isotopes. Of course, when a living organism dies, it ceases its metabolism of carbon, and the process of radioactive decay begins to deplete its $C^{14}$ content. The fraction of $C^{14}$ in the air remains roughly constant because new $C^{14}$ is being generated constantly by the action of cosmic rays on nitrogen atoms in the upper atmosphere, and this generation has long been in a steady-state equilibrium with the loss of $C^{14}$ through radioactive decay.

For our Stonehenge example, we take $t = 0$ as the time of "death" of the tree from which the charcoal was made. From Equation (8) we know that

$$k = \frac{\ln 2}{\tau} = \frac{\ln 2}{5700} \approx 0.0001216.$$

We are given that $N = (0.63)N_0$ now, so we solve the equation

$$(0.63)N_0 = N_0 e^{-kt}$$

with this value of $k$ and thus find that

$$t = -\frac{\ln(0.63)}{0.0001216} \approx 3800$$

years. Thus the sample is about 3800 years old, and if it has any connection with the builders of Stonehenge, our computation suggests that this observatory, monument, or temple—whichever it may be—dates from almost 1800 B.C.

**EXAMPLE 3** According to one cosmological theory, there were equal amounts of the two uranium isotopes $U^{235}$ and $U^{238}$ at the creation of the universe in the "big bang." At present there are 137.7 $U^{238}$ atoms for each atom of $U^{235}$. Using the half-lives

$$4.51 \text{ billion years for } U^{238},$$

$$0.71 \text{ billion years for } U^{235},$$

calculate the age of the universe.

**Solution**  Let $N_8(t)$ and $N_5(t)$ be the numbers of $U^{238}$ and $U^{235}$ atoms, respectively, at time $t$ (in billions of years after the creation of the universe). Then

$$N_8(t) = N_0 e^{-kt} \quad \text{and} \quad N_5(t) = N_0 e^{-ct}$$

where $N_0$ is the original number of atoms of each isotope. Also

$$k = \frac{\ln 2}{4.51} \quad \text{and} \quad c = \frac{\ln 2}{0.71},$$

a consequence of Equation (8). We divide the equation for $N_8$ by the equation for $N_5$, and find that when $t$ has the value corresponding to "now," then

$$137.7 = \frac{N_8}{N_5} = e^{(c-k)t}.$$

Finally, we solve this equation for

$$t = \frac{\ln 137.7}{[(1/0.71) - (1/4.51)] \ln 2} \approx 5.99.$$

Thus our estimate of the age of the universe is about 6 billion years, which is at least of the same order of magnitude as recent estimates of about 15 billion years.

### CONTINUOUSLY COMPOUNDED INTEREST

Consider a savings account that is opened with an initial deposit of $A_0$ dollars and which thenceforth earns interest at the annual rate $r$. If there are $A(t)$ dollars in the account at time $t$ and the interest is compounded at time $t + \Delta t$, this means that $rA(t) \Delta t$ dollars in interest are added to the account then. So

$$A(t + \Delta t) = A(t) + rA(t) \Delta t,$$

and thus

$$\frac{\Delta A}{\Delta t} = \frac{A(t + \Delta t) - A(t)}{\Delta t} = rA(t).$$

*Continuous* compounding of interest is the situation that results by taking the limit as $\Delta t \to 0$, so that

$$\frac{dA}{dt} = rA. \tag{9}$$

This is an exponential growth equation with solution

$$A(t) = A_0 e^{rt}. \tag{10}$$

For example, if $A_0 = \$1000$ is invested at $6\%$ annual interest so that $r = 0.06$, then Equation (10) gives

$$A(1) = 1000 e^{(0.06)(1)} = \$1061.84$$

for the amount present after one year if the interest is compounded continuously. Thus the "effective" interest rate is $6.184\%$. Most people are aware that the more often interest is compounded, the more rapidly their savings grow, but bank advertisements sometimes tend to overemphasize this advantage. For example, $6\%$ compounded *monthly* multiplies your investment by

$$1 + \frac{0.06}{12} = 1.005$$

at the end of each month, so an initial investment of $\$1000$ would grow in one year to

$$(1000)(1.005)^{12} = \$1061.68,$$

only 16 cents less than would be yielded by continuous compounding.

### *DRUG ELIMINATION

In many cases the amount $A(t)$ of a certain drug in the bloodstream, measured by the excess over the natural level of the drug, will decline at a rate proportional to the excess amount. That is,

$$\frac{dA}{dt} = -\lambda A, \quad \text{so that} \quad A(t) = A_0 e^{-\lambda t}. \tag{11}$$

The parameter $\lambda$ is called the *elimination constant* of the drug, and $T = 1/\lambda$ is called the *elimination time*.

**EXAMPLE 4** The elimination time for alcohol varies from one person to another. If a person's "sobering time" $T = 1/\lambda$ is 2.5 h, how long will it take for the excess bloodstream alcohol concentration to be reduced from $0.10\%$ to $0.02\%$?

*Solution* We assume that the normal concentration of alcohol in the blood is zero, so that any amount is the excess amount. In this problem, we have $\lambda = 1/2.5 = 0.4$, so Equation (10) yields

$$0.02 = (0.10)e^{-(0.4)t}.$$

Thus

$$t = -\frac{\ln(0.2)}{0.4} \approx 4.02 \quad \text{(hours)}.$$

## *SALES DECLINE

Marketing studies for certain products show that if advertising for a particular product is halted and other market conditions remain unchanged—we mean such things as number and promotion of competing products, their prices, and so on—then the sales of the unadvertised product will decline at a rate that is proportional at any time to the current sales $S$. That is,

$$\frac{dS}{dt} = -\lambda S, \quad \text{so that} \quad S(t) = S_0 e^{-\lambda t}. \tag{12}$$

As usual, $S_0$ denotes the initial value of the sales, which we take to be the sales in the last advertising month. If we take months as the natural units for time $t$, then $S(t)$ denotes sales $t$ months after advertising is halted, and $\lambda$ might be called the *sales decay constant*.

## *LINGUISTICS

Consider a basic list of $N_0$ words in use in a given language at time $t = 0$. Let $N(t)$ denote the number of these words that are still in use at time $t$—those that have neither disappeared from the language nor been replaced by noncognates. According to one theory in linguistics, the rate of decrease of $N$ is proportional to $N$. That is,

$$\frac{dN}{dt} = -\lambda N, \quad \text{so that} \quad N(t) = N_0 e^{-\lambda t}. \tag{13}$$

If $t$ is measured in millennia (standard in linguistics), then $k = e^{-\lambda}$ is the fraction of the words in the original list that survive in use for 1000 years.

## 7-5 PROBLEMS

**1** Suppose that $1000 is deposited in a savings account that pays 8% annual interest compounded continuously. At what rate (in dollars per year) is this account earning interest after 5 years? After 20 years?

**2** (Population growth) A certain city had a population of 25,000 in 1960 and a population of 30,000 in 1970. Assume that its population will continue to grow exponentially at a constant rate. What population can its city planners expect in the year 2000?

**3** (Population growth) In a certain culture of bacteria, the number of bacteria increased sixfold in 10 hours. How long did it take to double their number?

**4** (Radiocarbon dating) Carbon extracted from an ancient skull contained only $\frac{1}{6}$ as much radioactive $C^{14}$ as carbon extracted from present-day bone. How old is the skull?

**5** (Radiocarbon dating) Carbon taken from a purported relic of the time of Christ contained $4.6 \times 10^{10}$ atoms of $C^{14}$ per gram. Carbon extracted from a present-day specimen of the same substance contained $5.0 \times 10^{10}$ atoms of $C^{14}$ per gram. Compute the approximate age of the relic. What is your opinion as to its authenticity?

**6** An amount $A$ is invested for $t$ years at an annual interest rate $r$ that is compounded $n$ times over these years at equal intervals.
(a) Explain why the amount that has accrued after $t$ years is

$$A_{r,n} = A\left(1 + \frac{rt}{n}\right)^n.$$

(b) Conclude from the limit in Equation (18) of Section 7-3 that

$$\lim_{n \to \infty} A_{r,n} = Ae^{rt},$$

in agreement with Equation (10) in this section.

**7** If an investment of $A_0$ dollars returns $A_1$ dollars after one year, the **effective annual interest rate** $r$ is defined by means of the equation $A_1 = (1 + r)A_0$. Banks often advertise that they increase the effective interest rates on their customers' savings accounts by increasing the frequency of compounding. Calculate the effective annual interest rate if a 9% annual interest rate is compounded
(a) quarterly;
(b) monthly;

(c) weekly;

(d) daily;

(e) continuously.

8 (Continuously compounded interest) Upon the birth of their first child, a couple deposited $5000 in a savings account that draws 6% interest compounded continuously. The interest payments are allowed to accumulate. How much will the account contain when the child is ready to go to college at age 18?

9 (Continuously compounded interest) Suppose that you discover in your attic an overdue library book on which your great-great-grandfather owed a fine of 30¢ 100 years ago. If an overdue fine grows exponentially at a 5% annual rate compounded continuously, how much would you have to pay if you returned the book today?

10 (Drug elimination) Suppose that sodium pentobarbitol is used to anesthetize a dog; the dog is anesthetized when its bloodstream contains at least 45 mg of sodium pentobarbitol per kilogram of the dog's body weight. Suppose also that sodium pentobarbitol is eliminated exponentially from the dog's bloodstream, with a half-life of 5 h. What single dose should be administered in order to anesthetize a 50-kg dog for 1 h?

11 (Sales decline) Advertising of a certain product has been discontinued; the company plans to resume advertising when sales have declined to 75% of their initial rate. (This phenomenon actually occurred when the Sony Corporation first introduced home videotape recorders in the United States in 1976.) If after 1 week without advertising, sales have declined to 95% of their initial rate, when should the company expect to resume advertising?

12 (Linguistics) The English language evolves in such a way that 77% of all words disappear (or are replaced by noncognates) every 1000 years. Of a basic list of words used by Chaucer in A.D. 1400, what percentage should we expect to find still in use today?

13 (Half-life) The half-life of radioactive cobalt is 5.27 years. Suppose that a nuclear accident has left the level of cobalt radiation in a certain region at 100 times the level acceptable for human habitation. How long will it be before the region is again habitable? (Ignore the likely presence of other radioactive substances.)

14 Suppose that a mineral body formed in an ancient cataclysm—perhaps the formation of the earth itself—originally contained the uranium isotope $U^{238}$ (which has a half-life of $4.51 \times 10^9$ years) but no lead, the end product of the radioactive decay of $U^{238}$. If today the ratio of $U^{238}$ atoms to lead atoms in the mineral body is 0.9, when did the cataclysm occur?

15 A certain moon rock was found to contain equal numbers of potassium and argon atóms. Assume that all the argon is the result of radioactive decay of potassium (its half-life is about $1.28 \times 10^9$ years) and that 1 of every 9 potassium atom disintegrations yields an argon atom. What is the age of the rock, measured from the time it contained only potassium?

16 If a body is cooling in a medium with constant temperature $A$, then—according to Newton's law of cooling—the rate of change of the temperature $T$ of the body is proportional to $T - A$. A pitcher of buttermilk initially at $25°C$ is to be cooled by setting it out on the front porch, where the temperature is $0°C$. Suppose that the temperature of the buttermilk has dropped to $15°C$ after 20 min. When will it be at $5°C$?

17 When sugar is dissolved in water, the amount $A$ that remains undissolved after $t$ minutes satisfies the differential equation $dA/dt = -kA$ ($k > 0$). If 25% of the sugar dissolves in 1 min, how long does it take for half the sugar to dissolve?

18 The intensity of light $I$ at a depth $x$ meters below the surface of a lake satisfies the differential equation $dI/dx = -(1.4)I$.

(a) At what depth is the intensity half the intensity $I_0$ at the surface (where $x = 0$)?

(b) What is the intensity at a depth of 10 m (as a fraction of $I_0$)?

(c) At what depth will the intensity be $1/100$ of what it is at the surface?

19 The barometric pressure $p$ (in inches of mercury) at an altitude $x$ miles above sea level satisfies the differential equation $dp/dx = -(0.2)p$; $p(0) = 29.92$.

(a) Calculate the barometric pressure at 10,000 ft and again at 30,000 ft.

(b) Without prior conditioning, few people can survive when the pressure drops to less than 15 in. of mercury. How high is that?

## *7-6

## Linear First Order Differential Equations and Applications

A **first order differential equation** is one in which only the first derivative (and not higher derivatives) of the dependent variable appears. It is called **linear** if it can be written in the form

$$\frac{dx}{dt} = ax + b, \tag{1}$$

where $a$ and $b$ denote functions of the independent variable $t$. In this section we discuss applications of the special case in which the coefficients $a$ and $b$ are *constants*.

To solve Equation (1) when the coefficients are constant, we write it as

$$\frac{1}{ax + b}\frac{dx}{dt} = 1.$$

If $ax + b > 0$, this says that

$$\frac{1}{a}D_t(\ln(ax + b)) = D_t t,$$

and antidifferentiation gives

$$\ln(ax + b) = at + C.$$

Note that the same result is obtained formally if we first "separate the variables" in (1) and then integrate:

$$\frac{dx}{ax + b} = dt;$$

$$\int \frac{a\,dx}{ax + b} = \int a\,dt;$$

$$\ln(ax + b) = at + C.$$

Then exponentiation gives

$$ax + b = Ke^{at}$$

where $K = e^C$. When we substitute $t = 0$ and denote the resulting value of $x$ by $x_0$, we find that $K = ax_0 + b$. So

$$ax + b = (ax_0 + b)e^{at}.$$

Finally, we solve this equation for the solution $x = x(t)$ of Equation (1):

$$x(t) = \left(x_0 + \frac{b}{a}\right)e^{at} - \frac{b}{a}. \tag{2}$$

In this development, we have assumed that $ax + b > 0$, but this same formula gives the solution in the case $ax + b < 0$ as well (see Problem 17). In Problem 21 we outline a method by which Equation (1) may be solved when the coefficients $a$ and $b$ are functions of $t$ rather than constants, but Solution (2) for the constant-coefficient case will be sufficient for the following applications.

## POPULATION GROWTH WITH IMMIGRATION

Consider a population $P(t)$ with constant birth and death rates ($\beta$ and $\delta$, respectively), as in Section 7-5, but also with a constant immigration rate of $I$ persons per year entering the country. To account for the immigration, our derivation of Equation (2) in Section 7-5 must be amended as follows:

$$P(t + \Delta t) - P(t) = (\text{births}) - (\text{deaths}) + (\text{immigrants})$$
$$\approx \beta P(t)\,\Delta t - \delta P(t)\,\Delta t + I\,\Delta t,$$

so

$$\frac{P(t + \Delta t) - P(t)}{\Delta t} \approx (\beta - \delta)P(t) + I.$$

We take limits as $\Delta t \to 0$ and thus obtain the linear first order differential equation

$$\frac{dP}{dt} = kP + I \qquad (3)$$

with constant coefficients $k = \beta - \delta$ and $I$. According to Formula (2), the solution of (3) is

$$P(t) = P_0 e^{kt} + \frac{I}{k}(e^{kt} - 1). \qquad (4)$$

The first term on the right-hand side is the effect of natural population growth; the second term is the effect of immigration.

**EXAMPLE 1** Consider the U.S. population with $P_0 = 222$ million in 1980 ($t = 0$). Suppose that we ask about the effect of allowing immigration at the rate of half a million people per year for the next 20 years, assuming a natural growth rate of 1% annually, so that $k = 0.01$. Then

$$P_0 e^{kt} = 222 e^{(0.01)(20)} \approx 271.2 \quad \text{(million)},$$

and

$$\frac{I}{k}(e^{kt} - 1) = \frac{0.5}{0.01}(e^{(0.01)(20)} - 1) \approx 11.1 \quad \text{(million)}.$$

Thus the effect of the immigration would be to increase the U.S. population in the year 2000 from 271.2 million to 282.3 million.

## SAVINGS ACCOUNT WITH CONTINUOUS DEPOSITS

Consider the savings account of Section 7-5 containing $A_0$ dollars initially and earning interest at the annual rate $r$ compounded continuously. In addition, we now suppose that deposits are added to this account at the rate of $Q$ dollars per year. To simplify the mathematical model, we assume that these deposits are made continuously rather than (for instance) monthly. We may then regard the amount $A(t)$ in the account at time $t$ as a "population" of dollars, having a natural annual growth rate $r$ and with "immigration" (deposits) at the rate of $Q$ dollars annually. Then by merely changing the notation in Equations (3) and (4), we get the differential equation

$$\frac{dA}{dt} = rA + Q \qquad (5)$$

with solution

$$A(t) = A_0 e^{rt} + \frac{Q}{r}(e^{rt} - 1). \qquad (6)$$

**EXAMPLE 2** Suppose that you wish to arrange, at the time of her birth, for your daughter to have $20 thousand available for college expenses at age 18. You plan to do so by making frequent small—essentially continuous—deposits in a savings account, at the rate of $Q$ thousand dollars each year. This account accumulates 6% annual interest compounded continuously. What should $Q$ be so that you may achieve your goal?

**Solution** With $A_0 = 0$ and $r = 0.06$, we are asking for the value of $Q$ so that Equation (6) yields the result $A(18) = 20$. That is, we are to find $Q$ so that

$$20 = \frac{Q}{0.06}(e^{(0.06)(18)} - 1).$$

When we solve this equation, we find that $Q = 0.61707$. Thus you should deposit \$617.07 per year, or about \$51.42 per month, in order to have \$20,000 in the account after 18 years. You may wish to verify that your total deposits will be \$11,107.23 and that the total interest accumulated will be \$8892.77.

## COOLING AND HEATING

According to Newton's law of cooling (or heating!), the time rate of change of the temperature $T$ of a body is proportional to the difference between $T$ and the temperature $A$ of its surroundings, under the assumption that $A$ is constant. We may translate this law into the language of differential equations by writing

$$\frac{dT}{dt} = -k(T - A). \tag{7}$$

Here $k$ is a positive constant; the minus sign is needed to make $T'(t)$ negative when $T$ exceeds $A$.

From Equation (2) we obtain the solution of Equation (7):

$$T(t) = A + (T_0 - A)e^{-kt}. \tag{8}$$

**EXAMPLE 3** A 5-lb roast, initially at 50°F, is put into a 375°F oven when $t = 0$; it's found that the temperature of the roast $T(t)$ (since $T$ is a function of time $t$) is 125°F when $t = 75$ (min). When will the roast be medium rare, a temperature of 150°F?

**Solution** Though we could simply substitute $A = 375$ and $T_0 = 50$ in Equation (8), let us instead solve explicitly the differential equation

$$\frac{dT}{dt} = k(375 - T)$$

that we get from Equation (7). We separate the variables and integrate to obtain

$$\int \frac{dT}{375 - T} = \int k\, dt;$$

$$-\ln(375 - T) = kt + C.$$

When $t = 0$, $T = T_0 = 50$; substitution of these data now yields the value $C = -\ln 325$, so

$$-\ln(375 - T) = kt - \ln 325,$$

$$375 - T = 325e^{-kt},$$

and hence

$$T = 375 - 325e^{-kt}.$$

We know that $T = 125$ when $t = 75$, and using this information, we find that

$$k = \frac{1}{75} \ln \frac{325}{250} \approx 0.0035.$$

So all we need do is solve the equation

$$150 = 375 - 325e^{-kt}.$$

We find that $t$ is about 105, so the roast should remain in the oven for about another 30 minutes.

## DIFFUSION OF INFORMATION

Let $N(t)$ denote the number of people (in a fixed population $P$) who by time $t$ have heard a certain piece of information that is being spread by the mass media. Under certain common conditions, the time rate of increase of $N$ will be proportional to the number of people who have not yet heard the piece of information. That is,

$$\frac{dN}{dt} = k(P - N). \tag{9}$$

If $N(0) = 0$, the solution to Equation (9) is

$$N(t) = P(1 - e^{-kt}). \tag{10}$$

If $P$ and some value $N(t_1)$ are known, we can then solve for $k$ and thereby determine $N(t)$ for all $t$. Problem 15 illustrates this situation.

## ELIMINATION OF POLLUTANTS

In the following example, we envision a lake that has been polluted, perhaps by factories operating on its shores. Suppose that the pollution is halted, perhaps by a legal order, perhaps by improved technology. We ask how long it will take for natural processes to reduce the pollutant concentration in the lake to an acceptable level.

**EXAMPLE 4**   Consider a lake with a volume of 8 billion ft$^3$ and an initial pollutant concentration of 0.25%. Suppose that an inflowing river brings in 500 million ft$^3$ of water daily with a (low) pollutant concentration of 0.05%, and that an outflowing river also removes 500 million ft$^3$ of the lake water daily. We make the simplifying assumption that the water in the lake, including that removed by the second river, is perfectly mixed at all times. If so, how long will it take to reduce the pollutant concentration in the lake to 0.10%?

*Solution*   Let $x(t)$ denote the amount of pollutants in the lake after $t$ days, in millions of cubic feet. Then $x_0 = (0.0025)(8000) = 20$. We want to know when $x = (0.001)(8000) = 8$.

**EXAMPLE 3** Consider a solar power plant that would cost $1 billion if it could be built immediately and instantaneously. Suppose that the start of construction is delayed by 5 years because of government red tape and anti-trust suits filed against the consortium of sponsoring companies. Once construction has begun, it takes another 5 years to complete the power plant. Assume a 10% continuous annual inflation of construction costs during this 10-year period. What will be the actual cost of construction of the power plant?

*Solution* We assume that, during the 5-year period of construction, there is a constant rate of expenditures in terms of present-value dollars. This means that we use $g(t) = \frac{1}{5}$ (in billions of dollars per year). Then Formula (6) gives the total actual cost as

$$C = \int_5^{10} \frac{1}{5} e^{(0.1)t} \, dt = \left[ 2e^{(0.1)t} \right]_5^{10} = 2(e - e^{1/2}),$$

or about $2.14 billion. Thus the 10-year delay results in more than doubling the apparent cost of the power plant.

In Problems 1–4, assume that the known global reserves of natural gas are $1.2 \times 10^{15}$ ft$^3$ and that the current rate of consumption is $3 \times 10^{13}$ ft$^3$ per year.

**1** If the rate of consumption increases continuously at 5% per year, how long will the known global reserves of natural gas last?

**2** If the global reserves of natural gas are actually five times as large as now known, how long will they last with consumption increasing continuously at 5% per year?

**3** If through conservation measures the rate of increase of consumption is reduced to 2% per year, how long will the *known* global reserves of natural gas last?

**4** How much would the consumption rate have to be decreased in order for the known global reserves of natural gas to last for 60 years?

**5** Suppose that you have signed a contract under which you will be paid $10,000 per year (on a continuous basis) during the next 5 years.
(a) What is the present value at 6% continuous interest rate of the $50,000 you will be paid under this contract?
(b) What is the present value of the $10,000 that you will be paid during the fifth year of this contract?

**6** Suppose that you have signed a contract under which you will be paid a salary continuously for 10 years, starting at $10,000 per year and increasing steadily to $20,000 per year after 10 years. Thus your salary (in thousands of dollars) after $t$ years will be $f(t) = 10 + t$.
(a) What is the present value (at 6% continuous annual interest) of this contract?
(b) What will be the total amount of dollars you actually receive during the 10-year period? Use the integral formula

$$\int t e^{at} \, dt = \frac{e^{at}}{a^2} (at - 1) + C.$$

**7** Suppose that you have $100,000 in book royalties payable now, but for income-tax reasons you desire to be paid at a constant rate for 10 years starting 5 years hence. At what annual rate should you be paid in order that the present value of the amount you receive will be the fair sum of $100,000? Assume 8% continuous annual interest rate.

**8** Consider a nuclear carrier for which construction will require materials and services with current value of $1.5 billion. Suppose also that the annual rate of inflation is and remains at 7.5%.
(a) How much will the carrier cost if it is begun now and built over a 2-year period?
(b) Repeat Part (a), but assume a 3-year construction period.
(c) State carefully what assumptions you are making about the rate of construction in the other parts of this problem.
(d) How much will the carrier cost if its construction is delayed for 4 years for political reasons and then completed during the subsequent 2-year period?

**9** Suppose that a small but rapidly growing company had a total income last year of $1 million and that production costs were $0.9 million, so that net profit was $0.1 million. Suppose also that during the foreseeable future its total income increases linearly by $0.2 million annually, but that its production costs increase exponentially at a continuous annual rate of 12%.
(a) What will be the total profits of the company during the next 5 years?
(b) What will be its profits for the subsequent 5 years?

## Separable Differential Equations and Applications

A **separable** first order differential equation is one that can be written in the form

$$f(x) \frac{dx}{dt} = g(t), \tag{1}$$

so that the variables may be "separated" by writing $f(x) \, dx = g(t) \, dt$. It is easy to solve this sort of differential equation by writing

$$\int f(x) \, dx = \int g(t) \, dt + C, \tag{2}$$

provided that the antiderivatives $F(x) = D_x^{-1} f(x)$ and $G(t) = D_t^{-1} g(t)$ can be found. To see that (2) follows from (1), it is enough to note that (1) says that $D_t F(x(t)) = D_t G(t)$, which in turn implies that

$$F(x(t)) = G(t) + C. \tag{3}$$

When a particular value of $x$ is available, it may be used to determine the value of the constant of integration $C$. For instance, if $x_0 = x(0)$, then substitution of $t = 0$ into Equation (3) gives $C = F(x_0) - G(0)$. The problem of solving (1) subject to the initial condition $x(0) = x_0$ is called an **initial value problem,** and the resulting solution is called a **particular solution** of (1); in contrast, (3) is called the **general solution** of Equation (1).

**EXAMPLE 1**   Solve the initial value problem

$$\frac{dx}{dt} = 4xt^3, \qquad x(0) = 2.$$

*Solution*   From

$$\int \frac{dx}{x} = \int 4t^3 \, dt$$

we immediately obtain the general solution

$$\ln x = t^4 + C.$$

Substitution of $t = 0$ into *both* sides of the general solution gives $C = \ln 2$. Then we apply the natural exponential function to each side of the equation in order to solve for $x$; thus we find that

$$x = e^{t^4 + \ln 2} = 2e^{t^4}.$$

### LIMITED POPULATIONS

In Section 7-5 we saw that a population $P(t)$ with constant birth rate $\beta$ and constant death rate $\delta$ satisfies the differential equation

$$\frac{dP}{dt} = (\beta - \delta)P. \tag{4}$$

This equation also applies when $\beta$ and $\delta$ are variable. One important example is the case when $\beta$ is a linear decreasing function of the population $P$, so that $\beta = \beta_0 - \beta_1 P$ where $\beta_0$ and $\beta_1$ are positive constants. If the death rate

$\delta = \delta_0$ remains constant, then (4) takes the form

$$\frac{dP}{dt} = (\beta_0 - \beta_1 P - \delta_0)P,$$

$$\frac{dP}{dt} = kP(M - P), \tag{5}$$

where $k = \beta_1$ and $M = (\beta_0 - \delta_0)/\beta_1$. We assume that $\beta_0 > \delta_0$ so that $M > 0$.

Equation (5) is called the **logistic equation.** If we assume that $P < M$, it may be solved as follows.

$$\int \frac{dP}{P(M - P)} = \int k \, dt;$$

$$\frac{1}{M} \int \left( \frac{1}{P} + \frac{1}{M - P} \right) dP = \int k \, dt;$$

$$\ln\left( \frac{P}{M - P} \right) = kMt + C.$$

Exponentiation gives

$$\frac{P}{M - P} = Ae^{kMt}$$

where $A = e^C$. We substitute $t = 0$ into both sides of this last equation, and this gives $A = P_0/(M - P_0)$. So

$$\frac{P}{M - P} = \frac{P_0 e^{kMt}}{M - P_0}.$$

This equation is easily solved for

$$P(t) = \frac{MP_0}{P_0 + (M - P_0)e^{-kMt}}. \tag{6}$$

Whereas we assumed in deriving (6) that $P < M$, this restriction is unnecessary, because we can verify by direct substitution into Equation (5) that $P(t)$ as given by (6) satisfies the logistic equation whether $P < M$ or $P \geqq M$.

If the initial population satisfies $P_0 < M$, then (6) shows that $P(t) < M$ for all $t > 0$ and also that

$$\lim_{t \to \infty} P(t) = M. \tag{7}$$

Thus a population that satisfies the logistic equation is *not* like a naturally growing population; it does *not* grow without bound. Instead, the population approaches the finite *limiting population* $M$ as $t \to \infty$. But since

$$\frac{dP}{dt} = kP(M - P) > 0$$

in this case, we see that the population is steadily increasing. Moreover, differentiation with respect to $t$ gives

$$\frac{d^2P}{dt^2} = \left[ \frac{d}{dP}\left( \frac{dP}{dt} \right) \right]\left( \frac{dP}{dt} \right) = (kM - 2kP) \cdot kP(M - P).$$

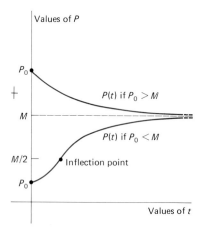

Values of P

$P_0$

$M$ — $P(t)$ if $P_0 > M$

$M/2$ — Inflection point

$P_0$

$P(t)$ if $P_0 < M$

Values of t

**7.24** Graphs of the two solutions of the logistic equation

| Year | Actual U.S. population (millions) | Calculated population |
|------|------|------|
| 1790 | 3.9 | 3.9 |
| 1800 | 5.3 | 5.3 |
| 1810 | 7.2 | 7.2 |
| 1820 | 9.6 | 9.7 |
| 1830 | 12.9 | 13.0 |
| 1840 | 17.1 | 17.4 |
| 1850 | 23.2 | 23.1 |
| 1860 | 31.4 | 30.2 |
| 1870 | 38.6 | 39.2 |
| 1880 | 50.2 | 49.9 |
| 1890 | 62.9 | 62.5 |
| 1900 | 76.0 | 76.6 |
| 1910 | 92.0 | 91.7 |
| 1920 | 106.5 | 107.1 |
| 1930 | 123.2 | 122.1 |

**7.25** Actual United States population data compared with data predicted by the logistic equation

So the graph of $P(t)$ has an inflection point where $P = M/2$. Therefore, the graph of $P$ has the shape of the lower curve in Fig. 7.21. The population increases at an increasing rate until $P = M/2$ and thereafter increases at a decreasing rate.

If $P_0 > M$, then a similar analysis shows that $P(t)$ is a steadily decreasing function with a graph like the upper curve in Fig. 7.24.

In 1845 the Belgian demographer Verhulst used the U.S. population data for 1790 through 1840 to predict the U.S. population through the year 1930, under his assumption that it would continue to satisfy the logistic equation. With $P_0 = 3.9$ (in millions), $M = 197.3$ (in millions), and $k = 0.00001589$, Equation (6) gives the remarkable results shown in the table of Fig. 7.25. Of course, the assumed limiting population of $M = 197.3$ has now been exceeded, so the U.S. population has *not* continued to satisfy the logistic equation in the decades following 1930.

**EXAMPLE 2** Suppose that in 1885 the population of a certain country was 50 million and was growing at the rate of 750,000 people per year at that time. Suppose also that in 1940 its population was 100 million and was then growing at the rate of 1 million per year. Assume that this population satisfies the logistic equation. Determine both the limiting population $M$ and the predicted population for the year 2000.

*Solution* We substitute the two given pairs of data into Equation (5) and find that

$$0.75 = 50k(M - 50),$$

$$1.00 = 100k(M - 100).$$

We solve simultaneously and find that $M = 200$ and $k = 0.0001$. Thus the limiting population of the country in question is 200 million. With these values of $M$ and $k$ and with $t = 0$ corresponding to the year 1940 (in which $P_0 = 100$), we find that, as a consequence of Equation (6), the population in the year 2000 will be

$$P = \frac{(100)(200)}{100 + (200 - 100)e^{-(0.0001)(200)(60)}},$$

or about 153.7 million people.

We describe next some situations that illustrate the varied circumstances in which the logistic equation is a satisfactory mathematical model.

1 *Limited Environment Situation.* A certain environment can support a population of at most $M$ individuals. It's then reasonable to expect the growth rate $\beta - \delta$ (the combined birth and death rates) to be proportional to $M - P$, since we may think of $M - P$ as the "potential" for further expansion. Then $\beta - \delta = k(M - P)$, so that

$$\frac{dP}{dt} = (\beta - \delta)P = kP(M - P).$$

The classic example of a limited environment situation is a fruit fly population in a closed container.

2 *Competition Situation.* If the birth rate $\beta$ is constant but the death rate $\delta$ is proportional to $P$, so that $\delta = \alpha P$, then

$$\frac{dP}{dt} = (\beta - \alpha P)P = kP(M - P).$$

**398**

This might be a reasonable working hypothesis in a study of a cannibalistic population, in which all deaths result from chance encounters between individuals. Of course, competition between individuals is not usually so deadly, nor its effects so immediate and so decisive.

**3** *Joint Proportion Situation.* Let $P(t)$ denote the number of individuals in a constant susceptible population $M$ that is infected with a certain contagious and incurable disease. The disease in question is spread by chance encounters. Then $P'(t)$ should be proportional to the product of the number $P$ of individuals having the disease and the number $M - P$ of those not having it, so that

$$\frac{dP}{dt} = kP(M - P).$$

Again we discover that the mathematical model is the logistic equation. The mathematical description of the spread of a rumor in a population of $M$ individuals is identical.

**EXAMPLE 3** Suppose that at time $t = 0$, half of a population of 100,000 persons have heard a certain rumor and that the number of those who have heard it is then increasing at the rate of 1000 persons per day. How long will it take for this rumor to spread to 80% of the population?

*Solution* Let us work in units of thousands of persons. Then we take $M = 100$ for the total fixed population. We substitute $M = 100$, $P'(0) = 1$, and $P_0 = 50$ into the logistic equation, and thereby obtain

$$1 = k(50)(100 - 50), \quad \text{so that} \quad k = 0.0004.$$

If $t$ denotes the number of days until 80 thousand people have heard the rumor, then Equation (6) gives

$$80 = \frac{(50)(100)}{50 + (100 - 50)e^{-(0.04)t}},$$

so that $t$ is approximately 34.66. Thus the rumor will have spread to 80% of the population in a little less than 35 days.

## CHEMICAL REACTIONS

Consider a chemical reaction of the sort symbolized by

$$mA + nB \longrightarrow pC, \tag{8}$$

in which $m$ molecules of reactant $A$ combine with $n$ molecules of reactant $B$ to form $p$ molecules of the product $C$ (plus, perhaps, other products in which we have no interest). The simplest assumption concerning the rate at which this reaction takes place is that the rate of increase of the concentration $[C]$ of the product $C$ is proportional to the product $[A] \cdot [B]$ of the concentrations of the reactants $A$ and $B$. That is,

$$\frac{d[C]}{dt} = k[A][B] \qquad (k > 0). \tag{9}$$

In such cases we say that Reaction (8) satisfies the **law of mass action.**

Concentrations of the components in a chemical reaction are generally specified in moles per unit volume. One **mole** (mol) of a substance consists of $6.023 \times 10^{23}$ molecules—the number of molecules present in an amount of the substance equal in mass (in grams) to its molecular weight. We consider only reactions taking place in a constant volume, which may then be taken as the unit of volume. The concentration of a substance is, in such a simple case, merely the number of moles of that substance present in the given volume.

Let $x(t)$ denote the concentration of the product $C$ at time $t$. Suppose that the initial concentrations of the reactants $A$ and $B$ are $a$ and $b$, respectively. Suppose also that $x(0) = 0$. Then the concentrations of $A$ and $B$ at time $t$ are

$$[A] = a - \frac{m}{p} x \quad \text{and} \quad [B] = b - \frac{n}{p} x,$$

because it takes $m$ molecules of $A$ and $n$ molecules of $B$ to form $p$ molecules of $C$. In this case, the law of mass action (9) for Reaction (8) becomes

$$\frac{dx}{dt} = k \left( a - \frac{m}{p} x \right) \left( b - \frac{n}{p} x \right). \tag{10}$$

**EXAMPLE 4**  Consider the chemical reaction

$$C_2H_4Br_2 + 3KI \longrightarrow KI_3 + \text{(other products)}$$

between ethylene bromide and potassium iodide. With $A = C_2H_4Br_2$, $B = KI$, and $C = KI_3$, we have $m = p = 1$ and $n = 3$. So Equation (10) becomes

$$\frac{dx}{dt} = k(a - x)(b - 3x). \tag{11}$$

Problems 16 and 17 deal with the solution of this separable differential equation.

## 7-8  PROBLEMS

In each of Problems 1–8, find the solution of the differential equation that satisfies the given initial condition.

1  $\dfrac{dx}{dt} = 2t - 1; \quad x(0) = 3.$

2  $\dfrac{dx}{dt} = 2x - 1; \quad x(0) = 1$

3  $2x \dfrac{dx}{dt} = x^2 + 1; \quad x(0) = 2$

4  $\dfrac{dx}{dt} = e^{t-x}; \quad x(0) = \ln 2$

5  $\dfrac{dx}{dt} = (x - 3)^2; \quad x(0) = -3$

6  $t \dfrac{dx}{dt} = x \ln x; \quad x(1) = e$

7  $\sqrt{t} \dfrac{dx}{dt} = x^2; \quad x(1) = 1$

8  $\sqrt{xt} \dfrac{dx}{dt} = 1; \quad x(1) = 4$

9  Suppose that when a certain lake is stocked with fish, the birth and death rates of the fish population $P(t)$ are each inversely proportional to $\sqrt{P}$, so that $dP/dt = k\sqrt{P}$ ($k$ is a constant).
(a) Show that $P(t) = (\frac{1}{2}kt + \sqrt{P_0})^2$.
(b) If $P_0 = 100$ and after 6 months there are 169 fish in the lake, how many will there be after 1 year?

10  Consider a prolific breed of rabbits for which the birth and death rates are each proportional to the square of the population $P$, so that $dP/dt = kP^2$ ($k > 0$).
(a) Show that

$$P(t) = \frac{P_0}{1 - kP_0 t}.$$

Note that $P(t) \to \infty$ as $t \to 1/kP_0$. This is "doomsday."
(b) Suppose that $P_0 = 2$ and that there are 4 rabbits after 3 months. When does doomsday occur?

Use the algebraic identity

$$\frac{1}{x+a} - \frac{1}{x+b} = \frac{b-a}{(x+a)(x+b)}$$

to find the solution $x(t)$ in Problems 11–13.

**11** $\dfrac{dx}{dt} = x^2 - 1;\quad x(0) = 2$

**12** $\dfrac{dx}{dt} = 4 - x^2;\quad x(0) = 6$

**13** $\dfrac{dx}{dt} = x^2 + 3x + 2;\quad x(0) = 0$

**14** The data in the table of Fig. 7.26 are given for a certain population $P(t)$ that satisfies the logistic equation.

| Year | P (in millions) |
|------|------|
| 1909 | 49.26 |
| 1910 | 50.00 |
| 1911 | 50.76 |
| ⋮ | ⋮ |
| 1939 | 99.02 |
| 1940 | 100.00 |
| 1941 | 101.02 |

**7.26** Population data for Problem 14

(a) What is the limiting population $M$? (*Suggestion:* Use the approximation

$$P'(t) \approx \frac{P(t+h) - P(t-h)}{2h}$$

with $h = 1$ to estimate the values of $dP/dt$ when $P = 50$ and when $P = 100$. Then substitute these values into the logistic equation, and solve for $k$ and $M$.)
(b) Use the values of $k$ and $M$ found in part (a) to determine when $P = 150$. (*Suggestion:* Take $t = 0$ to correspond to the year 1940.)

**15** A country contains 15,000 people who are susceptible to a spreading and contagious disease. At time $t = 0$, the number $N(t)$ of people who have the disease is 5000 and is then increasing at 500 cases per day. Assume that $N(t)$ satisfies the logistic equation. Determine how long it will take for another 5000 people to contract the disease.

**16** Suppose that $b = 3a$ in Example 4 (the reaction between ethylene bromide and potassium iodide), so that Equation (11) takes the form $dx/dt = 3k(a-x)^2$. Derive the solution

$$x(t) = a\left(1 - \frac{1}{3akt + 1}\right).$$

Conclude that $\lim\limits_{t\to\infty} x(t) = a$, and explain why this implies that *both* reactants are used up in the limit.

**17** If $b \neq 3a$ in Equation (11) of Example 4, use the suggestion for Problems 11–13 to derive the solution

$$x(t) = ab\,\frac{1 - e^{k(b-3a)t}}{3a - be^{k(b-3a)t}}.$$

Conclude that $\lim\limits_{t\to\infty} x(t) = a$ if $b > 3a$, while $\lim\limits_{t\to\infty} x(t) = b/3$ if $b < 3a$. Explain these results in terms of the reaction continuing until *one* of the reactants is used up.

**18** Consider the constant-volume gas reaction $A + B \longrightarrow 2C$. A typical example is the combination of hydrogen and chlorine to form the gas hydrogen chloride:

$$H_2 + Cl_2 \longrightarrow 2HCl.$$

Let $x(t)$ denote the concentration of the product $C$ at time $t$, and suppose that $x(0) = 0$. Let $a$ and $b$ denote the initial concentrations of the reactants $A$ and $B$.
(a) Show that the law of mass action for this reaction is $dx/dt = k(a - x/2)(b - x/2)$.
(b) Consider the case in which $a = b = 5$, and suppose that 5 mol of $C$ are produced in the first 2 min of the reaction. What is the value of $k$? How long will it take to produce 8 mol of $C$?
(c) If $a = 10$ and $b = 5$, how long will it take to produce 8 mol of $C$?

## CHAPTER 7 REVIEW:  Definitions, Concepts, Results

Use the list below as a guide to concepts you may need to review.

**1** The laws of exponents
**2** The laws of logarithms
**3** Definition of the natural logarithm
**4** The graph of $y = \ln x$
**5** Definition of the number $e$
**6** Definition of the natural exponential function
**7** The inverse function relationship between $\ln x$ and $e^x$
**8** The graphs of $y = e^x$ and $y = e^{-x}$

**9** Differentiation of $\ln u$ and $e^u$ where $u$ is a differentiable function of $x$
**10** The order of magnitude of $(\ln x)/x^k$ and $x^k/e^x$ as $x \to \infty$
**11** The number $e$ as a limit
**12** Definition of general exponential and logarithm functions
**13** Differentiation of $a^u$ and $\log_a u$
**14** Logarithmic differentiation
**15** Solution of the differential equation $dx/dt = kx$
**16** Natural population growth

**17** Radioactive decay and radiocarbon dating
**18** Solution of a linear first order differential equation with constant coefficients

**19** Solution of separable first order differential equations
**20** Evaluation of the constant of integration in an initial value problem

## MISCELLANEOUS PROBLEMS

Differentiate the given function $f(x)$ in Problems 1–24.

**1** $\ln 2\sqrt{x}$

**2** $e^{-2\sqrt{x}}$

**3** $\ln(x - e^x)$

**4** $10^{\sqrt{x}}$

**5** $\ln(2^x)$

**6** $\log_{10}(2^x)$

**7** $x^3 e^{-1/x^2}$

**8** $x(\ln x)^2$

**9** $(\ln x)[\ln(\ln x)]$

**10** $\exp(10^x)$

**11** $2^{\ln x}$

**12** $\ln\left(\dfrac{e^x + e^{-x}}{e^x - e^{-x}}\right)$

**13** $e^{(x+1)/(x-1)}$

**14** $\ln(\sqrt{1 + x}\sqrt[3]{2 + x^2})$

**15** $\ln[(x-1)/(3-4x^2)]^{3/2}$

**16** $(x \ln x)^{18}$

**17** $\exp\sqrt{1 + \ln x}$

**18** $\dfrac{x}{(\ln x)^2}$

**19** $\ln(\sqrt{x}3^x)$

**20** $(\ln x)^x$

**21** $x^{1/x}$

**22** $x^{\sqrt{\ln x}}$

**23** $(\ln x)^{\ln x}$

**24** $\ln(x^x)$

Evaluate the indefinite integrals in Problems 25–36.

**25** $\displaystyle\int \frac{dx}{1 - 2x}$

**26** $\displaystyle\int \frac{\sqrt{x}\,dx}{1 + x^{3/2}}$

**27** $\displaystyle\int \frac{(3 - x)\,dx}{1 + 6x - x^2}$

**28** $\displaystyle\int \frac{e^x - e^{-x}}{e^x + e^{-x}}\,dx$

**29** $\displaystyle\int \frac{e^x}{2 + e^x}\,dx$

**30** $\displaystyle\int \frac{e^{-1/x^2}}{x^3}\,dx$

**31** $\displaystyle\int \frac{10^{\sqrt{x}}}{\sqrt{x}}\,dx$

**32** $\displaystyle\int \frac{dx}{x(\ln x)^2}$

**33** $\displaystyle\int e^x\sqrt{1 + e^x}\,dx$

**34** $\displaystyle\int \frac{1}{x}\sqrt{1 + \ln x}\,dx$

**35** $\displaystyle\int 2^x 3^x\,dx$

**36** $\displaystyle\int \frac{dx}{x^{1/3}(1 + x^{2/3})}$

Solve the initial value problems in Problems 37–44.

**37** $\dfrac{dx}{dt} = 2t; \quad x(0) = 17$

**38** $\dfrac{dx}{dt} = 2x; \quad x(0) = 17$

**39** $\dfrac{dx}{dt} = e^t; \quad x(0) = 2$

**40** $\dfrac{dx}{dt} = e^x; \quad x(0) = 2$

**41** $\dfrac{dx}{dt} = 3x - 2; \quad x(0) = 3$

**42** $\dfrac{dx}{dt} = x^2 t^2; \quad x(0) = -1$

**43** $\dfrac{dx}{dt} = xe^t; \quad x(0) = e^2$

**44** $\dfrac{dx}{dt} = \sqrt{x}; \quad x(1) = 0$

Sketch the graphs of the curves given in Problems 45–49.

**45** $y = e^{-x}\sqrt{x}$

**46** $y = x - \ln x$

**47** $y = \sqrt{x} - \ln x$

**48** $y = x(\ln x)^2$

**49** $y = e^{-1/x}$

**50** Find the length of the curve $y = \frac{1}{2}x^2 - \frac{1}{4}\ln x$ from $x = 1$ to $x = e$.

**51** A grain warehouse holds $B$ bushels of grain, which is deteriorating in such a way that only $B \cdot 2^{-t/12}$ bushels will be salable after $t$ months. Meanwhile the market price is increasing linearly; after $t$ months it will be $2 + t/12$ dollars per bushel. After how many months should the grain be sold in order to maximize the revenue obtained?

**52** Suppose that you have borrowed \$1000 at 10% annual interest compounded continuously to plant timber on a tract of land and agree to repay this loan when the timber is cut and sold. If the cut timber can be sold after $t$ years for $800 \exp(\frac{1}{2}\sqrt{t})$ dollars, when should you cut and sell in order to maximize your profit?

**53** Suppose that blood samples from 1000 students are to be tested for a certain disease that is known to occur in 1% of the population. Each test costs \$5, so it would cost \$5000 to test the samples individually. Suppose, however, that "lots" made up of $x$ samples each are formed by pooling individual samples, and that these lots are tested first (for \$5 each). Only in case the test on a lot is positive—the probability of this is $1 - (0.99)^x$—will the $x$ samples used to make up this lot be tested individually.

(a) Show that the total expected number of tests is

$$f(x) = \frac{1000}{x}[(1)(0.99)^x + (x + 1)(1 - (0.99)^x)]$$

$$= 1000 + \frac{1000}{x} - 1000(0.99)^x \quad \text{if } x \geq 2.$$

(b) Show that the value of $x$ that minimizes $f(x)$ is a root of the equation

$$x = \frac{(0.99)^{-x/2}}{[\ln(100/99)]^{1/2}}.$$

Since the denominator above is approximately $1/10$, it may be convenient to solve instead the simpler equation $x = 10(0.99)^{-x/2}$. Whichever you choose, the method of repeated substitution (beginning with $x_0 = 10$) is very effective.

(c) From the results above, deduce the cost of using this batch method to test the original 1000 samples.

**54** Deduce from Problem 63 in Section 7-2 that

$$\lim_{x \to 0^+} x^x = 1.$$

**55** Show that

$$\lim_{x \to 0} \frac{\ln(1 + x)}{x} = 1$$

by considering the value of $D \ln x$ for $x = 1$. Thus show that $\ln(1 + x) \approx x$ if $x$ is very close to zero.

**56** Show that

$$\lim_{h \to 0} \frac{1}{h}(a^h - 1) = \ln a$$

by considering the definition of the derivative of $a^x$ at $x = 0$. Substitute $h = 1/n$ to obtain

$$\ln a = \lim_{n \to \infty} n(\sqrt[n]{a} - 1).$$

Approximate $\ln 2$ by taking $n = 1024 = 2^{10}$ and using only the square root key (10 times) on your calculator.

**57** Suppose that the fish population in a lake is attacked by disease at time $t = 0$, with the result that $dP/dt = -3\sqrt{P}$ thereafter ($t$ in weeks). How long will it take for all the fish in the lake to die, assuming that $P_0 = 900$?

**58** A race car sliding along a level surface is decelerated by frictional forces proportional to its speed. Initially it is decelerated at 2 m/s$^2$ and travels a total of 1800 m. What was its initial velocity? See Problem 18 in Section 7-6.

**59** A home mortgage of $120,000 is to be paid off continuously over a period of 25 years. Apply the result of Problem 13 in Section 7-6 to determine the monthly payment required if the annual interest rate is
(a) 8%;
(b) 12%.

**60** A powerboat weighs 32,000 lb, and its motor provides a thrust of 5000 lb. Assume that the water resistance is 100 lb for each foot per second of the boat's speed. Then

$$1000 \frac{dv}{dt} = 5000 - 100v.$$

If the boat starts from rest, what is the maximum velocity that it can attain?

**61** The temperature within a certain freezer is $-16°C$, and the room temperature is a constant $20°C$. At 11 P.M. the

power goes off, and at 6 A.M. the next morning the temperature in the freezer has risen to $-10°C$. At what time will the temperature in the freezer reach the critical value of $0°C$ if the power remains off?

**62** Suppose that the action of fluorocarbons depletes the ozone in the upper atmosphere by 0.25% annually, so that the amount $A$ of ozone in the atmosphere satisfies the differential equation

$$\frac{dA}{dt} = -\frac{1}{400} A \qquad (t \text{ in years}).$$

(a) What percentage of the original amount $A_0$ of atmospheric ozone will remain 25 years from now?
(b) How long will it take for the amount of atmospheric ozone to be reduced to half its initial amount?

**63** Suppose it is discovered experimentally that the amount $x$ in grams of a solute that dissolves in a certain solvent in $t$ seconds satisfies the logistic equation

$$\frac{dx}{dt} = (0.8)x - (0.0004)x^2.$$

(a) What is the maximum amount of the solute that will dissolve in this solvent?
(b) If $x = 50$ when $t = 0$, how long will it take for another 50 g to dissolve?

**64** A car starts from rest and travels along a straight road. Its engine provides a constant acceleration of $a$ feet per second per second. Air resistance and road friction cause a deceleration of $\rho$ feet per second per second for every foot per second of the car's velocity.
(a) Show that after $t$ seconds the car's velocity is

$$v = \frac{a}{\rho}(1 - e^{-\rho t}).$$

(b) If $a = 17.6$ ft/s$^2$ and $\rho = 0.1$, find $v$ when $t = 10$ (s), and also find the limiting velocity as $t \to \infty$. Give each answer in *miles per hour*.

**65** (a) In the notation of Problem 18 in Section 7-8, show that the law of mass action for the reaction

$$2A + B \longrightarrow C$$

takes the form $dx/dt = k(a - 2x)(b - x)$.
(b) Let $a = 10$ and $b = 5$, and suppose that 2 mol of $C$ are produced in the first 40 s. What is the value of $k$? How long will it take to produce 4 mol of $C$?
(c) Suppose that $a = b = 5$. How long will it take to produce 2 mol of $C$?

# 8

# Trigonometric and Hyperbolic Functions

## Introduction

The function $f$ is called an **algebraic function** provided that $y = f(x)$ satisfies an equation of the form

$$a_n(x)y^n + a_{n-1}(x)y^{n-1} + \cdots + a_1(x)y + a_0(x) = 0$$

where the coefficients $a_0(x), a_1(x), \ldots, a_n(x)$ are polynomials in $x$. For example, since the equation $y^2 - p(x) = 0$ has the above form, we see that the square root of the polynomial $p(x)$ is an algebraic function. The equation $q(x)y - p(x) = 0$ also has the necessary form, and so we see that a rational function (a quotient of polynomials) is also an algebraic function.

A function that is *not* algebraic is said to be **transcendental.** The natural logarithm function $\ln x$ and the exponential function $e^x$ are examples of transcendental functions. In this chapter we study the remaining transcendental functions of elementary character—the trigonometric and the hyperbolic functions. These functions have extensive scientific applications and also provide the basis for certain important methods of integration (these methods are discussed in Chapter 9).

## Trigonometric Functions and Limits

In elementary trigonometry, the six basic trigonometric functions of an acute angle $\theta$ in a right triangle are defined as ratios between pairs of sides of the triangle. As in Fig. 8.1,

$$\cos \theta = \frac{\text{adj}}{\text{hyp}}, \qquad \sin \theta = \frac{\text{opp}}{\text{hyp}}, \qquad \tan \theta = \frac{\text{opp}}{\text{adj}}, \tag{1}$$

$$\sec \theta = \frac{\text{hyp}}{\text{adj}}, \qquad \csc \theta = \frac{\text{hyp}}{\text{opp}}, \qquad \cot \theta = \frac{\text{adj}}{\text{opp}}.$$

8.1

We generalize these definitions to *directed* angles of arbitrary size in the following way. Suppose that the initial side of the angle $\theta$ is the positive $x$-axis, so its vertex is at the origin. The angle is **directed** if a direction of rotation from its initial side to its terminal side is specified. We call $\theta$ a **positive angle** if this rotation is counterclockwise and a **negative angle** if this rotation is clockwise.

Let $P(x, y)$ be the point at which the terminal side of $\theta$ intersects the *unit* circle $x^2 + y^2 = 1$. Then we define

$$\cos \theta = x, \qquad \sin \theta = y, \qquad \tan \theta = \frac{y}{x}, \tag{2}$$

$$\sec \theta = \frac{1}{x}, \qquad \csc \theta = \frac{1}{y}, \qquad \cot \theta = \frac{x}{y}.$$

Of course we assume that $x \neq 0$ in the case of $\tan \theta$ and $\sec \theta$ and that $y \neq 0$ in the case of $\cot \theta$ and $\csc \theta$. If the angle $\theta$ is positive and acute, then it

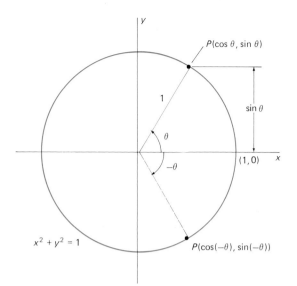

**8.2** Using the unit circle to define the trigonometric functions

is clear from Fig. 8.2 that the definitions in (2) agree with the right triangle definitions in (1) in terms of the coordinates of $P$. A glance at the figure also shows which of the functions are positive for angles in each of the four quadrants. The diagram in Fig. 8.3 summarizes this information.

In this section we discuss primarily the two most basic trigonometric functions, the sine and the cosine. From (2) we see immediately that

$$\tan \theta = \frac{\sin \theta}{\cos \theta}, \qquad \sec \theta = \frac{1}{\cos \theta},$$
$$\cot \theta = \frac{\cos \theta}{\sin \theta}, \qquad \csc \theta = \frac{1}{\sin \theta}. \tag{3}$$

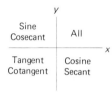

**8.3** The signs of the trigonometric functions

Next, we compare the angles $\theta$ and $-\theta$ in Fig. 8.4. We see that

$$\cos(-\theta) = \cos \theta \quad \text{and} \quad \sin(-\theta) = -\sin \theta. \tag{4}$$

Because $x = \cos \theta$ and $y = \sin \theta$ in (2), the equation $x^2 + y^2 = 1$ of the unit circle translates immediately into the **fundamental identity of trigonometry,**

$$\cos^2\theta + \sin^2\theta = 1. \tag{5}$$

In Problems 49 and 50 of this section, we outline derivations of the **addition formulas**

$$\sin(\alpha + \beta) = \sin \alpha \cos \beta + \cos \alpha \sin \beta, \tag{6}$$

$$\cos(\alpha + \beta) = \cos \alpha \cos \beta - \sin \alpha \sin \beta. \tag{7}$$

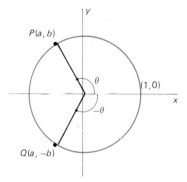

**8.4** Effect of replacing $\theta$ by $-\theta$ in sine and cosine

When we take $\alpha = \theta = \beta$ in (6) and (7), we obtain the **double-angle formulas**

$$\sin 2\theta = 2 \sin \theta \cos \theta, \tag{8}$$

$$\cos 2\theta = \cos^2\theta - \sin^2\theta \tag{9}$$

$$= 2 \cos^2\theta - 1 \tag{9a}$$

$$= 1 - 2 \sin^2\theta, \tag{9b}$$

where (9a) and (9b) are obtained from (9) by use of the fundamental identity in (5).

If we solve (9a) for $\cos^2\theta$ and (9b) for $\sin^2\theta$, we get the **half-angle formulas**

$$\cos^2\theta = \tfrac{1}{2}(1 + \cos 2\theta), \tag{10}$$

$$\sin^2\theta = \tfrac{1}{2}(1 - \cos 2\theta). \tag{11}$$

### RADIAN MEASURE

In elementary mathematics, angles frequently are measured in *degrees*, with $360°$ in one complete revolution. In calculus it is more convenient, and often imperative, to measure angles in *radians*. The **radian measure** of an angle is the length of the arc it subtends in (that is, the arc it cuts out of) the unit circle when the vertex of the angle is at the center of the circle.

Recall that the area $A$ and circumference (or perimeter) $C$ of a circle of radius $r$ are given by the formulas

$$A = \pi r^2 \quad \text{and} \quad C = 2\pi r$$

where the irrational number $\pi$ is approximately 3.14159. Because the circumference of the complete unit circle is $2\pi$ and its central angle is $360°$, it follows that

$$2\pi \text{ radians} = 360°,$$

$$180° = \pi \text{ radians} \approx 3.14159 \text{ radians}. \tag{12}$$

Using (12) we can easily convert back and forth between radians and degrees:

$$1 \text{ radian} = \frac{180°}{\pi} \approx 57°17'44.8'', \tag{12a}$$

$$1° = \frac{\pi}{180} \text{ radians} \approx 0.01745 \text{ radians}. \tag{12b}$$

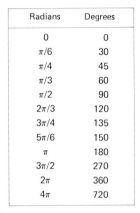

| Radians | Degrees |
|---------|---------|
| 0 | 0 |
| $\pi/6$ | 30 |
| $\pi/4$ | 45 |
| $\pi/3$ | 60 |
| $\pi/2$ | 90 |
| $2\pi/3$ | 120 |
| $3\pi/4$ | 135 |
| $5\pi/6$ | 150 |
| $\pi$ | 180 |
| $3\pi/2$ | 270 |
| $2\pi$ | 360 |
| $4\pi$ | 720 |

**8.5** Some radian-degree conversions

The table in Fig. 8.5 shows radian-degree conversions for some frequently occurring angles.

Now consider an angle of $\theta$ radians placed at the center of a circle of radius $r$, as in Fig. 8.6. Denote by $s$ the length of the arc subtended by $\theta$ and by $A$ the area of the sector of the circle bounded by this angle. Then the evident proportions

$$\frac{s}{2\pi r} = \frac{A}{\pi r^2} = \frac{\theta}{2\pi}$$

give the formulas

$$s = r\theta \quad (\theta \text{ in radians}) \tag{13}$$

and

$$A = \tfrac{1}{2}r^2\theta \quad (\theta \text{ in radians}). \tag{14}$$

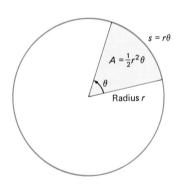

**8.6** Area of a sector and arc length for a circle

The definitions in (2) refer to trigonometric functions of *angles* rather than trigonometric functions of *numbers*. Suppose that $t$ is a real number. Then the number $\sin t$ is, *by definition*, the sine of an angle of $t$ radians—remembering that a positive angle is directed counterclockwise from the positive $x$-axis, while a negative angle is directed clockwise. Briefly, $\sin t$ is the sine of an angle of $t$ *radians*. The other trigonometric functions of the number $t$ have similar definitions. Hence, when we write $\sin t$, $\cos t$, and so on, with $t$ a real number, we *always* intend reference to an angle of $t$ *radians*.

When we need to refer to the sine of an angle of $t$ *degrees*, we will henceforth write $\sin t°$. The point is that $\sin t$ and $\sin t°$ are quite different functions of the variable $t$. For example, you will find that

$$\sin 1° \approx 0.0175 \quad \text{and} \quad \sin 30° = 0.5000$$

with your calculator set in degree mode. But when it is set radian mode, it will report that

$$\sin 1 \approx 0.8415 \quad \text{and} \quad \sin 30 \approx -0.9880.$$

The relationship between the functions $\sin t$ and $\sin t°$ is this:

$$\sin t° = \sin\left(\frac{\pi t}{180}\right). \tag{15}$$

The distinction extends even to programming languages. In FORTRAN, the function SIN is the radian sine function above, and you may need to write $\sin t°$ in the form SIND(T). In BASIC you must write SIN(PI*T/180) to get the correct value of the sine of an angle of $t$ degrees.

An angle of $2\pi$ radians corresponds to 1 revolution. This implies that the sine and cosine functions have **period** $2\pi$, meaning that

$$\begin{aligned} \sin(t + 2\pi) &= \sin t, \\ \cos(t + 2\pi) &= \cos t. \end{aligned} \tag{16}$$

It follows from (16) that

$$\sin(t + 2n\pi) = \sin t, \qquad \cos(t + 2n\pi) = \cos t \tag{17}$$

for any integer $n$. This periodicity of the sine and cosine functions is evident in their graphs, which are shown in Fig. 8.7.

From the definitions of $\sin t$ and $\cos t$, we see that

$$\sin 0 = 0, \qquad \sin \frac{\pi}{2} = 1, \qquad \sin \pi = 0,$$

$$\tag{18}$$

$$\cos 0 = 1, \qquad \cos \frac{\pi}{2} = 0, \qquad \cos \pi = -1.$$

8.7 Periodicity of $\sin t$ and $\cos t$

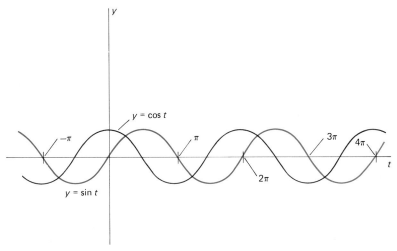

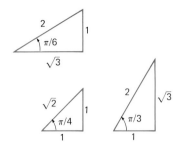

8.8 Familiar right triangles

The trigonometric functions of $\pi/6$, $\pi/4$, and $\pi/3$ (the radian equivalents of $30°$, $45°$, and $60°$, respectively) are easy to read from the well-known triangles of Fig. 8.8. For instance,

$$\sin \frac{\pi}{6} = \cos \frac{\pi}{3} = \frac{1}{2} = \frac{\sqrt{1}}{2},$$

$$\sin \frac{\pi}{4} = \cos \frac{\pi}{4} = \frac{1}{\sqrt{2}} = \frac{\sqrt{2}}{2}, \quad \text{and} \tag{19}$$

$$\sin \frac{\pi}{3} = \cos \frac{\pi}{6} = \frac{\sqrt{3}}{2}.$$

The following example illustrates the use of periodicity and the addition formulas in (6) and (7) to find sines and cosines of angles larger than $\pi/2$.

**EXAMPLE 1**

$$\cos \frac{2\pi}{3} = \cos\left(\pi - \frac{\pi}{3}\right)$$

$$= \cos \pi \cos \frac{\pi}{3} + \sin \pi \sin \frac{\pi}{3}$$

$$= (-1)\left(\frac{1}{2}\right) + (0)\left(\frac{\sqrt{3}}{2}\right) = -\frac{1}{2};$$

$$\sin \frac{5\pi}{4} = \sin\left(\pi + \frac{\pi}{4}\right)$$

$$= \sin \pi \cos \frac{\pi}{4} + \cos \pi \sin \frac{\pi}{4}$$

$$= (0)\left(\frac{\sqrt{2}}{2}\right) + (-1)\left(\frac{\sqrt{2}}{2}\right) = -\frac{\sqrt{2}}{2};$$

$$\sin \frac{17\pi}{6} = \sin\left(2\pi + \frac{5\pi}{6}\right) = \sin \frac{5\pi}{6}$$

$$= \sin\left(\pi - \frac{\pi}{6}\right)$$

$$= \sin \pi \cos \frac{\pi}{6} - \cos \pi \sin \frac{\pi}{6}$$

$$= (0)\left(\frac{\sqrt{3}}{2}\right) - (-1)\left(\frac{1}{2}\right) = \frac{1}{2}.$$

**EXAMPLE 2**  Find the solutions (if any) of the equation

$$\sin^2 x - 3 \cos^2 x + 2 = 0$$

that lie in the interval $[0, \pi]$.

**410**

***Solution*** Using the fundamental identity in (5), we substitute $\cos^2 x = 1 - \sin^2 x$ in the given equation to obtain

$$\sin^2 x - 3(1 - \sin^2 x) + 2 = 0;$$
$$4\sin^2 x - 1 = 0;$$
$$\sin x = \pm\tfrac{1}{2}.$$

Because $\sin x \geq 0$ for $x$ in $[0, \pi]$, $\sin x = -\tfrac{1}{2}$ is impossible. But $\sin x = \tfrac{1}{2}$ for $x = \pi/6$ and for $x = \pi - \pi/6 = 5\pi/6$. These are the solutions of the given equation in $[0, \pi]$.

## LIMITS AND CONTINUITY

We want to show that the sine and cosine functions are continuous. We begin with Fig. 8.9, which shows an angle $\theta$ with its vertex at the origin, its initial side along the positive $x$-axis, and its terminal side intersecting the unit circle at the point $P$. By the definition of the sine and cosine functions, the coordinates of $P$ are $P(\cos \theta, \sin \theta)$. It is geometrically obvious that, as $\theta \to 0$, the point $P(\cos \theta, \sin \theta)$ approaches the point $R(1, 0)$. Hence $\cos \theta \to 1$ and $\sin \theta \to 0$ as $\theta \to 0^+$. A similar picture gives the same results as $\theta \to 0^-$, so we see that

$$\lim_{\theta \to 0} \cos \theta = 1 \quad \text{and} \quad \lim_{\theta \to 0} \sin \theta = 0. \tag{20}$$

But $\cos 0 = 1$ and $\sin 0 = 0$, so these limits imply that the functions $\cos \theta$ and $\sin \theta$ are, by definition, continuous at the point $\theta = 0$. We can use this fact to prove that they are continuous everywhere.

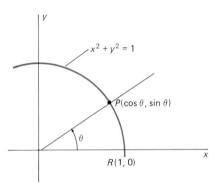

**8.9** An angle $\theta$

---

> **Theorem 1** *Continuity of Sine and Cosine*
> The functions $\sin x$ and $\cos x$ are continuous functions of $x$ on the whole real line.

---

***Proof*** We give the proof only for $\sin x$; the proof for $\cos x$ is similar (Problem 14). We want to show that $\lim_{x \to a} \sin x = \sin a$ for every real number $a$. If we write $x = a + h$, so that $h = x - a$, then $h \to 0$ as $x \to a$, so it will be sufficient if we show that

$$\lim_{h \to 0} \sin(a + h) = \sin a.$$

But the addition formula for the sine function yields

$$\lim_{h \to 0} \sin(a + h) = \lim_{h \to 0} (\sin a \cos h + \cos a \sin h)$$

$$= (\sin a)\left(\lim_{h \to 0} \cos h\right) + (\cos a)\left(\lim_{h \to 0} \sin h\right)$$

$$= \sin a$$

as desired; we used the limits in (20) in the last step. ∎

| $\theta$ | $\dfrac{\sin \theta}{\theta}$ | (Values rounded) |
|---|---|---|
| 1.0 | 0.84147 | |
| 0.1 | 0.99833 | |
| 0.01 | 0.99998 | |
| 0.001 | 1.00000 | |
| 0.0001 | 1.00000 | |

**8.10** The numerical data suggest that $\lim\limits_{\theta \to 0} \dfrac{\sin \theta}{\theta} = 1$.

The limit of $(\sin \theta)/\theta$ as $\theta \to 0$ plays a special role in the calculus of trigonometric functions. A calculator set in *radian mode* provides us with the numerical evidence shown in Fig. 8.10. This table strongly suggests that the limit of $(\sin \theta)/\theta$ is 1 as $\theta \to 0$. The next theorem replaces this suggestion with certainty.

---

**Theorem 2**    **The Basic Trigonometric Limit**

$$\lim_{\theta \to 0} \frac{\sin \theta}{\theta} = 1. \tag{21}$$

---

**Proof**    Figure 8.11 shows the angle $\theta$, the triangles $OPQ$ and $ORS$, and the circular sector $OPR$ that contains the triangle $OPQ$ and is contained in the triangle $ORS$. Hence

$$\text{area}(\triangle OPQ) < \text{area}(\text{sector } OPR) < \text{area}(\triangle ORS).$$

In terms of $\theta$, this means that

$$\frac{1}{2} \sin \theta \cos \theta < \frac{1}{2} \theta < \frac{1}{2} \tan \theta = \frac{\sin \theta}{2 \cos \theta}.$$

Here we use the standard formula for the area of a triangle to obtain the area of $\triangle OPQ$ and $\triangle ORS$. We also use the fact that the area of a circular sector in a circle of radius $r$ is $A = \frac{1}{2}r^2\theta$ if the sector is subtended by an angle of $\theta$ radians; here, we have $r = 1$. If $0 < \theta < \pi/2$, we can divide each member of the above inequality by $\frac{1}{2} \sin \theta$ to obtain

$$\cos \theta < \frac{\theta}{\sin \theta} < \frac{1}{\cos \theta}.$$

We take reciprocals, which reverses the inequalities, and thus

$$\cos \theta < \frac{\sin \theta}{\theta} < \frac{1}{\cos \theta}.$$

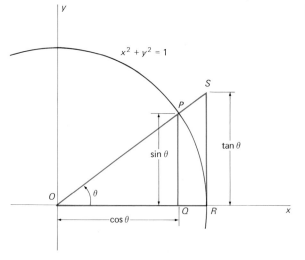

**8.11**    Aid to the proof of Theorem 2

CHAP. 8:   Trigonometric and Hyperbolic Functions

Now we apply the squeeze law of limits with

$$f(\theta) = \cos\theta, \quad g(\theta) = \frac{\sin\theta}{\theta}, \quad \text{and} \quad h(\theta) = \frac{1}{\cos\theta}.$$

Because it is clear from (20) that $f(\theta)$ and $h(\theta)$ approach 1 as $\theta \to 0^+$, so does $g(\theta) = (\sin\theta)/\theta$. This geometric argument shows that $(\sin\theta)/\theta \to 1$ for *positive* values of $\theta$ approaching zero. But the same result follows for negative values of $\theta$ because $\sin(-\theta) = -\sin\theta$. So we have proved (21). ∎

As in the following two examples, many other trigonometric limits can be reduced to the one in Theorem 2.

**EXAMPLE 3** Show that

$$\lim_{\theta \to 0} \frac{1 - \cos\theta}{\theta} = 0. \tag{22}$$

*Solution*

$$\lim_{\theta \to 0} \frac{1 - \cos\theta}{\theta} = \lim_{\theta \to 0} \frac{1 - \cos\theta}{\theta} \frac{1 + \cos\theta}{1 + \cos\theta}$$

$$= \lim_{\theta \to 0} \frac{\sin^2\theta}{\theta(1 + \cos\theta)}$$

$$= \left(\lim_{\theta \to 0} \frac{\sin\theta}{\theta}\right)\left(\lim_{\theta \to 0} \frac{\sin\theta}{1 + \cos\theta}\right)$$

$$= (1)\left(\frac{0}{1 + 1}\right) = 0.$$

Note that in the last step we used all the limits in Equations (20) and (21).

**EXAMPLE 4** Evaluate $\lim\limits_{x \to 0} \dfrac{\tan 3x}{x}$.

*Solution*

$$\lim_{x \to 0} \frac{\tan 3x}{x} = 3 \lim_{x \to 0} \frac{\tan 3x}{3x}$$

$$= 3 \lim_{\theta \to 0} \frac{\tan\theta}{\theta} \qquad (\theta = 3x)$$

$$= 3 \left(\lim_{\theta \to 0} \frac{\sin\theta}{\theta}\right)\left(\lim_{\theta \to 0} \frac{1}{\cos\theta}\right)$$

$$= (3)(1)\left(\frac{1}{1}\right) = 3.$$

We used the fact that $\tan\theta = (\sin\theta)/(\cos\theta)$ as well as some of the limits in (20) and (21).

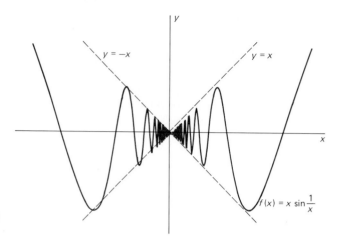

**8.12** The graph of an interesting function

**EXAMPLE 5** Figure 8.12 shows the graph of the function $f$ given by

$$f(x) = \begin{cases} x \sin \dfrac{1}{x} & \text{if} \quad x \neq 0; \\ 0 & \text{if} \quad x = 0. \end{cases}$$

As $x \to 0$, $1/x$ increases without bound, so because of the periodicity of the sine function, $f(x)$ oscillates infinitely often between the values $-x$ and $x$. Because $|\sin(1/x)| \leqq 1$ for all $x$, we see that

$$-|x| \leqq x \sin \frac{1}{x} \leqq |x|$$

for all $x$. Moreover, $|x| \to 0$ as $x \to 0$, so it follows from the squeeze law of limits that

$$\lim_{x \to 0} x \sin \frac{1}{x} = 0.$$

Thus $f$ is continuous at $x = 0$. In Problem 51 we ask you to show that $f$ is *not* differentiable at $x = 0$.

## 8-2 PROBLEMS

**1** Express the following angles in radian measure.

(a) $40°$;    (b) $-270°$;    (c) $315°$;    (d) $210°$;
(e) $-150°$

**2** Express the following angles, given in radian measure, in degrees.

(a) $\dfrac{\pi}{10}$;    (b) $\dfrac{2\pi}{5}$;    (c) $3\pi$;    (d) $\dfrac{15\pi}{4}$;    (e) $\dfrac{23\pi}{60}$

**3** Evaluate the six trigonometric functions of $x$ at the following values.

(a) $x = -\dfrac{\pi}{3}$;    (b) $x = \dfrac{3\pi}{4}$;

(c) $x = \dfrac{7\pi}{6}$;    (d) $x = \dfrac{5\pi}{3}$

**4** Find all numbers $x$ such that

(a) $\sin x = 0$;    (b) $\sin x = 1$;    (c) $\sin x = -1$.

**5** Find all numbers $x$ such that

(a) $\cos x = 0$;    (b) $\cos x = 1$;    (c) $\cos x = -1$.

**6** Find all numbers $x$ such that

(a) $\tan x = 0$;    (b) $\tan x = 1$;    (c) $\tan x = -1$.

**7** Suppose that $\tan x = \frac{3}{4}$ and that $\sin x$ is negative. Find the values of the other five trigonometric functions at $x$.

**8** Suppose that $\csc x = -\frac{5}{3}$ and that $\cos x$ is positive. Find the values of the other five trigonometric functions at $x$.

**9** Deduce from the fundamental identity $\cos^2\theta + \sin^2\theta = 1$ and the definitions of the other four trigonometric functions the identities

(a) $1 + \tan^2\theta = \sec^2\theta$;     (b) $1 + \cot^2\theta = \csc^2\theta$.

**10** Deduce from the addition formulas for the sine and cosine the addition formula for the tangent,

$$\tan(x + y) = \frac{\tan x + \tan y}{1 - \tan x \tan y}.$$

In Problems 11 and 12, use the method of Example 1 to find the indicated values.

**11** (a) $\sin\dfrac{5\pi}{6}$;     (b) $\cos\dfrac{7\pi}{6}$;

    (c) $\sin\dfrac{11\pi}{6}$;     (d) $\cos\dfrac{19\pi}{6}$.

**12** (a) $\sin\dfrac{2\pi}{3}$;     (b) $\cos\dfrac{4\pi}{3}$;

    (c) $\sin\dfrac{5\pi}{3}$;     (d) $\cos\dfrac{10\pi}{3}$.

**13** Suppose that $0 < \theta < \pi/2$. Show that

(a) $\sin(\pi \pm \theta) = \mp\sin\theta$;     (b) $\cos(\pi \pm \theta) = -\cos\theta$.

**14** Establish the other half of Theorem 1; that is, that $\cos x$ is continuous everywhere.

In each of Problems 15–20, find all solutions of the given equation that lie in the interval $[0, \pi]$.

**15** $3\sin^2 x - \cos^2 x = 2$

**16** $\sin^2 x = \cos^2 x$

**17** $2\cos^2 x + 3\sin x = 3$

**18** $2\sin^2 x + \cos x = 2$

**19** $8\sin^2 x \cos^2 x = 1$

**20** $\cos 2\theta - 3\cos\theta + 2 = 0$

Find the limits in Problems 21–45.

**21** $\displaystyle\lim_{\theta \to 0} \frac{\theta^2}{\sin\theta}$

**22** $\displaystyle\lim_{\theta \to 0} \frac{\sin^2\theta}{\theta^2}$

**23** $\displaystyle\lim_{\theta \to 0} \frac{1 - \cos\theta}{\theta^2}$

**24** $\displaystyle\lim_{\theta \to 0} \frac{\tan\theta}{\theta}$

**25** $\displaystyle\lim_{x \to 0} \frac{2x}{(\sin x) - x}$

**26** $\displaystyle\lim_{\theta \to 0} \frac{\sin(2\theta^2)}{\theta^2}$

**27** $\displaystyle\lim_{x \to 0} \frac{\sin 5x}{x}$

**28** $\displaystyle\lim_{x \to 0} \frac{\sin 2x}{x\cos 3x}$

**29** $\displaystyle\lim_{x \to 0} \frac{\sin x}{x^{1/2}}$

**30** $\displaystyle\lim_{x \to 0} \frac{1 - \cos 2x}{x}$

**31** $\displaystyle\lim_{x \to 0} \frac{1}{x}\sin\frac{x}{3}$

**32** $\displaystyle\lim_{x \to 0} \frac{(\sin 3x)^2}{x^2\cos x}$

**33** $\displaystyle\lim_{x \to 0} \frac{1 - \cos x}{\sin x}$

**34** $\displaystyle\lim_{x \to 0} \frac{\tan 3x}{\tan 5x}$

**35** $\displaystyle\lim_{x \to 0} x\sec x \csc x$

**36** $\displaystyle\lim_{\theta \to 0} \frac{\sin 2\theta}{\theta}$

**37** $\displaystyle\lim_{\theta \to 0} \frac{1 - \cos\theta}{\theta\sin\theta}$

**38** $\displaystyle\lim_{\theta \to 0} \frac{\sin^2\theta}{\theta}$

**39** $\displaystyle\lim_{x \to 0} \frac{\tan x}{x}$

**40** $\displaystyle\lim_{x \to 0} \frac{\tan 2x}{3x}$

**41** $\displaystyle\lim_{x \to 0} x\cot 3x$

**42** $\displaystyle\lim_{x \to 0} \frac{x - \tan x}{\sin x}$

**43** $\displaystyle\lim_{x \to 0} \frac{1}{x^2}\sin^2\!\left(\frac{x}{2}\right)$

**44** $\displaystyle\lim_{x \to 0} \frac{\sin 2x}{\sin 5x}$

**45** $\displaystyle\lim_{x \to 0} x^2\csc 2x \cot 2x$

Use the squeeze law of limits to find the limits in Problems 46–48.

**46** $\displaystyle\lim_{x \to 0} x^2\sin\frac{1}{x^2}$

**47** $\displaystyle\lim_{x \to 0} x^2\cos(x^{-1/3})$

**48** $\displaystyle\lim_{x \to 0^+} \sqrt{x}\,\sin\frac{1}{x}$

**49** The points $A(\cos\theta, -\sin\theta)$, $B(1, 0)$, $C(\cos\phi, \sin\phi)$, and $D(\cos(\theta + \phi), \sin(\theta + \phi))$ are shown in Fig. 8.13; all

**8.13** Deriving the cosine addition formula (Problem 49)

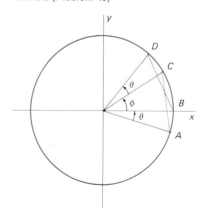

are points on the unit circle. Deduce from the fact that the line segments $AC$ and $BD$ have equal length (because they subtend the same angle, $\theta + \phi$) that

$$\cos(\theta + \phi) = \cos\theta\cos\phi - \sin\theta\sin\phi.$$

**50** (a) Use the triangles shown in Fig. 8.14 to deduce that

$$\sin\left(\theta + \frac{\pi}{2}\right) = \cos\theta \text{ and } \cos\left(\theta + \frac{\pi}{2}\right) = -\sin\theta.$$

(b) Use the results of Problem 49 and part (a) to derive the addition formula for the sine function.

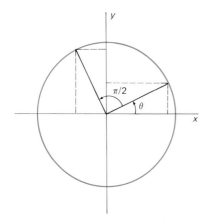

**8.14** The identities of Problem 50

**51** Show that the function $f(x) = x\sin(1/x)$ of Example 5 is *not* differentiable at $x = 0$. (*Suggestion:* Show that

whether $z = +1$ or $z = -1$, there are arbitrarily small values of $h$ such that $[f(h) - f(0)]/h = z$.)

**52** Let $f(x) = x^2\sin(1/x)$ for $x \neq 0$; $f(0) = 0$ (the graph of $f$ is shown in Fig. 8.15). Apply the definition of the derivative to show that $f$ is differentiable at $x = 0$ and that $f'(0) = 0$.

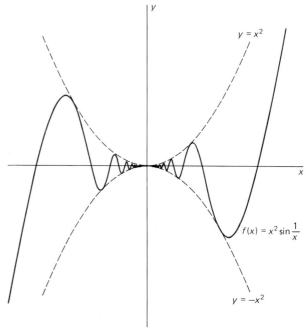

**8.15** Graph of the function of Problem 52

---

## Calculus of Sines and Cosines

In this section we begin our study of the calculus of trigonometric functions, concentrating mainly on the sine and cosine functions. Recall that $\sin x$ (or $\cos x$) means the sine (or cosine) of an angle of $x$ *radians*. The derivatives of the sine and cosine functions depend on the limits

$$\lim_{\theta \to 0} \frac{\sin\theta}{\theta} = 1, \quad \lim_{\theta \to 0} \frac{1 - \cos\theta}{\theta} = 0 \tag{1}$$

that we established in Section 8-2; see Theorem 2 and Example 3 there.

---

*Theorem*    *Derivatives of Sines and Cosines*

The functions $\sin x$ and $\cos x$ are differentiable everywhere, with

$$D\sin x = \cos x, \tag{2}$$

$$D\cos x = -\sin x. \tag{3}$$

---

**Proof** To differentiate $f(x) = \sin x$, we begin with the definition of the derivative,

$$f'(x) = \lim_{h \to 0} \frac{f(x + h) - f(x)}{h}$$

$$= \lim_{h \to 0} \frac{\sin(x + h) - \sin x}{h}.$$

Next we apply the addition formula for the sine and the limit laws to get

$$f'(x) = \lim_{h \to 0} \frac{(\sin x \cos h + \sin h \cos x) - \sin x}{h}$$

$$= \lim_{h \to 0} \left[ (\cos x) \frac{\sin h}{h} - (\sin x) \frac{1 - \cos h}{h} \right]$$

$$= (\cos x)\left( \lim_{h \to 0} \frac{\sin h}{h} \right) - (\sin x)\left( \lim_{h \to 0} \frac{1 - \cos h}{h} \right).$$

The limits in (1) now yield

$$f'(x) = (\cos x)(1) - (\sin x)(0) = \cos x,$$

which proves (2). The proof of (3) is quite similar (Problem 68). ■

The following four examples illustrate the application of (2) and (3) in conjunction with the general differentiation formulas of Chapter 3 to differentiate various combinations of trigonometric functions.

**EXAMPLE 1**

$$D(x^2 \sin x) = (Dx^2)(\sin x) + (x^2)(D \sin x)$$
$$= 2x \sin x + x^2 \cos x.$$

**EXAMPLE 2**

$$D(\cos^3 t) = D[(\cos t)^3]$$
$$= 3(\cos t)^2 D(\cos t)$$
$$= 3(\cos^2 t)(-\sin t) = -3 \cos^2 t \sin t.$$

**EXAMPLE 3** If $y = \dfrac{\cos x}{1 - \sin x}$, then

$$\frac{dy}{dx} = \frac{(D \cos x)(1 - \sin x) - (\cos x)D(1 - \sin x)}{(1 - \sin x)^2}$$

$$= \frac{(-\sin x)(1 - \sin x) - (\cos x)(-\cos x)}{(1 - \sin x)^2}$$

$$= \frac{-\sin x + \sin^2 x + \cos^2 x}{(1 - \sin x)^2}$$

$$= \frac{-\sin x + 1}{(1 - \sin x)^2};$$

$$\frac{dy}{dx} = \frac{1}{1 - \sin x}.$$

**EXAMPLE 4**  If $g(t) = (2 - 3 \cos t)^{3/2}$, then

$$g'(t) = \tfrac{3}{2}(2 - 3 \cos t)^{1/2}D(2 - 3 \cos t)$$
$$= \tfrac{3}{2}(2 - 3 \cos t)^{1/2}(3 \sin t);$$
$$g'(t) = \tfrac{9}{2}(2 - 3 \cos t)^{1/2}\sin t.$$

It is easy to differentiate the other four trigonometric functions because each of them is defined in terms of the sine and cosine. For instance, $\tan x = (\sin x)/(\cos x)$, so

$$D \tan x = \frac{(D \sin x)(\cos x) - (\sin x)(D \cos x)}{(\cos x)^2}$$

$$= \frac{(\cos x)(\cos x) - (\sin x)(-\sin x)}{\cos^2 x}$$

$$= \frac{\cos^2 x + \sin^2 x}{\cos^2 x} = \frac{1}{\cos^2 x};$$

$$D \tan x = \sec^2 x \tag{4}$$

because $\sec x = 1/\cos x$. In a similar way, we can derive the formulas

$$D \cot x = -\csc^2 x, \tag{5}$$

$$D \sec x = \sec x \tan x, \tag{6}$$

$$D \csc x = -\csc x \cot x. \tag{7}$$

Although we will have occasion to differentiate these other trigonometric functions, more extensive discussion of them is deferred to Section 8-4.

By now it is a familiar fact that the chain rule in the form

$$D_x g(u) = g'(u)\frac{du}{dx} \tag{8}$$

provides a chain rule version of each new differentiation formula that we learn. For the sine and cosine functions these chain rule versions are

$$D_x \sin u = (\cos u)\frac{du}{dx} \tag{9}$$

and

$$D_x \cos u = (-\sin u)\frac{du}{dx} \tag{10}$$

where $u$ denotes a differentiable function of $x$. The cases in which $u = kx$ (where $k$ is a constant) are worth noting specifically:

$$D_x \sin kx = k \cos kx \quad \text{and} \quad D_x \cos kx = -k \sin kx. \tag{11}$$

The formulas in (11) provide an explanation of why radian measure is more natural than degree measure. Recall from Equation (15) in Section 8-2 that the sine of an angle of $x$ degrees is given by

$$\sin x^\circ = \sin \frac{\pi x}{180}.$$

Hence the first formula in (11) yields

$$D \sin x° = \frac{\pi}{180} \cos \frac{\pi x}{180},$$

so

$$D \sin x° \approx (0.01745) \cos x°.$$

The necessity of using the approximate value 0.01745 here, and indeed its very presence, is one reason that radians instead of degrees are used in the calculus of trigonometric functions.

**EXAMPLE 5**

$$
\begin{aligned}
D(\sin^2 3x \cos^4 5x) &= 2(\sin 3x)(D \sin 3x) \cdot (\cos^4 5x) \\
&\quad + (\sin^2 3x) \cdot 4(\cos^3 5x)(D \cos 5x) \\
&= 2(\sin 3x)(3 \cos 3x) \cdot (\cos^4 5x) \\
&\quad + (\sin^2 3x) \cdot 4(\cos^3 5x)(-5 \sin 5x) \\
&= 6 \sin 3x \cos 3x \cos^4 5x - 20 \sin^2 3x \sin 5x \cos^3 5x.
\end{aligned}
$$

**EXAMPLE 6**   Differentiate $\cos \sqrt{x}$.

*Solution*   If $u = \sqrt{x}$, then $du/dx = 1/(2\sqrt{x})$, so (10) yields

$$D_x \cos \sqrt{x} = D_x \cos u = (-\sin u)\frac{du}{dx}$$

$$= -(\sin \sqrt{x})\frac{1}{2\sqrt{x}} = -\frac{\sin \sqrt{x}}{2\sqrt{x}}.$$

Alternatively, we can carry out this computation without introducing $u$, by writing

$$D \cos \sqrt{x} = (-\sin \sqrt{x})D(\sqrt{x}) = -\frac{\sin \sqrt{x}}{2\sqrt{x}}.$$

**EXAMPLE 7**   Differentiate

$$y = \sin^2(2x - 1)^{3/2} = [\sin(2x - 1)^{3/2}]^2.$$

*Solution*   Here, $y = u^2$ where $u = \sin(2x - 1)^{3/2}$, so

$$\frac{dy}{dx} = 2u\frac{du}{dx}$$

$$
\begin{aligned}
&= 2[\sin(2x - 1)^{3/2}]D_x[\sin(2x - 1)^{3/2}] \\
&= 2[\sin(2x - 1)^{3/2}][\cos(2x - 1)^{3/2}]D(2x - 1)^{3/2} \\
&= 2[\sin(2x - 1)^{3/2}][\cos(2x - 1)^{3/2}]\tfrac{3}{2}(2x - 1)^{1/2}(2) \\
&= 6(2x - 1)^{1/2}[\sin(2x - 1)^{3/2}][\cos(2x - 1)^{3/2}].
\end{aligned}
$$

The following three examples illustrate the applications of trigonometric functions to rate of change and maximum-minimum problems.

**EXAMPLE 8**  A rocket is launched vertically and is tracked by an observing station located on the ground 5 mi from the launch pad. Suppose that the elevation angle $\theta$ of the line of sight to the rocket is increasing at 3°/s when $\theta = 60°$. What is the velocity of the rocket at this instant?

*Solution*  First we convert the given data from degrees to radians. We are given that

$$\frac{d\theta}{dt} = 3\frac{\text{deg}}{\text{s}} \times \frac{\pi}{180}\frac{\text{rad}}{\text{deg}} = \frac{\pi}{60}\frac{\text{rad}}{\text{s}} \qquad \text{when } \theta = 60° = \frac{\pi}{3} \text{ rads.}$$

From Figure 8.16 we see that the height $y$ (in miles) of the rocket is

$$y = 5 \tan \theta = 5\frac{\sin \theta}{\cos \theta}.$$

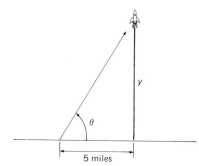

Hence its velocity is

$$\frac{dy}{dt} = \frac{dy}{d\theta}\frac{d\theta}{dt}$$

$$= 5\frac{(\cos \theta)(\cos \theta) - (\sin \theta)(-\sin \theta)}{\cos^2\theta} \cdot \frac{d\theta}{dt} = 5(\sec^2\theta)\frac{d\theta}{dt}.$$

Since $\sec(\pi/3) = 2$, the velocity of the rocket is

$$\frac{dy}{dt} = 5(2)^2\left(\frac{\pi}{60}\right) = \frac{\pi}{3}$$

**8.16**  Tracking an ascending rocket

mi/s, or about 3770 mi/h, at the time when $\theta = 60°$.

**EXAMPLE 9**  Find the maximal area of a rectangle inscribed in a semicircle of radius $R$, as shown in Fig. 8.17.

*Solution*  If we denote half the base of the rectangle by $x$ and its height by $y$, then its area is $A = 2xy$. We see in Fig. 8.17 that the right triangle has hypotenuse $R$, the radius of the circle. So

$$x = R \cos \theta, \qquad y = R \sin \theta. \tag{12}$$

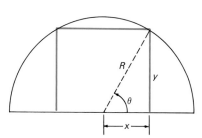

Each value of $\theta$ between 0 and $\pi/2$ corresponds to a possible inscribed rectangle; the values $\theta = 0$ and $\theta = \pi/2$ yield degenerate rectangles.

We substitute (12) in the formula $A = 2xy$ to obtain the area

$$A(\theta) = 2(R \cos \theta)(R \sin \theta)$$

$$= 2R^2(\cos \theta)(\sin \theta) \tag{13}$$

**8.17**  The rectangle of Example 9

as a function of $\theta$ on the closed interval $[0, \pi/2]$. To find the critical points, we differentiate:

$$\frac{dA}{d\theta} = 2R^2(-\sin \theta)(\sin \theta) + 2R^2(\cos \theta)(\cos \theta)$$

$$= 2R^2(\cos^2\theta - \sin^2\theta). \tag{14}$$

Thus we have a critical point only if

$$\cos^2\theta - \sin^2\theta = 0;$$
$$\sin^2\theta = \cos^2\theta;$$
$$\tan^2\theta = 1;$$
$$\tan\theta = \pm 1.$$

The only value of $\theta$ in $[0, \pi/2]$ such that $\tan\theta = \pm 1$ is $\theta = \pi/4$.

Upon evaluation of $A(\theta)$ at each of the possibilities $\theta = 0, \pi/4, \pi/2$ (the critical point and the two endpoints), we find that

$$A(0) = 0,$$

$$A\left(\frac{\pi}{4}\right) = 2R^2\left(\frac{1}{\sqrt{2}}\right)\left(\frac{1}{\sqrt{2}}\right) = R^2, \quad \longleftarrow \text{Maximum}$$

$$A\left(\frac{\pi}{2}\right) = 0.$$

Thus the area of the largest inscribed rectangle is $R^2$; its dimensions are $2x = R\sqrt{2}$ and $y = R/\sqrt{2}$.

**EXAMPLE 10**  Find the length of the longest rod that can be carried horizontally around the corner from a hall 2 m wide into one that is 4 m wide.

*Solution*  The desired length will actually be the *minimum* length $L$ of the dashed line in Fig. 8.18, which represents the rod being carried around the corner. From the two similar triangles in the figure we see that

$$L = L_1 + L_2 = 4\csc\theta + 2\sec\theta,$$

and thus

$$L = L(\theta) = \frac{4}{\sin\theta} + \frac{2}{\cos\theta}.$$

The domain of $L$ is the open interval $0 < \theta < \pi/2$. Clearly $L \to +\infty$ as either $\theta \to 0^+$ or $\theta \to (\pi/2)^-$.

We differentiate:

$$\frac{dL}{d\theta} = -\frac{4\cos\theta}{\sin^2\theta} + \frac{2\sin\theta}{\cos^2\theta} = \frac{2\sin^3\theta - 4\cos^3\theta}{\sin^2\theta\cos^2\theta},$$

so $dL/d\theta = 0$ when

$$2\sin^3\theta = 4\cos^3\theta,$$

$$\tan\theta = \sqrt[3]{2},$$

or $\theta \approx 0.90$ rads.

It's clear that $dL/d\theta < 0$ when $\theta < 0.90$, and that $dL/d\theta > 0$ when $\theta > 0.90$. So the graph of $L$ looks as indicated in Fig. 8.19. This means that the minimum value of $L$, and therefore the maximum length of the rod in question, is about

$$L(0.90) = \frac{4}{\sin(0.90)} + \frac{2}{\cos(0.90)};$$

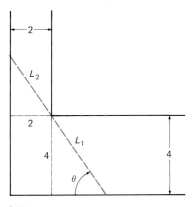

**8.18**  Carrying a rod around a corner

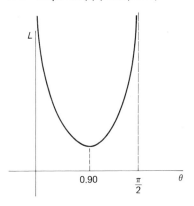

**8.19**  Graph of $L(\theta)$ (Example 10)

that is, approximately 8.32 m. This method of solution should be compared with that in Example 4 of Section 4-4, where we solved a similar problem without the use of trigonometric functions.

### INTEGRALS INVOLVING SINES AND COSINES

The integral versions of the chain rule differentiation formulas in (9) and (10) are

$$\int \cos u \, du = \sin u + C \tag{15}$$

and

$$\int \sin u \, du = -\cos u + C. \tag{16}$$

**EXAMPLE 11**  With $u = 2x$ and $du = 2 \, dx$, (16) gives

$$\int \sin 2x \, dx = \tfrac{1}{2} \int (\sin 2x)(2 \, dx)$$

$$= \tfrac{1}{2} \int \sin u \, du$$

$$= -\tfrac{1}{2} \cos u + C$$

$$= -\tfrac{1}{2} \cos 2x + C.$$

**EXAMPLE 12**  The substitution $u = x^{3/2}$, $du = \tfrac{3}{2}x^{1/2} \, dx$ yields

$$\int \sqrt{x} \, \cos x^{3/2} \, dx = \tfrac{2}{3} \int (\cos x^{3/2})(\tfrac{3}{2}x^{1/2} \, dx)$$

$$= \tfrac{2}{3} \int \cos u \, du$$

$$= \tfrac{2}{3} \sin u + C$$

$$= \tfrac{2}{3} \sin x^{3/2} + C.$$

In the following two examples we have integrals of *functions* of sines and cosines, rather than simply the forms in (15) and (16).

**EXAMPLE 13**  Evaluate $\int_0^{\pi/4} \sin^3 2t \, \cos 2t \, dt$.

*Solution*  We substitute

$$u = \sin 2t, \quad \text{so that} \quad du = 2 \cos 2t \, dt.$$

Then

$$u = 0 \quad \text{when} \quad t = 0; \quad u = 1 \quad \text{when} \quad t = \frac{\pi}{4}.$$

Hence

$$\int_0^{\pi/4} \sin^3 2t \, \cos 2t \, dt = \tfrac{1}{2} \int_0^1 u^3 \, du$$

$$= \tfrac{1}{2} \left[ \tfrac{1}{4} u^4 \right]_0^1 = \tfrac{1}{8}.$$

**EXAMPLE 14**  Evaluate $\displaystyle\int \frac{\sin t}{2 - \cos t}\, dt$.

**Solution**  We substitute

$$u = 2 - \cos t, \quad \text{so that} \quad du = \sin t\, dt.$$

Then

$$\int \frac{\sin t}{2 - \cos t}\, dt = \int \frac{du}{u}$$

$$= \ln|u| + C$$

$$= \ln|2 - \cos t| + C$$

$$= \ln(2 - \cos t) + C.$$

The last simplification is possible because $2 - \cos t > 0$ for all $t$.

## 8-3  PROBLEMS

Differentiate the functions given in Problems 1–20.

1  $f(x) = 3\sin^2 x$  

2  $f(x) = 2\cos^4 x$

3  $f(x) = x\cos x$  

4  $f(x) = \sqrt{x}\,\sin x$

5  $f(x) = \dfrac{\sin x}{x}$  

6  $f(x) = \dfrac{\cos x}{\sqrt{x}}$

7  $f(x) = \sin x \cos^2 x$  

8  $f(x) = \cos^3 x \sin^2 x$

9  $g(t) = (1 + \sin t)^4$  

10  $g(t) = (2 - \cos^2 t)^3$

11  $g(t) = \dfrac{1}{\sin t + \cos t}$  

12  $g(t) = \dfrac{\sin t}{1 + \cos t}$

13  $f(x) = 2x\sin x - 3x^2\cos x$

14  $f(x) = x^{1/2}\cos x - x^{-1/2}\sin x$

15  $f(x) = \cos 2x \sin 3x$  

16  $f(x) = \cos 5x \sin 7x$

17  $g(t) = t^3 \sin^2 2t$  

18  $g(t) = \sqrt{t}\,\cos^3 3t$

19  $g(t) = (\cos 3t + \cos 5t)^{5/2}$

20  $g(t) = \dfrac{1}{\sqrt{\sin^2 t + \sin^2 3t}}$

Find $dy/dx$ in Problems 21–40.

21  $y = \sin^2(\sqrt{x})$  

22  $y = \dfrac{\cos 2x}{x}$

23  $y = x^2\cos(3x^2 - 1)$  

24  $y = \sin^3(x^4)$

25  $y = (\sin 2x)(\cos 3x)$  

26  $y = \dfrac{x}{\sin 3x}$

27  $y = \dfrac{\cos 3x}{\sin 5x}$  

28  $y = \sqrt{\cos\sqrt{x}}$

29  $y = \sin^2(x^2)$  

30  $y = \cos^3(x^3)$

31  $y = \sin 2\sqrt{x}$  

32  $y = \cos 3x^{1/3}$

33  $y = x\sin x^2$  

34  $y = x^2\cos\left(\dfrac{1}{x}\right)$

35  $y = \sqrt{x}\,\sin\sqrt{x}$  

36  $y = (\sin x - \cos x)^2$

37  $y = \sqrt{x}(x - \cos x)^3$  

38  $y = \sqrt{x}\,\sin\sqrt{x + \sqrt{x}}$

39  $y = \cos(\sin x^2)$  

40  $y = \sin(1 + \sqrt{\sin x})$

Differentiate the functions given in Problems 41–50.

41  $f(x) = \cos(\ln x)$  

42  $f(x) = \ln(2\sin x)$

43  $f(x) = \ln(\cos x)$  

44  $f(x) = \sin(\ln 2x)$

45  $f(t) = t^2\ln(\cos t)$  

46  $g(t) = \sqrt{t}\,[\cos(\ln t)]^2$

47  $f(x) = \sin(2e^x)$  

48  $f(x) = e^x\cos 2x$

49  $g(t) = e^{\cos t}$  

50  $f(x) = \cos(e^x + e^{-x})$

Evaluate the integrals in Problems 51–66.

51  $\displaystyle\int \cos 3x\, dx$  

52  $\displaystyle\int \sin(5t + 7)\, dt$

53  $\displaystyle\int_0^1 \sin \pi x\, dx$  

54  $\displaystyle\int_0^{1/3} \cos 2\pi t\, dt$

55  $\displaystyle\int \sin\frac{t}{2}\, dt$  

56  $\displaystyle\int \frac{\cos\sqrt{x}}{\sqrt{x}}\, dx$

57  $\displaystyle\int_0^{\pi/6} \sin 2x \cos^3 2x\, dx$  

58  $\displaystyle\int_0^{\pi/2} \sin x \cos x\, dx$

59  $\displaystyle\int_0^{\pi/2} (1 + \sin t)^{3/2}\cos t\, dt$

60  $\displaystyle\int_0^{\sqrt{\pi}} x\sin\frac{x^2}{2}\, dx$  

61  $\displaystyle\int \frac{1}{t^2}\sin\frac{1}{t}\, dt$

62  $\displaystyle\int_0^{\pi/4} \frac{\sin t}{\sqrt{\cos t}}\, dt$  

63  $\displaystyle\int \frac{\sin 2x}{1 - \cos 2x}\, dx$

64  $\displaystyle\int \frac{\cos x}{1 + \sin x}\, dx$  

65  $\displaystyle\int (\sin 2x)e^{1 - \cos 2x}\, dx$

66  $\displaystyle\int (\cos x)e^{\sin x}\, dx$

**67** Derive the differentiation formulas in (5)–(7).

**68** Use the definition of the derivative to show directly that $D \cos x = -\sin x$.

**69** If a projectile is fired from ground level with initial velocity $v_0$ and inclination angle $\alpha$, then its horizontal range (ignoring air resistance) is $R = \frac{1}{16}v_0^2 \sin \alpha \cos \alpha$. What value of $\alpha$ maximizes $R$?

**70** A weather balloon is rising vertically and is being observed from a point on the ground 300 ft from the spot directly beneath the balloon. At what rate is the balloon rising when the angle between the ground and the observer's line of sight is $45°$ and is increasing at $1°/s$?

**71** A rocket is launched vertically upward from a point 2 mi west of an observer on the ground. What is the speed of the rocket when the angle of elevation (from the horizontal) of the observer's line of sight to the rocket is $50°$ and is increasing at $5°/s$?

**72** A plane flying at an altitude of 25,000 ft has an inoperative airspeed indicator. In order to determine his speed, the pilot sights a fixed point on the ground. At the moment when the angle of depression (from the horizontal) is $65°$, he observes that this angle is increasing at $1.5°/s$. What is the plane's speed?

**73** An observer on the ground sights an approaching plane that is flying at a constant speed and at an altitude of 20,000 feet. From her point of view, the plane's angle of elevation is increasing at $0.5°/s$ when the angle is $60°$. What is the plane's speed?

**74** (a) Verify by differentiation that

$$\int x \sin x \, dx = \sin x - x \cos x + C.$$

(b) Find the volume of the solid obtained by rotating about the $y$-axis the area under $y = \sin x$ from $x = 0$ to $x = \pi$.

**75** Use the integral formula in Problem 74 to find the volume of the solid generated by revolving the first-quadrant region bounded by $y = x$ and $y = \sin(\pi x/2)$ around the $y$-axis.

**76** Show that the length of one arch of the sine curve $y = \sin x$ is equal to half the circumference of the ellipse

$2x^2 + y^2 = 2$. (*Suggestion:* Substitute $x = \cos \theta$ into the arc length integral for the ellipse.)

**77** Use Simpson's approximation with $n = 6$ subintervals to estimate the length of the sine arch of Problem 76.

**78** Find the largest possible area $A$ of a rectangle inscribed in the unit circle by maximizing $A$ as a function of the angle $\theta$ indicated in Fig. 8.20.

**79** A water trough is to be made from a long strip of tin 6 ft wide by bending up at an angle $\theta$ a 2-ft strip on each side, as shown in Fig. 8.21. What should be the value of the angle $\theta$ to maximize the cross-sectional area, and thus the volume, of the trough?

**8.21** The water trough of Problem 79

**80** A circular area of radius 20 m is surrounded by a walkway, and a light is placed atop a lamp post at the center. At what height should the light be placed to illuminate the walkway most strongly? The intensity $I$ of illumination of a surface is given by $I = (k \sin \theta)/D^2$, where $D$ is the distance from the light source to the surface, $\theta$ is the angle at which light strikes the surface, and $k$ is a positive constant.

**81** Find the minimum possible volume $V$ of a cone in which a sphere of given radius $R$ is inscribed. Minimize $V$ as a function of the angle $\theta$ indicated in Fig. 8.22.

**8.22** Finding the smallest cone containing a fixed sphere

**82** A very long rectangular piece of paper is 20 cm wide. The bottom right-hand corner is folded along the crease

**8.20** A rectangle inscribed in the unit circle

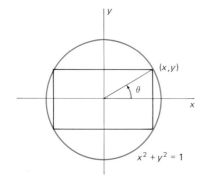

shown as a heavy line in Fig. 8.23, so that the corner just touches the left-hand side of the page. How should this be done so that the crease is as short as possible?

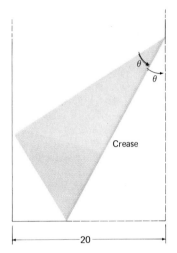

**8.23** Fold a piece of paper; make the crease of minimal length

**83** Find the maximum possible area $A$ of the trapezoid inscribed in a semicircle of radius 1, as shown in Fig. 8.24. Begin by expressing $A$ as a function of $\theta$.

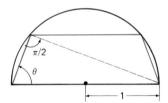

**8.24** A trapezoid inscribed in a circle

**8.25** A hexagonal beam cut from a circular log

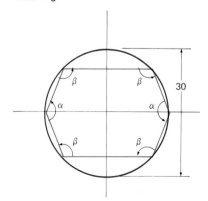

**84** A six-sided beam is to be cut from a circular log with diameter 30 cm, so that its cross section is as shown in Fig. 8.25; the beam is highly symmetrical, with only two different angles, as the figure shows. Show that the area of the cross section is maximal when it is a regular hexagon, with equal sides and angles (corresponding to $\alpha = \beta = 2\pi/3$). Note that $\alpha + 2\beta = 2\pi$. (Why?)

**85** Consider a circular arc of given length $s$ and with its end points on the $x$-axis. Show that the area $A$ bounded by this arc and the $x$-axis is maximal when the circular arc is in the shape of a semicircle. (*Suggestion:* Express $A$ in terms of the angle $\theta$ subtended by the arc at the center of the circle, as shown in Fig. 8.26. Show that $A$ is maximal when $\theta = \pi$.)

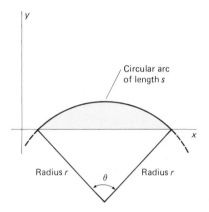

**8.26** Finding the maximum area bounded by a circular arc and its chord

**86** A hiker starting at a point $P$ on a straight road wants to get to a forest cabin that is 2 km from the point $Q$, which is 3 km down the road from $P$, as shown in Fig. 8.27. She can walk 8 km/h along the road, but only 3 km/h through the forest. She wants to minimize the time required to reach the cabin. How far down the road should she walk first before setting off through the forest straight for the cabin? (*Suggestion:* Use the angle between the road and the path she takes through the forest as an independent variable.)

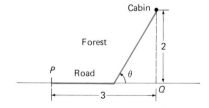

**8.27** Find the quickest path to the cabin in the forest.

**87** Use Newton's method to approximate each of the three solutions of the equation $x + 5 \cos x = 0$. Note that Fig. 8.28 will help in obtaining initial estimates of each solution.

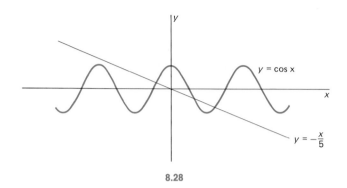

8.28

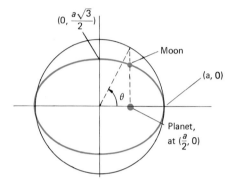

$\left(0, \dfrac{a\sqrt{3}}{2}\right)$

Moon

$(a, 0)$

$\theta$

Planet,
at $\left(\dfrac{a}{2}, 0\right)$

8.29   The elliptical orbit of Problem 88

**88** A moon of a certain planet has an elliptical orbit with eccentricity $\frac{1}{2}$, and its period of revolution about the planet is 100 days. If the moon is at the position $(a, 0)$ when $t = 0$, then, as illustrated in Fig. 8.29, its central angle $\theta$ after $t$ days is given by *Kepler's equation*

$$\frac{2\pi t}{100} = \theta - \frac{1}{2}\sin\theta.$$

Use Newton's method to solve for $\theta$ when $t = 17$ (days). Take $\theta_0 = 1.5$ (rad), and calculate the first two approximations $\theta_1$ and $\theta_2$. Express $\theta_2$ in degrees as well.

---

**8-4**

## Derivatives and Integrals of Other Trigonometric Functions

In Section 8-3 we saw that the derivatives of the sine and cosine functions are given by

$$D_x \sin u = (\cos u)\frac{du}{dx}, \qquad D_x \cos u = (-\sin u)\frac{du}{dx}, \tag{1}$$

where $u$ denotes a differentiable function of $x$. Recall that when differentiating and integrating trigonometric functions, we use radian measure of angles exclusively.

The other four trigonometric functions—the tangent, cotangent, secant, and cosecant—may be defined in terms of the sine and cosine:

$$\tan x = \frac{\sin x}{\cos x}, \qquad \cot x = \frac{\cos x}{\sin x},$$

$$\sec x = \frac{1}{\cos x}, \qquad \csc x = \frac{1}{\sin x}. \tag{2}$$

These formulas are valid except where the denominators are zero. Thus $\tan x$ and $\sec x$ are undefined when $x$ is an odd multiple of $\pi/2$, while $\cot x$ and $\csc x$ are undefined when $x$ is an integral multiple of $\pi$. The graphs of the six trigonometric functions appear in Fig. 8.30. There we show the sine and its reciprocal, the cosecant, in the same coordinate plane; we also pair the cosine with the secant and the tangent with the cotangent.

The graphs in Fig. 8.30 suggest that all six trigonometric functions are periodic, and this is indeed true. The sine and cosine functions each have period $2\pi$, and it follows from (2) that the other four must repeat their values in cycles of $2\pi$ as well. For example, $\tan(x + 2\pi) = \tan x$. But the **period** of the tangent and cotangent functions is the length of the shortest interval of repetition, so for those two functions the period is $\pi$, because $\tan(x + \pi) = \tan x$ for all (meaningful) values of $x$. This relation, too, is suggested in Fig. 8.30. The secant and cosecant functions have period $2\pi$.

If we begin with the derivatives in (1) and the definitions in (2), we can compute the derivatives of the latter four trigonometric functions. All we need is the quotient rule. The results (as listed in Section 8-3) are

$$D \tan x = \sec^2 x, \qquad D \cot x = -\csc^2 x,$$

$$D \sec x = \sec x \tan x, \qquad D \csc x = -\csc x \cot x.$$

When we combine these results with the chain rule, we get the four differentiation formulas

$$D_x \tan u = (\sec^2 u)\frac{du}{dx}, \tag{3}$$

$$D_x \cot u = (-\csc^2 u)\frac{du}{dx}, \tag{4}$$

$$D_x \sec u = (\sec u \tan u)\frac{du}{dx}, \tag{5}$$

$$D_x \csc u = (-\csc u \cot u)\frac{du}{dx}. \tag{6}$$

The patterns in (1) and in (3) through (6) make them easy to memorize. The formulas in (4) and (6) are the "cofunction analogues" of those in (3) and (5), respectively. Note also that the derivative formulas for the cofunctions are those involving minus signs.

**EXAMPLE 1** The following are some derivatives involving trigonometric functions.
(a) $D \tan(2x^3) = [\sec^2 2x^3]D_x(2x^3) = 6x^2 \sec^2 2x^3$.

(b) $D \cot^3(2x) = D (\cot 2x)^3 = [3(\cot 2x)^2]D_x(\cot 2x)$
$= [3 \cot^2(2x)][-\csc^2(2x)]D_x(2x)$
$= -6 \cot^2 2x \csc^2 2x$.

(c) $D \sec \sqrt{x} = [\sec \sqrt{x} \tan \sqrt{x}]D_x(\sqrt{x})$
$= \dfrac{\sec \sqrt{x} \tan \sqrt{x}}{2\sqrt{x}}$.

(d) $D \csc^{1/2} x = \dfrac{1}{2}[\csc x]^{-1/2}D_x(\csc x)$

$= \dfrac{-\csc x \cot x}{2(\csc x)^{1/2}} = -\dfrac{1}{2}\csc^{1/2}x \cot x$.

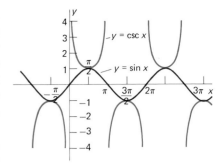

(a)

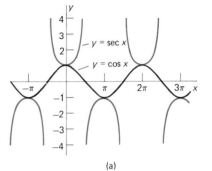

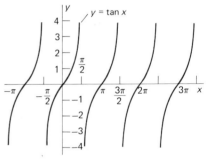

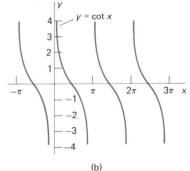

(b)

**8.30** Graphs of the six trigonometric functions

By now we know that to every derivative formula there corresponds an integral formula. The integral versions of (1) and (3) through (6) are these:

$$\int \cos u \, du = \sin u + C, \tag{7}$$

$$\int \sin u \, du = -\cos u + C, \tag{8}$$

$$\int \sec^2 u \, du = \tan u + C, \tag{9}$$

$$\int \csc^2 u \, du = -\cot u + C, \tag{10}$$

$$\int \sec u \tan u \, du = \sec u + C, \tag{11}$$

$$\int \csc u \cot u \, du = -\csc u + C. \tag{12}$$

**EXAMPLE 2**  Compute the antiderivative: $\int \sec^2 3x \, dx$.

**Solution**  With $u = 3x$, $du = 3 \, dx$, Formula (9) gives

$$\int \sec^2 3x \, dx = \tfrac{1}{3} \int \sec^2 u \, du$$

$$= \tfrac{1}{3} \tan u + C = \tfrac{1}{3} \tan 3x + C.$$

**EXAMPLE 3**  Find the antiderivative: $\displaystyle\int \frac{1}{\sqrt{x}} \csc \sqrt{x} \cot \sqrt{x} \, dx$.

**Solution**  With $u = \sqrt{x}$, $du = \dfrac{dx}{2\sqrt{x}}$, Formula (12) gives

$$\int \frac{\csc \sqrt{x} \cot \sqrt{x}}{\sqrt{x}} \, dx = 2 \int \csc u \cot u \, du$$

$$= -2 \csc u + C = -2 \csc \sqrt{x} + C.$$

To integrate $\tan x$, the substitution $u = \cos x$, $du = -\sin x \, dx$ gives

$$\int \tan x \, dx = \int \frac{\sin x \, dx}{\cos x} = -\int \frac{du}{u}$$

$$= -\ln|u| + C,$$

and thus

$$\int \tan x \, dx = -\ln|\cos x| + C = \ln|\sec x| + C. \tag{13}$$

We use here the fact that $|\sec x| = 1/|\cos x|$.

Similarly,

$$\int \cot x \, dx = \ln|\sin x| + C = -\ln|\csc x| + C. \tag{14}$$

The integration of sec $x$ may be accomplished by what amounts to an unmotivated trick:

$$\int \sec x \, dx = \int (\sec x) \frac{\tan x + \sec x}{\sec x + \tan x} \, dx$$

$$= \int \frac{\sec x \tan x + \sec^2 x}{\sec x + \tan x} \, dx$$

$$= \int \frac{du}{u} \qquad \text{(where } u = \sec x + \tan x)$$

$$= \ln|u| + C.$$

Therefore

$$\int \sec x \, dx = \ln|\sec x + \tan x| + C. \qquad (15)$$

Similarly,

$$\int \csc x \, dx = -\ln|\csc x + \cot x| + C. \qquad (16)$$

**EXAMPLE 4** The substitution $u = \frac{1}{2}x$, $du = \frac{1}{2} dx$ gives

$$\int_0^{\pi/2} \sec \frac{x}{2} \, dx = 2 \int_0^{\pi/4} \sec u \, du$$

$$= 2 \Big[ \ln|\sec u + \tan u| \Big]_0^{\pi/4} = 2 \ln(1 + \sqrt{2}).$$

To integrate $\sin^2 x$ and $\cos^2 x$ we use the half-angle identities (in Equations (10) and (11) in Section 8-2)

$$\sin^2 x = \frac{1}{2}(1 - \cos 2x), \qquad \cos^2 x = \frac{1}{2}(1 + \cos 2x). \qquad (17)$$

**EXAMPLE 5** Find $\int \sin^2 3x \, dx$.

*Solution* The first identity in (17)—with $3x$ in place of $x$—yields

$$\int \sin^2 3x \, dx = \int \frac{1}{2}(1 - \cos 6x) \, dx$$

$$= \frac{1}{2}(x - \frac{1}{6}\sin 6x) + C$$

$$= \frac{1}{12}(6x - \sin 6x) + C.$$

To integrate $\tan^2 x$ and $\cot^2 x$ we use the identities

$$1 + \tan^2 x = \sec^2 x, \qquad 1 + \cot^2 x = \csc^2 x. \qquad (18)$$

The first of these follows from the fundamental identity $\cos^2 x + \sin^2 x = 1$ upon division of both sides by $\cos^2 x$. To obtain the second formula in (18), divide both sides of the fundamental identity by $\sin^2 x$.

**EXAMPLE 6** Compute the antiderivative: $\int \cot^2 3x \, dx$.

**Solution** By using the second identity in (18) we obtain

$$\int \cot^2 3x \, dx = \int (\csc^2 3x - 1) \, dx$$

$$= \int (\csc^2 u - 1)(\tfrac{1}{3} \, du) \qquad \text{(with } u = 3x\text{)}$$

$$= \tfrac{1}{3}(-\cot u - u) + C$$

$$= -\tfrac{1}{3} \cot 3x - x + C.$$

## 8-4 PROBLEMS

Find the derivatives of the functions given in Problems 1–30.

1 $f(x) = \sin(2x + 3)$

2 $f(x) = \cos(x^2 - 1)$

3 $f(x) = \tan\left(\dfrac{2x}{3}\right)$

4 $f(x) = \cot \sqrt{x}$

5 $f(x) = \sec\left(\dfrac{1}{x}\right)$

6 $f(x) = \csc e^x$

7 $f(x) = \tan^3 4x$

8 $f(x) = \sec^2 \sqrt[3]{x}$

9 $f(x) = \ln(\sin x)$

10 $f(x) = e^{\cot 2x}$

11 $f(x) = \tan^2(\ln x)$

12 $f(x) = \sec(\csc x)$

13 $f(x) = \ln|\csc x + \cot x|$

14 $f(x) = \tan^3 x \sec^3 x$

15 $f(x) = \sin^3(\csc x)$

16 $f(x) = \dfrac{\cos^2 x}{1 - \sin x}$

17 $f(x) = \dfrac{\tan^2 x}{\sec x + 1}$

18 $f(x) = x^3 \tan^{3/2}\left(\dfrac{1}{x^2}\right)$

19 $f(x) = e^{\sin x} \sec x$

20 $f(x) = (\cot x) \ln|\cos x|$

21 $h(x) = \cos^2 x - \sin^2 x$

22 $f(x) = (1 + \sec^3 x)^{1/2}$

23 $g(t) = 2^{\sec 5t}$

24 $h(x) = (\sec x \cos x)^5$

25 $g(x) = (\sin x)^{1/3} - \sin(x^{1/3})$

26 $h(x) = \ln(\ln \sec x)$

27 $f(x) = \dfrac{1 + \tan x}{1 + \sec x}$

28 $g(x) = \sec(7 \ln x)$

29 $h(t) = \cot^3(1 + t^4)$

30 $f(x) = \cos(e^{-x \ln x})$

Evaluate the integrals in Problems 31–60.

31 $\displaystyle\int \sec^2 \dfrac{x}{2} \, dx$

32 $\displaystyle\int \csc^2 x \cot x \, dx$

33 $\displaystyle\int \cos^2 3x \, dx$

34 $\displaystyle\int \dfrac{\tan^2 \sqrt{x}}{\sqrt{x}} \, dx$

35 $\displaystyle\int \dfrac{dx}{\csc x}$

36 $\displaystyle\int \dfrac{dx}{\sin 2x \tan 2x}$

37 $\displaystyle\int \dfrac{\sec x \tan x}{1 + \sec x} \, dx$

38 $\displaystyle\int \tan 3x \sec^2 3x \, dx$

39 $\displaystyle\int x \sec(x^2) \tan(x^2) \, dx$

40 $\displaystyle\int \dfrac{\tan^2 x}{\sec x} \, dx$

41 $\displaystyle\int \dfrac{e^{\sin 2x}}{\sec 2x} \, dx$

42 $\displaystyle\int \dfrac{e^{\tan x}}{\cos^2 x} \, dx$

43 $\displaystyle\int \dfrac{\sin x}{\cos^2 x} \, dx$

44 $\displaystyle\int \dfrac{1}{x} \sec(\ln x) \, dx$

45 $\displaystyle\int \dfrac{e^x \sec^2 e^x}{\tan e^x} \, dx$

46 $\displaystyle\int \cos^2 2x \, dx$

47 $\displaystyle\int \sin^5 x \cos x \, dx$

48 $\displaystyle\int \tan^7 x \sec^2 x \, dx$

49 $\displaystyle\int \dfrac{\sin 2x}{\cos^5 2x} \, dx$

50 $\displaystyle\int \sin x \cos x \, dx$

51 $\displaystyle\int \sin^{10} x \cos x \, dx$

52 $\displaystyle\int e^{\tan x} \sec^2 x \, dx$

53 $\displaystyle\int \dfrac{\sin x}{(1 + \cos x)^5} \, dx$

54 $\displaystyle\int \tan^6 x \sec^2 x \, dx$

55 $\displaystyle\int e^{2x} \cos e^{2x} \, dx$

56 $\displaystyle\int (\cos x) \sec(\sin x) \, dx$

57 $\displaystyle\int 2^{\sec x} \sec x \tan x \, dx$

58 $\displaystyle\int \sec^2 5x \, dx$

59 $\displaystyle\int \dfrac{1}{x} \cos(\ln x) \, dx$

60 $\displaystyle\int \dfrac{\sin x - \cos x}{\cos x + \sin x} \, dx$

61 Find the critical points and inflection points of the function $f(x) = \tan x$, and verify that its graph looks as indicated in Fig. 8.30.

62 Repeat Problem 61, except use the function $g(x) = \sec x$.

63 Sketch the graph of $y = \sin x + \cos x$.

64 Sketch the graph of the function $f(x) = e^{-x} \cos x$ for $x \geqq 0$. This curve oscillates between the two curves $y = e^{-x}$ and $y = -e^{-x}$. Where are its local maxima and minima?

65 Find the area of the region between the curves $y = \tan^2 x$ and $y = \sec^2 x$ from $x = 0$ to $x = \pi/4$.

66 Find the volume generated by revolving the region under $y = \sec x$ from $x = 0$ to $x = \pi/4$ around the x-axis.

67 Find the volume generated by revolving the region under $y = \tan(\pi x^2/4)$ from $x = 0$ to $x = 1$ around the y-axis.

68 Find the length of the curve $y = \ln(\cos x)$ from $x = 0$ to $x = \pi/4$.

**430**

**69** Show by evaluating the integral in two different ways that

$$\int \cot x \csc^2 x \, dx = -\tfrac{1}{2} \cot^2 x + C_1 = -\tfrac{1}{2} \csc^2 x + C_2.$$

Are these two answers consistent with one another?

**70** Show by evaluating the integral in two different ways that

$$\int \sin x \cos x \, dx = \tfrac{1}{2} \sin^2 x + C_1 = -\tfrac{1}{2} \cos^2 x + C_2.$$

Are these two answers consistent with one another?

**71** Sketch the graphs of $y = 1/x$ and $y = \tan x$ on the same coordinate plane. Then use Newton's method to find the least two positive solutions of the equation

$$\frac{1}{x} = \tan x.$$

**72** Figure 8.31 shows a belt of total length $L$ stretched tightly around two pulleys with radii $R$ and $r$. The distance between the centers of the pulleys is $x$.

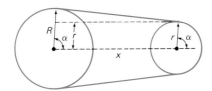

**8.31**  The belt and pulleys of Problem 72

(a) Show that $x = (R - r)\sec \alpha$ and that

$$L = 2\pi R + 2(R - r)[(\tan \alpha) - \alpha]. \tag{19}$$

(b) Find $x$ if $R = 3$ ft, $r = 1$ ft, and $L = 25$ ft. (*Suggestion:* First use Newton's method to solve Equation (19) for $\alpha$.)

**73** The equation $x + \tan x = 0$ is important in a variety of applications—for example, in the study of the diffusion of heat. It has a sequence $\alpha_1, \alpha_2, \alpha_3, \ldots$ of positive roots, with the $n$th one slightly larger than $(n - \tfrac{1}{2})\pi$. Use Newton's method to compute $\alpha_1$ and $\alpha_2$ to three-place accuracy.

**74** Let the function $g$ be defined as follows:

$$g(x) = \frac{x}{2} + x^2 \sin \frac{1}{x}$$

for $x \neq 0$; $g(0) = 0$.

(a) Apply the result of Problem 52 in Section 8-2 to show that $g'(0) = \tfrac{1}{2} > 0$.

(b) Perhaps with the aid of Fig. 8.15, sketch the graph of $g$ near $x = 0$. Is $g$ increasing on any open interval containing $x = 0$? (*Answer:* No.)

**75** Let the function $g$ be defined as follows:

$$g(x) = \frac{1}{4}x^2 + x^4 \sin \frac{1}{x}$$

for $x \neq 0$; $g(0) = 0$. First use the method of Problem 52 in Section 8-2 to show that $g'(0) = 0$ and $g''(0) = \tfrac{1}{2} > 0$. Then show that in spite of the behavior of $g''(x)$ at $x = 0$, the graph of $y = g(x)$ is neither bending upward nor bending downward on any interval containing $x = 0$.

---

**8-5**

# Inverse Trigonometric Functions

Recall from Section 2-3 that if the function $f$ is one-to-one on its domain of definition, then it has an inverse function $f^{-1}$. The domain of $f^{-1}$ is the range of values of $f$, and

$$f^{-1}(x) = y \quad \text{if and only if} \quad f(y) = x. \tag{1}$$

For example, the exponential function $e^x$ is increasing, and therefore one-to-one, on the whole real line, and it attains all positive values. Hence it has an inverse function, which we already know to be ln $x$, and this inverse function is defined for all $x > 0$.

Here we want to define the inverses of the trigonometric functions. We must, however, confront the fact that the trigonometric functions fail to be one-to-one because each has period $\pi$ or $2\pi$. For example, $\sin y = \tfrac{1}{2}$ if $y$ is either $\pi/6$ plus any multiple of $2\pi$ or $5\pi/6$ plus any multiple of $2\pi$. Consequently we *cannot* define $y = \sin^{-1}x$, the inverse of the sine function, by saying simply that $y$ is that value such that $\sin y = x$. There are *many* such values of $y$, and we must specify just which particular one of these is to be used.

We do this by suitably restricting the domain of the sine function. Since the function $\sin x$ is increasing on $[-\pi/2, \pi/2]$ and its range of values is $[-1, 1]$, for each $x$ in $[-1, 1]$ there is a *single* $y$ in $[-\pi/2, \pi/2]$ such that $\sin y = x$. This observation leads to the following definition of the **inverse sine** (or **arcsine**) **function**, denoted by $\sin^{-1}x$ or by arcsin $x$.

The **inverse sine function** is defined as follows:

$$y = \sin^{-1}x \quad \text{if and only if} \quad \sin y = x \qquad (2)$$

where $-1 \leq x \leq 1$ and $-\pi/2 \leq y \leq \pi/2$.

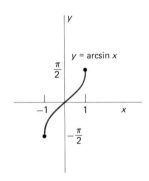

$\frac{\pi}{2}$

$y = \arcsin x$

$-\frac{\pi}{2}$

**8.32**  The graph of $y = \arcsin x$

Thus, if $x$ is between $-1$ and $+1$ (inclusive), then $\sin^{-1}x$ is that number $y$ between $-\pi/2$ and $\pi/2$ such that $\sin y = x$. Even more briefly, arcsin $x$ is the angle nearest zero whose sine is $x$. For instance,

$$\sin^{-1}1 = \frac{\pi}{2}, \qquad \sin^{-1}0 = 0, \qquad \sin^{-1}(-1) = -\frac{\pi}{2},$$

and $\sin^{-1}(2)$ does not exist.

Note that the symbol $-1$ in the notation $\sin^{-1}x$ is *not an exponent*—it does *not* mean $(\sin x)^{-1}$.

It follows from the reflection property of inverse functions (Section 2-3) that the graph of $y = \sin^{-1}x$ is the reflection in the line $y = x$ of the graph of $y = \sin x$, $-\pi/2 \leq x \leq \pi/2$, and therefore looks like the graph in Fig. 8.32.

It follows from Equation (2) that

$$\sin(\sin^{-1}x) = x \qquad \text{if } -1 \leq x \leq 1, \qquad (3a)$$

$$\sin^{-1}(\sin x) = x \qquad \text{if } -\frac{\pi}{2} \leq x \leq \frac{\pi}{2}. \qquad (3b)$$

Because the derivative of $\sin x$ is positive for $-\pi/2 < x < \pi/2$, it follows from Theorem 2 in Section 3-4 that $\sin^{-1}x$ is differentiable on $(-1, 1)$. We may therefore differentiate both sides of the identity in (3a), though we begin by writing it in the form

$$\sin y = x$$

where $y = \sin^{-1}x$. This gives

$$(\cos y)\frac{dy}{dx} = 1.$$

So,

$$\frac{dy}{dx} = \frac{1}{\cos y} = \frac{1}{\sqrt{1 - \sin^2 y}} = \frac{1}{\sqrt{1 - x^2}}.$$

We are correct in taking the positive square root in this computation because $\cos y > 0$ for $-\pi/2 < y < \pi/2$. Thus

$$D \sin^{-1}x = \frac{1}{\sqrt{1 - x^2}} \qquad (4)$$

provided that $-1 < x < 1$. When we combine this result with the chain rule, we get

$$D_x \sin^{-1}u = \frac{1}{\sqrt{1 - u^2}}\frac{du}{dx} \qquad (5)$$

if $u$ is a differentiable function with values in the interval $(-1, 1)$.

The definition of the inverse cosine function is similar, except that we begin by restricting the cosine function to the interval $[0, \pi]$, where it is a decreasing function. Thus the **inverse cosine** (or **arccosine**) **function** is defined by means of the rule

$$y = \cos^{-1}x \quad \text{if and only if} \quad \cos y = x \tag{6}$$

where $-1 \leq x \leq 1$ and $0 \leq y \leq \pi$. Thus $\cos^{-1}x$ is the angle in $[0, \pi]$ whose cosine is $x$. For instance,

$$\cos^{-1}1 = 0, \qquad \cos^{-1}0 = \frac{\pi}{2}, \qquad \cos^{-1}(-1) = \pi.$$

We may compute the derivative of $\cos^{-1}x$, also written arccos $x$, by differentiation of both sides of the identity

$$\cos(\cos^{-1}x) = x \qquad (-1 < x < 1).$$

The computations are similar to those for $D \sin^{-1}x$, and lead to the result

$$D \cos^{-1}x = -\frac{1}{\sqrt{1 - x^2}}.$$

And if $u$ denotes a differentiable function with values in $(-1, 1)$, the chain rule then gives

$$D_x \cos^{-1}u = -\frac{1}{\sqrt{1 - u^2}} \frac{du}{dx}. \tag{7}$$

The graph of $y = \cos^{-1}x$ is the reflection in the line $y = x$ of the graph of $y = \cos x$, $0 \leq x \leq \pi$, and is shown in Fig. 8.33.

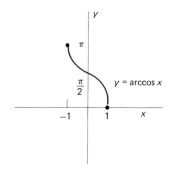

**8.33** The graph of $y = $ arccos $x$

**EXAMPLE 1** Suppose that $y = \sin^{-1}x^2$. Then (5) with $u = x^2$ gives

$$\frac{dy}{dx} = \frac{2x}{\sqrt{1 - x^4}}.$$

The tangent function is increasing on the *open* interval $(-\pi/2, \pi/2)$—it is not defined at the end points $-\pi/2$ and $\pi/2$—and its range of values is the whole real line. We may in consequence define the **inverse tangent** (or **arctangent**) **function**, denoted by $\tan^{-1}x$ or by arctan $x$, as the inverse of $y = \tan x$, $-\pi/2 < x < \pi/2$.

**8.34** The inverse tangent function has domain all real $x$

---

*Definition*

The **inverse tangent function** is defined as follows:

$$y = \tan^{-1}x \quad \text{if and only if} \quad \tan y = x \tag{8}$$

where $-\pi/2 < y < \pi/2$.

---

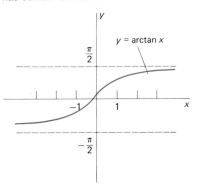

Since the tangent function attains all real values, $\tan^{-1}x$ is defined for all real numbers $x$; $\tan^{-1}x$ is that number $y$ between $-\pi/2$ and $\pi/2$ such that $\tan y = x$. Alternatively, arctan $x$ is the angle nearest zero whose tangent is $x$. The graph of $y = \tan^{-1}x$ is the reflection in the line $y = x$ of the graph of $y = \tan x$, $-\pi/2 < x < \pi/2$, and is shown in Fig. 8.34.

It follows from (8) that

$$\tan(\tan^{-1}x) = x \qquad \text{for all } x, \tag{9a}$$

$$\tan^{-1}(\tan x) = x \qquad \text{if } -\pi/2 < x < \pi/2. \tag{9b}$$

Because the derivative of $\tan x$ is positive for all $x$ in the interval $(-\pi/2, \pi/2)$, it follows from Theorem 2 in Section 3-4 that $\tan^{-1}x$ is differentiable for all $x$. We may therefore differentiate both sides of the identity in (9a). First, we write that identity in the form

$$\tan y = x$$

where $y = \tan^{-1}x$. Then

$$(\sec^2 y)\frac{dy}{dx} = 1.$$

So

$$\frac{dy}{dx} = \frac{1}{\sec^2 y} = \frac{1}{1 + \tan^2 y} = \frac{1}{1 + x^2}.$$

Thus

$$D \tan^{-1}x = \frac{1}{1 + x^2}, \tag{10}$$

and if $u$ is any differentiable function of $x$, then

$$D_x \tan^{-1}u = \frac{1}{1 + u^2}\frac{du}{dx}. \tag{11}$$

The definition of the inverse cotangent function is similar, except that we begin by restricting the cotangent function to the interval $(0, \pi)$, where it is a decreasing function attaining all real values. Thus the **inverse cotangent (or arccotangent) function** is defined as

$$y = \cot^{-1}x \quad \text{if and only if} \quad \cot y = x \tag{12}$$

where $x$ is any real number and $0 < y < \pi$. Then differentiation of the identity $\cot(\cot^{-1}x) = x$ leads, as in the derivation of (10), to

$$D \cot^{-1}x = -\frac{1}{1 + x^2}.$$

If $u$ is a differentiable function of $x$, then the chain rule gives

$$D_x \cot^{-1}u = -\frac{1}{1 + u^2}\frac{du}{dx}. \tag{13}$$

**8.35  The falling rock**

**EXAMPLE 2**   A mountain climber on one edge of a deep canyon 800 ft wide sees a large rock fall from the opposite edge at time $t = 0$. As he watches the rock plummeting downward, he notices that his eyes first move slowly, then faster, then more slowly again. Let $\theta$ denote the angle of depression of his line of sight below the horizontal. At what angle $\theta$ would the rock seem to be moving the most rapidly; that is, when would $d\theta/dt$ be maximal?

*Solution*   From our study of constant acceleration in Section 4-9, we know that the rock will fall $16t^2$ feet in the first $t$ seconds. We refer to Fig. 8.35 and

**434**

see that the value of $\theta$ after $t$ seconds will be

$$\theta = \theta(t) = \tan^{-1}\left(\frac{16t^2}{800}\right) = \tan^{-1}\left(\frac{t^2}{50}\right).$$

Hence

$$\frac{d\theta}{dt} = \frac{1}{1 + (t^2/50)^2} \cdot \frac{2t}{50} = \frac{100t}{t^4 + 2500}.$$

To find when $d\theta/dt$ is maximal, we find when *its* derivative is zero.

$$\frac{d}{dt}\left(\frac{d\theta}{dt}\right) = \frac{100(t^4 + 2500) - 100t(4t^3)}{(t^4 + 2500)^2}$$

$$= \frac{100(2500 - 3t^4)}{(t^4 + 2500)^2}.$$

So $d^2\theta/dt^2$ is zero when $3t^4 = 2500$; that is, when

$$t = \sqrt[4]{2500/3} \approx 5.37 \text{ s}.$$

This is the value of $t$ when $d\theta/dt$ is maximal, and at this time we have $t^2 = 50/\sqrt{3}$. So the angle at this time is

$$\theta = \arctan\left(\frac{1}{50} \cdot \frac{50}{\sqrt{3}}\right) = \arctan\left(\frac{1}{\sqrt{3}}\right) = \frac{\pi}{6}.$$

So the *apparent* speed of the falling rock is greatest when the climber's line of sight is 30° below the horizontal. The speed of the rock then is $32t$ with $t \approx 5.37$ and thus is about 172 ft/s.

Figure 8.36 shows that the secant function is increasing on each of the intervals $[0, \pi/2)$ and $(\pi/2, \pi]$. On the union of these two intervals the secant function attains all real values $y$ such that $|y| \geq 1$. We may therefore define the **inverse secant** (or **arcsecant**) **function,** denoted by $\sec^{-1}x$ or by arcsec $x$, by restricting the secant function to the union of the intervals $[0, \pi/2)$ and $(\pi/2, \pi]$.

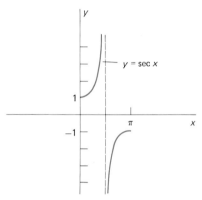

**8.36** Restriction of the secant function to the union of the intervals $[0, \pi/2)$ and $(\pi/2, \pi]$

---

*Definition*

The **inverse secant function** is defined as follows:

$$y = \sec^{-1}x \quad \text{if and only if} \quad \sec y = x \qquad (14)$$

where $|x| \geq 1$ and $0 \leq y \leq \pi$.

---

The graph of $y = \sec^{-1}x$ is the reflection in the line $y = x$ of the graph of $y = \sec x$, suitably restricted to the intervals $0 \leq x < \pi/2$ and $\pi/2 < x \leq \pi$. The graph appears in Fig. 8.37. It follows from the definition above that

$$\sec(\sec^{-1}x) = x \qquad \text{if } |x| \geq 1, \qquad (15a)$$

$$\sec^{-1}(\sec x) = x \qquad \text{for } x \text{ in } [0, \pi/2) \cup (\pi/2, \pi]. \qquad (15b)$$

Following the now familiar pattern, we find $D \sec^{-1}x$ by differentiating both sides of (15a) in the form

$$\sec y = x$$

**8.37** The graph of $y = \sec^{-1}x = \text{arcsec } x$

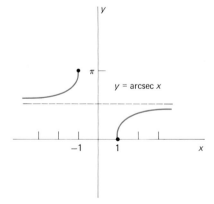

where $y = \sec^{-1} x$. This yields

$$(\sec y \tan y) \frac{dy}{dx} = 1,$$

and so

$$\frac{dy}{dx} = \frac{1}{\sec y \tan y} = \frac{1}{\pm x \sqrt{x^2 - 1}},$$

because $\tan y = \pm\sqrt{\sec^2 y - 1} = \pm\sqrt{x^2 - 1}$.

To obtain the correct choice of sign above, note what happens in the two cases $x > 1$ and $x < -1$. In the first case, $0 < y < \pi/2$ and $\tan y > 0$, so we choose the $+$ sign. If $x < -1$ then $\pi/2 < y < \pi$ and $\tan y < 0$, so we take the $-$ sign. Thus

$$D \sec^{-1} x = \frac{1}{|x|\sqrt{x^2 - 1}} \qquad (|x| > 1). \tag{16}$$

If $u$ is a differentiable function of $x$ with values that exceed 1 in magnitude, then by the chain rule we have

$$D_x \sec^{-1} u = \frac{1}{|u|\sqrt{u^2 - 1}} \frac{du}{dx}. \tag{17}$$

**EXAMPLE 3**   The function $\sec^{-1} e^x$ is defined if $x > 0$, for then $e^x > 1$. Then by (17)

$$D_x \sec^{-1} e^x = \frac{e^x}{|e^x|\sqrt{e^{2x} - 1}} = \frac{1}{\sqrt{e^{2x} - 1}}$$

because $|e^x| = e^x$ for all $x$.

---

The **inverse cosecant** (or **arccosecant**) **function** is the inverse of the function $y = \csc x$ where $x$ is restricted to the union of the intervals $[-\pi/2, 0)$ and $(0, \pi/2]$. Thus

$$y = \csc^{-1} x \quad \text{if and only if} \quad \csc y = x \tag{18}$$

where $|x| \geq 1$ and $-\pi/2 < y < \pi/2$. Its derivative formula, which has a derivation similar to that of (17) above, is

$$D_x \csc^{-1} u = -\frac{1}{|u|\sqrt{u^2 - 1}} \frac{du}{dx}. \tag{19}$$

It is noteworthy that the derivatives of the six inverse trigonometric functions are all simple *algebraic* functions. As a consequence, inverse trigonometric functions frequently appear when we integrate algebraic functions. Observe also that the derivatives of $\cos^{-1} x$, $\cot^{-1} x$, and $\csc^{-1} x$ differ only in sign from the derivatives of their respective cofunctions. For this reason only the arcsine, arctangent, and arcsecant functions are necessary for integration, and only these three are in common use. That is, one need commit to memory the integral formulas only for the latter three functions. They follow immediately from (5), (11), and (17), and may be written in the form below.

$$\int \frac{du}{\sqrt{1 - u^2}} = \sin^{-1}u + C, \tag{20}$$

$$\int \frac{du}{1 + u^2} = \tan^{-1}u + C, \tag{21}$$

$$\int \frac{du}{u\sqrt{u^2 - 1}} = \sec^{-1}|u| + C. \tag{22}$$

It is easy to verify that the absolute value on the right side in (22) follows from the one in (17). Remember also that, because $\sec^{-1}|u|$ is undefined unless $|u| \geq 1$, the definite integral

$$\int_a^b \frac{du}{u\sqrt{u^2 - 1}}$$

is meaningful only if the limits $a$ and $b$ are both at least 1 or both at most $-1$.

**EXAMPLE 4**   It follows immediately from Equation (21) that

$$\int_0^1 \frac{dx}{1 + x^2} = \left[\tan^{-1}x\right]_0^1$$

$$= \tan^{-1}1 - \tan^{-1}0 = \frac{\pi}{4}.$$

**EXAMPLE 5**   The substitution $u = x/2$, $du = \frac{1}{2}\,dx$ gives

$$\int \frac{dx}{\sqrt{4 - x^2}} = \int \frac{dx}{2\sqrt{1 - (x/2)^2}}$$

$$= \int \frac{du}{\sqrt{1 - u^2}} = \arcsin u + C$$

$$= \arcsin\left(\frac{x}{2}\right) + C.$$

**EXAMPLE 6**   The substitution $u = 3x$, $du = 3\,dx$ gives

$$\int \frac{dx}{1 + 9x^2} = \frac{1}{3}\int \frac{3\,dx}{1 + (3x)^2} = \frac{1}{3}\int \frac{du}{1 + u^2}$$

$$= \frac{1}{3}\tan^{-1}u + C = \frac{1}{3}\tan^{-1}3x + C.$$

**EXAMPLE 7**   The substitution $u = x\sqrt{2}$, $du = \sqrt{2}\,dx$ gives

$$\int_1^{\sqrt{2}} \frac{dx}{x\sqrt{2x^2 - 1}} = \int_{\sqrt{2}}^2 \frac{du}{u\sqrt{u^2 - 1}}$$

$$= \left[\sec^{-1}|u|\right]_{\sqrt{2}}^2 = \sec^{-1}2 - \sec^{-1}\sqrt{2}$$

$$= \frac{\pi}{3} - \frac{\pi}{4} = \frac{\pi}{12}.$$

## 8-5 PROBLEMS

Find the values indicated in each of Problems 1–4.

**1** (a) $\sin^{-1}(\frac{1}{2})$  (b) $\sin^{-1}(-\frac{1}{2})$

(c) $\sin^{-1}\left(\dfrac{\sqrt{2}}{2}\right)$  (d) $\sin^{-1}\left(-\dfrac{\sqrt{3}}{2}\right)$

**2** (a) $\cos^{-1}(\frac{1}{2})$  (b) $\cos^{-1}(-\frac{1}{2})$

(c) $\cos^{-1}\left(\dfrac{\sqrt{2}}{2}\right)$  (d) $\cos^{-1}\left(-\dfrac{\sqrt{3}}{2}\right)$

**3** (a) $\tan^{-1}0$  (b) $\tan^{-1}1$
(c) $\tan^{-1}(-1)$  (d) $\tan^{-1}\sqrt{3}$
**4** (a) $\sec^{-1}1$  (b) $\sec^{-1}(-1)$
(c) $\sec^{-1}2$  (d) $\sec^{-1}(-\sqrt{2})$

Differentiate the functions in Problems 5–30.

**5** $f(x) = \sin^{-1}(x^{100})$  **6** $f(x) = \arctan(e^x)$
**7** $f(x) = \sec^{-1}(\ln x)$  **8** $f(x) = \ln(\tan^{-1}x)$
**9** $f(x) = \arcsin(\tan x)$  **10** $f(x) = x\arctan x$
**11** $f(x) = \sin^{-1}e^x$  **12** $f(x) = \arctan\sqrt{x}$

**13** $f(x) = \cos^{-1}x + \sec^{-1}\left(\dfrac{1}{x}\right)$

**14** $f(x) = \cot^{-1}\left(\dfrac{1}{x^2}\right)$  **15** $f(x) = \csc^{-1}x^2$

**16** $f(x) = \arccos\left(\dfrac{1}{\sqrt{x}}\right)$  **17** $f(x) = \dfrac{1}{\arctan x}$

**18** $f(x) = (\arcsin x)^2$  **19** $f(x) = \tan^{-1}(\ln x)$
**20** $f(x) = \text{arcsec}\sqrt{x^2 + 1}$
**21** $f(x) = \tan^{-1}e^x + \cot^{-1}e^{-x}$
**22** $f(x) = \exp(\arcsin x)$  **23** $f(x) = \sin(\arctan x)$

**24** $f(x) = \sec(\sec^{-1}e^x)$  **25** $f(x) = \dfrac{e^x}{\arctan 3x}$

**26** $f(x) = (\sin^{-1}2x^2)^{-2}$  **27** $f(x) = \dfrac{\arctan x}{(1 + x^2)^2}$

**28** $f(x) = \arcsin\left(\dfrac{x}{a}\right)$  **29** $f(x) = \dfrac{1}{a}\tan^{-1}\left(\dfrac{x}{a}\right)$

**30** $f(x) = \dfrac{1}{a}\sec^{-1}\left(\dfrac{x}{a}\right)$

Evaluate or antidifferentiate, as appropriate, in Problems 31–55.

**31** $\displaystyle\int_0^1 \frac{dx}{1 + x^2}$  **32** $\displaystyle\int_0^{1/2} \frac{dx}{\sqrt{1 - x^2}}$

**33** $\displaystyle\int_{\sqrt{2}}^2 \frac{dx}{x\sqrt{x^2 - 1}}$  **34** $\displaystyle\int_{-2}^{-2/\sqrt{3}} \frac{dx}{x\sqrt{x^2 - 1}}$

**35** $\displaystyle\int_0^3 \frac{dx}{9 + x^2}$  **36** $\displaystyle\int_0^{\sqrt{12}} \frac{dx}{\sqrt{16 - x^2}}$

**37** $\displaystyle\int \frac{dx}{\sqrt{1 - 4x^2}}$  **38** $\displaystyle\int \frac{dx}{9x^2 + 4}$

**39** $\displaystyle\int \frac{dx}{x\sqrt{x^2 - 25}}$  **40** $\displaystyle\int \frac{dx}{x\sqrt{4x^2 - 9}}$

**41** $\displaystyle\int \frac{e^x\,dx}{1 + e^{2x}}$  **42** $\displaystyle\int \frac{x^2\,dx}{x^6 + 25}$

**43** $\displaystyle\int \frac{dx}{x\sqrt{x^6 - 25}}$  **44** $\displaystyle\int \frac{\sqrt{x}\,dx}{1 + x^3}$

**45** $\displaystyle\int \frac{dx}{\sqrt{x(1 - x)}}$  **46** $\displaystyle\int \frac{\sec x\tan x}{1 + \sec^2 x}\,dx$

**47** $\displaystyle\int \frac{x^{49}}{1 + x^{100}}\,dx$  **48** $\displaystyle\int \frac{x^4}{(1 - x^{10})^{1/2}}\,dx$

**49** $\displaystyle\int \frac{1}{x[1 + (\ln x)^2]}\,dx$  **50** $\displaystyle\int \frac{\arctan x}{1 + x^2}\,dx$

**51** $\displaystyle\int_0^1 \frac{1}{1 + (2x - 1)^2}\,dx$  **52** $\displaystyle\int_0^1 \frac{x^3}{1 + x^4}\,dx$

**53** $\displaystyle\int_1^e \frac{1}{x[1 - (\ln x)^2]^{1/2}}\,dx$

**54** $\displaystyle\int_1^2 \frac{dx}{x(x^2 - 1)^{1/2}}$

**55** $\displaystyle\int_1^3 \frac{dx}{2x^{1/2}(1 + x)}$  (*Suggestion:* Let $u = x^{1/2}$.)

**56** Conclude from the formula $D\cos^{-1}x = -D\sin^{-1}x$ that $\sin^{-1}x + \cos^{-1}x = \pi/2$ if $0 \leq x \leq 1$.

**57** Show that $D\sec^{-1}x = D\cos^{-1}(1/x)$ if $x \geq 1$, and conclude that $\sec^{-1}x = \cos^{-1}(1/x)$ if $x \geq 1$. This fact can be used to find arcsecants on a calculator that has the key for the arccosine function, usually written "inv cos" or "$\cos^{-1}$."

**58** (a) Deduce from the addition formula for tangents (Problem 10 in Section 8-2) that

$$\arctan x + \arctan y = \arctan\frac{x + y}{1 - xy}$$

provided that $xy < 1$.
(b) Apply part (a) to show that each of the following numbers is equal to $\pi/4$:
(i) $\arctan(\frac{1}{2}) + \arctan(\frac{1}{3})$
(ii) $2\arctan(\frac{1}{3}) + \arctan(\frac{1}{7})$

(iii) $\arctan(\frac{120}{119}) - \arctan(\frac{1}{239})$

(iv) $4 \arctan(\frac{1}{5}) - \arctan(\frac{1}{239})$

**59** A billboard *parallel* to a highway is to be 12 ft high, and its bottom will be 4 ft above the eye level of a passing motorist. How far from the highway should the billboard be placed in order to maximize the angle it subtends at the motorist's eyes?

**60** Use inverse trigonometric functions to prove that the vertical angle subtended by a rectangular painting on a wall is greatest when the painting is hung with its center at the level of the observer's eyes.

**61** Show that the circumference of a circle of radius $a$ is $2\pi a$ by finding the length of the circular arc $y = \sqrt{a^2 - x^2}$ from $x = 0$ to $x = a/\sqrt{2}$ and then multiplying by 8.

**62** Find the volume generated by revolving the area under $y = 1/(1 + x^4)$ from $x = 0$ to $x = 1$ around the $y$-axis.

**63** Show that the area of the unbounded region lying between the $x$-axis and the curve $y = 1/(1 + x^2)$ is finite by evaluating

$$\lim_{a \to \infty} \int_{-a}^{a} \frac{dx}{1 + x^2}.$$

**64** A building 250 ft high is equipped with an external elevator. The elevator starts at the top at time $t = 0$ and descends at the constant rate of 25 ft/s. You are watching the elevator from a window that is 100 ft above the ground, in a building 50 ft from the elevator. At what height does the elevator appear to you to be moving the fastest?

**65** Suppose that the function $f$ is defined for all $x$ with $|x| > 1$, and has the property that $f'(x) = 1/(x\sqrt{x^2 - 1})$ for all such $x$.

(a) Explain why there exist two constants $A$ and $B$ such that

$$f(x) = \operatorname{arcsec} x + A \qquad \text{if } x > 1;$$
$$f(x) = -\operatorname{arcsec} x + B \qquad \text{if } x < -1.$$

(b) Determine the values of $A$ and $B$ so that $f(\pm 2) = 1$. Then sketch the graph of $y = f(x)$.

**66** The arctangent is the only inverse trigonometric function included in many versions of BASIC and FORTRAN, so it is necessary in programming to express $\sin^{-1}x$ and $\sec^{-1}x$ in terms of the arctangent. Show each of the following.

(a) If $|x| < 1$, then $\sin^{-1}x = \tan^{-1}\left(\dfrac{x}{\sqrt{1 - x^2}}\right)$.

(b) If $x > 1$, then $\sec^{-1}x = \tan^{-1}(\sqrt{x^2 - 1})$.

(c) If $x < -1$, then $\sec^{-1}x = \pi - \tan^{-1}(\sqrt{x^2 - 1})$.

---

*8-6

## Periodic Phenomena and Simple Harmonic Motion

As we saw in Chapter 7, exponential functions are important in the study of natural growth and decay processes—population growth, radioactive decay, and the like. The trigonometric functions are equally important in the analysis of periodic phenomena in nature, such as the ebb and flow of the seasons and the tides, and the oscillatory motions that occur in vibrating mechanical systems. These applications are so remarkably diverse partly because of the truth of the following theorem.

> **Theorem** *The Equation* $x'' = -k^2 x$
>
> The function $x = f(t)$ satisfies the second order differential equation
>
> $$\frac{d^2x}{dt^2} = -k^2 x \qquad (k \text{ constant}) \qquad (1)$$
>
> if and only if there exist constants $A$ and $B$ such that
>
> $$f(t) = A \cos kt + B \sin kt. \qquad (2)$$

**Proof** Suppose first that $x = f(t)$ has the form given in Equation (2). Then

$$\frac{dx}{dt} = -kA \sin kt + kB \cos kt \qquad (3)$$

and

$$\frac{d^2x}{dt^2} = -k^2 A \cos kt - k^2 B \sin kt$$

$$= -k^2(A \cos kt + B \sin kt) = -k^2 x.$$

Thus the function in (2) satisfies the given differential equation in (1). This proves the *if* part of the theorem. Problem 21 is an outline of the *only if* part—the fact that if a function $x = f(t)$ satisfies the differential equation in (1), then it *must* be of the form in (2). ∎

If a particle moves along the $x$-axis with acceleration given by Equation (1), then the theorem implies that its position function $x = f(t)$ is of the form of (2). Motion of this type is called **simple harmonic motion.** The constants $A$ and $B$ may be determined in terms of the particle's initial position $x_0 = f(0)$ and initial velocity $v_0 = f'(0)$, as follows. Substitution of $t = 0$ into (2) gives $A = x_0$, and substitution of $t = 0$ into (3) gives $B = v_0/k$. Hence the position function of the particle is

$$x = x(t) = x_0 \cos kt + \frac{v_0}{k} \sin kt. \tag{4}$$

The expression on the right-hand side in Equation (2) can be rewritten in a way that makes it easier to visualize the graph $x = f(t)$. To do this, we choose constants $C$ and $\alpha$ as indicated in Fig. 8.38—that is, so that

$$C \sin k\alpha = -A \quad \text{and} \quad C \cos k\alpha = B. \tag{5}$$

Here is one way to find $C$ and $\alpha$. First, the sum of the squares of the left-hand sides above is $C^2$, and so $C^2 = A^2 + B^2$. When we form quotients using the two equations above, we find that $\tan k\alpha = -A/B$. And hence if $B > 0$, we take

$$C = \sqrt{A^2 + B^2} \quad \text{and} \quad \alpha = \frac{1}{k} \tan^{-1}(-A/B). \tag{6}$$

If $B < 0$, then the signs in (5) indicate that $k\alpha$ lies in the second or third quadrant, so we replace $\tan^{-1}(-A/B)$ with $\pi + \tan^{-1}(-A/B)$. Then substitution of the formulas in (5) in Equation (2) yields

$$x = (-C \sin k\alpha) \cos kt + (C \cos k\alpha) \sin kt$$
$$= C(\sin kt \cos k\alpha - \cos kt \sin k\alpha).$$

Thus

$$x = C \sin k(t - \alpha). \tag{7}$$

With $k\beta = k\alpha + \pi/2$, the addition formula for cosines gives the alternative form

$$x = C \cos k(t - \beta). \tag{8}$$

The graph of Equation (7) exhibits the same oscillatory behavior as the curve $x = \sin t$. But the effect of the coefficient $C$ is to stretch the graph in the $x$-direction, so that $x$ has maximum value $+C$ and minimum value $-C$.

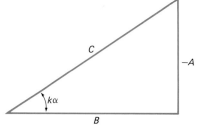

**8.38** The constants in Equation (5)

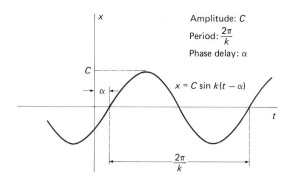

Amplitude: $C$

Period: $\dfrac{2\pi}{k}$

Phase delay: $\alpha$

$x = C \sin k(t - \alpha)$

$\dfrac{2\pi}{k}$

**8.39** Amplitude, period, and phase delay

Because of the constant $k$, the function $f(t) = C \sin k(t - \alpha)$ has period $2\pi/k$ instead of $2\pi$, since

$$f\left(t + \frac{2\pi}{k}\right) = C \sin k\left(t + \frac{2\pi}{k} - \alpha\right)$$

$$= C \sin(kt - k\alpha + 2\pi)$$

$$= C \sin k(t - \alpha) = f(t).$$

Finally, the effect of the constant $\alpha$ is to translate the graph $\alpha$ units to the right of the curve $x = C \sin kt$. When we assemble all this information, we see that the graph of $x(t)$ in (7) resembles the one shown in Fig. 8.39.

Simple harmonic motion with the position function in (7) may be described as having

| | |
|---|---|
| *Amplitude* | $C,$ |
| *Period* | $\dfrac{2\pi}{k},$ |
| *Phase delay* | $\alpha.$ |

(9)

**EXAMPLE 1**  Sketch the graph of $x = 3 \sin(2t - \pi/2)$.

***Solution***  We write $x = 3 \sin 2(t - \pi/4)$, which shows that the function $x$ has amplitude $C = 3$, period $2\pi/k = \pi$, and phase delay $\alpha = \pi/4$, so the graph looks like the one in Fig. 8.40.

**EXAMPLE 2**  Sketch the graph of $x = \cos 2t + \sin 2t$.

**8.40**  The graph for Example 1

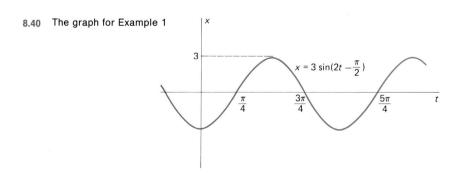

$x = 3 \sin\left(2t - \dfrac{\pi}{2}\right)$

$\dfrac{\pi}{4}$   $\dfrac{3\pi}{4}$   $\dfrac{5\pi}{4}$

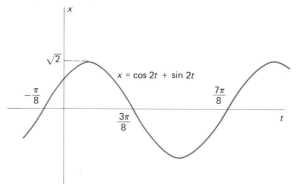

**8.41** The function of Example 2

**Solution** Here we have the form in (2) with $A = B = 1$ and $k = 2$. The relations in (6) give

$$C = \sqrt{2} \quad \text{and} \quad \alpha = \frac{1}{2}\tan^{-1}(-1) = -\frac{\pi}{8}.$$

Thus our curve is $x = \sqrt{2}\sin 2(t + \pi/8)$, with amplitude $C = \sqrt{2}$, period $2\pi/k = \pi$, and phase delay $\alpha = -\pi/8$. Hence its graph looks like the one in Fig. 8.41.

Some typical examples of simple harmonic motion are the motion of a mass attached to a vibrating spring, the motion of an air molecule in a sound wave, the motion of a point on a vibrating musical string (hence the term *harmonic*), and the motion of a piston in an engine.

### A RECIPROCATING PISTON

Consider the piston of Fig. 8.42, moving up and down on one end of a 6-ft shaft. The other end of the shaft is attached at $Q$ to a horizontal slotted arm fitted to a peg $P$ on the rim of a wheel of radius 2 ft. Suppose that the wheel begins with the point $P$ at $\theta = \pi/4$ when $t = 0$ and rotates counterclockwise at 5 rad/s. Then $\theta = 5t + \pi/4$ after $t$ seconds, so the $y$-coordinate of the point $Q$ after $t$ seconds is

$$y = 2\sin\theta = 2\sin\left(5t + \frac{\pi}{4}\right)$$

$$= 2\sin 5\left(t + \frac{\pi}{20}\right).$$

Thus the motion of the piston—essentially that of the point $Q$—is simple harmonic motion with amplitude $C = 2$ feet, period $2\pi/k = 2\pi/5$, and phase delay $\alpha = -\pi/20$.

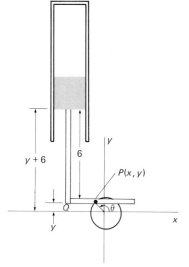

**8.42** The reciprocating piston

### MASS AND SPRING

Suppose that a mass $m$ rests on a frictionless horizontal plane, and is attached to a spring of natural (unstretched) length $L$. We assume that the spring obeys Hooke's law: If the spring is stretched an amount $x$, then it exerts a restorative force $F = -cx$, where $c$ is a positive constant (the *spring constant*). As shown in Fig. 8.43, the spring has one end attached to a fixed wall and the other to the mass $m$.

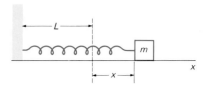

8.43  A simple mass-and-spring system

Now suppose that the mass is set in motion with position function $x(t)$, measured from the equilibrium position $x = 0$ (corresponding to an unstretched spring). Combining Newton's law $F = ma = mx''(t)$ and Hooke's law $F = -cx$, we find that

$$mx''(t) = -cx(t),$$

or

$$x''(t) = -k^2 x(t)$$

where $k = \sqrt{c/m}$. Thus the position function of the mass satisfies differential equation (1), and so by the theorem above, its motion is simple harmonic:

$$x(t) = A \cos kt + B \sin kt,$$

where the constants $A$ and $B$ are determined by the initial position $x_0$ and the initial velocity $v_0$ of the mass (as in Equation (4)). From (9) we see that the period of motion of the mass is

$$\frac{2\pi}{k} = 2\pi \sqrt{\frac{m}{c}}, \tag{10}$$

which depends only upon the mass $m$ and the spring constant $c$ and is independent of $x_0$ and $v_0$. The **frequency** $v$ of this simple harmonic motion—the number of cycles completed per second—is the reciprocal of the period:

$$v = \frac{k}{2\pi} = \frac{1}{2\pi} \sqrt{\frac{c}{m}}. \tag{11}$$

**EXAMPLE 3**  A mass-and-spring system is constructed as in the above discussion; the frequency turns out, by experiment, to be $v = 1/\pi$. The mass is then set into motion from the initial position $x_0 = 1$ with initial velocity $v_0 = 2$. Describe its subsequent motion.

*Solution*  By Equation (4) we know that

$$x(t) = \cos kt + \frac{2}{k} \sin kt.$$

From Equation (11) we see that $v = 1/\pi$ implies that $k = 2$, so

$$x(t) = \cos 2t + \sin 2t = \sqrt{2} \sin 2\left(t + \frac{\pi}{8}\right).$$

In the last step we used the result of Example 2; we see that the amplitude of the motion is $\sqrt{2}$ and the mass first reaches the position $x = \sqrt{2}$ when $t = \pi/8$ and first reaches the position $x = -\sqrt{2}$ when $t = 5\pi/8$.

**THE SIMPLE PENDULUM**

A simple pendulum consists of a mass $m$ swinging to and fro on the end of a string (or, better, a "massless rod") of length $L$, as in Fig. 8.44. We may specify the position of the mass at time $t$ by giving the counterclockwise

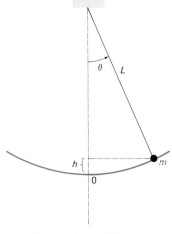

**8.44** The simple pendulum

angle $\theta = \theta(t)$ that the string or rod makes with the vertical. To analyze the motion of the mass $m$, we will apply the law of the *conservation of mechanical energy*, according to which the sum of the kinetic energy (K.E.) and the potential energy (P.E.) of $m$ remains constant.

The distance along the circular arc from 0 to $m$ is $s = L\theta$, so the speed of the mass is

$$v = \frac{ds}{dt} = L\frac{d\theta}{dt},$$

and its kinetic energy is

$$\text{K.E.} = \frac{1}{2}mv^2 = \frac{1}{2}m\left(\frac{ds}{dt}\right)^2 = \frac{1}{2}mL^2\left(\frac{d\theta}{dt}\right)^2.$$

We next choose as reference point the lowest point $O$ reached by the mass (see Fig. 8.44). Then its potential energy is the product of its weight $mg$ (where $g$ is gravitational acceleration) and its vertical height above $O$, $h = L(1 - \cos\theta)$, so that

$$\text{P.E.} = mgL(1 - \cos\theta).$$

The fact that the sum of K.E. and P.E. is a constant $C$ therefore gives

$$\frac{1}{2}mL^2\left(\frac{d\theta}{dt}\right)^2 + mgL(1 - \cos\theta) = C.$$

We differentiate both sides of this identity with respect to $t$ and find that

$$mL^2\left(\frac{d\theta}{dt}\right)\frac{d^2\theta}{dt^2} + mgL(\sin\theta)\frac{d\theta}{dt} = 0,$$

and so

$$\frac{d^2\theta}{dt^2} = -\frac{g}{L}\sin\theta \tag{12}$$

after cancellation of the common factor $mL^2(d\theta/dt)$.

Equation (12) is a bit too difficult to solve here. But recall from Section 8-2 that $(\sin\theta)/\theta \to 1$ as $\theta \to 0$, so $\theta$ and $\sin\theta$ are approximately equal if $\theta$ is sufficiently close to zero. In particular, examination of a table of sines or of some calculator values indicates that $\theta$ (in radians) and $\sin\theta$ agree to two decimal places when $\theta$ is at most $\pi/12$ (that is, $15°$). In a typical pendulum clock, for example, $\theta$ would certainly not exceed $15°$.

It therefore seems reasonable to simplify our mathematical model of the pendulum by replacing $\sin\theta$ by $\theta$ in Equation (12). This gives

$$\frac{d^2\theta}{dt^2} = -\frac{g}{L}\theta, \tag{13}$$

a differential equation of the form of (1) with $\theta$ instead of $x$ and with $k = \sqrt{g/L}$. Its solution is of the form

$$\theta = A\cos kt + B\sin kt = C\sin k(t - \alpha).$$

Thus the angular motion of the pendulum is simple harmonic, and by (9) its period $p$ is

$$p = \frac{2\pi}{k} = 2\pi\sqrt{\frac{L}{g}}. \tag{14}$$

CHAP. 8: Trigonometric and Hyperbolic Functions

Note that $p$ depends only upon the length $L$ of the pendulum and the gravitational acceleration $g$, but not upon either the mass $m$ of the pendulum bob or the initial conditions. For example, using $g = 32.2$ ft/s², we find that the period of oscillation of a 3-foot pendulum is

$$p = 2\pi\sqrt{\frac{3}{32.2}} \approx 1.92 \text{ s.}$$

Recall that the gravitational acceleration $g$ at a point on the earth's surface at distance $R$ from its center is $g = GM/R^2$, where $G$ is the gravitational constant and $M$ is the earth's mass. If this value of $g$ is substituted into Equation (14), the result is

$$p = 2\pi R\sqrt{\frac{L}{GM}}. \tag{15}$$

If the period of a pendulum of given length $L$ is first measured at sea level—where $R$ is known—and then measured atop a mountain, then Equation (15) may be used to compute the value of $R$ at the mountain top and thereby estimate its height. This method is only approximate, for it ignores the effect of the gravitational attraction of the mountain itself on the pendulum; it would be more accurate for finding the altitude of a dirigible gliding above level terrain.

**EXAMPLE 4**    A pendulum of length 49.2 in., located at a point at sea level where the radius of the earth is $R = 3956$ mi, has the same period $p$ as does a pendulum of length 49.0 in. in a hot-air balloon hovering directly overhead. What is the altitude of the balloon?

*Solution*    For the pendulum at sea level, Equation (15) gives

$$p = 2\pi R\sqrt{\frac{(49.2)(1/12)}{GM}}.$$

The factor $\frac{1}{12}$ is present to convert 49.2 in. into feet. For the pendulum in the balloon, the same equation gives

$$p = 2\pi(R + y)\sqrt{\frac{(49)(1/12)}{GM}}$$

where $y$ denotes the altitude of the balloon. When we equate these two expressions for $p$ and cancel common factors, we get

$$(R + y)\sqrt{49} = R\sqrt{49.2},$$

$$7(R + y) \approx (7.01427)R.$$

Hence the altitude of the balloon is

$$y \approx \frac{(0.01427)R}{7} = \frac{(0.01427)(3956)}{7} \approx 8.065 \text{ mi,}$$

or about 42,600 ft.

In 1672 the French astronomer Richer traveled to South America on a scientific mission. While there, he found that his pendulum clock, which had kept very accurate time in Paris, lost more than 2 min/day in French Guiana. Newton deduced from the behavior of Richer's clock that French Guiana must be farther than Paris from the center of the earth, and thereby Newton discovered the equatorial bulge of the earth. There are similar computations in the problems below.

## 8-6 PROBLEMS

In each of Problems 1–7 sketch the graph of the given function. Show at least one cycle.

**1** $x = 2 \sin \pi t$

**2** $x = 2 \cos 3\pi t$

**3** $x = \sin \pi(t - 1)$

**4** $x = 2 \cos(2t - \pi)$

**5** $x = 3 \sin \dfrac{\pi}{2}(4t - 1)$

**6** $x = \sin t + \sqrt{3} \cos t$

**7** $x = 3 \sin \pi t + 4 \cos \pi t$

**8** In a certain city the average temperature $f(t)$ on the $t$th day of the year (with $t = 1$ on January 1) is given by a formula of the form $f(t) = A + B \sin k(t - \alpha)$.
(a) What must $k$ be for the period of $f(t)$ to be 365 days?
(b) The minimum daily average temperature of 50°F occurs on January 20 and the maximum of 80°F occurs half a year later. What are the values of $A$, $B$, and $\alpha$?
(c) On what days of the year is the average temperature 60°F? Repeat for 70°F.

Problems 9–15 deal with a mass on a spring, undergoing simple harmonic motion with position function $x(t)$.

**9** What is the amplitude of the motion if its frequency is 2 hz (cycles per second), $x(0) = 1$, and $x'(0) = 0$?

**10** What is the amplitude of the motion if its frequency is $2/\pi$ hz, $x(0) = 0$, and $x'(0) = -4$?

**11** What is its amplitude if its period is $\pi$ seconds and $x(0) = x'(0) = 1$?

**12** What is the period of the motion if its amplitude is 4, $x(0) = 0$, and $x'(0) = 2$?

**13** What is the maximum velocity of the mass if the amplitude of the motion is 5 and its period is 2 s?

**14** What is the amplitude of the motion if its frequency is $1/\pi$ hz and the maximum velocity of the mass is 10?

**15** Find the period and amplitude of the motion if $x(0) = 3$, $x'(0) = 4$, and $x''(0) = -3$.

**16** Consider a piston as in Fig. 8.42, except that the piston shaft is 5 ft long and the diameter of the wheel is 3 ft. Suppose that the wheel begins at time $t = 0$ with $\theta = \pi$ and rotates counterclockwise at 5 revolutions per second. Express the displacement of the piston as a function of $t$. What is the maximum velocity of the piston?

**17** Two pendulums are of lengths $L_1$ and $L_2$, and—when located at the respective distances $R_1$ and $R_2$ from the center of the earth—have periods $p_1$ and $p_2$. Show that

$$\frac{p_1}{p_2} = \frac{R_1\sqrt{L_1}}{R_2\sqrt{L_2}}.$$

**18** A certain pendulum clock keeps perfect time in Paris, where the radius of the earth is $R = 3956$ mi. But this clock loses 2 min 40 s per day at a location on the equator. That is, the pendulum requires 24 h 2 min 40 s to complete the same number of periods that it completes in exactly 24 h in Paris. Apply the result of Problem 17 to find the distance of the equatorial point from the center of the earth, and thereby gauge the equatorial bulge of the earth. (*Remark:* This model of the situation is simplistic in that it assumes the earth to be a perfect sphere through Paris and also ignores the effects of centrifugal (spinning earth) forces on pendulums.)

**19** A pendulum of length 100.10 in., located at a point at sea level where the radius of the earth is $R = 3960$ mi, has the same period as does a pendulum of length 100.00 in. atop a nearby mountain. Use the result of Problem 17 to estimate the height of the mountain.

**20** Most grandfather clocks have pendulums with lengths that are adjustable. One such clock loses 10 min/day when the length of the pendulum is 30 in. With what length pendulum will this clock keep perfect time?

**21** This is an outline of the proof of the *only if* part of the theorem in this section.
(a) Suppose that $f(t)$ is twice differentiable and that $f''(t) = -k^2 f(t)$. Let the new function $g(t)$ be defined as follows:

$$g(t) = f(t) - f(0) \cos kt - \frac{f'(0)}{k} \sin kt.$$

Then show that $g''(t) = -k^2 g(t)$ and that $g(0) = g'(0) = 0$.
(b) Let $H(t) = [g'(t)]^2 + k^2[g(t)]^2$. Show that $H'(t) \equiv 0$. Conclude that $H(t) \equiv 0$. Why does this complete the proof?

**22** Suppose that the position function $x = x(t)$ satisfies the differential equation $x''(t) + k^2 x(t) = a$, where $a$ and $k$ are constants. Let $y(t) = x(t) - (a/k^2)$. Show that $y''(t) = -k^2 y(t)$. Conclude that

$$x(t) = \frac{a}{k^2} + A \cos kt + B \sin kt.$$

Thus $x(t)$ describes simple harmonic motion about the *equilibrium position* $x_e = a/k^2$.

**23** Consider a mass $m$ hanging on the end of a vertical spring, as in Fig. 8.45. Let $x(t)$ denote the distance of the mass *below* the position $x = 0$ when the spring is unstretched (as if the mass were not attached). Show that

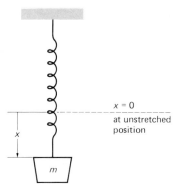

**8.45** Mass hanging from a spring, as in Problem 23

Newton's law $F = ma$ yields

$$mx''(t) + cx(t) = mg,$$

where $c$ is the spring constant. Conclude from the result of Problem 22 that the mass undergoes simple harmonic motion with period $p = 2\pi\sqrt{m/c}$ about the equilibrium position $x_e = mg/c$.

**24** Consider a floating cylindrical buoy with radius $r$, height $h$, and density $\rho \leq \frac{1}{2}$ (recall that the density of water is 1 gm/cm³). The buoy is initially suspended at rest with its bottom at the top surface of the water and is released at time $t = 0$. Thereafter it is acted upon by two forces: a downward gravitational force equal to its weight $mg = \rho\pi r^2 hg$ and an upward force of buoyancy equal to the weight $\pi r^2 xg$ of water displaced, where $x = x(t)$ is the depth of the buoy's bottom beneath the surface at time $t$. Conclude from Problem 22 that the buoy undergoes simple harmonic motion about the equilibrium position $x_e = \rho h$ with period $p = 2\pi\sqrt{\rho h/g}$. Compute $p$ and the amplitude of the motion if $\rho = 0.5$ gm/cm³, $h = 200$ cm, and $g = 980$ cm/s².

**8-7**

## Hyperbolic Functions

The **hyperbolic cosine** and the **hyperbolic sine** of the real number $x$ are denoted by $\cosh x$ and $\sinh x$ and are defined to be

$$\cosh x = \frac{e^x + e^{-x}}{2}, \qquad \sinh x = \frac{e^x - e^{-x}}{2}. \tag{1}$$

These particular combinations of familiar exponentials are useful in certain applications of calculus and are also helpful in evaluating certain integrals. The other four hyperbolic functions—the hyperbolic tangent, cotangent, secant, and cosecant—are defined in terms of $\cosh x$ and $\sinh x$ by analogy with trigonometry:

$$\tanh x = \frac{\sinh x}{\cosh x} = \frac{e^x - e^{-x}}{e^x + e^{-x}},$$

$$\coth x = \frac{\cosh x}{\sinh x} = \frac{e^x + e^{-x}}{e^x - e^{-x}} \qquad (x \neq 0); \tag{2}$$

$$\operatorname{sech} x = \frac{1}{\cosh x} = \frac{2}{e^x + e^{-x}},$$

$$\operatorname{csch} x = \frac{1}{\sinh x} = \frac{2}{e^x - e^{-x}} \qquad (x \neq 0). \tag{3}$$

The trigonometric terminology and notation for these hyperbolic functions stem from the fact that they satisfy a list of identities that, apart from an occasional difference of sign, much resemble the familiar trigonometric identities:

$$\cosh^2 x - \sinh^2 x = 1; \tag{4}$$

$$1 - \tanh^2 x = \operatorname{sech}^2 x; \tag{5}$$

$$\coth^2 x - 1 = \operatorname{csch}^2 x; \tag{6}$$

$$\sinh(x + y) = \sinh x \cosh y + \cosh x \sinh y; \tag{7}$$

$$\cosh(x + y) = \cosh x \cosh y + \sinh x \sinh y; \tag{8}$$

$$\sinh 2x = 2 \sinh x \cosh x; \tag{9}$$

$$\cosh 2x = \cosh^2 x + \sinh^2 x; \tag{10}$$

$$\cosh^2 x = \tfrac{1}{2}(\cosh 2x + 1); \tag{11}$$

$$\sinh^2 x = \tfrac{1}{2}(\cosh 2x - 1). \tag{12}$$

The identities in (4), (7), and (8) follow directly from the definitions of cosh $x$ and sinh $x$. For example,

$$\begin{aligned}
\cosh^2 x - \sinh^2 x &= \tfrac{1}{4}(e^x + e^{-x})^2 - \tfrac{1}{4}(e^x - e^{-x})^2 \\
&= \tfrac{1}{4}(e^{2x} + 2 + e^{-2x}) - \tfrac{1}{4}(e^{2x} - 2 + e^{-2x}) \\
&= 1.
\end{aligned}$$

The other identities above may be derived from (4), (7), and (8) in ways that parallel the derivations of the standard trigonometric identities.

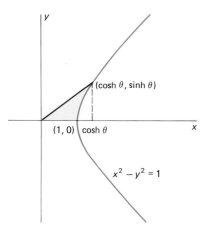

**8.46** Relation of the hyperbolic cosine and hyperbolic sine to the hyperbola $x^2 - y^2 = 1$.

The trigonometric functions are sometimes called *circular* functions because the point $(\cos \theta, \sin \theta)$ lies on the circle $x^2 + y^2 = 1$ for all $\theta$. Similarly, Identity (4) above tells us that the point $(\cosh \theta, \sinh \theta)$ lies on the hyperbola $x^2 - y^2 = 1$, and this is the reason for the name *hyperbolic* function (see Fig. 8.46).

The graphs of $y = \cosh x$ and $y = \sinh x$ are easily constructed: Add (for cosh) or subtract (for sinh) the ordinates of the graphs of $y = \tfrac{1}{2}e^x$ and $y = \tfrac{1}{2}e^{-x}$. The graphs of the other four hyperbolic functions can then be constructed by dividing ordinates. The graphs of all six are shown in Fig. 8.47.

A striking difference is apparent in these graphs, a difference between the hyperbolic functions and the ordinary trigonometric functions: None of the hyperbolic functions is periodic. They do, however, have even-odd properties like the circular functions. The two functions cosh and sech are even, because

$$\cosh(-x) = \cosh x \quad \text{and} \quad \operatorname{sech}(-x) = \operatorname{sech} x$$

for all $x$. The other four hyperbolic functions are odd:

$$\sinh(-x) = -\sinh x, \qquad \tanh(-x) = -\tanh x,$$

and so on.

The formulas for the derivatives of the hyperbolic functions parallel those for the trigonometric functions, with occasional sign differences. For example,

$$\begin{aligned}
D \cosh x &= D(\tfrac{1}{2}e^x + \tfrac{1}{2}e^{-x}) \\
&= \tfrac{1}{2}e^x - \tfrac{1}{2}e^{-x} = \sinh x.
\end{aligned}$$

The chain rule then gives

$$D_x \cosh u = (\sinh u)\frac{du}{dx} \tag{13}$$

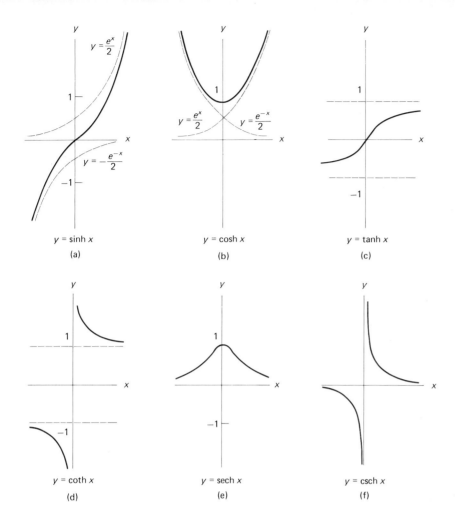

$y = \sinh x$

(a)

$y = \cosh x$

(b)

$y = \tanh x$

(c)

$y = \coth x$

(d)

$y = \operatorname{sech} x$

(e)

$y = \operatorname{csch} x$

(f)

**8.47** Graphs of the six hyperbolic functions

if $u$ is a differentiable function of $x$. The other five differentiation formulas are

$$D_x \sinh u = (\cosh u)\frac{du}{dx}, \tag{14}$$

$$D_x \tanh u = (\operatorname{sech}^2 u)\frac{du}{dx}, \tag{15}$$

$$D_x \coth u = -(\operatorname{csch}^2 u)\frac{du}{dx}, \tag{16}$$

$$D_x \operatorname{sech} u = (-\operatorname{sech} u \tanh u)\frac{du}{dx}, \tag{17}$$

$$D_x \operatorname{csch} u = (-\operatorname{csch} u \coth u)\frac{du}{dx}. \tag{18}$$

Formula (14) is derived just as (13) is. Then Formulas (15) through (18) follow from (13) and (14) with the aid of the quotient rule and identities (5) and (6).

As indicated in the following example, the differentiation of hyperbolic functions using the formulas in (13) through (18) is very similar to differentiation of trigonometric functions.

**EXAMPLE 1**

(a)  $D \cosh 2x = 2 \sinh 2x.$

(b)  $D \sinh^2 x = 2 \sinh x \cosh x.$

(c)  $D(x \tanh x) = \tanh x + x \, \text{sech}^2 x.$

(d)  $D \, \text{sech}(x^2) = -2x \, \text{sech} \, x^2 \tanh x^2.$

---

The antiderivative versions of the differentiation formulas in (14) through (18) are the following integral formulas.

$$\int \sinh u \, du = \cosh u + C, \tag{19}$$

$$\int \cosh u \, du = \sinh u + C, \tag{20}$$

$$\int \text{sech}^2 u \, du = \tanh u + C, \tag{21}$$

$$\int \text{csch}^2 u \, du = -\coth u + C, \tag{22}$$

$$\int \text{sech} \, u \tanh u \, du = -\text{sech} \, u + C, \tag{23}$$

$$\int \text{csch} \, u \coth u \, du = -\text{csch} \, u + C. \tag{24}$$

The integrals in the following example illustrate the fact that simple hyperbolic integrals may be treated in much the same way as simple trigonometric integrals (Sections 8-3 and 8-4).

**EXAMPLE 2**

(a)  $\int \sinh^2 x \, dx = \int \frac{1}{2}(\cosh 2x - 1) \, dx = \frac{1}{4} \sinh 2x - \frac{1}{2}x + C.$

(b)  $\int \cosh^2 x \sinh x \, dx = \frac{1}{3} \cosh^3 x + C.$

(c)  $\int \frac{\text{sech}^2 \sqrt{x}}{\sqrt{x}} \, dx = 2 \int (\text{sech}^2 \sqrt{x}) \frac{dx}{2\sqrt{x}} = 2 \tanh \sqrt{x} + C.$

(d)  $\int_0^1 \tanh^2 x \, dx = \int_0^1 (1 - \text{sech}^2 x) \, dx = \left[ x - \tanh x \right]_0^1$

$$= 1 - \tanh 1 = 1 - \frac{e^1 - e^{-1}}{e^1 + e^{-1}}$$

$$= \frac{2}{e^2 + 1} \approx 0.238.$$

(e)  $\int \coth x \, dx = \int \frac{\cosh x \, dx}{\sinh x} = \ln|\sinh x| + C.$

**DOWNWARD MOTION WITH RESISTANCE**

In Section 7-6 we discussed the vertical motion of a body of mass $m$ under the influence of a gravitational force $F_G = -mg$ and a resistance force—such as air resistance—that we assumed proportional to the velocity $v = dy/dt$ of the body. In the case of a heavy and rapidly moving body, it is often more accurate to assume that the force of air resistance is proportional to the *square* of the velocity:

$$F_R = \pm kv^2 \qquad (k \text{ constant}, \quad k > 0). \qquad (25)$$

Then Newton's law $F = ma$ gives

$$m\frac{dv}{dt} = F_G + F_R = -mg \pm kv^2,$$

so that

$$\frac{dv}{dt} = -g \pm \rho v^2 \qquad (26)$$

where $\rho = k/m > 0$.

The choice of plus or minus sign in (26) depends upon the direction of motion. For downward motion $F_R$ must be positive (directed upward, which we take as the positive direction), so we choose the plus sign:

$$\frac{dv}{dt} = -g + \rho v^2. \qquad (26a)$$

In the case of upward motion we would take the minus sign.

**EXAMPLE 3**   Suppose that a body is dropped at time $t = 0$ from a height $y_0$ above the ground, so that its height function $y(t)$ must satisfy Equation (26a) and the initial conditions $y(0) = y_0$ with $y'(0) = 0$. Verify that the function

$$y(t) = y_0 - \frac{1}{\rho}\ln(\cosh t\sqrt{\rho g}) \qquad (27)$$

meets these requirements.

*Solution*   We differentiate the function $y(t)$ of Equation (27), and thus we find that

$$v = \frac{dy}{dt} = -\frac{1}{\rho}\frac{\sinh t\sqrt{\rho g}}{\cosh t\sqrt{\rho g}}\sqrt{\rho g} = -\sqrt{g/\rho}\,\tanh t\sqrt{\rho g}. \qquad (28)$$

So,

$$v^2 = \frac{g}{\rho}\tanh^2 t\sqrt{\rho g}.$$

Another differentiation gives

$$\frac{dv}{dt} = -g\,\mathrm{sech}^2 t\sqrt{\rho g} = -g(1 - \tanh^2 t\sqrt{\rho g})$$

$$= -g + \rho \cdot \frac{g}{\rho}\tanh^2 t\sqrt{\rho g} = -g + \rho v^2.$$

This shows that the function of Equation (27) satisfies Equation (26a). The facts that $y(0) = y_0$ and $v(0) = 0$ follow from the facts that $\cosh 0 = 1$ and $\tanh 0 = 0$. Problem 30 in Section 8-8 will outline a systematic derivation of the solution in (27) from the differential equation in (26a).

**EXAMPLE 4**   Let us apply Equations (27) and (28) with $g = 32$ ft/s² and $\rho = 0.005$ in ft-lb-s units. After $t = 5$ seconds, the body has fallen a distance

$$\frac{1}{\rho} \ln(\cosh t\sqrt{\rho g}) = 200 \ln(\cosh 5\sqrt{0.16}) \approx 265$$

ft, and its speed is

$$\sqrt{\frac{g}{\rho}} \tanh t\sqrt{\rho g} = 80 \tanh 5\sqrt{0.16} \approx 77$$

ft/s. By contrast, without air resistance it would have fallen a distance of

$$\tfrac{1}{2}gt^2 = (16)(5)^2 = 400$$

ft in 5 s and would have speed

$$gt = (32)(5) = 160$$

ft/s.

## 8-7   PROBLEMS

Find the derivatives of the functions in Problems 1–14.

1  $f(x) = \cosh(3x - 2)$

2  $f(x) = \sinh\sqrt{x}$

3  $f(x) = x^2 \tanh\left(\dfrac{1}{x}\right)$

4  $f(x) = \operatorname{sech} e^{2x}$

5  $f(x) = \coth^3 4x$

6  $f(x) = \ln \sinh 3x$

7  $f(x) = e^{\operatorname{csch} x}$

8  $f(x) = \cosh \ln x$

9  $f(x) = \sin(\sinh x)$

10  $f(x) = \tan^{-1}(\tanh x)$

11  $f(x) = \sinh x^4$

12  $f(x) = \sinh^4 x$

13  $f(x) = \dfrac{1}{x + \tanh x}$

14  $f(x) = \cosh^2 x - \sinh^2 x$

Evaluate the integrals in Problems 15–28.

15  $\displaystyle\int x \sinh x^2 \, dx$

16  $\displaystyle\int \cosh^2 3u \, du$

17  $\displaystyle\int \tanh^2 3x \, dx$

18  $\displaystyle\int \frac{\operatorname{sech}\sqrt{x} \tanh\sqrt{x}}{\sqrt{x}} \, dx$

19  $\displaystyle\int \sinh^2 2x \cosh 2x \, dx$

20  $\displaystyle\int \tanh 3x \, dx$

21  $\displaystyle\int \frac{\sinh x}{\cosh^3 x} \, dx$

22  $\displaystyle\int \sinh^4 x \, dx$

23  $\displaystyle\int \coth x \operatorname{csch}^2 x \, dx$

24  $\displaystyle\int \operatorname{sech} x \, dx$

25  $\displaystyle\int \frac{\sinh x \, dx}{1 + \cosh x}$

26  $\displaystyle\int \frac{\sinh \ln x}{x} \, dx$

27  $\displaystyle\int \frac{dx}{(e^x + e^{-x})^2}$

28  $\displaystyle\int \frac{e^x + e^{-x}}{e^x - e^{-x}} \, dx$

29  Apply Definition (1) to prove Identity (7).

30  Derive Identities (5) and (6) from Identity (4).

31  Deduce Identities (10) and (11) from Identity (8).

32  Suppose that $A$ and $B$ are constants. Show that the function $x = A \cosh kt + B \sinh kt$ is a solution of the differential equation $x''(t) = k^2 x(t)$.

33  Find the length of the curve $y = \cosh x$ over the interval $[0, a]$.

34  Find the volume of the solid obtained by revolving around the $x$-axis the area under $y = \sinh x$ from $x = 0$ to $x = \pi$.

35  Show that the area $A(\theta)$ of the shaded sector in Fig. 8.46 is $\theta/2$. This corresponds to the fact that the area of the sector of the unit circle between the positive $x$-axis and the radius to the point $(\cos \theta, \sin \theta)$ is $\theta/2$. (*Suggestion:* Note first that

$$A(\theta) = \tfrac{1}{2} \cosh \theta \sinh \theta - \int_1^{\cosh \theta} \sqrt{x^2 - 1} \, dx.$$

Then use the fundamental theorem of calculus to show that $A'(\theta) = \tfrac{1}{2}$ for all $\theta$.)

36  (a) Verify that $\lim\limits_{x \to \infty} \tanh x = 1$, as indicated in Fig. 8.47.

(b) Deduce from the solution to Example 3 that a body falling from rest subject to a resistance force proportional to the square of its velocity (Equation (26a)) approaches the terminal speed $v_\tau = \sqrt{g/\rho}$.

A glance at Figure 8.47 in Section 8-7 shows that the hyperbolic sine and tangent functions are increasing on the whole real line. Thus each is one-to-one on the set of all real numbers. The hyperbolic cotangent and cosecant functions are also one-to-one where they are defined; each of the latter has domain the set of all real numbers other than zero. Thus each of these four hyperbolic functions has an inverse function whose domain is in each case the range of values of the original function:

**Inverse Hyperbolic Functions**

$$y = \sinh^{-1} x \quad \text{if } x = \sinh y, \quad \text{all } x; \tag{1}$$

$$y = \tanh^{-1} x \quad \text{if } x = \tanh y, \quad -1 < x < 1; \tag{2}$$

$$y = \coth^{-1} x \quad \text{if } x = \coth y, \quad |x| > 1; \tag{3}$$

$$y = \operatorname{csch}^{-1} x \quad \text{if } x = \operatorname{csch} y, \quad x \neq 0. \tag{4}$$

The hyperbolic cosine and secant functions are one-to-one on the half-line $x \geq 0$, where cosh is increasing and sech is decreasing. Hence we define

$$y = \cosh^{-1} x \quad \text{if} \quad x = \cosh y \tag{5}$$

where $x \geq 1$ and $y \geq 0$, and

$$y = \operatorname{sech}^{-1} x \quad \text{if} \quad x = \operatorname{sech} y \tag{6}$$

where $0 < x \leq 1$ and $y \geq 0$. The graphs of the six inverse hyperbolic functions may be obtained by reflecting through the line $y = x$ the graphs of the hyperbolic functions (restricted to the half-line $x \geq 0$ in the cases of cosh and sech). We show these graphs in Fig. 8.48.

The derivatives of the six inverse hyperbolic functions are these:

$$D \sinh^{-1} x = \frac{1}{\sqrt{x^2 + 1}}, \tag{7}$$

$$D \cosh^{-1} x = \frac{1}{\sqrt{x^2 - 1}}, \tag{8}$$

$$D \tanh^{-1} x = \frac{1}{1 - x^2}, \tag{9}$$

$$D \coth^{-1} x = \frac{1}{1 - x^2}, \tag{10}$$

$$D \operatorname{sech}^{-1} x = \frac{-1}{x\sqrt{1 - x^2}}, \tag{11}$$

$$D \operatorname{csch}^{-1} x = \frac{-1}{|x|\sqrt{1 + x^2}}. \tag{12}$$

One can derive these formulas by the standard method of finding the derivative of an inverse function given the derivative of the function itself.

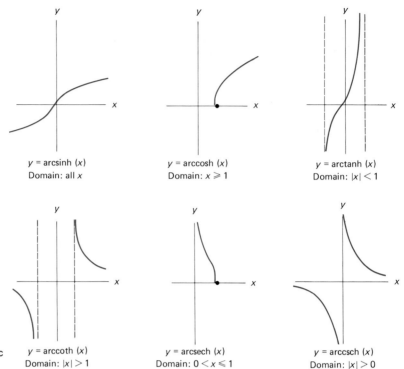

| $y = \text{arcsinh}\,(x)$ | $y = \text{arccosh}\,(x)$ | $y = \text{arctanh}\,(x)$ |
|---|---|---|
| Domain: all $x$ | Domain: $x \geqslant 1$ | Domain: $|x| < 1$ |

**8.48** The inverse hyperbolic functions

| $y = \text{arccoth}\,(x)$ | $y = \text{arcsech}\,(x)$ | $y = \text{arccsch}\,(x)$ |
|---|---|---|
| Domain: $|x| > 1$ | Domain: $0 < x \leqslant 1$ | Domain: $|x| > 0$ |

The only requirement is that the inverse function is known in advance to be differentiable. For example, to differentiate $\tanh^{-1}x$, we begin with the relation

$$\tanh(\tanh^{-1}x) = x$$

which follows from the definition in (2), and substitute $u = \tanh^{-1}x$. Then, because the equation above is actually an identity,

$$D_x \tanh u = Dx = 1,$$

so that

$$(\text{sech}^2 u)\,\frac{du}{dx} = 1.$$

Thus

$$D \tanh^{-1}x = \frac{du}{dx} = \frac{1}{\text{sech}^2 u} = \frac{1}{1 - \tanh^2 u}$$

$$= \frac{1}{1 - \tanh^2(\tanh^{-1}x)} = \frac{1}{1 - x^2}.$$

This establishes Formula (9). One can use similar methods to verify the other five derivative formulas above.

The hyperbolic functions are defined in terms of the natural exponential function $e^x$, so it is no surprise to find that their inverses may be expressed in terms of ln $x$. In fact,

$$\sinh^{-1}x = \ln(x + \sqrt{x^2 + 1}) \qquad \text{for all } x; \qquad (13)$$

$$\cosh^{-1}x = \ln(x + \sqrt{x^2 - 1}) \qquad \text{for all } x \geqq 1; \qquad (14)$$

$$\tanh^{-1}x = \frac{1}{2}\ln\left(\frac{1 + x}{1 - x}\right) \qquad \text{for } |x| < 1; \qquad (15)$$

$$\coth^{-1}x = \frac{1}{2}\ln\left(\frac{x + 1}{x - 1}\right) \qquad \text{for } |x| > 1; \qquad (16)$$

$$\text{sech}^{-1}x = \ln\left(\frac{1 + \sqrt{1 - x^2}}{x}\right) \qquad \text{if } 0 < x \leqq 1; \qquad (17)$$

$$\text{csch}^{-1}x = \ln\left(\frac{1}{x} + \frac{\sqrt{1 + x^2}}{|x|}\right) \qquad \text{if } x \neq 0. \qquad (18)$$

Each of these identities may be established by showing that each side has the same derivative, and that the two sides also agree for at least one value of $x$ in every interval in their respective domains. For example,

$$D \ln(x + \sqrt{x^2 + 1}) = \frac{1 + x(x^2 + 1)^{-1/2}}{x + (x^2 + 1)^{1/2}}$$

$$= \frac{1}{\sqrt{x^2 + 1}} = D \sinh^{-1}x.$$

So

$$\sinh^{-1}x = \ln(x + \sqrt{x^2 + 1}) + C.$$

But $\sinh^{-1}(0) = 0 = \ln(0 + \sqrt{0 + 1})$. This implies that $C = 0$ and thus establishes Formula (13). It is not quite so easy to show that $C = 0$ in the proofs of Formulas (16) and (18); see Problems 26 and 27.

The formulas in (13) through (18) may be used to calculate inverse hyperbolic function values. This is convenient if you own a calculator with a repertoire that does not include the inverse hyperbolic functions, or if you are programming in a language such as BASIC, most forms of which do not include these functions.

**EXAMPLE 1** Suppose that a body is dropped at time $t = 0$ from a height $y_0$ above the ground. Assume that it is acted on both by gravity and by air resistance proportional to $v^2$ (where $v$ is its velocity). By Example 3 in Section 8-5, its height $t$ seconds after release will be

$$y = y_0 - \frac{1}{\rho}\ln(\cosh t\sqrt{\rho g}).$$

Assume that $y_0 = 2000$ ft, that $g = 32$ ft/s$^2$, and that $\rho = 0.001$. How long will it take for this body to fall to the ground?

**Solution** If we set $y = 0$ in the above equation and then solve for $t$, we get

$$t = \frac{1}{\sqrt{\rho g}}\cosh^{-1}(e^{\rho y_0}).$$

Then Formula (14) gives

$$t = \frac{1}{\sqrt{\rho g}} \ln(e^{\rho y_0} + \sqrt{e^{2\rho y_0} - 1}).$$

We substitute the data given in the problem and find that

$$t = \frac{1}{\sqrt{0.032}} \ln(e^2 + \sqrt{e^4 - 1}) \approx 15.03$$

s—this is the time required for the body to fall to the ground.

---

The differentiation formulas in (7) through (12) may, in the usual way, be written as the following integral formulas.

$$\int \frac{du}{\sqrt{u^2 + 1}} = \sinh^{-1} u + C. \tag{19}$$

$$\int \frac{du}{\sqrt{u^2 - 1}} = \cosh^{-1} u + C. \tag{20}$$

$$\int \frac{du}{1 - u^2} = \begin{cases} \tanh^{-1} u + C & \text{if } |u| < 1; \tag{21a} \\ \coth^{-1} u + C & \text{if } |u| > 1; \tag{21b} \end{cases}$$

$$= \frac{1}{2} \ln \left| \frac{1 + u}{1 - u} \right| + C. \tag{21c}$$

$$\int \frac{du}{u\sqrt{1 - u^2}} = -\text{sech}^{-1} u + C. \tag{22}$$

$$\int \frac{du}{u\sqrt{1 + u^2}} = -\text{csch}^{-1} |u| + C. \tag{23}$$

The distinction between the two cases $|u| < 1$ and $|u| > 1$ in Formula (21) results from the fact that the inverse hyperbolic tangent is defined for $|x| < 1$, while the inverse hyperbolic cotangent is defined for $|x| > 1$.

**EXAMPLE 2**

$$\int \frac{dx}{\sqrt{4x^2 + 1}} = \frac{1}{2} \int \frac{du}{\sqrt{u^2 + 1}} \qquad [u = 2x, \quad dx = \tfrac{1}{2} \, du]$$

$$= \frac{1}{2} \sinh^{-1} 2x + C.$$

**EXAMPLE 3**

$$\int_0^{1/2} \frac{dx}{1 - x^2} = \left[ \tanh^{-1} x \right]_0^{1/2}$$

$$= \frac{1}{2} \left[ \ln \left| \frac{1 + x}{1 - x} \right| \right]_0^{1/2} = \frac{1}{2} \ln 3.$$

**EXAMPLE 4**

$$\int_2^5 \frac{dx}{1 - x^2} = \left[\coth^{-1}x\right]_2^5 = \frac{1}{2}\left[\ln\left|\frac{1 + x}{1 - x}\right|\right]_2^5$$

$$= \frac{1}{2}\left[\ln\left(\frac{6}{4}\right) - \ln 3\right] = -\frac{1}{2}\ln 2.$$

## THE HANGING CABLE

In Section 4-8 we investigated the shape of a flexible and inelastic cable supporting a load that is uniformly distributed horizontally, such as a supporting cable of a suspension bridge. We found that such a cable hangs in the shape of a parabola. Here, however, we investigate the shape of a cable that hangs of its own weight alone, which we assume is uniformly distributed along the cable, and therefore is not uniformly distributed horizontally.

Let $w$ denote the density of the cable, measured in pounds per foot of length. Figure 8.49 shows the cable with lowest point $P$ and with a coordinate system set up just as in Section 4-8.

Now consider a section $PQ$ of the cable of length $s$. We plan to obtain a differential equation for the shape $y = f(x)$ of the hanging cable by balancing horizontal and vertical components of the forces acting on $PQ$. These forces are:

$T_0$    the horizontal tension pulling on the cable at $P$;

$T$    the tangential tension pulling on the cable at $Q$;    and

$ws$    the force of gravity pulling downward on the section $PQ$.

When we equate the horizontal and vertical components, we find that

$$T \cos \phi = T_0 \quad \text{and} \quad T \sin \phi = ws. \tag{24}$$

As Fig. 8.49 shows, $\phi$ is the angle the tangent to the cable at $Q$ makes with the horizontal.

We divide the second equation above by the first and find that

$$\frac{dy}{dx} = \tan \phi = \frac{T \sin \phi}{T \cos \phi} = \frac{ws}{T_0}. \tag{25}$$

Since $s$ is a function of $x$, this differential equation is not as simple as we

**8.49** The hanging cable

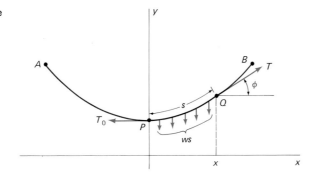

might hope. But after we differentiate both sides, we find that

$$\frac{d^2y}{dx^2} = \frac{w}{T_0}\frac{ds}{dx}.$$

We know from our study of arc length (in Section 6-4) that

$$\frac{ds}{dx} = \sqrt{1 + \left(\frac{dy}{dx}\right)^2},$$

and so

$$\frac{d^2y}{dx^2} = \frac{w}{T_0}\sqrt{1 + \left(\frac{dy}{dx}\right)^2}. \tag{26}$$

This is the differential equation that will give us the shape $y = y(x)$ of the hanging cable if we can manage to find the solution of the equation.

There is a standard method for solving a second order differential equation like the one above; it is used when the dependent variable $y$ does not appear explicitly. We substitute

$$p = \frac{dy}{dx} \quad \text{and} \quad \frac{dp}{dx} = \frac{d^2y}{dx^2}.$$

With Equation (26), this yields the simpler equation

$$\frac{dp}{dx} = \frac{w}{T_0}\sqrt{1 + p^2},$$

which we rewrite in the form

$$\frac{1}{\sqrt{1 + p^2}}\frac{dp}{dx} = \frac{w}{T_0}.$$

Since $D \sinh^{-1}x = 1/\sqrt{1 + x^2}$, integration of both sides of the above equation gives

$$\sinh^{-1}p = \frac{w}{T_0}x + C_1.$$

Now $p = 0$ when $x = 0$, because the cable has a horizontal tangent at its lowest point $P$. This tells us that $C_1 = \sinh^{-1}(0) = 0$, and thus that

$$\sinh^{-1}p = \frac{w}{T_0}x.$$

And so

$$\frac{dy}{dx} = p = \sinh\left(\frac{w}{T_0}x\right).$$

Finally, we integrate both sides of this last equation, and thus we find that

$$y = \frac{T_0}{w}\cosh\left(\frac{w}{T_0}x\right) + C_2.$$

If $y_0$ is the height of the point $P$ (above the $x$-axis), then

$$C_2 = y_0 - \frac{T_0}{w},$$

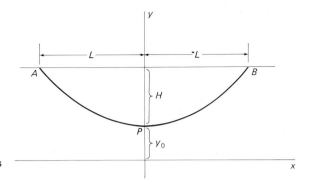

**8.50** A hanging cable of length $S$ and sag $H$, suspended from two points at distance $2L$ apart.

and so the shape of the hanging cable is given by

$$y = \frac{T_0}{w} \cosh\left(\frac{w}{T_0}x\right) + y_0 - \frac{T_0}{w}. \qquad (27)$$

We may choose the location of the $x$-axis so that $y_0 = T_0/w$, and then Equation (27) takes the simple form

$$y = \frac{T_0}{w} \cosh\left(\frac{wx}{T_0}\right), \qquad (28)$$

or, if we let $a = T_0/w$, the form

$$y = a \cosh\left(\frac{x}{a}\right).$$

This curve is frequently called a *catenary*, from the Latin word *catena* (chain).

Problems 31–35 deal with the relationship between the length $S$ of the cable, the distance $2L$ between the two points from which it is suspended (at equal heights), and the dip, or sag, $H$ of the cable at its midpoint. All these lengths are shown in Fig. 8.50.

## 8-8 PROBLEMS

Find the derivatives of the functions in Problems 1–10.

**1** $f(x) = \sinh^{-1}2x$

**2** $f(x) = \cosh^{-1}(x^2 + 1)$

**3** $f(x) = \tanh^{-1}\sqrt{x}$

**4** $f(x) = \coth^{-1}\sqrt{x^2 + 1}$

**5** $f(x) = \operatorname{sech}^{-1}\left(\frac{1}{x}\right)$

**6** $f(x) = \operatorname{csch}^{-1}e^x$

**7** $f(x) = (\sinh^{-1}x)^{3/2}$

**8** $f(x) = \sinh^{-1}(\ln x)$

**9** $f(x) = \ln(\tanh^{-1}x)$

**10** $f(x) = \dfrac{1}{\tanh^{-1}3x}$

Evaluate the integrals in Problems 11–20.

**11** $\displaystyle\int \frac{dx}{\sqrt{x^2 + 9}}$

**12** $\displaystyle\int \frac{dy}{\sqrt{4y^2 - 9}}$

**13** $\displaystyle\int_{1/2}^{1} \frac{dx}{4 - x^2}$

**14** $\displaystyle\int_{5}^{10} \frac{dx}{4 - x^2}$

**15** $\displaystyle\int \frac{dx}{x\sqrt{4 - 9x^2}}$

**16** $\displaystyle\int \frac{dx}{x\sqrt{x^2 + 25}}$

**17** $\displaystyle\int \frac{e^x\, dx}{\sqrt{e^{2x} + 1}}$

**18** $\displaystyle\int \frac{x\, dx}{\sqrt{x^4 - 1}}$

**19** $\displaystyle\int \frac{dx}{\sqrt{1 - e^{2x}}}$

**20** $\displaystyle\int \frac{\cos x\, dx}{\sqrt{1 + \sin^2 x}}$

**21** Establish Formula (7) for $D \sinh^{-1}x$.
**22** Establish Formula (11) for $D \operatorname{sech}^{-1}x$.
**23** Prove Formula (15) by differentiating both sides.

**24** Establish Formula (13) by solving the equation

$$x = \sinh y = \tfrac{1}{2}(e^y - e^{-y})$$

for $y$ in terms of $x$.

**25** Establish Formula (16) by solving the equation

$$x = \coth y = \frac{e^y + e^{-y}}{e^y - e^{-y}}$$

for $y$ in terms of $x$.

**26** (a) Differentiate both sides of Formula (16) to show that they differ by a constant $C$.
(b) Then prove that $C = 0$ by using the definition of $\coth x$ to show that $\coth^{-1} 2 = \tfrac{1}{2}\ln 3$.

**27** (a) Differentiate both sides of Formula (18) to show that they differ by a constant $C$.
(b) Then prove that $C = 0$ by using the definition of $\operatorname{csch} x$ to show that $\operatorname{csch}^{-1} 1 = \ln(1 + \sqrt{2})$.

**28** Suppose that the body of Example 1 is dropped from the top of a 500-ft building. After how long and with what speed does it strike the ground?

**29** Assume that $\rho = 0.075$ (in the formula of Example 1) for a paratrooper falling with open parachute. If she jumps from an altitude of 10,000 ft and opens her parachute immediately, after how long and with what speed will she land?

**30** (a) Starting with the initial value problem

$$\frac{dv}{dt} = -g + \rho v^2, \qquad v(0) = 0,$$

first separate variables, then use Formula (9) to derive the

solution

$$v = -\sqrt{\frac{g}{\rho}} \tanh t\sqrt{\rho g}.$$

(b) Integrate again to obtain Equation (27) in **Section 8-7**.
**31** Consider a hanging cable, of the sort to which **Equation (28)** applies. Show that the length of the section lying above the interval $[0, x]$ is

$$s = \frac{T_0}{w} \sinh\left(\frac{wx}{T_0}\right).$$

**32** Deduce from (28) that the sag of the cable at its middle is

$$H = \frac{T_0}{w}\left[\cosh\left(\frac{wL}{T_0}\right) - 1\right].$$

**33** Deduce from Equation (24) that the tension in the cable satisfies the equation $T^2 = T_0^2 + w^2 s^2$.
**34** Use the results of Problems 31 and 33 to show that the tension in the cable at the point $(x, y)$ is

$$T = T_0 \cosh\left(\frac{wx}{T_0}\right) = wy,$$

the weight of a cable of the same density and of length $y$.
**35** A high-voltage transmission line 205 ft long and weighing 1 lb/ft is strung between 2 towers 200 ft apart.
(a) Use the result of Problem 31 to find the minimum tension $T_0$. (*Suggestion:* Use Newton's method to solve for $u = 100/T_0$.)
(b) Use Problem 33 to find the maximum tension in the transmission line.
(c) Use Problem 32 to find the sag $H$ at the middle.

## CHAPTER 8 REVIEW:   Definitions and Formulas

Use the list below as a guide to concepts you may need to review.

**1** The derivatives of the six trigonometric functions, and the corresponding integral formulas
**2** The definitions of the six inverse trigonometric functions

**3** The derivatives of the inverse trigonometric functions
**4** The integral formulas corresponding to the derivatives of the inverse sine, tangent, and secant functions
**5** The definitions and derivatives of the hyperbolic functions
**6** The definitions of the inverse hyperbolic functions

## MISCELLANEOUS PROBLEMS

Differentiate the functions in Problems 1–20.

**1** $f(x) = \sin\sqrt{x}$

**2** $f(x) = x\cos\left(\dfrac{1}{x}\right)$

**3** $f(x) = \tan(\ln x)$

**4** $f(x) = \sec^2(3x + 1)$

**5** $f(x) = (\csc 2x + \cot 2x)^5$
**6** $f(x) = \sqrt{\tan x^2}$
**7** $f(x) = \ln(\sec^2 3x)$
**8** $f(x) = \cot e^{3x}$
**9** $f(x) = e^{\arctan x}$
**10** $f(x) = \ln(\sin^{-1} x)$
**11** $f(x) = \arcsin\sqrt{x}$
**12** $f(x) = x\sec^{-1} x^2$

13 $f(x) = \tan^{-1}(1 + x^2)$
14 $f(x) = \sin^{-1}\sqrt{1 - x^2}$
15 $f(x) = e^x \sinh e^x$
16 $f(x) = \ln \cosh x$
17 $f(x) = \tanh^2 3x + \text{sech}^2 3x$
18 $f(x) = \sinh^{-1}\sqrt{x^2 - 1}$
19 $f(x) = \cosh^{-1}\sqrt{x^2 + 1}$
20 $f(x) = \tanh^{-1}(1 - x^2)$

Evaluate the integrals in Problems 21–44.

21 $\int x \sec^2 x^2 \, dx$

22 $\int e^x \sec e^x \tan e^x \, dx$

23 $\int \csc^2 2x \cot 2x \, dx$

24 $\int \dfrac{\cot^2 \sqrt{x}}{\sqrt{x}} \, dx$

25 $\int \dfrac{\tan^2(1/x)}{x^2} \, dx$

26 $\int \tan^3 2x \sec^2 2x \, dx$

27 $\int x \tan(x^2 + 1) \, dx$

28 $\int \dfrac{e^x \, dx}{\sqrt{1 - e^x}}$

29 $\int \dfrac{e^x \, dx}{\sqrt{1 - e^{2x}}}$

30 $\int \dfrac{x \, dx}{1 + x^4}$

31 $\int \dfrac{dx}{\sqrt{9 - 4x^2}}$

32 $\int \dfrac{dx}{9 + 4x^2}$

33 $\int \dfrac{x^2 \, dx}{1 + x^6}$

34 $\int \dfrac{\cos x \, dx}{1 + \sin^2 x}$

35 $\int \dfrac{dx}{x\sqrt{4x^2 - 1}}$

36 $\int \dfrac{dx}{x\sqrt{x^4 - 1}}$

37 $\int \dfrac{dx}{\sqrt{e^{2x} - 1}}$

38 $\int x^2 \cosh x^3 \, dx$

39 $\int \dfrac{\sinh \sqrt{x}}{\sqrt{x}} \, dx$

40 $\int \text{sech}^2(3x - 2) \, dx$

41 $\int \dfrac{\tan^{-1} x}{1 + x^2} \, dx$

42 $\int \dfrac{dx}{\sqrt{4x^2 - 1}}$

43 $\int \dfrac{dx}{\sqrt{4x^2 + 9}}$

44 $\int \dfrac{x \, dx}{\sqrt{x^4 + 1}}$

45 Find the volume generated by revolving the region under $y = (1 - x^4)^{1/2}$ from $x = 0$ to $x = 1/\sqrt{2}$ around the $y$-axis.

46 Find the volume generated by revolving the region under $y = (x^4 + 1)^{-1/2}$ from $x = 0$ to $x = 1$ around the $y$-axis.

47 Sketch the graph of the curve $y = 5 \cos 2x + 12 \sin 2x$. Show at least one complete cycle.

48 A mass on the end of a spring oscillates vertically with a period of $2\pi$ seconds. At $t = 0$ the mass is 4 ft from the equilibrium point ($y = 0$) and is approaching it at 3 ft/s. What is the amplitude of the motion of this mass?

49 Use the formulas in Equations (14)–(17) of Section 8-8 to show each of the following.

(a) $\coth^{-1} x = \tanh^{-1}\left(\dfrac{1}{x}\right)$

(b) $\text{sech}^{-1} x = \cosh^{-1}\left(\dfrac{1}{x}\right)$

50 Show that $x''(t) = k^2 x(t)$ if

$$x(t) = A \cosh kt + B \sinh kt,$$

where $A$ and $B$ are constants. Determine $A$ and $B$ if
(a) $x(0) = 1, \quad x'(0) = 0$
(b) $x(0) = 0, \quad x'(0) = 1$

51 Use Newton's method to find the least positive solution of the equation $\cos x \cosh x = 1$. Begin by sketching the graphs $y = \cos x$ and $y = \text{sech } x$.

52 (a) Verify by differentiation that

$$\int \sec x \, dx = \sinh^{-1}(\tan x) + C.$$

(b) Show similarly that

$$\int \text{sech } x \, dx = \tan^{-1}(\sinh x) + C.$$

53 Deduce from Problems 31 and 32 in Section 8-8 that the length $S$, sag $H$, density $w$, and minimum tension $T_0$ of a hanging cable are related by the equation

$$4wH^2 - wS^2 + 8HT_0 = 0.$$

54 A 300-ft length of cable weighing 2 lb/ft is hung between two points at equal height. A 30-ft sag is then observed at the middle of the cable.
(a) Use the result of Problem 53 to find the minimum tension $T_0$.
(b) Find the distance $2L$ between the ends of the cable.
(c) Find the maximum tension in the cable.

55 A 400-ft wire weighing 2 lb/ft can withstand a tension of 5000 lb. Use the results of Problem 53 and Problem 33 in Section 8-8 to determine the minimum sag that can be allowed in hanging this wire between two points of equal height.

56 A cable is 105 ft long and weighs 2 lb/ft. It is hung with its ends at the same level, and the ends are 100 ft apart. Find the sag at its middle and the maximum and minimum tension in the cable.

**57** Assume that the earth is a solid sphere of uniform density, with mass $M$ and radius $R = 3960$ mi. For a particle of mass $m$ *within* the earth at distance $r$ from the center of the earth, the gravitational force attracting $m$ toward the center is $F_r = -GM_rm/r^2$, where $M_r$ is the mass of the part of the earth within a sphere of radius $r$.

(a) Show that $F_r = -(GMm/R^3)r$.

(b) Now suppose that a hole is drilled straight through the center of the earth, connecting two antipodal points on its surface. Let a particle of mass $m$ be dropped at time $t = 0$ into this hole with initial speed zero, and let $r(t)$ be its distance from the center of the earth at time $t$. Conclude from Newton's second law and part (a) that $r''(t) = -k^2r(t)$, where $k^2 = GM/R^3 = g/R$ and $g = 32.2$ feet per second per second.

(c) Deduce from part (b) that the particle undergoes simple harmonic motion back and forth between the ends of the hole, with a period of about 84 min.

(d) With what speed (in miles per hour) does the particle pass through the earth's center?

# 9

# Techniques of Integration

## Introduction

In the past three chapters we have seen that many geometric and physical quantities can be expressed as definite integrals. The fundamental theorem of calculus reduces the problem of calculating the definite integral $\int_a^b f(x)\,dx$ to that of finding an antiderivative $G(x)$ of $f(x)$; once this is accomplished, then

$$\int_a^b f(x)\,dx = \Big[ G(x) \Big]_a^b = G(b) - G(a).$$

But as yet we have relied largely on trial-and-error methods for finding the required antiderivative $G(x)$. Sometimes a knowledge of elementary derivative formulas, perhaps in combination with a simple substitution, will allow us to integrate a given function. This approach can, however, be inefficient and time-consuming, especially in view of the following surprising fact: There exist simple-looking integrals, such as

$$\int e^{-x^2}\,dx, \quad \int \frac{\sin x}{x}\,dx, \quad \text{and} \quad \int \sqrt{1 + x^4}\,dx,$$

that cannot be evaluated in terms of finite combinations of the familiar algebraic and elementary transcendental functions. For example, the antiderivative

$$H(x) = \int_0^x e^{-t^2}\,dt$$

of $e^{-x^2}$ has no finite expression in terms of elementary functions. Any attempt to find such an expression will, therefore, inevitably be unsuccessful.

The presence of such examples indicates that we cannot hope to reduce integration to a routine process like differentiation. In fact, finding antiderivatives is an art, depending upon experience and practice. Nevertheless, there are a number of techniques whose systematic use can substantially reduce our dependence upon chance and intuition alone. This chapter deals with some of these systematic techniques of integration.

## Integral Tables and Simple Substitutions

Integration would be a fairly simple matter if we had a list of integral formulas, an *integral table*, in which we could locate any integral that we ever needed to evaluate. But the diversity of integrals that we encounter in practice is too great for such an all-inclusive integral table to be practical. It is more sensible to print, or memorize, a short table of integrals of the sort seen frequently and to learn techniques by which the range of applicability of this short table can be extended. We may begin with the list of integrals in Fig. 9.1, which are familiar from earlier chapters; each is equivalent to one of the basic derivative formulas.

A table of approximately 110 integral formulas appears inside the covers of this book. Still more extensive integral tables are readily available. For example, the volume of *Standard Mathematical Tables* edited by Samuel M. Selby and published annually by the Chemical Rubber Company in

Cleveland, contains a list of over 700 integral formulas. But even such a lengthy table can be expected to include only a small fraction of the integrals we may need to evaluate. Thus it is necessary to learn techniques for deriving new formulas and for transforming a given integral either into one that's already familiar or into one appearing in an accessible table.

The principal such technique is the *method of substitution*, which we first considered in Section 5-6. Recall that if

$$\int f(u)\, du = F(u) + C,$$

then

$$\int f(g(x))g'(x)\, dx = F(g(x)) + C.$$

Thus the substitution $u = g(x)$, $du = g'(x)\, dx$ transforms the integral $\int f(g(x))g'(x)\, dx$ into the simpler integral $\int f(u)\, du$. The key to making this simplification lies in spotting the composition $f(g(x))$ in the given integrand. In order to convert this integrand into a function of $u$ alone, it is necessary that the remaining factor be a constant multiple of the derivative $g'(x)$ of the inside function $g(x)$. In this case we replace $f(g(x))$ by the simpler $f(u)$ and also $g'(x)\, dx$ by $du$. The preceding three chapters contain numerous illustrations of this method of substitution, and the problems for the present section provide an opportunity to review it.

**EXAMPLE 1** Find $\displaystyle\int \frac{1}{x}(1 + \ln x)^5\, dx$.

**Solution** We need to spot *both* the inner function $g(x)$ *and* its derivative $g'(x)$. If we choose $g(x) = 1 + \ln x$, then $g'(x) = 1/x$. Hence the given integral is of the form discussed above with $f(u) = u^5$, $u = 1 + \ln x$, and $du = dx/x$. Therefore,

$$\int \frac{1}{x}(1 + \ln x)^5\, dx = \int u^5\, du$$

$$= \frac{1}{6}u^6 + C = \frac{1}{6}(1 + \ln x)^6 + C.$$

**EXAMPLE 2** Find $\displaystyle\int \frac{x\, dx}{1 + x^4}$.

**Solution** Here it is not so clear what the inside function is. But comparison with the integral formula in (11) (in Fig. 9.1) suggests that we try the substitution $u = x^2$ and its consequent, $du = 2x\, dx$. We take advantage of the factor $x\, dx = \frac{1}{2}\, du$ that is available in the integrand and make these computations:

$$\int \frac{x\, dx}{1 + x^4} = \frac{1}{2}\int \frac{du}{1 + u^2}$$

$$= \frac{1}{2}\tan^{-1}u + C = \frac{1}{2}\tan^{-1}x^2 + C.$$

Note that the substitution $u = x^2$ would have been of little use had the integrand been either $1/(1 + x^4)$ or $x^2/(1 + x^4)$.

$$\int u^n\, du = \frac{u^{n+1}}{n+1} + C \quad [n \neq -1] \tag{1}$$

$$\int \frac{du}{u} = \ln|u| + C \tag{2}$$

$$\int e^u\, du = e^u + C \tag{3}$$

$$\int \cos u\, du = \sin u + C \tag{4}$$

$$\int \sin u\, du = -\cos u + C \tag{5}$$

$$\int \sec^2 u\, du = \tan u + C \tag{6}$$

$$\int \csc^2 u\, du = -\cot u + C \tag{7}$$

$$\int \sec u \tan u\, du = \sec u + C \tag{8}$$

$$\int \csc u \cot u\, du = -\csc u + C \tag{9}$$

$$\int \frac{du}{\sqrt{1 - u^2}} = \sin^{-1}u + C \tag{10}$$

$$\int \frac{du}{1 + u^2} = \tan^{-1}u + C \tag{11}$$

$$\int \frac{du}{u\sqrt{u^2 - 1}} = \sec^{-1}|u| + C \tag{12}$$

**9.1** A short table of integrals

Example 2 illustrates the device of using a substitution to convert a given integral into a familiar one. Often an integral that does not appear in any integral table may be transformed into one that does by using the techniques of this chapter. In the following example we employ an appropriate substitution to "reconcile" the given integral with the standard integral formula

$$\int \frac{u^2\, du}{\sqrt{a^2 - u^2}} = \frac{a^2}{2} \sin^{-1}\left(\frac{u}{a}\right) - \frac{u}{2}\sqrt{a^2 - u^2} + C,$$

which is Formula 56 (inside the front cover).

**EXAMPLE 3**  Find $\displaystyle\int \frac{\sin^2 x \cos x\, dx}{\sqrt{25 - 16 \sin^2 x}}$.

**Solution**  In order that $25 - 16 \sin^2 x$ be equal to $a^2 - u^2$, we take $a = 5$ and $u = 4 \sin x$, so that $du = 4 \cos x\, dx$. This gives

$$\int \frac{\sin^2 x \cos x\, dx}{\sqrt{25 - 16 \sin^2 x}} = \int \frac{(u/4)^2 (1/4)\, du}{\sqrt{25 - u^2}}$$

$$= \frac{1}{64} \int \frac{u^2\, du}{\sqrt{25 - u^2}}$$

$$= \frac{1}{64} \left[ \frac{25}{2} \sin^{-1}\left(\frac{u}{5}\right) - \frac{u}{2}\sqrt{25 - u^2} \right] + C$$

$$= \frac{25}{128} \sin^{-1}\left(\frac{4}{5} \sin x\right) - \frac{\sin x}{32}\sqrt{25 - 16 \sin^2 x} + C.$$

In Section 9-5 we will see how integral formulas like Formula 56, mentioned above, can be derived.

## 9-2  PROBLEMS

Evaluate the integrals in Problems 1–30.

**1** $\displaystyle\int (2 - 3x)^4\, dx$

**2** $\displaystyle\int \frac{dx}{(1 + 2x)^2}$

**3** $\displaystyle\int x^2 \sqrt{2x^3 - 4}\, dx$

**4** $\displaystyle\int \frac{5t\, dt}{5 + 2t^2}$

**5** $\displaystyle\int \frac{3x\, dx}{\sqrt[3]{2x^2 + 3}}$

**6** $\displaystyle\int x \sec^2 x^2\, dx$

**7** $\displaystyle\int \frac{\cot\sqrt{y}\,\csc\sqrt{y}}{\sqrt{y}}\, dy$

**8** $\displaystyle\int \sin \pi(2x + 1)\, dx$

**9** $\displaystyle\int (1 + \sin\theta)^5 \cos\theta\, d\theta$

**10** $\displaystyle\int \frac{\sin 2x}{4 + \cos 2x}\, dx$

**11** $\displaystyle\int e^{-\cot x} \csc^2 x\, dx$

**12** $\displaystyle\int \frac{e^{\sqrt{x+4}}}{\sqrt{x+4}}\, dx$

**13** $\displaystyle\int \frac{[\ln t]^{10}}{t}\, dt$

**14** $\displaystyle\int \frac{t\, dt}{\sqrt{1 - 9t^2}}$

**15** $\displaystyle\int \frac{dt}{\sqrt{1 - 9t^2}}$

**16** $\displaystyle\int \frac{e^{2x}}{1 + e^{2x}}\, dx$

**17** $\displaystyle\int \frac{e^{2x}}{1 + e^{4x}}\, dx$

**18** $\displaystyle\int \frac{e^{\tan^{-1} x}}{1 + x^2}\, dx$

**19** $\displaystyle\int \frac{3x\, dx}{\sqrt{1 - x^4}}$

**20** $\displaystyle\int \sin^3 2x \cos 2x\, dx$

**21** $\displaystyle\int \tan^4 3x \sec^2 3x\, dx$

**22** $\displaystyle\int \frac{dt}{1 + 4t^2}$

**23** $\displaystyle\int \frac{\cos\theta}{1 + \sin^2\theta}\, d\theta$

**24** $\displaystyle\int \frac{\sec^2\theta\, d\theta}{1 + \tan\theta}$

**25** $\displaystyle\int \frac{(1 + \sqrt{x})^4}{\sqrt{x}}\, dx$

**26** $\displaystyle\int t^{-1/3} \sqrt{t^{2/3} - 1}\, dt$

**466**

$$27 \int \frac{dt}{(1 + t^2)\arctan t}$$

$$28 \int \frac{\sec 2x \tan 2x \, dx}{(1 + \sec 2x)^{3/2}}$$

$$38 \int \frac{\cos x \, dx}{(\sin^2 x)\sqrt{1 + \sin^2 x}}$$

$$39 \int \frac{\sqrt{x^4 - 1}}{x} \, dx$$

$$29 \int \frac{dx}{\sqrt{e^{2x} - 1}}$$

$$30 \int \frac{x \, dx}{\sqrt{\exp(2x^2) - 1}}$$

$$40 \int \frac{e^{3x} \, dx}{\sqrt{25 + 16e^{2x}}}$$

In Problems 31–35, make the indicated substitution to evaluate the given integral.

$$41 \int \frac{(\ln x)^2}{x} \sqrt{1 + (\ln x)^2} \, dx$$

$$31 \int x^2 \sqrt{x - 2} \, dx, \quad u = x - 2$$

$$32 \int \frac{x^2 \, dx}{\sqrt{x + 3}}, \quad u = x + 3$$

$$42 \int x^8 \sqrt{4x^6 - 1} \, dx$$

43 The substitution $u = x^2$, $x = \sqrt{u}$, and $dx = du/2\sqrt{u}$ appears to lead to this result:

$$33 \int \frac{x \, dx}{\sqrt{2x + 3}}, \quad u = 2x + 3$$

$$\int_{-1}^{1} x^2 \, dx = \tfrac{1}{2} \int_{1}^{1} \sqrt{u} \, du = 0.$$

$$34 \int x \sqrt[3]{x - 1} \, dx, \quad u = x - 1$$

Do you believe this result? If not, why not?

44 Use the fact that $x^2 + 4x + 5 = (x + 2)^2 + 1$ to evaluate

$$35 \int \frac{x \, dx}{\sqrt[3]{x + 1}}, \quad u = x + 1$$

$$\int \frac{dx}{x^2 + 4x + 5}.$$

In each of Problems 36–42, evaluate the given integral by first transforming it into one of those listed in the integral table inside the covers of this book.

45 Use the fact that $1 - (x - 1)^2 = 2x - x^2$ to evaluate

$$\int \frac{dx}{\sqrt{2x - x^2}}.$$

$$36 \int x \sqrt{4 - x^4} \, dx$$

$$37 \int e^x \sqrt{9 + e^{2x}} \, dx$$

# Trigonometric Integrals

The substitution $u = \sin x$, $du = \cos x \, dx$ gives

$$\int \sin^3 x \cos x \, dx = \int u^3 \, du$$

$$= \tfrac{1}{4} u^4 + C = \tfrac{1}{4} \sin^4 x + C.$$

This type of substitution can be used to evaluate an integral of the form

$$\int \sin^m x \cos^n x \, dx \qquad (1)$$

in the first of the following two cases.

Case 1: At least one of the two numbers $m$ and $n$ is an *odd positive integer*. If so, the other may be any real number.

Case 2: Both $m$ and $n$ are *nonnegative even integers*.

Suppose, for example, that $m$ is an odd positive integer. Then we split off one $\sin x$ factor and use the identity $\sin^2 x = 1 - \cos^2 x$ to express the remaining factor $\sin^{m-1} x$ in terms of $\cos x$, as follows:

$$\int \sin^m x \cos^n x \, dx = \int \sin^{m-1} x \cos^n x \sin x \, dx$$

$$= \int (\sin^2 x)^{(m-1)/2} \cos^n x \sin x \, dx$$

$$= \int (1 - \cos^2 x)^{(m-1)/2} \cos^n x \sin x \, dx.$$

Now the substitution $u = \cos x$, $du = -\sin x \, dx$ yields

$$\int \sin^m x \cos^n x \, dx = -\int (1 - u^2)^{(m-1)/2} u^n \, du.$$

The exponent $(m - 1)/2$ is a nonnegative integer because $m$ is an odd positive integer. Thus the factor $(1 - u^2)^{(m-1)/2}$ of the integrand is a polynomial in the variable $u$, and so its product with $u^n$ is easy to integrate.

In essence, this method consists of peeling off one copy of $\sin x$ (if $m$ is odd) and then converting the remaining sines into cosines. If it is $n$ that is odd, then we can split off one copy of $\cos x$ and convert the remaining cosines into sines.

**EXAMPLE 1**

(a) $\displaystyle \int \sin^3 x \cos^2 x \, dx = \int (1 - \cos^2 x) \cos^2 x \sin x \, dx$

$$= \int (u^4 - u^2) \, du \qquad (u = \cos x)$$

$$= \tfrac{1}{5} u^5 - \tfrac{1}{3} u^3 + C$$

$$= \tfrac{1}{5} \cos^5 x - \tfrac{1}{3} \cos^3 x + C.$$

(b) $\displaystyle \int \cos^5 x \, dx = \int (1 - \sin^2 x)^2 \cos x \, dx$

$$= \int (1 - u^2)^2 \, du \qquad (u = \sin x)$$

$$= \int (1 - 2u^2 + u^4) \, du$$

$$= u - \tfrac{2}{3} u^3 + \tfrac{1}{5} u^5 + C$$

$$= \sin x - \tfrac{2}{3} \sin^3 x + \tfrac{1}{5} \sin^5 x + C.$$

---

In Case 2 of the sine-cosine integral in (1), with both $m$ and $n$ nonnegative even integers, we use the half-angle formulas

$$\sin^2 \theta = \tfrac{1}{2}(1 - \cos 2\theta) \tag{2a}$$

and

$$\cos^2 \theta = \tfrac{1}{2}(1 + \cos 2\theta) \tag{2b}$$

to halve the even powers of $\sin x$ and $\cos x$. Repetition of this process with the resulting powers of $\cos 2x$—if necessary—leads to integrals involving odd powers, and we have seen how to handle these in Case 1.

**EXAMPLE 2** Use of Formulas (2a) and (2b) gives

$$\int \sin^2 x \cos^2 x \, dx = \int \tfrac{1}{2}(1 - \cos 2x)\tfrac{1}{2}(1 + \cos 2x) \, dx$$

$$= \tfrac{1}{4} \int (1 - \cos^2 2x) \, dx$$

$$= \tfrac{1}{4} \int (1 - \tfrac{1}{2}(1 + \cos 4x)) \, dx$$

$$= \tfrac{1}{8} \int (1 - \cos 4x) \, dx$$

$$= \tfrac{1}{8} x - \tfrac{1}{32} \sin 4x + C.$$

In the third step we have used Formula (2b) with $\theta = 2x$.

**EXAMPLE 3**  Here we apply Formula (2b), first with $\theta = 3x$, then with $\theta = 6x$.

$$\int \cos^4 3x \, dx = \int \tfrac{1}{4}(1 + \cos 6x)^2 \, dx$$

$$= \tfrac{1}{4} \int (1 + 2 \cos 6x + \cos^2 6x) \, dx$$

$$= \tfrac{1}{4} \int (\tfrac{3}{2} + 2 \cos 6x + \tfrac{1}{2} \cos 12x) \, dx$$

$$= \tfrac{3}{8}x + \tfrac{1}{12} \sin 6x + \tfrac{1}{96} \sin 12x + C.$$

An integral of the form

$$\int \tan^m x \sec^n x \, dx \tag{3}$$

can be routinely evaluated in either of the following two cases.

> *Case 1:*  $m$ is an *odd positive integer.*
> *Case 2:*  $n$ is an *even positive integer.*

In Case 1, we split off the factor $\sec x \tan x$ to form, along with $dx$, the differential $\sec x \tan x \, dx$ of $\sec x$. We then use the identity $\tan^2 x = \sec^2 x - 1$ to convert the remaining even power of $\tan x$ into powers of $\sec x$. This prepares us for the substitution $u = \sec x$.

**EXAMPLE 4**

$$\int \tan^3 x \sec^3 x \, dx = \int (\sec^2 x - 1) \sec^2 x \sec x \tan x \, dx$$

$$= \int (u^4 - u^2) \, du \qquad (u = \sec x)$$

$$= \tfrac{1}{5}u^5 - \tfrac{1}{3}u^3 + C$$

$$= \tfrac{1}{5} \sec^5 x - \tfrac{1}{3} \sec^3 x + C.$$

To evaluate Integral (3) in Case 2, we split off $\sec^2 x$ to form, along with $dx$, the differential of $\tan x$. Use of the identity $\sec^2 x = 1 + \tan^2 x$ to convert the remaining even power of $\sec x$ to powers of $\tan x$ then prepares us for the substitution $u = \tan x$.

**EXAMPLE 5**

$$\int \sec^6 2x \, dx = \int (1 + \tan^2 2x)^2 \sec^2 2x \, dx$$

$$= \tfrac{1}{2} \int (1 + \tan^2 2x)^2 (2 \sec^2 2x) \, dx$$

$$= \tfrac{1}{2} \int (1 + u^2)^2 \, du \qquad (u = \tan 2x)$$

$$= \tfrac{1}{2} \int (1 + 2u^2 + u^4) \, du$$

$$= \tfrac{1}{2}u + \tfrac{1}{3}u^3 + \tfrac{1}{10}u^5 + C$$

$$= \tfrac{1}{2} \tan 2x + \tfrac{1}{3} \tan^3 2x + \tfrac{1}{10} \tan^5 2x + C.$$

Similar methods are effective with integrals of the form

$$\int \csc^m x \, \cot^n x \, dx.$$

The method of Case 1 succeeds with the integral $\int \tan^n x \, dx$ only when $n$ is an odd positive integer, but there is another approach that works equally well whether $n$ be even *or* odd. We split off the factor $\tan^2 x$ and replace it by $\sec^2 x - 1$. This gives

$$\int \tan^n x \, dx = \int (\tan^{n-2} x)(\sec^2 x - 1) \, dx$$

$$= \int \tan^{n-2} x \, \sec^2 x \, dx - \int \tan^{n-2} x \, dx.$$

We integrate what we can and find that

$$\int \tan^n x \, dx = \frac{\tan^{n-1} x}{n-1} - \int \tan^{n-2} x \, dx. \tag{4}$$

Equation (4) is our first example of a **reduction formula.** Its use reduces the original exponent from $n$ to $n - 2$. Repeated application of (4) leads eventually either to

$$\int \tan^2 x \, dx = \int (\sec^2 x - 1) \, dx$$

$$= \tan x - x + C$$

or to

$$\int \tan x \, dx = \int \frac{\sin x}{\cos x} \, dx$$

$$= -\ln|\cos x| + C.$$

**EXAMPLE 6**   Two applications of Formula (4) give

$$\int \tan^6 x \, dx = \tfrac{1}{5} \tan^5 x - \int \tan^4 x \, dx$$

$$= \tfrac{1}{5} \tan^5 x - \left( \tfrac{1}{3} \tan^3 x - \int \tan^2 x \, dx \right)$$

$$= \tfrac{1}{5} \tan^5 x - \tfrac{1}{3} \tan^3 x + \tan x - x + C.$$

Finally, in the case of a trigonometric integral involving tangents, cosecants, and so on, a last resort is to express the integrand in terms of sines and cosines. Simplification may then yield an integrable expression.

## 9-3   PROBLEMS

Evaluate the integrals in Problems 1–35.

1  $\int \sin^3 x \, dx$

2  $\int \sin^4 x \, dx$

3  $\int \sin^2 \theta \, \cos^3 \theta \, d\theta$

4  $\int \sin^3 t \, \cos^3 t \, dt$

5  $\int \cos^5 x \, dx$

6  $\int \dfrac{\sin t}{\cos^3 t} \, dt$

7  $\int \dfrac{\sin^3 x}{\sqrt{\cos x}} \, dx$

8  $\int \sin^3 3\phi \, \cos^4 3\phi \, d\phi$

9  $\int \sin^5 2z \, \cos^2 2z \, dz$

10  $\int \sin^{3/2} x \, \cos^3 x \, dx$

11  $\int \dfrac{\sin^4 4x}{\cos^2 4x} \, dx$

12  $\int \cos^6 4\theta \, d\theta$

13 $\int \sec^4 t \, dt$

14 $\int \tan^3 x \, dx$

15 $\int \cot^3 2x \, dx$

16 $\int \tan\theta \sec^4\theta \, d\theta$

17 $\int \tan^5 2x \sec^2 2x \, dx$

18 $\int \cot^3 x \csc^2 x \, dx$

19 $\int \csc^6 2t \, dt$

20 $\int \dfrac{\sec^4 t}{\tan^2 t} \, dt$

21 $\int \dfrac{\tan^3\theta}{\sec^4\theta} \, d\theta$

22 $\int \dfrac{\cot^3 x}{\csc^2 x} \, dx$

23 $\int \dfrac{\tan^3 t}{\sqrt{\sec t}} \, dt$

24 $\int \dfrac{1}{\cos^4 2x} \, dx$

25 $\int \dfrac{\cot\theta}{\csc^3\theta} \, d\theta$

26 $\int \sin^2 3\alpha \cos^2 3\alpha \, d\alpha$

27 $\int \cos^3 5t \, dt$

28 $\int \tan^4 x \, dx$

29 $\int \cot^4 3t \, dt$

30 $\int \tan^2 2t \sec^4 2t \, dt$

31 $\int \sin^5 2t \cos^{3/2} 2t \, dt$

32 $\int \cot^3\xi \csc^{3/2}\xi \, d\xi$

33 $\int \dfrac{\tan x + \sin x}{\sec x} \, dx$

34 $\int \dfrac{\cot x + \csc x}{\sin x} \, dx$

35 $\int \dfrac{\cot x + \csc^2 x}{1 - \cos^2 x} \, dx$

36 Derive a reduction formula, one analogous to that in Equation (4), for $\int \cot^n x \, dx$.

37 Find $\int \tan x \sec^4 x \, dx$ in two different ways. Then show that your two results are equivalent.

38 Find $\int \cot^3 x \, dx$ in two different ways. Then show that the results are equivalent.

Problems 39–42 are applications of the trigonometric identities

$$\sin\alpha \sin\beta = \tfrac{1}{2}[\cos(\alpha - \beta) - \cos(\alpha + \beta)],$$

$$\sin\alpha \cos\beta = \tfrac{1}{2}[\sin(\alpha - \beta) + \sin(\alpha + \beta)],$$

$$\cos\alpha \cos\beta = \tfrac{1}{2}[\cos(\alpha - \beta) + \cos(\alpha + \beta)].$$

39 Find $\int \sin 3x \cos 5x \, dx$.

40 Find $\int \sin 2x \sin 4x \, dx$.

41 Find $\int \cos x \cos 4x \, dx$.

42 Suppose that $m$ and $n$ are positive integers with $m \neq n$. Show that

(a) $\int_0^{2\pi} \sin mx \sin nx \, dx = 0$;

(b) $\int_0^{2\pi} \cos mx \sin nx \, dx = 0$;

(c) $\int_0^{2\pi} \cos mx \cos nx \, dx = 0$.

43 Substitute $\sec x \csc x = (\sec^2 x)/(\tan x)$ to derive the formula

$$\int \sec x \csc x \, dx = \ln|\tan x| + C.$$

44 Show that

$$\csc x = \frac{1}{2 \sin(x/2)\cos(x/2)}$$

and then apply the result of Problem 43 to derive the formula

$$\int \csc x \, dx = \ln|\tan(x/2)| + C.$$

45 Substitute $x = (\pi/2) - u$ in the integral formula of Problem 44 to show that

$$\int \sec x \, dx = \ln\left|\cot\left(\frac{\pi}{4} - \frac{x}{2}\right)\right| + C.$$

46 Use appropriate trigonometric identities to deduce from the result of Problem 45 that

$$\int \sec x \, dx = \ln|\sec x + \tan x| + C.$$

---

**9-4**

## Integration by Parts

One reason for transforming one given integral into another is to produce an integral that is easier to evaluate. There are two general ways to accomplish such a transformation—the first is integration by substitution and the second is *integration by parts*.

The formula for integration by parts is a simple consequence of the product rule for derivatives:

$$D_x(uv) = v\frac{du}{dx} + u\frac{dv}{dx}.$$

If we write this formula in the form

$$u(x)v'(x) = D_x[u(x)v(x)] - v(x)u'(x), \tag{1}$$

then antidifferentiation gives

$$\int u(x)v'(x)\,dx = u(x)v(x) - \int v(x)u'(x)\,dx. \qquad (2)$$

This is the formula for **integration by parts.** With $du = u'(x)\,dx$ and $dv = v'(x)\,dx$, Formula (2) becomes

$$\int u\,dv = uv - \int v\,du. \qquad (3)$$

To apply the integration by parts formula to a given integral, we must first factor its integrand into two "parts," $u$ and $dv$, the latter including the differential $dx$. We try to choose these parts in such a way that

  **1** The antiderivative $v = \int dv$ is easy to find, and
  **2** The new integral $\int v\,du$ is easier to compute than the original integral $\int u\,dv$.

An effective strategy is to choose for $dv$ the most complicated factor that can readily be integrated. The other part $u$ then gets differentiated (rather than integrated) to find $du$.

We begin with two examples in which we have little flexibility in choosing the parts $u$ and $dv$.

**EXAMPLE 1**  Find $\displaystyle\int \ln x\,dx$.

***Solution***  Here there is little alternative to the natural choice $u = \ln x$ and $dv = dx$. Then $du = dx/x$ and $v = x$. It is helpful to systematize the procedure of integration by parts by writing $u$, $dv$, $du$, and $v$ in an array like the one below:

$$u = \ln x \qquad dv = dx$$

$$du = \frac{dx}{x} \qquad v = x$$

The first line specifies one choice of $u$ and $dv$; the second line is computed from the first. Then Formula (3) gives

$$\int \ln x\,dx = x \ln x - \int dx = x \ln x - x + C.$$

---

COMMENT 1:  The constant of integration appears only at the last step. We know that, once we have found one antiderivative, any other may be obtained by adding a constant $C$ to the one we have found.

COMMENT 2:  In computing $v = \int dv$, we ordinarily take the constant of integration to be zero. Had we written $v = x + C_1$ in Example 1, the result would have been

$$\int \ln x\,dx = (x + C_1)\ln x - \int \left(1 + \frac{C_1}{x}\right)dx$$

$$= x \ln x + C_1 \ln x - (x + C_1 \ln x) + C$$

$$= x \ln x - x + C$$

as before, so introducing the extra constant $C_1$ makes no difference.

**EXAMPLE 2**   Find $\int \sin^{-1}x \, dx$.

**Solution**   Let

$$u = \sin^{-1}x \qquad dv = dx$$

$$du = \frac{dx}{\sqrt{1 - x^2}} \qquad v = x.$$

Then Formula (3) gives

$$\int \sin^{-1}x \, dx = x \sin^{-1}x - \int \frac{x \, dx}{\sqrt{1 - x^2}}$$

$$= x \sin^{-1}x + \sqrt{1 - x^2} + C.$$

**EXAMPLE 3**   Find $\int xe^{-x} \, dx$.

**Solution**   Here we have some flexibility. Suppose that we try

$$u = e^{-x} \qquad dv = x \, dx$$

$$du = -e^{-x} \, dx \qquad v = \tfrac{1}{2}x^2.$$

Then integration by parts gives

$$\int xe^{-x} \, dx = \tfrac{1}{2}x^2 e^{-x} + \tfrac{1}{2} \int x^2 e^{-x} \, dx.$$

The new integral on the right looks more troublesome than the one we started with! Let us begin anew with

$$u = x \qquad dv = e^{-x} \, dx$$

$$du = dx \qquad v = -e^{-x}.$$

Now integration by parts gives

$$\int xe^{-x} \, dx = -xe^{-x} + \int e^{-x} \, dx$$

$$= -xe^{-x} - e^{-x} + C.$$

---

Integration by parts can also be applied to definite integrals. We integrate Equation (1) from $x = a$ to $x = b$ and apply the fundamental theorem of calculus. This gives

$$\int_a^b u(x)v'(x) \, dx = \int_a^b D_x[u(x)v(x)] \, dx - \int_a^b v(x)u'(x) \, dx$$

$$= \Big[u(x)v(x)\Big]_a^b - \int_a^b v(x)u'(x) \, dx.$$

In the notation of (3) this is

$$\int_{x=a}^{x=b} u \, dv = \Big[uv\Big]_{x=a}^{x=b} - \int_{x=a}^{x=b} v \, du, \tag{4}$$

though we must not forget that $u$ and $v$ are functions of $x$. For example, with $u = x$ and $dv = e^{-x} \, dx$, as in Example 3, we obtain

$$\int_0^1 xe^{-x} \, dx = \Big[-xe^{-x}\Big]_0^1 + \int_0^1 e^{-x} \, dx$$

$$= -e^{-1} + \Big[-e^{-x}\Big]_0^1 = 1 - 2e^{-1}.$$

**EXAMPLE 4**  Find $\int x^2 e^{-x}\, dx$.

**Solution**  If we choose $u = x^2$, then $du = 2x\, dx$, so we will reduce the exponent of $x$ by this choice. With

$$u = x^2 \qquad dv = e^{-x}\, dx$$
$$du = 2x\, dx \qquad v = -e^{-x},$$

integration by parts gives

$$\int x^2 e^{-x}\, dx = -x^2 e^{-x} + 2 \int x e^{-x}\, dx.$$

We substitute the result $\int x e^{-x}\, dx = -x e^{-x} - e^{-x}$ of Example 3, and this yields

$$\int x^2 e^{-x}\, dx = -x^2 e^{-x} - 2x e^{-x} - 2e^{-x} + C$$
$$= -(x^2 + 2x + 2)e^{-x} + C.$$

In effect, we have annihilated the original $x^2$ factor by integrating by parts twice in succession (since Example 3 was itself an integration by parts).

**EXAMPLE 5**  Find $\int e^{2x} \sin 3x\, dx$.  Problem #171 (316)

**Solution**  This is another example in which repeated integration by parts succeeds, but with a different twist. Let

$$u = \sin 3x \qquad dv = e^{2x}\, dx$$
$$du = 3 \cos 3x\, dx \qquad v = \tfrac{1}{2} e^{2x}.$$

Then

$$\int e^{2x} \sin 3x\, dx = \tfrac{1}{2} e^{2x} \sin 3x - \tfrac{3}{2} \int e^{2x} \cos 3x\, dx.$$

At first it appears that little progress has been made, since the integral on the right seems just as difficult as the one on the left. We ignore this objection and try again. We apply integration by parts to the new integral. With

$$u = \cos 3x \qquad dv = e^{2x}\, dx$$
$$du = -3 \sin 3x\, dx \qquad v = \tfrac{1}{2} e^{2x},$$

we find that

$$\int e^{2x} \cos 3x\, dx = \tfrac{1}{2} e^{2x} \cos 3x + \tfrac{3}{2} \int e^{2x} \sin 3x\, dx.$$

When we substitute this information into the previous equation, we discover that

$$\int e^{2x} \sin 3x\, dx = \tfrac{1}{2} e^{2x} \sin 3x - \tfrac{3}{4} e^{2x} \cos 3x - \tfrac{9}{4} \int e^{2x} \sin 3x\, dx.$$

So we're back where we started. Or are we? In fact we are *not*, because we actually can *solve* the last equation for the desired integral! We move the right-hand integral above to the left-hand side of the equation; this gives

$$\tfrac{13}{4} \int e^{2x} \sin 3x\, dx = \tfrac{1}{4} e^{2x}(2 \sin 3x - 3 \cos 3x) + C_1,$$

and so

$$\int e^{2x} \sin 3x \, dx = \tfrac{1}{13} e^{2x}(2 \sin 3x - 3 \cos 3x) + C.$$

**EXAMPLE 6**   Find a reduction formula for $\int \sec^n x \, dx$.    *Problem # 185 (317)*

***Solution***   The idea is that $n$ is a (large) positive integer, and that we want to express the given integral in terms of a lower power of sec $x$. The easiest power of sec $x$ to integrate is $\sec^2 x$, so we let

$$u = \sec^{n-2} x \qquad\qquad dv = \sec^2 x \, dx$$
$$du = (n-2) \sec^{n-2} x \tan x \, dx \qquad v = \tan x.$$

This gives

$$\int \sec^n x \, dx = \sec^{n-2} x \tan x - (n-2) \int \sec^{n-2} x \tan^2 x \, dx$$

$$= \sec^{n-2} x \tan x - (n-2) \int (\sec^{n-2} x)(\sec^2 x - 1) \, dx.$$

Hence

$$\int \sec^n x \, dx = \sec^{n-2} x \tan x - (n-2) \int \sec^n x \, dx + (n-2) \int \sec^{n-2} x \, dx.$$

We solve this last equation for the desired integral and find that

$$\int \sec^n x \, dx = \frac{\sec^{n-2} x \tan x}{n-1} + \frac{n-2}{n-1} \int \sec^{n-2} x \, dx. \qquad (5)$$

This is the desired reduction formula. For example, if we take $n = 3$ in this formula, we find that

$$\int \sec^3 x \, dx = \tfrac{1}{2} \sec x \tan x + \tfrac{1}{2} \int \sec x \, dx$$

—— *Problem# 185 (317)*

$$= \tfrac{1}{2} \sec x \tan x + \tfrac{1}{2} \ln|\sec x + \tan x| + C. \qquad (6)$$

In the last step we used Equation (15) of Section 8-4:

$$\int \sec x \, dx = \ln|\sec x + \tan x| + C.$$

The reason for using the reduction formula in (5) is that repeated application must yield one of the two elementary integrals $\int \sec x \, dx$ and $\int \sec^2 x \, dx$. For instance, with $n = 4$ we get

$$\int \sec^4 x \, dx = \tfrac{1}{3} \sec^2 x \tan x + \tfrac{2}{3} \int \sec^2 x \, dx$$

—— *Problem 102 (315)*

$$= \tfrac{1}{3} \sec^2 x \tan x + \tfrac{2}{3} \tan x + C, \qquad (7)$$

and with $n = 5$ we get

$$\int \sec^5 x \, dx = \tfrac{1}{4} \sec^3 x \tan x + \tfrac{3}{4} \int \sec^3 x \, dx$$

$$= \tfrac{1}{4} \sec^3 x \tan x + \tfrac{3}{8} \sec x \tan x$$
$$+ \tfrac{3}{8} \ln|\sec x + \tan x| + C, \qquad (8)$$

using in the last step the formula in Equation (6).

# 9-4 PROBLEMS

Use integration by parts to compute the integrals in Problems 1–35.

**1** $\int xe^{2x}\, dx$  **2** $\int x^2 e^{2x}\, dx$

**3** $\int t \sin t\, dt$  **4** $\int t^2 \sin t\, dt$

**5** $\int x \cos 3x\, dx$  **6** $\int x \ln x\, dx$

**7** $\int x^3 \ln x\, dx$  **8** $\int e^{3z} \cos 3z\, dz$

**9** $\int \tan^{-1}x\, dx$  **10** $\int x \tan^{-1}x\, dx$

**11** $\int y^{1/2} \ln y\, dy$  **12** $\int x \sec^2 x\, dx$

**13** $\int (\ln t)^2\, dt$  **14** $\int t(\ln t)^2\, dt$

**15** $\int x\sqrt{x+3}\, dx$  **16** $\int x^3\sqrt{1-x^2}\, dx$

**17** $\int x^5\sqrt{x^3+1}\, dx$  **18** $\int \sin^2\theta\, d\theta$

**19** $\int \csc^3\theta\, d\theta$  **20** $\int \sin(\ln t)\, dt$

**21** $\int x^2 \arctan x\, dx$  **22** $\int \ln(1+x^2)\, dx$

**23** $\int \sec^{-1}\sqrt{x}\, dx$  **24** $\int x \tan^{-1}\sqrt{x}\, dx$

**25** $\int \tan^{-1}\sqrt{x}\, dx$  **26** $\int x^2\sin 4x\, dx$

**27** $\int x \csc^2 x\, dx$  **28** $\int \csc^3 x\, dx$

**29** $\int x^3\cos(x^2)\, dx$  **30** $\int e^{-3x}\sin 4x\, dx$

**31** $\int \dfrac{\ln x}{x^{3/2}}\, dx$  **32** $\int \dfrac{x^7\, dx}{(1+x^4)^{3/2}}$

**33** $\int x \cosh x\, dx$  **34** $\int e^x\cosh x\, dx$

**35** $\int x^2\sinh x\, dx$

**36** Use the method of cylindrical shells to find the volume generated by revolving the area under $y = \cos x$, where $0 \le x \le \pi/2$, around the $y$-axis.

**37** Find the volume of the solid obtained by revolving the area bounded by the graph of $y = \ln x$, the $x$-axis, and the vertical line $x = e$ about the $x$-axis.

**38** Find the centroid of the area described in Problem 37.

**39** Find the centroid of the solid of Problem 37.

**40** Find the centroid of the region between $y = e^{\sqrt{x}}$ and the $x$-axis over the interval $0 \le x \le 1$. (*Suggestion:* Substitute $u = \sqrt{x}$.)

Derive the reduction formulas given in Problems 41–46.

**41** $\int x^n e^x\, dx = x^n e^x - n\int x^{n-1}e^x\, dx$

**42** $\int x^n e^{-x^2}\, dx = -\dfrac{1}{2}x^{n-1}e^{-x^2} + \dfrac{n-1}{2}\int x^{n-2}e^{-x^2}\, dx$

**43** $\int (\ln x)^n\, dx = x(\ln x)^n - n\int (\ln x)^{n-1}\, dx$

**44** $\int x^n \cos x\, dx = x^n \sin x - n\int x^{n-1}\sin x\, dx$

**45** $\int \sin^n x\, dx = -\dfrac{\sin^{n-1}x \cos x}{n} + \dfrac{n-1}{n}\int \sin^{n-2}x\, dx$

**46** $\int \cos^n x\, dx = \dfrac{\cos^{n-1}x \sin x}{n} + \dfrac{n-1}{n}\int \cos^{n-2}x\, dx$

Use appropriate reduction formulas from the preceding list to evaluate the integrals in Problems 47–49.

**47** $\int_0^1 x^3 e^x\, dx$  **48** $\int_0^1 x^5\exp(-x^2)\, dx$

**49** $\int_1^e (\ln x)^3\, dx$

**50** Apply the reduction formula in Problem 45 to show that, for each positive integer $n$,

$$\int_0^{\pi/2} \sin^{2n}x\, dx = \frac{\pi}{2}\frac{1}{2}\frac{3}{4}\frac{5}{6}\cdots\frac{2n-1}{2n}$$

and

$$\int_0^{\pi/2} \sin^{2n+1}x\, dx = \frac{2}{3}\frac{4}{5}\frac{6}{7}\cdots\frac{2n}{2n+1}.$$

**51** Derive the formula

$$\int \ln(x+10)\, dx = (x+10)\ln(x+10) - x + C$$

in three different ways:

(a) By substituting $u = x + 10$ and applying the result of Example 1.

(b) By integrating by parts with $u = \ln(x+10)$ and $v = x$, noting that

$$\frac{x}{x+10} = 1 - \frac{10}{x+10}.$$

(c) By integrating by parts with $u = \ln(x+10)$ and $v = x + 10$.

**52** Derive the formula

$$\int x^3\tan^{-1}x\, dx = \tfrac{1}{4}(x^4-1)\tan^{-1}x - \tfrac{1}{12}x^3 + \tfrac{1}{4}x + C$$

by integrating by parts with $u = \tan^{-1}x$ and $v = \tfrac{1}{4}(x^4-1)$.

**53** Let

$$J_n = \int_0^1 x^n e^{-x}\,dx$$

for each integer $n \geq 0$.

(a) Show that

$$J_0 = 1 - \frac{1}{e}$$

and that

$$J_n = nJ_{n-1} - \frac{1}{e}$$

for $n \geq 1$.

(b) Deduce by mathematical induction that

$$J_n = n! - \frac{n!}{e}\sum_{k=0}^{n}\frac{1}{k!}$$

for each integer $n \geq 0$.

(c) Explain why $J_n \to 0$ as $n \to \infty$.

(d) Conclude that

$$e = \lim_{n \to \infty}\sum_{k=0}^{n}\frac{1}{k!}.$$

**54** Let $m$ and $n$ be positive integers. Derive the reduction formula

$$\int x^m (\ln x)^n\,dx = \frac{x^{m+1}}{m+1}(\ln x)^n - \frac{n}{m+1}\int x^m (\ln x)^{n-1}\,dx.$$

**55** According to an advertisement in the March 1984 issue of the *American Mathematical Monthly* for the computer algebra program MACSYMA®,

> An engineer working for a major aerospace company needed to evaluate the following integral dealing with turbulence and boundary layers:
>
> $$\int (k\ln x - 2x^3 + 3x^2 + b)^4\,dx.$$
>
> He had worked on this problem for more than three weeks with pencil and paper, always arriving at a different result. He was never sure which of the many results he had come upon was correct.
>
> In less than 10 seconds after entering the problem in the computer, MACSYMA gave him the correct answer, not in numerical terms, but in symbolic terms that gave him real insight into the physical nature of the problem.

(Reprinted with the permission of the Computer-Aided Mathematics Group, Symbolics, Inc., Eleven Cambridge Center, Cambridge, MA 02142.) Explain how you could use the reduction formula of Problem 54 to find the engineer's integral (but don't actually do it). Can you see any reason why it should take three weeks?

# Trigonometric Substitution

The method of *trigonometric substitution* is often effective in dealing with integrals when the integrands involve certain algebraic expressions such as $(a^2 - u^2)^{1/2}$, $(u^2 - a^2)^{3/2}$, and $1/(a^2 + u^2)^2$. There are three basic trigonometric substitutions:

| If the integral involves | then substitute | and use the identity |
|---|---|---|
| $a^2 - u^2$ | $u = a\sin\theta$ | $1 - \sin^2\theta = \cos^2\theta$ |
| $a^2 + u^2$ | $u = a\tan\theta$ | $1 + \tan^2\theta = \sec^2\theta$ |
| $u^2 - a^2$ | $u = a\sec\theta$ | $\sec^2\theta - 1 = \tan^2\theta$ |

By the substitution $u = a\sin\theta$ we mean, more precisely, the substitution

$$\theta = \sin^{-1}\frac{u}{a}, \qquad -\frac{\pi}{2} \leq \theta \leq \frac{\pi}{2},$$

where $|u| \leq a$. Suppose, for example, that an integral involves the expression $(a^2 - u)^{1/2}$. Then this substitution yields

$$(a^2 - u^2)^{1/2} = (a^2 - a^2\sin^2\theta)^{1/2}$$
$$= (a^2\cos^2\theta)^{1/2} = a\cos\theta.$$

We take the positive square root because $\cos\theta \geq 0$ for $-\pi/2 \leq \theta \leq \pi/2$. Thus the troublesome factor $(a^2 - u^2)^{1/2}$ becomes $a\cos\theta$, and meanwhile,

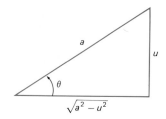

**9.2** The reference triangle for the substitution $u = a \sin \theta$

$du = a \cos \theta \, d\theta$. If the trigonometric integral that results from this substitution can be evaluated using the methods of Section 9-3, the result will normally involve $\theta = \sin^{-1}(u/a)$ and trigonometric functions of $\theta$. The final step will be to express the answer in terms of the original variable. For this purpose the values of the various trigonometric functions can be read from the right triangle shown in Fig. 9.2, which contains an angle $\theta$ such that $\sin \theta = u/a$ (if $u$ is negative, then $\theta$ is negative).

**EXAMPLE 1**   Evaluate $\displaystyle\int \frac{x^3 \, dx}{\sqrt{1 - x^2}}$ where $|x| < 1$.

**Solution**   Here $a = 1$ and $u = x$, so we substitute $x = \sin \theta$, $dx = \cos \theta \, d\theta$. This gives

$$\int \frac{x^3 \, dx}{\sqrt{1 - x^2}} = \int \frac{\sin^3\theta \cos \theta \, d\theta}{\sqrt{1 - \sin^2\theta}}$$

$$= \int \sin^3\theta \, d\theta = \int (\sin \theta)(1 - \cos^2\theta) \, d\theta$$

$$= \tfrac{1}{3}\cos^3\theta - \cos \theta + C.$$

Since $\cos \theta = \sqrt{1 - \sin^2\theta} = \sqrt{1 - x^2}$, our final answer is

$$\int \frac{x^3 \, dx}{\sqrt{1 - x^2}} = \tfrac{1}{3}(1 - x^2)^{3/2} - \sqrt{1 - x^2} + C.$$

**EXAMPLE 2**   Verify by using trigonometric substitution that the area

$$A = \int_0^a \sqrt{a^2 - x^2} \, dx$$

of the quarter-circle lying under $y = \sqrt{a^2 - x^2}$, $0 \leq x \leq a$, is $\tfrac{1}{4}\pi a^2$.

**Solution**   The substitution $x = a \sin \theta$, $dx = a \cos \theta \, d\theta$ gives

$$\int \sqrt{a^2 - x^2} \, dx = \int \sqrt{a^2 - a^2 \sin^2\theta} \, (a \cos \theta) \, d\theta$$

$$= \int a^2 \cos^2\theta \, d\theta = \frac{a^2}{2} \int (1 + \cos 2\theta) \, d\theta$$

$$= \frac{a^2}{2}\left(\theta + \frac{1}{2}\sin 2\theta\right) + C$$

$$= \frac{a^2}{2}(\theta + \sin \theta \cos \theta) + C.$$

(We used the identity $\sin 2\theta = 2 \sin \theta \cos \theta$ in the last step). Now $\sin \theta = x/a$ and (from Fig. 9.2, with $u = x$) $\cos \theta = (a^2 - x^2)^{1/2}/a$. Hence

$$A = \int_0^a \sqrt{a^2 - x^2} \, dx$$

$$= \frac{a^2}{2}\left[ \sin^{-1}\frac{x}{a} + \frac{x(a^2 - x^2)^{1/2}}{a^2} \right]_0^a$$

$$= \frac{a^2}{2}\frac{\pi}{2} = \frac{1}{4}\pi a^2.$$

By the substitution $u = a \tan \theta$ in an integral involving $a^2 + u^2$ is meant the substitution

$$\theta = \tan^{-1}\frac{u}{a}, \qquad -\frac{\pi}{2} < \theta < \frac{\pi}{2}.$$

Note that in this case

$$\sqrt{a^2 + u^2} = \sqrt{a^2 + a^2 \tan^2\theta}$$
$$= \sqrt{a^2 \sec^2\theta} = a \sec \theta$$

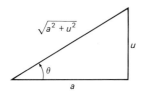

9.3   The substitution $u = a \tan \theta$

under the assumption that $a > 0$. We take the positive square root here because $\sec \theta > 0$ for $-\pi/2 < \theta < \pi/2$. The values of the various trigonometric functions of $\theta$ under this substitution can be read from the right triangle of Fig. 9.3, which shows a (positive or negative) acute angle $\theta$ such that $\tan \theta = u/a$.

**EXAMPLE 3**   Find $\displaystyle\int \frac{dx}{(4x^2 + 9)^2}$.

**Solution**   The factor $4x^2 + 9$ corresponds to $u^2 + a^2$ with $u = 2x$ and $a = 3$. Hence the substitution $u = a \tan \theta$ amounts to

$$2x = 3 \tan \theta, \qquad x = \tfrac{3}{2} \tan \theta, \quad \text{and} \quad dx = \tfrac{3}{2} \sec^2\theta \, d\theta.$$

This gives

$$\int \frac{dx}{(4x^2 + 9)^2} = \int \frac{\tfrac{3}{2} \sec^2\theta \, d\theta}{(9 \tan^2\theta + 9)^2}$$

$$= \frac{3}{2} \int \frac{\sec^2\theta \, d\theta}{(9 \sec^2\theta)^2} = \frac{1}{54} \int \frac{d\theta}{\sec^2\theta}$$

$$= \frac{1}{54} \int \cos^2\theta \, d\theta = \frac{1}{108} (\theta + \sin \theta \cos \theta) + C.$$

The actual integration in the last step is the same as in Example 2. Now $\theta = \tan^{-1}(2x/3)$, and the triangle of Fig. 9.4 gives

$$\sin \theta = \frac{2x}{\sqrt{4x^2 + 9}}, \qquad \cos \theta = \frac{3}{\sqrt{4x^2 + 9}}.$$

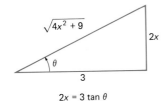

$2x = 3 \tan \theta$

9.4   The reference triangle for Example 3

Hence

$$\int \frac{dx}{(4x^2 + 9)^2} = \frac{1}{108}\left[ \tan^{-1}\left(\frac{2x}{3}\right) + \frac{2x}{\sqrt{4x^2 + 9}} \cdot \frac{3}{\sqrt{4x^2 + 9}} \right] + C$$

$$= \frac{1}{108} \tan^{-1}\left(\frac{2x}{3}\right) + \frac{x}{18(4x^2 + 9)} + C.$$

By the substitution $u = a \sec \theta$ in an integral involving $u^2 - a^2$ is meant the substitution

$$\theta = \sec^{-1}\frac{u}{a}, \qquad 0 \leqq \theta \leqq \pi,$$

where $|u| \geqq a$ (because of the domain and range of the inverse secant function). Then

$$\sqrt{u^2 - a^2} = \sqrt{a^2 \sec^2\theta - a^2}$$
$$= \sqrt{a^2 \tan^2\theta} = \pm a \tan \theta.$$

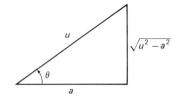

**9.5** The reference triangle for the substitution $u = a \sec \theta$

Here we must take the plus sign if $u > a$, so that $0 < \theta < \pi/2$ and $\tan \theta > 0$. If $u < -a$, so that $\pi/2 < \theta < \pi$ and $\tan \theta < 0$, we take the minus sign above. In either case the values of the various trigonometric functions of $\theta$ can be read from the right triangle of Fig. 9.5.

**EXAMPLE 4** Find $\displaystyle\int \frac{\sqrt{x^2 - 25}}{x}\, dx, \quad x > 5.$

**Solution** We substitute $x = 5 \sec \theta$, $dx = 5 \sec \theta \tan \theta\, d\theta$. Then

$$\sqrt{x^2 - 25} = \sqrt{25(\sec^2\theta - 1)} = +5 \tan \theta,$$

because $x > 5$ implies $0 < \theta < \pi/2$, so that $\tan \theta > 0$. Hence this substitution gives

$$\int \frac{\sqrt{x^2 - 25}}{x}\, dx = \int \frac{5 \tan \theta}{5 \sec \theta} (5 \sec \theta \tan \theta\, d\theta)$$

$$= 5 \int \tan^2\theta\, d\theta = 5 \int (\sec^2\theta - 1)\, d\theta$$

$$= 5 \tan \theta - 5\theta + C$$

$$= \sqrt{x^2 - 25} - 5 \sec^{-1}\left(\frac{x}{5}\right) + C.$$

Hyperbolic substitutions may be used in a similar way, and to the same effect, as trigonometric substitutions. The three basic hyperbolic substitutions—which are not ordinarily memorized—are listed here for reference.

| If the integral involves | use the substitution | and use the identity |
|---|---|---|
| $a^2 - u^2$ | $u = a \tanh \theta$ | $1 - \tanh^2\theta = \operatorname{sech}^2\theta$ |
| $a^2 + u^2$ | $u = a \sinh \theta$ | $1 + \sinh^2\theta = \cosh^2\theta$ |
| $u^2 - a^2$ | $u = a \cosh \theta$ | $\cosh^2\theta - 1 = \sinh^2\theta$ |

**EXAMPLE 5** Find $\displaystyle\int \frac{dx}{\sqrt{x^2 - 1}}, \quad x > 1.$

**Solution** For purpose of comparison, we evaluate this integral both by trigonometric substitution and by hyperbolic substitution. The trigonometric substitution

$$x = \sec \theta, \qquad dx = \sec \theta \tan \theta\, d\theta, \qquad \tan \theta = \sqrt{x^2 - 1}$$

gives

$$\int \frac{dx}{\sqrt{x^2 - 1}} = \int \frac{\sec \theta \tan \theta\, d\theta}{\tan \theta}$$

$$= \int \sec \theta\, d\theta$$

$$= \ln|\sec \theta + \tan \theta| + C \qquad \text{(Equation (15), Section 8-4)}$$

$$= \ln|x + \sqrt{x^2 - 1}| + C.$$

With the hyperbolic substitution $x = \cosh\theta$, $dx = \sinh\theta\,d\theta$, we have

$$\sqrt{x^2 - 1} = \sqrt{\cosh^2\theta - 1} = \sinh\theta.$$

We take the positive square root here because $x > 1$ implies that $\theta = \cosh^{-1}x > 0$ and thus that $\sinh\theta > 0$. Hence

$$\int \frac{dx}{\sqrt{x^2 - 1}} = \int \frac{\sinh\theta}{\sinh\theta}\,d\theta = \int d\theta$$

$$= \theta + C = \cosh^{-1}x + C.$$

Our two results appear to differ, but Equation (14) in Section 8-8 shows that they are equivalent.

## 9-5 PROBLEMS

Use trigonometric substitutions to evaluate the integrals in Problems 1–30.

1 $\displaystyle\int \frac{\sqrt{1 - x^2}}{x^2}\,dx$

2 $\displaystyle\int \frac{\sqrt{1 + x^2}}{x^2}\,dx$

3 $\displaystyle\int \frac{\sqrt{x^2 - 1}}{x^2}\,dx$

4 $\displaystyle\int x^3\sqrt{4 - x^2}\,dx$

5 $\displaystyle\int x^3\sqrt{9 + 4x^2}\,dx$

6 $\displaystyle\int \frac{x^3\,dx}{\sqrt{x^2 + 25}}$

7 $\displaystyle\int \frac{(1 - 4x^2)^{1/2}}{x}\,dx$

8 $\displaystyle\int \frac{dx}{\sqrt{1 + x^2}}$

9 $\displaystyle\int \frac{dx}{\sqrt{9 + 4x^2}}$

10 $\displaystyle\int \sqrt{1 + 4x^2}\,dx$

11 $\displaystyle\int \frac{x^2\,dx}{\sqrt{25 - x^2}}$

12 $\displaystyle\int \frac{x^3\,dx}{\sqrt{25 - x^2}}$

13 $\displaystyle\int \frac{x^2\,dx}{\sqrt{1 + x^2}}$

14 $\displaystyle\int \frac{x^3\,dx}{\sqrt{1 + x^2}}$

15 $\displaystyle\int \frac{x^2\,dx}{\sqrt{4 + 9x^2}}$

16 $\displaystyle\int (1 - x^2)^{3/2}\,dx$

17 $\displaystyle\int \frac{dx}{(1 + x^2)^{3/2}}$

18 $\displaystyle\int \frac{dx}{(4 - x^2)^2}$

19 $\displaystyle\int \frac{dx}{(4 - x^2)^3}$

20 $\displaystyle\int \frac{dx}{(4x^2 + 9)^3}$

21 $\displaystyle\int \sqrt{9 + 16x^2}\,dx$

22 $\displaystyle\int (9 + 16x^2)^{3/2}\,dx$

23 $\displaystyle\int \frac{1}{x}\sqrt{x^2 - 25}\,dx$

24 $\displaystyle\int \frac{\sqrt{9x^2 - 16}}{x}\,dx$

25 $\displaystyle\int x^2\sqrt{x^2 - 1}\,dx$

26 $\displaystyle\int \frac{x^2\,dx}{\sqrt{4x^2 - 9}}$

27 $\displaystyle\int \frac{dx}{(4x^2 - 1)^{3/2}}$

28 $\displaystyle\int \frac{dx}{x^2\sqrt{4x^2 - 9}}$

29 $\displaystyle\int \frac{\sqrt{x^2 - 5}}{x^2}\,dx$

30 $\displaystyle\int (4x^2 - 5)^{3/2}\,dx$

Use hyperbolic substitutions to evaluate the following integrals.

31 $\displaystyle\int \frac{dx}{\sqrt{25 + x^2}}$

32 $\displaystyle\int \sqrt{1 + x^2}\,dx$

33 $\displaystyle\int \frac{\sqrt{x^2 - 4}}{x^2}\,dx$

34 $\displaystyle\int \frac{dx}{\sqrt{1 + 9x^2}}$

35 $\displaystyle\int x^2\sqrt{1 + x^2}\,dx$

36 Compute the arc length of the parabola $y = x^2$ over the interval $0 \le x \le 1$.

37 Compute the area of the surface obtained by revolving the parabolic arc of Problem 36 around the $x$-axis.

38 Show that the length of one arch of the sine curve $y = \sin x$ is equal to half the circumference of the ellipse $x^2 + \frac{1}{2}y^2 = 1$. (*Suggestion:* Substitute $x = \cos\theta$ in the arc length integral for the ellipse.)

39 Compute the arc length of the curve $y = \ln x$ for $1 \le x \le 2$.

40 Compute the area of the surface obtained by revolving the curve of Problem 39 around the $y$-axis.

41 A torus is obtained by revolving the circle

$$(x - a)^2 + y^2 = b^2 \qquad (0 < a \le b)$$

around the $y$-axis. Show that its surface area is $4\pi^2 ab$.

42 Find the area under the curve $y = \sqrt{9 + x^2}$ from $x = 0$ to $x = 4$.

43 Find the area of the surface obtained by revolving the curve $y = \sin x$, $0 \le x \le \pi$, around the $x$-axis.

44 An ellipsoid of revolution is obtained by revolving the ellipse $x^2/a^2 + y^2/b^2 = 1$ around the $x$-axis. Suppose that $a > b$, and show then that the ellipsoid has surface area

$$A = 2\pi ab\left[\frac{b}{a} + \frac{a}{c}\sin^{-1}\left(\frac{c}{a}\right)\right]$$

where $c^2 = a^2 - b^2$. Assume that $a \approx b$, so that $c \approx 0$ and $\sin^{-1}(c/a) \approx c/a$. Conclude that $A \approx 4\pi a^2$.

**45** Suppose that $b > a$ for the ellipsoid of revolution of Problem 44. Show that its surface area is then

$$A = 2\pi ab \left[ \frac{b}{a} + \frac{a}{c} \ln \left( \frac{b+c}{a} \right) \right]$$

where $c^2 = b^2 - a^2$. Assume the fact that $\ln(1 + x) \approx x$ if $x \approx 0$, and thereby conclude that $A \approx 4\pi a^2$ if $a \approx b$.

**46** A road is to be built from the point $(2, 1)$ to the point $(5, 3)$, following the path of the parabola $y = 2\sqrt{x - 1} - 1$.

Calculate the length of this road (the units are in miles). (*Suggestion:* Substitute $x = \sec^2\theta$ in the arc length integral.)

**47** Suppose that the cost of the road in the previous problem is $\sqrt{x}$ million dollars per mile. Calculate its total cost.

**48** A kite is flying at a height of 500 ft at a horizontal distance of 100 ft from the string-holder on the ground. The kite string weighs 1/16 oz/ft and is hanging in the shape of the parabola $y = x^2/20$ joining the string-holder at $(0, 0)$ to the kite at $(100, 500)$. Calculate the work (in foot-pounds) done in lifting the kite string from the ground to its present position.

---

## Integrals Involving Quadratic Polynomials

An integral involving a square root or negative power of a quadratic polynomial $ax^2 + bx + c$ can often be simplified by the process of *completing the square*. For instance,

$$x^2 + 2x + 2 = (x + 1)^2 + 1,$$

and hence the substitution $u = x + 1$, $du = dx$ yields

$$\int \frac{dx}{x^2 + 2x + 2} = \int \frac{du}{u^2 + 1}$$

$$= \tan^{-1}u + C = \tan^{-1}(x + 1) + C.$$

In general, the object is to convert $ax^2 + bx + c$ into either a sum or a difference of two squares—either $u^2 \pm a^2$ or $a^2 - u^2$—so that the method of trigonometric substitution may then be used. To see how this works in practice, suppose first that $a = 1$, so that the quadratic in question is of the form $x^2 + bx + c$. The sum $x^2 + bx$ of the first two terms can be completed to a perfect square by adding $b^2/4$, the square of half the coefficient of $x$, and in turn subtracting $b^2/4$ from the constant term $c$. This gives

$$x^2 + bx + c = \left( x^2 + bx + \frac{b^2}{4} \right) + \left( c - \frac{b^2}{4} \right)$$

$$= \left( x + \frac{1}{2}b \right)^2 + \left( c - \frac{b^2}{4} \right).$$

With $u = x + \frac{1}{2}b$, our result above is of the form $u^2 + A^2$ or $u^2 - A^2$ (depending on the sign of $c - b^2/4$). If we factor out the coefficient $a$ to begin with, the general case is handled similarly:

$$ax^2 + bx + c = a \left( x^2 + \frac{b}{a}x + \frac{c}{a} \right).$$

**EXAMPLE 1** Find $\displaystyle\int \frac{dx}{9x^2 + 6x + 5}$.

**Solution**  Our first step is completion of the square.

$$9x^2 + 6x + 5 = 9(x^2 + \tfrac{2}{3}x) + 5$$
$$= 9(x^2 + \tfrac{2}{3}x + \tfrac{1}{9}) - 1 + 5$$
$$= 9(x + \tfrac{1}{3})^2 + 4$$
$$= (3x + 1)^2 + 2^2.$$

Hence

$$\int \frac{dx}{9x^2 + 6x + 5} = \int \frac{dx}{(3x + 1)^2 + 4}$$

$$= \frac{1}{3} \int \frac{du}{u^2 + 4} \qquad (u = 3x + 1)$$

$$= \frac{1}{6} \int \frac{\tfrac{1}{2}\,du}{(u/2)^2 + 1}$$

$$= \frac{1}{6} \int \frac{dv}{v^2 + 1} \qquad \left(v = \frac{1}{2}u\right)$$

$$= \frac{1}{6} \tan^{-1}v + C$$

$$= \frac{1}{6} \tan^{-1}\left(\frac{u}{2}\right) + C$$

$$= \frac{1}{6} \tan^{-1}\left(\frac{3x + 1}{2}\right) + C.$$

**EXAMPLE 2**  Find $\displaystyle\int \frac{dx}{\sqrt{9 + 16x - 4x^2}}$.

**Solution**  First we complete the square:

$$9 + 16x - 4x^2 = 9 - 4(x^2 - 4x)$$
$$= 9 - 4(x^2 - 4x + 4) + 16$$
$$= 25 - 4(x - 2)^2.$$

Hence

$$\int \frac{dx}{\sqrt{9 + 16x - 4x^2}} = \int \frac{dx}{\sqrt{25 - 4(x - 2)^2}}$$

$$= \frac{1}{5} \int \frac{dx}{\sqrt{1 - \tfrac{4}{25}(x - 2)^2}}$$

$$= \frac{1}{2} \int \frac{du}{\sqrt{1 - u^2}} \qquad \left(u = \frac{2(x - 2)}{5}\right)$$

$$= \frac{1}{2} \sin^{-1}u + C$$

$$= \frac{1}{2} \sin^{-1}\left(\frac{2(x - 2)}{5}\right) + C.$$

An alternative approach is to make the trigonometric substitution

$$2(x - 2) = 5 \sin \alpha, \qquad 2\,dx = 5 \cos \alpha \, d\alpha$$

immediately after completing the square. This yields

$$\int \frac{dx}{\sqrt{9 + 16x - 4x^2}} = \int \frac{dx}{\sqrt{25 - 4(x - 2)^2}}$$

$$= \int \frac{(\tfrac{5}{2})\cos \alpha \, d\alpha}{\sqrt{25 - 25 \sin^2\alpha}}$$

$$= \frac{1}{2} \int d\alpha = \frac{1}{2}\alpha + C$$

$$= \frac{1}{2} \arcsin \frac{2(x - 2)}{5} + C.$$

---

An integral involving a quadratic expression can sometimes be split into two simpler integrals. The next two examples illustrate this technique.

**EXAMPLE 3**  Find $\displaystyle\int \frac{(2x + 3)\,dx}{9x^2 + 6x + 5}.$

*Solution*  Because $D(9x^2 + 6x + 5) = 18x + 6$, this would be a simpler integral if the numerator $2x + 3$ were a constant multiple of $18x + 6$. Our strategy is to write

$$2x + 3 = A(18x + 6) + B,$$

so we can split the given integral into a sum of two integrals, one of which has numerator $18x + 6$ in its integrand. By matching coefficients, we find that $A = \frac{1}{9}$ and $B = \frac{7}{3}$. Hence

$$\int \frac{(2x + 3)\,dx}{9x^2 + 6x + 5} = \frac{1}{9} \int \frac{(18x + 6)\,dx}{9x^2 + 6x + 5} + \frac{7}{3} \int \frac{dx}{9x^2 + 6x + 5}.$$

The first integral on the right is a logarithm, and the second is given by Example 1. Thus

$$\int \frac{(2x + 3)\,dx}{9x^2 + 6x + 5} = \frac{1}{9} \ln(9x^2 + 6x + 5) + \frac{7}{18} \tan^{-1}\left(\frac{3x + 1}{2}\right) + C.$$

Alternatively, we first could complete the square in the denominator. The substitution $u = 3x + 1$, $x = \frac{1}{3}(u - 1)$, $dx = \frac{1}{3}\,du$ then gives

$$\int \frac{(2x + 3)\,dx}{(3x + 1)^2 + 4} = \int \frac{\left[\frac{2}{3}(u - 1) + 3\right]\frac{1}{3}\,du}{u^2 + 4}$$

$$= \frac{1}{9} \int \frac{2u\,du}{u^2 + 4} + \frac{7}{9} \int \frac{du}{u^2 + 4}$$

$$= \frac{1}{9} \ln(u^2 + 4) + \frac{7}{18} \tan^{-1}\left(\frac{u}{2}\right) + C$$

$$= \frac{1}{9} \ln(9x^2 + 6x + 5) + \frac{7}{18} \tan^{-1}\left(\frac{3x + 1}{2}\right) + C.$$

**484**

**EXAMPLE 4** Find $\displaystyle\int \frac{4-x}{(2x-x^2)^2}\,dx$ given $|x-1| < 1$.

**Solution** Because $D_x(2x-x^2) = 2 - 2x$, we first write

$$\int \frac{4-x}{(2x-x^2)^2}\,dx = \frac{1}{2}\int \frac{2-2x}{(2x-x^2)^2}\,dx + 3\int \frac{dx}{(2x-x^2)^2}$$

$$= -\frac{1}{2(2x-x^2)} + 3\int \frac{dx}{(2x-x^2)^2}.$$

To find the remaining integral, we complete the square:

$$2x - x^2 = 1 - (x-1)^2.$$

This suggests the substitution $x - 1 = \sin\theta$, so that

$$dx = \cos\theta\,d\theta \quad \text{and} \quad 1 - (x-1)^2 = \cos^2\theta.$$

This substitution gives

$$\int \frac{dx}{(2x-x^2)^2} = \int \frac{dx}{(1-(x-1)^2)^2}$$

$$= \int \frac{\cos\theta\,d\theta}{(\cos^2\theta)^2} = \int \sec^3\theta\,d\theta$$

$$= \frac{1}{2}\sec\theta\tan\theta + \frac{1}{2}\int \sec\theta\,d\theta$$

(by Formula (6) in Section 9-4)

$$= \frac{1}{2}\sec\theta\tan\theta + \frac{1}{2}\ln|\sec\theta + \tan\theta| + C$$

$$= \frac{1}{2}\left(\frac{x-1}{2x-x^2}\right) + \frac{1}{2}\ln\left|\frac{x}{\sqrt{2x-x^2}}\right| + C.$$

Here we have read the values of $\sec\theta$ and $\tan\theta$ in terms of $x$ from the right triangle in Fig. 9.6.

When we combine all our results above, we finally obtain the answer:

$$\int \frac{4-x}{(2x-x^2)^2}\,dx = -\frac{1}{2(2x-x^2)}$$

$$+ \frac{3(x-1)}{2(2x-x^2)} + \frac{3}{2}\ln\left|\frac{x}{\sqrt{2x-x^2}}\right| + C$$

$$= \frac{3x-4}{2(2x-x^2)} + \frac{3}{4}\ln\left|\frac{x^2}{2x-x^2}\right| + C.$$

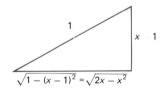

$$\sqrt{1-(x-1)^2} = \sqrt{2x-x^2}$$

**9.6** The reference triangle for Example 4

---

The method of Example 4 can be used to evaluate a general integral having the form

$$\int \frac{Ax + B}{(ax^2 + bx + c)^n}\,dx \tag{1}$$

where $n$ is a positive integer. By splitting such an integral into two simpler ones and by completing the square in the quadratic expression in the denominator, the problem of evaluating the integral in (1) can be reduced to that of computing

$$\int \frac{du}{(a^2 \pm u^2)^n}. \tag{2}$$

If the sign in (2) is the plus sign, then the substitution $u = a \tan \theta$ transforms the integral into the form

$$\int \cos^m \theta \, d\theta$$

(see Problem 29). This integral can be handled by the methods of Section 9-3 or by use of the reduction formula

$$\int \cos^k \theta \, d\theta = \frac{\cos^{k-1} \theta \sin \theta}{k} + \frac{k-1}{k} \int \cos^{k-2} \theta \, d\theta$$

of Problem 46 in Section 9-4.

If the sign in (2) is the minus sign, then the substitution $u = a \sin \theta$ transforms the integral into the form

$$\int \sec^m \theta \, d\theta$$

(see Problem 30). This integral may be evaluated with the aid of the reduction formula

$$\int \sec^k \theta \, d\theta = \frac{\sec^{k-2} \theta \tan \theta}{k-1} + \frac{k-2}{k-1} \int \sec^{k-2} \theta \, d\theta$$

(Equation (5) in Section 9-4).

## 9-6  PROBLEMS

Find the integrals in Problems 1–28.

**1** $\displaystyle\int \frac{dx}{x^2 + 4x + 5}$

**2** $\displaystyle\int \frac{(2x + 5) \, dx}{x^2 + 4x + 5}$

**3** $\displaystyle\int \frac{(5 - 3x) \, dx}{x^2 + 4x + 5}$

**4** $\displaystyle\int \frac{(x + 1) \, dx}{(x^2 + 4x + 5)^2}$

**5** $\displaystyle\int \frac{dx}{\sqrt{3 - 2x - x^2}}$

**6** $\displaystyle\int \frac{(x + 3) \, dx}{\sqrt{3 - 2x - x^2}}$

**7** $\displaystyle\int x\sqrt{3 - 2x - x^2} \, dx$

**8** $\displaystyle\int \frac{dx}{4x^2 + 4x - 3}$

**9** $\displaystyle\int \frac{(3x + 2) \, dx}{4x^2 + 4x - 3}$

**10** $\displaystyle\int \sqrt{4x^2 + 4x - 3} \, dx$

**11** $\displaystyle\int \frac{dx}{x^2 + 4x + 13}$

**12** $\displaystyle\int \frac{dx}{\sqrt{2x - x^2}}$

**13** $\displaystyle\int \frac{dx}{3 + 2x - x^2}$

**14** $\displaystyle\int x\sqrt{8 + 2x - x^2} \, dx$

**15** $\displaystyle\int \frac{2x - 5}{x^2 + 2x + 2} \, dx$

**16** $\displaystyle\int \frac{2x - 1}{4x^2 + 4x - 15} \, dx$

**17** $\displaystyle\int \frac{x \, dx}{\sqrt{5 + 12x - 9x^2}}$

**18** $\displaystyle\int (3x - 2)\sqrt{9x^2 + 12x + 8} \, dx$

**19** $\displaystyle\int (7 - 2x)\sqrt{9 + 16x - 4x^2} \, dx$

**20** $\displaystyle\int \frac{2x + 3}{\sqrt{x^2 + 2x + 5}} \, dx$

**21** $\displaystyle\int \frac{x + 4}{(6x - x^2)^{3/2}} \, dx$

**22** $\displaystyle\int \frac{x - 1}{(x^2 + 1)^2} \, dx$

**23** $\displaystyle\int \frac{2x + 3}{(4x^2 + 12x + 13)^2} \, dx$

**24** $\displaystyle\int \frac{x^3}{(1 - x^2)^4} \, dx$

**25** $\displaystyle\int \frac{3x-1}{x^2+x+1}\,dx$ 　　**26** $\displaystyle\int \frac{3x-1}{(x^2+x+1)^2}\,dx$

**27** $\displaystyle\int \frac{dx}{(x^2-4)^2}$ 　　**28** $\displaystyle\int (x-x^2)^{3/2}\,dx$

**29** Show that the substitution $u = a\tan\theta$ gives

$$\int \frac{du}{(a^2+u^2)^n} = \frac{1}{a^{2n-1}}\int \cos^{2n-2}\theta\,d\theta.$$

**30** Show that the substitution $u = a\sin\theta$ gives

$$\int \frac{du}{(a^2-u^2)^n} = \frac{1}{a^{2n-1}}\int \sec^{2n-1}\theta\,d\theta.$$

**31** A road is to be built joining the points (0, 0) and (3, 2), following the path of the circle

$$(4x+4)^2 + (4x-19)^2 = 377.$$

Find the length of this road, with distance measured in miles.

**32** Suppose that the road of Problem 31 costs $10/(1+x)$ million dollars per mile.
(a) Calculate its total cost.
(b) With the same cost per mile, calculate the total cost of a straight line road from (0, 0) to (3, 2). You should find that it is *more* than the cost of the *longer* circular road!

---

## Rational Functions and Partial Fractions

In this section we show that every rational function can be integrated in terms of elementary functions. Recall that a rational function $R(x)$ is one that can be expressed as a quotient of two polynomials. That is,

$$R(x) = \frac{P(x)}{Q(x)} \tag{1}$$

where $P(x)$ and $Q(x)$ are polynomials. The **method of partial fractions** involves decomposing $R(x)$ into a sum of terms:

$$R(x) = \frac{P(x)}{Q(x)} = p(x) + F_1(x) + F_2(x) + \cdots + F_k(x), \tag{2}$$

where $p(x)$ is a polynomial and each expression $F_i(x)$ is a fraction that can be integrated by the methods of earlier sections.

For example, one can verify (by finding a common denominator on the right) that

$$\frac{x^3-1}{x^3+x} = 1 - \frac{1}{x} + \frac{x-1}{x^2+1}. \tag{3}$$

It follows that

$$\int \frac{x^3-1}{x^3+x}\,dx = \int \left(1 - \frac{1}{x} + \frac{x}{x^2+1} - \frac{1}{x^2+1}\right)dx$$

$$= x - \ln|x| + \frac{1}{2}\ln|x^2+1| - \tan^{-1}x + C.$$

Of course, the key to this simple integration lies in finding the decomposition given in Equation (3). That such a decomposition exists and the technique of finding it is what the method of partial fractions is about.

According to a theorem that is proved in advanced algebra, every rational function can be written in the form of (2) with each of the $F_i(x)$ being either a fraction of the form

$$\frac{A}{(ax+b)^n} \tag{4}$$

or one of the form

$$\frac{Bx + C}{(ax^2 + bx + c)^n},$$  (5)

where the quadratic polynomial $ax^2 + bx + c$ is **irreducible,** meaning that it is not a product of linear factors with real coefficients. This is the same as saying that the equation $ax^2 + bx + c = 0$ has no real roots, and the quadratic formula tells us that this is the case exactly when $b^2 - 4ac < 0$.

Fractions of the forms in (4) and (5) are called **partial fractions,** and the sum in (2) is called the **partial-fraction decomposition** of $R(x)$. Thus (3) gives the partial-fraction decomposition of $(x^3 - 1)/(x^3 + x)$. A partial fraction of the form in (4) may be integrated immediately, and we saw in Section 9-6 how to integrate one of the form in (5).

The first step in finding the partial-fraction decomposition of $R(x)$ is finding the polynomial $p(x)$ in (2). It turns out that $p(x) \equiv 0$ provided that the degree of the numerator $P(x)$ is *less than* that of the denominator $Q(x)$; in this case the rational fraction $R(x) = P(x)/Q(x)$ is called **proper.** If $R(x)$ is not proper, then $p(x)$ may be found by "long division" of $Q(x)$ into $P(x)$, as in the following example.

**EXAMPLE 1**   Find $\displaystyle\int \frac{x^3 + x^2 + x - 1}{x^2 + 2x + 2}\,dx.$

*Solution*   Long division of denominator into numerator is carried out as follows.

$$
\begin{array}{r}
x - 1 \quad\quad p(x) \quad \text{(quotient)} \\
x^2 + 2x + 2 \overline{)\; x^3 + x^2 + x - 1} \\
\underline{x^3 + 2x^2 + 2x} \\
-x^2 - x - 1 \\
\underline{-x^2 - 2x - 2} \\
x + 1 \quad r(x) \quad \text{(remainder)}
\end{array}
$$

As in arithmetic,

$$\text{Fraction} = \text{quotient} + \frac{\text{remainder}}{\text{divisor}}.$$

Thus

$$\frac{x^3 + x^2 + x - 1}{x^2 + 2x + 2} = (x - 1) + \frac{x + 1}{x^2 + 2x + 2}.$$

And hence

$$\int \frac{x^3 + x^2 + x - 1}{x^2 + 2x + 2}\,dx = \int \left( x - 1 + \frac{x + 1}{x^2 + 2x + 2} \right) dx$$

$$= \tfrac{1}{2}x^2 - x + \tfrac{1}{2}\ln(x^2 + 2x + 2) + C.$$

By using long division as in Example 1, any rational function $R(x)$ can be written as a sum of a polynomial $p(x)$ and a *proper* rational function,

$$R(x) = p(x) + \frac{r(x)}{Q(x)}.$$

To see how to integrate an arbitrary rational function, we therefore need only see how to find the partial-fraction decomposition of a proper rational function.

To obtain such a decomposition, we first must factor the denominator $Q(x)$ into a product of linear factors (those of the form $ax + b$) and irreducible quadratic factors (those of the form $ax^2 + bx + c$ with $b^2 - 4ac < 0$). This is always possible in principle but may be quite difficult in practice. Once this factorization of $Q(x)$ has been found, the partial-fraction decomposition may be obtained by routine algebra (described below). Each linear or irreducible quadratic factor of $Q(x)$ leads to one or more partial fractions of the forms in (4) and (5).

## LINEAR FACTORS

Let $R(x) = P(x)/Q(x)$ be a *proper* rational fraction, and suppose that the linear factor $ax + b$ occurs $n$ times in the factorization of $Q(x)$. That is, $(ax + b)^n$ is the highest power of $ax + b$ that divides "evenly" into $Q(x)$. In this case we call $n$ the **multiplicity** of the factor $ax + b$.

---

**Rule 1  Linear Factor Partial Fractions**

The part of the partial-fraction decomposition of $R(x)$ corresponding to the linear factor $ax + b$ of multiplicity $n$ is a sum of $n$ partial fractions, having the form

$$\frac{A_1}{ax + b} + \frac{A_2}{(ax + b)^2} + \cdots + \frac{A_n}{(ax + b)^n}, \tag{6}$$

where $A_1, A_2, \ldots, A_n$ are constants.

---

If *all* the factors of $Q(x)$ are linear, then the partial-fraction decomposition of $R(x)$ is a sum of expressions like (6). The situation is especially simple if each of these linear factors is *nonrepeated*—that is, if each has multiplicity $n = 1$. In this case, the expression in (6) reduces to its first term, and the partial-fraction decomposition of $R(x)$ is a sum of such terms. The solutions in the following examples illustrate how the constant numerators can be determined.

**EXAMPLE 2**   Find $\displaystyle\int \frac{dx}{(2x + 1)(x - 2)}$.

*Solution*   The linear factors in the denominator are distinct, so we seek a partial-fraction decomposition of the form

$$\frac{1}{(2x + 1)(x - 2)} = \frac{A}{2x + 1} + \frac{B}{x - 2}.$$

To find the constants $A$ and $B$, we multiply both sides of this identity by the left-hand (common) denominator $(2x + 1)(x - 2)$ and get

$$1 = A(x - 2) + B(2x + 1) = (A + 2B)x + (-2A + B).$$

We next equate coefficients of $x$ and of 1 on the two sides; this yields the simultaneous equations

$$A + 2B = 0,$$
$$-2A + B = 1$$

that we readily solve for $A = -\frac{2}{5}$, $B = \frac{1}{5}$. Hence

$$\int \frac{dx}{(2x + 1)(x - 2)} = \int \left( \frac{-\frac{2}{5}}{2x + 1} + \frac{\frac{1}{5}}{x - 2} \right) dx$$

$$= -\frac{1}{5} \ln|2x + 1| + \frac{1}{5} \ln|x - 2| + C$$

$$= \frac{1}{5} \ln \left| \frac{x - 2}{2x + 1} \right| + C.$$

**EXAMPLE 3**  Find $\displaystyle\int \frac{4x^2 - 3x - 4}{x^3 + x^2 - 2x} \, dx$.

*Solution*  The rational function to be integrated is proper, so we immediately proceed to factor its denominator.

$$x^3 + x^2 - 2x = x(x^2 + x - 2)$$
$$= x(x - 1)(x + 2).$$

We are dealing with three nonrepeated linear factors, so the partial-fraction decomposition is of the form

$$\frac{4x^2 - 3x - 4}{x^3 + x^2 - 2x} = \frac{A}{x} + \frac{B}{x - 1} + \frac{C}{x + 2}.$$

To find the constants $A$, $B$, and $C$, we multiply both sides of this equation by the left-hand denominator $x(x - 1)(x + 2)$ and find thereby that

$$4x^2 - 3x - 4 = A(x - 1)(x + 2) + Bx(x + 2) + Cx(x - 1). \quad (7)$$

Then we collect coefficients of like powers on the right:

$$4x^2 - 3x - 4 = (A + B + C)x^2 + (A + 2B - C)x + (-2A).$$

Now two polynomials are equal only if the coefficients of corresponding powers of $x$ are the same, and so we may conclude that

$$A + B + C = 4,$$
$$A + 2B - C = -3,$$
$$-2A = -4.$$

We solve these simultaneous equations and thus find that $A = 2$, $B = -1$, and $C = 3$.

There is an alternative way of finding $A$, $B$, and $C$, one that is especially convenient in the case of nonrepeated linear factors. Substitute the values

$x = 0, 1,$ and $-2$ (the zeros of the linear factors of the denominator) in turn into Equation (7). Substitution of $x = 0$ into (7) immediately gives $-4 = -2A$, so that $A = 2$. Substitution of $x = 1$ into (7) gives $-3 = 3B$, and so $B = -1$. Substitution of $x = -2$ gives $18 = 6C$, so $C = 3$.

With these values $A = 2, B = -1,$ and $C = 3$, however obtained, we find that

$$\int \frac{4x^2 - 3x - 4}{x^3 + x^2 - 2x} \, dx = \int \left( \frac{2}{x} - \frac{1}{x-1} + \frac{3}{x+2} \right) dx$$

$$= 2 \ln|x| - \ln|x - 1| + 3 \ln|x + 2| + C.$$

Laws of logarithms allow us to write this antiderivative in the more compact form

$$\ln \left| \frac{x^2(x + 2)^3}{x - 1} \right| + C.$$

**EXAMPLE 4** Find $\displaystyle\int \frac{x^3 - 4x - 1}{x(x - 1)^3} \, dx.$

*Solution* Here we have a linear factor of multiplicity $n = 3$. According to Rule 1 above, the partial-fraction decomposition of the integrand has the form

$$\frac{x^3 - 4x - 1}{x(x - 1)^3} = \frac{A}{x} + \frac{B}{x-1} + \frac{C}{(x-1)^2} + \frac{D}{(x-1)^3}.$$

To find the constants $A, B, C,$ and $D$, we multiply both sides of this equation by $x(x - 1)^3$ and obtain

$$x^3 - 4x - 1 = A(x - 1)^3 + Bx(x - 1)^2 + Cx(x - 1) + Dx.$$

We multiply, then collect coefficients on the right-hand side; this yields

$$x^3 - 4x - 1 = (A + B)x^3 + (-3A - 2B + C)x^2$$
$$+ (3A + B - C + D)x - A.$$

Then we equate coefficients of like powers of $x$ on each side of this equation. We get the four equations

$$A + B = 1,$$
$$-3A - 2B + C = 0,$$
$$3A + B - C + D = -4,$$
$$-A = -1.$$

The last equation gives $A = 1$; then the first equation gives $B = 0$. Next, the second equation gives $C = 3$. Substitution of $A = 1, B = 0,$ and $C = 3$ into the third equation finally gives $D = -4$. Hence

$$\int \frac{x^3 - 4x - 1}{x(x - 1)^3} \, dx = \int \left( \frac{1}{x} + \frac{3}{(x-1)^2} - \frac{4}{(x-1)^3} \right) dx$$

$$= \ln|x| - \frac{3}{x - 1} + \frac{2}{(x - 1)^2} + C.$$

## QUADRATIC FACTORS

Suppose that $R(x) = P(x)/Q(x)$ is a proper rational fraction, and suppose that the irreducible linear factor $ax^2 + bx + c$ occurs $n$ times in the factorization of $Q(x)$. That is, $(ax^2 + bx + c)^n$ is the highest power of $ax^2 + bx + c$ that divides evenly into $Q(x)$.

---

**Rule 2** *Quadratic Factor Partial Fractions*

The part of the partial-fraction decomposition of $R(x)$ corresponding to the irreducible quadratic factor $ax^2 + bx + c$ of multiplicity $n$ is a sum of $n$ partial fractions, having the form

$$\frac{B_1 x + C_1}{ax^2 + bx + c} + \frac{B_2 x + C_2}{(ax^2 + bx + c)^2} + \cdots + \frac{B_n x + C_n}{(ax^2 + bx + c)^n}, \quad (8)$$

where $B_1, B_2, \ldots, B_n, C_1, C_2, \ldots, C_n$ are constants.

---

If $Q(x)$ has both linear and irreducible quadratic factors, then the partial-fraction decomposition of $R(x)$ is the sum of the expressions of the form in (6) corresponding to the linear factors, plus the sum of the expressions of the form in (8) corresponding to the quadratic factors. In the case of an irreducible quadratic factor of multiplicity $n = 1$, the expression (8) reduces to its first term, just as in the linear case.

**EXAMPLE 5** Find $\displaystyle\int \frac{5x^3 - 3x^2 + 2x - 1}{x^4 + x^2}\, dx.$

*Solution* The denominator $x^4 + x^2 = x^2(x^2 + 1)$ has both a quadratic factor and a repeated linear factor. The partial-fraction decomposition takes the form

$$\frac{5x^3 - 3x^2 + 2x - 1}{x^4 + x^2} = \frac{A}{x} + \frac{B}{x^2} + \frac{Cx + D}{x^2 + 1}.$$

We multiply both sides by $x^4 + x^2$ and obtain

$$5x^3 - 3x^2 + 2x - 1 = Ax(x^2 + 1) + B(x^2 + 1) + (Cx + D)x^2$$
$$= (A + C)x^3 + (B + D)x^2 + Ax + B.$$

As before, we equate coefficients of like powers of $x$ and obtain the four equations

$$A \quad\ + C \quad\quad = \quad 5,$$
$$B \quad\quad + D = -3,$$
$$A \quad\quad\quad\quad\quad = \quad 2,$$
$$B \quad\quad\quad\quad\quad = -1.$$

So $A = 2$, $B = -1$, $C = 3$, and $D = -2$. Thus

$$\int \frac{5x^3 - 3x^2 + 2x - 1}{x^4 + x^2} \, dx = \int \left( \frac{2}{x} - \frac{1}{x^2} + \frac{3x - 2}{x^2 + 1} \right) dx$$

$$= 2 \ln|x| + \frac{1}{x} + \frac{3}{2} \int \frac{2x \, dx}{x^2 + 1} - 2 \int \frac{dx}{x^2 + 1}$$

$$= 2 \ln|x| + \frac{1}{x} + \frac{3}{2} \ln(x^2 + 1) - 2 \tan^{-1}x + C.$$

**EXAMPLE 6**   Find $\displaystyle\int \frac{x^3 - 2x}{(x^2 + 2x + 2)^2} \, dx$.

*Solution*   The repeated quadratic factor $x^2 + 2x + 2$ is irreducible because

$$b^2 - 4ac = (2)^2 - 4(1)(2) = -4 < 0.$$

Hence the partial-fraction decomposition of the integrand, as given by Rule 2, is

$$\frac{x^3 - 2x}{(x^2 + 2x + 2)^2} = \frac{Ax + B}{x^2 + 2x + 2} + \frac{Cx + D}{(x^2 + 2x + 2)^2}.$$

We multiply each side by $(x^2 + 2x + 2)^2$ and find that

$$\begin{aligned} x^3 - 2x &= (Ax + B)(x^2 + 2x + 2) + (Cx + D) \\ &= Ax^3 + (2A + B)x^2 + (2A + 2B + C)x + (2B + D). \end{aligned}$$

And so

$$\begin{aligned} A &= 1, \\ 2A + B &= 0, \\ 2A + 2B + C &= -2, \\ 2B + D &= 0. \end{aligned}$$

We find that $A = 1$, $B = -2$, $C = 0$, and $D = 4$. So

$$\int \frac{x^3 - 2x}{(x^2 + 2x + 2)^2} \, dx = \int \frac{x - 2}{x^2 + 2x + 2} \, dx + 4 \int \frac{1}{(x^2 + 2x + 2)^2} \, dx$$

$$= \int \frac{x - 2}{(x + 1)^2 + 1} \, dx + 4 \int \frac{1}{[(x + 1)^2 + 1]^2} \, dx$$

$$= \int \frac{u - 3}{u^2 + 1} \, du + 4 \int \frac{1}{(u^2 + 1)^2} \, du \qquad (9)$$

where $u = x + 1$. The first integral here is simple to evaluate:

$$\int \frac{u - 3}{u^2 + 1} \, du = \frac{1}{2} \int \frac{2u}{u^2 + 1} \, du - 3 \int \frac{du}{u^2 + 1}$$

$$= \frac{1}{2} \ln(u^2 + 1) - 3 \tan^{-1}u + C$$

$$= \frac{1}{2} \ln(x^2 + 2x + 2) - 3 \tan^{-1}(x + 1) + C. \qquad (10)$$

For the second integral in (9), we use the substitution $u = \tan\theta$, so that $du = \sec^2\theta\, d\theta$. Then

$$4\int \frac{du}{(u^2 + 1)^2} = 4\int \frac{\sec^2\theta\, d\theta}{\sec^4\theta}$$

$$= 4\int \cos^2\theta\, d\theta = 2\int (1 + \cos 2\theta)\, d\theta$$

$$= 2(\theta + \sin\theta \cos\theta) + C$$

$$= 2\tan^{-1}u + \frac{2u}{u^2 + 1} + C$$

$$= 2\tan^{-1}(x + 1) + \frac{2(x + 1)}{x^2 + 2x + 2} + C. \qquad (11)$$

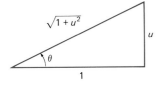

We have read the values of $\sin\theta$ and $\cos\theta$ in terms of $u$ from the right triangle of Fig. 9.7 and then resubstituted $x + 1$ for $u$. Finally, addition of our results in (10) and (11) gives the answer:

$$\int \frac{x^3 - 2x}{(x^2 + 2x + 2)^2}\, dx = \frac{1}{2}\ln(x^2 + 2x + 2) - \tan^{-1}(x + 1)$$

$$+ \frac{2(x + 1)}{x^2 + 2x + 2} + C.$$

**9.7** The reference triangle for Example 6

## 9-7  PROBLEMS

Find the integrals in Problems 1–35.

1. $\displaystyle\int \frac{x^2\, dx}{x + 1}$

2. $\displaystyle\int \frac{x^3\, dx}{2x - 1}$

3. $\displaystyle\int \frac{dx}{x^2 - 3x}$

4. $\displaystyle\int \frac{x\, dx}{x^2 + 4x}$

5. $\displaystyle\int \frac{dx}{x^2 + x - 6}$

6. $\displaystyle\int \frac{x^3\, dx}{x^2 + x - 6}$

7. $\displaystyle\int \frac{dx}{x^3 + 4x}$

8. $\displaystyle\int \frac{dx}{(x + 1)(x^2 + 1)}$

9. $\displaystyle\int \frac{x^4\, dx}{x^2 + 4}$

10. $\displaystyle\int \frac{dx}{(x^2 + 1)(x^2 + 4)}$

11. $\displaystyle\int \frac{x - 1}{x + 1}\, dx$

12. $\displaystyle\int \frac{2x^3 - 1}{x^2 + 1}\, dx$

13. $\displaystyle\int \frac{x^2 + 2x}{(x + 1)^2}\, dx$

14. $\displaystyle\int \frac{2x - 4}{x^2 - x}\, dx$

15. $\displaystyle\int \frac{dx}{x^2 - 4}$

16. $\displaystyle\int \frac{x^4\, dx}{x^2 + 4x + 4}$

17. $\displaystyle\int \frac{x + 10}{2x^2 + 5x - 3}\, dx$

18. $\displaystyle\int \frac{x + 1}{x^3 - x^2}\, dx$

19. $\displaystyle\int \frac{x^2 + 1}{x^3 + 2x^2 + x}\, dx$

20. $\displaystyle\int \frac{(x^2 + x)\, dx}{x^3 - x^2 - 2x}$

21. $\displaystyle\int \frac{4x^3 - 7x}{x^4 - 5x^2 + 4}\, dx$

22. $\displaystyle\int \frac{(2x^2 + 3)\, dx}{x^4 - 2x^2 + 1}$

23. $\displaystyle\int \frac{x^2\, dx}{(x + 2)^3}$

24. $\displaystyle\int \frac{(x^2 + x)\, dx}{(x^2 - 4)(x + 4)}$

25. $\displaystyle\int \frac{dx}{x^3 + x}$

26. $\displaystyle\int \frac{x^2\, dx}{x^2 + x + 1}$

27. $\displaystyle\int \frac{x + 4}{x^3 + 4x}\, dx$

28. $\displaystyle\int \frac{(x^2 + 1)\, dx}{x^3 + x^2 + x}$

29. $\displaystyle\int \frac{x\, dx}{(x + 1)(x^2 + 1)}$

30. $\displaystyle\int \frac{x^2 + 2}{(x^2 + 1)^2}\, dx$

31. $\displaystyle\int \frac{(x^2 - 10)\, dx}{2x^4 + 9x^2 + 4}$

32. $\displaystyle\int \frac{x^2\, dx}{x^4 - 1}$

33. $\displaystyle\int \frac{(3x + 1)\, dx}{(x^2 + 2x + 5)^2}$

34. $\displaystyle\int \frac{(x^2 + 4)\, dx}{(x^2 + 1)^2(x^2 + 2)}$

35. $\displaystyle\int \frac{x^4 + 3x^2 - 4x + 5}{(x - 1)^2(x^2 + 1)}\, dx$

In Problems 36–39, make a preliminary substitution before using the method of partial fractions.

**36** $\displaystyle\int \frac{\cos\theta\, d\theta}{\sin^2\theta - \sin\theta - 6}$

**37** $\displaystyle\int \frac{e^{4t}\, dt}{(e^{2t} - 1)^3}$

**38** $\displaystyle\int \frac{\sec^2 t\, dt}{\tan^3 t + \tan^2 t}$

**39** $\displaystyle\int \frac{(1 + \ln t)\, dt}{t(3 + 2\ln t)^2}$

In Problems 40 and 41, factor the denominator by first noting by inspection a root $r$ of the denominator and then employing long division by $x - r$. Finally, use the method of partial fractions to aid in finding the indicated antiderivative.

**40** $\displaystyle\int \frac{dx}{x^3 + 8}$

**41** $\displaystyle\int \frac{x^4 + 2x^2}{x^3 - 1}\, dx$

**42** (a) Find constants $a$ and $b$ such that

$$x^4 + 1 = (x^2 + ax + 1)(x^2 + bx + 1).$$

(b) Show that

$$\int_0^1 \frac{x^2 + 1}{x^4 + 1}\, dx = \frac{\pi}{2\sqrt{2}}.$$

**43** Factor $x^4 + x^2 + 1$ as in Problem 42, and then find

$$\int \frac{x^3 + 2x}{x^4 + x^2 + 1}\, dx.$$

**44** Find the volume generated by revolving the area bounded by the following curve around the $x$-axis:

$$y^2 = x^2 \frac{1 - x}{1 + x}, \qquad 0 \le x \le 1.$$

*9-8

In the following example we make a substitution that eliminates the radical. This is an example of a *rationalizing* substitution.

**Rationalizing Substitutions**

**EXAMPLE 1**   Find $\displaystyle\int x^2\sqrt{x + 1}\, dx$.

**Solution**   Let $u = x + 1$. Then $x = u - 1$ and $dx = du$. This substitution yields

$$\int x^2\sqrt{x + 1}\, dx = \int (u - 1)^2\sqrt{u}\, du$$

$$= \int (u^2 - 2u + 1)\sqrt{u}\, du$$

$$= \int (u^{5/2} - 2u^{3/2} + u^{1/2})\, du$$

$$= \tfrac{2}{7}u^{7/2} - \tfrac{4}{5}u^{5/2} + \tfrac{2}{3}u^{3/2} + C$$

$$= \tfrac{2}{7}(x + 1)^{7/2} - \tfrac{4}{5}(x + 1)^{5/2} + \tfrac{2}{3}(x + 1)^{3/2} + C.$$

The method of Example 1 succeeds with any integral of the form

$$\int p(x)\sqrt{ax + b}\, dx \tag{1}$$

where $p(x)$ is a polynomial. In fact, either the substitution $u = ax + b$ or the substitution $u = (ax + b)^{1/2}$ may be used. More generally, the substitution $u^n = f(x)$ may succeed when the integrand involves $\sqrt[n]{f(x)}$. It *always* succeeds in the case of an integral of the form

$$\int p(x)\sqrt[n]{\frac{ax + b}{cx + d}}\, dx, \tag{2}$$

where $p(x)$ is a polynomial. In particular, the substitution

$$u^n = \frac{ax + b}{cx + d}$$

converts (2) into the integral of a rational function of $u$ (see Problem 21), which can then be integrated by the method of partial fractions.

**EXAMPLE 2** Find $\displaystyle\int \frac{\sqrt{y}\,dy}{\sqrt{c-y}}$.

COMMENT   This integral arises in the solution of the problem of determining the shape of a wire joining two fixed points, down which a bead slides in the least possible time from the upper point $O$ to the lower point $P$. This is the famous *brachistochrone problem* (from the Greek *brachistos*, shortest, and *chronos*, time).

We take the $y$-axis pointing downward from $O$, as shown in Fig. 9.8. An analysis of the physics of the situation, together with an advanced technique from the calculus of variations, shows that the optimal shape $y = f(x)$ of the wire—the shape that gives the bead the least time of transit—satisfies the differential equation

$$y\left(1 + \left(\frac{dy}{dx}\right)^2\right) = c \qquad (c \text{ constant}),$$

so that

$$\frac{dy}{dx} = \frac{\sqrt{c-y}}{\sqrt{y}}.$$

Thus

$$x = x(y) = \int \frac{\sqrt{y}\,dy}{\sqrt{c-y}}.$$

So we can solve the brachistochrone problem if we can perform the above antidifferentiation, and thus obtain $x$ as a function of $y$; this will give us the formula for the inverse of the desired function $f$.

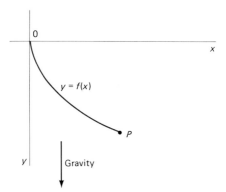

**9.8** The brachistochrone problem: Find the shape of the graph of $y = f(x)$ to minimize the time it takes a bead to slide down the graph from $O$ to $P$.

*Solution*   We use the substitution suggested above for an integral of the form (2). If $u^2 = y/(c - y)$, then

$$y = \frac{cu^2}{1 + u^2} \quad \text{and} \quad dy = \frac{2cu}{(1 + u^2)^2}\,du.$$

Hence

$$\begin{aligned}
x &= \int \frac{2cu^2}{(1 + u^2)^2}\,du \\
&= \int \frac{2c\tan^2\theta \sec^2\theta}{(1 + \tan^2\theta)^2}\,d\theta \qquad (u = \tan\theta) \\
&= 2c \int \frac{\tan^2\theta}{\sec^2\theta}\,d\theta = 2c \int \sin^2\theta\,d\theta.
\end{aligned}$$

**496**

And so

$$x = c(\theta - \sin\theta\cos\theta). \tag{3}$$

The constant of integration is zero because $y = u = \theta = 0$ when $x = 0$. Now, with the aid of Fig. 9.9, we retrace our steps.

$$x = \int \frac{\sqrt{y}}{\sqrt{c-y}}\, dy = c(\theta - \sin\theta\cos\theta)$$

$$= c\tan^{-1}u - \frac{cu}{u^2+1},$$

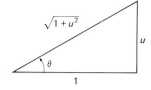

**9.9** The reference triangle for the substitution $u = \tan\theta$

and therefore

$$x = c\tan^{-1}\left(\frac{\sqrt{y}}{\sqrt{c-y}}\right) - \sqrt{cy - y^2}.$$

This *is* a solution, but its form tells us little about the actual shape of the curve of quickest descent. In Section 13-1 we will see what this curve looks like. For reference then, we record the fact that

$$y = \frac{cu^2}{1+u^2} = \frac{c\tan^2\theta}{1+\tan^2\theta} = c\sin^2\theta. \tag{4}$$

Equations (3) and (4) express $x$ and $y$ in terms of the substitution variable $\theta$. This is a "parametric" description of the curve of quickest descent.

---

In the case of an integral involving roots of the variable $x$, we may try the substitution $x = u^n$, with $n$ chosen to be the smallest integer that will rationalize all the roots.

**EXAMPLE 3**  Find $\displaystyle\int \frac{dx}{x^{1/2} + x^{1/3}}.$

*Solution*  The least common multiple of 2 and 3 (from the exponents $\frac{1}{2}$ and $\frac{1}{3}$) is 6, so we try the substitution $x = u^6$, $dx = 6u^5\, du$. This gives

$$\int \frac{dx}{x^{1/2} + x^{1/3}} = \int \frac{6u^5\, du}{u^3 + u^2} = 6\int \frac{u^3\, du}{u+1}$$

$$= 6\int \left(u^2 - u + 1 - \frac{1}{u+1}\right) du \qquad \text{(by long division)}$$

$$= 2u^3 - 3u^2 + 6u - 6\ln|u+1| + C$$

$$= 2x^{1/2} - 3x^{1/3} + 6x^{1/6} - 6\ln|1 + x^{1/6}| + C.$$

**RATIONAL FUNCTIONS OF sin $\theta$ AND cos $\theta$**

In Section 9-3 we saw how to evaluate certain integrals of the form

$$\int R(\sin\theta, \cos\theta)\, d\theta, \tag{5}$$

where $R(\sin\theta, \cos\theta)$ is a rational function of $\sin\theta$ and $\cos\theta$. That is, $R(\sin\theta, \cos\theta)$ is a quotient of polynomials in the two "variables" $\sin\theta$ and $\cos\theta$.

The special substitution

$$u = \tan \frac{\theta}{2} \tag{6}$$

can be used to find *any* integral of the Form (5).

In order to carry out the substitution in (6), we must express $\sin\theta$, $\cos\theta$, and $d\theta$ in terms of $u$ and $du$. Note first that

$$\theta = 2\tan^{-1}u, \quad \text{so that} \quad d\theta = \frac{2\,du}{1 + u^2}. \tag{7}$$

From the triangle in Fig. 9.10, we see that

$$\sin\frac{\theta}{2} = \frac{u}{\sqrt{1 + u^2}}, \quad \cos\frac{\theta}{2} = \frac{1}{\sqrt{1 + u^2}}.$$

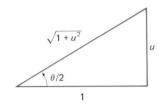

**9.10** The special rationalizing substitution $u = \tan(\theta/2)$

Hence

$$\sin\theta = 2\sin\frac{\theta}{2}\cos\frac{\theta}{2} = \frac{2u}{1 + u^2}, \tag{8}$$

$$\cos\theta = \cos^2\frac{\theta}{2} - \sin^2\frac{\theta}{2} = \frac{1 - u^2}{1 + u^2}. \tag{9}$$

It should be clear that these substitutions will convert the integral in (5) into an integral of a rational function of $u$. The latter can then be evaluated by the methods of Section 9-7.

In Section 8-4 we resorted to an unmotivated trick to show that

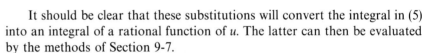

$$\int \sec\theta\,d\theta = \ln|\sec\theta + \tan\theta| + C.$$

We can now use the substitution $u = \tan(\theta/2)$ to integrate $\sec\theta$ more systematically.

**EXAMPLE 4**   Find $\int \sec\theta\,d\theta$.

**Solution**   Equations (7) through (9) give

$$\int \sec\theta\,d\theta = \int \frac{1}{\cos\theta}\,d\theta$$

$$= \int \frac{1 + u^2}{1 - u^2}\,\frac{2\,du}{1 + u^2}$$

$$= \int \frac{2\,du}{1 - u^2} = \int \left(\frac{1}{1 - u} + \frac{1}{1 + u}\right)du$$

$$= \ln\left|\frac{1 + u}{1 - u}\right| + C = \ln\left|\frac{1 + \tan(\theta/2)}{1 - \tan(\theta/2)}\right| + C.$$

Elementary trigonometric identities may now be used to show that this result is the same as our earlier one (see Problem 28).

**EXAMPLE 5** Suppose that $a > b > 0$. Find

$$\int \frac{d\theta}{a + b \cos \theta}.$$

**Solution** We make the substitutions of Equations (7) through (9) and thus find that

$$\int \frac{d\theta}{a + b \cos \theta} = \int \frac{1}{a + b(1 - u^2)/(1 + u^2)} \frac{2 \, du}{1 + u^2}$$

$$= \int \frac{2 \, du}{(a + b) + (a - b)u^2}$$

$$= \frac{2}{a - b} \int \frac{du}{c^2 + u^2} \quad \left( \text{where } c = \left( \frac{a + b}{a - b} \right)^{1/2} \right)$$

$$= \frac{2/c}{a - b} \tan^{-1}\left( \frac{u}{c} \right) + C.$$

Thus

$$\int \frac{d\theta}{a + b \cos \theta} = \frac{2}{\sqrt{a^2 - b^2}} \tan^{-1}\left( \sqrt{\frac{a - b}{a + b}} \tan \frac{\theta}{2} \right) + C. \qquad (10)$$

Find the integrals in Problems 1–20.

**1** $\int x^3 \sqrt{3x - 2} \, dx$

**2** $\int x^3 \sqrt[3]{x^2 + 1} \, dx$

**3** $\int \frac{1}{1 + \sqrt{x}} \, dx$

**4** $\int \frac{dx}{x^{1/2} - x^{1/4}}$

**5** $\int \frac{x^3 \, dx}{(x^2 - 1)^{4/3}}$

**6** $\int x^2 (x - 1)^{3/2} \, dx$

**7** $\int \frac{1 - \sqrt{x}}{1 + \sqrt[3]{x}} \, dx$

**8** $\int \frac{\sqrt[3]{x}}{1 + \sqrt{x}} \, dx$

**9** $\int \frac{x^5 \, dx}{\sqrt{x^3 + 1}}$

**10** $\int x^7 \sqrt[3]{x^4 + 1} \, dx$

**11** $\int \frac{dx}{1 + x^{2/3}}$

**12** $\int \sqrt{\frac{1 + x}{1 - x}} \, dx$

**13** $\int \frac{dx}{1 + \sqrt{x + 4}}$

**14** $\int \frac{\sqrt[3]{x} + 1}{x} \, dx$

**15** $\int \frac{d\theta}{1 + \sin \theta}$

**16** $\int \frac{d\theta}{(1 - \cos \theta)^2}$

**17** $\int \frac{d\theta}{\sin \theta + \cos \theta}$

**18** $\int \frac{d\phi}{\sin \phi + \cos \phi + 2}$

**19** $\int \frac{\sin \theta}{2 + \cos \theta} \, d\theta$

**20** $\int \frac{\sin \theta - \cos \theta}{\sin \theta + \cos \theta} \, d\theta$

**21** Show that if $p(x)$ is a polynomial, the substitution $u^n = (ax + b)/(cx + d)$ transforms the integral

$$\int p(x) \left( \frac{ax + b}{cx + d} \right)^{1/n} dx$$

into the integral of a rational function of $u$.

**22** Rework Example 2 using the trigonometric substitution $y = c \sin^2\theta$ instead of the algebraic substitution used in the text.

**23** Find the area bounded by the loop of the curve

$$y^2 = x^2(1 - x), \qquad 0 \le x \le 1.$$

**24** Find the area bounded by the loop of the curve

$$y^2 = x^2 \frac{1 - x}{1 + x}, \qquad 0 \le x \le 1.$$

**25** Find the length of the arc $y = 2\sqrt{x}, 0 \le x \le 1$. (*Suggestion:* Substitute $x = \tan^2\theta$ in the arc length integral.)

**26** Use the result of Example 5 to find

$$\int \frac{A + B \cos \theta}{a + b \cos \theta} \, d\theta \qquad (a > b > 0).$$

Do this by dividing the denominator into the numerator—think of the denominator as a polynomial in the variable $\cos \theta$.

**27** Show that

$$\int \frac{d\theta}{a + b \cos \theta} = \frac{1}{\sqrt{b^2 - a^2}}$$

$$\times \ln \left| \frac{\sqrt{b - a} \tan(\theta/2) + \sqrt{b + a}}{\sqrt{b - a} \tan(\theta/2) - \sqrt{b + a}} \right| + C$$

if $0 < a < b$.

**28** Use the trigonometric identity

$$\tan \frac{\theta}{2} = \sqrt{\frac{1 - \cos \theta}{1 + \cos \theta}}$$

to derive our earlier formula

$$\int \sec \theta \, d\theta = \ln|\sec \theta + \tan \theta| + C$$

from the solution to Example 4.

**29** (a) Use the method of Example 4 to show that

$$\int \csc \theta \, d\theta = \ln \left| \tan \frac{\theta}{2} \right| + C.$$

(b) Use trigonometric identities to derive the formula

$$\int \csc \theta \, d\theta = \ln|\csc \theta - \cot \theta| + C$$

from part (a).

## CHAPTER 9   Summary

When you confront the problem of evaluating a particular integral, you must first decide which of the several methods of this chapter to try. There are only two *general* methods of integration—integration by substitution (Section 9-2) and integration by parts (Section 9-4). These are the analogues for integration of the chain rule and product rule, respectively, for differentiation.

Look first at the given integral to see if you can spot a substitution that transforms it into an elementary or familiar integral or one likely to be found in an integral table. If the integrand is an unfamiliar product of two functions, one of which is easily differentiated and the other easily integrated, then an attempt to integrate by parts is indicated.

Beyond these two quite general methods, the chapter deals with a number of *special* methods. In the case of a conspicuously trigonometric integral, the simple "split-off" methods of Section 9-3 may succeed. Recall that reduction formulas (like Equation (5) and Problems 45 and 46 in Section 9-4) are available for integrating an integral power of a single trigonometric function. As a last resort, change to sines and cosines and remember that any rational function of $\sin \theta$ and $\cos \theta$ can be integrated using the substitution $u = \tan(\theta/2)$ of Section 9-8. The result of this substitution is a rational function of $u$.

Any integral of a rational function (quotient of two polynomials) can be evaluated by the method of partial fractions (Section 9-7). If the degree of the numerator is not less than that of the denominator—that is, if the rational function is not proper—first use long division to express it as the sum of a polynomial (easily integrated) and a proper rational fraction. Then decompose the latter into partial fractions. The partial fractions corresponding to linear factors are easily integrated, and those corresponding to irreducible quadratic factors can be integrated by completing the square and making (if necessary) a trigonometric substitution. As we explained in Section 9-6, the trigonometric integrals that result can always be evaluated.

In the case of an integral involving

$$\sqrt{ax^2 + bx + c},$$

you should first complete the square (Section 9-6) and then rationalize the integral by making an appropriate trigonometric substitution (Section 9-5); this will leave you with a trigonometric integral. Finally, an integral involving $\sqrt[n]{f(x)}$ is sometimes rationalized by the special substitution $u^n = f(x)$ of Section 9-8.

## MISCELLANEOUS PROBLEMS

Evaluate the integrals in Problems 1–100.

**1** $\int \dfrac{dx}{\sqrt{x}(1 + x)}$

**2** $\int \dfrac{\sec^2 t \, dt}{1 + \tan t}$

**3** $\int \sin x \sec x \, dx$

**4** $\int \dfrac{\csc x \cot x}{1 + \csc^2 x} \, dx$

**5** $\int \dfrac{\tan \theta}{\cos^2 \theta} \, d\theta$

**6** $\int \csc^4 x \, dx$

**7** $\int x \tan^2 x \, dx$

**8** $\int x^2 \cos^2 x \, dx$

**9** $\int x^5 \sqrt{2 - x^3} \, dx$

**10** $\int \dfrac{dx}{\sqrt{x^2 + 4}}$

11. $\displaystyle\int \frac{x^2\,dx}{\sqrt{25 + x^2}}$

12. $\displaystyle\int (\cos x)\sqrt{4 - \sin^2 x}\,dx$

51. $\displaystyle\int \frac{d\theta}{4 + 5\cos\theta}$

52. $\displaystyle\int \frac{(1 + x^{2/3})^{3/2}}{\sqrt[3]{x}}\,dx$

13. $\displaystyle\int \frac{dx}{x^2 - x + 1}$

14. $\displaystyle\int \sqrt{x^2 + x + 1}\,dx$

53. $\displaystyle\int \frac{(\sin^{-1}x)^2}{\sqrt{1 - x^2}}\,dx$

54. $\displaystyle\int \frac{dx}{x^{3/2}(1 + x^{1/3})}$

15. $\displaystyle\int \frac{5x + 31}{3x^2 - 4x + 11}\,dx$

16. $\displaystyle\int \frac{x^4 + 1}{x^2 + 2}\,dx$

55. $\displaystyle\int \tan^3 z\,dz$

56. $\displaystyle\int \sin^2\omega \cos^4\omega\,d\omega$

17. $\displaystyle\int \frac{d\theta}{5 + 4\cos\theta}$

18. $\displaystyle\int \frac{\sqrt{x}}{1 + x}\,dx$

57. $\displaystyle\int \frac{xe^{x^2}\,dx}{1 + e^{2x^2}}$

58. $\displaystyle\int \frac{\cos^3 x}{\sqrt{\sin x}}\,dx$

19. $\displaystyle\int \frac{\cos x\,dx}{\sqrt{4 - \sin^2 x}}$

20. $\displaystyle\int \frac{\cos 2x}{\cos x}\,dx$

59. $\displaystyle\int x^3 e^{-x^2}\,dx$

60. $\displaystyle\int \sin\sqrt{x}\,dx$

21. $\displaystyle\int \frac{\tan x}{\ln(\cos x)}\,dx$

22. $\displaystyle\int \frac{x^7\,dx}{\sqrt{1 - x^4}}$

61. $\displaystyle\int \frac{\arcsin x}{x^2}\,dx$

62. $\displaystyle\int \sqrt{x^2 - 9}\,dx$

63. $\displaystyle\int x^2\sqrt{1 - x^2}\,dx$

64. $\displaystyle\int x\sqrt{2x - x^2}\,dx$

23. $\displaystyle\int \ln(1 + x)\,dx$

24. $\displaystyle\int x\sec^{-1}x\,dx$

65. $\displaystyle\int \frac{(x - 2)\,dx}{4x^2 + 4x + 1}$

25. $\displaystyle\int \sqrt{x^2 + 9}\,dx$

26. $\displaystyle\int \frac{x^2\,dx}{\sqrt{4 - x^2}}$

66. $\displaystyle\int \frac{2x^2 - 5x - 1}{x^3 - 2x^2 - x + 2}\,dx$

27. $\displaystyle\int \sqrt{2x - x^2}\,dx$

28. $\displaystyle\int \frac{4x - 2}{x^3 - x}\,dx$

67. $\displaystyle\int \frac{e^{2x}\,dx}{e^{2x} - 1}$

68. $\displaystyle\int \frac{\cos x\,dx}{\sin^2 x - 3\sin x + 2}$

29. $\displaystyle\int \frac{x^4\,dx}{x^2 - 2}$

30. $\displaystyle\int \frac{\sec x \tan x\,dx}{\sec x + \sec^2 x}$

69. $\displaystyle\int \frac{2x^3 + 3x^2 + 4}{(x + 1)^4}\,dx$

70. $\displaystyle\int \frac{\sec^2 x\,dx}{\tan^2 x + 2\tan x + 2}$

31. $\displaystyle\int \frac{x\,dx}{(x^2 + 2x + 2)^2}$

32. $\displaystyle\int \frac{\sqrt[3]{x}\,dx}{\sqrt{x} + \sqrt[4]{x}}$

71. $\displaystyle\int \frac{x^3 + x^2 + 2x + 1}{x^4 + 2x^2 + 1}\,dx$

33. $\displaystyle\int \frac{d\theta}{1 + \cos 2\theta}$

34. $\displaystyle\int \frac{\sec x}{\tan x}\,dx$

72. $\displaystyle\int \frac{3 + \cos\theta}{2 - \cos\theta}\,d\theta$

73. $\displaystyle\int x^5\sqrt{x^3 - 1}\,dx$

35. $\displaystyle\int \sec^3 x \tan^3 x\,dx$

36. $\displaystyle\int x^2 \tan^{-1}x\,dx$

74. $\displaystyle\int \frac{d\theta}{2 + 2\cos\theta + \sin\theta}$

37. $\displaystyle\int x(\ln x)^3\,dx$

38. $\displaystyle\int \frac{dx}{x\sqrt{1 + x^2}}$

75. $\displaystyle\int \frac{\sqrt{1 + \sin x}}{\sec x}\,dx$

76. $\displaystyle\int \frac{dx}{x^{2/3}(1 + x^{2/3})}$

39. $\displaystyle\int e^x\sqrt{1 + e^{2x}}\,dx$

40. $\displaystyle\int \frac{x\,dx}{\sqrt{4x - x^2}}$

77. $\displaystyle\int \frac{\sin x}{\sin 2x}\,dx$

78. $\displaystyle\int \sqrt{1 + \cos t}\,dt$

41. $\displaystyle\int \frac{dx}{x^3\sqrt{x^2 - 9}}$

42. $\displaystyle\int \frac{x\,dx}{(7x + 1)^{17}}$

79. $\displaystyle\int \sqrt{1 + \sin t}\,dt$

80. $\displaystyle\int \frac{\sec^2 t\,dt}{1 - \tan^2 t}$

43. $\displaystyle\int \frac{4x^2 + x + 1}{4x^3 + x}\,dx$

44. $\displaystyle\int \frac{4x^3 - x + 1}{x^3 + 1}\,dx$

81. $\displaystyle\int \ln(x^2 + x + 1)\,dx$

82. $\displaystyle\int e^x\sin^{-1}(e^x)\,dx$

45. $\displaystyle\int \tan^2 x \sec x\,dx$

46. $\displaystyle\int \frac{x^2 + 2x + 2}{(x + 1)^3}\,dx$

83. $\displaystyle\int \frac{\arctan x\,dx}{x^2}$

84. $\displaystyle\int \frac{x^2\,dx}{\sqrt{x^2 - 25}}$

47. $\displaystyle\int \frac{x^4 + 2x + 2}{x^5 + x^4}\,dx$

48. $\displaystyle\int \frac{8x^2 - 4x + 7}{(x^2 + 1)(4x + 1)}\,dx$

85. $\displaystyle\int \frac{x^3\,dx}{(x^2 + 1)^2}$

86. $\displaystyle\int \frac{dx}{x\sqrt{6x - x^2}}$

49. $\displaystyle\int \frac{dx}{(x^3 - 1)^2}$

50. $\displaystyle\int \frac{x\,dx}{x^4 + 4x^2 + 8}$

87. $\displaystyle\int \frac{(3x + 2)\,dx}{(x^2 + 4)^{3/2}}$

88. $\displaystyle\int x^{3/2}\ln x\,dx$

**89** $\displaystyle\int \frac{(1 + \sin^2 x)^{1/2}}{\sec x \csc x}\, dx$

**90** $\displaystyle\int \frac{e^{\sqrt{\sin x}}\, dx}{\sec x \sqrt{\sin x}}$

**91** $\displaystyle\int x e^x \sin x\, dx$

**92** $\displaystyle\int x^2 \exp(x^{3/2})\, dx$

**93** $\displaystyle\int \frac{\tan^{-1} x}{(x - 1)^3}\, dx$

**94** $\displaystyle\int \ln(1 + \sqrt{x})\, dx$

**95** $\displaystyle\int \frac{(2x + 3)\, dx}{\sqrt{3 + 6x - 9x^2}}$

**96** $\displaystyle\int \frac{d\theta}{2 + 2 \sin \theta + \cos \theta}$

**97** $\displaystyle\int \frac{\sin^3 \theta\, d\theta}{\cos \theta - 1}$

**98** $\displaystyle\int x^{3/2} \tan^{-1}(x^{1/2})\, dx$

**99** $\displaystyle\int \sec^{-1}\sqrt{x}\, dx$

**100** $\displaystyle\int x \left(\frac{1 - x^2}{1 + x^2}\right)^{1/2} dx$

**101** Find the area of the surface generated by revolving the curve $y = \cosh x$, $0 \leq x \leq 1$, around the $x$-axis.

**102** Find the length of the curve $y = e^{-x}$, $0 \leq x \leq 1$.

**103** (a) Find the area $A_b$ of the surface generated by revolving the curve $y = e^{-x}$, $0 \leq x \leq b$, around the $x$-axis.
(b) Find $\lim\limits_{b \to \infty} A_b$.

**104** (a) Find the area $A_b$ of the surface generated by revolving the curve $y = 1/x$, $1 \leq x \leq b$, around the $x$-axis.
(b) Find $\lim\limits_{b \to \infty} A_b$.

**105** Find the area of the surface generated by revolving the curve

$$y = \sqrt{x^2 - 1}, \qquad 1 \leq x \leq 2,$$

around the $x$-axis.

**106** (a) Derive the reduction formula

$$\int x^m (\ln x)^n\, dx = \frac{x^{m+1}(\ln x)^n}{m + 1} - \frac{n}{m + 1} \int x^m (\ln x)^{n-1}\, dx.$$

(b) Find $\displaystyle\int_1^e x^3 (\ln x)^3\, dx$.

**107** Derive the reduction formula

$$\int \sin^m x \cos^n x\, dx = -\frac{\sin^{m-1} x \cos^{n+1} x}{m + n}$$
$$+ \frac{m - 1}{m + n} \int \sin^{m-2} x \cos^n x\, dx.$$

**108** Use the reduction formulas of Problem 107 and Problem 46 in Section 9-4 to evaluate

$$\int_0^{\pi/2} \sin^6 x \cos^5 x\, dx.$$

**109** Find the area bounded by the curve $y^2 = x^5(2 - x)$, $0 \leq x \leq 2$. (*Suggestion:* Substitute $x = 2 \sin^2 \theta$, and then use the result of Problem 50 in Section 9-4.)

**110** Show that

$$\int_0^1 \frac{t^4(1 - t)^4}{1 + t^2}\, dt = \frac{22}{7} - \pi.$$

**111** Evaluate $\displaystyle\int_0^1 t^4(1 - t)^4\, dt$, and then apply the result of Problem 110 to conclude that

$$\tfrac{22}{7} - \tfrac{1}{630} < \pi < \tfrac{22}{7} - \tfrac{1}{1260}.$$

Thus $3.1412 < \pi < 3.1421$.

**112** Find the length of the curve $y = \frac{4}{5}x^{5/4}$, $0 \leq x \leq 1$.

**113** Find the length of the curve $y = \frac{4}{3}x^{3/4}$, $1 \leq x \leq 4$.

**114** An initially empty water tank is shaped like a cone with vertical axis, vertex at the bottom, depth of 9 ft, and top radius 4.5 ft. Beginning at time $t = 0$, water is poured into this tank at 50 ft$^3$ per minute. Meanwhile, water leaks out a hole at the bottom at the rate of $10\sqrt{y}$ cubic feet per minute—consistent with Torricelli's law. How long does it take to fill the tank?

**115** (a) Evaluate $\displaystyle\int \frac{dx}{1 + e^x + e^{-x}}$.

(b) Explain why your substitution in (a) suffices to integrate any rational function of $e^x$.

**116** (a) The equation $x^3 + x + 1 = 0$ has one real root $r$. Use Newton's method to find it, accurate to at least two places.

(b) Use long division to find the irreducible quadratic factor of $x^3 + x + 1$.

(c) Use the factorization of part (b) to evaluate

$$\int_0^1 \frac{dx}{x^3 + x + 1}.$$

**117** Find $\displaystyle\int \frac{dx}{1 + e^x}$.

**118** The integral

$$\int \frac{1 + 2x^2}{x^5(1 + x^2)^3}\, dx = \int \frac{x + 2x^3}{(x^4 + x^2)^3}\, dx$$

would require solving eleven equations in eleven unknowns if the method of partial fractions were used. Use the substitution $u = x^4 + x^2$ to evaluate it much more simply.

**119** Find $\int \sqrt{\tan \theta}\, d\theta$. (*Suggestion:* First substitute $u = \tan \theta$. Then substitute $u = x^2$. Finally use the method of partial fractions; see Problem 42 in Section 9-7.)

# 10

Polar
Coordinates
and
Conic Sections

## Analytic Geometry and the Conic Sections

**Plane analytic geometry,** the main topic of this chapter, is the use of algebra and calculus to study the properties of curves in the plane. The ancient Greeks used deductive reasoning and the methods of axiomatic Euclidean geometry to study lines, circles, and the **conic sections** (parabolas, ellipses, and hyperbolas). The properties of conic sections have played an important role in diverse scientific applications since the 17th century, when Kepler discovered—and Newton explained—the fact that the orbits of planets and other bodies in the solar system are conic sections.

The French mathematicians Descartes and Fermat, working almost independently of one another, initiated analytic geometry in 1637. The central idea of analytic geometry is the correspondence between an equation $F(x, y) = 0$ and the set or **locus** (typically, a curve) of all those points $(x, y)$ in the plane with coordinates that satisfy this equation.

The idea is this: Given a geometric locus or curve, its properties can be derived algebraically or analytically from its defining equation $F(x, y) = 0$. For example, suppose that the equation of a given curve turns out to be the linear equation

$$Ax + By = C, \tag{1}$$

where $A$, $B$, and $C$ are constants with $B \neq 0$. This equation may be written in the form

$$y = mx + b \tag{2}$$

where $m = -A/B$ and $b = C/B$. But (2) is the slope-intercept equation (Section 1-4) of the straight line with slope $m$ and $y$-intercept $b$. Hence the given curve is this straight line. In the following example we use this approach to show that a certain geometrically described locus is a particular straight line.

**EXAMPLE 1** Show that the set of all points equally distant from the points $(1, 1)$ and $(5, 3)$ is the perpendicular bisector of the line segment joining these two points.

*Solution* The typical point $P(x, y)$ is equidistant from $(1, 1)$ and $(5, 3)$ if and only if

$$(x - 1)^2 + (y - 1)^2 = (x - 5)^2 + (y - 3)^2,$$
$$(x^2 - 2x + 1) + (y^2 - 2y + 1) = (x^2 - 10x + 25)$$
$$+ (y^2 - 6y + 9),$$
$$2x + y = 8. \tag{3}$$

Thus the given locus is the straight line in (3) with slope $-2$. The straight line through $(1, 1)$ and $(5, 3)$ has equation

$$y - 1 = \tfrac{1}{2}(x - 1) \tag{4}$$

and slope $\frac{1}{2}$. Because the products of the slopes of these two lines is $-1$, it follows (from Theorem 2 in Section 1-4) that they are perpendicular lines. If we solve Equations (3) and (4) simultaneously, we find that the intersection of these lines is, indeed, the midpoint $(3, 2)$ of the given line segment. Thus the locus described is the perpendicular bisector of this line segment.

---

The circle with center $(h, k)$ and radius $r$ is described geometrically as the set of all points $P(x, y)$ whose distance from $(h, k)$ is $r$. The distance formula then gives

$$(x - h)^2 + (y - k)^2 = r^2 \qquad (5)$$

as the equation of this circle. In particular, if $h = k = 0$, then (5) takes the simple form

$$x^2 + y^2 = r^2. \qquad (6)$$

We can see directly from this equation, without further reference to the definition, that a circle centered at the origin has the following symmetry properties:

- *Symmetry about the x-axis:* The equation of the curve is unaltered when $y$ is replaced by $-y$.
- *Symmetry about the y-axis:* The equation of the curve is unaltered when $x$ is replaced by $-x$.
- *Symmetry with respect to the origin:* The equation of the curve is unaltered when $x$ is replaced by $-x$ and $y$ is replaced by $-y$.
- *Symmetry about the 45°-line $y = x$:* The equation is unaltered when $x$ and $y$ are interchanged.

Equation (6) can be rewritten in each of the forms

$$y = \pm\sqrt{r^2 - x^2} \quad \text{and} \quad x = \pm\sqrt{r^2 - y^2}.$$

Since the expressions beneath the square roots must be nonnegative, these new equations show us that $|x| \leq r$ and $|y| \leq r$ for any point $P(x, y)$ of the circle of radius $r$ centered at the origin. (This is evident from the geometry.)

The relationship between Equations (5) and (6) is an illustration of the translation principle of Section 1-5. Imagine a translation (or "slide") of the plane that moves the point $(x, y)$ to the new position $(x + h, y + k)$. Under such a translation, a curve $C$ would be moved to a new curve. The equation of the new curve is easily obtained from the old equation—we simply replace $x$ by $x - h$ and $y$ by $y - k$. Conversely, we can recognize a translated circle from its equation: Any equation of the form

$$x^2 + y^2 + Ax + By + C = 0 \qquad (7)$$

can be rewritten in the form

$$(x - h)^2 + (y - k)^2 = p$$

by completing the square as in Example 1 of Section 1-5. Thus the graph of Equation (7) is either a circle (if $p > 0$), a single point (if $p = 0$), or empty (if $p < 0$). We use this approach in the following example to discover that the locus described is a particular circle.

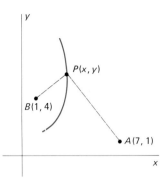

**10.1** The locus of Example 2

**EXAMPLE 2**   Determine the locus of a point $P(x, y)$ if its distance $|AP|$ from $A(7, 1)$ is twice its distance $|BP|$ from $B(1, 4)$.

**Solution**   The points $A$, $B$, and $P$ appear in Fig. 10.1, along with a curve through $P$ representing the given locus. Because

$$|AP|^2 = 4|BP|^2 \qquad \text{(from } |AP| = 2|BP|\text{)},$$

we get the equation

$$(x - 7)^2 + (y - 1)^2 = 4[(x - 1)^2 + (y - 4)^2].$$

Hence

$$3x^2 + 3y^2 + 6x - 30y + 18 = 0,$$
$$x^2 + y^2 + 2x - 10y = -6,$$

and so

$$(x + 1)^2 + (y - 5)^2 = 20.$$

Thus the locus is a circle with center $(-1, 5)$ and radius $r = \sqrt{20} = 2\sqrt{5}$.

### CONIC SECTIONS

The phrase *conic sections* stems from the fact that these are the curves in which a plane intersects a cone. The cone used is a right circular cone with two *nappes* extending infinitely far in both directions, as in Fig. 10.2. There are three types of conic sections, as illustrated in Fig. 10.3. If the cutting plane is parallel to some generator of the cone, then the curve of intersection is a *parabola*. Otherwise it is either a single closed curve—an *ellipse*—or a *hyperbola* with two *branches*.

In Section 14-6 we shall use the methods of three-dimensional analytic geometry to show that if an appropriate $xy$-coordinate system is set up in the intersecting plane, then the equations of the three conic sections take the following forms:

$$\text{Parabola:} \qquad y^2 = kx; \qquad (8)$$

$$\text{Ellipse:} \qquad \frac{x^2}{a^2} + \frac{y^2}{b^2} = 1; \qquad (9)$$

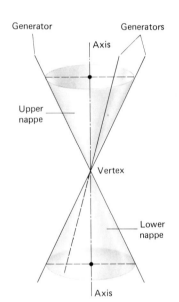

**10.2**   A cone of two nappes

  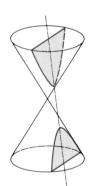

Ellipse    Parabola    Hyperbola

**10.3**   The conic sections

$$\text{Hyperbola:} \quad \frac{x^2}{a^2} - \frac{y^2}{b^2} = 1. \tag{10}$$

In Sections 10-4 through 10-6 we will discuss these conic sections on the basis of definitions that are "two-dimensional"—they will not require the three-dimensional setting of a cone and an intersecting plane. The following example illustrates one such approach to the conic sections.

**EXAMPLE 3**  Let $e$ be a given positive number (*not* to be confused with the natural logarithm base). Determine the locus of a point $P(x, y)$ if its distance from the fixed point $F(p, 0)$ is $e$ times its distance from the vertical line $L$ with equation $x = -p$. (See Fig. 10.4.)

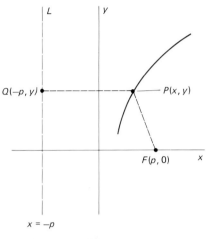

*Solution*  Let $PQ$ be the perpendicular from $P$ to the line $L$. Then the condition

$$\overline{PF} = e\overline{PQ}$$

takes the analytic form

$$\sqrt{(x - p)^2 + y^2} = e|x - (-p)|.$$

That is,

$$(x^2 - 2px + p^2) + y^2 = e^2(x^2 + 2px + p^2),$$

and so

$$x^2(1 - e^2) - 2p(1 + e^2)x + y^2 = -p^2(1 - e^2). \tag{11}$$

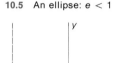
$x = -p$

**10.4**  A figure for Example 3

*Case 1:* $e = 1$.  Then (11) reduces to

$$y^2 = 4px. \tag{12}$$

We see upon comparison with Equation (8) that the locus of $P$ is a *parabola* if $e = 1$.

*Case 2:* $e < 1$.  Dividing Equation (11) by $1 - e^2$ yields

$$x^2 - 2p\frac{1 + e^2}{1 - e^2}x + \frac{y^2}{1 - e^2} = -p^2.$$

We now complete the square in $x$, and get

$$\left(x - p\frac{1 + e^2}{1 - e^2}\right)^2 + \frac{y^2}{1 - e^2} = p^2\left[\left(\frac{1 + e^2}{1 - e^2}\right)^2 - 1\right] = a^2.$$

**10.5**  An ellipse: $e < 1$

This equation takes the form

$$\frac{(x - h)^2}{a^2} + \frac{y^2}{b^2} = 1 \tag{13}$$

where

$$h = +p\frac{1 + e^2}{1 - e^2} \quad \text{and} \quad b^2 = a^2(1 - e^2). \tag{14}$$

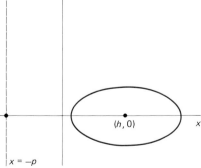

When we compare Equations (9) and (13), we see that if $e < 1$, then the locus of $P$ is an *ellipse* with $(0, 0)$ translated to $(h, 0)$, as illustrated in Fig. 10.5.

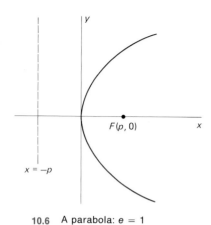

**10.6** A parabola: $e = 1$

*Case* 3: $e > 1$. In this case, Equation (11) reduces to a translated version of Equation (10), so the locus of $P$ is a *hyperbola*. The details, which are similar to those of Case 2, are left for Problem 35.

Thus the locus in Example 3 is a *parabola* if $e = 1$, an *ellipse* if $e < 1$, and a *hyperbola* if $e > 1$. The number $e$ is called the **eccentricity** of the conic section. The point $F(p, 0)$ is commonly called its **focus** in the parabolic case. Figure 10.6 shows the parabola of Case 1, and Fig. 10.7 illustrates the hyperbola of Case 3.

If we begin with Equations (8) through (10), we can derive the general characteristics of the three conic sections shown in Figs. 10.5 through 10.7. For example, in the case of the parabola with the equation in (8), we see that the curve passes through the origin, that $x \geq 0$ at each of its points, that $y \to \pm\infty$ as $x \to \infty$, and that the graph is symmetric about the $x$-axis (because the curve is unchanged when $y$ is replaced by $-y$).

In the case of the ellipse with the equation in (9), we see that the graph must be symmetric about both axes. At each point $(x, y)$ of the graph, we must have $|x| \leq a$ and $|y| \leq b$. The graph intersects the axes at the four points $(\pm a, 0)$ and $(0, \pm b)$.

Finally, the hyperbola with the equation in (10), or its alternative form

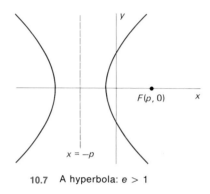

**10.7** A hyperbola: $e > 1$

$$y = \pm\frac{b}{a}\sqrt{x^2 - a^2},$$

is symmetric about both axes. It meets the $x$-axis at the two points $(\pm a, 0)$, has one branch consisting of points with $x \geq a$, and has another branch where $x \leq -a$. And $y \to \pm\infty$ as $x \to \pm\infty$.

## 10-1 PROBLEMS

In each of Problems 1–6, write an equation of the specified straight line.

**1** The line through the point $(1, -2)$ that is parallel to the line with equation $x + 2y = 5$.

**2** The line through the point $(-3, 2)$ that is perpendicular to the line with equation $3x - 4y = 7$.

**3** The line that is tangent to the circle $x^2 + y^2 = 25$ at the point $(3, -4)$.

**4** The line that is tangent to the curve $y^2 = x + 3$ at the point $(6, -3)$.

**5** The line that is perpendicular to the curve $x^2 + 2y^2 = 6$ at the point $(2, -1)$.

**6** The perpendicular bisector of the line segment with end points $(-3, 2)$ and $(5, -4)$.

In each of Problems 7–16, find the center and radius of the circle described in the given equation.

**7** $x^2 + 2x + y^2 = 4$
**8** $x^2 + y^2 - 4y = 5$
**9** $x^2 + y^2 - 4x + 6y = 3$
**10** $x^2 + y^2 + 8x - 6y = 0$
**11** $4x^2 + 4y^2 - 4x = 3$
**12** $4x^2 + 4y^2 + 12y = 7$
**13** $2x^2 + 2y^2 - 2x + 6y = 13$
**14** $9x^2 + 9y^2 - 12x = 5$

**15** $9x^2 + 9y^2 + 6x - 24y = 19$
**16** $36x^2 + 36y^2 - 48x - 108y = 47$

In each of Problems 17–20, show that the graph of the given equation consists either of a single point or of no points.

**17** $x^2 + y^2 - 6x - 4y + 13 = 0$
**18** $2x^2 + 2y^2 + 6x + 2y + 5 = 0$
**19** $x^2 + y^2 - 6x - 10y + 84 = 0$
**20** $9x^2 + 9y^2 - 6x + 6y + 11 = 0$

In each of Problems 21–24, write the equation of the specified circle.

**21** The circle with center $(-1, -2)$ that passes through the point $(2, 3)$.

**22** The circle with center $(2, -2)$ that is tangent to the line $y = x + 4$.

**23** The circle with center $(6, 6)$ that is tangent to the line $y = 2x - 4$.

**24** The circle that passes through the points $(4, 6)$, $(-2, -2)$, and $(5, -1)$. (*Suggestion:* Determine $A$, $B$, and $C$ so that each of the three given points satisfies the equation $x^2 + y^2 + Ax + By + C = 0$.)

In each of Problems 25–30, derive the equation of the set of all points $P(x, y)$ satisfying the given condition. Then sketch the graph of the equation.

**25** The point $P(x, y)$ is equally distant from the two points $(3, 2)$ and $(7, 4)$.

**26** The distance from $P$ to the point $(-2, 1)$ is one-half its distance from $(4, -2)$.

**27** The point $P$ is three times as far from the point $(-3, 2)$ as it is from the point $(5, 10)$.

**28** The distance of $P$ from the line $x = -3$ is equal to its distance from the point $(3, 0)$.

**29** The sum of the distances of $P$ from the points $(4, 0)$ and $(-4, 0)$ is 10.

**30** The sum of the distances of $P$ from the points $(0, 3)$ and $(0, -3)$ is 10.

**31** Find all lines through the point $(2, 1)$ that are tangent to the parabola $y = x^2$.

**32** Find all lines through the point $(-1, 2)$ that are normal to the parabola $y = x^2$.

**33** Find all lines that are normal to the curve $xy = 4$ and simultaneously parallel to the line $y = 4x$.

**34** Find all lines that are tangent to the curve $y = x^3$ and are also parallel to the line $3x - y = 5$.

**35** Suppose that $e > 1$. Show that Equation (11) of this section can be written in the form

$$\frac{(x - h)^2}{a^2} - \frac{y^2}{b^2} = 1,$$

thus showing that its graph is a hyperbola. Find $a$, $b$, and $h$ in terms of $p$ and $e$.

---

## 10-2

# Polar Coordinates

A familiar way to locate a point in the plane is by specifying its rectangular coordinates $(x, y)$—that is, by giving its abscissa $x$ and ordinate $y$ relative to given perpendicular axes. In some problems it is more convenient to locate a point by means of its **polar coordinates.** The polar coordinates of a point give its position relative to a fixed reference point $O$ (the **pole**) and a given ray beginning at $O$ (the **polar axis**).

For convenience, we begin with a given $xy$-coordinate system, then take the origin as the pole and the nonnegative $x$-axis as the polar axis. Given the pole $O$ and the polar axis, the point $P$ with **polar coordinates** $r$ and $\theta$, written as the ordered pair $(r, \theta)$, is located as follows. First find the terminal side of the angle $\theta$, given in radians, where $\theta$ is measured counterclockwise (if $\theta > 0$) from the positive $x$-axis (or polar axis) as its initial side. If $r \geq 0$, then $P$ is on this terminal side at the distance $r$ from the origin. If $r < 0$, then $P$ lies on the ray opposite the terminal side at the distance $|r| = -r$ from the pole. The **radial** coordinate $r$ can be described as the *directed* distance of $P$ from the pole along the terminal side of the angle $\theta$. Thus, if $r$ is positive, the point $P$ lies in the same quadrant as $\theta$, while if $r$ is negative, then $P$ lies in the opposite quadrant. If $r = 0$, the angle $\theta$ does not matter; the polar coordinates $(0, \theta)$ represent the origin whatever the **angular** coordinate $\theta$ might be. Of course, the origin or pole is the only point for which $r = 0$. The cases $r > 0$ and $r < 0$ are illustrated separately in Fig. 10.8.

Polar coordinates differ from rectangular coordinates in that any point has more than one representation in polar coordinates. For example, the

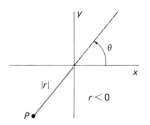

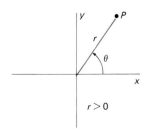

**10.8** The difference between the two cases $r > 0$ and $r < 0$.

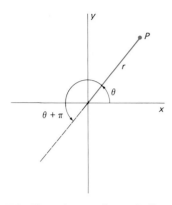

**10.9** The polar coordinates $(r, \theta)$ and $(-r, \theta + \pi)$ represent the same point $P$.

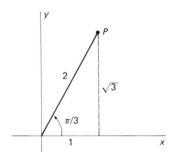

**10.10** The point $P$ can be described in many ways in polar coordinates.

| $\theta$ | $r$ |
|----------|------|
| 0 | 0.00 |
| $\pi/6$ | 1.00 |
| $\pi/3$ | 1.73 |
| $\pi/2$ | 2.00 |
| $2\pi/3$ | 1.73 |
| $5\pi/6$ | 1.00 |
| $\pi$ | 0.00 |
| $7\pi/6$ | −1.00 |
| $4\pi/3$ | −1.73 |
| $3\pi/2$ | −2.00 |
| $5\pi/3$ | −1.73 |
| $11\pi/6$ | −1.00 |
| $2\pi$ | 0.00 |
| | (data rounded) |

**10.11** Values of $r = 2 \sin \theta$

polar coordinates $(r, \theta)$ and $(-r, \theta + \pi)$ represent the same point $P$, as shown in Fig. 10.9. More generally, this same point $P$ has the polar coordinates $(r, \theta + n\pi)$ for any even integer $n$ and also the coordinates $(-r, \theta + n\pi)$ for any odd integer $n$. Thus the polar coordinate pairs

$$\left(2, \frac{\pi}{3}\right), \quad \left(-2, \frac{4\pi}{3}\right), \quad \left(2, \frac{7\pi}{3}\right), \quad \text{and} \quad \left(-2, \frac{-2\pi}{3}\right)$$

all represent the point $P$ shown in Fig. 10.10. The rectangular coordinates of $P$ are $(1, \sqrt{3})$.

Some curves have simpler equations in polar coordinates than in rectangular coordinates; this is an important reason for the usefulness of polar coordinates. The **graph** of an equation in the polar coordinate variables $r$ and $\theta$ is the set of all those points $P$ such that $P$ has a pair of polar coordinates $(r, \theta)$ satisfying the given equation. The graph of an equation $r = f(\theta)$ can be constructed by computing a table of values of $r$, then plotting the corresponding points $(r, \theta)$ on polar coordinate graph paper.

**EXAMPLE 1** Construct the graph of the equation $r = 2 \sin \theta$.

**Solution** Figure 10.11 shows a table of values of $r$ as a function of $\theta$. The corresponding points are plotted in Fig. 10.12, using the rays at multiples of $\pi/6$ and the circles (centered at the pole) of radii 1 and 2 to locate these points. A visual inspection of the smooth curve connecting these points suggests that it is a circle of radius 1. Let us assume for the moment that this is so. Note then that the point $P(r, \theta)$ moves *once around this circle counterclockwise* as $\theta$ increases from 0 to $\pi$, then moves around the circle *a second time* as $\theta$ increases from $\pi$ to $2\pi$. This is because the negative values of $r$ for $\theta$ between $\pi$ and $2\pi$ give (in this example) the same geometric points as the positive values of $r$ found when $\theta$ is between 0 and $\pi$.

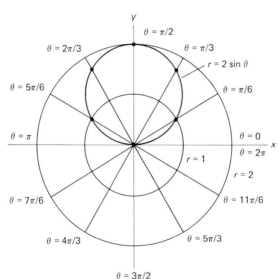

**10.12** The graph of the polar equation $r = 2 \sin \theta$

The verification that the graph of $r = 2 \sin \theta$ is the indicated circle illustrates the general procedure for transferring back and forth between polar and rectangular coordinates. For changing from polar to rectangular coordinates we use the basic relations

$$x = r \cos \theta, \qquad y = r \sin \theta \qquad (1)$$

that we read from the right triangle of Fig. 10.13. In the opposite direction we have

$$r^2 = x^2 + y^2, \qquad \theta = \tan^{-1}\left(\frac{y}{x}\right) \quad \text{if} \quad x \neq 0. \qquad (2)$$

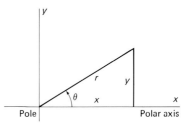

**10.13** Read Equations (1) and (2)—conversions between polar and rectangular coordinates–from this figure.

For example, the first of Equations (2) transforms the rectangular equation $x^2 + y^2 = a^2$ of a circle into $r^2 = a^2$, or simply $r = a$. To transform the equation $r = 2 \sin \theta$ of Example 1 into rectangular coordinates, we first multiply both sides by $r$ to get

$$r^2 = 2r \sin \theta.$$

Equations (1) and (2) now give

$$x^2 + y^2 = 2y.$$

Finally, after we complete the square, we have

$$x^2 + (y - 1)^2 = 1,$$

the rectangular coordinates equation of a circle with center $(0, 1)$ and radius 1. More generally, the graphs of the equations

$$r = 2a \sin \theta \quad \text{and} \quad r = 2a \cos \theta \qquad (3)$$

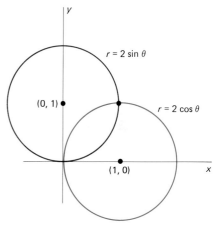

**10.14** The graphs of the circles with equations in (3)

are circles of radius $a$ centered, respectively, at the points $(0, a)$ and $(a, 0)$, as illustrated (with $a = 1$) in Fig. 10.14.

Substitution of Equations (1) transforms the rectangular equation $ax + by = c$ of a straight line into

$$ar \cos \theta + br \sin \theta = c.$$

Let us take $a = 1$ and $b = 0$. Then we see that the polar coordinates equation of the vertical line $x = c$ is $r = c \sec \theta$, as we can deduce directly from Fig. 10.15.

**10.15** Finding the polar equation of the vertical line $x = c$

**EXAMPLE 2** Sketch the graph of the equation

$$r = 2 + 2 \sin \theta.$$

**Solution** If we scan the second column of the table in Fig. 10.11, mentally adding 2 to each entry for $r$, we see that

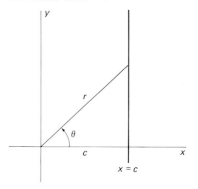

- $r$ increases from 2 to 4 as $\theta$ increases from 0 to $\pi/2$;
- $r$ decreases from 4 to 2 as $\theta$ increases from $\pi/2$ to $\pi$;
- $r$ decreases from 2 to 0 as $\theta$ increases from $\pi$ to $3\pi/2$; and
- $r$ increases from 0 to 2 as $\theta$ increases from $3\pi/2$ to $2\pi$.

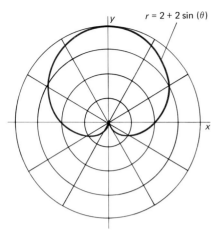

$r = 2 + 2 \sin (\theta)$

**10.16** A cardioid

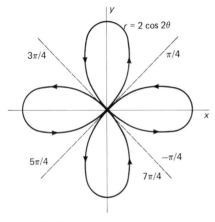

$r = 2 \cos 2\theta$

**10.17** A four-leaved rose

**10.18** The lemniscate
$r^2 = -4 \sin 2\theta$

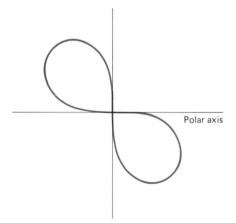

Polar axis

This information tells us that the graph resembles the curve shown in Fig. 10.16. This heart-shaped graph is called a **cardioid.** The graphs of the equations

$$r = a(1 \pm \sin \theta) \quad \text{and} \quad r = a(1 \pm \cos \theta)$$

are all cardioids, differing only in size (determined by $a$), axis of symmetry (whether horizontal or vertical), and the direction in which the cusp at the pole points.

**EXAMPLE 3**   Sketch the graph of the equation $r = 2 \cos 2\theta$.

***Solution***   Rather than constructing a table of values of $r$ as a function of $\theta$ and then plotting points, let us reason qualitatively. Note first that $r = 0$ if $\theta$ is an odd multiple of $\pi/4$. Also $|r| = 2$ if $\theta$ is an even multiple of $\pi/4$. We begin by plotting these points.

Then think of what happens as $\theta$ increases from one odd multiple of $\pi/4$ to the next. For instance, if $\theta$ goes from $\pi/4$ to $3\pi/4$, then $2\theta$ increases from $\pi/2$ to $3\pi/2$. So $r = 2 \cos 2\theta$ decreases from 0 to $-2$, then increases to 0. The graph for such values of $\theta$ forms a loop, starting at the origin and lying between the rays $\theta = 5\pi/4$ and $\theta = 7\pi/4$. Something similar happens between any two consecutive odd multiples of $\pi/4$, so the entire graph consists of four loops (Fig. 10.17). The arrows in the figure indicate the direction in which the point $P(r, \theta)$ moves along the curve as $\theta$ increases.

The curve of Example 3 is called a *four-leaved rose.* The equations $r = a \cos n\theta$ and $r = a \sin n\theta$ represent "roses" with $2n$ "leaves" or loops if $n$ is even and $n \geq 2$, and with $n$ loops if $n$ is odd and $n \geq 3$.

The four-leaved rose exhibits several types of symmetry. The following are some *sufficient* conditions for symmetry in polar coordinates.

- *For symmetry about the x-axis:* The equation is unaltered when $\theta$ is replaced by $-\theta$.
- *For symmetry about the y-axis:* The equation is unaltered when $\theta$ is replaced by $\pi - \theta$.
- *For symmetry with respect to the origin:* The equation is unchanged when $r$ is replaced by $-r$.

Since $\cos 2\theta = \cos(-2\theta) = \cos 2(\pi - \theta)$, the equation $r = 2 \cos 2\theta$ of the four-leaved rose satisfies the first two conditions above, and therefore its graph is symmetric about both the $x$-axis and the $y$-axis. Thus it is also symmetric about the origin. Nevertheless, this equation does *not* satisfy the third condition above, the condition for symmetry about the origin. This illustrates that while the conditions above are *sufficient* for the symmetries described, they are not *necessary* conditions.

**EXAMPLE 4**   Figure 10.18 shows the lemniscate with equation

$$r^2 = -4 \sin 2\theta.$$

To see why it has loops only in the second and fourth quadrants, we examine the table:

| $\theta$ | $2\theta$ | $-4\sin 2\theta$ |
|---|---|---|
| $0 < \theta < \dfrac{\pi}{2}$ | $0 < 2\theta < \pi$ | Negative |
| $\dfrac{\pi}{2} < \theta < \pi$ | $\pi < 2\theta < 2\pi$ | Positive |
| $\pi < \theta < \dfrac{3\pi}{2}$ | $2\pi < 2\theta < 3\pi$ | Negative |
| $\dfrac{3\pi}{2} < \theta < 2\pi$ | $3\pi < 2\theta < 4\pi$ | Positive |

When $\theta$ lies in either the first quadrant or the third quadrant, the quantity $-4\sin 2\theta$ is negative, so the equation $r^2 = -4\sin 2\theta$ cannot be satisfied for any real value of $r$.

Example 3 illustrates a peculiarity of graphs of polar equations, one that is caused by the fact that a single point has multiple representations in polar coordinates. For the point with polar coordinates $(2, \pi/2)$ clearly lies on the four-leaved rose, but these coordinates do *not* satisfy the equation $r = 2\cos 2\theta$. This means that a point may have one pair of polar coordinates that satisfy a given equation and others that do not. Hence we must be careful to understand this: The graph of a polar equation consists of all those points having *at least one* polar coordinate representation satisfying the given equation.

Another result of the multiplicity of polar coordinates is that the simultaneous solution of two polar equations does not always give all the points of intersection of their graphs. For instance, consider the circles $r = 2\sin\theta$ and $r = 2\cos\theta$ shown in Fig. 10.14. The origin is clearly a point of intersection of these two circles. Its polar representation $(0, \pi)$ satisfies the equation $r = 2\sin\theta$, and its representation $(0, \pi/2)$ satisfies the other equation $r = 2\cos\theta$. But the origin has no *single* polar representation that satisfies both equations simultaneously! If we think of $\theta$ as increasing uniformly with time, then the moving points on the two circles pass through the origin at different times. Hence the origin cannot be discovered as a point of intersection of the two circles by solving their equations algebraically.

As a consequence of the phenomenon illustrated by the above example, the only way we can be certain of finding *all* points of intersection of two curves in polar coordinates is by graphing both curves.

**EXAMPLE 5**    Find all points of intersection of the graphs of the equations $r = 1 + \sin\theta$ and $r^2 = 4\sin\theta$.

*Solution*    The graph of $r = 1 + \sin\theta$ is a scaled-down version of the cardioid of Example 2. In Problem 42 we ask you to show that the graph of $r^2 = 4\sin\theta$ is the figure-eight curve shown with the cardioid in Fig. 10.19. The figure shows four points of intersection, marked $A$, $B$, $C$, and $O$. Can we find all four points with algebra?

Given the two equations, we begin by eliminating $r$. Since

$$(1 + \sin\theta)^2 = r^2 = 4\sin\theta,$$

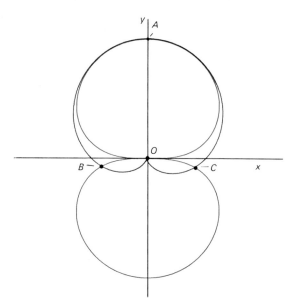

**10.19** The cardioid $r = 1 + \sin\theta$ and the figure eight $r^2 = 4\sin\theta$ meet in four points.

it follows that

$$\sin^2\theta - 2\sin\theta + 1 = 0,$$

$$(\sin\theta - 1)^2 = 0,$$

and thus that $\sin\theta = 1$. So $\theta$ must be an angle of the form $\pi/2 \pm 2n\pi$ where $n$ is an integer. All points on the cardioid and all points on the figure-eight curve are produced by letting $\theta$ increase from 0 to $2\pi$, so $\theta = \pi/2$ will produce all the solutions that we can possibly obtain by this method. The only such point is $A(2, \pi/2)$, and the other three points are not detected except by graphing the two equations.

## 10-2  PROBLEMS

**1** Plot the points with the following polar coordinates and then find the rectangular coordinates of each.
(a) $(1, \pi/4)$     (b) $(-2, 2\pi/3)$     (c) $(1, -\pi/3)$
(d) $(3, 3\pi/2)$     (e) $(2, 9\pi/4)$     (f) $(-2, -7\pi/6)$
(g) $(2, 5\pi/6)$

**2** Find two polar coordinate representations, one with $r > 0$ and the other with $r < 0$, for the points with these rectangular coordinates.
(a) $(-1, -1)$     (b) $(\sqrt{3}, -1)$     (c) $(2, 2)$
(d) $(-1, \sqrt{3})$     (e) $(\sqrt{2}, -\sqrt{2})$     (f) $(-3, \sqrt{3})$

In each of Problems 3–10, express the given rectangular equation in polar form.

**3** $x = 4$                **4** $y = 6$
**5** $x = 3y$            **6** $x^2 + y^2 = 25$
**7** $xy = 1$            **8** $x^2 - y^2 = 1$
**9** $y = x^2$          **10** $x + y = 4$

In each of Problems 11–18, express the given polar equation in rectangular coordinates.

**11** $r = 3$                **12** $\theta = 3\pi/4$
**13** $r = -5\cos\theta$      **14** $r = \sin 2\theta$
**15** $r = 1 - \cos 2\theta$    **16** $r = 2 + \sin\theta$
**17** $r = 3\sec\theta$         **18** $r^2 = \cos 2\theta$

For each of the curves described in Problems 19–28, write equations in both rectangular and polar coordinates.

**19** The vertical line through $(2, 0)$
**20** The horizontal line through $(1, 3)$
**21** The line with slope $-1$ through $(2, -1)$
**22** The line with slope $+1$ through $(4, 2)$
**23** The line through the points $(1, 3)$ and $(3, 5)$
**24** The circle with center $(3, 0)$ that passes through the origin
**25** The circle with center $(0, -4)$ that passes through the origin

**26** The circle with center $(3, 4)$ and radius 5

**27** The circle with center $(1, 1)$ that passes through the origin

**28** The circle with center $(5, -2)$ that passes through the point $(1, 1)$

Sketch the graphs of the polar equations in Problems 29–42, and indicate any symmetry about either axis or the origin.

**29** $r = 2 \cos \theta$

**30** $r = 2 \sin \theta + 2 \cos \theta$   (circle)

**31** $r = 1 + \cos \theta$   (cardioid)

**32** $r = 1 - \sin \theta$   (cardioid)

**33** $r = 2 + 4 \cos \theta$   (limaçon)

**34** $r = 4 + 2 \cos \theta$   (limaçon)

**35** $r^2 = 4 \sin 2\theta$   (lemniscate)

**36** $r^2 = 4 \cos 2\theta$   (lemniscate)

**37** $r = 2 \sin 2\theta$   (four-leaved rose)

**38** $r = 3 \sin 3\theta$   (three-leaved rose)

**39** $r = 3 \cos 3\theta$   (three-leaved rose)

**40** $r = 3\theta$   (spiral of Archimedes)

**41** $r = 2 \sin 5\theta$   (five-leaved rose)

**42** $r^2 = 4 \sin \theta$   (figure eight)

In Problems 43–48, find all points of intersection of the given curves.

**43** $r = 2$ and $r = \cos \theta$

**44** $r = \sin \theta$ and $r^2 = 3 \cos^2\theta$

**45** $r = \sin \theta$ and $r = \cos 2\theta$

**46** $r = 1 + \cos \theta$ and $r = 1 - \sin \theta$

**47** $r = 1 - \cos \theta$ and $r^2 = 4 \cos \theta$

**48** $r^2 = 4 \sin \theta$ and $r^2 = 4 \cos \theta$

**49** (a) The straight line $L$ passes through the point with polar coordinates $(p, \alpha)$ and is perpendicular to the line segment joining the pole and the point $(p, \alpha)$. Write the polar coordinates equation of $L$.

(b) Show that the rectangular coordinates equation of $L$ is $x \cos \alpha + y \sin \alpha = p$.

## *10-2   Optional Computer Application

The following program was written to calculate the data needed to plot the three-leaved rose   $r = 4 \sin 3\theta$   on rectangular coordinates graph paper. The polar coordinate equation appears (with   T   in place of $\theta$)   in line 30, and the $x$- and $y$-coordinates are calculated in line 40. A new point is calculated for each $10°$ increment in   $\theta$   from $0°$ to $360°$.

```
10    T = 0  :  H = 3.141592654/18.0
20    FOR  I = 0  TO  36
30            R = 4*SIN(3*T)
40            X = R*COS(T)  :  Y = R*SIN(T)
50            PRINT " X = "; X; " Y = "; Y
60            T = T + H
70    NEXT  I
80    END
```

**Exercise 1** Run the program and plot the points that it produces.

**Exercise 2** Alter the program to generate the data to plot the four-leaved rose $r = 3 \cos 2\theta$.

**Exercise 3** Alter the program to generate the data for plotting the figure-eight curve   $r^2 = 4 \sin 2\theta$.   You will need to take into account the fact that $4 \sin 2\theta$ is negative for certain values of $\theta$.

### 10-3

The graph of the polar coordinates equation $r = f(\theta)$ may bound an area, as does the cardioid $r = 2(1 + \sin \theta)$ of Fig. 10.16 (Section 10-2). To calculate this area, we may find it more convenient to work directly in polar coordinates, rather than change to rectangular coordinates.

## Area in Polar Coordinates

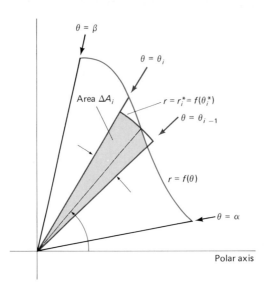

10.20 We obtain the area formula from Riemann sums.

To see how to set up an area integral in polar coordinates, we consider the region $R$ of Fig. 10.20. This region is bounded by the two radial lines $\theta = \alpha$ and $\theta = \beta$ and by the curve $r = f(\theta)$, $\alpha \leq \theta \leq \beta$. To approximate the area $A$ of $R$, we begin with a regular partition

$$\alpha = \theta_0 < \theta_1 < \theta_2 < \cdots < \theta_n = \beta$$

of the interval $[\alpha, \beta]$ into $n$ equal subintervals, each having length $\Delta\theta = (\beta - \alpha)/n$, and select a point $\theta_i^*$ in the $i$th subinterval $[\theta_{i-1}, \theta_i]$ for $i = 1, 2, \ldots, n$.

Let $\Delta A_i$ denote the area of the sector bounded by the lines $\theta = \theta_{i-1}$ and $\theta = \theta_i$ and the curve $f = f(\theta)$. We see from Fig. 10.20 that for small values of $\Delta\theta$, $\Delta A_i$ is approximately equal to the area of the *circular* sector with radius $r_i^* = f(\theta_i^*)$ and bounded by the same angle. That is,

$$\Delta A_i \approx \tfrac{1}{2}(r_i^*)^2 \, \Delta\theta = \tfrac{1}{2}[f(\theta_i^*)]^2 \, \Delta\theta.$$

We add the areas of these sectors for $i = 1, 2, \ldots, n$ and thereby find that

$$A = \sum_{i=1}^{n} \Delta A_i \approx \sum_{i=1}^{n} \tfrac{1}{2}[f(\theta_i^*)]^2 \, \Delta\theta.$$

The right-hand sum is a Riemann sum for the integral $\frac{1}{2}\int_{\alpha}^{\beta} [f(\theta)]^2 \, d\theta$. Hence the value of this integral is, assuming that $f$ is continuous, the limit of the sum above as $\Delta\theta \to 0$. We therefore conclude that *the area $A$ of the region $R$ bounded by the lines $\theta = \alpha$ and $\theta = \beta$ and the curve $r = f(\theta)$ is*

$$A = \tfrac{1}{2}\int_{\alpha}^{\beta} [f(\theta)]^2 \, d\theta. \tag{1}$$

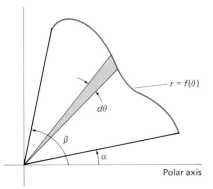

10.21 Heuristic derivation of the area formula in polar coordinates

The infinitesimal sector shown in Fig. 10.21, with radius $r$, central angle $d\theta$, and area $dA = \frac{1}{2}r^2 \, d\theta$, serves as a useful visual device for remembering Formula (1) in the abbreviated form

$$A = \tfrac{1}{2}\int_{\alpha}^{\beta} r^2 \, d\theta. \tag{2}$$

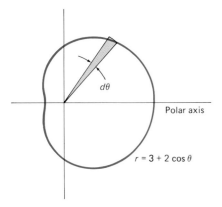

10.22 The limaçon of Example 1

**EXAMPLE 1** Find the area $A$ of the region bounded by the limaçon $r = 3 + 2\cos\theta$, $0 \leq \theta \leq 2\pi$, shown in Fig. 10.22.

**Solution**  We could apply the formula in (2) with $\alpha = 0$ and $\beta = 2\pi$. Here, instead, we will make use of symmetry: We will calculate the area of the upper half of the region and then double the result. Note that the infinitesimal sector shown in Fig. 10.22 sweeps out the upper half of the limaçon as $\theta$ increases from 0 to $\pi$. Hence

$$A = 2 \int_0^\pi \tfrac{1}{2} r^2 \, d\theta = \int_0^\pi (3 + 2 \cos \theta)^2 \, d\theta$$

$$= \int_0^\pi (9 + 12 \cos \theta + 4 \cos^2\theta) \, d\theta.$$

Because

$$4 \cos^2\theta = 4 \cdot \tfrac{1}{2}(1 + \cos 2\theta) = 2 + 2 \cos 2\theta,$$

we now get

$$A = \int_0^\pi (11 + 12 \cos \theta + 2 \cos 2\theta) \, d\theta$$

$$= \Big[ 11\theta + 12 \sin \theta + \sin 2\theta \Big]_0^\pi = 11\pi.$$

**EXAMPLE 2**  Find the area bounded by each loop of the limaçon with equation $r = 1 + 2 \cos \theta$. The limaçon is shown in Fig. 10.23.

**Solution**  The equation $1 + 2 \cos \theta = 0$ has two solutions for $\theta$ between 0 and $2\pi$: $\theta = 2\pi/3$ and $\theta = 4\pi/3$. The upper half of the outer loop of the limaçon corresponds to values of $\theta$ between 0 and $2\pi/3$, where $r$ is positive. Since the curve is symmetric about the $x$-axis, we can find the total area $A_1$ bounded by the outer loop by integrating from 0 to $2\pi/3$ and then doubling:

$$A_1 = 2 \int_0^{2\pi/3} \tfrac{1}{2}(1 + 2 \cos \theta)^2 \, d\theta$$

$$= \int_0^{2\pi/3} (1 + 4 \cos \theta + 4 \cos^2\theta) \, d\theta$$

$$= \int_0^{2\pi/3} (3 + 4 \cos \theta + 2 \cos 2\theta) \, d\theta$$

$$= \Big[ 3\theta + 4 \sin \theta + \sin 2\theta \Big]_0^{2\pi/3} = 2\pi + \tfrac{3}{2}\sqrt{3}.$$

The inner loop of the limaçon corresponds to values of $\theta$ between $2\pi/3$ and $4\pi/3$, where $r$ is negative. Hence the area bounded by the inner loop is

$$A_2 = \tfrac{1}{2} \int_{2\pi/3}^{4\pi/3} (1 + 2 \cos \theta)^2 \, d\theta$$

$$= \tfrac{1}{2} \Big[ 3\theta + 4 \sin \theta + \sin 2\theta \Big]_{2\pi/3}^{4\pi/3} = \pi - \tfrac{3}{2}\sqrt{3}.$$

The area of the region lying between the two loops of the limaçon is

$$A = A_1 - A_2 = (2\pi + \tfrac{3}{2}\sqrt{3}) - (\pi - \tfrac{3}{2}\sqrt{3}) = \pi + 3\sqrt{3}.$$

Now consider two curves $r = f(\theta)$ and $r = g(\theta)$, with $f(\theta) \geq g(\theta)$ for $\alpha \leq \theta \leq \beta$. Then the area of the region bounded by these two curves and the rays $\theta = \alpha$ and $\theta = \beta$ (shown in Fig. 10.24) may be found by subtracting the area bounded by the inner curve from that bounded by the outer curve.

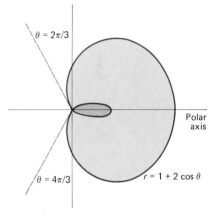

**10.23**  The limaçon of Example 2

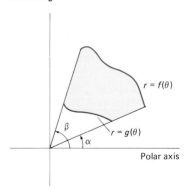

**10.24**  The area between the graphs of $f$ and $g$

That is, the area $A$ between the curves is given by

$$A = \frac{1}{2} \int_\alpha^\beta [f(\theta)]^2 \, d\theta - \frac{1}{2} \int_\alpha^\beta [g(\theta)]^2 \, d\theta$$

$$= \frac{1}{2} \int_\alpha^\beta ([f(\theta)]^2 - [g(\theta)]^2) \, d\theta. \tag{3}$$

With $r_2 = f(\theta)$ for the outer curve and $r_1 = g(\theta)$ for the inner curve, we get the abbreviated formula

$$A = \frac{1}{2} \int_\alpha^\beta (r_2^2 - r_1^2) \, d\theta \tag{4}$$

for the area of the region shown in Fig. 10.24.

**EXAMPLE 3**  Find the area $A$ of the region that lies within the limaçon $r = 1 + 2 \cos \theta$ and outside the circle $r = 2$.

*Solution*  The circle and the limaçon are shown in Fig. 10.25, where the area $A$ between them is shaded. The points of intersection of the circle and the limaçon are given by

$$1 + 2 \cos \theta = 2, \quad \text{so} \quad \cos \theta = \tfrac{1}{2},$$

and the figure shows that we should choose the solutions $\theta = \pm \pi/3$. These two values of $\theta$ are the needed limits of integration, and when we use Formula (3) we find that

$$A = \frac{1}{2} \int_{-\pi/3}^{\pi/3} [(1 + 2 \cos \theta)^2 - 2^2] \, d\theta$$

$$= \int_0^{\pi/3} (4 \cos \theta + 4 \cos^2\theta - 3) \, d\theta \quad \text{(symmetry about the polar axis)}$$

$$= \int_0^{\pi/3} (4 \cos \theta + 2 \cos 2\theta - 1) \, d\theta$$

$$= \Big[ 4 \sin \theta + \sin 2\theta - \theta \Big]_0^{\pi/3} = \frac{5}{2} \sqrt{3} - \frac{\pi}{3}.$$

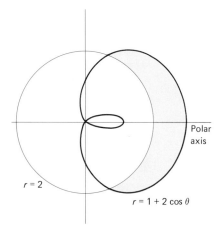

**10.25**   The region of Example 3

$r = 2$

$r = 1 + 2 \cos \theta$

Polar axis

## 10-3   PROBLEMS

In each of Problems 1–10, find the area bounded by the given curve.

**1** $r = 2 \cos \theta$
**2** $r = 4 \sin \theta$
**3** $r = 1 + \cos \theta$
**4** $r = 2 - 2 \sin \theta$
**5** $r = 2 - \cos \theta$
**6** $r = 3 + 2 \sin \theta$
**7** $r = -4 \cos \theta$
**8** $r = 5(1 + \sin \theta)$
**9** $r = 3 - \cos \theta$
**10** $r = 3 + \sin \theta + \cos \theta$

In each of Problems 11–18, find the area bounded by one loop of the given curve.

**11** $r = 2 \cos 2\theta$
**12** $r = 3 \sin 3\theta$
**13** $r = 2 \cos 4\theta$
**14** $r = \sin 5\theta$
**15** $r^2 = 4 \sin 2\theta$
**16** $r^2 = 4 \cos 2\theta$
**17** $r^2 = 4 \sin \theta$
**18** $r = 6 \cos 6\theta$

In each of Problems 19–28, find the area of the region described.

**19** Inside $r = 2 \sin \theta$ and outside $r = 1$.

**20** Inside both $r = 4 \cos \theta$ and $r = 2$.
**21** Inside both $r = \cos \theta$ and $r = \sqrt{3} \sin \theta$.
**22** Inside $r = 2 + \cos \theta$ and outside $r = 2$.
**23** Inside $r = 3 + 2 \sin \theta$ and outside $r = 4$.
**24** Inside $r^2 = 2 \cos 2\theta$ and outside $r = 1$.
**25** Inside both $r^2 = \cos 2\theta$ and $r^2 = \sin 2\theta$.
**26** Inside the large loop and outside the small loop of $r = 1 - 2 \sin \theta$.
**27** Inside $r = 2(1 + \cos \theta)$ and outside $r = 1$.
**28** Inside the figure-eight curve $r^2 = 4 \cos \theta$ and outside $r = 1 - \cos \theta$.
**29** Find the area of the circle $r = \sin \theta + \cos \theta$ by integration in polar coordinates. Check the answer by writing the equation of the circle in rectangular coordinates, finding its radius, and then using the familiar area formula.
**30** Find the area of the region that lies interior to all three circles $r = 1$, $r = 2 \cos \theta$, and $r = 2 \sin \theta$.

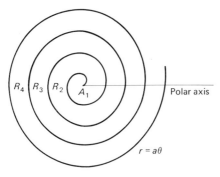

**10.26** The spiral of Archimedes

**31** The *spiral of Archimedes*, shown in Fig. 10.26, has the simple defining equation $r = a\theta$ ($a$ is a constant). Let $A_n$ denote the area bounded by the $n$th turn of the spiral, for $2\pi(n-1) \le \theta \le 2\pi n$, and the portion of the polar axis joining its end points. For each $n \ge 2$, let $R_n = A_n - A_{n-1}$ denote the area between the $(n-1)$st and the $n$th turns. Then derive the following results of Archimedes:
(a) $A_1 = \frac{1}{3}\pi(2\pi a)^2$,   (b) $A_2 = \frac{7}{12}\pi(4\pi a)^2$,
(c) $R_2 = 6A_1$,   (d) $R_{n+1} = nR_2$   for   $n \ge 2$.

**32** Two circles each have radius $a$, and each passes through the center of the other. Find the area of the region lying within both circles.

## 10-4

# The Parabola

The case $e = 1$ of Example 3 in Section 10-1 is motivation for this formal definition:

> **Definition**   *The Parabola*
>
> A **parabola** is the set of all points $P$ in the plane that are equally distant from a fixed point $F$ (called the **focus** of the parabola) and a fixed line $L$ (called its **directrix**) not containing $P$.

If the focus of the parabola is $F(p, 0)$ and its directrix is the vertical line $x = -p$, $p > 0$, then we know from Equation (12) of Section 10-1 that the equation of this parabola is

$$y^2 = 4px. \tag{1}$$

When we replace $x$ by $-x$, both in the equation and in the discussion that precedes it, we get the equation of a parabola with focus at $(-p, 0)$ and with directrix the vertical line $x = +p$. The new parabola has equation

$$y^2 = -4px. \tag{2}$$

The old and new parabolas appear in Fig. 10.27.

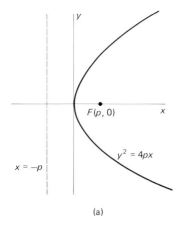

(a)

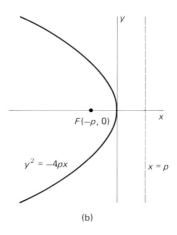

(b)

**10.27** Two parabolas with vertical directrices

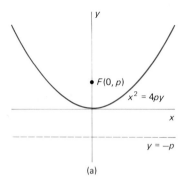

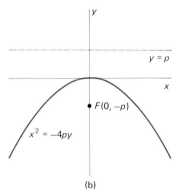

**10.28** Two parabolas with horizontal directrices

We could also interchange $x$ and $y$ in Equation (1). This would give the equation of a parabola that has focus at $(0, p)$ and directrix the (horizontal) line $y = -p$. This parabola opens upward. Its equation is

$$x^2 = 4py, \tag{3}$$

and it is shown in Fig. 10.28(a).

Finally, we replace $y$ with $-y$ in Equation (3). This gives the equation

$$x^2 = -4py \tag{4}$$

of a downward-opening parabola with focus $(0, -p)$ and directrix $y = p$. It appears in Fig. 10.28(b).

Each of the parabolas discussed above is symmetric about one of the coordinate axes. The line about which a parabola is symmetric is called its **axis.** The point on a parabola midway between its focus and its directrix is called its **vertex.** Note that the very definition of a parabola implies that the parabola contains its vertex. Each parabola we discussed in connection with Equations (1) through (4) has the origin as its vertex.

**EXAMPLE 1** Determine the focus, directrix, axis, and vertex of the parabola $x^2 = 12y$.

***Solution*** We write the given equation as $x^2 = 4(3)y$. In this form it matches Equation (3) with $p = 3$. Hence the given parabola has its focus at $(0, 3)$, and its directrix is the horizontal line $y = -3$. The $y$-axis is the axis of symmetry, and the parabola opens upward from its vertex at the origin.

Suppose that we begin with the parabola of Equation (1), and translate its vertex to the new point $(h, k)$. Then the translated parabola has equation

$$(y - k)^2 = 4p(x - h). \tag{1a}$$

The new parabola has focus $F(p + h, k)$ and its directrix is the vertical line $x = -p + h$, as indicated in Fig. 10.29. Its axis is the horizontal line $y = k$.

We can obtain the translates of the other three parabolas in (2) through (4) in the same way. If the vertex is moved from the origin to the point $(h, k)$, then the three equations take these forms:

$$(y - k)^2 = -4p(x - h), \tag{2a}$$

$$(x - h)^2 = 4p(y - k), \quad \text{and} \tag{3a}$$

$$(x - h)^2 = -4p(y - k). \tag{4a}$$

Now note that Equations (1a) and (2a) both take the general form

$$y^2 + Ax + By + C = 0 \quad (A \neq 0), \tag{5}$$

while Equations (3a) and (4a) take the general form

$$x^2 + Ax + By + C = 0 \quad (B \neq 0). \tag{6}$$

What is significant about these last two equations is what they have in common: Each is linear in one of the coordinate variables and quadratic in the other. In fact, *any* such equation can be reduced to one of the standard

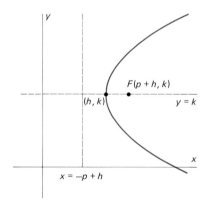

**10.29** A translation of the parabola $y^2 = 4px$

forms in (1a) through (4a) by completing the square in the coordinate variable that appears quadratically. This means that the graph of any equation of the form of either (5) or (6) is a parabola. The features of the parabola can be read from the standard form of the equation.

**EXAMPLE 2**  Determine the graph of the equation

$$4y^2 - 8x - 12y + 1 = 0.$$

*Solution*  This equation is linear in $x$ and quadratic in $y$. We divide through by the coefficient of $y^2$ and then collect all terms involving $y$ on one side of the equation:

$$y^2 - 3y = 2x - \tfrac{1}{4}.$$

Then we complete the square in the variable $y$ and thus find that

$$y^2 - 3y + \tfrac{9}{4} = 2x - \tfrac{1}{4} + \tfrac{9}{4} = 2x + 2.$$

The final step is to write the terms on the right involving $x$ in the form $4p(x - h)$:

$$(y - \tfrac{3}{2})^2 = 4(\tfrac{1}{2})(x + 1).$$

This equation has the form of Equation (1a), with $p = \tfrac{1}{2}, h = -1$, and $k = \tfrac{3}{2}$. Thus the graph is a parabola that opens to the right from the vertex at $(-1, \tfrac{3}{2})$. Its focus is at $(-\tfrac{1}{2}, \tfrac{3}{2})$, its directrix is the vertical line $x = -\tfrac{3}{2}$, and its axis is the horizontal line $y = \tfrac{3}{2}$, as indicated in Fig. 10.30.

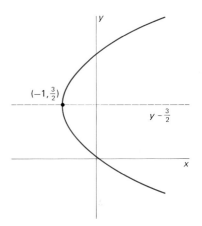

**10.30**  The parabola of Example 2

An important property of the parabola, known as its **reflection property,** has many practical applications. This property involves the fact that the tangent line to the parabola $y^2 = 4px$ at any point $P(x_0, y_0)$ makes equal angles with the horizontal line through $P$ and the line $PF$ from $P$ to the focus $F$ of the parabola. That is, $\alpha = \beta$ in Fig. 10.31. To verify this, note first that the

**10.31**  The reflection property of the parabola: $\alpha = \beta$.

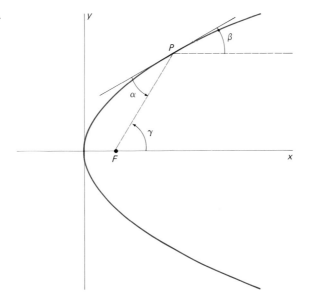

derivative of $y = 2\sqrt{px}$ is $dy/dx = \sqrt{p/x}$, so

$$\tan \beta = \sqrt{\frac{p}{x_0}}.$$

The slope of the line $PF$ is

$$\tan \gamma = \frac{y_0 - 0}{x_0 - p} = \frac{2\sqrt{px_0}}{x_0 - p}.$$

Now $\alpha = \gamma - \beta$ because $\gamma$ and $\alpha + \beta$ are supplementary to the same angle (see Fig. 10.31). It follows that

$$\begin{aligned}
\tan \alpha = \tan(\gamma - \beta) &= \frac{\tan \gamma - \tan \beta}{1 + \tan \gamma \tan \beta} \\
&= \frac{[2\sqrt{px_0}/(x_0 - p)] - (\sqrt{p}/\sqrt{x_0})}{1 + [2\sqrt{px_0}/(x_0 - p)](\sqrt{p}/\sqrt{x_0})} \cdot \frac{(x_0 - p)\sqrt{x_0}}{(x_0 - p)\sqrt{x_0}} \\
&= \frac{2\sqrt{px_0} - \sqrt{p}(x_0 - p)}{(x_0 - p)\sqrt{x_0} + 2\sqrt{px_0}\sqrt{p}} = \frac{\sqrt{p}(x_0 + p)}{\sqrt{x_0}(x_0 + p)} \\
&= \sqrt{\frac{p}{x_0}} = \tan \beta.
\end{aligned}$$

Since $\alpha$ and $\beta$ are acute angles, the fact that $\tan \alpha = \tan \beta$ implies that $\alpha = \beta$.

The reflection property is exploited in the design of parabolic mirrors. Such a mirror has the shape of the surface obtained by revolving a parabola around its axis of symmetry. Because of the law of reflection for light rays, a beam of incoming rays parallel to the axis will then be focused at the point $T$, as shown in Fig. 10.32. The reflection property is also exploited "in reverse"—rays emanating from the focus will be reflected in a beam parallel to the axis, thus keeping the light beam intense. Parabolic mirrors are used in visual and radio telescopes, radar antennas, searchlights, automobile headlights, microphone systems, satellite ground stations, and solar heating devices.

Galileo discovered in the early 1600s that the trajectory of a projectile fired from a gun is a parabola (under the assumption that air resistance is ignored). Suppose that the projectile is fired at time $t = 0$ from the origin with initial velocity $v_0$ at an angle $\alpha$ of inclination from the horizontal $x$-axis. Then the initial velocity splits into the components

$$v_{0x} = v_0 \cos \alpha \quad \text{and} \quad v_{0y} = v_0 \sin \alpha$$

as indicated in Fig. 10.33. The fact that the projectile continues to move horizontally with *constant* speed $v_{0x}$, together with Equation (8) in Section 4-9, implies that its $x$- and $y$-coordinates after $t$ seconds are

$$x = (v_0 \cos \alpha)t, \tag{7}$$

$$y = -\tfrac{1}{2}gt^2 + (v_0 \sin \alpha)t. \tag{8}$$

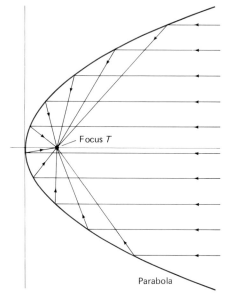

**10.32** Incident rays parallel to the axis reflect through the focus.

Focus $T$

Parabola

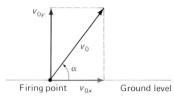

**10.33** Resolution of the initial velocity $v_0$ into horizontal and vertical components.

$v_{0y}$

$v_0$

$\alpha$

Firing point  $v_{0x}$  Ground level

**522**

By substituting $t = x/(v_0 \cos \alpha)$ from (7) into (8) and then completing the square, we can derive (as in Problem 24) an equation of the form

$$y - M = -4p(x - \tfrac{1}{2}R)^2. \tag{9}$$

Here,

$$M = \frac{v_0^2 \sin^2 \alpha}{2g} \tag{10}$$

is the maximum height attained by the projectile, and

$$R = \frac{v_0^2 \sin 2\alpha}{g} \tag{11}$$

is the **range,** or horizontal distance, it travels before returning to the ground. Thus the trajectory is the parabola shown in Fig. 10.34.

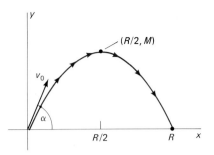

**10.34** The trajectory of the projectile, showing its maximum altitude and its range.

## 10-4 PROBLEMS

In each of Problems 1–5, find the equation and sketch the graph of the parabola with vertex $V$ and focus $F$.

1 $V(0, 0)$ and $F(3, 0)$
2 $V(0, 0)$ and $F(0, -2)$
3 $V(2, 3)$ and $F(2, 1)$
4 $V(-1, -1)$ and $F(-3, -1)$
5 $V(2, 3)$ and $F(0, 3)$

In each of Problems 6–10, find the equation and sketch the graph of the parabola with the given focus and directrix.

6 $F(1, 2)$;  $x = -1$      7 $F(0, -3)$;  $y = 0$
8 $F(1, -1)$;  $x = 3$      9 $F(0, 0)$;  $y = -2$
10 $F(-2, 1)$;  $x = -4$

In each of Problems 11–18, sketch the parabola with the given equation. Show and label its vertex, focus, axis, and directrix.

11 $y^2 = 12x$             12 $x^2 = -8y$
13 $y^2 = -6x$             14 $x^2 = 7y$
15 $x^2 - 4x - 4y = 0$
16 $y^2 - 2x + 6y + 15 = 0$
17 $4x^2 + 4x + 4y + 13 = 0$
18 $4y^2 - 12y + 9x = 0$
19 Show that the point of the parabola $y^2 = 4px$ that is closest to its focus is its vertex.
20 Find the equation of a parabola with vertical axis that passes through the points $(2, 3)$, $(4, 3)$, and $(6, -5)$.
21 Show that the equation of the tangent line to the parabola $y^2 = 4px$ at the point $(x_0, y_0)$ is

$$2px - y_0 y + 2px_0 = 0.$$

Conclude that this tangent line intersects the $x$-axis at the point $(-x_0, 0)$. This fact provides a quick method for constructing a tangent line to a parabola at a given point.
22 A comet has a parabolic orbit with the sun at its focus. When the comet is $100\sqrt{2}$ million miles from the sun, the

line from the sun to the comet makes an angle of $45°$ with the axis of the parabola, as shown in Fig. 10.35. What will be the minimum distance between the comet and the sun? (*Suggestion:* Write the equation of the parabola with the origin at the focus and then use the result of Problem 19.)

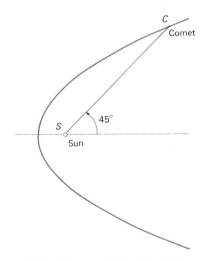

**10.35** The comet of Problem 22 in parabolic orbit about the sun.

23 Suppose that the angle of Problem 22 increases from $45°$ to $90°$ in 3 days. How much longer will be required for the comet to reach its point of closest approach to the sun? Assume that the line segment from the sun to the comet sweeps out area at a constant rate (Kepler's second law).
24 Use Equations (7) and (8) to derive Equation (9) with the values of $M$ and $R$ given by Equations (10) and (11).
25 Deduce from Equation (11) that, given a fixed initial velocity $v_0$, the maximum range of the projectile is $R_{max} = v_0^2/g$ and is attained when $\alpha = 45°$.

In Problems 26–28, assume that a projectile is fired from the origin with initial velocity $v_0 = 160$ ft/s and with initial angle of inclination $\alpha$. Use $g = 32$ ft/s$^2$.

**26** If $\alpha = 45°$, find the range $R$ of the projectile and the maximum height that it attains.

**27** For what value(s) of $\alpha$ is the range $R = 400$ ft?

**28** Find the range $R$ of the projectile and the length of time it is in the air if

(a) $\alpha = 30°$;

(b) $\alpha = 60°$.

## 10-5

### The Ellipse

An ellipse is a conic section with eccentricity $e$ less than 1, as in Example 3 of Section 10-1.

> **Definition**  *The Ellipse*
>
> Suppose that $0 < e < 1$, and let $F$ be a fixed point and $L$ a fixed line not containing $F$. Then the **ellipse** with **eccentricity** $e$, **focus** $F$, and **directrix** $L$ is the set of all points $P$ such that the distance $|PF|$ is $e$ times the (perpendicular) distance from $P$ to the line $L$.

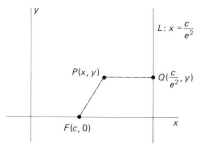

**10.36**  Ellipse: focus $F$, directrix $L$, eccentricity $e$

The equation of the ellipse is especially simple if $F$ is the point $(c, 0)$ on the $x$-axis and $L$ is the vertical line $x = c/e^2$. The case $c > 0$ is shown in Fig. 10.36. If $Q$ is the point $(c/e^2, y)$, then $PQ$ is the perpendicular from $P(x, y)$ to $L$. The condition $|PF| = e|PQ|$ gives

$$(x - c)^2 + y^2 = e^2 \left( x - \frac{c}{e^2} \right)^2,$$

$$x^2 - 2cx + c^2 + y^2 = e^2 x^2 - 2cx + \frac{c^2}{e^2},$$

$$x^2(1 - e^2) + y^2 = c^2 \left( \frac{1}{e^2} - 1 \right) = \frac{c^2}{e^2}(1 - e^2).$$

Thus

$$x^2(1 - e^2) + y^2 = a^2(1 - e^2)$$

where

$$a = \frac{c}{e}. \tag{1}$$

Division of both sides of the next-to-last equation above by $a^2(1 - e^2)$ gives

$$\frac{x^2}{a^2} + \frac{y^2}{a^2(1 - e^2)} = 1.$$

Finally, with the aid of the fact that $e < 1$, we may let

$$b^2 = a^2(1 - e^2) = a^2 - c^2. \tag{2}$$

Then the equation of the ellipse with focus $(c, 0)$ and directrix $x = c/e^2 = a/e$ takes the pleasant form

$$\frac{x^2}{a^2} + \frac{y^2}{b^2} = 1. \tag{3}$$

We see from Equation (3) that this ellipse is symmetric about both axes. Its $x$-intercepts are $(\pm a, 0)$, and its $y$-intercepts are $(0, \pm b)$. The points $(\pm a, 0)$ are called the **vertices** of the ellipse, and the line segment joining them is its **major axis.** The line segment joining its $y$-intercepts $(0, \pm b)$ is its **minor axis** (note from (2) that $b < a$). The alternative form

$$a^2 = b^2 + c^2 \qquad (4)$$

of (2) is the Pythagorean relation for the right triangle of Fig. 10.37. Indeed, visualization of this triangle is an excellent way to remember Equation (4). The numbers $a$ and $b$ are the lengths of the major and minor **semiaxes,** respectively.

Since $a = c/e$, the directrix of the ellipse in (3) is $x = a/e$. If we had begun instead with focus $(-c, 0)$ and directrix $x = -a/e$, we would still have got Equation (3), because only the squares of $a$ and $c$ are involved in the derivation. Thus the ellipse in (3) has *two* foci, $(c, 0)$ and $(-c, 0)$, and *two* directrices, $x = a/e$ and $x = -a/e$. These are shown in Fig. 10.38.

The larger the eccentricity $e < 1$ is, the more elongated is the ellipse. If $e = 0$ then Equation (2) gives $b = a$, so Equation (3) reduces to the equation of a circle of radius $a$. Thus a circle may be regarded as an ellipse with eccentricity zero. Compare the three cases shown in Fig. 10.39.

**EXAMPLE 1** Find the equation of the ellipse with foci $(\pm 3, 0)$ and vertices $(\pm 5, 0)$.

**Solution** We are given that $c = 3$ and $a = 5$, so Equation (2) gives $b = 4$. Thus Equation (3) gives

$$\frac{x^2}{25} + \frac{y^2}{16} = 1.$$

This ellipse is shown in Fig. 10.40.

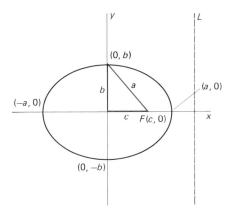

10.37　The parts of an ellipse

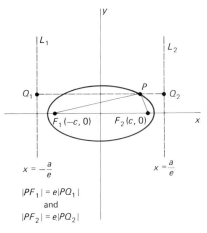

10.38　The ellipse as a conic section: Two foci, two directrices

If our ellipse has its two foci on the $y$-axis, for example $F_1(0, -c)$ and $F_2(0, c)$, then its equation is

$$\frac{x^2}{b^2} + \frac{y^2}{a^2} = 1, \qquad (5)$$

10.39　The relation between the eccentricity of an ellipse and its shape

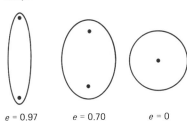

$e = 0.97 \qquad e = 0.70 \qquad e = 0$

10.40　The ellipse of Example 1

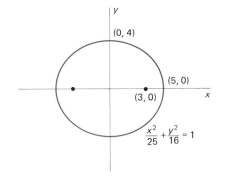

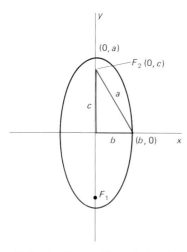

**10.41** An ellipse with vertical major axis

and it is still true that

$$a^2 = b^2 + c^2. \tag{4}$$

But now the major axis of length $2a$ is vertical, and the minor axis of length $2b$ is horizontal. The derivation of Equation (5) is similar to that of Equation (3); see Problem 23. Figure 10.41 shows the case of an ellipse with vertical major axis.

In practice, there is little chance of confusing Equations (3) and (5). The equation or given data will make it clear whether the major axis of the ellipse is horizontal or vertical. Just use the equation to read the ellipse's intercepts. The two that are farthest from the origin are the end points of the major axis, and the other two are the end points of the minor axis. The two foci lie on the major axis, each at distance $c$ from the ellipse's center—which will be at the origin if the ellipse's equation has the form of either of Equations (3) or (5).

**EXAMPLE 2**  Sketch the graph of the equation

$$\frac{x^2}{16} + \frac{y^2}{25} = 1.$$

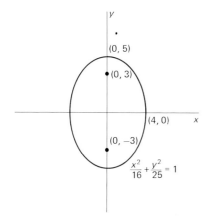

**10.42**  The ellipse of Example 2

***Solution***  The $x$-intercepts are $(\pm 4, 0)$; the $y$-intercepts are $(0, \pm 5)$. So the major axis is vertical. We take $a = 5$ and $b = 4$ in Equation (4) and find that $c = 3$. The foci are thus at $(0, \pm 3)$. Hence this ellipse has the appearance of the one shown in Fig. 10.42.

Any equation of the form

$$Ax^2 + Cy^2 + Dx + Ey + F = 0, \tag{6}$$

in which the coefficients $A$ and $C$ of the squared terms are *both nonzero* and have the *same sign*, may be reduced to the form

$$A(x - h)^2 + C(y - k)^2 = G$$

by completing the square in $x$ and $y$. We may assume that $A$ and $B$ are both positive. Then if $G < 0$, there are no points satisfying (6), and the graph is the empty set. If $G = 0$, there is exactly one point on the locus—the single point $(h, k)$. And if $G > 0$, we can divide both sides of the last equation by $G$ and get an equation that resembles one of the two below:

$$\frac{(x - h)^2}{a^2} + \frac{(y - k)^2}{b^2} = 1, \tag{7a}$$

$$\frac{(x - h)^2}{b^2} + \frac{(y - k)^2}{a^2} = 1. \tag{7b}$$

Which equation to choose? The one consistent with the condition $a \geq b > 0$. Finally note that either of Equations (7) is the equation of a translated ellipse. Thus, apart from the exceptional cases noted, the graph of Equation (6) is an ellipse if $AC > 0$.

**EXAMPLE 3**  Determine the graph of the equation

$$3x^2 + 5y^2 - 12x + 30y + 42 = 0.$$

CHAP. 10:  Polar Coordinates and Conic Sections

**Solution** We collect terms containing $x$, terms containing $y$, and complete the square in each variable. This gives

$$3(x^2 - 4x) + 5(y^2 + 6y) = -42,$$

$$3(x^2 - 4x + 4) + 5(y^2 + 6y + 9) = 15,$$

$$\frac{(x-2)^2}{5} + \frac{(y+3)^2}{3} = 1.$$

Thus the given equation is that of a translated ellipse. The center is at $(2, -3)$, the ellipse has horizontal major semiaxis of length $a = \sqrt{5}$, and the minor semiaxis has length $b = \sqrt{3}$. The distance from the center to each focus is $c = \sqrt{2}$, and the eccentricity is $e = c/a = \sqrt{\frac{2}{5}}$. This ellipse appears in Fig. 10.43.

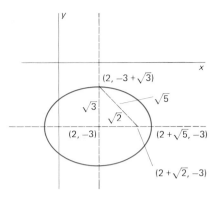

**10.43** The ellipse of Example 3

**EXAMPLE 4** The orbit of the earth is an ellipse with the sun at one focus. The planet's maximum distance from the sun is 94.56 million miles and its minimum distance is 91.45 million miles. What are the major and minor semiaxes of the earth's orbit, and what is its eccentricity?

**Solution** As Fig. 10.44 shows, we have

$$a + c = 94.56 \quad \text{and} \quad a - c = 91.45$$

with units in millions of miles. From these equations we conclude that $a = 93.00$, that $c = 1.56$, and then that

$$b = \sqrt{(93.00)^2 - (1.56)^2} \approx 92.99$$

million miles. Finally,

$$e = \frac{1.56}{93.00} \approx 0.017,$$

a number rather close to zero; this means that the earth's orbit is nearly circular. Indeed, the major and minor semiaxes are so nearly equal that, on any usual scale, the earth's orbit would appear to be a perfect circle. But the difference between uniform circular motion and the earth's actual motion has some important aspects, including the fact that the sun is 1.56 million miles off center, and—as we shall see in Chapter 13—the orbital speed of the earth is not constant.

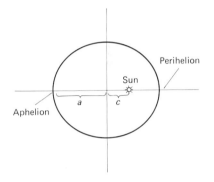

**10.44** The orbit of the earth with its eccentricity exaggerated

**EXAMPLE 5** One of the most famous of all comets is Halley's comet, named for Edmund Halley (1656–1742, a disciple of Newton). By studying the records of the paths of earlier comets, Halley deduced that the comet of 1682 was the same one that had been sighted in 1607, in 1531, in 1456, and in 1066 (an omen at the Battle of Hastings). In 1682, he predicted that the comet would return in 1759, in 1835, and in 1910; he was correct each time. The period of Halley's comet is about 76 years; it can vary a couple of years in either direction because of perturbations of its orbit by Jupiter. The orbit of the comet is an ellipse with the sun at one focus. In terms of astronomical units (1 A.U. is the earth's mean distance from the sun), the major and minor semiaxes of this elliptical orbit are 18.09 A.U. and 4.56 A.U., respectively. What are the maximum and minimum distances from the sun of Halley's comet?

**Solution** We are given that $a = 18.09$ (A.U.) and that $b = 4.56$ (A.U.), so

$$c = \sqrt{(18.09)^2 - (4.56)^2} \approx 17.51 \text{ A.U.}$$

Hence its maximum distance from the sun is $a + c \approx 35.60$ A.U., and its minimum distance is $a - c \approx 0.58$ A.U. The eccentricity of its orbit is

$$e = \frac{17.51}{18.09} \approx 0.97,$$

a very eccentric orbit.

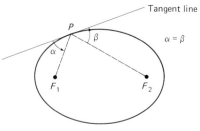

10.45 The reflection property: $\alpha = \beta$

The *reflection property* of the ellipse is the fact that at a point $P$ of an ellipse, the tangent line makes equal angles with the lines $PF_1$ and $PF_2$ from $P$ to the two foci of the ellipse (see Fig. 10.45). This property is the basis of the "whispering gallery" phenomenon, which can be observed (for instance) in the whispering gallery of the U.S. Senate. Suppose that the ceiling of a large room is shaped like half an ellipsoid of the sort obtained by revolving an ellipse about its major axis. Sound waves, like light waves, are reflected at equal angles of incidence and reflection. Thus if two diplomats are standing in quiet conversation near one focus of the ellipsoidal surface, a reporter standing near the other focus would be able to eavesdrop on their conversation even though standing 50 ft away, and even if the diplomats' conversation were inaudible to others in the same room.

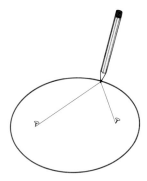

10.46 One way to draw an ellipse

An alternative definition of the ellipse with foci $F_1$ and $F_2$ and major axis of length $2a$ is this: It is the locus of a point $P$ such that the sum of the distances $|PF_1|$ and $|PF_2|$ is the constant $2a$ (Problem 26). This fact gives a convenient way of drawing the ellipse, making use of two tacks placed at $F_1$ and $F_2$, a string of length $2a$, and a pencil (see Fig. 10.46).

## 10-5 PROBLEMS

In each of Problems 1–15, find an equation of the ellipse specified.

1 Vertices ($\pm 4$, 0) and (0, $\pm 5$)
2 Foci ($\pm 5$, 0) and major semiaxis 13
3 Foci (0, $\pm 8$) and major semiaxis 17
4 Center (0, 0), vertical major axis 12, minor axis 8
5 Foci ($\pm 3$, 0), eccentricity 0.75
6 Foci (0, $\pm 4$), eccentricity $\frac{2}{3}$
7 Center (0, 0), horizontal major axis 20, eccentricity $\frac{1}{2}$
8 Center (0, 0), horizontal minor axis 10, eccentricity $\frac{1}{2}$
9 Foci ($\pm 2$, 0), directrices $x = \pm 8$
10 Foci (0, $\pm 4$), directrices $y = \pm 9$
11 Center (2, 3), horizontal axis 8, vertical axis 4
12 Center (1, $-2$), horizontal major axis 8, eccentricity $\frac{3}{4}$
13 Foci ($-2$, 1) and (4, 1), major axis 10
14 Foci ($-3$, 0) and ($-3$, 4), minor axis 6
15 Foci ($-2$, 2) and (4, 2), eccentricity $\frac{1}{3}$

Sketch the graphs of the equations in Problems 16–20. Indicate centers, foci, and lengths of axes.

16 $4x^2 + y^2 = 16$
17 $4x^2 + 9y^2 = 144$

18 $4x^2 + 9y^2 = 24x$
19 $9x^2 + 4y^2 - 32y + 28 = 0$
20 $2x^2 + 3y^2 + 12x - 24y + 60 = 0$
21 The orbit of the comet Kahoutek is an ellipse of extreme eccentricity $e = 0.999925$; the sun is at one focus of this ellipse. The minimum distance between Kahoutek and the sun is 0.13 A.U. What is the maximum distance between Kahoutek and the sun?
22 The orbit of the planet Mercury is an ellipse with eccentricity $e = 0.206$. Its maximum and minimum distances from the sun are 0.467 and 0.307 A.U., respectively. What are the major and minor semiaxes of Mercury's orbit? Would you describe this orbit as "nearly circular"?
23 Derive Equation (5) for an ellipse if its foci lie on the $y$-axis.
24 Show that the tangent line to the ellipse $x^2/a^2 + y^2/b^2 = 1$ at the point $P(x_0, y_0)$ of that ellipse has equation

$$\frac{xx_0}{a^2} + \frac{yy_0}{b^2} = 1.$$

**25** Use the result of Problem 24 to establish the reflection property of the ellipse. (*Suggestion:* Let $m$ be the slope of the normal line to the ellipse at $P(x_0, y_0)$, and $m_1$ and $m_2$ be the slopes of the lines $PF_1$ and $PF_2$, respectively. Show that

$$\frac{m - m_1}{1 + m_1 m} = \frac{m_2 - m}{1 + m_2 m},$$

and use the identity for $\tan(A - B)$.)

**26** Given $F_1(-c, 0)$ and $F_2(c, 0)$ and $a > c > 0$, show that the ellipse $x^2/a^2 + y^2/b^2 = 1$ (with $b^2 = a^2 - c^2$) is the locus of a point $P$ such that $|PF_1| + |PF_2| = 2a$.

**27** Find the equation of the ellipse with horizontal and vertical axes that passes through the points $(-1, 0)$, $(3, 0)$, $(0, 2)$ and $(0, -2)$.

**28** Derive an equation for the ellipse with foci $(3, -3)$, $(-3, 3)$, and major axis of length 10. Note that the foci of this ellipse lie on neither a vertical line nor a horizontal line.

---

## 10-6

A hyperbola is defined in the same way as is an ellipse, except that the eccentricity $e$ of a hyperbola is greater than 1.

# The Hyperbola

---

> **Definition**  *The Hyperbola*
>
> Suppose that $e > 1$, and let $F$ be a fixed point and $L$ a fixed line not containing $F$. Then the **hyperbola** with **eccentricity** $e$, **focus** $F$, and **directrix** $L$ is the set of all points $P$ such that the distance $|PF|$ is $e$ times the (perpendicular) distance from $P$ to the line $L$.

---

As with the ellipse, the equation of a hyperbola is simplest if $F$ is the point $(c, 0)$ on the $x$-axis and $L$ is the vertical line $x = c/e^2$. The case $c > 0$ is shown in Fig. 10.47. If $Q$ is the point $(c/e^2, y)$, then $PQ$ is the perpendicular from $P(x, y)$ to $L$. The condition $|PF| = e|PQ|$ gives

$$(x - c)^2 + y^2 = e^2\left(x - \frac{c}{e^2}\right)^2,$$

$$x^2 - 2cx + c^2 + y^2 = e^2 x^2 - 2cx + \frac{c^2}{e^2},$$

$$(e^2 - 1)x^2 - y^2 = c^2\left(1 - \frac{1}{e^2}\right) = \frac{c^2}{e^2}(e^2 - 1).$$

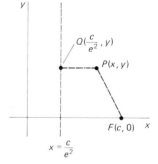

**10.47** The definition of the hyperbola

Thus

$$(e^2 - 1)x^2 - y^2 = a^2(e^2 - 1)$$

where

$$a = \frac{c}{e}. \tag{1}$$

Division of both sides of the next-to-last equation above by $a^2(e^2 - 1)$ gives

$$\frac{x^2}{a^2} - \frac{y^2}{a^2(e^2 - 1)} = 1.$$

To simplify this equation, we let

$$b^2 = a^2(e^2 - 1) = c^2 - a^2. \tag{2}$$

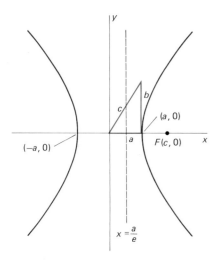

**10.48** A hyperbola has two branches.

This is permissible because $e > 1$. So the equation of the hyperbola with focus $(c, 0)$ and directrix $x = c/e^2 = a/e$ takes the form

$$\frac{x^2}{a^2} - \frac{y^2}{b^2} = 1. \tag{3}$$

The minus sign on the left is the only difference between the equation of a hyperbola and that of an ellipse. Of course, Equation (2) differs from the relation $a^2 = b^2 + c^2$ for the case of the ellipse.

The hyperbola of Equation (3) is clearly symmetric about both coordinate axes, and has $x$-intercepts $(\pm a, 0)$. But it has no $y$-intercepts! If we rewrite Equation (3) in the form

$$y = \pm \frac{b}{a} \sqrt{x^2 - a^2}, \tag{4}$$

then we see that there are points on the graph only if $|x| \geq a$. Hence the hyperbola has two **branches**, as shown in Fig. 10.48. We also see from Equation (4) that $|y| \to \infty$ as $|x| \to \infty$.

The $x$-intercepts $V_1(-a, 0)$ and $V_2(a, 0)$ are the **vertices** of the hyperbola, and the line segment joining them is its **transverse axis.** The line segment joining $W_1(0, -b)$ and $W_2(0, b)$ is its **conjugate axis.** The alternative form

$$c^2 = a^2 + b^2 \tag{5}$$

of Equation (2) is the Pythagorean relation for the right triangle of Fig. 10.48.

**10.49** The parts of a hyperbola

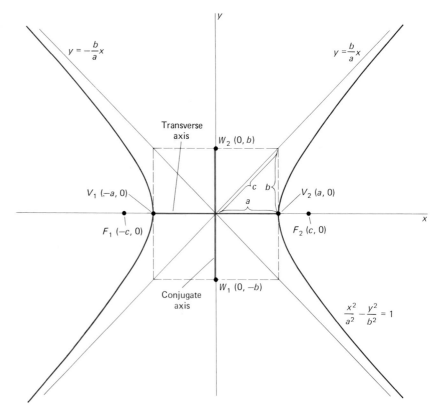

The lines $y = \pm bx/a$ that pass through the **center** $(0, 0)$ and the opposite vertices of the rectangle in Fig. 10.49 are **asymptotes** of the two branches of the hyperbola in both directions. That is, if

$$y_1 = \frac{bx}{a} \quad \text{and} \quad y_2 = \frac{b}{a}\sqrt{x^2 - a^2},$$

then

$$\lim_{x \to \infty} (y_1 - y_2) = 0 = \lim_{x \to -\infty} (y_1 - (-y_2)). \tag{6}$$

To verify the first limit, note that

$$\lim_{x \to +\infty} \frac{b}{a}(x - \sqrt{x^2 - a^2}) = \lim_{x \to +\infty} \frac{b}{a} \cdot \frac{(x - \sqrt{x^2 - a^2})(x + \sqrt{x^2 - a^2})}{x + \sqrt{x^2 - a^2}}$$

$$= \lim_{x \to +\infty} \frac{b}{a} \cdot \frac{a^2}{x + \sqrt{x^2 - a^2}} = 0.$$

Just as in the case of the ellipse, the hyperbola with focus $(c, 0)$ and directrix $x = a/e$ also has focus $(-c, 0)$ and directrix $x = -a/e$, as shown both in Fig. 10.49 and 10.50. Because $c = ae$ by (1), the foci $(\pm ae, 0)$ and the directrices $x = \pm a/e$ take the same forms in terms of $a$ and $e$ for both the hyperbola $(e > 1)$ and the ellipse $(e < 1)$.

If we interchange $x$ and $y$ in Equation (3), we obtain the equation

$$\frac{y^2}{a^2} - \frac{x^2}{b^2} = 1. \tag{7}$$

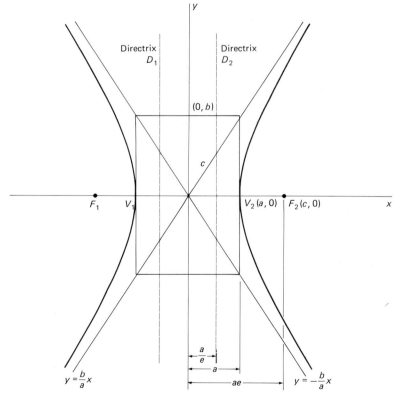

**10.50** The relations between the parts of a hyperbola

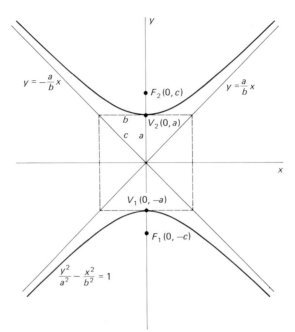

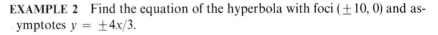

**10.51** The hyperbola of Equation (7) has horizontal directrices.

This hyperbola has foci at $(0, \pm c)$. They, and its transverse axis, lie on the $y$-axis. Its asymptotes are $y = \pm ax/b$, and its graph is like the one in Figure 10.51.

When we studied the ellipse, we saw that its orientation—whether the major axis is horizontal or vertical—is determined by the relative sizes of $a$ and $b$. In the case of the hyperbola, the situation is different, for the relative sizes of $a$ and $b$ make no such difference. The direction in which the hyperbola opens—horizontal as in Fig. 10.50 or vertical as in Fig. 10.51—is determined by the signs of the terms involving $x^2$ and $y^2$.

**EXAMPLE 1**   Sketch the graph of the hyperbola with equation

$$\frac{y^2}{9} - \frac{x^2}{16} = 1.$$

**10.52**   The hyperbola of Example 1

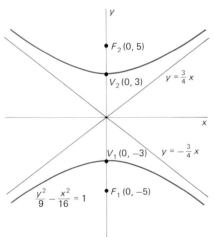

*Solution*   This is an equation of the form of (7), so the hyperbola opens vertically. Since $a = 3$ and $b = 4$, we find that $c = 5$ by using Equation (5): $c^2 = a^2 + b^2$. Thus the vertices are $(0, \pm 3)$, the foci are the two points $(0, \pm 5)$, and the asymptotes are the two lines $y = \pm 3x/4$. This hyperbola appears in Fig. 10.52.

**EXAMPLE 2**   Find the equation of the hyperbola with foci $(\pm 10, 0)$ and asymptotes $y = \pm 4x/3$.

*Solution*   Since $c = 10$, we have

$$a^2 + b^2 = 100 \quad \text{and} \quad \frac{b}{a} = \frac{4}{3}.$$

Thus $b = 8$ and $a = 6$, and the equation of the hyperbola is

$$\frac{x^2}{36} - \frac{y^2}{64} = 1.$$

As we noted in Section 10-5, any equation of the form

$$Ax^2 + Cy^2 + Dx + Ey + F = 0 \qquad (8)$$

with $A$ and $C$ nonzero can be reduced to the form

$$A(x - h)^2 + C(y - k)^2 = G$$

by completing the square in $x$ and $y$. Now suppose that the coefficients $A$ and $C$ of the quadratic terms have *opposite signs;* for example, suppose that $A = p^2$ and $C = -q^2$. The equation above becomes

$$p^2(x - h)^2 - q^2(y - k)^2 = G. \qquad (9)$$

If $G = 0$, then factorization of the difference of squares on the left yields the equations

$$p(x - h) + q(y - k) = 0,$$
$$p(x - h) - q(y - k) = 0$$

of two straight lines through $(h, k)$ with slopes $\pm p/q$. If $G \neq 0$, then division of Equation (9) by $G$ gives an equation that looks either like

$$\frac{(x - h)^2}{a^2} - \frac{(y - k)^2}{b^2} = 1 \qquad \text{(if } G > 0\text{)}$$

or like

$$\frac{(y - k)^2}{a^2} - \frac{(x - h)^2}{b^2} = 1 \qquad \text{(if } G < 0\text{)}.$$

Thus if $AC < 0$ in Equation (8), the graph is either a pair of intersecting straight lines or a hyperbola.

**EXAMPLE 3** Determine the graph of the equation

$$9x^2 - 4y^2 - 36x + 8y = 4.$$

*Solution* We collect the terms containing $x$, those containing $y$, and complete the square in each variable. We find that

$$9(x - 2)^2 - 4(y - 1)^2 = 36,$$

so

$$\frac{(x - 2)^2}{4} - \frac{(y - 1)^2}{9} = 1.$$

Hence the graph is a hyperbola with horizontal transverse axis and center $(2, 1)$. Since $a = 2$ and $b = 3$, $c = \sqrt{13}$; the vertices of the hyperbola are $(0, 1)$ and $(4, 1)$ and its foci are the two points $(2 \pm \sqrt{13}, 1)$. Its asymptotes

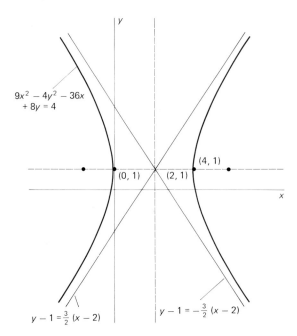

$9x^2 - 4y^2 - 36x + 8y = 4$

$(4, 1)$

$(0, 1)$  $(2, 1)$

$y - 1 = \frac{3}{2}(x - 2)$

$y - 1 = -\frac{3}{2}(x - 2)$

**10.53**  The hyperbola of Example 3, a translate of the hyperbola $x^2/4 - y^2/9 = 1$.

are the two lines

$$y - 1 = \pm\tfrac{3}{2}(x - 2),$$

translates of the asymptotes $y = \pm 3x/2$ of the hyperbola $x^2/4 - y^2/9 = 1$. Figure 10.53 shows the graph of the translated hyperbola.

The *reflection property* of the hyperbola takes the same form as for the ellipse. If $P$ is a point on a hyperbola, then the two lines $PF_1$ and $PF_2$ from $P$ to the two foci make equal angles with the tangent line at $P$. In Fig. 10.54 this means that $\alpha = \beta$.

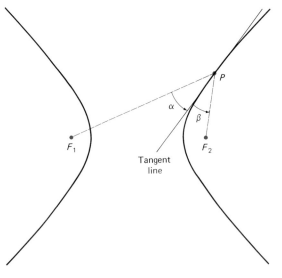

**10.54**  The reflection property of the hyperbola

$P$

$\alpha$

$\beta$

$F_1$

$F_2$

Tangent line

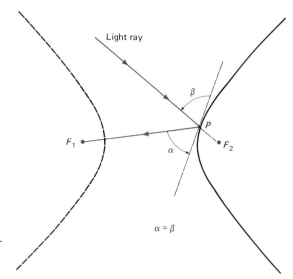

**10.55** How a hyperbolic mirror reflects a ray aimed at one focus: $\alpha = \beta$ again.

Light ray

$\beta$

$P$

$F_1$

$\alpha$

$F_2$

$\alpha = \beta$

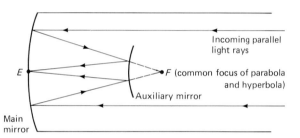

**10.56** One sort of reflecting telescope: Main mirror parabolic, auxiliary mirror hyperbolic

Incoming parallel light rays

$E$

$F$ (common focus of parabola and hyperbola)

Auxiliary mirror

Main mirror

For an application of this reflection property, consider a mirror shaped like one branch of a hyperbola and made reflective on its outer (convex) surface. Then an incoming light ray aimed toward one focus will be reflected toward the other focus, as shown in Fig. 10.55. Figure 10.56 indicates the design of a reflecting telescope that makes use of the reflection properties of the parabola and the hyperbola. The parallel incoming light rays first are reflected by the parabola toward its focus at $F$. Then they are intercepted by an auxiliary hyperbolic mirror with foci at $E$ and $F$ and reflected into the eyepiece located at $E$.

The following example illustrates how hyperbolas are sometimes used to determine the position of ships at sea.

**EXAMPLE 4**   A ship lies at sea due east of point $A$ on a long north-south coastline. Simultaneous signals are transmitted by a radio station at $A$ and by one at $B$ on the coast 200 miles south of $A$. The ship receives the signal from $A$ $500\,\mu s$ (microseconds) before it receives the one from $B$. Assume that the speed of radio signals is 980 ft/$\mu s$. How far out at sea is the ship?

*Solution*   The situation is diagrammed in Fig. 10.57.

The difference between the distances of the ship $S$ from $A$ and $B$ is

$$|SB| - |SA| = (500)(980)$$

**10.57** A navigation problem (Example 4)

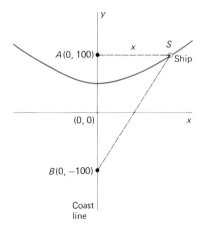

$y$

$A(0, 100)$

$x$

$S$

Ship

$(0, 0)$

$x$

$B(0, -100)$

Coast line

**535**

ft, or 92.8 mi. Thus (by Problem 24) the ship lies on a hyperbola with foci $A$ and $B$, and with $c = 100$, we have

$$a = \tfrac{1}{2}(92.8) = 46.4 \quad \text{and} \quad b = \sqrt{(100)^2 - (46.4)^2} \approx 88.6.$$

In the coordinate system of Figure 10.57, the hyperbola has equation

$$\frac{y^2}{(46.4)^2} - \frac{x^2}{(88.6)^2} = 1.$$

We substitute $y = 100$, because the ship is due east of $A$. We find that the ship's distance from the coastline is $x \approx 169.1$ miles.

## 10-6  PROBLEMS

In each of Problems 1–14, find an equation of the hyperbola described.

1  Foci $(\pm 4, 0)$ and vertices $(\pm 1, 0)$
2  Foci $(0, \pm 3)$ and vertices $(0, \pm 2)$

3  Foci $(\pm 5, 0)$ and asymptotes $y = \dfrac{\pm 3x}{4}$

4  Vertices $(\pm 3, 0)$ and asymptotes $y = \dfrac{\pm 3x}{4}$

5  Vertices $(0, \pm 5)$ and asymptotes $y = \pm x$
6  Vertices $(\pm 3, 0)$ and eccentricity $e = \tfrac{5}{3}$
7  Foci $(0, \pm 6)$ and eccentricity $e = 2$
8  Vertices $(\pm 4, 0)$ and passing through $(8, 3)$
9  Foci $(\pm 4, 0)$ and directrices $x = \pm 1$
10  Foci $(0, \pm 9)$ and directrices $y = \pm 4$
11  Center $(2, 2)$, horizontal transverse axis of length 6, and eccentricity 2
12  Center $(-1, 3)$, vertices at $(-4, 3)$ and $(2, 3)$, and foci at $(-6, 3)$ and $(4, 3)$
13  Center $(1, -2)$, vertices $(1, 1)$ and $(1, -5)$, and asymptotes $3x - 2y - 7 = 0$ and $3x + 2y + 1 = 0$
14  Focus $(8, -1)$ and asymptotes $3x - 4y - 13 = 0$ and $3x + 4y - 5 = 0$

Sketch the graph of the equation given in each of Problems 15–20; indicate centers, foci, and asymptotes.

15  $x^2 - y^2 - 2x + 4y = 4$
16  $x^2 - 2y^2 + 4x = 0$     17  $y^2 - 3x^2 - 6y = 0$
18  $x^2 - y^2 - 2x + 6y = 9$
19  $9x^2 - 4y^2 + 18x + 8y = 31$
20  $4y^2 - 9x^2 - 18x - 8y = 41$
21  Show that the graph of the equation

$$\frac{x^2}{15 - c} - \frac{y^2}{c - 6} = 1$$

is:
(a) a hyperbola with foci $(\pm 3, 0)$ if $6 < c < 15$;
(b) an ellipse if $c < 6$.
Identify the graph in the case $c > 15$.
22  Establish that the tangent line to the hyperbola $x^2/a^2 - y^2/b^2 = 1$ at the point $P(x_0, y_0)$ has equation

$$\frac{xx_0}{a^2} - \frac{yy_0}{b^2} = 1.$$

23  Use the result of Problem 22 to establish the reflection property of the hyperbola. (See the suggestion for Problem 25 in Section 10-5.)
24  Suppose that $0 < a < c$, and let $b = (c^2 - a^2)^{1/2}$. Show that the hyperbola $x^2/a^2 - y^2/b^2 = 1$ is the locus of a point $P$ such that the *difference* between the distances $|PF_1|$ and $|PF_2|$ is equal to $2a$, where $F_1$ and $F_2$ are the two foci of the hyperbola.
25  Derive an equation for the hyperbola with foci at $(\pm 5, \pm 5)$ and vertices at $(\pm 3/\sqrt{2}, \pm 3/\sqrt{2})$. Use the difference definition of a hyperbola implied by Problem 24.
26  Two radio signal stations at $A$ and $B$ lie on an east-west line with $A$ 100 mi west of $B$. A plane is flying west on a line 50 mi north of the line $AB$. Radio signals are sent (traveling at 980 ft/$\mu$s) simultaneously from $A$ and $B$, and the one sent from $B$ arrives 400 $\mu$s before the one from $A$. Where is the plane?
27  Two radio signal stations are located as in Problem 26 and transmit radio signals that travel at the same speed as in that problem. In this problem, however, it is known only that the plane is generally somewhere north of the line $AB$, that the signal sent from $B$ arrives 400 $\mu$s before the one from $A$, and that the signal sent from $A$ and reflected by the plane takes a total of 600 $\mu$s to reach $B$. Where is the plane?

In the previous three sections we have studied the second degree equation

$$Ax^2 + Cy^2 + Dx + Ey + F = 0, \qquad (1)$$

which contains no $xy$-term. We found that its graph is always a conic section, apart from exceptional cases of the following types:

| | |
|---|---|
| $2x^2 + 3y^2 = -1$ | (no locus), |
| $2x^2 + 3y^2 = 0$ | (one point), |
| $(2x - 1)^2 = 0$ | (a line), |
| $(2x - 1)^2 = 1$ | (two parallel lines), |
| $x^2 - y^2 = 0$ | (two intersecting lines). |

We may therefore say that the graph of Equation (1) is a conic section, possibly **degenerate.** If either $A$ or $C$ is zero (but not both), then the graph is a parabola. It is an ellipse if $AC > 0$ (by the discussion of Equation (6) in Section 10-5), a hyperbola if $AC < 0$ (by the discussion of Equation (8) in Section 10-6).

Let us assume that $AC \neq 0$. Then we can determine the particular conic section represented by Equation (1) by completing squares; that is, we write (1) in the form

$$A(x - h)^2 + B(y - k)^2 = G. \qquad (2)$$

This equation can be simplified further by a **translation of coordinates** to a new $\bar{x}\bar{y}$-coordinate system centered at the point $(h, k)$ in the old $xy$-system. The geometry of this change of coordinates is shown in Fig. 10.58. The relation between the old and the new coordinates is

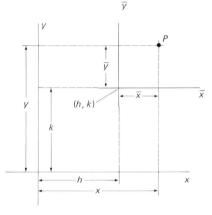

**10.58** A translation of coordinates

$$\begin{rmatrix} \bar{x} = x - h, \\ \bar{y} = y - k \end{rmatrix} \quad \text{or} \quad \begin{cases} x = \bar{x} + h, \\ y = \bar{y} + k. \end{cases} \qquad (3)$$

In the new $\bar{x}\bar{y}$-coordinate system, Equation (2) takes the simpler form

$$A\bar{x}^2 + C\bar{y}^2 = G, \qquad (2')$$

from which it is clear whether we have an ellipse, a hyperbola, or a degenerate case.

**10.59** A rotation of coordinates through the angle $\alpha$

We now turn to the general second-degree equation

$$Ax^2 + Bxy + Cy^2 + Dx + Ey + F = 0. \qquad (4)$$

Note the presence of the "cross-product," or $xy$-, term. In order to recognize its graph, we need to change to a new $x'y'$-coordinate system obtained by a **rotation of axes.**

We get the $x'y'$-axes from the $xy$-axes by a rotation through an angle $\alpha$ in the counterclockwise direction. In the notation of Fig. 10.59, we have

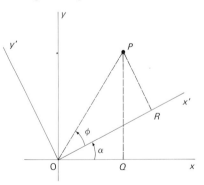

$$x = OQ = OP \cos(\phi + \alpha) \quad \text{and} \quad y = PQ = OP \sin(\phi + \alpha). \qquad (5)$$

Similarly,

$$x' = OR = OP \cos \phi \quad \text{and} \quad y' = PR = OP \sin \phi. \tag{6}$$

Recall the addition formulas

$$\cos(\phi + \alpha) = \cos \phi \cos \alpha - \sin \phi \sin \alpha,$$

$$\sin(\phi + \alpha) = \sin \phi \cos \alpha + \cos \phi \sin \alpha.$$

With the aid of these identities and the substitution of Equations (6) into Equations (5), we obtain this result:

---

**Equations for Rotation of Axes**

$$x = x' \cos \alpha - y' \sin \alpha,$$

$$y = x' \sin \alpha + y' \cos \alpha. \tag{7}$$

---

These equations express the old $xy$-coordinates of the point $P$ in terms of its new $x'y'$-coordinates and the rotation angle $\alpha$. The following example illustrates how Equations (7) may be used to transform the equation of a curve from $xy$-coordinates into the rotated $x'y'$-coordinates.

**EXAMPLE 1**   The $xy$-axes are rotated through an angle of $\alpha = 45°$. Find the equation of the curve $2xy = 1$ in the new coordinates $x'$, $y'$.

**Solution**   Since $\cos 45° = \sin 45° = 1/\sqrt{2}$, Equations (7) yield

$$x = \frac{x' - y'}{\sqrt{2}} \quad \text{and} \quad y = \frac{x' + y'}{\sqrt{2}}.$$

The original equation $2xy = 1$ thus becomes

$$(x')^2 - (y')^2 = 1.$$

So, in the $x'y'$-coordinate system, we have a hyperbola with $a = b = 1$, $c = \sqrt{2}$, and foci $(\pm\sqrt{2}, 0)$. In the original $xy$-coordinates, its foci are $(1, 1)$ and $(-1, -1)$, and its asymptotes are the $x$- and $y$-axes. This hyperbola is shown in Fig. 10.60. A hyperbola of this form, one which has equation $xy = k$, is called a **rectangular** hyperbola because its asymptotes are perpendicular.

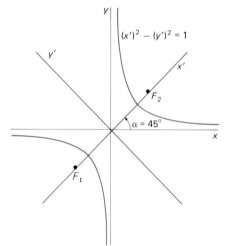

**10.60**  The graph of $2xy = 1$ is a hyperbola.

Example 1 suggests that the cross-product term $Bxy$ of Equation (4) may disappear upon rotation of the coordinate axes. One can, indeed, always choose an appropriate angle $\alpha$ of rotation so that, in the new coordinate system, there is no $xy$-term.

To determine the appropriate rotation angle, we substitute Equations (7) for $x$ and $y$ into the general second-degree equation in (4). We obtain the following new second-degree equation:

$$A'(x')^2 + B'x'y' + C'(y')^2 + D'x' + E'y' + F' = 0. \tag{8}$$

The new coefficients are given in terms of the old ones and the angle $\alpha$ by the following equations:

$$A' = A \cos^2\alpha + B \cos \alpha \sin \alpha + C \sin^2\alpha,$$

$$B' = B(\cos^2\alpha - \sin^2\alpha) + 2(C - A) \sin \alpha \cos \alpha,$$

$$C' = A \sin^2\alpha - B \sin \alpha \cos \alpha + C \cos^2\alpha,$$

$$D' = D \cos \alpha + E \sin \alpha, \tag{9}$$

$$E' = -D \sin \alpha + E \cos \alpha, \quad \text{and}$$

$$F' = F.$$

Now suppose that an equation of the form of (4) is given, with $B \neq 0$. We simply choose $\alpha$ so that $B' = 0$ in the above list of new coefficients. Then Equation (8) will have no cross-product term, and we can identify and sketch the curve with little trouble in the $x'y'$-coordinate system. But is it really easy to choose such an angle $\alpha$?

It is. We recall that

$$\cos 2\alpha = \cos^2\alpha - \sin^2\alpha \quad \text{and} \quad \sin 2\alpha = 2 \sin \alpha \cos \alpha.$$

So the above equation for $B'$ may be written

$$B' = B \cos 2\alpha + (C - A) \sin 2\alpha.$$

We can cause $B'$ to be zero by choosing $\alpha$ as that (unique) acute angle such that

$$\cot 2\alpha = \frac{A - C}{B}. \tag{10}$$

If we plan to use the equations in (9) to calculate the coefficients in the transformed Equation (8), we shall need the values of $\sin \alpha$ and $\cos \alpha$ that follow from Equation (10). It is sometimes convenient to calculate these values directly from $\cot 2\alpha$, as follows. From the triangle in Fig. 10.61, we can read the numerical value of $\cos 2\alpha$. Since the cosine and cotangent are both positive in the first quadrant and both negative in the second quadrant, we give $\cos 2\alpha$ the same sign as $\cot 2\alpha$. Then we use the half-angle formulas to get $\sin \alpha$ and $\cos \alpha$:

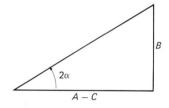

**10.61** Finding $\sin \alpha$ and $\cos \alpha$ given $\cot 2\alpha = (A - C)/B$.

$$\sin \alpha = \left(\frac{1 - \cos 2\alpha}{2}\right)^{1/2}, \qquad \cos \alpha = \left(\frac{1 + \cos 2\alpha}{2}\right)^{1/2}. \tag{11}$$

Once we have the values of $\sin \alpha$ and $\cos \alpha$, we can compute the coefficients in the resulting Equation (8) by means of Equations (9). Alternatively, it's frequently simpler to get Equation (8) directly by substituting Equations (7), with the numerical values of $\sin \alpha$ and $\cos \alpha$ obtained as above, into Equation (4). If we are using a calculator or computer that has an inverse tangent function provided, then we can calculate $\alpha$ more briefly by means of the formula

$$\alpha = \frac{\pi}{4} - \frac{1}{2} \tan^{-1}\left(\frac{A - C}{B}\right), \tag{12}$$

which follows from Equation (10) and the observation that $\cot^{-1} x = \pi/2 - \tan^{-1} x$.

**EXAMPLE 2**  Determine the graph of the equation

$$73x^2 - 72xy + 52y^2 - 30x - 40y - 75 = 0.$$

***Solution***  We begin with Equation (10) and find that $\cot 2\alpha = -\frac{7}{24}$, so that $\cos 2\alpha = -\frac{7}{25}$. Thus

$$\sin \alpha = \left(\frac{1 - (-\frac{7}{25})}{2}\right)^{1/2} = \frac{4}{5}, \qquad \cos \alpha = \left(\frac{1 + (-\frac{7}{25})}{2}\right)^{1/2} = \frac{3}{5}.$$

Then with $A = 73$, $B = -72$, $C = 52$, $D = -30$, $E = -40$, and $F = -75$, Equations (9) yield

| | |
|---|---|
| $A' = 25,$ | $D' = -50,$ |
| $B' = 0$  (this was our goal), | $E' = 0,$ |
| $C' = 100,$ | $F' = -75.$ |

Consequently the equation in the new $x'y'$-coordinate system, obtained by rotation through an angle of $\alpha = \arcsin(\frac{4}{5}) \approx 53.13°$, is

$$25(x')^2 + 100(y')^2 - 50x' = 75.$$

Alternatively, we could have obtained this equation by substituting

$$x = \tfrac{3}{5}x' - \tfrac{4}{5}y', \qquad y = \tfrac{4}{5}x' + \tfrac{3}{5}y'$$

into the original equation.

By completing the square in $x'$ we finally obtain

$$25(x' - 1)^2 + 100(y')^2 = 100,$$

which we put into the standard form

$$\frac{(x' - 1)^2}{4} + \frac{(y')^2}{1} = 1.$$

Thus the original curve is an ellipse with major semiaxis 2, minor semiaxis 1, and center $(1, 0)$ in the $x'y'$-coordinate system, as shown in Fig. 10.62.

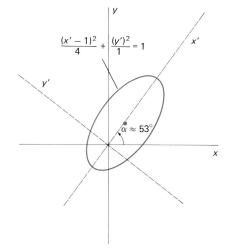

**10.62**  The ellipse of Example 2

### CLASSIFICATION OF CONICS

Example 2 illustrates the general procedure for finding the graph of a second degree equation. First, if there is a cross-product ($xy$-) term, rotate axes to eliminate it. Then translate axes (if necessary) to reduce the equation to the standard form of a parabola, ellipse, or hyperbola (or a degenerate case of one of these).

There *is* a test by which the nature of the curve may be discovered without actually carrying out the transformations described. This test stems from the fact (see Problem 14) that, whatever the angle $\alpha$ of rotation, the equations in (9) imply that

$$(B')^2 - 4A'C' = B^2 - 4AC. \tag{13}$$

Thus the **discriminant** $B^2 - 4AC$ is an *invariant* under any rotation of axes. If $\alpha$ is chosen so that $B' = 0$, then the left-hand side in Equation (13) is simply $-4A'C'$. Because $A'$ and $C'$ are the coefficients of the squared terms, our earlier discussion of Equation (1) now applies. It follows that the graph will be

1 a *parabola* if $B^2 - 4AC = 0$,
2 an *ellipse* if $B^2 - 4AC < 0$,     and
3 a *hyperbola* if $B^2 - 4AC > 0$.

Of course, degenerate cases may occur.

Here are some examples:

1 $x^2 + 2xy + y^2 = 1$ is a (degenerate) parabola,
2 $x^2 + xy + y^2 = 1$ is an ellipse,     and
3 $x^2 + 3xy + y^2 = 1$ is a hyperbola.

## 10-7   PROBLEMS

In each of Problems 1–12, first use the discriminant to classify the graph of the given equation. Then eliminate the cross-product term by an appropriate rotation of the coordinate axes. Finally, translate coordinates (if necessary) and sketch the curve in the original $xy$-coordinate system.

1 $x^2 + 2xy + y^2 = 2$     2 $x^2 + xy + y^2 = 3$
3 $x^2 + 3xy + y^2 = 5$     4 $5x^2 - 6xy + 5y^2 = 8$
5 $x^2 + 4xy - 2y^2 = 18$
6 $x^2 - 2xy + y^2 = 2x + 2y$
7 $x^2 - xy + y^2 - 2\sqrt{2}x + \sqrt{2}y = 0$
8 $41x^2 - 24xy + 34y^2 + 20x - 140y + 125 = 0$
9 $23x^2 - 72xy + 2y^2 + 140x + 20y = 75$
10 $9x^2 + 24xy + 16y^2 - 170x - 60y + 245 = 0$
11 $161x^2 + 480xy - 161y^2 - 510x - 272y = 0$
12 $144x^2 - 120xy + 25y^2 - 65x - 156y = 169$
13 Solve Equations (7) to show that the rotated coordinates $(x', y')$ are given in terms of the original coordinates by

$$x' = x \cos \alpha + y \sin \alpha, \qquad y' = -x \sin \alpha + y \cos \alpha.$$

14 Use the equations in (9) to verify Equation (13).
15 Show that the sum $A + C$ of the coefficients of $x^2$ and $y^2$ in Equation (4) is invariant under rotation. That is, show that $A' + C' = A + C$ for any rotation through an angle $\alpha$.

16 Use Equations (9) to show that any rotation of axes transforms the equation $x^2 + y^2 = r^2$ into the equation $(x')^2 + (y')^2 = r^2$.
17 Consider the equation

$$x^2 + Bxy - y^2 + Dx + Ey + F = 0.$$

Show that there is a rotation of axes such that $A' = 0 = C'$ in the resulting equation. (*Suggestion:* Find the angle $\alpha$ for which $A' = 0$. Then apply the result of Problem 15.) What can you conclude about the graph of the given equation?
18 Suppose that $B^2 - 4AC < 0$, so that the equation

$$Ax^2 + Bxy + Cy^2 = 1$$

represents an ellipse. Show that its area is

$$\pi ab = \frac{2\pi}{\sqrt{4AC - B^2}}$$

where $a$ and $b$ are the lengths of its semiaxes.
19 Show that the equation $27x^2 + 37xy + 17y^2 = 1$ represents an ellipse, and then find the points of the ellipse that are closest to and farthest from the origin.
20 Show that the equation $x^2 + 14xy + 49y^2 = 100$ represents a parabola (possibly degenerate), and then find the point of the parabola that is closest to the origin.

### *10-7   Optional Computer Application

The program below was written (in IBM-PC BASIC) to calculate the new coefficients that result when the axes are rotated to eliminate the $xy$-term in the equation

$$Ax^2 + Bxy + Cy^2 + Dx + Ey + F = 0.$$

```
100    REM—Program ROTATE
110    REM—Rotates the general second-degree curve
120    REM—   Ax∧2 + Bxy + Cy∧2 + Dx + Ey + F = 0
130    REM—to eliminate the cross-product term.
140    REM
150        INPUT  "  A, B, C  ";  A, B, C
160        INPUT  "  D, E  ";     D, E
170        INPUT  "  F  ";        F
180        PI = 3.141593
190    REM
200        IF  B = 0 THEN LET  ALPHA = 0
               ELSE LET  ALPHA = PI/4 − (ATN((A − C)/B))/2
210        COSA = COS(ALPHA)  :  SINA = SIN(ALPHA)
220    REM
230        A1 = A*COSA∧2 + B*COSA*SINA + C*SINA∧2
240        B1 = B*(COSA∧2 − SINA∧2) + 2*(C − A)*SINA*COSA
250        C1 = A*SINA∧2 − B*SINA*COSA + C*COSA∧2
260        D1 = D*COSA + E*SINA
270        E1 = −D*SINA + E*COSA
280    REM
290        PRINT  :  PRINT  "  ALPHA =  ";  180*ALPHA/PI;
                       "DEGREES"
300        PRINT  :  PRINT  "  NEW COEFFICIENTS:"
310        PRINT  "  A =  ";  A1
320        PRINT  "  B =  ";  B1
330        PRINT  "  C =  ";  C1
340        PRINT  "  D =  ";  D1
350        PRINT  "  E =  ";  E1
360        PRINT  "  F =  ";  F
370        END
```

Lines 150–170 call for the values of the coefficients A, B, C, D, E, and F to be entered. The angle $\alpha = $ ALPHA is then calculated in line 200 using the formula in Equation (12). Then the new coefficients are calculated in lines 230–270 using the rotation formulas in (9). A run of this program with the data of Example 2 produces the following output.

**RUN**

A, B, C?   73, −72, 52
D, E?   −30, −40
F?   −75

ALPHA = 53.13011 DEGREES

NEW COEFFICIENTS:
A = 25
B = 1.716614E-05
C = 100
D = −50
E = 5.722046E-06
F = −75

The "slightly nonzero" values for B and E result from round-off error.

**Exercise** Use this program—rewritten if necessary for a computer that accepts only single-character variables—to solve Problems 1–12 above.

# Conic Sections in Polar Coordinates

In order to investigate orbits of satellites, such as planets or comets orbiting the sun or natural or artificial moons orbiting a planet, we need the equations of the conic sections in polar coordinates. Recall the focus-directrix definition of a conic section with eccentricity $e$. It is the locus of a point $P$ whose distance from the focus is $e$ times its distance from the directrix. This locus is an ellipse if $e < 1$, a parabola if $e = 1$, and a hyperbola if $e > 1$. We shall find that all three conic sections have the same general equation in polar coordinates.

To derive the polar equation of a conic section, suppose that its focus is the pole (origin) and that its directrix is the vertical line $x = -p$ (with $p > 0$). In the notation of Fig. 10.63, the fact that $|OP| = e|PQ|$ tells us that

$$r = e(p + r \cos \theta),$$

so that

$$r = \frac{pe}{1 - e \cos \theta}.$$

If the directrix is the line $x = +p$, then a similar computation (Problem 22) gives the same result, except that the denominator $1 - e \cos \theta$ is replaced by $1 + e \cos \theta$.

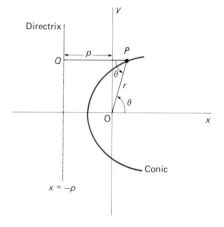

**10.63** A conic section: $|OP| = e|PQ|$.

---

**Polar Coordinates Equation of a Conic Section**

The polar equation of the conic section with eccentricity $e$, focus $O$, and directrix $x = \pm p$ is

$$r = \frac{pe}{1 \pm e \cos \theta}. \tag{1}$$

---

Here $p > 0$, and we take either the plus sign in both $x = \pm p$ and Equation (1), or the minus sign in both.

If the directrix of the conic section is the horizontal line $y = \pm p$, then a similar computation (Problem 23) yields the polar coordinates equation

$$r = \frac{pe}{1 \pm e \sin \theta}. \tag{2}$$

Now consider an ellipse given by Equation (1) for some choice of $e < 1$. Its vertices correspond to $\theta = 0$ and $\theta = \pi$. Hence it follows with the aid of Fig. 10.64 that the length $2a$ of its major axis is

$$2a = \frac{pe}{1 - e} + \frac{pe}{1 + e} = \frac{2pe}{1 - e^2}.$$

**10.64** Find the major axis of an ellipse given in polar form.

$$r_0 = \frac{pe}{1 - e},$$

$$r_1 = \frac{pe}{1 + e}.$$

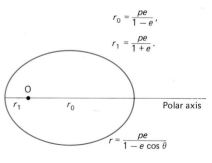

$$r = \frac{pe}{1 - e \cos \theta}$$

From this equation we obtain the important relation

$$pe = a(1 - e^2). \qquad (3)$$

**EXAMPLE 1**  Sketch the graph of the equation

$$r = \frac{16}{5 - 3\cos\theta}.$$

*Solution*  First we divide numerator and denominator by 5 and find that

$$r = \frac{\frac{16}{5}}{1 - \frac{3}{5}\cos\theta}.$$

Thus $e = \frac{3}{5}$ and $pe = \frac{16}{5}$. Equation (3) then implies that $a = 5$. Finally, $c = ae = 3$ and

$$b = \sqrt{a^2 - c^2} = 4.$$

So we have here an ellipse with major semiaxis $a = 5$, minor semiaxis $b = 4$, and center at $(3, 0)$ in rectangular coordinates. The ellipse is shown in Fig. 10.65.

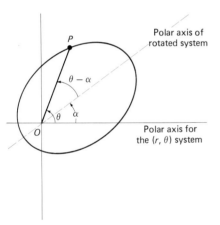

**10.65**  The ellipse of Example 1

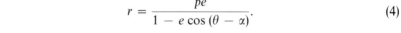

**10.66**  An ellipse rotated about one focus through the angle $\alpha$

In contrast with the rectangular coordinates situation discussed in Section 10-7, the polar coordinates equation of a rotated conic section is quite simple. Figure 10.66 shows an ellipse rotated through a counterclockwise angle $\alpha$ about its focus at the origin $O$. In the rotated system with polar coordinates $(r', \theta')$, the ellipse has equation

$$r' = \frac{pe}{1 - e\cos\theta'}.$$

Since $r' = r$ and $\theta' = \theta - \alpha$, the equation of the ellipse in the original (unrotated) polar coordinate system is

$$r = \frac{pe}{1 - e\cos(\theta - \alpha)}. \qquad (4)$$

With $e \geq 1$, Equation (4) represents a parabola or a hyperbola rotated through the angle $\alpha$.

If we combine Equations (1) and (3), we find that the equation of an ellipse with eccentricity $e$, major semiaxis $a$, and focus at the origin is

$$r = \frac{a(1 - e^2)}{1 \pm e\cos\theta}. \qquad (5)$$

The limiting form of this equation as $e \to 0$ is the equation $r = a$ of a circle. Since $p \to \infty$ as $e \to 0$ in Equation (3), we may thus regard any circle as an ellipse of eccentricity zero and with directrix $x = \pm\infty$.

If we begin with Equation (1) and let $e \to 1$, the limiting form of the equation is the equation of a parabola:

$$r = \frac{p}{1 \pm \cos\theta}.$$

Let us take $p = 1$ and compare values of $r$ for the parabola

$$r_{par} = \frac{1}{1 - \cos \theta}$$

and the ellipse

$$r_{ell} = \frac{0.999}{1 - (0.999) \cos \theta}.$$

The ellipse has eccentricity $e = 0.999$ and should be, in some sense, very like a parabola. The table in Fig. 10.67 shows the results of our comparison. The parabola and the ellipse almost coincide for $\theta > 30°$. This sort of approximation of an ellipse by a parabola is useful in studying comets with highly eccentric elliptical orbits. The way in which the two orbits would seem to coincide is illustrated in Fig. 10.68.

| $\theta$ (degrees) | $r_{par}$ | $r_{ell}$ |
|---|---|---|
| 180 | 0.500 | 0.500 |
| 135 | 0.586 | 0.585 |
| 90 | 1.000 | 0.999 |
| 60 | 2.000 | 1.998 |
| 45 | 3.414 | 3.403 |
| 30 | 7.464 | 7.409 |
| 0 | $\infty$ | 999.000 |

[data rounded]

10.67  Comparison between a parabola and an ellipse with eccentricity $e = 0.999$

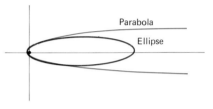

10.68  An eccentric ellipse resembles a parabola near their common focus.

**EXAMPLE 2**  A certain comet is known to have a highly eccentric elliptical orbit with the sun at one focus. Its position is described in polar coordinates, as shown in Fig. 10.69; the following measurements—made by radar—refer to the same figure:

$$r = 6 \text{ A.U.} \qquad \text{when } \theta = 60°,$$

$$r = 2 \text{ A.U.} \qquad \text{when } \theta = 90°.$$

Estimate the position of the comet at its point of closest approach to the sun.

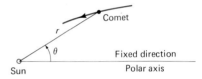

10.69  The comet of Example 2

**Solution**  First we write the equation of the orbit of the comet in the polar form suggested by Fig. 10.69. Because the orbit is highly eccentric ($e \approx 1$), we assume that near the sun it is approximately the parabola with polar equation

$$r = \frac{p}{1 - \cos(\theta - \alpha)}. \qquad (6)$$

This parabola has its axis rotated through the angle $\alpha$ from the polar axis.

We know that the vertex of a parabola is its point closest to the focus. Hence the minimum distance of the comet from the sun will be $r = p/2$ when $\theta = \pi + \alpha$. Our problem, then, is to determine $p$ and $\alpha$.

Substitution of the given data—$r = 6$ when $\theta = \pi/3$ and $r = 2$ when $\theta = \pi/2$—into Equation (6) gives us the two equations

$$6 = \frac{p}{1 - \cos(\pi/3 - \alpha)} \quad \text{and} \quad 2 = \frac{p}{1 - \cos(\pi/2 - \alpha)}.$$

We divide the first of these equations by the second. This eliminates $p$, and we obtain

$$3 = \frac{1 - \cos(\pi/2 - \alpha)}{1 - \cos(\pi/3 - \alpha)} = \frac{1 - \sin \alpha}{1 - \frac{1}{2} \cos \alpha - \frac{1}{2}\sqrt{3} \sin \alpha}.$$

We simplify this last equation and find that $\alpha$ is a root of the equation

$$f(\alpha) = (1.598) \sin \alpha + (1.5) \cos \alpha - 2 = 0.$$

We can approximate the value of $\alpha$ by using the iterative formula of Newton's method:

$$\alpha_{n+1} = \alpha_n - \frac{f(\alpha_n)}{f'(\alpha_n)}$$

$$= \alpha_n - \frac{(1.598)\sin\alpha_n + (1.5)\cos\alpha_n - 2}{(1.598)\cos\alpha_n - (1.5)\sin\alpha_n}.$$

The data of the problem suggest that $\alpha$ is a first-quadrant angle, so we use $\alpha_1 = 0.5$ (rad) as our first approximation. This leads to the sequence

$$\alpha_2 = 0.38, \qquad \alpha_3 = 0.39, \qquad \alpha_4 = \alpha_5 = 0.40,$$

and so $\alpha$ is about 0.40 rad, or about 23°. Then

$$p = 6\left[1 - \cos\left(\frac{\pi}{3} - 0.40\right)\right] \approx 1.23$$

astronomical units. The closest approach of the comet to the sun is thus approximately

$$\tfrac{1}{2}(1.23)(93) \approx 57.2$$

million miles, since 1 A.U. is about 93 million miles.

## 10-8 PROBLEMS

In each of Problems 1–10, identify and sketch the conic section with the given equation.

**1** $r = \dfrac{6}{1 + \cos\theta}$

**2** $r = \dfrac{6}{1 + 2\cos\theta}$

**3** $r = \dfrac{3}{1 - \sin\theta}$

**4** $r = \dfrac{8}{8 - 2\cos\theta}$

**5** $r = \dfrac{6}{2 - \sin\theta}$

**6** $r = \dfrac{12}{3 + 2\cos\theta}$

**7** $r = \dfrac{12}{2 - 3\cos\theta}$

**8** $r = \dfrac{12}{2 - 3\cos(\theta - \pi/3)}$

**9** $r = \dfrac{2}{1 - \sin\theta - \cos\theta}$

**10** $r = \dfrac{4}{2 - \sqrt{3}\cos\theta + \sin\theta}$

In each of Problems 11–15, find a polar equation of the given conic section with focus at the origin.

**11** Eccentricity 1, directrix $r = 3\sec\theta$

**12** Eccentricity 2, directrix $r = -4\sec\theta$

**13** Eccentricity $\frac{1}{3}$, directrix $r = 2\csc\theta$

**14** Eccentricity $\frac{2}{3}$, directrix $r = -3\csc\theta$

**15** Eccentricity 1, directrix $r(\cos\theta + \sin\theta) = 3$

In each of Problems 16–20, find an equation in rectangular coordinates for the graph of the indicated problem above.

**16** Problem 11

**17** Problem 12

**18** Problem 13

**19** Problem 14

**20** Problem 15

**21** Identify and sketch the graph of the equation

$$r = \frac{4}{2 - \sqrt{2}(\cos\theta + \sin\theta)}.$$

(*Suggestion:* Find $A$ and $\alpha$ so that

$$\cos\theta + \sin\theta = A\cos(\theta - \alpha).$$

Then refer to Equation (4).)

**22** Derive Equation (1) with the plus sign.

**23** Derive Equation (2) with the minus sign.

**24** A comet has a parabolic orbit with the sun at one focus. When the comet is 150 million miles from the sun, the sun-comet line makes an angle of 45° with the axis of the parabola. What will be the minimum distance between the comet and the sun?

**25** A satellite has an elliptical orbit with the center of the earth (take its radius to be 4000 mi) as one focus. The lowest point of its orbit is 500 mi above the North Pole, and the highest point is 5000 mi above the South Pole. What is the height of the satellite above the surface of the earth when the satellite crosses the equator?

**26** Find the closest approach to the sun of a comet as in Example 2 of this section; assume that $r = 2.5$ A.U. when $\theta = 45°$ and that $r = 1$ A.U. when $\theta = 90°$.

**27** An ellipse has semimajor axis $a$ and semiminor axis $b$. Derive the formula $A = \pi ab$ for its area by integration in polar coordinates.

**28** The orbit of a certain comet approaching the sun is the parabola $r = 1/(1 - \cos\theta)$. The units for $r$ are in as-tronomical units (recall that 1 A.U. is about 93 million miles). Suppose that it takes 15 days for the comet to travel from the position $\theta = 60°$ to the position $\theta = 90°$. How much longer will it require for the comet to reach its point of closest approach to the sun? Assume that the radius from the sun to the comet sweeps out area at a constant rate as the comet moves.

# CHAPTER 10 REVIEW:  Properties of Conic Sections

The parabola with focus $(p, 0)$ and directrix $x = -p$ has eccentricity $e = 1$ and equation $y^2 = 4px$. The table on the right compares the properties of an ellipse and a hyperbola, each with foci $(\pm c, 0)$ and major axis of length $2a$.

Use the list below as a guide to additional concepts that you may need to review.

1 Translation of coordinates
2 Equations and procedure for rotation of axes
3 Use of the discriminant to classify the graph of a second degree equation
4 The relationship between rectangular and polar coordinates
5 The graph of a polar coordinates equation
6 The equation of a conic section in polar coordinates
7 The area formula in polar coordinates

|  | *Ellipse* | *Hyperbola* |
|---|---|---|
| eccentricity | $e = \dfrac{c}{a} < 1$ | $e = \dfrac{c}{a} > 1$ |
| $a, b, c$ relation | $a^2 = b^2 + c^2$ | $c^2 = a^2 + b^2$ |
| equation | $\dfrac{x^2}{a^2} + \dfrac{y^2}{b^2} = 1$ | $\dfrac{x^2}{a^2} - \dfrac{y^2}{b^2} = 1$ |
| vertices | $(\pm a, 0)$ | $(\pm a, 0)$ |
| $y$-intercepts | $(0, \pm b)$ | none |
| directrices | $x = \pm\dfrac{a}{e}$ | $x = \pm\dfrac{a}{e}$ |
| asymptotes | none | $y = \pm\dfrac{b}{a}x$ |

# MISCELLANEOUS PROBLEMS

Sketch the graphs of the equations in Problems 1–30, labeling centers, foci, and vertices in the case of conic sections.

**1** $x^2 + y^2 - 2x + 2y = 2$
**2** $x^2 + y^2 = x + y$
**3** $x^2 + y^2 - 6x + 2y + 9 = 0$
**4** $y^2 = 4(x + y)$
**5** $x^2 = 8x - 2y - 20$
**6** $x^2 + 2y^2 - 2x + 8y + 8 = 0$
**7** $9x^2 + 4y^2 = 36x$
**8** $x^2 - y^2 = 2x - 2y - 1$
**9** $y^2 - 2x^2 = 4x + 2y + 3$
**10** $9y^2 - 4x^2 = 8x + 18y + 31$
**11** $x^2 + 2y^2 = 4x + 4y - 12$
**12** $x^2 + 2xy + y^2 + 1 = 0$
**13** $x^2 + 2xy - y^2 = 7$      **14** $xy + 8 = 0$
**15** $3x^2 - 2xy + 3y^2 = 4$    **16** $x^2 - 6xy + y^2 = 4$
**17** $9x^2 - 24xy + 16y^2 = 20x + 15y$
**18** $7x^2 + 48xy - 7y^2 + 25 = 0$
**19** $r = -2\cos\theta$      **20** $\cos\theta + \sin\theta = 0$.

**21** $r = \dfrac{1}{\sin\theta - \cos\theta}$      **22** $r\sin^2\theta = \cos\theta$.

**23** $r = 3\csc\theta$      **24** $r = 2(\cos\theta - 1)$
**25** $r^2 = 4\cos\theta$      **26** $r\theta = 1$

**27** $r = 3 - 2\sin\theta$      **28** $r = \dfrac{1}{1 + \cos\theta}$

**29** $r = \dfrac{4}{2 + \cos\theta}$      **30** $r = \dfrac{4}{1 - 2\cos\theta}$

In each of Problems 31–38, find the area of the region described.

**31** Inside both $r = 2\sin\theta$ and $r = 2\cos\theta$
**32** Inside $r^2 = 4\cos\theta$
**33** Inside $r = 3 - 2\sin\theta$ and outside $r = 4$
**34** Inside $r^2 = 2\sin 2\theta$ and outside $r = 2\sin\theta$
**35** Inside $r = 2\sin 2\theta$ and outside $r = \sqrt{2}$
**36** Inside $r = 3\cos\theta$ and outside $r = 1 + \cos\theta$
**37** Inside $r = 1 + \cos\theta$ and outside $r = \cos\theta$
**38** Between the loops of $r = 1 - 2\sin\theta$
**39** Find a polar coordinates equation of the circle that passes through the origin and is centered at the point with polar coordinates $(p, \alpha)$.
**40** Find the equation of the parabola with focus the origin and directrix the line $y = x + 4$. Recall from Chapter 3,

Miscellaneous Problem 65, that the distance from the point $(x, y)$ to the line $Ax + By + C = 0$ is

$$\frac{|Ax + By + C|}{\sqrt{A^2 + B^2}}.$$

**41** A *diameter* of an ellipse is a chord through its center. What are the maximum and minimum lengths of diameters of the ellipse with equation $x^2/a^2 + y^2/b^2 = 1$?

**42** Use calculus to prove that the ellipse $x^2/a^2 + y^2/b^2 = 1$ is normal to the coordinate axis at each of its four vertices.

**43** The parabolic arch of a bridge has base width $b$ and height $h$ at its center. Write its equation, choosing the origin on the ground at the left end of the arch.

**44** Use methods of calculus to find the points on the ellipse $x^2/a^2 + y^2/b^2 = 1$ that are nearest to and farthest from:
(a) The center $(0, 0)$;
(b) The focus $(c, 0)$.

**45** Consider a line segment $QR$ that contains a point $P$ such that $|QP| = a$ and $|PR| = b$. Suppose that $Q$ is constrained to move on the $y$-axis, while $R$ must remain on the $x$-axis. Show that the locus of $P$ is an ellipse.

**46** Suppose that $a > 0$ and that $F_1$ and $F_2$ are two fixed points in the plane with $|F_1F_2| > 2a$. Imagine a point $P$ that moves in such a way that $|PF_2| = 2a + |PF_1|$. Show that the locus of $P$ is one branch of a hyperbola with foci $F_1$ and $F_2$. Then—as a consequence—explain how to construct points on a hyperbola by drawing appropriate circles centered at its foci.

**47** Let $Q_1$ and $Q_2$ be two points on the parabola $y^2 = 4px$. Let $P$ be the point of the parabola at which the tangent line is parallel to $Q_1Q_2$. Show that the horizontal line through $P$ bisects the segment $Q_1Q_2$.

**48** Determine the locus of a point $P$ such that the product of its distances from the two fixed points $F_1(-a, 0)$ and $F_2(a, 0)$ is $a^2$.

**49** Find the eccentricity of the conic section

$$3x^2 - y^2 + 12x + 9 = 0$$

by writing its equation in polar coordinates.

**50** Find the area bounded by the loop of the *strophoid* $r = \sec \theta - 2 \cos \theta$ shown in Fig. 10.70.

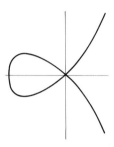

**10.70** Strophoid

**51** Find the area bounded by the loop of the *folium of Descartes* $x^3 + y^3 = 3xy$ shown in Fig. 3.28. (*Suggestion:* Change to polar coordinates, and then substitute $u = \tan \theta$ to evaluate the area integral.)

**52** Use the method of Problem 51 to find the area bounded by the first-quadrant loop (similar to the *folium*) of the curve $x^5 + y^5 = 5x^2y^2$.

**53** The asteroid Icarus has a period of revolution of 15 months in an elliptical orbit with the sun at one focus. Its distance from the sun ranges from 17 million to 183 million miles.
(a) Write the equation of its orbit in polar coordinates.
(b) Find how much time during one revolution Icarus spends in the portion of its orbit (nearest the sun) for which $-\pi/2 \leq \theta \leq \pi/2$. Assume Kepler's law of equal areas swept out in equal times by the sun-Icarus radius. (*Suggestion:* Use the substitution $u = \tan \theta/2$ of Section 9-8 to evaluate the integral $\int (1 + e \cos \theta)^{-2} \, d\theta$.)

# 11

# Indeterminate Forms, Taylor's Formula, and Improper Integrals

## Indeterminate Forms and L'Hôpital's Rule

An *indeterminate form* is a certain type of expression with a limit that is not evident by inspection. If

$$\lim_{x \to a} f(x) = 0 = \lim_{x \to a} g(x),$$

then we say that the quotient $f(x)/g(x)$ has the **indeterminate form** $0/0$ at $x = a$. For example, to differentiate the trigonometric functions (Section 8-3), we needed to know that

$$\lim_{x \to 0} \frac{\sin x}{x} = 1. \tag{1}$$

Here $f(x) = \sin x$ and $g(x) = x$. Because $\sin x$ and $x$ both approach zero as $x \to 0$, $(\sin x)/x$ has the indeterminate form $0/0$ at $x = 0$. Consequently, we had to use a special geometric argument to find the limit in (1)—see Theorem 2 of Section 8-2. Indeed, something of this sort happens whenever we compute a derivative, because the quotient

$$\frac{f(x) - f(a)}{x - a},$$

whose limit as $x \to a$ is the derivative $f'(a)$, has the indeterminate form $0/0$ at $x = a$.

Sometimes the limit of an indeterminate form can be found by a special algebraic manipulation or construction, as in our earlier computations of derivatives. Often, however, it is more convenient to apply a rule that appeared in the first calculus textbook ever published, in 1696, by the Marquis de l'Hôpital. L'Hôpital was a French nobleman who had hired the Swiss mathematician John Bernoulli as his calculus tutor, and "l'Hôpital's rule" is actually due to Bernoulli.

---

**Theorem 1** *L'Hôpital's Rule*

Suppose that the functions $f$ and $g$ are differentiable in a deleted neighborhood of the point $a$, and that $g'(x)$ is nonzero in that neighborhood. Suppose also that

$$\lim_{x \to a} f(x) = 0 = \lim_{x \to a} g(x).$$

Then

$$\lim_{x \to a} \frac{f(x)}{g(x)} = \lim_{x \to a} \frac{f'(x)}{g'(x)} \tag{2}$$

provided that the limit on the right either exists (as a finite number) or is $\pm\infty$.

---

In essence, the rule says that if $f(x)/g(x)$ has the indeterminate form $0/0$ at $x = a$, then—subject to a few mild restrictions—this quotient has the same limit at $x = a$ as does the quotient $f'(x)/g'(x)$ of *derivatives*. The proof appears at the end of this section.

**EXAMPLE 1** Find $\lim\limits_{x \to 0} \dfrac{e^x - 1}{\sin 2x}$.

*Solution* The above fraction has the indeterminate form $0/0$ at $x = 0$. The numerator and denominator are clearly differentiable in a deleted neighborhood of $x = 0$, and the derivative of the denominator is certainly nonzero if the neighborhood is small enough (if $0 < |x| < \pi/4$). So l'Hôpital's rule applies, and

$$\lim_{x \to 0} \frac{e^x - 1}{\sin 2x} = \lim_{x \to 0} \frac{e^x}{2 \cos 2x} = \frac{1}{2}.$$

If it turns out that the quotient $f'(x)/g'(x)$ is again indeterminate, then l'Hôpital's rule can be applied a second (or third, or ...) time, as in the following example. When the rule is applied repeatedly, however, the conditions for its applicability must be checked each time.

**EXAMPLE 2** Find $\lim\limits_{x \to 1} \dfrac{1 - x + \ln x}{1 + \cos \pi x}$.

*Solution*

$$\lim_{x \to 1} \frac{1 - x + \ln x}{1 + \cos \pi x}$$

$$= \lim_{x \to 1} \frac{-1 + 1/x}{-\pi \sin \pi x} \qquad \text{(still of the form 0/0)}$$

$$= \lim_{x \to 1} \frac{x - 1}{\pi x \sin \pi x} \qquad \text{(algebraic simplification)}$$

$$= \lim_{x \to 1} \frac{1}{\pi \sin \pi x + \pi^2 x \cos \pi x} \qquad \text{(l'Hôpital's rule again)}$$

$$= -\frac{1}{\pi^2} \qquad \text{(by inspection).}$$

Because the final limit exists, so do the previous ones; the existence of the right-hand limit in (2) implies the existence of the left-hand limit.

When you need to apply l'Hôpital's rule repeatedly in this way, you need only keep differentiating the numerator and denominator separately until at least one of them has a nonzero finite limit. At that point you can recognize the limit of the quotient by inspection, as in the final step of Example 2.

**EXAMPLE 3** Find $\lim\limits_{x \to 0} \dfrac{\sin x}{x + x^2}$.

***Solution***  If we simply applied l'Hôpital's rule twice in succession, the result would be the *incorrect* computation

$$\lim_{x \to 0} \frac{\sin x}{x + x^2} = \lim_{x \to 0} \frac{\cos x}{1 + 2x}$$

$$= \lim_{x \to 0} \frac{-\sin x}{2} = 0. \qquad \textbf{(WRONG!)}$$

The reason this answer is wrong is that $(\cos x)/(1 + 2x)$ is *not* an indeterminate form, so l'Hôpital's rule does not apply to it. The correct computation is

$$\lim_{x \to 0} \frac{\sin x}{x + x^2} = \lim_{x \to 0} \frac{\cos x}{1 + 2x}$$

$$= \frac{\lim\limits_{x \to 0} \cos x}{\lim\limits_{x \to 0} (1 + 2x)} = \frac{1}{1} = 1.$$

---

The point of Example 3 is to issue a warning: Verify the hypotheses of l'Hôpital's rule *before* applying it. It is an oversimplification to say that l'Hôpital's rule "works when you need it and doesn't work when you don't," but there is still much truth in this statement.

### INDETERMINATE FORMS INVOLVING ∞

L'Hôpital's rule has several variations. In addition to the fact that the limit in Equation (2) is allowed to be infinite, the real number $a$ in l'Hôpital's rule may be replaced by either $+\infty$ or $-\infty$. For example,

$$\lim_{x \to \infty} \frac{f(x)}{g(x)} = \lim_{x \to \infty} \frac{f'(x)}{g'(x)} \qquad (3)$$

provided the other hypotheses are satisfied. In particular, to use Equation (3) we must first verify that

$$\lim_{x \to \infty} f(x) = 0 = \lim_{x \to \infty} g(x)$$

and that the right-hand limit in Equation (3) exists. The proof of this version of l'Hôpital's rule is outlined in Problem 50.

L'Hôpital's rule may also be used when $f(x)/g(x)$ has the **indeterminate form** $\infty/\infty$. This means that

$$\lim_{x \to a} f(x) \quad \text{is either } +\infty \text{ or } -\infty,$$

and that

$$\lim_{x \to a} g(x) \quad \text{is either } +\infty \text{ or } -\infty.$$

The proof of this extension of the rule is more difficult, and will be omitted. For a proof, see, for example, A. E. Taylor and W. R. Mann, *Advanced Calculus*, third edition (New York: John Wiley, 1983), p. 107.

**EXAMPLE 4** Find $\displaystyle\lim_{x \to \infty} \frac{e^x}{x^2 + x}$.

***Solution*** The quotients $e^x/(x^2 + x)$ and $e^x/(2x + 1)$ each have the indeterminate form $\infty/\infty$ as $x \to \infty$, so two applications of l'Hôpital's rule yield

$$\lim_{x \to \infty} \frac{e^x}{x^2 + x} = \lim_{x \to \infty} \frac{e^x}{2x + 1} = \lim_{x \to \infty} \frac{e^x}{2} = +\infty.$$

Remember that l'Hôpital's rule "allows" the final result to be an infinite limit.

**EXAMPLE 5**

$$\lim_{x \to \infty} \frac{x^{1/3}}{\ln x} = \lim_{x \to \infty} \frac{\frac{1}{3}x^{-2/3}}{1/x} = \lim_{x \to \infty} \frac{1}{3}x^{1/3} = +\infty.$$

## CAUCHY'S MEAN VALUE THEOREM

In order to prove l'Hôpital's rule, we need a generalization of the mean value theorem due to the French mathematician Augustin-Louis Cauchy (1789–1857). He used it in the early nineteenth century to give rigorous proofs of several calculus results not previously established firmly.

---

**Theorem 2** *Cauchy's Mean Value Theorem*

Suppose that the functions $f$ and $g$ are continuous on the closed interval $[a, b]$ and differentiable on $(a, b)$. Then there exists a number $c$ in $(a, b)$ such that

$$[f(b) - f(a)]g'(c) = [g(b) - g(a)]f'(c). \tag{4}$$

---

REMARK 1 To see that this theorem is, indeed, a generalization of the (ordinary) mean value theorem, we take $g(x) = x$. Then $g'(x) = 1$, and the conclusion in (4) reduces to the fact that

$$f(b) - f(a) = (b - a)f'(c)$$

for some number $c$ in $(a, b)$.

REMARK 2 Equation (4) has a geometric interpretation like that of the ordinary mean value theorem. Let us think of the equations $x = g(t)$, $y = f(t)$ as describing the motion of a point $P(x, y)$ moving along a curve $C$ in the $xy$-plane (see Fig. 11.1) as $t$ increases from $a$ to $b$; $(g(t), f(t))$ is the location of the point $P$ at time $t$. Under the assumption that $g(a) \neq g(b)$, the slope of the line $L$ connecting the endpoints of the curve $C$ is

$$m = \frac{f(b) - f(a)}{g(b) - g(a)}. \tag{5}$$

On the other hand, if $g'(c) \neq 0$, then the chain rule gives

$$\frac{dy}{dx} = \frac{dy/dt}{dx/dt} = \frac{f'(c)}{g'(c)} \tag{6}$$

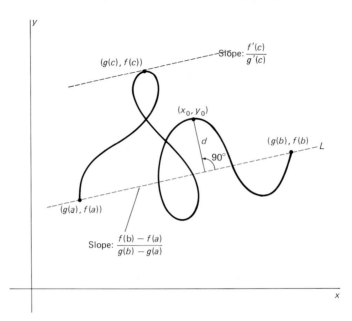

**11.1** The idea of Cauchy's mean value theorem

for the slope of the tangent line to the curve $C$ at the point $(g(c), f(c))$. But if $g(a) \neq g(b)$ and $g'(c) \neq 0$, then Equation (4) may be rewritten as

$$\frac{f(b) - f(a)}{g(b) - g(a)} = \frac{f'(c)}{g'(c)}, \tag{7}$$

so the two slopes in (5) and (6) are equal. Thus Cauchy's mean value theorem implies that (under our two assumptions) there is a point on the curve $C$ where the tangent line is *parallel* to the line joining the endpoints of $C$. Of course, this is exactly what the (ordinary) mean value theorem says for an explicitly defined curve $y = f(x)$. This geometric interpretation motivates the following proof of Cauchy's mean value theorem.

***Proof*** The line $L$ through the endpoints in Fig. 11.1 has point-slope equation

$$y - f(a) = \frac{f(b) - f(a)}{g(b) - g(a)} [x - g(a)],$$

which can be rewritten in the form $Ax + By + C = 0$ with

$$A = f(b) - f(a), \qquad B = -[g(b) - g(a)], \qquad \text{and}$$
$$C = f(a)[g(b) - g(a)] - g(a)[f(b) - f(a)]. \tag{8}$$

According to Miscellaneous Problem 65 at the end of Chapter 3, the (perpendicular) distance from the point $(x_0, y_0)$ to the line $L$ is

$$d = \frac{|Ax_0 + By_0 + C|}{\sqrt{A^2 + B^2}}.$$

Figure 11.1 suggests that the point $(g(c), f(c))$ will maximize this distance $d$ for points on the curve $C$.

We therefore are motivated to define the auxiliary function

$$\phi(t) = Ag(t) + Bf(t) + C, \tag{9}$$

CHAP. 11: Indeterminate Forms, Taylor's Formula, and Improper Integrals

Suppose that we want to calculate (or at least approximate closely) a specific value $f(x_0)$ of a given function $f$. It would suffice to find a polynomial $P(x)$ with a graph which is very close to that of $f$ on some interval containing $x_0$. For then we could use the value $P(x_0)$ as an approximation to the actual value of $f(x_0)$. Once we know how to find such an approximating polynomial $P(x)$, we can then ask how accurately $P(x_0)$ approximates the desired value $f(x_0)$.

The simplest example of polynomial approximation is the linear approximation

$$f(x) \approx f(a) + f'(a)(x - a)$$

obtained by writing $\Delta x = x - a$ in Equation (3) of Section 4-2. The graph of the first degree polynomial

$$P_1(x) = f(a) + f'(a)(x - a) \qquad (1)$$

is the tangent line to the curve $y = f(x)$ at the point $(a, f(a))$; see Fig. 11.3. Note that this first degree polynomial agrees with $f$ and with its first derivative at $x = a$; that is, $P_1(a) = f(a)$ and $P_1'(a) = f'(a)$.

For example, suppose that $f(x) = \ln x$ and that $a = 1$. Then $f(1) = 0$ and $f'(1) = 1$, so $P_1(x) = x - 1$. Hence we expect that $\ln x \approx x - 1$ for $x$ near 1. With $x = 1.1$, we find that

$$P_1(1.1) = 0.100000 \quad \text{while} \quad \ln(1.1) \approx 0.095310 \qquad \text{(rounded)}.$$

The error in this approximation is about 5%.

To better approximate $\ln x$ near $x = 1$, let us look for a second degree polynomial

$$P_2(x) = c_0 + c_1 x + c_2 x^2$$

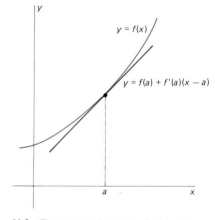

**11.3** The tangent line at $(a, f(a))$ is the best linear approximation to $f$ at $a$.

that has not only the same value and the same first derivative as does $f$ at $x = 1$ but also has the same second derivative there: $P_2''(1) = f''(1) = -1$. To satisfy these conditions, we must have

$$P_2(1) = c_2 + c_1 + c_0 = 0,$$
$$P_2'(1) = 2c_2 + c_1 = 1,$$
$$P_2''(1) = 2c_2 = -1.$$

When we solve these equations we find that $c_0 = -\frac{3}{2}$, $c_1 = 2$, and $c_2 = -\frac{1}{2}$, so

$$P_2(x) = -\tfrac{1}{2}x^2 + 2x - \tfrac{3}{2}.$$

With $x = 1.1$ we find that $P_2(1.1) = 0.095000$, which is accurate to three decimal places because $\ln(1.1) \approx 0.095310$. The graph of $y = -\frac{1}{2}x^2 + 2x - \frac{3}{2}$ is a parabola through $(1, 0)$ with the same slope *and curvature* there as $y = \ln x$; see Fig. 11.4.

The tangent line and the parabola used in these computations illustrate the general approach to polynomial approximation. In order to approximate the function $f(x)$ near $x = a$, we look for an $n$th-degree polynomial

$$P_n(x) = c_0 + c_1 x + c_2 x^2 + \cdots + c_n x^n$$

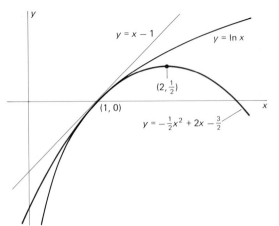

**11.4** The linear and parabolic approximations to $y = \ln x$ at the point $(1, 0)$

such that its value at $a$ and the values of its first $n$ derivatives at $a$ agree with the corresponding values of $f$. That is, we require that

$$P_n(a) = f(a),$$
$$P_n'(a) = f'(a),$$
$$P_n''(a) = f''(a), \qquad (2)$$
$$\vdots$$
$$P_n^{(n)}(a) = f^{(n)}(a).$$

These $n + 1$ conditions can be used to determine the values of the $n + 1$ coefficients $c_0, c_1, \ldots, c_n$.

The algebra involved is much simpler, however, if we begin with $P_n(x)$ expressed as an $n$th degree polynomial in powers of $x - a$ rather than in powers of $x$:

$$P_n(x) = b_0 + b_1(x - a) + b_2(x - a)^2 + \cdots + b_n(x - a)^n. \qquad (3)$$

Then substitution of $x = a$ in (3) yields $b_0 = P_n(a) = f(a)$ by the first condition in (2). Substitution of $x = a$ into

$$P_n'(x) = b_1 + 2b_2(x - a) + 3b_3(x - a)^2 + \cdots + nb_n(x - a)^{n-1}$$

yields $b_1 = P_n'(a) = f'(a)$ by the second condition in (2). Next, substitution of $x = a$ into

$$P_n''(x) = 2b_2 + 3 \cdot 2b_3(x - a) + \cdots + n(n - 1)b_n(x - a)^{n-2}$$

yields $2b_2 = P_n''(a) = f''(a)$, so that $b_2 = \frac{1}{2}f''(a)$.

We continue this process to find $b_3, b_4, \ldots, b_n$. In general, the constant term in the $k$th derivative $P_n^{(k)}(x)$ is $k!b_k$, since it is the $k$th derivative of the $k$th degree term $b_k(x - a)^k$ in $P_n(x)$:

$$P_n^{(k)}(x) = k!b_k + (\text{powers of } x - a).$$

So when we substitute $x = a$ into $P_n^{(k)}(x)$, we find that

$$k!b_k = P_n^{(k)}(a) = f^{(k)}(a),$$

and thus that

$$b_k = \frac{f^{(k)}(a)}{k!} \tag{4}$$

for $k = 1, 2, 3, \ldots, n$.

Indeed, Formula (4) also holds for $k = 0$ if we use the common convention that $0! = 1$ and agree that the zeroth derivative $g^{(0)}$ of the function $g$ is just $g$ itself. With such conventions, our computations establish the following theorem.

---

*Theorem 1*    *The nth Degree Taylor Polynomial*

Suppose that the first $n$ derivatives of the function $f(x)$ exist at $x = a$. Let $P_n(x)$ be the $n$th degree polynomial

$$P_n(x) = \sum_{k=0}^{n} \frac{f^{(k)}(a)}{k!} (x - a)^k \tag{5}$$

$$= f(a) + f'(a)(x - a)$$

$$+ \frac{f''(a)}{2!} (x - a)^2 + \cdots + \frac{f^{(n)}(a)}{n!} (x - a)^n.$$

Then the values of $P_n(x)$ and its first $n$ derivatives agree, at $x = a$, with the values of $f$ and its first $n$ derivatives there. That is, the equations in (2) hold.

---

The polynomial given in Equation (5) is called the **$n$th degree Taylor polynomial of the function $f$ at the point $x = a$.** Note that $P_n(x)$ is a polynomial in powers of $x - a$ rather than in powers of $x$. In order to use $P_n(x)$ effectively for the approximation of $f(x)$, for $x$ near $a$, we shall need to be able to compute the values of the derivatives $f(a), f'(a), f''(a)$, and so on, all the way to $f^{(n)}(a)$.

The line $y = P_1(x)$ is simply the tangent line to the curve $y = f(x)$ at the point $(a, f(a))$; thus $y = f(x)$ and $y = P_1(x)$ have the same slope at this point. Now recall from Section 4-6 that the second derivative $f''(a)$ measures the way the curve $y = f(x)$ is bending as it passes through $(a, f(a))$. Let us therefore call $f''(a)$ the "concavity" of $y = f(x)$ at $(a, f(a))$. Then, because $P_2''(a) = f''(a)$, we see that $y = P_2(x)$ has both the same slope and the same concavity at $(a, f(a))$ as does $y = f(x)$. The third degree Taylor polynomial $y = P_3(x)$ has not only the same slope and same concavity but [because $P_3'''(a) = f'''(a)$] also the same rate of change of concavity at $x = a$ as does $y = f(x)$. These observations suggest that the larger $n$ is, the more closely the $n$th degree Taylor polynomial $P_n(x)$ will approximate $f(x)$ for $x$ near $a$.

**EXAMPLE 1**    Find the $n$th degree Taylor polynomial of $f(x) = \ln x$ at $a = 1$.

*Solution*    The first few derivatives of $f(x) = \ln x$ are

$$f'(x) = \frac{1}{x}, \qquad f''(x) = -\frac{1}{x^2}, \qquad f'''(x) = \frac{2}{x^3},$$

$$f^{(4)}(x) = -\frac{3!}{x^4}, \qquad f^{(5)}(x) = \frac{4!}{x^5}.$$

The pattern should be clear:

$$f^{(k)}(x) = (-1)^{k-1}\frac{(k-1)!}{x^k} \qquad \text{for } k \geq 1.$$

Hence $f^{(k)}(1) = (-1)^{k-1}(k-1)!$, and so Equation (5) gives

$$P_n(x) = (x - 1) - \frac{1}{2}(x - 1)^2 + \frac{1}{3}(x - 1)^3 - \frac{1}{4}(x - 1)^4$$

$$+ \cdots + \frac{(-1)^{n-1}}{n}(x - 1)^n.$$

With $n = 2$ we obtain the quadratic polynomial

$$P_2(x) = (x - 1) - \tfrac{1}{2}(x - 1)^2 = -\tfrac{1}{2}x^2 + 2x - \tfrac{3}{2};$$

we saw its graph in Fig. 11.4. With the third degree Taylor polynomial

$$P_3(x) = (x - 1) - \tfrac{1}{2}(x - 1)^2 + \tfrac{1}{3}(x - 1)^3$$

we can go a step further in approximating $\ln(1.1)$—the value $P_3(1.1) = 0.095333\ldots$ is correct to four decimal places.

**EXAMPLE 2**  Find the $n$th degree Taylor polynomial for $f(x) = e^x$ at $a = 0$.

**Solution**  This is the easiest of all Taylor polynomials to compute because $f^{(k)}(x) = e^x$ for all $k \geq 0$. Hence $f^{(k)}(0) = 1$ for all $k \geq 0$, and so

$$P_n(x) = 1 + x + \frac{x^2}{2!} + \frac{x^3}{3!} + \cdots + \frac{x^n}{n!}.$$

The first few Taylor polynomials of the exponential function at $a = 0$ are, therefore,

$$P_0(x) = 1,$$
$$P_1(x) = 1 + x,$$
$$P_2(x) = 1 + x + \tfrac{1}{2}x^2,$$
$$P_3(x) = 1 + x + \tfrac{1}{2}x^2 + \tfrac{1}{6}x^3,$$
$$P_4(x) = 1 + x + \tfrac{1}{2}x^2 + \tfrac{1}{6}x^3 + \tfrac{1}{24}x^4,$$
$$P_5(x) = 1 + x + \tfrac{1}{2}x^2 + \tfrac{1}{6}x^3 + \tfrac{1}{24}x^4 + \tfrac{1}{120}x^5.$$

The table of Fig. 11.5 shows how these polynomials approximate $f(x) = e^x$ for $x = 0.1$ and for $x = 0.5$. Note that—at least for these two values of $x$—the closer $x$ is to $a = 0$, the more rapidly $P_n(x)$ appears to approach $f(x)$ as $n$ increases.

The closeness with which $P_n(x)$ approximates $f(x)$ is measured by the difference

$$R_n(x) = f(x) - P_n(x),$$

for which

$$f(x) = P_n(x) + R_n(x). \tag{6}$$

This difference $R_n(x)$ is called the **$n$th degree remainder for $f(x)$ at $x = a$.** It is the *error* made if the value $f(x)$ is replaced by the approximation $P_n(x)$.

| $n$ | $P_n(x)$ | $e^x$ | $e^x - P_n(x)$ |
|---|---|---|---|
| 0 | 1.000000 | 1.105171 | 0.105171 |
| 1 | 1.100000 | 1.105171 | 0.005171 |
| 2 | 1.105000 | 1.105171 | 0.000171 |
| 3 | 1.105167 | 1.105171 | 0.000004 |
| 4 | 1.105171 | 1.105171 | 0.000000085 |
| 5 | 1.105171 | 1.105171 | 0.000000001 |

x = 0.5

| $n$ | $P_n(x)$ | $e^x$ | $e^x - P_n(x)$ |
|---|---|---|---|
| 0 | 1.000000 | 1.648721 | 0.648721 |
| 1 | 1.500000 | 1.648721 | 0.148721 |
| 2 | 1.625000 | 1.648721 | 0.023721 |
| 3 | 1.645833 | 1.648721 | 0.002888 |
| 4 | 1.648438 | 1.648721 | 0.000284 |
| 5 | 1.648698 | 1.648721 | 0.000023 |

**11.5** Approximating $y = e^x$ with Taylor polynomials at $a = 0$

[all data rounded]

The theorem that lets us estimate the error or remainder $R_n(x)$ is called **Taylor's formula,** after Brook Taylor, a follower of Newton, who introduced "Taylor polynomials" in an article published in 1715. The particular expression for $R_n(x)$ that we give next is called the *Lagrange form* for the remainder because it first appeared in 1797 in a book written by the French mathematician Joseph Louis Lagrange (1736–1813).

---

**Theorem 2   Taylor's Formula**

Suppose that the $(n + 1)$st derivative of the function $f$ exists on an interval containing the points $a$ and $b$. Then

$$f(b) = f(a) + f'(a)(b - a) + \frac{f''(a)}{2!}(b - a)^2 + \frac{f^{(3)}(a)}{3!}(b - a)^3$$

$$+ \cdots + \frac{f^{(n)}(a)}{n!}(b - a)^n + \frac{f^{(n+1)}(z)}{(n + 1)!}(b - a)^{n+1} \qquad (7)$$

for some number $z$ between $a$ and $b$.

---

The proof of Taylor's formula is given at the end of this section. If we write $b = x$ in (7), we get the $n$th degree **Taylor's formula with remainder** at $x = a$,

$$f(x) = f(a) + f'(a)(x - a) + \frac{f''(a)}{2!}(x - a)^2 + \frac{f^{(3)}(a)}{3!}(x - a)^3$$

$$+ \cdots + \frac{f^{(n)}(a)}{n!}(x - a)^n + \frac{f^{(n+1)}(z)}{(n + 1)!}(x - a)^{n+1} \qquad (8)$$

where $z$ is some number between $a$ and $x$. Thus the $n$th degree remainder term is

$$R_n(x) = \frac{f^{(n+1)}(z)}{(n + 1)!}(x - a)^{n+1}, \qquad (9)$$

which is easy to remember—it's the same as the *last* term of $P_{n+1}(x)$, except that $f^{(n+1)}(a)$ is replaced by $f^{(n+1)}(z)$.

Ordinarily the exact value of $z$ is unknown. One effective way to skirt this difficulty is to estimate $f^{(n+1)}(z)$; what we usually seek is an overestimate, a number $M$ such that

$$\left| f^{(n+1)}(z) \right| \leq M$$

for all $z$ between $a$ and $x$. If we can find such a number $M$, then

$$\left| f(x) - P_n(x) \right| = \left| R_n(x) \right| \leq \frac{M|x-a|^{n+1}}{(n+1)!}. \tag{10}$$

The fact that $(n+1)!$ grows very rapidly as $n$ increases is often helpful in showing that $|R_n(x)|$ is small.

For a particular $x$, we may be able to show that $\lim\limits_{n \to \infty} R_n(x) = 0$. It will then follow that

$$f(x) = \lim_{n \to \infty} P_n(x).$$

In such a case, we can then approximate $f(x)$ with *any* desired degree of accuracy simply by choosing $n$ sufficiently large.

**EXAMPLE 3** Estimate the accuracy of the approximation $\ln(1.1) \approx 0.095333\ldots$ obtained in Example 1.

*Solution* Recall that if $f(x) = \ln x$, then

$$f^{(k)}(x) = (-1)^{k-1} \frac{(k-1)!}{x^k},$$

so that

$$f^{(k)}(1) = (-1)^{k-1}(k-1)!.$$

Hence the third degree Taylor formula *with remainder* at $a = 1$ is

$$\ln x = (x-1) - \frac{1}{2}(x-1)^2 + \frac{1}{3}(x-1)^3 + \frac{(-1)^3 3!}{4! z^4}(x-1)^4$$

for some $z$ between $a = 1$ and $x$. With $x = 1.1$, this gives

$$\ln(1.1) = 0.095333 - \frac{0.0001}{4z^4}$$

with $z$ between $a = 1$ and $x = 1.1$. The largest possible numerical value of the remainder term is obtained with $z = 1$,

$$\frac{0.0001}{4(1)^4} = 0.000025.$$

It follows that

$$0.095308 < \ln(1.1) < 0.095333,$$

so we can say that $\ln(1.1) = 0.0953$ to four-place accuracy.

**EXAMPLE 4** Use a Taylor polynomial for $f(x) = e^x$ at $a = 0$ (as in Example 2) to approximate the number $e$ to five-place accuracy.

**Solution** Since $f^{(k)}(x) = e^x$ for all $k$, the $n$th degree Taylor polynomial with remainder is

$$e^x = 1 + x + \frac{x^2}{2!} + \frac{x^3}{3!} + \cdots + \frac{x^n}{n!} + \frac{e^z}{(n+1)!} x^{n+1}$$

for some $z$ between $a = 0$ and $x$. With $x = 1$ we find that

$$e = 2 + \frac{1}{2!} + \frac{1}{3!} + \cdots + \frac{1}{n!} + \frac{e^z}{(n+1)!}$$

with $z$ between 0 and 1. Thus $e^0 < e^z < e^1$. Because we already know (Problem 68 in Section 7-2) that $e < 3$, it follows that $1 < e^z < 3$. We can therefore achieve at least five-place accuracy by choosing $n$ so large that $3/(n+1)! < 0.000005$; that is, so that

$$(n+1)! > \frac{3}{0.000005} = 600,000.$$

With either a table of factorials or a calculator, we find that the first factorial greater than 600,000 is $10! = 3,628,800$. We therefore take $n = 9$ and find that

$$e = 2 + \frac{1}{2!} + \frac{1}{3!} + \cdots + \frac{1}{9!} + R_9 = 2.7182815 + R_9$$

where

$$0 < R_9 = \frac{e^z}{10!} < \frac{3}{10!} = 0.000000827.$$

Thus $2.7182815 < e < 2.7182824$, and so the familiar $e = 2.71828$ is indeed the correct value rounded to five places.

---

Because of the factor $(x-a)^{n+1}$ in the remainder term $R_n(x)$, we see that (with $n$ fixed) the closer $x$ is to $a$, the better $P_n(x)$ approximates $f(x)$. To approximate $f(x)$ with the least labor, we naturally choose $a$ as the nearest point to $x$ at which we already know the value of $f$ and the required derivatives of $f$. For example, to compute the sine of $5°$ (which is $\pi/36$ in radians), we choose $a = 0$. But to compute the sine of $50°$ ($5\pi/18$ radians), it's better to use $a = \pi/4$.

**EXAMPLE 5** Show that the values of $\sin x$ for angles between $40°$ and $50°$ can be computed with three-place accuracy by using the approximation

$$\sin x \approx \frac{1}{\sqrt{2}} \left[ 1 + \left( x - \frac{\pi}{4} \right) - \frac{1}{2} \left( x - \frac{\pi}{4} \right)^2 \right].$$

**Solution** Let $f(x) = \sin x$. Since $f'(x) = \cos x$, $f''(x) = -\sin x$, $f^{(3)}(x) = -\cos x$, the second degree Taylor polynomial with remainder for $\sin x$ at $a = \pi/4$ is

$$\sin x = \sin \frac{\pi}{4} + \left[ \cos \frac{\pi}{4} \right] \left( x - \frac{\pi}{4} \right) - \frac{1}{2!} \left[ \sin \frac{\pi}{4} \right] \left( x - \frac{\pi}{4} \right)^2 + R_2(x)$$

$$= \frac{1}{\sqrt{2}} \left[ 1 + \left( x - \frac{\pi}{4} \right) - \frac{1}{2} \left( x - \frac{\pi}{4} \right)^2 \right] + R_2(x)$$

where

$$R_2(x) = -\frac{\cos z}{3!}\left(x - \frac{\pi}{4}\right)^3$$

for some number $z$ between $\pi/4$ and $x$. Now $|\cos z| \leq 1$ for any $z$, and

$$\left|x - \frac{\pi}{4}\right| \leq \frac{\pi}{36} < 0.1$$

for the angles in question. Hence

$$|R_2(x)| < \frac{(0.1)^3}{6} \approx 0.000167 < 0.0002.$$

Thus the given polynomial will indeed give three-place accuracy, exactly as desired. For example,

$$\sin 50° = \sin\left(\frac{\pi}{4} + \frac{\pi}{36}\right)$$

$$\approx \frac{1}{\sqrt{2}}\left[1 + \frac{\pi}{36} - \frac{1}{2}\left(\frac{\pi}{36}\right)^2\right] = 0.766$$

(to three-place accuracy).

### PROOF OF TAYLOR'S FORMULA

Several different proofs of Taylor's formula are known, but none of them seems very well motivated—each requires some "trick" to begin the proof. The trick we employ here is to begin by introducing *two* auxiliary functions $F(x)$ and $G(x)$ defined as follows:

$$F(x) = f(b) - f(x) - f'(x)(b - x) - \cdots - \frac{f^{(n)}(x)}{n!}(b - x)^n \qquad (11)$$

and

$$G(x) = \frac{(b - x)^{n+1}}{(n + 1)!}. \qquad (12)$$

Note first that $F(b) = G(b) = 0$. When we calculate $F'(x)$, we get

$$F'(x) = -f'(x) + \left[f'(x) - f^{(2)}(x)(b - x)\right]$$

$$+ \left[f^{(2)}(x)(b - x) - \frac{1}{2!}f^{(3)}(x)(b - x)^2\right]$$

$$+ \left[\frac{1}{2!}f^{(3)}(x)(b - x)^2 - \frac{1}{3!}f^{(4)}(x)(b - x)^3\right]$$

$$+ \cdots$$

$$+ \left[\frac{1}{(n - 1)!}f^{(n)}(x)(b - x)^{n-1} - \frac{1}{n!}f^{(n+1)}(x)(b - x)^n\right].$$

Upon careful examination of this result, we see that all terms but the final one cancel in pairs, so that

$$F'(x) = -\frac{f^{(n+1)}(x)}{n!}(b - x)^n. \qquad (13)$$

Differentiation in (12) immediately yields

$$G'(x) = -\frac{(b-x)^n}{n!}.\qquad(14)$$

Now we apply Cauchy's mean value theorem (Theorem 2 in Section 11-1); it tells us that

$$[F(b) - F(a)]G'(z) = [G(b) - G(a)]F'(z)\qquad(15)$$

for some $z$ in $(a, b)$. Because $F(b) = G(b) = 0$, while $G'(z) \neq 0$ by (14), it follows from Equation (15) that

$$F(a) = \frac{F'(z)}{G'(z)}\,G(a) = \frac{-f^{(n+1)}(z)(b-z)^n/n!}{-(b-z)^n/n!} \cdot \frac{(b-a)^{n+1}}{(n+1)!};$$

that is,

$$F(a) = \frac{f^{(n+1)}(z)}{(n+1)!}(b-a)^{n+1}.\qquad(16)$$

But substitution of $x = a$ in Equation (11) yields

$$F(a) = f(b) - f(a) - f'(a)(b-a) - \cdots - \frac{f^{(n)}(a)}{n!}(b-a)^n.\qquad(17)$$

Finally, we get Taylor's formula (Eq. (7)) when we equate the right-hand sides in (16) and (17). Note also that it makes no difference whether $a < b$ or $a > b$ because Cauchy's formula in (15) is unaffected by an interchange of $a$ and $b$. ∎

## 11-3 PROBLEMS

The case $a = 0$ of Taylor's formula with remainder is often called *Maclaurin's formula*, after the Scottish mathematician Colin Maclaurin, who used it as a basic tool in a calculus book he published in 1742. In Problems 1–10, find Maclaurin's formula with remainder for the given function and given value of $n$.

1  $f(x) = e^{-x};\quad n = 5$

2  $f(x) = \sin x;\quad n = 4$

3  $f(x) = \cos x;\quad n = 4$

4  $f(x) = \dfrac{1}{1-x};\quad n = 4$

5  $f(x) = \sqrt{1+x};\quad n = 3$

6  $f(x) = \ln(1+x);\quad n = 4$

7  $f(x) = \tan x;\quad n = 3$

8  $f(x) = \arctan x;\quad n = 2$

9  $f(x) = \sin^{-1}x;\quad n = 2$

10  $f(x) = x^3 - 3x^2 + 5x - 7;\quad n = 4$

In Problems 11–16, find the Taylor polynomial with remainder using the given values of $a$ and $n$.

11  $f(x) = e^x;\quad a = 1, n = 4$

12  $f(x) = \cos x;\quad a = \pi/4, n = 3$

13  $f(x) = \sin x;\quad a = \pi/6, n = 3$

14  $f(x) = \sqrt{x};\quad a = 100, n = 3$

15  $f(x) = 1/(x-4)^2;\quad a = 5, n = 5$

16  $f(x) = \tan x;\quad a = \pi/4, n = 4$

In Problems 17–20, determine the number of decimal places of accuracy the given approximation formula yields for $|x| \leq 0.1$.

17  $e^x \approx 1 + x + \frac{1}{2}x^2 + \frac{1}{6}x^3 + \frac{1}{24}x^4$

18  $\sin x \approx x - \dfrac{x^3}{6} + \dfrac{x^5}{120}$

19  $\ln(1+x) \approx x - \frac{1}{2}x^2 + \frac{1}{3}x^3 - \frac{1}{4}x^4$

20  $\sqrt{1+x} \approx 1 + \frac{1}{2}x - \frac{1}{8}x^2$

21  Show that the approximation in Problem 17 gives $e^x$ to within 0.001 if $|x| \leq 0.5$. Then compute $\sqrt[3]{e}$ accurate to two decimal places.

22  For what values of $x$ is the approximation $\sin x \approx x - \frac{1}{6}x^3$ accurate to five decimal places?

23  (a) Show that values of the cosine function for angles between 40° and 50° can be computed with five-place accuracy by means of the approximation

$$\cos x \approx \frac{\sqrt{2}}{2}\left[1 - \left(x - \frac{\pi}{4}\right) - \frac{1}{2}\left(x - \frac{\pi}{4}\right)^2 + \frac{1}{6}\left(x - \frac{\pi}{4}\right)^3\right].$$

(b) Show that this approximation yields eight-place accuracy for angles between $44°$ and $46°$.

In Problems 24–28, calculate the indicated number with the required accuracy using Taylor's formula for an appropriate function at the given value of $a$.

24 $\sin 10°$; $\quad a = 0$, four decimal places
25 $\cos 35°$; $\quad a = \pi/6$, four decimal places
26 $e^{1/4}$; $\quad a = 0$, four decimal places
27 $\sin 62°$; $\quad a = \pi/3$, six decimal places
28 $\sqrt{105}$; $\quad a = 100$, three decimal places
29 Show that the Taylor polynomial of degree $2n$ for

$f(x) = \cos x$ at $a = 0$ is

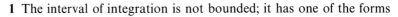

$$\sum_{k=0}^{n} (-1)^k \frac{x^{2k}}{(2k)!} = 1 - \frac{x^2}{2!} + \frac{x^4}{4!} - \cdots + (-1)^n \frac{x^{2n}}{(2n)!}.$$

30 Show that the Taylor polynomial of degree $2n + 1$ for $f(x) = \sin x$ at $a = 0$ is

$$\sum_{k=0}^{n} (-1)^k \frac{x^{2k+1}}{(2k+1)!} = x - \frac{x^3}{3!} + \frac{x^5}{5!}$$

$$- \cdots + (-1)^n \frac{x^{2n+1}}{(2n+1)!}.$$

---

## 11-4

## Improper Integrals

To show the existence of the definite integral, we have relied until now on the existence theorem stated in Section 5-3. This is the theorem that guarantees the existence of the definite integral $\int_a^b f(x)\,dx$ provided that the function $f$ is *continuous* on the closed (and *bounded*) interval $[a, b]$. Certain applications of calculus, however, lead naturally to the formulation of integrals in which either

1 The interval of integration is not bounded; it has one of the forms

$$[a, +\infty), \qquad (-\infty, a], \quad \text{or} \quad (-\infty, +\infty); \qquad \text{or}$$

2 The integrand has an infinite discontinuity at some point $c$:

$$\lim_{x \to c} f(x) = \pm\infty.$$

An example of Case (1) is the integral

$$\int_1^\infty \frac{dx}{x^2}.$$

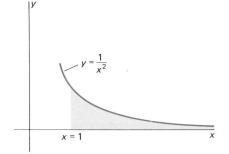

**11.6** The shaded area cannot be measured using our earlier techniques.

A geometric interpretation of this integral is the area of the unbounded region (shaded in Fig. 11.6) that lies between the curve $y = 1/x^2$ and the $x$-axis and to the right of the vertical line $x = 1$. An example of Case (2) is the integral

**11.7** Another area that must be measured with an improper integral.

$$\int_0^1 \frac{dx}{\sqrt{x}}.$$

This integral may be interpreted as the area of the unbounded region (shaded in Fig. 11.7) that lies under the curve $y = 1/\sqrt{x}$ from $x = 0$ to $x = 1$.

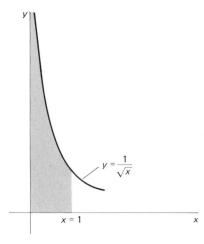

Such integrals are called **improper** integrals. The natural interpretation of an improper integral is the area of an unbounded region. The surprise is that such an area can nevertheless be finite, and this section is meant to show you how to find such areas—that is, how to evaluate improper integrals.

To see why improper integrals require special care, let us consider the integral

$$\int_{-1}^1 \frac{dx}{x^2},$$

CHAP. 11: Indeterminate Forms, Taylor's Formula, and Improper Integrals which is improper because its integrand $f(x) = 1/x^2$ is unbounded as $x \to 0$ and thus is *not* continuous at $x = 0$. If we blindly applied the fundamental theorem of calculus, we would get

$$\int_{-1}^{1} \frac{dx}{x^2} = \left[ -\frac{1}{x} \right]_{-1}^{1} = (-1) - (+1) = -2. \qquad \textbf{(WRONG!)}$$

The negative answer is obviously incorrect, because the area shown in Fig. 11.8 lies above the x-axis and hence cannot be negative. This simple example emphasizes that we cannot ignore the hypotheses—*continuous* function and *bounded* interval—of the fundamental theorem of calculus.

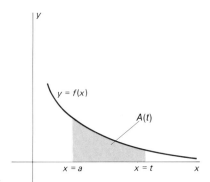

**11.8** The area under $y = 1/x^2$, $-1 \leq x \leq 1$

### INFINITE LIMITS OF INTEGRATION

Suppose that the function $f$ is continuous and nonnegative on the unbounded interval $[a, +\infty)$. Then, for any fixed $t > a$, the area $A(t)$ of the region under $y = f(x)$ from $x = a$ to $x = t$ (shaded in Fig. 11.9) is given by the (ordinary) definite integral

$$A(t) = \int_a^t f(x)\,dx.$$

Suppose now that we let $t \to \infty$, and find that the limit of $A(t)$ exists. Then we may regard this limit as the area of the unbounded region lying under $y = f(x)$ and over $[a, +\infty)$. For $f$ continuous on $[a, +\infty)$, we therefore *define*

$$\int_a^\infty f(x)\,dx = \lim_{t \to \infty} \int_a^t f(x)\,dx \qquad (1)$$

**11.9** The shaded area $A(t)$ exists provided $f$ is continuous.

provided that this limit exists (as a finite number). In this case we say that the improper integral on the left **converges;** otherwise, we say that it **diverges.** If $f(x)$ is *nonnegative* on $[a, +\infty)$, then the limit in Equation (1) either exists or is infinite, and in the latter case we write

$$\int_a^\infty f(x)\,dx = +\infty$$

and say that the above improper integral **diverges to infinity.**

If the function $f$ has both positive and negative values on $[a, +\infty)$, then the improper integral can diverge *by oscillation;* that is, without diverging to infinity. This occurs with $\int_0^\infty \sin x\,dx$, because it is easy to verify that $\int_0^t \sin x\,dx$ is 0 if $t$ is an even multiple of $\pi$ but is 2 if $t$ is an odd multiple of $\pi$. Thus $\int_0^t \sin x\,dx$ oscillates between 0 and 2 as $t \to \infty$, and so the limit in (1) does not exist.

We handle an infinite lower limit of integration similarly: We define

$$\int_{-\infty}^b f(x)\,dx = \lim_{t \to -\infty} \int_t^b f(x)\,dx \qquad (2)$$

provided the limit exists. If the function $f$ is continuous on the whole real line, we define

$$\int_{-\infty}^\infty f(x)\,dx = \int_{-\infty}^c f(x)\,dx + \int_c^\infty f(x)\,dx \qquad (3)$$

for any convenient choice of $c$, provided that both improper integrals on the right converge. Note that $\int_{-\infty}^{\infty} f(x)\, dx$ is *not* necessarily equal to $\lim\limits_{t \to \infty} \int_{-t}^{t} f(x)\, dx$ (see Problem 28).

It makes no difference what value of $c$ is used in (3) because, if $c < d$, then

$$\int_{-\infty}^{c} f(x)\, dx + \int_{c}^{\infty} f(x)\, dx = \int_{-\infty}^{c} f(x)\, dx + \int_{c}^{d} f(x)\, dx + \int_{d}^{\infty} f(x)\, dx$$

$$= \int_{-\infty}^{d} f(x)\, dx + \int_{d}^{\infty} f(x)\, dx,$$

under the assumption that the limits involved all exist.

**EXAMPLE 1**  Investigate the improper integrals:

$$\text{(a)} \ \int_{1}^{\infty} \frac{dx}{x^2}; \qquad \text{(b)} \ \int_{-\infty}^{0} \frac{dx}{\sqrt{1 - x}}.$$

*Solution*

(a) $\displaystyle \int_{1}^{\infty} \frac{dx}{x^2} = \lim_{t \to \infty} \int_{1}^{t} \frac{dx}{x^2}$

$$= \lim_{t \to \infty} \left[ -\frac{1}{x} \right]_{1}^{t} = \lim_{t \to \infty} \left( -\frac{1}{t} + 1 \right) = 1.$$

Thus this improper integral converges to 1, and this is the area of the region shown in Fig. 11.6.

(b) $\displaystyle \int_{-\infty}^{0} \frac{dx}{\sqrt{1 - x}} = \lim_{t \to -\infty} \int_{t}^{0} \frac{dx}{\sqrt{1 - x}}$

$$= \lim_{t \to -\infty} \left[ -2\sqrt{1 - x} \right]_{t}^{0} = \lim_{t \to -\infty} (2\sqrt{1 - t} - 2) = +\infty.$$

Thus the second improper integral of the example diverges to $+\infty$.

**EXAMPLE 2**  Investigate the improper integral

$$\int_{-\infty}^{\infty} \frac{1}{1 + x^2}\, dx.$$

*Solution*  The choice of $c = 0$ in Equation (3) gives

$$\int_{-\infty}^{\infty} \frac{dx}{1 + x^2} = \int_{-\infty}^{0} \frac{dx}{1 + x^2} + \int_{0}^{\infty} \frac{dx}{1 + x^2}$$

$$= \lim_{s \to -\infty} \int_{s}^{0} \frac{dx}{1 + x^2} + \lim_{t \to +\infty} \int_{0}^{t} \frac{dx}{1 + x^2}$$

$$= \lim_{s \to -\infty} \left[ \tan^{-1}x \right]_{s}^{0} + \lim_{t \to +\infty} \left[ \tan^{-1}x \right]_{0}^{t}$$

$$= \lim_{s \to -\infty} (-\tan^{-1}s) + \lim_{t \to +\infty} (\tan^{-1}t)$$

$$= \frac{\pi}{2} + \frac{\pi}{2} = \pi.$$

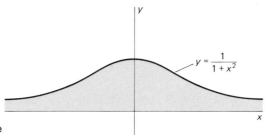

**11.10** The area measured by the integral of Example 2

The shaded area in Fig. 11.10 is a geometric interpretation of the integral of Example 2.

## INFINITE INTEGRANDS

Suppose that the function $f$ is continuous and nonnegative on $[a, b)$ but that $f(x) \to \infty$ as $x \to b^-$. The graph of such a function appears in Fig. 11.11. The area $A(t)$ of the region lying under $y = f(x)$ from $x = a$ to $x = t < b$ is the value of the (ordinary) definite integral

$$A(t) = \int_a^t f(x)\,dx.$$

If the limit of $A(t)$ as $t \to b^-$ exists, then this limit may be regarded as the area of the (unbounded) region under $y = f(x)$ from $x = a$ to $x = b$. For $f$ continuous on $[a, b)$, we therefore *define*

$$\int_a^b f(x)\,dx = \lim_{t \to b^-} \int_a^t f(x)\,dx, \tag{4}$$

provided that this limit exists (as a finite number), in which case we say that the improper integral on the left *converges*; otherwise we say that it *diverges*. If

$$\int_a^b f(x)\,dx = \lim_{t \to b^-} \int_a^t f(x)\,dx = \infty,$$

then we say that the improper integral *diverges to infinity*.

If $f$ is continuous on $(a, b]$ but the limit of $f(x)$ as $x \to a^+$ is infinite, then we *define*

$$\int_a^b f(x)\,dx = \lim_{t \to a^+} \int_t^b f(x)\,dx \tag{5}$$

provided that the limit exists. If $f$ is continuous at every point of $[a, b]$ except for the point $c$ in $(a, b)$ and one or both one-sided limits at $c$ are infinite, then we *define*

$$\int_a^b f(x)\,dx = \int_a^c f(x)\,dx + \int_c^b f(x)\,dx \tag{6}$$

provided that both improper integrals on the right converge.

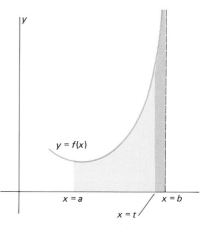

**11.11** An improper integral of the second type: $f(x) \to \infty$ as $x \to b^-$.

**EXAMPLE 3** Investigate the improper integrals:

$$\text{(a) } \int_0^1 \frac{dx}{\sqrt{x}}; \qquad \text{(b) } \int_1^2 \frac{dx}{(x-2)^2}.$$

*Solution*

(a) The integrand $1/\sqrt{x}$ becomes infinite as $x \to 0^+$, so

$$\int_0^1 \frac{dx}{\sqrt{x}} = \lim_{t \to 0^+} \int_t^1 \frac{dx}{\sqrt{x}}$$

$$= \lim_{t \to 0^+} \left[ 2\sqrt{x} \right]_t^1 = \lim_{t \to 0^+} 2(1 - \sqrt{t}) = 2.$$

Thus the area of the unbounded region shown in Fig. 11.7 is 2.

(b) Here the integrand becomes infinite as $x$ approaches the right-hand endpoint, so

$$\int_1^2 \frac{dx}{(x - 2)^2} = \lim_{t \to 2^-} \int_1^t \frac{dx}{(x - 2)^2}$$

$$= \lim_{t \to 2^-} \left[ -\frac{1}{x - 2} \right]_1^t = \lim_{t \to 2^-} \left( -1 - \frac{1}{t - 2} \right) = +\infty.$$

Hence this improper integral diverges to infinity. It follows that the improper integral

$$\int_1^3 \frac{dx}{(x - 2)^2} = \int_1^2 \frac{dx}{(x - 2)^2} + \int_2^3 \frac{dx}{(x - 2)^2}$$

also diverges, because not both of the right-hand improper integrals converge. (It turns out that the second one also diverges to $+\infty$.)

**EXAMPLE 4**  Investigate the improper integral

$$\int_0^2 \frac{dx}{(2x - 1)^{2/3}}.$$

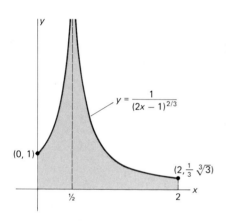

$$y = \frac{1}{(2x - 1)^{2/3}}$$

$(0, 1)$

$\left( 2, \frac{1}{3}\sqrt[3]{3} \right)$

$\frac{1}{2}$  $2$

**11.12**  The region of Example 4

*Solution*  This improper integral corresponds to the region shaded in Fig. 11.12. The integrand function has an infinite discontinuity at the point $c = \frac{1}{2}$ within the interval of integration, so we write

$$\int_0^2 \frac{dx}{(2x - 1)^{2/3}} = \int_0^{1/2} \frac{dx}{(2x - 1)^{2/3}} + \int_{1/2}^2 \frac{dx}{(2x - 1)^{2/3}}$$

and investigate separately the two improper integrals on the right. We find that

$$\int_0^{1/2} \frac{dx}{(2x - 1)^{2/3}} = \lim_{t \to (1/2)^-} \int_0^t \frac{dx}{(2x - 1)^{2/3}}$$

$$= \lim_{t \to (1/2)^-} \left[ \tfrac{3}{2}(2x - 1)^{1/3} \right]_0^t$$

$$= \lim_{t \to (1/2)^-} \tfrac{3}{2}[(2t - 1)^{1/3} - (-1)^{1/3}] = \tfrac{3}{2},$$

and

$$\int_{1/2}^2 \frac{dx}{(2x - 1)^{2/3}} = \lim_{t \to (1/2)^+} \int_t^2 \frac{dx}{(2x - 1)^{2/3}}$$

$$= \lim_{t \to (1/2)^+} \left[ \tfrac{3}{2}(2x - 1)^{1/3} \right]_t^2$$

$$= \lim_{t \to (1/2)^+} \tfrac{3}{2}[(3)^{1/3} - (2t - 1)^{1/3}] = \tfrac{3}{2}\sqrt[3]{3}.$$

Therefore,

$$\int_0^2 \frac{dx}{(2x-1)^{2/3}} = \frac{3}{2}(1 + \sqrt[3]{3}).$$

Special functions in advanced mathematics are frequently defined by means of improper integrals. An important example is the **gamma function** $\Gamma(t)$ that the prolific Swiss mathematician Leonhard Euler (1707–1783) introduced to "interpolate" the factorial function $n!$. The gamma function is defined for all real numbers $t > 0$ by

$$\Gamma(t) = \int_0^\infty x^{t-1} e^{-x}\, dx. \tag{7}$$

This definition gives a continuous function of $t$ such that

$$\Gamma(n+1) = n! \tag{8}$$

if $n$ is a positive integer (see Problems 29 and 30).

**EXAMPLE 5** Find the volume $V$ of the unbounded solid obtained by revolving around the $y$-axis the region under the curve $y = e^{-x^2}$, $x \geq 0$.

*Solution* The region in question is shown in Fig. 11.13. Let $V_t$ denote the volume generated by revolving the shaded part of this region between $x = 0$ and $x = t$. Then the method of cylindrical shells gives

$$V_t = \int_0^t 2\pi x e^{-x^2}\, dx.$$

We get the whole volume $V$ by letting $t \to \infty$:

$$V = \lim_{t\to\infty} V_t = \int_0^\infty 2\pi x e^{-x^2}\, dx$$
$$= \lim_{t\to\infty} \left[ -\pi e^{-x^2} \right]_0^t = \pi.$$

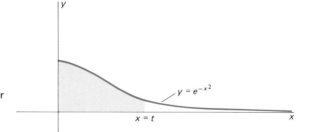

**11.13** *All* the region under the graph is rotated about the $y$-axis; this produces an unbounded solid.

$y = e^{-x^2}$

$x = t$

## *ESCAPE VELOCITY

We saw in Section 6-5 how to compute the work $W_r$ required to lift a body of mass $m$ from the surface of a planet (of mass $M$ and radius $R$) to a distance $r > R$ from the center of the planet. According to Equation (7) in that section, the answer is

$$W_r = \int_R^r \frac{GMm}{x^2}\, dx.$$

So the work required to move the mass $m$ "infinitely far" from the planet is

$$W = \lim_{r \to \infty} W_r = \int_R^\infty \frac{GMm}{x^2}\,dx = \lim_{r \to \infty}\left[-\frac{GMm}{x}\right]_R^r = \frac{GMm}{R}.$$

Suppose that the mass is projected straight upward from the planet's surface with initial velocity $v_0$, as in Jules Verne's novel *From the Earth to the Moon* (1865), in which a spaceship was fired from an immense cannon. Then the initial kinetic energy $\frac{1}{2}mv_0^2$ is available to supply this work—by conversion into potential energy. From the equation $\frac{1}{2}mv_0^2 = GMm/R$, we find that

$$v_0 = \sqrt{\frac{2GM}{R}}.$$

This provides an alternative derivation of the escape velocity formula (Equation (12) in Section 4-9).

### *PRESENT VALUE OF A PERPETUITY

Consider a "perpetual annuity," under which you and your heirs (and theirs, ad infinitum) will be paid $A$ dollars annually. By Equation (5) of Section 7-7, the **present value** of the amount you (and your heirs) receive between time $t = 0$ (the present) and time $t = T$ is

$$P_T = \int_0^T Ae^{-rt}\,dt.$$

In this formula, $r$ is the continuous interest rate. So the total present value of the perpetual annuity is

$$P = \lim_{T \to \infty} P_T = \int_0^\infty Ae^{-rt}\,dt = \lim_{T \to \infty}\left[-\frac{A}{r}e^{-rt}\right]_0^T = \frac{A}{r}.$$

Thus $A = rP$. For instance, at an interest rate of 8% ($r = 0.08$), you should be able to purchase for $P = (\$10{,}000)/(0.08) = \$125{,}000$ a perpetuity that pays you (and your heirs) an annual sum of $A = \$10{,}000$.

## 11-4  PROBLEMS

Determine whether or not the improper integrals in Problems 1–24 converge, and evaluate those that do converge.

1. $\displaystyle\int_4^\infty \frac{dx}{x^{3/2}}$

2. $\displaystyle\int_1^\infty \frac{dx}{x^{2/3}}$

3. $\displaystyle\int_0^4 \frac{dx}{x^{3/2}}$

4. $\displaystyle\int_0^8 \frac{dx}{x^{2/3}}$

5. $\displaystyle\int_1^\infty \frac{dx}{x + 1}$

6. $\displaystyle\int_3^\infty \frac{dx}{\sqrt{x + 1}}$

7. $\displaystyle\int_5^\infty \frac{dx}{(x - 1)^{3/2}}$

8. $\displaystyle\int_0^4 \frac{dx}{\sqrt{4 - x}}$

9. $\displaystyle\int_0^9 \frac{dx}{(9 - x)^{3/2}}$

10. $\displaystyle\int_0^3 \frac{dx}{(x - 3)^2}$

11. $\displaystyle\int_{-\infty}^{-2} \frac{dx}{(x + 1)^3}$

12. $\displaystyle\int_{-\infty}^0 \frac{dx}{\sqrt{4 - x}}$

13. $\displaystyle\int_{-1}^8 \frac{1}{x^{1/3}}\,dx$

14. $\displaystyle\int_{-4}^5 \frac{1}{(x + 4)^{2/3}}\,dx$

15. $\displaystyle\int_2^\infty \frac{dx}{(x - 1)^{1/3}}$

16. $\displaystyle\int_{-\infty}^\infty \frac{x\,dx}{(x^2 + 4)^{3/2}}$

17. $\displaystyle\int_{-\infty}^\infty \frac{x}{x^2 + 4}\,dx$

18. $\displaystyle\int_0^\infty e^{-(x + 1)}\,dx$

**19** $\displaystyle\int_0^1 \frac{e^{\sqrt{x}}}{\sqrt{x}}\,dx$ 

**20** $\displaystyle\int_0^2 \frac{x}{x^2-1}\,dx$

**21** $\displaystyle\int_1^\infty \frac{dx}{x\ln x}$ 

**22** $\displaystyle\int_0^\infty \sin^2 x\,dx$

**23** $\displaystyle\int_0^\infty xe^{-2x}\,dx$ 

**24** $\displaystyle\int_0^\infty e^{-x}\sin x\,dx$

Problems 25–27 deal with *Gabriel's horn*, the surface obtained by revolving the curve $y = 1/x$, $x \geq 1$, around the x-axis (see Fig. 11.14).

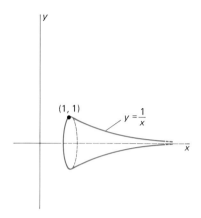

**11.14** Gabriel's horn

**25** Show that the area under the curve $y = 1/x$, $x \geq 1$, is infinite.

**26** Show that the volume of revolution enclosed by Gabriel's horn is finite, and compute it.

**27** Show that the surface area of Gabriel's horn is infinite. (*Suggestion:* If $A_t$ denotes the surface area from $x = 1$ to $x = t$, show that

$$A_t = \int_1^t \frac{2\pi}{x}\sqrt{1 + (1/x^4)}\,dx > 2\pi \ln t.)$$

Thus we could fill Gabriel's horn with a finite amount of paint (Problem 26), but no finite amount suffices to paint its surface.

**28** Show that

$$\int_{-\infty}^\infty \frac{1+x}{1+x^2}\,dx$$

diverges, but that

$$\lim_{t\to\infty} \int_{-t}^t \frac{1+x}{1+x^2}\,dx = \pi.$$

**29** Let $n$ be a fixed positive integer. Begin with Integral (7) defining the gamma function and integrate by parts to show that $\Gamma(n+1) = n\Gamma(n)$.

**30** (a) Show that $\Gamma(1) = 1$.

(b) Use the results of part (a) and Problem 29 to prove by mathematical induction that $\Gamma(n+1) = n!$.

**31** Use the substitution $x = e^{-u}$ and the fact that $\Gamma(n+1) = n!$ to show that, if $m$ and $n$ are fixed but arbitrary positive integers, then

$$\int_0^1 x^m (\ln x)^n\,dx = \frac{n!(-1)^n}{(m+1)^{n+1}}.$$

**32** Consider a perpetual annuity under which you and your heirs will be paid at the rate of $\$(10 + t)$ thousand per year $t$ years hence. Thus you will receive $20 thousand 10 years hence, your heir will receive $110 thousand 100 years hence, and so on. Show that the present value at an interest rate of 10% of this perpetuity is

$$P = \int_0^\infty (10 + t)e^{-t/10}\,dt,$$

and then evaluate this improper integral.

**33** A "semi-infinite" uniform rod occupies the nonnegative x-axis ($x \geq 0$) and has linear density $\rho$; that is, a segment of length $dx$ has mass $\rho\,dx$. Show that the force of gravitational attraction that the rod exerts on a point mass $m$ at $(-a, 0)$ is

$$F = \int_0^\infty \frac{Gm\rho\,dx}{(a+x)^2} = \frac{Gm\rho}{a}.$$

**34** A rod of linear density $\rho$ occupies the entire y-axis. A point mass $m$ is located at $(a, 0)$ on the x-axis, as indicated in Fig. 11.15. Show that the total (horizontal) gravitational force that the rod exerts on $m$ is

$$F = \int_{-\infty}^\infty \frac{Gm\rho \cos\theta}{r^2}\,dy = \frac{2Gm\rho}{a},$$

where $r^2 = a^2 + y^2$ and $\cos\theta = \dfrac{a}{r}$.

**11.15** Gravitational attraction exerted on a point mass by an infinite rod

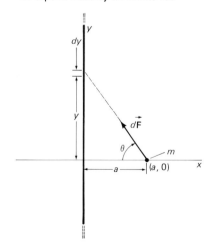

## Error Estimates for Numerical Integration Rules

In Section 5-8 we introduced several numerical integration rules for approximating the integral $\int_a^b f(x)\,dx$—the right- and left-end-point approximations, the midpoint and trapezoidal approximations, and Simpson's approximation. The latter two of these are the most commonly used. Here we give error estimates for the trapezoidal approximation and Simpson's approximation that are rather like the remainder term in Taylor's formula.

Let the points $a = x_0 < x_1 < \cdots < x_{n-1} < x_n = b$ partition the interval $[a, b]$ into $n$ equal subintervals, each with length $\Delta x = (b - a)/n$. Let $y_i = f(x_i)$ for $i = 0, 1, 2, \ldots, n$. Recall that the trapezoidal approximation $T_n$ and Simpson's approximation $S_n$ to $\int_a^b f(x)\,dx$ are defined as follows:

$$T_n = \frac{\Delta x}{2}(y_0 + 2y_1 + 2y_2 + \cdots + 2y_{n-1} + y_n) \tag{1}$$

and

$$S_n = \frac{\Delta x}{3}(y_0 + 4y_1 + 2y_2 + \cdots + 2y_{n-2} + 4y_{n-1} + y_n). \tag{2}$$

In the case of Simpson's approximation, the number $n$ of subintervals must be even.

For such approximations to be of real value, we must have some knowledge of the error involved. Let us denote these errors by $ET_n$ and $ES_n$, respectively; the error terms are defined by means of the equations

$$\int_a^b f(x)\,dx = T_n + ET_n \tag{3}$$

and

$$\int_a^b f(x)\,dx = S_n + ES_n. \tag{4}$$

The following two theorems give expressions for $ET_n$ and $ES_n$ that often allow us to estimate them effectively.

---

*Theorem 1   Trapezoidal Error Estimate*

If $f''$ is continuous on $[a, b]$, then

$$ET_n = -\frac{(b - a)^3}{12n^2} f''(z) \tag{5}$$

for some number $z$ in $[a, b]$. That is, for some such $z$,

$$\int_a^b f(x)\,dx = T_n - \frac{(b - a)^3}{12n^2} f''(z). \tag{6}$$

---

**Theorem 2**    *Simpson's Error Estimate*

If $f^{(4)}$ is continuous on $[a, b]$, then

$$ES_n = -\frac{(b-a)^5}{180n^4} f^{(4)}(z) \tag{7}$$

for some $z$ in $[a, b]$. That is, for some such $z$,

$$\int_a^b f(x)\, dx = S_n - \frac{(b-a)^5}{180n^4} f^{(4)}(z). \tag{8}$$

The proofs of Theorems 1 and 2 are based on the same general idea as the proof of Taylor's formula, that of approximating a function with a polynomial. As we indicated in Section 5-8, the trapezoidal approximation results from approximating $f(x)$ on each subinterval $[x_{i-1}, x_i]$ with a linear polynomial, while Simpson's approximation results from approximating $f(x)$ on each subinterval $[x_{2i-2}, x_{2i}]$ with a second degree polynomial. Detailed proofs of Theorems 1 and 2 may be found in most introductory numerical analysis books (for example, S. D. Conte and C. de Boor, *Elementary Numerical Analysis* (New York: McGraw-Hill, 1972) pp. 284–92).

If we let $M_2$ denote the maximum value of $|f''(z)|$ for $z$ in $[a, b]$ and let $M_4$ denote the maximum value of $|f^{(4)}(z)|$ there, then Theorems 1 and 2 yield the error estimates stated in Section 5-8:

$$\left| \int_a^b f(x)\, dx - T_n \right| \leq \frac{(b-a)^3}{12n^2} M_2 \tag{9}$$

and

$$\left| \int_a^b f(x)\, dx - S_n \right| \leq \frac{(b-a)^5}{180n^4} M_4. \tag{10}$$

If the value of either $M_2$ or $M_4$ is known, then we may attain any desired degree of accuracy by choosing $n$ sufficiently large.

**EXAMPLE 1**    Use the trapezoidal error estimate with $n = 10$ to find upper and lower bounds for the value of

$$\int_0^1 e^{-x^2}\, dx.$$

*Solution*    It is a sound idea to estimate the error first, so that we will know how many decimal places to retain in our computations. With $f(x) = e^{-x^2}$, we have

$$f'(x) = -2xe^{-x^2},$$

$$f''(x) = (4x^2 - 2)e^{-x^2}, \quad \text{and}$$

$$f^{(3)}(x) = 4x(3 - 2x^2)e^{-x^2}.$$

We have calculated the third derivative in order to find the maximum and minimum values of $f''(z)$ for $z$ in $[0, 1]$. Since $f^{(3)}(x) \geq 0$ for $0 \leq x \leq 1$, we see that $f''(z)$ lies between $f''(0) = -2$ and $f''(1) = 2/e \approx 0.736$. Thus

$$-0.736 < -f''(z) < 2.$$

We multiply each part of this inequality by

$$\frac{(b-a)^3}{12n^2} = \frac{1}{1200}$$

and find that

$$-0.00061 < ET_{10} < 0.00167.$$

This suggests that we compute $T_{10}$ to five decimal places. We obtain

$$T_{10} = \frac{0.1}{2}\left[e^{-(0.0)^2} + 2e^{-(0.1)^2} + \cdots + 2e^{-(0.9)^2} + e^{-(1.0)^2}\right]$$

$$= 0.74621$$

to five places. We get our upper and lower bounds by subtracting 0.00061 and adding 0.00167, respectively, to $T_{10}$. Thus we find that

$$0.74560 < \int_0^1 e^{-x^2}\, dx < 0.74788.$$

Since $|ET_{10}| < 0.00167$, a more crude estimate would be

$$\int_0^1 e^{-x^2}\, dx = 0.74621 \pm 0.00167.$$

Incidentally, to five-place accuracy, the true value of the integral is 0.74682.

**EXAMPLE 2**   We saw in Section 5-8 that

$$\frac{\pi}{4} = \int_0^1 \frac{dx}{1 + x^2}.$$

Use this fact and Simpson's approximation to approximate the number $\pi$ accurate to five decimal places.

*Solution*   If

$$\frac{\pi}{4} = \int_0^1 \frac{dx}{1 + x^2} = S_n + ES_n,$$

then

$$\pi = 4S_n + 4ES_n.$$

To bracket the value of $\pi$ in an interval of length less than 0.000005, we therefore need to choose $n$ so large that

$$|8ES_n| < 0.000005, \quad \text{and thus} \quad |ES_n| < 0.000000625.$$

To find the smallest suitable value of $n$, we first estimate $f^{(4)}(z)$ for $z$ in $[0, 1]$. We begin with $f(x) = 1/(1 + x^2)$; successive differentiation gives

$$f'(x) = \frac{-2x}{(1 + x^2)^2},$$

$$f''(x) = \frac{6x^2 - 2}{(1 + x^2)^3},$$

$$f^{(3)}(x) = \frac{24x(1 - x^2)}{(1 + x^2)^4},$$

$$f^{(4)}(x) = \frac{24(5x^4 - 10x^2 + 1)}{(1 + x^2)^5}, \quad \text{and}$$

$$f^{(5)}(x) = \frac{-240x(3x^4 - 10x^2 + 3)}{(1 + x^2)^6}.$$

The roots of $f^{(5)}(x) = 0$ are $x = 0$, $\pm 1/\sqrt{3}$, and $\pm\sqrt{3}$. We need check only the one of these that lies in the interval $(0, 1)$:

$$f^{(4)}(1/\sqrt{3}) = -10.125.$$

The only other critical points of $f^{(4)}(x)$ are the end points $x = 0$ and $x = 1$, and there we find that $f^{(4)}(0) = 24$, $f^{(4)}(1) = -3$. Thus

$$|f^{(4)}(z)| \leq 24$$

for $z$ in $[0, 1]$. We therefore must choose $n$ so that

$$\frac{(1)^5(24)}{180n^4} < 0.000000625,$$

which is equivalent to $n > 21.49$. Hence we take $n = 22$ and find that the maximum possible error is

$$|ES_{22}| \leq \frac{(1)(24)}{(180)(22)^4} \leq 0.00000057.$$

Simpson's approximation with $n = 22$ is

$$S_{22} = \frac{\frac{1}{22}}{3}\left(\frac{1}{1 + (0)^2} + \frac{4}{1 + (\frac{1}{22})^2} + \frac{2}{1 + (\frac{2}{22})^2}\right.$$

$$\left. + \cdots + \frac{2}{1 + (\frac{20}{22})^2} + \frac{4}{1 + (\frac{21}{22})^2} + \frac{1}{1 + (1)^2}\right),$$

so that

$$S_{22} \approx 0.78539816.$$

When we use our (high) estimate of $|ES_{22}|$ we write

$$S_{22} - 0.00000057 < \frac{\pi}{4} = \int_0^1 \frac{dx}{1 + x^2} < S_{22} + 0.00000057.$$

We replace $S_{22}$ by 0.78539816 and then multiply by 4; this finally gives

$$3.14159036 < \pi < 3.14159492.$$

Thus $\pi$ actually is 3.14159 to five decimal places.


## EXACT APPROXIMATIONS

The fourth derivative of a cubic (third degree) polynomial $P_3(x)$ is identically zero, so Simpson's approximation will be exact rather than a mere estimate. Thus, with $n = 2$,

$$\int_a^b P_3(x)\,dx = \frac{b - a}{6}\left[P_3(a) + 4P_3(m) + P_3(b)\right] \tag{11}$$

where $m = (a + b)/2$.

For a typical application of (11), suppose that the cross-sectional area function of a solid is a third degree polynomial (or one of lower degree). Then Equation (11) gives us an easy-to-apply formula for its volume:

$$V = \frac{H}{6}\left(A_0 + 4A_m + A_1\right). \tag{12}$$

Here, $H = b - a$ is the height of the solid, $A_0 = A(a)$ and $A_1 = A(b)$ are the areas of its ends, and $A_m = A(m)$ is the area of a cross section through its middle. Formula (12) is called the *prismoidal formula* for such solids (see Fig. 11.16).

For example, for a sphere with radius $r$, the prismoidal formula in (12) gives

$$V = \frac{2r}{6}(0 + 4\pi r^2 + 0) = \frac{4}{3}\pi r^3.$$

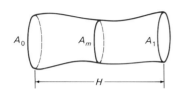

**11.16**  Using the prismoidal formula for volume

## 11-5  PROBLEMS

In Problems 1–5, use the trapezoidal error estimate, with the indicated number $n$ of subintervals, to find upper and lower bounds for the value of the integral. Then compare these with the exact value.

**1** $\displaystyle\int_0^{10} x\,dx; \quad n = 2$

**2** $\displaystyle\int_1^2 \sqrt{x}\,dx; \, n = 5$

**3** $\displaystyle\int_0^1 e^{-x}\,dx; \quad n = 10$

**4** $\displaystyle\int_1^2 \ln x\,dx; \quad n = 10$

**5** $\displaystyle\int_0^{\pi} \sin x\,dx; \quad n = 8$

In Problems 6–10, use Simpson's error estimate, with the indicated number $n$ of subintervals, to find upper and lower bounds for the value of the integral. Then compare these with the exact value.

**6** $\displaystyle\int_0^{10} x^3\,dx; \quad n = 2$

**7** $\displaystyle\int_1^2 x^{3/2}\,dx; \quad n = 6$

**8** $\displaystyle\int_0^1 e^{-x}\,dx; \quad n = 10$

**9** $\displaystyle\int_1^2 \ln x\,dx; \quad n = 10$

**10** $\displaystyle\int_0^{\pi} \sin x\,dx; \quad n = 8$

In Problems 11–14, determine a value of $n$ such that the trapezoidal approximation $T_n$ differs from the true value of the given integral by less than the indicated allowable error $E$.

**11** $\displaystyle\int_0^4 x^2\,dx; \quad E = 0.001$

**12** $\displaystyle\int_0^1 \sqrt{1 + x}\,dx; \quad E = 0.005$

**13** $\displaystyle\int_0^1 e^{-x}\,dx; \quad E = 0.005$

**14** $\displaystyle\int_0^{\pi} \sin x\,dx; \quad E = 0.0001$

In Problems 15–18, determine a value of $n$ such that Simpson's approximation $S_n$ differs from the true value of the given integral by no more than the maximum allowable error $E$ given in the problem.

**15** $\displaystyle\int_0^5 x^4\,dx; \quad E = 0.001$

**16** $\displaystyle\int_0^1 \frac{dx}{\sqrt{1 + x}}; \quad E = 0.005$

**17** $\displaystyle\int_1^2 \ln x\,dx; \quad E = 0.0005$

**18** $\displaystyle\int_0^1 e^{-x}\,dx; \quad E = 0.0001$

In Problems 19–22, use the prismoidal formula to calculate the exact volume of the solid described.

**19** A right circular cone with radius $r$ and height $h$.

**20** A frustum of a cone with height $h$ and base radii $r$ and $R$.

**21** The paraboloid obtained by revolving the parabola $y = \sqrt{x}$, $0 \leq x \leq a$, around the $x$-axis.

**22** The ellipsoid obtained by revolving the ellipse $x^2/a^2 + y^2/b^2 = 1$ around the $x$-axis.

**23** Use the trapezoidal rule to calculate $\ln 2 = \displaystyle\int_1^2 dx/x$ with error less than 0.002, thereby finding $\ln 2$ accurate to two decimal places.

**24** Repeat Problem 23, except use Simpson's approximation.

**25** Use the trapezoidal approximation to calculate

$$\int_0^1 \sqrt{1 + x^3}\, dx$$

with error less than 0.005. Begin by showing that if $f(x) = \sqrt{1 + x^3}$, then $|f''(x)| \leq 4$ for all $x$ in $[0, 1]$.

**26** Suppose that $P_4(x)$ is a fourth degree polynomial. Show that $ES_2$ is a computable *constant*. Hence explain how to use Simpson's error estimate with $n = 2$ to find the *exact value* of $\int_a^b P_4(x)\, dx$.

**27** Use Simpson's approximation with error estimate and $n = 2$ to compute the exact value of $\int_0^1 x^4\, dx$. (*Suggestion:* See Problem 26.)

**28** Use Simpson's approximation with error estimate and $n = 2$ to compute the exact volume of the solid obtained by revolving around the $x$-axis the area bounded by the $x$-axis and the parabola $y = 25 - x^2$.

# CHAPTER 11 REVIEW:   Definitions, Concepts, Results

Use the list below as a guide to concepts that you may need to review.

**1** L'Hôpital's rule and the indeterminate forms $0/0$, $\infty/\infty$, $0 \cdot \infty$, $\infty - \infty$, $0^0$, $\infty^0$, and $1^\infty$
**2** Cauchy's mean value theorem
**3** The $n$th degree Taylor polynomial of the function $f$ at the point $x = a$
**4** Taylor's formula with remainder
**5** Use of Taylor's formula to approximate values of functions

**6** Definition and evaluation of the improper integrals

$$\int_a^\infty f(x)\, dx \quad \text{and} \quad \int_{-\infty}^b f(x)\, dx$$

**7** Definition and evaluation of the improper integral $\int_a^b f(x)\, dx$, where $f$ has an infinite discontinuity at the point $c$ of $[a, b]$

**8** Error estimates for the trapezoidal approximation and Simpson's approximation

## MISCELLANEOUS PROBLEMS

Find the limits in Problems 1–15.

**1** $\displaystyle\lim_{x \to 2} \frac{x - 2}{x^2 - 4}$

**2** $\displaystyle\lim_{x \to 0} \frac{\sin 2x}{x}$

**3** $\displaystyle\lim_{x \to \pi} \frac{1 + \cos x}{(x - \pi)^2}$

**4** $\displaystyle\lim_{x \to 0} \frac{x - \sin x}{x^3}$

**5** $\displaystyle\lim_{t \to 0} \frac{t \tan^{-1} t - \sin^2 t}{t^6}$

**6** $\displaystyle\lim_{x \to \infty} \frac{\ln(\ln x)}{\ln x}$

**7** $\displaystyle\lim_{x \to 0} (\cot x) \ln(1 + x)$

**8** $\displaystyle\lim_{x \to 0^+} (e^{1/x} - 1) \tan x$

**9** $\displaystyle\lim_{x \to 0} \left( \frac{1}{x^2} - \frac{1}{1 - \cos x} \right)$

**10** $\displaystyle\lim_{x \to \infty} \left( \frac{x^2}{x + 2} - \frac{x^3}{x^2 + 3} \right)$

**11** $\displaystyle\lim_{x \to \infty} (\sqrt{x^2 - x + 1} - \sqrt{x})$

**12** $\displaystyle\lim_{x \to \infty} x^{1/x}$

**13** $\displaystyle\lim_{x \to \infty} (e^{2x} - 2x)^{1/x}$

**14** $\displaystyle\lim_{x \to \infty} (1 - e^{-x^2})^{1/x^2}$

**15** $\displaystyle\lim_{x \to \infty} x \left[ \left( 1 + \frac{1}{x} \right)^x - e \right]$   (*Suggestion:* Let $u = 1/x$ and take the limit as $u \to 0^+$.)

Determine whether or not the improper integrals in Problems 16–22 converge, and evaluate those that do converge.

**16** $\displaystyle\int_{-4}^0 \frac{dx}{\sqrt{x + 4}}$

**17** $\displaystyle\int_2^6 \frac{dx}{(x - 2)^{3/2}}$

**18** $\displaystyle\int_{-\infty}^\infty \frac{x\, dx}{\sqrt{x^2 + 9}}$

**19** $\displaystyle\int_e^\infty \frac{dx}{x(\ln x)^2}$

**20** $\displaystyle\int_{-\infty}^\infty \sin x\, dx$

**21** $\displaystyle\int_0^\infty x^2 e^{-x}\, dx$

**22** $\displaystyle\int_0^2 \frac{dx}{(x - 1)^{2/3}}$

In Problems 23–31, find the Taylor formula with remainder for $f(x)$ with the given values of $a$ and $n$.

**23** $f(x) = e^{x+1}$;   $a = 0, n = 5$
**24** $f(x) = xe^x$;   $a = 0, n = 3$
**25** $f(x) = e^{-x^2}$;   $a = 0, n = 3$
**26** $f(x) = \sqrt[3]{x - 8}$;   $a = 0, n = 3$
**27** $f(x) = \sqrt[3]{x}$;   $a = 125, n = 3$
**28** $f(x) = \tan^{-1} x$;   $a = 1, n = 3$
**29** $f(x) = x^{-1/2}$;   $a = 1, n = 4$
**30** $f(x) = (x + 2)^{3/2}$;   $a = 2, n = 3$
**31** $f(x) = x^4$;   $a = 1, n = 5$
**32** Use Taylor's formula to calculate $\sqrt[3]{120}$ accurate to three decimal places.

**33** Use Taylor's formula to calculate $\sqrt[10]{1000}$ accurate to five decimal places.

In Problems 34–35, determine the number of decimal places of accuracy the given approximation formula yields for $|x| \leq 1$.

**34** $\cos x \approx 1 - \frac{1}{2}x^2 + \frac{1}{24}x^4$

**35** $\tan^{-1}x \approx x - \frac{1}{3}x^3$

**36** Show that the approximation

$$\sin x \approx x - \frac{x^3}{3!} + \frac{x^5}{5!} - \frac{x^7}{7!} + \frac{x^9}{9!}$$

can be used to calculate $\sin x$ accurate to four decimal places for all values of $x$ that correspond to angles between $0°$ and $90°$.

**37** Show that the approximation of Problem 36 is accurate to 15 decimal places for values of $x$ that correspond to angles between $0°$ and $10°$.

**38** According to Problem 45 in Section 9-5, the surface area of the ellipsoid obtained by revolving the ellipse with equation $x^2/a^2 + y^2/b^2 = 1$ ($a < b$) around the $x$-axis is

$$A = 2\pi ab\left(\frac{b}{a} + \frac{a}{c}\ln\left(\frac{b+c}{a}\right)\right)$$

where $c^2 = b^2 - a^2$. Use l'Hôpital's rule to show that $\lim_{b \to a} A = 4\pi a^2$.

**39** Given: $\int_0^\infty e^{-x^2}\,dx = \frac{1}{2}\sqrt{\pi}$. Deduce that $\Gamma(\frac{1}{2}) = \sqrt{\pi}$.

**40** Recall that $\Gamma(x + 1) = x\Gamma(x)$ if $x > 0$. Suppose that $n$ is a positive integer. Use Problem 39 to establish that

$$\Gamma\left(n + \frac{1}{2}\right) = \frac{1 \cdot 3 \cdot 5 \cdots (2n-1)}{2^n}\sqrt{\pi}.$$

**41 (a)** Suppose that $k > 1$. Show by integration by parts that

$$\int_0^\infty x^k e^{-x^2}\,dx = \frac{k-1}{2}\int_0^\infty x^{k-2}e^{-x^2}\,dx.$$

**(b)** Suppose that $n$ is a positive integer. Show that

$$\int_0^\infty x^{n-1}e^{-x^2}\,dx = \frac{1}{2}\Gamma\left(\frac{n}{2}\right).$$

**42** Prove as follows that the number $e$ is irrational. First suppose to the contrary that $e = p/q$ where $p$ and $q$ are positive integers. Write

$$\frac{p}{q} = e = 1 + 1 + \frac{1}{2!} + \frac{1}{3!} + \cdots + \frac{1}{q!} + R_q$$

where $0 < R_q < 3/(q+1)!$. (Why?) Then show that multiplication of both sides of this equation by $q!$ would lead to the contradiction that one side of the result is an integer but the other side is not.

**43** In a letter of 1672 that he wrote to a person who wanted a quick method for gauging the content of a wine barrel (volume, not quality), Newton mentioned that if a symmetric barrel has the shape obtained by revolving a parabolic arc, then its volume is exactly

$$V = \pi H(\tfrac{2}{3}R^2 + \tfrac{1}{3}r^2 - \tfrac{2}{15}\delta^2).$$

In this formula, $H$ is the height of the barrel, $R$ is the radius of its midsection, $r$ is the radius of each end, and $\delta = R - r$. Use Simpson's approximation with error estimate to establish Newton's formula.

# 12

## Infinite Series

## 12-1

### Introduction

In the fifth century B.C., the Greek philosopher Zeno proposed the following paradox: In order for a runner to travel a given distance, the runner must first travel halfway, then half the remaining distance, then half the distance that still remains, and so on ad infinitum. But, Zeno argued, it is clearly impossible for a runner to accomplish these infinitely many steps in a finite period of time, so motion from one point to another must be impossible.

Zeno's paradox suggests the infinite subdivision of $[0, 1]$ indicated in Fig. 12.1. There is one subinterval of length $1/2^n$ for each integer $n = 1, 2, 3, \ldots$. If the length of the interval is the sum of the lengths of the subintervals into which it is divided, then it would appear that

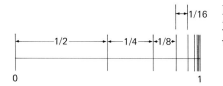

**12.1** Subdivision of an interval to illustrate Zeno's paradox.

$$1 = \frac{1}{2} + \frac{1}{4} + \frac{1}{8} + \frac{1}{16} + \cdots + \frac{1}{2^n} + \cdots,$$

with infinitely many terms somehow adding up to 1. On the other hand, the formal infinite sum

$$1 + 2 + 3 + \cdots + n + \cdots$$

of all the positive integers seems meaningless—it does not appear to add up to *any* (finite) value.

The question is this: What, if anything, is meant by the sum of an *infinite* collection of numbers? This chapter explores conditions under which an *infinite* sum

$$a_1 + a_2 + a_3 + \cdots + a_n + \cdots,$$

known as an *infinite series*, *is* meaningful. We shall discuss methods for computing the sum of an infinite series, and applications of the algebra and calculus of infinite series. Infinite series are important in science and mathematics because many functions either arise most naturally in the form of infinite series, or have infinite series representations (such as the Taylor series of Section 12-7) that are useful for numerical computations.

## 12-2

### Infinite Sequences

An (**infinite**) **sequence** of real numbers is a function whose domain of definition is the set of all positive integers. Thus if $s$ is a sequence, then to each positive integer $n$ there corresponds a real number $s(n)$. Ordinarily, a sequence is most conveniently described by listing its values in order, beginning with $s(1)$:

$$s(1), s(2), s(3), \ldots, s(n), \ldots.$$

With subscript notation rather than functional notation, we may write

$$s_1, s_2, s_3, \ldots, s_n, \ldots \tag{1}$$

for this list of values. The values in this list are the **terms** of the sequence; $s_1$ is the first term, $s_2$ the second term, $s_n$ the **nth term.**

We use the notation $\{s_n\}_1^\infty$, or simply $\{s_n\}$, as abbreviation for the **ordered** list in (1), and we may refer to the sequence by saying simply "the sequence $\{s_n\}$." When a particular sequence is so described, the $n$th term $s_n$ is generally (though not always) given by a formula in terms of its subscript $n$. In this case, listing the first few terms of the sequence often helps us to see it more concretely.

**EXAMPLE 1** The following table lists explicitly the first four terms of each of several sequences.

| $\{s_n\}_1^\infty$ | $s_1, s_2, s_3, s_4, \ldots$ |
|---|---|
| $\left\{\dfrac{1}{n}\right\}_1^\infty$ | $1, \dfrac{1}{2}, \dfrac{1}{3}, \dfrac{1}{4}, \ldots$ |
| $\left\{\dfrac{1}{10^n}\right\}_1^\infty$ | $0.1, 0.01, 0.001, 0.0001, \ldots$ |
| $\left\{\dfrac{1}{n!}\right\}_1^\infty$ | $1, \dfrac{1}{2}, \dfrac{1}{6}, \dfrac{1}{24}, \ldots$ |
| $\left\{\dfrac{(-1)^n}{n(n+1)}\right\}_1^\infty$ | $-\dfrac{1}{2}, \dfrac{1}{6}, -\dfrac{1}{12}, \dfrac{1}{20}, \ldots$ |
| $\left\{\sin \dfrac{n\pi}{2}\right\}_1^\infty$ | $1, 0, -1, 0, \ldots$ |
| $\{1 + (-1)^n\}_1^\infty$ | $0, 2, 0, 2, \ldots$ |

**EXAMPLE 2** The Fibonacci sequence $\{F_n\}$ is defined as follows:

$$F_1 = 1, \qquad F_2 = 1, \quad \text{and} \quad F_{n+1} = F_{n-1} + F_n \qquad \text{for } n \geqq 2.$$

The first ten terms of the Fibonacci sequence are

$$1, 1, 2, 3, 5, 8, 13, 21, 34, 55.$$

This is a *recursively defined* sequence—after the initial values are given, each term is defined in terms of its predecessors.

---

The limit of a sequence is defined in much the same way as the limit of an ordinary function (Section 2.1).

> *Definition*   *Limit of a Sequence*
>
> We say that the sequence $\{s_n\}$ **converges** to the real number $L$, or has **limit** $L$, and we write
>
> $$\lim_{n \to \infty} s_n = L, \tag{2}$$
>
> provided that $s_n$ can be made as close to $L$ as we please by choosing $n$ sufficiently large. That is, given any number $\varepsilon > 0$, there exists an integer $N$ such that
>
> $$|s_n - L| < \varepsilon \qquad \text{for all} \quad n \geqq N. \tag{3}$$

**EXAMPLE 3** Show that $\lim\limits_{n \to \infty} \dfrac{1}{n} = 0$.

*Solution*  We need to show this: To each positive number $\varepsilon$, there corresponds an integer $N$ such that, for all $n \geq N$,

$$\left| \frac{1}{n} - 0 \right| = \frac{1}{n} < \varepsilon.$$

It suffices to pick any fixed integer $N > 1/\varepsilon$. Then $n \geq N$ implies $1/n \leq 1/N < \varepsilon$, as desired.

---

The limit laws stated in Section 2-1 for limits of functions have natural analogues for limits of sequences. Their proofs are based on techniques similar to those used in Section 2-5.

---

**Theorem 1**   *Limit Laws for Sequences*

If the limits $\lim\limits_{n \to \infty} a_n = A$ and $\lim\limits_{n \to \infty} b_n = B$ exist (so that $A$ and $B$ are real numbers), then:

(i) $\lim\limits_{n \to \infty} ca_n = cA$      ($c$ any real number);

(ii) $\lim\limits_{n \to \infty} (a_n + b_n) = \cdot A + B$;

(iii) $\lim\limits_{n \to \infty} a_n b_n = AB$;

(iv) $\lim\limits_{n \to \infty} \dfrac{a_n}{b_n} = \dfrac{A}{B}.$

In this last case we must assume also that $B \neq 0$ and that $b_n \neq 0$ for all sufficiently large $n$.

---

**Theorem 2**   *Substitution Law for Sequences*

If $\lim\limits_{n \to \infty} a_n = A$ and the function $f(x)$ is continuous at $x = A$, then

$$\lim\limits_{n \to \infty} f(a_n) = f(A).$$

---

**Theorem 3**   *Squeeze Law for Sequences*

If $a_n \leq b_n \leq c_n$ for all $n$ and

$$\lim\limits_{n \to \infty} a_n = L = \lim\limits_{n \to \infty} c_n,$$

then $\lim\limits_{n \to \infty} b_n = L$ also.

---

These theorems can be used to compute limits of many sequences formally, without recourse to the definition. For example, if $k$ is a positive integer and $c$ is a constant, then Example 3 and the product law (Theorem 1(iii)) give

$$\lim_{n \to \infty} \frac{c}{n^k} = c \cdot 0 \cdot 0 \cdots 0 = 0.$$

**EXAMPLE 4**  Show that $\lim\limits_{n \to \infty} \dfrac{(-1)^n \cos n}{n^2} = 0$.

*Solution*  This result follows from the squeeze law and the fact that $1/n^2 \to 0$ as $n \to \infty$, because

$$-\frac{1}{n^2} \le \frac{(-1)^n \cos n}{n^2} \le \frac{1}{n^2}.$$

**EXAMPLE 5**  Show that if $a > 0$, then $\lim\limits_{n \to \infty} \sqrt[n]{a} = 1$.

*Solution*  We apply the substitution law with $f(x) = a^x$ and $A = 0$. Since $1/n \to 0$ as $n \to \infty$ and $f$ is continuous at $x = 0$, this gives

$$\lim_{n \to \infty} a^{1/n} = a^0 = 1.$$

**EXAMPLE 6**  The limit laws and the continuity of $\sqrt{x}$ at $x = 4$ yield

$$\lim_{n \to \infty} \sqrt{\frac{4n - 1}{n + 1}} = \sqrt{\lim_{n \to \infty} \frac{4 - (1/n)}{1 + (1/n)}} = \sqrt{4} = 2.$$

**EXAMPLE 7**  Show that if $|r| < 1$, then $\lim\limits_{n \to \infty} r^n = 0$.

*Solution*  Since $|r^n| = |(-r)^n|$, we may assume that $0 < r < 1$. Then $1/r = 1 + a$ with $a > 0$, so the binomial formula yields

$$\frac{1}{r^n} = (1 + a)^n = 1 + na + \text{(positive terms)} > 1 + na,$$

so

$$0 < r^n < \frac{1}{1 + na}.$$

Since $1/(1 + na) \to 0$ as $n \to \infty$, the squeeze law implies that $r^n \to 0$ as $n \to \infty$.

---

Let $f(x)$ be a function defined for every real number $x \ge 1$, and $\{a_n\}$ a sequence such that $f(n) = a_n$ for every positive integer $n$. Then it follows from the definitions of limits of functions and sequences that

$$\text{if} \quad \lim_{x \to \infty} f(x) = L \quad \text{then} \quad \lim_{n \to \infty} a_n = L. \tag{4}$$

Note that the converse of the above statement is generally false. For example,

$$\lim_{n \to \infty} \sin \pi n = 0 \quad \text{but} \quad \lim_{x \to \infty} \sin \pi x \quad \text{does not exist.}$$

Because of (4) we can use **l'Hôpital's rule for sequences:** If $a_n = f(n)$, $b_n = g(n)$ and $f(x)/g(x)$ has the indeterminate form $\infty/\infty$ as $x \to \infty$, then

$$\lim_{n \to \infty} \frac{a_n}{b_n} = \lim_{x \to \infty} \frac{f(x)}{g(x)} = \lim_{x \to \infty} \frac{f'(x)}{g'(x)}, \qquad (5)$$

provided that the right-hand limit exists.

**EXAMPLE 8**  Show that $\displaystyle\lim_{n \to \infty} \frac{\ln n}{n} = 0$.

*Solution*  The function $(\ln x)/x$ is defined for all $x \geq 1$ and agrees with the given sequence when $x = n$, an integer. Since $(\ln x)/x$ has the indeterminate form $\infty/\infty$ as $x \to \infty$, l'Hôpital's rule gives

$$\lim_{n \to \infty} \frac{\ln n}{n} = \lim_{x \to \infty} \frac{\ln x}{x} = \lim_{x \to \infty} \frac{1/x}{1} = 0.$$

**EXAMPLE 9**  Show that $\displaystyle\lim_{n \to \infty} \sqrt[n]{n} = 1$.

*Solution*  First we note that

$$\ln \sqrt[n]{n} = \frac{\ln n}{n} \to 0 \text{ as } n \to \infty$$

by Example 8. By the substitution law with $f(x) = e^x$, this gives

$$\lim_{n \to \infty} \sqrt[n]{n} = \lim_{n \to \infty} \exp(\ln \sqrt[n]{n}) = e^0 = 1.$$

**EXAMPLE 10**  Find $\displaystyle\lim_{n \to \infty} \frac{3n^3}{e^{2n}}$.

*Solution*  We apply l'Hôpital's rule repeatedly, though we must be careful at each intermediate step to verify that we still have an indeterminate form. Thus we find that

$$\lim_{n \to \infty} \frac{3n^3}{e^{2n}} = \lim_{x \to \infty} \frac{3x^3}{e^{2x}} = \lim_{x \to \infty} \frac{9x^2}{2e^{2x}}$$

$$= \lim_{x \to \infty} \frac{18x}{4e^{2x}} = \lim_{x \to \infty} \frac{18}{8e^{2x}} = 0.$$

## BOUNDED MONOTONE SEQUENCES

The set of all *rational* numbers has by itself all the most familiar elementary algebraic properties of the entire real number system. In order to guarantee the existence of irrational numbers, we must assume in addition a "completeness property" of the real numbers. Otherwise the real line might have "holes" where the irrational numbers ought to be. One way of stating this completeness property involves the convergence of an important type of sequence.

The sequence $\{a_n\}_1^\infty$ is called **monotone increasing** if

$$a_1 \leq a_2 \leq a_3 \leq \cdots \leq a_n \leq \cdots$$

and **monotone decreasing** if

$$a_1 \geqq a_2 \geqq a_3 \geqq \cdots \geqq a_n \geqq \cdots.$$

It is **monotone** if it is *either* monotone increasing *or* monotone decreasing. The sequence $\{a_n\}$ is **bounded** if there is a number $M$ such that $|a_n| \leqq M$ for all $n$. The following assertion may be taken as an axiom for the real number system.

---

**Bounded Monotone Sequence Property**

Every bounded monotone infinite sequence converges (that is, has a finite limit).

---

Suppose, for example, that the monotone increasing sequence $\{a_n\}_1^\infty$ is bounded above by a number $M$, meaning that $a_n \leqq M$ for all $n$. Since it is also bounded below (by $a_1$, for instance), the bounded monotone sequence property implies that $\lim_{n \to \infty} a_n = A$ for some real number $A \leqq M$. If the monotone increasing sequence $\{a_n\}$ is *not* bounded above (by any number), then it follows that $\lim_{n \to \infty} a_n = \infty$ (Problem 38). These two alternatives are illustrated in Fig. 12.2.

**12.2** A bounded increasing sequence and an increasing sequence that is not bounded above.

**EXAMPLE 11** Investigate the sequence $\{a_n\}$ defined recursively by

$$a_1 = \sqrt{2}, \qquad a_{n+1} = \sqrt{2 + a_n} \qquad \text{for} \quad n \geqq 1. \tag{6}$$

*Solution* The first four terms of $\{a_n\}$ are

$$\sqrt{2}, \ \sqrt{2 + \sqrt{2}}, \ \sqrt{2 + \sqrt{2 + \sqrt{2}}}, \ \sqrt{2 + \sqrt{2 + \sqrt{2 + \sqrt{2}}}}.$$

If $\lim_{n \to \infty} a_n$ exists, then this limit would seem to be the natural interpretation of the infinite expression

$$\sqrt{2 + \sqrt{2 + \sqrt{2 + \sqrt{2 + \cdots}}}}.$$

We will apply the bounded monotone sequence property to show that this limit does indeed exist. Note first that $a_1 < 2$. If we assume inductively that $a_n < 2$ for some $n \geqq 1$, then it follows that

$$(a_{n+1})^2 = 2 + a_n < 4, \quad \text{so} \quad a_{n+1} < 2.$$

Thus the sequence $\{a_n\}$ is bounded above by the number 2. We observe next that

$$(a_{n+1})^2 - (a_n)^2 = (2 + a_n) - (a_n)^2$$
$$= (2 - a_n)(1 + a_n) > 0$$

because $2 - a_n$ and $1 + a_n$ are both positive. Because $a_{n+1}$ and $a_n$ are positive, it follows that $a_{n+1} > a_n$ for $n \geqq 1$. Thus $\{a_n\}$ is a monotone increasing sequence.

Therefore, the bounded monotone sequence property tells us that the sequence $\{a_n\}$ has a limit $A$. It does not tell us what the number $A$ is. But now that we know that $A = \lim\limits_{n \to \infty} a_n$ exists, we can write

$$\lim_{n \to \infty} a_{n+1} = \lim_{n \to \infty} \sqrt{2 + a_n},$$

$$A = \sqrt{2 + A},$$

and thus

$$A^2 = 2 + A.$$

The roots of this equation are $-1$ and $2$. It is clear that $A > 0$, so we finally conclude that

$$\lim_{n \to \infty} a_n = 2.$$

To indicate what the bounded monotone sequence property has to do with the "completeness" of the real numbers, we outline in Problem 42 a proof (using this property) of the existence of the number $\sqrt{2}$. In Problems 43 and 44, we outline a proof of the equivalence of the bounded monotone sequence property and another common statement of the completeness of the real numbers—the *least upper bound property*.

## 12-2  PROBLEMS

In Problems 1–35, determine whether or not the sequence $\{a_n\}$ converges, and find its limit if it does converge.

**1** $a_n = \dfrac{2n}{5n - 3}$

**2** $a_n = \dfrac{1 - n^2}{2 + 3n^2}$

**3** $a_n = \dfrac{n^3 - n + 7}{2n^3 + n^2}$

**4** $a_n = \dfrac{n^3}{10n^2 + 1}$

**5** $a_n = 1 + \left(\frac{9}{10}\right)^n$

**6** $a_n = 2 - \left(-\frac{1}{2}\right)^n$

**7** $a_n = 1 + (-1)^n$

**8** $a_n = \dfrac{1 + (-1)^n}{\sqrt{n}}$

**9** $a_n = \dfrac{1 + (-1)^n \sqrt{n}}{\left(\frac{3}{2}\right)^n}$

**10** $a_n = \dfrac{\sin n}{3^n}$

**11** $a_n = \dfrac{\sin^2 n}{\sqrt{n}}$

**12** $a_n = \sqrt{\dfrac{2 + \cos n}{n}}$

**13** $a_n = n \sin \pi n$

**14** $a_n = n \cos \pi n$

**15** $a_n = \pi^{-(\sin n)/n}$

**16** $a_n = 2^{\cos n\pi}$

**17** $a_n = \dfrac{\ln n}{\sqrt{n}}$

**18** $a_n = \dfrac{\ln 2n}{\ln 3n}$

**19** $a_n = \dfrac{(\ln n)^2}{n}$

**20** $a_n = n \sin\left(\dfrac{1}{n}\right)$

**21** $a_n = \dfrac{\tan^{-1} n}{n}$

**22** $a_n = \dfrac{n^3}{e^{n/10}}$

**23** $a_n = \dfrac{2^n + 1}{e^n}$

**24** $a_n = \dfrac{\sinh n}{\cosh n}$

**25** $a_n = \left(1 + \dfrac{1}{n}\right)^n$

**26** $a_n = \sqrt[n]{2n + 5}$

**27** $a_n = \left(\dfrac{n - 1}{n + 1}\right)^n$

**28** $a_n = (0.001)^{-1/n}$

**29** $a_n = \sqrt[n]{2^{n+1}}$

**30** $a_n = \left(1 - \dfrac{2}{n^2}\right)^n$

**31** $a_n = \left(\dfrac{2}{n}\right)^{3/n}$

**32** $a_n = (-1)^n \sqrt[n]{n^2 + 1}$

**33** $a_n = \left(\dfrac{2 - n^2}{3 + n^2}\right)^n$

**34** $a_n = \dfrac{\left(\frac{2}{3}\right)^n}{1 - \sqrt[n]{n}}$

**35** $a_n = \dfrac{\left(\frac{2}{3}\right)^n}{\left(\frac{1}{2}\right)^n + \left(\frac{9}{10}\right)^n}$

**36** Suppose that $\lim\limits_{n \to \infty} a_n = A$. Prove that $\lim\limits_{n \to \infty} |a_n| = |A|$.

**37** Prove that $\{(-1)^n a_n\}$ diverges if $\lim\limits_{n \to \infty} a_n = A \neq 0$.

**38** Suppose that $\{a_n\}$ is a monotone increasing sequence that is not bounded. Prove that $\lim\limits_{n \to \infty} a_n = \infty$.

**39** Suppose that $A > 0$. Given $x_1$ arbitrary, define the sequence $\{x_n\}$ recursively by

$$x_{n+1} = \frac{1}{2}\left(x_n + \frac{A}{x_n}\right).$$

Prove that if $L = \lim\limits_{n \to \infty} x_n$ exists, then $L = \sqrt{A}$.

**40** Let $\{F_n\}$ be the Fibonacci sequence of Example 2. Assume that $\tau = \lim\limits_{n \to \infty} (F_{n+1}/F_n)$ exists; show that $\tau = \frac{1}{2}(1 + \sqrt{5})$. (*Suggestion:* Substitute $F_{n+1} = F_n + F_{n-1}$ in the equation $\tau = \lim\limits_{n \to \infty} (F_{n+1}/F_n)$. Deduce that $\tau$ satisfies the equation $\tau^2 = \tau + 1$, and note that $\tau > 0$.)

**41** Let the sequence $\{a_n\}$ be defined recursively by

$$a_1 = 2, \qquad a_{n+1} = \tfrac{1}{2}(a_n + 4) \qquad \text{for} \quad n \geq 1.$$

(a) Show by induction on $n$ that $a_n < 4$ for each $n$ and that $\{a_n\}$ is a monotone increasing sequence.
(b) Find the limit of this sequence.

**42** For each $n = 1, 2, 3, \ldots$, let $a_n$ be the largest multiple of $1/10^n$ such that $a_n^2 \leq 2$.
(a) Prove that $\{a_n\}$ is a bounded and monotone increasing sequence, so that $A = \lim\limits_{n \to \infty} a_n$ exists.
(b) Show that $A^2 > 2$ implies that $a_n^2 > 2$ for $n$ sufficiently large.

(c) Show that $A^2 < 2$ implies that $a_n^2 < B$ for some $B < 2$ and all sufficiently large $n$.
(d) Conclude that $A^2 = 2$.

Problems 43 and 44 deal with the **least upper bound property** of the real numbers: If a nonempty set $S$ of real numbers has an upper bound, then it has a least upper bound. The number $M$ is an **upper bound** for $S$ if $x \leq M$ for all $x$ in $S$. The upper bound $L$ of $S$ is a **least upper bound** of $S$ if no number smaller than $L$ is an upper bound for $S$.

**43** Show that the least upper bound property implies the bounded monotone sequence property. (*Suggestion:* If $\{a_n\}$ is a bounded and monotone increasing sequence, and $A$ is the least upper bound of $\{a_n\}$, show that $\lim\limits_{n \to \infty} a_n = A$.)

**44** Show that the bounded monotone sequence property implies the least upper bound property. (*Suggestion:* For each $n$, let $a_n$ be the least multiple of $1/10^n$ that is an upper bound of the set $S$. Show that $\{a_n\}$ is a bounded and monotone decreasing sequence and then that $A = \lim\limits_{n \to \infty} a_n$ is a least upper bound for $S$.)

## *12-2 Optional Computer Application

Whether the $n$th term $a_n$ of a sequence is defined explicitly in terms of $n$ or recursively in terms of the previous term $a_{n-1}$, it is a simple matter to program a computer to calculate and display the successive terms $a_1, a_2, a_3, \ldots$. The following program was written to investigate the sequence $\{a_n\}$ defined recursively in Example 11 by $a_1 = \sqrt{2}, a_{n+1} = \sqrt{2 + a_n}$ for $n \geq 1$.

```
10    A = SQR(2)  :  N = 1
20    PRINT N; " : "; A
30    B = A  :  N = N + 1
40    A = SQR(2 + B)
50    IF  ABS(A − B) > 1.0E-09  THEN GOTO 20
60    PRINT "  LIMIT: "; A
70    END
```

Lines 20–50 constitute a loop that calculates the new term $A = a_n$ in terms of the previous term $B = a_{n-1}$. Note that the recursive formula that determines the sequence appears in line 40.

Line 50 directs that further terms be calculated until two successive terms differ by at most $10^{-9}$, at which point the apparent value of the limit is displayed. In this example (as well as almost any similar example) the stopping condition in line 50 will be satisfied without the limit actually having been reached. (In fact, it is easy to give examples in which the stopping condition is satisfied even though the limit does not exist!) Therefore, you should be cautious in believing the computer merely because it prints what *you* told it to print.

Observe that only lines 10 and 40 refer to the definition of the sequence $\{a_n\}$, so only those two lines need to be altered in order to investigate a different sequence.

**Exercise 1** Enter the program and run it to confirm the value $2$ of the limit found in Example 11.

**Exercise 2** Suppose that the sequence $\{a_n\}$ is defined as follows: $a_1 = \sqrt{3}$ and $a_{n+1} = \sqrt{3 + a_n}$ for $n \geq 1$. Alter the program and then run it to verify that

$$\tfrac{1}{2}(1 + \sqrt{13}) = \sqrt{3 + \sqrt{3 + \sqrt{3 + \sqrt{3 + \cdots}}}}.$$

**Exercise 3** Suppose that the sequence $\{a_n\}$ is defined as follows:

$$a_1 = 1 \quad \text{and} \quad a_{n+1} = \frac{1}{1 + a_n} \quad \text{for} \quad n \geq 1.$$

Alter the program and then run it to verify that

$$\tfrac{1}{2}(-1 + \sqrt{5}) = \cfrac{1}{1 + \cfrac{1}{1 + \cfrac{1}{1 + \cdots}}}.$$

**Exercise 4** Write a program that calculates and displays the successive terms of the Fibonacci sequence $\{F_n\}$ of Example 2.

**Exercise 5** Write a program to verify that

$$\lim_{n \to \infty} \frac{F_{n+1}}{F_n} = \frac{1}{2}(1 + \sqrt{5})$$

where $\{F_n\}$ is the Fibonacci sequence.

---

### 12-3

## Convergence of Infinite Series

An **infinite series** is an expression of the form

$$\sum_{n=1}^{\infty} a_n = a_1 + a_2 + a_3 + \cdots + a_n + \cdots \tag{1}$$

where $\{a_n\}$ is an infinite sequence of real numbers. The number $a_n$ is called the **nth term** of the series. The symbol $\sum_{n=1}^{\infty} a_n$ is simply an abbreviation—compact notation—for the right-hand side in (1). An example of an infinite series is the series

$$\sum_{n=1}^{\infty} \frac{1}{2^n} = \frac{1}{2} + \frac{1}{4} + \frac{1}{8} + \cdots + \frac{1}{2^n} + \cdots$$

that was mentioned in Section 12-1. The $n$th term of this infinite series is $a_n = 1/2^n$.

To say what is meant by such an infinite sum, we introduce the partial sums of the infinite series (1). The **nth partial sum** $S_n$ of the series is the sum of its first $n$ terms:

$$S_n = a_1 + a_2 + a_3 + \cdots + a_n. \tag{2}$$

Thus each infinite series has associated with it an infinite **sequence of partial sums**

$$S_1, S_2, S_3, \ldots, S_n, \ldots.$$

We define the sum of the infinite series to be the limit of its sequence of partial sums, provided that this limit exists.

---

**Definition** *Sum of an Infinite Series*

We say that the infinite series

$$\sum_{n=1}^{\infty} a_n \text{ converges (or is convergent)}$$

with **sum** $S$ provided that the limit of its sequence of partial sums,

$$S = \lim_{n \to \infty} S_n, \tag{3}$$

exists (and is finite). Otherwise, we say that the series **diverges** (or is **divergent**). If a series diverges then it has no sum.

---

Thus an infinite sum is a limit of finite sums,

$$S = \sum_{n=1}^{\infty} a_n = \lim_{N \to \infty} \sum_{n=1}^{N} a_n,$$

provided that this limit exists.

**EXAMPLE 1** Show that the series

$$\sum_{n=1}^{\infty} (\tfrac{1}{2})^n = \tfrac{1}{2} + \tfrac{1}{4} + \tfrac{1}{8} + \tfrac{1}{16} + \cdots$$

converges, and find its sum.

**Solution** The first four partial sums are

$$S_1 = \tfrac{1}{2}, \qquad S_2 = \tfrac{3}{4}, \qquad S_3 = \tfrac{7}{8}, \qquad S_4 = \tfrac{15}{16}.$$

It seems likely that $S_n = (2^n - 1)/2^n$, and indeed this follows easily by induction on $n$, since

$$S_{n+1} = S_n + \frac{1}{2^{n+1}} = \frac{2^n - 1}{2^n} + \frac{1}{2^{n+1}}$$

$$= \frac{2^{n+1} - 2 + 1}{2^{n+1}} = \frac{2^{n+1} - 1}{2^{n+1}}.$$

Hence the sum of the given series is

$$S = \lim_{n \to \infty} S_n = \lim_{n \to \infty} \frac{2^n - 1}{2^n} = \lim_{n \to \infty} \left( 1 - \frac{1}{2^n} \right) = 1.$$

**EXAMPLE 2** Show that the series

$$\sum_{n=1}^{\infty} (-1)^{n+1} = 1 - 1 + 1 - 1 + 1 - 1 + \cdots$$

diverges.

*Solution*   The sequence of partial sums of the given series is

$$1, 0, 1, 0, 1, 0, 1, \ldots,$$

and this sequence has no limit. Therefore the series diverges.

**EXAMPLE 3**   Show that the infinite series

$$\sum_{n=1}^{\infty} \frac{1}{4n^2 - 1} = \frac{1}{3} + \frac{1}{15} + \frac{1}{35} + \frac{1}{63} + \cdots$$

converges, and find its sum.

*Solution*   We need a formula for the $n$th partial sum $S_n$ so that we can evaluate its limit as $n \to \infty$. To find such a formula, we begin with the observation that

$$a_n = \frac{1}{4n^2 - 1} = \frac{1}{2}\left(\frac{1}{2n - 1} - \frac{1}{2n + 1}\right).$$

It follows that

$$S_n = \frac{1}{2}\left(1 - \frac{1}{3}\right) + \frac{1}{2}\left(\frac{1}{3} - \frac{1}{5}\right) + \frac{1}{2}\left(\frac{1}{5} - \frac{1}{7}\right)$$

$$+ \cdots + \frac{1}{2}\left(\frac{1}{2n - 1} - \frac{1}{2n + 1}\right)$$

$$= \frac{1}{2}\left(1 - \frac{1}{2n + 1}\right) = \frac{n}{2n + 1}.$$

And hence

$$\sum_{n=1}^{\infty} \frac{1}{4n^2 - 1} = \lim_{n \to \infty} \frac{n}{2n + 1} = \frac{1}{2}.$$

The sum for $S_n$ in Example 3 is called a *telescoping* sum and provides us with a way to find the sums of certain series. The series in Examples 1 and 2 are examples of a much more common sort of series, the *geometric series*.

---

*Definition*   *Geometric Series*

The series $\displaystyle\sum_{n=0}^{\infty} a_n$ is said to be a **geometric series** if each term after the first is a fixed multiple of the term before it; that is, if there is a number $r$—called the **ratio** of the series—such that

$$a_{n+1} = ra_n$$

for all $n \geq 0$.

---

Thus every geometric series takes the form

$$a_0 + ra_0 + r^2 a_0 + r^3 a_0 + \cdots = \sum_{n=0}^{\infty} r^n a_0. \qquad (4)$$

Note that it is convenient to begin the summation at $n = 0$, and thus we regard the sum

$$S_n = a_0(1 + r + r^2 + \cdots + r^n)$$

of the first $n + 1$ terms as the $n$th partial sum of the series.

**EXAMPLE 4** The infinite series

$$\sum_{n=0}^{\infty} \frac{2}{3^n} = 2 + \frac{2}{3} + \frac{2}{9} + \cdots + \frac{2}{3^n} + \cdots$$

is a geometric series with first term $a_0 = 2$ and ratio $r = \frac{1}{3}$.

---

**Theorem 1** *Sum of a Geometric Series*

If $|r| < 1$ then the geometric series in (4) converges, and its sum is

$$S = \sum_{n=0}^{\infty} r^n a_0 = \frac{a_0}{1 - r}. \qquad (5)$$

If $|r| \geq 1$ and $a_0 \neq 0$, then the geometric series diverges.

---

**Proof** If $r = 1$, then $S_n = (n + 1)a_0$, so the series certainly diverges. If $r = -1$, then it diverges by an argument like the one in Example 2. So we suppose that $r \neq 1$ and $r \neq -1$. Then the elementary identity

$$1 + r + r^2 + \cdots + r^n = \frac{1 - r^{n+1}}{1 - r}$$

follows after multiplying each side by $1 - r$. Hence

$$S_n = a_0(1 + r + r^2 + \cdots + r^n)$$

$$= a_0\left(\frac{1}{1 - r} - \frac{r^{n+1}}{1 - r}\right).$$

If $|r| < 1$, then $r^{n+1} \to 0$ as $n \to \infty$, by Example 7 in Section 12-2. So in this case

$$S = \lim_{n \to \infty} a_0\left(\frac{1}{1 - r} - \frac{r^{n+1}}{1 - r}\right) = \frac{a_0}{1 - r}.$$

If $|r| > 1$, then $\lim_{n \to \infty} r^{n+1}$ does not exist, and so $\lim_{n \to \infty} S_n$ does not exist. This establishes the theorem. ∎

**EXAMPLE 5** With $a_0 = 1$ and $r = -\frac{1}{2}$, we find that

$$1 - \frac{1}{2} + \frac{1}{4} - \frac{1}{8} + \cdots = \sum_{n=0}^{\infty} \left(-\frac{1}{2}\right)^n$$

$$= \frac{1}{1 - (-\frac{1}{2})} = \frac{2}{3}.$$

---

The following theorem says that the operations of addition and multiplication by a constant can be carried out termwise in the case of *convergent* series. Since the sum of an infinite series is the limit of its sequence of partial sums, this theorem follows immediately from the limit laws for sequences (Theorem 1 in Section 12-2).

---

*Theorem 2  Termwise Addition and Multiplication*

If the series $A = \sum a_n$ and $B = \sum b_n$ converge to the indicated sums and $c$ is a constant, then the series $\sum (a_n + b_n)$ and $\sum ca_n$ also converge, with sums

(i) $\sum (a_n + b_n) = A + B$;

(ii) $\sum ca_n = cA$.

---

The geometric series formula in (5) may be used to find the rational number represented by a given repeating infinite decimal.

**EXAMPLE 6**

$$0.55555 \cdots = \frac{5}{10} + \frac{5}{100} + \frac{5}{1000} + \cdots$$

$$= \sum_{n=0}^{\infty} \frac{5}{10} \left( \frac{1}{10} \right)^n = \frac{\frac{5}{10}}{1 - \left( \frac{1}{10} \right)} = \frac{5}{9}.$$

In a more complicated example, we may wish to use the "termwise algebra" of Theorem 2:

$$0.728\ 28\ 28\ 28 \ldots = \frac{7}{10} + \frac{28}{10^3} + \frac{28}{10^5} + \frac{28}{10^7} + \cdots$$

$$= \frac{7}{10} + \frac{28}{10^3} \left( 1 + \frac{1}{10^2} + \frac{1}{10^4} + \cdots \right)$$

$$= \frac{7}{10} + \frac{28}{1000} \sum_{n=0}^{\infty} \left( \frac{1}{10^2} \right)^n = \frac{7}{10} + \frac{28}{1000} \left( \frac{1}{1 - \frac{1}{100}} \right)$$

$$= \frac{7}{10} + \frac{28}{1000} \left( \frac{100}{99} \right) = \frac{7}{10} + \frac{28}{990} = \frac{721}{990}.$$

It should be apparent that this technique can be generalized to show that every repeating infinite decimal represents a rational number; consequently, the decimal expansions of irrational numbers such as $\pi$, $e$, and $\sqrt{2}$ must be nonrepeating as well as infinite. Conversely, if $p$ and $q$ are integers with $q \neq 0$, then long division of $q$ into $p$ yields a repeating decimal expansion for the rational number $p/q$ because such a division can yield at each stage only $q$ possible different remainders.

**EXAMPLE 7**  Suppose that Peter, Paul, and Mary toss a six-sided die in turn until one of them wins by getting the first six. Calculate the probability that it is Mary who wins.

*Solution*  We assume that the die is "fair," so the probability that Mary gets a six on the first round is $(\frac{5}{6})(\frac{5}{6})(\frac{1}{6})$—the product of the probabilities that Peter and Paul each roll something else and Mary rolls a six. The probability that she gets the first six on the second round is $(\frac{5}{6})^3(\frac{5}{6})(\frac{5}{6})(\frac{1}{6})$—the product of the probability that no one rolls a six on the first round, $(\frac{5}{6})^3$, and the probability that only Mary rolls a six on the second round. Mary's probability $p$ of getting the first six in the game is the sum of her probabilities of getting it on the first round, on the second round, on the third round, and so on. Hence

$$p = \left(\frac{5}{6}\right)\left(\frac{5}{6}\right)\left(\frac{1}{6}\right) + \left(\frac{5}{6}\right)^3\left(\frac{5}{6}\right)\left(\frac{5}{6}\right)\left(\frac{1}{6}\right)$$

$$+ \left(\frac{5}{6}\right)^6\left(\frac{5}{6}\right)\left(\frac{5}{6}\right)\left(\frac{1}{6}\right) + \cdots$$

$$= \left(\frac{5}{6}\right)\left(\frac{5}{6}\right)\left(\frac{1}{6}\right) \sum_{n=0}^{\infty} \left[\left(\frac{5}{6}\right)^3\right]^n$$

$$= \frac{25}{216} \cdot \frac{1}{1 - (\frac{5}{6})^3} = \frac{25}{91}.$$

Observe that because she tosses last in sequence, Mary has less than the fair probability $\frac{1}{3}$ of getting the first six and thus winning the game. In Problem 53 we ask you to calculate Peter's and Paul's probabilities of rolling the first six.

---

The following theorem is often useful in showing that a given series does *not* converge.

> **Theorem 3   The nth Term Test for Divergence**
> If $\lim\limits_{n \to \infty} a_n \neq 0$, then the infinite series $\sum a_n$ diverges.

Here the hypothesis $\lim\limits_{n \to \infty} a_n \neq 0$ is to have its most liberal interpretation— that the limit does not exist *or*, should it exist, is not equal to zero.

**Proof**   We want to show under the stated hypothesis that the series $\sum a_n$ diverges. It suffices to show that, *if* the series $\sum a_n$ does converge, then $\lim\limits_{n \to \infty} a_n = 0$. So suppose that $\sum a_n$ converges with sum $S = \lim\limits_{n \to \infty} S_n$, where

$$S_n = a_1 + a_2 + a_3 + \cdots + a_n.$$

We note that $a_n = S_n - S_{n-1}$, so that

$$\lim_{n \to \infty} a_n = \lim_{n \to \infty} (S_n - S_{n-1})$$

$$= \lim_{n \to \infty} S_n - \lim_{n \to \infty} S_{n-1} = S - S = 0.$$

Consequently, if $\lim\limits_{n \to \infty} a_n \neq 0$, then the series $\sum a_n$ diverges.   ∎

**EXAMPLE 8**   The series

$$\sum_{n=1}^{\infty} (-1)^{n-1} n^2 = 1 - 4 + 9 - 16 + 25 - \cdots$$

diverges because $\lim\limits_{n \to \infty} a_n$ does not exist, while the series

$$\sum_{n=1}^{\infty} \frac{n}{3n+1} = \frac{1}{4} + \frac{2}{7} + \frac{3}{10} + \frac{4}{13} + \cdots$$

diverges because

$$\lim_{n \to \infty} \frac{n}{3n+1} = \frac{1}{3} \neq 0.$$

WARNING  The converse of Theorem 3 is false. The condition $\lim\limits_{n \to \infty} a_n = 0$ is necessary *but not sufficient* for convergence of the series $\sum a_n$. That is, a series may satisfy the condition $\lim\limits_{n \to \infty} a_n = 0$ and yet diverge. An important example of a divergent series with terms that approach zero is the *harmonic series*

$$\sum_{n=1}^{\infty} \frac{1}{n} = 1 + \frac{1}{2} + \frac{1}{3} + \frac{1}{4} + \frac{1}{5} + \cdots. \tag{6}$$

> **Theorem 4**  *The harmonic series diverges.*

***Proof***  Since each term of the harmonic series is positive, its sequence of partial sums $\{S_n\}$ is monotone increasing. We shall prove that $\lim\limits_{n \to \infty} S_n = \infty$, and thus that the harmonic series diverges, by showing that there are arbitrarily large partial sums. Consider the following grouping of terms:

$$\sum_{n=1}^{\infty} \frac{1}{n} = 1 + \frac{1}{2} + \left( \frac{1}{3} + \frac{1}{4} \right) + \left( \frac{1}{5} + \cdots + \frac{1}{8} \right)$$

$$+ \left( \frac{1}{9} + \cdots + \frac{1}{16} \right) + \left( \frac{1}{17} + \cdots + \frac{1}{32} \right) + \cdots$$

$$= S_2 + (S_4 - S_2) + (S_8 - S_4) + (S_{16} - S_8) + (S_{32} - S_{16}) + \cdots.$$

The sum of each group of terms within parentheses is greater than $\frac{1}{2}$ because

$$S_{2n} - S_n = \frac{1}{n+1} + \frac{1}{n+2} + \cdots + \frac{1}{2n}$$

$$> \frac{1}{2n} + \frac{1}{2n} + \cdots + \frac{1}{2n} = \frac{1}{2}.$$

Hence

$$S_{2^k} = S_2 + (S_4 - S_2) + (S_8 - S_4) + \cdots + (S_{2^k} - S_{2^{k-1}})$$
$$> \tfrac{3}{2} + \tfrac{1}{2} + \tfrac{1}{2} + \cdots + \tfrac{1}{2} \quad (k \text{ terms})$$
$$= \tfrac{3}{2} + \tfrac{1}{2}(k-1) = 1 + \tfrac{1}{2}k.$$

If $n > 2^k$, then $S_n > S_{2^k} > 1 + \frac{1}{2}k$. Since $1 + \frac{1}{2}k$ can be made arbitrarily large, it follows that

$$\lim_{n \to \infty} S_n = \infty.$$

Since its sequence of partial sums has no (finite) limit, the harmonic series therefore diverges. ∎

If the sequence of partial sums of the series $\sum a_n$ diverges to infinity, then we say that the series itself **diverges to infinity,** and we write

$$\sum a_n = \infty.$$

The series $\sum (-1)^{n+1}$ of Example 2 is one that diverges but does not diverge to infinity. In the nineteenth century it was common to say that such a series was divergent by oscillation; today we say merely that it diverges.

Our proof of Theorem 4 shows that $\sum 1/n = \infty$. But the partial sums of the harmonic series diverge to infinity very slowly. If $N_A$ denotes the smallest integer such that

$$\sum_{n=1}^{N_A} \frac{1}{n} \geqq A,$$

then it is known that

$$N_5 = 83; \qquad \text{(This can be verified with the aid of a}$$
$$N_{10} = 12{,}367; \qquad \text{programmable calculator.)}$$
$$N_{20} = 272{,}400{,}600;$$
$$N_{100} \approx 1.5 \times 10^{43};$$
$$N_{1000} \approx 1.1 \times 10^{434}.$$

Thus you would need to add more than a quarter of a billion terms of the harmonic series to get a partial sum over 20. At this point the next few terms would each be approximately $0.000000004 = 4 \times 10^{-9}$. The number of terms you'd have to add to reach 1000 is far larger than the estimated number of elementary particles in the entire universe ($10^{80}$). If you enjoy such large numbers, see the article *Partial Sums of Infinite Series, and How They Grow* by R. P. Boas, Jr., in the American Mathematical Monthly, 84 (1977), 237–48.

The following theorem says that if two infinite series have the same terms from some point on, then either both converge or both diverge. The proof is left for Problem 43.

---

*Theorem 5   Series That Are Eventually the Same*

If there exists a positive integer $k$ such that $a_n = b_n$ for all $n > k$, then the series $\sum a_n$ and $\sum b_n$ either both converge or both diverge.

---

It follows that a *finite* number of terms can be changed, deleted from, or adjoined to an infinite series without altering its convergence or divergence (although the *sum* of a convergent series will be changed by such alterations). In particular, taking $b_n = 0$ for $n \leq k$ and $b_n = a_n$ for $n > k$, we see that the series $\sum_{n=1}^{\infty} a_n$ and the series $\sum_{n=k+1}^{\infty} a_n$ obtained by deleting its first $k$ terms either both converge or both diverge.

## *SAFE DRUG DOSAGE

As a typical application of Formula (5) for the sum of a geometric series, we use it to determine the minimum permissible time between successive doses of a certain drug administered to a patient. Suppose that the patient receives a fixed dose every $t_0$ hours, and that a single dose of this particular drug results in an immediate bloodstream concentration of $C_0$. As we saw in Section 7-5, the concentration $t$ hours after administration of a single dose is $C_0 e^{-kt}$, where $k$ is the elimination constant of the drug. The concentration resulting from $n$ successive doses, immediately before taking the $(n + 1)$st dose, is then

$$C_0 e^{-kt_0} + C_0 e^{-2kt_0} + \cdots + C_0 e^{-nkt_0}.$$

It is therefore natural to expect the *residual concentration* of this drug—resulting from an extended regime of dosage at the prescribed level—to be

$$R = \sum_{n=1}^{\infty} C_0 e^{-nkt_0} = C_0 e^{-kt_0} \sum_{n=0}^{\infty} (e^{-kt_0})^n.$$

We take $r = e^{-kt_0}$ in Formula (5), and find the residual concentration to be

$$R = \frac{C_0 e^{-kt_0}}{1 - e^{-kt_0}}. \tag{7}$$

This will be the expected bloodstream concentration before each new dose.

Now let $C_S$ denote the highest concentration that is safe (or otherwise desirable). In order that the concentration immediately *after* each dose not exceed $C_S$, we require that

$$C_0 + R = C_0 \left( 1 + \frac{e^{-kt_0}}{1 - e^{-kt_0}} \right) \leqq C_S.$$

After we solve for $t_0$, we find that this requirement amounts to

$$t_0 \geqq \frac{1}{k} \ln\left( \frac{C_S}{C_S - C_0} \right). \tag{8}$$

The quantity on the right in (8) is, then, the minimum permissible time between successive doses.

**EXAMPLE 9** Suppose that the drug under consideration is alcohol, and that the contemplated dose of 1 oz produces a bloodstream concentration of $C_0 = 0.05\%$ (a volume concentration). How often can this dose be taken if the bloodstream concentration is not to exceed $C_S = 0.15\%$? Assume an elimination constant that corresponds to half of the blood alcohol being eliminated in 3 h.

***Solution*** We are given that $C_0 e^{-3k} = \frac{1}{2} C_0$, so the elimination constant is $k = \frac{1}{3} \ln 2$. From (8) we therefore obtain

$$t_0 = \frac{3}{\ln 2} \ln\left( \frac{0.15}{0.15 - 0.05} \right) \approx 1.755 \text{ h},$$

or about 1 h 45 min.

In Problems 1–29, determine whether each given infinite series converges or diverges. If it converges, find its sum.

**1** $1 + \frac{1}{3} + \frac{1}{9} + \cdots + (\frac{1}{3})^n + \cdots$

**2** $1 + e^{-1} + e^{-2} + \cdots + e^{-n} + \cdots$

**3** $1 + 3 + 5 + 7 + \cdots + (2n - 1) + \cdots$

**4** $\dfrac{1}{2} + \dfrac{1}{\sqrt{2}} + \dfrac{1}{\sqrt[3]{2}} + \cdots + \dfrac{1}{\sqrt[n]{2}} + \cdots$

**5** $1 - 2 + 4 - 8 + \cdots + (-2)^n + \cdots$

**6** $1 - \frac{1}{4} + \frac{1}{16} - \cdots + (-\frac{1}{4})^n + \cdots$

**7** $4 + \dfrac{4}{3} + \dfrac{4}{9} + \dfrac{4}{27} + \cdots + \dfrac{4}{3^n} + \cdots$

**8** $\dfrac{1}{3} + \dfrac{2}{9} + \dfrac{4}{27} + \cdots + \dfrac{2^{n-1}}{3^n} + \cdots$

**9** $1 + (1.01) + (1.01)^2 + \cdots + (1.01)^n + \cdots$

**10** $1 + \dfrac{1}{\sqrt{2}} + \dfrac{1}{\sqrt[3]{3}} + \cdots + \dfrac{1}{\sqrt[n]{n}} + \cdots$

**11** $\displaystyle\sum_{n=0}^{\infty} \frac{(-1)^n n}{n + 1}$

**12** $\displaystyle\sum_{n=1}^{\infty} \left(\frac{e}{10}\right)^n$

**13** $\displaystyle\sum_{n=0}^{\infty} (-1)^n \left(\frac{3}{e}\right)^n$

**14** $\displaystyle\sum_{n=0}^{\infty} \frac{3^n - 2^n}{4^n}$

**15** $\displaystyle\sum_{n=1}^{\infty} (\sqrt{2})^{1-n}$

**16** $\displaystyle\sum_{n=1}^{\infty} \left(\frac{2}{n} - \frac{1}{2^n}\right)$

**17** $\displaystyle\sum_{n=1}^{\infty} \frac{n}{10n + 17}$

**18** $\displaystyle\sum_{n=1}^{\infty} \frac{\sqrt{n}}{\ln(n + 1)}$

**19** $\displaystyle\sum_{n=1}^{\infty} (5^{-n} - 7^{-n})$

**20** $\displaystyle\sum_{n=0}^{\infty} \frac{1}{1 + (\frac{9}{10})^n}$

**21** $\displaystyle\sum_{n=1}^{\infty} \left(\frac{e}{\pi}\right)^n$

**22** $\displaystyle\sum_{n=1}^{\infty} \left(\frac{\pi}{e}\right)^n$

**23** $\displaystyle\sum_{n=0}^{\infty} \left(\frac{100}{99}\right)^n$

**24** $\displaystyle\sum_{n=0}^{\infty} \left(\frac{99}{100}\right)^n$

**25** $\displaystyle\sum_{n=0}^{\infty} \frac{1 + 2^n + 3^n}{5^n}$

**26** $\displaystyle\sum_{n=0}^{\infty} \frac{1 + 2^n + 5^n}{3^n}$

**27** $\displaystyle\sum_{n=0}^{\infty} \frac{7(5)^n + 3(11)^n}{13^n}$

**28** $\displaystyle\sum_{n=1}^{\infty} \sqrt[n]{2}$

**29** $\displaystyle\sum_{n=1}^{\infty} [(\tfrac{7}{11})^n - (\tfrac{3}{5})^n]$

**30** Use the method of Example 6 to verify that:
(a) $0.666666666\ldots = \frac{2}{3}$;
(b) $0.111111111\ldots = \frac{1}{9}$;
(c) $0.249999999\ldots = \frac{1}{4}$;
(d) $0.999999999\ldots = 1$.

In Problems 31–35, find the rational number represented by the given repeating decimal.

**31** $0.474747\ldots$      **32** $0.252525\ldots$

**33** $0.123123123\ldots$      **34** $0.337733773377\ldots$

**35** $3.141591415914159\ldots$

In each of Problems 36–40, use the method of Example 3 to find a formula for the $n$th partial sum $S_n$, and then compute the sum of the given infinite series.

**36** $\displaystyle\sum_{n=1}^{\infty} \frac{1}{n(n + 1)}$

**37** $\displaystyle\sum_{n=1}^{\infty} \ln\left(\frac{n + 1}{n}\right)$

**38** $\displaystyle\sum_{n=0}^{\infty} \frac{4}{16n^2 - 8n - 3}$

**39** $\displaystyle\sum_{n=2}^{\infty} \frac{2}{n^2 - 1}$

**40** $\dfrac{1}{1 \cdot 3} - \dfrac{1}{2 \cdot 4} + \dfrac{1}{3 \cdot 5} - \dfrac{1}{4 \cdot 6} + \cdots$

**41** Prove: If $\sum a_n$ diverges and $c$ is a nonzero constant, then $\sum c a_n$ diverges.

**42** Suppose that $\sum a_n$ converges and that $\sum b_n$ diverges. Prove that $\sum (a_n + b_n)$ diverges.

**43** Let $S_n$ and $T_n$ denote the $n$th partial sums of $\sum a_n$ and $\sum b_n$, respectively. Suppose that $a_n = b_n$ for all $n > k$. Show that $S_n - T_n = S_k - T_k$ for $n > k$. Hence prove Theorem 5.

**44** Suppose that $0 < x \le 1$. Integrate both sides of the identity

$$\frac{1}{1 + t} = 1 - t + t^2 - \cdots + (-1)^n t^n + \frac{(-1)^{n+1} t^{n+1}}{1 + t}$$

from $t = 0$ to $t = x$ to show that

$$\ln(1 + x) = x - \frac{x^2}{2} + \frac{x^3}{3} - \cdots + (-1)^n \frac{x^{n+1}}{n + 1} + R_n$$

where $\lim\limits_{n \to \infty} R_n = 0$. Hence conclude that

$$\ln(1 + x) = \sum_{n=1}^{\infty} (-1)^{n+1} \frac{x^n}{n}$$

if $0 < x \le 1$.

**45** Criticize the following "proof" that $2 = 1$. Substitution of $x = 1$ into the result of Problem 44 gives the fact that

$$\ln 2 = 1 - \tfrac{1}{2} + \tfrac{1}{3} - \tfrac{1}{4} + \cdots.$$

If

$$S = 1 + \tfrac{1}{2} + \tfrac{1}{3} + \tfrac{1}{4} + \cdots,$$

then

$$\ln 2 = S - 2(\tfrac{1}{2} + \tfrac{1}{4} + \tfrac{1}{6} + \tfrac{1}{8} + \cdots) = S - S = 0.$$

Hence $2 = e^{\ln 2} = e^0 = 1$.

**46** A ball has *bounce coefficient* $r < 1$ if, when it is dropped from height $h$, it bounces back to a height of $rh$. Suppose that such a ball is dropped from the initial height $a$ and subsequently bounces infinitely many times. Use a geometric series to show that the total up-and-down distance it travels in all its bouncing is

$$D = a\,\frac{1 + r}{1 - r}.$$

**47** A ball with bounce coefficient $r = 0.64$ (see Problem 46) is dropped from an initial height of $a = 4$ ft. Use a geometric series to compute the total time required for it to complete its infinitely many bounces. The time required for a ball to drop $h$ feet (from rest) is $\sqrt{2h/g}$ seconds where $g = 32$ ft/s$^2$.

**48** Suppose that the government spends $1 billion and that each recipient spends 90% of the dollars he or she receives. In turn, the secondary recipients spend 90% of the dollars they receive, and so on. How much total spending results from the original injection of $1 billion into the economy?

**49** A tank initially contains a mass $M_0$ of air. Each stroke of a vacuum pump removes 5% of the air in the container. Compute

(a) the mass $M_n$ of air remaining in the tank after $n$ strokes of the pump;

(b) $\lim\limits_{n \to \infty} M_n$.

**50** It requires 1 day for half of a dose of a certain drug to be eliminated from the bloodstream. The maximum safe concentration of the drug is three times that resulting from a single dose. How often can doses of this drug be taken?

**51** A pane of glass of a certain material reflects half the incident light, absorbs a fourth, and transmits a fourth. A window is made of two panes of this glass separated by a small space, as shown in Fig. 12.3. How much of the incident light $I$ is transmitted by the double window?

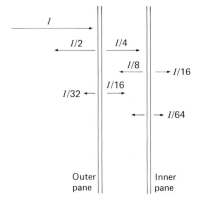

**12.3** Diagram of the double-pane window of Problem 51.

**52** If Peter, Paul, and Mary toss a fair coin in turn until one of them wins by getting the first head, calculate for each the probability that he or she wins the game. Check your answers by verifying that the sum of the three probabilities is 1.

**53** In Example 7, calculate Peter's and Paul's probabilities of getting the first six.

---

## 12-4

## The Integral Test

Given an infinite series $\sum a_n$, it is the exception rather than the rule when a nice formula for its $n$th partial sum $S_n$ can be found and used directly to determine whether the series converges or diverges. There are, however, several *convergence tests* that involve the *terms* of an infinite series rather than its partial sums. Such a test will (when successful) tell us whether or not the series converges. Once we know that a series $\sum a_n$ does converge, it is then a separate matter actually to find its sum $S$. It may be necessary to approximate $S$ by adding up sufficiently many terms; in this case we shall need to know how many terms are required for the desired accuracy.

In this section and the following one, we concentrate our attention on **positive term series**—that is, those with terms that are all positive. If $a_n > 0$ for all $n$, then

$$S_1 < S_2 < S_3 < \cdots < S_n < \cdots,$$

so the sequence of partial sums $\{S_n\}$ is monotone increasing. Hence there are just two possibilities. If the sequence $\{S_n\}$ is *bounded*—there exists a number $M$ such that $S_n \leq M$ for all $n$—then the bounded monotone sequence property (Section 12-2) implies that $S = \lim\limits_{n \to \infty} S_n$ exists, so the series $\sum a_n$ converges. Otherwise it diverges to infinity (by Problem 38 in Section 12-2).

A similar alternative holds for improper integrals. Suppose that the function $f$ is continuous and positive-valued for $x \geq 1$. Then it follows (Problem 35) that the improper integral

$$\int_1^\infty f(x)\,dx = \lim_{b \to \infty} \int_1^b f(x)\,dx \tag{1}$$

either converges (the limit is a real number) or diverges to infinity (the limit is $+\infty$). This analogy between positive-term series and improper integrals of positive functions is the key to the **integral test.** We compare the behavior of the series $\sum a_n$ with that of the improper integral in (1), where $f$ is an appropriately selected function. (Among other things, we require that $f(n) = a_n$ for all $n \geq 1$. For example, in the case of the harmonic series $\sum 1/n$, we would take $f(x) = 1/x$.)

---

**Theorem 1  The Integral Test**

Suppose that $\sum a_n$ is a positive-term series and that $f$ is a positive-valued, decreasing, continuous function for $x \geq 1$. If $f(n) = a_n$ for all integers $n \geq 1$, then the series and the improper integral

$$\sum_{n=1}^{\infty} a_n \quad \text{and} \quad \int_1^\infty f(x)\,dx$$

either both converge or both diverge.

---

**Proof**  Since $f$ is a decreasing function, the rectangular polygon with area

$$S_n = a_1 + a_2 + \cdots + a_n$$

shown in Fig. 12.4 contains the region under $y = f(x)$ from $x = 1$ to $x = n + 1$. Hence

$$\int_1^{n+1} f(x)\,dx \leq S_n. \tag{2}$$

**12.4**  Underestimating the partial sums with an integral

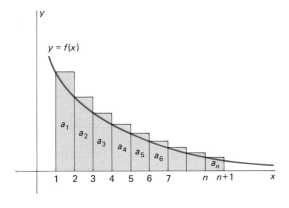

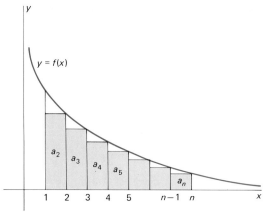

12.5 Overestimating the partial sums with an integral

Similarly, the rectangular polygon with area

$$S_n - a_1 = a_2 + a_3 + \cdots + a_n$$

shown in Fig. 12.5 is contained in the region under $y = f(x)$ from $x = 1$ to $x = n$. Hence

$$S_n - a_1 \leqq \int_1^n f(x)\,dx. \tag{3}$$

Suppose first that the improper integral $\int_1^\infty f(x)\,dx$ diverges (to infinity). Then

$$\lim_{n \to \infty} \int_1^{n+1} f(x)\,dx = \infty,$$

so it follows from (2) that $\lim_{n \to \infty} S_n = \infty$ also, and hence the infinite series $\sum a_n$ likewise diverges.

On the other hand, suppose that the improper integral $\int_1^\infty f(x)\,dx$ converges, with (finite) value $I$. Then (3) implies that

$$S_n \leqq a_1 + \int_1^n f(x)\,dx \leqq a_1 + I,$$

so the monotone increasing sequence $\{S_n\}_1^\infty$ is bounded. Thus the infinite series $\sum a_n = \lim_{n \to \infty} S_n$ converges also. Hence we have shown that the infinite series and the improper integral either both converge or both diverge. ∎

**EXAMPLE 1**  The integral test gives another proof that the harmonic series

$$\sum_{n=1}^\infty \frac{1}{n} = 1 + \frac{1}{2} + \frac{1}{3} + \cdots + \frac{1}{n} + \cdots$$

diverges. The function $f(x) = 1/x$ is positive, continuous, and decreasing for $x \geqq 1$, and

$$\int_1^\infty \frac{dx}{x} = \lim_{b \to \infty} \int_1^b \frac{dx}{x} = \lim_{b \to \infty} \left[ \ln x \right]_1^b$$

$$= \lim_{b \to \infty} (\ln b - \ln 1) = +\infty.$$

Thus the improper integral diverges and, therefore, so does the harmonic series.

The harmonic series is the case $p = 1$ of the **p-series**

$$\sum_{n=1}^{\infty} \frac{1}{n^p} = 1 + \frac{1}{2^p} + \frac{1}{3^p} + \cdots + \frac{1}{n^p} + \cdots. \tag{4}$$

Whether the $p$-series converges or diverges depends upon the value of $p$.

**EXAMPLE 2** Show that the $p$-series converges if $p > 1$ but diverges if $0 < p < 1$.

**Solution** If $p > 0$ but $p \neq 1$, then the function $f(x) = 1/x^p$ satisfies the conditions of the integral test, and

$$\int_1^{\infty} \frac{dx}{x^p} = \lim_{b \to \infty} \int_1^b \frac{dx}{x^p} = \lim_{b \to \infty} \left[ \frac{-1}{(p-1)x^{p-1}} \right]_1^b$$

$$= \lim_{b \to \infty} \frac{1}{p-1} \left( 1 - \frac{1}{b^{p-1}} \right).$$

If $p > 1$, then

$$\int_1^{\infty} \frac{dx}{x^p} = \frac{1}{p-1} < \infty,$$

so the integral and the series both converge. But if $0 < p < 1$, then

$$\int_1^{\infty} \frac{dx}{x^p} = \lim_{b \to \infty} \frac{1}{1-p} (b^{1-p} - 1) = \infty,$$

and in this case the integral and the series both diverge.

For example, the series

$$\sum_{n=1}^{\infty} \frac{1}{n^2} = 1 + \frac{1}{2^2} + \frac{1}{3^2} + \frac{1}{4^2} + \cdots + \frac{1}{n^2} + \cdots$$

converges ($p = 2$), while the series

$$\sum_{n=1}^{\infty} \frac{1}{\sqrt{n}} = 1 + \frac{1}{\sqrt{2}} + \frac{1}{\sqrt{3}} + \frac{1}{\sqrt{4}} + \cdots + \frac{1}{\sqrt{n}} + \cdots$$

diverges ($p = \frac{1}{2}$).

Now suppose that the positive-term series $\sum a_n$ converges by the integral test, and we wish to approximate its sum by adding up sufficiently many of its terms. The difference between the sum $S$ and the $n$th partial sum $S_n$ is the **remainder**

$$R_n = S - S_n = a_{n+1} + a_{n+2} + a_{n+3} + \cdots. \tag{5}$$

This remainder is the error made when the actual sum $S$ is estimated by using in its place the partial sum $S_n$.

**Theorem 2** *Integral Test Remainder Estimate*

Suppose that the infinite series and improper integral

$$\sum_{n=1}^{\infty} a_n \quad \text{and} \quad \int_1^{\infty} f(x)\, dx$$

satisfy the hypotheses of the integral test, and that both converge. Then

$$\int_{n+1}^{\infty} f(x)\, dx \leq R_n \leq \int_n^{\infty} f(x)\, dx \tag{6}$$

where $R_n$ is the remainder given in (5).

**Proof** From Fig. 12.6 we see that

$$\int_k^{k+1} f(x)\, dx \leq a_k \leq \int_{k-1}^{k} f(x)\, dx$$

for $k = n + 1, n + 2, \ldots$. We add these inequalities for all such values of $k$, and the result is (6), since

$$\sum_{k=n+1}^{\infty} \int_k^{k+1} f(x)\, dx = \int_{n+1}^{\infty} f(x)\, dx$$

and

$$\sum_{k=n+1}^{\infty} \int_{k-1}^{k} f(x)\, dx = \int_n^{\infty} f(x)\, dx. \qquad \blacksquare$$

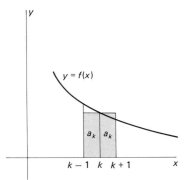

**12.6** Establishing the integral test remainder estimate

In Section 12-8 we will see that the exact sum of the $p$-series with $p = 2$ is $\pi^2/6$, thus giving the beautiful formula

$$\frac{\pi^2}{6} = 1 + \frac{1}{2^2} + \frac{1}{3^2} + \frac{1}{4^2} + \cdots. \tag{7}$$

**EXAMPLE 3** Approximate the number $\pi$ by applying the integral test remainder estimate with $n = 50$ to the series in Equation (7).

**Solution** The sum of the first 50 terms in (7) is, to the accuracy shown,

$$\sum_{n=1}^{50} \frac{1}{n^2} = 1.6251327.$$

You can add these 50 terms one by one on a pocket calculator in a few minutes, but this is precisely the sort of thing a computer or programmable calculator is really good for.

Because

$$\int_a^{\infty} \frac{dx}{x^2} = \lim_{b \to \infty} \left[ -\frac{1}{x} \right]_a^b = \frac{1}{a},$$

the integral test remainder estimate with $f(x) = 1/x^2$ gives $\frac{1}{51} \leq R_{50} \leq \frac{1}{50}$, so

$$\frac{1}{51} + 1.6251327 \leq \frac{\pi^2}{6} \leq \frac{1}{50} + 1.6251327.$$

We multiply through by 6, extract the square root, and round to four-place accuracy. The result is that

$$3.1414 < \pi < 3.1418.$$

With $n = 200$, a programmable calculator gives

$$\sum_{n=1}^{200} \frac{1}{n^2} = 1.6399465,$$

so

$$\frac{1}{201} + 1.6399465 \leq \frac{\pi^2}{6} \leq \frac{1}{200} + 1.6399465.$$

This leads to the inequality $3.14158 < \pi < 3.14161$, and it follows that $\pi = 3.1416$ rounded to four decimal places.

**EXAMPLE 4** Show that the series

$$\sum_{n=2}^{\infty} \frac{1}{n(\ln n)^2}$$

converges, and find how many terms you would need to add to find its sum accurate to within 0.01.

*Solution* We take $f(x) = 1/x(\ln x)^2$ and find that

$$\int_2^\infty \frac{dx}{x(\ln x)^2} = \lim_{b \to \infty} \left[ -\frac{1}{\ln x} \right]_2^b = \frac{1}{\ln 2} < \infty,$$

so the series does converge by the integral test. We want to choose $n$ sufficiently large that $R_n < 0.01$. The right-hand inequality in (6) gives

$$R_n \leq \int_n^\infty \frac{dx}{x(\ln x)^2} = \frac{1}{\ln n}.$$

Hence we need

$$\frac{1}{\ln n} \leq 0.01,$$

$$\ln n \geq 100,$$

and thus

$$n \geq e^{100} \approx 2.7 \times 10^{43}.$$

This is a far larger number of terms than any conceivable computer could add within the expected lifetime of the universe. But accuracy to within 0.05 would require only that $n \geq e^{20} \approx 4.85 \times 10^8$, fewer than half a billion terms—well within the range of a modern computer.

In Problems 1–28, use the integral test to test the given series for convergence.

1. $\displaystyle\sum_{n=1}^{\infty} \frac{n}{n^2 + 1}$

2. $\displaystyle\sum_{n=1}^{\infty} \frac{n}{e^{n^2}}$

3. $\displaystyle\sum_{n=1}^{\infty} \frac{1}{\sqrt{n + 1}}$

4. $\displaystyle\sum_{n=1}^{\infty} \frac{1}{(n + 1)^{4/3}}$

5. $\displaystyle\sum_{n=1}^{\infty} \frac{1}{n^2 + 1}$

6. $\displaystyle\sum_{n=1}^{\infty} \frac{1}{n(n + 1)}$

7. $\displaystyle\sum_{n=2}^{\infty} \frac{1}{n \ln n}$

8. $\displaystyle\sum_{n=1}^{\infty} \frac{\ln n}{n}$

9. $\displaystyle\sum_{n=1}^{\infty} \frac{1}{2^n}$

10. $\displaystyle\sum_{n=1}^{\infty} \frac{n}{e^n}$

**11** $\displaystyle\sum_{n=1}^{\infty} \frac{n^2}{e^n}$  **12** $\displaystyle\sum_{n=1}^{\infty} \frac{1}{17n - 13}$

**13** $\displaystyle\sum_{n=1}^{\infty} \frac{\ln n}{n^2}$  **14** $\displaystyle\sum_{n=1}^{\infty} \frac{n + 1}{n^2}$

**15** $\displaystyle\sum_{n=1}^{\infty} \frac{n}{n^4 + 1}$  **16** $\displaystyle\sum_{n=1}^{\infty} \frac{1}{n^3 + n}$

**17** $\displaystyle\sum_{n=1}^{\infty} \frac{1}{n(n + 1)(n + 2)}$  **18** $\displaystyle\sum_{n=1}^{\infty} \ln\left(\frac{n + 1}{n}\right)$

**19** $\displaystyle\sum_{n=1}^{\infty} \ln\left(1 + \frac{1}{n^2}\right)$  **20** $\displaystyle\sum_{n=1}^{\infty} \frac{\sqrt[n]{2}}{n^2}$

**21** $\displaystyle\sum_{n=1}^{\infty} \frac{n}{4n^2 + 5}$  **22** $\displaystyle\sum_{n=1}^{\infty} \frac{n}{(4n^2 + 5)^{3/2}}$

**23** $\displaystyle\sum_{n=2}^{\infty} \frac{1}{n\sqrt{\ln n}}$  **24** $\displaystyle\sum_{n=2}^{\infty} \frac{1}{n(\ln n)^2}$

**25** $\displaystyle\sum_{n=1}^{\infty} \frac{1}{4n^2 + 9}$  **26** $\displaystyle\sum_{n=1}^{\infty} \frac{n + 1}{n + 100}$

**27** $\displaystyle\sum_{n=1}^{\infty} \frac{n}{n^4 + 2n^2 + 1}$  **28** $\displaystyle\sum_{n=1}^{\infty} \frac{1}{(n + 1)^3}$

In each of Problems 29–31, tell why the integral test does *not* apply to the given infinite series.

**29** $\displaystyle\sum_{n=1}^{\infty} \frac{(-1)^n}{n}$  **30** $\displaystyle\sum_{n=1}^{\infty} e^{-n}\sin n$

**31** $\displaystyle\sum_{n=1}^{\infty} \frac{2 + \sin n}{n^2}$

In each of Problems 32–34, find the least positive integer $n$ such that the remainder $R_n$ in Theorem 2 is less than $E$.

**32** $\displaystyle\sum_{n=1}^{\infty} \frac{1}{n^2}$;  $E = 0.0001$  **33** $\displaystyle\sum_{n=1}^{\infty} \frac{1}{n^3}$;  $E = 0.00005$

**34** $\displaystyle\sum_{n=1}^{\infty} \frac{1}{n^6}$;  $E = 2 \times 10^{-11}$

**35** Suppose that the function $f$ is continuous and positive-valued for $x \geq 1$. Let $b_n = \displaystyle\int_1^n f(x)\, dx$ for $n = 1, 2, 3, \ldots$.
(a) Suppose that the monotone increasing sequence $\{b_n\}$ is bounded, so that $B = \lim_{n \to \infty} b_n$ exists. Prove that $\displaystyle\int_1^{\infty} f(x)\, dx = B$.

(b) Prove that if the sequence $\{b_n\}$ is not bounded, then $\displaystyle\int_1^{\infty} f(x)\, dx = \infty$.

**36** Show that the series

$$\sum_{n=2}^{\infty} \frac{1}{n(\ln n)^p}$$

converges if $p > 1$ and diverges if $p \leq 1$.

**37** Use the integral test remainder estimate to find $\displaystyle\sum_{n=1}^{\infty} 1/n^5$ accurate to three decimal places.

**38** Using the integral test remainder estimate, how many terms are needed to approximate $\displaystyle\sum_{n=1}^{\infty} 1/n^{3/2}$ with two-place accuracy?

**39** Deduce from Inequalities (2) and (3) with $f(x) = 1/x$ that

$$\ln n \leq 1 + \frac{1}{2} + \frac{1}{3} + \cdots + \frac{1}{n} \leq 1 + \ln n.$$

If a computer adds 1 million terms of the harmonic series $\sum 1/n$ per second, how long will it take for the partial sum to reach 50?

**40** (a) Let

$$c_n = \left(1 + \frac{1}{2} + \cdots + \frac{1}{n}\right) - \ln n.$$

Deduce from Problem 39 that $0 \leq c_n \leq 1$ for all $n$.
(b) Note that

$$\int_n^{n+1} \frac{1}{x}\, dx \geq \frac{1}{n + 1};$$

conclude that the sequence $\{c_n\}$ is monotone decreasing. Therefore, the sequence $\{c_n\}$ converges. The number

$$\gamma = \lim_{n \to \infty}\left(1 + \frac{1}{2} + \cdots + \frac{1}{n} - \ln n\right) \approx 0.57722$$

is known as **Euler's constant.**

**41** It is known that

$$\sum_{n=1}^{\infty} \frac{1}{n^4} = \frac{\pi^4}{90}.$$

Use the integral test remainder estimate and the first ten terms of this series to show that $\pi = 3.1416$ rounded to four decimal places.

## *12-4 Optional Computer Application

The program below was written to approximate the sum of the series

$$\sum_{n=1}^{\infty} \frac{1}{n^p}$$

that (by Example 2) converges if $p > 1$.

```
10      INPUT  "  P  ";  P
20      INPUT  "  E  ";  E
30      N = 0  :  S = 0
40      N = N + 1
50      T = 1/(N↑P)
60      S = S + T
70      IF  T > E  THEN GOTO  40
80      PRINT  "  SUM  :  ";  S
90      END
```

In lines 10 and 20 we enter the exponent P and our error tolerance E. At each stage in the computation, S denotes the current partial sum and T the next term to be added; T is calculated in line 50 and added in line 60. Line 70 directs that if T > E, then another term is to be added.

**Exercise 1** Check the computation in Example 3 by running this program with $P = 2$ and $E = 4 \times 10^{-4}$.

**Exercise 2** Apply this program to the series in Problem 41 to calculate $\pi$ accurate to six places to the right of the decimal.

**Exercise 3** Alter the program to approximate the sum of the series in Problem 36.

---

## Comparison Tests for Positive-Term Series

With the integral test we attempt to determine whether or not an infinite series converges by comparing it with an improper integral. The methods of this section involve comparing the terms of the *positive-term* series $\sum a_n$ with those of another series $\sum b_n$ whose convergence or divergence is already known. We have already developed two families of *reference series* for the role of the known series $\sum b_n$; these are the geometric series of Section 12-3 and the *p*-series of Section 12-4. They are well adapted for our new purposes because their convergence or divergence is quite easy to determine: The geometric series $\sum r^n$ converges if $|r| < 1$ and diverges if $|r| \geq 1$, and the *p*-series $\sum 1/n^p$ converges for $p > 1$ and diverges if $0 < p \leq 1$.

Let $\sum a_n$ and $\sum b_n$ be positive-term series. Then we say that the series $\sum b_n$ **dominates** the series $\sum a_n$ provided that $a_n \leq b_n$ for all $n$. The following theorem says that the positive-term series $\sum a_n$ converges if it is dominated by a convergent series, and diverges if it dominates a divergent series.

---

> *Theorem 1   Comparison Test*
> Suppose that $\sum a_n$ and $\sum b_n$ are positive-term series. Then
>
> (i) $\sum a_n$ converges if $\sum b_n$ converges and $a_n \leq b_n$ for all $n$;
> (ii) $\sum a_n$ diverges if $\sum b_n$ diverges and $a_n \geq b_n$ for all $n$.

---

***Proof*** Denote the *n*th partial sums of the series $\sum a_n$ and $\sum b_n$ by $S_n$ and $T_n$, respectively. Then $\{S_n\}$ and $\{T_n\}$ are monotone increasing sequences. To prove part (i), suppose that $\sum b_n$ converges, so that $T = \lim_{n \to \infty} T_n$ exists ($T$ is a real number). Then the fact that $a_n \leq b_n$ for all $n$ implies that $S_n \leq T_n \leq T$ for all $n$. Thus the sequence $\{S_n\}$ of partial sums of $\sum a_n$ is bounded and monotone increasing and therefore converges. Thus $\sum a_n$ converges.

Part (ii) is just a restatement of part (i). If the series $\sum a_n$ converged, then the fact that $\sum a_n$ dominates $\sum b_n$ would imply—by part (i) with $a_n$ and $b_n$ interchanged—that $\sum b_n$ converged. But $\sum b_n$ diverges, so it follows that $\sum a_n$ must also diverge. ∎

We know by Theorem 5 in Section 12-3 that the convergence or divergence of an infinite series is not affected by the addition or deletion of finitely many terms. Consequently the conditions $a_n \leq b_n$ and $a_n \geq b_n$ in the two parts of the comparison test need only be assumed to hold for all $n \geq k$, where $k$ is some fixed positive integer. Thus we can say that the series $\sum a_n$ converges if it is "eventually dominated" by a convergent series.

**EXAMPLE 1**   Since

$$\frac{1}{n(n+1)(n+2)} < \frac{1}{n^3}$$

for all $n \geq 1$, the series

$$\sum_{n=1}^{\infty} \frac{1}{n(n+1)(n+2)} = \frac{1}{1 \cdot 2 \cdot 3} + \frac{1}{2 \cdot 3 \cdot 4} + \frac{1}{3 \cdot 4 \cdot 5} + \cdots$$

is dominated by the series $\sum 1/n^3$, which is a convergent $p$-series with $p = 3$. Hence the series $\sum 1/n(n+1)(n+2)$ converges by part (i) of the comparison test.

**EXAMPLE 2**   Since

$$\frac{1}{\sqrt{2n-1}} > \frac{1}{\sqrt{2n}}$$

for all $n \geq 1$, the series

$$\sum_{n=1}^{\infty} \frac{1}{\sqrt{2n-1}} = 1 + \frac{1}{\sqrt{3}} + \frac{1}{\sqrt{5}} + \frac{1}{\sqrt{7}} + \cdots$$

dominates the series

$$\sum_{n=1}^{\infty} \frac{1}{\sqrt{2n}} = \frac{1}{\sqrt{2}} \sum_{n=1}^{\infty} \frac{1}{n^{1/2}}.$$

But $\sum 1/n^{1/2}$ is a divergent $p$-series with $p = \frac{1}{2}$, and a constant multiple of a divergent series diverges, so part (ii) of the comparison test shows that the series $\sum 1/\sqrt{2n-1}$ diverges.

**EXAMPLE 3**   Test the series

$$\sum_{n=0}^{\infty} \frac{1}{n!} = 1 + \frac{1}{1!} + \frac{1}{2!} + \frac{1}{3!} + \cdots$$

for convergence.

**Solution** We note first that if $n \geq 1$ then

$$n! = n(n-1)(n-2) \cdots 3 \cdot 2 \cdot 1$$
$$\geq 2 \cdot 2 \cdot 2 \cdots 2 \cdot 2 \cdot 1; \qquad \text{(same number of factors)}$$

that is, $n! \geq 2^{n-1}$ for $n \geq 1$. Thus

$$\frac{1}{n!} \leq \frac{1}{2^{n-1}} \qquad \text{for} \quad n \geq 1,$$

so the series $\sum\limits_{n=0}^{\infty} \dfrac{1}{n!}$ is dominated by the series

$$1 + \sum_{n=1}^{\infty} \frac{1}{2^{n-1}} = 1 + \sum_{n=0}^{\infty} \frac{1}{2^n},$$

which is a convergent geometric series. Hence the series $\sum\limits_{n=0}^{\infty} \dfrac{1}{n!}$ converges.
In Section 12-7 we will see that the sum of this series is the number $e$, so that

$$e = 1 + \frac{1}{1!} + \frac{1}{2!} + \frac{1}{3!} + \cdots + \frac{1}{n!} + \cdots.$$

Indeed, this series provides the simplest way of showing that

$$e \approx 2.71828\,18284\,59045\,23536.$$

---

Suppose that $\sum a_n$ is a series such that $a_n \to 0$ as $n \to \infty$, so that (in connection with the $n$th term divergence test of Section 12-3) $\sum a_n$ has at least a chance of converging. How do we choose an appropriate series $\sum b_n$ to compare it with? A good idea is to pick $b_n$ as a *simple* function of $n$, simpler than $a_n$, such that $a_n$ and $b_n$ approach zero at the same rate as $n \to \infty$. If the formula for $a_n$ is a fraction, we can try discarding all but the terms of largest magnitude in its numerator and denominator. For example, if

$$a_n = \frac{3n^2 + n}{n^4 + \sqrt{n}},$$

then we reason that $n$ is small in comparison with $3n^2$ and $\sqrt{n}$ is small in comparison with $n^4$ when $n$ is quite large. This suggests that we pick $b_n = 3n^2/n^4 = 3/n^2$. The series $\sum 3/n^2$ converges ($p = 2$), but when we attempt to compare $\sum a_n$ and $\sum b_n$, we find that $a_n \geq b_n$ (rather than $a_n \leq b_n$). Consequently the comparison test does not apply immediately—the fact that $\sum a_n$ dominates a convergent series does *not* imply that $\sum a_n$ itself converges. The following theorem provides a convenient way of handling such a situation.

---

**Theorem 2  Limit Comparison Test**

Suppose that $\sum a_n$ and $\sum b_n$ are positive-term series. If the limit

$$L = \lim_{n \to \infty} \frac{a_n}{b_n}$$

exists and $0 < L < \infty$, then either both series converge or both series diverge.

---

***Proof*** Choose two fixed positive numbers $P$ and $Q$ such that $P < L < Q$. Then $P < a_n/b_n < Q$ for all $n$ sufficiently large, and so

$$Pb_n < a_n < Qb_n$$

for all $n$ sufficiently large. If $\sum b_n$ converges, then $\sum a_n$ is eventually dominated by the convergent series $Q \sum b_n$, so part (i) of the comparison test implies that $\sum a_n$ also converges. If $\sum b_n$ diverges, then $\sum a_n$ eventually dominates the divergent series $P \sum b_n$, so part (ii) of the comparison test implies that $\sum a_n$ also diverges. Thus the convergence of either series implies the convergence of the other. ∎

**EXAMPLE 4** With

$$a_n = \frac{3n^2 + n}{n^4 + \sqrt{n}} \quad \text{and} \quad b_n = \frac{1}{n^2}$$

(motivated by the discussion preceding Theorem 2), we find that

$$\lim_{n \to \infty} \frac{a_n}{b_n} = \lim_{n \to \infty} \frac{3n^4 + n^3}{n^4 + \sqrt{n}} = \lim_{n \to \infty} \frac{3 + 1/n}{1 + n^{-7/2}} = 3.$$

Since $\sum 1/n^2$ is a convergent $p$-series ($p = 2$), the limit comparison test tells us that the series

$$\sum_{n=1}^{\infty} \frac{3n^2 + n}{n^4 + \sqrt{n}}$$

also converges.

**EXAMPLE 5** Test for convergence: $\displaystyle\sum_{n=1}^{\infty} \frac{1}{2n + \ln n}$.

***Solution*** Because $\lim_{n \to \infty} (\ln n)/n = 0$ by Example 8 in Section 12-2, we note that $\ln n$ is small in comparison with $2n$ when $n$ is quite large. We therefore take $a_n = 1/(2n + \ln n)$ and—ignoring the constant coefficient 2—we take $b_n = 1/n$. Then we find that

$$\lim_{n \to \infty} \frac{a_n}{b_n} = \lim_{n \to \infty} \frac{n}{2n + \ln n} = \lim_{n \to \infty} \frac{1}{2 + (\ln n)/n} = \frac{1}{2}.$$

Since the harmonic series $\sum 1/n$ diverges, it follows that the given series $\sum a_n$ also diverges.

---

It is important to realize that if $L = \lim(a_n/b_n)$ is either 0 or $\infty$, then the limit comparison test does not apply, and no conclusion is possible. For instance, if $a_n = 1/n^2$ and $b_n = 1/n$, then $\lim(a_n/b_n) = 0$. But in this case $\sum a_n$ converges while $\sum b_n$ diverges.

We close our discussion of positive-term series with the observation that the sum of a convergent *positive*-term series is not altered by grouping or rearranging its terms. For example, let $\sum a_n$ be a convergent positive-term series and consider

$$\sum_{n=1}^{\infty} b_n = (a_1 + a_2 + a_3) + (a_4) + (a_5 + a_6) + \cdots.$$

That is, the new series has $b_1 = a_1 + a_2 + a_3$, $b_2 = a_4$, $b_3 = a_5 + a_6$, and so on. Then every partial sum $T_n$ of $\sum b_n$ is equal to some partial sum $S_{n_i}$ of $\sum a_n$. Since $\{S_n\}$ is a monotone increasing sequence with limit $S = \sum a_n$, it follows easily that $\{T_n\}$ is a monotone increasing sequence with the same limit. Thus $\sum b_n = S$ also. The argument is more subtle if terms of $\sum a_n$ are "moved out of place," as in

$$\sum_{n=1}^{\infty} c_n = (a_1 + a_3 + a_2) + (a_5 + a_7 + a_4) + (a_9 + a_{11} + a_6) + \cdots,$$

but the same conclusion holds: Any rearrangement of a convergent *positive-term* series also converges, and to the same sum.

Similarly, it is easy to prove that any grouping or rearrangement of a divergent positive-term series also diverges. But these observations all fail in the case of infinite series having both positive and negative terms. For example, the series $\sum (-1)^n$ diverges, but it has the convergent regrouping

$$(-1 + 1) + (-1 + 1) + (-1 + 1) + \cdots = 0 + 0 + 0 + \cdots = 0.$$

It follows from Problem 44 in Section 12-3 that

$$\ln 2 = 1 - \tfrac{1}{2} + \tfrac{1}{3} - \tfrac{1}{4} + \tfrac{1}{5} - \cdots,$$

but the rearrangement

$$1 + \tfrac{1}{3} - \tfrac{1}{2} + \tfrac{1}{5} + \tfrac{1}{7} - \tfrac{1}{4} + \tfrac{1}{9} + \tfrac{1}{11} - \tfrac{1}{6} + \cdots$$

converges to $\tfrac{3}{2} \ln 2$. The above series for $\ln 2$ even has rearrangements that diverge.

## 12-5 PROBLEMS

Use comparison tests to determine whether the infinite series in Problems 1–35 converge or diverge.

1. $\displaystyle\sum_{n=1}^{\infty} \frac{1}{n^2 + n + 1}$

2. $\displaystyle\sum_{n=1}^{\infty} \frac{n^3 + 1}{n^4 + 2}$

3. $\displaystyle\sum_{n=1}^{\infty} \frac{1}{n + \sqrt{n}}$

4. $\displaystyle\sum_{n=1}^{\infty} \frac{1}{n + n^{3/2}}$

5. $\displaystyle\sum_{n=1}^{\infty} \frac{1}{1 + 3^n}$

6. $\displaystyle\sum_{n=1}^{\infty} \frac{10n^2}{n^4 + 1}$

7. $\displaystyle\sum_{n=2}^{\infty} \frac{10n^2}{n^3 - 1}$

8. $\displaystyle\sum_{n=1}^{\infty} \frac{n^2 - n}{n^4 + 2}$

9. $\displaystyle\sum_{n=1}^{\infty} \frac{1}{\sqrt{37n^3 + 3}}$

10. $\displaystyle\sum_{n=1}^{\infty} \frac{1}{\sqrt{n^2 + 1}}$

11. $\displaystyle\sum_{n=1}^{\infty} \frac{\sqrt{n}}{n^2 + n}$

12. $\displaystyle\sum_{n=1}^{\infty} \frac{1}{3 + 5^n}$

13. $\displaystyle\sum_{n=2}^{\infty} \frac{1}{\ln n}$

14. $\displaystyle\sum_{n=2}^{\infty} \frac{1}{n - \ln n}$

15. $\displaystyle\sum_{n=1}^{\infty} \frac{\sin^2 n}{n^2 + 1}$

16. $\displaystyle\sum_{n=1}^{\infty} \frac{\cos^2 n}{3^n}$

17. $\displaystyle\sum_{n=1}^{\infty} \frac{n + 2^n}{n + 3^n}$

18. $\displaystyle\sum_{n=0}^{\infty} \frac{1}{2^n + 3^n}$

19. $\displaystyle\sum_{n=2}^{\infty} \frac{1}{n^2 \ln n}$

20. $\displaystyle\sum_{n=1}^{\infty} \frac{1}{n^{1 + \sqrt{n}}}$

21. $\displaystyle\sum_{n=1}^{\infty} \frac{\ln n}{n^2}$

22. $\displaystyle\sum_{n=1}^{\infty} \frac{\tan^{-1} n}{n}$

23. $\displaystyle\sum_{n=1}^{\infty} \frac{\sin^2(1/n)}{n^2}$

24. $\displaystyle\sum_{n=1}^{\infty} \frac{e^{1/n}}{n}$

25. $\displaystyle\sum_{n=2}^{\infty} \frac{\ln n}{e^n}$

26. $\displaystyle\sum_{n=1}^{\infty} \frac{n^2 + 2}{n^3 + 3n}$

**27** $\displaystyle\sum_{n=1}^{\infty} \frac{n^{3/2}}{n^2 + 4}$

**28** $\displaystyle\sum_{n=1}^{\infty} \frac{1}{n \cdot 2^n}$

**29** $\displaystyle\sum_{n=1}^{\infty} \frac{3}{4 + \sqrt{n}}$

**30** $\displaystyle\sum_{n=1}^{\infty} \frac{n^2 + 1}{e^n(n + 1)^2}$

**31** $\displaystyle\sum_{n=1}^{\infty} \frac{2n^2 - 1}{n^2 \cdot 3^n}$

**32** $\displaystyle\sum_{n=1}^{\infty} \frac{1}{\sqrt[3]{2n^4 + 1}}$

**33** $\displaystyle\sum_{n=1}^{\infty} \frac{2 + \sin n}{n^2}$

**34** $\displaystyle\sum_{n=1}^{\infty} \frac{\ln n}{n^3}$

**35** $\displaystyle\sum_{n=1}^{\infty} \frac{(n + 1)^n}{n^{n+1}}$

$\left(\textit{Suggestion: } \displaystyle\lim_{n \to \infty} \left(1 + \frac{1}{n}\right)^n = e.\right)$

**36** (a) Show that $\ln n < n^{1/8}$ for all $n$ sufficiently large.
(b) Conclude that $\sum 1/(\ln n)^8$ diverges.
**37** Prove that $\sum (a_n/n)$ converges if $\sum a_n$ is a convergent positive-term series.

**38** Suppose that $\sum a_n$ is a convergent positive-term series and $\{c_n\}$ is a sequence of positive numbers such that $\lim\limits_{n \to \infty} c_n = 0$. Prove that $\sum a_n c_n$ converges.

**39** Use the result of Problem 38 to prove that if $\sum a_n$ and $\sum b_n$ are convergent positive-term series, then $\sum a_n b_n$ converges.

**40** Show that the series

$$\sum_{n=1}^{\infty} \frac{1}{1 + 2 + 3 + \cdots + n}$$

converges.

**41** Use the result of Problem 40 in Section 12-4 to show that the series

$$\sum_{n=1}^{\infty} \left(1 + \frac{1}{2} + \frac{1}{3} + \cdots + \frac{1}{n}\right)^{-1}$$

diverges.

---

## 12-6

## Alternating Series and Absolute Convergence

In Sections 12-4 and 12-5, we concentrated on positive-term series. Now we discuss infinite series that have both positive and negative terms. An important example is a series with terms that are alternately positive and negative. An **alternating series** is an infinite series of the form

$$\sum_{n=1}^{\infty} (-1)^{n+1} a_n = a_1 - a_2 + a_3 - a_4 + \cdots \tag{1}$$

or of the form $\displaystyle\sum_{n=1}^{\infty} (-1)^n a_n$, where $a_n > 0$ for all $n$. For example, the series

$$\sum_{n=1}^{\infty} \frac{(-1)^{n+1}}{n} = 1 - \frac{1}{2} + \frac{1}{3} - \frac{1}{4} + \cdots$$

is an alternating series. The following theorem shows that this series converges because the sequence of absolute values of its terms is monotone decreasing with limit zero.

---

> ### Theorem 1   Alternating Series Test
> If $a_n \geqq a_{n+1} > 0$ for all $n$ and $\lim\limits_{n \to \infty} a_n = 0$, then the alternating series in (1) converges.

---

**Proof**   We consider first the "even" partial sums $S_2, S_4, S_6, \ldots, S_{2n}$. We may write

$$S_{2n} = (a_1 - a_2) + (a_3 - a_4) + \cdots + (a_{2n-1} - a_{2n}).$$

Since $a_k - a_{k+1} \geqq 0$ for all $k$, we see that the sequence $\{S_{2n}\}$ is monotone increasing. Also, since

$$S_{2n} = a_1 - (a_2 - a_3) - \cdots - (a_{2n-2} - a_{2n-1}) - a_{2n},$$

we see that $S_{2n} \leq a_1$ for all $n$, so the monotone increasing sequence $\{S_{2n}\}$ is bounded above. Hence the limit

$$S = \lim_{n \to \infty} S_{2n}$$

exists by the bounded monotone sequence property of Section 12-2. And it follows that

$$0 < S < a_1$$

because the even partial sums $S_{2n}$ are positive and bounded by $a_1$. It remains only to see that the "odd" partial sums $S_1, S_3, S_5, \ldots$ also converge to $S$. But $S_{2n+1} = S_{2n} + a_{2n+1}$ and $\lim_{n \to \infty} a_{2n+1} = 0$, so

$$\lim_{n \to \infty} S_{2n+1} = \lim_{n \to \infty} S_{2n} + \lim_{n \to \infty} a_{2n+1} = S.$$

Thus $\lim_{n \to \infty} S_n = S$, and therefore the series converges. ∎

**EXAMPLE 1**   The series

$$\sum_{n=1}^{\infty} \frac{(-1)^{n+1}}{2n-1} = 1 - \frac{1}{3} + \frac{1}{5} - \frac{1}{7} + \cdots$$

satisfies the conditions of Theorem 1 and therefore converges. The alternating series test does not tell us what its sum is, but we will see in Section 12-7 that the sum of this series is $\pi/4$.

**EXAMPLE 2**   The series

$$\sum_{n=1}^{\infty} (-1)^{n+1} \frac{n}{2n-1} = 1 - \frac{2}{3} + \frac{3}{5} - \frac{4}{7} + \cdots$$

is an alternating series, and

$$a_n = \frac{n}{2n-1} > \frac{n+1}{2n+1} = a_{n+1}$$

for all $n$ because $n(2n+1) > (2n-1)(n+1) = 2n^2 + n - 1$ if $n \geq 1$. But $\lim_{n \to \infty} a_n = \frac{1}{2} \neq 0$, so the alternating series test does not apply. In fact, this series diverges by the $n$th term divergence test.

---

If a series converges by the alternating series test, the following theorem shows how to approximate its sum with any desired degree of accuracy—*if* you have some method (a computer?) for adding enough of its terms.

---

**Theorem 2**   *Alternating Series Error Estimate*

Suppose that the series $\sum (-1)^{n+1} a_n$ satisfies the conditions of the alternating series test and therefore converges. Let $S$ denote its sum. Denote by $R_n = S - S_n$ the error made in replacing $S$ with the $n$th partial sum $S_n$. Then $R_n$ has the same sign as the next term $(-1)^{n+2} a_{n+1}$, and

$$0 < |R_n| < a_{n+1}. \tag{2}$$

---

**Proof**  We note that

$$R_n = S - S_n$$
$$= (-1)^{n+2}a_{n+1} + (-1)^{n+3}a_{n+2} + (-1)^{n+4}a_{n+3} + \cdots$$
$$= (-1)^n(a_{n+1} - a_{n+2} + a_{n+3} - \cdots).$$

So

$$|R_n| = a_{n+1} - a_{n+2} + a_{n+3} - \cdots$$

is an alternating series with first term $a_{n+1}$. Hence Inequality (2) follows from the fact that the sum of a convergent alternating series lies between zero and its first term, as we saw in the proof of Theorem 1.  ∎

**EXAMPLE 3**  In Section 12-7 we will see that

$$e^x = \sum_{n=0}^{\infty} \frac{x^n}{n!}$$

and thus that

$$\frac{1}{e} = e^{-1} = 1 - 1 + \frac{1}{2!} - \frac{1}{3!} + \frac{1}{4!} - \cdots.$$

Use this alternating series to compute $e^{-1}$ accurate to four decimal places.

**Solution**  We want $|R_n| < 1/(n+1)! \leq 0.00005$. The least value of $n$ for which this inequality holds is $n = 7$. Then

$$e^{-1} = 1 - 1 + \frac{1}{2!} - \frac{1}{3!} + \frac{1}{4!} - \frac{1}{5!} + \frac{1}{6!} - \frac{1}{7!} + R_7$$

$$= 0.367857 + R_7.$$

(We are carrying six places because we want four-place accuracy in the final answer.) Now Inequality (2) gives

$$0 < R_7 < \frac{1}{8!} < 0.000025.$$

Thus

$$0.367857 < e^{-1} < 0.367882.$$

If we take reciprocals, we find that

$$2.718263 < e < 2.718448.$$

Thus $e^{-1} = 0.3679$ rounded to four places and also $e = 2.718$ rounded to three places.

**ABSOLUTE CONVERGENCE**

The series

$$\sum_{n=1}^{\infty} \frac{(-1)^{n+1}}{n} = 1 - \frac{1}{2} + \frac{1}{3} - \frac{1}{4} + \cdots$$

converges, but if we simply add the absolute values of its terms, we get the *divergent* series

$$1 + \tfrac{1}{2} + \tfrac{1}{3} + \tfrac{1}{4} + \cdots .$$

On the other hand, the series

$$\sum_{n=0}^{\infty} \frac{(-1)^n}{2^n} = 1 - \frac{1}{2} + \frac{1}{4} - \frac{1}{8} + \cdots$$

has the property that the associated positive-term series

$$1 + \tfrac{1}{2} + \tfrac{1}{4} + \tfrac{1}{8} + \tfrac{1}{16} + \cdots$$

converges. Our next theorem tells us this: If a series of *positive* terms converges, then we may insert minus signs in front of any of the terms—every other one, for instance—and the resulting series must also converge.

---

**Theorem 3  Absolute Convergence Implies Convergence**
If the series $\sum |a_n|$ converges, then so does the series $\sum a_n$.

---

**Proof**  Suppose that the series $\sum |a_n|$ converges. Note that

$$0 \le a_n + |a_n| \le 2|a_n|$$

for all $n$. Let $b_n = a_n + |a_n|$, and then it follows from the comparison test that the positive-term series $\sum b_n$ converges; it is dominated by the convergent series $\sum 2|a_n|$. Since it is easily verified that the termwise difference of two convergent series also converges, we now see that the series

$$\sum a_n = \sum (b_n - |a_n|) = \sum b_n - \sum |a_n|$$

converges. ∎

Thus we have another convergence test: Given the series $\sum a_n$, test $\sum |a_n|$ for convergence. If the latter series converges, then so does the former. This phenomenon motivates us to make the following definition.

---

**Definition  Absolute Convergence**
The series $\sum a_n$ is said to **converge absolutely** (and is called **absolutely convergent**) provided that the series

$$\sum |a_n| = |a_1| + |a_2| + |a_3| + \cdots + |a_n| + \cdots$$

converges.

---

Thus Theorem 3 says simply that *every absolutely convergent series is convergent.* The two examples preceding Theorem 3 show that a convergent series may either converge absolutely or fail to do so:

$$1 - \tfrac{1}{2} + \tfrac{1}{4} - \tfrac{1}{8} + \tfrac{1}{16} - \cdots$$

is an absolutely convergent series because

$$1 + \tfrac{1}{2} + \tfrac{1}{4} + \tfrac{1}{8} + \tfrac{1}{16} + \cdots$$

converges, while

$$1 - \tfrac{1}{2} + \tfrac{1}{3} - \tfrac{1}{4} + \tfrac{1}{5} - \cdots$$

is an example of a series that, though it converges, does *not* converge absolutely. A series that converges but does not converge absolutely is said to be **conditionally convergent.**

Consequently the terms *absolutely convergent, conditionally convergent,* and *divergent* are simultaneously all-inclusive and mutually exclusive.

Note that there is some advantage in the application of Theorem 3, since to apply it we test the *positive*-term series $\sum |a_n|$ for convergence; we have a variety of tests, such as comparison tests or the integral test, designed for use on positive-term series.

Note also that absolute convergence of the series $\sum a_n$ means convergence of *another* series $\sum |a_n|$, and the two sums generally will differ. For example, with $a_n = (-\tfrac{1}{3})^n$ the formula for the sum of a geometric series gives

$$\sum_{n=0}^{\infty} a_n = \sum_{n=0}^{\infty} (-\tfrac{1}{3})^n = \frac{1}{1 - (-\tfrac{1}{3})} = \frac{3}{4}$$

while

$$\sum_{n=0}^{\infty} |a_n| = \sum_{n=0}^{\infty} (\tfrac{1}{3})^n = \frac{1}{1 - (\tfrac{1}{3})} = \frac{3}{2}.$$

**EXAMPLE 4**  Discuss the convergence of the series

$$\sum_{n=1}^{\infty} \frac{\cos n}{n^2} = \cos 1 + \frac{\cos 2}{4} + \frac{\cos 3}{9} + \cdots.$$

*Solution*  Let $a_n = (\cos n)/n^2$. Then

$$|a_n| = \frac{|\cos n|}{n^2} \leqq \frac{1}{n^2}$$

for all $n \geqq 1$. Hence the positive-term series $\sum |a_n|$ converges by the comparison test, since it is dominated by the convergent *p*-series $\sum (1/n^2)$. Thus the given series is absolutely convergent, and therefore (by Theorem 3) it converges.

One reason for the importance of absolute convergence is the fact (proved in advanced calculus) that the terms of an absolutely convergent series may be regrouped or rearranged without changing the sum of the series. As we remarked at the end of Section 12-5, this is *not* true of conditionally convergent series.

**THE RATIO AND ROOT TESTS**

Each of our next two convergence tests involves a way of measuring the rate of growth or decrease of the sequence $\{a_n\}$ of terms in order to determine whether the series $\sum a_n$ diverges or converges absolutely.

> **Theorem 4   The Ratio Test**
>
> Suppose that the limit
>
> $$\rho = \lim_{n \to \infty} \left| \frac{a_{n+1}}{a_n} \right| \qquad (3)$$
>
> either exists or is infinite. Then the infinite series $\sum a_n$ with nonzero terms
>
> (i) Converges absolutely if $\rho < 1$;
> (ii) Diverges if $\rho > 1$.
>
> If $\rho = 1$, the ratio test is inconclusive.

**Proof**   If $\rho < 1$, pick a (fixed) number $r$ with $\rho < r < 1$. Then (3) implies that there exists an integer $N$ such that $|a_{n+1}| \leq r|a_n|$ for all $n \geq N$. It follows that

$$|a_{N+1}| \leq r|a_N|,$$
$$|a_{N+2}| \leq r|a_{N+1}| \leq r^2|a_N|,$$
$$|a_{N+3}| \leq r|a_{N+2}| \leq r^3|a_N|,$$

and—in general—that

$$|a_{N+k}| \leq r^k|a_N| \qquad \text{for} \quad k \geq 0.$$

Hence the series

$$|a_N| + |a_{N+1}| + |a_{N+2}| + \cdots$$

is dominated by the geometric series

$$|a_N|(1 + r + r^2 + \cdots),$$

and the latter converges because $r < 1$. Thus the series $\sum |a_n|$ converges, and so the series $\sum a_n$ converges absolutely.

If $\rho > 1$, then (3) implies that there exists a positive integer $N$ such that $|a_{n+1}| > |a_n|$ for all $n \geq N$. It follows that $|a_n| > |a_N| > 0$ for all $n > N$. Thus the sequence $\{a_n\}$ cannot approach zero as $n \to \infty$, and consequently—by the $n$th term divergence test—the series $\sum a_n$ diverges.  ∎

To see that $\sum a_n$ may either converge or diverge if $\rho = 1$, consider the divergent series $\sum (1/n)$ and the convergent series $\sum (1/n^2)$, for both of which $\rho = 1$:

$$\lim_{n \to \infty} \left| \frac{1/(n+1)}{1/n} \right| = \lim_{n \to \infty} \frac{n}{n+1} = 1,$$

$$\lim_{n \to \infty} \left| \frac{1/(n+1)^2}{1/n^2} \right| = \lim_{n \to \infty} \frac{n^2}{(n+1)^2} = 1.$$

**EXAMPLE 5**   Consider the series

$$\sum_{n=1}^{\infty} \frac{(-1)^n 2^n}{n!} = -2 + \frac{4}{2!} - \frac{8}{3!} + \frac{16}{4!} - \cdots.$$

Then

$$\rho = \lim_{n \to \infty} \left| \frac{a_{n+1}}{a_n} \right| = \lim_{n \to \infty} \left| \frac{(-1)^{n+1} 2^{n+1}/(n+1)!}{(-1)^n 2^n/n!} \right|$$

$$= \lim_{n \to \infty} \frac{2}{n+1} = 0.$$

Since $\rho < 1$, the series converges absolutely.

**EXAMPLE 6**   Test the following series for convergence.

$$\text{(a)} \sum_{n=1}^{\infty} \frac{n}{2^n} \qquad \text{(b)} \sum_{n=1}^{\infty} \frac{3^n}{n^2}$$

**Solution**   (a) We have

$$\rho = \lim_{n \to \infty} \left| \frac{a_{n+1}}{a_n} \right| = \lim_{n \to \infty} \frac{(n+1)/2^{n+1}}{n/2^n} = \lim_{n \to \infty} \frac{n+1}{2n} = \frac{1}{2}.$$

Since $\rho < 1$, this series converges absolutely.
(b) Here we have

$$\rho = \lim_{n \to \infty} \left| \frac{a_{n+1}}{a_n} \right| = \lim_{n \to \infty} \frac{3^{n+1}/(n+1)^2}{3^n/n^2}$$

$$= \lim_{n \to \infty} \frac{3n^2}{(n+1)^2} = 3.$$

Since $\rho > 1$, this series diverges.

---

**Theorem 5   The Root Test**

Suppose that the limit

$$\rho = \lim_{n \to \infty} \sqrt[n]{|a_n|} \qquad (4)$$

either exists or is infinite. Then the infinite series $\sum a_n$

(i) Converges absolutely if $\rho < 1$;
(ii) Diverges if $\rho > 1$.

If $\rho = 1$, the root test is inconclusive.

---

**Proof**   If $\rho < 1$, pick a (fixed) number $r$ such that $\rho < r < 1$. Then $|a_n|^{1/n} < r$, and hence $|a_n| < r^n$, for $n$ sufficiently large. Thus the series $\sum |a_n|$ is eventually dominated by the convergent geometric series $\sum r^n$. Therefore, $\sum |a_n|$ converges, and so the series $\sum a_n$ converges absolutely.
    If $\rho > 1$, then $|a_n|^{1/n} > 1$, and hence $|a_n| > 1$, for $n$ sufficiently large. Therefore the $n$th term divergence test implies that the series $\sum a_n$ diverges. ∎

The ratio test is generally simpler to apply than the root test and therefore is ordinarily the one to try first. But there are certain series for which the root test succeeds while the ratio test fails.

**EXAMPLE 7**   Consider the series

$$\sum_{n=0}^{\infty} \frac{1}{2^{n+(-1)^n}} = \frac{1}{2} + \frac{1}{1} + \frac{1}{8} + \frac{1}{4} + \frac{1}{32} + \frac{1}{16} + \cdots .$$

Then $a_{n+1}/a_n = \frac{1}{2}$ if $n$ is even while $a_{n+1}/a_n = \frac{1}{8}$ if $n$ is odd. So the limit required for the ratio test does not exist. But

$$\lim_{n \to \infty} |a_n|^{1/n} = \lim_{n \to \infty} \left( \frac{1}{2^{n+(-1)^n}} \right)^{1/n}$$

$$= \lim_{n \to \infty} \frac{1}{2} \left( \frac{1}{2^{(-1)^n/n}} \right) = \frac{1}{2},$$

so the given series converges by the root test. (Its convergence also follows from the fact that it is a rearrangement of the positive-term geometric series $\sum 1/2^n$.)

## 12-6   PROBLEMS

Determine whether or not the alternating series in Problems 1–10 converge.

1  $\displaystyle\sum_{n=1}^{\infty} \frac{(-1)^{n+1}}{n^2}$

2  $\displaystyle\sum_{n=1}^{\infty} \frac{(-1)^{n+1}n}{3n+2}$

3  $\displaystyle\sum_{n=1}^{\infty} (-1)^n \frac{n}{n^2+1}$

4  $\displaystyle\sum_{n=2}^{\infty} (-1)^n \frac{n}{\ln n}$

5  $\displaystyle\sum_{n=2}^{\infty} (-1)^n \frac{1}{\ln n}$

6  $\displaystyle\sum_{n=1}^{\infty} \frac{(-1)^n}{\sqrt{2n+1}}$

7  $\displaystyle\sum_{n=0}^{\infty} \frac{(-1)^n}{n!}$

8  $\displaystyle\sum_{n=1}^{\infty} (-1)^n \frac{(1.01)^n}{n^4+1}$

9  $\displaystyle\sum_{n=1}^{\infty} \frac{(-1)^n}{n^{1/n}}$

10  $\displaystyle\sum_{n=1}^{\infty} (-1)^n \frac{n!}{(2n)!}$

Determine whether the series in Problems 11–32 converge absolutely, converge conditionally, or diverge.

11  $\displaystyle\sum_{n=1}^{\infty} \frac{(-1)^{n+1}}{2^n}$

12  $\displaystyle\sum_{n=1}^{\infty} \frac{n}{n^2+1}$

13  $\displaystyle\sum_{n=1}^{\infty} \frac{(-1)^n \ln n}{n}$

14  $\displaystyle\sum_{n=1}^{\infty} \frac{1}{n^n}$

15  $\displaystyle\sum_{n=1}^{\infty} \left( \frac{10}{n} \right)^n$

16  $\displaystyle\sum_{n=1}^{\infty} \frac{3^n}{n!n}$

17  $\displaystyle\sum_{n=0}^{\infty} \frac{(-10)^n}{n!}$

18  $\displaystyle\sum_{n=1}^{\infty} (-1)^n \frac{n!}{n^n}$

19  $\displaystyle\sum_{n=1}^{\infty} (-1)^n \left( \frac{n}{n+1} \right)^n$

20  $\displaystyle\sum_{n=1}^{\infty} \frac{n!n^2}{(2n)!}$

21  $\displaystyle\sum_{n=1}^{\infty} \left( \frac{\ln n}{n} \right)^n$

22  $\displaystyle\sum_{n=0}^{\infty} (-1)^n \frac{2^{3n}}{7^n}$

23  $\displaystyle\sum_{n=0}^{\infty} (-1)^n (\sqrt{n+1} - \sqrt{n})$

24  $\displaystyle\sum_{n=1}^{\infty} n(\tfrac{3}{4})^n$

25  $\displaystyle\sum_{n=2}^{\infty} \left[ \ln\left( \frac{1}{n} \right) \right]^n$

26  $\displaystyle\sum_{n=0}^{\infty} \frac{(n!)^2}{(2n)!}$

27  $\displaystyle\sum_{n=1}^{\infty} (-1)^n \frac{3^n}{n(2^n+1)}$

28  $\displaystyle\sum_{n=1}^{\infty} (-1)^n \frac{\tan^{-1} n}{n}$

29  $\displaystyle\sum_{n=1}^{\infty} (-1)^n \frac{n!}{1 \cdot 3 \cdot 5 \cdots (2n-1)}$

30  $\displaystyle\sum_{n=1}^{\infty} \frac{1 \cdot 3 \cdot 5 \cdots (2n-1)}{1 \cdot 4 \cdot 7 \cdots (3n-2)}$

31  $\displaystyle\sum_{n=0}^{\infty} \frac{(n+2)!}{3^n(n!)^2}$

32  $\displaystyle\sum_{n=1}^{\infty} (-1)^n \frac{n^n}{3^{n^2}}$

In each of Problems 33–36, find the least positive integer $n$ such that $|R_n| = |S - S_n| < 0.0005$, so that the $n$th partial sum $S_n$ of the given alternating series approximates its sum $S$ accurate to three decimal places.

33  $\displaystyle\sum_{n=1}^{\infty} \frac{(-1)^{n+1}}{n}$

34  $\displaystyle\sum_{n=1}^{\infty} \frac{(-1)^{n+1}}{n^2}$

$35 \quad \displaystyle\sum_{n=0}^{\infty} \frac{(-1)^n}{3^n}$

$36 \quad \displaystyle\sum_{n=1}^{\infty} \frac{(-1)^n}{n^n}$

In each of Problems 37–40, approximate the sum of the given series accurate to three decimal places.

$37 \quad e^{-1/2} = \displaystyle\sum_{n=0}^{\infty} \frac{(-1)^n}{n!2^n}$

$38 \quad \cos 1 = \displaystyle\sum_{n=0}^{\infty} \frac{(-1)^n}{(2n)!}$

$39 \quad \ln(1.1) = \displaystyle\sum_{n=1}^{\infty} (-1)^{n+1} \frac{(0.1)^n}{n}$

$40 \quad \displaystyle\sum_{n=1}^{\infty} \frac{(-1)^{n+1}}{n^5}$

41 Approximate the sum of the series

$$\frac{\pi^2}{12} = \sum_{n=1}^{\infty} \frac{(-1)^{n+1}}{n^2}$$

with error less than 0.01. Use the corresponding partial sum and error estimate to verify that $3.13 < \pi < 3.15$.

42 Show that $\sum |a_n|$ diverges if the series $\sum a_n$ diverges.

43 Give an example of two convergent series $\sum a_n$ and $\sum b_n$ such that $\sum a_n b_n$ diverges.

44 (a) Suppose that $r$ is a (fixed) number with $|r| < 1$. Use the ratio test to show that the series

$$\sum_{n=0}^{\infty} nr^n$$

converges. Let $S$ denote its sum.
(b) Show that

$$(1 - r)S = \sum_{n=1}^{\infty} r^n.$$

Show how to conclude that

$$\sum_{n=0}^{\infty} nr^n = \frac{r}{(1 - r)^2}.$$

## *12-6   Optional Computer Application

Here we include a program that uses the series

$$\frac{1}{e} = 1 - \frac{1}{1!} + \frac{1}{2!} - \frac{1}{3!} + \cdots$$

to calculate the value of $e$ more precisely.

```
10    N  =  0  :  T = 1  :  S = 1
20    N  =  N + 1
30    T  =  -T/N
40    S  =  S + T
50    IF  ABS(T) > 1.0E-10  THEN GOTO  20
60    PRINT  "  N  = "; N; ":"
70    PRINT  "  SUM : "; S
80    PRINT  "  E  = "; 1/S
90    END
```

The data in line 10 correspond to the initial term $a_0 = 1$. During the loop in lines 20–50, T denotes the Nth term and S the Nth partial sum. Note in line 30 that we use the fact that $a_n = -a_{n-1}/n$ instead of calculating the factorial afresh for each term. Line 50 directs that the computation stop when we reach a term that has absolute value not exceeding $10^{-10}$. The output of this program is

```
N  = 14:
SUM : 0.3678794412
E  = 2.718281828
```

Thus we find the value of $e$ accurate to nine decimal places.

Up to this point we have concentrated on infinite series with *constant* terms. **Applications of**
Much of the practical importance of infinite series, however, derives from **Taylor's Formula**
the fact that many functions have useful representations as infinite series
with *variable* terms. A number of such series representations can be derived
from Taylor's formula (Section 11-3).

Suppose that $f$ is a function with continuous derivatives of *all* orders
in a neighborhood of the point $a$ and that $n$ is an arbitrary positive integer.
Taylor's formula then implies that

$$f(x) = P_n(x) + R_n(x), \tag{1}$$

where

$$P_n(x) = f(a) + f'(a)(x - a) + f''(a)\frac{(x - a)^2}{2!}$$

$$+ \cdots + f^{(n)}(a)\frac{(x - a)^n}{n!} \tag{2}$$

and

$$R_n(x) = \frac{f^{(n+1)}(z)}{(n + 1)!}(x - a)^{n+1} \tag{3}$$

where $z$ is some number between $a$ and $x$.

Now suppose that, for some particular *fixed* value of $x$, we can show
that

$$\lim_{n \to \infty} R_n(x) = 0. \tag{4}$$

Then it follows from Equation (1) that

$$f(x) = \lim_{n \to \infty} P_n(x) = \lim_{n \to \infty} \left( \sum_{k=0}^{n} \frac{f^{(k)}(a)}{k!}(x - a)^k \right), \tag{5}$$

and thus

$$f(x) = \sum_{k=0}^{\infty} \frac{f^{(k)}(a)}{k!}(x - a)^k$$

$$= f(a) + f'(a)(x - a) + \frac{f''(a)}{2!}(x - a)^2 + \cdots$$

$$+ \frac{f^{(n)}(a)}{n!}(x - a)^n + \cdots. \tag{6}$$

The infinite series in Equation (6) is called the **Taylor series** of the function
$f$ at $x = a$. Note that the $n$th degree Taylor polynomial $P_n(x)$ is the sum of
the first $n + 1$ terms of the Taylor series and thus is the $(n + 1)$st term in
the associated sequence of partial sums of the series in (6).

Now suppose that Equation (4) holds—that, for a particular value of $x$,
$R_n(x) \to 0$ as $n \to \infty$. Then, by Equation (5), we can compute the value of

$f(x)$ with any desired accuracy by adding enough terms of the Taylor series of $f$ at $a$. For example, from the problems and examples of Section 11-3, we know the following Taylor formulas for the exponential and trigonometric functions:

$$e^x = 1 + x + \frac{x^2}{2!} + \frac{x^3}{3!} + \cdots + \frac{x^n}{n!} + \frac{e^z}{(n+1)!} x^{n+1}.$$

$$\cos x = 1 - \frac{x^2}{2!} + \frac{x^4}{4!} - \cdots + (-1)^n \frac{x^{2n}}{(2n)!} + (-1)^{n+1} \frac{\cos z}{(2n+2)!} x^{2n+2}.$$

$$\sin x = x - \frac{x^3}{3!} + \frac{x^5}{5!} - \cdots + (-1)^n \frac{x^{2n+1}}{(2n+1)!} + (-1)^{n+1} \frac{\cos z}{(2n+3)!} x^{2n+3}.$$

In each case, $z$ is a number between 0 and $x$.

Since $z$ is between 0 and $x$, it follows that $0 < e^z \leq e^{|x|}$ in Taylor's formula for $e^x$. In the formulas for the sine and cosine functions, $0 \leq |\cos z| \leq 1$. Therefore, the fact that

$$\lim_{n \to \infty} \frac{x^n}{n!} = 0$$

for all $x$ (see Problem 23) implies that $\lim_{n \to \infty} R_n(x) = 0$ in all three cases above. This gives the following Taylor series:

$$e^x = \sum_{n=0}^{\infty} \frac{x^n}{n!} = 1 + x + \frac{x^2}{2!} + \frac{x^3}{3!} + \frac{x^4}{4!} + \cdots, \tag{7}$$

$$\cos x = \sum_{n=0}^{\infty} (-1)^n \frac{x^{2n}}{(2n)!} = 1 - \frac{x^2}{2!} + \frac{x^4}{4!} - \frac{x^6}{6!} + \cdots, \tag{8}$$

$$\sin x = \sum_{n=0}^{\infty} (-1)^n \frac{x^{2n+1}}{(2n+1)!} = x - \frac{x^3}{3!} + \frac{x^5}{5!} - \frac{x^7}{7!} + \cdots. \tag{9}$$

Our discussion of these and similar infinite series representations of familiar functions is continued in Section 12-8. The remainder of this section is devoted to several additional applications of Taylor's formula.

### *THE NUMBER $\pi$

In Section 5-2 we described how Archimedes used polygons inscribed in and circumscribed about the unit circle to show that $3\frac{10}{71} < \pi < 3\frac{1}{7}$. With the aid of large economic computers, $\pi$ has been computed to as many as ten million decimal places. We describe now some of the methods that have been used for such computations. (For a chronicle of mankind's perennial fascination with the number $\pi$, see Howard Eves, *An Introduction to the History of Mathematics*, 4th ed. (Boston: Allyn & Bacon, 1976), pp. 96–102.)

We begin with the elementary algebraic identity

$$\frac{1}{1+x} = 1 - x + x^2 - x^3 + \cdots + (-1)^{k-1} x^{k-1} + \frac{(-1)^k x^k}{1+x}, \tag{10}$$

which can be verified by multiplying both sides by $1 + x$. We substitute $t^2$

for $x$ and $n + 1$ for $k$, and thus find that

$$\frac{1}{1 + t^2} = 1 - t^2 + t^4 - \cdots + (-1)^n t^{2n} + \frac{(-1)^{n+1} t^{2n+2}}{1 + t^2}.$$

Since $D_t \tan^{-1} t = 1/(1 + t^2)$, integration of both sides of this last equation from $t = 0$ to $t = x$ gives

$$\tan^{-1} x = x - \frac{x^3}{3} + \frac{x^5}{5} - \cdots + (-1)^n \frac{x^{2n+1}}{2n + 1} + R_{2n+1} \qquad (11)$$

where

$$|R_{2n+1}| = \left| \int_0^x \frac{t^{2n+2}}{1 + t^2} \, dt \right| \leqq \left| \int_0^x t^{2n+2} \, dt \right| = \frac{|x|^{2n+3}}{2n + 3}. \qquad (12)$$

This estimate of the error makes it clear that

$$\lim_{n \to \infty} R_{2n+1} = 0$$

if $|x| \leqq 1$. Hence we obtain the Taylor series for the inverse tangent function:

$$\tan^{-1} x = \sum_{n=0}^{\infty} (-1)^n \frac{x^{2n+1}}{2n + 1} = x - \frac{x^3}{3} + \frac{x^5}{5} - \frac{x^7}{7} + \cdots, \qquad (13)$$

valid for $-1 \leqq x \leqq 1$.

If we substitute $x = 1$ into Equation (13), we obtain *Leibniz's series*

$$\frac{\pi}{4} = 1 - \frac{1}{3} + \frac{1}{5} - \frac{1}{7} + \cdots.$$

Though this is a beautiful series, it is not an effective way to compute $\pi$. But the error estimate in (12) shows that Formula (11) is effective for the calculation of $\tan^{-1} x$ if $|x|$ is small. For example, if $x = \frac{1}{5}$, the fact that

$$\frac{1}{(9)(5)^9} \approx 0.000000057$$

implies that the approximation

$$\alpha = \tan^{-1}(\tfrac{1}{5}) \approx \tfrac{1}{5} - \tfrac{1}{3}(\tfrac{1}{5})^3 + \tfrac{1}{5}(\tfrac{1}{5})^5 - \tfrac{1}{7}(\tfrac{1}{5})^7$$

is accurate to six decimal places.

Let us begin with $\alpha = \tan^{-1}(\tfrac{1}{5})$. The addition formula for the tangent function can be used to show (Problem 14) that

$$\tan\left(\frac{\pi}{4} - 4\alpha\right) = -\frac{1}{239}.$$

Hence

$$\frac{\pi}{4} = 4 \tan^{-1}\left(\frac{1}{5}\right) - \tan^{-1}\left(\frac{1}{239}\right). \qquad (14)$$

In 1706 John Machin used Formula (14) to calculate the first 100 decimal places of $\pi$; in Problem 22 we ask you to use it to show that $\pi = 3.14159$ to five decimal places. In 1844 the lightning calculator Zacharias Dase, a

German, computed the first 200 decimal places of $\pi$ using the related formula

$$\frac{\pi}{4} = \tan^{-1}\left(\frac{1}{2}\right) + \tan^{-1}\left(\frac{1}{5}\right) + \tan^{-1}\left(\frac{1}{8}\right), \tag{15}$$

and you might enjoy verifying this formula (see Problem 15). A recent computation of 1 million decimal places of $\pi$ used the formulas

$$\frac{\pi}{4} = 12 \tan^{-1}\left(\frac{1}{18}\right) + 8 \tan^{-1}\left(\frac{1}{57}\right) - 5 \tan^{-1}\left(\frac{1}{239}\right)$$

$$= 6 \tan^{-1}\left(\frac{1}{8}\right) + 2 \tan^{-1}\left(\frac{1}{57}\right) + \tan^{-1}\left(\frac{1}{239}\right).$$

For derivations of these and similar formulas, with further discussion of computations of the number $\pi$, see the article *An Algorithm for the Calculation of $\pi$* by George Miel in the American Mathematical Monthly, 86 (1979), pp. 694–697. Although no practical application is likely to require more than ten or twelve decimal places of $\pi$, these computations provide dramatic evidence of the power of Taylor's formula.

## NUMERICAL INTEGRATION WITH TAYLOR POLYNOMIALS

As an alternative to the use of the trapezoidal rule or Simpson's approximation to compute $\int_a^b f(x)\,dx$, it is sometimes more convenient to substitute for the integrand $f(x)$ the Taylor formula for $f$ at $a$:

$$f(x) = P_n(x) + \frac{f^{(n+1)}(z)}{(n+1)!}(x-a)^{n+1}$$

(for some number $z$ between $a$ and $x$). Then

$$\int_a^b f(x)\,dx = \int_a^b P_n(x)\,dx + E_n$$

with

$$|E_n| = \left| \int_a^b f^{(n+1)}(z)\frac{(x-a)^{n+1}}{(n+1)!}\,dx \right| \leq \frac{M(b-a)^{n+2}}{(n+2)!}, \tag{16}$$

where $M$ is the maximum of $|f^{(n+1)}(x)|$ for $x$ in $[a, b]$. Finding the integral of the polynomial $P_n(x)$ is simple, and (16) provides an upper bound on the error.

**EXAMPLE 1** Substitute the eighth degree Taylor polynomial of $\sin x$ to compute

$$\int_0^1 \frac{\sin x}{x}\,dx.$$

*Solution* The eighth degree Taylor polynomial with remainder for $\sin x$ at $a = 0$ is

$$\sin x = x - \frac{x^3}{3!} + \frac{x^5}{5!} - \frac{x^7}{7!} + \frac{x^9}{9!}\cos z$$

**628**

with $z$ between 0 and $x$. We divide both sides by $x$ and then integrate:

$$\int_0^1 \frac{\sin x}{x}\, dx = \int_0^1 \left(1 - \frac{x^2}{3!} + \frac{x^4}{5!} - \frac{x^6}{7!}\right) dx + E$$

$$= 1 - \frac{1}{3!3} + \frac{1}{5!5} - \frac{1}{7!7} + E \approx 0.946083 + E.$$

Now

$$E = \int_0^1 \frac{x^8}{9!} \cos z\, dx \leqq \int_0^1 \frac{x^8}{9!}\, dx = \frac{1}{9!9} \approx 0.0000003.$$

Therefore, we find that

$$\int_0^1 \frac{\sin x}{x}\, dx = 0.946083$$

to six decimal places, with considerably less work involved than would be necessary to achieve such accuracy with Simpson's approximation.

There's a bonus. The original integral is actually, in some sense, improper: The integrand is an indeterminate form at the endpoint $x = 0$. Notice how easily the Taylor series method avoids this potential difficulty.

### TAYLOR'S FORMULA AND THE SECOND DERIVATIVE TEST

Here is a demonstration of the theoretical power of Taylor's formula: We use it to give a quick proof of the second derivative maximum-minimum test for a function that has a continuous second derivative. Suppose that $f'(a) = 0$ but that $f''(a) \neq 0$. Then the second degree Taylor formula for $f$ at $a$ is

$$f(x) = f(a) + \tfrac{1}{2}f''(z)(x - a)^2$$

with $z$ somewhere between $a$ and $x$. Therefore, $f(x) - f(a)$ has the same sign as $f''(z)$. If $x$ is sufficiently close to $a$ to insure that $f''(z)$ has the same sign as $f''(a)$, it follows that $f(x) > f(a)$ if $f''(a) > 0$, while $f(x) < f(a)$ if $f''(a) < 0$. Therefore, $f(a)$ is a local minimum value if $f''(a) > 0$; it is a local maximum value if $f''(a) < 0$.

## 12-7  PROBLEMS

In each of Problems 1–13, find the Taylor series (Equation (6)) of the given function $f(x)$ at the indicated point $a$.

1 $f(x) = \ln(1 + x);\quad a = 0$

2 $f(x) = \dfrac{1}{1 - x};\quad a = 0$

3 $f(x) = e^{-x};\quad a = 0$

4 $f(x) = \cosh x;\quad a = 0$

5 $f(x) = \ln x;\quad a = 1$

6 $f(x) = \sin x;\quad a = \dfrac{\pi}{2}$

7 $f(x) = \cos x;\quad a = \dfrac{\pi}{4}$

8 $f(x) = e^{2x};\quad a = 0$

9 $f(x) = \sinh x;\quad a = 0$

10 $f(x) = \dfrac{1}{(1 - x)^2};\quad a = 0$

11 $f(x) = \dfrac{1}{x};\quad a = 1$

12 $f(x) = \cos x;\quad a = \dfrac{\pi}{2}$

13 $f(x) = \sin x;\quad a = \dfrac{\pi}{4}$

14 Beginning with $\alpha = \tan^{-1}(\tfrac{1}{5})$, use the addition formula

$$\tan(A + B) = \frac{\tan A + \tan B}{1 - \tan A \tan B}$$

to show in turn that:

(a) $\tan 2\alpha = \frac{5}{12}$;

(b) $\tan 4\alpha = \frac{120}{119}$; and

(c) $\tan(\pi/4 - 4\alpha) = -\frac{1}{239}$.

**15** Apply the addition formula for the tangent function to verify the formula in Equation (15).

In Problems 16–21, substitute an appropriate Taylor formula with remainder for part or all of the integrand in order to compute the given integral accurate to the indicated number of decimal places.

**16** $\int_0^1 e^{-x^2}\,dx$;  two decimal places

**17** $\int_0^1 \frac{1 - e^{-x}}{x}\,dx$;  two decimal places

**18** $\int_0^{1/2} \sin x^2\,dx$;  five decimal places

**19** $\int_0^{1/2} \frac{\sin x}{\sqrt{x}}\,dx$;  five decimal places

**20** $\int_0^1 e^{\sqrt{x}}\,dx$;  two decimal places

**21** $\int_0^{1/2} \frac{\tan^{-1}x}{x}\,dx$;  three decimal places

**22** Use Formulas (11), (12), and (14) to show that $\pi = 3.14159$ to five decimal places. (*Suggestion:* Compute $\arctan(\frac{1}{5})$ and $\arctan(\frac{1}{239})$ each with error less than $0.5 \times 10^{-7}$. You can do this by applying Formula (11) with $n = 4$ and with $n = 1$. Carry out your computations to seven decimal places and keep track of errors.)

**23** Prove that $\lim_{n \to \infty} x^n/n! = 0$ if $x$ is a fixed real number. (*Suggestion:* Choose an integer $k$ with $k > 2|x|$, and let $L = |x|^k/k!$. Then show that $|x|^n/n! < L/2^{n-k}$ if $n > k$.)

**24** (a) Suppose that $t$ is a positive real number but is otherwise arbitrary. Integrate both sides of (10) from $x = 0$ to $x = t$ to show that

$$\ln(1 + t) = t - \frac{t^2}{2} + \frac{t^3}{3} - \cdots + (-1)^{k-1}\frac{t^k}{k} + R_k$$

where $|R_k| \leq \dfrac{t^{k+1}}{k+1}$.

(b) Conclude that

$$\ln(1 + t) = t - \frac{t^2}{2} + \frac{t^3}{3} - \cdots = \sum_{k=1}^{\infty} (-1)^{k-1}\frac{t^k}{k}$$

if $0 \leq t \leq 1$.

**25** Use the result of Problem 24(a) to calculate $\ln(\frac{4}{3})$ accurate to three decimal places.

**26** Suppose that $f$ is a function with a continuous third derivative. If $f'(a) = f''(a) = 0$ but $f^{(3)}(a) \neq 0$, prove that $f(a)$ is *neither* a local maximum nor a local minimum value. Begin by writing the third degree Taylor expansion of $f$ at $x = a$.

**27** Differentiate both sides of the identity

$$\frac{1}{1 - x} = 1 + x + x^2 + \cdots + x^n + \frac{x^{n+1}}{1 - x}$$

to obtain the result

$$\frac{1}{(1 - x)^2} = 1 + 2x + 3x^2 + \cdots + nx^{n-1} + R_n$$

where, assuming that $|x| < 1$, $R_n \to 0$ as $n \to \infty$. Thus if $-1 < x < 1$, then

$$\frac{1}{(1 - x)^2} = \sum_{n=0}^{\infty} (n + 1)x^n.$$

**28** Substitute $t^2$ for $x$ in the first identity of Problem 27; then integrate from $t = 0$ to $t = x$ to obtain the result that

$$\tanh^{-1}x = x + \frac{x^3}{3} + \cdots + \frac{x^{2n+1}}{2n + 1} + R_n$$

where $R_n \to 0$ as $n \to \infty$ (under the assumption that $|x| < 1$). Thus if $-1 < x < 1$, then

$$\tanh^{-1}x = \sum_{n=0}^{\infty} \frac{x^{2n+1}}{2n + 1}.$$

**29** Use the formula $\ln 2 = \ln \frac{5}{4} + 2 \ln \frac{6}{5} + \ln \frac{10}{9}$ to calculate $\ln 2$ accurate to three decimal places. See Problem 24(a) for the Taylor's formula for $\ln(1 + x)$.

## *12-7  Optional Computer Application

The program listed below uses the Taylor series

$$\sin x = x - \frac{x^3}{3!} + \frac{x^5}{5!} - \cdots$$

to compute  $\sin x$.  Line 10 asks us to enter a value for  $x$  (in radians). Line 20 sets the initial term  T  and the initial partial sum  S  each equal to  $x$.  To understand line 40, note that the term involving  $x^{n+2}$  in the sine series is obtained from the term involving $x^n$ by multiplying it by

$$-\frac{x^2}{(n+1)(n+2)}.$$

Line 70 specifies that we continue to add terms until we arrive at one whose absolute value does not exceed $10^{-10}$, at which point the value of $\sin x$ is printed. The series is alternating, and its terms are monotonically decreasing toward zero for $n$ sufficiently large, so the error should be less than $10^{-10}$ (that is, if roundoff error is negligible).

```
10    INPUT  " ANGLE IN RADIANS "; X
20    N = 1 : T = X : S = X
30    PRINT  S
40    T = −T*X*X/((N + 1)*(N + 2))
50    S = S + T
60    N = N + 2
70    IF  ABS(T) > 1.0E−10 THEN GOTO  30
80    PRINT  " SINE: "; S
90    END
```

For instance, to compute the sine of $10°$, we run this program with input

$$x = \frac{\pi}{18} \approx 0.1745329252.$$

We get the following output:

```
0.1745329252
0.1736468290
0.1736481786
0.1736481777
   SINE:  0.1736481777
```

The answer is equal to $\sin 10°$ accurate to all ten decimal places.

**Exercise 1** Alter the program so that it accepts input in degrees followed by an internal conversion to radians.

**Exercise 2** Alter *just a line or two* of the program to get one that computes values of $\cos x$.

**Exercise 3** Write a program to compute $e^x$ given the input $x$.

**Exercise 4** Use the series in Problem 24 to write a program to compute $\ln(1 + x)$ given $0 < x < 1$.

**Exercise 5** Use the series in Equation (13) to construct a program that computes values of $\tan^{-1}x$.

---

## 12-8

## Power Series

The most important infinite series representations of functions are those whose terms are constant multiples of (successive) integral powers of the independent variable $x$—that is, series that resemble "infinite polynomials."

For example, from Theorem 1 of Section 12-3 (with $x$ in place of $r$), we know that the geometric series

$$\frac{1}{1-x} = 1 + x + x^2 + \cdots + x^n + \cdots \tag{1}$$

represents (that is, *converges to*) the function $f(x) = 1/(1-x)$ when $|x| < 1$. In Section 12-7 we derived the Taylor series

$$e^x = \sum_{n=0}^{\infty} \frac{x^n}{n!} = 1 + x + \frac{x^2}{2!} + \frac{x^3}{3!} + \cdots, \tag{2}$$

$$\cos x = \sum_{n=0}^{\infty} (-1)^n \frac{x^{2n}}{(2n)!} = 1 - \frac{x^2}{2!} + \frac{x^4}{4!} - \cdots, \tag{3}$$

and

$$\sin x = \sum_{n=0}^{\infty} (-1)^n \frac{x^{2n+1}}{(2n+1)!} = x - \frac{x^3}{3!} + \frac{x^5}{5!} - \cdots. \tag{4}$$

There we used Taylor's formula with remainder to show that these series converge (for all $x$) to the functions $e^x$, $\cos x$, and $\sin x$, respectively. Thus we did not then require the convergence tests of this chapter.

The infinite series above all have the form

$$\sum_{n=0}^{\infty} a_n x^n = a_0 + a_1 x + a_2 x^2 + \cdots + a_n x^n + \cdots \tag{5}$$

with constant *coefficients* $a_0, a_1, a_2, \ldots$. An infinite series of this form is called a **power series in** (powers of) $x$. In order that the first terms of the two sides of Equation (5) agree, we adopt here the convention that $x^0 = 1$ even if $x = 0$.

The power series (5) obviously converges when $x = 0$. In general, it will converge for some nonzero values of $x$ and diverge for others. Because of the way in which powers of $x$ are involved, the ratio test is particularly effective in deciding for which values of $x$ a power series converges. Assume that the limit

$$\rho = \lim_{n \to \infty} \left| \frac{a_{n+1}}{a_n} \right| \tag{6}$$

exists. This is the limit that we need if we want to apply the ratio test to the series $\sum a_n$ of constants. To apply the ratio test to the power series of (5), we write $u_n = a_n x^n$ and compute the limit

$$\lim_{n \to \infty} \left| \frac{u_{n+1}}{u_n} \right| = \lim_{n \to \infty} \left| \frac{a_{n+1} x^{n+1}}{a_n x^n} \right| = \rho |x|. \tag{7}$$

If $\rho = 0$, we see that $\sum a_n x^n$ converges absolutely for all $x$. If $\rho = \infty$, we see that $\sum a_n x^n$ diverges for all $x \neq 0$. If $\rho$ is a positive real number, we see from (7) that $\sum a_n x^n$ converges absolutely for all $x$ such that $\rho |x| < 1$; that is, when

$$|x| < r = \frac{1}{\rho} = \lim_{n \to \infty} \left| \frac{a_n}{a_{n+1}} \right|. \tag{8}$$

In this case the ratio test also implies that $\sum a_n x^n$ diverges if $|x| > r$, but the test is inconclusive when $x = \pm r$. We therefore have proved the following theorem, under the additional hypothesis that the limit in (6) exists. In Problems 43 and 44, we outline a proof that does not require this additional hypothesis.

---

**Theorem 1   Convergence of Power Series**

If $\sum a_n x^n$ is a power series, then either

(i) The series converges absolutely for all $x$;   or

(ii) The series converges only when $x = 0$;   or

(iii) There exists a number $r > 0$ such that $\sum a_n x^n$ converges absolutely if $|x| < r$ and diverges if $|x| > r$.

---

The number $r$ of Case (iii) is called the **radius of convergence** of the power series $\sum a_n x^n$. We shall write $r = \infty$ in Case (i) and $r = 0$ in Case (ii). The set of all real numbers $x$ for which the series converges is called its **interval of convergence**. If $0 < r < \infty$, then the interval of convergence is one of the intervals

$$(-r, r), \quad (-r, r], \quad [-r, r), \quad \text{or} \quad [-r, r].$$

When we substitute either of the end points $x = \pm r$ into the series $\sum a_n x^n$, we obtain an infinite series with constant terms whose convergence must be determined separately. Since these will be numerical series, the earlier tests of this chapter are appropriate.

**EXAMPLE 1**   Find the interval of convergence of the series

$$\sum_{n=1}^{\infty} \frac{x^n}{n \cdot 3^n}.$$

**Solution**   With $u_n = x^n/(n \cdot 3^n)$ we find that

$$\lim_{n \to \infty} \left| \frac{u_{n+1}}{u_n} \right| = \lim_{n \to \infty} \left| \frac{x^{n+1}/([n+1] \cdot 3^{n+1})}{x^n/(n \cdot 3^n)} \right|$$

$$= \lim_{n \to \infty} \frac{n|x|}{3(n+1)} = \frac{|x|}{3}.$$

Now $|x|/3 < 1$ provided that $|x| < 3$, so the ratio test implies that the given series converges absolutely if $|x| < 3$ and diverges if $|x| > 3$. When $x = 3$ we have the divergent harmonic series $\sum (1/n)$, and when $x = -3$ we have the convergent alternating series $\sum (-1)^n/n$. Thus the interval of convergence of the given power series is $[-3, 3)$.

**EXAMPLE 2**   Find the interval of convergence of

$$\sum_{n=0}^{\infty} \frac{2^n x^n}{n!}.$$

**Solution** With $u_n = 2^n x^n/n!$, we find that

$$\lim_{n \to \infty} \left| \frac{u_{n+1}}{u_n} \right| = \lim_{n \to \infty} \left| \frac{2^{n+1} x^{n+1}/(n+1)!}{2^n x^n/n!} \right|$$

$$= \lim_{n \to \infty} \frac{2|x|}{n+1} = 0$$

for all $x$. Hence the ratio test implies that the power series converges for all $x$, and its interval of convergence is $(-\infty, +\infty)$, the whole real line.

**EXAMPLE 3** Find the interval of convergence of the series

$$\sum_{n=0}^{\infty} n^n x^n.$$

**Solution** With $u_n = n^n x^n$, we find that

$$\lim_{n \to \infty} \left| \frac{u_{n+1}}{u_n} \right| = \lim_{n \to \infty} \left| \frac{(n+1)^{n+1} x^{n+1}}{n^n x^n} \right|$$

$$= \lim_{n \to \infty} (n+1) \left( 1 + \frac{1}{n} \right)^n |x| = +\infty$$

for all $x \ne 0$, because $\lim_{n \to \infty} (1 + 1/n)^n = e$. Thus the given series diverges for all $x \ne 0$, and its interval of convergence consists of the single point $x = 0$.

**EXAMPLE 4** Use the ratio test to verify that the Taylor series for $\cos x$ in (3) above converges for all $x$.

**Solution** With $u_n = (-1)^n x^{2n}/(2n)!$ we find that

$$\lim_{n \to \infty} \left| \frac{u_{n+1}}{u_n} \right| = \lim_{n \to \infty} \left| \frac{(-1)^{n+1} x^{2n+2}/(2n+2)!}{(-1)^n x^{2n}/(2n)!} \right|$$

$$= \lim_{n \to \infty} \frac{x^2}{(2n+1)(2n+2)} = 0$$

for all $x$, so the series converges for all $x$. Note, however, that the ratio test tells us only that the series converges to *some* number, *not* necessarily the particular number $\cos x$. The argument of Section 12-7 (using the Taylor formula with remainder) is required to establish that the sum of the series is actually $\cos x$.

---

An infinite series of the form

$$\sum_{n=0}^{\infty} a_n (x-c)^n = a_0 + a_1(x-c) + a_2(x-c)^2 + \cdots \tag{9}$$

(where $c$ is a constant) is called a **power series in** (powers of) $x - c$. By the same reasoning that leads to Theorem 1, with $x^n$ replaced by $(x-c)^n$ throughout, we conclude that either

1 The series (9) converges absolutely for all $x$;    or
2 The series converges only when $x - c = 0$—that is, when $x = c$;    or
3 There exists a number $r > 0$ such that $\sum a_n(x - c)^n$ converges absolutely if $|x - c| < r$ and diverges if $|x - c| > r$.

As in the case of a power series with $c = 0$, the number $r$ is called its **radius of convergence,** and the **interval of convergence** of the series $\sum a_n(x - c)^n$ is the set of all numbers $x$ for which it converges. As before, when $0 < r < \infty$ the convergence of the series at the endpoints $x = c - r$ and $x = c + r$ of its interval of convergence must be checked separately.

**EXAMPLE 5** Determine the interval of convergence of the series

$$\sum_{n=1}^{\infty} (-1)^n \frac{(x - 2)^n}{(4^n)\sqrt{n}}.$$

*Solution*   We let $u_n = (-1)^n(x - 2)^n/[(4^n)\sqrt{n}]$. Then

$$\lim_{n \to \infty} \left| \frac{u_{n+1}}{u_n} \right| = \lim_{n \to \infty} \left| \frac{(-1)^{n+1}(x - 2)^{n+1}/[(4^{n+1})\sqrt{n + 1}]}{(-1)^n(x - 2)^n/[(4^n)\sqrt{n}]} \right|$$

$$= \lim_{n \to \infty} \frac{|x - 2|}{4} \left( \frac{n}{n + 1} \right)^{1/2} = \frac{|x - 2|}{4}.$$

Hence the given series converges when $|x - 2|/4 < 1$; that is, when $|x - 2| < 4$, so the radius of convergence is $r = 4$. Since $c = 2$, the series converges when $-2 < x < 6$ and diverges if either $x < -2$ or $x > 6$. When $x = -2$ the series reduces to the divergent $p$-series $\sum (1/\sqrt{n})$, and when $x = 6$ it reduces to the convergent alternating series $\sum (-1)^n/\sqrt{n}$. Thus the interval of convergence of the given power series is $(-2, 6]$.

## POWER SERIES REPRESENTATION OF FUNCTIONS

Power series provide us with an important tool for computing (or approximating) values of functions. Suppose that the series $\sum a_n x^n$ converges to the value $f(x)$; that is,

$$f(x) = a_0 + a_1 x + a_2 x^2 + \cdots + a_n x^n + \cdots$$

for each $x$ in the interval of convergence of the power series. Then we call $\sum a_n x^n$ a **power series representation** of $f(x)$. For example, the geometric series $\sum x^n$ in (1) is a power series representation of the function $f(x) = 1/(1 - x)$ on the interval $(-1, 1)$.

In Section 12-7 we saw how Taylor's formula with remainder can often be used to find a power series representation of a given function. Recall that the $n$th degree Taylor's formula for $f(x)$ at $x = a$ is

$$f(x) = f(a) + f'(a)(x - a) + \frac{f''(a)}{2!} (x - a)^2 + \cdots$$

$$+ \frac{f^{(n)}(a)}{n!} (x - a)^n + R_n(x). \tag{10}$$

The remainder $R_n(x)$ is given by

$$R_n(x) = \frac{f^{(n+1)}(c)}{(n+1)!} (x-a)^{n+1}$$

where $c$ is some number between $a$ and $x$. If we let $n \to +\infty$ in Formula (10), we obtain the following theorem.

---

**Theorem 2** *Taylor Series Representation*

Suppose that the function $f$ has derivatives of all orders on some interval containing $a$ and also that

$$\lim_{n \to \infty} R_n(x) = 0 \qquad (11)$$

for each $x$ in that interval. Then

$$f(x) = \sum_{n=0}^{\infty} \frac{f^{(n)}(a)}{n!} (x-a)^n \qquad (12)$$

for each $x$ in the interval.

---

The power series in (12) is the **Taylor series** of the function $f$ **at** $x = a$ (or *in powers of* $x - a$, or *with center a*). If $a = 0$ we obtain the power series

$$f(x) = \sum_{n=0}^{\infty} \frac{f^{(n)}(0)}{n!} x^n$$

$$= f(0) + f'(0)x + \frac{f''(0)}{2!} x^2 + \cdots, \qquad (13)$$

which is often called the **Maclaurin series** of $f$. Thus the power series in Equations (2) through (4) at the beginning of this section are the Maclaurin series of the functions $e^x$, $\cos x$, and $\sin x$, respectively.

Upon replacing $x$ by $-x$ in the Maclaurin series for $e^x$, we obtain

$$e^{-x} = 1 - x + \frac{x^2}{2!} - \frac{x^3}{3!} + \cdots + (-1)^n \frac{x^n}{n!} + \cdots.$$

Let us add the series for $e^x$ and $e^{-x}$. This gives

$$\cosh x = \frac{e^x + e^{-x}}{2}$$

$$= \frac{1}{2}\left(1 + x + \frac{x^2}{2!} + \frac{x^3}{3!} + \frac{x^4}{4!} + \cdots\right)$$

$$+ \frac{1}{2}\left(1 - x + \frac{x^2}{2!} - \frac{x^3}{3!} + \frac{x^4}{4!} - \cdots\right),$$

so that

$$\cosh x = 1 + \frac{x^2}{2!} + \frac{x^4}{4!} + \frac{x^6}{6!} + \cdots.$$

Similarly,

$$\sinh x = x + \frac{x^3}{3!} + \frac{x^5}{5!} + \frac{x^7}{7!} + \cdots.$$

Note the resemblance with the series for $\sin x$ and $\cos x$.

Upon replacement of $x$ by $-x^2$ in the series for $e^x$, we obtain

$$e^{-x^2} = \sum_{n=0}^{\infty} (-1)^n \frac{x^{2n}}{n!} = 1 - x^2 + \frac{x^4}{2!} - \frac{x^6}{3!} + \cdots.$$

Because this power series converges to $e^{-x^2}$ for all $x$, it must be the Maclaurin series for $e^{-x^2}$ (see Problem 40). Think how tedious it would be to compute the derivatives of $e^{-x^2}$ needed to write its Maclaurin series directly from (13).

## THE BINOMIAL SERIES

The following example gives one of the most famous and useful of all series, the *binomial series*, which was discovered by Newton in the 1660s. It is the infinite series generalization of the (finite) binomial formula of elementary algebra.

**EXAMPLE 6** Suppose that $\alpha$ is a nonzero real number. Show that the Maclaurin series of $f(x) = (1 + x)^\alpha$ is

$$(1 + x)^\alpha = 1 + \sum_{n=1}^{\infty} \frac{\alpha(\alpha - 1)(\alpha - 2) \cdots (\alpha - n + 1)}{n!} x^n$$

$$= 1 + \alpha x + \frac{\alpha(\alpha - 1)}{2!} x^2 + \frac{\alpha(\alpha - 1)(\alpha - 2)}{3!} x^3 + \cdots. \quad (14)$$

Also determine the interval of convergence of this **binomial series.**

*Solution* To derive the series itself, we simply list the derivatives of $f(x) = (1 + x)^\alpha$:

$$f(x) = (1 + x)^\alpha,$$
$$f'(x) = \alpha(1 + x)^{\alpha - 1},$$
$$f''(x) = \alpha(\alpha - 1)(1 + x)^{\alpha - 2},$$
$$f^{(3)}(x) = \alpha(\alpha - 1)(\alpha - 2)(1 + x)^{\alpha - 3},$$
$$\vdots$$
$$f^{(n)}(x) = \alpha(\alpha - 1)(\alpha - 2) \cdots (\alpha - n + 1)(1 + x)^{\alpha - n}.$$

Thus

$$f^{(n)}(0) = \alpha(\alpha - 1)(\alpha - 2) \cdots (\alpha - n + 1).$$

Substitution of this value of $f^{(n)}(0)$ into the Maclaurin series formula of (13) gives the binomial series in (14).

To determine the interval of convergence of the binomial series, we let

$$u_n = \frac{\alpha(\alpha - 1)(\alpha - 2) \cdots (\alpha - n + 1)}{n!} x^n.$$

We find that

$$\lim_{n \to \infty} \left| \frac{u_{n+1}}{u_n} \right| = \lim_{n \to \infty} \left| \frac{\alpha(\alpha-1)(\alpha-2)\cdots(\alpha-n)x^{n+1}/(n+1)!}{\alpha(\alpha-1)(\alpha-2)\cdots(\alpha-n+1)x^n/n!} \right|$$

$$= \lim_{n \to \infty} \left| \frac{(\alpha-n)x}{n+1} \right| = |x|.$$

Hence the ratio test shows that the binomial series converges absolutely if $|x| < 1$ and diverges if $|x| > 1$. Its convergence at the end points $x = \pm 1$ depends upon the value of $\alpha$; we shall not pursue this. Problem 41 outlines a proof that the sum of the binomial series actually is $(1 + x)^{\alpha}$ if $|x| < 1$.

If $\alpha = k$, a positive integer, then the coefficient of $x^n$ in (14) vanishes for $n > k$, and the binomial series reduces to the binomial formula

$$(1 + x)^k = \sum_{n=0}^{k} \frac{k!}{n!(k-n)!} x^n.$$

Otherwise (14) is an infinite series. For example, with $\alpha = \frac{1}{2}$, we obtain

$$\sqrt{1+x} = 1 + \frac{1}{2}x + \frac{(\frac{1}{2})(-\frac{1}{2})}{2!}x^2 + \frac{(\frac{1}{2})(-\frac{1}{2})(-\frac{3}{2})}{3!}x^3$$

$$+ \frac{(\frac{1}{2})(-\frac{1}{2})(-\frac{3}{2})(-\frac{5}{2})}{4!}x^4 + \cdots$$

$$= 1 + \frac{1}{2}x - \frac{1}{8}x^2 + \frac{1}{16}x^3 - \frac{5}{128}x^4 + \cdots. \tag{15}$$

If we replace $x$ by $-x$ and take $\alpha = -\frac{1}{2}$, we get the series

$$\frac{1}{\sqrt{1-x}} = 1 + \frac{(-\frac{1}{2})}{1!}(-x) + \frac{(-\frac{1}{2})(-\frac{3}{2})}{2!}(-x)^2$$

$$+ \cdots + \frac{1 \cdot 3 \cdot 5 \cdots (2n-1)}{n! \cdot 2^n}x^n + \cdots,$$

which in summation notation takes the form

$$\frac{1}{\sqrt{1-x}} = 1 + \sum_{n=1}^{\infty} \frac{1 \cdot 3 \cdot 5 \cdots (2n-1)}{2 \cdot 4 \cdot 6 \cdots (2n)}x^n. \tag{16}$$

Sometimes it is inconvenient to compute the repeated derivatives of a function in order to find its Taylor series. Another common method of deriving new power series is by the differentiation and integration of known power series. Suppose that a power series representation of the function $f(x)$ is known. Then the following theorem (we leave its proof to advanced calculus) says that the function $f(x)$ may be differentiated by separately differentiating the individual terms in its power series. That is, the power series obtained by termwise differentiation converges to the derivative $f'(x)$. Similarly, a function can be integrated by termwise integration of its power series.

> **Theorem 3** *Termwise Differentiation and Integration*
>
> Suppose that the function $f$ has a power series representation
>
> $$f(x) = \sum_{n=0}^{\infty} a_n x^n = a_0 + a_1 x + a_2 x^2 + a_3 x^3 + \cdots$$
>
> with nonzero radius of convergence $r$. Then $f$ is differentiable on $(-r, r)$ and
>
> $$f'(x) = \sum_{n=1}^{\infty} n a_n x^{n-1} = a_1 + 2a_2 x + 3a_3 x^2 + \cdots. \qquad (17)$$
>
> Also
>
> $$\int_0^x f(t)\, dt = \sum_{n=0}^{\infty} \frac{a_n x^{n+1}}{n+1} = a_0 x + \frac{a_1 x^2}{2} + \frac{a_2 x^3}{3} + \cdots \qquad (18)$$
>
> for each $x$ in $(-r, r)$. Moreover, the power series in (17) and (18) have the same radius of convergence $r$.

REMARK    Although the proof of Theorem 3 is omitted, we observe that the radius of convergence of the series in (17) is

$$r = \lim_{n \to \infty} \left| \frac{n a_n}{(n+1) a_{n+1}} \right|$$

$$= \left( \lim_{n \to \infty} \frac{n}{n+1} \right) \left( \lim_{n \to \infty} \left| \frac{a_n}{a_{n+1}} \right| \right)$$

$$= \lim_{n \to \infty} \left| \frac{a_n}{a_{n+1}} \right|.$$

Thus (by the formula in (8)) the power series for $f(x)$ and the power series for $f'(x)$ have the same radius of convergence.

**EXAMPLE 7**    Termwise differentiation of the geometric series for $f(x) = 1/(1-x)$ yields

$$\frac{1}{(1-x)^2} = D_x\left(\frac{1}{1-x}\right)$$

$$= D_x(1 + x + x^2 + x^3 + \cdots)$$

$$= 1 + 2x + 3x^2 + 4x^3 + \cdots.$$

Thus

$$\frac{1}{(1-x)^2} = \sum_{n=1}^{\infty} n x^{n-1} = \sum_{n=0}^{\infty} (n+1) x^n.$$

The series converges to $1/(1-x)^2$ if $-1 < x < 1$.

**EXAMPLE 8**    Replacement of $x$ by $-t$ in the geometric series above gives

$$\frac{1}{1+t} = 1 - t + t^2 - t^3 + \cdots + (-1)^n t^n + \cdots.$$

Since $D_t \ln(1 + t) = 1/(1 + t)$, termwise integration from $t = 0$ to $t = x$ now gives

$$\ln(1 + x) = \int_0^x \frac{dt}{1 + t}$$

$$= \int_0^x (1 - t + t^2 - \cdots + (-1)^n t^n + \cdots)\, dt;$$

$$\ln(1 + x) = x - \frac{x^2}{2} + \frac{x^3}{3} - \frac{x^4}{4} + \cdots + (-1)^{n-1} \frac{x^n}{n} + \cdots \quad (19)$$

if $|x| < 1$.

**EXAMPLE 9** Find a power series representation for the arctangent function.

**Solution** Since $D_t \tan^{-1} t = 1/(1 + t^2)$, termwise integration of the series

$$\frac{1}{1 + t^2} = 1 - t^2 + t^4 - t^6 + t^8 - \cdots$$

gives

$$\tan^{-1} x = \int_0^x \frac{dt}{1 + t^2} = \int_0^x (1 - t^2 + t^4 - t^6 + t^8 - \cdots)\, dt.$$

Therefore,

$$\tan^{-1} x = x - \frac{x^3}{3} + \frac{x^5}{5} - \frac{x^7}{7} + \frac{x^9}{9} - \cdots \quad (20)$$

if $-1 < x < 1$.

**EXAMPLE 10** Find a power series representation for the arcsine function.

**Solution** First we substitute $t^2$ for $x$ in Equation (16). This yields

$$\frac{1}{\sqrt{1 - t^2}} = 1 + \sum_{n=1}^{\infty} \frac{1 \cdot 3 \cdot 5 \cdots (2n - 1)}{2 \cdot 4 \cdot 6 \cdots (2n)} t^{2n}$$

if $|t| < 1$. Since $D_t \sin^{-1} t = 1/(1 - t^2)^{1/2}$, termwise integration of this series from $t = 0$ to $t = x$ gives

$$\sin^{-1} x = \int_0^x \frac{dt}{\sqrt{1 - t^2}}$$

$$= x + \sum_{n=1}^{\infty} \frac{1 \cdot 3 \cdot 5 \cdots (2n - 1)}{2 \cdot 4 \cdot 6 \cdots (2n)} \cdot \frac{x^{2n+1}}{2n + 1} \quad (21)$$

if $|x| < 1$. Problem 42 shows how to use this series for $\sin^{-1} x$ to derive the series

$$\frac{\pi^2}{6} = 1 + \frac{1}{2^2} + \frac{1}{3^2} + \frac{1}{4^2} + \cdots + \frac{1}{n^2} + \cdots$$

that was used in Example 3 of Section 12-4 to approximate the number $\pi$.

---

Theorem 3 has the important consequence that, if the two power series $\sum a_n x^n$ and $\sum b_n x^n$ both converge and $\sum a_n x^n = \sum b_n x^n$ for all $|x| < r\, (r > 0)$, then $a_n = b_n$ for all $n$. In particular, the Taylor series of a function is its unique power series representation (if any). See Problem 40.

Find the interval of convergence of each of the power series in Problems 1–20.

$1 \displaystyle\sum_{n=1}^{\infty} \frac{x^n}{n}$

$2 \displaystyle\sum_{n=0}^{\infty} \frac{(-1)^n x^n}{n^2 + 1}$

$3 \displaystyle\sum_{n=1}^{\infty} (-1)^n n^2 x^n$

$4 \displaystyle\sum_{n=1}^{\infty} n! x^n$

$5 \displaystyle\sum_{n=1}^{\infty} \frac{(-1)^n x^{2n}}{2n - 1}$

$6 \displaystyle\sum_{n=1}^{\infty} \frac{n x^n}{5^n}$

$7 \displaystyle\sum_{n=0}^{\infty} (5x - 3)^n$

$8 \displaystyle\sum_{n=1}^{\infty} \frac{(2x - 1)^n}{n^4 + 16}$

$9 \displaystyle\sum_{n=1}^{\infty} \frac{2^n (x - 3)^n}{n^2}$

$10 \displaystyle\sum_{n=1}^{\infty} \frac{n!}{n^n} x^n$  (Do not test the endpoints; it diverges at each.)

$11 \displaystyle\sum_{n=1}^{\infty} \frac{(2n)!}{n!} x^n$

$12 \displaystyle\sum_{n=1}^{\infty} \frac{1 \cdot 3 \cdot 5 \cdots (2n + 1)}{n!} x^n$  (Do not test the endpoints; it diverges at each.)

$13 \displaystyle\sum_{n=1}^{\infty} \frac{n^3 (x + 1)^n}{3^n}$

$14 \displaystyle\sum_{n=1}^{\infty} (-1)^n \frac{(x - 2)^n}{n^2}$

$15 \displaystyle\sum_{n=1}^{\infty} \frac{(3 - x)^n}{n^3}$

$16 \displaystyle\sum_{n=1}^{\infty} (-1)^n \frac{10^n}{n!} (x - 10)^n$

$17 \displaystyle\sum_{n=1}^{\infty} \frac{n!}{2^n} (x - 5)^n$

$18 \displaystyle\sum_{n=1}^{\infty} \frac{(-1)^n}{n \cdot 10^n} (x - 2)^n$

$19 \displaystyle\sum_{n=0}^{\infty} x^{(2n)}$

$20 \displaystyle\sum_{n=0}^{\infty} \left( \frac{x^2 + 1}{5} \right)^n$

In each of Problems 21–30, use power series established in this section to find a power series representation of the given function. Then determine the radius of convergence of the resulting series.

$21 \ f(x) = x^2 e^{-3x}$

$22 \ f(x) = \dfrac{1}{10 + x}$

$23 \ f(x) = \sin x^2$
$24 \ f(x) = \cos^2 x = \frac{1}{2}(1 + \cos 2x)$
$25 \ f(x) = \sqrt[3]{1 - x}$

$26 \ f(x) = (1 + x^2)^{3/2}$

$27 \ f(x) = (1 + x)^{-3}$

$28 \ f(x) = \dfrac{1}{\sqrt{9 + x^3}}$

$29 \ f(x) = \dfrac{\ln(1 + x)}{x}$

$30 \ f(x) = \dfrac{x - \tan^{-1} x}{x^3}$

In each of Problems 31–36, find a power series representation for the given function $f(x)$ using termwise integration.

$31 \ f(x) = \displaystyle\int_0^x \sin t^3 \, dt$

$32 \ f(x) = \displaystyle\int_0^x \frac{\sin t}{t} \, dt$

$33 \ f(x) = \displaystyle\int_0^x e^{-t^3} \, dt$

$34 \ f(x) = \displaystyle\int_0^x \frac{\tan^{-1} t}{t} \, dt$

$35 \ f(x) = \displaystyle\int_0^x \frac{1 - e^{-t^2}}{t^2} \, dt$

$36 \ \tanh^{-1} x = \displaystyle\int_0^x \frac{dt}{1 - t^2}$

37 Deduce from the arctangent series (Example 9) that

$$\pi = \frac{6}{\sqrt{3}} \sum_{n=0}^{\infty} \frac{(-1)^n}{2n + 1} \left( \frac{1}{3} \right)^n.$$

Then use this alternating series to show that $\pi = 3.14$ accurate to two decimal places.

38 Substitute the Maclaurin series for $\sin x$, and then assume the validity of termwise integration of the resulting series, to derive the formula

$$\int_0^{\infty} e^{-t} \sin xt \, dt = \frac{x}{1 + x^2} \quad (|x| < 1).$$

Use the fact that $\int_0^{\infty} t^n e^{-t} \, dt = \Gamma(n + 1) = n!$ (Section 11-4).

39 (a) Deduce from the Maclaurin series for $e^t$ that

$$\frac{1}{x^x} = \sum_{n=0}^{\infty} \frac{(-1)^n}{n!} (x \ln x)^n.$$

(b) Assuming the validity of termwise integration of the series in (a), use the integral formula of Problem 31 in Section 11-4 to conclude that

$$\int_0^1 \frac{dx}{x^x} = \sum_{n=1}^{\infty} \frac{1}{n^n}.$$

40 Suppose that $f(x)$ is represented by the power series $\sum_{n=0}^{\infty} a_n x^n$ for all $x$ in some neighborhood of $x = 0$. Show by repeated termwise differentiation of the series, substituting $x = 0$ each time, that $a_n = f^{(n)}(0)/n!$. Thus the only power series in $x$ that represents a given function at and near $x = 0$ is its Maclaurin series.

41 (a) Consider the binomial series

$$f(x) = \sum_{n=0}^{\infty} \frac{\alpha(\alpha - 1)(\alpha - 2) \cdots (\alpha - n + 1)}{n!} x^n,$$

which converges (to *something*) if $|x| < 1$. Compute the derivative $f'(x)$ by termwise differentiation, and show that it satisfies the differential equation $(1 + x)f'(x) = \alpha f(x)$, so that

$$\frac{f'(x)}{f(x)} = \frac{\alpha}{1 + x}.$$

(b) Antidifferentiate both sides of the last equation in part (a) to obtain $f(x) = C(1 + x)^{\alpha}$ for some constant $C$. Finally, show that $C = 1$. Thus the binomial series converges to $(1 + x)^{\alpha}$ if $|x| < 1$.

**42** (a) Show by direct integration that

$$\int_0^1 \frac{\sin^{-1}x}{\sqrt{1 - x^2}}\, dx = \frac{\pi^2}{8}.$$

(b) Use the result of Problem 50 in Section 9-4 to show that

$$\int_0^1 \frac{x^{2n+1}}{\sqrt{1 - x^2}}\, dx = \frac{2 \cdot 4 \cdot 6 \cdots (2n)}{1 \cdot 3 \cdot 5 \cdots (2n + 1)}.$$

(c) Substitute the series of Example 10 for $\sin^{-1}x$ into the integral of part (a), and then use the integral of part (b) to integrate termwise. Conclude that

$$\int_0^1 \frac{\sin^{-1}x}{\sqrt{1 - x^2}}\, dx = 1 + \frac{1}{3^2} + \frac{1}{5^2} + \cdots.$$

(d) Note that

$$\sum_{n=1}^{\infty} \frac{1}{n^2} = \sum_{n=1}^{\infty} \frac{1}{(2n - 1)^2} + \sum_{n=1}^{\infty} \frac{1}{(2n)^2}.$$

Use this information and parts (a) and (c) to show that

$$\sum_{n=1}^{\infty} \frac{1}{n^2} = \frac{\pi^2}{6}.$$

**43** Prove that, if the power series $\sum a_n x^n$ converges for $x = x_0 \neq 0$, then it converges absolutely for all $x$ such that $|x| < |x_0|$. (*Suggestion:* Conclude from the fact that $\lim_{n \to \infty} a_n x_0^n = 0$ that $|a_n x^n| \leq |x/x_0|^n$ for $n$ sufficiently large. Thus the series $\sum |a_n x^n|$ is eventually dominated by the geometric series $\sum |x/x_0|^n$, which converges if $|x| < |x_0|$.)

**44** Suppose that the power series $\sum a_n x^n$ converges for some but not all nonzero values of $x$. Let $S$ be the set of all real numbers $x$ for which the series converges absolutely. (a) Conclude from Problem 43 that the set $S$ is bounded above.

(b) Let $r$ be the least upper bound of the set $S$ (see Problem 44 in Section 12-2). Then show that $\sum a_n x^n$ converges absolutely if $|x| < r$ and diverges if $|x| > r$. Explain why this proves Theorem 1 without the additional hypothesis that $\lim_{n \to \infty} |a_{n+1}/a_n|$ exists.

---

## 12-9

### Power Series Computations

In Section 12-7 we discussed the use of Taylor polynomials to approximate numerical values of functions and integrals. Power series are often used for these same purposes. In the case of an *alternating* power series, it may be simpler to work with the alternating series error estimate (Theorem 2 in Section 12-6) than with a Taylor formula remainder term.

**EXAMPLE 1**   Use the binomial series

$$\sqrt{1 + x} = 1 + \tfrac{1}{2}x - \tfrac{1}{8}x^2 + \tfrac{1}{16}x^3 - \tfrac{5}{128}x^4 + \cdots$$

to approximate $\sqrt{105}$ and estimate the accuracy.

**Solution**   The series above is, after the first term, an alternating series. Hence

$$\sqrt{105} = \sqrt{100 + 5} = 10\sqrt{1 + 0.05}$$
$$= 10[1 + \tfrac{1}{2}(0.05) - \tfrac{1}{8}(0.05)^2 + \tfrac{1}{16}(0.05)^3 + E]$$
$$\approx 10(1.02469531 + E) = 10.2469531 + 10E$$

where $E$ is negative and

$$|E| < \frac{5}{128}(0.05)^4 < 0.0000003.$$

Consequently

$$10.246950 < \sqrt{105} < 10.246954,$$

and so $\sqrt{105} = 10.24695$ to five decimal places.

---

Suppose that we had been asked in advance to approximate $\sqrt{105}$ accurate to five decimal places. A convenient way to do this is to continue writing terms of the series until it is clear that they have become too small in magnitude to affect the fifth decimal place. A good rule of thumb is to use in the computations two more decimal places than required. Thus we use seven decimal places in this case and get

$$\sqrt{105} = 10(1 + 0.05)^{1/2}$$
$$\approx 10(1 + 0.025 - 0.0003125 + 0.0000078 - 0.0000002 + \cdots)$$
$$\approx 10.2469510 \approx 10.24695.$$

**EXAMPLE 2** Approximate

$$\int_0^1 \frac{1 - \cos x}{x^2}\, dx$$

accurate to five decimal places.

*Solution*  We replace $\cos x$ by its Maclaurin series and get

$$\int_0^1 \frac{1 - \cos x}{x^2}\, dx = \int_0^1 \frac{1}{x^2}\left(\frac{x^2}{2!} - \frac{x^4}{4!} + \frac{x^6}{6!} - \cdots\right) dx$$
$$= \int_0^1 \left(\frac{1}{2!} - \frac{x^2}{4!} + \frac{x^4}{6!} - \frac{x^6}{8!} + \cdots\right) dx$$
$$= \frac{1}{2!} - \frac{1}{4!3} + \frac{1}{6!5} - \frac{1}{8!7} + \cdots.$$

Since this last series is an alternating series, it follows that

$$\int_0^1 \frac{1 - \cos x}{x^2}\, dx = \frac{1}{2!} - \frac{1}{4!3} + \frac{1}{6!5} + E$$
$$\approx 0.486389 + E,$$

where $E$ is negative and

$$|E| < \frac{1}{8!7} < 0.000004.$$

Therefore,

$$\int_0^1 \frac{1 - \cos x}{x^2}\, dx \approx 0.48639$$

rounded to five decimal places.

**EXAMPLE 3**  The binomial series with $\alpha = \frac{1}{3}$ gives

$$(1 + x^2)^{1/3} = 1 + \tfrac{1}{3}x^2 - \tfrac{1}{9}x^4 + \tfrac{5}{81}x^6 - \tfrac{10}{243}x^8 + \cdots,$$

which alternates after its first term. Use the first five terms of this series to estimate the value of

$$\int_0^{1/2} \sqrt[3]{1 + x^2}\, dx.$$

**Solution**   Termwise integration of the above series yields

$$\int_0^{1/2} \sqrt[3]{1 + x^2}\, dx = \int_0^{1/2} \left(1 + \frac{x^2}{3} - \frac{x^4}{9} + \frac{5x^6}{81} - \frac{10x^8}{243} + \cdots \right) dx$$

$$= \left[ x + \frac{1}{9}x^3 - \frac{1}{45}x^5 + \frac{5}{567}x^7 - \frac{10}{2187}x^9 + \cdots \right]_0^{1/2}$$

$$= \frac{1}{2} + \frac{1}{9}\left(\frac{1}{2}\right)^3 - \frac{1}{45}\left(\frac{1}{2}\right)^5 + \frac{5}{567}\left(\frac{1}{2}\right)^7$$

$$- \frac{10}{2187}\left(\frac{1}{2}\right)^9 + \cdots.$$

Since this last series is an alternating series, it follows that

$$\int_0^{1/2} \sqrt[3]{1 + x^2}\, dx = \tfrac{1}{2} + \tfrac{1}{9}(\tfrac{1}{2})^3 - \tfrac{1}{45}(\tfrac{1}{2})^5 + \tfrac{5}{567}(\tfrac{1}{2})^7 + E$$

$$\approx 0.513263 + E$$

where $E$ is negative and

$$|E| < \tfrac{10}{2187}(\tfrac{1}{2})^9 < 0.000009.$$

Therefore,

$$0.513254 < \int_0^{1/2} \sqrt[3]{1 + x^2}\, dx < 0.513263,$$

so

$$\int_0^{1/2} \sqrt[3]{1 + x^2}\, dx = 0.51326 \pm 0.00001.$$

## THE ALGEBRA OF POWER SERIES

The following theorem, which we state without proof, says that power series may be added and multiplied much like polynomials. The guiding principle is that of collecting coefficients of like powers of $x$.

---

**Theorem 1   Adding and Multiplying Power Series**

Let $\sum a_n x^n$ and $\sum b_n x^n$ be power series with nonzero radii of convergence. Then

$$\sum_{n=0}^{\infty} a_n x^n + \sum_{n=0}^{\infty} b_n x^n = \sum_{n=0}^{\infty} (a_n + b_n) x^n \tag{1}$$

and

$$\left( \sum_{n=0}^{\infty} a_n x^n \right)\left( \sum_{n=0}^{\infty} b_n x^n \right) = \sum_{n=0}^{\infty} c_n x^n$$

$$= a_0 b_0 + (a_0 b_1 + a_1 b_0)x$$

$$+ (a_0 b_2 + a_1 b_1 + a_2 b_0)x^2 + \cdots \tag{2}$$

where

$$c_n = a_0 b_n + a_1 b_{n-1} + \cdots + a_{n-1} b_1 + a_n b_0. \tag{3}$$

Series (1) and (2) converge for any $x$ that lies interior to the intervals of convergence of both $\sum a_n x^n$ and $\sum b_n x^n$.

---

Thus if $\sum a_n x^n$ and $\sum b_n x^n$ are power series representations of the functions $f(x)$ and $g(x)$, respectively, then the product power series $\sum c_n x^n$ found by "ordinary multiplication" and collection of terms is a power series representation of the function $f(x)g(x)$. This fact can also be used to divide one power series by another, *provided* it is known in advance that the quotient has a power series representation.

**EXAMPLE 4** Assume that the tangent function has a power series representation $\tan x = \sum a_n x^n$. Use the Maclaurin series for $\sin x$ and $\cos x$ to find $a_0, a_1, a_2,$ and $a_3$.

*Solution* We multiply series to obtain

$$\sin x = \tan x \cos x$$

$$= (a_0 + a_1 x + a_2 x^2 + a_3 x^3 + \cdots)\left(1 - \frac{x^2}{2} + \frac{x^4}{24} - \cdots\right)$$

$$= a_0 + a_1 x + \left(a_2 - \frac{1}{2}a_0\right)x^2 + \left(a_3 - \frac{1}{2}a_1\right)x^3 + \cdots.$$

Since

$$\sin x = x - \tfrac{1}{6}x^3 + \tfrac{1}{120}x^5 - \cdots,$$

comparison of coefficients gives the equations

$$a_0 = 0,$$
$$a_1 = 1,$$
$$-\tfrac{1}{2}a_0 + a_2 = 0,$$
$$-\tfrac{1}{2}a_1 + a_3 = -\tfrac{1}{6}.$$

Then we find that $a_0 = 0$, $a_1 = 1$, $a_2 = 0$, and $a_3 = \tfrac{1}{3}$. So

$$\tan x = x + \tfrac{1}{3}x^3 + \cdots.$$

Things are not always as they first appear; the continuation of this series is

$$\tan x = x + \tfrac{1}{3}x^3 + \tfrac{2}{15}x^5 + \tfrac{17}{315}x^7 + \cdots.$$

For the general form of the $n$th coefficient $a_n$, see page 204 of K. Knopp's *Theory and Application of Infinite Series* (Hafner, 1971). You may check that the first few terms given above agree with the result of the ordinary long division indicated below.

$$\left(1 - \tfrac{1}{2}x^2 + \tfrac{1}{24}x^4 - \cdots\right) \overline{)\, x - \tfrac{1}{6}x^3 + \tfrac{1}{120}x^5 - \cdots} \qquad \frac{x + \tfrac{1}{3}x^3 + \tfrac{2}{15}x^5 + \cdots}{}$$

## POWER SERIES AND INDETERMINATE FORMS

According to Theorem 3 in Section 12-8, a power series is differentiable and therefore continuous within its interval of convergence. It follows that

$$\lim_{x \to c} \sum_{n=0}^{\infty} a_n(x - c)^n = a_0. \qquad (4)$$

The following two examples illustrate the use of this fact to find the limit of an indeterminate form $f(x)/g(x)$ by first substituting power series representations for $f(x)$ and $g(x)$.

**EXAMPLE 5**   Find $\lim\limits_{x \to 0} \dfrac{\sin x - \tan^{-1}x}{x^2 \ln(1 + x)}$.

**Solution**   The power series of Equations (4), (19), and (20) in Section 12-8 give

$$\sin x - \tan^{-1}x = (x - \tfrac{1}{6}x^3 + \tfrac{1}{120}x^5 - \cdots) - (x - \tfrac{1}{3}x^3 + \tfrac{1}{5}x^5 - \cdots)$$
$$= \tfrac{1}{6}x^3 - \tfrac{23}{120}x^5 + \cdots$$

and

$$x^2 \ln(1 + x) = x^2(x - \tfrac{1}{2}x^2 + \tfrac{1}{3}x^3 - \cdots)$$
$$= x^3 - \tfrac{1}{2}x^4 + \tfrac{1}{3}x^5 - \cdots .$$

Hence

$$\lim_{x \to 0} \frac{\sin x - \tan^{-1}x}{x^2 \ln(1 + x)} = \lim_{x \to 0} \frac{\tfrac{1}{6}x^3 - \tfrac{23}{120}x^5 + \cdots}{x^3 - \tfrac{1}{2}x^4 + \cdots}$$
$$= \lim_{x \to 0} \frac{\tfrac{1}{6} - \tfrac{23}{120}x^2 + \cdots}{1 - \tfrac{1}{2}x + \cdots} = \frac{1}{6}.$$

**EXAMPLE 6**   Find $\lim\limits_{x \to 1} \dfrac{\ln x}{x - 1}$.

**Solution**   We first replace $x$ by $x - 1$ in the power series for $\ln(1 + x)$ used in Example 5. This gives us

$$\ln x = (x - 1) - \tfrac{1}{2}(x - 1)^2 + \tfrac{1}{3}(x - 1)^3 - \cdots .$$

Hence

$$\lim_{x \to 1} \frac{\ln x}{x - 1} = \lim_{x \to 1} \frac{(x - 1) - \tfrac{1}{2}(x - 1)^2 + \tfrac{1}{3}(x - 1)^3 - \cdots}{x - 1}$$
$$= \lim_{x \to 1} \left[ 1 - \tfrac{1}{2}(x - 1) + \tfrac{1}{3}(x - 1)^2 - \cdots \right] = 1.$$

The method of Examples 5 and 6 provides a useful alternative to l'Hôpital's rule, especially when repeated differentiation of numerator and denominator is inconvenient or too time-consuming.

## 12-9   PROBLEMS

In each of Problems 1–10, use an infinite series to approximate the indicated number accurate to three decimal places.

1  $\sqrt[3]{65}$

2  $\sqrt[4]{630}$

3  $\sin(0.5)$

4  $e^{-0.2}$

5  $\tan^{-1}(0.5)$

6  $\ln(1.1)$

7  $\sin\left(\dfrac{\pi}{10}\right)$

8  $\cos\left(\dfrac{\pi}{20}\right)$

9  $\sin 10°$

10  $\cos 5°$

In each of Problems 11–20, use an infinite series to approximate the value of the given integral accurate to three decimal places.

11  $\displaystyle\int_0^1 \frac{\sin x}{x}\, dx$

12  $\displaystyle\int_0^1 \frac{\sin x}{\sqrt{x}}\, dx$

13  $\displaystyle\int_0^{1/2} \frac{\tan^{-1}x}{x}\, dx$

14  $\displaystyle\int_0^1 \sin x^2\, dx$

$15 \quad \int_0^{0.1} \dfrac{\ln(1 + x)}{x} \, dx$

$16 \quad \int_0^{1/2} \dfrac{dx}{\sqrt{1 + x^4}}$

$17 \quad \int_0^{1/2} \dfrac{1 - e^{-x}}{x} \, dx$

$18 \quad \int_0^{1/2} \sqrt{1 + x^3} \, dx$

$19 \quad \int_0^1 \exp(-x^2) \, dx$

$20 \quad \int_0^{1/2} \dfrac{dx}{1 + x^5}$

In each of Problems 21–26, use power series to evaluate the given limit.

$21 \quad \lim\limits_{x \to 0} \dfrac{1 + x - e^x}{x^2}$

$22 \quad \lim\limits_{x \to 0} \dfrac{x - \sin x}{x^3 \cos x}$

$23 \quad \lim\limits_{x \to 0} \dfrac{1 - \cos x}{x(e^x - 1)}$

$24 \quad \lim\limits_{x \to 0} \dfrac{e^x - e^{-x} - 2x}{x - \arctan x}$

$25 \quad \lim\limits_{x \to 0} \left( \dfrac{1}{x} - \dfrac{1}{\sin x} \right)$

$26 \quad \lim\limits_{x \to 1} \dfrac{\ln(x^2)}{x - 1}$

27 Derive the geometric series by long division of $1 - x$ into 1.

28 Derive the series for $\tan x$ listed in Example 4 by long division of the Taylor series of $\cos x$ into the Taylor series of $\sin x$.

29 Derive the geometric series representation of $1/(1 - x)$ by finding $a_0, a_1, a_2, \ldots$ such that

$$(1 - x)(a_0 + a_1 x + a_2 x^2 + a_3 x^3 + \cdots) = 1.$$

30 Derive the first five coefficients in the binomial series for $\sqrt{1 + x}$ by finding $a_0, a_1, a_2, a_3,$ and $a_4$ such that

$$(a_0 + a_1 x + a_2 x^2 + a_3 x^3 + a_4 x^4 + \cdots)^2 = 1 + x.$$

31 Use the method of Example 4 to find the coefficients $a_0, a_1, a_2, a_3,$ and $a_4$ in the series

$$\sec x = \dfrac{1}{\cos x} = \sum_{n=0}^{\infty} a_n x^n.$$

32 Multiply the geometric series for $1/(1 - x)$ and the series for $\ln(1 - x)$ to show that, if $|x| < 1$, then

$$-\dfrac{1}{1 - x} \ln(1 - x) = x + \left(1 + \dfrac{1}{2}\right) x^2 + \left(1 + \dfrac{1}{2} + \dfrac{1}{3}\right) x^3$$

$$+ \left(1 + \dfrac{1}{2} + \dfrac{1}{3} + \dfrac{1}{4}\right) x^4 + \cdots.$$

33 Take the logarithmic series

$$\ln(1 + x) = x - \dfrac{x^2}{2} + \dfrac{x^3}{3} - \dfrac{x^4}{4} + \cdots$$

as known. Find the first four coefficients in the series for $e^x$ by finding $a_0, a_1, a_2,$ and $a_3$ so that

$$1 + x = e^{\ln(1 + x)} = \sum_{n=0}^{\infty} a_n \left(x - \dfrac{x^2}{2} + \dfrac{x^3}{3} - \dfrac{x^4}{4} + \cdots\right)^n.$$

This is exactly how the power series for $e^x$ was first discovered (by Newton)!

34 Use the method of Example 4 to show that

$$\dfrac{x}{\sin x} = 1 + \dfrac{1}{6} x^2 + \dfrac{7}{360} x^4 + \cdots.$$

35 Show that long division of power series gives

$$\dfrac{2 + x}{1 + x + x^2} = 2 - x - x^2 + 2x^3 - x^4 - x^5$$

$$+ 2x^6 - x^7 - x^8 + \cdots.$$

Show that the radius of convergence of this series is $r = 1$.

## *12-9  Optional Computer Application

If you worked Problem 11 of this section, you found that

$$\int_0^1 \dfrac{\sin x}{x} \, dx = 1 - \dfrac{1}{3!3} + \dfrac{1}{5!5} - \dfrac{1}{7!7} + \cdots.$$

To write a program to sum this series, we note that

$$\dfrac{1}{(n + 2)!(n + 2)} = \dfrac{n}{(n + 1)(n + 2)^2} \cdot \dfrac{1}{n!n}.$$

This observation enables us (see line 20 of the program) to "convert" each term  T of the series into the following term.

```
10    N = 1  :  T = 1  :  S = 1
20    T = −T*N/((N + 1)*(N + 2)*(N + 2))
30    S = S + T
40    N = N + 2
50    IF ABS(T) > 1.0E−06 THEN GOTO 20
60    PRINT " ANSWER: "; S
70    END
```

To alter this program to sum a different alternating series, we need only change the initial values in line 10, the conversion from each term to its successor in line 20, and perhaps the shift of indices in line 40. For exercises, try this with any of Problems 1–20 of this section.

## *12-10

### Additional Applications

Let us consider the ellipse with equation

$$\frac{x^2}{a^2} + \frac{y^2}{b^2} = 1 \qquad (a > b). \tag{1}$$

As in Section 10-5, its eccentricity $\varepsilon$ is defined to be

$$\varepsilon = \frac{1}{a}\sqrt{a^2 - b^2}, \tag{2}$$

whence

$$b = a\sqrt{1 - \varepsilon^2}. \tag{3}$$

A rather simple integration (Problem 43 in Section 5-7) yields $\pi ab$ for the area of the region bounded by this ellipse. To find its perimeter, we use the parametrization

$$x = a\cos\theta, \qquad y = b\sin\theta, \qquad 0 \leq \theta \leq 2\pi. \tag{4}$$

Then the arc length element is

$$\begin{aligned}
ds &= \sqrt{(dx)^2 + (dy)^2} \\
&= \sqrt{a^2\sin^2\theta + b^2\cos^2\theta}\, d\theta \\
&= \sqrt{a^2\sin^2\theta + a^2(1 - \varepsilon^2)\cos^2\theta}\, d\theta
\end{aligned}$$

(using Equation (3)). We simplify to obtain

$$ds = a\sqrt{1 - \varepsilon^2\cos^2\theta}\, d\theta.$$

Consequently the perimeter $p$ is given by

$$p = 4a\int_0^{\pi/2}\sqrt{1 - \varepsilon^2\cos^2\theta}\, d\theta. \tag{5}$$

Naturally enough, this integral is called an *elliptic integral*.

This integral is known to be nonelementary if $0 < \varepsilon < 1$. Therefore, to evaluate it, we expand the integrand using the binomial series

$$\sqrt{1 - x} = 1 - \tfrac{1}{2}x - \tfrac{1}{8}x^2 - \tfrac{1}{16}x^3 - \tfrac{5}{128}x^4 - \cdots \tag{6}$$

of Formula (15) in Section 12-8. This yields

$$\begin{aligned}
p = 4a\int_0^{\pi/2} \Big[ &1 - \tfrac{1}{2}\varepsilon^2\cos^2\theta - \tfrac{1}{8}\varepsilon^4\cos^4\theta \\
&- \tfrac{1}{16}\varepsilon^6\cos^6\theta - \tfrac{5}{128}\varepsilon^8\cos^8\theta - \cdots \Big]\, d\theta.
\end{aligned} \tag{7}$$

When we integrate termwise, applying the integral formula

$$\int_0^{\pi/2}\cos^{2n}\theta\, d\theta = \frac{1\cdot 3\cdot 5\cdots(2n-1)}{2\cdot 4\cdot 6\cdots(2n)}\cdot\frac{\pi}{2} \tag{8}$$

from the endpapers, the result—after some arithmetic that we ask you to verify in Problem 1—is the power series

$$p = 2\pi a\left(1 - \frac{1}{4}\varepsilon^2 - \frac{3}{64}\varepsilon^4 - \frac{5}{256}\varepsilon^6 - \frac{175}{16{,}384}\varepsilon^8 - \cdots\right) \tag{9}$$

that gives the perimeter $p$ of the ellipse in terms of its major semiaxis $a$ and its eccentricity $\varepsilon$.

Note that when $\varepsilon = 0$, Equation (9) reduces to the formula $p = 2\pi a$ for the circumference of a circle of radius $a$. When $\varepsilon$ is fairly small, the series converges very rapidly. For instance, if $\varepsilon = 0.1$, then the last term listed in (9) is $(175)(0.1)^8/16{,}384 \approx 10^{-10}$.

**EXAMPLE 1** Ignoring the effects of the sun and the other planets, the orbit of the moon is almost a perfect ellipse with the earth at one focus. Assume that $a = 238{,}857$ mi (exactly) and that $\varepsilon = 0.0549$ (exactly) for this ellipse. Then when we sum the series in (9) using a computer that carries 17 decimal digits, we get

$$p = 1{,}499{,}651.30945658 \text{ mi} \tag{10}$$

(eight digits to the right of the decimal are accurate) for the perimeter of the moon's orbit.

## AN APPROXIMATION FORMULA

The *arithmetic mean* of the major and minor semiaxes of an ellipse is

$$A = \tfrac{1}{2}(a + b) \tag{11}$$

while their *root-square mean* is

$$R = \sqrt{\tfrac{1}{2}(a^2 + b^2)}. \tag{12}$$

The simple approximation

$$p \approx \pi(A + R) \tag{13}$$

is often used to calculate the perimeter of an ellipse.

We can use the binomial series in (6) to check the accuracy of the approximation in (13). First we write

$$2\pi A = \pi(a + b) = \pi a\left(1 + \frac{b}{a}\right),$$

and thus

$$2\pi A = \pi a(1 + \sqrt{1 - \varepsilon^2}). \tag{14}$$

In Problem 2 we ask you to use the binomial series to verify that

$$2\pi A = 2\pi a(1 - \tfrac{1}{4}\varepsilon^2 - \tfrac{1}{16}\varepsilon^4 - \tfrac{1}{32}\varepsilon^6 - \tfrac{5}{256}\varepsilon^8 - \cdots). \tag{15}$$

Note that the first two terms agree with those of the series in (9).

Next we write

$$2\pi R = 2\pi\sqrt{\tfrac{1}{2}(a^2 + b^2)} = 2\pi\sqrt{\tfrac{1}{2}[a^2 + a^2(1 - \varepsilon^2)]},$$

and so

$$2\pi R = 2\pi a(1 - \tfrac{1}{2}\varepsilon^2)^{1/2}. \tag{16}$$

In Problem 3 we ask you to verify that

$$2\pi R = 2\pi a(1 - \tfrac{1}{4}\varepsilon^2 - \tfrac{1}{32}\varepsilon^4 - \tfrac{1}{128}\varepsilon^6 - \tfrac{5}{2048}\varepsilon^8 - \cdots). \qquad (17)$$

Finally, something wonderful happens when we average the two series in (15) and (17). In Problem 4 we ask you to show that

$$\pi(A + R) = 2\pi a(1 - \tfrac{1}{4}\varepsilon^2 - \tfrac{3}{64}\varepsilon^4 - \tfrac{5}{256}\varepsilon^6 - \tfrac{180}{16,384}\varepsilon^8 - \cdots). \qquad (18)$$

Note that the exact series for $p$ in Equation (9) and the approximate series for $p$ in Equation (18) agree through the terms in $\varepsilon^6$. Thus

$$p = \pi(A + R) + 2\pi a\left(\frac{5}{16,384}\varepsilon^8 + \cdots\right). \qquad (19)$$

If $\varepsilon$ is close to zero then the difference between the exact value of $p$ and the simple approximation $\pi(A + R)$ is exceedingly small.

**EXAMPLE 2**   Using the lunar orbital parameters given in Example 1, we first calculate $b$ using the formula in (3) and then the means $A$ and $R$ defined in (11) and (13). The result (with eight digits correct to the right of the decimal) is

$$p \approx \pi(A + R) = 1,499,651.30945654 \text{ mi.} \qquad (20)$$

Upon comparison of the results in (10) and (20), we see that the error in the approximation $p \approx \pi(A + R)$ is

$$4 \times 10^{-8} \text{ mi} \approx \frac{1}{400} \text{ in.,}$$

which is roughly the thickness of a single page of this book!

### THE SIMPLE PENDULUM AGAIN

In Section 8-6 we introduced the *approximate* formula

$$T \approx 2\pi \sqrt{\frac{L}{g}} \qquad (21)$$

for the period $T$ of oscillation of the simple pendulum shown in Fig. 12.7. In Problems 11 and 12 we outline a derivation of the *exact* formula

$$T = 4\sqrt{\frac{L}{g}} \int_0^{\pi/2} \frac{d\phi}{\sqrt{1 - k^2\sin^2\phi}}, \qquad (22)$$

where $\alpha$ is the initial angle from which the pendulum is released from rest and

$$k = \sin\frac{\alpha}{2}. \qquad (23)$$

In Problem 13 we ask you to expand the integrand in (22) using the binomial series

$$\frac{1}{\sqrt{1 - x}} = \sum_{n=0}^{\infty} \frac{1 \cdot 3 \cdot 5 \cdots (2n - 1)}{2 \cdot 4 \cdot 6 \cdots (2n)} x^n \qquad (24)$$

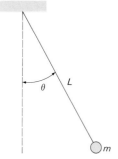

**12.7**   The simple pendulum

of Section 12-8, and then integrate term by term (as in our derivation of the series in (9)) to obtain the series

$$T = 2\pi\left(\frac{L}{g}\right)^{1/2}\left[1 + \left(\frac{1}{2}\right)^2 k^2 + \left(\frac{1\cdot 3}{2\cdot 4}\right)^2 k^4 + \left(\frac{1\cdot 3\cdot 5}{2\cdot 4\cdot 6}\right)^2 k^6 + \cdots\right] \quad (25)$$

for the exact value of the period $T$ in powers of $k$.

Let us denote by $S(k)$ the infinite series factor in Equation (25). We find that our earlier estimate $T \approx 2\pi(L/g)^{1/2}$ can now be replaced by

$$T = 2\pi\left(\frac{L}{g}\right)^{1/2} S(k). \quad (26)$$

We can check the approximation $T \approx 2\pi(L/g)^{1/2}$ by verifying that $S(k)$ is near 1 for very small initial angles $\alpha$. We take $\alpha = 10°$, so that $k = \sin 5°$. Then

$$S(k) \geqq 1 + \tfrac{1}{4}k^2 \approx 1.00190.$$

To get an upper bound for $S(k)$, we note that the coefficients in the series for $S(k)$ are all at most $\tfrac{1}{4}$, so

$$S(k) \leqq 1 + \frac{1}{4}k^2 + \frac{1}{4}k^4 + \frac{1}{4}k^6 + \cdots$$

$$= 1 + \frac{1}{4}k^2(1 + k^2 + k^4 + k^6 + \cdots)$$

$$= 1 + \frac{1}{4}k^2\left(\frac{1}{1 - k^2}\right) = \frac{4 - 3k^2}{4 - 4k^2}.$$

Hence

$$S(k) \leqq \frac{4 - 3k^2}{4 - 4k^2} \approx 1.00191$$

when $k = \sin 5°$.

Thus $S(k) = 1.0019$ to four decimal places, so the approximation $T \approx 2\pi(L/g)^{1/2}$ is about 0.19% low. In a well-designed pendulum clock, the lessening of the angular amplitude due to friction tends to compensate for this increased period and helps the clock to keep time more accurately.

## 12-10 PROBLEMS

1 Integrate termwise in Equation (7) to derive the series in (9) for the perimeter of an ellipse.

2 Verify that Equation (15) is the result of a binomial series expansion in (14).

3 Verify that Equation (17) is the result of a binomial series expansion in (16).

4 Derive the series in (18) from the series in (15) and (17).

5 Consider two pendulum clocks with identical working mechanisms. The period of oscillation of the pendulum of one clock is 0.19% greater than the period of the other. If the two clocks initially agree, by how much will they disagree when the slower clock indicates that exactly one week has elapsed?

6 Figure 12.8 shows a parabolic suspension cable for a bridge of total span $2S$ and height $H$ at each end.

12.8 The parabolic supporting cable of a suspension bridge

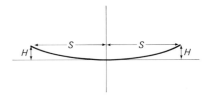

(a) Show that the total length $L$ of the cable is

$$L = 2 \int_0^S \left(1 + \frac{4H^2}{S^4} x^2\right)^{1/2} dx.$$

(b) Then use the binomial series to show that

$$L = 2S\left(1 + \frac{2H^2}{3S^2} - \frac{2H^4}{5S^4} + \frac{4H^6}{7S^6} - \cdots\right).$$

**7** A single-span suspension bridge across the Strait of Messina (5 mi wide) between Italy and Sicily is currently in the planning stage. The plans include suspension towers 1250 ft high at each end. Use the series of Problem 6 to estimate the length of the suspension cable for this proposed bridge.

**8** Consider the power series

$$S(x) = \sum_{n=0}^{\infty} a_n x^n,$$

with positive and nonincreasing coefficients; thus $a_n \geq a_{n+1} > 0$ for all $n \geq 0$. Let

$$S_{n-1}(x) = a_0 + a_1 x + a_2 x^2 + \cdots + a_{n-1}x^{n-1}.$$

Use the geometric series to show that, if $0 < x < 1$, then

$$S_{n-1}(x) + a_n x^n < S(x) < S_{n-1}(x) + \frac{a_n x^n}{1 - x}.$$

Explain how this result may be used to compute $S(x)$ with any desired degree of accuracy.

**9** Apply the result of Problem 8 to show that for a simple pendulum with initial angle $\alpha = 90°$, $S(k) = 1.18$ to two decimal places. Thus the actual period of such a pendulum would be 18% larger than $2\pi(L/g)^{1/2}$.

**10** Use the result of Problem 8 and the first five terms of the Maclaurin series for $e^x$ to compute $e^{1/3}$ accurate to three decimal places.

**11** Suppose the same simple pendulum of Fig. 12.7 is released from rest at initial angle $\alpha$. Then, as explained in Section 8-6, conservation of mechanical energy yields the equation

$$\frac{1}{2} mL^2\left(\frac{d\theta}{dt}\right)^2 + mgL(1 - \cos\theta) = mgL(1 - \cos\alpha).$$

Solve for $dt/d\theta$, and then integrate over a quarter-period $(0 \leq \theta \leq \pi/2)$ to obtain the period integral

$$T = 4\left(\frac{L}{2g}\right)^{1/2} \int_0^\alpha \frac{d\theta}{\sqrt{\cos\theta - \cos\alpha}}.$$

**12** First substitute $\cos\theta = 1 - 2\sin^2(\theta/2)$ into the integral of Problem 11, and next substitute $u = (1/k)\sin(\theta/2)$, where $k = \sin(\alpha/2)$, to deduce that the period $T$ of oscillation of the pendulum is

$$T = 4\left(\frac{L}{g}\right)^{1/2} \int_0^{\pi/2} \frac{d\phi}{\sqrt{1 - k^2 \sin^2\phi}}.$$

**13** Derive the series in (25) from Equations (22)–(24).

## CHAPTER 12 REVIEW: Definitions, Concepts, Results

Use the list below as a guide to concepts that you may need to review.

**1** Limit of a sequence (definition)
**2** The limit laws for sequences
**3** The bounded monotone sequence property
**4** Sum of an infinite series (definition)
**5** Sum of a geometric series (formula)
**6** The $n$th term test for divergence
**7** Divergence of the harmonic series
**8** The integral test
**9** Convergence of $p$-series
**10** The comparison and limit comparison tests
**11** The alternating series test

**12** Absolute convergence (definition *and* the fact that it implies convergence)
**13** The ratio test
**14** The root test
**15** Power series; radius and interval of convergence
**16** Taylor series of the elementary transcendental functions
**17** The binomial series
**18** Termwise differentiation and integration of power series
**19** Use of power series to approximate values of functions and integrals
**20** Product of two power series
**21** Use of power series to evaluate indeterminate forms

## MISCELLANEOUS PROBLEMS

In Problems 1–15, determine whether or not the sequence $\{a_n\}$ converges, and find the limit if it does converge.

**1** $a_n = \dfrac{n^2 + 1}{n^2 + 4}$

**2** $a_n = \dfrac{8n - 7}{7n - 8}$

**3** $a_n = 10 - (0.99)^n$

**4** $a_n = n \sin \pi n$

**5** $a_n = \dfrac{1 + (-1)^n \sqrt{n}}{n + 1}$

**6** $a_n = \sqrt{\dfrac{1 + (-\frac{1}{2})^n}{n + 1}}$

$7\ a_n = \dfrac{\sin 2n}{n}$

$8\ a_n = 2^{-(\ln n)/n}$

$9\ a_n = (-1)^{\sin(n\pi/2)}$

$10\ a_n = \dfrac{(\ln n)^3}{n^2}$

$11\ a_n = \dfrac{1}{n}\sin\dfrac{1}{n}$

$12\ a_n = \dfrac{n - e^n}{n + e^n}$

$13\ a_n = \dfrac{\sinh n}{n}$

$14\ a_n = \left(1 + \dfrac{2}{n}\right)^{2n}$

$15\ a_n = \sqrt[n]{2n^2 + 1}$

Determine whether each infinite series in Problems 16–30 converges or diverges.

$16\ \displaystyle\sum_{n=1}^{\infty} \dfrac{(n^2)!}{n^n}$

$17\ \displaystyle\sum_{n=1}^{\infty} (-1)^n \dfrac{\ln n}{n^2}$

$18\ \displaystyle\sum_{n=0}^{\infty} \dfrac{3^n}{2^n + 4^n}$

$19\ \displaystyle\sum_{n=0}^{\infty} \dfrac{n!}{e^{n^2}}$

$20\ \displaystyle\sum_{n=1}^{\infty} \dfrac{\sin(1/n)}{n^{3/2}}$

$21\ \displaystyle\sum_{n=0}^{\infty} \dfrac{(-2)^n}{3^n + 1}$

$22\ \displaystyle\sum_{n=1}^{\infty} 2^{-(2/n^2)}$

$23\ \displaystyle\sum_{n=2}^{\infty} \dfrac{(-1)^n n}{(\ln n)^3}$

$24\ \displaystyle\sum_{n=1}^{\infty} \dfrac{(-1)^n}{\sqrt[n]{10}}$

$25\ \displaystyle\sum_{n=1}^{\infty} \dfrac{\sqrt{n} + \sqrt[3]{n}}{n^2 + n^3}$

$26\ \displaystyle\sum_{n=1}^{\infty} (-1)^n \dfrac{1}{n^{[1 + (1/n)]}}$

$27\ \displaystyle\sum_{n=1}^{\infty} (-1)^n \dfrac{\tan^{-1} n}{\sqrt{n}}$

$28\ \displaystyle\sum_{n=1}^{\infty} n \sin\dfrac{1}{n}$

$29\ \displaystyle\sum_{n=3}^{\infty} \dfrac{1}{n(\ln n)(\ln \ln n)}$

$30\ \displaystyle\sum_{n=3}^{\infty} \dfrac{1}{n(\ln n)(\ln \ln n)^2}$

Find the interval of convergence of each of the power series in Problems 31–40.

$31\ \displaystyle\sum_{n=0}^{\infty} \dfrac{2^n x^n}{n!}$

$32\ \displaystyle\sum_{n=0}^{\infty} \dfrac{(3x)^n}{2^{n+1}}$

$33\ \displaystyle\sum_{n=1}^{\infty} \dfrac{(x - 1)^n}{n(3^n)}$

$34\ \displaystyle\sum_{n=0}^{\infty} \dfrac{(2x - 3)^n}{4^n}$

$35\ \displaystyle\sum_{n=1}^{\infty} \dfrac{(-1)^n x^n}{4n^2 - 1}$

$36\ \displaystyle\sum_{n=0}^{\infty} \dfrac{(2x - 1)^n}{n^2 + 1}$

$37\ \displaystyle\sum_{n=0}^{\infty} \dfrac{n! x^{2n}}{10^n}$

$38\ \displaystyle\sum_{n=2}^{\infty} \dfrac{x^n}{\ln n}$

$39\ \displaystyle\sum_{n=0}^{\infty} \dfrac{1 + (-1)^n}{2(n!)} x^n$

$40\ \displaystyle\sum_{n=1}^{\infty} \left(1 + \dfrac{1}{n}\right)^n (x - 1)^n$

Find the set of values of $x$ for which each series in Problems 41–43 converges.

$41\ \displaystyle\sum_{n=1}^{\infty} (x - n)^n$

$42\ \displaystyle\sum_{n=1}^{\infty} (\ln x)^n$

$43\ \displaystyle\sum_{n=0}^{\infty} \dfrac{e^{nx}}{n!}$

**44** Find the rational number that has repeated decimal expansion $2.7\ 1828\ 1828\ 1828\cdots$.

**45** Criticize the following argument:

If $x = 1 + 2 + 4 + 8 + \cdots$, then
$1 + 2x = 1 + 2 + 4 + 8 + \cdots$, so $2x + 1 = x$,
and hence $x = -1$.

**46** Prove that $\sum a_n^2$ converges if $\sum a_n$ is a convergent positive-term series.

**47** Let the sequence $\{a_n\}$ be defined recursively by

$$a_1 = 1, \qquad a_{n+1} = 1 + \dfrac{1}{1 + a_n} \qquad \text{if} \quad n \geq 1.$$

The limit of the sequence $\{a_n\}$ is the value of the *continued fraction*

$$1 + \cfrac{1}{2 + \cfrac{1}{2 + \cfrac{1}{2 + \cdots}}}.$$

Assuming that $A = \lim_{n \to \infty} a_n$ exists, show that $A = \sqrt{2}$.

**48** Let $\{F_n\}_{n=1}^{\infty}$ be the Fibonacci sequence of Example 2 in Section 12-2.
(a) Show that $0 < F_n \leq 2^n$ for all $n \geq 1$, and hence conclude that the power series $F(x) = \displaystyle\sum_{n=1}^{\infty} F_n x^n$ converges if $|x| < \frac{1}{2}$.
(b) Show that $(1 - x - x^2)F(x) = x$, so that $F(x) = x/(1 - x - x^2)$.

**49** We say that the "infinite product" indicated by

$$\prod_{n=1}^{\infty} (1 + a_n) = (1 + a_1)(1 + a_2)(1 + a_3)\cdots$$

converges provided that the infinite series

$$S = \sum_{n=1}^{\infty} \ln(1 + a_n)$$

converges, in which case the value of the infinite product is $e^S$. Use the integral test to show that

$$\prod_{n=1}^{\infty} \left(1 + \dfrac{1}{n}\right)$$

diverges.

**50** (See Problem 49.) Show that the infinite product

$$\prod_{n=1}^{\infty} \left(1 + \dfrac{1}{n^2}\right)$$

converges, and use the integral test remainder estimate to approximate its value. The actual value of this infinite product is known to be $(\sinh \pi)/\pi \approx 3.67608$.

In each of Problems 51–55, use infinite series to approximate the indicated number accurate to three decimal places.

**51** $\sqrt[3]{1.5}$

**52** $\ln(1.2)$

**53** $\int_0^1 e^{-x^2}\,dx$

**54** $\int_0^{1/2} \sqrt[3]{1 + x^4}\,dx$

**55** $\int_0^{1/2} \dfrac{1 - e^{-x}}{x}\,dx$

**56** Substitute the Maclaurin series for $\sin x$ into that for $e^x$ to obtain

$$e^{\sin x} = 1 + x + \tfrac{1}{2}x^2 - \tfrac{1}{8}x^4 + \cdots.$$

**57** Substitute the Maclaurin series for the cosine and then integrate termwise to derive the formula

$$\int_0^\infty e^{-t^2} \cos 2xt\,dt = \frac{\sqrt{\pi}}{2} e^{-x^2}.$$

Use the formula for $\int_0^\infty t^{2n} e^{-t^2}\,dt$ given in Miscellaneous Problem 41 in Chapter 11. It should be noted that the validity of this improper termwise integration is subject to verification.

**58** Show that

$$\tanh^{-1} x = \int_0^x \frac{dt}{1 - t^2} = \sum_{n=0}^\infty \frac{x^{2n+1}}{2n + 1}$$

if $|x| < 1$.

**59** Show that

$$\sinh^{-1} x = \int_0^x \frac{dt}{(1 + t^2)^{1/2}}$$
$$= \sum_{n=0}^\infty (-1)^n \frac{1 \cdot 3 \cdot 5 \cdots (2n - 1)}{2 \cdot 4 \cdot 6 \cdots (2n)} \frac{x^{2n+1}}{2n + 1}$$

if $|x| < 1$.

**60** Suppose that $\tan y = \sum a_n y^n$. Determine $a_0, a_1, a_2,$ and $a_3$ by substituting the inverse tangent series (Equation (20) in Section 12-7) into the equation

$$x = \tan(\tan^{-1} x) = \sum a_n(\tan^{-1} x)^n.$$

**61** According to *Stirling's series*, the value of $n!$ for $n$ large is given to a close approximation by

$$n! \approx \sqrt{2\pi n}\left(\frac{n}{e}\right)^n e^{\mu(n)}$$

where

$$\mu(n) = \frac{1}{12n} - \frac{1}{360n^3} + \frac{1}{1260n^5}.$$

Substitute $\mu(n)$ into Maclaurin's series for $e^x$ to show that

$$e^{\mu(n)} = 1 + \frac{1}{12n} + \frac{1}{288n^2} - \frac{139}{51840n^3} + \cdots.$$

Can you show that the next term is $-571/(2,488,320n^4)$?

**62** Define

$$T(n) = \int_0^{\pi/4} \tan^n x\,dx$$

for $n \geq 0$.

(a) Show by "reduction" of the integral that

$$T(n + 2) = \frac{1}{n + 1} - T(n)$$

for $n \geq 0$.

(b) Conclude that $T(n) \to 0$ as $n \to \infty$.

(c) Show that $T_0 = \pi/4$ and that $T_1 = \tfrac{1}{2}\ln 2$.

(d) Show by induction on $n$ that

$$T(2n) = (-1)^{n+1}\left(1 - \frac{1}{3} + \frac{1}{5} - \cdots \pm \frac{1}{2n - 1} - \frac{\pi}{4}\right).$$

(e) Conclude from (b) and (d) that

$$1 - \frac{1}{3} + \frac{1}{5} - \cdots = \frac{\pi}{4}.$$

(f) Show by induction on $n$ that

$$T(2n + 1) = \frac{1}{2}(-1)^n(1 - \frac{1}{2} + \frac{1}{3} - \cdots \pm \frac{1}{n} - \ln 2).$$

(g) Conclude from (b) and (f) that

$$1 - \frac{1}{2} + \frac{1}{3} - \frac{1}{4} + \cdots = \ln 2.$$

# 13

# Parametric Curves and Vectors in the Plane

## Parametric Curves

Until now we have encountered *curves* mainly as graphs of equations. An equation of the form $y = f(x)$ or one of the form $x = g(y)$ determines a curve by giving one of the coordinate variables explicitly as a function of the other. An equation of the form $F(x, y) = 0$ may also determine a curve, but here each variable is given implicitly as a function of the other.

Another important sort of curve is the trajectory of a point moving in the plane. The motion of the point can be described by giving its position $(x(t), y(t))$ at time $t$. Such a description involves expressing both the rectangular coordinate variables $x$ and $y$ as functions of a third variable, or **parameter,** $t$, rather than as functions of one another. In this context a *parameter* is an independent variable (not a constant, as is sometimes meant in popular usage). This approach motivates the following definition.

---

*Definition*   *Parametric Curve*

A **parametric curve** $C$ in the plane is a pair of functions

$$x = f(t), \qquad y = g(t), \tag{1}$$

that give $x$ and $y$ as continuous functions of the real number $t$ (the parameter) in some interval $I$.

---

Each value of the parameter $t$ gives a point $(f(t), g(t))$, and the set of all such points is the **graph** of the curve $C$. Often the distinction between the curve—the pair of **coordinate functions** $f$ and $g$—and the graph is not made. Therefore we shall sometimes refer interchangeably to the curve and to its graph. The two equations in (1) are called the **parametric equations** of the curve.

In most cases the interval $I$ will be a closed interval of the form $[a, b]$, and in such cases the two points $P_a(f(a), g(a))$ and $P_b(f(b), g(b))$ are called the **endpoints** of the curve $C$. If these two points coincide, then we say that the curve $C$ is **closed.** If no distinct pair of values of $t$, except possibly for the values $t = a$ and $t = b$, give rise to the same point on the graph of the curve, then the curve $C$ is non-self-intersecting, and we say that the curve is **simple.** These concepts are illustrated in Fig. 13.1.

**13.1** Parametric curves may be simple or not, closed or not.

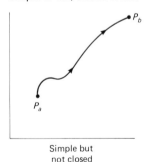

Simple but
not closed

Neither closed
nor simple

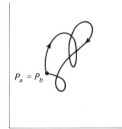

Closed but
not simple

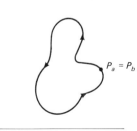

Both simple
and closed

The graph of a parametric curve may be sketched by plotting enough points to indicate its likely shape. Sometimes one can eliminate the parameter $t$ and thus obtain an equation in $x$ and $y$. This equation may give us more information about the shape of the curve.

**EXAMPLE 1**   Determine the graph of the curve

$$x = \cos t, \qquad y = \sin t, \qquad 0 \leq t \leq 2\pi.$$

**Solution**   Figure 13.2 shows a table of values of $x$ and $y$ that correspond to multiples of $\pi/4$ for the parameter $t$. These values give the eight points $(\pm 1, 0)$, $(0, \pm 1)$, and $(\pm 1/\sqrt{2}, \pm 1/\sqrt{2})$, all of which lie on the unit circle. This suggests that the graph is, in fact, the unit circle. To verify this, we note that the fundamental identity of trigonometry gives

$$x^2 + y^2 = \cos^2 t + \sin^2 t = 1,$$

so every point of the graph lies on the unit circle $x^2 + y^2 = 1$. Conversely, the point of the circle with angular (polar) coordinate $t$ is the point $(\cos t, \sin t)$ of the graph. Thus the graph is precisely the unit circle.

| $t$ | $x$ | $y$ |
|---|---|---|
| 0 | 1 | 0 |
| $\pi/4$ | $1/\sqrt{2}$ | $1/\sqrt{2}$ |
| $\pi/2$ | 0 | 1 |
| $3\pi/4$ | $-1/\sqrt{2}$ | $1/\sqrt{2}$ |
| $\pi$ | $-1$ | 0 |
| $5\pi/4$ | $-1/\sqrt{2}$ | $-1/\sqrt{2}$ |
| $3\pi/2$ | 0 | $-1$ |
| $7\pi/4$ | $1/\sqrt{2}$ | $-1/\sqrt{2}$ |
| $2\pi$ | 1 | 0 |

What is lost in the above process is the information about how the graph is produced as $t$ goes from 0 to $2\pi$. But this is easy to determine by inspection. As $t$ increases from 0 to $2\pi$, the point $(\cos t, \sin t)$ starts at $(1, 0)$ and travels counterclockwise around the circle, ending at $(1, 0)$ when $t = 2\pi$. The direction of motion is indicated by the arrow in Fig. 13.2.

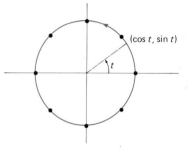

**13.2**   Table of values for Example 1 and the graph

A given figure in the plane may be the graph of different curves. Speaking more loosely, a given curve may have different **parametrizations.** For example, the graph of the curve

$$x = \frac{1 - t^2}{1 + t^2}, \qquad y = \frac{2t}{1 + t^2}, \qquad -\infty < t < \infty$$

also lies on the unit circle, because we find that $x^2 + y^2 = 1$ here, too. If $t$ begins at 0 and increases (decreases), then the point $P(x(t), y(t))$ begins at $(1, 0)$ and travels along the upper (lower) half of the circle. As $t \to \pm\infty$, the point $P$ approaches the point $(-1, 0)$. Thus the graph consists of the unit circle with the single point $(-1, 0)$ deleted. A slight modification of Example 1,

$$x = \cos t, \qquad y = \sin t, \qquad -\pi < t < \pi,$$

is a different parametrization of this graph.

**EXAMPLE 2**   Eliminate the parameter to determine the graph of the parametric curve

$$x = t - 1, \qquad y = 2t^2 - 4t + 1, \qquad 0 \leq t \leq 2.$$

**Solution**   We substitute $t = x + 1$ (from the equation for $x$) into the equation for $y$. This yields

$$y = 2(x + 1)^2 - 4(x + 1) + 1 = 2x^2 - 1$$

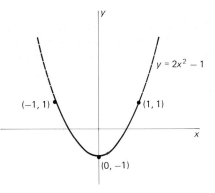

**13.3** The curve of Example 2 is part of a parabola.

for $-1 \leq x \leq 1$. Thus the graph of the given curve is a portion of the parabola $y = 2x^2 - 1$, as shown in Fig. 13.3. As $t$ increases from 0 to 2, the point $(t - 1, 2t^2 - 4t + 1)$ travels along the parabola from $(-1, 1)$ to $(1, 1)$.

---

The same parabolic arc can be reparametrized with

$$x = \sin t, \qquad y = 2 \sin^2 t - 1.$$

Now, as $t$ increases, the point $(\sin t, 2 \sin^2 t - 1)$ travels back and forth along the parabola between the two points $(-1, 1)$ and $(1, 1)$, rather like the bob of a pendulum.

Examples 1 and 2 are parametric curves in which we can eliminate the parameter and thus obtain an explicit equation $y = f(x)$. Conversely, any explicit curve $y = f(x)$ can be viewed as a parametric curve by writing

$$x = t, \qquad y = f(t),$$

with $t$ running through the values in the original domain of $f$. For example, the straight line

$$y - y_0 = m(x - x_0)$$

with slope $m$ and passing through the point $(x_0, y_0)$ is parametrized by

$$x = t, \qquad y = y_0 + m(t - t_0).$$

An even simpler parametrization is obtained with $x - x_0 = t$. This gives

$$x = x_0 + t, \qquad y = y_0 + mt.$$

The use of parametric equations $x = x(t)$ and $y = y(t)$ to describe a curve is most advantageous when elimination of the parameter is either impossible or would lead to an equation $y = f(x)$ considerably more complicated than the original parametric equations. This often happens when the curve is a geometric locus or the path of a point moving under specified conditions.

**EXAMPLE 3**  The curve traced by a point $P$ on the edge of a rolling circle is called a **cycloid**. The circle is to roll along a straight line without slipping or stopping. You will see a cycloid if you watch a patch of bright paint on the tire of a bicycle crossing your path from left to right. Find parametric equations for the cycloid if the line along which the circle rolls is the $x$-axis, the circle is above the $x$-axis but always tangent to it, and the point $P$ begins at the origin.

*Solution*  Evidently the cycloid consists of a series of arches. As parameter $t$ we take the angle (in radians) through which the circle has turned since it began with $P$ at the origin. This is the angle $TCP$ in Fig. 13.4.

The distance the circle has rolled is $|OT|$, so this is also the length of the circumference subtended by the angle $TCP$. Thus $|OT| = at$ if $a$ is the radius of the circle, and so the center $C$ of the rolling circle has coordinates $(at, a)$ at time $t$. The right triangle $CPQ$ of Fig. 13.4 provides us with the relations

$$at - x = a \sin t \quad \text{and} \quad a - y = a \cos t.$$

CHAP. 13:  Parametric Curves and Vectors in the Plane

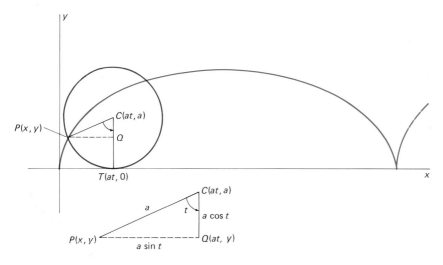

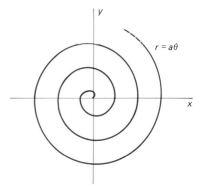

13.4 The cycloid

Therefore the parametric equations of the cycloid, the path of the moving point $P$, are

$$x = a(t - \sin t), \qquad y = a(1 - \cos t). \qquad (2)$$

In Example 2 of Section 9-8 (Equations (3) and (4) of that section), we obtained the parametrization

$$x = c(\theta - \sin \theta \cos \theta), \qquad y = c \sin^2 \theta \qquad (3)$$

of the brachistochrone—the curve of quickest descent. But the substitution $t = 2\theta$, $a = c/2$, together with the identities $\sin 2\theta = 2 \sin \theta \cos \theta$ and $\sin^2 \theta = \frac{1}{2}(1 - \cos 2\theta)$, transform Equations (3) into Equations (2). In June 1696 John Bernoulli proposed the brachistochrone problem as a public challenge, with a 6-month deadline (later extended to Easter 1697 at Leibniz's request). Newton received Bernoulli's challenge on January 29, 1697 (new style), probably after a hard day at the office, and the next day communicated his solution—the curve of minimal descent time is an arc of an inverted cycloid—to the Royal Society of London.

A curve given in polar coordinates by the equation $r = f(\theta)$ can be viewed as a parametric curve with parameter $\theta$. To see this, we recall that the equations $x = r \cos \theta$ and $y = r \sin \theta$ allow us to change from polar to rectangular coordinates. We replace $r$ with $f(\theta)$, and this gives the parametric equations

$$x = f(\theta) \cos \theta, \qquad y = f(\theta) \sin \theta, \qquad (4)$$

expressing $x$ and $y$ in terms of the parameter $\theta$.

For instance, the spiral of Archimedes shown in Fig. 13.5 has the polar coordinates equation $r = a\theta$. The equations in (4) give the spiral the parametrization

$$x = a\theta \cos \theta, \qquad y = a\theta \sin \theta.$$

13.5 The spiral of Archimedes

### TANGENT LINES TO PARAMETRIC CURVES

The parametric curve $x = f(t)$, $y = g(t)$ is called **smooth** if the derivatives $f'(t)$ and $g'(t)$ are continuous and never simultaneously zero. In some neighborhood of each point of its graph, a smooth parametric curve can be described either in the form $y = F(x)$ or in the form $x = G(y)$. To see why this is so, suppose (for example) that $f'(t) > 0$ on the interval $I$. Then $f(t)$ is an increasing function on $I$, and therefore has an inverse function $t = \phi(x)$ there. Substitution of $t = \phi(x)$ into the equation $y = g(t)$ then gives

$$y = g(\phi(x)) = F(x).$$

We can use the chain rule to compute the slope $dy/dx$ of the tangent line to a smooth parametric curve. Differentiation of $y = F(x)$ with respect to $t$ yields

$$\frac{dy}{dt} = \frac{dy}{dx}\frac{dx}{dt},$$

so

$$\frac{dy}{dx} = \frac{dy/dt}{dx/dt} = \frac{g'(t)}{f'(t)} \tag{5}$$

at any point where $f'(t) \neq 0$. The tangent line is vertical at a point where $f'(t) = 0$ but $g'(t) \neq 0$.

Equation (5) gives $y' = dy/dx$ as a function of $t$. Another differentiation with respect to $t$, again using the chain rule, results in the formula

$$\frac{dy'}{dt} = \frac{dy'}{dx}\frac{dx}{dt}.$$

So

$$\frac{d^2y}{dx^2} = \frac{dy'}{dx} = \frac{dy'/dt}{dx/dt}. \tag{6}$$

**EXAMPLE 4**  Calculate $dy/dx$ and $d^2y/dx^2$ for the cycloid with the parametric equations in (2).

*Solution*  We begin with

$$x = a(t - \sin t), \qquad y = a(1 - \cos t). \tag{2}$$

Then Equation (5) gives

$$\frac{dy}{dx} = \frac{dy/dt}{dx/dt} = \frac{a \sin t}{a(1 - \cos t)} = \frac{\sin t}{1 - \cos t}. \tag{7}$$

This derivative is zero when $t$ is an odd multiple of $\pi$, so the tangent line is horizontal at the midpoint of each arch of the cycloid. The endpoints of the arches correspond to even multiples of $\pi$, where both the numerator and the denominator of (7) vanish. These are isolated points (called *cusps*) at which the cycloid fails to be a smooth curve.

Next, Equation (6) yields

$$\frac{d^2y}{dx^2} = \frac{(\cos t)(1 - \cos t) - (\sin t)(\sin t)}{(1 - \cos t)^2} \div a(1 - \cos t)$$

$$= -\frac{1}{a(1 - \cos t)^2}.$$

Since $d^2y/dx^2 < 0$ for all $t$ (except for the isolated even multiples of $\pi$), this shows that each arch of the cycloid is concave downward, as shown in Fig. 13.4.

The slope $dy/dx$ can also be computed in terms of polar coordinates. Given a polar coordinates curve $r = f(\theta)$, we use the parametrization

$$x = f(\theta) \cos \theta, \qquad y = f(\theta) \sin \theta$$

of Equations (4). Then Equation (5), with $\theta$ in place of $t$, gives

$$\frac{dy}{dx} = \frac{f'(\theta) \sin \theta + f(\theta) \cos \theta}{f'(\theta) \cos \theta - f(\theta) \sin \theta} \qquad (8)$$

or, alternatively,

$$\frac{dy}{dx} = \frac{r' \sin \theta + r \cos \theta}{r' \cos \theta - r \sin \theta} \qquad \left( \text{where } r' = \frac{dr}{d\theta} \right) \qquad (9)$$

Formula (9) has the following useful consequence. Let $\psi$ denote the angle between the tangent line at $P$ and the radius $OP$ (extended) from the origin, as in Fig. 13.6. Then

$$\cot \psi = \frac{1}{r} \frac{dr}{d\theta}. \qquad (10)$$

In Problem 26 we indicate how Formula (10) can be derived from Formula (9).

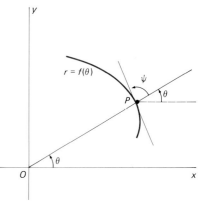

**13.6** The interpretation of the angle $\psi$ (see the formula in (10))

**EXAMPLE 5** Consider the *logarithmic spiral* with polar equation $r = e^\theta$. Show that $\psi = 45°$ at every point of the spiral, and write the equation of its tangent line at the point $(e^{\pi/2}, \pi/2)$.

**Solution** Since $dr/d\theta = e^\theta$, Equation (10) tells us that $\cot \psi = e^\theta/e^\theta = 1$. Thus $\psi = 45°$. When $\theta = \pi/2$, Equation (9) gives

$$\frac{dy}{dx} = \frac{e^{\pi/2} \sin(\pi/2) + e^{\pi/2} \cos(\pi/2)}{e^{\pi/2} \cos(\pi/2) - e^{\pi/2} \sin(\pi/2)} = -1.$$

Since $x - 0$ and $y = e^{\pi/2}$ when $\theta = \pi/2$, it follows that the equation of the desired tangent line is

$$y - e^{\pi/2} = -x, \quad \text{or} \quad x + y = e^{\pi/2}.$$

The line and the spiral appear in Fig. 13.7.

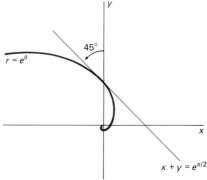

**13.7** The angle $\psi$ is always 45° for the logarithmic spiral.

## 13-1 PROBLEMS

In Problems 1–12, eliminate the parameter and then sketch the curve.

1. $x = t + 1, y = 2t - 1$
2. $x = t^2 + 1, y = 2t^2 - 1$
3. $x = t^2, y = t^3$
4. $x = \sqrt{t}, y = 3t - 2$
5. $x = t + 1, y = 2t^2 - t - 1$
6. $x = t^2 + 3t, y = t - 2$
7. $x = e^t, y = 4e^{2t}$
8. $x = 2e^t, y = 2e^{-t}$
9. $x = 5 \cos t, y = 3 \sin t$
10. $x = 2 \cosh t, y = 3 \sinh t$
11. $x = \sec t, y = \tan t$
12. $x = \cos 2t, y = \sin t$

In Problems 13–17, write the equation of the tangent line to the given parametric curve at the point corresponding to the given value of $t$.

**13** $x = 2t^2 + 1, y = 3t^3 + 2; \quad t = 1$
**14** $x = \cos^3 t, y = \sin^3 t; \quad t = \pi/4$
**15** $x = t \sin t, y = t \cos t; \quad t = \pi/2$
**16** $x = e^t, y = e^{-t}; \quad t = 0$

**17** $x = \dfrac{3t}{1 + t^3}, y = \dfrac{3t^2}{1 + t^3}; \quad t = 1$

In each of Problems 18–21, find the angle $\psi$ between the radius $OP$ and the tangent line at the point $P$ that corresponds to the given value of $\theta$.

**18** $r = 1/\theta, \theta = 1$      **19** $r = e^{\theta\sqrt{3}}, \theta = \pi/2$
**20** $r = 1 - \cos\theta, \theta = \pi/3$      **21** $r = \sin 3\theta, \quad \theta = \pi/6$
**22** Eliminate $t$ to determine the graph of the curve $x = t^2$, $y = t^3 - 3t$, $-2 \leq t \leq 2$. Note that $t = \sqrt{x}$ if $t > 0$, while $t = -\sqrt{x}$ if $t < 0$. Your sketch should contain a loop and should be symmetric about the $x$-axis.
**23** The curve $C$ is determined by the parametric equations $x = e^{-t}$, $y = e^{2t}$. Calculate $dy/dx$ and $d^2y/dx^2$ directly from these parametric equations. Conclude that $C$ is concave upward at every point. Then sketch $C$.
**24** The graph of the folium of Descartes with rectangular equation $x^3 + y^3 = 3xy$ appears in Fig. 13.8. Parametrize its loop as follows: Let $P$ be the point of intersection of the line $y = tx$ with the loop; then solve for the coordinates $x$ and $y$ of $P$ in terms of $t$.

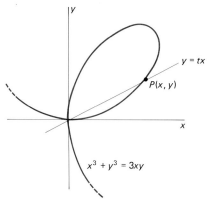

**13.8**    The loop of the folium of Descartes
(Problem 24)

**25** Parametrize the parabola $y^2 = 4px$ by expressing $x$ and $y$ as functions of the slope $m$ of the tangent line at the point $P(x, y)$ of the parabola.
**26** Let $P$ be a point of the curve with polar equation $r = f(\theta)$, and let $\psi$ be the angle between the extended radius $OP$ and the tangent line at $P$. Let $\alpha$ be the angle of inclination of this tangent line, measured counterclockwise from the horizontal. Then $\psi = \alpha - \theta$. Verify Formula

(10) by substituting $\tan\alpha = dy/dx$ from Formula (9) and $\tan\theta = y/x = (\sin\theta)/(\cos\theta)$ into the identity

$$\cot\psi = \frac{1}{\tan(\alpha - \theta)} = \frac{1 + \tan\alpha\tan\theta}{\tan\alpha - \tan\theta}.$$

**27** Let $P_0$ be the highest point of the circle of Fig. 13.4— the circle that generates the cycloid of Example 3. Show that the line through $P_0$ and the point $P$ of the cycloid (the point $P$ is shown in Fig. 13.4) is tangent to the cycloid at $P$. This fact gives a geometric construction of the tangent line to the cycloid.
**28** A circle of radius $b$ rolls without slipping inside a circle of radius $a > b$. The path of a point $P$ fixed on the circumference of the rolling circle is called a *hypocycloid*. Let $P$ begin its journey at $A(a, 0)$, shown in Fig. 13.9, and let $t$ be the angle $AOC$, where $O$ is the center of the large circle and $C$ the center of the rolling circle. Show that the coordinates of $P$ are given by the parametric equations

$$x = (a - b)\cos t + b\cos\left(\frac{a - b}{b}t\right),$$

$$y = (a - b)\sin t - b\sin\left(\frac{a - b}{b}t\right).$$

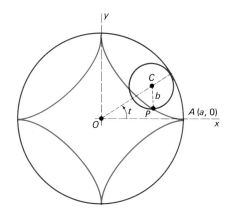

**13.9**    The hypocycloid of Problem 28

**29** If $b = a/4$ in Problem 28, show that the parametric equations of the hypocycloid reduce to

$$x = a\cos^3 t, \qquad y = a\sin^3 t.$$

**30** (a) Show that the hypocycloid $x = \cos^3 t$, $y = \sin^3 t$ is the graph of the equation $x^{2/3} + y^{2/3} = 1$.
(b) Find the points of the hypocycloid where its tangent line is either horizontal or vertical and the intervals on which it is concave upward and those on which it is concave downward.
(c) Sketch this hypocycloid.

**31** Consider a point $P$ on the spiral of Archimedes, the curve shown in Fig. 13.10 with polar equation $r = a\theta$.

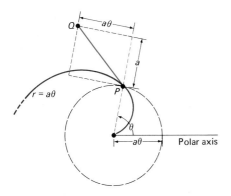

**13.10** The segment $PQ$ is tangent to the spiral (a result of Archimedes).

Archimedes viewed the path of $P$ as compounded of two motions, one with speed $a$ directly away from the origin $O$ and the other a circular motion with unit angular speed around $O$. This suggests Archimedes' result that the line $PQ$ in the figure is tangent to the spiral at $P$. Prove that this is indeed true.

**32** (a) Deduce from Equation (7) that if $t$ is not a multiple of $2\pi$, then the slope of the tangent line at the corresponding point of the cycloid is $\cot(t/2)$.

(b) Conclude that at the cusp point of the cycloid where $t$ is a multiple of $2\pi$, the cycloid has a vertical tangent line.

**33** A *loxodrome* is a curve $r = f(\theta)$ such that the tangent line and the radius $OP$ in Fig. 13.6 make a constant angle. Use Equation (10) to show that every loxodrome is of the form $r = Ae^{k\theta}$ where $A$ and $k$ are constants. Thus every loxodrome is a logarithmic spiral similar to the one considered in Example 5.

---

**13-2**

In Chapter 6 we discussed the computation of a variety of geometric quantities associated with the graph $y = f(x)$ of a nonnegative function on the interval $[a, b]$. These included:

## Integral Computations with Parametric Curves

Area under the curve:
$$A = \int_a^b y\, dx. \tag{1}$$

Volume of revolution around the $x$-axis:
$$V_x = \int_a^b \pi y^2\, dx. \tag{2a}$$

Volume of revolution around the $y$-axis:
$$V_y = \int_a^b 2\pi x y\, dx. \tag{2b}$$

Arc length of the curve:
$$s = \int_0^s ds = \int_a^b [1 + (y')^2]^{1/2}\, dx. \tag{3}$$

Area of surface of revolution about $x$-axis:
$$A_x = \int_a^b 2\pi y\, ds. \tag{4a}$$

Area of surface of revolution about $y$-axis:
$$A_y = \int_a^b 2\pi x\, ds. \tag{4b}$$

Of course we substitute $y = f(x)$ in each of these integrals before integrating from $x = a$ to $x = b$.

We now want to compute these same quantities for a smooth parametric curve

$$x = f(t), \qquad y = g(t), \qquad \alpha \leq t \leq \beta.$$

To insure that this parametric curve looks like the graph of a function defined on some interval $[a, b]$ of the $x$-axis, we assume that $f(t)$ *is either an increasing*

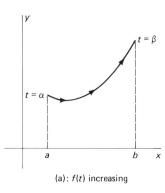

(a): $f(t)$ increasing

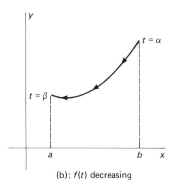

(b): $f(t)$ decreasing

**13.11** Tracing a parametrized curve.

*or a decreasing function of t on* $[\alpha, \beta]$, *and that* $g(t) \geqq 0$. If $a = f(\alpha)$ and $b = f(\beta)$, then the curve will be traced from left to right as $t$ increases, while if $a = f(\beta)$ and $b = f(\alpha)$, it is traced from right to left. (See Fig. 13.11.)

Any one of the quantities $A$, $V_x$, $V_y$, $s$, $A_x$, and $A_y$ may then be computed by making the substitutions

$$x = f(t), \qquad y = g(t),$$
$$dx = f'(t)\,dt, \qquad dy = g'(t)\,dt, \qquad \text{and} \qquad (5)$$
$$ds = \sqrt{[f'(t)]^2 + [g'(t)]^2}\,dt$$

in the appropriate one of the integral formulas in (1) through (4) above. In the case of Formulas (1) and (2), which involve $dx$, we then integrate from $t = \alpha$ to $t = \beta$ in the case shown in part (a) of Fig. 13.11, and from $t = \beta$ to $t = \alpha$ in the case shown in part (b). Thus the proper choice of limits on $t$ corresponds to traversing the curve *from left to right*. For example,

$$A = \int_\alpha^\beta g(t)f'(t)\,dt \quad \text{if} \quad f(\alpha) < f(\beta),$$

while

$$A = \int_\beta^\alpha g(t)f'(t)\,dt \quad \text{if} \quad f(\beta) < f(\alpha).$$

The validity of this method of evaluating the integrals in (1) and (2) follows from the theorem on integration by substitution in Section 5-6.

In the case of the formulas in (3) and (4), which involve $ds$, we integrate from $t = \alpha$ to $t = \beta$ in either case. To see why this is so, recall from Equation (5) in Section 13-1 that $dy/dx = g'(t)/f'(t)$ if $f'(t) \neq 0$ on $[\alpha, \beta]$. Hence

$$s = \int_a^b \sqrt{1 + \left(\frac{dy}{dx}\right)^2}\,dx$$
$$= \int_{f^{-1}(a)}^{f^{-1}(b)} \sqrt{1 + \left[\frac{g'(t)}{f'(t)}\right]^2}\,f'(t)\,dt.$$

Since $f'(t) > 0$ if $f(\alpha) = a$ and $f(\beta) = b$, while $f'(t) < 0$ if $f(\alpha) = b$ and $f(\beta) = a$, it follows that

$$s = \int_\alpha^\beta \sqrt{1 + \left[\frac{g'(t)}{f'(t)}\right]^2}\,|f'(t)|\,dt$$

and so

$$s = \int_\alpha^\beta \sqrt{[f'(t)]^2 + [g'(t)]^2}\,dt$$
$$= \int_\alpha^\beta \sqrt{\left(\frac{dx}{dt}\right)^2 + \left(\frac{dy}{dt}\right)^2}\,dt. \qquad (6)$$

This formula, derived under the assumption that $f'(t) \neq 0$ on $[\alpha, \beta]$, may be taken as the *definition* of arc length for an arbitrary smooth parametric curve. Similarly, the area of a surface of revolution is defined for smooth parametric curves as the result of first making substitutions (5) in Formula (4a) or (4b) and then integrating from $t = \alpha$ to $t = \beta$.

The infinitesimal triangle of Fig. 13.12 serves as a convenient device for remembering the substitution

$$ds = \sqrt{[f'(t)]^2 + [g'(t)]^2}\, dt.$$

When the Pythagorean theorem is applied to this right triangle, it leads to the symbolic manipulation

$$ds = \sqrt{(dx)^2 + (dy)^2}$$
$$= \sqrt{\left(\frac{dx}{dt}\right)^2 + \left(\frac{dy}{dt}\right)^2}\, dt$$
$$= \sqrt{[f'(t)]^2 + [g'(t)]^2}\, dt. \tag{7}$$

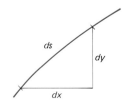

**13.12** Nearly a right triangle for $dx$ and $dy$ close to zero.

**EXAMPLE 1** Use the parametrization $x = a \cos t$, $y = a \sin t$ of the circle with center $O$ and radius $a$ to find the volume $V$ and surface area $A$ of the sphere obtained by revolving this circle around the $x$-axis.

**Solution** Half of the sphere is obtained by revolving the first quadrant of the circle, which is shown in Fig. 13.13. The left-to-right direction along the curve is from $t = \pi/2$ to $t = 0$, so Formula (2a) gives

$$V = 2\pi \int_{\pi/2}^{0} (a \sin t)^2(-a \sin t \, dt)$$
$$= 2\pi a^3 \int_{0}^{\pi/2} (1 - \cos^2 t) \sin t \, dt$$
$$= 2\pi a^3 \left[ -\cos t + \tfrac{1}{3}\cos^3 t \right]_{0}^{\pi/2} = \tfrac{4}{3}\pi a^3.$$

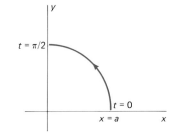

**13.13** The quarter circle of Example 1

The arc length differential for the parametrized curve is

$$ds = \sqrt{(-a \sin t)^2 + (a \cos t)^2}\, dt = a \, dt.$$

Hence Formula (4a) gives

$$A = 2 \int_{0}^{\pi/2} 2\pi(a \sin t)(a \, dt) = 4\pi a^2 \int_{0}^{\pi/2} \sin t \, dt$$
$$= 4\pi a^2 \left[ -\cos t \right]_{0}^{\pi/2} = 4\pi a^2.$$

Of course the results of Example 1 are familiar. By contrast, the following example requires the new methods of this section.

**EXAMPLE 2** Find the area under and the arc length of the cycloidal arch

$$x = a(t - \sin t), \qquad y = a(1 - \cos t), \qquad 0 \le t \le 2\pi.$$

**Solution** We use the identity $1 - \cos t = 2\sin^2(t/2)$ and the result of Problem 50 in Section 9-4,

$$\int_{0}^{\pi} \sin^{2n}u \, du = \pi \frac{1}{2}\frac{3}{4}\frac{5}{6} \cdots \frac{2n-1}{2n}$$

Since $dx = a(1 - \cos t)\, dt$ and the left-to-right direction along the curve is from $t = 0$ to $t = 2\pi$, Formula (1) gives

$$A = \int_0^{2\pi} a(1 - \cos t) \cdot a(1 - \cos t)\, dt$$

$$= a^2 \int_0^{2\pi} (1 - \cos t)^2\, dt = 4a^2 \int_0^{2\pi} \sin^4 \frac{t}{2}\, dt$$

$$= 8a^2 \int_0^{\pi} \sin^4 u\, du \qquad \left(u = \frac{t}{2}\right)$$

$$= 8a^2 \cdot \pi \cdot \frac{1}{2} \cdot \frac{3}{4} = 3\pi a^2$$

for the area under one arch of the cycloid. The arc length differential is

$$ds = \sqrt{a^2(1 - \cos t)^2 + (a \sin t)^2}\, dt$$

$$= a\sqrt{2(1 - \cos t)}\, dt = 2a \sin\left(\frac{t}{2}\right) dt,$$

so Formula (3) gives

$$s = \int_0^{2\pi} 2a \sin \frac{t}{2}\, dt = \left[-4a \cos \frac{t}{2}\right]_0^{2\pi} = 8a$$

for the length of one arch of the cycloid.

## PARAMETRIC POLAR COORDINATES

Suppose that a parametric curve is determined by giving its polar coordinates

$$r = r(t), \qquad \theta = \theta(t), \qquad \alpha \leqq t \leqq \beta$$

as functions of the parameter $t$. Then this curve is described in rectangular coordinates by means of the parametric equations

$$x(t) = r(t) \cos \theta(t), \qquad y(t) = r(t) \sin \theta(t), \qquad \alpha \leqq t \leqq \beta,$$

giving $x$ and $y$ as functions of $t$. These latter parametric equations may then be used in integral formulas (1) through (4).

To compute $ds$, we first calculate the derivatives

$$\frac{dx}{dt} = (\cos \theta) \frac{dr}{dt} - (r \sin \theta) \frac{d\theta}{dt},$$

$$\frac{dy}{dt} = (\sin \theta) \frac{dr}{dt} + (r \cos \theta) \frac{d\theta}{dt}.$$

Upon substituting these expressions for $dx/dt$ and $dy/dt$ into (7) and making algebraic simplifications, we find that the arc length differential in parametric polar coordinates is

$$ds = \sqrt{\left(\frac{dr}{dt}\right)^2 + \left(r \frac{d\theta}{dt}\right)^2}\, dt. \tag{8}$$

In the case of a curve with explicit polar coordinate equation $r = f(\theta)$, we may use $\theta$ rather than $t$ as the parameter. Then (8) takes the simpler form

$$ds = \sqrt{\left(\frac{dr}{d\theta}\right)^2 + r^2}\; d\theta. \tag{9}$$

**EXAMPLE 3** Find the perimeter $s$ of the cardioid $r = 1 + \cos\theta$; also find the surface area $A$ generated by revolving it around the $x$-axis.

*Solution* This cardioid is shown in Fig. 13.14. Formula (9) gives

$$ds = \sqrt{(-\sin\theta)^2 + (1 + \cos\theta)^2}\; d\theta$$

$$= \sqrt{2(1 + \cos\theta)}\; d\theta = \sqrt{4\cos^2\left(\frac{\theta}{2}\right)}\; d\theta.$$

Hence $ds = 2\cos(\theta/2)\, d\theta$ on the upper half of the cardioid, where $0 \le \theta \le \pi$ and $\cos(\theta/2) \ge 0$. Therefore,

$$s = 2\int_0^\pi 2\cos\frac{\theta}{2}\, d\theta = 8\left[\sin\frac{\theta}{2}\right]_0^\pi = 8.$$

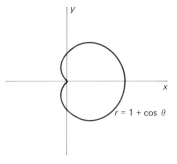

**13.14** The cardioid of Example 3

The surface area of revolution is

$$A = \int 2\pi y\, ds = \int_{\theta=0}^\pi 2\pi(r\sin\theta)\, ds$$

$$= \int_0^\pi 2\pi(1 + \cos\theta)(\sin\theta)\cdot 2\cos\frac{\theta}{2}\, d\theta$$

$$= 16\pi\int_0^\pi \cos^4\frac{\theta}{2}\sin\frac{\theta}{2}\, d\theta$$

$$= 16\pi\left[-\frac{2}{5}\cos^5\frac{\theta}{2}\right]_0^\pi = \frac{32\pi}{5}.$$

## 13-2  PROBLEMS

In each of Problems 1–6, find the area of the region that lies between the given parametric curve and the $x$-axis.

1 $x = t^3,\ y = 2t^2 + 1;\ -1 \le t \le 1$
2 $x = e^{3t},\ y = e^{-t};\ 0 \le t \le \ln 2$
3 $x = \cos t,\ y = \sin^2 t;\ 0 \le t \le \pi$
4 $x = 2 - 3t,\ y = e^{2t};\ 0 \le t \le 1$
5 $x = \cos t,\ y = e^t;\ 0 \le t \le \pi$
6 $x = 1 - e^t,\ y = 2t + 1;\ 0 \le t \le 1$

In Problems 7–10, find the volume obtained by revolving the region indicated in the given problem above around the $x$-axis.

7 Problem 1          8 Problem 2
9 Problem 3         10 Problem 5

In Problems 11–16, find the arc length of the given curve.

11 $x = 2t,\ y = \frac{2}{3}t^{3/2};\ 5 \le t \le 12$
12 $x = \frac{1}{2}t^2,\ y = \frac{1}{3}t^3;\ 0 \le t \le 1$

13 $x = \sin t - \cos t,\ y - \sin t + \cos t;\ \dfrac{\pi}{4} \le t \le \dfrac{\pi}{2}$

14 $x = e^t\sin t,\ y = e^t\cos t;\ 0 \le t \le \pi$
15 $r = e^{\theta/2};\ 0 \le \theta \le 4\pi$          16 $r = \theta;\ 2\pi \le \theta \le 4\pi$

In Problems 17–22, find the area of the surface of revolution generated by revolving the given curve around the indicated axis.

17 $x = 1 - t,\ y = 2\sqrt{t},\ 1 \le t \le 4;$   the $x$-axis

18 $x = 2t^2 + \dfrac{1}{t},\ y = 8t^{1/2},\ 1 \le t \le 2;$   the $x$-axis

19 $x = t^3,\ y = 2t + 3,\ -1 \le t \le 1;$   the $y$-axis
20 $x = 2t + 1,\ y = t^2 + t,\ 0 \le t \le 3;$   the $y$-axis
21 $r = 4\sin\theta;$   the $x$-axis

22 $r = e^\theta,\ 0 \le \theta \le \dfrac{\pi}{2};$   the $y$-axis

**23** Find the volume generated by revolving the region under the cycloidal arch of Example 2 around the $x$-axis.

**24** Find the area of the surface generated by revolving the cycloidal arch of Example 2 around the $x$-axis.

**25** Use the parametrization $x = a \cos t$, $y = b \sin t$ to find
(a) the area bounded by the ellipse $x^2/a^2 + y^2/b^2 = 1$;
(b) the volume of the ellipsoid generated by revolving this ellipse around the $x$-axis.

**26** Find the area bounded by the loop of the parametric curve $x = t^2$, $y = t^3 - 3t$ of Problem 22 in Section 13-1.

**27** Use the parametrization $x = t \cos t$, $y = t \sin t$ of the Archimedean spiral to find the arc length of the first full turn of this spiral (corresponding to $0 \le t \le 2\pi$).

**28** The circle $(x - b)^2 + y^2 = a^2$ with radius $a < b$ and center $(b, 0)$ can be parametrized by

$$x = b + a \cos t, \qquad y = a \sin t, \qquad 0 \le t \le 2\pi.$$

Find the surface area of the torus obtained by revolving this circle around the $y$-axis.

**29** The *astroid* (or four-cusped hypocycloid) with equation $x^{2/3} + y^{2/3} = 1$ arose in connection with Problems 28–30 of Section 13-1 and is shown in Fig. 13.9 at the end of that section. It has the parametrization

$$x = a \cos^3 t, \qquad y = a \sin^3 t, \qquad 0 \le t \le 2\pi.$$

Find the area of the region bounded by the astroid.

**30** Find the total length of the astroid of Problem 29.

**31** Find the area of the surface obtained by revolving the astroid of Problem 29 around the $x$-axis.

**32** Find the area of the surface generated by revolving the lemniscate $r^2 = 2a^2 \cos 2\theta$ around the $y$-axis. (*Suggestion:* Use Formula (9) and note that $r\, dr = -2a^2 \sin 2\theta\, d\theta$.)

**33** Figure 13.15 shows the graph of the parametric curve

$$x = t^2\sqrt{3}, \qquad y = 3t - \tfrac{1}{3}t^3;$$

the shaded region is bounded by the part $-3 \le t \le 3$ of the curve. Find its area.

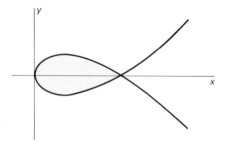

**13.15** The parametric curve of Problems 33–36

**34** Find the arc length of the loop in Problem 33.

**35** Find the volume of the solid obtained by revolving the shaded region in Fig. 13.15 around the $x$-axis.

**36** Find the surface area of revolution generated by revolving the loop of Fig. 13.15 around the $x$-axis.

**37** (a) With reference to Problem 24 and Fig. 13.8 in Section 13-1, show that the arc length of the first quadrant loop of the folium of Descartes is

$$s = 6 \int_0^1 \frac{(1 + 4t^2 - 4t^3 - 4t^5 + 4t^6 + t^8)^{1/2}}{(1 + t^3)^2}\, dt.$$

(b) If you have a programmable calculator or computer, apply Simpson's approximation to obtain $s \approx 4.9175$.

---

## Vectors in the Plane

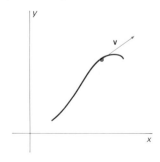

**13.16** A velocity vector may be represented by an arrow.

A physical quantity such as length, temperature, or mass can be specified in terms of a single real number, its magnitude. Such a quantity is called a **scalar.** Other physical entities, such as force and velocity, possess both magnitude and direction; these entities are called **vector quantities,** or simply **vectors.**

For example, to specify the velocity of a moving point in the plane, we must give both the rate at which it moves (its speed) and the direction of that motion. The combination is the **velocity vector** of the moving point. It is convenient to represent this velocity vector by an arrow, located at the current position of the moving point on its trajectory, as shown in Fig. 13.16.

Although the arrow or directed line segment carries the desired information—both magnitude (its length) and direction—it is a pictorial object rather than a mathematical object. We shall see that the following formal definition of a vector captures the essence of magnitude combined with direction.

### Definition of Vector

A **vector v** in the Cartesian plane is an ordered pair of real numbers, of the form $\langle a, b \rangle$. We write $\mathbf{v} = \langle a, b \rangle$ and call $a$ and $b$ the **components** of the vector **v**.

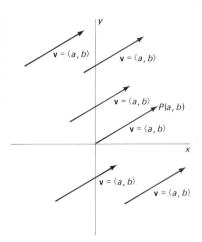

**13.17** All these arrows represent the same vector $\mathbf{v} = \langle a, b \rangle$.

The directed line segment $\overrightarrow{OP}$ from the origin $O$ to the point $P(a, b)$ is one geometric representation of the vector **v**. For this reason, the vector $\mathbf{v} = \langle a, b \rangle$ is called the **position vector** of the point $P(a, b)$. In fact, the relationship between $\mathbf{v} = \langle a, b \rangle$ and $P(a, b)$ is so close that, in certain contexts, it is convenient to confuse the two deliberately—to regard **v** and $P$ as the same mathematical object. In other contexts the distinction between the two is important; in particular, *any* directed line segment with the same direction and magnitude (length) as $\overrightarrow{OP}$ is a representation of **v**, as Fig. 13.17 suggests. What is important about the vector **v** is usually not *where* it is, but how long it is and which way it points.

The magnitude associated with the vector $\mathbf{v} = \langle a, b \rangle$, called its **length,** is denoted by $v = |\mathbf{v}|$ and is defined to be

$$v = |\mathbf{v}| = |\langle a, b \rangle| = \sqrt{a^2 + b^2}. \tag{1}$$

For instance, the length of the vector $\mathbf{v} = \langle 1, -2 \rangle$ is

$$|\mathbf{v}| = |\langle 1, -2 \rangle| = \sqrt{(1)^2 + (-2)^2} = \sqrt{5}.$$

The notation $v = |\mathbf{v}|$ is used because the length of a vector is in some ways analogous to the absolute value of a real number.

The only vector with length zero is the **zero vector** with both components zero, denoted by $\mathbf{0} = \langle 0, 0 \rangle$. The zero vector is also unique in having no specific direction.

The operations of addition and multiplication of real numbers have analogues for vectors. We shall define each of these operations of "vector algebra" in terms of components of vectors and then give a geometric interpretation in terms of arrows.

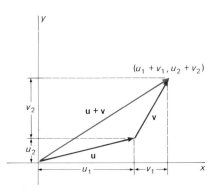

**13.18** The triangle law, a geometric interpretation of vector addition.

**13.19** The parallelogram law for vector addition.

### Definition *Addition of Vectors*

The **sum u + v** of the two vectors $\mathbf{u} = \langle u_1, u_2 \rangle$ and $\mathbf{v} = \langle v_1, v_2 \rangle$ is the vector

$$\mathbf{u} + \mathbf{v} = \langle u_1 + v_1, u_2 + v_2 \rangle. \tag{2}$$

For instance, the sum of the vectors $\mathbf{u} = \langle 4, 3 \rangle$ and $\mathbf{v} = \langle -5, 2 \rangle$ is

$$\langle 4, 3 \rangle + \langle -5, 2 \rangle = \langle 4 + (-5), 3 + 2 \rangle = \langle -1, 5 \rangle.$$

Thus we add vectors by adding corresponding components—that is, by *componentwise addition*. The geometric interpretation of vector addition is the **triangle law of addition,** illustrated in Fig. 13.18, where the labeled lengths indicate why this interpretation is valid. An equivalent interpretation is the **parallelogram law of addition** illustrated in Fig. 13.19.

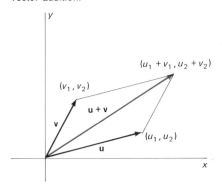

It is natural to write $2\mathbf{u} = \mathbf{u} + \mathbf{u}$. If $\mathbf{u} = \langle u_1, u_2 \rangle$, then

$$2\mathbf{u} = \mathbf{u} + \mathbf{u} = \langle u_1, u_2 \rangle + \langle u_1, u_2 \rangle = \langle 2u_1, 2u_2 \rangle.$$

This suggests that multiplication of a vector by a scalar (real number) can also be defined in a componentwise manner.

---

**Definition**   *Multiplication of a Vector by a Scalar*
If $\mathbf{u} = \langle u_1, u_2 \rangle$ and $c$ is a real number, then the **scalar multiple** $c\mathbf{u}$ is the vector

$$c\mathbf{u} = \langle cu_1, cu_2 \rangle. \qquad (3)$$

---

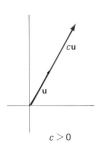

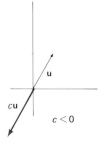

**13.20**  The vector $c\mathbf{u}$ may have the same direction as $\mathbf{u}$ or the opposite direction.

Note that

$$|c\mathbf{u}| = \sqrt{(cu_1)^2 + (cu_2)^2} = |c|\sqrt{(u_1)^2 + (u_2)^2} = |c| \cdot |\mathbf{u}|.$$

Thus the length of $c\mathbf{u}$ is $|c|$ times that of $\mathbf{u}$.

The negative of the vector $\mathbf{u}$ is the vector

$$-\mathbf{u} = (-1)\mathbf{u} = \langle -u_1, -u_2 \rangle$$

with the same length as $\mathbf{u}$ but the opposite direction. We say that the two nonzero vectors $\mathbf{u}$ and $\mathbf{v}$ have

**1** The **same direction** if $\mathbf{u} = c\mathbf{v}$ for some $c > 0$;
**2** **Opposite directions** if $\mathbf{u} = c\mathbf{v}$ for some $c < 0$.

The geometric interpretation of scalar multiplication is that $c\mathbf{u}$ is the vector with length $|c| \cdot |\mathbf{u}|$, with the same direction as $\mathbf{u}$ if $c > 0$ but the opposite direction if $c < 0$. See Fig. 13.20.

The **difference** $\mathbf{u} - \mathbf{v}$ of the vectors $\mathbf{u} = \langle u_1, u_2 \rangle$ and $\mathbf{v} = \langle v_1, v_2 \rangle$ is defined to be

$$\mathbf{u} - \mathbf{v} = \mathbf{u} + (-\mathbf{v}) = \langle u_1 - v_1, u_2 - v_2 \rangle. \qquad (4)$$

If we think of $\langle u_1, u_2 \rangle$ and $\langle v_1, v_2 \rangle$ as position vectors of the points $P$ and $Q$, respectively, then $\mathbf{u} - \mathbf{v}$ may be represented by the arrow $\overrightarrow{QP}$ from $Q$ to $P$. We may therefore write

$$\mathbf{u} - \mathbf{v} = \overrightarrow{OP} - \overrightarrow{OQ} = \overrightarrow{QP},$$

**13.21**  Geometric interpretation of the difference $\mathbf{u} - \mathbf{v}$

as illustrated in Fig. 13.21.

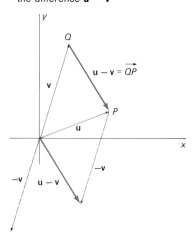

**EXAMPLE 1**  Suppose that $\mathbf{u} = \langle 4, -3 \rangle$ and $\mathbf{v} = \langle -2, 3 \rangle$. Find $|\mathbf{u}|$ and the vectors $\mathbf{u} + \mathbf{v}$, $\mathbf{u} - \mathbf{v}$, $3\mathbf{u}$, $-2\mathbf{v}$, and $2\mathbf{u} + 4\mathbf{v}$.

**Solution**

$$|\mathbf{u}| = \sqrt{4^2 + (-3)^2} = \sqrt{25} = 5.$$

$$\mathbf{u} + \mathbf{v} = \langle 4 + (-2), (-3) + 3 \rangle = \langle 2, 0 \rangle.$$

$$\mathbf{u} - \mathbf{v} = \langle 4 - (-2), (-3) - 3 \rangle = \langle 6, -6 \rangle.$$

$$3\mathbf{u} = \langle 3 \cdot (4), 3 \cdot (-3) \rangle = \langle 12, -9 \rangle.$$

$$-2\mathbf{v} = \langle -2 \cdot (-2), -2 \cdot (3) \rangle = \langle 4, -6 \rangle.$$

$$2\mathbf{u} + 4\mathbf{v} = \langle 2 \cdot (4) + 4 \cdot (-2), 2 \cdot (-3) + 4 \cdot (3) \rangle = \langle 0, 6 \rangle.$$

---

> **Theorem** *Test for Perpendicular Vectors*
>
> The two nonzero vectors $\mathbf{a} = \langle a_1, a_2 \rangle$ and $\mathbf{b} = \langle b_1, b_2 \rangle$ are perpendicular if and only if
>
> $$a_1 b_1 + a_2 b_2 = 0. \qquad (10)$$

Note that it is consistent with (10) to regard the zero vector $\mathbf{0} = \langle 0, 0 \rangle$ as perpendicular to every vector.

The particular combination of the components of $\mathbf{a}$ and $\mathbf{b}$ in (10) has important applications and is called the *dot product* of $\mathbf{a}$ and $\mathbf{b}$.

> **Definition** *The Dot Product of Two Vectors*
>
> The **dot product** $\mathbf{a} \cdot \mathbf{b}$ of the two vectors $\mathbf{a} = \langle a_1, a_2 \rangle$ and $\mathbf{b} = \langle b_1, b_2 \rangle$ is the real number
>
> $$\mathbf{a} \cdot \mathbf{b} = a_1 b_1 + a_2 b_2. \qquad (11)$$

Note that the dot product (sometimes called the *scalar product*) of two vectors is a *scalar*. For example,

$$\langle 2, -3 \rangle \cdot \langle -4, 2 \rangle = (2)(-4) + (-3)(2) = -14.$$

The following algebraic properties of the dot product are easy to verify by componentwise calculations.

$$
\begin{aligned}
&\text{(i)} && \mathbf{a} \cdot \mathbf{a} = |\mathbf{a}|^2; \\
&\text{(ii)} && \mathbf{a} \cdot \mathbf{b} = \mathbf{b} \cdot \mathbf{a}; \\
&\text{(iii)} && \mathbf{a} \cdot (\mathbf{b} + \mathbf{c}) = \mathbf{a} \cdot \mathbf{b} + \mathbf{a} \cdot \mathbf{c}; \\
&\text{(iv)} && (r\mathbf{a}) \cdot \mathbf{b} = r(\mathbf{a} \cdot \mathbf{b}) = \mathbf{a} \cdot (r\mathbf{b}).
\end{aligned}
\qquad (12)
$$

For example,

$$\mathbf{a} \cdot \mathbf{a} = (a_1)^2 + (a_2)^2 = |\mathbf{a}|^2$$

and

$$
\begin{aligned}
\mathbf{a} \cdot (\mathbf{b} + \mathbf{c}) &= \langle a_1, a_2 \rangle \cdot \langle b_1 + c_1, b_2 + c_2 \rangle \\
&= a_1(b_1 + c_1) + a_2(b_2 + c_2) \\
&= (a_1 b_1 + a_2 b_2) + (a_1 c_1 + a_2 c_2) = \mathbf{a} \cdot \mathbf{b} + \mathbf{a} \cdot \mathbf{c}.
\end{aligned}
$$

Since $\mathbf{i} \cdot \mathbf{i} = \mathbf{j} \cdot \mathbf{j} = 1$ and $\mathbf{i} \cdot \mathbf{j} = \mathbf{j} \cdot \mathbf{i} = 0$, application of Properties (12) gives

$$
\begin{aligned}
\mathbf{a} \cdot \mathbf{b} &= (a_1 \mathbf{i} + a_2 \mathbf{j}) \cdot (b_1 \mathbf{i} + b_2 \mathbf{j}) \\
&= (a_1 b_1) \mathbf{i} \cdot \mathbf{i} + (a_1 b_2 + a_2 b_1) \mathbf{i} \cdot \mathbf{j} + (a_2 b_2) \mathbf{j} \cdot \mathbf{j} \\
&= a_1 b_1 + a_2 b_2,
\end{aligned}
$$

in accord with the definition of $\mathbf{a} \cdot \mathbf{b}$. Thus dot products of vectors expressed in terms of $\mathbf{i}$ and $\mathbf{j}$ can be computed by "ordinary" algebra.

The test for perpendicularity is best remembered in terms of the dot product.

> **Corollary** *Test for Perpendicular Vectors*
> The two nonzero vectors **a** and **b** are perpendicular if and only if **a** · **b** = 0.

**EXAMPLE 3** Determine whether or not the following pairs of vectors are perpendicular.

$$\text{(a) } \langle 3, 4 \rangle \text{ and } \langle 8, -6 \rangle \qquad \text{(b) } \langle 2, 3 \rangle \text{ and } \langle 4, -3 \rangle$$

*Solution* (a) $\langle 3, 4 \rangle \cdot \langle 8, -6 \rangle = 24 - 24 = 0$, so these two vectors *are* perpendicular.
(b) $\langle 2, 3 \rangle \cdot \langle 4, -3 \rangle = 8 - 9 = -1 \neq 0$, so $\langle 2, 3 \rangle$ and $\langle 4, -3 \rangle$ are *not* perpendicular.

In Section 14-1, we shall see that $\mathbf{a} \cdot \mathbf{b} = |\mathbf{a}| \, |\mathbf{b}| \cos \theta$, where $\theta$ is the angle between the vectors **a** and **b**. The case $\theta = \pi/2$ gives the corollary above.

## 13-3 PROBLEMS

In each of Problems 1–8 find $|\mathbf{a}|$, $|-2\mathbf{b}|$, $|\mathbf{a} - \mathbf{b}|$, $\mathbf{a} + \mathbf{b}$, and $3\mathbf{a} - 2\mathbf{b}$. Also determine whether or not **a** and **b** are perpendicular.

1 $\mathbf{a} = \langle 1, -2 \rangle$, $\mathbf{b} = \langle -3, 2 \rangle$
2 $\mathbf{a} = \langle 3, 4 \rangle$, $\mathbf{b} = \langle -4, 3 \rangle$
3 $\mathbf{a} = \langle -2, -2 \rangle$, $\mathbf{b} = \langle -3, -4 \rangle$
4 $\mathbf{a} = -2\langle 4, 7 \rangle$, $\mathbf{b} = -3\langle -4, -2 \rangle$
5 $\mathbf{a} = \mathbf{i} + 3\mathbf{j}$, $\mathbf{b} = 2\mathbf{i} - 5\mathbf{j}$
6 $\mathbf{a} = 2\mathbf{i} - 5\mathbf{j}$, $\mathbf{b} = \mathbf{i} - 6\mathbf{j}$
7 $\mathbf{a} = 4\mathbf{i}$, $\mathbf{b} = -7\mathbf{j}$
8 $\mathbf{a} = -\mathbf{i} - \mathbf{j}$, $\mathbf{b} = 2\mathbf{i} + 2\mathbf{j}$

In each of Problems 9–12, find a unit vector **u** with the same direction as the given vector **a**. Express **u** in terms of **i** and **j**. Also find a unit vector **v** with the direction opposite that of **a**.

9 $\mathbf{a} = \langle -3, -4 \rangle$
10 $\mathbf{a} = \langle 5, -12 \rangle$
11 $\mathbf{a} = 8\mathbf{i} + 15\mathbf{j}$
12 $\mathbf{a} = 7\mathbf{i} - 24\mathbf{j}$

In each of Problems 13–16, find the vector **a**, expressed in terms of **i** and **j**, that is represented by the arrow $\overrightarrow{PQ}$ in the plane.

13 $P = (3, 2)$, $Q = (3, -2)$
14 $P = (-3, 5)$, $Q = (-3, 6)$
15 $P = (-4, 7)$, $Q = (4, -7)$
16 $P = (1, -1)$, $Q = (-4, -1)$

In each of Problems 17–20, determine whether or not the given vectors **a** and **b** are perpendicular.

17 $\mathbf{a} = \langle 6, 0 \rangle$, $\mathbf{b} = \langle 0, -7 \rangle$
18 $\mathbf{a} = 3\mathbf{j}$, $\mathbf{b} = 3\mathbf{i} - \mathbf{j}$
19 $\mathbf{a} = 2\mathbf{i} - \mathbf{j}$, $\mathbf{b} = 4\mathbf{j} + 8\mathbf{i}$
20 $\mathbf{a} = 8\mathbf{i} + 10\mathbf{j}$, $\mathbf{b} = 15\mathbf{i} - 12\mathbf{j}$

21 Find a vector that has the same direction as $5\mathbf{i} - 7\mathbf{j}$ and
(a) three times its length;
(b) one-third its length.
22 Find a vector that has the opposite direction from $-3\mathbf{i} + 5\mathbf{j}$ and
(a) four times its length;
(b) one-fourth its length.
23 Find a vector of length 5 with
(a) the same direction as $7\mathbf{i} - 3\mathbf{j}$;
(b) the opposite direction from $8\mathbf{i} + 5\mathbf{j}$.
24 For what numbers $c$ are the vectors $\langle c, 2 \rangle$ and $\langle c, -8 \rangle$ perpendicular?
25 For what numbers $c$ are the vectors $2c\mathbf{i} - 4\mathbf{j}$ and $3\mathbf{i} + c\mathbf{j}$ perpendicular?
26 Given the three points $A(2, 3)$, $B(-5, 7)$, and $C(1, -5)$, verify by direct computation of the vectors and their sum that $\overrightarrow{AB} + \overrightarrow{BC} + \overrightarrow{CA} = \mathbf{0}$.

In each of Problems 27–32, give a componentwise proof of the indicated property of vector algebra. Take $\mathbf{a} = \langle a_1, a_2 \rangle$ and $\mathbf{b} = \langle b_1, b_2 \rangle$ throughout.

27 $\mathbf{a} + (\mathbf{b} + \mathbf{c}) = (\mathbf{a} + \mathbf{b}) + \mathbf{c}$
28 $(r + s)\mathbf{a} = r\mathbf{a} + s\mathbf{a}$
29 $(rs)\mathbf{a} = r(s\mathbf{a})$
30 $\mathbf{a} \cdot \mathbf{b} = \mathbf{b} \cdot \mathbf{a}$
31 $(r\mathbf{a}) \cdot \mathbf{b} = r(\mathbf{a} \cdot \mathbf{b})$
32 If $\mathbf{a} + \mathbf{b} = \mathbf{a}$, then $\mathbf{b} = \mathbf{0}$

In Problems 33–35, assume the following fact: If an airplane flies with velocity vector $\mathbf{v}_a$ relative to the air and the velocity vector of the wind is **w**, then the velocity vector of the plane relative to the ground is

$$\mathbf{v}_g = \mathbf{v}_a + \mathbf{w}.$$

The **apparent velocity** vector is $\mathbf{v}_a$, while $\mathbf{v}_g$ is called the **true velocity** vector.

**33** Suppose that the wind is blowing from the northeast at 50 mi/h, and the pilot wishes to fly due east at 500 mi/h. What should the plane's apparent velocity vector be?

**34** Repeat Problem 33 with the phrase *due east* replaced by *due west*.

**35** Repeat Problem 33 in the case that the pilot wishes to fly northwest at 500 mi/h.

**36** In the triangle $ABC$, let $M_1$ and $M_2$ be the midpoints of $AB$ and $AC$, respectively. Show that $\overrightarrow{M_1 M_2} = \frac{1}{2}\overrightarrow{BC}$. Conclude that the line segment joining the midpoints of two sides of a triangle is parallel to the third side.

**37** Prove that the diagonals of a parallelogram $ABCD$ bisect each other. (*Suggestion:* If $M_1$ and $M_2$ denote the midpoints of $AC$ and $BD$, respectively, show that $\overrightarrow{OM_1} = \overrightarrow{OM_2}$.)

**38** Use vectors to prove that the midpoints of the four sides of an arbitrary quadrilateral are the vertices of a parallelogram.

**39** Show that the vector $\mathbf{n} = a\mathbf{i} + b\mathbf{j}$ is perpendicular to the line with equation $ax + by + c = 0$. (*Suggestion:* If $P_1(x, y)$ and $P_2(\bar{x}, \bar{y})$ are two points on the line, show that $\mathbf{n} \cdot \overrightarrow{P_1 P_2} = 0$.)

**40** Figure 13.24 shows the vector $\mathbf{a}_\perp$ that is obtained by rotating the vector $\mathbf{a} = a_1\mathbf{i} + a_2\mathbf{j}$ through a counterclockwise angle of $90°$. Show that

$$\mathbf{a}_\perp = -a_2\mathbf{i} + a_1\mathbf{j}.$$

(*Suggestion:* Begin by writing

$$\mathbf{a} = (r \cos \theta)\mathbf{i} + (r \sin \theta)\mathbf{j}.)$$

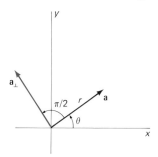

**13.24** Rotate $\mathbf{a}$ $90°$ counterclockwise to obtain $\mathbf{a}_\perp$.

## 13-4

## Motion and Vector-Valued Functions

We now use vectors to study the motion of a point in the plane. If the coordinates of the moving point at time $t$ are given by the parametric equations $x = f(t)$, $y = g(t)$, then the vector

$$\mathbf{r}(t) = f(t)\mathbf{i} + g(t)\mathbf{j}, \quad \text{or} \quad \mathbf{r} = x\mathbf{i} + y\mathbf{j}, \tag{1}$$

is called the **position vector** of the point. Equation (1) determines a **vector-valued function** $\mathbf{r} = \mathbf{r}(t)$ that associates with the number $t$ the vector $\mathbf{r}(t)$. In the bracket notation for vectors, a vector-valued function is an ordered pair of real-valued functions: $\mathbf{r}(t) = \langle f(t), g(t) \rangle$.

Much of the calculus of (ordinary) real-valued functions applies to vector-valued functions. To begin with, the **limit** of a vector-valued function $\mathbf{r}$ is defined as follows:

$$\lim_{t \to a} \mathbf{r}(t) = \left\langle \lim_{t \to a} f(t), \lim_{t \to a} g(t) \right\rangle$$

$$= \mathbf{i} \left( \lim_{t \to a} f(t) \right) + \mathbf{j} \left( \lim_{t \to a} g(t) \right), \tag{2}$$

provided that the limits in the latter two expressions exist. Thus we take limits of vector-valued functions by taking limits of their component functions.

We say that $\mathbf{r} = \mathbf{r}(t)$ is **continuous** at the number $a$ provided that

$$\lim_{t \to a} \mathbf{r}(t) = \mathbf{r}(a).$$

This amounts to saying that $\mathbf{r}$ is continuous at $a$ if and only if its component functions $f$ and $g$ are continuous at $a$.

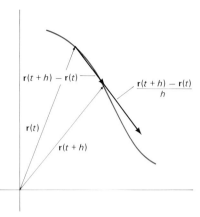

**13.25** Geometry of the derivative of a vector-valued function.

The derivative $\mathbf{r}'(t)$ of the vector-valued function $\mathbf{r}(t)$ is defined in much the same way as the derivative of a real-valued function. Specifically,

$$\mathbf{r}'(t) = \lim_{h \to 0} \frac{\mathbf{r}(t+h) - \mathbf{r}(t)}{h}, \qquad (3)$$

provided that this limit exists. A geometric note: Fig. 13.25 suggests that $\mathbf{r}'(t)$ will be tangent to the curve swept out by $\mathbf{r}$ if $\mathbf{r}'(t)$ is attached at the point of evaluation. Note also that because $|\mathbf{r}(t+h) - \mathbf{r}(t)|$ is the distance from the point with position vector $\mathbf{r}(t)$ to the point with position vector $\mathbf{r}(t+h)$, the quotient

$$\frac{|\mathbf{r}(t+h) - \mathbf{r}(t)|}{h}$$

is equal to the average speed of a particle that travels from $\mathbf{r}(t)$ to $\mathbf{r}(t+h)$ in time $h$. Consequently the limit in (3) yields both the direction of motion and the instantaneous speed of a particle moving with position vector $\mathbf{r}(t)$ along a curve.

Our next theorem tells us the simple *but important* fact that $\mathbf{r}'(t)$ can be calculated by componentwise differentiation. We shall also denote derivatives by

$$\mathbf{r}'(t) = D_t\mathbf{r}(t) = \frac{d\mathbf{r}}{dt}.$$

---

**Theorem 1** *Componentwise Differentiation*

Suppose that

$$\mathbf{r}(t) = \langle f(t), g(t) \rangle = f(t)\mathbf{i} + g(t)\mathbf{j},$$

where both $f$ and $g$ are differentiable functions. Then

$$\mathbf{r}'(t) = \langle f'(t), g'(t) \rangle = f'(t)\mathbf{i} + g'(t)\mathbf{j}. \qquad (4)$$

That is, if $\mathbf{r} = x\mathbf{i} + y\mathbf{j}$, then

$$\frac{d\mathbf{r}}{dt} = \frac{dx}{dt}\mathbf{i} + \frac{dy}{dt}\mathbf{j}.$$

---

**Proof**   We simply take the limit in Equation (3) by taking limits of the components. We find that

$$\mathbf{r}'(t) = \lim_{h \to 0} \frac{1}{h}[\mathbf{r}(t+h) - \mathbf{r}(t)]$$

$$= \lim_{h \to 0} \frac{1}{h}[f(t+h)\mathbf{i} + g(t+h)\mathbf{j} - f(t)\mathbf{i} - g(t)\mathbf{j}]$$

$$= \left(\lim_{h \to 0} \frac{f(t+h) - f(t)}{h}\right)\mathbf{i} + \left(\lim_{h \to 0} \frac{g(t+h) - g(t)}{h}\right)\mathbf{j}$$

$$= f'(t)\mathbf{i} + g'(t)\mathbf{j}. \qquad \blacksquare$$

In Equation (5) of Section 13-1, we saw that the tangent line to the parametric curve $x = f(t)$, $y = g(t)$ has slope $g'(t)/f'(t)$ and therefore is parallel to the line through $(0, 0)$ and $(f'(t), g'(t))$. Hence, by Theorem 1, the derivative vector $\mathbf{r}'(t) = \langle f'(t), g'(t) \rangle$ may be visualized as a tangent vector to the curve at the point $(f(t), g(t))$. (See Fig. 13.26.)

Our next theorem tells us that the formulas for computing derivatives of sums and products of vector-valued functions are formally similar to those for real-valued functions.

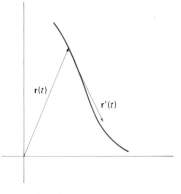

**13.26** The derivative vector is tangent to the curve at the point of evaluation.

---

**Theorem 2    Differentiation Formulas**

Let $\mathbf{u}(t)$ and $\mathbf{v}(t)$ be differentiable vector-valued functions. Let $h(t)$ be a differentiable real-valued function, and let $c$ be a (constant) scalar. Then

(i) $D_t[\mathbf{u}(t) + \mathbf{v}(t)] = \mathbf{u}'(t) + \mathbf{v}'(t)$;

(ii) $D_t[c\mathbf{u}(t)] = c\mathbf{u}'(t)$;

(iii) $D_t[h(t)\mathbf{u}(t)] = h'(t)\mathbf{u}(t) + h(t)\mathbf{u}'(t)$;    and

(iv) $D_t[\mathbf{u}(t) \cdot \mathbf{v}(t)] = \mathbf{u}'(t) \cdot \mathbf{v}(t) + \mathbf{u}(t) \cdot \mathbf{v}'(t)$.

---

**Proof**    We shall prove part (iv) and leave the other parts as exercises. If

$$\mathbf{u}(t) = f_1(t)\mathbf{i} + f_2(t)\mathbf{j} \quad \text{and} \quad \mathbf{v}(t) = g_1(t)\mathbf{i} + g_2(t)\mathbf{j},$$

then

$$\mathbf{u}(t) \cdot \mathbf{v}(t) = f_1(t)g_1(t) + f_2(t)g_2(t).$$

Hence the ordinary product rule gives

$$\begin{aligned}
D_t[\mathbf{u}(t) \cdot \mathbf{v}(t)] &= D_t[f_1(t)g_1(t) + f_2(t)g_2(t)] \\
&= [f_1'(t)g_1(t) + f_2'(t)g_2(t)] + [f_1(t)g_1'(t) + f_2(t)g_2'(t)] \\
&= \mathbf{u}'(t) \cdot \mathbf{v}(t) + \mathbf{u}(t) \cdot \mathbf{v}'(t). \qquad \blacksquare
\end{aligned}$$

Now we can discuss the motion of a point with position vector $\mathbf{r}(t) = f(t)\mathbf{i} + g(t)\mathbf{j}$. Its **velocity vector** $\mathbf{v}(t)$ and **acceleration vector** $\mathbf{a}(t)$ are defined by

$$\left. \begin{aligned}
\mathbf{v}(t) &= \mathbf{r}'(t) = f'(t)\mathbf{i} + g'(t)\mathbf{j}, \\
\mathbf{v} &= \frac{d\mathbf{r}}{dt} = \frac{dx}{dt}\mathbf{i} + \frac{dy}{dt}\mathbf{j};
\end{aligned} \right\} \tag{5}$$

$$\left. \begin{aligned}
\mathbf{a}(t) &= \mathbf{v}'(t) = f''(t)\mathbf{i} + g''(t)\mathbf{j}, \\
\mathbf{a} &= \frac{d^2\mathbf{r}}{dt^2} = \frac{d^2x}{dt^2}\mathbf{i} + \frac{d^2y}{dt^2}\mathbf{j}.
\end{aligned} \right\} \tag{6}$$

The moving point also has a **speed** $v(t)$ and **scalar acceleration** $a(t)$. These are just the lengths of the corresponding velocity and acceleration vectors; that is,

$$v(t) = |\mathbf{v}(t)| = \sqrt{\left(\frac{dx}{dt}\right)^2 + \left(\frac{dy}{dt}\right)^2} \tag{7}$$

and

$$a(t) = |\mathbf{a}(t)| = \sqrt{\left(\frac{d^2x}{dt^2}\right)^2 + \left(\frac{d^2y}{dt^2}\right)^2}. \tag{8}$$

Do *not* assume that the scalar acceleration $a$ and the derivative $dv/dt$ of the speed are equal. Though this holds part of the time for one-dimensional motion, it is almost *never* true for two-dimensional motion.

**EXAMPLE 1**   A moving particle has position vector

$$\mathbf{r}(t) = t\mathbf{i} + t^2\mathbf{j}.$$

Find the velocity and acceleration vectors and the speed and the scalar acceleration when $t = 2$.

*Solution*   Since $\mathbf{r}(2) = 2\mathbf{i} + 4\mathbf{j}$, the particle is at the point $(2, 4)$ on the parabola $y = x^2$ when $t = 2$. Its velocity vector is

$$\mathbf{v}(t) = \mathbf{i} + 2t\mathbf{j}.$$

So $\mathbf{v}(2) = \mathbf{i} + 4\mathbf{j}$ and $v(2) = |\mathbf{v}(2)| = \sqrt{17}$. Note that the *velocity* $\mathbf{v}(2)$ is a *vector*, while the *speed* $v(2)$ is a *scalar*.

The acceleration vector is $\mathbf{a} = 2\mathbf{j}$, so the acceleration of the particle is the (constant) vector $2\mathbf{j}$. In particular, when $t = 2$, its acceleration is $2\mathbf{j}$ and its scalar acceleration is $a = 2$. Figure 13.27 shows the trajectory of the particle with the vectors $\mathbf{v}(2)$ and $\mathbf{a}(2)$ attached at the point corresponding to $t = 2$.

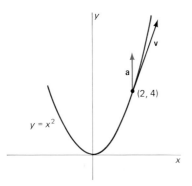

**13.27**   The velocity and acceleration vectors at $t = 2$ (Example 1)

Integrals of vector-valued functions are defined by analogy with the definition of an integral of a real-valued function:

$$\int_a^b \mathbf{r}(t)\,dt = \lim_{\Delta t \to 0} \sum_{i=1}^n \mathbf{r}(t_i^*)\,\Delta t, \tag{9}$$

where $t_i^*$ is a point of the $i$th subinterval of a subdivision of $[a, b]$ into $n$ equal subintervals each of length $\Delta t = (b - a)/n$.

If $\mathbf{r}(t) = f(t)\mathbf{i} + g(t)\mathbf{j}$ is continuous, then—by taking limits componentwise—we get

$$\int_a^b \mathbf{r}(t)\,dt = \lim_{\Delta t \to 0} \sum_{i=1}^n \mathbf{r}(t_i^*)\,\Delta t$$

$$= \mathbf{i}\left(\lim_{\Delta t \to 0} \sum_{i=1}^n f(t_i^*)\,\Delta t\right) + \mathbf{j}\left(\lim_{\Delta t \to 0} \sum_{i=1}^n g(t_i^*)\,\Delta t\right).$$

This gives the result that

$$\int_a^b \mathbf{r}(t)\,dt = \mathbf{i}\left(\int_a^b f(t)\,dt\right) + \mathbf{j}\left(\int_a^b g(t)\,dt\right). \tag{10}$$

Thus *a vector-valued function may be integrated componentwise.*

Now suppose that $\mathbf{R}(t)$ is an *antiderivative* of $\mathbf{r}(t)$, meaning that $\mathbf{R}'(t) = \mathbf{r}(t)$. That is, if $\mathbf{R}(t) = F(t)\mathbf{i} + G(t)\mathbf{j}$, then

$$\mathbf{R}'(t) = F'(t)\mathbf{i} + G'(t)\mathbf{j}$$
$$= f(t)\mathbf{i} + g(t)\mathbf{j} = \mathbf{r}(t).$$

Then componentwise integration yields

$$\int_a^b \mathbf{r}(t)\, dt = \mathbf{i}\left(\int_a^b f(t)\, dt\right) + \mathbf{j}\left(\int_a^b g(t)\, dt\right)$$

$$= \mathbf{i}\left(\left[F(t)\right]_a^b\right) + \mathbf{j}\left(\left[G(t)\right]_a^b\right)$$

$$= (F(b)\mathbf{i} + G(b)\mathbf{j}) - (F(a)\mathbf{i} + G(a)\mathbf{j}).$$

Thus the *fundamental theorem of calculus* for vector-valued functions takes the form

$$\int_a^b \mathbf{r}(t)\, dt = \left[\mathbf{R}(t)\right]_a^b = \mathbf{R}(b) - \mathbf{R}(a) \tag{11}$$

where $\mathbf{R}'(t) = \mathbf{r}(t)$.

Vector integration is the basis for one method of navigation. If a submarine is cruising beneath the icecap at the North Pole, and thus can use neither visual nor radio methods to determine its position, here is an alternative: Build a sensitive gyroscope-accelerometer combination and install it in the submarine. The device continuously measures the sub's acceleration vector, starting at the time $t = 0$ when its position $\mathbf{r}(0)$ and velocity $\mathbf{v}(0)$ are known. Since $\mathbf{v}'(t) = \mathbf{a}(t)$, Equation (10) gives

$$\int_0^t \mathbf{a}(t)\, dt = \left[\mathbf{v}(t)\right]_0^t = \mathbf{v}(t) - \mathbf{v}(0),$$

so that

$$\mathbf{v}(t) = \mathbf{v}(0) + \int_0^t \mathbf{a}(t)\, dt.$$

Thus the velocity at every time $t$ is known. Similarly, because $\mathbf{r}'(t) = \mathbf{v}(t)$, a second integration gives

$$\mathbf{r}(t) = \mathbf{r}(0) + \int_0^t \mathbf{v}(t)\, dt$$

for the position at every time $t$. On-board computers can be programmed to carry out these integrations (perhaps using Simpson's approximation), and thus continuously provide the crew with the submarine's position and velocity.

Indefinite integrals of vector-valued functions may also be computed. If $\mathbf{R}'(t) = \mathbf{r}(t)$, then every antiderivative of $\mathbf{r}(t)$ is of the form $\mathbf{R}(t) + \mathbf{C}$ for some constant vector $\mathbf{C}$. We therefore write

$$\int \mathbf{r}(t)\, dt = \mathbf{R}(t) + \mathbf{C} \quad \text{if} \quad \mathbf{R}'(t) = \mathbf{r}(t), \tag{12}$$

on the basis of a componentwise computation similar to the one leading to Equation (11).

**EXAMPLE 2** Suppose that a moving point has initial position $\mathbf{r}(0) = 2\mathbf{i}$, initial velocity $\mathbf{v}(0) = \mathbf{i} - \mathbf{j}$, and acceleration $\mathbf{a}(t) = 2\mathbf{i} + 6t\mathbf{j}$. Find its position and velocity at time $t$.

*Solution* Because $\mathbf{a}(t) = \mathbf{v}'(t)$, Equation (12) gives

$$\mathbf{v}(t) = \int \mathbf{a}(t)\, dt + \mathbf{C} = \int (2\mathbf{i} + 6t\mathbf{j})\, dt + \mathbf{C},$$

$$= 2t\mathbf{i} + 3t^2\mathbf{j} + \mathbf{C}.$$

To evaluate $\mathbf{C}$, we use the fact that $\mathbf{v}(0)$ is known; we substitute $t = 0$ into both sides of the last equation and find that $\mathbf{C} = \mathbf{v}(0) = \mathbf{i} - \mathbf{j}$. So

$$\mathbf{v}(t) = (2t\mathbf{i} + 3t^2\mathbf{j}) + (\mathbf{i} - \mathbf{j}) = (2t + 1)\mathbf{i} + (3t^2 - 1)\mathbf{j}.$$

Next,

$$\mathbf{r}(t) = \int \mathbf{v}(t)\, dt + \mathbf{C}^* = (t^2 + t)\mathbf{i} + (t^3 - t)\mathbf{j} + \mathbf{C}^*.$$

Again we substitute $t = 0$, and find that $\mathbf{C}^* = \mathbf{r}(0) = 2\mathbf{i}$. Hence

$$\mathbf{r}(t) = (t^2 + t + 2)\mathbf{i} + (t^3 - t)\mathbf{j}$$

is the position vector of the moving point.

## 13-4  PROBLEMS

In Problems 1–8, find the values of $\mathbf{r}'(t)$ and $\mathbf{r}''(t)$ for the indicated value of $t$.

1  $\mathbf{r}(t) = 3\mathbf{i} - 2\mathbf{j}$;  $t = 1$
2  $\mathbf{r}(t) = t^2\mathbf{i} - t^3\mathbf{j}$;  $t = 2$
3  $\mathbf{r}(t) = e^{2t}\mathbf{i} + e^{-t}\mathbf{j}$;  $t = 0$
4  $\mathbf{r}(t) = \mathbf{i} \cos t + \mathbf{j} \sin t$;  $t = \pi/4$
5  $\mathbf{r}(t) = 3\mathbf{i} \cos 2\pi t + 3\mathbf{j} \sin 2\pi t$;  $t = \frac{3}{4}$
6  $\mathbf{r}(t) = 5\mathbf{i} \cos t + 4\mathbf{j} \sin t$;  $t = \pi$
7  $\mathbf{r}(t) = \mathbf{i} \sec t + \mathbf{j} \tan t$;  $t = 0$
8  $\mathbf{r}(t) = (2t + 3)\mathbf{i} + (6t - 5)\mathbf{j}$;  $t = 10$

Calculate the integrals in Problems 9–12.

9  $\displaystyle\int_0^{\pi/4} (\mathbf{i} \sin t + 2\mathbf{j} \cos t)\, dt$

10  $\displaystyle\int_1^e \left(\frac{1}{t}\mathbf{i} - \mathbf{j}\right) dt$

11  $\displaystyle\int_0^2 t^2(1 + t^3)^{3/2}\mathbf{i}\, dt$    12  $\displaystyle\int_0^1 (\mathbf{i}e^t - \mathbf{j}te^{-t^2})\, dt$

In Problems 13–15, apply Theorem 2 to compute the derivative $D_t[\mathbf{u}(t) \cdot \mathbf{v}(t)]$.

13  $\mathbf{u}(t) = 3t\mathbf{i} - \mathbf{j}$, $\mathbf{v}(t) = 2\mathbf{i} - 5t\mathbf{j}$
14  $\mathbf{u}(t) = t\mathbf{i} + t^2\mathbf{j}$, $\mathbf{v}(t) = t^2\mathbf{i} - t\mathbf{j}$
15  $\mathbf{u}(t) = \mathbf{i} \cos t + \mathbf{j} \sin t$, $\mathbf{v}(t) = \mathbf{i} \sin t - \mathbf{j} \cos t$

In Problems 16–20, find $\mathbf{v}(t)$ and $\mathbf{r}(t)$ corresponding to the given values of the acceleration $\mathbf{a}(t)$, initial velocity $\mathbf{v}_0$, and initial position $\mathbf{r}_0$.

16  $\mathbf{a} = \mathbf{0}$, $\mathbf{r}_0 = 2\mathbf{i} + 3\mathbf{j}$, $\mathbf{v}_0 = -2\mathbf{j}$
17  $\mathbf{a} = 2\mathbf{j}$, $\mathbf{r}_0 = \mathbf{0}$, $\mathbf{v}_0 = \mathbf{i}$
18  $\mathbf{a} = \mathbf{i} - \mathbf{j}$, $\mathbf{v}_0 = \mathbf{i} + \mathbf{j}$, $\mathbf{r}_0 = \mathbf{j}$
19  $\mathbf{a} = t\mathbf{i} + t^2\mathbf{j}$, $\mathbf{r}_0 = \mathbf{i}$, $\mathbf{v}_0 = \mathbf{0}$
20  $\mathbf{a} = -\mathbf{i} \sin t + \mathbf{j} \cos t$, $\mathbf{r}_0 = \mathbf{i}$, $\mathbf{v}_0 = \mathbf{j}$
21  Suppose that the vector-valued functions $\mathbf{u}(t)$ and $\mathbf{v}(t)$ both have limits as $t \to a$. Prove that

(a)  $\displaystyle\lim_{t \to a} (\mathbf{u}(t) + \mathbf{v}(t)) = \lim_{t \to a} \mathbf{u}(t) + \lim_{t \to a} \mathbf{v}(t)$;

(b)  $\displaystyle\lim_{t \to a} (\mathbf{u}(t) \cdot \mathbf{v}(t)) = \left(\lim_{t \to a} \mathbf{u}(t)\right) \cdot \left(\lim_{t \to a} \mathbf{v}(t)\right)$.

22  Prove part (i) of Theorem 2.
23  Prove part (ii) of Theorem 2.
24  Suppose that both the vector-valued function $\mathbf{r}(t)$ and the real-valued function $h(t)$ are differentiable. Deduce the chain rule for vector-valued functions,

$$D_t[\mathbf{r}(h(t))] = h'(t)\mathbf{r}'(h(t)),$$

in componentwise fashion from the ordinary chain rule.
25  A point moves with constant speed, so that its velocity vector $\mathbf{v}$ satisfies the condition $|\mathbf{v}|^2 = \mathbf{v} \cdot \mathbf{v} = C$ (a constant). Prove that the velocity and acceleration vectors for the point are always perpendicular to each other.
26  A point moves on a circle with center at the origin. Use the dot product to show that the position and velocity vectors of the moving point are always perpendicular.
27  A point moves on the hyperbola $x^2 - y^2 = 1$, with position vector

$$\mathbf{r}(t) = \mathbf{i} \cosh \omega t + \mathbf{j} \sinh \omega t$$

(the number $\omega$ is a constant). Show that $\mathbf{a}(t) = c\mathbf{r}(t)$ where $c$ is a positive constant. What sort of external force would produce this sort of motion?
28  Suppose that a point moves on the ellipse $x^2/a^2 + y^2/b^2 = 1$ with position vector $\mathbf{r}(t) = \mathbf{i}a \cos \omega t + \mathbf{j}b \sin \omega t$, $\omega$ a constant. Show that $\mathbf{a}(t) = c\mathbf{r}(t)$ where $c$ is a negative constant. To what sort of external force $\mathbf{F}(t)$ does this motion correspond?
29  A point moves in the plane with constant acceleration vector $\mathbf{a} = a\mathbf{j}$. Show that its path is a parabola or a straight line.
30  Suppose that a particle is subject to no force, so its acceleration vector $\mathbf{a}(t)$ is identically zero. Prove that the particle travels with constant speed along a straight line (Newton's first law of motion).

## Projectiles and Uniform Circular Motion

Suppose that a projectile is launched from the point $(x_0, y_0)$, with $y_0$ denoting its initial height above the surface of the earth. Let $\alpha$ be the angle of inclination from the horizontal of its initial velocity vector $\mathbf{v}_0$, as in Fig. 13.28. Then its initial position vector is

$$\mathbf{r}_0 = x_0\mathbf{i} + y_0\mathbf{j}, \tag{1a}$$

and from Fig. 13.28 we see that

$$\mathbf{v}_0 = (v_0 \cos \alpha)\mathbf{i} + (v_0 \sin \alpha)\mathbf{j} \tag{1b}$$

where $v_0 = |\mathbf{v}_0|$ is the initial speed of the projectile.

We suppose that the motion takes place sufficiently close to the surface that we may assume the earth is flat and gravity perfectly uniform. Then, if we also ignore air resistance, the acceleration of the projectile is

$$\mathbf{a} = \frac{d\mathbf{v}}{dt} = -g\mathbf{j}$$

where $g \approx 32$ ft/s$^2$. Antidifferentiation gives

$$\mathbf{v} = -gt\mathbf{j} + \mathbf{C}_1.$$

Put $t = 0$ into both sides of this last equation. This shows that $\mathbf{C}_1 = \mathbf{v}_0$, and thus that

$$\mathbf{v} = \frac{d\mathbf{r}}{dt} = -gt\mathbf{j} + \mathbf{v}_0.$$

Another antidifferentiation gives

$$\mathbf{r} = -\tfrac{1}{2}gt^2\mathbf{j} + \mathbf{v}_0 t + \mathbf{C}_2.$$

Now substitution of $t = 0$ gives $\mathbf{C}_2 = \mathbf{r}_0$, so the position vector of the projectile at time $t$ is

$$\mathbf{r}(t) = -\tfrac{1}{2}gt^2\mathbf{j} + \mathbf{v}_0 t + \mathbf{r}_0. \tag{2}$$

Equations (1) now give

$$\mathbf{r}(t) = [(v_0 \cos \alpha)t + x_0]\mathbf{i} + [-\tfrac{1}{2}gt^2 + (v_0 \sin \alpha)t + y_0]\mathbf{j},$$

so the parametric equations of the trajectory of the projectile are

$$x = (v_0 \cos \alpha)t + x_0, \tag{3}$$

$$y = -\tfrac{1}{2}gt^2 + (v_0 \sin \alpha)t + y_0. \tag{4}$$

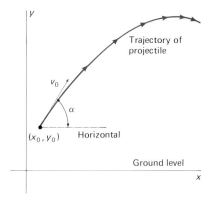

**13.28** Trajectory of a projectile launched at the angle $\alpha$

**EXAMPLE 1** An airplane is flying horizontally at an altitude of 1600 ft, in order to pass directly over snowbound cattle on the ground. Its speed is a constant 150 mi/h (220 ft/s). At what angle of sight $\phi$ (between the horizontal and the direct line to the target) should a bale of hay be released in order to hit the target?

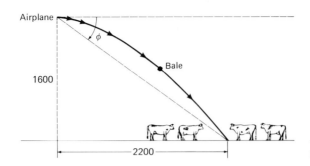

**13.29** Trajectory of the hay bale (Example 1)

*Solution* See Fig. 13.29. We take $x_0 = 0$ where the bale of hay is released at time $t = 0$. Then $y_0 = 1600$ (ft), $v_0 = 220$ (ft/s), and $\alpha = 0$. Then Equations (3) and (4) give

$$x = 220t, \qquad y = -16t^2 + 1600.$$

From the second of these equations we find that $t = 10$ (s) when the bale of hay hits the ground ($y = 0$). It has then traveled a horizontal distance of

$$x = (220)(10) = 2200$$

ft. Hence the required angle of sight is

$$\phi = \tan^{-1}\left(\frac{1600}{2200}\right) \approx 36°.$$

## UNIFORM CIRCULAR MOTION

Consider a point that moves around the circle with center $(0, 0)$ and radius $r$ at a constant angular speed of $\omega$ radians per second. If its initial position is the point $(r, 0)$, then its position vector at time $t$ is

$$\mathbf{r} = \mathbf{i}r \cos \omega t + \mathbf{j}r \sin \omega t. \tag{5}$$

We differentiate to find the velocity vector:

$$\mathbf{v} = -\mathbf{i}r\omega \sin \omega t + \mathbf{j}r\omega \cos \omega t.$$

Note that $\mathbf{v} \cdot \mathbf{r} = 0$, so that $\mathbf{v}$ is tangent to the circle. This is shown in Fig. 13.30. The speed of the moving point is

$$v = |\mathbf{v}| = r\omega. \tag{6}$$

A second differentiation gives the acceleration vector

$$\frac{d\mathbf{v}}{dt} = \mathbf{a} = -\mathbf{i}r\omega^2 \cos \omega t - \mathbf{j}r\omega^2 \sin \omega t,$$

and we note that $\mathbf{a}$ is a scalar multiple of $\mathbf{r}$; indeed,

$$\mathbf{a} = -\omega^2\mathbf{r}.$$

Thus the acceleration vector, when we think of it as located at the endpoint of $\mathbf{r}$, always points back toward the origin. The scalar acceleration is

$$a = |\mathbf{a}| = \omega^2|\mathbf{r}| = r\omega^2, \tag{7}$$

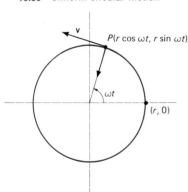

**13.30** Uniform circular motion

CHAP. 13: Parametric Curves and Vectors in the Plane

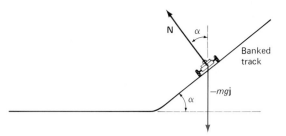

**13.31** Racing on a circular banked track.

proportional to the radius of the circle and to the *square* of the angular velocity.

If the moving point is a particle of mass $m$, then the **central** force (one directed towards the origin) required to produce this motion is

$$F = ma = mr\omega^2 = \frac{mv^2}{r}, \tag{8}$$

since $v = r\omega$. In the examples below and in the problems we give some of the many applications of Formula (8).

**EXAMPLE 2** A circular racetrack has a radius of $\frac{1}{4}$ mi. At what angle $\alpha$ must the track be banked in order that a car can travel around it at 150 mi/h (220 ft/s) with no danger of skidding? The idea is to assume that the track is coated with oil and water, so there's no friction between the tires and the surface of the track.

*Solution* The forces involved are shown in Fig. 13.31. The two forces acting on the car are the gravitational force $mg$ and a normal force $\mathbf{N}$ exerted on the car by the track. The vertical component of $\mathbf{N}$ must balance the gravitational force:

$$|\mathbf{N}| \cos \alpha = mg.$$

The horizontal component of $\mathbf{N}$ supplies the force of Equation (8):

$$|\mathbf{N}| \sin \alpha = \frac{mv^2}{r}.$$

If we divide the second of these two equations by the first, we eliminate $|\mathbf{N}|$. Thus we discover that

$$\tan \alpha = \frac{v^2}{gr}. \tag{9}$$

Now we use the data of the problem—that $v = 220$ (ft/s) and $r = 1320$ (the number of feet in a quarter mile). We take $g = 32$, and Equation (9) tells us that $\alpha$ is about 49°. Indeed, the *rated speed* of a circular curve in a road banked at angle $\alpha$ is the speed $v$ given by Equation (9).

Suppose now that our particle of mass $m$ is a natural or an artificial satellite in a circular orbit around a body of mass $M$. The latter might be either the sun or a planet. According to Newton's law of gravitation, the force that $M$ exerts on $m$ has the scalar value

$$F = G\frac{Mm}{r^2} \qquad (G \text{ is a constant}).$$

We equate this value of $F$ with that in Equation (8), which leads to

$$F = \frac{GMm}{r^2} = \frac{mv^2}{r}. \qquad (10)$$

If the satellite has period of revolution $T$, then its (constant) speed is $v = 2\pi r/T$. A consequence of Equation (10) is

$$T^2 = kr^3 \qquad (11)$$

where $k$ is the constant $4\pi^2/GM$. Thus for satellites in circular orbits about the *same* body, *the square of the period of revolution is proportional to the cube of the radius of the orbit*. This is Kepler's third law for the case of uniform circular motion.

**EXAMPLE 3**  A communications relay satellite is to be placed in a circular orbit about the earth and to have a period of revolution of 24 h. This is a *synchronous* orbit, in which the satellite appears to be stationary in the sky. Assume that the earth's natural moon has a period of 27.32 days in a circular orbit of radius 238,850 mi. What should be the radius of the satellite's orbit?

*Solution*  Equation (11), when applied to the moon, yields $(27.32)^2 = k(238,850)^3$. For the stationary satellite that has period $T = 1$ (day), it yields $(1)^2 = kr^3$, where $r$ is the radius of the synchronous orbit. To eliminate $k$, we divide the second of these equations by the first, and we find that

$$r^3 = \frac{(238,850)^3}{(27.32)^2}.$$

Thus $r$ is approximately 26,330 mi. The radius of the earth is about 3960 mi, so the satellite will be 22,370 mi above the surface.

## 13-5  PROBLEMS

Problems 1–6 deal with a projectile fired from the origin (so that $x_0 = y_0 = 0$) with initial speed $v_0$ and initial angle of inclination $\alpha$. The *range R* of the projectile is the horizontal distance it travels before returning to the ground.

**1**  If $\alpha = 45°$, what value of $v_0$ gives a range of 1 mi?

**2**  If $\alpha = 60°$ and $R = 1$ mi, what is the maximum height attained by the projectile?

**3**  Deduce from Equations (3) and (4) the fact that $R = \frac{1}{16}v_0^2 \sin \alpha \cos \alpha$.

**4**  Given the initial speed $v_0$, find the angle $\alpha$ that maximizes the range $R$. Use the result of Problem 3.

**5**  Suppose that $v_0 = 160$ ft/s. Find the maximum height $y_m$ and the range $R$ of the projectile if
(a) $\alpha = 30°$;

(b) $\alpha = 45°$;
(c) $\alpha = 60°$.

**6**  The projectile of Problem 5 is to be fired at a target 600 ft away, and there is a hill 300 ft high midway between the gun site and this target. At what initial angle of inclination should the projectile be fired?

**7**  A projectile is to be fired horizontally from the top of a 400-ft cliff at a target 1 mi from the base of the cliff. What should be the initial velocity of the projectile?

**8**  A bomb is dropped (initial speed zero) from a helicopter hovering at a height of 800 ft. A projectile is fired from a gun located on the ground 800 ft to the west of the point directly beneath the helicopter. The intent is for the projectile to intercept the bomb at a height of exactly 400 ft.

If the projectile is fired at the same instant that the bomb is dropped, what should be its initial velocity and angle of inclination?

9 Suppose, more realistically, that the projectile of Problem 8 is fired 1 s after the bomb is dropped. What should be its initial velocity and angle of inclination?

In Problems 10–14, use the value $G = 6.673 \times 10^{-11}$ for the gravitational constant. This is its approximate value in mks units.

10 The acceleration of gravity at the surface of the earth is 32.17 ft/s², and the radius of the earth is 3960 mi. Calculate the mass of the earth in kilograms.

11 Assume that the earth's orbit around the sun is circular (this is very nearly true) with a radius of $1.496 \times 10^{11}$ m. Assume also that the period of the earth is 365.26 days. Calculate the mass of the sun and its ratio to the mass of the earth.

12 Given the fact that Jupiter's period of (almost) circular revolution around the sun is 11.86 years, calculate the distance of Jupiter from the sun.

13 Given the period 27.32 days of the moon's almost circular motion around the earth, compute the distance (in miles) to the moon.

14 Jupiter's moon Ganymede has a period of revolution of 7.166 (Earth) days in an orbit with a radius of 1,070,000 km. What is the mass of Jupiter, both in kilograms and as a multiple of the mass of the earth?

15 Suppose that the circular race track of Example 2 has a radius of 1 mi. At what angle should it be banked in order to allow cars to travel at 200 mi/h?

16 A circular racetrack is like the one of Example 2 except that it is *not* banked. Instead, the necessary centripetal force $mr\omega^2$ is supplied by a force of resistance to sliding that is equal to one-fourth of the weight $W = mg$ of the car. How fast can a car travel (without skidding) around this track?

17 At carnivals one sometimes sees a motorcyclist riding around the interior of a vertical cylinder. The motorcycle is acted upon by three forces, as shown in Fig. 13.32:

- A normal force $N = mr\omega^2$ that supplies the necessary centripetal acceleration,
- The weight $W = mg$ of the motorcycle, and
- A force $R = kN = W$ of sliding resistance.

If the radius of the cylinder is $r = 25$ ft and the proportionality constant $k$ is 0.4, what should the speed $v$ of the motorcycle be?

**13.32** A daring motorcyclist

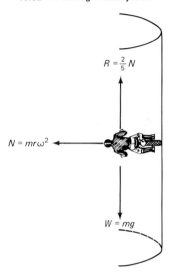

$N = mr\omega^2$

$R = \frac{2}{5}N$

$W = mg$

---

# Curvature and Acceleration

The word *curvature* has an intuitive meaning that we need to make precise. Most people would agree that a straight line does not curve at all, while a circle of small radius is more curved than a circle of large radius (see Fig. 13.33). This judgment may be based on a feeling that curvature is "rate of change of direction." The direction of a curve is determined by its velocity vector, so you would expect the idea of curvature to have something to do with the rate at which the velocity vector is turning.

Let $\mathbf{r}(t) = x(t)\mathbf{i} + y(t)\mathbf{j}$, $a \leq t \leq b$, be the position vector of a smooth curve with nonzero velocity vector $\mathbf{v}(t) = \mathbf{r}'(t)$. The curve's **unit tangent vector** at the point $\mathbf{r}(t)$ is the unit vector

$$\mathbf{T}(t) = \frac{\mathbf{v}(t)}{|\mathbf{v}(t)|} = \frac{\mathbf{v}(t)}{v(t)} \tag{1}$$

where $v(t) = |\mathbf{v}(t)|$ is the speed. Now denote by $\phi$ the angle of inclination of

**13.33** The intuitive idea of curvature

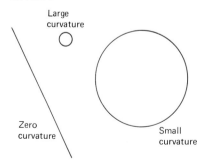

Large curvature

Zero curvature

Small curvature

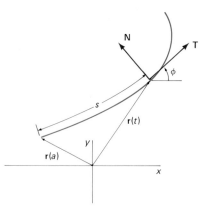

13.34 The unit tangent vector **T**

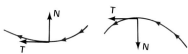

13.35 Direction of the principal unit normal vector **N**

**T**, measured counterclockwise from the positive $x$-axis, as in Fig. 13.34. Then

$$\mathbf{T} = \mathbf{i} \cos \phi + \mathbf{j} \sin \phi. \qquad (2)$$

The **principal unit normal vector** to the curve is the unit vector

$$\mathbf{N} = \pm \mathbf{T}_\perp$$

$$= \pm \left[ \mathbf{i} \cos\left( \phi + \frac{\pi}{2} \right) + \mathbf{j} \sin\left( \phi + \frac{\pi}{2} \right) \right],$$

so that

$$\mathbf{N} = \pm(-\mathbf{i} \sin \phi + \mathbf{j} \cos \phi). \qquad (3)$$

Here, $\mathbf{T}_\perp$ is the vector obtained by rotating **T** through a counterclockwise angle of $90°$. We choose the sign in (3) so that **N** points in the direction in which the curve is bending. Specifically, if **T** points to the right (its **i**-component is positive), then we take the plus sign in (3) if the curve is concave upward and the minus sign if it is concave downward. If **T** points to the left, the choice of signs is reversed. The relationship between **T**, **N**, and the direction of curvature is shown in Fig. 13.35.

**EXAMPLE 1**  The parabola $y = x^2$ can be parametrized by $x = t$, $y = t^2$. Find **T** and **N** at the point $(1, 1)$.

**Solution**  The position vector is $\mathbf{r}(t) = t\mathbf{i} + t^2\mathbf{j}$, so $\mathbf{v}(t) = \mathbf{i} + 2t\mathbf{j}$. The speed is $v(t) = (1 + 4t^2)^{1/2}$, so Equation (1) yields

$$\mathbf{T}(t) = \frac{\mathbf{i} + 2t\mathbf{j}}{\sqrt{1 + 4t^2}}.$$

By substituting $t = 1$, we find that the unit tangent vector at $(1, 1)$ is

$$\mathbf{T} = \frac{1}{\sqrt{5}}\mathbf{i} + \frac{2}{\sqrt{5}}\mathbf{j}.$$

Since the parabola is concave upward at $(1, 1)$, the plus sign is the correct one to choose in Equation (3), and the principal unit normal vector is thus

$$\mathbf{N} = -\frac{2}{\sqrt{5}}\mathbf{i} + \frac{1}{\sqrt{5}}\mathbf{j}.$$

The **curvature** at a point of a curve, denoted by the lowercase Greek letter kappa, is defined to be

$$\kappa = \left| \frac{d\phi}{ds} \right|, \qquad (4)$$

the absolute value of the rate of change of the angle $\phi$ with respect to arc length $s$. We measure $s$ along the curve from its initial point $\mathbf{r}(a)$ to the point $\mathbf{r}(t)$, so that

$$s = \int_a^t \sqrt{[x'(u)]^2 + [y'(u)]^2} \, du = \int_a^t v(u) \, du \qquad (5)$$

by the formula in Equation (6) in Section 13-2. The fundamental theorem of calculus then implies that

$$v = \frac{ds}{dt}.$$ (6)

We define the curvature $\kappa$ in terms of $d\phi/ds$ rather than $d\phi/dt$ because the latter depends not only upon the shape of the curve but also upon the speed of the moving point $\mathbf{r}(t)$. For a straight line the angle $\phi$ is a constant, so the curvature given by Equation (4) is zero. If you imagine a point moving with constant speed along a curve, the curvature is greatest at points where $\phi$ changes most rapidly, such as the points $P$ and $R$ on the curve of Fig. 13.36. The curvature is least at points such as $Q$ and $S$, where $\phi$ is changing the least rapidly.

We need to derive a formula that is effective in computing the curvature of a *smooth* parametric curve $x = x(t)$, $y = y(t)$. First we note that

$$\phi = \tan^{-1}\left(\frac{dy}{dx}\right) = \tan^{-1}\left(\frac{y'(t)}{x'(t)}\right)$$

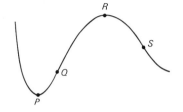

13.36 The curvature is large at $P$ and $R$, small at $Q$ and $S$.

provided $x'(t) \neq 0$. Hence

$$\frac{d\phi}{dt} = \frac{y''x' - y'x''}{(x')^2} \div \left(1 + \left(\frac{y'}{x'}\right)^2\right) = \frac{x'y'' - x''y'}{(x')^2 + (y')^2}.$$

Since $v = ds/dt > 0$, Equation (4) gives

$$\kappa = \left|\frac{d\phi}{ds}\right| = \left|\frac{d\phi}{dt}\frac{dt}{ds}\right| = \frac{1}{v}\left|\frac{d\phi}{dt}\right|;$$

thus

$$\kappa = \frac{|x'y'' - x''y'|}{[(x')^2 + (y')^2]^{3/2}} = \frac{|x'y'' - x''y'|}{v^3}.$$ (7)

At a point where $x'(t) = 0$, we know that $y'(t) \neq 0$ because the curve is smooth. Thus we will obtain the same result if we begin with the equation $\phi = \cot^{-1}(x'/y')$.

An explicitly described curve $y = f(x)$ may be regarded as a parametric curve $x = x$, $y = f(x)$. Then $x' = 1$ and $x'' = 0$, so Equation (7)—with $x$ in place of $t$ as the parameter—becomes

$$\kappa = \frac{|y''|}{[1 + (y')^2]^{3/2}} = \frac{|d^2y/dx^2|}{[1 + (dy/dx)^2]^{3/2}}.$$ (8)

**EXAMPLE 2** Show that the curvature at each point of a circle of radius $a$ is $\kappa = 1/a$.

**Solution** With the familiar parametrization $x = a\cos t$, $y = a\sin t$ of such a circle centered at the origin, we have

$$x' = -a\sin t, \qquad y' = a\cos t,$$

and

$$x'' = -a\cos t, \qquad y'' = -a\sin t.$$

Hence (7) gives

$$\kappa = \frac{|(-a\sin t)(-a\sin t) - (-a\cos t)(a\cos t)|}{[(-a\sin t)^2 + (a\cos t)^2]^{3/2}} = \frac{a^2}{a^3} = \frac{1}{a}.$$

Alternatively, we could have used Formula (8); our point of departure would be the equation $x^2 + y^2 = a^2$ of the same circle, and we would compute $y'$ and $y''$ by implicit differentiation (see Problem 21).

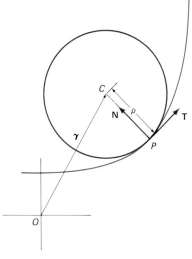

**13.37**  Osculating circle, radius of curvature, center of curvature

Suppose that $P$ is a point on a parametrized curve where $\kappa \neq 0$. Consider the circle that is tangent to the curve at the point $P$ and which has the same curvature there. The center of the circle is to lie on the concave side of the curve; that is, on the side toward which the normal vector $\mathbf{N}$ points. This circle is called the **osculating circle** (or **circle of curvature**) of the curve at the given point because it touches the curve so closely there (*osculum* is the Latin word for *kiss*). Let $\rho$ be the radius of the osculating circle and let $\gamma$ be the position vector of its center; $\gamma = \overrightarrow{OC}$ where $C$ is the center of the osculating circle. Then $\rho$ is called the **radius of curvature** of the curve at the point $P$, and $\gamma$ is called the (vector) **center of curvature** of the curve at $P$. (See Fig. 13.37.)

Example 2 implies that the radius of curvature is

$$\rho = \frac{1}{\kappa}, \tag{9}$$

and the fact that $|\mathbf{N}| = 1$ implies that the position vector of the center of curvature is

$$\gamma = \mathbf{r} + \rho\mathbf{N} \qquad (\mathbf{r} = \overrightarrow{OP}). \tag{10}$$

**EXAMPLE 3**  Determine the radius of curvature and the center of curvature of the parabola $y = x^2$ at the point $(1, 1)$.

*Solution*  By Equation (8),

$$\kappa = \frac{|2|}{(1 + 4x^2)^{3/2}},$$

so $\kappa = 2/(5\sqrt{5})$ and $\rho = (5\sqrt{5})/2$ at the point $(1, 1)$.

Now for the vector center of curvature: In Example 1 we found the unit tangent and normal vectors to the same curve at the same point:

$$\mathbf{T} = \frac{1}{\sqrt{5}}\mathbf{i} + \frac{2}{\sqrt{5}}\mathbf{j} \quad \text{and} \quad \mathbf{N} = -\frac{2}{\sqrt{5}}\mathbf{i} + \frac{1}{\sqrt{5}}\mathbf{j}.$$

Hence Equation (10) gives the center of curvature as

$$\gamma = \langle 1, 1 \rangle + \frac{5\sqrt{5}}{2}\left\langle -\frac{2}{\sqrt{5}}, \frac{1}{\sqrt{5}} \right\rangle = \left\langle -4, \frac{7}{2} \right\rangle.$$

The equation of the osculating circle to the parabola at $(1, 1)$ is, therefore,

$$(x + 4)^2 + (y - \tfrac{7}{2})^2 = \rho^2 = \tfrac{125}{4}.$$

The parabola and this osculating circle are shown in Fig. 13.38.

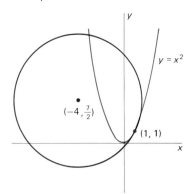

**13.38**  The osculating circle of Example 3

CHAP. 13:  Parametric Curves and Vectors in the Plane

There is an important relationship between the curvature $\kappa$, the unit normal vector $\mathbf{N}$, and the rate of change of the unit tangent vector $\mathbf{T}$ with respect to arc length. We begin with Equation (2), and differentiate each side with respect to arc length $s$. This gives

$$\frac{d\mathbf{T}}{ds} = (-\mathbf{i}\sin\phi + \mathbf{j}\cos\phi)\frac{d\phi}{ds} = (\pm\mathbf{N})(\pm\kappa).$$

To determine the correct sign, we first consider the case in which $\mathbf{T}$ points to the right. From our analysis of the signs in Equation (3) it follows that

$$-\mathbf{i}\sin\phi + \mathbf{j}\cos\phi = +\mathbf{N}$$

if the curve is concave upward, while

$$-\mathbf{i}\sin\phi + \mathbf{j}\cos\phi = -\mathbf{N}$$

if it is concave downward. But the definition of $\kappa$ shows that $d\phi/ds = +\kappa > 0$ if the curve is concave upward, while $d\phi/ds = -\kappa < 0$ if it is concave downward. Thus we take both plus signs in the one case and both minus signs in the other. In either case we obtain

$$\frac{d\mathbf{T}}{ds} = \kappa\mathbf{N}. \tag{11}$$

The result of the analysis is the same if $\mathbf{T}$ points to the left.

## NORMAL AND TANGENTIAL COMPONENTS OF ACCELERATION

We may apply Equation (11) to analyze the meaning of the acceleration vector of a moving particle with velocity vector $\mathbf{v}$ and speed $v$. Then Equation (1) gives $\mathbf{v} = v\mathbf{T}$, so the acceleration vector of the particle is

$$\mathbf{a} = \frac{d\mathbf{v}}{dt} = \frac{dv}{dt}\mathbf{T} + v\frac{d\mathbf{T}}{dt}$$

$$= \frac{dv}{dt}\mathbf{T} + v\frac{d\mathbf{T}}{ds}\frac{ds}{dt}.$$

Since $ds/dt = v$, Equation (11) gives us the formula

$$\mathbf{a} = \frac{dv}{dt}\mathbf{T} + \kappa v^2\mathbf{N}. \tag{12}$$

Since $\mathbf{T}$ and $\mathbf{N}$ are unit vectors tangent and normal to the curve, respectively, Equation (12) provides a *decomposition of the acceleration vector* into its components tangent and normal to the trajectory. Such a decomposition is illustrated in Fig. 13.39. The **tangential component**

$$a_T = \frac{dv}{dt} \tag{13}$$

is the particle's rate of change of speed, while the **normal component**

$$a_N = \kappa v^2 = \frac{v^2}{\rho} \tag{14}$$

measures the rate of change of its direction of motion.

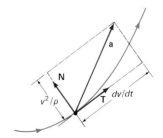

**13.39** Resolution of acceleration **a** into its tangential and normal components

The results of Section 13-5 on uniform circular motion (motion with constant speed $v$ around a circle of radius $r$) follow directly from Eq. (12). Because $v$ is constant, $dv/dt = 0$, and so $\mathbf{a} = \kappa v^2 \mathbf{N}$. But $\mathbf{N}$ is a unit vector directed toward the center of the circle, and $\kappa = 1/r$ by Example 2 of this section. Hence the centripetal acceleration is

$$|\mathbf{a}| = \frac{v^2}{r} = r\omega^2,$$

where $\omega = v/r$ is the angular speed.

As an application of Equation (12), think of a train moving along a straight track with constant speed $v$, so that $a_T = 0 = a_N$ (the latter because $\kappa = 0$ for a straight line). Suppose that at time $t = 0$, the train enters a circular curve of radius $\rho$. At that instant, it will be *suddenly* subjected to a normal acceleration of magnitude $v^2/\rho$. A passenger in the train will experience a sudden jerk. If $v$ is large, the stresses may be so great as to damage the track or to derail the train. It is for exactly this reason that railroads are not built with curves shaped like arcs of circles, but with *approach curves* in which the curvature, and hence the normal acceleration, builds up smoothly.

**EXAMPLE 4** A particle moves with parametric equations

$$x = \tfrac{3}{2}t^2, \qquad y = \tfrac{4}{3}t^3.$$

Find the tangential and normal components of its acceleration vector when $t = 1$.

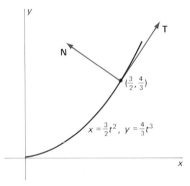

**13.40** The moving particle of Example 4

*Solution* The trajectory, $\mathbf{N}$, and $\mathbf{T}$ appear in Fig. 13.40; $\mathbf{N}$ and $\mathbf{T}$ are shown attached at the point of evaluation, at which $t = 1$. The particle has position vector $\mathbf{r} = \tfrac{3}{2}t^2\mathbf{i} + \tfrac{4}{3}t^3\mathbf{j}$, and velocity vector $\mathbf{v} = 3t\mathbf{i} + 4t^2\mathbf{j}$. Hence its speed is

$$v = \sqrt{9t^2 + 16t^4},$$

from which we calculate

$$a_T = \frac{dv}{dt} = \frac{9t + 32t^3}{\sqrt{9t^2 + 16t^4}}.$$

Thus $v = 5$ and $a_T = \tfrac{41}{5}$ when $t = 1$.

Since $x' = 3t$, $y' = 4t^2$, $x'' = 3$, and $y'' = 8t$, we can use Equation (7) to find the curvature at $t = 1$:

$$\kappa = \frac{|x'y'' - x''y'|}{v^3} = \frac{|(3)(8) - (3)(4)|}{(5)^3} = \frac{12}{125}.$$

Hence

$$a_N = \kappa v^2 = \left(\tfrac{12}{125}\right)(5)^2 = \tfrac{12}{5}$$

when $t = 1$. As a check (Problem 22), you might compute $\mathbf{T}$ and $\mathbf{N}$ when $t = 1$ and verify that

$$\tfrac{41}{5}\mathbf{T} + \tfrac{12}{5}\mathbf{N} = \mathbf{a} = 3\mathbf{i} + 8\mathbf{j}.$$

In Problems 1–6, find the curvature of the given curve at the indicated point.

1  $y = x^3$;  at $(0, 0)$
2  $y = x^3$;  at $(-1, -1)$
3  $y = \cos x$;  at $(0, 1)$
4  $x = t - 1, y = t^2 + 3t + 2$;  where $t = 2$
5  $x = 5 \cos t, y = 4 \sin t$;  where $t = \pi/4$
6  $x = 5 \cosh t, y = 3 \sinh t$;  where $t = 0$

In Problems 7–10, find the point or points of the given curve at which the curvature is a maximum.

7  $y = e^x$
8  $y = \ln x$
9  $x = 5 \cos t, y = 3 \sin t$
10  $xy = 1$

For each of the curves in Problems 11–15, find the unit tangent and normal vectors at the indicated point.

11  $y = x^3$;  at $(-1, -1)$
12  $x = t^3, y = t^2$;  at $(-1, 1)$
13  $x = 3 \sin 2t, y = 4 \cos 2t$;  where $t = \pi/6$
14  $x = t - \sin t, y = 1 - \cos t$;  where $t = \pi/2$
15  $x = \cos^3 t, y = \sin^3 t$;  where $t = 3\pi/4$

The position vector of a moving particle is given in each of Problems 16–20. Find the tangential and normal components of the acceleration vector.

16  $\mathbf{r}(t) = 3\mathbf{i} \sin \pi t + 3\mathbf{j} \cos \pi t$
17  $\mathbf{r}(t) = (2t + 1)\mathbf{i} + (3t^2 - 1)\mathbf{j}$
18  $\mathbf{r}(t) = \mathbf{i} \cosh 3t + \mathbf{j} \sinh 3t$
19  $\mathbf{r}(t) = \mathbf{i}t \cos t + \mathbf{j}t \sin t$
20  $\mathbf{r}(t) = \mathbf{i}e^t \sin t + \mathbf{j}e^t \cos t$
21  Use Formula (8) to compute the curvature of the circle with equation $x^2 + y^2 = a^2$.
22  Verify the equation $\frac{41}{5}\mathbf{T} + \frac{12}{5}\mathbf{N} = 3\mathbf{i} + 8\mathbf{j}$ mentioned at the end of Example 4.

In each of Problems 23–25, find the equation of the osculating circle for the given curve at the indicated point.

23  $y = 1 - x^2$;  at $(0, 1)$
24  $y = e^x$;  at $(0, 1)$
25  $xy = 1$;  at $(1, 1)$
26  (a) Deduce from Equation (11) that

$$\mathbf{N} = \frac{d\mathbf{T}/dt}{|d\mathbf{T}/dt|}.$$

(b) Use the above formula to compute $\mathbf{N}$ for the circle $x = a \sin \omega t, y = a \cos \omega t$.

27  A particle moves under the influence of a force that is always perpendicular to its direction of motion. Show that the speed of the particle must be constant.
28  Deduce from Equation (12) that

$$\kappa = \frac{\sqrt{a^2 - a_T^2}}{v^2} = \frac{[(x'')^2 + (y'')^2 - (v')^2]^{1/2}}{(x')^2 + (y')^2}$$

where each prime denotes differentiation with respect to $t$.
29  Apply the formula of Problem 28 to calculate the curvature of the curve

$$x = \cos t + t \sin t, \qquad y = \sin t - t \cos t.$$

30  Find the curvature and center of curvature of the folium of Descartes $x^3 + y^3 = 3xy$ at the point $(\frac{3}{2}, \frac{3}{2})$. Begin by calculating $dy/dx$ and $d^2y/dx^2$ by implicit differentiation.
31  Determine the constants $A, B, C, D, E,$ and $F$ so that the curve

$$y = Ax^5 + Bx^4 + Cx^3 + Dx^2 + Ex + F$$

does, simultaneously, *all* the following:

- Joins the two points $(0, 0)$ and $(1, 1)$;
- Has slope 0 at $(0, 0)$ and slope 1 at $(1, 1)$;
- Has curvature 0 at both $(0, 0)$ and $(1, 1)$.

The curve in question is shown in color in Fig. 13.41. Why would this be a good curve to join the railway tracks, which are shown in black in the figure?

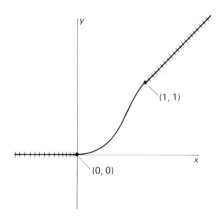

**13.41**  Connecting two railroad tracks (Problem 31)

32  Show that the point on the curve $y = e^x$ where the curvature $\kappa$ is maximal is the point where $y = 1/\sqrt{5}$.

## Orbits of Planets and Satellites

Ancient Greek astronomers and mathematicians developed an elaborate mathematical model to account for the complicated motions of the sun, moon, and six planets (then known) as viewed from Earth. A combination of uniform circular motions was used to describe the motion of each body about the Earth—if the Earth is placed at the origin, then each body *does* orbit the Earth.

In this system, it was typical for a planet *P* to travel uniformly around a small circle (the *epicycle*) with center *C*, which in turn traveled uniformly around a circle centered at the Earth, labeled *E* in Fig. 13.42. The radii of the circles and the angular speeds of *P* and *C* around them were chosen to match the observed motion of the planet as closely as possible. For greater accuracy, one could use secondary circles. In fact, several circles were required for each body in the solar system. The theory of epicycles reached its definitive form in Ptolemy's *Almagest* of the second century A.D.

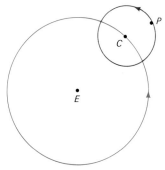

**13.42** The small circle is the epicycle.

In 1543 Copernicus altered Ptolemy's approach by placing the center of each primary circle at the sun rather than at the Earth. This change was of much greater philosophical than mathematical importance. For, contrary to popular belief, this *heliocentric system* was *not* simpler than Ptolemy's geocentric system; indeed, Copernicus's system actually required more circles.

It was Johann Kepler (1571–1630) who finally got rid of all these circles. On the basis of a detailed analysis of a lifetime of planetary observations by the Danish astronomer Tycho Brahe, Kepler stated the following three propositions, now known as **Kepler's laws of planetary motion:**

1 The orbit of each planet is an ellipse with the sun at one focus.
2 The radius vector from the sun to a planet sweeps out area at a constant rate.
3 The *square* of the period of revolution of a planet is proportional to the *cube* of the major semiaxis of its elliptical orbit.

In his *Principia Mathematica* (1687) Newton showed that Kepler's laws follow from the basic principles of mechanics ($F = ma$, and so on) and the inverse-square law of gravitational attraction. His success in using mathematics to explain natural phenomena ("I now demonstrate the frame of the System of the World") inspired confidence that the universe could be understood, and perhaps even mastered. This new confidence permanently altered humanity's perception of itself and of its place in the scheme of things.

In this section we show how Kepler's laws can be derived as indicated above. To begin with, set up a coordinate system in which the sun is located at the origin in the plane of the planet's motion. Let $r = r(t)$ and $\theta = \theta(t)$ be the polar coordinates at time $t$ of the planet as it moves in its orbit about the sun. We want first to split the planet's position, velocity, and acceleration vectors **r**, **v**, and **a** into *radial* and *transverse* components. To do so, we introduce at each point $(r, \theta)$ of the plane (the origin excepted) the *unit* vectors

$$\mathbf{u}_r = \mathbf{i} \cos \theta + \mathbf{j} \sin \theta, \qquad \mathbf{u}_\theta = -\mathbf{i} \sin \theta + \mathbf{j} \cos \theta. \qquad (1)$$

If we substitute $\theta = \theta(t)$, then $\mathbf{u}_r$ and $\mathbf{u}_\theta$ become functions of $t$. The **radial** unit vector $\mathbf{u}_r$ always points directly away from the origin; the **transverse** unit vector $\mathbf{u}_\theta$ is obtained from $\mathbf{u}_r$ by a $90°$ counterclockwise rotation, as shown in Fig. 13.43.

In Problem 6 we ask you to verify, by componentwise differentiation of Equations (1), that

$$\frac{d\mathbf{u}_r}{dt} = \mathbf{u}_\theta \frac{d\theta}{dt} \qquad \text{and} \qquad \frac{d\mathbf{u}_\theta}{dt} = -\mathbf{u}_r \frac{d\theta}{dt}. \qquad (2)$$

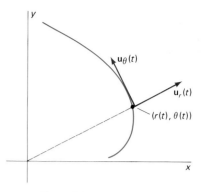

**13.43** The radial and transverse unit vectors $\mathbf{u}_r$ and $\mathbf{u}_\theta$

The position vector $\mathbf{r}$ points directly away from the origin and has length $|\mathbf{r}| = r$, so

$$\mathbf{r} = r\mathbf{u}_r. \qquad (3)$$

Differentiation of both sides of Equation (3) with respect to $t$ gives the result

$$\mathbf{v} = \frac{d\mathbf{r}}{dt} = \mathbf{u}_r \frac{dr}{dt} + r \frac{d\mathbf{u}_r}{dt}.$$

We use the first equation in (2) and find that the planet's velocity vector is

$$\mathbf{v} = \mathbf{u}_r \frac{dr}{dt} + r \frac{d\theta}{dt} \mathbf{u}_\theta. \qquad (4)$$

Thus we have expressed the velocity $\mathbf{v}$ in terms of the radial vector $\mathbf{u}_r$ and the transverse vector $\mathbf{u}_\theta$.

We differentiate this last equation, and find that

$$\mathbf{a} = \frac{d\mathbf{v}}{dt} = \left( \mathbf{u}_r \frac{d^2r}{dt^2} + \frac{dr}{dt} \frac{d\mathbf{u}_r}{dt} \right) + \left( \frac{dr}{dt} \frac{d\theta}{dt} \mathbf{u}_\theta + r \frac{d^2\theta}{dt^2} \mathbf{u}_\theta + r \frac{d\theta}{dt} \frac{d\mathbf{u}_\theta}{dt} \right).$$

Then, by using Equations (2) and collecting the coefficients of $\mathbf{u}_r$ and $\mathbf{u}_\theta$ (Problem 7), we obtain the decomposition

$$\mathbf{a} = \left[ \frac{d^2r}{dt^2} - r \left( \frac{d\theta}{dt} \right)^2 \right] \mathbf{u}_r + \left[ \frac{1}{r} \frac{d}{dt} \left( r^2 \frac{d\theta}{dt} \right) \right] \mathbf{u}_\theta \qquad (5)$$

of the acceleration vector into its radial and transverse components.

Let $M$ denote the mass of the sun and $m$ the mass of the orbiting planet. The inverse-square law of gravitation in vector form is

$$\mathbf{F} = m\mathbf{a} = -\frac{GMm}{r^2} \mathbf{u}_r,$$

so the acceleration of the planet *also* is given by

$$\mathbf{a} = -\frac{\mu}{r^2} \mathbf{u}_r \qquad (6)$$

where $\mu = GM$. We equate the transverse components of (5) and (6), and thus obtain

$$\frac{1}{r}\frac{d}{dt}\left(r^2\frac{d\theta}{dt}\right) = 0.$$

We antidifferentiate both sides and find that

$$r^2\frac{d\theta}{dt} = h \qquad (h \text{ is a constant}). \tag{7}$$

We know from Section 10-3 that if $A(t)$ denotes the area swept out by the planet's radius vector from time 0 to time $t$ (see Fig. 13.44), then

$$A(t) = \int \frac{1}{2}r^2\,d\theta = \int_0^t \frac{1}{2}r^2\frac{d\theta}{dt}\,dt.$$

Now we apply the fundamental theorem of calculus, which yields

$$\frac{dA}{dt} = \frac{1}{2}r^2\frac{d\theta}{dt}. \tag{8}$$

When we compare Equations (7) and (8), we see that

$$\frac{dA}{dt} = \frac{h}{2}. \tag{9}$$

Since $h/2$ is a constant, we have derived Kepler's second law: The radius vector from sun to planet sweeps out area at a constant rate.

Now we derive Kepler's first law. Choose coordinate axes so that at time $t = 0$ the planet is on the polar axis and is at its closest point of approach to the sun, with initial position vector $\mathbf{r}_0$ and initial velocity vector $\mathbf{v}_0$, as in Fig. 13.45. By Equation (4),

$$\mathbf{v}_0 = r_0\,\theta'(0)\mathbf{u}_\theta = v_0\mathbf{j} \tag{10}$$

because, when $t = 0$, $\mathbf{u}_\theta = \mathbf{j}$ and $dr/dt = 0$ (because $r$ is minimal then).

From Equation (6) we have

$$\frac{d\mathbf{v}}{dt} = -\frac{\mu}{r^2}\mathbf{u}_r = -\frac{\mu}{r^2}\left(-\frac{1}{d\theta/dt}\frac{d\mathbf{u}_\theta}{dt}\right)$$

$$= \frac{\mu}{r^2\,d\theta/dt}\frac{d\mathbf{u}_\theta}{dt} = \frac{\mu}{h}\frac{d\mathbf{u}_\theta}{dt},$$

by using the second equation in (2) in the first step and then Equation (7). Next we antidifferentiate and find that

$$\mathbf{v} = \frac{\mu}{h}\mathbf{u}_\theta + \mathbf{C}.$$

To find $\mathbf{C}$, we substitute $t = 0$; this yields $\mathbf{u}_\theta = \mathbf{j}$ and $\mathbf{v} = v_0\mathbf{j}$. We get

$$\mathbf{C} = v_0\mathbf{j} - \frac{\mu}{h}\mathbf{j} = \left(v_0 - \frac{\mu}{h}\right)\mathbf{j}.$$

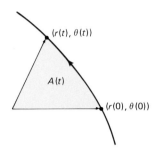

**13.44** Area swept out by the radius vector.

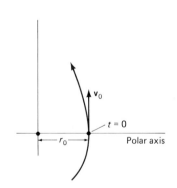

**13.45** Setup of the coordinate system for the derivation of Kepler's first law

Consequently

$$\mathbf{v} = \frac{\mu}{h}\mathbf{u}_\theta + \left(v_0 - \frac{\mu}{h}\right)\mathbf{j}. \qquad (11)$$

Now we take the dot product of each side of Equation (11) with the unit vector $\mathbf{u}_\theta$. Remember that

$$\mathbf{u}_\theta \cdot \mathbf{v} = r\frac{d\theta}{dt} \quad \text{and} \quad \mathbf{u}_\theta \cdot \mathbf{j} = \cos\theta,$$

consequences of Equations (4) and (1), respectively. The result is

$$r\frac{d\theta}{dt} = \frac{\mu}{h} + \left(v_0 - \frac{\mu}{h}\right)\cos\theta.$$

Since $r(d\theta/dt) = h/r$ by Equation (7), we have

$$\frac{h}{r} = \frac{\mu}{h} + \left(v_0 - \frac{\mu}{h}\right)\cos\theta.$$

We solve for $r$, and find that

$$r = \frac{h^2/\mu}{1 + [(v_0 h/\mu) - 1]\cos\theta}.$$

Next, from Equations (7) and (10) we see that

$$h = r_0 v_0, \qquad (12)$$

so finally we may write the equation of the planet's orbit:

$$r = \frac{(r_0 v_0)^2/\mu}{1 + (r_0 v_0^2/\mu - 1)\cos\theta} = \frac{pe}{1 + e\cos\theta}. \qquad (13)$$

From Equation (1) in Section 10-8, we see that the orbit of the planet is a conic section with eccentricity

$$e = \frac{r_0 v_0^2}{GM} - 1. \qquad (14)$$

Because the nature of a conic section is determined by its eccentricity, we see that the planet's orbit is:

- A circle if $r_0 v_0^2 = GM$,
- An ellipse if $GM < r_0 v_0^2 < 2GM$,
- A parabola if $r_0 v_0^2 = 2GM$,
- A hyperbola if $r_0 v_0^2 > 2GM$.

$$(15)$$

This comprehensive description of the situation is a generalization of Kepler's first law. Of course, the elliptical case is the one that holds for planets in the solar system.

Let us consider further the case in which the orbit is an ellipse with major semiaxis $a$ and minor semiaxis $b$. (The case in which the ellipse is actually a circle, with $a = b$, will be a special case of this discussion.) By Equation (3) of Section 10-8, the constant

$$pe = \frac{h^2}{GM}$$

used in (13) satisfies the equations

$$pe = a(1 - e^2) = a\left(1 - \frac{a^2 - b^2}{a^2}\right) = \frac{b^2}{a}. \tag{16}$$

We equate these two expressions for $pe$ and find that

$$h^2 = GM\frac{b^2}{a}.$$

Now let $T$ denote the period of revolution of the planet in its elliptical orbit. Then from Equation (9) we see that the area of the ellipse is $A = \frac{1}{2}hT = \pi ab$, and thus that

$$T^2 = \frac{4\pi^2 a^2 b^2}{h^2} = \frac{4\pi^2 a^2 b^2}{GMb^2/a},$$

so that

$$T^2 = \gamma a^3 \tag{17}$$

where $\gamma = 4\pi^2/GM$. This is Kepler's third law.

**EXAMPLE 1**  The period of revolution of Mercury in its elliptical orbit about the sun is 87.97 days, while that of Earth is 365.26 days. Compute the major semiaxis (in astronomical units) of the orbit of Mercury.

*Solution*  The major semiaxis of Earth's orbit is, by definition, 1 A.U. So Equation (17) gives us the value of the constant $\gamma = (365.26)^2$ (day$^2$/au$^3$). Hence the major semiaxis of the orbit of Mercury is

$$a = \left(\frac{T^2}{\gamma}\right)^{1/3} = \left(\frac{(87.97)^2}{(365.26)^2}\right)^{1/3} \approx 0.387 \text{ A.U.}$$

As yet we have considered only planets in orbit about the sun. But Kepler's laws and the equations of this section apply to bodies in orbit about any common central mass, so long as they move solely under the influence of its gravitational attraction. Examples include satellites (artificial or natural) orbiting the Earth or the moons of Jupiter. All we need to know is the value of the constant $\mu = GM$ for the central body. For Earth, we can compute $\mu$ by beginning with the values

$$g = \frac{GM}{R^2} = 32.15$$

ft/sec$^2$ for surface gravitational acceleration and the radius $R = 3960$ (mi) of Earth—which we assume to be spherical. Then

$$\mu = GM = R^2 g = (3960)^2 \left(\frac{32.15}{5280}\right) \approx 95{,}500 \text{ mi}^3/\text{s}^2.$$

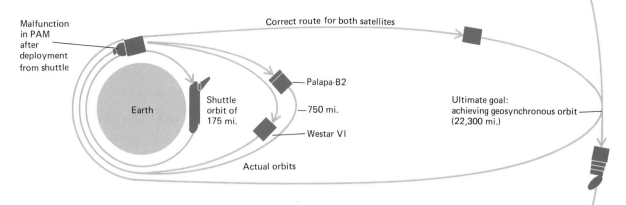

Figure 13.46 shows an attempted launch of a Westar VI geosynchronous satellite from a space shuttle in February 1984. The space shuttle was traveling in a circular orbit 175 mi above the earth's surface, or at an approximate distance $r_0 = 4130$ mi from the earth's center. The plan (which failed on this attempt but has succeeded on others) was as follows: After deployment of the satellite from the shuttle's cargo bay, its rockets were to be fired to increase its (tangential) velocity enough to place it in an elliptical orbit with apogee (farthest distance) of 26,330 mi—the radius of a geosynchronous earth orbit (Example 3 in Section 13-5). Upon reaching this distance, its rockets were to be fired again to increase its velocity enough to insert it into a *circular* orbit with radius $r_1 = 26,330$ mi. We use these figures in the following example.

**EXAMPLE 2**   A rocket is in a circular orbit about the earth with orbital radius $r_0 = 4130$ mi. Its thrusters, when activated, provide a tangential acceleration of 0.1 mi/s$^2$. We want to insert the rocket into an elliptical orbit with perigee (closest distance) $r_0 = 4130$ mi and apogee $r_1 = 26,330$ mi. How long must the thrusters burn to accomplish this?

***Solution***   The problem is to determine by how much the speed $v_0$ of the rocket, when at the perigee of the desired elliptical orbit (see Fig. 13.47), exceeds the speed $v_c$ of the rocket in its original circular orbit.

From our study of uniform circular motion (Section 13-5), we know that the acceleration in a circular orbit with radius $r_0$ is

$$a = \frac{v_c^2}{r_0} = \frac{GM}{r_0^2},$$

so the rocket's original speed—in the circular orbit—is

$$v_c = \sqrt{\frac{GM}{r_0}} = \sqrt{\frac{95,500}{4130}} \approx 4.81$$

mi/s.

We may use Equation (12), that $h = r_0 v_0$, to find $v_0$ once we know $h$. And we can obtain $h$ from Equation (16):

$$pe = \frac{h^2}{GM} = a(1 - e^2),$$

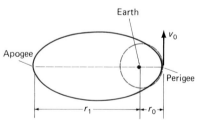

**13.47**   The transfer orbit of Example 2

after we find $a$ and $e$ for the desired ellipse. Recall that each focus of an ellipse is at distance $ae$ from its center. Thus

$$r_0 = a(1 - e) = 4130 \quad \text{and} \quad r_1 = a(1 + e) = 26{,}330$$

(the units are miles). After we solve these simultaneous equations for $a = 15{,}230$ (mi) and $e = 0.729$, the previous equation gives

$$h = \sqrt{(95{,}500)(15{,}230)(1 - [0.729]^2)} \approx 26{,}106.$$

Thus insertion into the elliptical orbit requires an initial speed of

$$v_0 = \frac{h}{r_0} = \frac{26{,}106}{4130} \approx 6.32$$

mi/s. The thrusters must be activated for long enough to increase the rocket's speed by $v_0 - v_c$, or approximately 1.51 mi/s. With an acceleration of 0.1 mi/s$^2$, this will require a burn time of 15.1 s.

---

The ellipse of Example 2 is a **transfer orbit** between the original circular orbit and the orbit of a "stationary satellite" at $r = 26{,}330$ mi. In Problem 13 we ask you to continue Example 2 to determine how to complete the transfer to the stationary satellite's circular orbit.

NOTE  In November of 1984, during the second flight of the space shuttle Discovery, astronauts Joseph Allen, Anna Fisher, and Dale Gardner successfully retrieved the Westar VI that had been improperly deployed in February as well as another errant satellite, the Palapa.

## 13-7  PROBLEMS

For each of the parametric curves in Problems 1–5, express its velocity and acceleration vectors in terms of $\mathbf{u}_r$ and $\mathbf{u}_\theta$; that is, in the form $A(t)\mathbf{u}_r + B(t)\mathbf{u}_\theta$.

1 $r = a$, $\theta = t$
2 $r = 2\cos t$, $\theta = t$
3 $r = t$, $\theta = t$
4 $r = e^t$, $\theta = 2t$
5 $r = 3\sin 4t$, $\theta = 2t$
6 Derive both equations in (2) by differentiation of Equations (1).
7 Derive Equation (5) by differentiating Equation (4).
8 Consider a body in an elliptical orbit with major and minor semiaxes $a$ and $b$ and period of revolution $T$.
(a) Deduce from Equation (4) that $v = r(d\theta/dt)$ when the body is nearest to and farthest from its focus.
(b) Then apply Kepler's second law to conclude that $v = 2\pi ab/rT$ at the body's nearest and farthest points.

In each of Problems 9–12, apply the equation of Part (b) of Problem 8 to compute the speed (in miles per second) of the given body at the nearest and farthest points of its orbit. Use the value 1 A.U. = 92,956,000 mi for the major semiaxis of Earth's orbit.

9 The planet Mercury: $a = 0.387$ A.U., $e = 0.206$, $T = 87.97$ days
10 The planet Earth: $e = 0.0167$, $T = 365.26$ days
11 The moon: $a = 238{,}900$ miles, $e = 0.055$, $T = 27.32$ days
12 An earth satellite: $a = 10{,}000$ miles, $e = 0.5$
13 Continue Example 2 as follows.
(a) Apply Kepler's third law to find how long it will take the rocket to travel from perigee to apogee in the elliptical transfer orbit.
(b) Then compute, as in Example 2, how long the thrusters should burn at apogee to insert the rocket into a circular orbit with a radius of 26,330 miles.

The eccentricity of the earth's elliptical orbit about the sun is approximately $e = 0.017$. Taking its major semiaxis as $a = 93$ million miles, we calculate

$$b = (93)\sqrt{1 - (0.017)^2} \approx 92.99,$$

$$c = (93)(0.017) \approx 1.58.$$

Thus the orbit of the earth is very nearly a perfect circle with radius $a = 93$ million miles, but the sun is located about $1.58$ million miles off-center. For the planet Mercury, the corresponding approximate values are $a = 36$, $b = 35.23$, $c = 7.42$, and $e = 0.206$. Thus Mercury's orbit is also quite circular, but the sun is somewhat more off-center. Mercury's period of revolution about the sun is approximately 88 days, so the earth's period is $365/88 \approx 4.15$ times that of Mercury.

For the sake of a simplified model based on uniform circular motion, let us assume that both Earth and Mercury travel in circular orbits about the sun, with radii $a = 93$ and $b = 36$, respectively. The angular speed of Mercury will be about $k = 4.15$ times that of Earth. Here, however, we want to look at Mercury's motion from the viewpoint of Earth, as did the ancient Greek astronomers. Hence we place Earth at the origin, with the sun traveling in a circular orbit about Earth (as in the Ptolemaic theory of epicycles) and with Mercury in turn orbiting the sun. This model of the motion of Mercury (about Earth) is illustrated in Fig. 13.48.

If we add the coordinates of the sun relative to Earth to the coordinates of Mercury relative to the sun, as indicated in Fig. 13.48, we then get the parametric equations

$$x = a \cos t + b \cos kt,$$

$$y = a \sin t + b \sin kt$$

of Mercury's orbit about Earth. The curve these parametric equations describe is called an *epitrochoid*.

The program shown in Fig. 13.49 plots this epitrochoid on the monitor of an IBM Personal Computer. The effect of the **WINDOW** command in line 180 is to

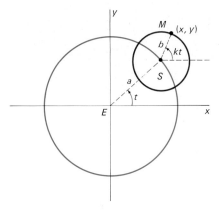

**13.48**   The circles that generate an epitrochoid.

**13.49**   Program MERCURY

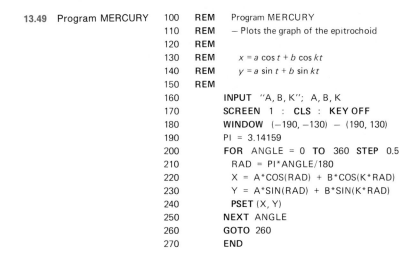

```
100   REM     Program MERCURY
110   REM     — Plots the graph of the epitrochoid
120   REM
130   REM     x = a cos t + b cos kt
140   REM     y = a sin t + b sin kt
150   REM
160         INPUT  "A, B, K";  A, B, K
170         SCREEN  1  :  CLS  :  KEY OFF
180         WINDOW  (−190, −130)  −  (190, 130)
190         PI  =  3.14159
200         FOR  ANGLE  =  0  TO  360  STEP  0.5
210           RAD  =  PI * ANGLE/180
220           X  =  A * COS(RAD)  +  B * COS(K * RAD)
230           Y  =  A * SIN(RAD)  +  B * SIN(K * RAD)
240           PSET (X, Y)
250         NEXT  ANGLE
260         GOTO  260
270         END
```

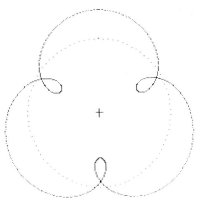

**13.50** The epitrochoid with $k = 4$ (program output)

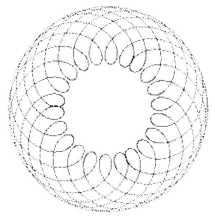

**13.51** The epitrochoid with $k = 4.15$ (program output)

establish $xy$-coordinates on the monitor screen, with $(-190, -130)$ and $(190, 130)$ being the lower-left-hand and the upper-right-hand corners of the screen, respectively.

When the values $a = 93$, $b = 36$, and $k = 4$ are used, we get the epitrochoid shown in Fig. 13.50. With the value $k = 4.15$ that is more accurate for Mercury, the curve no longer "closes" after one revolution. Instead, when $7\frac{1}{3}$ revolutions are plotted, we get the more exotic epitrochoid shown in Fig. 13.51.

## CHAPTER 13 REVIEW: Definitions and Concepts

Use the list below as a guide to concepts that you may need to review.

**1** Definition of a parametric curve; of a smooth parametric curve

**2** The slope of the tangent line to a smooth parametric curve (both in rectangular and in polar coordinates)

**3** Integral computations with parametric curves (Formulas (1) through (4) in Section 13-2)

**4** Arc length of a parametric curve

**5** Vectors: their definition, length, addition, multiplication by scalars, and dot product

**6** Test for perpendicular vectors

**7** Vector-valued functions, velocity and acceleration vectors

**8** Componentwise differentiation and integration of vector-valued functions

**9** The equations of motion of a projectile

**10** The speed and the scalar acceleration of a particle in uniform circular motion

**11** Definitions of unit tangent vector and principal unit normal vector

**12** Definition and computation of curvature of an explicit or parametric curve

**13** Normal and tangential components of acceleration

**14** Kepler's three laws of planetary motion

**15** The radial and transverse unit polar vectors

**16** Polar decomposition of velocity and acceleration vectors

**17** Outline of the derivation of Kepler's laws from Newton's law of gravitation

## MISCELLANEOUS PROBLEMS

In Problems 1–5, eliminate the parameter and sketch the curve.

**1** $x = 2t^3 - 1, y = 2t^3 + 1$
**2** $x = \cosh t, y = \sinh t$
**3** $x = 2 + \cos t, y = 1 - \sin t$
**4** $x = \cos^4 t, y = \sin^4 t$
**5** $x = 1 + t^2, y = t^3$

In Problems 6–10, write the equation of the tangent line to the given curve at the indicated point.

**6** $x = t^2, y = t^3; t = 1$
**7** $x = 3 \sin t, y = 4 \cos t; \quad t = \pi/4$
**8** $x = e^t, y = e^{-t}; \quad t = 0$

**9** $r = \theta; \quad \theta = \pi/2$

**10** $r = 1 + \sin \theta; \quad \theta = \pi/3$

In each of Problems 11–14, find the area of the region between the given curve and the *x*-axis.

**11** $x = 2t + 1, \; y = t^2 + 3; \quad -1 \le t \le 2$

**12** $x = e^t, \; y = e^{-t}; \quad 0 \le t \le 10$

**13** $x = 3 \sin t, \; y = 4 \cos t; \quad 0 \le t \le \pi/2$

**14** $x = \cosh t, \; y = \sinh t; \quad 0 \le t \le 1$

In each of Problems 15–19, find the arc length of the given curve.

**15** $x = t^2, \; y = t^3; \quad 0 \le t \le 1$

**16** $x = \ln(\cos t), \; y = t; \quad 0 \le t \le \pi/4$

**17** $x = 2t, \; y = t^3 + 1/3t; \quad 1 \le t \le 2$

**18** $r = \sin \theta; \quad 0 \le \theta \le \pi$

**19** $r = \sin^3(\theta/3); \quad 0 \le \theta \le \pi$

In each of the Problems 20–24, find the area of the surface generated by revolving the given curve around the *x*-axis.

**20** $x = t^2 + 1, \; y = 3t; \quad 0 \le t \le 2$

**21** $x = 4t^{1/2}, \; y = t^3/3 + 1/2t^2; \quad 1 \le t \le 4$

**22** $r = 4 \cos \theta$

**23** $r = e^{\theta/2}; \quad 0 \le \theta \le \pi$

**24** $x = e^t \cos t, \; y = e^t \sin t; \quad 0 \le t \le \pi/2$

**25** Consider the rolling circle of radius $a$ that was used to generate the cycloid in Example 3 of Section 13-1. Suppose that this circle is the rim of a disk, and let $Q$ be a point of this disk at distance $b < a$ from its center. Find parametric equations for the curve traced by $Q$ as the circle rolls along the *x*-axis; assume that $Q$ begins at the point $(0, a - b)$. Sketch this curve, which is called a *trochoid*.

**26** If the smaller circle of Problem 28 in Section 13-1 rolls around the *outside* of the larger circle, the path of the point $P$ is called an *epicycloid*. Show that it has parametric equations

$$x = (a + b) \cos t - b \cos\left(\frac{a + b}{b} t\right),$$

$$y = (a + b) \sin t - b \sin\left(\frac{a + b}{b} t\right).$$

**27** Suppose that $b = a$ in Problem 26. Show that the epicycloid is then the cardioid $r = 2a(1 - \cos \theta)$ translated $a$ units to the right.

**28** Find the area of the surface generated by revolving the lemniscate $r^2 = 2a^2 \cos 2\theta$ around the *x*-axis.

**29** Find the volume generated by revolving around the *y*-axis the area under the cycloid

$$x = a(t - \sin t), \qquad y = a(1 - \cos t), \qquad 0 \le t \le 2\pi.$$

**30** Show that the length of one arch of the hypocycloid of Problem 28 in Section 13-1 is $s = 8b(a - b)/a$.

**31** Let $ABC$ be an isosceles triangle with $|AB| = |AC|$. Let $M$ be the midpoint of $BC$. Use the dot product to show that $AM$ and $BC$ are perpendicular. (*Suggestion:* Show first that $\overrightarrow{BC} = \overrightarrow{AC} - \overrightarrow{AB}$ and $\overrightarrow{AM} = \frac{1}{2}(\overrightarrow{AC} + \overrightarrow{AB})$.)

**32** Use the dot product to show that the diagonals of a rhombus (a parallelogram with all four sides equal) are perpendicular to each other.

**33** The acceleration of a certain particle is $\mathbf{a} = \mathbf{i} \sin t - \mathbf{j} \cos t$. Assume that the particle begins at time $t = 0$ at the point $(0, 1)$, and has initial velocity $\mathbf{v}_0 = -\mathbf{i}$. Show that its path is a circle.

**34** A particle moves in an attracting central force field, with force proportional to the distance from the origin. This implies that the particle's acceleration is given by $\mathbf{a} = -\omega^2 \mathbf{r}$, where $\mathbf{r}$ is the position vector of the particle. Assume that the particle's initial position is $\mathbf{r}_0 = p\mathbf{i}$ and that its initial velocity is $\mathbf{v}_0 = q\omega\mathbf{j}$. Show that the trajectory of the particle is the ellipse $x^2/p^2 + y^2/q^2 = 1$. (*Suggestion:* Apply the theorem of Section 8-4.)

**35** At time $t = 0$, a target is 160 ft from a gun and is moving directly away from it with a constant speed of 80 ft/s. If the muzzle velocity of the gun is 320 ft/s, at what angle of elevation $\alpha$ should it be fired in order to strike the moving target?

**36** Suppose that a gun with muzzle velocity $v_0$ is located at the foot of a hill with a 30° slope. At what angle of elevation (from the horizontal) should the gun be fired in order to maximize its range, as measured up the hill?

**37** Find the points on the curve $y = \sin x$ where the curvature is maximal and those where it is minimal.

**38** The right branch of the hyperbola $x^2 - y^2 = 1$ is parametrized by $x = \cosh t, \; y = \sinh t$. Find the point where its curvature is maximal.

**39** Find the vectors $\mathbf{N}$ and $\mathbf{T}$ at the point of $x = t \cos t$, $y = t \sin t$ where $t = \pi/2$.

**40** Find the points on the ellipse $x^2/a^2 + y^2/b^2 = 1$ (with $a > b$) where the curvature is maximal and those where it is minimal.

**41** Suppose that the curve $r = f(\theta)$ is given in polar coordinates. Put $r' = f'(\theta)$ and $r'' = f''(\theta)$. Show that its curvature is given by

$$\kappa = \frac{|r^2 + 2(r')^2 - rr''|}{(r^2 + (r')^2)^{3/2}}.$$

**42** Use the formula of Problem 41 to calculate the curvature $\kappa(\theta)$ at the point $(r, \theta)$ of the Archimedean spiral $r = \theta$. Then show that $\kappa(\theta) \to 0$ as $\theta \to \infty$.

**43** A railway curve is to join two straight tracks, one extending westward from $(-1, -1)$, the other extending eastward from $(1, 1)$. Determine $A$, $B$, and $C$ so that the curve $y = Ax + Bx^3 + Cx^5$ joins $(-1, -1)$ and $(1, 1)$ and so that the slope and curvature of this connecting curve are zero at both its end points.

**44** Assume that a body has an elliptical orbit $r = pe/(1 + e \cos \theta)$ and satisfies Kepler's second law in the

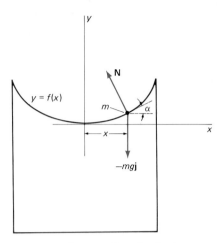

**13.52**  Cross section of water in a rotating bucket

form $r^2(d\theta/dt) = h$ (constant). Then deduce from Equation (5) in Section 13-7 that its acceleration vector **a** is equal to $(k/r^2)\mathbf{u}_r$ for some constant $k$. This shows that Newton's inverse-square law of gravitation follows from Kepler's first and second laws.

**45** Consider the water surface in a cylindrical bucket that is rotating with angular speed $\omega$. Assume that the water has reached steady state with all the water particles rotating at the same angular speed as the bucket. A water particle of mass $m$ at the water surface is subject to two forces, shown in Fig. 13.52: a force **N** perpendicular to the surface and the weight $mg$ of the water particle. The vertical components balance because of the steady-state situation, and the horizontal component of **N** supplies the necessary centripetal force $mx\omega^2$. Deduce that a vertical cross section $y = f(x)$ of the water surface through the axis of the bucket is a parabola.

# 14

# Vectors, Curves, and Surfaces in Space

## Rectangular Coordinates and Three-Dimensional Vectors

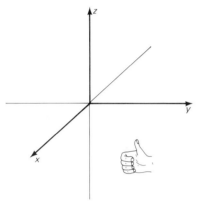

**14.1** The right-handed rectangular coordinate system

In the first thirteen chapters we have discussed many aspects of the calculus of functions of a *single* variable. The geometry of such functions is two-dimensional because the graph of a function of a single variable is a curve in the plane. Most of the remainder of this book deals with the calculus of functions of *several* (two or more) independent variables. The geometry of functions of two variables is three-dimensional because the graphs of such functions are generally surfaces in space.

Rectangular coordinates in the plane may be generalized in a natural way to rectangular coordinates in space. A point in space is determined by giving its location relative to three mutually perpendicular **coordinate axes** passing through the origin $O$. We shall always draw the $x$-, $y$-, and $z$-axes as shown in Fig. 14.1, with arrows indicating the positive direction along each axis. With this configuration of axes, our rectangular coordinate system is said to be **right-handed:** If the curled fingers of the right hand point in the direction of a $90°$ rotation from the positive $x$-axis to the positive $y$-axis, then the thumb points in the direction of the positive $z$-axis. If the $x$- and $y$-axes were interchanged, then we would have a left-handed coordinate system. These two coordinate systems are different in that it is impossible to bring one into coincidence with the other by means of rotations and translations. Similarly, the L- and D-alanine molecules shown in Fig. 14.2 are different; you can digest the left-handed ("levo") version but not the right-handed ("dextro") version. In this book we shall use right-handed coordinate systems exclusively and always draw the $x$-, $y$-, and $z$-axes with the orientation shown in Fig. 14.1.

The three coordinate axes taken in pairs determine three **coordinate planes:**

- the (horizontal) $xy$-plane, where $z = 0$;
- the (vertical) $yz$-plane, where $x = 0$; and
- the (vertical) $xz$-plane, where $y = 0$.

**14.2** The stereoisomers of the amino acid alanine are physically and biologically different.

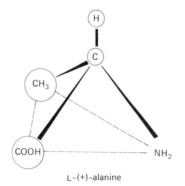

L-(+)-alanine

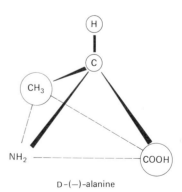

D-(−)-alanine

The point $P$ in space is said to have **rectangular coordinates** $(x, y, z)$ if (see Fig. 14.3):

- $x$ is its signed distance from the $yz$-plane;
- $y$ is its signed distance from the $xz$-plane;   and
- $z$ is its signed distance from the $xy$-plane.

In this case we may describe the location of the point $P$ by simply calling it "the point $P(x, y, z)$." There is a one-to-one correspondence between ordered triples $(x, y, z)$ of real numbers and points $P$ in space; this correspondence is called a **rectangular coordinate system** in space. In Fig. 14.4 the point $P$ is located in the **first octant**—the eighth of space in which all three rectangular coordinates are positive.

If we apply the Pythagorean theorem to the right triangles $P_1QR$ and $P_1RP_2$ in Fig. 14.5, we get

$$|P_1P_2|^2 = |RP_2|^2 + |P_1R|^2 = |RP_2|^2 + |QR|^2 + |P_1Q|^2$$
$$= (x_1 - x_2)^2 + (y_1 - y_2)^2 + (z_1 - z_2)^2.$$

Thus the **distance formula** for the **distance** $|P_1P_2|$ between the points $P_1$ and $P_2$ is

$$|P_1P_2| = \sqrt{(x_1 - x_2)^2 + (y_1 - y_2)^2 + (z_1 - z_2)^2}. \qquad (1)$$

For example, the distance between the points $P_1(1, 3, -2)$ and $P_2(4, -3, 1)$ is

$$|P_1P_2| = \sqrt{(4 - 1)^2 + (-3 - 3)^2 + (1 + 2)^2}$$
$$= \sqrt{54} \approx 7.34847.$$

In Problem 43 we ask you to apply the distance formula in (1) to show that the **midpoint** $M$ of the line segment joining $P_1(x_1, y_1, z_1)$ and $P_2(x_2, y_2, z_2)$ is

$$M(\tfrac{1}{2}[x_1 + x_2], \tfrac{1}{2}[y_1 + y_2], \tfrac{1}{2}[z_1 + z_2]). \qquad (2)$$

The **graph** of an equation in three variables $x$, $y$, and $z$ is the set of all points in space with rectangular coordinates that satisfy the equation. In general, the graph of an equation in three variables will be a *two-dimensional*

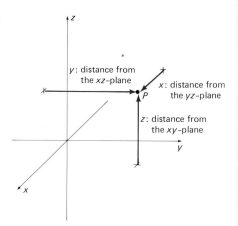

**14.3**   Locating the point $P$ with rectangular coordinates

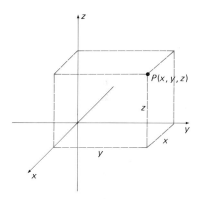

**14.4**   "Completing the box" to show $P$ with the illusion of the third dimension

**14.5**   The distance between $P_1$ and $P_2$ is the length of the long diagonal of the box.

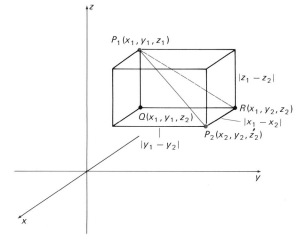

*surface* in $\mathcal{R}^3$ (three-dimensional space with rectangular coordinates). For example, let $C(h, k, l)$ be a fixed point. Then the graph of the equation

$$(x - h)^2 + (y - k)^2 + (z - l)^2 = r^2 \qquad (3)$$

is the set of all points $P(x, y, z)$ at distance $r$ from the fixed point $C$. This means that Equation (3) is the equation of **sphere with radius $r$ and center $C(h, k, l)$**. Moreover, given an equation of the form

$$x^2 + y^2 + z^2 + Ax + By + Cz + D = 0,$$

we can attempt—by completing the square in each variable—to write it in the form of Equation (3), and thereby show that its graph is a sphere.

**EXAMPLE 1** Determine the graph of the equation

$$x^2 + y^2 + z^2 + 4x + 2y - 6z - 2 = 0.$$

*Solution* We complete the squares, and the equation takes the form

$$(x^2 + 4x + 4) + (y^2 + 2y + 1) + (z^2 - 6z + 9) = 16.$$

That is,

$$(x + 2)^2 + (y + 1)^2 + (z - 3)^2 = 4^2.$$

Thus the given equation has as its graph a sphere with radius 4 and center $(-2, -1, 3)$.

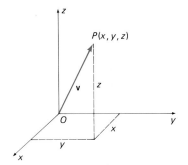

**14.6** The segment $OP$ is a realization of the vector $\mathbf{v} = \overrightarrow{OP}$.

## VECTORS IN SPACE

The discussion of vectors in the plane in Section 13-3 may be repeated almost verbatim for vectors in space. The major difference is that a vector in space has three components rather than two. The vector determined by the point $P(x, y, z)$ is its **position vector** $\mathbf{v} = \overrightarrow{OP} = \langle x, y, z \rangle$, which is represented pictorially (Fig. 14.6) by the arrow from the origin $O$ to $P$ (or by any parallel translate of this arrow). The distance formula in (1) gives

$$|\mathbf{v}| = \sqrt{x^2 + y^2 + z^2} \qquad (4)$$

for the **length** of the vector $\mathbf{v} = \langle x, y, z \rangle$.

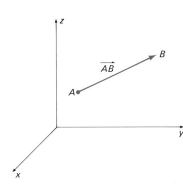

**14.7** The segment $AB$ is an instance of the vector $\overrightarrow{AB}$.

The vector $\overrightarrow{AB}$ represented by the arrow (Fig. 14.7) from $A(a_1, a_2, a_3)$ to $B(b_1, b_2, b_3)$ is defined to be

$$\overrightarrow{AB} = \langle b_1 - a_1, b_2 - a_2, b_3 - a_3 \rangle,$$

and its length is simply the distance between the two points $A$ and $B$.

We define addition and scalar multiplication of vectors exactly as in Section 13-3, taking into account that our vectors now have three components instead of two: The **sum** of the vectors $\mathbf{a} = \langle a_1, a_2, a_3 \rangle$ and $\mathbf{b} = \langle b_1, b_2, b_3 \rangle$ is the vector

$$\mathbf{a} + \mathbf{b} = \langle a_1 + b_1, a_2 + b_2, a_3 + b_3 \rangle. \qquad (5)$$

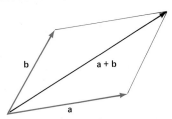

**14.8** The parallelogram law for addition of vectors

Because $\mathbf{a}$ and $\mathbf{b}$ lie in a plane (though perhaps not the $xy$-plane) if their initial points coincide, addition of vectors obeys the same **parallelogram law** as in the two-dimensional case (see Fig. 14.8).

If $c$ is a real number, then the **scalar multiple** $c\mathbf{a}$ is the vector

$$c\mathbf{a} = \langle ca_1, ca_2, ca_3 \rangle. \tag{6}$$

The length of $c\mathbf{a}$ is $|c|$ times that of $\mathbf{a}$, and $c\mathbf{a}$ has the same direction as $\mathbf{a}$ if $c > 0$ but the opposite direction if $c < 0$. The following algebraic properties of vector addition and scalar multiplication are easy to establish; they follow from computations with components, just as in Section 13-3.

$$\mathbf{a} + \mathbf{b} = \mathbf{b} + \mathbf{a},$$
$$\mathbf{a} + (\mathbf{b} + \mathbf{c}) = (\mathbf{a} + \mathbf{b}) + \mathbf{c},$$
$$r(\mathbf{a} + \mathbf{b}) = r\mathbf{a} + r\mathbf{b}, \tag{7}$$
$$(r + s)\mathbf{a} = r\mathbf{a} + s\mathbf{a},$$
$$(rs)\mathbf{a} = r(s\mathbf{a}) = s(r\mathbf{a}).$$

**EXAMPLE 2**  If $\mathbf{a} = \langle 3, 4, 12 \rangle$ and $\mathbf{b} = \langle -4, 3, 0 \rangle$, then

$$\mathbf{a} + \mathbf{b} = \langle (3) + (-4), (4) + (3), (12) + (0) \rangle = \langle -1, 7, 12 \rangle,$$
$$|\mathbf{a}| = \sqrt{3^2 + 4^2 + 12^2} = \sqrt{169} = 13,$$
$$2\mathbf{a} = \langle (2)(3), (2)(4), (2)(12) \rangle = \langle 6, 8, 24 \rangle,$$

and

$$2\mathbf{a} - 3\mathbf{b} = \langle 6 + 12, 8 - 9, 24 - 0 \rangle = \langle 18, -1, 24 \rangle.$$

A **unit vector** is one with length 1. Any space vector can be expressed in terms of the three **basic unit vectors**

$$\mathbf{i} = \langle 1, 0, 0 \rangle, \qquad \mathbf{j} = \langle 0, 1, 0 \rangle, \qquad \mathbf{k} = \langle 0, 0, 1 \rangle.$$

When located with their initial points at the origin, these basic unit vectors form a right-handed triple of vectors pointing in the positive directions along the three coordinate axes (as shown in Fig. 14.9).

Any space vector $\mathbf{a} = \langle a_1, a_2, a_3 \rangle$ can be written as

$$\mathbf{a} = a_1\mathbf{i} + a_2\mathbf{j} + a_3\mathbf{k}$$

in terms of the basic unit vectors. As in the two-dimensional case, the usefulness of this representation is that algebraic operations involving vectors may be carried out simply by collecting coefficients of $\mathbf{i}$, $\mathbf{j}$, and $\mathbf{k}$. For example,

$$\mathbf{a} + \mathbf{b} = (a_1\mathbf{i} + a_2\mathbf{j} + a_3\mathbf{k}) + (b_1\mathbf{i} + b_2\mathbf{j} + b_3\mathbf{k})$$
$$= (a_1 + b_1)\mathbf{i} + (a_2 + b_2)\mathbf{j} + (a_3 + b_3)\mathbf{k}.$$

The **dot product** of the two vectors

$$\mathbf{a} = a_1\mathbf{i} + a_2\mathbf{j} + a_3\mathbf{k} \quad \text{and} \quad \mathbf{b} = b_1\mathbf{i} + b_2\mathbf{j} + b_3\mathbf{k}$$

is also defined almost exactly as before: Multiply corresponding components, then add the results. Thus

$$\mathbf{a} \cdot \mathbf{b} = a_1 b_1 + a_2 b_2 + a_3 b_3. \tag{8}$$

If $a_3 = 0 = b_3$, then we may think of $\mathbf{a}$ and $\mathbf{b}$ as vectors in the $xy$-plane. Then the above definition reduces to the one given in Section 13-3 for the

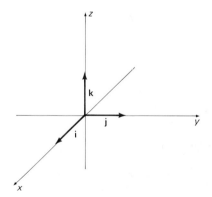

**14.9**  The basic unit vectors $\mathbf{i}$, $\mathbf{j}$, and $\mathbf{k}$

dot product of plane vectors. The three-dimensional dot product has the same list of properties as the two-dimensional dot product, and all those below can be established routinely by working with components.

$$\text{(i)} \qquad \mathbf{a} \cdot \mathbf{a} = |\mathbf{a}|^2,$$

$$\text{(ii)} \qquad \mathbf{a} \cdot \mathbf{b} = \mathbf{b} \cdot \mathbf{a},$$

$$\text{(iii)} \ \ \mathbf{a} \cdot (\mathbf{b} + \mathbf{c}) = \mathbf{a} \cdot \mathbf{b} + \mathbf{a} \cdot \mathbf{c}, \tag{9}$$

$$\text{(iv)} \qquad (r\mathbf{a}) \cdot \mathbf{b} = r(\mathbf{a} \cdot \mathbf{b}) = \mathbf{a} \cdot (r\mathbf{b}).$$

**EXAMPLE 3** If $\mathbf{a} = \langle 3, 4, 12 \rangle$ and $\mathbf{b} = \langle -4, 3, 0 \rangle$, then

$$\mathbf{a} \cdot \mathbf{b} = (3)(-4) + (4)(3) + (12)(0) = -12 + 12 + 0 = 0.$$

If $\mathbf{c} = \langle 4, 5, -3 \rangle$ then

$$\mathbf{a} \cdot \mathbf{c} = (3)(4) + (4)(5) + (12)(-3) = 12 + 20 - 36 = -4.$$

It is important to remember that the dot product of two *vectors* is a *scalar*—that is, a real number. For this reason the dot product is sometimes called the *scalar product*.

The significance of the dot product resides in its geometric interpretation. Let the vectors $\mathbf{a}$ and $\mathbf{b}$ be represented by the position vectors $\overrightarrow{OP}$ and $\overrightarrow{OQ}$, respectively. Then the angle $\theta$ between $\mathbf{a}$ and $\mathbf{b}$ is the angle at $O$ in the triangle $OPQ$ of Fig. 14.10. We say that $\mathbf{a}$ and $\mathbf{b}$ are **parallel** if $\theta = 0$ or if $\theta = \pi$ and that $\mathbf{a}$ and $\mathbf{b}$ are **perpendicular** if $\theta = \pi/2$. For convenience, we regard the zero vector $\mathbf{0}$ as both parallel to *and* perpendicular to *every* vector.

---

> **Theorem** *Interpretation of the Dot Product*
>
> If $\theta$ is the angle between the vectors $\mathbf{a}$ and $\mathbf{b}$, then
>
> $$\mathbf{a} \cdot \mathbf{b} = |\mathbf{a}| \, |\mathbf{b}| \cos \theta. \tag{10}$$

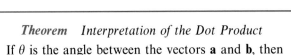

**Proof** If either $\mathbf{a} = \mathbf{0}$ or $\mathbf{b} = \mathbf{0}$, then Equation (10) follows immediately. If the vectors $\mathbf{a}$ and $\mathbf{b}$ are parallel, then $\mathbf{b} = t\mathbf{a}$ with either $t > 0$ and $\theta = 0$ or $t < 0$ and $\theta = \pi$. In either case, both sides of (10) reduce to $t|\mathbf{a}|^2$.

So we turn to the general case in which the vectors $\mathbf{a} = \overrightarrow{OP}$ and $\mathbf{b} = \overrightarrow{OQ}$ are nonzero and nonparallel. Then

$$
\begin{aligned}
|\overrightarrow{PQ}|^2 = |\mathbf{a} - \mathbf{b}|^2 &= (\mathbf{a} - \mathbf{b}) \cdot (\mathbf{a} - \mathbf{b}) \\
&= \mathbf{a} \cdot \mathbf{a} - \mathbf{a} \cdot \mathbf{b} - \mathbf{b} \cdot \mathbf{a} + \mathbf{b} \cdot \mathbf{b} \\
&= |\mathbf{a}|^2 + |\mathbf{b}|^2 - 2\mathbf{a} \cdot \mathbf{b}.
\end{aligned}
$$

On the other hand, $c = |\overrightarrow{PQ}|$ is the side of the triangle $OPQ$ (Fig. 14.10) opposite the angle $\theta$ included by the sides $a = |\mathbf{a}|$ and $b = |\mathbf{b}|$. Hence the law of cosines yields

$$
\begin{aligned}
|\overrightarrow{PQ}|^2 = c^2 &= a^2 + b^2 - 2ab \cos \theta \\
&= |\mathbf{a}|^2 + |\mathbf{b}|^2 - 2|\mathbf{a}| \, |\mathbf{b}| \cos \theta.
\end{aligned}
$$

Finally, comparison of these two expressions for $|\overrightarrow{PQ}|^2$ gives Equation (10). ∎

**14.10** The angle $\theta$ between the vectors $\mathbf{a}$ and $\mathbf{b}$

This theorem tells us that the angle between the nonzero vectors **a** and **b** is given by

$$\cos \theta = \frac{\mathbf{a} \cdot \mathbf{b}}{|\mathbf{a}| \, |\mathbf{b}|}. \tag{11}$$

Note that this immediately implies the perpendicularity test of Section 13-3: *The two vectors* **a** *and* **b** *are perpendicular* ($\theta = \pi/2$) *if and only if* $\mathbf{a} \cdot \mathbf{b} = 0$. For instance, the vectors **a** and **b** of Example 3 are perpendicular because we found that $\mathbf{a} \cdot \mathbf{b} = 0$.

**EXAMPLE 4**   Find the angles in the triangle shown in Fig. 14.11 that has its vertices at $A(2, -1, 0)$, $B(5, -4, 3)$, and $C(1, -3, 2)$.

*Solution*   We apply Equation (10) with $\theta = \angle A$, $\mathbf{a} = \overrightarrow{AB} = \langle 3, -3, 3 \rangle$, and $\mathbf{b} = \overrightarrow{AC} = \langle -1, -2, 2 \rangle$. This yields

$$\angle A = \cos^{-1} \left( \frac{\langle 3, -3, 3 \rangle \cdot \langle -1, -2, 2 \rangle}{\sqrt{27}\sqrt{9}} \right)$$

$$= \cos^{-1} \left( \frac{9}{\sqrt{27}\sqrt{9}} \right) \approx 54.74°.$$

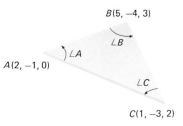

14.11   The triangle of Example 4

Similarly,

$$\angle B = \cos^{-1} \left( \frac{\overrightarrow{BA} \cdot \overrightarrow{BC}}{|\overrightarrow{BA}| \, |\overrightarrow{BC}|} \right) = \cos^{-1} \left( \frac{\langle -3, 3, -3 \rangle \cdot \langle -4, 1, -1 \rangle}{\sqrt{27}\sqrt{18}} \right)$$

$$= \cos^{-1} \left( \frac{18}{\sqrt{27}\sqrt{18}} \right) \approx 35.26°.$$

Then $\angle C = 180° - \angle A - \angle B = 90°$. As a check, note that

$$\overrightarrow{CA} \cdot \overrightarrow{CB} = \langle 1, 2, -2 \rangle \cdot \langle 4, -1, 1 \rangle = 0.$$

So angle $C$ is, indeed, a right angle.

---

The **direction angles** of the nonzero vector $\mathbf{a} = \langle a_1, a_2, a_3 \rangle$ are the angles $\alpha$, $\beta$, and $\gamma$ that it makes with the vectors **i**, **j**, and **k**, respectively, as shown in Fig. 14.12. The cosines of these angles, $\cos \alpha$, $\cos \beta$, and $\cos \gamma$, are called the **direction cosines** of the vector **a**. When we replace **b** in Equation (11) by **i**, **j**, and **k** in turn, we find that

$$\cos \alpha = \frac{\mathbf{a} \cdot \mathbf{i}}{|\mathbf{a}| \, |\mathbf{i}|} = \frac{a_1}{|\mathbf{a}|},$$

$$\cos \beta = \frac{\mathbf{a} \cdot \mathbf{j}}{|\mathbf{a}| \, |\mathbf{j}|} = \frac{a_2}{|\mathbf{a}|}, \qquad \text{and} \tag{12}$$

$$\cos \gamma = \frac{\mathbf{a} \cdot \mathbf{k}}{|\mathbf{a}| \, |\mathbf{k}|} = \frac{a_3}{|\mathbf{a}|}.$$

14.12   The direction angles of the vector **a**

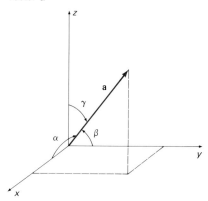

That is, the direction cosines of **a** are the components of the **unit vector** $\mathbf{a}/|\mathbf{a}|$ with the same direction as **a**. Consequently

$$\cos^2 \alpha + \cos^2 \beta + \cos^2 \gamma = 1. \tag{13}$$

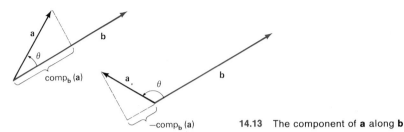

**14.13** The component of **a** along **b**

**EXAMPLE 5**  Find the direction angles of the vector $\mathbf{a} = 2\mathbf{i} + 3\mathbf{j} - \mathbf{k}$.

**Solution**  Since $|\mathbf{a}| = \sqrt{14}$, Equations (12) give $\alpha = \cos^{-1}(2/\sqrt{14}) \approx 57.69°$, $\beta = \cos^{-1}(3/\sqrt{14}) \approx 36.70°$, and $\gamma = \cos^{-1}(-1/\sqrt{14}) \approx 105.50°$.

Sometimes we need to find the component of one vector **a** in the direction of another (nonzero) vector **b**. Think of the two vectors as located with the same initial point, as in Fig. 14.13. Then the (scalar) **component of a along b**, denoted by $\mathrm{comp}_\mathbf{b}\mathbf{a}$, is numerically the length of the perpendicular projection of **a** onto the straight line determined by **b**. It is positive if the angle $\theta$ between **a** and **b** is acute (so that **a** and **b** point in the same general direction) and negative if $\theta > \pi/2$. Thus $\mathrm{comp}_\mathbf{b}\mathbf{a} = |\mathbf{a}| \cos \theta$ in either case. Equation (10) then gives

$$\mathrm{comp}_\mathbf{b}\mathbf{a} = \frac{|\mathbf{a}|\,|\mathbf{b}| \cos \theta}{|\mathbf{b}|} = \frac{\mathbf{a} \cdot \mathbf{b}}{|\mathbf{b}|}. \tag{14}$$

There is no real need to memorize this formula, for—in practice—we can always read $\mathrm{comp}_\mathbf{b}\mathbf{a} = |\mathbf{a}| \cos \theta$ from the figure and then apply (10) to eliminate $\cos \theta$. Note that $\mathrm{comp}_\mathbf{b}\mathbf{a}$ is a scalar, not a vector.

**EXAMPLE 6**  Given $\mathbf{a} = \langle 4, -5, 3 \rangle$ and $\mathbf{b} = \langle 2, 1, -2 \rangle$, write **a** as the sum of a vector $\mathbf{a}_\parallel$ parallel to **b** and a vector $\mathbf{a}_\perp$ perpendicular to **b**.

**Solution**  Our method of solution is motivated by the diagram of Fig. 14.14. We take

$$\mathbf{a}_\parallel = (\mathrm{comp}_\mathbf{b}\mathbf{a}) \frac{\mathbf{b}}{|\mathbf{b}|} = \frac{\mathbf{a} \cdot \mathbf{b}}{|\mathbf{b}|^2}\mathbf{b} = \frac{8 - 5 - 6}{9}\mathbf{b}$$

$$= -\tfrac{1}{3}\langle 2, 1, -2 \rangle = \langle -\tfrac{2}{3}, -\tfrac{1}{3}, \tfrac{2}{3} \rangle,$$

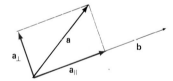

**14.14**  Construction of $\mathbf{a}_\parallel$ and $\mathbf{a}_\perp$ (see Example 6)

and

$$\mathbf{a}_\perp = \mathbf{a} - \mathbf{a}_\parallel = \langle 4, -5, 3 \rangle - \langle -\tfrac{2}{3}, -\tfrac{1}{3}, \tfrac{2}{3} \rangle = \langle \tfrac{14}{3}, -\tfrac{14}{3}, \tfrac{7}{3} \rangle.$$

The diagram makes our choice of $\mathbf{a}_\parallel$ plausible, and we have deliberately chosen $\mathbf{a}_\perp$ so that $\mathbf{a} = \mathbf{a}_\perp + \mathbf{a}_\parallel$. To verify that the vector $\mathbf{a}_\parallel$ is, indeed, parallel to **b**, we simply note that it is a scalar multiple of **b**. To verify that $\mathbf{a}_\perp$ is perpendicular to **b**, we compute the dot product:

$$\mathbf{a}_\perp \cdot \mathbf{b} = \tfrac{28}{3} - \tfrac{14}{3} - \tfrac{14}{3} = 0.$$

Thus $\mathbf{a}_\parallel$ and $\mathbf{a}_\perp$ have the required properties.

One important application of vector components is to the definition and computation of work. Recall that the work $W$ done by a constant force $F$ exerted along the line of motion in moving a particle a distance $d$ is given by $W = Fd$. But what if the force is a constant vector $\mathbf{F}$ pointing in some direction other than the line of motion, as when a child pulls a sled against the resistance of friction (see Fig. 14.15). Suppose that $\mathbf{F}$ moves a particle along the line segment from $P$ to $Q$, and let $\mathbf{u} = \overrightarrow{PQ}$. Then the **work** $W$ done by $\mathbf{F}$ in moving the particle along the line from $P$ to $Q$ is *by definition* the product of the component of $\mathbf{F}$ along $\mathbf{u} = \overrightarrow{PQ}$ and the distance moved:

$$W = (\text{comp}_\mathbf{u}\mathbf{F})|\mathbf{u}|. \tag{15}$$

**EXAMPLE 7** Show that the work done by a constant force $\mathbf{F}$ in moving a particle along the line segment from $P$ to $Q$ is

$$W = \mathbf{F} \cdot \mathbf{D} \tag{16}$$

where $\mathbf{D} = \overrightarrow{PQ}$ is the displacement vector. This formula is the natural vector generalization of the scalar formula $W = Fd$.

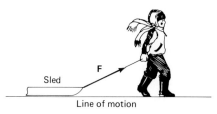

**14.15** The vector force **F** is constant but acts at an angle to the line of motion.

*Solution* We combine Formulas (14) and (15). This gives

$$W = (\text{comp}_\mathbf{D}\mathbf{F})|\mathbf{D}| = \frac{\mathbf{F} \cdot \mathbf{D}}{|\mathbf{D}|}|\mathbf{D}| = \mathbf{F} \cdot \mathbf{D}.$$

## 14-1 PROBLEMS

In Problems 1–5 find: (a) $2\mathbf{a} + \mathbf{b}$; (b) $3\mathbf{a} - 4\mathbf{b}$; (c) $\mathbf{a} \cdot \mathbf{b}$; (d) $|\mathbf{a} - \mathbf{b}|$; (e) $\mathbf{a}/|\mathbf{a}|$.

**1** $\mathbf{a} = \langle 2, 5, -4 \rangle$ and $\mathbf{b} = \langle 1, -2, -3 \rangle$
**2** $\mathbf{a} = \langle -1, 0, 2 \rangle$ and $\mathbf{b} = \langle 3, 4, -5 \rangle$
**3** $\mathbf{a} = \mathbf{i} + \mathbf{j} + \mathbf{k}$ and $\mathbf{b} = \mathbf{j} - \mathbf{k}$
**4** $\mathbf{a} = 2\mathbf{i} - 3\mathbf{j} + 5\mathbf{k}$ and $\mathbf{b} = 5\mathbf{i} + 3\mathbf{j} - 7\mathbf{k}$
**5** $\mathbf{a} = 2\mathbf{i} - \mathbf{j}$ and $\mathbf{b} = \mathbf{j} - 3\mathbf{k}$
**6–10** Find the angle between the vectors $\mathbf{a}$ and $\mathbf{b}$ in Problems 1–5.
**11–15** Find $\text{comp}_\mathbf{b}\mathbf{a}$ and $\text{comp}_\mathbf{a}\mathbf{b}$ for the vectors $\mathbf{a}$ and $\mathbf{b}$ given in Problems 1–5.

In Problems 16–20, write the equation of the indicated sphere.

**16** Center (3, 1, 2), radius 5
**17** Center $(-2, 1, -5)$, radius $\sqrt{7}$
**18** One diameter the segment joining $(3, 5, -3)$ and $(7, 3, 1)$
**19** Center $(4, 5, -2)$, passing through the point $(1, 0, 0)$
**20** Center $(3, -4, 3)$, tangent to the $xz$-plane

In Problems 21–23, find the center and radius of the sphere having the given equation.

**21** $x^2 + y^2 + z^2 + 4x - 6y = 0$
**22** $x^2 + y^2 + z^2 - 8x - 9y + 10z + 40 = 0$
**23** $3x^2 + 3y^2 + 3z^2 - 18z - 48 = 0$

In Problems 24–30, describe the graph of the given equation in geometric terms.

**24** $x = 0$      **25** $z = 10$
**26** $xy = 0$      **27** $xyz = 0$
**28** $x^2 + y^2 + z^2 + 7 = 0$
**29** $x^2 + y^2 + z^2 - 2x + 1 = 0$
**30** $x^2 + y^2 + z^2 - 6x + 8y + 25 = 0$

In Problems 31–33, find the direction angles of the vector $\overrightarrow{PQ}$.

**31** $P(1, -1, 0)$ and $Q(3, 4, 5)$
**32** $P(2, -3, 5)$ and $Q(1, 0, -1)$
**33** $P(-1, -2, -3)$ and $Q(5, 6, 7)$

In Problems 34 and 35 find the work $W$ done by the force $\mathbf{F}$ in moving a particle from $P$ to $Q$.

**34** $\mathbf{F} = \mathbf{i} - \mathbf{k}$;   $P(0, 0, 0)$, $Q(3, 1, 0)$
**35** $\mathbf{F} = 2\mathbf{i} - 3\mathbf{j} + 5\mathbf{k}$;   $P(5, 3, -4)$, $Q(-1, -2, 5)$
**36** Prove the **Cauchy-Schwarz inequality**: $|\mathbf{a} \cdot \mathbf{b}| \le |\mathbf{a}|\,|\mathbf{b}|$ for all pairs of vectors $\mathbf{a}$ and $\mathbf{b}$.
**37** Given two arbitrary vectors $\mathbf{a}$ and $\mathbf{b}$, prove that they must satisfy the **triangle inequality**

$$|\mathbf{a} + \mathbf{b}| \le |\mathbf{a}| + |\mathbf{b}|.$$

(*Suggestion:* Square both sides.)

**38** Show that, if **a** and **b** are two arbitrary vectors, then

$$|\mathbf{a} - \mathbf{b}| \geq |\mathbf{a}| - |\mathbf{b}|.$$

(*Suggestion:* Write $\mathbf{a} = (\mathbf{a} - \mathbf{b}) + \mathbf{b}$; then apply the triangle inequality (Problem 37).)

**39** Find the area of the triangle with vertices $A(1, 1, 1)$, $B(3, -2, 3)$, and $C(3, 4, 6)$.

**40** Find the three angles of the triangle of Problem 39.

**41** Find the angle between any longest diagonal of a cube and any edge it meets.

**42** Show that the three points $P(0, -2, 4)$, $Q(1, -3, 5)$, and $R(4, -6, 8)$ lie on a single straight line. (*Suggestion:* Compare the vectors $\overrightarrow{PQ}$ and $\overrightarrow{PR}$.)

**43** Show that the point $M$ given in Formula (2) is indeed the midpoint of the line segment $P_1 P_2$. (*Note:* You must show *both* that $M$ is equally distant from $P_1$ and $P_2$ *and* that $M$ lies on the line segment $P_1 P_2$.)

**44** Given vectors **a** and **b**, let $a = |\mathbf{a}|$ and $b = |\mathbf{b}|$. Show that the vector $\mathbf{c} = (b\mathbf{a} + a\mathbf{b})/(a + b)$ bisects the angle between **a** and **b**.

**45** Let **a**, **b**, and **c** be three vectors in the $xy$-plane with **a** and **b** nonzero and not parallel. Show that there exist scalars $\alpha$ and $\beta$ such that $\mathbf{c} = \alpha\mathbf{a} + \beta\mathbf{b}$. Begin by expressing **a**, **b**, and **c** in terms of **i** and **j**.

**46** Let $ax + by + c = 0$ be the equation of the line $L$ in the $xy$-plane with normal vector $\mathbf{n} = \langle a, b \rangle$. Let $P_0(x_0, y_0)$ be a point on this line and $P_1(x_1, y_1)$ a point not on $L$. Show that the (perpendicular) distance $d$ from $P_1$ to $L$ is

$$d = \frac{|\mathbf{n} \cdot \overrightarrow{P_0 P_1}|}{|\mathbf{n}|} = \frac{|ax_1 + by_1 + c|}{\sqrt{a^2 + b^2}}.$$

**47** Given the two points $A(3, -2, 4)$ and $B(5, 7, -1)$, write an equation in $x$, $y$, and $z$ that says this: The point $P(x, y, z)$ is equally distant from the points $A$ and $B$. Then simplify this equation and give a geometric description of the set of all such points $P(x, y, z)$.

**48** Given the fixed point $A(1, 3, 5)$, the point $P(x, y, z)$, and the vector $\mathbf{n} = \mathbf{i} - \mathbf{j} + 2\mathbf{k}$, use the dot product to help you write an equation in $x$, $y$, and $z$ that says this: **n** and $\overrightarrow{AP}$ are perpendicular. Then simplify this equation and give a geometric description of the set of all such points $P(x, y, z)$.

**49** Show that the points $(0, 0, 0)$, $(1, 1, 0)$, $(1, 0, 1)$, and $(0, 1, 1)$ are the vertices of a regular tetrahedron by showing that each of the six edges has length $\sqrt{2}$. Then use the dot product to find the angle between any two edges.

**50** The methane molecule $CH_4$ is arranged with the four hydrogen atoms at the vertices of a regular tetrahedron and with the carbon atom at its centroid. Suppose that the axes and scale are chosen so that the tetrahedron is the one of Problem 49, which has centroid at $(\frac{1}{2}, \frac{1}{2}, \frac{1}{2})$. Find the *bond angle* between the lines from the carbon atom to two of the hydrogen atoms.

---

## 14-2

## The Vector Product of Two Vectors

We often need to find a vector that is perpendicular to each of two space vectors **a** and **b**. A routine way of doing this is provided by the **vector product**, or **cross product, a × b** of the vectors **a** and **b**. This vector product is quite unlike the dot product $\mathbf{a} \cdot \mathbf{b}$ in that $\mathbf{a} \cdot \mathbf{b}$ is a scalar, while $\mathbf{a} \times \mathbf{b}$ is a vector.

The vector product of the vectors $\mathbf{a} = \langle a_1, a_2, a_3 \rangle$ and $\mathbf{b} = \langle b_1, b_2, b_3 \rangle$ can be defined by the formula

$$\mathbf{a} \times \mathbf{b} = \langle a_2 b_3 - a_3 b_2, a_3 b_1 - a_1 b_3, a_1 b_2 - a_2 b_1 \rangle. \tag{1}$$

Though this formula seems unmotivated, it has a redeeming feature: The product $\mathbf{a} \times \mathbf{b}$ is perpendicular both to **a** and **b**.

---

> **Theorem 1** *Perpendicularity of the Vector Product*
>
> The vector $\mathbf{a} \times \mathbf{b}$ is perpendicular both to **a** and to **b**.

---

**Proof** We show that $\mathbf{a} \times \mathbf{b}$ is perpendicular to **a** by showing that the dot product of **a** and $\mathbf{a} \times \mathbf{b}$ is zero. With the components as in Formula (1), we find that

$$\mathbf{a} \cdot (\mathbf{a} \times \mathbf{b}) = a_1(a_2 b_3 - a_3 b_2) + a_2(a_3 b_1 - a_1 b_3) + a_3(a_1 b_2 - a_2 b_1)$$
$$= a_1 a_2 b_3 - a_1 a_3 b_2 + a_2 a_3 b_1 - a_2 a_1 b_3 + a_3 a_1 b_2 - a_3 a_2 b_1$$
$$= 0.$$

A similar computation shows that $\mathbf{b} \cdot (\mathbf{a} \times \mathbf{b}) = 0$, so that $\mathbf{a} \times \mathbf{b}$ is perpendicular to the vector $\mathbf{b}$ as well. ∎

Formula (1) need not be memorized, because there is an alternative version involving determinants that is easy to remember. Recall that a determinant of order two is defined by

$$\begin{vmatrix} a_1 & a_2 \\ b_1 & b_2 \end{vmatrix} = a_1 b_2 - a_2 b_1. \tag{2}$$

For example,

$$\begin{vmatrix} 2 & -1 \\ 3 & 4 \end{vmatrix} = (2)(4) - (-1)(3) = 11.$$

A determinant of order three can be defined in terms of determinants of order two:

$$\begin{vmatrix} a_1 & a_2 & a_3 \\ b_1 & b_2 & b_3 \\ c_1 & c_2 & c_3 \end{vmatrix} = +a_1 \begin{vmatrix} b_2 & b_3 \\ c_2 & c_3 \end{vmatrix} - a_2 \begin{vmatrix} b_1 & b_3 \\ c_1 & c_3 \end{vmatrix} + a_3 \begin{vmatrix} b_1 & b_2 \\ c_1 & c_2 \end{vmatrix}. \tag{3}$$

Note that each element $a_i$ of the first row is multiplied by the 2-by-2 "subdeterminant" obtained by deleting the row *and* column that contain $a_i$. Note also in (3) that signs are attached to the $a_i$ in accord with the checkerboard pattern

$$\begin{vmatrix} + & - & + \\ - & + & - \\ + & - & + \end{vmatrix}.$$

The formula in (3) is an expansion of the 3-by-3 determinant along its first row. It also can be expanded along any other row or column. For instance, its expansion along its second column is

$$\begin{vmatrix} a_1 & a_2 & a_3 \\ b_1 & b_2 & b_3 \\ c_1 & c_2 & c_3 \end{vmatrix} = -a_2 \begin{vmatrix} b_1 & b_3 \\ c_1 & c_3 \end{vmatrix} + b_2 \begin{vmatrix} a_1 & a_3 \\ c_1 & c_3 \end{vmatrix} - c_2 \begin{vmatrix} a_1 & a_3 \\ b_1 & b_3 \end{vmatrix}.$$

In linear algebra it is proved that all such expansions yield the same value for the determinant.

Although a determinant of order three can be expanded along any row or column, we shall use only expansions along the first row, as in (3). For example,

$$\begin{vmatrix} 1 & 3 & -2 \\ 2 & -1 & 4 \\ -3 & 7 & 5 \end{vmatrix} = (1) \begin{vmatrix} -1 & 4 \\ 7 & 5 \end{vmatrix} - (3) \begin{vmatrix} 2 & 4 \\ -3 & 5 \end{vmatrix} + (-2) \begin{vmatrix} 2 & -1 \\ -3 & 7 \end{vmatrix}$$

$$= (1)[-5 - 28] + (-3)[10 + 12] + (-2)[14 - 3]$$

$$= -33 - 66 - 22 = -121.$$

Formula (1) for the vector product of the vectors $\mathbf{a} = a_1 \mathbf{i} + a_2 \mathbf{j} + a_3 \mathbf{k}$ and $\mathbf{b} = b_1 \mathbf{i} + b_2 \mathbf{j} + b_3 \mathbf{k}$ is equivalent to

$$\mathbf{a} \times \mathbf{b} = \begin{vmatrix} a_2 & a_3 \\ b_2 & b_3 \end{vmatrix} \mathbf{i} - \begin{vmatrix} a_1 & a_3 \\ b_1 & b_3 \end{vmatrix} \mathbf{j} + \begin{vmatrix} a_1 & a_2 \\ b_1 & b_2 \end{vmatrix} \mathbf{k}. \tag{4}$$

This is easy to verify by expanding the 2 by 2 determinants of the right-hand side, and noting that the three components of the right-hand side of Formula (1) result. Motivated by Formula (4), we write

$$\mathbf{a} \times \mathbf{b} = \begin{vmatrix} \mathbf{i} & \mathbf{j} & \mathbf{k} \\ a_1 & a_2 & a_3 \\ b_1 & b_2 & b_3 \end{vmatrix} \tag{5}$$

The "symbolic determinant" in this equation is to be evaluated by expanding along its first row, just as in Equation (3) and just as though it were an ordinary determinant with real-number entries. The result of this expansion is the right-hand side of Formula (4). Note that the components of the *first* vector **a** in **a** × **b** constitute the *second* row of the 3 by 3 determinant, while the components of the *second* vector **b** constitute the *third* row of the determinant. The order in which the vectors **a** and **b** are written is important, because we shall see presently that **a** × **b** is generally *not* equal to **b** × **a**: The vector product is *not commutative*.

Formula (5) for the vector product is the form that is most convenient for computational purposes. For example, if

$$\mathbf{a} = 3\mathbf{i} - \mathbf{j} + 2\mathbf{k} \quad \text{and} \quad \mathbf{b} = 2\mathbf{i} + 2\mathbf{j} - \mathbf{k}$$

then

$$\mathbf{a} \times \mathbf{b} = \begin{vmatrix} \mathbf{i} & \mathbf{j} & \mathbf{k} \\ 3 & -1 & 2 \\ 2 & 2 & -1 \end{vmatrix}$$

$$= \begin{vmatrix} -1 & 2 \\ 2 & -1 \end{vmatrix} \mathbf{i} - \begin{vmatrix} 3 & 2 \\ 2 & -1 \end{vmatrix} \mathbf{j} + \begin{vmatrix} 3 & -1 \\ 2 & 2 \end{vmatrix} \mathbf{k}$$

$$= (1 - 4)\mathbf{i} - (-3 - 4)\mathbf{j} + (6 - [-2])\mathbf{k};$$

thus

$$\mathbf{a} \times \mathbf{b} = -3\mathbf{i} + 7\mathbf{j} + 8\mathbf{k}.$$

You might pause to verify (using the dot product) that the vector $-3\mathbf{i} + 7\mathbf{j} + 8\mathbf{k}$ is perpendicular both to **a** and to **b**.

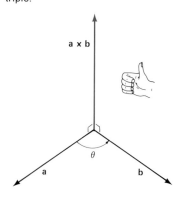

**14.16** The vectors **a**, **b**, and **a** × **b** form—in that order—a right-handed triple.

If the vectors **a** and **b** are located with the same initial point, then Theorem 1 tells us that **a** × **b** is normal to the plane determined by **a** and **b**, as indicated in Fig. 14.16. There are still two possible directions for **a** × **b**, but it turns out that if **a** × **b** ≠ **0**, then the triple **a**, **b**, **a** × **b** is a **right-handed** triple in exactly the same sense as the triple **i**, **j**, **k**. Thus if the thumb of your right hand points in the direction of **a** × **b**, your fingers curl in the direction of a rotation (less than 180°) from **a** to **b**.

Once the direction of **a** × **b** has been established, the vector product can be described in completely geometric terms by telling what the length |**a** × **b**| of the vector **a** × **b** is. This is given by the formula

$$|\mathbf{a} \times \mathbf{b}|^2 = |\mathbf{a}|^2 |\mathbf{b}|^2 - (\mathbf{a} \cdot \mathbf{b})^2. \tag{6}$$

This vector identity can be verified routinely (though tediously) by writing $\mathbf{a} = \langle a_1, a_2, a_3 \rangle$ and $\mathbf{b} = \langle b_1, b_2, b_3 \rangle$, computing both sides of Equation

(6), and then verifying that the results are equal. That is,

$$(a_2b_3 - a_3b_2)^2 + (a_1b_3 - a_3b_1)^2 + (a_1b_2 - a_2b_1)^2$$
$$= (a_1^2 + a_2^2 + a_3^2)(b_1^2 + b_2^2 + b_3^2) - (a_1b_1 + a_2b_2 + a_3b_3)^2.$$

Formula (6) tells us what $|\mathbf{a} \times \mathbf{b}|$ is, but the following theorem reveals the *geometric significance* of the length of the vector product.

---

**Theorem 2   Length of the Vector Product**

Let $\theta$ be the angle between the nonzero vectors $\mathbf{a}$ and $\mathbf{b}$ (measured so that $0 \le \theta \le \pi$). Then

$$|\mathbf{a} \times \mathbf{b}| = |\mathbf{a}|\,|\mathbf{b}|\sin\theta. \tag{7}$$

---

**Proof**   We begin with Equation (6), and use the fact that $\mathbf{a} \cdot \mathbf{b} = |\mathbf{a}|\,|\mathbf{b}|\cos\theta$. Thus

$$\begin{aligned}
|\mathbf{a} \times \mathbf{b}|^2 &= |\mathbf{a}|^2|\mathbf{b}|^2 - (\mathbf{a} \cdot \mathbf{b})^2 \\
&= |\mathbf{a}|^2|\mathbf{b}|^2 - (|\mathbf{a}|\,|\mathbf{b}|\cos\theta)^2 \\
&= |\mathbf{a}|^2|\mathbf{b}|^2[1 - \cos^2\theta] \\
&= |\mathbf{a}|^2|\mathbf{b}|^2\sin^2\theta.
\end{aligned}$$

Equation (7) now follows after we take the positive square root of both sides. (This is the correct root on the right-hand side because $\sin\theta \ge 0$ for $0 \le \theta \le \pi$.)   ∎

---

**Corollary   Parallel Vectors**

Two nonzero vectors $\mathbf{a}$ and $\mathbf{b}$ are parallel ($\theta = 0$ or $\theta = \pi$) if and only if $\mathbf{a} \times \mathbf{b} = \mathbf{0}$.

---

In particular, the cross product of any vector with itself is the zero vector. Also, Formula (1) shows immediately that the cross product of any vector with the zero vector is the zero vector again. Thus

$$\mathbf{a} \times \mathbf{a} = \mathbf{a} \times \mathbf{0} = \mathbf{0} \times \mathbf{a} = \mathbf{0} \tag{8}$$

for any vector $\mathbf{a}$.

Equation (7) has an important geometric interpretation. Suppose that $\mathbf{a}$ and $\mathbf{b}$ are represented by adjacent sides of a parallelogram $PQRS$, with $\mathbf{a} = \overrightarrow{PQ}$ and $\mathbf{b} = \overrightarrow{PS}$, as indicated in Fig. 14.17. The parallelogram then has base of length $|\mathbf{a}|$ and height $|\mathbf{b}|\sin\theta$, so its area is

$$A = |\mathbf{a}|\,|\mathbf{b}|\sin\theta = |\mathbf{a} \times \mathbf{b}|. \tag{9}$$

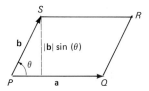

**14.17**   The area of the parallelogram $PQRS$ is $|\mathbf{a} \times \mathbf{b}|$.

Thus *the length of the vector product* $\mathbf{a} \times \mathbf{b}$ *is the same, numerically, as the area of the parallelogram determined by* $\mathbf{a}$ *and* $\mathbf{b}$. It follows that the area of the triangle $PQS$—since it has half the area of the parallelogram—is

$$\tfrac{1}{2}A = \tfrac{1}{2}|\mathbf{a} \times \mathbf{b}| = \tfrac{1}{2}|\overrightarrow{PQ} \times \overrightarrow{PS}|. \tag{10}$$

Formula (10) gives a quick way to compute the area of a triangle—even one in space—without the need of finding any of its angles.

**EXAMPLE 1** Find the area of the triangle with vertices $A(3, 0, -1)$, $B(4, 2, 5)$, and $C(7, -2, 4)$.

**Solution** $\overrightarrow{AB} = \langle 1, 2, 6 \rangle$ and $\overrightarrow{AC} = \langle 4, -2, 5 \rangle$, so

$$\overrightarrow{AB} \times \overrightarrow{AC} = \begin{vmatrix} \mathbf{i} & \mathbf{j} & \mathbf{k} \\ 1 & 2 & 6 \\ 4 & -2 & 5 \end{vmatrix} = 22\mathbf{i} + 19\mathbf{j} - 10\mathbf{k}.$$

So by Equation (10) the area of triangle $ABC$ is

$$\tfrac{1}{2}\sqrt{(22)^2 + (19)^2 + (-10)^2} = \tfrac{1}{2}\sqrt{945} \approx 15.37.$$

Now let $\mathbf{u}$, $\mathbf{v}$, $\mathbf{w}$ be a right-handed triple of mutually perpendicular unit vectors. The angle between any two of these is $\theta = \pi/2$, so $|\mathbf{u}| = |\mathbf{v}| = |\mathbf{w}| = \sin\theta = 1$. Thus it follows from (7) that $\mathbf{u} \times \mathbf{v} = \mathbf{w}$. When we apply this observation to the basic unit vectors $\mathbf{i}$, $\mathbf{j}$, and $\mathbf{k}$, we see that

$$\mathbf{i} \times \mathbf{j} = \mathbf{k}, \quad \mathbf{j} \times \mathbf{k} = \mathbf{i}, \quad \text{and} \quad \mathbf{k} \times \mathbf{i} = \mathbf{j}. \tag{11a}$$

But

$$\mathbf{j} \times \mathbf{i} = -\mathbf{k}, \quad \mathbf{k} \times \mathbf{j} = -\mathbf{i}, \quad \text{and} \quad \mathbf{i} \times \mathbf{k} = -\mathbf{j}. \tag{11b}$$

These relations, together with the fact that

$$\mathbf{i} \times \mathbf{i} = \mathbf{j} \times \mathbf{j} = \mathbf{k} \times \mathbf{k} = \mathbf{0}, \tag{11c}$$

also follow directly from the original definition of the cross product (in the form of Equation (5)). The products in (11a) are easily remembered in terms of the sequence

$$\mathbf{i}, \mathbf{j}, \mathbf{k}, \mathbf{i}, \mathbf{j}, \mathbf{k}, \ldots.$$

The product of any two consecutive unit vectors—in the order they appear here—equals the next one in the sequence.

Note that $\mathbf{i} \times \mathbf{j} \neq \mathbf{j} \times \mathbf{i}$. *The vector product is not commutative.* Instead, it is *anticommutative:* $\mathbf{a} \times \mathbf{b} = -\mathbf{b} \times \mathbf{a}$. This is the first part of the following theorem.

---

**Theorem 3** *Algebraic Properties of Vector Products*

If $\mathbf{a}$, $\mathbf{b}$, and $\mathbf{c}$ are vectors and $k$ is a real number, then

(i) $\quad \mathbf{a} \times \mathbf{b} = -(\mathbf{b} \times \mathbf{a});$ \hfill (12)

(ii) $\quad (k\mathbf{a}) \times \mathbf{b} = \mathbf{a} \times (k\mathbf{b}) = k(\mathbf{a} \times \mathbf{b});$ \hfill (13)

(iii) $\mathbf{a} \times (\mathbf{b} + \mathbf{c}) = (\mathbf{a} \times \mathbf{b}) + (\mathbf{a} \times \mathbf{c});$ \hfill (14)

(iv) $\quad \mathbf{a} \cdot (\mathbf{b} \times \mathbf{c}) = (\mathbf{a} \times \mathbf{b}) \cdot \mathbf{c};$ and \hfill (15)

(v) $\mathbf{a} \times (\mathbf{b} \times \mathbf{c}) = (\mathbf{a} \cdot \mathbf{c})\mathbf{b} - (\mathbf{a} \cdot \mathbf{b})\mathbf{c}.$ \hfill (16)

---

The proofs of Properties (12) through (15) are straightforward applications of the definition of the vector product in terms of components. See Problem 20 for an outline of the proof of (16).

Cross products of vectors expressed in terms of the basic unit vectors $\mathbf{i}$, $\mathbf{j}$, and $\mathbf{k}$ can be found by means of computations that closely resemble those of ordinary algebra. We simply apply the algebraic properties summarized in Theorem 3 together with the relations in (11) giving the various products of the basic unit vectors. Care must be taken to preserve the order of the factors, since vector multiplication is not commutative—although, of course, one should not hesitate to use Property (12). For example,

$$
\begin{aligned}
(\mathbf{i} - 2\mathbf{j} + 3\mathbf{k}) \times (3\mathbf{i} + 2\mathbf{j} - 4\mathbf{k}) = {}& 3(\mathbf{i} \times \mathbf{i}) + 2(\mathbf{i} \times \mathbf{j}) - 4(\mathbf{i} \times \mathbf{k}) \\
& - 6(\mathbf{j} \times \mathbf{i}) - 4(\mathbf{j} \times \mathbf{j}) + 8(\mathbf{j} \times \mathbf{k}) \\
& + 9(\mathbf{k} \times \mathbf{i}) + 6(\mathbf{k} \times \mathbf{j}) - 12(\mathbf{k} \times \mathbf{k}) \\
= {}& 3(\mathbf{0}) + 2\mathbf{k} - 4(-\mathbf{j}) - 6(-\mathbf{k}) - 4(\mathbf{0}) \\
& + 8\mathbf{i} + 9\mathbf{j} + 6(-\mathbf{i}) - 12(\mathbf{0}) \\
= {}& 2\mathbf{i} + 13\mathbf{j} + 8\mathbf{k}.
\end{aligned}
$$

### SCALAR TRIPLE PRODUCTS

Let us examine the product $\mathbf{a} \cdot (\mathbf{b} \times \mathbf{c})$ appearing in Equation (15). Note first that the expression would not make sense were the parentheses around $\mathbf{a} \cdot \mathbf{b}$, since $\mathbf{a} \cdot \mathbf{b}$ is a scalar, and thus we could not form the cross product of $\mathbf{a} \cdot \mathbf{b}$ with the vector $\mathbf{c}$. This means that we may omit the parentheses; the expression $\mathbf{a} \cdot \mathbf{b} \times \mathbf{c}$ is not ambiguous. The dot product of the vectors $\mathbf{a}$ and $\mathbf{b} \times \mathbf{c}$ is a real number, called the **scalar triple product** of the vectors $\mathbf{a}$, $\mathbf{b}$, and $\mathbf{c}$. Equation (15) says that the operations $\cdot$ (dot) and $\times$ (cross) can be interchanged without affecting the value of the expression:

$$
\mathbf{a} \cdot \mathbf{b} \times \mathbf{c} = \mathbf{a} \times \mathbf{b} \cdot \mathbf{c}
$$

for all vectors $\mathbf{a}$, $\mathbf{b}$, and $\mathbf{c}$.

To compute the scalar triple product in terms of components, write $\mathbf{a} = \langle a_1, a_2, a_3 \rangle$, $\mathbf{b} = \langle b_1, b_2, b_3 \rangle$, and $\mathbf{c} = \langle c_1, c_2, c_3 \rangle$. Then

$$
\mathbf{b} \times \mathbf{c} = (b_2 c_3 - b_3 c_2)\mathbf{i} - (b_1 c_3 - b_3 c_1)\mathbf{j} + (b_1 c_2 - b_2 c_1)\mathbf{k},
$$

so

$$
\mathbf{a} \cdot (\mathbf{b} \times \mathbf{c}) = a_1(b_2 c_3 - b_3 c_2) - a_2(b_1 c_3 - b_3 c_1) + a_3(b_1 c_2 - b_2 c_1).
$$

But the expression on the right is the value of the following 3 by 3 determinant:

$$
\mathbf{a} \cdot \mathbf{b} \times \mathbf{c} = \begin{vmatrix} a_1 & a_2 & a_3 \\ b_1 & b_2 & b_3 \\ c_1 & c_2 & c_3 \end{vmatrix}. \tag{17}
$$

This is the quickest way to compute the scalar triple product. For instance, if

$$
\mathbf{a} = 2\mathbf{i} - 3\mathbf{k}, \qquad \mathbf{b} = \mathbf{i} + \mathbf{j} + \mathbf{k}, \quad \text{and} \quad \mathbf{c} = 4\mathbf{j} - \mathbf{k},
$$

then

$$
\begin{aligned}
\mathbf{a} \cdot \mathbf{b} \times \mathbf{c} &= \begin{vmatrix} 2 & 0 & -3 \\ 1 & 1 & 1 \\ 0 & 4 & -1 \end{vmatrix} \\
&= +(2)\begin{vmatrix} 1 & 1 \\ 4 & -1 \end{vmatrix} - (0)\begin{vmatrix} 1 & 1 \\ 0 & -1 \end{vmatrix} + (-3)\begin{vmatrix} 1 & 1 \\ 0 & 4 \end{vmatrix} \\
&= +(2)(-5) + (-3)(4) = -22.
\end{aligned}
$$

The importance of the scalar triple product for applications depends upon the following geometric interpretation. Let $\mathbf{a}$, $\mathbf{b}$, and $\mathbf{c}$ be three vectors

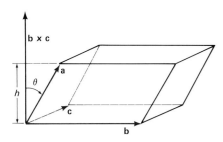

**14.18** The volume of the parallelepiped is $|\mathbf{a} \cdot \mathbf{b} \times \mathbf{c}|$.

with the same initial point. Figure 14.18 shows the parallelepiped determined by these vectors—that is, with the vectors as adjacent edges. If the vectors **a**, **b**, and **c** are coplanar (lie in a single plane) then the parallelepiped is *degenerate;* its volume is zero. Theorem 4 holds whether or not the three vectors are coplanar, but it is of most interest when they are not.

---

**Theorem 4**  *Scalar Triple Products and Volume*

The volume $V$ of the parallelepiped determined by the vectors **a**, **b**, and **c** is the absolute value of the scalar triple product $\mathbf{a} \cdot \mathbf{b} \times \mathbf{c}$; that is,

$$V = |\mathbf{a} \cdot \mathbf{b} \times \mathbf{c}|. \tag{18}$$

---

**Proof**  If the three vectors are coplanar, then **a** and $\mathbf{b} \times \mathbf{c}$ are perpendicular, so $V = |\mathbf{a} \cdot \mathbf{b} \times \mathbf{c}| = 0$. Assume they are not coplanar. By Equation (9) the area of the base (determined by **b** and **c**) of the parallelepiped is $A = |\mathbf{b} \times \mathbf{c}|$.

Now let $\alpha$ be the *acute* angle between **a** and the line through $\mathbf{b} \times \mathbf{c}$ that is perpendicular to the base. Then the height of the parallelepiped is $h = |\mathbf{a}| \cos \alpha$. If $\theta$ is the angle between the vectors **a** and $\mathbf{b} \times \mathbf{c}$, then either $\theta = \alpha$ or $\theta = \pi - \alpha$. Hence $\cos \alpha = |\cos \theta|$, so

$$V = Ah = |\mathbf{b} \times \mathbf{c}||\mathbf{a}| \cos \alpha = |\mathbf{a}||\mathbf{b} \times \mathbf{c}||\cos \theta| = |\mathbf{a} \cdot \mathbf{b} \times \mathbf{c}|.$$

Thus we have verified Equation (18). ∎

**EXAMPLE 2**  Use the scalar triple product to show that the points $A(1, -1, 2)$, $B(2, 0, 1)$, $C(3, 2, 0)$, and $D(5, 4, -2)$ are coplanar.

**Solution**  It is enough to show that the vectors $\overrightarrow{AB} = \langle 1, 1, -1 \rangle$, $\overrightarrow{AC} = \langle 2, 3, -2 \rangle$, and $\overrightarrow{AD} = \langle 4, 5, -4 \rangle$ are coplanar. But their scalar triple product is

$$\begin{vmatrix} 1 & 1 & -1 \\ 2 & 3 & -2 \\ 4 & 5 & -4 \end{vmatrix} = (1)(-2) - (1)(0) + (-1)(-2) = 0,$$

so Theorem 4 guarantees us that the parallelepiped determined by these three vectors has volume zero. Hence the four given points are coplanar.

---

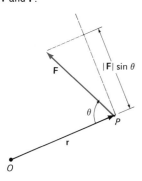

**14.19** The torque vector $\tau$ is normal to both **r** and **F**.

The vector product occurs quite naturally in many scientific applications. For example, suppose that a body in space is free to rotate about the fixed point $O$. If a force **F** acts at the point $P$ of the body, its effect is to cause rotation of the body. This effect is measured by the **torque vector** $\tau$ defined by the relation

$$\tau = \mathbf{r} \times \mathbf{F}$$

where $\mathbf{r} = \overrightarrow{OP}$. The straight line through $O$ determined by $\tau$ is the axis of rotation, and the length $|\tau| = |\mathbf{r}||\mathbf{F}| \sin \theta$ (see Fig. 14.19) is the **moment** of the force **F** about this axis.

Another example is the force exerted on a moving charged particle by a magnetic field. This force is important in the cyclotron and in the television picture tube; controlling the paths of the ions is accomplished through the interplay of electric and magnetic fields. In such circumstances, the force **F** on the particle due to a magnetic field depends upon three things: the charge

CHAP. 14:  Vectors, Curves, and Surfaces in Space

$q$ of the particle, its velocity vector $\mathbf{v}$, and the magnetic field vector $\mathbf{B}$ at the instantaneous location of the particle. And it turns out that

$$\mathbf{F} = (q\mathbf{v}) \times \mathbf{B}.$$

Find $\mathbf{a} \times \mathbf{b}$ in Problems 1–4.

**1**  $\mathbf{a} = \langle 5, -1, -2 \rangle, \mathbf{b} = \langle -3, 2, 4 \rangle$
**2**  $\mathbf{a} = \langle 3, -2, 0 \rangle, \mathbf{b} = \langle 0, 3, -2 \rangle$
**3**  $\mathbf{a} = \mathbf{i} - \mathbf{j} + 3\mathbf{k}, \mathbf{b} = -2\mathbf{i} + 3\mathbf{j} + \mathbf{k}$
**4**  $\mathbf{a} = 4\mathbf{i} + 2\mathbf{j} - 2\mathbf{k}, \mathbf{b} = 2\mathbf{i} - 5\mathbf{j} + 5\mathbf{k}$
**5**  Apply Equation (5) to verify Equations (11a).
**6**  Apply Equation (5) to verify Equations (11b).
**7**  Show that the vector product is not associative by calculating and comparing both $\mathbf{a} \times (\mathbf{b} \times \mathbf{c})$ and $(\mathbf{a} \times \mathbf{b}) \times \mathbf{c}$ with

$$\mathbf{a} = \mathbf{i}, \quad \mathbf{b} = \mathbf{i} + \mathbf{j}, \quad \text{and} \quad \mathbf{c} = \mathbf{i} + \mathbf{j} + \mathbf{k}.$$

**8**  Find nonzero vectors $\mathbf{a}$, $\mathbf{b}$, and $\mathbf{c}$ such that $\mathbf{a} \times \mathbf{b} = \mathbf{a} \times \mathbf{c}$ but $\mathbf{b} \neq \mathbf{c}$.
**9**  Suppose that the three vectors $\mathbf{a}$, $\mathbf{b}$, and $\mathbf{c}$ are mutually perpendicular. Show that $\mathbf{a} \times (\mathbf{b} \times \mathbf{c}) = \mathbf{0}$.
**10**  Find the area of the triangle with vertices $P(1, 1, 0)$, $Q(1, 0, 1)$, and $R(0, 1, 1)$.
**11**  Find the area of the triangle with vertices $P(1, 3, -2)$, $Q(2, 4, 5)$ and $R(-3, -2, 2)$.
**12**  Find the volume of the parallelepiped with adjacent edges $OP$, $OQ$, and $OR$, where $P$, $Q$, and $R$ are the points given in Problem 10.
**13**  (a) Find the volume of the parallelepiped with adjacent edges $OP$, $OQ$, and $OR$, where $P$, $Q$, and $R$ are the points of Problem 11.
(b) Find the volume of the tetrahedron with vertices $O$, $P$, $Q$, and $R$.
**14**  Find a unit vector $\mathbf{n}$ perpendicular to the plane through the three points $P$, $Q$, and $R$ of Problem 11. Then find the distance from the origin to this plane by computing $\mathbf{n} \cdot \overrightarrow{OP}$.
**15**  Apply Formula (5) to verify Equation (12), the anticommutativity of the vector product.
**16**  Apply Formula (17) to verify Identity (15) for scalar triple products.
**17**  Suppose that $P$ and $Q$ are points on a line $L$ in space. Let $A$ be a point not on $L$. Calculate in two ways the area of the triangle $APQ$ to show that the perpendicular distance from $A$ to the line $L$ is $d = |\overrightarrow{AP} \times \overrightarrow{AQ}|/|\overrightarrow{PQ}|$. Then use this formula to compute the distance from the point $A(1, 0, 1)$ to the line through the two points $P(2, 3, 1)$ and $Q(-3, 1, 4)$.
**18**  Suppose that $A$ is a point not on the plane determined by the three points $P$, $Q$, and $R$. Calculate in two ways the volume of the tetrahedron $APQR$ to show that the perpendicular distance from $A$ to this plane is

$$d = \frac{|\overrightarrow{AP} \cdot \overrightarrow{AQ} \times \overrightarrow{AR}|}{|\overrightarrow{PQ} \times \overrightarrow{PR}|}.$$

Use this formula to compute the distance from the point $A(1, 0, 1)$ to the plane through the points $P(2, 3, 1)$, $Q(3, -1, 4)$, and $R(0, 0, 2)$.
**19**  Suppose that $P_1$ and $Q_1$ are two points on the line $L_1$, and that $P_2$ and $Q_2$ are two points on the line $L_2$. If the lines $L_1$ and $L_2$ are not parallel, then the perpendicular distance $d$ between them is the projection of $\overrightarrow{P_1P_2}$ onto a vector $\mathbf{n}$ that is perpendicular both to $\overrightarrow{P_1Q_1}$ and to $\overrightarrow{P_2Q_2}$. Show that

$$d = \frac{|\overrightarrow{P_1P_2} \cdot \overrightarrow{P_1Q_1} \times \overrightarrow{P_2Q_2}|}{|\overrightarrow{P_1Q_1} \times \overrightarrow{P_2Q_2}|}.$$

**20**  Use the following method to establish that the **vector triple product** $(\mathbf{a} \times \mathbf{b}) \times \mathbf{c}$ is equal to $(\mathbf{a} \cdot \mathbf{c})\mathbf{b} - (\mathbf{b} \cdot \mathbf{c})\mathbf{a}$.
(a) Let $\mathbf{I}$ be a unit vector in the direction of $\mathbf{a}$, and let $\mathbf{J}$ be a unit vector perpendicular to $\mathbf{I}$ and parallel to the plane of $\mathbf{a}$ and $\mathbf{b}$. Let $\mathbf{K} = \mathbf{I} \times \mathbf{J}$. Explain why there are scalars $a_1, b_1, b_2, c_1, c_2,$ and $c_3$ such that

$$\mathbf{a} = a_1\mathbf{I},$$
$$\mathbf{b} = b_1\mathbf{I} + b_2\mathbf{J},$$
$$\mathbf{c} = c_1\mathbf{I} + c_2\mathbf{J} + c_3\mathbf{K}.$$

(b) Now show that

$$(\mathbf{a} \times \mathbf{b}) \times \mathbf{c} = -a_1b_2c_2\mathbf{I} + a_1b_2c_1\mathbf{J}.$$

(c) Finally, substitute for $\mathbf{I}$ and $\mathbf{J}$ in terms of $\mathbf{a}$ and $\mathbf{b}$.
**21**  By permutation of the vectors $\mathbf{a}$, $\mathbf{b}$, and $\mathbf{c}$, deduce from Problem 20 that

$$\mathbf{a} \times (\mathbf{b} \times \mathbf{c}) = (\mathbf{a} \cdot \mathbf{c})\mathbf{b} - (\mathbf{a} \cdot \mathbf{b})\mathbf{c}$$

[Formula (16)].
**22**  Deduce from the orthogonality properties of the vector product that the vector $(\mathbf{a} \times \mathbf{b}) \times (\mathbf{c} \times \mathbf{d})$ can be written both in the form $r_1\mathbf{a} + r_2\mathbf{b}$ and in the form $s_1\mathbf{c} + s_2\mathbf{d}$.
**23**  Consider the triangle in the $xy$-plane that has vertices $(x_1, y_1, 0)$, $(x_2, y_2, 0)$, and $(x_3, y_3, 0)$. Use the cross product to show that the area of this triangle is *half* the *absolute value* of the determinant

$$\begin{vmatrix} 1 & 1 & 1 \\ x_1 & x_2 & x_3 \\ y_1 & y_2 & y_3 \end{vmatrix}.$$

## Lines and Planes in Space

A straight line in space is determined by any two points $P_0$ and $P_1$ on it. Alternatively, a line in space can be specified by giving a point $P_0$ on it *and* a vector, such as $\overrightarrow{P_0P_1}$, that determines the direction of the line.

To investigate equations describing lines in space, let us begin with a straight line $L$ that passes through the point $P_0(x_0, y_0, z_0)$ and is parallel to the vector $\mathbf{v} = a\mathbf{i} + b\mathbf{j} + c\mathbf{k}$, as in Fig. 14.20. Then another point $P(x, y, z)$ lies on the line $L$ if and only if the vectors $\mathbf{v}$ and $\overrightarrow{P_0P}$ are parallel, in which case

$$\overrightarrow{P_0P} = t\mathbf{v} \tag{1}$$

for some real number $t$. If $\mathbf{r}_0 = \overrightarrow{OP_0}$ and $\mathbf{r} = \overrightarrow{OP}$ are the position vectors of the points $P_0$ and $P$, respectively, then $\overrightarrow{P_0P} = \mathbf{r} - \mathbf{r}_0$. Hence (1) gives the *vector equation*

$$\mathbf{r} = \mathbf{r}_0 + t\mathbf{v} \tag{2}$$

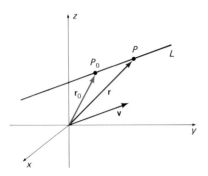

**14.20** Finding the equation of the line $L$ through the point $P_0$ parallel to the vector $\mathbf{v}$.

describing the line $L$. As indicated in Fig. 14.20, $\mathbf{r}$ is the position vector of an *arbitrary* point $P$ on the line $L$, and Equation (2) gives $\mathbf{r}$ in terms of the parameter $t$, the position vector $\mathbf{r}_0$ of a *fixed* point $P_0$ on $L$, and the fixed vector $\mathbf{v}$ that determines the direction of $L$.

The left-hand and right-hand sides are equal in Equation (2), and each side is a vector. So corresponding components are also equal. When we write the resulting equations, we get a scalar description of the line $L$. Since $\mathbf{r}_0 = \langle x_0, y_0, z_0 \rangle$ and $\mathbf{r} = \langle x, y, z \rangle$, Equation (2) thereby yields the three scalar equations

$$\left. \begin{aligned} x &= x_0 + at, \\ y &= y_0 + bt, \\ z &= z_0 + ct. \end{aligned} \right\} \tag{3}$$

These are **parametric equations** of the line $L$ through the point $(x_0, y_0, z_0)$ parallel to the vector $\mathbf{v} = \langle a, b, c \rangle$.

**EXAMPLE 1** Write parametric equations of the line $L$ that passes through the points $P_1(1, 2, 2)$ and $P_2(3, -1, 3)$.

*Solution* The line $L$ is parallel to the vector

$$\mathbf{v} = \overrightarrow{P_1P_2} = 2\mathbf{i} - 3\mathbf{j} + \mathbf{k},$$

so we take $a = 2, b = -3$, and $c = 1$. With $P_1$ as the fixed point, the equations in (3) give

$$x = 1 + 2t, \qquad y = 2 - 3t, \qquad z = 2 + t$$

as parametric equations of $L$. In contrast, with $P_2$ as the fixed point and with

$$-2\mathbf{v} = -4\mathbf{i} + 6\mathbf{j} - 2\mathbf{k}$$

as the direction vector, the equations in (3) yield the parametric equations

$$x = 3 - 4t, \qquad y = -1 + 6t, \qquad z = 3 - 2t.$$

Thus the parametric equations of a line are not unique.

---

Given two straight lines $L_1$ and $L_2$ with parametric equations

$$x = x_1 + a_1 t, \qquad y = y_1 + b_1 t, \qquad z = z_1 + c_1 t \qquad (4)$$

and

$$x = x_2 + a_2 s, \qquad y = y_2 + b_2 s, \qquad z = z_2 + c_2 s, \qquad (5)$$

respectively, we can see at a glance whether or not $L_1$ and $L_2$ are parallel. Since $L_1$ is parallel to $\mathbf{v}_1 = \langle a_1, b_1, c_1 \rangle$ and $L_2$ is parallel to $\mathbf{v}_2 = \langle a_2, b_2, c_2 \rangle$, it follows that the lines $L_1$ and $L_2$ are parallel if and only if the vectors $\mathbf{v}_1$ and $\mathbf{v}_2$ are scalar multiples of each other. If the two lines are not parallel, we can attempt to find a point of intersection by solving the equations

$$x_1 + a_1 t = x_2 + a_2 s \quad \text{and} \quad y_1 + b_1 t = y_2 + b_2 s$$

simultaneously for $s$ and $t$. If these values satisfy the equation $z_1 + c_1 t = z_2 + c_2 s$, then we have found a point of intersection (whose coordinates are obtained by substitution of the resulting value of $t$ in Equations (4) or that of $s$ in Equations (5)). Otherwise, the two lines $L_1$ and $L_2$ do not intersect; two nonparallel and nonintersecting lines in space are called **skew** lines.

**EXAMPLE 2** The line $L_1$ with parametric equations

$$x = 1 + 2t, \qquad y = 2 - 3t, \qquad z = 2 + t$$

passes through the point $P_1(1, 2, 2)$ and is parallel to the vector $\mathbf{v}_1 = \langle 2, -3, 1 \rangle$. The line $L_2$ with parametric equations

$$x = 3 + 4t, \qquad y = 1 - 6t, \qquad z = 5 + 2t$$

passes through the point $P_2(3, 1, 5)$ and is parallel to the vector $\mathbf{v}_2 = \langle 4, -6, 2 \rangle$. Because $\mathbf{v}_2 = 2\mathbf{v}_1$, we see that $L_1$ and $L_2$ are parallel.

But are $L_1$ and $L_2$ actually different lines, or are we perhaps dealing with two different parametrizations of the same line? To answer this question, we note that $\overrightarrow{P_1 P_2} = \langle 2, -1, 3 \rangle$ is not parallel to $\mathbf{v}_1 = \langle 2, -3, 1 \rangle$. Thus the point $P_2$ does not lie on the line $L_1$, and hence the lines $L_1$ and $L_2$ are indeed distinct.

---

If the coefficients $a$, $b$, and $c$ in Equations (3) are all nonzero, then we can eliminate the parameter by equating the three expressions obtained by solving for $t$. This gives

$$\frac{x - x_0}{a} = \frac{y - y_0}{b} = \frac{z - z_0}{c}. \qquad (6)$$

These are called the **symmetric equations** of the line $L$. If one of $a$ or $b$ or $c$ *is* zero, this means that $L$ lies in a plane parallel to one of the coordinate planes, in which case the line does not have symmetric equations. For example, if $c = 0$, then $L$ lies in the horizontal plane $z = z_0$. Of course, it is still possible to write equations for $L$ not involving the parameter; if $c = 0$ while

$a$ and $b$ are nonzero, we could describe the line $L$ as the simultaneous solution of the equations

$$\frac{x - x_0}{a} = \frac{y - y_0}{b}, \qquad z = z_0.$$

**EXAMPLE 3** Find both parametric equations and symmetric equations of the line $L$ through the points $P_0(3, 1, -2)$ and $P_1(4, -1, 1)$. Also find the points in which $L$ intersects the three coordinate planes.

***Solution*** The line $L$ is parallel to the vector $\mathbf{v} = \overrightarrow{P_0P_1} = \langle 1, -2, 3 \rangle$, so we take $a = 1$, $b = -2$, and $c = 3$. Equations (3) then give the parametric equations

$$x = 3 + t, \qquad y = 1 - 2t, \qquad z = -2 + 3t$$

of $L$, while Equations (6) give the symmetric equations

$$\frac{x - 3}{1} = \frac{y - 1}{-2} = \frac{z + 2}{3}.$$

To find the point at which $L$ intersects the $xy$-plane, we set $z = 0$ in the symmetric equations. This gives

$$\frac{x - 3}{1} = \frac{y - 1}{-2} = \frac{2}{3},$$

and so $x = \frac{11}{3}$, $y = -\frac{1}{3}$. Thus $L$ meets the $xy$-plane at the point $(\frac{11}{3}, -\frac{1}{3}, 0)$. Similarly, $x = 0$ gives $(0, 7, -11)$ for the point where $L$ meets the $yz$-plane, and $y = 0$ gives $(\frac{7}{2}, 0, -\frac{1}{2})$ for its intersection with the $xz$-plane.

---

A **plane** $\mathscr{P}$ in space is determined by a point $P_0(x_0, y_0, z_0)$ through which $\mathscr{P}$ passes and a line through $P_0$ that is normal to $\mathscr{P}$. Alternatively, we may be given $P_0$ on $\mathscr{P}$ and a normal vector $\mathbf{n} = \langle a, b, c \rangle$ to the plane $\mathscr{P}$. Then the point $P(x, y, z)$ lies on the plane $\mathscr{P}$ if and only if the vectors $\mathbf{n}$ and $\overrightarrow{P_0P}$ are perpendicular (see Fig. 14.21), in which case $\mathbf{n} \cdot \overrightarrow{P_0P} = 0$. We write $\overrightarrow{P_0P} = \mathbf{r} - \mathbf{r}_0$, where $\mathbf{r}$ and $\mathbf{r}_0$ are the position vectors $\mathbf{r} = \overrightarrow{OP}$ and $\mathbf{r}_0 = \overrightarrow{OP_0}$ of the points $P$ and $P_0$, respectively. Thus we obtain a **vector equation**

$$\mathbf{n} \cdot (\mathbf{r} - \mathbf{r}_0) = 0 \tag{7}$$

of the plane $\mathscr{P}$.

If we substitute $\mathbf{n} = \langle a, b, c \rangle$, $\mathbf{r} = \langle x, y, z \rangle$, and $\mathbf{r}_0 = \langle x_0, y_0, z_0 \rangle$ in Equation (7), we thereby obtain a **scalar equation**

$$a(x - x_0) + b(y - y_0) + c(z - z_0) = 0 \tag{8}$$

**of the plane through $P_0(x_0, y_0, z_0)$ with normal vector $\mathbf{n} = \langle a, b, c \rangle$.**

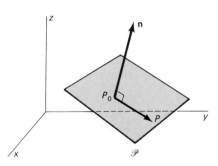

**14.21** Because $\mathbf{n}$ is normal to $\mathscr{P}$, it follows that $\mathbf{n}$ is normal to $\overrightarrow{P_0P}$ for all points $P$ in $\mathscr{P}$.

**EXAMPLE 4** An equation of the plane through $P_0(-1, 5, 2)$ with normal vector $\mathbf{n} = \langle 1, -3, 2 \rangle$ is

$$(1)(x + 1) + (-3)(y - 5) + (2)(z - 2) = 0;$$

that is,

$$x - 3y + 2z = -12.$$

Note that the coefficients of $x$, $y$, and $z$ in the last equation are the components of the normal vector. This is always the case, for Equation (8) can be written in the form

$$ax + by + cz = d \qquad (9)$$

where $d = ax_0 + by_0 + cz_0$. Conversely, every *linear equation* in $x$, $y$, and $z$ of the form in (9) represents a plane in space provided that the coefficients $a$, $b$, and $c$ are not all zero. For if $c \neq 0$ (for instance), we can pick $x_0$ and $y_0$ arbitrarily and solve the equation $ax_0 + by_0 + cz_0 = d$ for $z_0$. With these values, Equation (9) takes the form

$$ax + by + cz = ax_0 + by_0 + cz_0$$

or

$$a(x - x_0) + b(y - y_0) + c(z - z_0) = 0,$$

so this equation represents the plane through $(x_0, y_0, z_0)$ with normal vector $\langle a, b, c \rangle$.

**EXAMPLE 5**  Find an equation for the plane through the three points $P(2, 4, -3)$, $Q(3, 7, -1)$, and $R(4, 3, 0)$.

*Solution*  We want to use Equation (8), so we first need a vector $\mathbf{n}$ that is normal to the plane in question. One easy way to obtain such a normal vector is through the cross product: Let

$$\mathbf{n} = \overrightarrow{PQ} \times \overrightarrow{PR} = \begin{vmatrix} \mathbf{i} & \mathbf{j} & \mathbf{k} \\ 1 & 3 & 2 \\ 2 & -1 & 3 \end{vmatrix} = 11\mathbf{i} + \mathbf{j} - 7\mathbf{k}.$$

Because $\overrightarrow{PQ}$ and $\overrightarrow{PR}$ are in the plane, their cross product $\mathbf{n}$ is normal to the plane, as indicated in Fig. 14.22. Hence our plane has equation

$$11(x - 2) + (y - 4) - 7(z + 3) = 0.$$

After simplifications, we write it as

$$11x + y - 7z = 47.$$

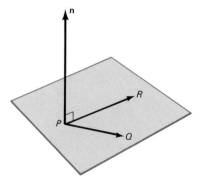

**14.22**  The normal vector $\mathbf{n}$ as a cross product

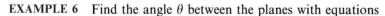

Two planes with normal vectors $\mathbf{n}$ and $\mathbf{m}$ are called **parallel** provided that $\mathbf{n}$ and $\mathbf{m}$ are parallel. Otherwise the two planes meet in a straight line. We define the angle between the two planes to be the angle between their normal vectors $\mathbf{n}$ and $\mathbf{m}$, as in Fig. 14.23.

**14.23**  Vectors $\mathbf{m}$ and $\mathbf{n}$ normal to the planes $\mathscr{P}$ and $\mathscr{Q}$, respectively

**EXAMPLE 6**  Find the angle $\theta$ between the planes with equations

$$2x + 3y - z = -3 \quad \text{and} \quad 4x + 5y + z = 1.$$

Then write symmetric equations of their line of intersection $L$.

*Solution*  The vectors $\mathbf{n} = \langle 2, 3, -1 \rangle$ and $\mathbf{m} = \langle 4, 5, 1 \rangle$ are normal to the two planes, so

$$\cos \theta = \frac{\mathbf{n} \cdot \mathbf{m}}{|\mathbf{n}| \, |\mathbf{m}|} = \frac{22}{\sqrt{14}\sqrt{42}}.$$

Hence $\theta = \cos^{-1}(11\sqrt{3}/21) \approx 24.87°$.

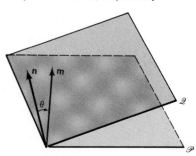

To determine the line $L$ of intersection of the two planes, we need first to find a point $P_0$ lying on $L$. We can do this by substituting an arbitrarily chosen value of $x$ into the equations of the given planes and then solving the resulting two equations for $y$ and $z$. With $x = 1$ we get the equations

$$2 + 3y - z = -3,$$

$$4 + 5y + z = 1.$$

The common solution is $y = -1$, $z = 2$; thus the point $P_0(1, -1, 2)$ lies on the line $L$.

Next we need a vector $\mathbf{v}$ parallel to $L$. The normal vectors $\mathbf{n}$ and $\mathbf{m}$ to the two planes are both perpendicular to $L$, so their cross product will be parallel to $L$. Hence we may choose

$$\mathbf{v} = \mathbf{n} \times \mathbf{m} = \begin{vmatrix} \mathbf{i} & \mathbf{j} & \mathbf{k} \\ 2 & 3 & -1 \\ 4 & 5 & 1 \end{vmatrix} = 8\mathbf{i} - 6\mathbf{j} - 2\mathbf{k}.$$

From (6) we now find the symmetric equations

$$\frac{x - 1}{8} = \frac{y + 1}{-6} = \frac{z - 2}{-2}$$

of the line of intersection of the two given planes.

In conclusion, we note that the symmetric equations of a line $L$ exhibit the line as an intersection of planes. For we can rewrite Equations (6) in the form

$$b(x - x_0) - a(y - y_0) = 0,$$

$$c(x - x_0) - a(z - z_0) = 0, \quad \text{and} \quad (10)$$

$$c(y - y_0) - b(z - z_0) = 0.$$

These are the equations of three planes that intersect in the line $L$. The first has normal vector $\langle b, -a, 0 \rangle$, a vector parallel to the $xy$-plane. So the first plane is perpendicular to the $xy$-plane. The second plane is perpendicular to the $xz$-plane, and the third is perpendicular to the $yz$-plane.

Equations (10) are symmetric equations of the line through $P_0(x_0, y_0, z_0)$ parallel to $\mathbf{v} = \langle a, b, c \rangle$. They enjoy the advantage over Equations (6) of being meaningful whether or not the components $a$, $b$, and $c$ of $\mathbf{v}$ are all nonzero. They have a special form, though, if one of these components is zero. If (for instance) $a = 0$, then the first two equations in (10) take the form $x = x_0$. The line is then the intersection of the two planes $x = x_0$ and $c(y - y_0) = b(z - z_0)$.

## 14-3 PROBLEMS

In Problems 1–3, write parametric equations of the straight line through the point $P$ that is parallel to the vector $\mathbf{v}$.

1  $P(0, 0, 0)$,  $\mathbf{v} = \mathbf{i} + 2\mathbf{j} + 3\mathbf{k}$
2  $P(3, -4, 5)$,  $\mathbf{v} = -2\mathbf{i} + 7\mathbf{j} + 3\mathbf{k}$
3  $P(4, 13, -3)$,  $\mathbf{v} = 2\mathbf{i} - 3\mathbf{k}$

In Problems 4–6, write parametric equations of the straight line that passes through the points $P_1$ and $P_2$.

4  $P_1(0, 0, 0)$ and $P_2(-6, 3, 5)$
5  $P_1(3, 5, 7)$ and $P_2(6, -8, 10)$
6  $P_1(3, 5, 7)$ and $P_2(6, 5, 4)$

In Problems 7–10, write an equation of the plane with normal vector **n** that passes through the point $P$.

7  $P(0, 0, 0)$,  $\mathbf{n} = \mathbf{i} + 2\mathbf{j} + 3\mathbf{k}$
8  $P(3, -4, 5)$,  $\mathbf{n} = -2\mathbf{i} + 7\mathbf{j} + 3\mathbf{k}$
9  $P(5, 12, 13)$,  $\mathbf{n} = \mathbf{i} - \mathbf{k}$   10  $P(5, 12, 13)$,  $\mathbf{n} = \mathbf{j}$

In Problems 11–16, write both parametric and symmetric equations for the indicated straight line.

11  Through $P(2, 3, -4)$ and parallel to $\mathbf{v} = \langle 1, -1, -2 \rangle$
12  Through $P(2, 5, -7)$ and $Q(4, 3, 8)$
13  Through $P(1, 1, 1)$ and perpendicular to the $xy$-plane
14  Through the origin and perpendicular to the plane with equation $x + y + z = 1$
15  Through $P(2, -3, 4)$ and perpendicular to the plane with equation $2x - y + 3z = 4$
16  Through $P(2, -1, 5)$ and parallel to the line with parametric equations $x = 3t$, $y = 2 + t$, $z = 2 - t$

In Problems 17–24, write an equation of the indicated plane.

17  Through $P(5, 7, -6)$ parallel to the $xz$-plane
18  Through $P(1, 0, -1)$ with normal vector $\mathbf{n} = \langle 2, 2, -1 \rangle$
19  Through $P(10, 4, -3)$ with normal vector $\mathbf{n} = \langle 7, 11, 0 \rangle$
20  Through $P(1, -3, 2)$ with normal vector $\mathbf{n} = \overrightarrow{OP}$
21  Through the origin and parallel to the plane with equation $3x + 4y = z + 10$
22  Through $P(5, 1, 4)$ and parallel to the plane with equation $x + y - 2z = 0$
23  Through the origin and the points $P(1, 1, 1)$ and $Q(1, -1, 3)$
24  Through the points $A(1, 0, -1)$, $B(3, 3, 2)$, and $C(4, 5, -1)$

In Problems 25–27, find the angle between the given planes.

25  $x = 10$ and $x + y + z = 0$
26  $2x - y + z = 5$ and $x + y - z = 1$
27  $x - y - 2z = 1$ and $x - y - 2z = 5$
28  Find parametric equations of the line of intersection of the planes $2x + y + z = 4$ and $3x - y + z = 3$.
29  Write symmetric equations for the line through $P(3, 3, 1)$ that is parallel to the line of Problem 28.
30  Find an equation of the plane through $P(3, 3, 1)$ that is perpendicular to the planes $x + y = 2z$ and $2x + z = 10$.
31  Find an equation of the plane through $(1, 1, 1)$ that intersects the $xy$-plane in the same line as does the plane $3x + 2y - z = 6$.
32  Find an equation for the plane through the point

$P(1, 3, -2)$ and the line of intersection of the planes
$$x - y + z = 1 \quad \text{and} \quad x + y - z = 1.$$

33  Find an equation of the plane through the points $P(1, 0, -1)$ and $Q(2, 1, 0)$ that is also parallel to the line of intersection of the planes $x + y + z = 5$ and $3x - y = 4$.
34  Show that the lines $x - 1 = \frac{1}{2}(y + 1) = z - 2$ and $x - 2 = \frac{1}{3}(y - 2) = \frac{1}{2}(z - 4)$ intersect. Find an equation of the (only) plane containing them both.
35  Show that the line of intersection of the planes
$$x + 2y - z = 2 \quad \text{and} \quad 3x + 2y + 2z = 7$$
is parallel to the line $x = 1 + 6t$, $y = 3 - 5t$, $z = 2 - 4t$. Find an equation of the plane determined by these two lines.
36  Show that the perpendicular distance $D$ from the point $P_0(x_0, y_0, z_0)$ to the plane $ax + by + cz = d$ is
$$D = \frac{|ax_0 + by_0 + cz_0 - d|}{\sqrt{a^2 + b^2 + c^2}}.$$

(*Suggestion:* The line through $P_0$ perpendicular to the given plane has parametric equations $x = x_0 + at$, $y = y_0 + bt$, and $z = z_0 + ct$. Let $P_1(x_1, y_1, z_1)$ be the point of this line, corresponding to $t = t_1$, at which it intersects the given plane. Solve for $t_1$, and then compute $D = |\overrightarrow{P_0P_1}|$.)

In Problems 37 and 38, use the formula of Problem 36 to find the distance between the given point and the given plane.

37  The origin and the plane $x + y + z = 10$.
38  The point $P(5, 12, -13)$ and the plane with equation $3x + 4y + 5z = 12$.
39  Show that any two skew lines $L_1$ and $L_2$ lie in parallel planes.
40  Use the formula of Problem 36 to show that the perpendicular distance $D$ between the two parallel planes $ax + by + cz + d_1 = 0$ and $ax + by + cz + d_2 = 0$ is
$$D = \frac{|d_1 - d_2|}{\sqrt{a^2 + b^2 + c^2}}.$$

41  The line $L_1$ is described by the equations
$$x - 1 = 2y + 2; \qquad z = 4.$$
The line $L_2$ passes through the points $P(2, 1, -3)$ and $Q(0, 8, 4)$.
(a)  Show that $L_1$ and $L_2$ are skew lines.
(b)  Use the results of Problems 39 and 40 to find the perpendicular distance between $L_1$ and $L_2$.

---

## 14-4

## Curves and Motion in Space

In Sections 13-4 and 13-6, we used vector-valued functions to discuss curves and motion in the plane. Much of that discussion applies, with only minor changes, to curves and motion in three-dimensional space. The principal difference is that our vectors now have three components rather than two.

Think of a point that moves along a curve in space. Its position at time $t$ can be described by *parametric equations*

$$x = f(t), \qquad y = g(t), \qquad z = h(t)$$

that give its coordinates at time $t$. Alternatively, the location of the point can be given by its **position vector**

$$\mathbf{r} = f(t)\mathbf{i} + g(t)\mathbf{j} + h(t)\mathbf{k} = x\mathbf{i} + y\mathbf{j} + z\mathbf{k} \qquad (1)$$

shown in Fig. 14.24.

A three-dimensional vector-valued function like $\mathbf{r}(t)$ can be differentiated and integrated in a componentwise manner, just like a two-dimensional vector-valued function (see Theorem 1 in Section 13-4). Thus the **velocity vector** $\mathbf{v} = \mathbf{v}(t)$ of our moving point at time $t$ is given by

$$\mathbf{v}(t) = \mathbf{r}'(t) = f'(t)\mathbf{i} + g'(t)\mathbf{j} + h'(t)\mathbf{k},$$
$$\mathbf{v} = \frac{d\mathbf{r}}{dt} = \frac{dx}{dt}\mathbf{i} + \frac{dy}{dt}\mathbf{j} + \frac{dz}{dt}\mathbf{k}. \qquad (2)$$

Its **acceleration** vector $\mathbf{a} = \mathbf{a}(t)$ is given by

$$\mathbf{a}(t) = \mathbf{v}'(t) = f''(t)\mathbf{i} + g''(t)\mathbf{j} + h''(t)\mathbf{k},$$
$$\mathbf{a} = \frac{d\mathbf{v}}{dt} = \frac{d^2x}{dt^2}\mathbf{i} + \frac{d^2y}{dt^2}\mathbf{j} + \frac{d^2z}{dt^2}\mathbf{k}. \qquad (3)$$

The **speed** $v(t)$ and **scalar acceleration** $a(t)$ of the moving point are the lengths of its velocity and acceleration vectors, respectively:

$$v(t) = |\mathbf{v}(t)| = \sqrt{\left(\frac{dx}{dt}\right)^2 + \left(\frac{dy}{dt}\right)^2 + \left(\frac{dz}{dt}\right)^2} \qquad (4)$$

and

$$a(t) = |\mathbf{a}(t)| = \sqrt{\left(\frac{d^2x}{dt^2}\right)^2 + \left(\frac{d^2y}{dt^2}\right)^2 + \left(\frac{d^2z}{dt^2}\right)^2}. \qquad (5)$$

**EXAMPLE 1**   The parametric equations of a moving point are

$$x = a \cos \omega t, \qquad y = a \sin \omega t, \qquad z = bt,$$

where $a$, $b$, and $\omega$ are positive. Describe the path of the point in geometric terms. Compute its velocity, speed, and acceleration at time $t$.

*Solution*   Because

$$x^2 + y^2 = a^2\cos^2\omega t + a^2\sin^2\omega t = a^2,$$

the path of the moving point lies on the **cylinder** that stands above and below the circle $x^2 + y^2 = a^2$ in the $xy$-plane, extending infinitely far in both directions, as indicated in Fig. 14.25. By our discussion of uniform circular motion in Section 13-5, the given parametric equations for $x$ and $y$ tell us that the projection of the point into the $xy$-plane moves counterclockwise around the circle $x^2 + y^2 = a^2$ with angular speed $\omega$. Meanwhile, because $z = bt$, the point itself is rising with vertical speed $b$. Its path on the cylinder is a spiral called a **helix** (also shown in Fig. 14.25).

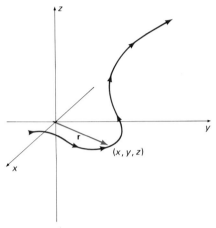

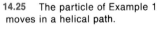

**14.24** The position vector $\mathbf{r} = \langle x, y, z \rangle$ of a moving particle in space

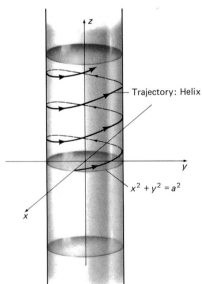

**14.25** The particle of Example 1 moves in a helical path.

The derivative of the position vector

$$\mathbf{r}(t) = \mathbf{i}a \cos \omega t + \mathbf{j}a \sin \omega t + \mathbf{k}bt$$

of the moving point is its velocity vector

$$\mathbf{v}(t) = -\mathbf{i}a\omega \sin \omega t + \mathbf{j}a\omega \cos \omega t + b\mathbf{k}. \tag{6}$$

Another differentiation gives its acceleration vector

$$\begin{aligned}
\mathbf{a}(t) &= -\mathbf{i}a\omega^2 \cos \omega t - \mathbf{j}a\omega^2 \sin \omega t \\
&= -a\omega^2(\mathbf{i} \cos \omega t + \mathbf{j} \sin \omega t). \tag{7}
\end{aligned}$$

The speed of the moving point is a constant, for

$$v(t) = |\mathbf{v}(t)| = \sqrt{a^2\omega^2 + b^2}.$$

Note that the acceleration vector is a horizontal vector of length $a\omega^2$. Moreover, if we think of $\mathbf{a}(t)$ as attached to the moving point at the time $t$ of evaluation—so that the initial point of $\mathbf{a}(t)$ is the terminal point of $\mathbf{r}(t)$—then $\mathbf{a}(t)$ points directly toward the point $(0, 0, bt)$ on the $z$-axis.

---

The helix of Example 1 is a typical trajectory of a charged particle in a constant magnetic field. Such a particle must satisfy both Newton's law $\mathbf{F} = m\mathbf{a}$ and the magnetic force law $\mathbf{F} = q\mathbf{v} \times \mathbf{B}$ mentioned in Section 14-2. Hence its velocity and acceleration vectors must satisfy the equation

$$q\mathbf{v} \times \mathbf{B} = m\mathbf{a}. \tag{8}$$

If the constant magnetic field is vertical, $\mathbf{B} = B\mathbf{k}$, then with the velocity vector of Equation (6), we find that

$$q\mathbf{v} \times \mathbf{B} = q \begin{vmatrix} \mathbf{i} & \mathbf{j} & \mathbf{k} \\ -a\omega \sin \omega t & a\omega \cos \omega t & b \\ 0 & 0 & B \end{vmatrix}$$

$$= qa\omega B(\mathbf{i} \cos \omega t + \mathbf{j} \sin \omega t).$$

The acceleration vector of Equation (7) gives

$$m\mathbf{a} = -ma\omega^2(\mathbf{i} \cos \omega t + \mathbf{j} \sin \omega t).$$

When we compare these results, we see that the helix of Example 1 satisfies Equation (8) provided that

$$qa\omega B = -ma\omega^2; \quad \text{that is,} \quad \omega = -\frac{qB}{m}.$$

For example, this equation would determine the angular speed $\omega$ for the helical trajectory of electrons ($q < 0$) in a cathode-ray tube placed in a constant magnetic field parallel to the axis of the tube (see Fig. 14.26).

Differentiation of three-dimensional vector-valued functions satisfies the same formal properties that we listed for two-dimensional vector-valued functions in Theorem 2 of Section 13-4:

$$\left.\begin{aligned}
&\text{(i)} \quad D_t[\mathbf{u}(t) + \mathbf{v}(t)] = \mathbf{u}'(t) + \mathbf{v}'(t), \\
&\text{(ii)} \quad D_t[h(t)\mathbf{u}(t)] = h'(t)\mathbf{u}(t) + h(t)\mathbf{u}'(t), \\
&\text{(iii)} \quad D_t[\mathbf{u}(t) \cdot \mathbf{v}(t)] = \mathbf{u}'(t) \cdot \mathbf{v}(t) + \mathbf{u}(t) \cdot \mathbf{v}'(t).
\end{aligned}\right\} \tag{9}$$

**14.26** A spiraling electron in a cathode-ray tube

Here, $\mathbf{u}(t)$ and $\mathbf{v}(t)$ are vector-valued functions with differentiable component functions, and $h(t)$ is a differentiable real-valued function. In addition, the expected product rule for the derivative of a cross product also holds:

$$\text{(iv)} \quad D_t[\mathbf{u}(t) \times \mathbf{v}(t)] = \mathbf{u}'(t) \times \mathbf{v}(t) + \mathbf{u}(t) \times \mathbf{v}'(t). \tag{10}$$

Note that the order of the factors in Equation (10) *must* be preserved because the cross product is not commutative. Each of the properties in (9) and (10) can be verified routinely by componentwise differentiation, as in Section 13-4.

**EXAMPLE 2** If $\mathbf{u}(t) = 3\mathbf{j} + 4t\mathbf{k}$ and $\mathbf{v}(t) = 5t\mathbf{i} - 4\mathbf{k}$, then

$$\mathbf{u}'(t) = 4\mathbf{k} \quad \text{and} \quad \mathbf{v}'(t) = 5\mathbf{i}.$$

Hence Equation (10) yields

$$\begin{aligned}
D_t[\mathbf{u}(t) \times \mathbf{v}(t)] &= 4\mathbf{k} \times (5t\mathbf{i} - 4\mathbf{k}) + (3\mathbf{j} + 4t\mathbf{k}) \times 5\mathbf{i} \\
&= 20t(\mathbf{k} \times \mathbf{i}) - 16(\mathbf{k} \times \mathbf{k}) + 15(\mathbf{j} \times \mathbf{i}) + 20t(\mathbf{k} \times \mathbf{i}) \\
&= 20t(\mathbf{j}) - 16(\mathbf{0}) + 15(-\mathbf{k}) + 20t(\mathbf{j}) = 40t\mathbf{j} - 15\mathbf{k},
\end{aligned}$$

without our having to calculate the cross product $\mathbf{u}(t) \times \mathbf{v}(t)$.

**EXAMPLE 3** A ball is thrown northward into the air from the origin in $xyz$-space (where the $xy$-plane represents the ground) with initial velocity vector

$$\mathbf{v}_0 = \mathbf{v}(0) = 80\mathbf{j} + 80\mathbf{k}. \tag{11}$$

In addition to gravitational acceleration, the spin of the ball causes an eastward acceleration of 2 ft/s², so the acceleration vector produced by the combination of gravity and the spin is

$$\mathbf{a}(t) = 2\mathbf{i} - 32\mathbf{k}. \tag{12}$$

First find the velocity vector $\mathbf{v}(t)$ of the ball, as well as its position vector $\mathbf{r}(t)$. Then determine where and with what speed it hits the ground. See Fig. 14.27.

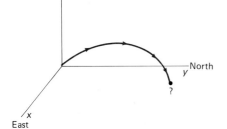

**14.27** The trajectory of the ball of Example 3

*Solution* When we antidifferentiate $\mathbf{a}(t)$ we get

$$\mathbf{v}(t) = 2t\mathbf{i} - 32t\mathbf{k} + \mathbf{C}_1.$$

We substitute $t = 0$ to find that $\mathbf{C}_1 = \mathbf{v}_0 = 80\mathbf{j} + 80\mathbf{k}$, so

$$\mathbf{v}(t) = 2t\mathbf{i} + 80\mathbf{j} + (80 - 32t)\mathbf{k}.$$

Another antidifferentiation yields

$$\mathbf{r}(t) = t^2\mathbf{i} + 80t\mathbf{j} + (80t - 16t^2)\mathbf{k} + \mathbf{C}_2,$$

and substitution of $t = 0$ gives $\mathbf{C}_2 = \mathbf{r}(0) = \mathbf{0}$. Hence the position vector of the ball is

$$\mathbf{r}(t) = t^2\mathbf{i} + 80t\mathbf{j} + (80t - 16t^2)\mathbf{k}.$$

The ball hits the ground when $z = 80t - 16t^2 = 0$, thus when $t = 5$. Its position vector then is

$$\mathbf{r}(5) = (5)^2\mathbf{i} + (80)(5)\mathbf{j} = 25\mathbf{i} + 400\mathbf{j},$$

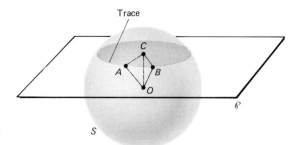

**14.33** The intersection of the sphere $S$ and the plane $\mathscr{P}$ is a circle.

The simplest example of a surface is a plane with linear equation $Ax + By + Cz + D = 0$. This section is devoted to examples of other simple surfaces that frequently appear in multivariable calculus.

In order to sketch a surface $S$, it is often helpful to examine its intersections with various planes. The **trace** of the surface $S$ in the plane $\mathscr{P}$ is the intersection of $\mathscr{P}$ and $S$. For example, if $S$ is a sphere, then the methods of elementary geometry may be used to verify that the trace of $S$ in the plane $\mathscr{P}$ is a circle (see Fig. 14.33), provided that $\mathscr{P}$ intersects the sphere but is not merely tangent to it (Problem 49). When we want to visualize a specified surface in space, it often suffices to examine its traces in the coordinate planes and possibly in a few planes parallel to them.

**EXAMPLE 1** Consider the plane with equation

$$3x + 2y + 2z = 6.$$

We find its trace in the $xy$-plane by setting $z = 0$; the equation then reduces to the equation $3x + 2y = 6$ of a line in the $xy$-plane. Similarly, when we set $y = 0$ we get the line $3x + 2z = 6$ as the trace of our plane in the $xz$-plane. To find its trace in the $yz$-plane we set $x = 0$; this yields the line $y + z = 3$. Figure 14.34 shows the portions of these three trace lines that lie in the first octant. Together they give us a good picture of how the plane $3x + 2y + 2z = 6$ is situated in space.

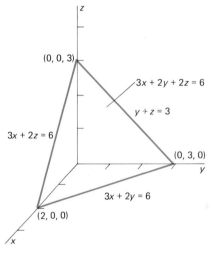

**14.34** Traces of the plane $3x + 2y + 2z = 6$ in the coordinate planes

**14.35** A right circular cylinder

Let $C$ be a curve in a plane, and let $L$ be a line not parallel to the plane. Then the set of all points on all lines parallel to $L$ that intersect $C$ is called a **cylinder.** This is a generalization of the familiar right circular cylinder, for which the curve $C$ is a circle and the line $L$ is perpendicular to the plane of the circle. Figure 14.35 shows this cylinder when $C$ is the circle $x^2 + y^2 = a^2$ in the $xy$-plane. The trace of this cylinder in any horizontal plane $z = c$ is a circle of radius $a$ and with center the point $(0, 0, c)$ on the $z$-axis. Thus the point $(x, y, c)$ lies on the cylinder if and only if $x^2 + y^2 = a^2$. Hence this cylinder is the graph of the equation $x^2 + y^2 = a^2$, considered as an equation in *three* variables.

The fact that the variable $z$ does not appear explicitly in the equation $x^2 + y^2 = a^2$ means that given any point $(x_0, y_0, 0)$ on the *circle* $x^2 + y^2 = a^2$ in the $xy$-plane, the point $(x_0, y_0, z)$ lies on the cylinder for any and all values of $z$. The set of all such points is the vertical line through the point $(x_0, y_0, 0)$. Thus the *cylinder* $x^2 + y^2 = a^2$ in space is the union of all vertical

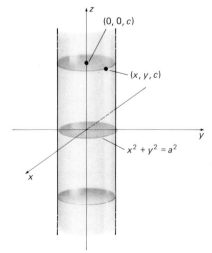

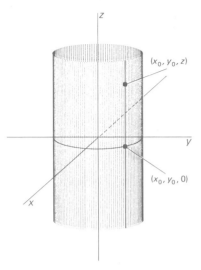

**14.36** The cylinder $x^2 + y^2 = a^2$; its rulings are parallel to the z-axis

lines through points of the *circle* $x^2 + y^2 = a^2$ in the plane, as illustrated in Fig. 14.36.

**EXAMPLE 2** Consider the sphere of radius 5 with equation

$$x^2 + y^2 + z^2 = 25.$$

If $0 < |z_0| < 5$, then the cylinder

$$x^2 + y^2 = 25 - (z_0)^2$$

intersects the sphere in two circles with radius $r_0 = \sqrt{25 - (z_0)^2}$ each; these circles are the traces of the sphere in the horizontal planes $z = +z_0$ and $z = -z_0$. Figure 14.37 shows the sphere presented as the union of such circles (plus the two isolated points $(0, 0, -5)$ and $(0, 0, 5)$). For instance, $z = 3$ gives the circle with equation $x^2 + y^2 = 16$ in the horizontal plane $z = 3$.

Given a general cylinder generated by a plane curve $C$ and a line $L$ as in the definition above, the lines on the cylinder parallel to the line $L$ are called **rulings** of the cylinder. Thus the rulings of the cylinder $x^2 + y^2 = a^2$ are vertical lines (parallel to the z-axis), as described above.

If the curve $C$ in the $xy$-plane has equation

$$f(x, y) = 0, \tag{2}$$

then the cylinder through $C$ with vertical rulings has the same equation. This is so because the point $P(x, y, z)$ lies on the cylinder if and only if the point $Q(x, y, 0)$ lies on the curve $C$. Similarly, the graph of an equation $g(x, z) = 0$ is a cylinder with rulings parallel to the $y$-axis, and the graph of an equation $h(y, z) = 0$ is a cylinder with rulings parallel to the $x$-axis. Thus the graph in space of an equation involving only two of the three coordinate variables is always a cylinder; its rulings are parallel to the axis corresponding to the *missing* variable.

**14.37** A sphere as a union of circles (and two points).

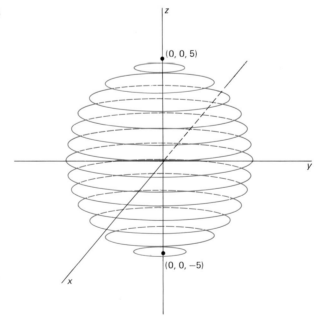

**EXAMPLE 3** The graph of the equation $4y^2 + 9z^2 = 36$ is the **elliptic cylinder** shown in Fig. 14.38. Its rulings are parallel to the $x$-axis, and its trace in every plane perpendicular to the $x$-axis is an ellipse with semiaxes of lengths 3 and 2 (just like the pictured ellipse $y^2/9 + z^2/4 = 1$ in the $yz$-plane).

**EXAMPLE 4** The graph of the equation $z = 4 - x^2$ is the **parabolic cylinder** shown in Fig. 14.39. Its rulings are parallel to the $y$-axis, and its trace in every plane perpendicular to the $y$-axis is a parabola that is a parallel translate of the parabola $z = 4 - x^2$ in the $xz$-plane.

Another way to use a plane curve $C$ to generate a surface is to revolve the curve (in space) around a line $L$ in its plane. This gives a **surface of revolution** with **axis** $L$. For example, Fig. 14.40 shows the surface generated by revolving the curve $f(x, y) = 0$ in the first quadrant of the $xy$-plane around

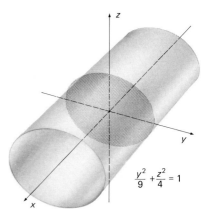

$$\frac{y^2}{9} + \frac{z^2}{4} = 1$$

**14.38** An elliptic cylinder

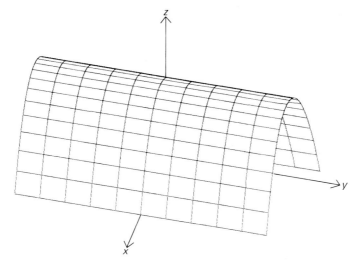

**14.39** The parabolic cylinder $z = 4 - x^2$

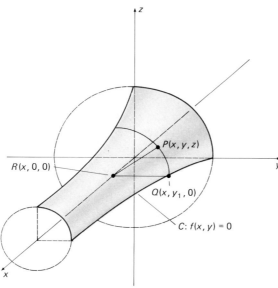

**14.40** The surface generated by rotating $C$ around the $x$-axis. (For clarity, only a quarter of the surface is shown.)

$R(x, 0, 0)$
$P(x, y, z)$
$Q(x, y_1, 0)$
$C: f(x, y) = 0$

the $x$-axis. The point $P(x, y, z)$ lies on the surface of revolution if and only if the point $Q(x, y_1, 0)$ lies on the curve, where

$$y_1 = |RQ| = |RP| = \sqrt{y^2 + z^2}.$$

Thus it is necessary that $f(x, y_1) = 0$, so the equation of the indicated surface of revolution around the $x$-axis is

$$f(x, \sqrt{y^2 + z^2}) = 0. \tag{3}$$

The equations of surfaces of revolution around other coordinate axes are obtained similarly. If the first quadrant curve $f(x, y) = 0$ above is revolved around the $y$-axis, we replace $x$ by $(x^2 + z^2)^{1/2}$ to get the equation $f((x^2 + z^2)^{1/2}, y) = 0$ of the resulting surface of revolution. If the curve $g(y, z) = 0$ in the first quadrant of the $yz$-plane is revolved around the $z$-axis, we replace $y$ by $(x^2 + y^2)^{1/2}$. Thus the equation of the resulting surface of revolution around the $z$-axis is $g((x^2 + y^2)^{1/2}, z) = 0$. These assertions are easily verified with the aid of diagrams similar to Fig. 14.40.

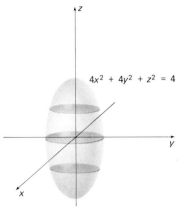

**EXAMPLE 5** Find an equation of the **ellipsoid of revolution** generated by revolving the ellipse $4y^2 + z^2 = 4$ around the $z$-axis (see Fig. 14.41).

**Solution** We replace $y$ by $(x^2 + y^2)^{1/2}$ in the given equation. This gives us $4x^2 + 4y^2 + z^2 = 4$ as an equation of the ellipsoid.

**EXAMPLE 6** Determine the graph of the equation $z^2 = x^2 + y^2$.

**Solution** First we rewrite the given equation in the form $z = \pm(x^2 + y^2)^{1/2}$. Thus the surface is symmetric about the $xy$-plane, and the upper half has equation $z = +(x^2 + y^2)^{1/2}$. This last equation is obtained from the simple equation $z = y$ by replacing $y$ by $(x^2 + y^2)^{1/2}$. Thus the upper half of the surface is obtained by revolving the line $z = y$ (for $y \geq 0$) around the $z$-axis. Thus we have the **cone** shown in Fig. 14.42. Its upper half has equation $z = +(x^2 + y^2)^{1/2}$, and its lower half has equation $z = -(x^2 + y^2)^{1/2}$. The whole cone $z^2 = x^2 + y^2$ is obtained by revolving the whole line $z = y$ around the $z$-axis.

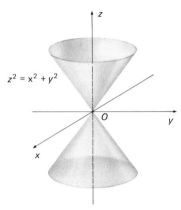

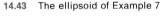

**14.41** The ellipsoid of revolution of Example 5

**14.42** The cone of Example 6

## QUADRIC SURFACES

Cones, spheres, circular and parabolic cylinders, and ellipsoids of revolution are all examples of surfaces with graphs that are second degree equations in $x$, $y$, and $z$. The graph of a second degree equation in three variables is called a **quadric surface.** We discuss here some important special cases of the equation

$$Ax^2 + By^2 + Cz^2 + Dx + Ey + Fz + G = 0. \tag{4}$$

This is a somewhat special second-degree equation in that it contains no terms involving the products $xy$, $xz$, or $yz$.

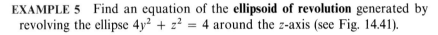

**14.43** The ellipsoid of Example 7

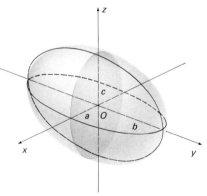

**EXAMPLE 7** The **ellipsoid**

$$\frac{x^2}{a^2} + \frac{y^2}{b^2} + \frac{z^2}{c^2} = 1 \tag{5}$$

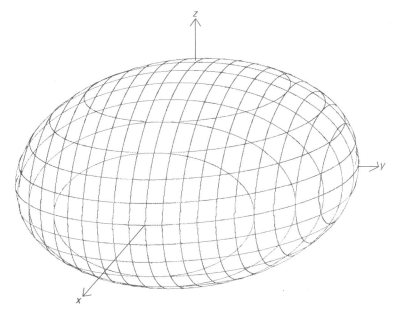

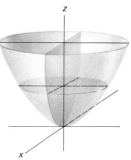

14.44 The traces of the ellipsoid $x^2/a^2 + y^2/b^2 + z^2/c^2 = 1$

is symmetric about each of the three coordinate planes and has intercepts $(\pm a, 0, 0), (0, \pm b, 0)$, and $(0, 0, \pm c)$ on the three coordinate axes. If $P(x, y, z)$ is a point of this ellipsoid, then $|x| \leq a, |y| \leq b$, and $|z| \leq c$. Each trace in a plane parallel to one of the three coordinate axes is either a single point or an ellipse. For example, if $-c < z_0 < c$, then Equation (5) can be reduced to

$$\frac{x^2}{a^2} + \frac{y^2}{b^2} = 1 - \frac{z_0^2}{c^2} > 0,$$

which is the equation of an ellipse with semiaxes $(a/c)(c^2 - z_0^2)^{1/2}$ and $(b/c)(c^2 - z_0^2)^{1/2}$. Figure 14.43 shows this ellipsoid with its semiaxes $a$, $b$, and $c$ labeled. Figure 14.44 shows its trace ellipses in planes parallel to the three coordinate planes.

14.45 An elliptic paraboloid

14.46 Trace parabolas of a circular paraboloid

**EXAMPLE 8** The **elliptic paraboloid**

$$\frac{x^2}{a^2} + \frac{y^2}{b^2} = \frac{z}{c} \tag{6}$$

is shown in Fig. 14.45. Its trace in the horizontal plane $z = z_0 > 0$ is the ellipse $x^2/a^2 + y^2/b^2 = z_0/c$ with semiaxes $a\sqrt{z_0/c}$ and $b\sqrt{z_0/c}$. Its trace in any vertical plane is a parabola. For instance, its trace in the plane $y = y_0$ has equation $x^2/a^2 + y_0^2/b^2 = z/c$, which can be rewritten in the form $z - z_1 = k(x - x_1)^2$ by taking $z_1 = cy_0^2/b^2$ and $x_1 = 0$. The paraboloid opens upward if $c > 0$ and downward if $c < 0$. If $a = b$ the paraboloid is circular. Figure 14.46 shows the traces of a circular paraboloid in planes parallel to the $xz$- and $yz$-planes.

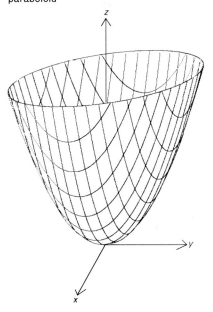

**EXAMPLE 9** The **elliptic cone**

$$\frac{x^2}{a^2} + \frac{y^2}{b^2} = \frac{z^2}{c^2} \tag{7}$$

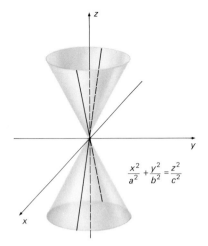

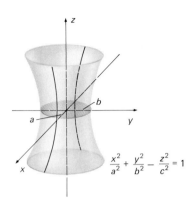

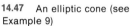

**14.47** An elliptic cone (see Example 9)

**14.48** A hyperboloid of one sheet (Example 10)

is shown in Fig. 14.47. Its trace in the horizontal plane $z = z_0 \neq 0$ is an ellipse with semiaxes $a|z_0|/c$ and $b|z_0|/c$.

**EXAMPLE 10** The **hyperboloid of one sheet** with equation

$$\frac{x^2}{a^2} + \frac{y^2}{b^2} - \frac{z^2}{c^2} = 1 \tag{8}$$

is shown in Fig. 14.48. Its trace in the horizontal plane $z = z_0$ is the ellipse $x^2/a^2 + y^2/b^2 = 1 + z_0^2/c^2 > 0$. Its trace in a vertical plane is a hyperbola except when the vertical plane intersects the $xy$-plane in a line tangent to the ellipse $x^2/a^2 + y^2/b^2 = 1$. In this special case, the trace is a degenerate hyperbola consisting of two intersecting lines. Figure 14.49 shows the traces (in planes parallel to the coordinate planes) of a circular ($a = b$) hyperboloid of one sheet.

**14.49** A circular hyperboloid of one sheet. Its traces in horizontal planes are circles; its traces in vertical planes are hyperbolas.

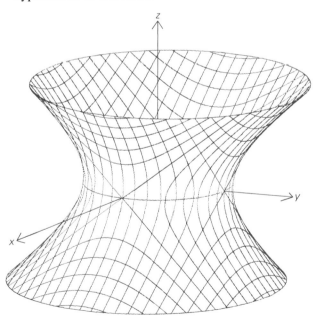

The graphs of the equations

$$\frac{y^2}{b^2} + \frac{z^2}{c^2} - \frac{x^2}{a^2} = 1 \quad \text{and} \quad \frac{x^2}{a^2} + \frac{z^2}{c^2} - \frac{y^2}{b^2} = 1$$

are also hyperboloids of one sheet, opening along the $x$- and $y$-axes, respectively.

**EXAMPLE 11**  The **hyperboloid of two sheets** with equation

$$\frac{z^2}{c^2} - \frac{x^2}{a^2} - \frac{y^2}{b^2} = 1 \tag{9}$$

consists of two connected pieces or sheets (Fig. 14.50). The two sheets open along the positive and negative $z$-axis and intersect it at the points $(0, 0, \pm c)$. The trace of this hyperboloid in a horizontal plane $z = z_0$ with $|z_0| > c$ is the ellipse

$$\frac{x^2}{a^2} + \frac{y^2}{b^2} = \frac{z_0^2}{c^2} - 1 > 0.$$

Its trace in any vertical plane is a nondegenerate hyperbola. Figure 14.51 shows traces of a circular hyperboloid of two sheets.

The graphs of the equations

$$\frac{x^2}{a^2} - \frac{y^2}{b^2} - \frac{z^2}{c^2} = 1 \quad \text{and} \quad \frac{y^2}{b^2} - \frac{x^2}{a^2} - \frac{z^2}{c^2} = 1$$

are also hyperboloids of two sheets, opening along the $x$- and $y$-axes, respectively. Note that when the equation of a hyperboloid is written in standard form with $+1$ on the right-hand side (as in Equations (8) and (9)), the number of sheets is equal to the number of negative terms on the left-hand side.

**EXAMPLE 12**  The **hyperbolic paraboloid**

$$\frac{y^2}{b^2} - \frac{x^2}{a^2} = \frac{z}{c} \qquad (c > 0) \tag{10}$$

is saddle-shaped, as indicated in Fig. 14.52. Its trace in the horizontal plane $z = z_0$ is a hyperbola (or two intersecting lines if $z_0 = 0$). Its trace in a vertical plane parallel to the $xz$-plane is a parabola that opens downward, while its trace in a vertical plane parallel to the $yz$-plane is a parabola that

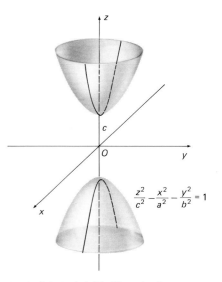

**14.50**  A hyperboloid of two sheets (see Example 11)

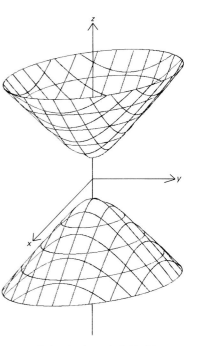

**14.51**  A circular hyperboloid of two sheets. Its (nondegenerate) traces in horizontal planes are circles; its traces in vertical planes are hyperbolas.

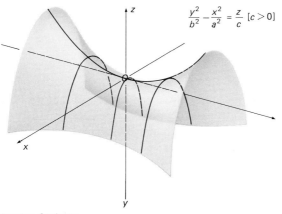

**14.52**  The hyperbolic paraboloid is a saddle-shaped surface.

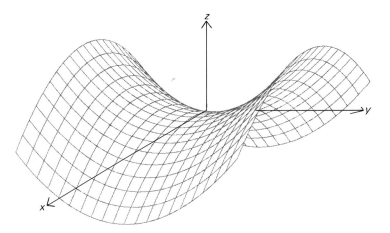

**14.53** The vertical traces of the hyperbolic paraboloid $z = y^2 - x^2$

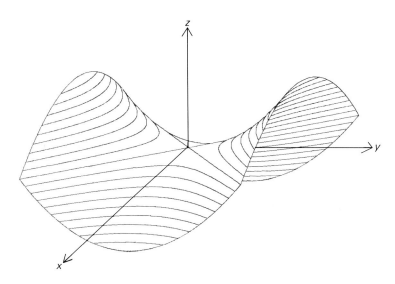

**14.54** The horizontal traces of the hyperbolic paraboloid $z = y^2 - x^2$

opens upward. In particular, the trace of the hyperbolic paraboloid in the $xz$-plane is a parabola opening downward from the origin, while its trace in the $yz$-plane is a parabola opening upward from the origin. Thus the origin looks like a local maximum from one direction but like a local minimum from another direction. Such a point on a surface is called a **saddle point.**

Figure 14.53 shows the parabolic traces in vertical planes of the hyperbolic paraboloid $z = y^2 - x^2$. Figure 14.54 shows its hyperbolic traces in horizontal planes.

### *CONIC SECTIONS AS SECTIONS OF A CONE

The parabola, ellipse, and hyperbola that we studied in Chapter 10 were originally introduced by the ancient Greek mathematicians as plane sections (or traces) of a right circular cone. Here we show that the intersection of a plane and a cone is, indeed, one of the three conic sections as defined in Chapter 10.

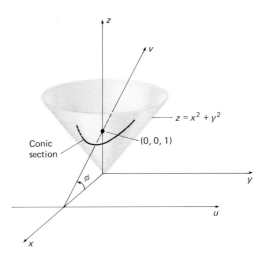

**14.55** Finding an equation for a conic section

Figure 14.55 shows the cone with equation $z = (x^2 + y^2)^{1/2}$ and its intersection with a plane $\mathcal{P}$ that passes through the point $(0, 0, 1)$ and the line $x = c > 0$ in the $xy$-plane. The equation of $\mathcal{P}$ is

$$z = 1 - \frac{x}{c}. \tag{11}$$

The angle between $\mathcal{P}$ and the $xy$-plane is $\phi = \tan^{-1}(1/c)$. We want to show that the conic section obtained by intersecting the cone and the plane is:

> A parabola if $\phi = 45°$    $(c = 1)$,
>
> An ellipse if $\phi < 45°$    $(c > 1)$,    and
>
> A hyperbola if $\phi > 45°$    $(c < 1)$.

We begin by introducing $uv$-coordinates in the plane $\mathcal{P}$ as follows. The $u$-coordinate of the point $(x, y, z)$ of $\mathcal{P}$ is $u = y$. The $v$-coordinate of the same point is its perpendicular distance from the line $x = c$. This explains the $u$- and $v$-axes indicated in Fig. 14.55. Figure 14.56 shows the cross section in the plane $y = 0$ exhibiting the relation between $v$, $x$, and $z$. We see that

$$z = v \sin \phi = \frac{v}{\sqrt{1 + c^2}}. \tag{12}$$

Equations (11) and (12) give

$$x = c(1 - z) = c\left(1 - \frac{v}{\sqrt{1 + c^2}}\right). \tag{13}$$

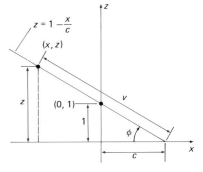

**14.56** Computing coordinates in the $uv$-plane

We had $z^2 = x^2 + y^2$ for the equation of the cone. We make the following substitutions in this equation: Replace $y$ by $u$, and replace $z$ and $x$ by the expressions on the right-hand sides of Equations (12) and (13), respectively. This yields

$$\frac{v^2}{1 + c^2} = c^2\left(1 - \frac{v}{\sqrt{1 + c^2}}\right)^2 + u^2.$$

After simplification, this last equation takes the form

$$u^2 + \frac{c^2 - 1}{c^2 + 1} v^2 - \frac{2c^2}{\sqrt{1 + c^2}} v + c^2 = 0. \tag{14}$$

This is the equation of the curve of intersection in the $uv$-plane. We proceed to examine the three cases for the angle $\phi$.

First suppose that $\phi = 45°$. Then $c = 1$, so that Equation (14) contains a $u^2$-term, a $v$-term, and a constant term. So the curve is a parabola; see Equation (6) of Section 10-4.

Next suppose that $\phi < 45°$. Then $c > 1$, and the coefficients of $u^2$ and $v^2$ in Equation (14) are both positive. So the curve is an ellipse; see Equation (6) of Section 10-5.

Finally, if $\phi > 45°$, then $c < 1$, and the coefficients of $u^2$ and $v^2$ in Equation (14) have different signs. So the curve is a hyperbola; see Equation (8) in Section 10-6.

## 14-6 PROBLEMS

Describe and sketch the graphs of the equations given in Problems 1–30.

1 $3x + 2y + 10z = 20$
2 $3x + 2y = 30$
3 $x^2 + y^2 = 9$
4 $y^2 = x^2 - 9$
5 $xy = 4$
6 $z = 4x^2 + 4y^2$
7 $z^2 = 4x^2 + y^2$
8 $4x^2 + 9y^2 = 36$
9 $z = 4 - x^2 - y^2$
10 $y^2 + z^2 = 1$
11 $2z = x^2 + y^2$
12 $x = 1 + y^2 + z^2$
13 $z^2 = 4(x^2 + y^2)$
14 $y^2 = 4x$
15 $x^2 = 4z + 8$
16 $x = 9 - z^2$
17 $4x^2 + y^2 = 4$
18 $x^2 + z^2 = 4$
19 $x^2 = 4y^2 + 9z^2$
20 $x^2 - 4y^2 = z$
21 $x^2 + y^2 + 4z = 0$
22 $x = \sin y$
23 $x = 2y^2 - z^2$
24 $x^2 + 4y^2 + 2z^2 = 4$
25 $x^2 + y^2 - 9z^2 = 9$
26 $x^2 - y^2 - 9z^2 = 9$
27 $y = 4x^2 + 9z^2$
28 $y^2 + 4x^2 - 9z^2 = 36$
29 $y^2 - 9x^2 - 4z^2 = 36$
30 $x^2 + 9y^2 + 4z^2 = 36$

Each of Problems 31–40 gives the equation of a curve in one of the coordinate planes. Write an equation for the surface generated by revolving this curve about the indicated axis. Then sketch the surface.

31 $x = 2z^2$;   the $x$-axis
32 $4x^2 + 9y^2 = 36$;   the $y$-axis
33 $y^2 - z^2 = 1$;   the $z$-axis
34 $z = 4 - x^2$;   the $z$-axis
35 $y^2 = 4x$;   the $x$-axis
36 $yz = 1$;   the $z$-axis
37 $z = e^{-x^2}$;   the $z$-axis
38 $(y - z)^2 + z^2 = 1$;   the $z$-axis
39 The line $z = 2x$;   the $z$-axis
40 The line $z = 2x$;   the $x$-axis

In each of Problems 41–47, describe the traces of the given surface in planes of the indicated type.

41 $x^2 + 4y^2 = 4$;   in horizontal planes (those parallel to the $xy$-plane)
42 $x^2 + 4y^2 + 4z^2 = 4$;   in horizontal planes
43 $x^2 + 4y^2 + 4z^2 = 4$;   in planes parallel to the $yz$-plane
44 $z = 4x^2 + 9y^2$;   in horizontal planes
45 $z = 4x^2 + 9y^2$;   in planes parallel to the $xz$-plane
46 $z = xy$;   in horizontal planes
47 $z = xy$;   in vertical planes through the $z$-axis
48 Identify the surface $z = xy$ by making a suitable rotation of axes in the $xy$-plane (as in Section 10-7).
49 Show that the triangles $OAC$ and $OBC$ in Fig. 14.33 are congruent, and thereby conclude that the trace of a sphere in an intersecting plane is a circle.
50 Show that the projection into the $yz$-plane of the curve of intersection of the surfaces $x = 1 - y^2$ and $x = y^2 + z^2$ is an ellipse.
51 Show that the projection into the $xy$-plane of the intersection of the plane $z = 2y$ and the paraboloid $z = x^2 + y^2$ is a circle.
52 Show that the projection into the $zx$-plane of the intersection of the paraboloids $y = 2x^2 + 3z^2$ and $y = 5 - 3x^2 - 2z^2$ is a circle.
53 Show that the projection into the $xy$-plane of the intersection of the plane $x + y + z = 1$ and the ellipsoid $x^2 + 4y^2 + 4z^2 = 4$ is an ellipse.
54 Show that the curve of intersection of the plane $z = ky$ and the cylinder $x^2 + y^2 = 1$ is an ellipse. (*Suggestion:* Introduce $uv$-coordinates in the plane $z = ky$ as follows: Let the $u$-axis be the original $x$-axis, and let the $v$-axis be the line $z = ky$, $x = 0$.)

Rectangular coordinates provide only one of several useful ways of describing points, curves, and surfaces in space. In this section we discuss two additional coordinate systems in three-dimensional space. Each is a generalization of polar coordinates in the plane.

Recall that the relationship between the rectangular coordinates $(x, y)$ and the polar coordinates $(r, \theta)$ of a point in the plane is given by the equations

$$x = r \cos \theta, \qquad y = r \sin \theta \tag{1}$$

and

$$r^2 = x^2 + y^2, \qquad \tan \theta = \frac{y}{x} \quad \text{if} \quad x \neq 0. \tag{2}$$

These relationships may be read directly from the triangle of Fig. 14.57.

The cylindrical coordinates $(r, \theta, z)$ of a point $P$ in space are a perfectly natural hybrid of its polar and rectangular coordinates. We use the polar coordinates $(r, \theta)$ of the point in the plane with rectangular coordinates $(x, y)$ and use the same $z$-coordinate as in rectangular coordinates. (The cylindrical coordinates of a point $P$ in space are illustrated in Fig. 14.58.) This means that the relation between the rectangular coordinates $(x, y, z)$ of the point $P$ and its cylindrical coordinates $(r, \theta, z)$ is obtained by simply adjoining the identity $z = z$ to Equations (1) and (2):

$$x = r \cos \theta, \qquad y = r \sin \theta, \qquad z = z \tag{3}$$

and

$$r^2 = x^2 + y^2, \qquad \theta = \tan^{-1}\left(\frac{y}{x}\right), \qquad z = z. \tag{4}$$

These equations may be used to convert from rectangular to cylindrical coordinates and vice versa. The given table lists the rectangular and corresponding cylindrical coordinates of some typical points in space.

## Cylindrical and Spherical Coordinates*

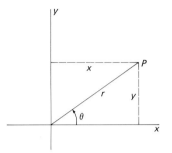

**14.57** The relation between rectangular and polar coordinates in the $xy$-plane.

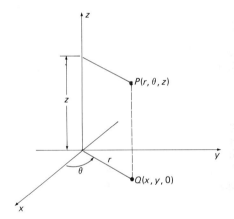

**14.58** Finding the cylindrical coordinates of the point $P$

| $(x, y, z)$ | $(r, \theta, z)$ | $(x, y, z)$ | $(r, \theta, z)$ |
|---|---|---|---|
| $(1, 0, 0)$ | $(1, 0, 0)$ | $(1, -1, 2)$ | $\left(\sqrt{2}, \frac{7\pi}{4}, 2\right)$ |
| $(-1, 0, 0)$ | $(1, \pi, 0)$ | $(-1, 1, 2)$ | $\left(\sqrt{2}, \frac{3\pi}{4}, 2\right)$ |
| $(0, 2, 3)$ | $\left(2, \frac{\pi}{2}, 3\right)$ | $(-1, -1, -2)$ | $\left(\sqrt{2}, \frac{5\pi}{4}, -2\right)$ |
| $(1, 1, 2)$ | $\left(\sqrt{2}, \frac{\pi}{4}, 2\right)$ | $(0, -3, -3)$ | $\left(3, \frac{3\pi}{2}, -3\right)$ |

The term **cylindrical coordinates** arises from the fact that the graph in space of the equation $r = c$ ($c$ a constant) is a cylinder of radius $c$ symmetric

---

* The material in this section will not be required until Section 16-7 and thus may be deferred until just before that section is covered.

about the $z$-axis. This suggests that cylindrical coordinates be used when working problems involving circular symmetry about the $z$-axis; cylindrical coordinates exhibit a special simplicity in such cases. For instance, the sphere $x^2 + y^2 + z^2 = a^2$ and the cone $z^2 = x^2 + y^2$ have cylindrical coordinate equations $r^2 + z^2 = a^2$ and $z^2 = r^2$, respectively. It follows from our discussion of surfaces of revolution in Section 14-6 that if the curve $f(y, z) = 0$ in the $yz$-plane is revolved around the $z$-axis, then the cylindrical coordinate equation of the surface generated is $f(r, z) = 0$.

**EXAMPLE 1**   If the parabola $z = y^2$ is revolved around the $z$-axis, then we get the cylindrical coordinate equation of the paraboloid thus generated quite simply: We replace $y$ with $r$. Thus the paraboloid has cylindrical equation

$$z = r^2.$$

Similarly, if the ellipse $y^2/9 + z^2/4 = 1$ is revolved around the $z$-axis, the cylindrical coordinate equation of the resulting ellipsoid is

$$\frac{r^2}{9} + \frac{z^2}{4} = 1.$$

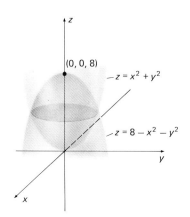

**14.59**  The two paraboloids of Example 2

**EXAMPLE 2**   Sketch the region that is bounded by the graphs of the cylindrical coordinate equations $z = r^2$ and $z = 8 - r^2$.

**Solution**   We first substitute $r^2 = x^2 + y^2$ from Equation (4) in the given equations. Thus the two surfaces of the example have rectangular coordinate equations $z = x^2 + y^2$ and $z = 8 - x^2 - y^2$. Their graphs are the two paraboloids shown in Fig. 14.59. The region in question is bounded above by the paraboloid $z = 8 - x^2 - y^2$ and below by the paraboloid $z = x^2 + y^2$.

---

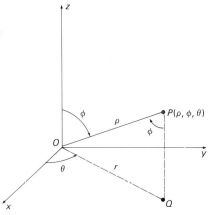

**14.60**  Finding the spherical coordinates of the point $P$

**14.61**   The two nappes of a 45° cone; $\phi = \pi/2$ is the $xy$-plane

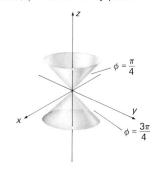

Figure 14.60 shows the **spherical coordinates** of the point $P$ in space. The first spherical coordinate $\rho$ is simply the distance $\rho = |OP|$ from the origin $O$ to $P$. The second spherical coordinate $\phi$ is the angle between $\overline{OP}$ and the positive $z$-axis; thus $0 \le \phi \le \pi$. Finally, $\theta$ is the familiar angle of cylindrical coordinates, the angular polar coordinate of the projection $Q$ of the point $P$ into the $xy$-plane. Both angles $\phi$ and $\theta$ are always measured in *radians*.

The term *spherical coordinates* is used because the graph of the equation $\rho = c$ (constant) is a sphere of radius $c$ centered at the origin. Note also that $\phi = c$ (constant) describes (one nappe of) a cone if $0 < c < \pi/2$ or $\pi/2 < c < \pi$. The spherical coordinate equation of the $xy$-plane is $\phi = \pi/2$. (See Fig. 14.61).

From the right triangle $OPQ$ of Fig. 14.60, we see that

$$r = \rho \sin \phi, \qquad z = \rho \cos \phi. \qquad (5)$$

Indeed, these equations are most easily remembered by visualizing this triangle. Substitution of Equations (5) into Equations (3) gives the equations

$$x = \rho \sin \phi \cos \theta,$$
$$y = \rho \sin \phi \sin \theta, \qquad (6)$$
$$z = \rho \cos \phi.$$

CHAP. 14:   Vectors, Curves, and Surfaces in Space

These three equations give the relationship between rectangular and spherical coordinates. Also useful is the formula

$$\rho^2 = x^2 + y^2 + z^2, \tag{7}$$

a consequence of the distance formula.

It is important to observe the order in which the spherical coordinates $(\rho, \phi, \theta)$ of a point $P$ are written—first the distance $\rho$ of $P$ from the origin, then the angle $\phi$ down from the positive $z$-axis, and last the counterclockwise angle $\theta$ around from the positive $x$-axis. The given table lists the rectangular coordinates and corresponding spherical coordinates of some typical points in space.

| $(x, y, z)$ | $(\rho, \phi, \theta)$ | $(x, y, z)$ | $(\rho, \phi, \theta)$ |
|---|---|---|---|
| $(1, 0, 0)$ | $\left(1, \dfrac{\pi}{2}, 0\right)$ | $(1, 1, \sqrt{2})$ | $\left(2, \dfrac{\pi}{4}, \dfrac{\pi}{4}\right)$ |
| $(0, 1, 0)$ | $\left(1, \dfrac{\pi}{2}, \dfrac{\pi}{2}\right)$ | $(-1, -1, \sqrt{2})$ | $\left(2, \dfrac{\pi}{4}, \dfrac{5\pi}{4}\right)$ |
| $(0, 0, 1)$ | $(1, 0, 0)$ | $(1, -1, -\sqrt{2})$ | $\left(2, \dfrac{3\pi}{4}, \dfrac{7\pi}{4}\right)$ |
| $(0, 0, -1)$ | $(1, \pi, 0)$ | $(1, 1, \sqrt{6})$ | $\left(2\sqrt{2}, \dfrac{\pi}{6}, \dfrac{\pi}{4}\right)$ |

Given the rectangular coordinates $(x, y, z)$ of the point $P$, one systematic method for finding the spherical coordinates of $P$ goes as follows. First we find its cylindrical coordinates $r$ and $\theta$ with the aid of the triangle in Fig. 14.62(a). Then we find $\rho$ and $\phi$ using the triangle in Fig. 14.62(b).

**EXAMPLE 3**  Find a spherical coordinate equation for the paraboloid $z = x^2 + y^2$.

***Solution***  We substitute $z = \rho \cos \phi$ from (5) and $x^2 + y^2 = r^2 = \rho^2 \sin^2\phi$ from (6). This gives $\rho \cos \phi = \rho^2 \sin^2\phi$. Cancellation of $\rho$ gives $\cos \phi = \rho \sin^2\phi$, or $\rho = \csc \phi \cot \phi$ as the spherical coordinate equation of the paraboloid. We get the whole paraboloid by using $\phi$ in the range $0 < \phi \leq \pi/2$. Note that $\phi = \pi/2$ gives the point $\rho = 0$ that might otherwise have been lost by canceling $\rho$.

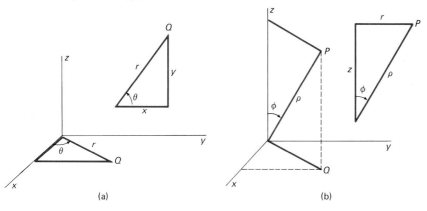

**14.62**  Triangles used in finding spherical coordinates.

(a)

(b)

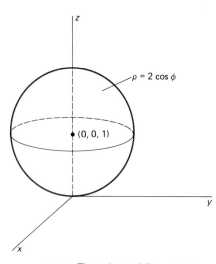

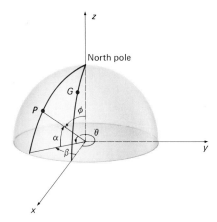

**EXAMPLE 4** Determine the graph of the spherical coordinate equation $\rho = 2 \cos \phi$.

**Solution** Multiplication by $\rho$ gives

$$\rho^2 = 2\rho \cos \phi;$$

then substitution of $\rho^2 = x^2 + y^2 + z^2$ and $z = \rho \cos \theta$ yields

$$x^2 + y^2 + z^2 = 2z$$

as the rectangular coordinate equation of the graph. Completion of the square in $z$ now gives

$$x^2 + y^2 + (z - 1)^2 = 1,$$

so the graph is a sphere with center $(0, 0, 1)$ and radius 1. It is tangent to the $xy$-plane at the origin (see Fig. 14.63).

$\rho = 2 \cos \phi$

$(0, 0, 1)$

**14.63** The sphere of Example 4

**EXAMPLE 5** Determine the graph of the spherical coordinate equation $\rho = \sin \phi \sin \theta$.

**Solution** We first multiply through by $\rho$ and get $\rho^2 = \rho \sin \phi \sin \theta$. We then use Equations (6) and (7) and find that $x^2 + y^2 + z^2 = y$. This is the rectangular coordinate equation of a sphere with center $(0, \frac{1}{2}, 0)$ and with radius $\frac{1}{2}$.

**\*LATITUDE AND LONGITUDE**

A **great circle** of a spherical surface is a circle formed by the intersection of the surface with a plane through the center of the sphere it bounds. Thus a great circle of a spherical surface is a circle in the surface of the same radius as the sphere and therefore is a circle of maximum possible circumference lying on the sphere. It is easy to see that any two points on a spherical surface lie on a great circle (uniquely determined unless the two points are antipodal). In the calculus of variations it is shown that the shortest distance between two such points—measured along the curved surface—is the shorter of the two arcs of the great circle containing them. The surprise is that the *shortest* distance is found by using the *largest* circle.

The spherical coordinates $\phi$ and $\theta$ are closely related to the latitude and longitude of points on the earth's surface. Assume that the earth is a sphere with radius $\rho = 3960$ mi. We begin with the **prime meridian** (a **meridian** is a north-south great semicircle) through Greenwich, just outside London. This is the point marked $G$ in Fig. 14.64.

**14.64** The relation between latitude, longitude, and spherical coordinates

We take the $z$-axis through the North Pole and the $x$-axis through the point where the prime meridian intersects the equator. The **latitude** $\alpha$ and (west) **longitude** $\beta$ of a point $P$ in the Northern Hemisphere are given by the equations

$$\alpha = 90° - \phi° \quad \text{and} \quad \beta = 360° - \theta°, \tag{8}$$

where $\phi°$ and $\theta°$ are the angular spherical coordinates, measured in *degrees*, of $P$. (That is, $\phi°$ and $\theta°$ denote the degree equivalents of the angles $\theta$ and $\phi$, respectively, which are always measured in radians unless otherwise specified.)

Thus the latitude $\alpha$ is measured northward from the equator, and the longitude $\beta$ is measured westward from the prime meridian.

**EXAMPLE 6** Find the great-circle distance between New York (latitude 40.75° north, longitude 74°) and London (latitude 51.5° north, longitude 0°). See Fig. 14.65.

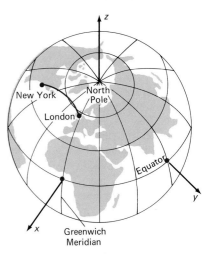

**14.65** The great circle arc between New York and London

***Solution*** From Equations (8) we find that $\phi° = 49.25°$, $\theta° = 286°$ for New York, while $\phi° = 38.5°$, $\theta° = 360°$ (or 0°) for London. Hence the spherical coordinates of New York are $\phi = (49.25/180)\pi$, $\theta = (286/180)\pi$, while those of London are $\phi = (38.5/180)\pi$, $\theta = 0$. With these values of $\phi$ and $\theta$ and with $\rho = 3960$ (mi), Equations (6) give the rectangular coordinates

New York: $P_1(826.90, -2883.74, 2584.93)$

and

London: $P_2(2465.16, 0.0, 3099.13)$.

The angle $\gamma$ between the radius vectors $\overrightarrow{OP_1}$ and $\overrightarrow{OP_2}$ in Fig. 14.66 is given by

$$\cos \gamma = \frac{\overrightarrow{OP_1} \cdot \overrightarrow{OP_2}}{|\overrightarrow{OP_1}||\overrightarrow{OP_2}|}$$

$$= \frac{(826.90)(2465.16) + (-2883.74)(0.0) + (2584.93)(3099.13)}{(3960)^2}$$

$$\approx 0.641.$$

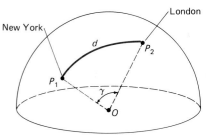

**14.66** Finding the great circle distance $d$ from New York to London

Thus $\gamma$ is approximately 0.875 rad. Hence the great circle distance between New York and London is close to $d = (3960)(0.875) \approx 3465$ mi.

## 14-7 PROBLEMS

In Problems 1–10 find both the cylindrical coordinates and the spherical coordinates of the point $P$ having the given rectangular coordinates.

1 $P(0, 0, 5)$   2 $P(0, 0, -3)$
3 $P(1, 1, 0)$   4 $P(2, -2, 0)$
5 $P(1, 1, 1)$   6 $P(-1, 1, -1)$
7 $P(2, 1, -2)$   8 $P(-2, -1, -2)$
9 $P(3, 4, 12)$   10 $P(-2, 4, -12)$

In each of Problems 11–24, describe the graph of the given equation.

11 $r = 5$   12 $\theta = \dfrac{3\pi}{4}$

13 $\theta = \dfrac{\pi}{4}$   14 $\rho = 5$

15 $\phi = \dfrac{\pi}{6}$   16 $\phi = \dfrac{5\pi}{6}$

17 $\phi = \dfrac{\pi}{2}$   18 $\phi = \pi$

19 $r = 2 \sin \theta$   20 $\rho = 2 \sin \phi$
21 $\cos \theta + \sin \theta = 0$   22 $z = 10 - 3r^2$
23 $\rho \cos \phi = 1$   24 $\rho = \cot \phi$

In each of Problems 25–30, convert the given equation both to cylindrical and to spherical coordinates.

25 $x^2 + y^2 + z^2 = 25$   26 $x^2 + y^2 = 2x$
27 $x + y + z = 1$   28 $x + y = 4$
29 $x^2 + y^2 + z^2 = x + y + z$
30 $z = x^2 - y^2$

31 The parabola $z = x^2$, $y = 0$ is rotated about the $z$-axis. Write a cylindrical coordinate equation for the surface generated.

32 The hyperbola $y^2 - z^2 = 1$, $x = 0$ is rotated about the $z$-axis. Write a cylindrical coordinate equation for the surface generated.

33 A sphere of radius 2 is centered at the origin. A hole of radius 1 is drilled through the sphere, with the axis of the hole lying on the $z$-axis. Describe the solid region that remains, in
(a) cylindrical coordinates;
(b) spherical coordinates.

**34** Find the great-circle distance from Atlanta (latitude 33.75°, longitude 84.40°) to San Francisco (latitude 37.78°, longitude 122.42°).

**35** Find the great-circle distance from Fairbanks (latitude 64.80°, longitude 147.85) to Leningrad (latitude 59.91°, longitude 30.43° *east* of Greenwich; in our earlier notation, we would say that its longitude is 329.57°).

**36** Because Fairbanks and Leningrad are at almost the same latitude, a plane could fly from one to the other roughly along the 62nd parallel of latitude. Accurately estimate the length of such a trip.

**37** In flying the great-circle route from Fairbanks to Leningrad, how close to the North Pole would a plane come?

**38** A right circular cone of radius $R$ and height $H$ is located with its vertex at the origin and its axis coincident with part of the nonnegative $z$-axis. Describe the solid cone in cylindrical coordinates.

**39** Describe the cone of Problem 38 in spherical coordinates.

## CHAPTER 14 REVIEW: Definitions, Concepts, Results

Use the list below as a guide to concepts that you may need to review.

**1** Properties of addition of vectors in space, and of multiplication of vectors by scalars

**2** The dot (scalar) product of vectors—definition and geometric interpretation

**3** Use of the dot product to test perpendicularity of vectors

**4** The cross (vector) product of vectors—definition and geometric interpretation

**5** The scalar triple product of vectors—definition and geometric interpretation

**6** The parametric and symmetric equations of the straight line through a given point and parallel to a given vector

**7** Equation of the plane through a given point normal to a given vector

**8** The velocity and acceleration vectors of a parametric space curve

**9** Arc length of a parametric space curve

**10** The curvature, unit tangent vector, and principal unit normal vector of a space curve

**11** Tangential and normal components of the acceleration vector of a parametric space curve

**12** Equations of cylinders and of surfaces of revolution

**13** The standard examples of quadric surfaces

**14** Definition of the cylindrical coordinate and spherical coordinate systems, and the equations relating cylindrical and spherical coordinates with rectangular coordinates

## MISCELLANEOUS PROBLEMS

**1** Suppose that $M$ is the midpoint of the line segment $PQ$ in space and that $A$ is another point. Show that $AM = \frac{1}{2}(\overrightarrow{AP} + AQ)$.

**2** Let $\mathbf{a}$ and $\mathbf{b}$ be nonzero vectors, and let

$$\mathbf{a}_{\parallel} = (\text{comp}_b\mathbf{a})\,\frac{\mathbf{b}}{|\mathbf{b}|} \quad \text{and} \quad \mathbf{a}_{\perp} = \mathbf{a} - \mathbf{a}_{\parallel}.$$

Show that $\mathbf{a}_{\perp}$ is perpendicular to $\mathbf{b}$.

**3** Let $P$ and $Q$ be different points in space. Show that the point $R$ lies on the line through $P$ and $Q$ *if and only if* there exist numbers $a$ and $b$ such that $a + b = 1$ and $\overrightarrow{OR} = a\overrightarrow{OP} + b\overrightarrow{OQ}$. Conclude that $\mathbf{r}(t) = t\overrightarrow{OP} + (1 - t)\overrightarrow{OQ}$ is a parametric equation of this line.

**4** Conclude from the result of Problem 3 that the points $P, Q,$ and $R$ are collinear if and only if there exist numbers $a, b,$ and $c,$ not all zero, such that $a + b + c = 0$ and $a\overrightarrow{OP} + b\overrightarrow{OQ} + c\overrightarrow{OR} = \mathbf{0}$.

**5** Let $P(x_0, y_0), Q(x_1, y_1),$ and $R(x_2, y_2)$ be points in the $xy$-plane. Use the cross product to show that the area of

the triangle $PQR$ is

$$A = \tfrac{1}{2}|(x_1 - x_0)(y_2 - y_0) - (x_2 - x_0)(y_1 - y_0)|.$$

**6** Write both symmetric and parametric equations for the line through $P(1, -1, 0)$ parallel to $\mathbf{v} = \langle 2, -1, 3 \rangle$.

**7** Write both symmetric and parametric equations of the line through $P_1(1, -1, 2)$ and $P_2(3, 2, -1)$.

**8** Write an equation of the plane through $P(3, -5, 1)$ with normal vector $\mathbf{n} = \mathbf{i} + \mathbf{j}$.

**9** Show that the lines with symmetric equations

$$x - 1 = 2(y + 1) = 3(z - 2)$$

and

$$x - 3 = 2(y - 1) = 3(z + 1)$$

are parallel. Then write an equation of the plane through these two lines.

**10** Let the lines $L_1$ and $L_2$ have symmetric equations

$$\frac{x - x_i}{a_i} = \frac{y - y_i}{b_i} = \frac{z - z_i}{c_i}$$

for $i = 1, 2$. Show that $L_1$ and $L_2$ are skew lines if and only if

$$\begin{vmatrix} x_1 - x_2 & y_1 - y_2 & z_1 - z_2 \\ a_1 & b_1 & c_1 \\ a_2 & b_2 & c_2 \end{vmatrix} \neq 0.$$

**11** Given the four points $A(2, 3, 2)$, $B(4, 1, 0)$, $C(-1, 2, 0)$, and $D(5, 4, -2)$, find an equation of the plane through $A$ and $B$ parallel to the line through $C$ and $D$.

**12** Given the points $A$, $B$, $C$ and $D$ of Problem 11, find points $P$ on the line $AB$ and $Q$ on the line $CD$ such that the line $PQ$ is perpendicular to both $AB$ and $CD$. What is the perpendicular distance $d$ between the lines $AB$ and $CD$?

**13** Let $P_0(x_0, y_0, z_0)$ be a point of the plane with equation

$$ax + by + cz + d = 0.$$

By projecting $\overrightarrow{OP_0}$ onto the normal vector $\mathbf{n} = \langle a, b, c \rangle$, show that the distance $D$ from the origin to this plane is

$$D = \frac{|d|}{\sqrt{a^2 + b^2 + c^2}}.$$

**14** Show that the distance $D$ from the point $P_1(x_1, y_1, z_1)$ to the plane $ax + by + cz + d = 0$ is equal to the distance from the origin to the plane with equation

$$a(x + x_1) + b(y + y_1) + c(z + z_1) + d = 0.$$

Hence conclude from the result of Problem 13 that

$$D = \frac{|ax_1 + by_1 + cz_1 + d|}{\sqrt{a^2 + b^2 + c^2}}.$$

**15** Find the perpendicular distance between the parallel planes $2x - y + 2z = 4$ and $2x - y + 2z = 13$.

**16** Write an equation of the plane through the point $(1, 1, 1)$ that is normal to the twisted cubic $x = t$, $y = t^2$, $z = t^3$ at this point.

**17** A particle moves in space with parametric equations $x = t$, $y = t^2$, $z = \frac{4}{3}t^{3/2}$. Find the curvature of its trajectory and the tangential and normal components of its acceleration when $t = 1$.

**18** The **osculating plane** to a space curve at a point $P$ of that curve is the plane through $P$ that is parallel to the curve's unit tangent and principal unit normal vectors at $P$. Write an equation of the osculating plane to the curve of Problem 17 at the point $(1, 1, \frac{4}{3})$.

**19** Show that the equation of the plane through the point $P_0(x_0, y_0, z_0)$ and parallel to the vectors $\mathbf{v}_1 = \langle a_1, b_1, c_1 \rangle$ and $\mathbf{v}_2 = \langle a_2, b_2, c_2 \rangle$ can be written in the form

$$\begin{vmatrix} x - x_0 & y - y_0 & z - z_0 \\ a_1 & b_1 & c_1 \\ a_2 & b_2 & c_2 \end{vmatrix} = 0.$$

**20** Deduce from Problem 19 that the equation of the osculating plane (Problem 18) to the parametric curve

$\mathbf{r}(t)$ at the point $\mathbf{r}(t_0)$ can be written in the form

$$(\mathbf{R} - \mathbf{r}(t_0)) \cdot \mathbf{r}'(t_0) \times \mathbf{r}''(t_0) = 0$$

where $\mathbf{R} = \langle x, y, z \rangle$. Note first that the vectors $\mathbf{T}$ and $\mathbf{N}$ are coplanar with $\mathbf{r}'(t)$ and $\mathbf{r}''(t)$.

**21** Use the result of Problem 20 to write an equation of the osculating plane to the twisted cubic $x = t$, $y = t^2$, $z = t^3$ at the point $(1, 1, 1)$.

**22** Let a parametric space curve be described by equations $r = r(t)$, $\theta = \theta(t)$, $z = z(t)$ giving the cylindrical coordinates of a moving point on the curve for $a \leq t \leq b$. Use the equations relating rectangular and cylindrical coordinates to show that the arc length of the curve is

$$s = \int_a^b \left[ \left( \frac{dr}{dt} \right)^2 + \left( r \frac{d\theta}{dt} \right)^2 + \left( \frac{dz}{dt} \right)^2 \right]^{1/2} dt.$$

**23** A point moves on the *unit* sphere $\rho = 1$ with its spherical angular coordinates at time $t$ given by $\phi = \phi(t)$, $\theta = \theta(t)$, $a \leq t \leq b$. Use the equations relating rectangular and spherical coordinates to show that the arc length of its path is

$$s = \int_a^b \left[ \left( \frac{d\phi}{dt} \right)^2 + (\sin^2 \phi) \left( \frac{d\theta}{dt} \right)^2 \right]^{1/2} dt.$$

**24** The cross product $\mathbf{B} = \mathbf{T} \times \mathbf{N}$ of the unit tangent vector and the principal unit normal vector is the *unit binormal vector* $\mathbf{B}$ of a curve.
(a) Differentiate $\mathbf{B} \cdot \mathbf{T} = 0$ to show that $\mathbf{T}$ is perpendicular to $d\mathbf{B}/ds$.
(b) Differentiate $\mathbf{B} \cdot \mathbf{B} = 1$ to show that $\mathbf{B}$ is perpendicular to $d\mathbf{B}/ds$.
(c) Conclude from (a) and (b) that $d\mathbf{B}/ds = -\tau\mathbf{N}$ for some number $\tau$, called the **torsion** of the curve. The torsion of a curve measures the amount it twists at each point in space

**25** Show that the torsion of the helix of Example 1 in Section 14-4 is constant by showing that its value is

$$\tau = \frac{b\omega}{a^2\omega^2 + b^2}.$$

**26** Deduce from the definition of torsion (Problem 24) that $\tau \equiv 0$ for any curve such that $\mathbf{r}(t)$ lies in a fixed plane.

**27** Write an equation in spherical coordinates of the sphere with radius 1 and center $x = 0 = y$, $z = 1$.

**28** Let $C$ be the circle in the $yz$-plane with radius 1 and center $y = 1$, $z = 0$. Write equations in both rectangular and cylindrical coordinates of the surface obtained by revolving $C$ around the $z$-axis.

**29** Let $C$ be the curve in the $yz$-plane with equation $(y^2 + z^2)^2 = 2(z^2 - y^2)$. Write an equation in spherical coordinates of the surface obtained by revolving this curve around the $z$-axis. Then sketch this surface. (*Suggestion:* Remember that $r^2 = 2 \cos 2\theta$ is the polar equation of a figure-eight curve.)

**30** Let $A$ be the area of a parallelogram $PQRS$ in space determined by the vectors $\mathbf{a} = \overrightarrow{PQ}$ and $\mathbf{b} = \overrightarrow{PS}$. Let $A'$ be the area of the perpendicular projection of *PQRS* into a plane that makes an acute angle $\gamma$ with the plane of *PQRS*. Assuming that $A' = A \cos \gamma$ in this situation, show that the areas of the perpendicular projections of the parallelogram *PQRS* into the three coordinate planes are

$$|\mathbf{i} \cdot \mathbf{a} \times \mathbf{b}|, \quad |\mathbf{j} \cdot \mathbf{a} \times \mathbf{b}|, \quad \text{and} \quad |\mathbf{k} \cdot \mathbf{a} \times \mathbf{b}|.$$

Conclude that the square of the area of a parallelogram in space is equal to the sum of the squares of its perpendicular projections into the three coordinate planes.

**31** Take $\mathbf{a} = \langle a_1, a_2, a_3 \rangle$ and $\mathbf{b} = \langle b_1, b_2, b_3 \rangle$ in Problem 30. Show that

$$A^2 = \begin{vmatrix} a_2 & a_3 \\ b_2 & b_3 \end{vmatrix}^2 + \begin{vmatrix} a_3 & a_1 \\ b_3 & b_1 \end{vmatrix}^2 + \begin{vmatrix} a_1 & a_2 \\ b_1 & b_2 \end{vmatrix}^2.$$

**32** Let *C* be a curve in a plane $\mathscr{P}$ that is not parallel to the *z*-axis. Suppose that the projection of *C* into the *xy*-plane is an ellipse. Introduce *uv*-coordinates in the plane $\mathscr{P}$ to prove that the curve *C* is itself an ellipse.

**33** Conclude from Problem 32 that the intersection of a nonvertical plane and an elliptic cylinder with vertical axis is an ellipse.

**34** Use the result of Problem 32 to show that the intersection of the plane $z = Ax + By$ and the paraboloid $z = a^2 x^2 + b^2 y^2$ is either empty, a single point, or an ellipse.

**35** Use the result of Problem 32 to show that the intersection of the plane $z = Ax + By$ and the ellipsoid $x^2/a^2 + y^2/b^2 + z^2/c^2 = 1$ is either empty, a single point, or an ellipse.

# 15

# Partial
# Differentiation

## Introduction

In this and the following two chapters, we turn our attention to the calculus of functions of more than one variable. Many real-world functions depend upon two or more variables. For example:

- In physical chemistry the ideal gas law $PV = nRT$ (where $n$ and $R$ are constants) is used to express any one of the variables $P$, $V$, and $T$ as a function of the other two.
- The altitude above sea level at a particular location on the earth's surface depends upon its latitude and longitude.
- A manufacturer's profit depends upon sales, overhead costs, the cost of each raw product that is used, and perhaps upon additional variables.
- The amount of useable energy a solar panel can gather depends upon its efficiency, its angle of inclination to the sun's rays, the angle of elevation of the sun above the horizon, and perhaps upon other factors.

A typical application may call for us to find an extreme value of a function of several variables. For example, suppose that we want to minimize the cost of making a rectangular box with a volume of 48 cubic feet, given that its front and back cost $1/ft^2$, its top and bottom cost $2/ft^2$, and its two ends cost $3/ft^2$. Figure 15.1 shows such a box with length $x$, width $y$, and height $z$. Under the conditions given, its total cost will be

$$C = 2xz + 4xy + 6yz \quad \text{(dollars)}.$$

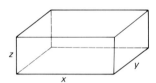

**15.1** Minimizing the total cost of a box

But $x$, $y$, and $z$ are not independent variables, because the box has fixed volume $V = xyz = 48$. We eliminate $z$ from the first formula by using the second formula; since $z = 48/xy$, we find that the cost we want to minimize is given by

$$C = 4xy + \frac{288}{x} + \frac{96}{y}.$$

Because neither of the variables $x$ and $y$ can be expressed in terms of the other, the single-variable maximum-minimum techniques of Chapter 3 cannot be applied here. We need new optimization techniques applicable to functions of two or more independent variables. In Section 15-5 we shall return to this problem.

The problem of optimization is just a typical example. In this chapter, we shall see that the main ingredients of single-variable differential calculus—limits, derivatives and rates of change, chain-rule computations, and maximum-minimum techniques—can all be generalized to functions of two or more variables.

## Functions of Several Variables

Recall that a real-valued *function* is a rule or correspondence $f$ that associates a unique real number with each element of a set $D$. The set $D$ is called the *domain* of definition of $f$. The domain $D$ has always been a subset of the real line for the functions of a single variable that we have studied up to this point. If $D$ is a subset of the plane, then $f$ is a function of *two* variables—for, given a point $P$ in $D$, we naturally associate with $P$ its rectangular co-ordinates $(x, y)$.

---

**Definition** *Functions of Two or Three Variables*

A **function of two variables,** defined on the **domain** $D$ in the plane, is a rule $f$ that associates with each point $(x, y)$ in $D$ a real number $f(x, y)$. A **function of three variables,** defined on the **domain** $D$ in space, is a rule that associates with each point $(x, y, z)$ in $D$ a real number $f(x, y, z)$.

---

A function $f$ of two (or three) variables is often defined by giving a formula that specifies $f(x, y)$ in terms of $x$ and $y$ (or $f(x, y, z)$ in terms of $x$, $y$, and $z$). In case the domain $D$ of $f$ is not explicitly specified, it is to be understood that $D$ consists of all points for which the given formula is meaningful. For example, the domain of the function $f$ with formula $f(x, y) = \sqrt{25 - x^2 - y^2}$ is the set of all $(x, y)$ such that $25 - x^2 - y^2 \geq 0$; that is, the circular disk $x^2 + y^2 \leq 25$ of radius 5 centered at the origin. Similarly, the function $g$ defined as

$$g(x, y, z) = \frac{x + y + z}{\sqrt{x^2 + y^2 + z^2}}$$

is defined at all points of space where $x^2 + y^2 + z^2 \neq 0$; that is, everywhere except at the origin $(0, 0, 0)$.

**EXAMPLE 1**  Find the domain of definition of the function

$$f(x, y) = \frac{y}{\sqrt{x - y^2}}. \tag{1}$$

Also find the points $(x, y)$ at which $f(x, y) = \pm 1$.

*Solution*  In order that $f(x, y)$ be defined, it is necessary that the radicand $x - y^2$ be positive; that is, that $y^2 < x$. Hence the domain of $f$ is the set of points lying strictly to the right of the parabola $x = y^2$—the domain is shown shaded in Fig. 15.2. The parabola itself is shown dotted because it is not included in the domain; any point for which $x = y^2$ would entail division by zero in (1).

The function $f(x, y)$ has value $\pm 1$ wherever

$$\frac{y}{\sqrt{x - y^2}} = \pm 1;$$

that is, $y^2 = x - y^2$, so that $x = 2y^2$. Thus $f(x, y) = \pm 1$ at each point of the parabola $x = 2y^2$ (other than its vertex $(0, 0)$, which is not included in the domain of $f$).

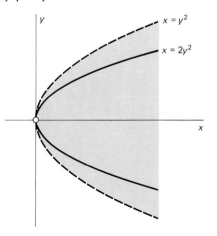

**15.2**  The domain of $f(x, y) = y/\sqrt{x - y^2}$

---

In a geometric, physical, or economic situation, a function typically re-sults from expressing one descriptive variable in terms of others. Thus in Section 15-1, we saw that the cost $C$ of the box discussed there was given by the formula

$$C = C(x, y) = 4xy + \frac{288}{x} + \frac{96}{y}$$

in terms of its length $x$ and width $y$. The value $C$ of this function is a variable that depends upon the values of $x$ and $y$. Hence we call $C$ a **dependent variable,** while $x$ and $y$ are **independent variables.** If the temperature $T$ at the point $(x, y, z)$ is given by some formula $T = h(x, y, z)$, then the dependent variable $T$ is a function of the independent variables $x$, $y$, and $z$.

A function of four or more variables can be defined by giving a formula involving the appropriate number of independent variables. For example, if an amount $A$ of heat is released at the origin at time $t = 0$ in a medium with diffusivity $k$, then—under appropriate conditions—it turns out that the temperature at the point $(x, y, z)$ at the time $t > 0$ is given by

$$T(x, y, z, t) = \frac{A}{(4\pi kt)^{3/2}} e^{-(x^2 + y^2 + z^2)/4kt}.$$

This formula gives the temperature $T$ as a function of the four independent variables $x$, $y$, $z$, and $t$.

We shall see that the main differences between single-variable and multivariable calculus show up with examples involving only two independent variables. Hence most of our results will be stated in terms of functions of two variables. Many of these results will, however, readily generalize by analogy to the case of three or more independent variables.

### GRAPHS AND LEVEL CURVES

A function $f$ of two variables $x$ and $y$ has the property that we can visualize how it "works" in terms of its graph. The **graph** of $f$ is the graph of the equation $z = f(x, y)$; that is, the set of all points in space with coordinates $(x, y, z)$ that satisfy the equation $z = f(x, y)$. (See Fig. 15.3.)

You saw several examples of such graphs in Chapter 14. For example, the graph of the function $f(x, y) = x^2 + y^2$ is the paraboloid $z = x^2 + y^2$

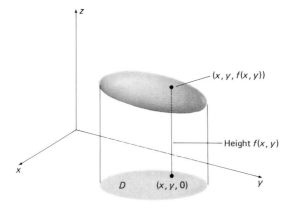

**15.3** The graph of a function of two variables is frequently a surface "over" the domain of the function.

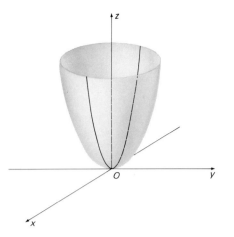

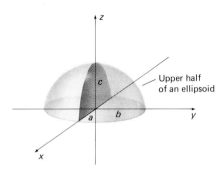

**15.4** The paraboloid is (part of) the graph of the function $f(x, y) = x^2 + y^2$.

**15.5** The upper half of an ellipsoid is the graph of a function of two variables.

shown in Fig. 15.4. The graph of the function

$$g(x, y) = c\sqrt{1 - \frac{x^2}{a^2} - \frac{y^2}{b^2}}$$

is the *upper half* of the ellipsoid $x^2/a^2 + y^2/b^2 + z^2/c^2 = 1$ and is shown in Fig. 15.5. Generally speaking, the graph of a function of two variables is a surface that lies above (or below, or both) its domain $D$ in the $xy$-plane.

The intersection of the horizontal plane $z = k$ with the surface $z = f(x, y)$ is called the **contour curve** of height $k$ on the surface, as in Fig. 15.6. The vertical projection of this contour curve into the $xy$-plane is the **level curve** $f(x, y) = k$ of the function $f$. The level curves of $f$ are simply the sets on which the value of $f$ is constant. On a topographic map, like the one in Fig. 15.7, the level curves are the curves of constant height above sea level.

**15.6** A contour curve and the corresponding level curve

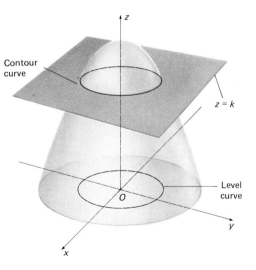

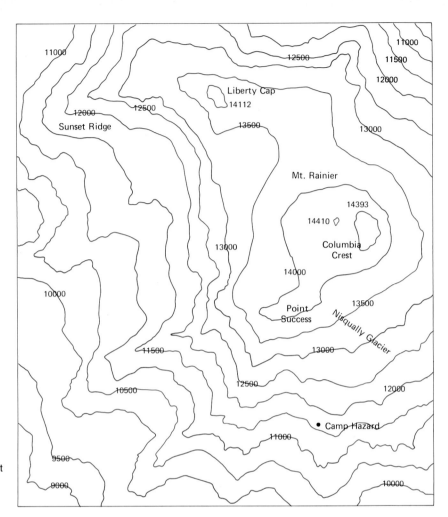

**15.7** Area around Mt. Rainier, Washington, showing level curves at 500-ft intervals (Adapted from U.S. Geological Survey Map N4645-W12145/7.5 (1971))

**15.8** Contour curves and level curves for a hill

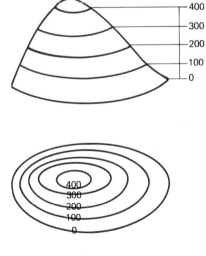

**15.9** Contour curves on $z = 25 - x^2 - y^2$

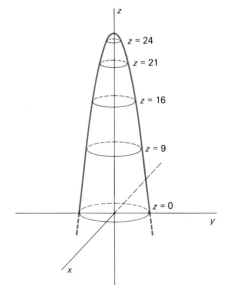

CHAP. 15:  Partial Differentiation

Level curves afford us a two-dimensional way of representing a three-dimensional surface $z = f(x, y)$, just as the two-dimensional map in Fig. 15.7 represents the three-dimensional mountain. We do this by drawing typical level curves of $z = f(x, y)$ in the $xy$-plane, labeling each with the corresponding (constant) value of $z$. Figure 15.8 illustrates this process for a simple hill.

**EXAMPLE 2**  Figure 15.9 shows some typical contour curves on the paraboloid $z = 25 - x^2 - y^2$. Figure 15.10 shows the corresponding level curves.

**EXAMPLE 3**  Figure 15.11 shows contour curves on the hyperbolic paraboloid $z = y^2 - x^2$. Figure 15.12 shows the corresponding level curves of the function $f(x, y) = y^2 - x^2$. If $z = k > 0$, then $y^2 - x^2 = k$ is a hyperbola opening along the $y$-axis, while if $k < 0$, it opens along the $x$-axis. The level curve with $k = 0$ consists of the two straight lines $y = x$ and $y = -x$.

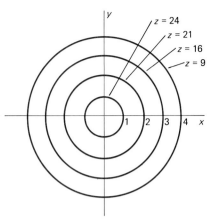

**15.10**  Level curves of $f(x, y) = 25 - x^2 - y^2$.

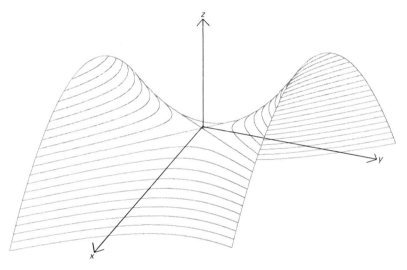

**15.11**  Contour curves on $z = y^2 - x^2$

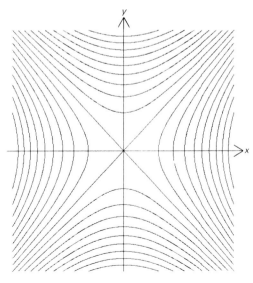

**15.12**  Level curves of $f(x, y) = y^2 - x^2$

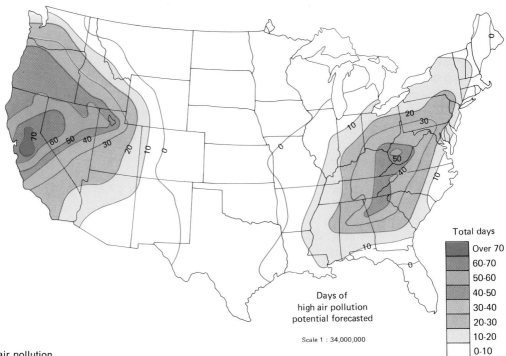

15.13 Days of high air pollution potential forecast in the United States (from National Atlas of the United States, U.S. Department of the Interior, 1970)

The graph of a function $f(x, y, z)$ of three variables cannot be drawn in three dimensions, but we can readily visualize its **level surfaces** of the form $f(x, y, z) = k$. For example, the level surfaces of the function $f(x, y, z) = x^2 + y^2 + z^2$ are spheres centered at the origin. Thus the level surfaces of $f$ are the sets in space on which the value $f(x, y, z)$ is constant.

If $f$ is a temperature function, then its level curves or surfaces are called **isotherms.** A weather map often includes level curves of the atmospheric pressure; these are called **isobars.** Even though you may be able to construct the graph of a function of two variables, that graph might be so complicated that information about the function (or the situation it describes) is obscure. Frequently the level curves themselves give more information—as in weather maps. For example, Fig. 15.13 shows level curves for the annual numbers of days of *high* air pollution forecast at different localities in the United States. The scale of this figure does not show local variations caused by individual cities. But a glance indicates that western Colorado, southern Georgia, and central Illinois all expect about the same number (10, in this case) of high-pollution days each year.

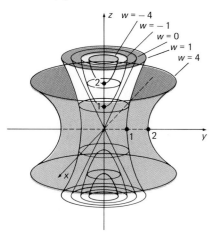

15.14 Some level surfaces of the function $w = f(x, y, z) = x^2 + y^2 - z^2$

**EXAMPLE 4** Figure 15.14 shows some level surfaces of the function

$$f(x, y, z) = x^2 + y^2 - z^2. \qquad (2)$$

If $k > 0$, then the graph of $x^2 + y^2 - z^2 = k$ is a hyperboloid of one sheet, while if $k < 0$, it is a hyperboloid of two sheets. The cone $x^2 + y^2 - z^2 = 0$ lies between these two types of hyperboloids.

**EXAMPLE 5**  The surface

$$z = \sin\sqrt{x^2 + y^2} \qquad (3)$$

is symmetrical with respect to the $z$-axis, because Equation (3) reduces to the equation $z = \sin r$ (see Fig. 15.15) in terms of the radial coordinate $r = \sqrt{x^2 + y^2}$ that measures perpendicular distance from the $z$-axis. The *surface* $z = \sin r$ is generated by revolving the curve $z = \sin x$ around the $z$-axis. Hence its level curves are circles centered at the origin in the $xy$-plane. For instance, $z = 0$ if $r$ is an integral multiple of $\pi$, while $z = \pm 1$ if $r$ is any odd multiple of $\pi/2$. Figure 15.16 shows traces of this surface in planes parallel to the $yz$-plane. The "hat effect" was achieved by plotting $(x, y, z)$ for those points $(x, y)$ lying within an appropriate ellipse in the $xy$-plane.

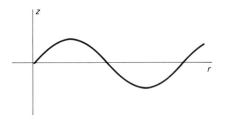

15.15  The curve $z = \sin r$

Given an arbitrary function $f(x, y)$, it can be quite a challenge to construct a picture of the surface $z = f(x, y)$. The following example illustrates some special techniques that may be useful. Additional surface-sketching techniques will appear in the remainder of the chapter.

**EXAMPLE 6**  Investigate the graph of the function

$$f(x, y) = \tfrac{3}{4}y^2 + \tfrac{1}{24}y^3 - \tfrac{1}{32}y^4 - x^2. \qquad (4)$$

**Solution**  The key feature in (4) is that the right-hand side is a *sum* of a function of $x$ and a function of $y$. If we set $x = 0$, we get the curve

$$z = \tfrac{3}{4}y^2 + \tfrac{1}{24}y^3 - \tfrac{1}{32}y^4 \qquad (5)$$

in which the surface $z = f(x, y)$ intersects the $yz$-plane. On the other hand, if we set $y = y_0$ in (4), we get

$$z = (\tfrac{3}{4}y_0^2 + \tfrac{1}{24}y_0^3 - \tfrac{1}{32}y_0^4) - x^2;$$

that is,

$$z = k - x^2, \qquad (6)$$

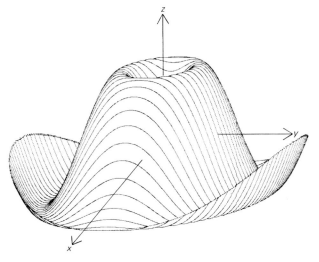

15.16  The hat surface
$z = \sin\sqrt{x^2 + y^2}$

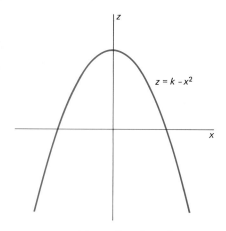

z = k - x²

**15.17** Intersection of $z = f(x, y)$ and the plane $y = y_0$

which is the equation of a parabola in the $xz$-plane. Hence the trace of $z = f(x, y)$ in each plane $y = y_0$ is a parabola of the form in (6); see Fig. 15.17.

We can use the techniques of Section 4-5 to sketch the curve in (5). Calculating the derivative of $z$ with respect to $y$, we get

$$\frac{dz}{dy} = \frac{3}{2}y + \frac{1}{8}y^2 - \frac{1}{8}y^3$$

$$= -\frac{1}{8}y(y^2 - y - 12)$$

$$= -\frac{1}{8}y(y + 3)(y - 4).$$

Hence the critical points are $y = -3$, $y = 0$, and $y = 4$. The corresponding values of $z$ are $f(0, -3) \approx 3.09$, $f(0, 0) = 0$, and $f(0, 4) \approx 6.67$. Because $z \to -\infty$ as $y \to \pm\infty$, it follows readily that the graph of (5) looks like the one shown in Fig. 15.18.

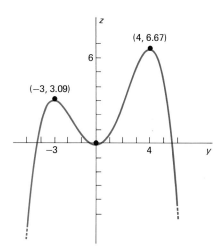

**15.18** The curve $z = \frac{3}{4}y^2 + \frac{1}{24}y^3 - \frac{1}{32}y^4$

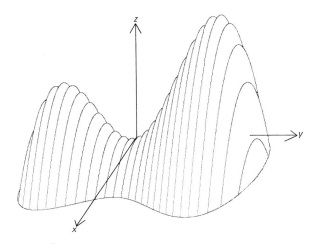

**15.19** Trace parabolas of $z = f(x, y)$

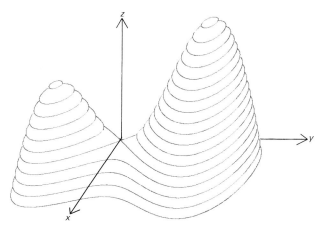

**15.20** Contour curves on $z = f(x, y)$

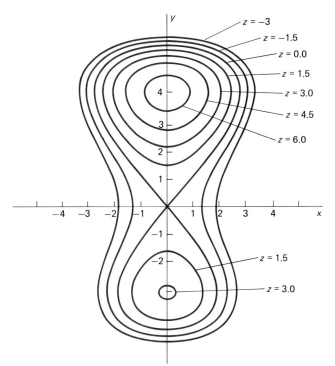

**15.21** Level curves of the function
$$f(x, y) = \tfrac{3}{4} y^2 + \tfrac{1}{24} y^3 - \tfrac{1}{32} y^4 - x^2$$

Now we can see what the surface $z = f(x, y)$ looks like. Each vertical plane $y = y_0$ intersects the curve in (5) in a single point, and this point is the vertex of a parabola that opens downward like the one in (6); this parabola is the intersection of the plane and the surface. Thus the surface $z = f(x, y)$ is generated by translating the vertex of such a parabola along the curve $z = \tfrac{3}{4} y^2 + \tfrac{1}{24} y^3 - \tfrac{1}{32} y^4$, as indicated in Fig. 15.19.

Figure 15.20 shows some typical contour curves on this surface. Note that it resembles two peaks separated by a mountain pass. To check this figure, we programmed a microcomputer to plot typical level curves of the function $f(x, y)$; the result is shown in Fig. 15.21. The nested level curves around the points $(0, -3)$ and $(0, 4)$ are indicative of the local maxima of $z = f(x, y)$, and the level figure-eight curve through $(0, 0)$ is indicative of the *saddle point* we see in Figs. 15.18 and 15.19. Local extrema and saddle points of functions of two variables are discussed in Sections 15-5 and 15-10.

## 15-2  PROBLEMS

In each of Problems 1–10, state the largest possible domain of definition of the given function $f$.

**1** $f(x, y) = e^{-x^2 - y^2}$

**2** $f(x, y) = \ln(x^2 - y^2 - 1)$

**3** $f(x, y) = \dfrac{x + y}{x - y}$

**4** $f(x, y) = \sqrt{4 - x^2 - y^2}$

**5** $f(x, y) = \dfrac{1 + \sin xy}{xy}$

**6** $f(x, y) = \dfrac{1 + \sin xy}{x^2 + y^2}$

**7** $f(x, y) = \dfrac{xy}{x^2 - y^2}$

**8** $f(x, y, z) = \dfrac{1}{\sqrt{z - x^2 - y^2}}$

**9** $f(x, y, z) = \exp\left(\dfrac{1}{x^2 + y^2 + z^2}\right)$

**10** $f(x, y, z) = \ln(xyz)$

In each of Problems 11–20, describe the graph of the function $f$.

**11** $f(x, y) = 10$        **12** $f(x, y) = x$

**13** $f(x, y) = x + y$       **14** $f(x, y) = \sqrt{x^2 + y^2}$

**15** $f(x, y) = x^2 + y^2$     **16** $f(x, y) = 4 - x^2 - y^2$

**17** $f(x, y) = \sqrt{4 - x^2 - y^2}$

**18** $f(x, y) = 16 - y^2$

**19** $f(x, y) = 10 - \sqrt{x^2 + y^2}$

**20** $f(x, y) = -\sqrt{36 - 4x^2 - 9y^2}$

In each of Problems 21–30, sketch some typical level curves of the function $f$.

**21** $f(x, y) = x - y$       **22** $f(x, y) = x^2 - y^2$

**23** $f(x, y) = x^2 + 4y^2$     **24** $f(x, y) = y - x^2$

**25** $f(x, y) = y - x^3$      **26** $f(x, y) = y - \cos x$

**27** $f(x, y) = x^2 + y^2 - 4x$

**28** $f(x, y) = x^2 + y^2 - 6x + 4y + 7$

**29** $f(x, y) = \exp(-x^2 - y^2)$

**30** $f(x, y) = \dfrac{1}{1 + x^2 + y^2}$

In each of Problems 31–36, describe the level surfaces of the function $f$.

**31** $f(x, y, z) = x^2 + y^2 - z$

**32** $f(x, y, z) = z + \sqrt{x^2 + y^2}$

**33** $f(x, y, z) = x^2 + y^2 + z^2 - 4x - 2y - 6z$

**34** $f(x, y, z) = z^2 - x^2 - y^2$

**35** $f(x, y, z) = x^2 + 4y^2 - 4x - 8y + 17$

**36** $f(x, y, z) = x^2 + z^2 + 25$

In each of Problems 37–40, the function $f(x, y)$ is the sum of a function of $x$ and a function of $y$. Use the method of Example 6 to construct a sketch of the surface $z = f(x, y)$.

**37** $f(x, y) = y^3 - x^2$

**38** $f(x, y) = y^4 + x^2$

**39** $f(x, y) = y^4 - 2y^2 + x^2$

**40** $f(x, y) = 2y^3 - 3y^2 - 12y + x^2$

**41** Figures 15.22 through 15.27 show the graphs of six functions $z = f(x, y)$. Figures 15.28 through 15.33 show the level curves of the same six functions, but not in the same order. Match each surface with its level curves.

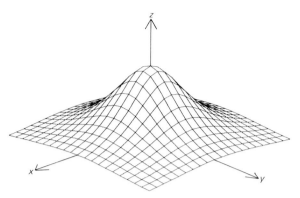

**15.22**    $z = \dfrac{1}{1 + x^2 + y^2}$, $|x| \leq 2$, $|y| \leq 2$

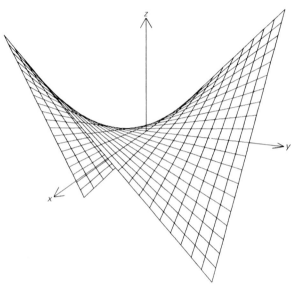

**15.23**    $z = -xy$, $|x| \leq 1$, $|y| \leq 1$

**15.24**    $z = r^2 \exp(-r^2) \cos^2(\tfrac{3}{2}\theta)$, $|x| \leq 3$, $|y| \leq 3$

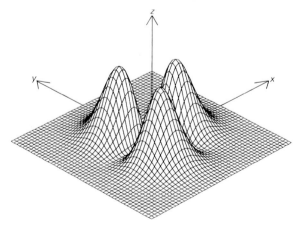

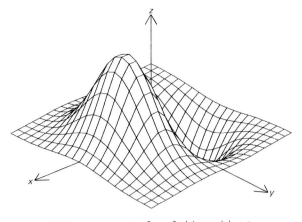

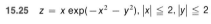

**15.25** $z = x \exp(-x^2 - y^2), |x| \leq 2, |y| \leq 2$

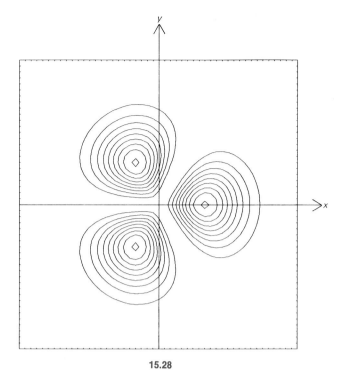

**15.28**

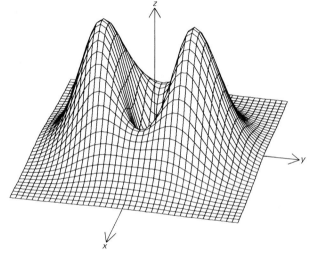

**15.26** $z = 3(x^2 + 3y^2) \exp(-x^2 - y^2), |x| \leq 2.5, |y| \leq 2.5$

**15.27** $z = xy \exp(-\frac{1}{2}[x^2 + y^2]), |x| \leq 3.5, |y| \leq 3.5$

**15.29**

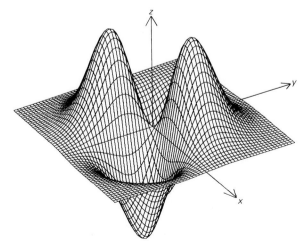

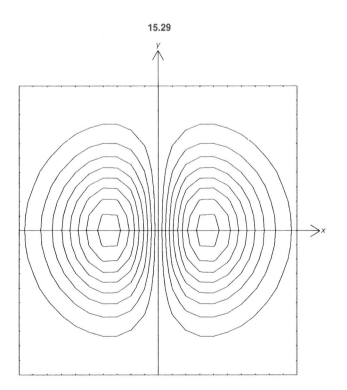

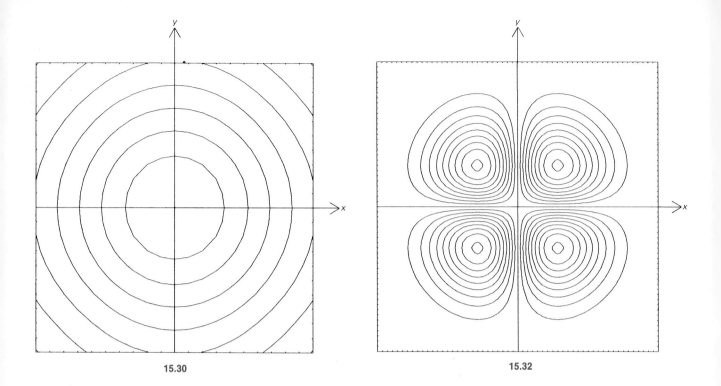

**15.30**

**15.32**

**15.31**

**15.33**

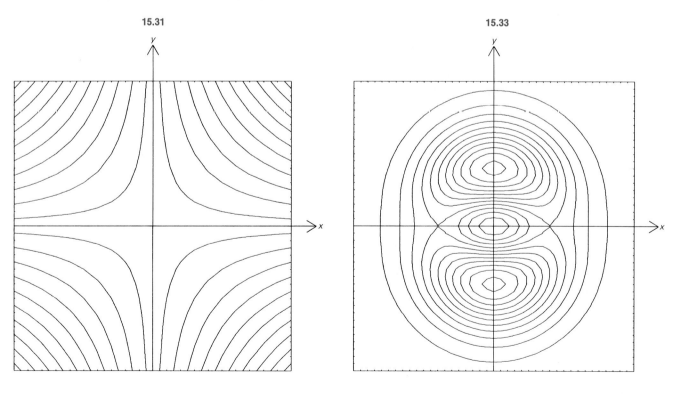

CHAP. 15:   Partial Differentiation

The special form of the equation

$$z = \tfrac{3}{2}y^2 + \tfrac{1}{24}y^3 - \tfrac{1}{32}y^4 - x^2 = f(x, y) \tag{7}$$

makes it a simple matter to write a program to plot level curves in the $xy$-plane. The reason is that we can solve (7) explicitly for $x$:

$$x = \pm(\tfrac{3}{2}y^2 + \tfrac{1}{24}y^3 - \tfrac{1}{32}y^4 - z)^{1/2}. \tag{8}$$

Now suppose that we want to plot the level curve $z = k$ on a $y$-interval subdivided by the equally spaced points

$$y_0 < y_1 < y_2 < \cdots < y_n.$$

When we substitute $y = y_i$ and $z = k$ in (8), we get two, one, or no values of $x$, depending on the sign of the quantity whose square root is indicated in (8). If there are two real roots $\pm x_i$, then we have found two points $(\pm x_i, y_i)$ on the level curve $z = k$.

Figure 15.34 shows an IBM-PC program that implements this procedure. Line 120 defines the function $F(Y, Z)$ that appears as the radicand in (8). Line 130 sets the screen for graphics. Line 140 establishes $xy$-coordinates on the screen with $(-8.5, -6)$ and $(8.5, 6)$ as the lower-left-hand and upper-right-hand corners, respectively. Lines 150–170 draw the two axes and the unit interval ticks along them. In line 180 we see that we want to plot the level curves for $Z = -3, -1.5, 0, 1.5, 3,$ 4.5, and 6—in steps of 0.1 from $Y = -6$ to $Y = 6$ (line 190). The values of $X$ are calculated in lines 200–220.

This program took about 40 s to produce the "dot-plot" shown in Fig. 15.35, and Fig. 15.21 was produced by joining the dots with smooth curves. The speed of this program was a consequence of the explicit solution for $x$ in (8). A more general program that essentially uses the method of bisection to solve Equation (7) numerically required 56 min to produce the same dot-plot. Sophisticated computer graphics involves so much number-crunching that—according to an article in the March 1984 issue of BYTE Magazine—motion picture companies are now using supercomputers such as the CRAY XMP for this purpose.

```
100   REM       Program LEVEL
110   REM
120   DEF  FNF(Y, Z) = 3*Y*Y/4 + Y*Y*Y/24 - Y*Y*Y*Y/32 - Z
130   KEY OFF  :  CLS  :  SCREEN 1
140   WINDOW  (-8.5, -6) - (8.5, 6)
150   LINE  (-6, 0) - (6, 0)  :  LINE (0, -6) - (0, 6)
160   FOR  I = -5 TO 5  :  LINE (I, -0.1) - (I, 0.1)  :  NEXT I
170   FOR  I = -5 TO 5  :  LINE (-0.1, I) - (0.1, I)  :  NEXT I
180   FOR  Z = -3 TO 6 STEP 1.5
190     FOR  Y = -6 TO 6 STEP 0.1
200       A = FNF(Y, Z)
210       IF  A = 0 THEN PSET (0, Y)
220       IF  A > 0 THEN X = SQR(A)  :  PSET (X, Y)  :  PSET (-X, Y)
230     NEXT  Y
240   NEXT Z
250   GOTO 250
260   END
```

15.34  Program to plot level curves

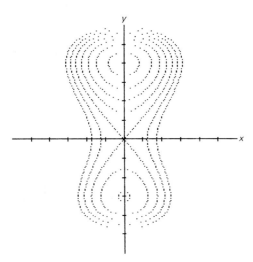

15.35 Dot-plot produced by the program in Fig. 15.34

## 15-3

### Limits and Continuity

We need limits of functions of several variables for the same reasons that we needed limits of functions of a single variable—so that we can discuss slopes and rates of change. Both the definition and the basic properties of limits of functions of several variables are essentially the same as those we stated in Section 2-1 for functions of a single variable. For simplicity, we shall state them here only for functions of two variables $x$ and $y$; for a function of three variables, the pair $(x, y)$ should be replaced by the triple $(x, y, z)$.

For a function $f$ of two variables, we ask what number (if any) the values $f(x, y)$ are approaching as $(x, y)$ approaches the fixed point $(a, b)$ in the plane. For a function $f$ of three variables, we ask what number (if any) the values $f(x, y, z)$ are approaching as $(x, y, z)$ approaches the fixed point $(a, b, c)$ in space.

| $x$ | $y$ | $f(x, y) = xy$ (rounded) |
|-----|-----|--------------------------|
| 2.2 | 2.5 | 5.50000 |
| 1.98 | 3.05 | 6.03900 |
| 2.002 | 2.995 | 5.99599 |
| 1.9998 | 3.0005 | 6.00040 |
| 2.00002 | 2.99995 | 5.99996 |
| 1.999998 | 3.000005 | 6.00000 |
| ↓ | ↓ | ↓ |
| 2 | 3 | 6 |

15.36 The numerical data of Example 1

**EXAMPLE 1** The numerical data in the table of Fig. 15.36 suggest that the values of the function $f(x, y) = xy$ are approaching 6 as $x \to 2$ and $y \to 3$ simultaneously; that is, as the point $(x, y)$ approaches the point $(2, 3)$. It therefore is natural to write

$$\lim_{(x, y)\to(2, 3)} xy = 6.$$

Our intuitive idea of the limit of a function of two variables is this. We say that the number $L$ is the *limit* of the function $f(x, y)$ as $(x, y)$ approaches the point $(a, b)$ and write

$$\lim_{(x,y)\to(a,b)} f(x, y) = L, \tag{1}$$

provided that the number $f(x, y)$ can be made as close as we please to $L$ if the point $(x, y)$ is chosen sufficiently close to—though not equal to—the point $(a, b)$.

To make this idea precise, we must specify how close to $L$—within $\varepsilon > 0$, say—we want $f(x, y)$ to be and then how close to $(a, b)$ the point $(x, y)$ must be to accomplish this. We think of the point $(x, y)$ as being close to $(a, b)$ provided that it lies within a small square (Fig. 15.37) with center $(a, b)$ and edge length $2\delta$, where $\delta > 0$ is small. The point $(x, y)$ lies within this square if and only if both

$$|x - a| < \delta \quad \text{and} \quad |y - b| < \delta. \tag{2}$$

This observation serves as motivation for the formal definition, with two additional conditions. First, we define the limit of $f(x, y)$ as $(x, y) \to (a, b)$ *only* under the condition that the domain of definition of $f$ contains points $(x, y) \neq (a, b)$ lying arbitrarily close to $(a, b)$—within *every* square of the sort in Fig. 15.37 and thus within any and every preassigned positive distance of $(a, b)$. Hence we do not speak of the limit of $f$ at an isolated point of its domain $D$. Finally, we do *not* require that $f$ be defined at the point $(a, b)$ itself, so we deliberately exclude the possibility that $(x, y) = (a, b)$.

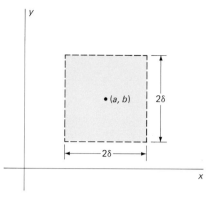

**15.37** The square $|x - a| < \delta$, $|y - b| < \delta$

---

**Definition** *The Limit of $f(x, y)$*

We say that the **limit of $f(x, y)$ as $(x, y)$ approaches $(a, b)$** is $L$ provided that for every number $\varepsilon > 0$, there exists a number $\delta > 0$ with the following property: If $(x, y)$ is a point of the domain of $f$ other than $(a, b)$ such that both

$$|x - a| < \delta \quad \text{and} \quad |y - b| < \delta, \tag{2}$$

then it follows that

$$|f(x, y) - L| < \varepsilon. \tag{3}$$

---

We ordinarily shall rely upon continuity rather than the formal definition of the limit to evaluate limits of functions of several variables. We say that $f$ is **continuous at the point** $(a, b)$ provided that $f(a, b)$ exists and $f(x, y)$ approaches $f(a, b)$ as $(x, y)$ approaches $(a, b)$; that is,

$$\lim_{(x,y) \to (a,b)} f(x, y) = f(a, b).$$

Thus $f$ is continuous at $(a, b)$ if it is defined there and its limit there is equal to its value there. The function $f$ is **continuous on the set** $D$ if it is continuous at each point of $D$.

**EXAMPLE 2** Let $D$ be the circular disk consisting of the points $(x, y)$ such that $x^2 + y^2 \leq 1$, and let $f(x, y) = 1$ at each point of $D$. Then the limit of $f(x, y)$ at each point of $D$ is obviously 1, so $f$ is continuous on $D$. But let the new function $g(x, y)$ be defined on the entire plane $\mathscr{R}^2$ as follows:

$$g(x, y) = \begin{cases} f(x, y) & \text{if} \quad (x, y) \text{ is in } D; \\ 0 & \text{otherwise.} \end{cases}$$

Then $g$ is *not* continuous on $\mathscr{R}^2$. For instance, the limit of $g(x, y)$ as $(x, y) \to (1, 0)$ does not exist because there exist both points within $D$ arbitrarily close to $(1, 0)$ at which $g$ has value 1 and points outside $D$ arbitrarily close to $(1, 0)$ at which $g$ has value 0. Thus $g(x, y)$ cannot approach any single value as $(x, y) \to (1, 0)$. Because $g$ has no limit at $(1, 0)$, it cannot be continuous there.

---

The limit laws of Section 2-1 have natural analogues for functions of several variables. If

$$\lim_{(x,y)\to(a,b)} f(x, y) = L \quad \text{and} \quad \lim_{(x,y)\to(a,b)} g(x, y) = M, \tag{4}$$

then the sum, product, and quotient laws say that

$$\lim_{(x,y)\to(a,b)} [f(x, y) + g(x, y)] = L + M, \tag{5}$$

$$\lim_{(x,y)\to(a,b)} [f(x, y)g(x, y)] = LM, \quad \text{and} \tag{6}$$

$$\lim_{(x,y)\to(a,b)} \frac{f(x, y)}{g(x, y)} = \frac{L}{M} \quad \text{if} \quad M \neq 0. \tag{7}$$

**EXAMPLE 3** Show that $\displaystyle\lim_{(x,y)\to(a,b)} xy = ab.$

*Solution* We take $f(x, y) = x$ and $g(x, y) = y$. Then it follows from the definition of limits that

$$\lim_{(x,y)\to(a,b)} f(x, y) = a \quad \text{and} \quad \lim_{(x,y)\to(a,b)} g(x, y) = b.$$

Hence the product law gives

$$\lim_{(x,y)\to(a,b)} xy = \lim_{(x,y)\to(a,b)} f(x, y)g(x, y)$$

$$= \left( \lim_{(x,y)\to(a,b)} f(x, y) \right) \left( \lim_{(x,y)\to(a,b)} g(x, y) \right) = ab.$$

---

More generally, suppose that $P(x, y)$ is a polynomial in the two variables $x$ and $y$, so that it can be written in the form

$$P(x, y) = \sum c_{ij} x^i y^j.$$

Then the sum and product laws imply that

$$\lim_{(x,y)\to(a,b)} P(x, y) = P(a, b).$$

An immediate but important consequence is that every polynomial in two (or more) variables is a continuous function.

Just as in the single-variable case, any composition of continuous multivariable functions is also a continuous function. For instance, suppose that the functions $f$ and $g$ are each continuous at $(a, b)$, and that $h$ is continuous at the point $(f(a, b), g(a, b))$. Then the composite function $H(x, y) = h(f(x, y), g(x, y))$ is also continuous at $(a, b)$. As a consequence, any finite combination involving sums, products, quotients, and compositions of the familiar elementary functions is continuous, except possibly at points where a denominator is zero or where the formula for the function is otherwise meaningless. This general rule suffices for the evaluation of most limits that we shall encounter.

**EXAMPLE 4** By application of the limit laws we get

$$\lim_{(x,y)\to(1,2)}\left[e^{xy}\sin\frac{\pi y}{4}+xy\ln\sqrt{y-x}\right]$$

$$=\lim_{(x,y)\to(1,2)}e^{xy}\sin\frac{\pi y}{4}+\lim_{(x,y)\to(1,2)}xy\ln\sqrt{y-x}$$

$$=\left(\lim_{(x,y)\to(1,2)}e^{xy}\right)\left(\lim_{(x,y)\to(1,2)}\sin\frac{\pi y}{4}\right)$$

$$+\left(\lim_{(x,y)\to(1,2)}xy\right)\left(\lim_{(x,y)\to(1,2)}\ln\sqrt{y-x}\right)$$

$$=(e^2)(1)+(2)(\ln 1)=e^2.$$

The following two examples illustrate techniques that sometimes are successful in handling cases with denominators that approach zero.

**EXAMPLE 5** Show that $\displaystyle\lim_{(x,y)\to(0,0)}\frac{xy}{\sqrt{x^2+y^2}}=0.$

*Solution* Let $(r,\theta)$ be the polar coordinates of the point $(x,y)$. Then $x=r\cos\theta$ and $y=r\sin\theta$, so

$$\frac{xy}{\sqrt{x^2+y^2}}=\frac{(r\cos\theta)(r\sin\theta)}{\sqrt{r^2\cos^2\theta+r^2\sin^2\theta}}$$

$$=r\cos\theta\sin\theta\qquad\text{for}\quad r>0.$$

Since $r=\sqrt{x^2+y^2}$, it is clear that $r\to 0$ as $x$ and $y$ both approach zero. It therefore follows that

$$\lim_{(x,y)\to(0,0)}\frac{xy}{\sqrt{x^2+y^2}}=\lim_{r\to0}r\cos\theta\sin\theta=0,$$

since $|\cos\theta\sin\theta|\leq 1$ for all values of $\theta$.

**EXAMPLE 6** Show that $\displaystyle\lim_{(x,y)\to(0,0)}\frac{xy}{x^2+y^2}$ docs not cxist.

*Solution* Our plan is to show that $f(x,y)=xy/(x^2+y^2)$ approaches different values as $(x,y)$ approaches $(0,0)$ from different directions. Suppose that $(x,y)$ approaches $(0,0)$ along the line of slope $m$ through the origin. On this line we have $y=mx$. So on this line

$$f(x,y)=\frac{x(mx)}{x^2+m^2x^2}=\frac{m}{1+m^2}$$

if $x\neq 0$. If we take $m=1$, we see that $f(x,y)=\frac{1}{2}$ at every point of the line $y=x$ other than $(0,0)$. If we take $m=-1$, we see that $f(x,y)=-\frac{1}{2}$ at every point of the line $y=-x$ other than $(0,0)$. Thus $f(x,y)$ approaches two different values as $(x,y)$ approaches $(0,0)$ along these two lines, shown in Fig. 15.38. Hence $f(x,y)$ cannot approach any *single* value as $(x,y)$ approaches $(0,0)$, and this implies that the limit in question cannot exist.

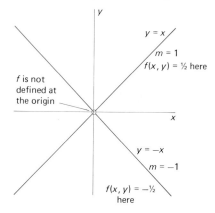

15.38 The function $f$ of Example 6 takes on both values $+\frac{1}{2}$ and $-\frac{1}{2}$ at points arbitrarily close to the origin.

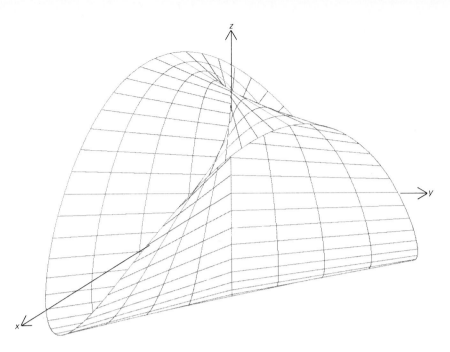

**15.39** The graph of $f(x, y) = xy/(x^2 + y^2)$

Figure 15.39 shows a computer-generated graph of the function $f(x, y) = xy/(x^2 + y^2)$. It consists of linear rays along each of which $\theta$ is constant. For each number $z$ between $-\frac{1}{2}$ and $\frac{1}{2}$ (inclusive), there are rays along which $f(x, y)$ has the constant value $z$. Hence we can make $f(x, y)$ approach any number we please in $\left[-\frac{1}{2}, \frac{1}{2}\right]$ by letting $(x, y)$ approach $(0, 0)$ from the appropriate direction.

In order for $\lim_{(x,y) \to (a,b)} f(x, y)$ to exist, $f(x, y)$ must approach $L$ for *any* mode of approach of $(x, y)$ to $(a, b)$. In Problem 27 below, we give an example of a function $f$ such that $f(x, y) \to 0$ as $(x, y) \to (0, 0)$ along any straight line through the origin, but $f(x, y) \to 1$ as $(x, y)$ approaches the origin along the parabola $y = x^2$. Fortunately, many important applications, including those we discuss in the remainder of this chapter, involve only functions that exhibit no such exotic behavior.

## 15-3 PROBLEMS

Use the limit laws and consequences of continuity to evaluate the limits in Problems 1–15.

**1** $\lim_{(x,y) \to (0,0)} (7 - x^2 + 5xy)$

**2** $\lim_{(x,y) \to (1,-2)} (3x^2 - 4xy + 5y^2)$

**3** $\lim_{(x,y) \to (1,-1)} e^{-xy}$

**4** $\lim_{(x,y) \to (0,0)} \dfrac{x + y}{1 + xy}$

**5** $\lim_{(x,y) \to (0,0)} \dfrac{5 - x^2}{3 + x + y}$

**6** $\lim_{(x,y) \to (2,3)} \dfrac{9 - x^2}{1 + xy}$

**7** $\lim_{(x,y) \to (0,0)} \ln \sqrt{1 - x^2 - y^2}$

**8** $\lim_{(x,y) \to (2,-1)} \ln \dfrac{1 + x + 2y}{3y^2 - x}$

**9** $\lim_{(x,y) \to (0,0)} \dfrac{e^{xy} \sin xy}{xy}$

**10** $\lim_{(x,y) \to (0,0)} \exp\left(\dfrac{-1}{x^2 + y^2}\right)$

**11** $\lim_{(x,y,z) \to (1,1,1)} \dfrac{x^2 + y^2 + z^2}{1 - x - y - z}$

**12** $\displaystyle\lim_{(x,y,z)\to(1,1,1)} (x + y + z)\ln xyz$

**13** $\displaystyle\lim_{(x,y,z,)\to(1,1,0)} \frac{xy - z}{\cos xyz}$

**14** $\displaystyle\lim_{(x,y,z)\to(2,-1,3)} \frac{x + y + z}{x^2 + y^2 + z^2}$

**15** $\displaystyle\lim_{(x,y,z)\to(2,8,1)} \sqrt{xy} \tan\frac{3\pi z}{4}$

In each of Problems 16–20, evaluate the limits

$$\lim_{h\to 0} \frac{f(x + h, y) - f(x, y)}{h} \quad\text{and}\quad \lim_{k\to 0} \frac{f(x, y + k) - f(x, y)}{k}.$$

**16** $f(x, y) = x + y$       **17** $f(x, y) = xy$
**18** $f(x, y) = x^2 + y^2$       **19** $f(x, y) = xy^2 - 2$
**20** $f(x, y) = x^2y^3 - 10$

In Problems 21–23, use the method of Example 5 to verify the given limit.

**21** $\displaystyle\lim_{(x,y)\to(0,0)} \frac{x^2 - y^2}{\sqrt{x^2 + y^2}} = 0$

**22** $\displaystyle\lim_{(x,y)\to(0,0)} \frac{x^3 - y^3}{x^2 + y^2} = 0$

**23** $\displaystyle\lim_{(x,y)\to(0,0)} \frac{x^4 + y^4}{(x^2 + y^2)^{3/2}} = 0$

**24** Use the method of Example 6 to show that

$$\lim_{(x,y)\to(0,0)} \frac{x^2 - y^2}{x^2 + y^2}$$

does not exist.

**25** Substitute spherical coordinates $x = \rho \sin\phi \cos\theta$, $y = \rho \sin\phi \sin\theta$, $z = \rho \cos\phi$ to show that

$$\lim_{(x,y,z)\to(0,0,0)} \frac{xyz}{x^2 + y^2 + z^2} = 0.$$

**26** Determine whether or not

$$\lim_{(x,y,z)\to(0,0,0)} \frac{xy + xz + yz}{x^2 + y^2 + z^2}$$

exists.

**27** Let $f(x, y) = 2x^2y/(x^4 + y^2)$. Show that
(a) $f(x, y) \to 0$ as $(x, y) \to (0, 0)$ along *any* (every) straight line through the origin; but that
(b) $f(x, y) \to 1$ as $(x, y) \to (0, 0)$ along the parabola $y = x^2$. Hence conclude that the limit of $f(x, y)$ as $(x, y) \to (0, 0)$ does not exist.

**28** Suppose that $f(x, y) = (x - y)/(x^3 - y)$ except at points of the curve $y = x^3$; we *define* $f(x, y)$ to be 1 at those points. Show that $f$ is not continuous at the point $(1, 1)$. Evaluate the limits of $f(x, y)$ as $(x, y) \to (1, 1)$ along the vertical line $x = 1$ and along the horizontal line $y = 1$. (*Suggestion:* Recall that $a^3 - b^3 = (a - b)(a^2 + ab + b^2)$.)

**29** Locate and identify the extrema (local or global, maximum or minimum) of the function $f(x, y) = x^2 - x + y^2 + 2y + 1$. (*Note:* The point of this problem is that you do *not* need calculus to work it.)

**30** Sketch enough level curves of the function $h(x, y) = y - x^2$ to show that the function $h$ has *no* extreme values— no local or global maxima or minima.

---

## 15-4

## Partial Derivatives

Suppose that $y = f(x)$ is a function of *one* real variable. Its first derivative

$$\frac{dy}{dx} = D_xf(x) = \lim_{h\to 0} \frac{f(x + h) - f(x)}{h} \tag{1}$$

can be interpreted as the instantaneous rate of change of $y$ with respect to $x$. For a function $z = f(x, y)$ of two variables, we need a similar understanding of the rate at which $z$ changes as $x$ and $y$ vary (either singly or simultaneously). To reach this more complicated concept, we adopt a "divide and conquer" strategy.

First we hold $y$ fixed and let $x$ vary. The rate of change of $z$ with respect to $x$ is then denoted by $\partial z/\partial x$ and has the value

$$\frac{\partial z}{\partial x} = \lim_{h\to 0} \frac{f(x + h, y) - f(x, y)}{h}. \tag{2}$$

The value of this limit (assuming it exists) is called the **partial derivative of** $f$ **with respect to** $x$. In like manner, we may hold $x$ fixed and let $y$ vary. The rate of change of $z$ with respect to $y$ is then the **partial derivative of** $f$ **with**

**respect to** $y$, defined to be

$$\frac{\partial z}{\partial y} = \lim_{k \to 0} \frac{f(x, y + k) - f(x, y)}{k} \tag{3}$$

for all $(x, y)$ for which this limit exists. Note the symbol $\partial$ that is used instead of $d$ to denote the partial derivatives of a function of two variables. A function of three or more independent variables has a partial derivative (defined similarly) with respect to each of its independent variables. Some other commonly used notations for partial derivatives are

$$\frac{\partial z}{\partial x} = \frac{\partial f}{\partial x} = f_x(x, y) = D_x f(x, y) = D_1 f(x, y), \tag{4}$$

$$\frac{\partial z}{\partial y} = \frac{\partial f}{\partial y} = f_y(x, y) = D_y f(x, y) = D_2 f(x, y). \tag{5}$$

Note that if the symbol $y$ in Equation (2) is deleted, the result is the limit of Equation (1). This means that $\partial z/\partial x$ can be calculated as an "ordinary" derivative with respect to $x$, simply regarding $y$ as a constant during the process of differentiation. Similarly, we can compute $\partial z/\partial y$ as an ordinary derivative, thinking of $y$ as the *only* variable and treating $x$ as a constant during the computation.

**EXAMPLE 1**   Compute the partial derivatives $\partial f/\partial x$ and $\partial f/\partial y$ of the function $f(x, y) = x^2 + 2xy^2 - y^3$.

*Solution*   To compute the partial of $f$ with respect to $x$, we regard $y$ as a constant. Then we differentiate normally and find that

$$\frac{\partial f}{\partial x} = 2x + 2y^2.$$

When we regard $x$ as a constant and differentiate with respect to $y$, we find that

$$\frac{\partial f}{\partial y} = 4xy - 3y^2.$$

To get an intuitive feel for the meaning of partial derivatives, we can think of $f(x, y)$ as the temperature at the point $(x, y)$. Then $f_x(x, y)$ is the instantaneous rate of change of temperature at $(x, y)$ per unit increase in $x$ (with $y$ held constant), while $f_y(x, y)$ is the rate of change of temperature per unit increase in $y$ (with $x$ held constant). For example, with the temperature function $f(x, y) = x^2 + 2xy^2 - y^3$ of Example 1, the rate of change of temperature at the point $(1, -1)$ is $+4°$ per unit distance in the positive $x$-direction and $-7°$ per unit distance in the positive $y$-direction.

**EXAMPLE 2**   Find $\partial z/\partial x$ and $\partial z/\partial y$ if $z = (x^2 + y^2)e^{-xy}$.

*Solution*   Because $\partial z/\partial x$ is calculated as if it were an ordinary derivative with respect to $x$ while $y$ is held constant, we can use the product rule. This gives

$$\frac{\partial z}{\partial x} = (2x)(e^{-xy}) + (x^2 + y^2)(-ye^{-xy})$$

$$= (2x - x^2 y - y^3)e^{-xy}.$$

Because $x$ and $y$ appear symmetrically in the expression $z = (x^2 + y^2)e^{-xy}$, we get $\partial z/\partial y$ when we interchange $x$ and $y$ in the expression for $\partial z/\partial x$:

$$\frac{\partial z}{\partial y} = (2y - xy^2 - x^3)e^{-xy}.$$

You should check this result by differentiating $z$ with respect to $y$.

**EXAMPLE 3**  The volume $V$ (in cubic centimeters) of 1 mole of an ideal gas is given by

$$V = \frac{(82.06)T}{P},$$

where $P$ is the pressure (in atmospheres) and $T$ is the absolute temperature (in degrees Kelvin ($^\circ$K), where $^\circ$K $= {}^\circ$C $+ 273$). Find the rates of change of the volume of 1 mol of an ideal gas with respect to pressure and with respect to temperature with $T = 300^\circ$K and $P = 5$ atm.

*Solution*   The partial derivatives of $V$ with respect to its two variables are

$$\frac{\partial V}{\partial P} = -\frac{(82.06)T}{P^2} \quad \text{and} \quad \frac{\partial V}{\partial T} = \frac{82.06}{P}.$$

With $T = 300^\circ$K and $P = 5$ atm, we have the two values $\partial V/\partial P = -984.72$ cm$^3$/atm and $\partial V/\partial T = 16.41$ cm$^3$/$^\circ$K. These partial derivatives allow us to estimate the effect of a change in temperature or in pressure on the volume $V$ of gas in question, as follows. We are given $T = 300^\circ$K and $P = 5$ atm, so the volume of gas we are dealing with is $V = (82.06)(300)/5 = 4923.60$ cm$^3$. We would expect an increase in pressure of one atmosphere (with temperature held constant) to decrease the volume of gas by roughly 1 liter (1000 cm$^3$). An increase in temperature of $1^\circ$K (or $1^\circ$C) would, with pressure held constant, increase the volume by about 16 cm$^3$.

## GEOMETRIC INTERPRETATION
## OF PARTIAL DERIVATIVES

The partial derivatives $f_x$ and $f_y$ are the slopes of tangent lines to certain curves on the surface $z = f(x, y)$. Consider the point $P(a, b, f(a, b))$ on this surface. It lies directly above the point $Q(a, b, 0)$ in the $xy$-plane, as shown in Fig. 15.40. The vertical plane $y = b$ parallel to the $xz$-plane intersects the surface in the curve $z = f(x, b)$ through the point $P$. Along this curve, $x$ varies

**15.40**  Geometric interpretation of $f_x(a, b)$ as the slope of a tangent to a *curve*

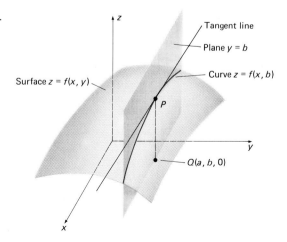

Tangent line

Plane $y = b$

Curve $z = f(x, b)$

Surface $z = f(x, y)$

$P$

$Q(a, b, 0)$

but $y$ is constant; $y$ is always equal to $b$ because the curve lies in the vertical plane with equation $y = b$.

Let us project this curve and its tangent line at $P$ into the $xz$-plane. We get exactly the same curve and tangent line, for this is a normal projection. But because we are now working in the $xz$-plane, we can actually "forget" the presence of $y$. In effect, as indicated in Fig. 15.41, we are dealing with $z$ as a function of the *single* variable $x$, and the projected curve is the graph of this function. Hence the slope of the tangent line to the original curve at the point $P(a, b, f(a, b))$ is equal to the slope of the tangent line of Fig. 15.41. But by familiar single-variable calculus, this slope is

$$\lim_{h \to 0} \frac{f(a + h, b) - f(a, b)}{h} = f_x(a, b).$$

*Thus $f_x(a, b)$ is the slope of the tangent line at $P$ to the curve formed by intersecting the plane $y = b$ parallel to the xz-plane with the surface $z = f(x, y)$.*

We proceed in much the same way to get a geometric interpretation of the partial derivative of $f$ with respect to $y$. As Fig. 15.42 suggests, *$f_y(a, b)$ is the slope of a line tangent to the curve of intersection of the surface $z = f(x, y)$ with the plane $x = a$ parallel to the yz-plane.*

The two tangent lines we have just found determine a unique plane through the point $P(a, b, f(a, b))$. In Section 15-7 we will see that if the partial derivatives $f_x$ and $f_y$ are continuous functions of $x$ and $y$, then this plane contains the tangent line at $P$ to *every* smooth curve on the surface $z = f(x, y)$ that passes through $P$. This plane is therefore (by definition) the tangent plane to the surface at $P$.

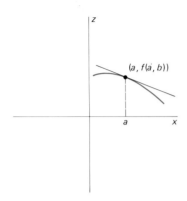

**15.41** The projection of the curve and tangent line of Fig. 15.40 into the $xz$-plane

---

*Definition*  *Tangent Plane to $z = f(x, y)$*

Suppose that the function $f(x, y)$ has continuous partial derivatives on a rectangle in the $xy$-plane containing $(a, b)$ in its interior. Then the **tangent plane** to the surface $z = f(x, y)$ at the point $P(a, b, f(a, b))$ is the plane through $P$ that contains the tangent lines to the two curves

$$z = f(x, b), \qquad y = b \tag{6}$$

and

$$z = f(a, y), \qquad x = a. \tag{7}$$

---

**15.42** The partial derivative $f_y$ is also the slope of a line tangent to a curve in the surface $z = f(x, y)$.

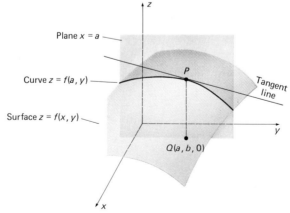

CHAP. 15:  Partial Differentiation

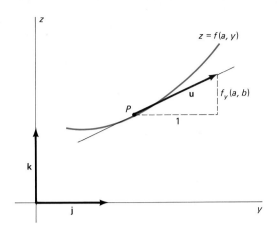

**15.43** The curve $z = f(a, y)$ in the plane $x = a$

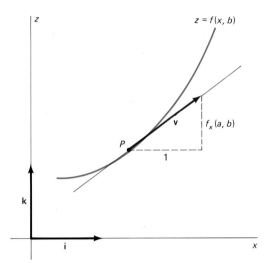

**15.44** The curve $z = f(x, b)$ in the plane $y = b$

In order to write the equation of this tangent plane, all we need is a vector **n** normal to the plane. One way to get such a vector is to find the cross product of tangent vectors to the curves in (6) and (7). Figures 15.43 and 15.44 show these two curves. As we saw above, the curve in (7) has slope $f_y(a, b)$, so we can take

$$\mathbf{u} = \mathbf{j} + \mathbf{k}f_y(a, b) \tag{8}$$

as its tangent vector at $P$. The curve in (6) has slope $f_x(a, b)$, so we can take

$$\mathbf{v} = \mathbf{i} + \mathbf{k}f_x(a, b) \tag{9}$$

as its tangent vector. Using these two tangent vectors, we obtain the normal vector

$$\mathbf{n} = \mathbf{u} \times \mathbf{v} = \begin{vmatrix} \mathbf{i} & \mathbf{j} & \mathbf{k} \\ 0 & 1 & f_y(a, b) \\ 1 & 0 & f_x(a, b) \end{vmatrix},$$

so that

$$\mathbf{n} = \mathbf{i}f_x(a, b) + \mathbf{j}f_y(a, b) - \mathbf{k}. \tag{10}$$

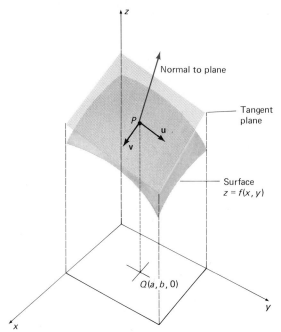

**15.45** A normal vector to the tangent plane can be found by forming the cross product of two tangent vectors.

Note that

$$\mathbf{n} = \left\langle \frac{\partial z}{\partial x}, \frac{\partial z}{\partial y}, -1 \right\rangle \tag{11}$$

is a downward-pointing vector (because of its negative $z$-component); its negative $-\mathbf{n}$ is the normal vector shown in Fig. 15.45.

Finally we use the normal vector $\mathbf{n}$ of Equation (10) to find the equation of the tangent plane to the surface $z = f(x, y)$ at the point $(a, b, f(a, b))$. This equation is

$$f_x(a, b)(x - a) + f_y(a, b)(y - b) - [z - f(a, b)] = 0. \tag{12}$$

An equivalent form of this equation is

$$z - c = \frac{\partial z}{\partial x}(x - a) + \frac{\partial z}{\partial y}(y - b) \tag{13}$$

where $c = f(a, b)$ and we remember that the partial derivatives $\partial z/\partial x$ and $\partial z/\partial y$ are evaluated at the point $(a, b)$.

**EXAMPLE 4**  Write an equation of the tangent plane to the paraboloid $z = x^2 + y^2$ at the point $(2, -1, 5)$.

*Solution*  We begin by computing $\partial z/\partial x = 2x$ and $\partial z/\partial y = 2y$. The values of these partial derivatives when $x = 2$ and $y = -1$ are 4 and $-2$, respectively, so Equation (12) gives

$$4(x - 2) - 2(y + 1) - (z - 5) = 0$$

—when simplified, $4x - 2y - z = 5$—as an equation of the indicated tangent plane.

## HIGHER PARTIAL DERIVATIVES

The **first order** partial derivatives $f_x$ and $f_y$ are themselves functions of $x$ and $y$, so they may be differentiated with respect to $x$ and $y$. The partial derivatives of $f_x(x, y)$ and $f_y(x, y)$ are called the **second order partial derivatives** of $f$. There are four of them because there are four possibilities in the order of differentiation:

$$(f_x)_x = f_{xx} = \frac{\partial f_x}{\partial x} = \frac{\partial}{\partial x}\left(\frac{\partial f}{\partial x}\right) = \frac{\partial^2 f}{\partial x^2},$$

$$(f_x)_y = f_{xy} = \frac{\partial f_x}{\partial y} = \frac{\partial}{\partial y}\left(\frac{\partial f}{\partial x}\right) = \frac{\partial^2 f}{\partial y\,\partial x},$$

$$(f_y)_x = f_{yx} = \frac{\partial f_y}{\partial x} = \frac{\partial}{\partial x}\left(\frac{\partial f}{\partial y}\right) = \frac{\partial^2 f}{\partial x\,\partial y},$$

$$(f_y)_y = f_{yy} = \frac{\partial f_y}{\partial y} = \frac{\partial}{\partial y}\left(\frac{\partial f}{\partial y}\right) = \frac{\partial^2 f}{\partial y^2}.$$

If we write $z = f(x, y)$, we can replace each occurrence of the symbol $f$ above by $z$.

Note that $f_{xy}$ is the second order partial derivative of $f$ with respect to $x$ first, then $y$; $f_{yx}$ is the result of differentiating with respect to $x$ and $y$ in the reverse order. It is proved in advanced calculus that these two "mixed" second partial derivatives are equal if they are continuous. More precisely, if $f_{xy}$ and $f_{yx}$ are continuous on a circular disk centered at the point $(a, b)$, then

$$f_{xy}(a, b) = f_{yx}(a, b). \tag{14}$$

Because most functions of interest to us will have second order partial derivatives that are continuous everywhere they are defined, we will ordinarily need deal only with three distinct second partial derivatives rather than with four. Similarly, if $f(x, y, z)$ is a function of three variables with continuous second order partial derivatives, then

$$\frac{\partial^2 f}{\partial x\,\partial y} = \frac{\partial^2 f}{\partial y\,\partial x}, \qquad \frac{\partial^2 f}{\partial x\,\partial z} = \frac{\partial^2 f}{\partial z\,\partial x}, \quad \text{and} \quad \frac{\partial^2 f}{\partial y\,\partial z} = \frac{\partial^2 f}{\partial z\,\partial y}.$$

Third order and higher order partial derivatives are defined similarly, and the order in which the differentiations are performed is unimportant so long as all derivatives involved are continuous. For instance, the distinct third order partial derivatives of a function $z = f(x, y)$ are

$$f_{xxx} = \frac{\partial}{\partial x}\left(\frac{\partial^2 f}{\partial x^2}\right) = \frac{\partial^3 f}{\partial x^3}, \qquad f_{xxy} = \frac{\partial}{\partial y}\left(\frac{\partial^2 f}{\partial x^2}\right) = \frac{\partial^3 f}{\partial y\,\partial x^2},$$

$$f_{xyy} = \frac{\partial}{\partial y}\left(\frac{\partial^2 f}{\partial y\,\partial x}\right) = \frac{\partial^3 f}{\partial y^2\,\partial x}, \qquad f_{yyy} = \frac{\partial}{\partial y}\left(\frac{\partial^2 f}{\partial y^2}\right) = \frac{\partial^3 f}{\partial y^3}.$$

**EXAMPLE 5**  Show that the partial derivatives of third order and higher of the function $f(x, y) = x^2 + 2xy^2 - y^3$ are constant.

*Solution* We find that

$$f_x(x, y) = 2x + 2y^2 \quad \text{and} \quad f_y(x, y) = 4xy - 3y^2.$$

So

$$f_{xx}(x, y) = 2, \quad f_{xy}(x, y) = f_{yx}(x, y) = 4y, \quad \text{and} \quad f_{yy}(x, y) = 4x - 6y.$$

And finally,

$$f_{xxx}(x, y) = 0, \qquad f_{xxy}(x, y) = 0,$$
$$f_{xyy}(x, y) = 4, \quad \text{and} \quad f_{yyy}(x, y) = -6.$$

The function $f$ is a polynomial, so all its partial derivatives are polynomials and are therefore continuous. Hence there is no need to check any other third order partial derivatives; each equals one of the last four above. Since the third order partial derivatives are all constant, all higher partial derivatives of $f$ are zero.

## 15-4 PROBLEMS

Compute the first order partial derivatives of the functions in Problems 1–20.

**1** $f(x, y) = x^4 - x^3y + x^2y^2 - xy^3 + y^4$
**2** $f(x, y) = x \sin y$
**3** $f(x, y) = e^x(\cos y - \sin y)$

**4** $f(x, y) = x^2e^{xy}$      **5** $f(x, y) = \dfrac{x + y}{x - y}$

**6** $f(x, y) = \dfrac{xy}{x^2 + y^2}$      **7** $f(x, y) = \ln(x^2 + y^2)$

**8** $f(x, y) = (x - y)^{14}$      **9** $f(x, y) = x^y$
**10** $f(x, y) = \tan^{-1}xy$      **11** $f(x, y, z) = x^2y^3z^4$
**12** $f(x, y, z) = x^2 + y^3 + z^4$
**13** $f(x, y, z) = e^{xyz}$
**14** $f(x, y, z) = x^4 - 16yz$    **15** $f(x, y, z) = x^2e^y \ln z$
**16** $f(u, v) = (2u^2 + 3v^2)e^{-u^2 - v^2}$

**17** $f(r, s) = \dfrac{r^2 - s^2}{r^2 + s^2}$

**18** $f(u, v) = e^{uv}(\cos uv + \sin uv)$
**19** $f(u, v, w) = ue^v + ve^w + we^u$
**20** $f(r, s, t) = (1 - r^2 - s^2 - t^2)e^{-rst}$

Verify that $z_{xy} = z_{yx}$ in Problems 21–30.

**21** $z = x^2 - 4xy + 3y^2$
**22** $z = 2x^3 + 5x^2y - 6y^2 + xy^4$
**23** $z = x^2e^{-y^2}$      **24** $z = xye^{-xy}$
**25** $z = \ln(x + y)$      **26** $z = (x^3 + y^3)^{10}$
**27** $z = e^{-3x}\cos y$      **28** $z = (x + y)\sec xy$
**29** $z = x^2\cosh(1/y^2)$      **30** $z = \sin xy + \tan^{-1}xy$

In each of Problems 31–40, find an equation of the tangent plane to the given surface $z = f(x, y)$ at the indicated point $P$.

**31** $z = x^2 + y^2$;   $P = (3, 4, 25)$
**32** $z = \sqrt{25 - x^2 - y^2}$;   $P = (4, -3, 0)$

**33** $z = \sin \dfrac{\pi xy}{2}$;   $P = (3, 5, -1)$

**34** $z = \dfrac{4}{\pi} \tan^{-1}xy$;   $P = (1, 1, 1)$

**35** $z = x^3 - y^3$;   $P = (3, 2, 19)$
**36** $z = 3x + 4y$;   $P = (1, 1, 7)$
**37** $z = xy$;   $P = (1, -1, -1)$
**38** $z = e^{-x^2 - y^2}$;   $P = (0, 0, 1)$
**39** $z = x^2 - 4y^2$;   $P = (5, 2, 9)$
**40** $z = \sqrt{x^2 + y^2}$;   $P = (3, -4, 5)$
**41** Verify that the mixed second order partial derivatives $f_{xy}$ and $f_{yx}$ are indeed equal if $f(x, y) - x^my^n$ where $m$ and $n$ are positive integers.
**42** Suppose that $z = e^{x+y}$. Show that the result of differentiating $z$ $m$ times with respect to $x$ and $n$ times with respect to $y$ is

$$\frac{\partial^{m+n}z}{\partial x^m \, \partial y^n} = e^{x+y}.$$

**43** Let $f(x, y, z) = e^{xyz}$. Calculate the six distinct second order partial derivatives of $f$ and also the third order partial derivative $f_{xyz}$.
**44** Suppose that $g(x, y) = \sin xy$. Verify that $g_{xy} = g_{yx}$ and that $g_{xxy} = g_{xyx} = g_{yxx}$.
**45** In physics it is shown that the temperature $u(x, y)$ at the point $x$ and the time $t$ of a long insulated rod lying along the $x$-axis satisfies the *one-dimensional heat equation*

$$\frac{\partial u}{\partial t} = k \frac{\partial^2 u}{\partial x^2} \qquad (k \text{ is a constant}).$$

Show that the function

$$u = u(x, y, t) = e^{-n^2kt}\sin nx$$

satisfies the heat equation for any choice of the constant $n$.

**46** The *two-dimensional heat equation* for an insulated plane is

$$\frac{\partial u}{\partial t} = k\left(\frac{\partial^2 u}{\partial x^2} + \frac{\partial^2 u}{\partial y^2}\right).$$

Show that the function

$$u = u(x, y, t) = e^{-(m^2 + n^2)kt}\sin mx \cos ny$$

satisfies this equation for any choice of constants $m$ and $n$.

**47** A string is stretched along the $x$-axis, fixed at each end, and then set in vibration. In physics it is shown that the displacement $y = y(x, t)$ of the point of the string at location $x$ at time $t$ satisfies the *one-dimensional wave equation*

$$\frac{\partial^2 y}{\partial t^2} = a^2 \frac{\partial^2 y}{\partial x^2}$$

where the constant $a$ depends upon the density and tension of the string. Show that the following functions each satisfy the wave equation above.

(a) $y = \sin(x + at)$
(b) $y = \cosh(3[x - at])$
(c) $y = \sin kx \cos kat$   ($k$ is a constant)

**48** A steady-state temperature function $u = u(x, y)$ for a thin flat plate satisfies *Laplace's equation*

$$\frac{\partial^2 u}{\partial x^2} + \frac{\partial^2 u}{\partial y^2} = 0.$$

Determine which of the following functions satisfies Laplace's equation.

(a) $u = \ln(\sqrt{x^2 + y^2})$
(b) $u = (x^2 + y^2)^{1/2}$
(c) $u = \arctan(y/x)$
(d) $u = e^{-x}\sin y$

**49** The **ideal gas law** $PV = nRT$ ($n$ is the number of moles of gas, $R$ is a constant) determines each of the three variables $P$, $V$, and $T$ (pressure, volume, and absolute temperature, respectively) as functions of the other two. Show that

$$\frac{\partial P}{\partial V} \frac{\partial V}{\partial T} \frac{\partial T}{\partial P} = -1.$$

**50** It is geometrically clear that every tangent plane to the cone $z^2 = x^2 + y^2$ passes through the origin. Show this by methods of calculus.

**51** There is only one point at which the tangent plane to the surface $z = x^2 + 2xy + 2y^2 - 6x + 8y$ is horizontal. Find it.

**52** Show that the plane tangent to the paraboloid with equation $z = x^2 + y^2$ at the point $(a, b, c)$ intersects the

$xy$-plane in the line with equation $2ax + 2by = a^2 + b^2$. Then apply the formula for the distance from a point to a line to conclude that this line is tangent to the circle with equation $4x^2 + 4y^2 = a^2 + b^2$.

**53** According to the van der Waals equation, 1 mol of a gas satisfies the equation

$$\left(P + \frac{a}{V^2}\right)(V - b) = (82.06)T$$

where $P$, $V$, and $T$ are as in Example 2. For carbon dioxide, $a = 3.59 \times 10^6$ and $b = 42.7$, and $V$ is 25,600 cm$^3$ when $P$ is 1 atm and $T = 313°$K.

(a) Compute $\partial V/\partial P$ by differentiating the above equation with $T$ held constant. Then estimate the change in volume that would result from an increase of 0.1 atm in pressure with $T$ at $313°$K.

(b) Compute $\partial V/\partial T$ by differentiating the above equation with $P$ held constant. Then estimate the change in volume that would result from an increase of $1°$K in temperature with $P$ held at 1 atm.

**54** A *minimal surface* is one that has the least surface area of all surfaces with the same boundary. Figure 15.46 shows Scherk's minimal surface with equation

$$z = \ln(\cos x) - \ln(\cos y).$$

It is known that a minimal surface $z = f(x, y)$ satisfies the partial differential equation

$$(1 + z_y^2)z_{xx} - 2z_x z_y z_{xy} + (1 + z_x^2)z_{yy} = 0.$$

Verify this in the case of Scherk's surface.

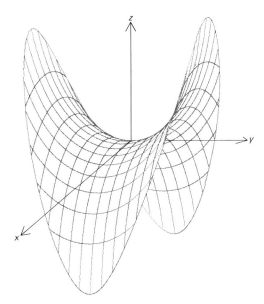

**15.46**   Scherk's minimal surface

## Maxima and Minima of Functions of Several Variables

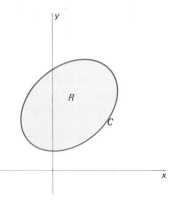

**15.47** A bounded plane region $R$ with boundary the simple closed curve $C$

The single-variable maximum-minimum techniques of Section 3-5 generalize in a natural manner to functions of several variables. We consider first a function $f$ of two variables. Suppose that we are interested in the extreme values attained by $f$ on a plane region $R$ consisting of the points on and within a simple closed curve $C$, as in Fig. 15.47. We say that the function $f$ attains its **absolute** (or **global**) **maximum value** $M$ on $R$ at the point $(a, b)$ of $R$ provided that

$$f(x, y) \leqq M = f(a, b)$$

for all points $(x, y)$ of $R$. Similarly, $f$ attains its **absolute** (or **global**) **minimum value** $m$ on $R$ at the point $(c, d)$ provided that $f(x, y) \geqq m = f(c, d)$ for all points $(x, y)$ of $R$. The following theorem, proved in advanced calculus courses, guarantees the existence of absolute maximum and minimum values in many situations of practical interest.

---

*Theorem 1  Existence of Extreme Values*

Suppose that the function $f$ is continuous on the region $R$ which consists of the points on and within a simple closed curve $C$ in the plane. Then $f$ attains an absolute maximum value at some point $(a, b)$ of $R$ and attains an absolute minimum value at some point $(c, d)$ of $R$.

---

But how do we *find* these extrema? We are interested mainly in the case in which the function $f$ attains its absolute maximum (or minimum) value at an interior point of $R$. The point $(a, b)$ of $R$ is called an **interior point** of $R$ provided that some circular disk centered at $(a, b)$ lies wholly within $R$. The interior points of a region $R$ of the sort described in Theorem 1 are precisely those that do *not* lie on the boundary curve $C$. Although this fact seems intuitively plausible, it is by no means obvious; we omit its proof.

An absolute extreme value attained by the function $f$ at an *interior* point of $R$ is necessarily a local extreme value. We say that $f(a, b)$ is a **local maximum value** of $f$ if there is a circular disk $D$ centered at $(a, b)$ such that $f$ is defined on $D$ and $f(x, y) \leqq f(a, b)$ for all points $(x, y)$ of $D$. If the inequality is reversed, then $f(a, b)$ is a **local minimum value** of the function $f$.

**EXAMPLE 1**  Figure 15.48 shows the graph of a certain function $f(x, y)$ that is defined on a region $R$ in the $xy$-plane bounded by the simple closed curve $C$. We can think of the surface $z = f(x, y)$ as formed by starting with a stretched elastic membrane with fixed boundary $C$ and then poking two fingers upward and two fingers downward. The two fingers in each direction have different lengths. Looking at the four extreme values of $f(x, y)$ that result, we see

 **1** A local maximum that is not an absolute maximum,
 **2** A local maximum that is also an absolute maximum,
 **3** A local minimum that is not an absolute minimum,      and
 **4** A local minimum that is also a global minimum.

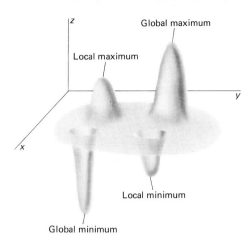

**15.48** Local extrema contrasted with global extrema

In Theorem 1 of Section 3-5, we saw that if the function $f(x)$ has a local maximum or local minimum value at a point $x = c$ where it is differentiable, then $f'(c) = 0$. We now shall show that an analogous event occurs in the two-dimensional situation: If $f(a, b)$ is either a local maximum value or a local minimum value of the function $f(x, y)$, then $f_x(a, b)$ and $f_y(a, b)$ are both zero, provided that these two partial derivatives both exist at the point $(a, b)$.

Suppose, for example, that $f(a, b)$ is a local maximum value of $f(x, y)$ and that the partial derivatives $f_x(a, b)$ and $f_y(a, b)$ both exist. We look at cross sections of the graph of $z = f(x, y)$ in the same way as when we defined these partial derivatives in Section 15-4. Let

$$G(x) = f(x, b) \quad \text{and} \quad H(y) = f(a, y).$$

Because $f$ is defined on a circular disk centered at $(a, b)$, it follows that $G(x)$ is defined on some open interval containing the point $x = a$ and that $H(y)$ is defined on some open interval containing the point $y = b$. Since $f$ has a local maximum at $(a, b)$, it follows that $G(x)$ has a local maximum at $x = a$ and that $H(y)$ has a local maximum at $y = b$. The single-variable maximum-minimum result cited above therefore implies that $G'(a) = 0$ and that $H'(b) = 0$. But

$$G'(a) = \lim_{h \to 0} \frac{G(a + h) - G(a)}{h} = \lim_{h \to 0} \frac{f(a + h, b) - f(a, b)}{h} = f_x(a, b)$$

and

$$H'(b) = \lim_{k \to 0} \frac{H(b + k) - H(b)}{k} = \lim_{k \to 0} \frac{f(a, b + k) - f(a, b)}{k} = f_y(a, b).$$

Hence we conclude that $f_x(a, b) = 0 = f_y(a, b)$. A similar argument gives the same conclusion if $f(a, b)$ is a local minimum value of $f$. This discussion establishes the following theorem.

---

**Theorem 2** *Necessary Conditions for Local Extrema*

Suppose that $f(x, y)$ attains a local maximum value or a local minimum value at the point $(a, b)$ and that the partial derivatives $f_x(a, b)$ and $f_y(a, b)$ both exist. Then

$$f_x(a, b) = 0 = f_y(a, b). \tag{1}$$

---

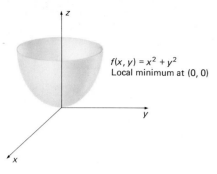

$f(x, y) = x^2 + y^2$
Local minimum at (0, 0)

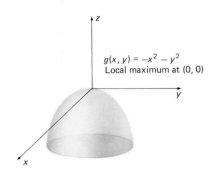

$g(x, y) = -x^2 - y^2$
Local maximum at (0, 0)

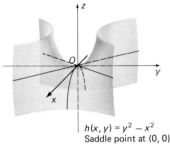

$h(x, y) = y^2 - x^2$
Saddle point at (0, 0)

**15.49** When both partial derivatives are zero, there may be a maximum, a minimum, or neither.

Equations (1) imply that the tangent plane to the surface $z = f(x, y)$ must be horizontal at any local maximum or local minimum point $(a, b, f(a, b))$, just as in the single-variable case (in which the tangent line is horizontal at any local maximum or minimum point).

**EXAMPLE 2** Consider the three familiar surfaces

$$z = f(x, y) = x^2 + y^2,$$
$$z = g(x, y) = -x^2 - y^2, \quad \text{and}$$
$$z = h(x, y) = y^2 - x^2$$

shown in Fig. 15.49. In each case $\partial z/\partial x = \pm 2x$ and $\partial z/\partial y = \pm 2y$. Thus both partial derivatives vanish only at the origin $(0, 0)$. It is clear from the figure that $f(x, y) = x^2 + y^2$ has a local minimum at $(0, 0)$; indeed, because a square cannot be negative, it is plain that $z = x^2 + y^2$ has the global minimum value 0 at $(0, 0)$. Similarly, $g(x, y) = -x^2 - y^2$ has a local maximum at $(0, 0)$, while $h(x, y) = y^2 - x^2$ has neither a local minimum nor a local maximum there—the origin is a *saddle point* of $h$. This example shows that a point $(a, b)$, where

$$\frac{\partial z}{\partial x} = 0 = \frac{\partial z}{\partial y}$$

may correspond to either a local maximum, a local minimum, or neither. Thus the necessary condition in (1) is *not* a sufficient condition for a local extremum.

**EXAMPLE 3** Find all points on the surface

$$z = \tfrac{3}{4}y^2 + \tfrac{1}{24}y^3 - \tfrac{1}{32}y^4 - x^2$$

where the tangent plane is horizontal.

***Solution*** We first calculate the partial derivatives $\partial z/\partial x$ and $\partial z/\partial y$:

$$\frac{\partial z}{\partial x} = -2x,$$

$$\frac{\partial z}{\partial y} = \frac{3}{2}y + \frac{1}{8}y^2 - \frac{1}{8}y^3$$

$$= -\frac{1}{8}y(y^2 - y - 12)$$

$$= -\frac{1}{8}y(y + 3)(y - 4).$$

We next equate both $\partial z/\partial x$ and $\partial z/\partial y$ to zero. This yields

$$-2x = 0$$

and

$$-\frac{1}{8}y(y + 3)(y - 4) = 0.$$

Simultaneous solution of these equations yields exactly three points where both partial derivatives vanish—$(0, -3)$, $(0, 0)$, and $(0, 4)$. The three corre-

ditions that $df$ be a good approximation to the increment $\Delta f$ when $\Delta x$ and $\Delta y$ are small.

---

**Theorem** *Linear Approximation*

Suppose that $f(x, y)$ has continuous first order partial derivatives on a rectangular region with horizontal and vertical sides and containing the points $P(a, b)$ and $Q(a + \Delta x, b + \Delta y)$. Let $\Delta f$ be the increment of Equation (1). Then

$$\Delta f = f_x(a, b)\, \Delta x + f_y(a, b)\, \Delta y + \varepsilon_1\, \Delta x + \varepsilon_2\, \Delta y \qquad (8)$$

where $\varepsilon_1$ and $\varepsilon_2$ are functions of $\Delta x$ and $\Delta y$ that approach zero as $\Delta x \to 0$ and $\Delta y \to 0$.

---

**Proof**  If $R$ is the point $(a + \Delta x, b)$ indicated in Fig. 15.57, then

$$\Delta f = f(Q) - f(P) = [f(R) - f(P)] + [f(Q) - f(R)]$$
$$= [f(a + \Delta x, b) - f(a, b)] + [f(a + \Delta x, b + \Delta y) - f(a + \Delta x, b)].$$
$$(9)$$

We consider separately the two terms on the right in Equation (9). For the first term, we define the single-variable function

$$g(x) = f(x, b) \quad \text{for } x \text{ in } [a, a + \Delta x].$$

Then the mean value theorem gives

$$f(a + \Delta x, b) - f(a, b) = g(a + \Delta x) - g(a)$$
$$= g'(X)\, \Delta x = f_x(X, b)\, \Delta x$$

for some number $X$ in the open interval $(a, a + \Delta x)$.

For the second term on the right in Equation (9), we define the single-variable function

$$h(y) = f(a + \Delta x, y) \quad \text{for } y \text{ in } [b, b + \Delta y].$$

The mean value theorem now yields

$$f(a + \Delta x, b + \Delta y) - f(a + \Delta x, b) = h(b + \Delta y) - h(b)$$
$$= h'(Y)\, \Delta y = f_y(a + \Delta x, Y)\, \Delta y$$

for some number $Y$ in $(b, b + \Delta y)$.

When we substitute these two results into Equation (9), we find that

$$\Delta f = f_x(X, b)\, \Delta x + f_y(a + \Delta x, Y)\, \Delta y$$
$$= [f_x(a, b) + f_x(X, b) - f_x(a, b)]\, \Delta x$$
$$\quad + [f_y(a, b) + f_y(a + \Delta x, Y) - f_y(a, b)]\, \Delta y.$$

So

$$\Delta f = f_x(a, b)\, \Delta x + f_y(a, b)\, \Delta y + \varepsilon_1\, \Delta x + \varepsilon_2\, \Delta y,$$

where

$$\varepsilon_1 = f_x(X, b) - f_x(a, b), \qquad \varepsilon_2 = f_y(a + \Delta x, Y) - f_y(a, b).$$

Finally, since the points $(X, b)$ and $(a + \Delta x, Y)$ both approach $(a, b)$ as $\Delta x \to 0$ and $\Delta y \to 0$, it follows from the continuity of $f_x$ and $f_y$ that $\varepsilon_1$ and

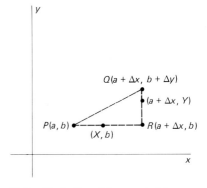

**15.57**  Illustration of the proof of the linear approximation theorem

$\varepsilon_2$ both approach zero as $\Delta x$ and $\Delta y$ approach zero. This completes the proof. ∎

The function of Problem 39 illustrates the fact that a function $f(x, y)$ of two variables can have partial derivatives at a point without even being continuous there. Thus the mere existence of partial derivatives means much less for a function of two (or more) variables than does differentiability for a single-variable function. But it follows from the linear approximation theorem above that a function with partial derivatives that are *continuous* at every point within a circle is itself continuous within that circle (see Problem 40).

A function $f$ of two variables is called **differentiable** at the point $(a, b)$ if $f_x(a, b)$ and $f_y(a, b)$ both exist *and also* there exist functions $\varepsilon_1$ and $\varepsilon_2$ of $\Delta x$ and $\Delta y$ that approach zero as $\Delta x$ and $\Delta y$ do and such that Equation (8) holds. We shall have no need of this concept of differentiability, since it will always suffice for us to assume that our functions have continuous partial derivatives.

The basic theorem above can be generalized in a natural way to functions of three or more variables. For example, if $w = f(x, y, z)$, the analogue of Equation (8) is

$$\Delta f = f_x(a, b, c)\,\Delta x + f_y(a, b, c)\,\Delta y + f_z(a, b, c)\,\Delta z$$
$$+ \varepsilon_1\,\Delta x + \varepsilon_2\,\Delta y + \varepsilon_3\,\Delta z,$$

where $\varepsilon_1$, $\varepsilon_2$, and $\varepsilon_3$ all approach zero as $\Delta x$, $\Delta y$, and $\Delta z$ approach zero. The proof for the three-variable case is much like the one above for two variables.

## 15-6 PROBLEMS

Find the differential $dw$ in Problems 1–16.

1 $w = 3x^2 + 4xy - 2y^3$    2 $w = e^{-x^2-y^2}$

3 $w = \sqrt{1 + x^2 + y^2}$    4 $w = xye^{x+y}$

5 $w = \arctan\left(\dfrac{y}{x}\right)$    6 $w = xz^2 - yx^2 + zy^2$

7 $w = \ln(x^2 + y^2 + z^2)$    8 $w = \sin xyz$

9 $w = x\tan yz$    10 $w = xye^{uv}$

11 $w = e^{-xyz}$    12 $w = \ln(1 + rs)$

13 $w = u^2 e^{-v^2}$    14 $w = \dfrac{s+t}{s-t}$

15 $w = \sqrt{x^2 + y^2 + z^2}$    16 $w = pqre^{-(p^2+q^2+r^2)}$

In each of Problems 17–23, use differentials to approximate $\Delta f = f(Q) - f(P)$.

17 $f(x, y) = \sqrt{x^2 + y^2}$;   $P(3, 4)$, $Q(2.97, 4.04)$

18 $f(x, y) = \sqrt{x^2 - y^2}$;   $P(13, 5)$, $Q(13.2, 4.9)$

19 $f(x, y) = \dfrac{1}{1 + x + y}$;   $P(3, 6)$, $Q(3.02, 6.05)$

20 $f(x, y, z) = \sqrt{xyz}$;   $P(1, 3, 3)$, $Q(0.9, 2.9, 3.1)$

21 $f(x, y, z) = \sqrt{x^2 + y^2 + z^2}$;   $P(3, 4, 12)$, $Q(3.03, 3.96, 12.05)$

22 $f(x, y, z) = \dfrac{xyz}{x + y + z}$;   $P(2, 3, 5)$, $Q(1.98, 3.03, 4.97)$

23 $f(x, y, z) = e^{-xyz}$;   $P(1, 0, -2)$, $Q(1.02, 0.03, -2.02)$

In each of Problems 24–29, use differentials to approximate the indicated number.

24 $\sqrt{26}\,\sqrt[3]{28}\,\sqrt[4]{17}$    25 $(\sqrt{15} + \sqrt{99})^2$

26 $\dfrac{\sqrt[3]{25}}{\sqrt[5]{30}}$    27 $e^{0.4} = e^{(1.1)^2 - (0.9)^2}$

28 The $y$-coordinate of the point $P$ near $(1, 2)$ on the curve $2x^3 + 2y^3 = 9xy$, if the $x$-coordinate of $P$ is 1.1.

29 The $x$-coordinate of the point $P$ near $(2, 4)$ on the curve $4x^4 + 4y^4 = 17x^2y^2$, if the $y$-coordinate of $P$ is 3.9.

30 The base and height of a rectangle are measured as 10 cm and 15 cm, respectively, with a possible error of as much as 0.1 cm in each. Use differentials to estimate the maximum resulting error in computing the area of the rectangle.

**31** The base radius $r$ and height $h$ of a right circular cone are measured as 5 in. and 10 in., respectively. There is a possible error of as much as $\frac{1}{16}$ in. in each measurement. Use differentials to estimate the maximum resulting error that might occur in computing the volume of the cone. (*Note:* The volume $V$ is given by $V = \frac{1}{3}\pi r^2 h$.)

**32** The dimensions of a closed rectangular box are found by measurement to be 10 cm, 15 cm, and 20 cm, but there is a possible error of 0.1 cm in each. Use differentials to estimate the maximum resulting error in computing the total surface area of the box.

**33** A surveyor wants to find the area in acres (1 acre is 43,560 ft$^2$) of a certain field. She measures two adjacent sides, finding them to be $a = 500$ ft and $b = 700$ ft, with a possible error of as much as 1 ft in each. She finds the angle between these two sides to be $\theta = 30°$, with a possible error of 0.25°. The field is triangular, so its area is given by $A = \frac{1}{2}ab \sin \theta$. Use differentials to estimate the maximum resulting error, in acres, in computing the area of the field by this formula.

**34** Use differentials to estimate the change in the volume of the gas of Example 3 if its pressure is decreased from 5 atm to 4.9 atm and its temperature is decreased from 300°K to 280°K.

**35** The period of a pendulum of length $L$ is given (approximately) by the formula $T = 2\pi\sqrt{L/g}$. Estimate the change in the period of a pendulum if its length is increased from 2 ft to 2 ft 1 in. and if it is simultaneously moved from a location where $g$ is exactly 32 ft/s$^2$ to one where $g = 32.2$ ft/s$^2$.

**36** Show that, for the pendulum of Problem 35, the relative error in the determination of $T$ is half the dif-

ference of the relative errors in measuring $L$ and $g$; that is, that

$$\frac{dT}{T} = \frac{1}{2}\left(\frac{dL}{L} - \frac{dg}{g}\right).$$

**37** The range of a projectile fired (in vacuum) with initial velocity $v_0$ and inclination angle $\alpha$ from the horizontal is $R = \frac{1}{32}v_0^2 \sin 2\alpha$. Use differentials to approximate the change in range if $v_0$ is increased from 400 ft/s to 410 ft/s and $\alpha$ is increased from 30° to 31°.

**38** A horizontal beam is supported at both ends and supports a uniform load. The deflection, or sag, at its midpoint is given by $S = k/wh^3$, where $w$ and $h$ are the width and height, respectively, of the beam, and $k$ is a constant that depends on the length and composition of the beam and the amount of the load. Show that

$$dS = -S\left(\frac{dw}{w} + \frac{3\,dh}{h}\right).$$

If $S$ is 1 in. when $w$ is 2 in. and $h$ is 4 in., approximate the sag when $w = 2.1$ in. and $h = 4.1$ in. Compare your approximation with the actual value you compute from the formula $S = k/wh^3$.

**39** Let the function $f$ be defined on the whole $xy$-plane by $f(x, y) = 1$ if $x = y \neq 0$, while $f(x, y) = 0$ otherwise. Show that

(a) $f$ is not continuous at $(0, 0)$; but

(b) both first partial derivatives $f_x$ and $f_y$ exist at $(0, 0)$.

**40** Deduce from Equation (8) that under the hypotheses of the linear approximation theorem, $\Delta f \to 0$ as $\Delta x \to 0$ and $\Delta y \to 0$. What does this imply about the continuity of $f$ at the point $(a, b)$?

---

## 15-7

## The Chain Rule

The single-variable chain rule expresses the derivative of a composite function $f(g(t))$ in terms of the derivatives of $f$ and $g$:

$$D_t f(g(t)) = f'(g(t))g'(t). \tag{1}$$

With $w = f(x)$ and $x = g(t)$, the chain rule says that

$$\frac{dw}{dt} = \frac{dw}{dx}\frac{dx}{dt}. \tag{2}$$

The simplest multivariable chain rule situation involves a function $w = f(x, y)$ where $x$ and $y$ are each functions of the same single variable $t$: $x = g(t)$ and $y = h(t)$. The composite function $f(g(t), h(t))$ is then a single-variable function of $t$, and the following theorem expresses its derivative in terms of the partial derivatives of $f$ and the ordinary derivatives of $g$ and $h$. We assume that the stated hypotheses hold on suitable domains such that the composite function is defined.

> **Theorem 1**    *The Chain Rule*
>
> Suppose that $w = f(x, y)$ has continuous first order partial derivatives, and that $x = g(t)$ and $y = h(t)$ are differentiable functions. Then $w$ is a differentiable function of $t$ and
>
> $$\frac{dw}{dt} = \frac{\partial w}{\partial x}\frac{dx}{dt} + \frac{\partial w}{\partial y}\frac{dy}{dt}. \qquad (3)$$

The variable notation of Formula (3) ordinarily will be more useful than functional notation. It must be remembered, however, that the partial derivatives in (3) are to be evaluated at the point $(g(t), h(t))$, so what (3) actually says is that

$$D_t[f(g(t), h(t))] = f_x(g(t), h(t))g'(t) + f_y(g(t), h(t))h'(t). \qquad (4)$$

**EXAMPLE 1**    Suppose that

$$w = e^{xy}, \qquad x = t^2, \quad \text{and} \quad y = t^3.$$

Then

$$\frac{\partial w}{\partial x} = ye^{xy}, \qquad \frac{\partial w}{\partial y} = xe^{xy}, \qquad \text{and}$$

$$\frac{dx}{dt} = 2t, \qquad \frac{dy}{dt} = 3t^2.$$

So the formula in (3) yields

$$\frac{dw}{dt} = \frac{\partial w}{\partial x}\frac{dx}{dt} + \frac{\partial w}{\partial y}\frac{dy}{dt}$$

$$= (ye^{xy})(2t) + (xe^{xy})(3t^2)$$

$$= (t^3e^{t^5})(2t) + (t^2e^{t^5})(3t^2) = 5t^4e^{t^5}.$$

Had our purpose not been to illustrate the chain rule, we could have obtained the same result more simply by writing

$$w = e^{xy} = e^{(t^2)(t^3)} = e^{t^5}$$

and then differentiating $w$ as a (single-variable) function of $t$.

---

***Proof of the Chain Rule***    We choose a point $t_0$ at which we want to compute $dw/dt$ and write

$$a = g(t_0), \qquad b = h(t_0).$$

Let

$$\Delta x = g(t_0 + \Delta t) - g(t_0), \qquad \Delta y = h(t_0 + \Delta t) - h(t_0).$$

Then

$$g(t_0 + \Delta t) = a + \Delta x \quad \text{and} \quad h(t_0 + \Delta t) = b + \Delta y.$$

If

$$\Delta w = f(g(t_0 + \Delta t), h(t_0 + \Delta t)) - f(g(t_0), h(t_0))$$

$$= f(a + \Delta x, b + \Delta y) - f(a, b),$$

then what we want to compute is

$$\frac{dw}{dt} = \lim_{\Delta t \to 0} \frac{\Delta w}{\Delta t}.$$

The linear approximation theorem of Section 15-6 gives

$$\Delta w = f_x(a, b)\, \Delta x + f_y(a, b)\, \Delta y + \varepsilon_1\, \Delta x + \varepsilon_2\, \Delta y$$

where $\varepsilon_1$ and $\varepsilon_2$ approach zero as $\Delta x \to 0$ and $\Delta y \to 0$. We note that $\Delta x$ and $\Delta y$ approach zero as $\Delta t \to 0$ because the derivatives

$$\frac{dx}{dt} = \lim_{\Delta t \to 0} \frac{\Delta x}{\Delta t} \quad \text{and} \quad \frac{dy}{dt} = \lim_{\Delta t \to 0} \frac{\Delta y}{\Delta t}$$

both exist. Therefore,

$$\begin{aligned}
\frac{dw}{dt} &= \lim_{\Delta t \to 0} \frac{\Delta w}{\Delta t} \\
&= \lim_{\Delta t \to 0} \left[ f_x(a, b) \frac{\Delta x}{\Delta t} + f_y(a, b) \frac{\Delta y}{\Delta t} + \varepsilon_1 \frac{\Delta x}{\Delta t} + \varepsilon_2 \frac{\Delta y}{\Delta t} \right] \\
&= f_x(a, b) \frac{dx}{dt} + f_y(a, b) \frac{dy}{dt} + 0 \frac{dx}{dt} + 0 \frac{dy}{dt};
\end{aligned}$$

hence

$$\frac{dw}{dt} = \frac{\partial w}{\partial x} \frac{dx}{dt} + \frac{\partial w}{\partial y} \frac{dy}{dt}.$$

Thus we have established Formula (3), writing $\partial w/\partial x$ and $\partial w/\partial y$ for the partial derivatives $f_x(a, b)$ and $f_y(a, b)$, respectively, in the final step. ∎

In the situation of Theorem 1, we may refer to $w$ as the **dependent** variable, $x$ and $y$ as **intermediate** variables, and $t$ as the **independent** variable. Then note that the right-hand side of Equation (3) has two terms, one for each intermediate variable, each like the right-hand side of the single-variable chain rule of Equation (2). If there are more than two intermediate variables, then there is still one term on the right-hand side for each intermediate variable. For example, if $w = f(x, y, z)$, with $x$, $y$, and $z$ each a function of $t$, then the chain rule takes the form

$$\frac{dw}{dt} = \frac{\partial w}{\partial x} \frac{dx}{dt} + \frac{\partial w}{\partial y} \frac{dy}{dt} + \frac{\partial w}{\partial z} \frac{dz}{dt}. \tag{5}$$

The proof of (5) is essentially the same as the proof of (3); it uses the linear approximation theorem for three variables rather than for two variables.

**EXAMPLE 2** Find $dw/dt$ if $w = x^2 + ze^y + \sin xz$ and $x = t$, $y = t^2$, $z = t^3$.

*Solution* Equation (5) gives

$$\begin{aligned}
\frac{dw}{dt} &= \frac{\partial w}{\partial x} \frac{dx}{dt} + \frac{\partial w}{\partial y} \frac{dy}{dt} + \frac{\partial w}{\partial z} \frac{dz}{dt} \\
&= (2x + z \cos xz)(1) + (ze^y)(2t) + (e^y + x \cos xz)(3t^2) \\
&= 2t + (3t^2 + 2t^4)e^{t^2} + 4t^3 \cos t^4.
\end{aligned}$$

In this example we could check the result given by the chain rule by first writing $w$ as a function of $t$ and then computing the ordinary single-variable derivative with respect to $t$.

There may be several independent variables as well as several intermediate variables. For example, if $w = f(x, y, z)$, where $x = g(u, v)$, $y = h(u, v)$, and $z = k(u, v)$, so that

$$w = f(x, y, z) = f(g(u, v), h(u, v), k(u, v)),$$

then we have the three intermediate variables $x$, $y$, and $z$ and the two independent variables $u$ and $v$. In this case we would want to compute the *partial* derivatives $\partial w/\partial u$ and $\partial w/\partial v$ of the composite function. The general chain rule below says that each partial derivative of the dependent variable $w$ is given by a chain rule formula like (3) or (5) above. The only difference is that the derivatives with respect to the independent variables will be partial derivatives. For instance,

$$\frac{\partial w}{\partial u} = \frac{\partial w}{\partial x}\frac{\partial x}{\partial u} + \frac{\partial w}{\partial y}\frac{\partial y}{\partial u} + \frac{\partial w}{\partial z}\frac{\partial z}{\partial u}.$$

The "molecular model" in Fig. 15.58 illustrates this formula. The "atom" at the top represents the independent variable $w$. The atoms at the next level represent the intermediate variables $x$, $y$, and $z$. The atoms at the bottom represent the independent variables $u$ and $v$. Each "bond" in the diagram represents a partial derivative involving the two variables represented by the atoms joined by that bond. Finally, note that the formula displayed before this paragraph expresses $\partial w/\partial u$ as the sum of the products of the partial derivatives taken along every path from $w$ to $u$. Similarly, the sum of the products of the partial derivatives along every path from $w$ to $v$ yields the correct formula

$$\frac{\partial w}{\partial v} = \frac{\partial w}{\partial x}\frac{\partial x}{\partial v} + \frac{\partial w}{\partial y}\frac{\partial y}{\partial v} + \frac{\partial w}{\partial z}\frac{\partial z}{\partial v}.$$

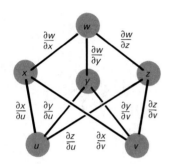

**15.58** Diagram for $w = w(x, y, z)$, where $x = x(u, v)$, $y = y(u, v)$, $z = z(u, v)$

The following theorem describes the most general such situation.

---

**Theorem 2  The General Chain Rule**

Suppose that $w$ is a function of the variables $x_1, x_2, x_3, \ldots, x_m$, and that each of these is a function of the variables $t_1, t_2, \ldots, t_n$. If all these functions have continuous first order partial derivatives, then

$$\frac{\partial w}{\partial t_i} = \frac{\partial w}{\partial x_1}\frac{\partial x_1}{\partial t_i} + \frac{\partial w}{\partial x_2}\frac{\partial x_2}{\partial t_i} + \cdots + \frac{\partial w}{\partial x_m}\frac{\partial x_m}{\partial t_i} \qquad (6)$$

for each $i$, $1 \leq i \leq n$.

---

Thus there is a formula in (6) for *each* of the independent variables $t_1, t_2, \ldots, t_n$, and the right-hand side of each such formula contains one typical chain rule term for each of the intermediate variables $x_1, x_2, \ldots, x_m$.

**EXAMPLE 3**  Suppose that

$$z = f(u, v), \qquad u = 2x + y, \qquad v = 3x - 2y.$$

Given the values $\partial z/\partial u = 3$ and $\partial z/\partial v = -2$ at the point $u = 3$, $v = 1$, find the values $\partial z/\partial x$ and $\partial z/\partial y$ at the corresponding point $x = 1$, $y = 1$.

*Solution* The relationships between the variables are shown in Fig. 15.59. The chain rule gives

$$\frac{\partial z}{\partial x} = \frac{\partial z}{\partial u}\frac{\partial u}{\partial x} + \frac{\partial z}{\partial v}\frac{\partial v}{\partial x}$$

$$= (3)(2) + (-2)(3) = 0$$

and

$$\frac{\partial z}{\partial y} = \frac{\partial z}{\partial u}\frac{\partial u}{\partial y} + \frac{\partial z}{\partial v}\frac{\partial v}{\partial y}$$

$$= (3)(1) + (-2)(-2) = 7$$

at the indicated point $(x, y) = (1, 1)$.

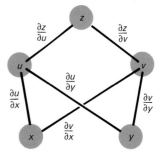

**15.59** Diagram for $z = z(u, v)$, where $u = u(x, y)$ and $v = v(x, y)$

**EXAMPLE 4** Let $w = f(x, y)$, where $x$ and $y$ are given in polar coordinates by the equations $x = r \cos \theta$ and $y = r \sin \theta$. Calculate

$$\frac{\partial w}{\partial r}, \quad \frac{\partial w}{\partial \theta}, \quad \text{and} \quad \frac{\partial^2 w}{\partial r^2}$$

in terms of $r$ and $\theta$ and the partial derivatives of $w$ with respect to $x$ and $y$. (See Fig. 15.60.)

*Solution* Here $x$ and $y$ are intermediate variables, while the independent variables are $r$ and $\theta$. First note that

$$\frac{\partial x}{\partial r} = \cos \theta, \qquad \frac{\partial y}{\partial r} = \sin \theta,$$

$$\frac{\partial x}{\partial \theta} = -r \sin \theta, \quad \text{and} \quad \frac{\partial y}{\partial \theta} = r \cos \theta.$$

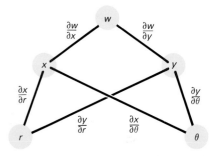

**15.60** Diagram for $w = w(x, y)$, where $x = x(r, \theta)$ and $y = y(r, \theta)$

Then

$$\frac{\partial w}{\partial r} = \frac{\partial w}{\partial x}\frac{\partial x}{\partial r} + \frac{\partial w}{\partial y}\frac{\partial y}{\partial r} = \frac{\partial w}{\partial x}\cos \theta + \frac{\partial w}{\partial y}\sin \theta \qquad (7a)$$

and

$$\frac{\partial w}{\partial \theta} = \frac{\partial w}{\partial x}\frac{\partial x}{\partial \theta} + \frac{\partial w}{\partial y}\frac{\partial y}{\partial \theta} = -r\frac{\partial w}{\partial x}\sin \theta + r\frac{\partial w}{\partial y}\cos \theta. \qquad (7b)$$

Next,

$$\frac{\partial^2 w}{\partial r^2} = \frac{\partial}{\partial r}\left(\frac{\partial w}{\partial r}\right)$$

$$= \frac{\partial}{\partial r}\left(\frac{\partial w}{\partial x}\cos \theta + \frac{\partial w}{\partial y}\sin \theta\right)$$

$$= \frac{\partial w_x}{\partial r}\cos \theta + \frac{\partial w_y}{\partial r}\sin \theta$$

where $w_x = \partial w/\partial x$ and $w_y = \partial w/\partial y$. Applying the formula in (7a) to calculate $\partial w_x/\partial r$ and $\partial w_y/\partial\theta$, we get

$$\frac{\partial^2 w}{\partial r^2} = \left(\frac{\partial w_x}{\partial x}\frac{\partial x}{\partial r} + \frac{\partial w_x}{\partial y}\frac{\partial y}{\partial r}\right)\cos\theta + \left(\frac{\partial w_y}{\partial x}\frac{\partial x}{\partial r} + \frac{\partial w_y}{\partial y}\frac{\partial y}{\partial r}\right)\sin\theta$$

$$= \left(\frac{\partial^2 w}{\partial x^2}\cos\theta + \frac{\partial^2 w}{\partial y\,\partial x}\sin\theta\right)\cos\theta + \left(\frac{\partial^2 w}{\partial x\,\partial y}\cos\theta + \frac{\partial^2 w}{\partial y^2}\sin\theta\right)\sin\theta.$$

Finally, because $w_{yx} = w_{xy}$, we get

$$\frac{\partial^2 w}{\partial r^2} = \frac{\partial^2 w}{\partial x^2}\cos^2\theta + 2\frac{\partial^2 w}{\partial x\,\partial y}\cos\theta\sin\theta + \frac{\partial^2 w}{\partial y^2}\sin^2\theta. \tag{8}$$

**EXAMPLE 5** Suppose that $w = f(u, v, x, y)$, where $u$ and $v$ are functions of $x$ and $y$. Here $x$ and $y$ appear to play dual roles as both intermediate and independent variables. The chain rule yields

$$\frac{\partial w}{\partial x} = \frac{\partial f}{\partial u}\frac{\partial u}{\partial x} + \frac{\partial f}{\partial v}\frac{\partial v}{\partial x} + \frac{\partial f}{\partial x}\frac{\partial x}{\partial x} + \frac{\partial f}{\partial y}\frac{\partial y}{\partial x}$$

$$= \frac{\partial f}{\partial u}\frac{\partial u}{\partial x} + \frac{\partial f}{\partial v}\frac{\partial v}{\partial x} + \frac{\partial f}{\partial x}$$

because $\partial x/\partial x = 1$ and $\partial y/\partial x = 0$. Similarly,

$$\frac{\partial w}{\partial y} = \frac{\partial f}{\partial u}\frac{\partial u}{\partial y} + \frac{\partial f}{\partial v}\frac{\partial v}{\partial y} + \frac{\partial f}{\partial y}.$$

Note that these results are consistent with the paths from $w$ to $x$ and from $w$ to $y$ in the molecular model shown in Fig. 15.61.

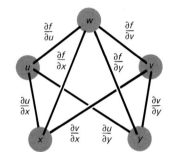

**15.61** Diagram for $w = f(u, v, x, y)$, where $u = u(x, y)$ and $v = v(x, y)$.

**EXAMPLE 6** Consider a parametric curve $x = x(t), y = y(t), z = z(t)$ lying on the surface $z = f(x, y)$ in space. Recall that if

$$\mathbf{T} = \left\langle\frac{dx}{dt}, \frac{dy}{dt}, \frac{dz}{dt}\right\rangle \quad \text{and} \quad \mathbf{N} = \left\langle\frac{\partial z}{\partial x}, \frac{\partial z}{\partial y}, -1\right\rangle,$$

then $\mathbf{T}$ is tangent to the curve and $\mathbf{N}$ is normal to the surface. Show that $\mathbf{T}$ and $\mathbf{N}$ are everywhere perpendicular.

*Solution* The chain rule of Equation (3) tells us that

$$\frac{dz}{dt} = \frac{\partial z}{\partial x}\frac{dx}{dt} + \frac{\partial z}{\partial y}\frac{dy}{dt}.$$

But this equation is equivalent to the vector equation

$$\left\langle\frac{\partial z}{\partial x}, \frac{\partial z}{\partial y}, -1\right\rangle \cdot \left\langle\frac{dx}{dt}, \frac{dy}{dt}, \frac{dz}{dt}\right\rangle = 0.$$

Thus $\mathbf{N} \cdot \mathbf{T} = 0$, and so $\mathbf{N}$ and $\mathbf{T}$ are indeed perpendicular.

> **Theorem 3   Implicit Partial Differentiation**
>
> Suppose that the function $F(x, y, z)$ has continuous first partial derivatives, and that the equation $F(x, y, z) = 0$ implicitly defines a function $z = f(x, y)$ that has continuous first partial derivatives. Then
>
> $$\frac{\partial z}{\partial x} = -\frac{F_x}{F_z} \quad \text{and} \quad \frac{\partial z}{\partial y} = -\frac{F_y}{F_z} \tag{9}$$
>
> wherever $F_z = \partial F/\partial z \neq 0$.

**Proof**   Because $w = F(x, y, f(x, y))$ is identically zero, differentiation with respect to $x$ yields

$$0 = \frac{\partial w}{\partial x} = \frac{\partial F}{\partial x}\frac{\partial x}{\partial x} + \frac{\partial F}{\partial y}\frac{\partial y}{\partial x} + \frac{\partial F}{\partial z}\frac{\partial z}{\partial x}$$

$$= (F_x)(1) + (F_y)(0) + (F_z)\left(\frac{\partial z}{\partial x}\right);$$

and so

$$F_x + F_z\frac{\partial z}{\partial x} = 0.$$

This gives the first formula in (9); the second is obtained similarly by differentiating $w$ with respect to $y$.  ∎

REMARK   In a specific example it usually is simpler to differentiate the equation $F(x, y, f(x, y)) = 0$ implicitly than to apply the formulas in (9).

**EXAMPLE 7**   Find the tangent plane at the point $(1, 3, 2)$ to the surface with equation

$$z^3 + xz - y^2 = 1.$$

**Solution**   Partial differentiation of the given equation with respect to $x$ and with respect to $y$ yields the equations

$$3z^2\frac{\partial z}{\partial x} + z + x\frac{\partial z}{\partial x} = 0 \quad \text{and}$$

$$3z^2\frac{\partial z}{\partial y} + x\frac{\partial z}{\partial y} - 2y = 0.$$

When we substitute $x = 1$, $y = 3$, and $z = 2$, we find that $\partial z/\partial x = -\frac{2}{13}$ and $\partial z/\partial y = \frac{6}{13}$. Hence the equation of the desired tangent plane is

$$z - 2 = (-\tfrac{2}{13})(x - 1) + (\tfrac{6}{13})(y - 3);$$

that is,

$$2x - 6y + 13z = 10.$$

## *AN APPLICATION TO PHYSICAL CHEMISTRY

We close this section with an example that shows how we can use chain rule computations to *actually do something*. Consider a quantity of liquid which has pressure $P$, volume $V$, and temperature $T$ that satisfy a "state equation" of the form

$$F(P, V, T) = 0, \qquad (10)$$

analogous to the state equation $PV - nRT = 0$ of an ideal gas. We shall not need to know precisely what the function $F$ is, but we shall assume that Equation (10) can be solved for each variable in terms of the other two. Let us write

$$P = g(V, T) \quad \text{and} \quad V = h(P, T). \qquad (11)$$

The partial derivatives $\partial P/\partial V$, $\partial P/\partial T$, $\partial V/\partial P$, and $\partial V/\partial T$ of $g$ and $h$ are important in physical chemistry.

Suppose that $V$ and $T$ are changed to $V + \Delta V$ and $T + \Delta T$, while $P$ is held constant. Then the fractional change in volume per unit change in temperature is $(\Delta V/V)/\Delta T$. The limit of this expression as $\Delta T \to 0$ gives the *thermal expansivity*

$$\alpha = \frac{1}{V}\frac{\partial V}{\partial T} \qquad (12)$$

of our liquid. Similarly, if $V$ and $P$ change while $T$ remains constant, the limit as $\Delta P \to 0$ of the negative of the fractional change in volume per unit change in pressure is the *isothermal compressivity* of our liquid,

$$\beta = -\frac{1}{V}\frac{\partial V}{\partial P}. \qquad (13)$$

We may differentiate the equation $F(P, V, T) = 0$ with respect to $P$ and $T$ as in Theorem 3. This yields the results

$$\frac{\partial V}{\partial P} = -\frac{F_P}{F_V} \quad \text{and} \quad \frac{\partial V}{\partial T} = -\frac{F_T}{F_V}. \qquad (14)$$

But when we differentiate Equation (10) with respect to $V$ and $T$, we find that

$$\frac{\partial P}{\partial V} = -\frac{F_V}{F_P} \quad \text{and} \quad \frac{\partial P}{\partial T} = -\frac{F_T}{F_P}. \qquad (15)$$

We finally substitute Equations (14) into the second of Equations (15) and obtain

$$\frac{\partial P}{\partial T} = -\frac{F_T}{F_P} = -\frac{-F_T/F_V}{-F_P/F_V} = -\frac{\partial V/\partial T}{\partial V/\partial P} = \frac{(1/V)(\partial V/\partial T)}{(-1/V)(\partial V/\partial P)};$$

consequently

$$\frac{\partial P}{\partial T} = \frac{\alpha}{\beta}. \qquad (16)$$

The following example shows that this last equation has significant practical consequences.

**EXAMPLE 8** The thermal expansivity and isothermal compressivity of liquid mercury are $\alpha = 1.8 \times 10^{-4}$ and $\beta = 3.9 \times 10^{-6}$, respectively, in liter-atm-°C units. Suppose that a thermometer bulb is exactly filled with mercury at 50°C. If the bulb can withstand an internal pressure of only 200 atm, can it be heated to 55°C without breaking?

**Solution**  By Equation (16), we find that

$$\frac{\partial P}{\partial T} = \frac{\alpha}{\beta} = \frac{1.8 \times 10^{-4}}{3.9 \times 10^{-6}} \approx 46.$$

Thus each increase of 1° in temperature will generate an additional internal pressure of about 46 atm. Hence if you heat the bulb to 55°C, you will generate a pressure of 230 atm and will, therefore, break the bulb.

## 15-7  PROBLEMS

In Problems 1–4, find $dw/dt$ both by using the chain rule *and* by expressing $w$ explicitly as a function of $t$ before differentiation.

**1** $w = e^{-x^2 - y^2}, \; x = t, \; y = t^{1/2}$

**2** $w = \dfrac{1}{u^2 + v^2}, \; u = \cos 2t, \; v = \sin 2t$

**3** $w = \sin xyz, \; x = t, \; y = t^2, \; z = t^3$

**4** $w = \ln(u + v + z), \; u = \cos^2 t, \; v = \sin^2 t, \; z = t^2$

In Problems 5–8, find $\partial w/\partial s$ and $\partial w/\partial t$.

**5** $w = \ln(x^2 + y^2 + z^2), \; x = s - t, \; y = s + t, \; z = 2\sqrt{st}$

**6** $w = pq \sin r, \; p = 2s + t, \; q = s - t, \; r = st$

**7** $w = (u^2 + v^2 + z^2)^{1/2}, \quad u = 3e^t \sin s, \quad v = 3e^t \cos s, \; z = 4e^t$

**8** $w = yz + zx + xy, \; x = s^2 - t^2, \; y = s^2 + t^2, \; z = s^2 t^2$

In Problems 9–10, find $\partial r/\partial x$, $\partial r/\partial y$, and $\partial r/\partial z$.

**9** $r = e^{u+v+w}, \; u = yz, \; v = xz, \; w = xy$

**10** $r = uvw - u^2 - v^2 - w^2, \; u = y + z, \; v = x + z, \; w = x + y$

In Problems 11–15, find $\partial z/\partial x$ and $\partial z/\partial y$ as functions of $x$, $y$, and $z$, assuming that $z = f(x, y)$ satisfies the given equation.

**11** $x^{2/3} + y^{2/3} + z^{2/3} = 1$

**12** $x^3 + y^3 + z^3 = xyz$

**13** $xe^{xy} + ye^{zx} + ze^{xy} = 3$

**14** $z^5 + xy^2 + yz = 5$

**15** $\dfrac{x^2}{a^2} + \dfrac{y^2}{b^2} + \dfrac{z^2}{c^2} = 1$

In Problems 16–19, use the method of Example 5 to find $\partial w/\partial x$ and $\partial w/\partial y$ as functions of $x$ and $y$.

**16** $w = u^2 + v^2 + x^2 + y^2, \; u = x - y, \; v = x + y$

**17** $w = \sqrt{uvxy}, \; u = \sqrt{x - y}, \; v = \sqrt{x + y}$

**18** $w = xy \ln(u + v), \; u = (x^2 + y^2)^{1/3}, \; v = (x^3 + y^3)^{1/2}$

**19** $w = uv - xy, \; u = \dfrac{x}{x^2 + y^2}, \; v = \dfrac{y}{x^2 + y^2}$

In Problems 20–23, write an equation for the tangent plane at the point $P$ to the surface with the given equation.

**20** $x^2 + y^2 + z^2 = 9; \quad P(1, 2, 2)$
**21** $x^2 + 2y^2 + 2z^2 = 14; \quad P(2, 1, -2)$
**22** $x^3 + y^3 + z^3 = 5xyz; \quad P(2, 1, 1)$
**23** $z^3 + (x + y)z^2 + x^2 + y^2 = 13; \quad P(2, 2, 1)$

**24** Suppose that $y = g(x, z)$ satisfies the equation $F(x, y, z) = 0$ and that $F_y \neq 0$. Show that

$$\frac{\partial y}{\partial x} = -\frac{\partial F/\partial x}{\partial F/\partial y}.$$

**25** Suppose that $x = h(y, z)$ satisfies the equation $F(x, y, z) = 0$ and that $F_x \neq 0$. Show that

$$\frac{\partial x}{\partial y} = -\frac{\partial F/\partial y}{\partial F/\partial x}.$$

**26** Differentiate the equation $F(V, P, T) = 0$ with respect to $P$ and $T$ to establish the equations in (14) of this section.

**27** Differentiate the equation $F(V, P, T) = 0$ with respect to $V$ and $T$ to establish the equations in (15) of this section.

**28** Suppose that $w = f(x, y)$, $x = r \cos \theta$ and $y = r \sin \theta$. Show that

$$\left(\frac{\partial w}{\partial x}\right)^2 + \left(\frac{\partial w}{\partial y}\right)^2 = \left(\frac{\partial w}{\partial r}\right)^2 + \frac{1}{r^2}\left(\frac{\partial w}{\partial \theta}\right)^2.$$

**29** Suppose that $w = f(u)$ and that $u = x + y$. Show that $\partial w/\partial x = \partial w/\partial y$.

**30** Suppose that $w = f(u)$ and that $u = x - y$. Show that $\partial w/\partial x = -\partial w/\partial y$ and that

$$\frac{\partial^2 w}{\partial x^2} = \frac{\partial^2 w}{\partial y^2} = -\frac{\partial^2 w}{\partial x \, \partial y}.$$

**31** Suppose that $w = f(x, y)$ where $x = u + v$ and $y = u - v$. Show that

$$\frac{\partial^2 w}{\partial x^2} - \frac{\partial^2 w}{\partial y^2} = \frac{\partial^2 w}{\partial u \, \partial v}.$$

**32** Assume that $w = f(x, y)$ where $x = 2u + v$ and $y = u - v$. Show that

$$5\frac{\partial^2 w}{\partial x^2} + 2\frac{\partial^2 w}{\partial x \, \partial y} + 2\frac{\partial^2 w}{\partial y^2} = \frac{\partial^2 w}{\partial u^2} + \frac{\partial^2 w}{\partial v^2}.$$

**33** Suppose that $w = f(x, y)$, $x = r \cos \theta$, and $y = r \sin \theta$. Show that

$$\frac{\partial^2 w}{\partial x^2} + \frac{\partial^2 w}{\partial y^2} = \frac{\partial^2 w}{\partial r^2} + \frac{1}{r}\frac{\partial w}{\partial r} + \frac{1}{r^2}\frac{\partial^2 w}{\partial \theta^2}.$$

(*Suggestion:* First find $\partial^2 w/\partial \theta^2$ by the method of Example 4. Then combine the result with Equations (7) and (8) of this section.)

**34** Suppose that $w = (1/r)f(t - r/a)$ and that $r = \sqrt{x^2 + y^2 + z^2}$. Show that

$$\frac{\partial^2 w}{\partial x^2} + \frac{\partial^2 w}{\partial y^2} + \frac{\partial^2 w}{\partial z^2} = \frac{1}{a^2}\frac{\partial^2 w}{\partial t^2}.$$

**35** Suppose that $w = f(r)$ and that $r = \sqrt{x^2 + y^2 + z^2}$. Show that

$$\frac{\partial^2 w}{\partial x^2} + \frac{\partial^2 w}{\partial y^2} + \frac{\partial^2 w}{\partial z^2} = \frac{d^2 w}{dr^2} + \frac{2}{r}\frac{dw}{dr}.$$

**36** Suppose that $w = f(u) + g(v)$, that $u = x - at$, and that $v = x + at$. Show that

$$\frac{\partial^2 w}{\partial t^2} = a^2 \frac{\partial^2 w}{\partial x^2}.$$

**37** Assume that $w = f(u, v)$ where $u = x + y$ and $v = x - y$. Show that

$$\frac{\partial w}{\partial x}\frac{\partial w}{\partial y} = \left(\frac{\partial w}{\partial u}\right)^2 - \left(\frac{\partial w}{\partial v}\right)^2.$$

**38** Given: $w = f(x, y)$, $x = e^u \cos v$, and $y = e^u \sin v$. Show that

$$\left(\frac{\partial w}{\partial x}\right)^2 + \left(\frac{\partial w}{\partial y}\right)^2 = e^{-2u}\left[\left(\frac{\partial w}{\partial u}\right)^2 + \left(\frac{\partial w}{\partial v}\right)^2\right].$$

**39** Assume that $w = f(x, y)$ and that there is a constant $\alpha$ such that $x = u \cos \alpha - v \sin \alpha$ and $y = u \sin \alpha + v \cos \alpha$. Show that

$$\left(\frac{\partial w}{\partial u}\right)^2 + \left(\frac{\partial w}{\partial v}\right)^2 = \left(\frac{\partial w}{\partial x}\right)^2 + \left(\frac{\partial w}{\partial y}\right)^2.$$

**40** Suppose that $w = f(u)$ where

$$u = \frac{x^2 - y^2}{x^2 + y^2}.$$

Show that $xw_x + yw_y = 0$.

Suppose that the equation $F(x, y, z) = 0$ defines implicitly the three functions $z = f(x, y)$, $y = g(x, z)$, and $x = h(y, z)$. To keep track of the various partial derivatives, the notation

$$\left(\frac{\partial z}{\partial x}\right)_y = \frac{\partial f}{\partial x}, \qquad \left(\frac{\partial z}{\partial y}\right)_x = \frac{\partial f}{\partial y}, \tag{17a}$$

$$\left(\frac{\partial y}{\partial x}\right)_z = \frac{\partial g}{\partial x}, \qquad \left(\frac{\partial y}{\partial z}\right)_x = \frac{\partial g}{\partial z}, \tag{17b}$$

$$\left(\frac{\partial x}{\partial y}\right)_z = \frac{\partial h}{\partial y}, \qquad \left(\frac{\partial x}{\partial z}\right)_y = \frac{\partial h}{\partial z} \tag{17c}$$

is used. In short, the general symbol $(\partial w/\partial u)_v$ denotes the derivative of $w$ with respect to $u$, with $w$ regarded as a function of the independent variables $u$ and $v$.

**41** Using the notation in (17), show that

$$\left(\frac{\partial x}{\partial y}\right)_z \left(\frac{\partial y}{\partial z}\right)_x \left(\frac{\partial z}{\partial x}\right)_y = -1.$$

(*Suggestion:* Find the three partial derivatives on the right-hand side in terms of $F_x$, $F_y$, and $F_z$.)

**42** Verify the result in Problem 41 for the equation

$$F(x, y, z) = x^2 + y^2 + z^2 - 1 = 0.$$

**43** Verify the result in Problem 41 (with $P$, $V$, and $T$ in place of $x$, $y$, and $z$) for the equation

$$F(P, V, T) = PV - nRT = 0$$

($n$ and $R$ constant) that expresses the ideal gas law.

---

**15-8**

### Directional Derivatives and the Gradient Vector

We know that the partial derivatives $f_x(x, y, z)$, $f_y(x, y, z)$, and $f_z(x, y, z)$ give the rates of change of $w = f(x, y, z)$ at the point $P(x, y, z)$ in the $x$-, $y$-, and $z$-directions, respectively. Now we want to say what is meant by the rate of change of $w$ in the direction of an *arbitrary unit* vector $\mathbf{u} = \langle a, b, c \rangle$.

Let $Q$ be a point on the ray in the direction of $\mathbf{u}$ from the point $P$ (see Fig. 15.62). The **average rate of change of $w$ with respect to distance**

**between** $P$ and $Q$ is

$$\frac{f(Q) - f(P)}{|\overrightarrow{PQ}|} = \frac{\Delta w}{\text{distance } PQ}.$$

We want to find the *instantaneous* rate of change of $w$ in the direction $\mathbf{u}$ at the point $P$. So we take the limit of this average rate of change as $Q$ approaches $P$ along the ray:

$$\lim_{Q \to P} \frac{f(Q) - f(P)}{|\overrightarrow{PQ}|}. \qquad (1)$$

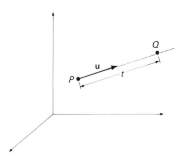

**15.62** First step in computing the rate of change of $f(x, y, z)$ in the direction of the unit vector $\mathbf{u}$

The value of this limit is called the **directional derivative of $f$ at $P(x, y, z)$ in the direction $\mathbf{u}$**, and we denote it by $D_{\mathbf{u}}f(x, y, z)$.

To make the limit in (1) more precise, let $t = |\overrightarrow{PQ}|$. Then $\overrightarrow{PQ} = t\mathbf{u} = \langle at, bt, ct \rangle$, so $f(Q) = f(x + at, y + bt, z + ct)$. Now we can rewrite the limit in (1) as

$$D_{\mathbf{u}}f(x, y, z) = \lim_{t \to 0} \frac{f(x + at, y + bt, z + ct) - f(x, y, z)}{t}. \qquad (2)$$

This is the formal definition of the directional derivative $D_{\mathbf{u}}f(x, y, z)$. Note that the limit in (2) is equivalent to

$$D_{\mathbf{u}}f(x, y, z) = D_t f(x + at, y + bt, z + ct)\big|_{t=0}, \qquad (2')$$

where the derivative on the right is a single-variable derivative. This form of the definition can be used to compute directional derivatives.

**EXAMPLE 1** Find $D_{\mathbf{u}}f(x, y, z)$ given $\mathbf{u} = \langle a, b, c \rangle$ is a unit vector and $f(x, y, z) = x^2 + y^2 + z^2$.

*Solution* Applying (2'), we get

$$D_{\mathbf{u}}f(x, y, z) = D_t \left[ (x + at)^2 + (y + bt)^2 + (z + ct)^2 \right]\big|_{t=0}$$
$$= \left[ 2a(x + at) + 2b(y + bt) + 2c(z + ct) \right]\big|_{t=0},$$

and thus

$$D_{\mathbf{u}}f(x, y, z) = 2ax + 2by + 2cz.$$

For instance, if $\mathbf{u} = \frac{1}{3}(\mathbf{i} + 2\mathbf{j} + 2\mathbf{k})$, then $a = \frac{1}{3}$, $b = \frac{2}{3}$, and $c = \frac{2}{3}$, so

$$D_{\mathbf{u}}f(2, -1, 3) = (2)(\tfrac{1}{3})(2) + (2)(\tfrac{2}{3})(-1) + (2)(\tfrac{2}{3})(3) = 4.$$

For a physical interpretation of the final result in Example 1, suppose that $f(x, y, z) = x^2 + y^2 + z^2$ is the temperature at the point $(x, y, z)$, with temperature in degrees Celsius and distance in centimeters. Then the rate of change of temperature at the point $P(2, -1, 3)$ in the direction of the vector $\mathbf{u} = \langle \frac{1}{3}, \frac{2}{3}, \frac{2}{3} \rangle$ is $4°/\text{cm}$. If an insect flew through the point $P$ with velocity vector $\mathbf{u}$, it would experience a time rate of change of temperature of $4°/\text{s}$.

**THE GRADIENT VECTOR**

To ease the labor of computing directional derivatives, we use a form of the chain rule that involves the **gradient vector** $\nabla f$ of the function $f$, which is defined to be

$$\nabla f(x, y, z) = \mathbf{i} f_x(x, y, z) + \mathbf{j} f_y(x, y, z) + \mathbf{k} f_z(x, y, z). \qquad (3)$$

We also write

$$\nabla f = \left\langle \frac{\partial f}{\partial x}, \frac{\partial f}{\partial y}, \frac{\partial f}{\partial z} \right\rangle = \frac{\partial f}{\partial x}\mathbf{i} + \frac{\partial f}{\partial y}\mathbf{j} + \frac{\partial f}{\partial z}\mathbf{k}.$$

**EXAMPLE 2** If $f(x, y, z) = x^2 + yz - 2xy - z^2$, then the definition of the gradient vector in (3) yields

$$\nabla f(x, y, z) = \frac{\partial f}{\partial x}\mathbf{i} + \frac{\partial f}{\partial y}\mathbf{j} + \frac{\partial f}{\partial z}\mathbf{k}$$

$$= (2x - 2y)\mathbf{i} + (z - 2x)\mathbf{j} + (y - 2z)\mathbf{k}.$$

For instance, the value of $\nabla f$ at the point $(2, 1, 3)$ is

$$\nabla f(2, 1, 3) = 2\mathbf{i} - \mathbf{j} - 5\mathbf{k}.$$

---

Suppose now that the first order partial derivatives of $f$ are continuous and that

$$\mathbf{r}(t) = x(t)\mathbf{i} + y(t)\mathbf{j} + z(t)\mathbf{k}$$

is a differentiable vector-valued function. Then $f(\mathbf{r}(t)) = f(x(t), y(t), z(t))$ is a differentiable function of $t$, and its derivative is

$$D_t f(\mathbf{r}(t)) = \nabla f(\mathbf{r}(t)) \cdot \mathbf{r}'(t). \tag{4}$$

The operation on the right-hand side above is the *dot* product, because the gradient of $f$ and the derivative of $\mathbf{r}$ are both *vector*-valued functions.

The rule of Equation (4) is the **vector chain rule;** it is a new way of writing the chain rule of Section 15-7. To see why this is so, recall that

$$\mathbf{r}'(t) = \frac{d\mathbf{r}}{dt} = \left\langle \frac{dx}{dt}, \frac{dy}{dt}, \frac{dz}{dt} \right\rangle.$$

Then

$$D_t f(\mathbf{r}(t)) = D_t f(x(t), y(t), z(t))$$

$$= \frac{\partial f}{\partial x}\frac{dx}{dt} + \frac{\partial f}{\partial y}\frac{dy}{dt} + \frac{\partial f}{\partial z}\frac{dz}{dt} = \nabla f \cdot \frac{d\mathbf{r}}{dt}.$$

Now we are ready to apply the gradient vector to compute directional derivatives. Suppose that $\mathbf{r}(t) = \langle x + at, y + bt, z + ct \rangle$, so that

$$\mathbf{r}(0) = \langle x, y, z \rangle \quad \text{and} \quad \mathbf{r}'(0) = \langle a, b, c \rangle = \mathbf{u}.$$

Then, by (2), we find that

$$D_\mathbf{u} f(x, y, z) = \lim_{t \to 0} \frac{f(\mathbf{r}(t)) - f(\mathbf{r}(0))}{t}$$

$$= D_t f(\mathbf{r}(t)) \quad \text{at} \quad t = 0$$

$$= \nabla f(r(0)) \cdot \mathbf{r}'(0) \qquad \text{[by Equation (4)]}.$$

Therefore,

$$D_\mathbf{u} f(x, y, z) = \nabla f(x, y, z) \cdot \mathbf{u}. \tag{5}$$

This formula provides the quickest and easiest way of computing the directional derivative $D_\mathbf{u} f$; all we need is knowledge of the partial derivatives of $f$. In terms of the components $a$, $b$, and $c$ of the unit vector $\mathbf{u}$, Formula

(5) says simply that

$$D_{\mathbf{u}}f = a\frac{\partial f}{\partial x} + b\frac{\partial f}{\partial y} + c\frac{\partial f}{\partial z}. \tag{6}$$

If $a = \cos \alpha$, $b = \cos \beta$, and $c = \cos \gamma$ are the direction cosines of $\mathbf{u}$, then

$$D_{\mathbf{u}}f = \frac{\partial f}{\partial x}\cos \alpha + \frac{\partial f}{\partial y}\cos \beta + \frac{\partial f}{\partial z}\cos \gamma. \tag{7}$$

**EXAMPLE 3** Compute the directional derivative of $f(x, y, z) = x^2 - 2yz$ in the direction of the vector $\mathbf{v} = \langle 2, 2, 1 \rangle$. At what rate is $f$ increasing in the direction of $\mathbf{v}$ at the point $(1, 2, 3)$?

*Solution* Since $\mathbf{v}$ is not a unit vector, we must replace it by a unit vector having the same direction before using the formulas above. So we take

$$\mathbf{u} = \frac{\mathbf{v}}{|\mathbf{v}|} = \left\langle \frac{2}{3}, \frac{2}{3}, \frac{1}{3} \right\rangle.$$

Then Formula (5) gives

$$D_{\mathbf{u}}f(x, y, z) = \langle 2x, -2z, -2y \rangle \cdot \langle \tfrac{2}{3}, \tfrac{2}{3}, \tfrac{1}{3} \rangle = \tfrac{1}{3}(4x - 2y - 4z).$$

And at the point $(1, 2, 3)$, we find that

$$D_{\mathbf{u}}f(1, 2, 3) = \tfrac{1}{3}(4 - 4 - 12) = -4.$$

Thus $f$ is *decreasing* at the rate of 4 units per unit distance moved in the direction of $\mathbf{u}$ (or $\mathbf{v}$) at $(1, 2, 3)$.

The gradient vector $\nabla f$ has an important interpretation that involves the *maximal* directional derivative of $f$. If $\phi$ is the angle between $\nabla f$ at the point $P$ and the unit vector $\mathbf{u}$ (Fig. 15.63), then the formula in Equation (5) gives

$$D_{\mathbf{u}}f(P) = \nabla f(P) \cdot \mathbf{u} = |\nabla f(P)| \cos \phi$$

because $|\mathbf{u}| = 1$. The maximum value of $\cos \phi$ is 1, and this occurs when $\phi = 0$. This is so when $\mathbf{u}$ is the particular unit vector $\nabla f(P)/|\nabla f(P)|$ that points in the direction of the gradient vector. In this case the formula above yields

$$D_{\mathbf{u}}f(P) = |\nabla f(P)|,$$

so that the value of the directional derivative is the length of the gradient vector. We have therefore proved the following theorem.

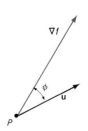

**15.63** The angle $\phi$ between $\nabla f$ and the unit vector $\mathbf{u}$

---

**Theorem 1** *Significance of the Gradient Vector*

The maximum value of the directional derivative $D_{\mathbf{u}}f(P)$ is obtained when $\mathbf{u}$ is the unit vector in the direction of the gradient vector $\nabla f(P)$; that is, when $\mathbf{u} = \nabla f(P)/|\nabla f(P)|$. The value of this maximum directional derivative is $|\nabla f(P)|$, the length of the gradient vector.

---

Thus *the gradient vector $\nabla f$ points in the direction in which the function $f$ increases most rapidly, and its length is the rate of increase of $f$ (with respect to distance) in that direction.*

**EXAMPLE 4** Suppose that the temperature $W$ (in degrees Celsius) at the point $(x, y, z)$ in space is given by $W = 50 + xyz$.
(a) Find the rate of change of temperature with respect to distance (in feet) at the point $P(3, 4, 1)$ in the direction of the vector $\mathbf{v} = \mathbf{i} + 2\mathbf{j} + 2\mathbf{k}$.
(b) Find the maximal directional derivative $D_\mathbf{u} W$ at the point $P(3, 4, 1)$ and the direction $\mathbf{u}$ in which that maximum occurs.

***Solution*** (a) $\nabla W = yz\mathbf{i} + xz\mathbf{j} + xy\mathbf{k}$, so $\nabla W(P) = 4\mathbf{i} + 3\mathbf{j} + 12\mathbf{k}$. The *unit* vector in the direction of $\mathbf{v}$ is

$$\mathbf{u} = \frac{\mathbf{v}}{|\mathbf{v}|} = \frac{1}{3}\mathbf{i} + \frac{2}{3}\mathbf{j} + \frac{2}{3}\mathbf{k}.$$

Hence the desired directional derivative is

$$D_\mathbf{u} W(3, 4, 1) = \nabla W(3, 4, 1) \cdot \mathbf{u}$$
$$= (4)(\tfrac{1}{3}) + (3)(\tfrac{2}{3}) + (12)(\tfrac{2}{3}) = \tfrac{34}{3}.$$

Thus the rate of change of temperature in the direction of the vector $\mathbf{v}$ is $11\tfrac{1}{3}°\text{C/ft}$.
(b) According to Theorem 1, the maximal directional derivative of $W$ at $P$ is

$$|\nabla W(P)| = |4\mathbf{i} + 3\mathbf{j} + 12\mathbf{k}| = 13°\text{C/ft}.$$

This is the directional derivative of $W$ in the direction of the unit vector

$$\mathbf{u} = \frac{\nabla W(P)}{|\nabla W(P)|} = \frac{4\mathbf{i} + 3\mathbf{j} + 12\mathbf{k}}{13}.$$

---

Thus far, we only have discussed directional derivatives of functions of three variables. For a function of two variables the formulas are analogous:

$$\nabla f(x, y) = \left\langle \frac{\partial f}{\partial x}, \frac{\partial f}{\partial y} \right\rangle = \frac{\partial f}{\partial x}\mathbf{i} + \frac{\partial f}{\partial y}\mathbf{j} \tag{8}$$

and

$$D_\mathbf{u} f(x, y) = \nabla f(x, y) \cdot \mathbf{u} = a\frac{\partial f}{\partial x} + b\frac{\partial f}{\partial y} \tag{9}$$

where $\mathbf{u} = \langle a, b \rangle$ is a unit vector. If $\alpha$ is the inclination angle of $\mathbf{u}$, then $a = \cos \alpha$ and $b = \sin \alpha$, so Equation (9) takes the form

$$D_\mathbf{u} f(x, y) = \frac{\partial f}{\partial x}\cos \alpha + \frac{\partial f}{\partial y}\sin \alpha. \tag{10}$$

As in Example 4, change in temperature provides a tangible interpretation of the three-dimensional directional derivative. In the two-dimensional case, it is helpful to think of the directional derivative $D_\mathbf{u} f(x, y)$ as the slope of a curve on the surface $z = f(x, y)$. If $\mathbf{u} = \langle a, b \rangle$ is a (two-dimensional) unit vector, then the quotient

$$\frac{f(x + at, y + bt) - f(x, y)}{t} = \frac{\text{vertical rise}}{\text{horizontal run}}$$

(c) $\nabla\left(\dfrac{u}{v}\right) = \dfrac{v\,\nabla u - u\,\nabla v}{v^2}$

(d) $\nabla u^n = nu^{n-1}\,\nabla u$

**32** Suppose that $f$ is a function of the three independent variables $x$, $y$, and $z$. Show that $D_{\mathbf{i}}f = f_x$, $D_{\mathbf{j}}f = f_y$, and $D_{\mathbf{k}}f = f_z$.

**33** Show that the equation of the line tangent to the conic section $Ax^2 + Bxy + Cy^2 = D$ at the point $(x_0, y_0)$ is

$$(Ax_0)x + \tfrac{1}{2}B(y_0x + x_0y) + (Cy_0)y = D.$$

**34** Show that the equation of the tangent plane to the quadric surface $Ax^2 + By^2 + Cz^2 = D$ at the point $(x_0, y_0, z_0)$ is

$$(Ax_0)x + (By_0)y + (Cz_0)z = D.$$

**35** Assume that the temperature at the point $(x, y, z)$ in space is given by the function $W = 50 + xyz$, as in Example 4. You are standing at the point $P(3, 4, 1)$ in a mountain valley shaped like the surface $z = x^2 - 2y$. Suppose that you start climbing with unit speed and a northeast compass heading. What initial rate of change of temperature do you observe?

**36** Suppose that the temperature at the point $(x, y, z)$ in space is given by the formula $W = 100 - x^2 - y^2 - z^2$.
(a) Find the rate of change of temperature at $P(3, -4, 5)$ in the direction of the vector $\mathbf{v} = 3\mathbf{i} - 4\mathbf{j} + 12\mathbf{k}$.

(b) In what direction does $W$ increase most rapidly at $P$? What is the value of the maximal directional derivative?

**37** You are standing at the point $(30, 20, 5)$ on a hill with the shape of the surface

$$z = 100 \exp\left[-\dfrac{(x^2 + 3y^2)}{701}\right].$$

(a) In what direction (that is, with what compass heading) should you go in order to climb most steeply? At what angle from the horizontal will you initially be climbing?
(b) If instead of climbing as in part (a), you head directly west (the direction of decreasing $x$), then at what angle will you be climbing initially?

**38** Find an equation for the plane tangent to the paraboloid $z = 2x^2 + 3y^2$ and, simultaneously, parallel to the plane $4x - 3y - z = 10$.

**39** The cone $z^2 = x^2 + y^2$ and the plane $2x + 3y + 4z + 2 = 0$ intersect in an ellipse. Write an equation of the plane normal to this ellipse at the point $(3, 4, -5)$.

**40** It is geometrically apparent that the highest and lowest points of the ellipse of Problem 39 are those points where its tangent line is horizontal. Find those points.

**41** Show that the sphere $x^2 + y^2 + z^2 = r^2$ and the elliptical cone $z^2 = a^2x^2 + b^2y^2$ are orthogonal (that is, have perpendicular tangent planes) at every point of their intersection.

---

### 15-9

# Lagrange Multipliers and Constrained Maximum-Minimum Problems

In Section 15-5 we discussed the problem of finding the maximum and minimum values attained by a function $f(x, y)$ at points of the plane region $R$, in the simple case in which $R$ consists of the points on and within the simple closed curve $C$. We saw that any local maximum or minimum in the *interior* of $R$ occurs at a point where $f_x = 0 = f_y$. In this section we discuss the very different matter of finding the maximum and minimum values attained by $f$ at points of the *boundary* curve $C$.

If the curve $C$ is the graph of the equation $g(x, y) = 0$, then our task is to maximize or minimize the function $f(x, y)$ subject to the **constraint** or **side condition**

$$g(x, y) = 0. \tag{1}$$

We could in principle try to solve this constraint equation for $y = \phi(x)$, and then maximize or minimize the single-variable function $f(x, \phi(x))$ by the standard method of finding where its derivative is zero. But what if it is impractical or impossible to solve Equation (1) explicitly for $y$ in terms of $x$? An alternative approach that does not require that we first solve this equation is the **method of Lagrange multipliers.** It is named for its discoverer, the French mathematician Joseph Louis Lagrange (1736–1813). The method is based on the following theorem.

Let $f(x, y)$ and $g(x, y)$ be functions with continuous first order partial derivatives. If the maximum (or minimum) value of $f$ subject to the condition

$$g(x, y) = 0 \tag{1}$$

occurs at a point $P$ where $\nabla g(P) \neq \mathbf{0}$, then

$$\nabla f(P) = \lambda \, \nabla g(P) \tag{2}$$

for some constant $\lambda$.

***Proof*** By the implicit function theorem mentioned in Section 15-8, the fact that $\nabla g(P) \neq \mathbf{0}$ allows us to represent the curve $g(x, y) = 0$ near $P$ by a parametric curve $\mathbf{r}(t)$ and in such fashion that $\mathbf{r}$ has a nonzero tangent vector near $P$; thus $\mathbf{r}'(t) \neq \mathbf{0}$ (see Fig. 15.70). Let $t_0$ be the value of $t$ such that $\mathbf{r}(t_0) = \overrightarrow{OP}$. If $f(x, y)$ attains its maximum value at $P$, then the composite function $f(\mathbf{r}(t))$ attains its maximum value at $t = t_0$, and so

$$D_t f(\mathbf{r}(t)) = \nabla f(\mathbf{r}(t_0)) \cdot \mathbf{r}'(t_0)$$
$$= \nabla f(P) \cdot \mathbf{r}'(t_0) = 0. \tag{3}$$

Here we have used the vector chain rule—Equation (4) of Section 15-8.

Because $\mathbf{r}(t)$ lies on the curve $g(x, y) = 0$, the composite function $g(\mathbf{r}(t))$ is a constant function. Therefore,

$$D_t g(\mathbf{r}(t)) = \nabla g(\mathbf{r}(t_0)) \cdot \mathbf{r}'(t_0)$$
$$= \nabla g(P) \cdot \mathbf{r}'(t_0) = 0. \tag{4}$$

Equations (3) and (4) tell us that the vectors $\nabla f(P)$ and $\nabla g(P)$ are both perpendicular to the tangent vector $\mathbf{r}'(t_0)$. Hence $\nabla f(P)$ must be a scalar multiple of $\nabla g(P)$, and this is what Equation (2) says. Thus we have proved Theorem 1. ∎

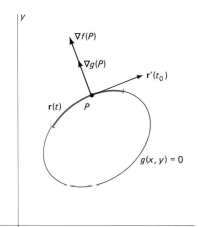

**15.70** The conclusion of Theorem 1 illustrated

The Lagrange multiplier method based on Theorem 1 goes like this. Suppose that we want to maximize (or minimize) $z = f(x, y)$ subject to the constraint or side condition $g(x, y) = 0$. Equation (1) and the two scalar components of Equation (2) yield the three equations

$$g(x, y) = 0, \tag{1}$$

$$f_x(x, y) = \lambda g_x(x, y), \quad \text{and} \tag{2a}$$

$$f_y(x, y) = \lambda g_y(x, y). \tag{2b}$$

Thus we have three equations which we can attempt to solve for the three unknowns $x$, $y$, and $\lambda$. The points $(x, y)$ that we find (assuming that our efforts are successful) are the only possible locations for the extrema of $f$ subject to the constraint $g(x, y) = 0$. The associated values of $\lambda$, called **Lagrange multipliers,** may come out in the wash but are usually not of much interest to us. Finally we calculate the value $f(x, y)$ at each of the solution points $(x, y)$ so as to spot its maximum and minimum values.

We must bear in mind the additional possibility that the maximum or minimum (or both) values might occur at a point where $g_x = 0 = g_y$. The Lagrange multiplier method may fail to locate these exceptional points, but they can usually be recognized as points at which the graph $g(x, y) = 0$ fails to be a smooth curve.

**EXAMPLE 1**    Let us return to the sawmill problem of Example 5 in Section 3-6: to maximize the cross-sectional area of a rectangular beam cut from a circular log with radius 1 ft. We want to show that the optimal beam has a square cross section.

*Solution*    With the coordinate system indicated in Fig. 15.71, we want to maximize the area $A = f(x, y) = 4xy$ subject to the constraint

$$g(x, y) = x^2 + y^2 - 1 = 0.$$

Equations (2a) and (2b) take the forms

$$4y = 2\lambda x \quad \text{and} \quad 4x = 2\lambda y.$$

It is clear that neither $x = 0$ nor $y = 0$ gives the maximum area. Hence we may divide the first equation by $4x$ and the second by $4y$ to obtain

$$\frac{\lambda}{2} = \frac{y}{x} = \frac{x}{y}.$$

We forget $\lambda$, and see that $x^2 = y^2$. We're looking for a solution point in the first quadrant, so we conclude that $x = 1/\sqrt{2} = y$ gives the maximum. This corresponds to a square beam of edge $\sqrt{2}$ feet, which has cross-sectional area of 2 square feet—about 64% of the total cross-sectional area $\pi$ of the log.

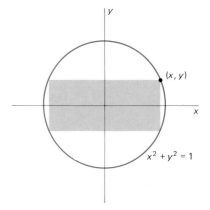

15.71    Cutting a rectangular beam from a circular log

---

Note that $f(x, y) = 4xy$ attains its maximum value of 2 at both $(1/\sqrt{2}, 1/\sqrt{2})$ and $(-1/\sqrt{2}, -1/\sqrt{2})$ and its minimum value of $-2$ at both $(-1/\sqrt{2}, 1/\sqrt{2})$ and $(1/\sqrt{2}, -1/\sqrt{2})$. The Lagrange multiplier method actually locates all four of these points for us.

**EXAMPLE 2**    After the square beam of Example 1 has been cut from our circular log of radius 1 ft, let us cut four planks from the remaining pieces, each of dimensions $u$ by $2v$, as shown in Fig. 15.72. How should this be

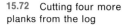

15.72    Cutting four more planks from the log

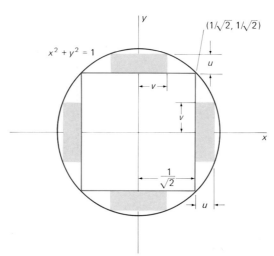

done in order to maximize the combined cross-sectional area of all four planks, thereby using what might otherwise be scrap lumber as efficiently as possible?

*Solution*  Since the point $(u + 1/\sqrt{2}, v)$ lies on the unit circle, we want to maximize the function $f(u, v) = 8uv$ subject to the condition

$$g(u, v) = \left(u + \frac{1}{\sqrt{2}}\right)^2 + v^2 - 1 = 0.$$

The conditions $f_u = \lambda g_u$ and $f_v = \lambda g_v$ give us the equations

$$8v = 2\lambda\left(u + \frac{1}{\sqrt{2}}\right) \quad \text{and} \quad 8u = 2\lambda v.$$

We solve each of these equations for $\lambda/4$ and find that

$$\frac{\lambda}{4} = \frac{v}{u + 1/\sqrt{2}} = \frac{u}{v}.$$

Thus $v^2 = u(u + 1/\sqrt{2})$.

Finally we apply the one equation that we've not used yet—the constraint equation $g(u, v) = 0$. We substitute our last equation into that equation and get the quadratic equation

$$\left(u + \frac{1}{\sqrt{2}}\right)^2 + u\left(u + \frac{1}{\sqrt{2}}\right) - 1 = 2u^2 + \frac{3}{\sqrt{2}}u - \frac{1}{2} = 0.$$

The only positive root of this equation is $u = 0.199$ (to three-place accuracy), and this in turn tells us that $v = 0.424$. Thus our planks should be 0.199 ft thick and 0.848 ft wide. Their combined cross-sectional area will be $f(0.199, 0.424) \approx 0.673$ ft$^2$, or about 21% of the original log.

## LAGRANGE MULTIPLIERS IN THREE DIMENSIONS

Now suppose that $f(x, y, z)$ and $g(x, y, z)$ have continuous first order partial derivatives, and that we want to find the points of the *surface*

$$g(x, y, z) = 0 \tag{5}$$

at which the function $f(x, y, z)$ attains its maximum and minimum values. With functions of three rather than two variables, Theorem 1 holds precisely as we stated it above. We leave the details to Problem 33, but an argument similar to the proof of Theorem 1 shows that, at a maximum-minimum point $P$ of $f(x, y, z)$ on the surface in (5), the gradient vectors $\nabla f(P)$ and $\nabla g(P)$ are both normal vectors to the surface, as indicated in Fig. 15.73. It follows that

$$\nabla f(P) = \lambda \nabla g(P) \tag{6}$$

for some scalar $\lambda$. This vector equation corresponds to three scalar equations, and so we can attempt to solve simultaneously the four equations

$$g(x, y, z) = 0, \tag{5}$$

$$f_x(x, y, z) = \lambda g_x(x, y, z), \tag{6a}$$

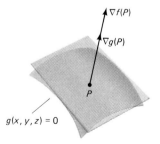

**15.73**  The natural generalization of Theorem 1 also holds for functions of three variables.

$\nabla f(P)$

$\nabla g(P)$

$P$

$g(x, y, z) = 0$

**820**

$$f_y(x, y, z) = \lambda g_y(x, y, z), \quad \text{and} \quad \text{(6b)}$$

$$f_z(x, y, z) = \lambda g_z(x, y, z) \quad \text{(6c)}$$

for the four unknowns $x$, $y$, $z$, and $\lambda$. If successful, we then evaluate $f(x, y, z)$ at each of the solution points $(x, y, z)$ to see at which it attains its maximum and its minimum values. Thus the Lagrange multiplier method with one side condition is essentially the same in dimension three as in dimension two.

**EXAMPLE 3**   Find the maximum volume of a rectangular box inscribed in the ellipsoid $x^2/a^2 + y^2/b^2 + z^2/c^2 = 1$ with its faces parallel to the coordinate planes.

*Solution*   Let $(x, y, z)$ be the vertex of our box that lies in the first octant (where $x$, $y$, and $z$ are all positive). We want to maximize the volume $V = f(x, y, z) = 8xyz$ subject to the constraint

$$g(x, y, z) = \frac{x^2}{a^2} + \frac{y^2}{b^2} + \frac{z^2}{c^2} - 1 = 0.$$

Equations (6a), (6b), and (6c) give

$$8yz = \frac{2\lambda x}{a^2}, \qquad 8xz = \frac{2\lambda y}{b^2}, \qquad 8xy = \frac{2\lambda z}{c^2}.$$

Part of the art of mathematics lies in pausing for a moment to find an elegant way to solve a problem, rather than rushing in headlong with brute force methods. Here, if we multiply the first equation by $x$, the second by $y$, and the third by $z$, we find that

$$2\lambda \frac{x^2}{a^2} = 2\lambda \frac{y^2}{b^2} = 2\lambda \frac{z^2}{c^2} = 8xyz.$$

Now $\lambda \neq 0$ because (at maximum volume) $x$, $y$, and $z$ are nonzero. We conclude that

$$\frac{x^2}{a^2} = \frac{y^2}{b^2} = \frac{z^2}{c^2}.$$

The sum of the last three expressions is 1—that is the constraint condition for this problem—and so each must be equal to $\frac{1}{3}$. All three of $x$, $y$, and $z$ are positive, and so

$$x = \frac{a}{\sqrt{3}}, \qquad y = \frac{b}{\sqrt{3}}, \quad \text{and} \quad z = \frac{c}{\sqrt{3}}.$$

Therefore the box of maximum volume has volume $V = 8abc/3\sqrt{3}$.

---

Check the ratio of box volume to ellipsoid volume ($4\pi abc/3$) to see if the above answer is plausible.

**PROBLEMS WITH TWO CONSTRAINTS**

Suppose that we want to find the maximum and minimum values of the function $f(x, y, z)$ at points of the curve of intersection of the two surfaces

$$g(x, y, z) = 0 \quad \text{and} \quad h(x, y, z) = 0. \quad \text{(7)}$$

This is a maximum-minimum problem with *two* constraints or side conditions. The Lagrange multiplier method for such situations is based upon the following theorem.

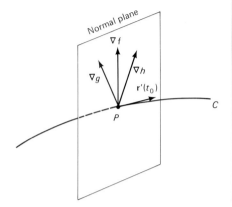

**15.74** The relation between the gradient vectors in the proof of Theorem 2

---

> **Theorem 2**   *Lagrange Multipliers (two constraints)*
>
> Let $f(x, y, z)$, $g(x, y, z)$, and $h(x, y, z)$ be functions with continuous first order partial derivatives. If the maximum (or minimum) value of $f$ subject to the two conditions
>
> $$g(x, y, z) = 0 \quad \text{and} \quad h(x, y, z) = 0 \tag{7}$$
>
> occurs at a point $P$ where the vectors $\nabla g(P)$ and $\nabla h(P)$ are nonzero and nonparallel, then
>
> $$\nabla f(P) = \lambda_1 \nabla g(P) + \lambda_2 \nabla h(P) \tag{8}$$
>
> for some two constants $\lambda_1$ and $\lambda_2$.

---

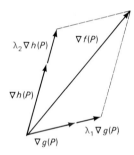

**15.75** Geometry of the equation $\nabla f(P) = \lambda_1 \nabla g(P) + \lambda_2 \nabla h(P)$

**15.76** The ellipse of Example 4

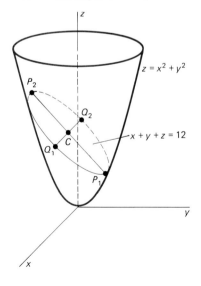

*Outline of Proof*   By an appropriate version of the implicit function theorem, the curve $C$ of intersection of the two surfaces (Fig. 15.74) may be represented near $P$ by a parametric curve $\mathbf{r}(t)$ with nonzero tangent vector $\mathbf{r}'(t)$. Let $t_0$ be the value of $t$ such that $\mathbf{r}(t_0) = P$. We compute the derivatives at $t_0$ of the composite functions $f(\mathbf{r}(t))$, $g(\mathbf{r}(t))$, and $h(\mathbf{r}(t))$. We find—exactly as in the proof of Theorem 1—that

$$\nabla f(P) \cdot \mathbf{r}'(t_0) = 0, \qquad \nabla g(P) \cdot \mathbf{r}'(t_0) = 0, \quad \text{and} \quad \nabla h(P) \cdot \mathbf{r}'(t_0) = 0.$$

These three equations say that all three gradient vectors are perpendicular to the curve $C$ at $P$ and thus all lie in a single plane, the normal plane to the curve $C$ at the point $P$.

Now $\nabla g(P)$ and $\nabla h(P)$ are nonzero and nonparallel, so $\nabla f(P)$ is the sum of its projections onto $\nabla g(P)$ and $\nabla h(P)$ (see Problem 45 of Section 14-1). As illustrated in Fig. 15.75, this fact implies Equation (8). ∎

In examples we prefer to avoid subscripts by writing $\lambda$ and $\mu$ for the Lagrange multipliers $\lambda_1$ and $\lambda_2$ in the statement of Theorem 2. The equations in (7) and the three scalar components of the vector equation in (8) then give us the five simultaneous equations

$$g(x, y, z) = 0, \tag{7a}$$

$$h(x, y, z) = 0, \tag{7b}$$

$$f_x(x, y, z) = \lambda g_x(x, y, z) + \mu h_x(x, y, z), \tag{8a}$$

$$f_y(x, y, z) = \lambda g_y(x, y, z) + \mu h_y(x, y, z), \quad \text{and} \tag{8b}$$

$$f_z(x, y, z) = \lambda g_z(x, y, z) + \mu h_z(x, y, z). \tag{8c}$$

**EXAMPLE 4**   The plane $x + y + z = 12$ intersects the paraboloid $z = x^2 + y^2$ in an ellipse, as shown in Fig. 15.76. Find the highest and lowest points on this ellipse, and thereby determine its semiaxes.

*Solution*   The height of a point $(x, y, z)$ is $z$, so we want to find the maximum and minimum values of

$$f(x, y, z) = z \tag{9}$$

subject to the conditions

$$g(x, y, z) = x + y + z - 12 = 0 \qquad (10)$$

and

$$h(x, y, z) = x^2 + y^2 - z = 0. \qquad (11)$$

The conditions in (8a) through (8c) yield

$$0 = \lambda + 2\mu x, \qquad (12a)$$

$$0 = \lambda + 2\mu y, \qquad \text{and} \qquad (12b)$$

$$1 = \lambda - \mu. \qquad (12c)$$

If $\mu$ were zero, then Equation (12a) would imply that $\lambda = 0$, which contradicts Equation (12c). Hence $\mu \neq 0$, and therefore the equation

$$2\mu x = -\lambda = 2\mu y$$

implies that $x = y$. Substitution of $x = y$ in Equation (11) gives $z = 2x^2$, and then Equation (10) yields

$$2x^2 + 2x - 12 = 0, \qquad (13)$$

$$x^2 + x - 6 = 0,$$

$$(x + 3)(x - 2) = 0.$$

Thus we obtain the two solutions $x = -3$ and $x = 2$. Because $y = x$ and $z = 2x^2$, the corresponding points of the ellipse are $P_1(2, 2, 8)$ and $P_2(-3, -3, 18)$. Obviously $P_1$ is the lowest point and $P_2$ is the highest.

The center of the ellipse is the midpoint $C(-\frac{1}{2}, -\frac{1}{2}, 13)$. Substitution of $z = 13$ in Equations (10) and (11) yields the equations

$$x + y + 1 = 0,$$

$$x^2 + y^2 = 13.$$

Substitution of $y = -1 - x$ in $x^2 + y^2 = 13$ then gives the same quadratic equation as in (13). This leads to the two points $Q_1(2, -3, 13)$ and $Q_2(-3, 2, 13)$ at height $z = 13$ on the ellipse.

Finally, the ellipse has major axis $P_1 P_2$ and minor axis $Q_1 Q_2$, so the lengths of its major and minor semiaxes are

$$a = \tfrac{1}{2}\sqrt{(-3 - 2)^2 + (-3 - 2)^2 + (18 - 8)^2} = \tfrac{5}{2}\sqrt{6}$$

and

$$b = \tfrac{1}{2}\sqrt{(2 + 3)^2 + (-3 - 2)^2 + (13 - 13)^2} = \tfrac{5}{2}\sqrt{2}.$$

## ECONOMIC APPLICATIONS

The method of Lagrange multipliers is applicable to economic problems in which total production (or profit, or what-have-you) is to be maximized subject to the constraint of fixed available resources. For example, let $P$ be the number of units of a certain product that is being manufactured. Then $P$ may be given in terms of the number of units $x$ of labor and $y$ of capital utilized. In particular, if

$$P = f(x, y) = kx^\alpha y^\beta \qquad (\alpha + \beta = 1),$$

then this function—known as the Cobb-Douglas production function—has the convenient property that $f(cx, cy) = cf(x, y)$. Thus doubling (or tripling) each of the resources utilized will double (or triple) production.

If the cost of each unit of labor is $A$ dollars, that of capital is $B$ dollars, and a total of $C$ dollars is available, then we want to maximize the production $f(x, y)$ subject to the constraint

$$g(x, y) = Ax + By - C = 0.$$

In Problem 27 we ask you to show that production is maximized when $x = \alpha C/A$ and $y = \beta C/B$. In Problem 28 we ask you to minimize the cost function $Ax + By$ subject to the constraint of fixed production.

### *CLOSE TO THE WIND

How is it possible for a sailboat to tack in such a way to make northward progress against a wind blowing directly from the north? If indeed this is possible, how should the skipper set the sail to maximize the northward component of the boat's velocity?

Figure 15.77 depicts a sailboat with its sail set in the direction **s** and its rudder set so that it will sail in the direction **u**. We assume (unrealistically, but for the sake of simplicity) that the wind force vector **F** is independent of the angles $\alpha$ and $\beta$ shown, and that the effective thrust of the wind on the sail is the component

$$|\mathbf{F}_n| = |\mathbf{F}| \sin \alpha \tag{14}$$

of **F** normal to **s**. The effective force pushing the boat in the direction **u** is then the component

$$|\mathbf{F}_n| \sin \beta = |\mathbf{F}| \sin \alpha \sin \beta \tag{15}$$

of $\mathbf{F}_n$ in the direction of **u**. Finally, the northward component of this effective force is

$$|\mathbf{F}| \sin \alpha \sin \beta \cos(\alpha + \beta) = |\mathbf{F}| \sin \alpha \sin \beta \sin \gamma, \tag{16}$$

where $\gamma = (\pi/2) - (\alpha + \beta)$.

It follows that the boat's rate of northward progress is maximized by maximizing the function

$$f(\alpha, \beta, \gamma) = \sin \alpha \sin \beta \sin \gamma \tag{17}$$

subject to the constraint

$$g(\alpha, \beta, \gamma) = \alpha + \beta + \gamma - \frac{\pi}{2} = 0. \tag{18}$$

The Lagrange multiplier equations for this problem are

$$\cos \alpha \sin \beta \sin \gamma = \lambda, \tag{19a}$$

$$\sin \alpha \cos \beta \sin \gamma = \lambda, \tag{19b}$$

$$\sin \alpha \sin \beta \cos \gamma = \lambda. \tag{19c}$$

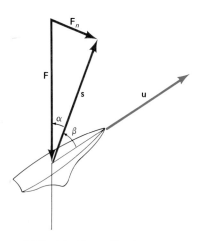

**15.77** A tacking sailboat

---

* This is the title of the glass sculpture on the cover of the first edition. The discussion here is based on the more detailed exposition in Ivan Niven's *Maxima and Minima Without Calculus*, The Mathematical Association of America, 1981, pages 106–110.

Equations (19a) and (19b) imply that

$$\cos \alpha \sin \beta = \sin \alpha \cos \beta;$$

$$\tan \alpha = \tan \beta;$$

$$\alpha = \beta.$$

Similarly, Equations (19b) and (19c) imply that $\beta = \gamma$. Because $\alpha + \beta + \gamma = \pi/2$ (90°), it follows that $\alpha = \beta = \gamma = 30°$

Thus the northward component of the boat's velocity is maximized if we set a course 60° from due north—either 60° east of north or 60° west of north—and set the sail so that it bisects the angle between the wind direction and the course direction.

---

## 15-9 PROBLEMS

In each of Problems 1–10, find the maximum and minimum values—if any—of the given function $f$ subject to the given constraint or constraints.

1 $f(x, y) = x^2 - y^2;\quad x^2 + y^2 = 4$
2 $f(x, y) = x^2 + y^2;\quad 2x + 3y = 6$
3 $f(x, y) = xy;\quad 4x^2 + 9y^2 = 36$
4 $f(x, y) = 4x^2 + 9y^2;\quad x^2 + y^2 = 1$
5 $f(x, y, z) = x^2 + y^2 + z^2;\quad 3x + 2y + z = 6$
6 $f(x, y, z) = 3x + 2y + z;\quad x^2 + y^2 + z^2 = 1$
7 $f(x, y, z) = x + y + z;\quad x^2 + 4y^2 + 9z^2 = 36$
8 $f(x, y, z) = xyz;\quad x^2 + y^2 + z^2 = 1$
9 $f(x, y, z) = x^2 + y^2 + z^2;\quad x + y + z = 1$ and $x + 2y + 3z = 6$
10 $f(x, y, z) = z;\quad x^2 + y^2 = 1$ and $2x + 2y + z = 5$

In each of Problems 11–20, use Lagrange multipliers to solve the indicated problem.

11 Section 15-5, Problem 27
12 Section 15-5, Problem 28
13 Section 15-5, Problem 29
14 Section 15-5, Problem 30
15 Section 15-5, Problem 31
16 Section 15-5, Problem 32
17 Section 15-5, Problem 34
18 Section 15-5, Problem 37
19 Section 15-5, Problem 38
20 Section 15-5, Problem 39
21 Find the point(s) of the surface $z = xy + 5$ closest to the origin. (*Suggestion:* Minimize the *square* of the distance.)
22 A triangle with sides $x$, $y$, and $z$ has fixed perimeter $2s = x + y + z$. Its area $A$ is given by Heron's formula:

$$A^2 = s(s - x)(s - y)(s - z).$$

Use the method of Lagrange multipliers to show that, among all triangles with the given perimeter, the one of largest area is the equilateral one.

23 Use the method of Lagrange multipliers to show that, of all triangles inscribed in the unit circle, the one of greatest area is equilateral. (*Suggestion:* Use Fig. 15.78 and the fact that the area of a triangle with sides $a$ and $b$ and included angle $\theta$ is given by the formula $A = \frac{1}{2} ab \sin \theta$.)

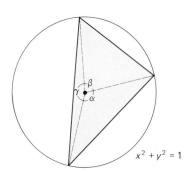

**15.78** A triangle inscribed in a circle (see Problem 23)

24 Find the points on the rotated ellipse $x^2 + xy + y^2 = 3$ that are nearest to and farthest from the origin. (*Suggestion:* Write the Lagrange multiplier equations in the form

$$ax + by = 0 \quad \text{and} \quad cx + dy = 0.$$

These equations have a nontrivial solution *only if*

$$ad - bc = 0.$$

Use this fact to solve first for $\lambda$.)
25 Use the method of Problem 24 to find the points of the rotated hyperbola $x^2 + 12xy + 6y^2 = 130$ that are closest to the origin.
26 Find the points of the ellipse $4x^2 + 9y^2 = 36$ that are closest to, and those farthest from, the point $(1, 1)$.

**27** Show that the production function $P = P(x, y) = kx^\alpha y^\beta$ (where $\alpha + \beta = 1$) is maximized subject to fixed costs $Ax + By = C$ by $x = \alpha C/A$, $y = \beta C/B$.

**28** Show that the cost function $C = C(x, y) = Ax + By$ is minimized subject to fixed production $P = kx^\alpha y^\beta$ (where $\alpha + \beta = 1$) by

$$x = \frac{P}{k}\left(\frac{\alpha B}{\beta A}\right)^\beta, \qquad y = \frac{P}{k}\left(\frac{\beta A}{\alpha B}\right)^\alpha.$$

**29** Find the highest and lowest points on the ellipse of intersection of the cylinder $x^2 + y^2 = 1$ and the plane $2x + y - z = 4$.

**30** Apply the method of Example 4 to find the highest and lowest points on the ellipse of intersection of the cone $z^2 = x^2 + y^2$ and the plane $x + 2y + 3z = 3$.

**31** Find the points on the ellipse of Problem 30 that are nearest to, and those farthest from, the origin.

**32** The ice tray shown in Fig. 15.79 is to be made of material that costs $1\not\!c/\text{in}^2$. Minimize the cost function $f(x, y, z) = xy + 3xz + 7yz$ subject to the constraints that each of the 12 compartments is to be square and the total volume (ignoring partitions) is to be 12 in.³

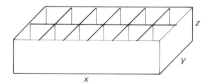

**15.79** The ice tray of Problem 32

**33** Prove Theorem 1 for functions of three variables by showing that each of the vectors $\nabla f(P)$ and $\nabla g(P)$ is perpendicular at $P$ to every curve on the surface $g(x, y, z) = 0$.

---

## 15-10

### The Second Derivative Test for Functions of Two Variables

We saw in Section 15-5 that in order for the differentiable function $f(x, y)$ to have either a local minimum or a local maximum at the point $P(a, b)$, it is a *necessary* condition that $P$ be a *critical point* of $f$; that is, that

$$f_x(a, b) = 0 = f_y(a, b).$$

In this section we give *sufficient* conditions that $f$ have a local extremum at a critical point. The criterion stated below involves the second order partial derivatives of $f$ at $(a, b)$, and plays the role of the single-variable second derivative test (Section 4-6) for functions of two variables. To simplify the statement of this result, we shall use the following abbreviations:

$$A = f_{xx}(a, b), \qquad B = f_{xy}(a, b), \qquad C = f_{yy}(a, b), \tag{1}$$

and

$$\Delta = AC - B^2 = f_{xx}(a, b)f_{yy}(a, b) - [f_{xy}(a, b)]^2. \tag{2}$$

We will prove the following theorem at the end of this section.

> **Theorem** *Sufficient Conditions for Local Extrema*
>
> Let $(a, b)$ be a critical point of the function $f(x, y)$, and suppose that $f$ has continuous first order and second order partial derivatives in some circular disk centered at $(a, b)$. Then:
>
> (i) If $\Delta > 0$ and $A > 0$, then $f$ has a local minimum at $(a, b)$.
> (ii) If $\Delta > 0$ and $A < 0$, then $f$ has a local maximum at $(a, b)$.
> (iii) If $\Delta < 0$, then $f$ has neither a local minimum nor a local maximum at $(a, b)$; instead, it has a saddle point.

Thus $f$ has *either* a local maximum *or* a local minimum at the critical point $(a, b)$ provided that the **discriminant** $\Delta = AC - B^2$ is *positive*. In this case, $A = f_{xx}(a, b)$ plays the role of the second derivative of a single-variable function: There is a local minimum at $(a, b)$ if $A > 0$ and a local maximum if $A < 0$.

the line $y = mx + b$ that lies directly above or below the $i$th empirical point $P_i(x_i, y_i)$. We define the **deviation** between the $i$th point and the line to be

$$d_i = y_i - (mx_i + b) \tag{2}$$

for $1 \leq i \leq n$.

Our first inclination might be to choose the line so that the sum $\sum d_i$ of the deviations is as small as possible. But in this case positive deviations could cancel negative deviations, so that $\sum d_i$ is small while none of the points lies very close to the line. We avoid this by minimizing the sum of the *squares* of the deviations given in (2). Doing so is called the **least squares method** of fitting a straight line to empirical data; we choose $m$ and $b$ in Equation (1) so as to minimize the function

$$f(m, b) = d_1^2 + d_2^2 + d_3^2 + \cdots + d_n^2 = \sum_{i=1}^{n} [y_i - (mx_i + b)]^2. \tag{3}$$

The function $f$ of Equation (3) is a function of the variables $m$ and $b$; the numbers $x_i$ and $y_i$ are all constants. The partial derivatives of $f$ are

$$\frac{\partial f}{\partial m} = \sum_{i=1}^{n} 2[y_i - (mx_i + b)](-x_i)$$

$$= 2m \sum_{i=1}^{n} x_i^2 + 2b \sum_{i=1}^{n} x_i - 2 \sum_{i=1}^{n} x_i y_i$$

and

$$\frac{\partial f}{\partial b} = \sum_{i=1}^{n} 2[y_i - (mx_i + b)](-1)$$

$$= 2m \sum_{i=1}^{n} x_i + 2b \sum_{i=1}^{n} 1 - 2 \sum_{i=1}^{n} y_i.$$

When we set both partial derivatives equal to zero, we get the pair of *linear* equations

$$m \sum_{i=1}^{n} x_i + nb = \sum_{i=1}^{n} y_i, \tag{4}$$

$$m \sum_{i=1}^{n} x_i^2 + b \sum_{i=1}^{n} x_i = \sum_{i=1}^{n} x_i y_i. \tag{5}$$

We can now solve these two linear equations simultaneously for $m$ and $b$. It can be shown (see Problem 16) that the resulting line $y = mx + b$ always minimizes the sum of the squares of the deviations. Hence we need not worry in each individual problem whether we have found a maximum, a minimum, or neither.

**EXAMPLE 1**   As a step toward determining the selling price that would maximize its profits, a perfume company wished to express its weekly sales $y$ (in thousands of bottles) as a linear function of the price $x$ (in dollars per bottle). In order to do this, it test-sold the perfume in four similar cities, with the following results:

|   | City 1 | City 2 | City 3 | City 4 |
|---|--------|--------|--------|--------|
| $x$ | 6.25 | 6.75 | 8.00 | 8.75 |
| $y$ | 6.03 | 5.62 | 4.78 | 4.34 |

If the company has fixed weekly costs of $25,000, plus $1.25 per bottle, what selling price will maximize its weekly profit?

**Solution**  To find the sales-price relation $y = mx + b$, we first compute the coefficients in Equations (4) and (5). We find that

$$\sum x_i = 6.25 + 6.75 + 8.00 + 8.75 = 29.75,$$

$$\sum x_i^2 = (6.25)^2 + (6.75)^2 + (8.00)^2 + (8.75)^2 = 225.19,$$

$$\sum y_i = 6.03 + 5.62 + 4.78 + 4.34 = 20.77, \quad \text{and}$$

$$\sum x_i y_i = (6.25)(6.03) + (6.75)(5.62) + (8.00)(4.78) + (8.75)(4.34) = 151.84$$

(we carry two-place accuracy since the data are no better than this). Since $n = 4$ in this example, we then solve the equations [from (4) and (5)]

$$29.75m + \quad 4b = \quad 20.77,$$

$$225.19m + 29.75b = 151.84$$

and find that $m = -0.67$ and $b = 10.19$. Thus each increase of $1 in price will reduce weekly sales by 670 bottles, because the sales-price relation is

$$y = -0.67x + 10.19 \quad \text{(thousands)}.$$

The company's weekly production costs for $y$ thousand bottles are $C = 25 + 1.25y$ (in thousands of dollars). So its weekly profits (still in thousands of dollars) are

$$P = xy - C = x(-0.67x + 10.19) - 25 - (1.25)(-0.67x + 10.19),$$

and so

$$P = P(x) = -0.67x^2 + 11.03x - 37.74.$$

Now $dP/dx = -1.34x + 11.03$, and $dP/dx = 0$ when $x = 8.23$. Thus $8.23 is the optimal selling price for each bottle of perfume. To complete the analysis, it follows that

$$y = 4.68 \text{ (thousand) bottles will be the weekly sales,} \quad \text{and}$$

$$P = 7.66 \text{ (thousand) dollars will be the weekly profit,}$$

with a profit of $1.64 on each bottle.

---

You can use the method of least squares to fit empirical data to non-linear equations or to equations with more than one independent variable. To fit the data $(x_1, y_1), (x_2, y_2), \ldots, (x_n, y_n)$ to the second degree equation

$$y = Ax^2 + Bx + C,$$

you need only to minimize the function of three variables

$$g(A, B, C) = \sum_{i=1}^{n} [y_i - (Ax_i^2 + Bx_i + C)]^2.$$

To fit the data $(x_1, y_1, z_1), (x_2, y_2, z_2), \ldots, (x_n, y_n, z_n)$ to the linear equation

$$z = Ax + By + C,$$

**836**

you need only to minimize the function

$$g(A, B, C) = \sum_{i=1}^{n} [z_i - (Ax_i + By_i + C)]^2.$$

Each case would call for solving the *three* simultaneous equations

$$\frac{\partial g}{\partial A} = \frac{\partial g}{\partial B} = \frac{\partial g}{\partial C} = 0.$$

The great German mathematician Carl Friedrich Gauss (1777–1855) invented the method of least squares when he was 18 years old. A short time later, he used it to determine the orbit of the first-discovered asteroid Ceres, initially observed on the first day of the nineteenth century, but lost from sight a few weeks later. Ceres was later rediscovered in the position that Gauss had predicted from the limited number of original observations.

## COMPETITION AND COLLUSION

Here we give an economic application of maximum-minimum techniques. Suppose that companies 1 and 2 make competitive products, with the weekly sales of each product being determined jointly by the price of the product and the price of the competing product. For simplicity, let us suppose that the cost of production of each product is the same, $1 per unit.

The management of each company does some market research like that described in Example 1. They find that the weekly sales $m$ and $n$ of products 1 and 2 (in thousands of units) are given in terms of their respective prices $x$ and $y$ by the equations

$$m = 10 - 2x + y, \tag{6}$$

$$n = 15 + 2x - 3y. \tag{7}$$

Because

$$\frac{\partial m}{\partial x} = -2, \qquad \frac{\partial n}{\partial x} = +2,$$

$$\frac{\partial m}{\partial y} = +1, \quad \text{and} \quad \frac{\partial n}{\partial y} = -3,$$

we see that our model is at least qualitatively plausible. For an increase of $1 in the price of product 1 decreases its weekly sales by 2 (thousand) units, but increases the sales of product 2 by 2 (thousand) units. An increase of $1 in the price of product 2 decreases its weekly sales by 3 (thousand) units but increases the sales of product 1 by 1 (thousand) units. This is part of the basic nature of competition: Each price change in either product has opposite effects on their respective sales, as one's gain is the other's loss.

Suppose first that the two companies act independently—that is, in *competition*—to maximize their individual weekly profits. From the point of view of company 1, its own selling price $x$ is a *variable* under its control, but the selling price $y$ of product 2 is *fixed* by its competitor. The weekly profit of company 1 is then a function of $x$, $P = P(x)$, given by

$$P = mx - m = (10 - 2x + y)(x - 1);$$

$$P(x) = -2x^2 + 12x + xy - y - 10. \tag{8}$$

If the management of company 1 knew in advance the selling price $y$ of product 2, then it would determine its own selling price $x$ by solving the equation

$$P'(x) = -4x + y + 12 = 0 \qquad (9)$$

for the purpose of maximizing its own profit.

From the point of view of company 2, the roles of the selling prices $x$ and $y$ are reversed. Its own selling price $y$ is a *variable*, while $x$ is *fixed*. The weekly profit of company 2 is then a function of $y$, $Q = Q(y)$, given by

$$Q = ny - n = (15 + 2x - 3y)(y - 1);$$
$$Q(y) = -3y^2 + 18y + 2xy - 2x - 15. \qquad (10)$$

If the management of company 2 knew in advance its competitor's selling price $x$, then it would determine its own selling price $y$ by solving the equation

$$Q'(y) = 2x - 6y + 18 = 0. \qquad (11)$$

If the management of each company credits its competitor with a knowledge of elementary calculus, then each will solve the simultaneous linear equations in (9) and (11) and get

$$x = \tfrac{45}{11} \approx \$4.09, \qquad y = \tfrac{48}{11} \approx \$4.36.$$

Equations (6) and (7) will then give their weekly sales

$$m = 6.182, \qquad n = 10.091$$

(in thousands of units). Equations (8) and (10) finally give their respective weekly profits

$$P = 19.108, \qquad Q = 33.942$$

in thousands of dollars.

Now suppose instead that the managements of the two companies act in *collusion*. They enter into a price-fixing agreement (legal or otherwise) by which they plan to maximize their total weekly profit. (We suppose that they will divide the resulting profit in an equitable way, but the details of this intriguing problem are, for the moment, not the issue.)

The total weekly profit $T = T(x, y)$ of the two companies will be given by

$$T = (\text{revenue}) - (\text{cost}) = (mx + ny) - (m + n);$$
$$T(x, y) = 10x + 17y - 2x^2 + 3xy - 3y^2 - 25. \qquad (12)$$

To maximize $T$ they solve the two equations

$$\frac{\partial T}{\partial x} = 10 - 4x + 3y = 0,$$

$$\frac{\partial T}{\partial y} = 17 + 3x - 6y = 0$$

for $x = \$7.40$, $y = \$6.53$. These values give a total weekly profit of $11,072 by the first company and $56,461 by the second. The total weekly profit of the two companies is $67,533, an increase of $14,483 over their combined weekly profits when they acted independently. Figure 15.89 shows a comparison of the two situations—competition versus collusion.

```
                    Competition model

     Price 1:  $4.09     Company 1 profit/week:  $19,108

     Price 2:  $4.36     Company 2 profit/week:  $33,942


                         Total weekly profit:   $53,050
```

```
                      Collusion model

     Price 1:  $7.40     Company 1 profit/week:  $11,072

     Price 2:  $6.53     Company 2 profit/week:  $56,461


                         Total weekly profit:   $67,533
```

**15.89** Comparison of the competition model with the collusion model

## 15-11  PROBLEMS

In Problems 1–3, find the straight line $y = mx + b$ that best fits the given data points $(x_i, y_i)$.

**1** $(-1, 1.9)$, $(1, 8.1)$, $(4, 16.8)$

**2** $(-1.73, 4.61)$, $(2.25, -3.23)$, $(5.67, -2.13)$

**3** $(1.4, -11)$, $(2.7, 11)$, $(3.6, 26)$, $(5.5, 58)$

**4** In order to find a linear relation $y = mx + b$ between the price $x$ per box (in cents) and the weekly sales $y$ (in thousands of cases), a soap manufacturer test-sold soap in three cities at different prices, with the following results.

| $x$ | 79 | 89 | 99 |
|---|---|---|---|
| $y$ | 110 | 95 | 85 |

What weekly sales should be expected if the price is set at 69¢ per box?

**5** The per capita consumption of cigarettes in 1930 and the lung cancer death rate (deaths per million males) for 1950 in the Scandinavian countries were as follows:

| Country | Cigarette Consumption | Deaths from Lung Cancer |
|---|---|---|
| Denmark | 350 | 165 |
| Finland | 1100 | 350 |
| Norway | 250 | 95 |
| Sweden | 300 | 120 |

Fit these data to a straight line to estimate the 1950 male lung cancer death rate for Australia, in which the 1930 per capita cigarette consumption was 470. (The actual death rate was 170.) What lung cancer death rate might you expect in a country in which no cigarettes at all are smoked?

**6** The total cost $y$ (dollars) for printing $x$ copies of a certain booklet is known for three lot sizes; here are the details:

| $x$ | 700 | 2700 | 3700 |
|---|---|---|---|
| $y$ | 2700 | 4200 | 5000 |

Express the total cost as the sum of a fixed cost and an additional cost per book printed. What, then, should 2000 copies cost?

**7** The systolic blood pressure $p$ (in millimeters of mercury) of a healthy child is essentially a linear function $p = m(\ln w) + b$ of the *logarithm* of his or her weight $w$ (in pounds). The numbers $m$ and $b$ are constants. Determine them from the following experimental data.

| $w$ | 41 | 67 | 78 | 93 | 125 |
|---|---|---|---|---|---|
| $p$ | 89 | 100 | 103 | 107 | 110 |

What should the systolic blood pressure of a healthy 100-pound child be?

**8** Consider the sum of squares

$$g(A, B, C) = \sum_{i=1}^{n} [z_i - (Ax_i + By_i + C)]^2$$

that you would use to fit the data $(x_1, y_1, z_1), (x_2, y_2, z_2), \ldots, (x_n, y_n, z_n)$ to the linear equation $z = Ax + By + C$. Show that the conditions $g_A = g_B = g_C = 0$ reduce to the simultaneous equations

$$A \sum x_i + B \sum y_i + nC = \sum z_i,$$
$$A \sum x_i^2 + B \sum x_i y_i + C \sum x_i = \sum x_i z_i,$$
$$A \sum x_i y_i + B \sum y_i^2 + C \sum y_i = \sum y_i z_i.$$

**9** A company wishes to determine the weekly sales $z$ (in thousands of units) as a linear function $z = Ax + By + C$ of the price $x$ (in dollars) of its own product and the price $y$ of a competitor's product. Given the data below, use the equations of Problem 8 to find $A$, $B$, and $C$. What sales should be expected if $x$ is \$7.50 and $y$ is \$6.50?

| $x$ | $y$ | $z$ |
|---|---|---|
| 5 | 6 | 7.1 |
| 7 | 5 | 13.8 |
| 6 | 7 | 6.2 |
| 6 | 6 | 8.9 |

**10** A company manufactures a product and sells it in two different countries. In country A, it can sell $1000 - 10x$ units at $x$ dollars each, and in country B it can sell $2000 - 15y$ units at $y$ dollars each. The production costs for the units are \$30,000 (fixed cost) plus \$20 for each unit. What prices $x$ and $y$ will maximize the total profit of the company?

**11** A company manufactures and sells two competing products (I and II) that cost \$15 and \$25 per unit, respectively, to produce. The total revenue from making $x$ units of product I and $y$ units of product II is

$$55x + 45y - (0.02)xy - (0.15)x^2 - (0.05)y^2.$$

Find the values of $x$ and $y$ that will maximize the total profit (revenue minus cost) of the company.

**12** The owner of an appliance store charges $x$ dollars per videotape recorder and $y$ dollars per service policy for each such machine. As a result he can sell

$$27,500 - 30x - 5y \quad \text{videotape recorders}$$

and

$$6500 - 5x - 20y \quad \text{service policies.}$$

What should be the values of $x$ and $y$ to maximize his total revenue?

**13** A farmer can raise sheep, hogs, and cattle. She has space for 80 sheep or 120 hogs or 60 cattle or any combination using the same amount of space; that is, 8 sheep use as much space as 12 hogs or 6 cattle. The anticipated profits per animal are \$10 per sheep, \$8 per hog, and \$20 for each head of cattle. State law requires that a farmer raise as many hogs as sheep and cattle combined. How does the farmer maximize her profit?

**14** The profit of two firms depends in each case upon the production of both. Say the first firm has profit $P$, the second $Q$, the first has production $x$, and the second $y$.

These quantities are thus related:

$$P = 48x - 2x^2 - 3y^2 + 1000;$$

$$Q = 60y - 3y^2 - 2x^2 + 2000.$$

(a) If each firm acts to maximize its profit independently (but while crediting the competition with intelligence), find $P$, $Q$, $x$, $y$, and $T = P + Q$.
(b) If the firms form an agreement to maximize $T = P + Q$ and act accordingly, find $x$, $y$, and $T$.

**15** Three firms—Ajax Products (AP), Behemoth Quicksilver (BQ), and Conglomerate Resources (CR)—produce products in quantities $A$, $B$, and $C$, respectively. The profits that accrue to each obey the following laws:

| | |
|---|---|
| AP: | $P = 1000A - A^2 - 2AB.$ |
| BQ: | $Q = 2000B - 2B^2 - 4BC.$ |
| CR: | $R = 1500C - 3C^2 - 6AC.$ |

(a) If each firm acts independently to maximize its profit, what will the profits be?
(b) If firms AP and CR join to maximize their total profit while BQ continues to act alone, what effects will this have? Give a *complete* answer to this part. Assume that the fact of the merger of AP and CR *is* known to the management of BQ.

**16** Apply the maximum-minimum test of Section 15-10 to prove that the method of least squares always gives a minimum. That is, show that the function

$$f(m, b) = \sum_{i=1}^{n} \left[ y_i - (mx_i + b) \right]^2$$

has a single critical point and that it is a local minimum. You may use the fact that

$$n \sum (x_i)^2 > \left( \sum x_i \right)^2$$

unless the numbers $x_1, x_2, \ldots, x_n$ are all equal (see Miscellaneous Problem 37).

# CHAPTER 15 REVIEW:  Definitions, Concepts, Results

Use the list below as a guide to concepts that you may need to review.

**1** Graphs and level curves of functions of two variables
**2** Limits and continuity for functions of two and three variables
**3** Partial derivatives—definition and computation
**4** Geometric interpretation of partial derivatives and the tangent plane to the surface $z = f(x, y)$
**5** Absolute and local maxima and minima
**6** Necessary conditions for a local extremum
**7** The method of least squares
**8** Increments and differentials for functions of two and three variables

**9** The linear approximation theorem
**10** The chain rule for functions of several variables
**11** Directional derivatives—definition and computation
**12** The gradient vector and the vector chain rule
**13** Significance of the length and direction of the gradient vector
**14** The gradient vector as a normal vector; tangent plane to a surface $F(x, y, z) = 0$
**15** Constrained maximum-minimum problems and the Lagrange multiplier method
**16** Sufficient conditions for a local extremum of a function of two variables

**1** Use the method of Example 5 in Section 15-3 to show that

$$\lim_{(x,y)\to(0,0)} \frac{x^2y^2}{x^2 + y^2} = 0.$$

**2** Use spherical coordinates to show that

$$\lim_{(x,y,z)\to(0,0,0)} \frac{x^3 + y^3 - z^3}{x^2 + y^2 + z^2} = 0.$$

**3** Suppose that $g(x, y) = xy/(x^2 + y^2)$ unless $x = 0 = y$; let $g(0, 0)$ be defined to be zero. Show that $g$ is not continuous at $(0, 0)$.

**4** Compute $g_x(0, 0)$ and $g_y(0, 0)$ for the function of Problem 3.

**5** Find a function $f(x, y)$ such that

$$f_x(x, y) = 2xy^3 + e^x \sin y$$

and

$$f_y(x, y) = 3x^2y^2 + e^x \cos y + 1.$$

**6** Prove that there is *no* function with continuous second order partial derivatives such that $f_x(x, y) = 6xy^2$ and $f_y(x, y) = 8x^2y$.

**7** Find the points on the paraboloid $z = x^2 + y^2$ at which the normal line passes through the point $(0, 0, 1)$.

**8** Write an equation of the tangent plane to the surface $\sin xy + \sin yz + \sin xz = 1$ at the point $(1, \pi/2, 0)$.

**9** Prove that every normal line to the cone with equation $z = (x^2 + y^2)^{1/2}$ intersects the $z$-axis.

**10** Show that the function

$$u(x, t) = (4\pi kt)^{-1/2} \exp\left(\frac{-x^2}{4kt}\right)$$

satisfies the one-dimensional heat equation of Problem 45 in Section 15-4.

**11** Show that the function

$$u(x, y, t) = (4\pi kt)^{-1} \exp\left(\frac{-[x^2 + y^2]}{4kt}\right)$$

satisfies the two-dimensional heat equation of Problem 46 in Section 15-4.

**12** Let $f(x, y) = xy(x^2 - y^2)/(x^2 + y^2)$ unless $x = 0 = y$, in which case we have $f(0, 0) = 0$. Show that the second-order partial derivatives $f_{xx}$, $f_{xy}$, $f_{yx}$, and $f_{yy}$ all exist at $(0, 0)$ but that $f_{xy}(0, 0) \neq f_{yx}(0, 0)$.

**13** Define the partial derivatives $\mathbf{r}_x$ and $\mathbf{r}_y$ of the vector-valued function $\mathbf{r}(x, y) = x\mathbf{i} + y\mathbf{j} + f(x, y)\mathbf{k}$ by component-wise partial differentiation. Then show that the vector $\mathbf{r}_x \times \mathbf{r}_y$ is normal to the surface $z = f(x, y)$.

**14** An open-topped rectangular box is to have a total surface area of 300 in$^2$. Find the dimensions that maximize its volume.

**15** A rectangular shipping crate is to have a volume of 60 ft$^3$. Its sides cost \$1/ft$^2$, its top costs \$2/ft$^2$, and its bottom costs \$3/ft$^2$. What dimensions will minimize the cost of the box?

**16** A pyramid is bounded by the three coordinate planes and the tangent plane to the surface $xyz = 1$ at a point in the first octant. Find the volume of this pyramid (it is independent of the point of tangency).

**17** The total resistance $R$ of two resistances $R_1$ and $R_2$ connected in parallel is given by the formula

$$\frac{1}{R} = \frac{1}{R_1} + \frac{1}{R_2}.$$

Suppose that $R_1$ and $R_2$ are measured to be 300 and 600 $\Omega$, respectively, with a maximum error of 1% in each measurement. Use differentials to estimate the maximum error (in ohms) in the calculated value of $R$.

**18** Consider the gas of Problem 53 in Section 15-4, a gas satisfying the van der Waals equation. Use differentials to approximate the change in its volume if $P$ is increased from 1 atm to 1.1 atm and $T$ is decreased from 313°K to 303°K.

**19** The semiaxes $a$, $b$, and $c$ of an ellipsoid with volume $V = \frac{4}{3}\pi abc$ are each measured with a maximum percentage error of 1%. Use differentials to estimate the maximum percentage error in the calculated value of $V$.

**20** Two spheres have radii $a$ and $b$ and the distance between their centers is $c < a + b$, so that the spheres meet in a common circle. Let $P$ be a point on this circle and let $\mathcal{P}_1$ and $\mathcal{P}_2$ be the tangent planes at $P$ to the two spheres. Find the angle between $\mathcal{P}_1$ and $\mathcal{P}_2$ in terms of $a$, $b$, and $c$. (*Note:* The angle between two planes in space is by definition the angle between their normal vectors.)

**21** Find every point on the surface of the ellipsoid $x^2 + 4y^2 + 9z^2 = 16$ at which the normal line at the point passes through the center $(0, 0, 0)$ of the ellipsoid.

**22** Suppose that

$$F(x) = \int_{g(x)}^{h(x)} f(t)\, dt.$$

Show that $F'(x) = f(h(x))h'(x) - f(g(x))g'(x)$. (*Suggestion:* Write $w = \int_u^v f(t)\, dt$ with $u = g(x)$ and $v = h(x)$.)

**23** Suppose that $\mathbf{a}$, $\mathbf{b}$, and $\mathbf{c}$ are mutually perpendicular unit vectors, and that $f$ is a function of the three independent variables $x$, $y$, and $z$. Show that

$$\nabla f = (D_\mathbf{a} f)\mathbf{a} + (D_\mathbf{b} f)\mathbf{b} + (D_\mathbf{c} f)\mathbf{c}.$$

**24** Let $\mathbf{R} = \langle \cos\theta, \sin\theta, 0 \rangle$ and $\mathbf{\Theta} = \langle -\sin\theta, \cos\theta, 0 \rangle$

be the polar coordinate unit vectors. Given $f(x, y, z) = w(r, \theta, z)$, show that

$$D_{\mathbf{R}}f = \frac{\partial w}{\partial r} \quad \text{and} \quad D_{\mathbf{\Theta}}f = \frac{1}{r}\frac{\partial w}{\partial \theta}.$$

Then conclude from Problem 23 that the gradient vector is given in cylindrical coordinates by

$$\nabla f = \frac{\partial w}{\partial r}\mathbf{R} + \frac{1}{r}\frac{\partial w}{\partial \theta}\mathbf{\Theta} + \frac{\partial w}{\partial z}\mathbf{k}.$$

**25** Suppose that you are standing at the point with co-ordinates $(-100, -100, 430)$ on the hill of Example 5 in Section 15-8.
(a) In what (horizontal) direction should you move in order to climb the most steeply?
(b) What will be your resulting rate of climb (rise/run)?
**26** Suppose that the blood concentration in the water at the point $(x, y)$ is given by $f(x, y) = A\exp(-k[x^2 + 2y^2])$ where $A$ and $k$ are positive constants. A shark always swims in the direction of $\nabla f$. (Why?) Show that its path is a parabola $y = cx^2$. (*Suggestion:* Show that the condition that $\langle dx/dt, dy/dt \rangle$ be a multiple of $\nabla f$ implies that $x'/x = y'/2y$. Then antidifferentiate this equation.)
**27** Consider a tangent plane to the surface $x^{2/3} + y^{2/3} + z^{2/3} = 1$. Show that the sum of the squares of the $x$-, $y$-, and $z$-intercepts of this plane is 1.
**28** Find the points on the ellipse $x^2/a^2 + y^2/b^2 = 1$ (with $a \neq b$) where the normal line passes through the origin.
**29** (a) Show that the origin $(0, 0)$ is a critical point of the function $f$ of Problem 12.
(b) Show that $f$ does not have a local extremum at $(0, 0)$.
**30** Find the point of the surface $z = xy + 1$ that is closest to the origin.
**31** Use the method of Problem 24 in Section 15-9 to find the semiaxes of the rotated ellipse

$$73x^2 + 72xy + 52y^2 = 100.$$

**32** Use the Lagrange multiplier method to show that the longest chord of the sphere $x^2 + y^2 + z^2 = 1$ has length 2. (*Suggestion:* There is no loss of generality in assuming that $(1, 0, 0)$ is one end point of the chord.)
**33** Use the method of Lagrange multipliers, the law of cosines, and Fig. 15.78 to find the triangle of minimum perimeter inscribed in the unit circle.

**34** When a current $I$ enters two resistances $R_1$ and $R_2$ connected in parallel, it splits into two currents $I_1$ and $I_2$ to minimize $R_1I_1^2 + R_2I_2^2$ (the total power). Express $I_1$ and $I_2$ in terms of $R_1$, $R_2$, and $I$.
**35** Use the method of Lagrange multipliers to find the points of the ellipse $x^2 + 2y^2 = 1$ that are closest to and farthest from the line $x + y = 2$. (*Suggestion:* Let $f(x, y, u, v)$ denote the square of the distance between the point $(x, y)$ of the ellipse and the point $(u, v)$ of the line.)
**36** (a) Show that the maximum value of $f(x, y, z) = x + y + z$ at points of the sphere $x^2 + y^2 + z^2 = a^2$ is $a\sqrt{3}$.
(b) Conclude from the result of (a) that

$$(x + y + z)^2 \leq 3(x^2 + y^2 + z^2)$$

for any three numbers $x$, $y$, and $z$.
**37** Generalize the method of Problem 36 to show that

$$\left(\sum_{i=1}^{n} x_i\right)^2 \leq n\sum_{i=1}^{n} x_i^2$$

for any $n$ real numbers $x_1, x_2, \ldots, x_n$.
**38** Find the minimum and maximum values of $f(x, y) = xy - x - y$ at points on and within the triangle with vertices $(0, 0)$, $(0, 1)$, and $(3, 0)$.
**39** Find the maximum and minimum values of $f(x, y, z) = x^2 - yz$ at points of the sphere $x^2 + y^2 + z^2 = 1$.
**40** Find the maximum and minimum values of $f(x, y) = x^2y^2$ at points of the ellipse $x^2 + 4y^2 = 24$.

Locate and classify the critical points (local maxima, local minima, saddle points, and other points at which the tangent plane is horizontal) of the functions in Problems 41–50.

**41** $f(x, y) = x^3y - 3xy + y^2$
**42** $f(x, y) = x^2 + xy + y^2 - 6x + 2$
**43** $f(x, y) = x^3 - 6xy + y^3$
**44** $f(x, y) = x^2y + xy^2 + x + y$
**45** $f(x, y) = x^3y^2(1 - x - y)$
**46** $f(x, y) = x^4 - 2x^2 + y^2 + 4y + 3$
**47** $f(x, y) = e^{xy} - 2xy$
**48** $f(x, y) = x^3 - y^3 + x^2 + y^2$
**49** $f(x, y) = (x - y)(xy - 1)$
**50** $f(x, y) = (2x^2 + y^2)e^{-x^2 - y^2}$

**842**

# 16

# Multiple Integrals

## Double Integrals

This chapter is devoted to integrals of functions of two and three variables. Such integrals are called **multiple integrals.** The applications of multiple integrals include computations of area, volume, mass, and surface area in a wider variety of situations than can be handled with the single integral of Chapters 5 and 6.

The simplest sort of multiple integral is the *double integral*

$$\iint\limits_{R} f(x, y) \, dA$$

of a continuous function $f(x, y)$ over the *rectangle*

$$R = [a, b] \times [c, d] = \{(x, y) | a \leq x \leq b, c \leq y \leq d\}$$

in the $xy$-plane. Just as the definition of the single integral is motivated by the problem of computing areas, the definition of the double integral is motivated by the problem of computing the volume $V$ of the solid of Fig. 16.1: a solid that is bounded above by the graph $z = f(x, y)$ of the nonnegative function $f$ and that lies above the rectangle $R$ in the $xy$-plane.

To define the double integral $\iint_{R} f(x, y) \, dA$ which has value $V$, we begin with an *approximation* to $V$. To obtain this approximation, our first step is to construct a **partition** $\mathscr{P}$ of $R$ into subrectangles $R_1, R_2, \ldots, R_k$ determined by the partitions

$$a = x_0 < x_1 < x_2 < \cdots < x_m = b$$

of $[a, b]$ and

$$c = y_0 < y_1 < y_2 < \cdots < y_n = d$$

of $[c, d]$. Such a partition of $R$ into $k = mn$ subrectangles is shown in Fig. 16.2. The order in which these rectangles are labeled makes no difference.

**16.1** We will use a double integral to compute the volume $V$.

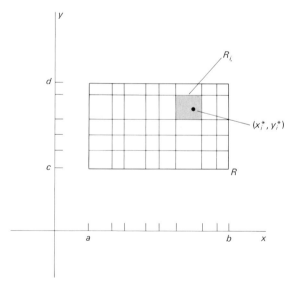

**16.2** A partition $\mathscr{P}$ of the rectangle $R$

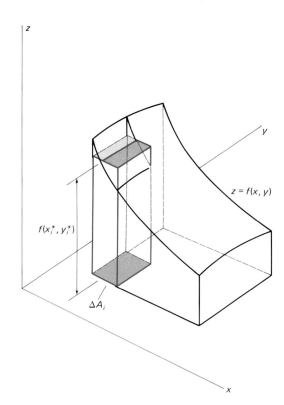

**16.3** Approximating the volume under the surface by summing volumes of towers with rectangular bases

Next, we choose an arbitrary point $(x_i^*, y_i^*)$ of the $i$th subrectangle $R_i$ for each $i$ ($1 \le i \le k$). The collection of points $S = \{(x_i^*, y_i^*)\}_{i=1}^k$ is called a **selection** for the partition $\mathscr{P} = \{R_i\}_{i=1}^k$. As a measure of the size of the subrectangles of the partition $\mathscr{P}$, we define its **mesh** $|\mathscr{P}|$ to be the maximum of the lengths of the diagonals of the rectangles $\{R_i\}$.

Figure 16.3 shows the rectangular column with base the subrectangle $R_i$ and height the value $f(x_i^*, y_i^*)$ of $f$ at the selected point $(x_i^*, y_i^*)$ of $R_i$. If $\Delta A_i$ denotes the area of $R_i$, then the volume of this column is $f(x_i^*, y_i^*)\Delta A_i$. The sum of the volumes of all such columns is the **Riemann sum**

$$\sum_{i=1}^k f(x_i^*, y_i^*)\, \Delta A_i, \tag{1}$$

an approximation to the volume $V$ of the solid region that lies over the rectangle $R$ and under the graph $z = f(x, y)$.

We would expect to get the exact volume $V$ by taking the limit of the Riemann sum in (1) as the mesh $|\mathscr{P}|$ of the partition $\mathscr{P}$ approaches zero. We therefore define the **(double) integral** of the function $f$ over the rectangle $R$ by

$$\iint_R f(x, y)\, dA = \lim_{|\mathscr{P}| \to 0} \sum_{i=1}^k f(x_i^*, y_i^*)\, \Delta A_i, \tag{2}$$

provided that this limit (we discuss its meaning in more detail below) exists. In advanced calculus it is proved that the limit in (2) *does* exist if $f$ is continuous on $R$. In order to motivate the introduction of the Riemann sum in (1), we assumed that $f$ was nonnegative on $R$, but Equation (2) serves to define the double integral whether or not $f$ is nonnegative.

The direct evaluation of the limit in Equation (2) is generally even less practical than the direct evaluation of the limit we used in Section 5-3 to define the single-variable integral. In practice, we shall calculate double integrals over rectangles by means of the **iterated integrals** that appear in the following theorem.

---

**Theorem** *Iterated Double Integrals*

Suppose that $f(x, y)$ is continuous on the rectangle $R = [a, b] \times [c, d]$. Then

$$\iint_R f(x, y) \, dA = \int_a^b \left( \int_c^d f(x, y) \, dy \right) dx = \int_c^d \left( \int_a^b f(x, y) \, dx \right) dy. \quad (3)$$

---

This theorem tells us how to compute a double integral by means of two successive (or *iterated*) single-variable integrations, each of which can be carried out using the fundamental theorem of calculus (provided that the function $f$ is nice enough).

The meaning of the parentheses in the iterated integral

$$\int_a^b \int_c^d f(x, y) \, dy \, dx = \int_a^b \left( \int_c^d f(x, y) \, dy \right) dx \quad (4)$$

is this: First we hold $x$ constant and integrate with respect to $y$, from $y = c$ to $y = d$. The result of this first integration is the **partial integral of $f$ with respect to $y$**, denoted by

$$\int_c^d f(x, y) \, dy,$$

and it is a function of $x$ alone. The final step is to integrate this latter function with respect to $x$, from $x = a$ to $x = b$.

Similarly, the iterated integral

$$\int_c^d \int_a^b f(x, y) \, dx \, dy = \int_c^d \left( \int_a^b f(x, y) \, dx \right) dy \quad (5)$$

is calculated by first integrating from $a$ to $b$ with respect to $x$ (while holding $y$ fixed) and then integrating the result from $c$ to $d$ with respect to $y$. Note that the order of integration (either first with respect to $x$ and then with respect to $y$ or the reverse) is determined by the order in which the differentials $dx$ and $dy$ appear in the iterated integrals in (4) and (5). We always work "from the inside out." Theorem 1 guarantees that the value obtained is independent of the order of integration provided that $f$ is continuous.

**EXAMPLE 1** Compute the iterated integrals in (4) and (5) for the function $f(x, y) = 4x^3 + 6xy^2$ on the rectangle $R = [1, 3] \times [-2, 1]$.

**Solution**

$$\int_1^3 \left( \int_{-2}^1 (4x^3 + 6xy^2)\,dy \right) dx = \int_1^3 \left[ 4x^3 y + 2xy^3 \right]_{-2}^1 dx$$

$$= \int_1^3 \left[ (4x^3 + 2x) - (-8x^3 - 16x) \right] dx$$

$$= \int_1^3 (12x^3 + 18x)\,dx$$

$$= \left[ 3x^4 + 9x^2 \right]_1^3 = 312.$$

On the other hand,

$$\int_{-2}^1 \left( \int_1^3 (4x^3 + 6xy^2)\,dx \right) dy = \int_{-2}^1 \left[ x^4 + 3x^2 y^2 \right]_1^3 dy$$

$$= \int_{-2}^1 \left[ (81 + 27y^2) - (1 + 3y^2) \right] dy$$

$$= \int_{-2}^1 (80 + 24y^2)\,dy$$

$$= \left[ 80y + 8y^3 \right]_{-2}^1 = 312.$$

When we note that iterated double integrals are always evaluated from the inside out, it becomes clear that the parentheses appearing on the right-hand sides in Equations (4) and (5) are unnecessary. They are therefore generally omitted, as in the two examples that follow. When $dy\,dx$ appears in the integrand we integrate first with respect to $y$, while the appearance of $dx\,dy$ tells us to integrate first with respect to $x$.

**EXAMPLE 2**

$$\int_0^\pi \int_0^{\pi/2} \cos x \cos y \, dy \, dx = \int_0^\pi \left[ \cos x \sin y \right]_{y=0}^{\pi/2} dx$$

$$= \int_0^\pi \cos x \, dx = \left[ \sin x \right]_0^\pi = 0.$$

**EXAMPLE 3**

$$\int_0^1 \int_0^{\pi/2} (e^y + \sin x)\,dx\,dy = \int_0^1 \left[ xe^y - \cos x \right]_{x=0}^{\pi/2} dy$$

$$= \int_0^1 \left( \frac{1}{2} \pi e^y + 1 \right) dy$$

$$= \left[ \frac{1}{2} \pi e^y + y \right]_0^1 = \frac{\pi(e-1)}{2} + 1.$$

An outline of the proof of the theorem above exhibits an instructive relationship between iterated integrals and the method of cross sections (for computing volumes) discussed in Section 6-2. First, we subdivide $[a, b]$ into $n$ equal subintervals each with length $\Delta x = (b - a)/n$, and we also subdivide $[c, d]$ into $n$ equal subintervals each with length $\Delta y = (d - c)/n$. This gives $n^2$ subrectangles, each of which has area $\Delta A = \Delta x \Delta y$. Pick a point $x_i^*$ in $[x_{i-1}, x_i]$ for each $i$, $1 \leq i \leq n$. Then the average value theorem for single integrals (Section 5-5) gives a point $y_{ij}^*$ in $[y_{j-1}, y_j]$ such that

$$\int_{y_{j-1}}^{y_j} f(x_i^*, y)\, dy = f(x_i^*, y_{ij}^*)\, \Delta y.$$

This gives our selected point $(x_i^*, y_{ij}^*)$ in the subrectangle $[x_{i-1}, x_i] \times [y_{j-1}, y_j]$. Then

$$\iint_R f(x, y)\, dA \approx \sum_{i,j=1}^n f(x_i^*, y_{ij}^*)\, \Delta A$$

$$= \sum_{i=1}^n \sum_{j=1}^n f(x_i^*, y_{ij}^*)\, \Delta y\, \Delta x$$

$$= \sum_{i=1}^n \left( \sum_{j=1}^n \int_{y_{j-1}}^{y_j} f(x_i^*, y)\, dy \right) \Delta x$$

$$= \sum_{i=1}^n \left( \int_c^d f(x_i^*, y)\, dy \right) \Delta x = \sum_{i=1}^n A(x_i^*)\, \Delta x$$

where

$$A(x) = \int_c^d f(x, y)\, dy.$$

This last sum is a Riemann sum for the integral $\int_a^b A(x)\, dx$, so the result of our computation is that

$$\iint_R f(x, y)\, dA \approx \sum_{i=1}^n A(x_i^*)\, \Delta x$$

$$\approx \int_a^b A(x)\, dx = \int_a^b \left( \int_c^d f(x, y)\, dy \right) dx.$$

This outline can be converted in a complete proof of the theorem by showing that the approximations above become equalities when we take limits as $n \to \infty$.

In case the function $f$ is nonnegative on $R$, the function $A(x)$ introduced above gives the area of the vertical cross section perpendicular to the $x$-axis shown in Fig. 16.4. Thus the iterated integral in (4) expresses the volume $V$ as the integral from $x = a$ to $x = b$ of the cross-sectional area function $A(x)$. Similarly, the iterated integral in (5) expresses $V$ as the integral from $y = c$ to $y = d$ of the function

$$A(y) = \int_a^b f(x, y)\, dx,$$

which gives the area of a vertical cross section in a plane perpendicular to the $y$-axis. (Although it is suggestive to use $A(y)$ here, note that $A(y)$ and $A(x)$ are not the same function.)

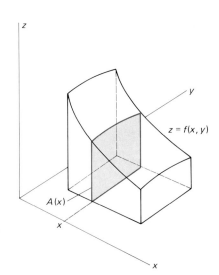

**16.4** $A(x) = \int_c^d f(x, y)\, dy$ is the area of a cross section.

## 16-1 PROBLEMS

Evaluate the iterated double integrals in Problems 1–20.

1. $\int_0^2 \int_0^4 (3x + 4y)\, dx\, dy$

2. $\int_0^3 \int_0^2 x^2 y\, dx\, dy$

3. $\int_{-1}^2 \int_1^3 (2x - 7y)\, dy\, dx$

4. $\int_{-2}^1 \int_2^4 x^2 y^3\, dy\, dx$

5. $\int_0^3 \int_0^3 (xy + 7x + y)\, dx\, dy$

6. $\int_0^2 \int_2^4 (x^2 y^2 - 17)\, dx\, dy$

7. $\int_{-1}^2 \int_{-1}^2 (2xy^2 - 3x^2 y)\, dy\, dx$

**8** $\displaystyle\int_1^3 \int_{-3}^{-1} (x^3y - xy^3)\, dy\, dx$

**9** $\displaystyle\int_0^{\pi/2} \int_0^{\pi/2} (\sin x \cos y)\, dx\, dy$

**10** $\displaystyle\int_0^{\pi/2} \int_0^{\pi/2} (\cos x \sin y)\, dy\, dx$

**11** $\displaystyle\int_0^1 \int_0^1 xe^y\, dy\, dx$

**12** $\displaystyle\int_0^1 \int_{-2}^2 x^2 e^y\, dx\, dy$

**13** $\displaystyle\int_0^1 \int_0^\pi e^x \sin y\, dy\, dx$

**14** $\displaystyle\int_0^1 \int_0^1 e^{x+y}\, dx\, dy$

**15** $\displaystyle\int_0^\pi \int_0^\pi (xy + \sin x)\, dx\, dy$

**16** $\displaystyle\int_0^{\pi/2} \int_0^{\pi/2} (y-1)\cos x\, dx\, dy$

**17** $\displaystyle\int_0^{\pi/2} \int_1^e \frac{\sin y}{x}\, dx\, dy$

**18** $\displaystyle\int_1^e \int_1^e \frac{dy\, dx}{xy}$

**19** $\displaystyle\int_0^1 \int_0^1 \left( \frac{1}{x+1} + \frac{1}{y+1} \right) dx\, dy$

**20** $\displaystyle\int_1^2 \int_1^3 \left( \frac{x}{y} + \frac{y}{x} \right) dy\, dx$

In Problems 21–24, verify that the values of $\iint_R f(x, y)\, dA$ given by the iterated integrals in (4) and (5) are indeed equal.

**21** $f(x, y) = 2xy - 3y^2; \quad R = [-1, 1] \times [-2, 2]$

**22** $f(x, y) = \sin x \cos y; \quad R = [0, \pi] \times \left[ -\dfrac{\pi}{2}, \dfrac{\pi}{2} \right]$

**23** $f(x, y) = \sqrt{x + y}; \quad R = [0, 1] \times [0, 2]$

**24** $f(x, y) = e^{x+y}; \quad R = [0, \ln 2] \times [0, \ln 3]$

**25** Show that

$$\lim_{n \to \infty} \int_0^1 \int_0^1 x^n y^n\, dx\, dy = 0.$$

## *16-1 Optional Computer Application

The program below was written to compute values of the Riemann sum

$$\sum_{i=1}^n \left( \sum_{j=1}^n f(x_i^*, y_j^*) \right) \Delta y\, \Delta x \tag{6}$$

with $f(x, y) = 4x^3 + 6xy^2$ corresponding to Example 1 in this section. Our approach is to subdivide the interval $a \le x \le b$ into $n$ subintervals all having the same length $h = \Delta x = (b - a)/n$, and the interval $c \le y \le d$ into $n$ subintervals all having the same length $k = \Delta y = (d - c)/n$. Then for each $i = 1, 2, 3, \ldots, n$ we select $x_i^*$ to be the midpoint of the $i$th subinterval $[x_{i-1}, x_i]$, and for each $j = 1, 2, 3, \ldots, n$, we select $y_j^*$ to be the midpoint of the $j$th subinterval $[y_{j-1}, y_j]$.

```
100     A = 1    : B = 3
110     C = -2 : D = 1
120     INPUT "N: "; N
130     H = (B - A)/N  : K = (D - C)/N
140     X = A + H/2 : S = 0
150     FOR I = 1 TO N
160         Y = C + K/2
170         FOR J = 1 TO N
180             F = 4*X*X*X + 6*X*Y*Y
190             S = S + F
200             Y = Y + K
210         NEXT J
220         X = X + H
230     NEXT I
240     S = S*H*K
250     PRINT "RIEMANN SUM: "; S
260     END
```

Lines 100–110 specify the limits of the iterated integrals in Example 1. The integer $n$ is entered by the operator of the program at line 120, and then line 130 calculates $h = \Delta x$ and $k = \Delta y$. The inner **FOR-NEXT** loop at lines 170–210 computes the inner sum in Equation (6), and the outer **FOR-NEXT** loop in lines 150–230 computes the outer sum.

When this program was executed with N = 10, N = 20, and N = 100 in turn, we obtained the values 310.98, 311.75, and 311.99, respectively, in apparent agreement with the exact value 312 obtained for the integral in Example 1.

---

## 16-2

## Double Integrals Over More General Regions

Now we want to define and compute double integrals over regions more general than rectangles. Let the function $f$ be defined on the plane region $R$, and suppose that $R$ is **bounded**—that is, suppose that $R$ lies within some rectangle $S$. To define the (double) integral of $f$ over the region $R$, we begin with a partition $\mathcal{Q}$ of the rectangle $S$ into subrectangles. Some of the rectangles of $\mathcal{Q}$ will lie inside $R$, some will lie outside $R$, and some will lie partly inside and partly outside. We consider the collection $\mathcal{P} = \{R_1, R_2, \ldots, R_k\}$ of all those subrectangles of $\mathcal{Q}$ that lie *completely within* the region $R$. This collection $\mathcal{P}$ is called the **inner partition** of the region $R$ determined by the partition $\mathcal{Q}$ of the rectangle $S$ (see Fig. 16.5). By the **mesh** $|\mathcal{P}|$ of the inner partition $\mathcal{P}$ is meant the mesh of the partition $\mathcal{Q}$ that determines $\mathcal{P}$. (Note that $|\mathcal{P}|$ depends not only upon $\mathcal{P}$ but upon $\mathcal{Q}$ as well.)

Using the inner partition $\mathcal{P}$ of the region $R$, we can proceed in much the same way as before. By choosing an arbitrary point $(x_i^*, y_i^*)$ in the $i$th subrectangle $R_i$ of $\mathcal{P}$ for $i = 1, 2, 3, \ldots, k$, we obtain a **selection** for the inner partition $\mathcal{P}$. Let us denote by $\Delta A_i$ the area of $R_i$. Then this selection gives the **Riemann sum**

$$\sum_{i=1}^{k} f(x_i^*, y_i^*) \, \Delta A_i$$

associated with the inner partition $\mathcal{P}$. In case $f$ is nonnegative on $R$, this Riemann sum approximates the volume of the three-dimensional region that lies under the surface $z = f(x, y)$ and above the region $R$ in the $xy$-plane. We therefore define the double integral of $f$ over the region $R$ by taking the limit

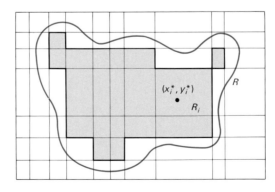

**16.5** The rectangular partition of $S$ produces an associated inner partition (shown shaded) of the region $R$.

of this Riemann sum as the mesh $|\mathcal{P}|$ approaches zero. Thus

$$\iint\limits_{R} f(x, y)\, dA = \lim_{|\mathcal{P}| \to 0} \sum_{i=1}^{k} f(x_i^*, y_i^*)\, \Delta A_i \qquad (1)$$

provided that this limit exists in the sense of the following definition.

---

*Definition*    *The Double Integral*

The **(double) integral** of the bounded function $f$ over the plane region $R$ is the number

$$I = \iint\limits_{R} f(x, y)\, dA$$

provided that, for every $\varepsilon > 0$, there exists a number $\delta > 0$ such that

$$\left| \sum_{i=1}^{k} f(x_i^*, y_i^*)\, \Delta A_i - I \right| < \varepsilon$$

for every inner partition $\mathcal{P} = \{R_1, R_2, R_3, \dots, R_k\}$ of $R$ having mesh $|\mathcal{P}| < \delta$ and every selection of points $(x_i^*, y_i^*)$ in $R_i$ $(i = 1, 2, 3, \dots, k)$.

---

Thus the meaning of the limit in (1) is that the Riemann sum can be made arbitrarily close to $I = \iint_R f(x, y)\, dA$ by choosing the mesh of the inner partition $\mathcal{P}$ sufficiently small.

Note that, if $R$ is a rectangle and we choose $S = R$ (so that an inner partition of $R$ is simply a partition of $R$), then the definition above reduces to our earlier definition of a double integral over a rectangle. In advanced calculus it is proved that the double integral of the function $f$ over the bounded plane region $R$ exists provided that $f$ is continuous on $R$ and the *boundary* of $R$ is reasonably nice. In particular, it suffices for the boundary of $R$ to consist of finitely many piecewise smooth simple closed curves (that is, each boundary curve consists of finitely many smooth arcs).

For certain common types of regions, we can evaluate double integrals by using iterated integrals, in much the same way as when the region is a rectangle. The region $R$ is called **vertically simple** if it is described by means of the inequalities

$$a \leq x \leq b, \qquad g_1(x) \leq y \leq g_2(x), \qquad (2)$$

where $g_1$ and $g_2$ are continuous on $[a, b]$. Such a region appears in Fig. 16.6. The region $R$ is called **horizontally simple** if it is described by the inequalities

$$c \leq y \leq d, \qquad h_1(y) \leq x \leq h_2(y), \qquad (3)$$

where $h_1$ and $h_2$ are continuous on $[c, d]$. The region of Fig. 16.7 is horizontally simple.

The following theorem tells us how to compute by iterated integration a double integral over a region $R$ that is either vertically simple or horizontally simple.

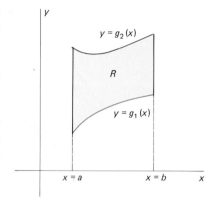

**16.6**   A vertically simple region $R$

**16.7**   A horizontally simple region $R$

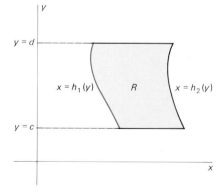

**Theorem** *General Iterated Double Integrals*

Suppose that $f(x, y)$ is continuous on the region $R$. If $R$ is the vertically simple region given in (2), then

$$\iint_R f(x, y)\, dA = \int_a^b \int_{g_1(x)}^{g_2(x)} f(x, y)\, dy\, dx. \tag{4}$$

If $R$ is the horizontally simple region given in (3), then

$$\iint_R f(x, y)\, dA = \int_c^d \int_{h_1(y)}^{h_2(y)} f(x, y)\, dx\, dy. \tag{5}$$

This theorem includes the theorem of Section 16-1 as a special case (when $R$ is a rectangle), and it can be proved by a generalization of the argument we outlined there.

**EXAMPLE 1**   Compute in two different ways the integral

$$\iint_R xy^2\, dA,$$

where $R$ is the first-quadrant region bounded by the two curves $y = x^2$ and $y = x^3$.

**Solution**   *Always sketch the region $R$ of integration before attempting to evaluate a double integral.* The region $R$ bounded by $y = x^2$ and $y = x^3$ is shown in Fig. 16.8. This region is vertically simple with $a = 0$, $b = 1$, $g_1(x) = x^3$, and $g_2(x) = x^2$. Therefore (4) yields

$$\iint_R xy^2\, dA = \int_0^1 \int_{x^3}^{x^2} xy^2\, dy\, dx = \int_0^1 \left[\tfrac{1}{3}xy^3\right]_{x^3}^{x^2} dx$$

$$= \int_0^1 (\tfrac{1}{3}x^7 - \tfrac{1}{3}x^{10})\, dx = \tfrac{1}{24} - \tfrac{1}{33} = \tfrac{1}{88}.$$

The region $R$ is also horizontally simple with $c = 0$, $d = 1$, $h_1(y) = y^{1/2}$, and $h_2(y) = y^{1/3}$, so we can reverse the order of integration, and integrate first with respect to $x$. Then Equation (5) gives

$$\iint_R xy^2\, dA = \int_0^1 \int_{y^{1/2}}^{y^{1/3}} xy^2\, dx\, dy = \int_0^1 \left[\tfrac{1}{2}x^2 y^2\right]_{y^{1/2}}^{y^{1/3}} dy$$

$$= \int_0^1 (\tfrac{1}{2}y^{8/3} - \tfrac{1}{2}y^3)\, dy = \tfrac{1}{88}.$$

**EXAMPLE 2**   Evaluate $\iint_R (6x + 2y^2)\, dA$, where $R$ is the region bounded by the parabola $x = y^2$ and the straight line $x + y = 2$.

**Solution**   The region $R$ appears in Fig. 16.9. It is both horizontally and vertically simple. If we wished to integrate first with respect to $y$ and then with respect to $x$, we would need to evaluate two integrals:

$$\iint_R f(x, y)\, dA = \int_0^1 \int_{-\sqrt{x}}^{\sqrt{x}} (6x + 2y^2)\, dy\, dx + \int_1^4 \int_{-\sqrt{x}}^{2-x} (6x + 2y^2)\, dy\, dx.$$

The reason is that the "top" function $g_2(x)$ changes its formula at the point $(1, 1)$ (see the sketch of the region $R$).

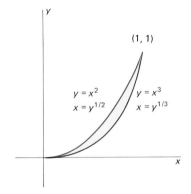

**16.8**   The region $R$ of Example 1

**16.9**   The region of Example 2

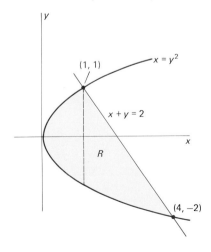

To avoid this extra work, we prefer to integrate in the opposite order:

$$
\begin{aligned}
\iint\limits_{R} (6x + 2y^2)\, dA &= \int_{-2}^{1} \int_{y^2}^{2-y} (6x + 2y^2)\, dx\, dy \\
&= \int_{-2}^{1} \left[ 3x^2 + 2xy^2 \right]_{y^2}^{2-y} dy \\
&= \int_{-2}^{1} \left[ 3(2 - y)^2 + 2(2 - y)y^2 - 3(y^2)^2 - 2(y^2)y^2 \right] dy \\
&= \int_{-2}^{1} (12 - 12y + 7y^2 - 2y^3 - 5y^4)\, dy \\
&= \left[ 12y - 6y^2 + \tfrac{7}{3}y^3 - \tfrac{1}{2}y^4 - y^5 \right]_{-2}^{1} = \tfrac{99}{2}.
\end{aligned}
$$

Example 2 indicates that even when the region $R$ is both vertically and horizontally simple, it may be simpler to integrate in one order rather than the other because of the shape of $R$. We naturally prefer the easier route. The choice of the preferable order of integration may also be influenced by the nature of the function $f(x, y)$. For it may be difficult—or even impossible—to compute a given iterated integral, but easy to do *after reversing the order of integration.* The following example shows that the key to reversing the order of integration is this: Find (and sketch) the region $R$ over which the integration is performed.

**EXAMPLE 3** Evaluate

$$
\int_{0}^{2} \int_{y/2}^{1} y e^{x^3}\, dx\, dy.
$$

***Solution*** We cannot integrate first with respect to $x$, as indicated, because it happens that $e^{x^3}$ has no elementary antiderivative. So we decide to try to find the value of the above integral by reversing the order. To do this, we first sketch the region of integration specified by the limits in the given iterated double integral.

This region $R$ is determined by the inequalities

$$
\frac{y}{2} \le x \le 1 \quad \text{and} \quad 0 \le y \le 2.
$$

Thus all points $(x, y)$ of $R$ lie between the horizontal lines $y = 0$ and $y = 2$, and also between the two graphs $x = y/2$ and $x = 1$. We draw the four straight lines $y = 0$, $y = 2$, $x = y/2$, and $x = 1$ and find that the region of integration is the triangle that appears in Fig. 16.10.

Integrating first with respect to $y$, from $g_1(x) = 0$ to $g_2(x) = 2x$, we obtain

$$
\begin{aligned}
\int_{0}^{2} \int_{y/2}^{1} y e^{x^3}\, dx\, dy &= \int_{0}^{1} \int_{0}^{2x} y e^{x^3}\, dy\, dx \\
&= \int_{0}^{1} \left[ \tfrac{1}{2}y^2 \right]_{0}^{2x} e^{x^3}\, dx = \int_{0}^{1} 2x^2 e^{x^3}\, dx \\
&= \left[ \tfrac{2}{3}e^{x^3} \right]_{0}^{1} = \tfrac{2}{3}(e - 1).
\end{aligned}
$$

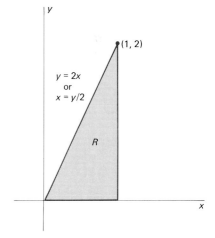

**16.10** The region $R$ of Example 3

We conclude this section by listing some useful formal properties of double integrals. Let $c$ be a constant and $f$ and $g$ continuous functions on a region $R$ on which $f(x, y)$ attains a minimum value $m$ and a maximum value $M$. Let $a(R)$ denote the area of the region $R$. If the indicated integrals all exist, then

$$\iint_R cf(x, y)\, dA = c \iint_R f(x, y)\, dA. \tag{6}$$

$$\iint_R [f(x, y) + g(x, y)]\, dA = \iint_R f(x, y)\, dA + \iint_R g(x, y)\, dA. \tag{7}$$

$$m \cdot a(R) \leq \iint_R f(x, y)\, dA \leq M \cdot a(R). \tag{8}$$

$$\iint_R f(x, y)\, dA = \iint_{R_1} f(x, y)\, dA + \iint_{R_2} f(x, y)\, dA. \tag{9}$$

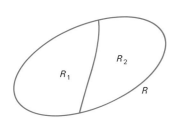

**16.11** The regions of the formula in Equation (9)

In Equation (9), $R_1$ and $R_2$ are two nonoverlapping regions (regions with disjoint interiors) with union $R$, as in Fig. 16.11. In Problems 25 through 28, we indicate proofs of the properties in (6) through (9) for the special case in which $R$ is a rectangle.

The property in Equation (9) enables us to evaluate $\iint_R f(x, y)\, dA$ when the region $R$ is neither horizontally simple nor vertically simple but can be decomposed into finitely many simple regions $R_1, R_2, R_3, \ldots, R_n$. Then we may write

$$\iint_R f(x, y)\, dA = \sum_{i=1}^{n} \iint_{R_i} f(x, y)\, dA$$

and use the theorem of this section to evaluate each of the integrals on the right-hand side by iterated integration.

## 16-2 PROBLEMS

Evaluate the iterated integrals in Problems 1–14.

**1** $\int_0^1 \int_0^x (1 + x)\, dy\, dx$    **2** $\int_0^2 \int_0^{2x} (1 + y)\, dy\, dx$

**3** $\int_0^1 \int_y^1 (x + y)\, dx\, dy$    **4** $\int_0^2 \int_{y/2}^1 (x + y)\, dx\, dy$

**5** $\int_0^1 \int_0^{x^2} xy\, dy\, dx$    **6** $\int_0^1 \int_y^{\sqrt{y}} (x + y)\, dx\, dy$

**7** $\int_0^1 \int_x^{\sqrt{x}} (2x - y)\, dy\, dx$

**8** $\int_0^2 \int_{-\sqrt{2y}}^{\sqrt{2y}} (3x + 2y)\, dx\, dy$

**9** $\int_0^1 \int_{x^4}^x (y - x)\, dy\, dx$

**10** $\int_{-1}^2 \int_{-y}^{y+2} (x + 2y^2)\, dx\, dy$

**11** $\int_0^1 \int_0^{x^3} e^{y/x}\, dy\, dx$    **12** $\int_0^\pi \int_0^{\sin x} y\, dy\, dx$

**13** $\int_0^3 \int_0^y \sqrt{y^2 + 16}\, dx\, dy$    **14** $\int_1^{e^2} \int_0^{1/y} e^{xy}\, dx\, dy$

In Problems 15–24, reverse the order of integration and then evaluate the resulting integral. Also sketch each region of integration.

**15** $\int_{-2}^2 \int_{x^2}^4 x^2 y\, dy\, dx$    **16** $\int_0^1 \int_{x^4}^x (x - 1)\, dy\, dx$

**17** $\int_{-1}^3 \int_{x^2}^{2x+3} x\, dy\, dx$    **18** $\int_{-2}^2 \int_{y^2-4}^{4-y^2} y\, dx\, dy$

**19** $\int_0^2 \int_{2x}^{4x-x^2} 1\, dy\, dx$    **20** $\int_0^1 \int_y^1 e^{-x^2}\, dx\, dy$

**21** $\int_0^\pi \int_x^\pi \dfrac{\sin y}{y}\, dy\, dx$    **22** $\int_0^{\sqrt{\pi}} \int_y^{\sqrt{\pi}} \sin x^2\, dx\, dy$

**23** $\int_0^1 \int_y^1 \dfrac{dx\, dy}{1 + x^4}$    **24** $\int_0^1 \int_{\tan^{-1}y}^{\pi/4} \sec x\, dx\, dy$

**25** Use Riemann sums to prove (6) for the case in which $R$ is a rectangle.

**26** Use iterated integrals and familiar properties of single integrals to prove (7) for the case in which $R$ is a rectangle.

**27** Use Riemann sums to prove (8) when $R$ is a rectangle.

**28** Use iterated integrals and familiar properties of single integrals to prove (9) when the right edge of the rectangle $R_1$ is the left edge of the rectangle $R_2$.

**29** Use Riemann sums to show that

$$\iint_R f(x, y)\, dA \leq \iint_R g(x, y)\, dA$$

if $f(x, y) \leq g(x, y)$ at each point of the rectangle $R$.

**30** Suppose that the continuous function $f$ is integrable on the plane region $R$ and that $f$ attains a minimum value $m$ and a maximum value $M$ at points of $R$. Assume also

that $R$ is *connected* in the following sense: For any two points $(x_0, y_0)$ and $(x_1, y_1)$ of $R$, there is a continuous parametric curve $\mathbf{r}(t)$ in $R$ with $\mathbf{r}(0) = \langle x_0, y_0 \rangle$ and $\mathbf{r}(1) = \langle x_1, y_1 \rangle$. Then deduce the *average value property* of double integrals from (8):

$$\iint_R f(x, y)\, dA = f(\hat{x}, \hat{y}) \cdot \text{area}(R)$$

for some point $(\hat{x}, \hat{y})$ of $R$. [*Suggestion:* If $m = f(x_0, y_0)$ and $M = f(x_1, y_1)$, then you may apply the intermediate value property of the function $f(\mathbf{r}(t))$.]

---

## 16-3

## Area and Volume by Double Integration

In Section 16-2 our definition of $\iint_R f(x, y)\, dA$ was *motivated* by the problem of computing the volume of the solid

$$T = \{(x, y, z) | (x, y) \in R \quad \text{and} \quad 0 \leq z \leq f(x, y)\}$$

that lies under the surface $z = f(x, y)$ and above the region $R$ in the $xy$-plane. Such a solid $T$ appears in Fig. 16.12. Despite this geometric motivation, the actual definition of the double integral as a limit of Riemann sums does not depend upon the concept of volume. We may, therefore, turn matters around and use the double integral to *define* volume.

---

> *Definition*    Volume under $z = f(x, y)$
>
> Suppose that the function $f$ is continuous and nonnegative on the bounded plane region $R$. Then the **volume** $V$ of the solid that lies under the surface $z = f(x, y)$ and above the region $R$ is defined to be
>
> $$V = \iint_R f(x, y)\, dA, \qquad (1)$$
>
> provided this integral exists.

---

It is of interest to note the connection between this definition and the cross-sections approach to volume that we discussed in Section 6-2. If, for

**16.12** A solid region with vertical sides and base $R$ in the $xy$-plane

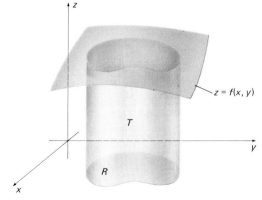

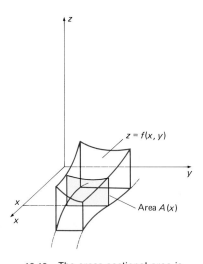

**16.13** The cross-sectional area is

$$A(x) = \int_{g_1(x)}^{g_2(x)} f(x, y)\, dy.$$

instance, the region $R$ is vertically simple, then the volume integral in (1) takes the form

$$V = \int_a^b \int_{g_1(x)}^{g_2(x)} f(x, y)\, dy\, dx$$

in terms of iterated integrals. The inner integral

$$A(x) = \int_{g_1(x)}^{g_2(x)} f(x, y)\, dy$$

is simply the area of the cross section of the solid region $T$ in a plane perpendicular to the $x$-axis (see Fig. 16.13). Thus

$$V = \int_a^b A(x)\, dx,$$

and so in this case the formula in (1) reduces to "volume is the integral of cross-sectional area."

**EXAMPLE 1**    Find the volume of the solid that lies under the surface $z = 1 + xy$ and above the rectangle $R$ in the $xy$-plane consisting of those points $(x, y)$ for which $0 \le x \le 2$ and $0 \le y \le 1$.

*Solution*    Here $f(x, y) = 1 + xy$, so the formula in (1) yields

$$\begin{aligned}
V &= \int_0^2 \int_0^1 (1 + xy)\, dy\, dx \\
&= \int_0^2 \left[ y + \tfrac{1}{2}xy^2 \right]_{y=0}^{1} dx \\
&= \int_0^2 (1 + \tfrac{1}{2}x)\, dx \\
&= \left[ x + \tfrac{1}{4}x^2 \right]_0^2 = 3.
\end{aligned}$$

A three-dimensional region $T$ is typically described in terms of the surfaces that bound it. The first step in applying the formula in Equation (1) to compute its volume $V$ is to determine the region $R$ in the $xy$-plane over which $T$ lies. The second step is to determine the appropriate order of iterated integration. This may be done in the following way.

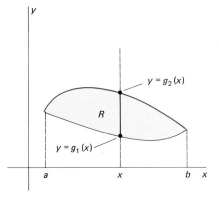

**16.14** A vertically simple region

If each vertical line in the $xy$-plane meets $R$ in a *single* line segment, then $R$ is vertically simple, and you may integrate first with respect to $y$. The limits on $y$ will be the $y$-coordinates $g_1(x)$ and $g_2(x)$ of the end points of this line segment (as indicated in Fig. 16.14). The limits on $x$ will be the end points $a$ and $b$ of the interval on the $x$-axis onto which $R$ projects. The theorem of Section 16-2 then gives

$$V = \iint_R f(x, y)\, dA = \int_a^b \int_{g_1(x)}^{g_2(x)} f(x, y)\, dy\, dx. \tag{2}$$

Alternatively:

> If each horizontal line in the $xy$-plane meets $R$ in a *single* line segment, then $R$ is horizontally simple, and you may integrate with respect to $x$ first. In this case,
>
> $$V = \iint\limits_{R} f(x, y)\, dA = \int_{c}^{d} \int_{h_1(y)}^{h_2(y)} f(x, y)\, dx\, dy. \tag{3}$$
>
> As indicated in Fig. 16.15, $h_1(y)$ and $h_2(y)$ are the $x$-coordinates of the end points of this horizontal line segment, and $c$ and $d$ are the end points of the corresponding interval on the $y$-axis.

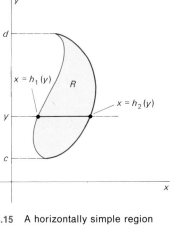

If the region $R$ is both vertically simple and horizontally simple, then you have the pleasant option of choosing the order of integration that will lead to the simpler subsequent computations. If $R$ is neither vertically simple nor horizontally simple, then you must first subdivide $R$ into simple regions before proceeding with iterated integration.

**16.15** A horizontally simple region

**EXAMPLE 2** Compute the area of the region $R$ in the $xy$-plane that is bounded by the two parabolas $y^2 = x/2$ and $y^2 = x - 4$.

*Solution* The region $R$ is shown in Fig. 16.16. It is horizontally simple but not vertically simple, so we integrate first with respect to $x$. We read from Fig. 16.16 that $h_1(y) = 2y^2$ and $h_2(y) = y^2 + 4$. So $R$ has area

$$
\begin{aligned}
a(R) &= \int_{-2}^{2} \int_{2y^2}^{y^2+4} 1\, dx\, dy = \int_{-2}^{2} \Big[ x \Big]_{2y^2}^{y^2+4}\, dy \\
&= \int_{-2}^{2} (4 - y^2)\, dy = \Big[ 4y - \tfrac{1}{3}y^3 \Big]_{-2}^{2} = \tfrac{32}{3}.
\end{aligned}
$$

Suppose now that the solid region $T$ lies above the plane region $R$, as before, but *between* the surfaces $z = f_1(x, y)$ and $z = f_2(x, y)$, where $f_1(x, y) \leqq f_2(x, y)$ for all $(x, y)$ in $R$ (see Fig. 16.17). Then we get the volume $V$ of $T$ by subtracting the volume under $z = f_1(x, y)$ from the volume under $z = f_2(x, y)$, so

$$V = \iint\limits_{R} \left[ f_2(x, y) - f_1(x, y) \right] dA. \tag{4}$$

**16.16** The region $R$ of Example 2

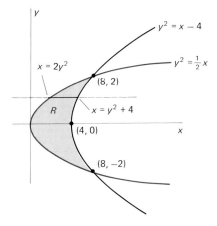

**16.17** The solid $T$ has vertical sides and is bounded above and below by surfaces.

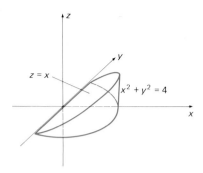

16.18   The wedge of Example 3

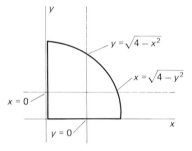

16.19   *Half* of the base *R* of the wedge

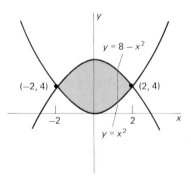

16.20   The region *R* of Example 4

More briefly,

$$V = \iint\limits_{R} (z_{\text{top}} - z_{\text{bot}}) \, dA$$

where $z_{\text{top}} = f_2(x, y)$ describes the top surface and $z_{\text{bot}} = f_1(x, y)$ the bottom surface of $T$. This is a natural generalization of the formula for the area of the plane region between the curves $y = f_1(x)$ and $y = f_2(x)$ over the interval $[a, b]$.

The special case $f(x, y) \equiv 1$ in the formula of Equation (1) gives the area

$$A = a(R) = \iint\limits_{R} 1 \, dA = \iint\limits_{R} dA \qquad (5)$$

of the plane region $R$. In this case the solid region $T$ is a flat-topped "mesa"— that is, a solid cylinder with base area $R$ and height 1. And the volume of any cylinder—not necessarily circular—is the product of its height and the area of its base.

**EXAMPLE 3**   Find the volume of the wedge-shaped solid $T$ that lies above the $xy$-plane, under the plane $z = x$, and within the cylinder $x^2 + y^2 = 4$ (as shown in Fig. 16.18).

**Solution**   The base region $R$ is a semicircle of radius 2, but by symmetry we may integrate over the first-quadrant quarter-circle alone and then double the result. A sketch of the quarter-circle (see Fig. 16.19) helps establish the limits of integration. We could integrate in either order, but integration with respect to $x$ first gives a slightly simpler computation:

$$V = 2 \int_0^2 \int_0^{\sqrt{4-y^2}} x \, dx \, dy = 2 \int_0^2 \left[ \tfrac{1}{2} x^2 \right]_0^{\sqrt{4-y^2}} dy$$

$$= \int_0^2 (4 - y^2) \, dy = \left[ 4y - \tfrac{1}{3} y^3 \right]_0^2 = \tfrac{16}{3}.$$

You should integrate in the other order and then compare the results.

**EXAMPLE 4**   Find the volume $V$ of the solid $T$ bounded by the parabolic cylinders $z = x^2$, $z = 2x^2$, $y = x^2$, and $y = 8 - x^2$.

**Solution**   We think of $T$ as lying between the surfaces $z = x^2$ and $z = 2x^2$ and above the region $R$ in the $xy$-plane that is bounded by the parabolas $y = x^2$ and $y = 8 - x^2$. These parabolas intersect at the points $(-2, 4)$ and $(2, 4)$, and they and $R$ are shown in Fig. 16.20.

The figure indicates that we should integrate first with respect to $y$, for otherwise we would need two integrals. Integrating $2x^2 - x^2 = x^2$ first from $y = x^2$ to $y = 8 - x^2$ and then from $x = -2$ to $x = 2$, we get

$$V = \int_{-2}^2 \int_{x^2}^{8-x^2} x^2 \, dy \, dx = \int_{-2}^2 \left[ x^2 y \right]_{x^2}^{8-x^2} dx$$

$$= \int_{-2}^2 (8x^2 - 2x^4) \, dx = 2 \left[ \tfrac{8}{3} x^3 - \tfrac{2}{5} x^5 \right]_0^2 = \tfrac{256}{15}.$$

## 16-3   PROBLEMS

In each of Problems 1–10, use double integration to find the area of the region in the $xy$-plane that is bounded by the given curves.

1  $y = x$, $y^2 = x$
2  $y = x$, $y = x^4$

3  $y = x^2$, $y = 2x + 3$
4  $y = 2x + 3$, $y = 6x - x^2$
5  $y = x^2$, $x + y = 2$, $y = 0$
6  $y = (x - 1)^2$, $y = (x + 1)^2$, $y = 0$

**7** $y = x^2 + 1$, $y = 2x^2 - 3$
**8** $y = x^2 + 1$, $y = 9 - x^2$
**9** $y = x$, $y = 2x$, $xy = 2$

**10** $y = x^2$, $y = \dfrac{2}{1 + x^2}$

In each of Problems 11–26, find the volume of the solid that lies under the surface $z = f(x, y)$ and above the region in the $xy$-plane that is bounded by the given curves.

**11** $z = 1 + x + y$;   $x = 0$, $x = 1$, $y = 0$, $y = 1$
**12** $z = 2x + 3y$;   $x = 0$, $x = 3$, $y = 0$, $y = 2$
**13** $z = y + e^x$;   $x = 0$, $x = 1$, $y = 0$, $y = 2$
**14** $z = 3 + \cos x + \cos y$;   $x = 0$, $x = \pi$, $y = 0$, $y = \pi$
**15** $z = x + y$;   $x = 0$, $y = 0$, $x + y = 1$
**16** $z = 3x + 2y$;   $x = 0$, $y = 0$, $x + 2y = 4$
**17** $z = 1 + x + y$;   $x = 1$, $y = 0$, $y = x^2$
**18** $z = 2x + y$;   $x = 0$, $y = 1$, $y = x^2$
**19** $z = x^2$;   $y = x^2$, $y = 1$
**20** $z = y^2$;   $x = y^2$, $x = 4$
**21** $z = x^2 + y^2$;   $x = 0$, $y = 0$, $x = 1$, $y = 2$
**22** $z = 1 + x^2 + y^2$;   $y = x$, $y = 2 - x^2$
**23** $z = 9 - x - y$;   $y = 0$, $x = 3$, $y = 2x/3$
**24** $z = 10 + y - x^2$;   $y = x^2$, $x = y^2$
**25** $z = 4x^2 + y^2$;   $x = 0$, $y = 0$, $2x + y = 2$
**26** $z = 2x + 3y$;   $y = x^2$, $y = x^3$

**27** Use double integration to find the volume of the tetrahedron in the first octant that is bounded by the coordinate planes and the plane with equation $x/a + y/b + z/c = 1$. The numbers $a$, $b$, and $c$ are positive constants.
**28** Suppose that $h > a > 0$. Show that the volume of the solid bounded by the cylinder $x^2 + y^2 = a^2$, the plane $z = 0$, and the plane $z = x + h$ is $\pi a^2 h$.
**29** Find the volume of the first octant part of the solid bounded by the cylinders $x^2 + y^2 = 1$ and $y^2 + z^2 = 1$. (*Suggestion:* One order of integration is considerably easier than the other.)
**30** Find the areas of the two regions bounded by the parabola $y = x^2$ and the curve $y(2x - 7) = -9$, a translated rectangular hyperbola. (*Suggestion:* $x = -1$ is a root of the cubic equation you will need to solve.)

In the following problems you may consult Chapter 9 or the integral table inside the covers of this book to find such antiderivatives as

$$\int \sqrt{a^2 - u^2}\, du \quad \text{and} \quad \int (a^2 - u^2)^{3/2}\, du.$$

**31** Find the volume of a sphere of radius $a$ by double integration.
**32** Use double integration to find the formula $V = V(a, b, c)$ for the volume of an ellipsoid with semiaxes $a$, $b$, and $c$.
**33** Find the volume of the solid bounded by the $xy$-plane and the paraboloid $z = 25 - x^2 - y^2$ by evaluating a double integral.
**34** Find the volume of the solid bounded by the paraboloids $z = x^2 + 2y^2$ and $z = 12 - 2x^2 - y^2$.
**35** Find the volume removed when a vertical square hole of edge length $R$ is cut directly through the center of a long horizontal cylinder of radius $R$.
**36** Find the volume of the solid bounded by the two surfaces $z = x^2 + 3y^2$ and $z = 4 - y^2$.

---

## 16-4

# Double Integrals in Polar Coordinates

A double integral may be easier to evaluate after it has been "transformed" from rectangular $xy$-coordinates to polar $r\theta$-coordinates. In particular, this is likely to be the case when the region $R$ of integration is a *polar rectangle*. A **polar rectangle** is a region described in polar coordinates by the inequalities

$$a \leqq r \leqq b, \qquad \alpha \leqq \theta \leqq \beta. \tag{1}$$

This polar rectangle is shown in Fig. 16.21. If $a = 0$, it is a sector of a circular disk of radius $b$. If $0 < a < b$, $\alpha = 0$, and $\beta = 2\pi$, it is an annular ring of inner radius $a$ and outer radius $b$. Since the area of a circular sector with radius $r$ and central angle $\theta$ is $\frac{1}{2}r^2\theta$, the area of the polar rectangle (1) is

$$
\begin{aligned}
A &= \tfrac{1}{2}b^2(\beta - \alpha) - \tfrac{1}{2}a^2(\beta - \alpha) \\
&= \tfrac{1}{2}(a + b)(b - a)(\beta - \alpha) = \bar{r}\,\Delta r\,\Delta\theta,
\end{aligned}
\tag{2}
$$

where $\Delta r = b - a$, $\Delta\theta = \beta - \alpha$, and $\bar{r} = \frac{1}{2}(a + b)$ is the *average radius* of the polar rectangle.

Now suppose that we want to compute the double integral $\iint_R f(x, y)\, dA$, where $R$ is the polar rectangle in (1). In Section 16-1 we defined the double

**16.21** A "polar rectangle"

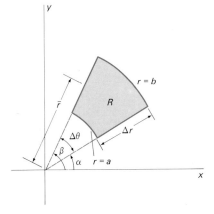

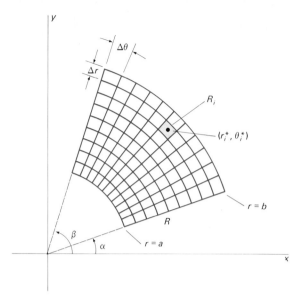

**16.22**  A polar partition of the polar rectangle $R$

integral as a limit of Riemann sums associated with partitions consisting of ordinary rectangles. It can also be defined in terms of *polar partitions*, consisting of polar rectangles. We begin with a partition

$$a = r_0 < r_1 < r_2 < \cdots < r_m = b$$

of $[a, b]$ into $m$ equal subintervals of length $\Delta r = (b - a)/m$, and a partition

$$\alpha = \theta_0 < \theta_1 < \theta_2 < \cdots < \theta_n = \beta$$

of $[\alpha, \beta]$ into $n$ equal subintervals of length $\Delta\theta = (\beta - \alpha)/n$. This gives the **polar partition** $\mathscr{P}$ of $R$ into the $k = mn$ polar rectangles $R_1, R_2, R_3, \ldots, R_k$ indicated in Fig. 16.22. The mesh $|\mathscr{P}|$ of this polar partition is the maximum of the lengths of the diagonals of its polar subrectangles.

Let the center point of $R_i$ have polar coordinates $(r_i^*, \theta_i^*)$, where $r_i^*$ is the average radius of $R_i$; then the rectangular coordinates of this point are $x_i^* = r_i^* \cos \theta_i^*$ and $y_i^* = r_i^* \sin \theta_i^*$. Therefore, the Riemann sum for the function $f(x, y)$ associated with the polar partition $\mathscr{P}$ is $\sum_{i=1}^{k} f(x_i^*, y_i^*) \, \Delta A_i$, where $\Delta A_i = r_i^* \, \Delta r \, \Delta\theta$ is the area of the polar rectangle $R_i$ (in part a consequence of Formula (2)). When we express this Riemann sum in polar coordinates, we obtain

$$\sum_{i=1}^{k} f(x_i^*, y_i^*) \, \Delta A_i = \sum_{i=1}^{k} f(r_i^* \cos \theta_i^*, r_i^* \sin \theta_i^*) \, r_i^* \, \Delta r \, \Delta\theta$$

$$= \sum_{i=1}^{k} g(r_i^*, \theta_i^*) \, \Delta r \, \Delta\theta,$$

where $g(r, \theta) = f(r \cos \theta, r \sin \theta) \, r$. This last sum is simply a Riemann sum for the double integral

$$\int_{\alpha}^{\beta} \int_{a}^{b} g(r, \theta) \, dr \, d\theta = \int_{\alpha}^{\beta} \int_{a}^{b} f(r \cos \theta, r \sin \theta) \, r \, dr \, d\theta,$$

CHAP. 16:   Multiple Integrals

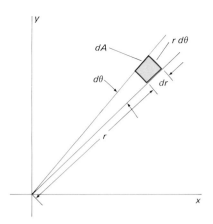

**16.23** The dimensions of the small polar rectangle suggest that $dA = r\, dr\, d\theta$.

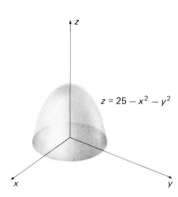

**16.24** The solid of Example 1

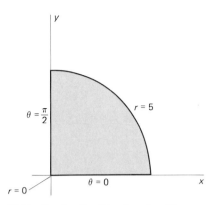

**16.25** One-fourth of the domain of the integral of Example 1

so it follows finally that

$$\iint_R f(x, y)\, dA = \lim_{|\mathcal{P}| \to 0} \sum_{i=1}^{k} f(x_i^*, y_i^*)\, \Delta A_i$$

$$= \lim_{\Delta r, \Delta\theta \to 0} \sum_{i=1}^{k} g(r_i^*, \theta_i^*)\, \Delta r\, \Delta\theta = \int_\alpha^\beta \int_a^b g(r, \theta)\, dr\, d\theta.$$

That is,

$$\iint_R f(x, y)\, dA = \int_\alpha^\beta \int_a^b f(r\cos\theta, r\sin\theta)\, r\, dr\, d\theta. \tag{3}$$

Thus we formally transform a double integral over a polar rectangle (1) into polar coordinates by replacing $x$ by $r \cos \theta$, $y$ by $r \sin \theta$, $dA = dx\, dy$ by $r\, dr\, d\theta$, and inserting the appropriate limits on $r$ and $\theta$. In particular, *note the "extra" $r$ on the right-hand side in Formula (3)*. It may be remembered by visualizing the "infinitesimal polar rectangle" of Fig. 16.23, with "area" $dA = r\, dr\, d\theta$ (formally).

**EXAMPLE 1** Find the volume of the solid shown in Fig. 16.24. This is the figure bounded below by the $xy$-plane and above by the paraboloid $z = 25 - x^2 - y^2$.

*Solution* The paraboloid intersects the $xy$-plane in the circle $x^2 + y^2 = 25$. We can compute the volume of the solid by integrating over the quarter of that circle in the first quadrant (see Fig. 16.25) and then multiplying the result by 4. Thus

$$V = 4 \int_0^5 \int_0^{\sqrt{25 - x^2}} (25 - x^2 - y^2)\, dy\, dx.$$

There is no difficulty in performing the integration with respect to $y$, but then we are confronted with the integrals

$$\int \sqrt{25 - x^2}\, dx, \quad \int x^2\sqrt{25 - x^2}\, dx, \quad \text{and} \quad \int (25 - x^2)^{3/2}\, dx.$$

Let us instead transform the integral into polar coordinates. Since $25 - x^2 - y^2 = 25 - r^2$ and since the first quadrant of our circular disk

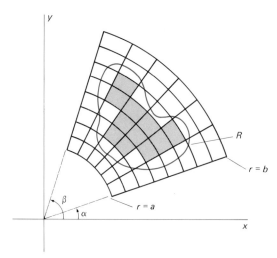

**16.26** A "polar inner partition" of the region $R$

is described by $0 \leq r \leq 5$, $0 \leq \theta \leq \pi/2$, Formula (3) yields

$$V = 4 \int_0^{\pi/2} \int_0^5 (25 - r^2) r \, dr \, d\theta$$

$$= 4 \int_0^{\pi/2} \left[ \frac{25}{2} r^2 - \frac{1}{4} r^4 \right]_0^5 d\theta$$

$$= 4 \left( \frac{625}{4} \right) \left( \frac{\pi}{2} \right) = \frac{625\pi}{2}.$$

If $R$ is a more general region, then the double integral $\iint_R f(x, y) \, dA$ can be transformed into polar coordinates by expressing it as a limit of Riemann sums associated with "polar inner partitions" of the sort indicated in Fig. 16.26. Instead of giving the detailed derivation—a generalization of the above derivation of the formula in (3)—we shall simply give the results in two special cases of practical importance. These correspond to the two types of plane regions that play the same role in polar coordinates that horizontally simple and vertically simple regions play in rectangular coordinates.

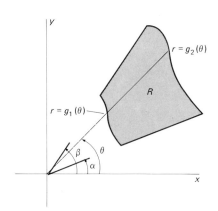

**16.27** A radially simple region $R$

Figure 16.27 shows a *radially simple* region $R$ consisting of those points that have polar coordinates satisfying the inequalities

$$\alpha \leq \theta \leq \beta, \qquad g_1(\theta) \leq r \leq g_2(\theta).$$

In this case, the formula

$$\iint_R f(x, y) \, dA = \int_\alpha^\beta \int_{g_1(\theta)}^{g_2(\theta)} f(r \cos \theta, r \sin \theta) \, r \, dr \, d\theta \qquad (4)$$

gives the evaluation in polar coordinates of a double integral over $R$ (under the usual assumption that the indicated integrals exist). Note that we integrate first with respect to $r$, with the limits $g_1(\theta)$ and $g_2(\theta)$ being the $r$-coordinates of the end points of a typical radial segment in $R$, as indicated in Fig. 16.27.

Figure 16.28 indicates how the iterated integral on the right-hand side in (4) can be set up in a formal way. A typical area element $dA = r \, dr \, d\theta$

**16.28** Integrating first with respect to $r$, then with respect to $\theta$

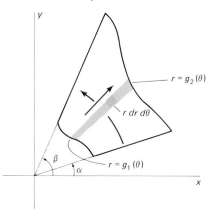

**862**

is first swept radially from $r = g_1(\theta)$ to $r = g_2(\theta)$. The resulting strip is then rotated from $\theta = \alpha$ to $\theta = \beta$ to sweep out the region $R$.

Figure 16.29 shows an *angularly simple* region $R$ described in polar coordinates by the inequalities

$$a \leq r \leq b, \qquad h_1(r) \leq \theta \leq h_2(r).$$

In this case

$$\iint\limits_{R} f(x, y)\, dA = \int_a^b \int_{h_1(r)}^{h_2(r)} f(r \cos \theta, r \sin \theta)\, r\, d\theta\, dr. \qquad (5)$$

Here we integrate first with respect to $\theta$, with the limits $h_1(r)$ and $h_2(r)$ being the $\theta$-coordinates of the endpoints of a typical circular arc in $R$, as indicated in Fig. 16.29. Figure 16.30 indicates how the iterated integral in (5) can be set up in a formal manner.

Observe that the formulas in Equations (3), (4), and (5) for the evaluation of a double integral in polar coordinates all take the form

$$\iint\limits_{R} f(x, y)\, dA = \iint\limits_{S} f(r \cos \theta, r \sin \theta)\, r\, dr\, d\theta. \qquad (6)$$

The letter $S$ on the right-hand side corresponds to choosing appropriate limits on $r$ and $\theta$ (once the order of integration has been determined) so that the region $R$ is swept out in the manner of Figs. 16.28 and 16.30. With $f(x, y) \equiv 1$, the formula in (6) reduces to the formula

$$A = a(R) = \iint\limits_{S} r\, dr\, d\theta \qquad (7)$$

for computing the area of $R$ by double integration in polar coordinates. Note again that the letter $S$ on the right-hand side refers not to a new region in the $xy$-plane, but to a new description—in terms of polar coordinates—of the same region $R$.

**EXAMPLE 2** The region $R$ bounded by the circle $(x - a)^2 + y^2 = a^2$ is shown in Fig. 16.31. Here is one way to write the equation of its boundary in polar coordinates:

$$(x^2 - 2ax + a^2) + y^2 = a^2;$$

$$x^2 + y^2 = 2ax;$$

$$r^2 = 2ar \cos \theta;$$

$$r = 2a \cos \theta.$$

Thus the region $R$ is described in polar coordinates by the inequalities $0 \leq r \leq 2a \cos \theta$, $-\pi/2 \leq \theta \leq \pi/2$. Hence the formula in (7) yields

$$A = \int_{-\pi/2}^{\pi/2} \int_0^{2a \cos \theta} r\, dr\, d\theta = \int_{-\pi/2}^{\pi/2} \left[ \tfrac{1}{2} r^2 \right]_0^{2a \cos \theta} d\theta$$

$$= \int_{-\pi/2}^{\pi/2} 2a^2 \cos^2\theta\, d\theta = \int_{-\pi/2}^{\pi/2} a^2 (1 + \cos 2\theta)\, d\theta$$

$$= a^2 \left[ \theta + \tfrac{1}{2} \sin 2\theta \right]_{-\pi/2}^{\pi/2} = \pi a^2.$$

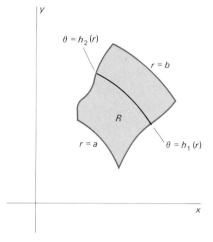

**16.29** An angularly simple region

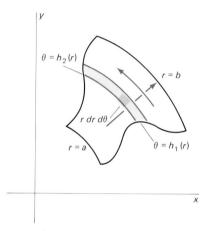

**16.30** Integrating first with respect to $\theta$, last with respect to $r$

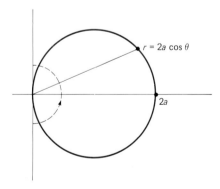

**16.31** The circle of Example 2

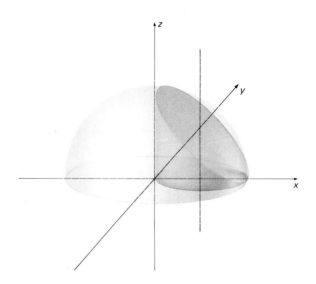

**16.32** The *upper half* of the sphere-with-hole of Example 3

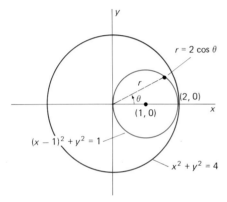

**16.33** The small circle is the domain of the integral for Example 3

**16.34** The solid of Example 4

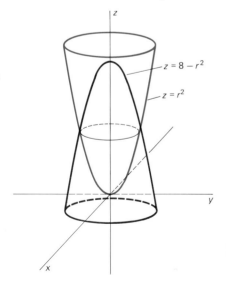

**EXAMPLE 3** Find the volume of the solid region that is interior to both the sphere $x^2 + y^2 + z^2 = 4$ of radius 2 and the cylinder $(x - 1)^2 + y^2 = 1$. This is the volume of material removed when an off-center hole of radius 1 is bored just tangent to a diameter all the way through a sphere of radius 2. The upper hemisphere is shown in Fig. 16.32.

**Solution** We need to integrate the function $f(x, y) = (4 - x^2 - y^2)^{1/2}$ over the disk $R$ bounded by the circle with center $(1, 0)$ and radius 1, shown in Fig. 16.33. The desired volume is twice that of the part above the $xy$-plane:

$$V = 2 \iint_R \sqrt{4 - x^2 - y^2} \, dA.$$

But this integral would be troublesome to evaluate in rectangular coordinates, so we change to polar coordinates.

The unit circle in Fig. 16.33 is familiar from Example 2; its polar coordinate equation is $r = 2 \cos \theta$. Therefore, the region $R$ is described by the inequalities

$$0 \le r \le 2 \cos \theta, \qquad -\pi/2 \le \theta \le \pi/2.$$

We shall integrate over only the upper half of $R$, taking advantage of the symmetry of the sphere-with-hole; this involves doubling for a *second* time the integral we write. So—using Formula (4)—we find that

$$V = 4 \int_0^{\pi/2} \int_0^{2 \cos \theta} \sqrt{4 - r^2} \, r \, dr \, d\theta$$

$$= 4 \int_0^{\pi/2} \left[ -\tfrac{1}{3}(4 - r^2)^{3/2} \right]_0^{2 \cos \theta} d\theta = \tfrac{32}{3} \int_0^{\pi/2} (1 - \sin^3 \theta) \, d\theta.$$

Now we see from Formula (113) inside the back cover that $\int_0^{\pi/2} \sin^3 \theta \, d\theta = \tfrac{2}{3}$, and so $V = 16\pi/3 - \tfrac{64}{9} \approx 9.64405$.

**EXAMPLE 4** Find the volume of the solid that is bounded above by the paraboloid $z = 8 - r^2$ and below by the paraboloid $z = r^2$. (See Fig. 16.34.)

*Solution* The curve of intersection of the two paraboloids is found by simultaneous solution of the equations of the two surfaces. We eliminate $z$ to obtain

$$r^2 = 8 - r^2; \quad \text{that is,} \quad r^2 = 4.$$

Hence the solid lies above the circular disk $r \leq 2$, and so its volume is

$$V = \iint_{r \leq 2} (z_{\text{top}} - z_{\text{bot}}) \, dA = \int_0^{2\pi} \int_0^2 [(8 - r^2) - r^2] r \, dr \, d\theta$$

$$= \int_0^{2\pi} \int_0^2 (8r - 2r^3) \, dr \, d\theta = 2\pi \Big[ 4r^2 - \tfrac{1}{2} r^4 \Big]_0^2 = 16\pi.$$

**EXAMPLE 5** Here we apply a standard polar coordinates technique to show that

$$I = \int_0^\infty e^{-x^2} \, dx = \frac{\sqrt{\pi}}{2}. \tag{8}$$

This important improper integral converges because

$$\int_0^b e^{-x^2} \, dx \leq \int_0^b e^{-x} \, dx \leq \int_0^\infty e^{-x} \, dx = 1.$$

(The first inequality is valid because $e^{-x^2} \leq e^{-x}$ for $x \geq 1$.) It follows that $\int_1^b e^{-x^2} \, dx$ is a bounded, increasing function of $b$.

*Solution* Let $V_b$ denote the volume of the region that lies under the surface $z = e^{-x^2 - y^2}$ and above the square with vertices $(\pm b, \pm b)$ in the $xy$-plane. Then

$$V_b = \int_{-b}^b \int_{-b}^b e^{-x^2 - y^2} \, dx \, dy = \int_{-b}^b e^{-y^2} \left( \int_{-b}^b e^{-x^2} \, dx \right) dy$$

$$= \left( \int_{-b}^b e^{-x^2} \, dx \right) \left( \int_{-b}^b e^{-y^2} \, dy \right) = \left( \int_{-b}^b e^{-x^2} \, dx \right)^2 = 4 \left( \int_0^b e^{-x^2} \, dx \right)^2.$$

It follows that the volume under $e^{-x^2 - y^2}$ above the entire $xy$-plane is

$$V = \lim_{b \to \infty} V_b = \lim_{b \to \infty} 4 \left( \int_0^b e^{-x^2} \, dx \right)^2$$

$$= 4 \left( \int_0^\infty e^{-x^2} \, dx \right)^2 = 4I^2.$$

Now we compute $V$ by another method—the use of polar coordinates. We take the limit, as $b \to \infty$, of the volume under $z = e^{-x^2 - y^2} = e^{-r^2}$ above the circular disk with center $(0, 0)$ and radius $b$. This disk is described by $0 \leq r \leq b$, $0 \leq \theta \leq 2\pi$, so we obtain

$$V = \lim_{b \to \infty} \int_0^{2\pi} \int_0^b e^{-r^2} r \, dr \, d\theta = \lim_{b \to \infty} \int_0^{2\pi} \Big[ -\tfrac{1}{2} e^{-r^2} \Big]_0^b \, d\theta$$

$$= \lim_{b \to \infty} \int_0^{2\pi} \tfrac{1}{2} (1 - e^{-b^2}) \, d\theta = \lim_{b \to \infty} \pi (1 - e^{-b^2}) = \pi.$$

We equate our two values of $V$, and it follows that $4I^2 = \pi$. Therefore, $I = \tfrac{1}{2}\sqrt{\pi}$, as desired.

In Problems 1–7, find the indicated area by double integration in polar coordinates.

1 The area bounded by the circle $r = 1$

2 The area bounded by the circle $r = 3 \sin \theta$

3 The area bounded by the cardioid $r = 1 + \cos \theta$

4 The area bounded by one loop of $r = 2 \cos 2\theta$

5 The area within both the circles $r = 1$ and $r = 2 \sin \theta$

6 The area within $r = 2 + \cos \theta$ and outside the circle $r = 2$

7 The area within the smaller loop of $r = 1 - 2 \sin \theta$

In each of Problems 8–12, find by double integration in polar coordinates the volume of the solid that lies under the given surface and over the plane region $R$ bounded by the given curve.

8 $z = x^2 + y^2$; $r = 3$

9 $z = \sqrt{x^2 + y^2}$; $r = 2$

10 $z = x^2 + y^2$; $r = 2 \cos \theta$

11 $z = 10 + 2x + 3y$; $r = \sin \theta$

12 $z = a^2 - x^2 - y^2$; $r = a$

In Problems 13–18, evaluate the given integral by first changing to polar coordinates.

13 $\int_0^1 \int_0^{\sqrt{1-y^2}} \dfrac{dx\,dy}{1 + x^2 + y^2}$

14 $\int_0^1 \int_0^{\sqrt{1-x^2}} \dfrac{dy\,dx}{\sqrt{4 - x^2 - y^2}}$

15 $\int_0^2 \int_0^{\sqrt{4-x^2}} (x^2 + y^2)^{3/2}\,dy\,dx$

16 $\int_0^1 \int_x^1 x^2\,dy\,dx$

17 $\int_0^1 \int_0^{\sqrt{1-y^2}} \sin(x^2 + y^2)\,dx\,dy$

18 $\int_1^2 \int_0^{\sqrt{2x-x^2}} \dfrac{dy\,dx}{\sqrt{x^2 + y^2}}$

In each of Problems 19–22, find the volume of the solid that is bounded above and below by the given surfaces $z = f_1(x, y)$ and $z = f_2(x, y)$ and lies over the plane region $R$ bounded by the given curve $r = g(\theta)$.

19 $z = 1$; $z = 3 + x + y$; $r = 1$

20 $z = 2 + x$; $z = 4 + 2x$; $r = 2$

21 $z = 0$; $z = 3 + x + y$; $r = 2 \sin \theta$

22 $z = 0$; $z = 1 + x$; $r = 1 + \cos \theta$

Solve Problems 23–32 by double integration in polar coordinates.

23 Problem 31, Section 16-3

24 Problem 34, Section 16-3

25 Problem 28, Section 16-3

26 Find the volume of the wedge-shaped solid described in Example 3 of Section 16-3.

27 Find the volume bounded by the paraboloids $z = x^2 + y^2$ and $z = 4 - 3x^2 - 3y^2$.

28 Find the volume bounded by the paraboloids $z = x^2 + y^2$ and $z = 2x^2 + 2y^2 - 1$.

29 Find the volume of the "ice cream cone" bounded by the sphere $x^2 + y^2 + z^2 = a^2$ and the cone $z = \sqrt{x^2 + y^2}$.

30 Find the volume bounded by the paraboloid $z = r^2$, the cylinder $r = 2a \sin \theta$, and the plane $z = 0$.

31 Find the volume that lies under the paraboloid $z = r^2$ and above one loop of the lemniscate with equation $r^2 = 2 \sin \theta$.

32 Find the volume that lies inside both the cylinder $x^2 + y^2 = 4$ and the ellipsoid $2x^2 + 2y^2 + z^2 = 18$.

33 If $0 < h < a$, then the plane $z = a - h$ cuts off a spherical segment of height $h$ and radius $b$ from the sphere $x^2 + y^2 + z^2 = a^2$.

(a) Show that $b^2 = 2ah - h^2$.

(b) Show that the volume of this spherical segment is

$$V = \frac{\pi h}{6}(3b^2 + h^2).$$

34 Show by the method of Example 5 that

$$\int_0^\infty \int_0^\infty \frac{dx\,dy}{(1 + x^2 + y^2)^2} = \frac{\pi}{4}.$$

35 Find the volume of the solid torus obtained by revolving the disk $r \leqq a$ about the line $x = b > a$. (Suggestion: If the area element $dA = r\,dr\,d\theta$ is revolved around the line, the volume generated is $dV = 2\pi(b - x)\,dA$. Express everything in polar coordinates.)

## 16-5

### Applications of Double Integrals

The double integral can be used to find the mass $M$ and the centroid $(\bar{x}, \bar{y})$ of a thin plate or plane *lamina* that occupies a bounded region $R$ in the $xy$-plane. We suppose that the density of the lamina (in units of mass per unit *area*) at the point $(x, y)$ is given by the continuous function $\rho(x, y)$.

Let $\mathscr{P} = \{R_1, R_2, R_3, \ldots, R_n\}$ be an inner partition of $R$ and choose a point $(x_i^*, y_i^*)$ in each subrectangle $R_i$ (see Fig. 16.35). Then the mass of the piece of the lamina occupying $R_i$ will be approximately $\rho(x_i^*, y_i^*) \Delta A_i$, where $\Delta A_i$ denotes the area $a(R_i)$ of $R_i$. Hence the mass of the entire lamina is given approximately by

$$M \approx \sum_{i=1}^{n} \rho(x_i^*, y_i^*) \Delta A_i.$$

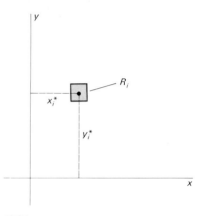

Recall from Section 6-6 that the moment of a particle about the $x$-axis is the product of its mass and its directed distance from the $x$-axis; its moment about the $y$-axis is the product of its mass and its directed distance from the $y$-axis. So the moments $M_x$ and $M_y$ of our lamina about the $x$- and $y$-axes, respectively, are given approximately by

$$M_x \approx \sum_{i=1}^{n} y_i^* \rho(x_i^*, y_i^*) \Delta A_i \quad \text{and} \quad M_y \approx \sum_{i=1}^{n} x_i^* \rho(x_i^*, y_i^*) \Delta A_i.$$

**16.35** The moments about the coordinate axes of a partition subrectangle

As the mesh $|\mathscr{P}|$ of the inner partition $\mathscr{P}$ approaches zero, these Riemann sums approach the corresponding double integrals over $R$, so we may *define* the **mass** $M$ and the **moments** $M_x$ and $M_y$ by means of the formulas

$$M = \iint_R \rho(x, y) \, dA, \tag{1}$$

$$M_x = \iint_R y\rho(x, y) \, dA, \quad \text{and} \tag{2}$$

$$M_y = \iint_R x\rho(x, y) \, dA. \tag{3}$$

These formulas can be abbreviated as

$$M = \iint_R dM, \quad M_x = \iint_R y \, dM, \quad M_y = \iint_R x \, dM$$

in terms of the element of mass $dM = \rho \, dA$.

As in Section 6-6, the coordinates $(\bar{x}, \bar{y})$ of the **centroid** or center of mass of the lamina are given by the formulas

$$\bar{x} = \frac{M_y}{M} = \frac{1}{M} \iint_R x\rho(x, y) \, dA, \tag{4}$$

$$\bar{y} = \frac{M_x}{M} = \frac{1}{M} \iint_R y\rho(x, y) \, dA. \tag{5}$$

**16.36** The lamina of Example 1

These formulas may be remembered in the form

$$\bar{x} = \frac{1}{M} \iint_R x \, dM, \quad \bar{y} = \frac{1}{M} \iint_R y \, dM.$$

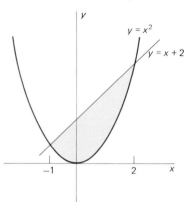

Thus $\bar{x}$ and $\bar{y}$ are the *average values* of $x$ and $y$ *with respect to mass* in the region $R$. The centroid $(\bar{x}, \bar{y})$ is the point of the lamina where it would balance horizontally if placed on the point of an icepick.

**EXAMPLE 1** A lamina occupies the region bounded by the line $y = x + 2$ and the parabola $y = x^2$, as in Fig. 16.36. The density of the lamina at the point $P(x, y)$ is proportional to the square of the distance of $P$ from the $y$-axis—thus $\rho(x, y) = kx^2$ (where $k$ is a positive constant). Find the mass and centroid of this lamina.

**Solution**   The line and the parabola intersect in the two points $(-1, 1)$ and $(2, 4)$, so Formula (1) gives

$$M = \int_{-1}^{2} \int_{x^2}^{x+2} kx^2 \, dy \, dx = k \int_{-1}^{2} \left[ x^2 y \right]_{x^2}^{x+2} dx$$

$$= k \int_{-1}^{2} (x^3 + 2x^2 - x^4) \, dx = \frac{63k}{20}.$$

Then the formulas in (4) and (5) give

$$\bar{x} = \frac{20}{63k} \int_{-1}^{2} \int_{x^2}^{x+2} kx^3 \, dy \, dx = \frac{20}{63} \int_{-1}^{2} \left[ x^3 y \right]_{x^2}^{x+2} dx$$

$$= \frac{20}{63} \int_{-1}^{2} (x^4 + 2x^3 - x^5) \, dx = \frac{20}{63} \cdot \frac{18}{5} = \frac{8}{7};$$

$$\bar{y} = \frac{20}{63k} \int_{-1}^{2} \int_{x^2}^{x+2} kx^2 y \, dy \, dx = \frac{20}{63} \int_{-1}^{2} \left[ \frac{1}{2} x^2 y^2 \right]_{x^2}^{x+2} dx$$

$$= \frac{10}{63} \int_{-1}^{2} (x^4 + 4x^3 + 4x^2 - x^6) \, dx = \frac{10}{63} \cdot \frac{531}{35} = \frac{118}{49}.$$

Thus the lamina of our example has mass $63k/20$ and its centroid is located at the point $(\frac{8}{7}, \frac{118}{49})$.

**EXAMPLE 2**   A lamina is shaped like the first-quadrant quarter-circle of radius $a$ shown in Fig. 16.37. Its density is proportional to distance from the origin; that is, its density at $(x, y)$ is $\rho(x, y) = k\sqrt{x^2 + y^2} = kr$ (where $k$ is positive constant). Find its mass and centroid.

**Solution**   First we change to polar coordinates—the shape of the boundary of the lamina suggests that this will make the problem easier. Then Formula (1) yields the result

$$M = \iint_R \rho \, dA = \int_0^{\pi/2} \int_0^a kr^2 \, dr \, d\theta$$

$$= k \int_0^{\pi/2} \left[ \tfrac{1}{3} r^3 \right]_0^a d\theta = k \int_0^{\pi/2} \tfrac{1}{3} a^3 \, d\theta = \frac{k\pi a^3}{6}.$$

By symmetry of the lamina and its density function, the centroid lies on the line $y = x$. So Formula (5) gives

$$\bar{x} = \bar{y} = \frac{1}{M} \iint_R y\rho \, dA = \frac{6}{k\pi a^3} \int_0^{\pi/2} \int_0^a kr^3 \sin\theta \, dr \, d\theta$$

$$= \frac{6}{\pi a^3} \int_0^{\pi/2} \left[ \frac{1}{4} r^4 \sin\theta \right]_0^a d\theta = \frac{6}{\pi a^3} \frac{a^4}{4} \int_0^{\pi/2} \sin\theta \, d\theta = \frac{3a}{2\pi}.$$

Thus the given lamina has mass $\frac{1}{6} k\pi a^3$ and its centroid is located at the point $(3a/2\pi, 3a/2\pi)$.

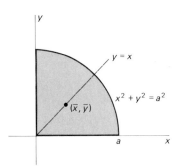

**16.37**  Finding mass and centroid—see Example 2.

An important physical concept, the *moment of inertia*, is defined this way: Consider a finite system of particles with masses $m_1, m_2, m_3, \ldots, m_n$, all revolving in uniform circular motion with angular speed $\omega$ (in radians per second) about a given fixed axis. Let $r_i$ denote the (perpendicular) distance of the $i$th particle from this axis. Then the linear speed of this particle is $v_i = r_i\omega$. Hence the total kinetic energy of this system of particles will be

$$\text{K.E.} = \sum_{i=1}^{n} \tfrac{1}{2}m_i v_i^2 = \sum_{i=1}^{n} \tfrac{1}{2}m_i r_i^2 \omega^2,$$

so that

$$\text{K.E.} = \tfrac{1}{2}I\omega^2 \tag{6}$$

where $I$ denotes the **moment of inertia** of the system of particles about the given axis and is defined by

$$I = \sum_{i=1}^{n} m_i r_i^2. \tag{7}$$

Because linear kinetic energy has the formula $\text{K.E.} = \tfrac{1}{2}mv^2$, the result of the formula in (6) suggests that moment of inertia is the rotational analogue of mass.

The **radius of gyration** of the system is defined by means of the equation

$$\hat{r} = \sqrt{\frac{I}{M}}, \tag{8}$$

where $M = m_1 + m_2 + \cdots + m_n$. Then

$$I = M\hat{r}^2, \quad \text{so that} \quad \text{K.E.} = \tfrac{1}{2}M(\hat{r}\omega)^2.$$

Thus the kinetic energy of the rotating system is the same as that of a single particle of mass $M$ revolving at the distance $\hat{r}$ from the given axis.

Formula (7) gives the definition of the moment of inertia about a given axis of a *discrete* system of mass particles. We want to generalize this formula, to define the moment of inertia of a *continuous* distribution of mass in the form of a plane lamina occupying the region $R$ in the $xy$-plane. The moment of inertia of this lamina *about the z-axis* is called its **polar moment of inertia** and is denoted by $I_0$.

To define $I_0$, we consider our original inner partition $\mathscr{P} = \{R_1, R_2, R_3, \ldots, R_n\}$ of the region $R$. Denote by $r_i^*$ the distance from the origin (or from the $z$-axis) of the selected point $(x_i^*, y_i^*)$ in $R_i$. Then, by analogy with Formula (7), $I_0$ is given approximately by

$$I_0 \approx \sum_{i=1}^{n} (r_i^*)^2 \rho(x_i^*, y_i^*)\,\Delta A_i.$$

We take the limit as $|\mathscr{P}| \to 0$ and obtain the integral formula

$$I_0 = \iint_R r^2 \rho(x, y)\,dA = \iint_R (x^2 + y^2)\rho(x, y)\,dA, \tag{9}$$

which serves to define the polar moment of inertia $I_0$ of the lamina about the origin. In short,

$$I_0 = \iint_R r^2\,dM = \iint_R (x^2 + y^2)\,dM.$$

It follows that

$$I_0 = I_x + I_y,$$

where

$$I_x = \iint_R y^2 \, dM = \iint_R y^2 \rho \, dA \qquad (10)$$

and

$$I_y = \iint_R x^2 \, dM = \iint_R x^2 \rho \, dA. \qquad (11)$$

Here $I_x$ is the lamina's **moment of inertia about the x-axis,** while $I_y$ is its **moment of inertia about the y-axis.**

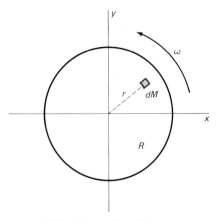

**16.38** The rotating disk

An important application of moments of inertia involves *kinetic energy of rotation.* Consider a circular disk that is revolving about its center (the origin) with angular speed $\omega$ radians per second. A mass element $dM$ at distance $r$ from the origin is moving with (linear) velocity $r = r\omega$ (see Fig. 16.38), so the kinetic energy of this mass element is

$$\tfrac{1}{2}(dM)v^2 = \tfrac{1}{2}\omega^2 r^2 \, dM.$$

Summing (by integration) over the whole disk, we find that its kinetic energy due to rotation with angular speed $\omega$ is

$$\text{K.E.}_{\text{rot}} = \iint_R \tfrac{1}{2}\omega^2 r^2 \, dM = \tfrac{1}{2}\omega^2 \iint_R r^2 \, dM;$$

that is,

$$\text{K.E.}_{\text{rot}} = \tfrac{1}{2} I_0 \omega^2. \qquad (12)$$

Note the analogy between Equations (6) and (12).

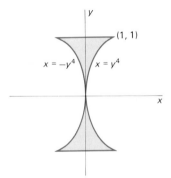

**16.39** The lamina of Example 3

**EXAMPLE 3**   Compute $I_x$ for a lamina of constant density $\rho = 1$ occupying the region bounded by the curves $x = \pm y^4$, $-1 \leq y \leq 1$. (See Fig. 16.39.)

*Solution*   The formula in Equation (10) gives

$$I_x = \int_{-1}^{1} \int_{-y^4}^{y^4} y^2 \, dx \, dy$$

$$= \int_{-1}^{1} \left[ xy^2 \right]_{-y^4}^{y^4} dy = \int_{-1}^{1} 2y^6 \, dy = \tfrac{4}{7}.$$

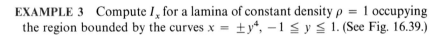

The region of Example 3 resembles the cross section of an I beam. It is known that the stiffness, or resistance to bending, of a horizontal beam is proportional to the moment of inertia of its cross section with respect to a horizontal axis through the centroid of the cross section. Let us compare our I beam with a rectangular beam of equal height 2 and equal area

$$A = \int_{-1}^{1} \int_{-y^4}^{y^4} 1 \, dx \, dy = \tfrac{4}{5}.$$

The cross section of such a rectangular beam is shown in Fig. 16.40. Its width will be $\frac{2}{5}$, and the moment of inertia of its cross section will be

$$I_x = \int_{-1}^{1} \int_{-1/5}^{1/5} y^2 \, dx \, dy = \frac{4}{15}.$$

Since the ratio of $\frac{4}{7}$ to $\frac{4}{15}$ is $\frac{15}{7}$, we see that the I beam is more than twice as strong as a rectangular beam of the same cross-sectional area. This is why I beams frequently are used in construction.

**EXAMPLE 4**  Find the polar moment of inertia of a circular lamina of radius $a$ and constant density $\rho$ centered at the origin.

*Solution*  Formula (9) gives

$$I_0 = \iint\limits_{x^2+y^2 \le a^2} r^2 \rho \, dA$$

$$= \int_0^{2\pi} \int_0^a \rho r^3 \, dr \, d\theta = \frac{\rho \pi a^4}{2} = \frac{1}{2} M a^2,$$

where $M = \rho \pi a^2$ is the mass of the circular lamina.

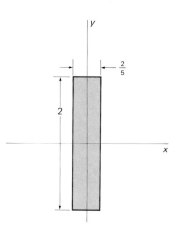

**16.40**  A rectangular beam for comparison with the I beam of Example 3

To show the moment of inertia in action, we apply the result of Example 4 to show that a Hula-Hoop will roll down an incline more slowly than a quarter will. You can easily perform an experiment to verify this assertion.

Consider a circular object of mass $M$ and polar moment of inertia $I$. Suppose that it starts from rest and rolls down an incline under the influence of gravity. Suppose that the incline has vertical height $h$ and makes an angle $\alpha$ with the horizontal, as shown in Fig. 16.41.

At any instant $t$ after the start at time $t = 0$, the linear speed $v$ and angular speed $\omega$ of the rolling mass $M$ are related by $v = a\omega$, where $a$ is the radius of $M$. After a vertical descent of $h$ down the incline, the rolling mass has lost potential energy $Mgh$ ($g$ is the acceleration of gravity). But it arrives at the bottom with translational kinetic energy

$$\text{K.E.}_{\text{tr}} = \frac{1}{2} M v^2$$

and—by Equation (12)—with rotational kinetic energy

$$\text{K.E.}_{\text{rot}} = \frac{1}{2} I \omega^2 = \frac{I v^2}{2a^2},$$

**16.41**  A circular object rolling down an incline

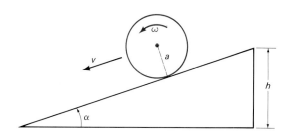

where $v$ now denotes the velocity of $M$ upon reaching the bottom of the incline.

Next we apply the law of conservation of mechanical energy. In this situation it implies that

$$\text{K.E.}_{\text{tr}} + \text{K.E.}_{\text{rot}} + \text{P.E.} = C \text{ (a constant)}.$$

Hence the potential energy lost must equal the kinetic energy gained. Therefore,

$$Mgh = \frac{Mv^2}{2} + \frac{Iv^2}{2a^2} \tag{13}$$

where $h$ is the total vertical height of the incline. Of course, these calculations are independent of $h$, so Equation (13) also gives the relationship between the velocity $v$ and the vertical descent $h$ of $M$ at any time during its journey.

If we know the polar moment of inertia $I$ of the circular object, we can use the formula in (13) to calculate the velocity the object acquires when it has descended the vertical height $h$ down the incline.

Now for the Hula-Hoop and the quarter. First, a quarter has polar moment $I = \frac{1}{2}Ma^2$ by Example 4. So the formula in (13) yields

$$Mgh = \frac{Mv^2}{2} + \frac{Ma^2v^2}{4a^2} = \frac{3}{4}Mv^2.$$

We cancel $M$ and solve for $v$:

$$v = 2\sqrt{\frac{gh}{3}}.$$

For instance, $v \approx 46.2$ ft/s if $h = 50$ ft and $g \approx 32$ ft/s$^2$.

The Hula-Hoop can be thought of as a circular object with mass $M$ all concentrated in its rim at distance $a$ from its center, so that $I = Ma^2$. We substitute this formula in (13), and we find that

$$Mgh = \frac{Mv^2}{2} + \frac{Ma^2v^2}{2a^2} = Mv^2.$$

So $v = \sqrt{gh}$. If, as before, we take $h = 50$ ft, then we find that $v$ will be approximately 40 ft/s. This establishes our assertion about which of the two rolls faster, because $h$ is actually arbitrary.

Finally, the **radius of gyration** of a lamina about an axis is defined by Equation (8), with $I$ being its moment of inertia about that axis. For example, the radii of gyration $\hat{x}$ and $\hat{y}$ about the $y$-axis and the $x$-axis, respectively, are given by

$$\hat{x} = \left(\frac{I_y}{M}\right)^{1/2} \quad \text{and} \quad \hat{y} = \left(\frac{I_x}{M}\right)^{1/2}. \tag{14}$$

Now suppose that this lamina lies in the right half-plane $x > 0$ and is symmetric about the $x$-axis. If it represents the "face" of a racquet whose handle (considered weightless) extends along the $x$-axis from the origin to the face, then the point $(\hat{x}, 0)$ is a plausible candidate for the tennis player's "sweet spot."

**EXAMPLE 5** Suppose that a lamina of uniform density $\rho$ occupies the region $R$ in the $xy$-plane, and that a mass $m$ is located at the point $(0, 0, b)$ on the $z$-axis. From Fig. 16.42 we see that the $z$-component $dF_z$ of the force of gravitational attraction between $m$ and an infinitesimal piece $dA$ of the lamina is

$$dF_z = \frac{Gm\rho\, dA}{w^2} \cos\alpha = \frac{Gm\rho b\, dA}{w^3}$$

where $w = \sqrt{r^2 + b^2}$. It follows that the vertical component $F_z$ of the force of gravitational attraction between the point mass and the lamina is

$$F_z = \iint\limits_R \frac{Gm\rho b\, dA}{(r^2 + b^2)^{3/2}}.$$

In Problem 38 we ask you to compute this integral in the case in which $R$ is a circular disk.

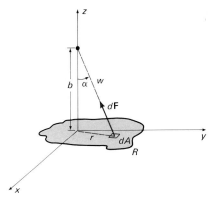

**16.42** Computing the gravitational attraction between a point mass and a lamina

---

In each of Problems 1–20, find the mass and centroid of a plane lamina with the indicated shape and density.

**1** The triangular region bounded by $x = 0$, $y = 0$, and $x + y = 1$, with $\rho(x, y) = xy$.

**2** The triangular region of Problem 1, with $\rho(x, y) = x^2$.

**3** The region bounded by $y = 0$ and $y = 4 - x^2$, with $\rho(x, y) = y$.

**4** The region bounded by $x = 0$ and $x = 9 - y^2$, with $\rho(x, y) = x^2$.

**5** The region bounded by the parabolas $y = x^2$ and $x = y^2$, with $\rho(x, y) = xy$.

**6** The region of Problem 5, with $\rho(x, y) = x^2 + y^2$.

**7** The region bounded by the parabolas $y = x^2$ and $y = 2 - x^2$, with $\rho(x, y) = y$.

**8** The region bounded by $x = 0$, $x = e$, and $y = \ln x$ for $1 \leq x \leq e$, with $\rho(x, y) = 1$.

**9** The region bounded by $y = 0$ and $y = \sin x$ for $0 \leq x \leq \pi$, with $\rho(x, y) = x$.

**10** The region bounded by $y = 0$, $x = -1$, $x = 1$, and $y = \exp(-x^2)$, with $\rho(x, y) = |xy|$.

**11** The square with vertices $(0, 0)$, $(0, a)$, $(a, a)$, $(a, 0)$; $\rho = x + y$.

**12** The triangular region bounded by the coordinate axes and the line $x + y = a$; $\rho = x^2 + y^2$.

**13** The region bounded by $y = x^2$ and $y = 4$; $\rho = y$.

**14** The region bounded by $y = x^2$ and $y = 2x + 3$; $\rho = x^2$.

**15** The region between the $x$-axis and $y = \sin x$ over the interval $0 \leq x \leq \pi$; $\rho = x$.

**16** The semicircular region $x^2 + y^2 \leq a^2$, $y \geq 0$; $\rho = y$.

**17** The semicircular region $x^2 + y^2 \leq a^2$, $y \geq 0$; $\rho = r$.

**18** The region bounded by the cardioid $r = 1 + \cos\theta$; $\rho = r$.

**19** The region within the circle $r = 2 \sin\theta$ and outside the circle $r = 1$; $\rho = y$.

**20** The region within the limaçon $r = 1 + 2 \cos\theta$ and outside the circle $r = 2$; $\rho = r$.

In each of Problems 21–25, find the polar moment of inertia $I_0$ of the indicated lamina.

**21** The region bounded by the circle $r = a$; $\rho = r^n$, where $n$ is a fixed positive integer.

**22** The lamina of Problem 16.

**23** The disk bounded by $r = 2 \cos\theta$; $\rho = k$ (a constant).

**24** The lamina of Problem 19.

**25** The region bounded by the right-hand loop of the lemniscate $r^2 = \cos 2\theta$; $\rho = r^2$.

In each of Problems 26–30, find the radii of gyration $\hat{x}$ and $\hat{y}$ of the indicated lamina about the coordinate axes.

**26** The lamina of Problem 11

**27** The lamina of Problem 13

**28** The lamina of Problem 14

**29** The lamina of Problem 17

**30** The lamina of Problem 23

**31** Show that the moment of inertia of a uniform rectangular lamina with dimensions $a$ and $b$ about a perpendicular axis through its centroid is given by

$$I = \tfrac{1}{12}\rho(a^3 b + ab^3).$$

**32** Show that the moment of inertia of a uniform plane lamina in the $xy$-plane with centroid $(0, 0)$, about an axis perpendicular to the $xy$-plane at $(x_0, y_0)$, is

$$I = I_0 + M(x_0^2 + y_0^2).$$

**33** Suppose that a plane lamina consists of two nonoverlapping laminae. Show that its polar moment of inertia is the sum of theirs. Use this fact, together with the results of Problems 31 and 32, to find the polar moment of inertia of the $T$-shaped lamina of constant density $\rho = k$ shown in Fig. 16.43.

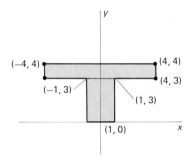

**16.43** One lamina made of two simple ones—see Problem 33.

**34** Find the polar moment of inertia of a uniform elliptical lamina with semiaxes $a$ and $b$ and centered at the origin.

**35** A homogeneous washer-shaped lamina is bounded by the circles $r = a$ and $r = b > a$.
(a) Compute its polar moment of inertia.
(b) Suppose that $a = 1$ ft and $b = 2$ ft. Compute the speed with which this washer will arrive at the bottom of the incline shown in Fig. 16.41.

**36** A racquet consists of a uniform lamina that occupies the region within the right-hand loop of $r^2 = \cos 2\theta$ on the end of a handle (which we will assume for simplicity to be of negligible mass) corresponding to the interval $-1 \leq x \leq 0$, as in Fig. 16.44. Find the radius of gyration of the racquet about the line $x = -1$. Where is its sweet spot?

**37** Find the centroid of the unbounded region under the graph of $y = e^{-x}$, $x \geq 0$.

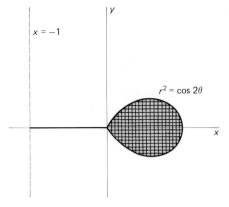

**16.44** The racquet of Problem 36

**38** (a) Suppose that the lamina of Example 5 is a uniform circular disk of radius $a$ centered at the origin. Show by integration in polar coordinates that

$$F_z = 2\pi Gm\rho \left[ 1 - \left( 1 + \frac{a^2}{b^2} \right)^{-1/2} \right].$$

(b) Suppose that $b$ is very large in comparison with $a$. Use the binomial series (Section 12-8, Example 6) to show that $F_z \approx GmM/b^2$ where $M = \pi\rho a^2$ is the mass of the disk.
(c) Note that $F_z \to 2\pi Gm\rho$ as $a \to \infty$. Thus the gravitational field of an infinite plane with uniform mass density is *constant*—independent of distance from the plane.

**39** Consider a solid sphere of radius $R$ centered at the origin with constant density $\rho$ and with mass $M = \frac{4}{3}\pi\rho R^3$. A point mass $m$ is located at the point $(0, 0, D)$, where $D > R$. Think of the slice of the sphere between $z$ and $z + dz$ as a circular lamina with density $\rho\, dz$ and radius $a = (R^2 - z^2)^{1/2}$. The vertical force it exerts on $m$ is then given by the formula of Problem 38(a) with $b = D - z$. Integrate from $z = -R$ to $z = R$ to show that the total force exerted by the solid sphere on the point mass is

$$F = G\frac{Mm}{D^2}.$$

Thus the solid sphere acts gravitationally as if all its mass were concentrated at its center.

## 16-6

### Triple Integrals

The definition of the triple integral is merely the three-dimensional version of the definition of the double integral in Section 16-2. Let $f(x, y, z)$ be continuous on the bounded space region $T$, and suppose that $T$ lies inside the rectangular block $R$ determined by the inequalities $a \leq x \leq b$, $c \leq y \leq d$, and $p \leq z \leq q$. We subdivide $[a, b]$ into equal subintervals of length $\Delta x$, $[c, d]$

into equal subintervals of length $\Delta y$, and $[p, q]$ into equal subintervals of length $\Delta z$. This generates a partition of $R$ into smaller rectangular blocks (as in Fig. 16.45), each with volume $\Delta V = \Delta x \, \Delta y \, \Delta z$. Let $\mathscr{P} = \{T_1, T_2, T_3, \ldots, T_n\}$ be the collection of these smaller blocks that lie wholly within $T$. Then $\mathscr{P}$ is called an **inner partition** of the region $T$. The **mesh** $|\mathscr{P}|$ of $\mathscr{P}$ is the length of a longest diagonal of any of the blocks $T_i$. If $(x_i^*, y_i^*, z_i^*)$ is an arbitrarily selected point of $T_i$ (for each $i = 1, 2, 3, \ldots, n$), then the **Riemann sum**

$$\sum_{i=1}^{n} f(x_i^*, y_i^*, z_i^*) \, \Delta V$$

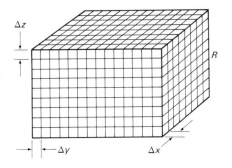

**16.45** A partition of a large block into many small ones

is an approximation to the triple integral of $f$ over the region $T$.

For instance, if $T$ is a solid body with density function $f$, then the above sum approximates its total mass. We define the **triple integral of $f$ over $T$** by means of the equation

$$\iiint_T f(x, y, z) \, dV = \lim_{|\mathscr{P}| \to 0} \sum_{i=1}^{n} f(x_i^*, y_i^*, z_i^*) \, \Delta V. \tag{1}$$

It is proved in advanced calculus that the limit of the Riemann sums exists as the mesh $|\mathscr{P}|$ approaches zero, provided that $f$ is continuous on $T$ and the boundary of the region $T$ is reasonably nice. For instance, it suffices for the boundary of $T$ to consist of finitely many pieces of smooth surfaces.

Just like double integrals, triple integrals are ordinarily computed by means of iterated integrals. If the region of integration is a rectangular block as in Fig. 16.45, then we can integrate in any order we wish.

**EXAMPLE 1** If $f(x, y, z) = xy + yz$ and $T$ consists of those points $(x, y, z)$ in space satisfying the inequalities $-1 \leq x \leq 1, 2 \leq y \leq 3$, and $0 \leq z \leq 1$, then

$$\begin{aligned}
\iiint_T f(x, y, z) \, dV &= \int_{-1}^{1} \int_{2}^{3} \int_{0}^{1} (xy + yz) \, dz \, dy \, dx \\
&= \int_{-1}^{1} \int_{2}^{3} \left[ xyz + \tfrac{1}{2}yz^2 \right]_{z=0}^{1} dy \, dx \\
&= \int_{-1}^{1} \int_{2}^{3} (xy + \tfrac{1}{2}y) \, dy \, dx \\
&= \int_{-1}^{1} \left[ \tfrac{1}{2}xy^2 + \tfrac{1}{4}y^2 \right]_{y=2}^{3} dx \\
&= \int_{-1}^{1} (\tfrac{5}{2}x + \tfrac{5}{4}) \, dx \\
&= \left[ \tfrac{5}{4}x^2 + \tfrac{5}{4}x \right]_{-1}^{1} = \tfrac{5}{2}.
\end{aligned}$$

The applications of double integrals that we saw in earlier sections generalize immediately to triple integrals. If $T$ is a solid body with density function $\rho(x, y, z)$, then its **mass** $M$ is given by

$$M = \iiint_T \rho \, dV. \tag{2}$$

The case $\rho \equiv 1$ gives the **volume**

$$V = \iiint_T dV$$

of $T$. The **moments** of $T$ about the three coordinate planes are defined to be

$$M_{yz} = \iiint_T x\rho \, dV, \qquad M_{xz} = \iiint_T y\rho \, dV, \qquad M_{xy} = \iiint_T z\rho \, dV. \qquad (3)$$

The coordinates of the **centroid** of $M$ are given by

$$\bar{x} = \frac{M_{yz}}{M} = \frac{1}{M} \iiint_T x\rho \, dV, \qquad\qquad (4a)$$

$$\bar{y} = \frac{M_{xz}}{M} = \frac{1}{M} \iiint_T y\rho \, dV, \qquad \text{and} \qquad (4b)$$

$$\bar{z} = \frac{M_{xy}}{M} = \frac{1}{M} \iiint_T z\rho \, dV. \qquad\qquad (4c)$$

The **moments of inertia** of $T$ about the three coordinate axes are defined to be

$$I_x = \iiint_T (y^2 + z^2)\rho \, dV, \qquad\qquad (5a)$$

$$I_y = \iiint_T (x^2 + z^2)\rho \, dV, \qquad \text{and} \qquad (5b)$$

$$I_z = \iiint_T (x^2 + y^2)\rho \, dV. \qquad\qquad (5c)$$

As indicated previously, triple integrals are almost always evaluated by iterated single integration. Suppose that the nice region $T$ is **z-simple,** in that each line parallel to the $z$-axis intersects $T$ (if at all) in a single line segment. In effect, this means that $T$ can be described by the inequalities

$$g_1(x, y) \leqq z \leqq g_2(x, y), \qquad (x, y) \text{ in } R,$$

where $R$ is the vertical projection of $T$ into the $xy$-plane. Then

$$\iiint_T f(x, y, z)\,dV = \iint_R \left( \int_{g_1(x,y)}^{g_2(x,y)} f(x, y, z)\,dx \right) dA. \qquad (6)$$

In Formula (6), we take $dA = dx\,dy$ or $dy\,dx$, depending upon the preferable order of integration over the set $R$. The limits $g_1(x, y)$ and $g_2(x, y)$ for $z$ will be the $z$-coordinates of the end points of the line segment in which the vertical line at $(x, y)$ meets $T$.

If the region $R$ has the description

$$\alpha(y) \leqq x \leqq \beta(y), \qquad c \leqq y \leqq d,$$

and we integrate *last* with respect to $y$, then

$$\iiint_T f(x, y, z)\,dV = \int_c^d \int_{\alpha(y)}^{\beta(y)} \int_{g_1(x,y)}^{g_2(x,y)} f(x, y, z)\,dz\,dx\,dy.$$

**27** Find the moment of inertia about the $y$-axis of the solid of Problem 24.

**28** Find the moment of inertia about the $z$-axis of the solid cylinder $x^2 + y^2 \leqq R^2, 0 \leqq z \leqq H$.

**29** Find the moment of inertia about the $z$-axis of the solid bounded by $x = 0, y = 0, z = 0,$ and $x + y + z = 1.$

**30** Find the moment of inertia about the $z$-axis of the cube with vertices $(\pm\frac{1}{2}, \frac{7}{2} \pm \frac{1}{2}, \pm\frac{1}{2})$.

In the following problems the indicated solid has uniform density $\rho \equiv 1$ unless otherwise indicated.

**31** Find the moment of inertia about one of its edges of a cube with edge length $a$.

**32** The first octant cube with edge $a$, faces parallel to the coordinate planes, and opposite vertices $(0, 0, 0)$ and $(a, a, a)$, has density at $P(x, y, z)$ proportional to the square of the distance from $P$ to the origin. Find the coordinates of its centroid.

**33** Find the moment of inertia about the $z$-axis of the cube of Problem 32.

**34** The cube bounded by the coordinate planes and the planes $x = 1, y = 1,$ and $z = 1$ has density $\rho = kz$ at the point $P(x, y, z)$ ($k$ is a positive constant). Find its centroid.

**35** Find the moment of inertia about the $z$-axis of the cube of Problem 34.

**36** Find the moment of inertia about a diameter of a solid sphere of radius $a$.

**37** Find the centroid of the first octant region that is interior to the two cylinders $x^2 + z^2 = 1$ and $y^2 + z^2 = 1$.

**38** Find the moment of inertia about the $z$-axis of the solid of Problem 37.

**39** Find the volume bounded by the ellliptic paraboloids $z = 2x^2 + y^2$ and $z = 12 - x^2 - 2y^2$. Note that this solid projects onto a circular disk in the $xy$-plane.

**40** Find the volume bounded by the elliptic paraboloid $y = x^2 + 4z^2$ and the plane $y = 2x + 3$.

**41** Find the volume of the elliptical cone bounded by $z = \sqrt{x^2 + 4y^2}$ and $z = 1$. (*Suggestion:* Integrate first with respect to $x$.)

**42** Find the volume of the region that is bounded by the paraboloid $x = y^2 + 2z^2$ and the parabolic cylinder $x = 2 - y^2$.

---

**16-7**

**Integration in Cylindrical and Spherical Coordinates**

Suppose that $f(x, y, z)$ is a continuous function defined on the $z$-simple region $T$, which—because it is $z$-simple—can be described by

$$g_1(x, y) \leqq z \leqq g_2(x, y) \qquad \text{for } (x, y) \text{ in } R.$$

We saw in Section 16-6 that

$$\iiint_T f(x, y, z)\, dV = \iint_R \left( \int_{g_1(x,y)}^{g_2(x,y)} f(x, y, z)\, dz \right) dA. \qquad (1)$$

If the region $R$ is described more naturally in polar coordinates than in rectangular coordinates, then it is likely that the integration over the plane region $R$ will be simpler if it is carried out in polar coordinates.

We first express the inner partial integral in (1) in terms of $r$ and $\theta$ by writing

$$\int_{g_1(x,y)}^{g_2(x,y)} f(x, y, z)\, dz = \int_{G_1(r,\theta)}^{G_2(r,\theta)} F(r, \theta, z)\, dz \qquad (2)$$

where

$$F(r, \theta, z) = f(r \cos \theta, r \sin \theta, z) \qquad (3a)$$

and

$$G_i(r, \theta) = g_i(r \cos \theta, r \sin \theta) \qquad \text{for } i = 1, 2. \qquad (3b)$$

Substitution of (2) into (1) with $dA = r\, dr\, d\theta$ gives

$$\iiint_T f(x, y, z)\, dV = \iint_S \left( \int_{G_1(r,\theta)}^{G_2(r,\theta)} F(r, \theta, z)\, dz \right) r\, dr\, d\theta \qquad (4)$$

where $F$, $G_1$, and $G_2$ are the functions given in (3) and $S$ represents the appropriate limits on $r$ and $\theta$ needed to describe the plane region $R$ in polar

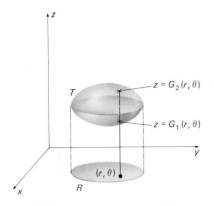

**16.54** The limits on *z* in a triple integral in cylindrical coordinates are determined by the lower and upper surfaces.

coordinates (as discussed in Section 16-4). The limits on *z* are simply the *z*-coordinates (in terms of *r* and *θ*) of a typical line segment joining the lower and upper boundary surfaces of *T*, as indicated in Fig. 16.54.

Thus the general formula for **triple integration in cylindrical coordinates** is

$$\iiint_T f(x, y, z)\, dV = \iiint f(r \cos \theta, r \sin \theta, z)\, r\, dz\, dr\, d\theta, \tag{5}$$

with limits on *z*, *r*, and *θ* appropriate to describe the space region *T* in cylindrical coordinates. Before integrating, the variables *x* and *y* are replaced by *r* cos *θ* and *r* sin *θ*, respectively, while *z* is left unchanged. The cylindrical coordinates volume element

$$dV = r\, dz\, dr\, d\theta$$

may be regarded formally as the product of *dz* and the polar coordinates area element $dA = r\, dr\, d\theta$. It is a consequence of the formula $\Delta V = \bar{r}\, \Delta z\, \Delta r\, \Delta \theta$ for the volume of the *cylindrical block* shown in Fig. 16.55.

Integration in cylindrical coordinates is particularly useful for computations associated with solids of revolution. The solid should be placed so that the axis of revolution is the *z*-axis.

**EXAMPLE 1** Find the centroid of the first-octant part of the solid ball bounded by the sphere $r^2 + z^2 = a^2$.

***Solution*** The volume of the first octant of the solid ball is $V = \frac{1}{8}(\frac{4}{3}\pi r^3) = \frac{1}{6}\pi a^3$. Because $\bar{x} = \bar{y} = \bar{z}$ by symmetry, we need only calculate

$$\bar{z} = \frac{1}{V} \iiint z\, dV = \frac{6}{\pi a^3} \int_0^{\pi/2} \int_0^a \int_0^{\sqrt{a^2 - r^2}} zr\, dz\, dr\, d\theta$$

$$= \frac{6}{\pi a^3} \int_0^{\pi/2} \int_0^a \frac{r}{2}(a^2 - r^2)\, dr\, d\theta = \frac{3}{\pi a^3} \int_0^{\pi/2} \left[ \frac{1}{2} a^2 r^2 - \frac{1}{4} r^4 \right]_0^a d\theta$$

$$= \frac{3}{\pi a^3} \cdot \frac{\pi}{2} \cdot \frac{a^4}{4} = \frac{3}{8} a.$$

Thus the centroid is located at (3*a*/8, 3*a*/8, 3*a*/8).

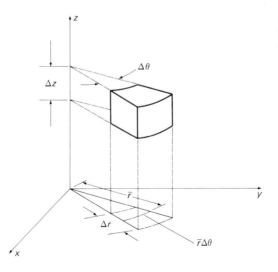

**16.55** The volume of the cylindrical block is $\Delta V = \bar{r}\, \Delta z\, \Delta r\, \Delta \theta$.

**EXAMPLE 2** Find the volume and centroid of the solid $T$ that is bounded by the paraboloid $z = b(x^2 + y^2)$ $(b > 0)$ and the plane $z = h$ $(h > 0)$.

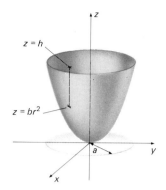

**Solution** Figure 16.56 makes it clear that we get the radius of the circular top by equating $z = b(x^2 + y^2) = br^2$ and $z = h$. This gives $a = \sqrt{h/b}$ for the radius of the circle over which the solid lies. Hence the formula in (4), with $f(x, y, z) \equiv 1$, gives the volume:

$$V = \iiint_T dV = \int_0^{2\pi} \int_0^a \int_{br^2}^h r \, dz \, dr \, d\theta$$

$$= \int_0^{2\pi} \int_0^a (hr - br^3) \, dr \, d\theta$$

$$= 2\pi \left( \frac{1}{2} ha^2 - \frac{1}{4} ba^4 \right) = \frac{\pi h^2}{2b} = \frac{1}{2} \pi a^2 h$$

**16.56** The paraboloid of Example 2

(because $a^2 = h/b$).

By symmetry, the centroid of $T$ lies on the $z$-axis, so all that remains is the computation of $\bar{z}$:

$$\bar{z} = \frac{1}{V} \iiint_T z \, dV = \frac{2}{\pi a^2 h} \int_0^{2\pi} \int_0^a \int_{br^2}^h rz \, dz \, dr \, d\theta$$

$$= \frac{2}{\pi a^2 h} \int_0^{2\pi} \int_0^a \left( \frac{1}{2} h^2 r - \frac{1}{2} b^2 r^5 \right) dr \, d\theta$$

$$= \frac{4}{a^2 h} \left( \frac{1}{4} h^2 a^2 - \frac{1}{12} b^2 a^6 \right) = \frac{2}{3} h,$$

again using $a^2 = h/b$. Note that this answer is dimensionally correct; $\bar{z}$ *must* be in units of length.

---

The results of Example 2 can be summarized as follows: The volume of a right circular paraboloid is *half* that of the circumscribed cylinder (see Fig. 16.57), and its centroid lies on its axis of symmetry *two-thirds* of the way from the "vertex" $(0, 0, 0)$ to the base (top).

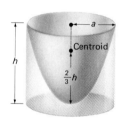

**16.57** Volume and centroid of a right circular paraboloid in terms of the circumscribed cylinder

### SPHERICAL COORDINATES

When the boundary surfaces of the region $T$ of integration are spheres, cones, or other surfaces with simple descriptions in spherical coordinates, it is generally advantageous to transform the integral into spherical coordinates. Recall from Section 14-7 that the relationship between spherical coordinates $(\rho, \phi, \theta)$ (shown in Fig. 16.58) and rectangular coordinates $(x, y, z)$ is given by the equations

**16.58** The spherical coordinates $(\rho, \phi, \theta)$ of the point $P$

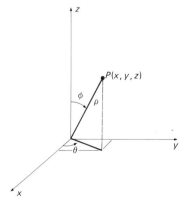

$$x = \rho \sin \phi \cos \theta, \qquad y = \rho \sin \phi \sin \theta, \qquad z = \rho \cos \phi. \qquad (6)$$

Suppose, for example, that $T$ is the **spherical block** determined by the inequalities

$$\rho_1 \leq \rho \leq \rho_2 = \rho_1 + \Delta\rho,$$

$$\phi_1 \leq \phi \leq \phi_2 = \phi_1 + \Delta\phi, \qquad (7)$$

$$\theta_1 \leq \theta \leq \theta_2 = \theta_1 + \Delta\theta.$$

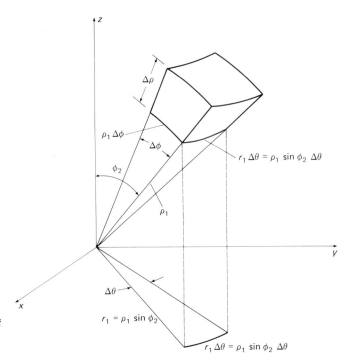

**16.59** The approximate volume of the spherical block is $\rho_1^2\sin\phi_2\,\Delta\rho\,\Delta\phi\,\Delta\theta$.

As indicated by the dimensions labeled in Fig. 16.59, this spherical block is (if $\Delta\rho$, $\Delta\phi$, and $\Delta\theta$ are small) *approximately* a rectangular block with dimensions $\Delta\rho$, $\rho_1\,\Delta\phi$, and $\rho_1\sin\phi_2\,\Delta\theta$. Thus its volume is approximately $\rho_1^2\sin\phi_2\,\Delta\rho\,\Delta\phi\,\Delta\theta$. It can be shown (see Problem 19 in Section 16-8) that the *exact* volume of the spherical block described in (7) is

$$\Delta V = \hat{\rho}^2\sin\hat{\phi}\,\Delta\rho\,\Delta\phi\,\Delta\theta \tag{8}$$

for appropriate numbers $\hat{\rho}$ and $\hat{\phi}$ such that $\rho_1 < \hat{\rho} < \rho_2$ and $\phi_1 < \hat{\phi} < \phi_2$.

Now suppose that we partition each of the intervals $[\rho_1, \rho_2]$, $[\phi_1, \phi_2]$, and $[\theta_1, \theta_2]$ into $n$ equal subintervals with lengths

$$\Delta\rho_i = \frac{\rho_2 - \rho_1}{n}, \qquad \Delta\phi_i = \frac{\phi_2 - \phi_1}{n}, \quad \text{and} \quad \Delta\theta_i = \frac{\theta_2 - \theta_1}{n},$$

respectively. This produces a **spherical partition** $\mathscr{P}$ of the spherical block $T$ into $k = n^3$ smaller spherical blocks $T_1, T_2, T_3, \ldots, T_k$. By Formula (8) there exists a point $(\hat{\rho}_i, \hat{\phi}_i, \hat{\theta}_i)$ of the spherical block $T_i$ such that its volume is $\Delta V_i = \hat{\rho}_i^2\sin\hat{\theta}_i\,\Delta\rho_i\,\Delta\phi_i\,\Delta\theta_i$. The mesh $|\mathscr{P}|$ of $\mathscr{P}$ is the length of the longest diagonal of any of the small spherical blocks $T_1, T_2, T_3, \ldots, T_k$.

If $(x_i^*, y_i^*, z_i^*)$ are the rectangular coordinates of the point with spherical coordinates $(\hat{\rho}_i, \hat{\phi}_i, \hat{\theta}_i)$, then the definition of the triple integral as a limit of Riemann sums as the mesh $|\mathscr{P}|$ approaches zero gives

$$\iiint_T f(x, y, z)\,dV = \lim_{|\mathscr{P}|\to 0}\sum_{i=1}^{n} f(x_i^*, y_i^*\,z_i^*)\,\Delta V_i$$

$$= \lim_{|\mathscr{P}|\to 0}\sum_{i=1}^{k} F(\hat{\rho}_i, \hat{\phi}_i, \hat{\theta}_i)\,\hat{\rho}_i^2\sin\hat{\phi}_i\,\Delta\rho_i\,\Delta\phi_i\,\Delta\theta_i, \tag{9}$$

where

$$F(\rho, \phi, \theta) = f(\rho\sin\phi\cos\theta,\ \rho\sin\phi\sin\theta,\ \rho\cos\phi). \tag{10}$$

**884**

But the sum in limit (9) is simply a Riemann sum for the triple integral

$$\int_{\theta_1}^{\theta_2} \int_{\phi_1}^{\phi_2} \int_{\rho_1}^{\rho_2} F(\rho, \phi, \theta)\, \rho^2 \sin \phi \, d\rho \, d\phi \, d\theta.$$

It therefore follows that

$$\iiint_T f(x, y, z)\, dV = \int_{\theta_1}^{\theta_2} \int_{\phi_1}^{\phi_2} \int_{\rho_1}^{\rho_2} F(\rho, \phi, \theta)\, \rho^2 \sin \phi \, d\rho \, d\phi \, d\theta. \quad (11)$$

Thus we transform the integral $\iiint_T f(x, y, z)\, dV$ into spherical coordinates by replacing the rectangular coordinate variables $x, y, z$, by their expressions in (6) in terms of the spherical coordinate variables $\rho, \phi, \theta$ and formally writing

$$dV = \rho^2 \sin \phi \, d\rho \, d\phi \, d\theta$$

for the volume element in spherical coordinates.

More generally, we can transform the triple integral $\iiint_T f(x, y, z)\, dV$ into spherical coordinates whenever the region $T$ is **centrally simple;** that is, whenever it has a spherical coordinates description of the form

$$G_1(\phi, \theta) \leq \rho \leq G_2(\phi, \theta), \qquad \phi_1 \leq \phi \leq \phi_2, \qquad \theta_1 \leq \theta \leq \theta_2. \quad (12)$$

If so, then

$$\iiint_T f(x, y, z)\, dV = \int_{\theta_1}^{\theta_2} \int_{\phi_1}^{\phi_2} \int_{G_1(\phi,\theta)}^{G_2(\phi,\theta)} F(\rho, \phi, \theta)\, \rho^2 \sin \phi \, d\rho \, d\phi \, d\theta. \quad (13)$$

The limits on $\rho$ are simply the $\rho$-coordinates (in terms of $\phi$ and $\theta$) of the end points of a typical radial segment joining the "inner" and "outer" parts of the boundary of $T$ (see Fig. 16.60). Thus the general formula for **triple integration in spherical coordinates** is

$$\iiint_T f(x, y, z)\, dV$$

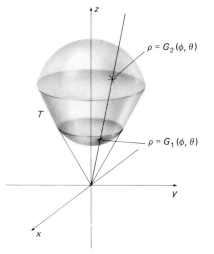

16.60   A centrally simple region

$$= \iiint f(\rho \sin \phi \cos \theta, \rho \sin \phi \sin \theta, \rho \cos \phi)\, \rho^2 \sin \phi \, d\rho \, d\phi \, d\theta \quad (14)$$

with limits on $\rho, \phi, \theta$ appropriate to describe the region $T$ in spherical coordinates.

**EXAMPLE 3** A solid ball $T$ with constant density $\delta$ is bounded by the sphere $\rho = a$. Use spherical coordinates to compute its volume $V$ and its moment of inertia $I_z$ about the $z$-axis.

*Solution* The points of the ball are described by the inequalities

$$0 \leq \rho \leq a, \qquad 0 \leq \phi \leq \pi, \qquad 0 \leq \theta \leq 2\pi.$$

We take $f = F \equiv 1$ in Formula (11) and thereby obtain

$$V = \iiint_T dV = \int_0^{2\pi} \int_0^{\pi} \int_0^a \rho^2 \sin \phi \, d\rho \, d\phi \, d\theta$$

$$= \frac{a^3}{3} \int_0^{2\pi} \int_0^{\pi} \sin \phi \, d\phi \, d\theta = \frac{a^3}{3} \int_0^{2\pi} \left[ -\cos \phi \right]_0^{\pi} d\theta$$

$$= \frac{2a^3}{3} \int_0^{2\pi} d\theta = \frac{4}{3}\pi a^3.$$

The distance from the typical point $(\rho, \phi, \theta)$ to the $z$-axis is $r = \rho \sin \phi$, so the moment of inertia of the sphere about that axis is

$$I_z = \iiint_T r^2 \delta \, dV = \int_0^{2\pi} \int_0^\pi \int_0^a \delta \rho^4 \sin^3\phi \, d\rho \, d\phi \, d\theta$$

$$= \frac{\delta a^5}{5} \int_0^{2\pi} \int_0^\pi \sin^3\phi \, d\phi \, d\theta = \frac{2\pi \delta a^5}{5} \int_0^\pi \sin^3\phi \, d\phi$$

$$= \frac{2\pi \delta a^5}{5} \cdot 2 \cdot \frac{2}{3} = \frac{2}{5} Ma^2$$

where $M = \frac{4}{3}\pi a^3 \delta$ is the mass of the ball. In Problem 32 we ask you to apply this result to continue the discussion associated with Equation (13) in Section 16-5 and show that a marble rolls down an inclined plane faster than a quarter does.

**EXAMPLE 4** Find the volume and centroid of the "ice cream cone" bounded by the cone $\phi = \pi/6$ and the sphere $\rho = 2a \cos \phi$ of radius $a$ and tangent to the $xy$-plane at the origin. The sphere and the part of the cone within it are shown in Fig. 16.61.

**Solution** The ice cream cone is described by the inequalities

$$0 \le \theta \le 2\pi, \qquad 0 \le \phi \le \frac{\pi}{6}, \qquad 0 \le \rho \le 2a \cos \phi.$$

Using Formula (13) to compute its volume, we get

$$V = \int_0^{2\pi} \int_0^{\pi/6} \int_0^{2a \cos \phi} \rho^2 \sin \phi \, d\rho \, d\phi \, d\theta$$

$$= \frac{8}{3} a^3 \int_0^{2\pi} \int_0^{\pi/6} \cos^3\phi \sin \phi \, d\phi \, d\theta$$

$$= \frac{16}{3} \pi a^3 \left[ -\frac{1}{4} \cos^4\phi \right]_0^{\pi/6} = \frac{7\pi a^3}{12}.$$

Now for the centroid. It is clear by symmetry that $\bar{x} = 0 = \bar{y}$. Since $z = \rho \cos \phi$, the moment of the ice cream cone about the $xy$-plane is

$$M_{xy} = \iiint z \, dV$$

$$= \int_0^{2\pi} \int_0^{\pi/6} \int_0^{2a \cos \phi} \rho^3 \cos \phi \sin \phi \, d\rho \, d\phi \, d\theta$$

$$= 4a^4 \int_0^{2\pi} \int_0^{\pi/6} \cos^5\phi \sin \phi \, d\phi \, d\theta = 8\pi a^4 \left[ -\frac{1}{6} \cos^6\phi \right]_0^{\pi/6} = \frac{37\pi a^4}{48}.$$

Hence the $z$-coordinate of the centroid is $\bar{z} = M_{xy}/V = 37a/28$.

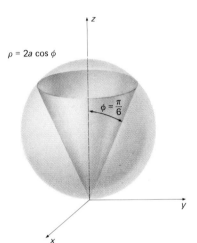

$\rho = 2a \cos \phi$

$\phi = \dfrac{\pi}{6}$

**16.61** The ice cream cone of Example 4 is the part of the cone that lies within the sphere.

## 16-7 PROBLEMS

Solve Problems 1–20 by triple integration in cylindrical coordinates.

**1** Find the volume of the solid that is bounded above by the plane $z = 4$ and below by the paraboloid $z = r^2$.

**2** Find the centroid of the solid described in Problem 1.
**3** Derive the formula $V = 4\pi a^3/3$ for the volume of a sphere of radius $a$.
**4** Find the moment of inertia about the $z$-axis of the solid

sphere of Problem 3 under the assumption that it has unit density.

**5** Find the volume of the region that lies inside both the sphere $x^2 + y^2 + z^2 = 4$ and the cylinder $x^2 + y^2 = 1$.

**6** Find the centroid of the portion $z \geq 0$ of the region of Problem 5.

**7** Find the mass of the cylinder $r \leq a$, $0 \leq z \leq h$, if its density at $(x, y, z)$ is $z$.

**8** Find the centroid of the cylinder described in Problem 7.

**9** Find the moment of inertia about the $z$-axis of the cylinder of Problem 7.

**10** Find the volume of the region that lies inside both the sphere $x^2 + y^2 + z^2 = 4$ and the cylinder $x^2 + y^2 - 2x = 0$.

**11** Find the volume and centroid of the region bounded by the plane $z = 0$ and the paraboloid $z = 9 - x^2 - y^2$.

**12** Find the volume and centroid of the region bounded by the paraboloids $z = x^2 + y^2$ and $z = 12 - 2x^2 - 2y^2$.

**13** Find the volume of the region bounded by the paraboloids $z = 2x^2 + y^2$ and $z = 12 - x^2 - 2y^2$.

**14** Find the volume of the region that is bounded below by the paraboloid $z = x^2 + y^2$ and above by the plane $z = 2x$.

**15** Find the volume of the region that is bounded above by the spherical surface $x^2 + y^2 + z^2 = 2$ and below by the paraboloid $z = x^2 + y^2$.

**16** A homogeneous solid cylinder has mass $M$ and radius $a$. Show that its moment of inertia about its axis of symmetry is $\frac{1}{2}Ma^2$.

**17** Find the moment of inertia $I$ of a homogeneous solid cylinder about a diameter of its base. Express $I$ in terms of the radius $a$, the height $h$, and the (constant) density $\delta$ of the cylinder.

**18** Find the centroid of a homogeneous solid hemisphere of radius $a$.

**19** Find the volume of the region bounded by the plane $z = 1$ and the cone $z = r$.

**20** Show that the centroid of a homogeneous solid circular cone lies on its axis three-quarters of the way from its vertex to its base.

Solve Problems 21–30 by triple integration in spherical coordinates.

**21** Find the centroid of a homogeneous solid hemisphere of radius $a$.

**22** Find the mass and centroid of a solid hemisphere of radius $a$ if its density $\delta$ is proportional to distance $z$ from its base—so that $\delta = kz$ ($k$ a positive constant).

**23** Solve Problem 19 by triple integration in spherical coordinates.

**24** Solve Problem 20 by triple integration in spherical coordinates.

**25** Find the volume and centroid of the solid that lies within the sphere $\rho = a$ and above the cone $r = z$.

**26** Find the moment of inertia $I_z$ of the solid of Problem 25 under the assumption that it has constant density $\delta$.

**27** Find the moment of inertia about a tangent line of a solid homogeneous sphere of radius $a$ and total mass $M$.

**28** A spherical shell of mass $M$ is bounded by the spheres $\rho = a$ and $\rho = 2a$, and its density function is $\delta = \rho^2$. Find its moment of inertia about a diameter.

**29** Describe the surface $\rho = 2a \sin \phi$, and compute the volume of the region it bounds.

**30** Describe the surface $\rho = 1 + \cos \phi$, and compute the volume of the region it bounds.

**31** Find the moment of inertia about the $x$-axis of the region that lies inside both the cylinder $r = a$ and the sphere $\rho = 2a$.

**32** Substitute the result of Example 3 of this section in Equation (13) of Section 16-5 and thereby find the speed with which a homogeneous solid spherical ball will reach the bottom of the inclined plane of Fig. 16.41.

**33** A homogeneous spherical shell is bounded by the concentric spheres $\rho = a$ and $\rho = b > a$.
(a) Compute its moment of inertia about the $z$-axis.
(b) Assume that $a = 1$ ft and $b = 2$ ft; find the speed with which it will arrive at the bottom of the inclined plane of Fig. 16.41.

**34** Consider a homogeneous spherical ball of radius $a$ centered at the origin, with density $\delta$ and mass $M = \frac{4}{3}\pi a^3 \delta$. Show that the gravitational force $\mathbf{F}$ exerted by this ball upon a point mass $m$ located at the point $(0, 0, c)$, with $c > a$ (Fig. 16.62), is the same as though all the mass of

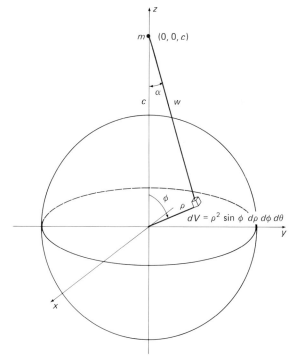

**16.62** The point-and-mass system of Problem 34

the ball were concentrated at its center $(0, 0, 0)$. That is, show that $|\mathbf{F}| = GMm/c^2$. (*Suggestion:* By symmetry you may assume that the force is vertical, so that $\mathbf{F} = F_z\mathbf{k}$. Set up the integral

$$F_z = -\int_0^{2\pi} \int_0^a \int_0^\pi \frac{GM\delta \cos \alpha}{w^2} \rho^2 \sin \phi \, d\phi \, d\rho \, d\theta.$$

Change the first variable of integration from $\phi$ to $w$ by using the law of cosines:

$$w^2 = \rho^2 + c^2 - 2\rho c \cos \phi.$$

Then $2w \, dw = 2\rho c \sin \phi \, d\phi$ and $w \cos \alpha + \rho \cos \phi = c$. (*Why?*))

**35** Consider now the spherical shell $a \leq \rho \leq b$ with uniform density $\delta$. Show that this shell exerts *no* force on a point mass $m$ located at the point $(0, 0, c)$ *inside* it—that is, with $c < a$. The computation will be the same as in Problem 34 except for the limits of integration on $\rho$ and $w$.

---

## 16-8

### Surface Area

Until now our concept of a surface has been the graph $z = f(x, y)$ of a function of two variables. Occasionally we have seen such a surface defined implicitly by an equation of the form $F(x, y, z) = 0$. Now we want to introduce the more precise concept of a parametric surface—the two-dimensional analogue of a parametric curve.

A **parametric surface** $S$ is the image of a function or transformation $\mathbf{r}$ defined on a region $R$ in the $uv$-plane and with values in $xyz$-space. The **image** under $\mathbf{r}$ of each point $(u, v)$ in $R$ is the point in $xyz$-space with position vector

$$\mathbf{r}(u, v) = \langle x(u, v), y(u, v), z(u, v) \rangle. \tag{1}$$

We shall assume throughout this section that the component functions of $\mathbf{r}$ have continuous partial derivatives with respect to $u$ and $v$, and also that the vectors

$$\mathbf{r}_u = \frac{\partial \mathbf{r}}{\partial u} = \langle x_u, y_u, z_u \rangle = \frac{\partial x}{\partial u}\mathbf{i} + \frac{\partial y}{\partial u}\mathbf{j} + \frac{\partial z}{\partial u}\mathbf{k} \tag{2}$$

and

$$\mathbf{r}_v = \frac{\partial \mathbf{r}}{\partial v} = \langle x_v, y_v, z_v \rangle = \frac{\partial x}{\partial v}\mathbf{i} + \frac{\partial y}{\partial v}\mathbf{j} + \frac{\partial z}{\partial v}\mathbf{k} \tag{3}$$

are nonzero and nonparallel at each interior point of $R$.

We call the variables $u$ and $v$ the *parameters* for the surface $S$ (analogous to the single parameter $t$ for a parametric curve). The graph $z = f(x, y)$ of a function may be regarded as a parametric surface with parameters $x$ and $y$. In this case the transformation $\mathbf{r}$ from the $xy$-plane to $xyz$-space has the component functions

$$x = x, \qquad y = y, \qquad z = f(x, y). \tag{4}$$

Similarly, a surface given in cylindrical coordinates as the graph of $z = g(r, \theta)$ may be regarded as a parametric surface with parameters $r$ and $\theta$; the transformation $\mathbf{r}$ from the $r\theta$-plane to $xyz$-space is then given by

$$x = r \cos \theta, \qquad y = r \sin \theta, \qquad z = g(r, \theta). \tag{5}$$

A surface given in spherical coordinates by $\rho = h(\phi, \theta)$ may be regarded as a parametric surface with parameters $\phi$ and $\theta$, and the corresponding transformation from the $\phi\theta$-plane to $xyz$-space is then given by

$$x = h(\phi, \theta) \sin \phi \cos \theta, \qquad y = h(\phi, \theta) \sin \phi \sin \theta, \qquad z = h(\phi, \theta) \cos \phi. \tag{6}$$

The concept of a parametric surface lets us treat all these cases, and others, with the same techniques.

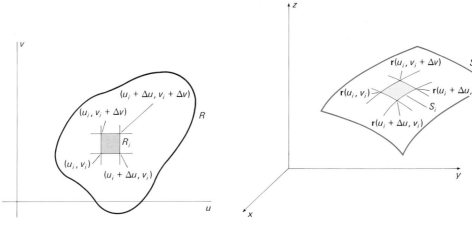

**16.63** The rectangle $R_i$ in the $uv$-plane

**16.64** The image of $R_i$ is a curvilinear figure.

Now we want to define the *surface area* of the general parametric surface given by Equation (1). We begin with an inner partition of the region $R$—the domain of $\mathbf{r}$ in the $uv$-plane—into rectangles $R_1, R_2, R_3, \ldots, R_n$, each with dimensions $\Delta u$ and $\Delta v$. Let $(u_i, v_i)$ be the lower left-hand corner of $R_i$ (as in Fig. 16.63). The image $S_i$ of $R_i$ under $\mathbf{r}$ will not generally be a rectangle in $xyz$-space; it will look more like a *curvilinear figure* on the image surface $S$, with $\mathbf{r}(u_i, v_i)$ as one "vertex." (See Fig. 16.64.) Let $\Delta S_i$ denote the area of this curvilinear figure $S_i$.

The parametric curves $\mathbf{r}(u, v_i)$ and $\mathbf{r}(u_i, v)$—with parameters $u$ and $v$, respectively—lie on the surface $S$ and meet at the point $\mathbf{r}(u_i, v_i)$. At this point of intersection, these two curves have the tangent vectors $\mathbf{r}_u(u_i, v_i)$ and $\mathbf{r}_v(u_i, v_i)$ shown in Fig. 16.65. Hence their cross product

$$\mathbf{N}(u_i, v_i) = \mathbf{r}_u(u_i, v_i) \times \mathbf{r}_v(u_i, v_i) \tag{7}$$

is a normal vector to $S$ at the point $\mathbf{r}(u_i, v_i)$.

Now suppose that $\Delta u$ and $\Delta v$ are both small. Then the area $\Delta S_i$ of the curvilinear figure $S_i$ will be approximately equal to the area $\Delta P_i$ of the parallelogram with adjacent sides $\mathbf{r}_u(u_i, v_i) \Delta u$ and $\mathbf{r}_v(u_i, v_i) \Delta v$ (see Fig. 16.66).

**16.66** The area of the actual parallelogram $P_i$ is an approximation to the area of the curvilinear figure $S_i$.

**16.65** The normal $\mathbf{N}$ to the surface at $\mathbf{r}(u_i, v_i)$

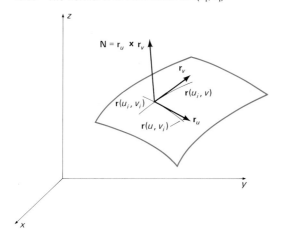

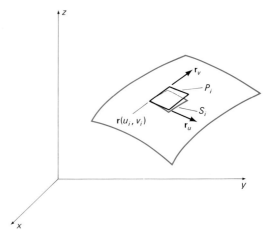

But the area of this parallelogram is

$$\Delta P_i = \left| \mathbf{r}_u(u_i, v_i) \, \Delta u \times \mathbf{r}_v(u_i, v_i) \, \Delta v \right| = \left| \mathbf{N}(u_i, v_i) \right| \Delta u \, \Delta v.$$

This means that the area of the surface $S$ is given approximately by

$$\text{Area } (S) = \sum_{i=1}^{n} \Delta S_i \approx \sum_{i=1}^{n} \Delta P_i,$$

so that

$$\text{Area } (S) \approx \sum_{i=1}^{n} \left| \mathbf{N}(u_i, v_i) \right| \Delta u \, \Delta v.$$

But this last sum is a Riemann sum for the double integral $\iint_R \left| \mathbf{N}(u, v) \right| du \, dv$. We are therefore motivated to *define* the **surface area** of the parametric surface $S$ by

$$A = \text{Area } (S) = \iint_R \left| \mathbf{N}(u, v) \right| du \, dv = \iint_R \left| \frac{\partial \mathbf{r}}{\partial u} \times \frac{\partial \mathbf{r}}{\partial v} \right| du \, dv. \tag{8}$$

In the case of the surface $z = f(x, y)$, for $(x, y)$ in the region $R$ in the $xy$-plane, the component functions of $\mathbf{r}$ are given by Equations (4) with parameters $x$ and $y$ (in place of $u$ and $v$). Then

$$\mathbf{N} = \frac{\partial \mathbf{r}}{\partial x} \times \frac{\partial \mathbf{r}}{\partial y} = \begin{vmatrix} \mathbf{i} & \mathbf{j} & \mathbf{k} \\ 1 & 0 & \dfrac{\partial f}{\partial x} \\ 0 & 1 & \dfrac{\partial f}{\partial y} \end{vmatrix} = -\frac{\partial f}{\partial x} \mathbf{i} - \frac{\partial f}{\partial y} \mathbf{j} + \mathbf{k},$$

So Formula (8) becomes

$$A = \text{Area } (S) = \iint_R \sqrt{1 + \left( \frac{\partial f}{\partial x} \right)^2 + \left( \frac{\partial f}{\partial y} \right)^2} \, dx \, dy$$

$$= \iint_R \sqrt{1 + z_x^2 + z_y^2} \, dx \, dy. \tag{9}$$

**EXAMPLE 1** Find the area of the ellipse cut from the plane $z = 2x + 2y + 1$ by the cylinder $x^2 + y^2 = 1$.

***Solution*** Here, $R$ is the unit circle in the $xy$-plane with area $\iint_R 1 \, dx \, dy = \pi$, so Formula (9) gives

$$A = \iint_R \sqrt{1 + \left( \frac{\partial z}{\partial x} \right)^2 + \left( \frac{\partial z}{\partial y} \right)^2} \, dx \, dy$$

$$= \iint_R \sqrt{1 + (2)^2 + (2)^2} \, dx \, dy = \iint_R 3 \, dx \, dy = 3\pi.$$

Now consider a cylindrical coordinates surface $z = g(r, \theta)$ parametrized by Equations (5), for $(r, \theta)$ in a region $R$ of the $r\theta$-plane. Then the normal vector is

$$\mathbf{N} = \frac{\partial \mathbf{r}}{\partial r} \times \frac{\partial \mathbf{r}}{\partial \theta} = \begin{vmatrix} \mathbf{i} & \mathbf{j} & \mathbf{k} \\ \cos\theta & \sin\theta & \dfrac{\partial z}{\partial r} \\ -r\sin\theta & r\cos\theta & \dfrac{\partial z}{\partial \theta} \end{vmatrix}$$

$$= \mathbf{i}\left(\frac{\partial z}{\partial\theta}\sin\theta - r\frac{\partial z}{\partial r}\cos\theta\right) - \mathbf{j}\left(\frac{\partial z}{\partial\theta}\cos\theta + r\frac{\partial z}{\partial r}\sin\theta\right) + r\mathbf{k}.$$

After some simplification, we find that

$$|\mathbf{N}| = \sqrt{r^2 + r^2\left(\frac{\partial z}{\partial r}\right)^2 + \left(\frac{\partial z}{\partial\theta}\right)^2}.$$

Thus Formula (8) yields the formula

$$A = \iint_R \sqrt{r^2 + r^2\left(\frac{\partial z}{\partial r}\right)^2 + \left(\frac{\partial z}{\partial\theta}\right)^2}\, dr\, d\theta \tag{10}$$

for surface area in cylindrical coordinates.

**EXAMPLE 2**   Find the surface area cut from the paraboloid $z = r^2$ by the cylinder $r = 1$.

*Solution*   Formula (10) gives

$$A = \int_0^{2\pi}\int_0^1 \sqrt{r^2 + r^2(2r)^2}\, dr\, d\theta = 2\pi\int_0^1 r\sqrt{1 + 4r^2}\, dr$$

$$= 2\pi\left[\frac{2}{3}\cdot\frac{1}{8}(1 + 4r^2)^{3/2}\right]_0^1 = \frac{\pi}{6}(5\sqrt{5} - 1) \approx 5.3304.$$

In Example 2, you would get the same result if you first wrote $z = x^2 + y^2$, used Equation (9), which gives

$$A = \iint_R \sqrt{1 + 4x^2 + 4y^2}\, dx\, dy,$$

and then changed to polar coordinates. In the following example it would be less convenient to begin with rectangular coordinates.

**EXAMPLE 3**   Find the area of the *spiral ramp* $z = \theta, 0 \le r \le 1, 0 \le \theta \le \pi$: the upper surface of the solid shown in Fig. 16.67.

*Solution*   Formula (10) gives

$$A = \int_0^{\pi}\int_0^1 \sqrt{r^2 + 1}\, dr\, d\theta = \frac{\pi}{2}[\sqrt{2} + \ln(1 + \sqrt{2})] \approx 3.6059;$$

we avoided a trigonometric substitution by using the table of integrals inside the front cover.

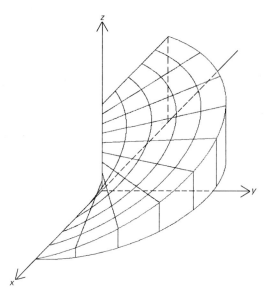

**16.67** The spiral ramp of Example 3

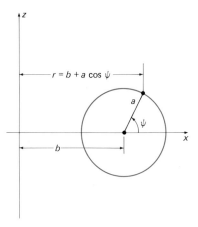

**16.68** The circle that generates the torus of Example 4

**EXAMPLE 4** Find the surface area of the torus generated by revolving the circle $(x - b)^2 + z^2 = a^2$ $(a < b)$ in the $xz$-plane around the $z$-axis.

***Solution*** With the ordinary polar coordinate $\theta$ and the angle $\psi$ of Fig. 16.68, the torus is described for $0 \leq \theta \leq 2\pi$ and $0 \leq \psi \leq 2\pi$ by the parametric equations

$$x = r \cos \theta = (b + a \cos \psi) \cos \theta,$$

$$y = r \sin \theta = (b + a \cos \psi) \sin \theta,$$

$$z = a \sin \psi.$$

When we compute $\mathbf{N} = \mathbf{r}_\theta \times \mathbf{r}_\psi$ and simplify, we find that

$$|\mathbf{N}| = a(b + a \cos \psi).$$

Hence the general surface area formula (Formula (8)) gives

$$A = \int_0^{2\pi} \int_0^{2\pi} a(b + a \cos \psi) \, d\theta \, d\psi$$

$$= 2\pi a \left[ b\psi + a \sin \psi \right]_0^{2\pi} = 4\pi^2 ab.$$

We obtained the same result in Section 6-6 with the aid of Pappus's theorem.

## 16-8 PROBLEMS

**1** Find the area of the portion of the plane $z = x + 3y$ that lies inside the elliptical cylinder with equation $x^2/4 + y^2/9 = 1$.

**2** Find the area of the region in the plane $y = 1 + 2x + 2y$ that lies directly above the region in the $xy$-plane bounded by the parabolas $y = x^2$ and $x = y^2$.

**3** Find the area of the part of the paraboloid $z = 9 - x^2 - y^2$ that lies above the plane $z = 5$.

**4** Find the area of the part of the surface $z = \frac{1}{2}x^2$ that lies directly above the triangle in the $xy$-plane with vertices at $(0, 0)$, $(1, 0)$, and $(1, 1)$.

**5** Find the area of the surface that is the graph of $z = x + y^2$ for $0 \leq x \leq 1, 0 \leq y \leq 2$.

**6** Find the area of that part of the surface of Problem 5 that lies above the triangle in the $xy$-plane with vertices at $(0, 0)$, $(0, 1)$, and $(1, 1)$.

**7** Find by integration the area of the part of the plane $2x + 3y + z = 6$ that lies in the first octant.

**8** Find the area of the ellipse that is cut from the plane of Problem 7 by the cylinder $x^2 + y^2 = 2$.

**9** Find the area that is cut from the saddle-shaped surface $z = xy$ by the cylinder $x^2 + y^2 = 1$.

**10** Find the area that is cut from the surface $z = x^2 - y^2$ by the cylinder $x^2 + y^2 = 4$.

**11** Find the surface area of the part of the paraboloid $z = 16 - x^2 - y^2$ that lies above the $xy$-plane.

**12** Show by integration that the surface area of the conical surface $z = br$ between the planes $z = 0$ and $z = h = ab$ is given by $A = \pi aL$, where $L$ is the slant height $\sqrt{a^2 + h^2}$ and $a$ is the radius of the base of the cone.

**13** Let the part of the cylinder $x^2 + y^2 = a^2$ between the planes $z = 0$ and $z = h$ be parametrized by $x = a \cos \theta$, $y = a \sin \theta$, $z = z$. Apply the formula in Equation (8) to show that its surface area is $A = 2\pi ah$.

**14** Consider the meridianal zone of height $h = c - b$ lying on the sphere $r^2 + z^2 = a^2$ between the planes $z = b$ and $z = c$, where $0 \leq b < c \leq a$. Apply the formula in Equation (10) to show that the area of this zone is $A = 2\pi ah$.

**15** Find the area of the part of the cylinder $x^2 + z^2 = a^2$ that lies within the cylinder $r^2 = x^2 + y^2 = a^2$.

**16** Find the area of the part of the sphere $r^2 + z^2 = a^2$ that lies within the cylinder $r = a \sin \theta$.

**17** (a) Apply the formula in Equation (8) to show that the surface area of the surface $y = f(x, z)$, for $(x, z)$ in the region $R$ of the $xz$-plane, is given by

$$A = \iint_R \sqrt{1 + \left(\frac{\partial f}{\partial x}\right)^2 + \left(\frac{\partial f}{\partial z}\right)^2} \; dx \; dz.$$

(b) State and derive a similar formula for the area of the surface $x = f(y, z)$, $(y, z)$ in $R$.

**18** Suppose that $R$ is a region in the $\phi\theta$-plane. Consider the part of the sphere $\rho = a$ that corresponds to $(\phi, \theta)$ in $R$, parametrized by Equations (6) with $h(\phi, \theta) = a$. Apply (8) to show that the surface area of this part of the sphere is $A = \iint_R a^2 \sin \phi \; d\phi \; d\theta$.

**19** (a) Consider the "spherical rectangle" defined by $\rho = a$, $\phi_1 \leq \phi \leq \phi_2 = \phi_1 + \Delta\phi$, $\theta_1 \leq \theta \leq \theta_2 = \theta_1 + \Delta\theta$. Apply the formula of Problem 18 and the average value property (see Problem 30 in Section 16-2) to show that the area of this spherical rectangle is $A = a^2 \sin \hat{\phi} \, \Delta\phi \, \Delta\theta$ for some $\hat{\phi}$ in $(\phi_1, \phi_2)$.

(b) Conclude from the result of part (a) that the volume of the spherical block defined by $\rho_1 \leq \rho \leq \rho_2 = \rho_1 + \Delta\rho$ and $\phi_1 \leq \phi \leq \phi_2, \theta_1 \leq \theta \leq \theta_2$ is

$$\Delta V = \tfrac{1}{3}(\rho_2^3 - \rho_1^3) \sin \hat{\phi} \, \Delta\phi \, \Delta\theta.$$

Finally, derive the formula in Equation (8) of Section 16-7 by applying the mean value theorem to the function $f(\rho) = \rho^3$ on the interval $[\rho_1, \rho_2]$.

**20** Describe the surface $z = 2a \sin \phi$. Why is it called a pinched torus? It is parametrized by the equation in (6) with $h(\phi, \theta) = 2a \sin \phi$. Show that its surface area is $A = 4\pi^2 a^2$.

**21** The surface of revolution obtained when one revolves the curve $x = f(z)$, $a \leq z \leq b$ around the $z$-axis is parametrized in terms of $\theta$ ($0 \leq \theta \leq 2\pi$) and $z$ ($a \leq z \leq b$) by $x = f(z) \cos \theta$, $y = f(z) \sin \theta$, $z = z$. From (8) derive the surface area formula

$$A = \int_0^{2\pi} \int_a^b f(z) \sqrt{1 + [f'(z)]^2} \; dz \; d\theta.$$

Note that this agrees with surface area of revolution as defined in Section 6-4.

**22** Apply the formula of Problem 18 in both parts of this problem.

(a) Verify the formula $A = 4\pi a^2$ for the surface area of a sphere of radius $a$.

(b) Find the area of that part of a sphere of radius $a$ that lies within the cone $\phi = \pi/6$.

**23** Apply the result of Problem 21 to verify the formula $a = 2\pi rh$ for the lateral surface area of a right circular cylinder of radius $r$ and height $h$.

**24** Apply the formula in Equation (9) to verify the formula $A = 2\pi rh$ for the lateral surface area of the cylinder $x^2 + z^2 = r^2, 0 \leq y \leq h$ of radius $r$ and height $h$.

---

**\*16-9**

# Change of Variables in Multiple Integrals

We have seen in preceding sections that it is easier to evaluate certain multiple integrals by transforming them into polar or spherical coordinates. The technique of changing coordinate systems in order to evaluate a multiple integral is the multivariable analogue of substitution in a single integral. Recall from Section 5-6 that if $x = g(u)$, then

$$\int_a^b f(x) \, dx = \int_c^d f(g(u))g'(u) \, du \qquad (1)$$

where $a = g(c)$ and $b = g(d)$. The method of substitution involves a "change of variables" that is tailored to the evaluation of a given integral.

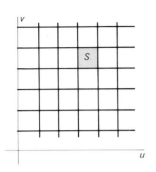

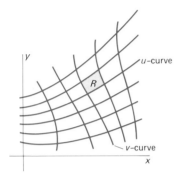

**16.69** The transformation $T$ turns the rectangle $S$ into the curvilinear figure $R$.

Suppose that we want to evaluate the double integral $\iint_R F(x, y)\, dx\, dy$. A change of variables for this integral is determined by a **transformation** $T$ from the $uv$-plane to the $xy$-plane; that is, a function that associates with the point $(u, v)$ a point $T(u, v) = (x, y)$ given by equations of the form

$$x = f(u, v), \qquad y = g(u, v). \tag{2}$$

The point $(x, y)$ is called the **image** of the point $(u, v)$ under the transformation $T$. If no two different points in the $uv$-plane have the same image point in the $xy$-plane, then the transformation $T$ is said to be **one-to-one.** In this case it may be possible to solve Equations (2) for $u$ and $v$ in terms of $x$ and $y$ and thus obtain the equations

$$u = h(x, y), \qquad v = k(x, y) \tag{3}$$

of the **inverse transformation** $T^{-1}$ from the $xy$-plane to the $uv$-plane.

It is often convenient to visualize the transformation $T$ geometrically in terms of its $u$-curves and $v$-curves. The $u$-**curves** of $T$ are the images in the $xy$-plane of horizontal lines in the $uv$-plane; the $v$-**curves** are the images of vertical lines in the $uv$-plane. Note that the image under $T$ of a rectangle that is bounded by horizontal and vertical lines in the $uv$-plane is a *curvilinear figure* bounded by $u$-curves and $v$-curves in the $xy$-plane. This phenomenon is shown in Fig. 16.69. If we know the equations in (3) of the inverse transformation, then we can find the $u$-curves and the $v$-curves quite simply by writing the equations

$$k(x, y) = C_1 \quad \text{and} \quad h(x, y) = C_2,$$

respectively, where $C_1$ and $C_2$ are constants.

**EXAMPLE 1**  Determine the $u$-curves and the $v$-curves of the transformation $T$ whose inverse $T^{-1}$ is specified by the equations $u = xy$, $v = x^2 - y^2$.

*Solution*  The $v$-curves are the rectangular hyperbolas

$$xy = u = C_1 \qquad \text{(constant)},$$

while the $u$-curves are the hyperbolas

$$x^2 - y^2 = v = C_2 \qquad \text{(constant)}.$$

These two familiar families of hyperbolas are shown in Fig. 16.70.

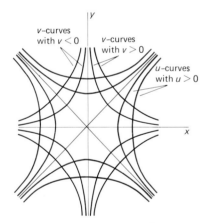

**16.70**  The $u$-curves and $v$-curves of Example 1

Now we describe the change of variables in a double integral that corresponds to the transformation $T$ specified by Equations (2). Let the region $R$ in the $xy$-plane be the image under $T$ of the region $S$ in the $uv$-plane. Let $F(x, y)$ be continuous on $R$. Let $\{S_1, S_2, S_3, \ldots, S_n\}$ be an inner partition of $S$ into rectangles each having dimensions $\Delta u$ by $\Delta v$. Each rectangle $S_i$ is transformed by $T$ into a curvilinear figure $R_i$ in the $xy$-plane, as shown in

Fig. 16.71. The images $\{R_1, R_2, R_3, \ldots, R_n\}$ under $T$ of the $S_i$ then constitute an inner partition of the region $R$ (into curvilinear figures rather than rectangles).

Let $(u_i^*, v_i^*)$ be the lower left-hand corner point of $S_i$ and write $(x_i^*, y_i^*) = (f(u_i^*, v_i^*), g(u_i^*, v_i^*))$ for its image under $T$. The $u$-curve through $(x_i^*, y_i^*)$ has velocity vector

$$\mathbf{t}_u = \mathbf{i} f_u(u_i^*, v_i^*) + \mathbf{j} g_u(u_i^*, v_i^*) = \frac{\partial x}{\partial u}\mathbf{i} + \frac{\partial y}{\partial u}\mathbf{j},$$

while the $v$-curve through $(x_i^*, y_i^*)$ has velocity vector

$$\mathbf{t}_v = \mathbf{i} f_v(u_i^*, v_i^*) + \mathbf{j} g_v(u_i^*, v_i^*) = \frac{\partial x}{\partial v}\mathbf{i} + \frac{\partial y}{\partial v}\mathbf{j}.$$

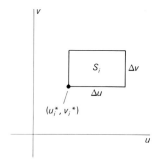

Thus we can approximate the curvilinear figure $R_i$ by a parallelogram $P_i$ with edges that are "copies" of the vectors $\mathbf{t}_u\,\Delta u$ and $\mathbf{t}_v\,\Delta v$. These edges and the approximating parallelogram $P_i$ also appear in Fig. 16.71.

Now the area $\Delta A_i$ of $R_i$ is also approximated by the area of the parallelogram $P_i$, and we can compute the latter. Indeed,

$$\Delta A_i \approx \text{area}\,(P_i) = |\mathbf{t}_u\,\Delta u \times \mathbf{t}_v\,\Delta v| = |\mathbf{t}_u \times \mathbf{t}_v|\,\Delta u\,\Delta v.$$

But

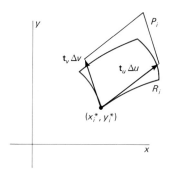

$$\mathbf{t}_u \times \mathbf{t}_v = \begin{vmatrix} \mathbf{i} & \mathbf{j} & \mathbf{k} \\ \dfrac{\partial x}{\partial u} & \dfrac{\partial y}{\partial u} & 0 \\ \dfrac{\partial x}{\partial v} & \dfrac{\partial y}{\partial v} & 0 \end{vmatrix} = \begin{vmatrix} \dfrac{\partial x}{\partial u} & \dfrac{\partial x}{\partial v} \\ \dfrac{\partial y}{\partial u} & \dfrac{\partial y}{\partial v} \end{vmatrix}\mathbf{k}.$$

16.71 The effect of the transformation $T$; we estimate the area of $R_i = T(S_i)$ by computing the area of $P_i$.

The two-by-two determinant above is called the **Jacobian** of the transformation $T$, after the German mathematician Carl Jacobi (1804–1851), who first investigated general changes of variables in multiple integrals. The Jacobian of the transformation $T$ is a function of $u$ and $v$, and we denote it by $J_T = J_T(u, v)$. Thus

$$J_T(u, v) = \begin{vmatrix} f_u(u, v) & f_v(u, v) \\ g_u(u, v) & g_v(u, v) \end{vmatrix}. \tag{4}$$

A common and particularly suggestive notation for the Jacobian is

$$J_T = \frac{\partial(x, y)}{\partial(u, v)}.$$

The computation above shows that the area $\Delta A_i$ of $R_i$ is given approximately by

$$\Delta A_i \approx |J_T(u_i^*, v_i^*)|\,\Delta u\,\Delta v.$$

Therefore, when we set up Riemann sums for approximating double integrals, we find that

$$\iint_R F(x, y)\,dx\,dy \approx \sum_{i=1}^n F(x_i^*, y_i^*)\,\Delta A_i$$

$$\approx \sum_{i=1}^n F(f(u_i^*, v_i^*), g(u_i^*, v_i^*))|J_T(u_i^*, v_i^*)|\,\Delta u\,\Delta v$$

$$\approx \iint_S F(f(u, v), g(u, v))|J_T(u, v)|\,du\,dv.$$

This discussion is, in fact, an outline of a proof of the following general **change of variables** theorem. We assume that $T$ transforms the bounded region $S$ in the $uv$-plane into the bounded region $R$ in the $xy$-plane, and that $T$ is one-to-one from the interior of $S$ to the interior of $R$. Suppose that the function $F(x, y)$ and the first order partial derivatives of the component functions of $T$ are continuous functions. Finally, in order to assure the existence of the indicated double integrals, we assume that the boundary of each of the regions $S$ and $R$ consists of finitely many piecewise-smooth simple closed curves.

---

**Theorem** *Change of Variables*

If the transformation $T$ with component functions $x = f(u, v)$, $y = g(u, v)$ satisfies the above conditions, then

$$\iint_R F(x, y) \, dx \, dy = \iint_S F(f(u, v), g(u, v)) |J_T(u, v)| \, du \, dv. \qquad (5)$$

---

If we write $G(u, v) = F(f(u, v), g(u, v))$, then the change of variables formula in (5) becomes

$$\iint_R F(x, y) \, dx \, dy = \iint_S G(u, v) \left| \frac{\partial(x, y)}{\partial(u, v)} \right| du \, dv. \qquad (5a)$$

Thus we formally transform $\iint_R F(x, y) \, dA$ by replacing the variables $x$ and $y$ by $f(u, v)$ and $g(u, v)$, respectively, and writing

$$dA = \left| \frac{\partial(x, y)}{\partial(u, v)} \right| du \, dv$$

for the area element in terms of $u$ and $v$. Note the analogy between (5a) and the single-variable formula (1). In fact, if $g'(x) \neq 0$ on $[c, d]$, and we denote by $\alpha$ the smaller and by $\beta$ the larger of the two limits $c$ and $d$ in (1), then it becomes

$$\int_a^b f(x) \, dx = \int_\alpha^\beta f(g(u)) |g'(u)| \, du. \qquad (1a)$$

Thus the Jacobian in (5a) plays the role of the derivative $g'(u)$ in Formula (1a).

For example, suppose that the transformation $T$ from the $r\theta$-plane to the $xy$-plane is determined by the polar coordinates equations

$$x = f(r, \theta) = r \cos \theta, \qquad y = g(r, \theta) = r \sin \theta.$$

Then the Jacobian of $T$ is

$$\frac{\partial(x, y)}{\partial(r, \theta)} = \begin{vmatrix} \cos \theta & -r \sin \theta \\ \sin \theta & r \cos \theta \end{vmatrix} = r > 0,$$

so Equation (5) or (5a) reduces to the familiar formula

$$\iint_R F(x, y) \, dx \, dy = \iint_S F(r \cos \theta, r \sin \theta) \, r \, dr \, d\theta.$$

Given a particular double integral $\iint_R f(x, y) \, dx \, dy$, how do we find a *productive* change of variables? One standard approach is to choose a transformation $T$ such that the boundary of $R$ consists of $u$-curves and $v$-curves.

In case it is more convenient to express $u$ and $v$ in terms of $x$ and $y$, we can first compute $\partial(u, v)/\partial(x, y)$ explicitly and then find the needed Jacobian $\partial(x, y)/\partial(u, v)$ from the formula

$$\frac{\partial(x, y)}{\partial(u, v)} \cdot \frac{\partial(u, v)}{\partial(x, y)} = 1. \tag{6}$$

The formula in (6) is a consequence of the chain rule (see Problem 18).

**EXAMPLE 2** Suppose that $R$ is the plane region that is bounded by the hyperbolas

$$xy = 1, \quad xy = 3 \quad \text{and} \quad x^2 - y^2 = 1, \quad x^2 - y^2 = 4.$$

Find the polar moment of inertia $I_0 = \iint_R (x^2 + y^2)\, dx\, dy$ of this region.

**Solution** The hyperbolas bounding $R$ are $v$-curves and $u$-curves if $u = xy$ and $v = x^2 - y^2$, as in Example 1. We can most easily write $x^2 + y^2$ in terms of $u$ and $v$ by first noting that

$$4u^2 + v^2 = 4x^2y^2 + (x^2 - y^2)^2 = (x^2 + y^2)^2,$$

so that $x^2 + y^2 = \sqrt{4u^2 + v^2}$. Now

$$\frac{\partial(u, v)}{\partial(x, y)} = \begin{vmatrix} y & x \\ 2x & -2y \end{vmatrix} = -2(x^2 + y^2).$$

Hence Equation (6) gives

$$\frac{\partial(x, y)}{\partial(u, v)} = -\frac{1}{2(x^2 + y^2)} = -\frac{1}{2\sqrt{4u^2 + v^2}}.$$

We are now ready to apply the theorem, with the regions $S$ and $R$ as shown in Fig. 16.72. With $F(x, y) = x^2 + y^2$, the formula in (5a) gives

$$I_0 = \iint_R (x^2 + y^2)\, dx\, dy = \int_1^4 \int_1^3 \sqrt{4u^2 + v^2} \cdot \frac{1}{2\sqrt{4u^2 + v^2}}\, du\, dv$$

$$= \int_1^4 \int_1^3 \frac{1}{2}\, du\, dv = 3.$$

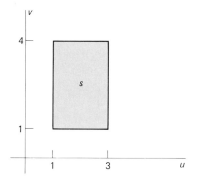

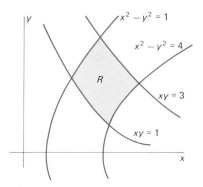

**16.72** The transformation $T$ and the new region $S$ constructed in Example 2

**16.73** Finding the area of the region $R$ (Example 3)

Our next example has an important application as its motivation. Consider an engine with an operating cycle that consists of alternate expansion and compression of the gas in a piston. During a cycle the point $(P, V)$ giving the pressure and volume of this gas traces a closed curve in the $PV$-plane. The work done by the engine—ignoring friction and related losses—is then equal (in appropriate units) to the *area enclosed by this curve*, called the *indicator diagram* of the engine. In an ideal Carnot engine, the indicator diagram consists of two *isothermals* $xy = a$, $xy = b$ and two *adiabatics* $xy^\gamma = c$, $xy^\gamma = d$, where $\gamma$ is the heat capacity ratio of the working gas in the piston. A typical value is $\gamma = 1.4$.

**EXAMPLE 3** Find the area of the region $R$ bounded by the curves $xy = 1$, $xy = 3$, and $xy^{1.4} = 1$, $xy^{1.4} = 2$ (see Fig. 16.73).

**Solution** To force the given curves to be $u$-curves and $v$-curves, we define our change of variables transformation by means of the equations $u = xy$

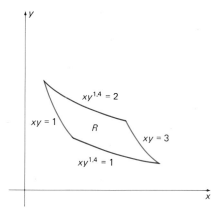

and $v = xy^{1.4}$. Then

$$\frac{\partial(u, v)}{\partial(x, y)} = \begin{vmatrix} y & x \\ y^{1.4} & (1.4)xy^{0.4} \end{vmatrix} = (0.4)xy^{1.4} = (0.4)v.$$

So

$$\frac{\partial(x, y)}{\partial(u, v)} = \frac{1}{\partial(u, v)/\partial(x, y)} = \frac{2.5}{v}.$$

Consequently, the change of variables theorem gives the formula

$$A = \iint_R 1 \, dx \, dy = \int_1^2 \int_1^3 \frac{2.5}{v} \, du \, dv = 5 \ln 2.$$

---

The change of variables formula for triple integrals is similar to Formula (5). Let $S$ and $R$ be regions that correspond under the one-to-one transformation $T$ from $uvw$-space to $xyz$-space, where the coordinate functions that comprise $T$ are

$$x = f(u, v, w), \qquad y = g(u, v, w), \qquad z = h(u, v, w). \tag{7}$$

The Jacobian of $T$ is

$$J_T(u, v, w) = \frac{\partial(x, y, z)}{\partial(u, v, w)} = \begin{vmatrix} \dfrac{\partial x}{\partial u} & \dfrac{\partial x}{\partial v} & \dfrac{\partial x}{\partial w} \\[2mm] \dfrac{\partial y}{\partial u} & \dfrac{\partial y}{\partial v} & \dfrac{\partial y}{\partial w} \\[2mm] \dfrac{\partial z}{\partial u} & \dfrac{\partial z}{\partial v} & \dfrac{\partial z}{\partial w} \end{vmatrix}. \tag{8}$$

Then the change of variables formula for triple integrals is

$$\iiint_R F(x, y, z) \, dx \, dy \, dz = \iiint_S G(u, v, w) \left| \frac{\partial(x, y, z)}{\partial(u, v, w)} \right| du \, dv \, dw, \tag{9}$$

where $G(u, v, w) = F(f(u, v, w), g(u, v, w), h(u, v, w))$ is the function obtained from $F(x, y, z)$ by expressing the variables $x$, $y$, and $z$ in terms of $u$, $v$, and $w$.

For example, if $T$ is the spherical coordinates transformation given by

$$x = \rho \sin \phi \cos \theta, \qquad y = \rho \sin \phi \sin \theta, \qquad z = \rho \cos \phi,$$

then the Jacobian of $T$ is

$$\frac{\partial(x, y, z)}{\partial(\rho, \phi, \theta)} = \begin{vmatrix} \sin \phi \cos \theta & \rho \cos \phi \cos \theta & -\rho \sin \phi \sin \theta \\ \sin \phi \sin \theta & \rho \cos \phi \sin \theta & \rho \sin \phi \cos \theta \\ \cos \phi & -\rho \sin \phi & 0 \end{vmatrix} = \rho^2 \sin \phi.$$

Thus (9) reduces to the familiar formula

$$\iiint_R F(x, y, z) \, dx \, dy \, dz = \iiint_S G(\rho, \phi, \theta) \rho^2 \sin \phi \, d\rho \, d\phi \, d\theta$$

because $\rho^2 \sin \phi \geq 0$ for $\phi$ in $[0, \pi]$.

**EXAMPLE 4** Find the volume of the solid torus $R$ that is obtained by revolving around the $z$-axis the circular disk $(x - b)^2 + z^2 \leq a^2$, $a < b$, in the $xz$-plane.

**898**

**Solution** This is the torus of Example 4 in Section 16-8. Let us write $u$ for the ordinary polar coordinate angle $\theta$, $v$ for the angle $\psi$ of Fig. 16.68, and $w$ for distance from the center of the circular disk described by the above inequality. We then define the transformation $T$ by means of the equations

$$x = (b + w\cos v)\cos u, \qquad y = (b + w\cos v)\sin u, \qquad z = w\sin v.$$

Then the solid torus $R$ is the image under $T$ of the region $S$ in $uvw$-space described by the inequalities $0 \le u \le 2\pi,\ 0 \le v \le 2\pi,\ 0 \le w \le a$. By a routine computation, we find that the Jacobian of $T$ is

$$\frac{\partial(x,\,y,\,z)}{\partial(u,\,v,\,w)} = w(b + w\cos v).$$

Hence Formula (9) with $F(x, y, z) \equiv 1$ gives

$$V = \iiint_R dx\, dy\, dz = \int_0^{2\pi}\int_0^{2\pi}\int_0^a (bw + w^2\cos v)\, dw\, du\, dv$$

$$= 2\pi \int_0^{2\pi} (\tfrac{1}{2}a^2 b + \tfrac{1}{3}a^3\cos v)\, dv = 2\pi^2 a^2 b,$$

which agrees with the value $V = (2\pi b)(\pi a^2)$ given by Pappus's theorem (Section 6-6).

---

## 16-9 PROBLEMS

In each of Problems 1–6, solve for $x$ and $y$ in terms of $u$ and $v$ and then compute the Jacobian $\partial(x, y)/\partial(u, v)$.

**1** $u = x + y,\quad v = x - y$

**2** $u = x - 2y,\quad v = 3x + y$

**3** $u = xy,\quad v = \dfrac{y}{x}$

**4** $u = 2(x^2 + y^2),\quad v = 2(x^2 - y^2)$

**5** $u = x + 2y^2,\quad v = x - 2y^2$

**6** $u = \dfrac{2x}{x^2 + y^2},\quad v = \dfrac{-2y}{x^2 + y^2}$

**7** Let $R$ be the parallelogram bounded by the lines $x + y = 1$, $x + y = 2$ and $2x - 3y = 2$, $2x - 3y = 5$. Substitute $u = x + y$, $v = 2x - 3y$ to find the area $A = \iint_R dx\, dy$ of $R$.

**8** Substitute $u = xy$, $v = y/x$ to find the area of the first quadrant region bounded by the lines $y = x$, $y = 2x$ and the hyperbolas $xy = 1$, $xy = 2$.

**9** Substitute $u = xy$, $v = xy^3$ to find the area of the region in the first quadrant bounded by the curves $xy = 2$, $xy = 4$ and $xy^3 = 3$, $xy^3 = 6$.

**10** Find the area of the region in the first quadrant bounded by the curves $y = x^2$, $y = 2x^2$ and $x = y^2$, $x = 4y^2$. (*Suggestion:* Let $y = ux^2$ and $x = vy^2$.)

**11** Use the method of Problem 10 to find the area of the region in the first quadrant bounded by the curves $y = x^3$, $y = 2x^3$ and $x = y^3$, $x = 4y^3$.

**12** Let $R$ be the region in the first quadrant bounded by the circles $x^2 + y^2 = 2x$, $x^2 + y^2 = 6x$ and the circles $x^2 + y^2 = 2y$, $x^2 + y^2 = 8y$. Use the transformation

$u = 2x/(x^2 + y^2)$, $v = 2y/(x^2 + y^2)$ to evaluate the integral $\iint_R (x^2 + y^2)^{-2}\, dx\, dy$.

**13** Use elliptical coordinates $x = 3r\cos\theta$, $y = 2r\sin\theta$ to find the volume of the region that is bounded by the $xy$-plane, the paraboloid $z = x^2 + y^2$, and the elliptical cylinder $x^2/9 + y^2/4 = 1$.

**14** Let $R$ be the solid ellipsoid with outer boundary surface $x^2/a^2 + y^2/b^2 + z^2/c^2 = 1$. Use the transformation $x = au$, $y = bv$, $z = cw$ to show that the volume of this ellipsoid is $V = \iiint_R 1\, dx\, dy\, dz = \tfrac{4}{3}\pi abc$.

**15** Find the volume of the region in the first octant that is bounded by the hyperbolic cylinders $xy = 1$, $xy = 4$; $xz = 1$, $xz = 9$; $yz = 4$, $yz = 9$. (*Suggestion:* Let $u = xy$, $v = xz$, $w = yz$, and note that $uvw = x^2 y^2 z^2$.)

**16** Use the transformation $x = (r/t)\cos\theta$, $y = (r/t)\sin\theta$, $z = r^2$ to find the volume of the region $R$ that lies between the paraboloids $z = x^2 + y^2$, $z = 4(x^2 + y^2)$ and also between the planes $z = 1$, $z = 4$.

**17** Let $R$ be the rotated elliptical region bounded by the graph of $x^2 + xy + y^2 = 3$. Let $x = u + v$ and $y = u - v$. Show that

$$\iint_R e^{-(x^2 + xy + y^2)}\, dx\, dy = 2\iint_S e^{-(3u^2 + v^2)}\, du\, dv.$$

Then substitute $u = r\cos\theta$, $v = \sqrt{3}(r\sin\theta)$ to evaluate the latter integral.

**18** Derive Relation (6) between the Jacobians of a transformation and its inverse from the chain rule and the following property of determinants:

$$\begin{vmatrix} a_1 & b_1 \\ c_1 & d_1 \end{vmatrix} \cdot \begin{vmatrix} a_2 & b_2 \\ c_2 & d_2 \end{vmatrix} = \begin{vmatrix} a_1 a_2 + b_1 c_2 & a_1 b_2 + b_1 d_2 \\ a_2 c_1 + c_2 d_1 & b_2 c_1 + d_1 d_2 \end{vmatrix}.$$

**19** Change to spherical coordinates to show that, for $k > 0$,

$$\int_{-\infty}^{+\infty} \int_{-\infty}^{+\infty} \int_{-\infty}^{+\infty} (x^2 + y^2 + z^2)^{1/2} e^{-k(x^2+y^2+z^2)} \, dx \, dy \, dz$$
$$= \frac{2\pi}{k^2}.$$

**20** Let $R$ be the solid ellipsoid with constant density $\delta$ and boundary surface $x^2/a^2 + y^2/b^2 + z^2/c^2 = 1$. Use ellipsoidal coordinates $x = a\rho \sin \phi \cos \theta$, $y = b\rho \sin \phi \sin \theta$, $z = c\rho \cos \phi$ to show that its mass is $M = \frac{4}{3}\pi\delta abc$.

**21** Show that the moment of inertia of the ellipsoid of Problem 20 about the $z$-axis is $I_z = \frac{1}{5}M(a^2 + b^2)$.

## CHAPTER 16 REVIEW: Definitions, Concepts, Results

Use the list below as a guide to concepts that you may need to review.

**1** Definition of the double integral as a limit of Riemann sums

**2** Evaluation of double integrals by iterated single integrals

**3** Use of the double integral to find the volume between two surfaces above a given plane region

**4** Transformation of the double integral $\iint_R f(x, y) \, dA$ into polar coordinates

**5** Application of double integrals to find mass, centroids, and moments of inertia of plane laminae

**6** Definition of the triple integral as a limit of Riemann sums

**7** Evaluation of triple integrals by iterated single integrals

**8** Application of triple integrals to find volume, mass, centroids, and moments of inertia

**9** Transformation of the triple integral $\iiint_T f(x, y, z) \, dV$ into cylindrical and spherical coordinates

**10** The surface area of a parametric surface

**11** The area of a surface $z = f(x, y)$ for $(x, y)$ in the plane region $R$

**12** The Jacobian of a transformation of coordinates

**13** The transformation of a double or triple integral corresponding to a given change of variables

## MISCELLANEOUS PROBLEMS

In each of Problems 1–5, evaluate the given integral by first reversing the order of integration.

**1** $\int_0^1 \int_{y^{1/3}}^1 \frac{dx \, dy}{\sqrt{1 + x^2}}$

**2** $\int_0^1 \int_y^1 \frac{\sin x}{x} \, dx \, dy$

**3** $\int_0^1 \int_x^1 e^{-y^2} \, dy \, dx$

**4** $\int_0^8 \int_{x^{2/3}}^4 x \cos y^4 \, dy \, dx$

**5** $\int_0^4 \int_{\sqrt{y}}^2 \frac{ye^{x^2}}{x^3} \, dx \, dy$

**6** The double integral $\int_0^\infty \int_x^\infty y^{-1}e^{-y} \, dy \, dx$ is an improper integral over the unbounded region in the first quadrant bounded by the lines $y = x$ and $x = 0$. Assuming the validity of reversing the order of integration, evaluate this integral by integrating first with respect to $x$.

**7** Find the volume of the solid $T$ that lies under the paraboloid $z = x^2 + y^2$ and over the triangle $R$ in the $xy$-plane having vertices $(0, 0, 0)$, $(1, 1, 0)$, and $(2, 0, 0)$.

**8** Find by integration in cylindrical coordinates the volume bounded by the paraboloids $z = 2x^2 + 2y^2$ and $z = 48 - x^2 - y^2$.

**9** Use integration in spherical coordinates to find the volume and centroid of the solid region that is inside the

sphere $\rho = 3$, under the cone $\phi = \pi/3$, and above the $xy$-plane $\phi = \pi/2$.

**10** Find the volume of the solid bounded by the elliptic paraboloids $z = x^2 + 3y^2$ and $z = 8 - x^2 - 5y^2$.

**11** Find the volume bounded by the paraboloid $y = x^2 + 3z^2$ and the parabolic cylinder $y = 4 - z^2$.

**12** Find the volume of the region that is bounded by the parabolic cylinders $z = x^2$, $z = 2 - x^2$, and the planes $y = 0$, $y + z = 4$.

**13** Find the volume of the region bounded by the elliptical cylinder $y^2 + 4z^2 = 4$ and the planes $x = 0$, $x = y + 2$.

**14** Show that the volume of the solid bounded by the elliptic cylinder $x^2/a^2 + y^2/b^2 = 1$ and the planes $z = 0$, $z = h + x$ (where $h > a > 0$) is $V = \pi abh$.

**15** Let $R$ be the first-quadrant region bounded by the line $y = x$ and the curve $x^4 + x^2y^2 = y^2$. Use polar coordinates to evaluate the integral $\iint_R (1 + x^2 + y^2)^{-2} \, dA$.

In each of Problems 16–20, find the mass and centroid of a plane lamina having the indicated shape and density $\rho$.

**16** The region bounded by $y = x^2$ and $x = y^2$; $\rho = x^2 + y^2$.

**17** The region bounded by $y^2 = \frac{1}{2}x$ and $y^2 = x - 4$; $\rho = y^2$.

**18** The region between $y = \ln x$ and the $x$-axis over the interval $1 \leq x \leq 2$; $\rho = 1/x$.

**19** The circle bounded by $r = 2 \cos \theta$; $\rho = k$ (a constant).

**20** The region of Problem 19; $\rho = r$.

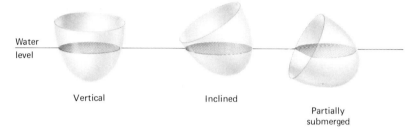

Water level

Vertical       Inclined

Partially submerged

**16.74** How a uniform solid paraboloid might float

**21** Find the centroid of the region in the $xy$-plane bounded by the $x$-axis and the parabola $y = 4 - x^2$.

**22** Find the volume of the solid that lies under the parabolic cylinder $z = x^2$ and over the triangle in the $xy$-plane bounded by the $x$-axis, the $y$-axis, and the line $x + y = 1$.

**23** Use cylindrical coordinates to find the volume of the ice cream cone bounded above by the sphere $x^2 + y^2 + z^2 = 5$ and below by the cone $z = 2(x^2 + y^2)^{1/2}$.

**24** Find the volume and centroid of the ice cream cone that is bounded above by the sphere $\rho = a$ and below by the cone $\phi = \pi/3$.

**25** Find the moment of inertia about its natural axis of a homogeneous solid circular cone with mass $M$ and base radius $a$.

**26** Find the mass of the first octant of the ball $\rho \leq a$ if its density function is $\delta = xyz$.

**27** Find the moment of inertia about the $x$-axis of the homogeneous solid ellipsoid bounded by $x^2/a^2 + y^2/b^2 + z^2/c^2 = 1$.

**28** Find the volume of the region in the first octant that is bounded by the sphere $\rho = a$, the cylinder $r = a$, the plane $z = a$, the $xz$-plane, and the $yz$-plane.

**29** Find the moment of inertia about the $z$-axis of the homogeneous region that lies inside both the sphere $\rho = 2$ and the cylinder $r = 2 \cos \theta$.

In each of Problems 30–32, a volume is generated by revolving a plane region $R$ around an axis. To find this volume, set up a *double* integral over $R$ by revolving an area element $dA$ around the indicated axis to generate a volume element $dV$.

**30** Find the volume of the solid that is obtained by revolving around the $y$-axis the area within the circle $r = 2a \cos \theta$.

**31** Find the volume of the solid obtained by revolving the area enclosed by the cardioid $r = 1 + \cos \theta$ around the $x$-axis.

**32** Find the volume of the solid torus that is obtained by revolving the circle $r \leq a$ about the line $x = -b, |b| \geq a > 0$.

**33** This problem deals with the oblique segment of a paraboloid discussed in Example 3 of Section 16-6 (shown in Fig. 16.52).

(a) Show that its centroid is the point $C(0, \frac{1}{2}, \frac{7}{4})$.

(b) Show that the center of the elliptical upper "base" of the solid paraboloid is the point $Q(0, \frac{1}{2}, \frac{5}{2})$.

(c) Verify that the point $V(0, \frac{1}{2}, \frac{1}{4})$ is the point where the tangent plane to the paraboloid is parallel to the upper base. The point $V$ is called the *vertex* of the oblique segment, and the line segment $VQ$ is its *principal axis*.

(d) Show that $C$ lies on the principal axis *two-thirds* of the way from the vertex to the upper base. Archimedes showed that this is true for any segment cut off from a paraboloid by a plane. This was a key step in his determination of the equilibrium position of a floating right circular paraboloid, in terms of its density and dimensions. The possible positions are shown in Fig. 16.74. The principles he introduced for the solution of this problem are still important in naval architecture.

Problems 34–40 deal with average distance. The **average distance** $\bar{d}$ of the point $(x_0, y_0)$ from the points of the plane region $R$ with area $A$ is defined to be

$$\bar{d} = \frac{1}{A} \iint_R [(x - x_0)^2 + (y - y_0)^2]^{1/2} \, dA.$$

The average distance of a point $(x_0, y_0, z_0)$ from the points of a space region is defined similarly.

**34** Show that the average distance of the points of a disk of radius $a$ from its center is $2a/3$.

**35** Show that the average distance of the points of a disk of radius $a$ from a fixed point on its boundary is $\bar{d} = 32a/9\pi$.

**36** A circle of radius 1 is interior to and tangent to a circle of radius 2. Find the average distance of the point of tangency from the points that lie between the two circles.

**37** Show that the average distance of the points of a spherical ball of radius $a$ from its center is $3a/4$.

**38** Show that the average distance of the points of a spherical ball of radius $a$ from a fixed point on its surface is $6a/5$.

**39** A sphere of radius 1 is interior to and tangent to a sphere of radius 2. Find the average distance of the point of tangency from the set of all points between the two spheres.

**40** A right circular cone has radius $R$ and height $H$. Find the average distance of the points of the cone from its vertex.

**41** Find the surface area of the part of the paraboloid $z = 10 - r^2$ that lies between the two planes $z = 1$ and $z = 6$.

**42** Find the surface area of the part of the surface $z = y^2 - x^2$ that is inside the cylinder $x^2 + y^2 = 4$.

**43** Deduce from the formula of Problem 18 in Section 16-8 that the surface area of the zone on the sphere $\rho = a$ between the planes $z = z_1$ and $z = z_2$ (where $-a \leqq z_1 < z_2 \leqq a$) is $A = 2\pi ah$ where $h = z_2 - z_1$.

**44** Find the surface area of the part of the sphere $\rho = 2$ that is inside the cylinder $x^2 + y^2 = 2x$.

**45** A square hole with side length 2 is cut through a cone of height and base radius 2; the center line of the hole is the axis of symmetry of the cone. Find the area of the surface removed from the cone.

**46** Numerically approximate the surface area of the part of the parabolic cylinder $z = \frac{1}{2}x^2$ that lies inside the cylinder $x^2 + y^2 = 1$.

**47** A "fence" of variable height $h(t)$ stands above the plane curve $(x(t), y(t))$. Thus the fence has the parametrization $x = x(t)$, $y = y(t)$, $z = z$ for $a \leqq t \leqq b$, $0 \leqq z \leqq h(t)$. Apply the formula in Equation (8) of Section 16-8 to show that the area of the fence is

$$A = \int_a^b \int_0^{h(t)} \left[ \left(\frac{dx}{dt}\right)^2 + \left(\frac{dy}{dt}\right)^2 \right]^{1/2} dz\, dt.$$

**48** Apply the formula of Problem 47 to compute the area of the part of the cylinder $r = a \sin \theta$ that lies inside the sphere $r^2 + z^2 = a^2$.

**49** Find the polar moment of inertia of the first-quadrant region of constant density $\delta$ that is bounded by the hyperbolas $xy = 1$, $xy = 3$ and $x^2 - y^2 = 1$, $x^2 - y^2 = 4$.

**50** Substitute $u = x - y$, $v = x + y$ to evaluate

$$\iint_R \exp\left(\frac{x-y}{x+y}\right) dx\, dy,$$

where $R$ is bounded by the coordinate axes and the line $x + y = 1$.

**51** Use ellipsoidal coordinates $x = a\rho \sin\phi \cos\theta$, $y = b\rho \sin\phi \sin\theta$, $z = c\rho \cos\phi$ to find the mass of the solid ellipsoid $x^2/a^2 + y^2/b^2 + z^2/c^2 = 1$ if its density at the point $(x, y, z)$ is $\delta = 1 - x^2/a^2 - y^2/b^2 - z^2/c^2$.

**52** Let $R$ be the first-quadrant region bounded by the lemniscates $r^2 = 3 \cos 2\theta$, $r^2 = 4 \cos 2\theta$, and $r^2 = 3 \sin 2\theta$, $r^2 = 4 \sin 2\theta$ (see Fig. 16.75). Show that its area is $A = (10 - 7\sqrt{2})/4$. (*Suggestion:* Define the transformation $T$ from the $uv$-plane to the $r\theta$-plane by $r^2 = u^{1/2}\cos 2\theta$, $r^2 = v^{1/2}\sin 2\theta$. Show first that

$$r^4 = \frac{uv}{u+v}, \qquad \theta = \frac{1}{2}\tan^{-1}\frac{u^{1/2}}{v^{1/2}}.$$

Then show that $\partial(r, \theta)/\partial(u, v) = -1/[16r(u+v)^{3/2}]$.)

**16.75** The region $R$ of Problem 52

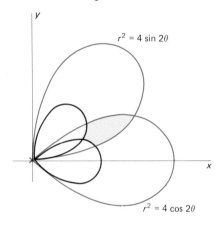

# 17

# Vector Analysis

## Vector Fields

This chapter is devoted to topics in the calculus of vector fields of importance in science and engineering. A **vector field** defined on a region $T$ in space is a vector-valued function $\mathbf{F}$ that associates with each point $(x, y, z)$ of $T$ a vector

$$\mathbf{F}(x, y, z) = \mathbf{i}P(x, y, z) + \mathbf{j}Q(x, y, z) + \mathbf{k}R(x, y, z). \tag{1}$$

We may more briefly describe the vector field $\mathbf{F}$ in terms of its *component functions* $P$, $Q$, and $R$ by writing $\mathbf{F} = \langle P, Q, R \rangle$. Note that $P$, $Q$, and $R$ are scalar (real-valued) functions.

A **vector field** in the plane is similar except that neither $z$-components nor $z$-coordinates are involved. Thus a vector field on the plane region $R$ is a vector-valued function $\mathbf{F}$ that associates with each point $(x, y)$ of $R$ a vector

$$\mathbf{F}(x, y) = \mathbf{i}P(x, y) + \mathbf{j}Q(x, y). \tag{2}$$

It is useful to be able to visualize a given vector field $\mathbf{F}$. One common way is to sketch a collection of typical vectors $\mathbf{F}(x, y)$ each represented by an arrow of length $|\mathbf{F}(x, y)|$ and placed with $(x, y)$ as its initial point. This procedure is illustrated in the following example and the accompanying figure.

**EXAMPLE 1**   Describe the vector field $\mathbf{F}(x, y) = x\mathbf{i} + y\mathbf{j}$.

*Solution*   For each point $(x, y)$ in the plane, $\mathbf{F}(x, y)$ is simply its position vector; it points directly away from the origin and has length

$$|\mathbf{F}(x, y)| = |x\mathbf{i} + y\mathbf{j}| = \sqrt{x^2 + y^2} = r$$

**17.1**  The vector field $F(x, y) = x\mathbf{i} + y\mathbf{j}$

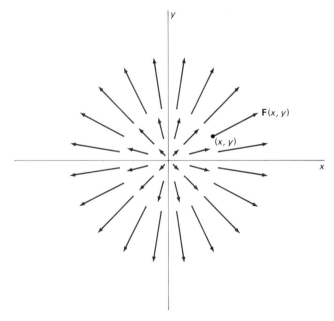

equal to the distance from the origin to $(x, y)$. Figure 17.1 shows some typical vectors representing this vector field.

Among the most important vector fields in applications are *velocity vector fields*. Imagine the steady flow of a fluid, such as the air in the wind or the water in a river. By a *steady flow* is meant one such that the velocity vector $\mathbf{v}(x, y, z)$ of the fluid flowing through each point $(x, y, z)$ is independent of time, so the pattern of the flow remains constant. Then $\mathbf{v}(x, y, z)$ is the **velocity vector field** of the fluid flow.

**EXAMPLE 2**  Suppose the horizontal $xy$-plane to be covered with a thin sheet of water that is revolving—rather like a whirlpool—about the origin with constant angular speed $\omega$ radians per second in the counterclockwise direction. Describe the associated velocity vector field.

*Solution*  In this case we obviously have a two-dimensional vector field $\mathbf{v}(x, y)$. At each point $(x, y)$ the water is moving tangential to the circle of radius $r = \sqrt{x^2 + y^2}$ with speed $v = r\omega$. We note that the vector field

$$\mathbf{v}(x, y) = \omega(-y\mathbf{i} + x\mathbf{j}) \tag{3}$$

has length $r\omega$, points in a generally counterclockwise direction, and that

$$\mathbf{v} \cdot \mathbf{r} = \omega(-y\mathbf{i} + x\mathbf{j}) \cdot (x\mathbf{i} + y\mathbf{j}) = 0,$$

so $\mathbf{v}$ is tangent to the circle mentioned above. The velocity field determined by Equation (3) is illustrated in Fig. 17.2.

Equally important in physical applications are *force fields*. Suppose that some circumstance (perhaps gravitational or electrical in character) causes a force $\mathbf{F}(x, y, z)$ to act on a particle when it is placed at the point $(x, y, z)$. Then we have a force field $\mathbf{F}$. The following example deals with what is perhaps the most common force field experienced by human beings.

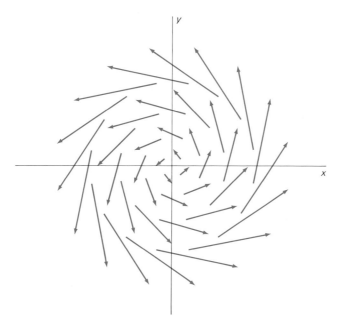

17.2  The velocity vector field $\mathbf{v}(x, y) = \omega(-y\mathbf{i} + x\mathbf{j})$, drawn with $\omega = 1$

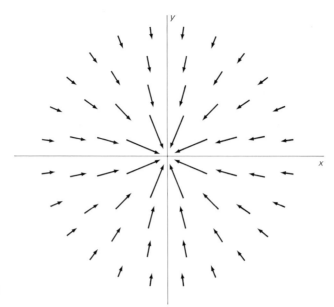

**17.3** An inverse-square-law force field

**EXAMPLE 3** Suppose that a mass $M$ is fixed at the origin. When a particle of unit mass is placed at the point $(x, y, z)$ other than the origin, it is subjected to a force $\mathbf{F}(x, y, z)$ of gravitational attraction directed toward the mass $M$ at the origin. By the inverse-square law of gravitation, the magnitude of $\mathbf{F}$ is $F = GM/r^2$ where $r = \sqrt{x^2 + y^2 + z^2}$ is the length of the position vector $\mathbf{r} = x\mathbf{i} + y\mathbf{j} + z\mathbf{k}$. It follows immediately that

$$\mathbf{F}(x, y, z) = -\frac{k\mathbf{r}}{r^3} \tag{4}$$

where $k = GM$, because this vector has both the correct magnitude and the correct direction (toward the origin, for $\mathbf{F}$ is a multiple of $-\mathbf{r}$). A force field of the form in (4) is called an *inverse-square* force field. Note that $\mathbf{F}(x, y, z)$ is not defined at the origin, and that $|\mathbf{F}| \to \infty$ as $r \to 0$. Figure 17.3 illustrates an inverse-square force field.

### THE GRADIENT VECTOR FIELD

In Section 15-8 we introduced the gradient vector of the real-valued function $f(x, y, z)$. It is the vector $\nabla f$ defined as follows:

$$\nabla f = \mathbf{i}\frac{\partial f}{\partial x} + \mathbf{j}\frac{\partial f}{\partial y} + \mathbf{k}\frac{\partial f}{\partial z}. \tag{5}$$

The partial derivatives on the right-hand side in (5) are evaluated at the point $(x, y, z)$. Thus $\nabla f(x, y, z)$ is a vector field—it is the **gradient vector field** of the function $f$. According to Theorem 1 in Section 15-8, the vector $\nabla f(x, y, z)$ points in the direction in which the maximal directional derivative of $f$ at $(x, y, z)$ is obtained. For instance, if $f(x, y, z)$ is the temperature at the point $(x, y, z)$, then one should go in the direction of $\nabla f(x, y, z)$ in order to get warmer the most quickly.

The notation in Equation (5) suggests the formal expression

$$\nabla = \mathbf{i}\frac{\partial}{\partial x} + \mathbf{j}\frac{\partial}{\partial y} + \mathbf{k}\frac{\partial}{\partial z}. \tag{6}$$

It is fruitful to think of $\mathbf{V}$ as a *vector differential operator*—$\mathbf{V}$ is the operation which, when applied to the scalar function $f$, yields its gradient vector field $\mathbf{V}f$. This operation behaves in several familiar and important ways like the operation of (single-variable) differentiation. For a familiar example of this, recall that in Chapter 15 we found the critical points of a function of several variables to be those points at which $\mathbf{V}f(x, y, z) = \mathbf{0}$ and those at which $\mathbf{V}f(x, y, z)$ does not exist. As a computationally useful instance, suppose that $f$ and $g$ are functions and $a$ and $b$ are constants. It then follows readily from (5) and from the linearity of partial differentiation that

$$\mathbf{V}(af + bg) = a\,\mathbf{V}f + b\,\mathbf{V}g. \tag{7}$$

Thus the gradient operator is *linear*. It also satisfies the product rule of the following example.

**EXAMPLE 4**   Given the differentiable functions $f(x, y, z)$ and $g(x, y, z)$, show that

$$\mathbf{V}(fg) = f\,\mathbf{V}g + g\,\mathbf{V}f. \tag{8}$$

*Solution*   We apply the definition in (5) and the product rule for partial derivatives. Thus

$$\mathbf{V}(fg) = \mathbf{i}\,\frac{\partial(fg)}{\partial x} + \mathbf{j}\,\frac{\partial(fg)}{\partial y} + \mathbf{k}\,\frac{\partial(fg)}{\partial z}$$

$$= \mathbf{i}\left(f\,\frac{\partial g}{\partial x} + g\,\frac{\partial f}{\partial x}\right) + \mathbf{j}\left(f\,\frac{\partial g}{\partial y} + g\,\frac{\partial f}{\partial y}\right) + \mathbf{k}\left(f\,\frac{\partial g}{\partial z} + g\,\frac{\partial f}{\partial z}\right)$$

$$= f\left(\mathbf{i}\,\frac{\partial g}{\partial x} + \mathbf{j}\,\frac{\partial g}{\partial y} + \mathbf{k}\,\frac{\partial g}{\partial z}\right) + g\left(\mathbf{i}\,\frac{\partial f}{\partial x} + \mathbf{j}\,\frac{\partial f}{\partial y} + \mathbf{k}\,\frac{\partial f}{\partial z}\right)$$

$$= f\,\mathbf{V}g + g\,\mathbf{V}f,$$

as desired.

## THE DIVERGENCE OF A VECTOR FIELD

Suppose now that

$$\mathbf{F}(x, y, z) = \mathbf{i}P(x, y, z) + \mathbf{j}Q(x, y, z) + \mathbf{k}R(x, y, z)$$

with differentiable component functions $P$, $Q$, and $R$. Then the **divergence** of $\mathbf{F}$ is the scalar function div $\mathbf{F}$ defined as follows:

$$\text{div } \mathbf{F} = \mathbf{V} \cdot \mathbf{F} = \frac{\partial P}{\partial x} + \frac{\partial Q}{\partial y} + \frac{\partial R}{\partial z}. \tag{9}$$

Of course *div* is an abbreviation for divergence, and the alternative notation $\mathbf{V} \cdot \mathbf{F}$ is consistent with the formal expression for $\mathbf{V}$ in Equation (6). That is,

$$\mathbf{V} \cdot \mathbf{F} = \left\langle \frac{\partial}{\partial x}, \frac{\partial}{\partial y}, \frac{\partial}{\partial z} \right\rangle \cdot \langle P, Q, R \rangle = \frac{\partial P}{\partial x} + \frac{\partial Q}{\partial y} + \frac{\partial R}{\partial z}.$$

In Section 17-6 we will see that if $\mathbf{v}$ is the velocity vector field of a steady fluid flow, then the value of div $\mathbf{v}$ at a point $(x, y, z)$ is essentially the net rate per unit volume at which fluid mass is flowing away (or "diverging") from the point $(x, y, z)$.

**EXAMPLE 5** If the vector field **F** is given by

$$\mathbf{F}(x, y, z) = (xe^y)\mathbf{i} + (z \sin y)\mathbf{j} + (xy \ln z)\mathbf{k},$$

then $P(x, y, z) = xe^y$, $Q(x, y, z) = z \sin y$, and $R(x, y, z) = xy \ln z$. Hence (9) yields

$$\text{div } \mathbf{F} = \frac{\partial}{\partial x}(xe^y) + \frac{\partial}{\partial y}(z \sin y) + \frac{\partial}{\partial z}(xy \ln z) = e^y + z \cos y + \frac{xy}{z}.$$

For instance, the value of div **F** at the point $(-3, 0, 2)$ is

$$\mathbf{V} \cdot \mathbf{F}(-3, 0, 2) = e^0 + 2 \cos 0 + 0 = 3.$$

---

The analogues for divergence of Equations (7) and (8) are the formulas

$$\mathbf{V} \cdot (a\mathbf{F} + b\mathbf{G}) = a\,\mathbf{V} \cdot \mathbf{F} + b\,\mathbf{V} \cdot \mathbf{G} \tag{10}$$

and

$$\mathbf{V} \cdot (f\mathbf{G}) = (f)(\mathbf{V} \cdot \mathbf{G}) + (\nabla f) \cdot \mathbf{G}; \tag{11}$$

we ask you to verify these in the problems. Note that the formula in (11)—where $f$ is a scalar function and **G** is a vector field—is consistent in that $f$ and $\mathbf{V} \cdot \mathbf{G}$ are scalar functions, while $\nabla f$ and **G** are vector fields, so the sum on the right-hand side is a scalar function.

### THE CURL OF A VECTOR FIELD

The **curl** of the vector field $\mathbf{F} = P\mathbf{i} + Q\mathbf{j} + R\mathbf{k}$ is the vector field curl **F** with the following definition:

$$\text{curl } \mathbf{F} = \mathbf{V} \times \mathbf{F} = \begin{vmatrix} \mathbf{i} & \mathbf{j} & \mathbf{k} \\ \dfrac{\partial}{\partial x} & \dfrac{\partial}{\partial y} & \dfrac{\partial}{\partial z} \\ P & Q & R \end{vmatrix}. \tag{12}$$

Evaluation of the formal determinant expression in (12) yields

$$\text{curl } \mathbf{F} = \mathbf{i}\left(\frac{\partial R}{\partial y} - \frac{\partial Q}{\partial z}\right) + \mathbf{j}\left(\frac{\partial P}{\partial z} - \frac{\partial R}{\partial x}\right) + \mathbf{k}\left(\frac{\partial Q}{\partial x} - \frac{\partial P}{\partial y}\right). \tag{13}$$

Although you may wish to memorize this complicated formula, we recommend—because you generally will find it simpler—that in practice you set up and evaluate directly the formal determinant in (12). Our next example shows how easy this is.

**EXAMPLE 6** For the vector field **F** of Example 5, the formula in (12) yields

$$\text{curl } \mathbf{F} = \begin{vmatrix} \mathbf{i} & \mathbf{j} & \mathbf{k} \\ \dfrac{\partial}{\partial x} & \dfrac{\partial}{\partial y} & \dfrac{\partial}{\partial z} \\ xe^y & z \sin y & xy \ln z \end{vmatrix}$$

$$= \mathbf{i}(x \ln z - \sin y) + \mathbf{j}(-y \ln z) + \mathbf{k}(-xe^y).$$

**908**

For instance, the value of curl $\mathbf{F}$ at the point $(3, \pi/2, e)$ is

$$\mathbf{V} \times \mathbf{F}\left(3, \frac{\pi}{2}, e\right) = 2\mathbf{i} - \frac{\pi}{2}\mathbf{j} - 3e^{\pi/2}\mathbf{k}.$$

In Section 17-7 we will see that if $\mathbf{v}$ is the velocity vector of a fluid flow, then the value of the vector curl $\mathbf{v}$ (where it is nonzero) at the point $(x, y, z)$ determines the axis through $(x, y, z)$ about which the fluid is rotating (or whirling or "curling").

The analogues for curl of Equations (10) and (11) are the formulas

$$\mathbf{V} \times (a\mathbf{F} + b\mathbf{G}) = a(\mathbf{V} \times \mathbf{F}) + b(\mathbf{V} \times \mathbf{G}) \tag{14}$$

and

$$\mathbf{V} \times (f\mathbf{G}) = (f)(\mathbf{V} \times \mathbf{G}) + (\nabla f) \times \mathbf{G} \tag{15}$$

that we ask you to verify in the problems.

**EXAMPLE 7** If the function $f(x, y, z)$ has continuous second order partial derivatives, show that

$$\operatorname{curl}(\operatorname{grad} f) = \mathbf{0}. \tag{16}$$

**Solution** Direct computation yields

$$\mathbf{V} \times \mathbf{V}f = \begin{vmatrix} \mathbf{i} & \mathbf{j} & \mathbf{k} \\ \dfrac{\partial}{\partial x} & \dfrac{\partial}{\partial y} & \dfrac{\partial}{\partial z} \\ \dfrac{\partial f}{\partial x} & \dfrac{\partial f}{\partial y} & \dfrac{\partial f}{\partial z} \end{vmatrix}$$

$$= \mathbf{i}\left(\frac{\partial^2 f}{\partial y\, \partial z} - \frac{\partial^2 f}{\partial z\, \partial y}\right) + \mathbf{j}\left(\frac{\partial^2 f}{\partial z\, \partial x} - \frac{\partial^2 f}{\partial x\, \partial z}\right) + \mathbf{k}\left(\frac{\partial^2 f}{\partial x\, \partial y} - \frac{\partial^2 f}{\partial y\, \partial x}\right);$$

therefore,

$$\mathbf{V} \times \mathbf{V}f = \mathbf{0}$$

because of the equality of continuous mixed second order partial derivatives.

In Section 17-2 we define line integrals, which are used (for example) to compute the work done by a force field in moving a particle along a curved path. In Section 17-5 we discuss surface integrals, which are used (for example) to compute the rate at which a fluid with a known velocity vector field is moving across a given surface. The three basic integral theorems of vector analysis—Green's theorem (Section 17-4), the divergence theorem (Section 17-6), and Stokes' theorem (Section 17-7)—play much the same role for line and surface integrals that the fundamental theorem of calculus plays for ordinary single-variable integrals. In advanced calculus you will see that all three of these theorems are special cases of a very general and yet surprisingly simple theorem.

In each of Problems 1–10, illustrate the given vector field $\mathbf{F}$ by sketching several typical vectors in the field.

1  $\mathbf{F}(x, y) = \mathbf{i} + \mathbf{j}$
2  $\mathbf{F}(x, y) = 3\mathbf{i} - 2\mathbf{j}$
3  $\mathbf{F}(x, y) = x\mathbf{i} - y\mathbf{j}$
4  $\mathbf{F}(x, y) = 2\mathbf{i} + x\mathbf{j}$
5  $\mathbf{F}(x, y) = (x^2 + y^2)^{1/2}(x\mathbf{i} + y\mathbf{j})$
6  $\mathbf{F}(x, y) = (x^2 + y^2)^{-1/2}(x\mathbf{i} + y\mathbf{j})$
7  $\mathbf{F}(x, y, z) = \mathbf{j} + \mathbf{k}$
8  $\mathbf{F}(x, y, z) = \mathbf{i} + \mathbf{j} - \mathbf{k}$
9  $\mathbf{F}(x, y, z) = -x\mathbf{i} - y\mathbf{j}$
10  $\mathbf{F}(x, y, z) = x\mathbf{i} + y\mathbf{j} + z\mathbf{k}$

In each of Problems 11–20, calculate the divergence and curl of the given vector field $\mathbf{F}$.

11  $\mathbf{F}(x, y, z) = x\mathbf{i} + y\mathbf{j} + z\mathbf{k}$
12  $\mathbf{F}(x, y, z) = 3x\mathbf{i} - 2y\mathbf{j} - 4z\mathbf{k}$
13  $\mathbf{F}(x, y, z) = yz\mathbf{i} + xz\mathbf{j} + xy\mathbf{k}$
14  $\mathbf{F}(x, y, z) = x^2\mathbf{i} + y^2\mathbf{j} + z^2\mathbf{k}$
15  $\mathbf{F}(x, y, z) = xy^2\mathbf{i} + yz^2\mathbf{j} + zx^2\mathbf{k}$
16  $\mathbf{F}(x, y, z) = (2x - y)\mathbf{i} + (3y - 2z)\mathbf{j} + (7z - 3x)\mathbf{k}$
17  $\mathbf{F}(x, y, z) = (y^2 + z^2)\mathbf{i} + (x^2 + z^2)\mathbf{j} + (x^2 + y^2)\mathbf{k}$
18  $\mathbf{F}(x, y, z) = (e^{xz}\sin y)\mathbf{j} + (e^{xy}\cos z)\mathbf{k}$
19  $\mathbf{F}(x, y, z) = (x + \sin yz)\mathbf{i} + (y + \sin xz)\mathbf{j} + (z + \sin xy)\mathbf{k}$
20  $\mathbf{F}(x, y, z) = (x^2 e^{-z})\mathbf{i} + (y^3\ln x)\mathbf{j} + (z \cosh y)\mathbf{k}$

Apply the definitions of gradient, divergence, and curl to establish the identities in Problems 21–27, where $a$ and $b$ denote constants, $f$ and $g$ denote differentiable scalar functions, and $\mathbf{F}$ and $\mathbf{G}$ denote differentiable vector fields.

21  $\nabla(af + bg) = a\,\nabla f + b\,\nabla g$
22  $\nabla \cdot (a\mathbf{F} + b\mathbf{G}) = a\,\nabla \cdot \mathbf{F} + b\,\nabla \cdot \mathbf{G}$
23  $\nabla \times (a\mathbf{F} + b\mathbf{G}) = a(\nabla \times \mathbf{F}) + b(\nabla \times \mathbf{G})$
24  $\nabla \cdot (f\mathbf{G}) = (f)(\nabla \cdot \mathbf{G}) + (\nabla f) \cdot \mathbf{G}$
25  $\nabla \times (f\mathbf{G}) = (f)(\nabla \times \mathbf{G}) + (\nabla f) \times \mathbf{G}$
26  $\nabla(f/g) = (g\,\nabla f - f\,\nabla g)/(g^2)$
27  $\nabla \cdot (\mathbf{F} \times \mathbf{G}) = \mathbf{G} \cdot (\nabla \times \mathbf{F}) - \mathbf{F} \cdot (\nabla \times \mathbf{G})$

Establish the identities in Problems 28–30 under the assumption that the scalar functions $f$ and $g$ and the vector field $\mathbf{F}$ are twice differentiable.

28  $\operatorname{div}(\operatorname{curl} \mathbf{F}) = 0$
29  $\operatorname{div}(\nabla\, fg) = f\,\operatorname{div}(\nabla g) + g\,\operatorname{div}(\nabla f) + 2(\nabla f) \cdot (\nabla g)$
30  $\operatorname{div}(\nabla f \times \nabla g) = 0$

Verify the identities in Problems 31–40, in which $\mathbf{a}$ is a constant vector, $\mathbf{r} = x\mathbf{i} + y\mathbf{j} + z\mathbf{k}$, and $r = |\mathbf{r}|$. Problems 33 and 34 imply that the divergence and curl of an inverse-square vector field both vanish identically.

31  $\nabla \cdot \mathbf{r} = 3$ and $\nabla \times \mathbf{r} = \mathbf{0}$
32  $\nabla \cdot (\mathbf{a} \times \mathbf{r}) = 0$ and $\nabla \times (\mathbf{a} \times \mathbf{r}) = 2\mathbf{a}$

33  $\nabla \cdot \dfrac{\mathbf{r}}{r^3} = 0$
34  $\nabla \times \dfrac{\mathbf{r}}{r^3} = \mathbf{0}$

35  $\nabla(r) = \dfrac{\mathbf{r}}{r}$
36  $\nabla\!\left(\dfrac{1}{r}\right) = -\dfrac{\mathbf{r}}{r^3}$

37  $\nabla \cdot (r\mathbf{r}) = 4r$
38  $\nabla \cdot (\nabla r) = 0$

39  $\nabla(\ln r) = \dfrac{\mathbf{r}}{r^2}$
40  $\nabla(r^{10}) = 10r^8\mathbf{r}$

---

## 17-2

### Line Integrals

17.4  A wire of variable density in the shape of the smooth curve $C$

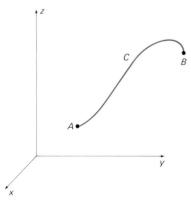

To motivate the definition of the line integral, we imagine a thin wire shaped like the smooth curve $C$ with end points $A$ and $B$, as in Fig. 17.4. Suppose that the wire has variable density given at the point $(x, y, z)$ by the known continuous function $f(x, y, z)$, in units such as grams per (linear) centimeter. Let

$$x = x(t), \quad y = y(t), \quad z = z(t), \qquad t \text{ in } [a, b], \tag{1}$$

be a smooth parametrization of the curve $C$, with $t = a$ corresponding to the initial point $A$ of the curve and $t = b$ to its terminal point $B$.

In order to *approximate* the total mass $M$ of our curved wire, we begin with a partition

$$a = t_0 < t_1 < t_2 < \cdots < t_{n-1} < t_n = b$$

of $[a, b]$ into $n$ equal subintervals, each with length $\Delta t = (b - a)/n$. These subdivision points of $[a, b]$ produce, via our parametrization, a physical subdivision of the wire into short curve segments, as shown in Fig. 17.5. We

CHAP. 17:  Vector Analysis

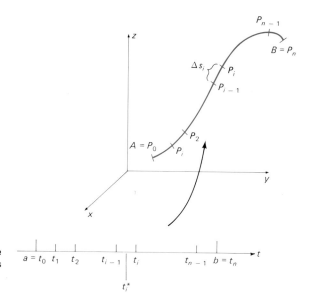

**17.5** The partition of the interval [a, b] determines a related partition of the curve C into short arcs.

let $P_i = (x(t_i), y(t_i), z(t_i))$ (for $i = 0, 1, 2, \ldots, n$) be the resulting subdivision points of C. Note that $A = P_0$ and that $B = P_n$.

From our study of arc length in Sections 13-2 and 14-4, we know that the arc length $\Delta s_i$ of the segment of C from $P_{i-1}$ to $P_i$ is

$$\Delta s_i = \int_{t_{i-1}}^{t_i} \sqrt{[x'(t)]^2 + [y'(t)]^2 + [z'(t)]^2} \, dt$$
$$= \sqrt{[x'(t_i^*)]^2 + [y'(t_i^*)]^2 + [z'(t_i^*)]^2} \, \Delta t \qquad (2)$$

for some number $t_i^*$ in the interval $[t_{i-1}, t_i]$—a consequence of the average value theorem for integrals in Section 5-5.

If we multiply the density at the point $(x_i^*, y_i^*, z_i^*)$ by the length $\Delta s_i$ of the segment of C containing it, we obtain an estimate of the mass of that segment of C. So, after we sum over all the segments, we have an estimate of the total mass of the wire:

$$M \approx \sum_{i=1}^{n} f(x(t_i^*), y(t_i^*), z(t_i^*)) \, \Delta s_i.$$

The limit of this sum as $\Delta t \to 0$ should be the actual mass M. This is our motivation for the definition of the line integral of the function $f$ along the curve C, denoted by $\int_C f(x, y, z) \, ds$.

---

*Definition*  *Line Integral with Respect to Arc Length*

Suppose that the function $f(x, y, z)$ is continuous at each point of the smooth parametric curve C from A to B, as given in (1). Then the **line integral of $f$ along C from A to B with respect to arc length** is defined to be

$$\int_C f(x, y, z) \, ds = \lim_{\Delta t \to 0} \sum_{i=1}^{n} f(x(t_i^*), y(t_i^*), z(t_i^*)) \, \Delta s_i. \qquad (3)$$

---

When we substitute (2) into (3), we recognize the result as the limit of a Riemann sum; therefore

$$\int_C f(x, y, z)\, ds = \int_a^b f(x(t), y(t), z(t))([x'(t)]^2 + [y'(t)]^2 + [z'(t)]^2)^{1/2}\, dt. \quad (4)$$

Thus we may evaluate the line integral $\int_C f(x, y, z)\, ds$ by expressing everything in terms of the parameter $t$, including the symbolic arc length element

$$ds = \sqrt{[x'(t)]^2 + [y'(t)]^2 + [z'(t)]^2}\, dt.$$

The result—the right-hand side of Equation (4)—is an **ordinary integral with respect to the single real variable** $t$.

A curve $C$ that lies in the $xy$-plane may be regarded as a space curve for which $z \equiv 0$. In this case we suppress the variable $z$ in (4) and write

$$\int_C f(x, y)\, ds = \int_a^b f(x(t), y(t))\sqrt{|x'(t)|^2 + |y'(t)|^2}\, dt. \quad (5)$$

**EXAMPLE 1**  Evaluate the line integral $\int_C xy\, ds$, where $C$ is the first-quadrant quarter-circle parametrized by $x = \cos t$, $y = \sin t$, $0 \le t \le \pi/2$.

*Solution*  Here

$$ds = \sqrt{(-\sin t)^2 + (\cos t)^2}\, dt = dt,$$

so the formula in (5) yields

$$\int_C xy\, ds = \int_0^{\pi/2} \cos t \sin t\, dt$$

$$= \left[ \frac{1}{2} \sin^2 t \right]_0^{\pi/2} = \frac{1}{2}.$$

Let us now return to our physical wire, and denote its density function by the more usual $\rho(x, y, z)$. The mass of a small piece of length $\Delta s$ is $\Delta m = \rho\, \Delta s$, so we write

$$dm = \rho(x, y, z)\, ds$$

for its (symbolic) element of mass. Then the **mass** $M$ of the wire and its **moments** $M_{yz}$, $M_{xz}$, and $M_{xy}$ about the respective coordinate planes are given by the line integrals

$$M = \int_C dm = \int_C \rho\, ds, \qquad M_{yz} = \int_C x\, dm,$$

$$M_{xz} = \int_C y\, dm, \qquad M_{xy} = \int_C z\, dm. \quad (6)$$

The **centroid** of the wire is the point $(\bar{x}, \bar{y}, \bar{z})$ where

$$\bar{x} = \frac{M_{yz}}{M}, \qquad \bar{y} = \frac{M_{xz}}{M}, \quad \text{and} \quad \bar{z} = \frac{M_{xy}}{M}.$$

Note the analogy with Equations (3) and (4) in Section 16-5. The **moment of inertia** of the wire about a given axis is

$$I = \int_C w^2\, dm, \quad (7)$$

where $w = w(x, y, z)$ denotes the perpendicular distance from the point $(x, y, z)$ to the axis in question.

**EXAMPLE 2** Find the centroid of a wire with density $\rho = kz$ if it has the shape of the helix $C$ with parametrization

$$x = 3 \cos t, \quad y = 3 \sin t, \quad z = 4t, \quad 0 \leq t \leq \pi.$$

**Solution** The mass element of the wire is

$$dm = \rho \, ds = kz \, ds$$
$$= 4kt \sqrt{(-3 \sin t)^2 + (3 \cos t)^2 + (4)^2} \, dt = 20kt \, dt.$$

Hence Formulas (6) yield

$$M = \int_C \rho \, ds = \int_0^\pi 20kt \, dt = 10k\pi^2;$$

$$M_{yz} = \int_C \rho x \, ds = \int_0^\pi 60kt \cos t \, dt$$
$$= 60k \Big[ \cos t + t \sin t \Big]_0^\pi = -120k;$$

$$M_{xz} = \int_C \rho y \, ds = \int_0^\pi 60kt \sin t \, dt$$
$$= 60k \Big[ \sin t - t \cos t \Big]_0^\pi = 60k\pi;$$

and

$$M_{xy} = \int_C \rho z \, ds = \int_0^\pi 80kt^2 \, dt = \frac{80k\pi^3}{3}.$$

Hence

$$\bar{x} = \frac{-120k}{10k\pi^2} \approx -1.22, \qquad \bar{y} = \frac{60k\pi}{10k\pi^2} \approx 1.91, \qquad \bar{z} = \frac{80k\pi^3}{3(10k\pi^2)} \approx 8.38.$$

So the centroid of the wire is approximately the point $(-1.22, 1.91, 8.38)$.

## LINE INTEGRALS WITH RESPECT TO COORDINATE VARIABLES

A different type of line integral is obtained by replacing $\Delta s_i$ in (3) by

$$\Delta x_i = x(t_i) - x(t_{i-1}) = x'(t_i^*) \, \Delta t.$$

The **line of integral $f$ along $C$ with respect to** $x$ is defined to be

$$\int_C f(x, y, z) \, dx = \lim_{\Delta t \to 0} \sum_{i=1}^n f(x(t_i^*), y(t_i^*), z(t_i^*)) \, \Delta x_i;$$

$$\int_C f(x, y, z) \, dx = \int_a^b f(x(t), y(t), z(t)) x'(t) \, dt. \qquad (8a)$$

Similarly, the line integrals of $f$ along $C$ **with respect to $y$** and **with respect to $z$** are given by

$$\int_C f(x, y, z) \, dy = \int_a^b f(x(t), y(t), z(t)) y'(t) \, dt \qquad (8b)$$

and

$$\int_C f(x, y, z) \, dz = \int_a^b f(x(t), y(t), z(t)) z'(t) \, dt. \qquad (8c)$$

The three integrals in (8) often occur together. If $P$, $Q$, and $R$ are continuous functions of the variables $x$, $y$, and $z$, then we write (indeed, *define*)

$$\int_C P\,dx + Q\,dy + R\,dz = \int_C P\,dx + \int_C Q\,dy + \int_C R\,dz. \qquad (9)$$

The line integrals in (8) and (9) are evaluated by expressing $x$, $y$, $z$, $dx$, $dy$, and $dz$ in terms of $t$ as determined by a suitable parametrization of the curve $C$; the result is an ordinary single-variable integral. For instance, if $C$ is a parametric plane curve parametrized over the interval $a \leqq t \leqq b$, then

$$\int_C P\,dx + Q\,dy = \int_a^b \left[ P(x(t), y(t))x'(t) + Q(x(t), y(t))y'(t) \right]\,dt.$$

**EXAMPLE 3**  Evaluate the line integral $\int_C y\,dx + z\,dy + x\,dz$ where $C$ is the parametric curve $x = t$, $y = t^2$, $z = t^3$, $0 \leqq t \leqq 1$.

*Solution*  Because $dx = dt$, $dy = 2t\,dt$, and $dz = 3t^2\,dt$, substitution in terms of $t$ yields

$$\int_C y\,dx + z\,dy + x\,dz = \int_0^1 t^2\,(dt) + t^3(2t\,dt) + t(3t^2\,dt)$$

$$= \int_0^1 (t^2 + 3t^3 + 2t^4)\,dt$$

$$= \left[ \frac{1}{3}t^3 + \frac{3}{4}t^4 + \frac{2}{5}t^5 \right]_0^1 = \frac{89}{60}.$$

There is an important difference between the line integral in (4) with respect to arc length $s$ and the line integral in (9) with respect to $x$, $y$, and $z$. Suppose that the *orientation* of the curve $C$ (the direction in which it is traced as $t$ increases) is reversed. Then, because of the terms $x'(t)$, $y'(t)$, and $z'(t)$ in (8), the *sign* of the line integral (9) is changed. But this reversal of orientation does *not* change the value of the line integral in (4). We express this by writing

$$\int_{-C} f\,ds = \int_C f\,ds, \qquad (10)$$

17.6  The three arcs of Example 4

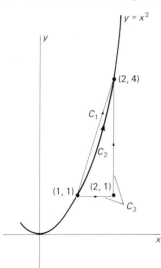

in contrast with the formula

$$\int_{-C} P\,dx + Q\,dy + R\,dz = -\int_C P\,dx + Q\,dy + R\,dz. \qquad (11)$$

Here the symbol $-C$ denotes the curve $C$ with its orientation reversed (from $B$ to $A$, rather than from $A$ to $B$). It is proved in advanced calculus that for either type of line integral, two one-to-one parametrizations of the curve $C$ that *agree in orientation* will give the same value.

If the curve $C$ consists of finitely many smooth curves joined at corner points, we say that $C$ is **piecewise smooth**. The value of a line integral along $C$ is then defined to be the sum of its values along the smooth segments of $C$.

**EXAMPLE 4**  Evaluate the line integral

$$\int_C y\,dx + 2x\,dy$$

for $C$ each of the three curves $C_1$, $C_2$, and $C_3$ from $A(1, 1)$ to $B(2, 4)$ shown in Fig. 17.6.

**Solution** The straight line segment $C_1$ from $A$ to $B$ can be parametrized by $x = 1 + t$, $y = 1 + 3t$, $0 \leq t \leq 1$. So

$$\int_{C_1} y \, dx + 2x \, dy = \int_0^1 (1 + 3t)(dt) + 2(1 + t)(3 \, dt)$$

$$= \int_0^1 (7 + 9t) \, dt = \tfrac{23}{2}.$$

Next, the arc $C_2$ of the parabola $y = x^2$ from $A$ to $B$ has the parametrization $x = x$, $y = x^2$, $1 \leq x \leq 2$, so

$$\int_{C_2} y \, dx + 2x \, dy = \int_1^2 (x^2)(dx) + 2(x)(2x \, dx) = \int_1^2 5x^2 \, dx = \tfrac{35}{3}.$$

Finally, along the straight line segment from $(1, 1)$ to $(2, 1)$ we have $y = 1$ and $dy = 0$. Along the vertical segment from $(2, 1)$ to $(2, 4)$ we have $x = 2$ and $dx = 0$. Therefore,

$$\int_{C_3} y \, dx + 2x \, dy = \int_1^2 [(1)(dx) + (2x)(0)] + \int_1^4 [(y)(0) + (4)(dy)]$$

$$= \int_1^2 1 \, dx + \int_1^4 4 \, dy = 13.$$

---

Example 4 also shows that we may obtain different values for the line integral from $A$ to $B$ if we evaluate it along different curves from $A$ to $B$. In Section 17-3 we give a sufficient condition for the line integral $\int_C P \, dx + Q \, dy + R \, dz$ to have the same value for *all* smooth curves $C$ from $A$ to $B$, and thus for the line integral to be *independent of path*.

### LINE INTEGRALS AND WORK

Suppose now that $\mathbf{F} = P\mathbf{i} + Q\mathbf{j} + R\mathbf{k}$ is a force field that is defined on a region containing the curve $C$ from $A$ to $B$. Suppose also that our parametrization

$$x = x(t), \qquad y = y(t), \qquad z = z(t), \qquad t \text{ in } [a, b]$$

of $C$ has a *nonzero* velocity vector

$$\mathbf{v} = \mathbf{i}\frac{dx}{dt} + \mathbf{j}\frac{dy}{dt} + \mathbf{k}\frac{dz}{dt}.$$

The speed associated with this velocity vector is

$$v = |\mathbf{v}| = \sqrt{\left(\frac{dx}{dt}\right)^2 + \left(\frac{dy}{dt}\right)^2 + \left(\frac{dz}{dt}\right)^2}.$$

Recall that the *unit tangent vector* to the curve $C$ is

$$\mathbf{T} = \frac{\mathbf{v}}{v} = \frac{1}{v}\left(\frac{dx}{dt}\mathbf{i} + \frac{dy}{dt}\mathbf{j} + \frac{dz}{dt}\mathbf{k}\right).$$

We want to approximate the work $W$ done by the force field $\mathbf{F}$ in moving a particle along the curve $C$ from $A$ to $B$. Subdivide $C$ as indicated in Fig. 17.7. Think of $\mathbf{F}$ moving the particle from $P_{i-1}$ to $P_i$, two consecutive division points of $C$. The work $\Delta W_i$ done is approximately the product of the distance $\Delta s_i$ from $P_{i-1}$ to $P_i$ (measured along $C$) and the tangential component $\mathbf{F} \cdot \mathbf{T}$ of the force $\mathbf{F}$ at a typical point $(x(t_i^*), y(t_i^*), z(t_i^*))$ between $P_{i-1}$

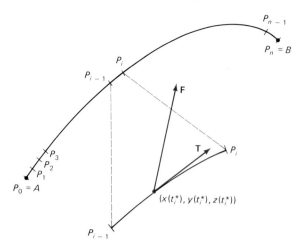

17.7 The component of **F** along $C$ from $P_{i-1}$ to $P_i$ is **F · T**.

and $P_i$. Thus

$$\Delta W_i \approx \mathbf{F}(x(t_i^*), y(t_i^*), z(t_i^*)) \cdot \mathbf{T}(t_i^*) \, \Delta s_i,$$

so the total work $W$ is given approximately by

$$W \approx \sum_{i=1}^{n} \mathbf{F}(x(t_i^*), y(t_i^*), z(t_i^*)) \cdot \mathbf{T}(t_i^*) \, \Delta s_i.$$

This approximation suggests that we define the **work** $W$ as

$$W = \int_C \mathbf{F} \cdot \mathbf{T} \, ds. \tag{12}$$

Thus *work is the integral with respect to arc length of the tangential component of the force.* Intuitively, we may regard $dW = \mathbf{F} \cdot \mathbf{T} \, ds$ as the infinitesimal element of work done by the tangential component $\mathbf{F} \cdot \mathbf{T}$ of the force in moving the particle along the arc length element $ds$. The line integral in (12) is then the "sum" of all these infinitesimal elements of work.

It is customary to write formally

$$\mathbf{r} = x\mathbf{i} + y\mathbf{j} + z\mathbf{k}, \qquad d\mathbf{r} = \mathbf{i} \, dx + \mathbf{j} \, dy + \mathbf{k} \, dz,$$

and

$$\mathbf{T} \, ds = \left( \frac{dx}{ds} \mathbf{i} + \frac{dy}{ds} \mathbf{j} + \frac{dz}{ds} \mathbf{k} \right) ds = d\mathbf{r}.$$

With this notation, (12) takes the form

$$W = \int_C \mathbf{F} \cdot d\mathbf{r} \tag{13}$$

that is common in physics and engineering texts.

To evaluate the line integral $\int_C \mathbf{F} \cdot \mathbf{T} \, ds$, we express its integrand in terms of the parameter $t$, as usual. Thus

$$
\begin{aligned}
W &= \int_C \mathbf{F} \cdot \mathbf{T} \, ds = \int_a^b (P\mathbf{i} + Q\mathbf{j} + R\mathbf{k}) \cdot \frac{1}{v} \left( \frac{dx}{dt} \mathbf{i} + \frac{dy}{dt} \mathbf{j} + \frac{dz}{dt} \mathbf{k} \right) v \, dt \\
&= \int_a^b \left( P \frac{dx}{dt} + Q \frac{dy}{dt} + R \frac{dz}{dt} \right) dt
\end{aligned}
$$

and so

$$W = \int_C P\,dx + Q\,dy + R\,dz. \qquad (14)$$

This computation yields an important relation between the two types of line integrals we have defined in this section.

---

**Theorem**  *Equivalent Line Integrals*

Suppose that the vector field $\mathbf{F} = P\mathbf{i} + Q\mathbf{j} + R\mathbf{k}$ has continuous component functions and that $\mathbf{T}$ is the unit tangent to the smooth curve $C$. Then

$$\int_C \mathbf{F} \cdot \mathbf{T}\,ds = \int_C P\,dx + Q\,dy + R\,dz. \qquad (15)$$

---

Note that, if the orientation of the curve $C$ is reversed, then the sign of the right-hand integral is changed according to Equation (11), while the sign of the left-hand integral is changed because $\mathbf{T}$ is replaced by $-\mathbf{T}$.

**EXAMPLE 5**  The work done by the force field $F = y\mathbf{i} + z\mathbf{j} + x\mathbf{k}$ in moving a particle from $(0, 0, 0)$ to $(1, 1, 1)$ along the twisted cubic $x = t,\ y = t^2,\ z = t^3$ is given by the line integral

$$W = \int_C \mathbf{F} \cdot \mathbf{T}\,ds = \int_C y\,dx + z\,dy + x\,dz,$$

and we computed the value of this integral in Example 3. Hence $W = \frac{89}{60}$.

**EXAMPLE 6**  Find the work done by the inverse-square law force field

$$\mathbf{F}(x, y, z) = \frac{k\mathbf{r}}{r^3} = \frac{k(x\mathbf{i} + y\mathbf{j} + z\mathbf{k})}{(x^2 + y^2 + z^2)^{3/2}}$$

in moving a particle along the straight line segment $C$ from $(0, 4, 0)$ to $(0, 4, 3)$.

***Solution***  Along $C$ we have $x = 0,\ y = 4$, and $z = z$. We choose $z$ as the parameter. Since $dx = 0 = dy$, Formula (14) gives

$$W = \int_C \frac{k(x\,dx + y\,dy + z\,dz)}{(x^2 + y^2 + z^2)^{3/2}}$$

$$= \int_0^3 \frac{kz\,dz}{(16 + z^2)^{3/2}} = \left[\frac{-k}{(16 + z^2)^{1/2}}\right]_0^3 = \frac{k}{20}.$$

## 17-2  PROBLEMS

In Problems 1–5, evaluate the line integrals $\int_C f(x, y)\,ds$, $\int_C f(x, y)\,dx$, and $\int_C f(x, y)\,dy$ along the indicated parametric curve.

**1** $f(x, y) = x^2 + y^2;\quad x = 4t - 1,\ y = 3t + 1$, $-1 \le t \le 1$

**2** $f(x, y) = x;\quad x = t,\ y = t^2,\ 0 \le t \le 1$

**3** $f(x, y) = x + y;\quad x = e^t + 1,\ y = e^t - 1$, $0 \le t \le \ln 2$

**4** $f(x, y) = 2x - y;\quad x = \sin t,\ y = \cos t$, $0 \le t \le \pi/2$

**5** $f(x, y) = xy;\quad x = 3t,\ y = t^4,\ 0 \le t \le 1$

**6** Evaluate $\int_C xy\,dx + (x + y)\,dy$ where $C$ is the part of the graph of $y = x^2$ from $(-1, 1)$ to $(2, 4)$.

**7** Evaluate $\int_C y^2\,dx + x\,dy$ where $C$ is the part of the graph of $x = y^3$ from $(-1, -1)$ to $(1, 1)$.

**8** Evaluate $\int_C y\sqrt{x}\,dx + x\sqrt{x}\,dy$ where $C$ is the part of the graph of $y^2 = x^3$ from $(1, 1)$ to $(4, 8)$.

9 Evaluate the line integral $\int_C x^2 y\, dx + xy^3\, dy$ where $C$ consists of the line segments from $(-1, 1)$ to $(2, 1)$ and from $(2, 1)$ to $(2, 5)$.

10 Evaluate $\int_C (x + 2y)\, dx + (2x - y)\, dy$ where $C$ consists of the line segments from $(3, 2)$ to $(3, -1)$ and from $(3, -1)$ to $(-2, -1)$.

In Problems 11–15, evaluate the line integral $\int_C \mathbf{F} \cdot \mathbf{T}\, ds$ along the indicated curve $C$.

11 $\mathbf{F} = z\mathbf{i} + x\mathbf{j} - y\mathbf{k}$; $x = t$, $y = t^2$, $z = t^3$, $t$ in $[0, 1]$.

12 $\mathbf{F} = yz\mathbf{i} + xz\mathbf{j} + xy\mathbf{k}$; $C$ is the straight line segment from $(2, -1, 3)$ to $(4, 2, -1)$.

13 $\mathbf{F} = y\mathbf{i} - x\mathbf{j} + z\mathbf{k}$; $x = \sin t$, $y = \cos t$, $z = 2t$, $0 \leq t \leq \pi$.

14 $\mathbf{F} = (2x + 3y)\mathbf{i} + (3x + 2y)\mathbf{j} + 3z^2\mathbf{k}$; $C$ is the path from $(0, 0, 0)$ to $(4, 2, 3)$ that consists of three line segments parallel to the $x$-axis, the $y$-axis, and the $z$-axis, in that order.

15 $\mathbf{F} = yz^2\mathbf{i} + xz^2\mathbf{j} + 2xyz\mathbf{k}$; $C$ is the path from $(-1, 2, -2)$ to $(1, 5, 2)$ consisting of three line segments parallel to the $z$-axis, the $x$-axis, and the $y$-axis, in that order.

16 Find $\int_C xyz\, ds$ if $C$ is the line segment from $(1, -1, 2)$ to $(3, 2, 5)$.

17 Find $\int_C (2x + 9xy)\, ds$ given that $C$ is the curve $x = t$, $y = t^2$, $z = t^3$, $0 \leq t \leq 1$.

18 Evaluate $\int_C xy\, ds$ where $C$ is the elliptical helix $x = 4\cos t$, $y = 9\sin t$, $z = 7t$, $0 \leq t \leq 5\pi/2$.

19 Find the centroid of a uniform thin wire shaped like the semicircle $x^2 + y^2 = a^2$, $y \geq 0$.

20 Find the moments of inertia about the $x$- and $y$-axes of the wire of Problem 19.

21 Find the mass and centroid of a wire with constant density $\rho = k$ and shaped like the helix $x = 3\cos t$, $y = 3\sin t$, $z = 4t$, $t$ in $[0, 2\pi]$.

22 Find the moment of inertia about the $z$-axis of the wire of Example 1 of this section.

23 A wire shaped like the first-quadrant portion of the circle $x^2 + y^2 = a^2$ has density $\rho = kxy$ at the point $(x, y)$. Find its mass, its centroid, and its moment of inertia about each coordinate axis.

24 Find the work done by the inverse-square force field of Example 6 in moving a particle from $(1, 0, 0)$ to $(0, 3, 4)$. Integrate first along the line segment from $(1, 0, 0)$ to $(5, 0, 0)$ and then along a path on the sphere $x^2 + y^2 + z^2 = 25$, and note that the second integral is automatically zero (why?).

25 Imagine an infinitely long and uniformly charged wire that coincides with the $z$-axis. The electric force that it exerts on a unit charge at the point $(x, y) \neq (0, 0)$ in the $xy$-plane is $\mathbf{F} = k(x\mathbf{i} + y\mathbf{j})/(x^2 + y^2)$. Find the work done by $\mathbf{F}$ in moving a unit charge along the straight line segment from:
(a) $(1, 0)$ to $(1, 1)$;
(b) $(1, 1)$ to $(0, 1)$.

26 Show that if $\mathbf{F}$ is a *constant* force field, then it does zero work on a particle that moves once uniformly counterclockwise around the unit circle in the $xy$-plane.

27 Show that if $\mathbf{F} = k\mathbf{r} = k(x\mathbf{i} + y\mathbf{j})$, then $\mathbf{F}$ does zero work on a particle that moves once uniformly counterclockwise around the unit circle in the $xy$-plane.

28 Find the work done by the force field $\mathbf{F} = -y\mathbf{i} + x\mathbf{j}$ in moving a particle counterclockwise once around the unit circle in the $xy$-plane.

29 Let $C$ be a curve on the unit sphere $x^2 + y^2 + z^2 = 1$. Explain why the inverse-square force field of Example 6 does zero work in moving a particle along $C$.

In each of Problems 30–32, the given curve $C$ joins the points $P$ and $Q$ in the $xy$-plane. The point $P$ represents the top of a ten-story building and $Q$ is a point on the ground 100 ft from the base of the building. A 150-lb person slides down a frictionless slide shaped like the curve $C$ from $P$ to $Q$ under the influence of gravitational force $\mathbf{F} = -150\mathbf{j}$. In each problem show that $\mathbf{F}$ does the same amount of work on the person, $W = 15{,}000$ ft-lb, as though he or she dropped straight down to the ground.

30 $C$ is the straight line segment $y = x$ from $P(100, 100)$ to $Q(0, 0)$.

31 $C$ is the circular arc $x = 100\sin t$, $y = 100\cos t$ from $P(0, 100)$ to $Q(100, 0)$.

32 $C$ is the parabolic arc $y = \frac{1}{100}x^2$ from $P(100, 100)$ to $Q(0, 0)$.

---

## 17-3

## Independence of Path

Let $\mathbf{F} = P\mathbf{i} + Q\mathbf{j} + R\mathbf{k}$ be a vector field with continuous component functions. By the theorem in Section 17-2, we know that

$$\int_C \mathbf{F} \cdot \mathbf{T}\, ds = \int_C P\, dx + Q\, dy + R\, dz \tag{1}$$

for any piecewise smooth curve $C$. Thus the two sides of Equation (1) are two different ways of writing the same line integral. In this section we discuss the question whether this line integral has the *same value* for any *two* curves with the same end points (the same initial point and the same terminal point).

> **Definition   Independence of Path**
> The line integral in (1) is said to be **independent of the path in the region**
> $D$ provided that given any two points $A$ and $B$ of $D$, the integral has
> the same value along every piecewise smooth curve or **path** in $D$ from
> $A$ to $B$. In this case, we may write
>
> $$\int_C \mathbf{F} \cdot \mathbf{T} \, ds = \int_A^B \mathbf{F} \cdot \mathbf{T} \, ds \qquad (2)$$
>
> because the value of the integral depends only on the points $A$ and $B$
> and not on the particular choice of the path $C$ joining them.

For a tangible interpretation of independence of path, let us think of
walking along the curve $C$ from point $A$ to point $B$ in the plane where a
wind with velocity vector field $\mathbf{w}(x, y)$ is blowing. Suppose that when we are
at $(x, y)$ the wind exerts a force $\mathbf{F} = k\mathbf{w}(x, y)$ on us, $k$ being a constant that
depends on our size and shape (and perhaps on other factors as well). Then,
by the formula in Equation (12) of Section 17-2, the amount of work the
wind does on us as we walk along $C$ is given by

$$W = \int_C \mathbf{F} \cdot \mathbf{T} \, ds = k \int_C \mathbf{w} \cdot \mathbf{T} \, ds. \qquad (3)$$

This is the wind's contribution to our trip from $A$ to $B$. In this context, the
question of independence of path is the question whether the wind's work
$W$ depends on *which* path from point $A$ to point $B$ we choose.

**EXAMPLE 1**   Suppose that there is a steady wind blowing toward the
northeast with $\mathbf{w} = 10\mathbf{i} + 10\mathbf{j}$ in fps units, so its speed is $|\mathbf{w}| = 10\sqrt{2} \approx$
14 ft/s—about 10 mi/h. Assume that $k = \frac{1}{2}$, so the wind exerts $\frac{1}{2}$ lb of force
for each foot per second of its velocity. Then $\mathbf{F} = 5\mathbf{i} + 5\mathbf{j}$, so the formula
in (3) yields

$$W = \int_C \langle 5, 5 \rangle \cdot \mathbf{T} \, ds = \int_C 5 \, dx + 5 \, dy \qquad (4)$$

for the work done on us by the wind as we walk along $C$.

For instance, if $C$ is the straight line path $x = 10t$, $y = 10t$, $0 \le t \le 1$
from $(0, 0)$ to $(10, 10)$, then (4) gives

$$W = \int_0^1 (5)(10 \, dt) + (5)(10 \, dt) = 100 \int_0^1 1 \, dt = 100$$

ft-lb of work.

On the other hand, if $C$ is the parabolic path $y = \frac{1}{10}x^2$, $0 \le x \le 10$
from the same initial point $(0, 0)$ to the same terminal point $(10, 10)$, then
(4) yields

$$W = \int_0^{10} (5)(dx) + (5)\left( \frac{1}{5} x \, dx \right)$$

$$= \int_0^{10} (5 + x) \, dx = \left[ 5x + \frac{1}{2} x^2 \right]_0^{10} = 100$$

ft-lb of work, the same as before. Indeed, it follows from Theorem 1 below
that the line integral in (4) is independent of path, so the wind does 100 ft-lb
of work along any path from $(0, 0)$ to $(10, 10)$.

**EXAMPLE 2** Now suppose that $\mathbf{w} = -2y\mathbf{i} + 2x\mathbf{j}$. This wind is blowing counterclockwise around the origin, as in a hurricane with its eye at the origin. With $k = \frac{1}{2}$ as before, $\mathbf{F} = -y\mathbf{i} + x\mathbf{j}$, so the work integral is

$$W = \int_C \mathbf{F} \cdot \mathbf{T}\, ds = \int_C -y\, dx + x\, dy. \qquad (5)$$

If we walk from $(10, 0)$ to $(-10, 0)$ along the straight path $C_1$ (through the eye of the hurricane!), then—as illustrated in Fig. 17.8—the wind is always perpendicular to our unit tangent vector $\mathbf{T}$. Hence $\mathbf{F} \cdot \mathbf{T} = 0$, and therefore

$$W = \int_{C_1} \mathbf{F} \cdot \mathbf{T}\, ds = \int_{C_1} -y\, dx + x\, dy = 0.$$

On the other hand, if we walk along the semicircular path $C_2$ shown in Fig. 17.8, then $\mathbf{w}$ remains tangent to our path, so $\mathbf{F} \cdot \mathbf{T} = |\mathbf{F}| = 10$ at each point. Hence

$$W = \int_{C_2} -y\, dx + x\, dy$$

$$= \int_{C_2} \mathbf{F} \cdot \mathbf{T}\, ds = (10)(10\pi) = 100\pi.$$

The fact that we get different values along different paths from $(10,0)$ to $(-10, 0)$ shows that the line integral in (5) is *not* independent of path.

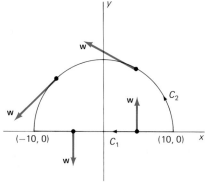

**17.8** Around and through the eye of the hurricane.

The following theorem tells when a given line integral is independent of path and when it is not.

---

> **Theorem 1** *Independence of Path*
>
> The line integral $\int_C \mathbf{F} \cdot \mathbf{T}\, ds$ is independent of path in the plane region $D$ if and only if $\mathbf{F} = \nabla f$ for some function $f$ defined on $D$.

---

***Proof*** Suppose first that $\mathbf{F} = \nabla f = \langle \partial f/\partial x, \partial f/\partial y, \partial f/\partial z \rangle$ and that $C$ is a path from $A$ to $B$ parametrized as usual with parameter $t$ in $[a, b]$. Then, by Equation (1),

$$\int_C \mathbf{F} \cdot \mathbf{T}\, ds = \int_C \frac{\partial f}{\partial x}\, dx + \frac{\partial f}{\partial y}\, dy + \frac{\partial f}{\partial z}\, dz$$

$$= \int_a^b \left( \frac{\partial f}{\partial x} \frac{dx}{dt} + \frac{\partial f}{\partial y} \frac{dy}{dt} + \frac{\partial f}{\partial z} \frac{dz}{dt} \right) dt$$

$$= \int_a^b D_t(f(x(t), y(t), z(t)))\, dt$$

$$= f(x(b), y(b), z(b)) - f(x(a), y(a), z(a)).$$

And so

$$\int_C \mathbf{F} \cdot \mathbf{T}\, ds = f(B) - f(A). \qquad (6)$$

The last step follows from the fundamental theorem of calculus. Equation (6) shows that the value of the line integral depends only upon the points $A$ and $B$ and is therefore independent of the choice of the particular path $C$. This proves the *if* part of Theorem 1.

**920**

To outline the *only if* part, we suppose that the line integral is independent of path in $D$. Choose a *fixed* point $A_0 = A_0(x_0, y_0, z_0)$ in $D$, and let $B = B(x, y, z)$ be an arbitrary point in $D$. Given any path $C$ from $A_0$ to $B$ in $D$, we *define* the function $f$ by means of the equation

$$f(x, y, z) = \int_C \mathbf{F} \cdot \mathbf{T} \, ds = \int_{(x_0, y_0, z_0)}^{(x, y, z)} \mathbf{F} \cdot \mathbf{T} \, ds. \tag{7}$$

Because of our hypothesis of independence of path, the resulting value of $f(x, y, z)$ depends only upon $(x, y, z)$ and not upon the particular path $C$ used. We shall omit the verification that $\nabla f = \mathbf{F}$. (See Problem 19 in Section 17-7.) ∎

As an application of Theorem 1, we see that the vector field $\mathbf{F} = -y\mathbf{i} + x\mathbf{j}$ of Example 2 is not the gradient of any scalar function $f$ because $\mathbf{F}$ is not independent of path. More precisely, $\mathbf{F}$ is not independent of path in any region that either includes or encloses the origin.

---

**Definition**  *Conservative Fields and Potential Functions*

The vector field $\mathbf{F}$ defined on a region $D$ is called **conservative** provided that there exists a scalar function $f$ defined on $D$ such that

$$\mathbf{F} = \nabla f \tag{8}$$

at each point of $D$. In this case $f$ is called a **potential function** for the vector field $\mathbf{F}$.*

---

Note that the formula in (6) has the form of a "fundamental theorem of calculus" for line integrals, with the potential function $f$ playing the role of an antiderivative. If the line integral $\int_C \mathbf{F} \cdot \mathbf{T} \, ds$ is known to be independent of path, then Theorem 1 guarantees that the vector field $\mathbf{F}$ is conservative and that the formula in (7) yields a potential function for $\mathbf{F}$.

**EXAMPLE 3**    Find a potential function for the conservative vector field

$$\mathbf{F}(x, y) = (6xy - y^3)\mathbf{i} + (4y + 3x^2 - 3xy^2)\mathbf{j}. \tag{9}$$

*Solution*    Because we are given the information that $\mathbf{F}$ is a conservative field, the line integral $\int_C \mathbf{F} \cdot \mathbf{T} \, ds$ is independent of path by Theorem 1. Therefore, we may apply the formula in Equation (7) to find a scalar potential function for $\mathbf{F}$. Let $C$ be the straight line path from $A(0, 0)$ to $B(x_1, y_1)$ parametrized by $x = x_1 t$, $y = y_1 t$, $0 \leq t \leq 1$. Then Equation (7) yields

$$f(x_1, y_1) = \int_A^B \mathbf{F} \cdot \mathbf{T} \, ds$$

$$= \int_A^B (6xy - y^3) \, dx + (4y + 3x^2 - 3xy^2) \, dy$$

$$= \int_0^1 (6x_1 y_1 t^2 - y_1^3 t^3)(x_1 \, dt) + (4y_1 t + 3x_1^2 t^2 - 3x_1 y_1^2 t^3)(y_1 \, dt)$$

$$= \int_0^1 (4y_1^2 t + 9x_1^2 y_1 t^2 - 4x_1 y_1^3 t^3) \, dt$$

$$= \left[ 2y_1^2 t^2 + 3x_1^2 y_1 t^3 - x_1 y_1^3 t^4 \right]_0^1 = 2y_1^2 + 3x_1^2 y_1 - x_1 y_1^3.$$

---

* In some physical applications the scalar function $f$ is called a *potential function* for the vector field $\mathbf{F}$ provided that $\mathbf{F} = -\nabla f$.

We may, of course, delete the subscripts because $(x_1, y_1)$ is an arbitrary point of the plane. Thus we obtain the potential function

$$f(x, y) = 2y^2 + 3x^2y - xy^3$$

for the vector field $\mathbf{F}$ in (9). As a check, we can differentiate $f$ to obtain

$$\frac{\partial f}{\partial x} = 6xy - y^3, \qquad \frac{\partial f}{\partial y} = 4y + 3x^2 - 3xy^2.$$

---

But how did we know in advance that the vector field $\mathbf{F}$ in Example 3 was conservative? The answer is provided by the following theorem, which is proved in Section 1.7 of Edwards and Penney, *Elementary Differential Equations with Applications* (Englewood Cliffs, N.J.: Prentice-Hall, 1985).

---

*Theorem 2*    *Conservative Fields and Potential Functions*

Suppose that the functions $P(x, y)$ and $Q(x, y)$ are continuous and have continuous first order partial derivatives in the open rectangle $R = \{(x, y) | a < x < b, c < y < d\}$. Then the vector field $\mathbf{F} = P\mathbf{i} + Q\mathbf{j}$ is conservative in $R$—and hence has a potential function $f(x, y)$ defined on $R$—if and only if

$$\frac{\partial P}{\partial y} = \frac{\partial Q}{\partial x} \tag{10}$$

at each point of $R$.

---

Observe that the vector field $\mathbf{F}$ in (9), where $P(x, y) = 6xy - y^3$ and $Q(x, y) = 4y + 3x^2 - 3xy^2$, satisfies the criterion in (10) because

$$\frac{\partial P}{\partial y} = 6x - 3y^2 = \frac{\partial Q}{\partial x}.$$

When this sufficient condition for the existence of a potential function is satisfied, the method illustrated in the following example is usually an easier way to find a potential function than the evaluation of the line integral in (7)—the method we used in Example 3.

**EXAMPLE 4**   Given

$$P(x, y) = 6xy - y^3, \qquad Q(x, y) = 4y + 3x^2 - 3xy^2$$

satisfying the condition $\partial P/\partial y = \partial Q/\partial x$, find a potential function $f(x, y)$ such that

$$\frac{\partial f}{\partial x} = 6xy - y^3 \quad \text{and} \quad \frac{\partial f}{\partial y} = 4y + 3x^2 - 3xy^2. \tag{11}$$

*Solution*   Upon integrating the first of these two equations with respect to $x$, we get

$$f(x, y) = 3x^2y - xy^3 + \xi(y), \tag{12}$$

where $\xi(y)$ is an "arbitrary function" of $y$ alone—it acts as a "constant of integration" with respect to $x$ because its derivative with respect to $x$ is zero. We next determine $\xi(y)$ by imposing the second condition in (11):

$$\frac{\partial f}{\partial y} = 3x^2 - 3xy^2 + \xi'(y) = 4y + 3x^2 - 3xy^2.$$

It follows that $\xi'(y) = 4y$, so $\xi(y) = 2y^2 + C$. When we set $C = 0$ and substitute the result in (12), we get the same potential function

$$f(x, y) = 3x^2y - xy^3 + 2y^2$$

that we found by entirely different methods in Example 3.

### CONSERVATIVE FORCE FIELDS

Given a conservative force field $\mathbf{F}$, it is customary in physics to introduce a minus sign and write $\mathbf{F} = -\nabla V$. Then $V(x, y, z)$ is called the **potential energy** at the point $(x, y, z)$. With $f = -V$ in Equation (6), we have

$$W = \int_A^B \mathbf{F} \cdot \mathbf{T}\, ds = V(A) - V(B), \tag{13}$$

and this means that the work done by $\mathbf{F}$ in moving a particle from $A$ to $B$ is equal to the *decrease* in potential energy.

For example, a brief computation shows that

$$\nabla\left(-\frac{k}{\sqrt{x^2 + y^2 + z^2}}\right) = \frac{k(x\mathbf{i} + y\mathbf{j} + z\mathbf{k})}{(x^2 + y^2 + z^2)^{3/2}} = \mathbf{F}$$

for the inverse-square force field of Example 6 in Section 17-2. With $V = k/(x^2 + y^2 + z^2)^{1/2}$, Equation (16) then gives

$$\int_{(0,4,0)}^{(0,4,3)} \mathbf{F} \cdot \mathbf{T}\, ds = \frac{k}{(0^2 + 4^2 + 0^2)^{1/2}} - \frac{k}{(0^2 + 4^2 + 3^2)^{1/2}} = \frac{k}{20},$$

as we also found by direct integration.

Here is the reason why the expression *conservative field* is used. Suppose that a particle of mass $m$ moves from $A$ to $B$ under the influence of the conservative force $\mathbf{F}$, with position vector $\mathbf{r}(t)$, $a \leq t \leq b$. Then Newton's law $\mathbf{F}(\mathbf{r}(t)) = m\mathbf{r}''(t) = m\mathbf{v}'(t)$ gives

$$\int_A^B \mathbf{F} \cdot \mathbf{T}\, ds = \int_a^b m\mathbf{v}'(t) \cdot \frac{\mathbf{v}(t)}{v}\, v\, dt$$

$$= \int_a^b mD_t[\tfrac{1}{2}\mathbf{v}(t) \cdot \mathbf{v}(t)]\, dt = \left[\frac{1}{2}m[\mathbf{v}(t)]^2\right]_a^b.$$

Thus with the abbreviations $v_A$ for $v(a)$ and $v_B$ for $v(b)$, we see that

$$\int_A^B \mathbf{F} \cdot \mathbf{T}\, ds = \frac{1}{2}m(v_B)^2 - \frac{1}{2}m(v_A)^2. \tag{14}$$

By equating the right-hand sides in Equations (13) and (14), we get the formula

$$\tfrac{1}{2}m(v_A)^2 + V(A) = \tfrac{1}{2}m(v_B)^2 + V(B). \tag{15}$$

This is the law of **conservation of mechanical energy** for a particle moving under the influence of a *conservative* force field—the sum of its kinetic energy and potential energy remains constant.

Apply the method of Example 4 to find a potential function for each of the vector fields in Problems 1–12.

**1** $\mathbf{F}(x, y) = (2x + 3y)\mathbf{i} + (3x + 2y)\mathbf{j}$

**2** $\mathbf{F}(x, y) = (4x - y)\mathbf{i} + (6y - x)\mathbf{j}$

**3** $\mathbf{F}(x, y) = (3x^2 + 2y^2)\mathbf{i} + (4xy + 6y^2)\mathbf{j}$

**4** $\mathbf{F}(x, y) = (2xy^2 + 3x^2)\mathbf{i} + (2x^2y + 4y^3)\mathbf{j}$

**5** $\mathbf{F}(x, y) = \left(x^3 + \dfrac{y}{x}\right)\mathbf{i} + (y^2 \ln x)\mathbf{j}$

**6** $\mathbf{F}(x, y) = (1 + ye^{xy})\mathbf{i} + (2y + xe^{xy})\mathbf{j}$

**7** $\mathbf{F}(x, y) = (\cos x + \ln y)\mathbf{i} + \left(\dfrac{x}{y} + e^y\right)\mathbf{j}$

**8** $\mathbf{F}(x, y) = (x + \tan^{-1}y)\mathbf{i} + \dfrac{x + y}{1 + y^2}\mathbf{j}$

**9** $\mathbf{F}(x, y) = (3x^2y^3 + y^4)\mathbf{i} + (3x^3y^2 + y^4 + 4xy^3)\mathbf{j}$

**10** $\mathbf{F}(x, y) = (e^x \sin y + \tan y)\mathbf{i} + (e^x \cos y + x \sec^2 y)\mathbf{j}$

**11** $\mathbf{F}(x, y) = \left(\dfrac{2x}{y} - \dfrac{3y^2}{x^4}\right)\mathbf{i} + \left(\dfrac{2y}{x^3} - \dfrac{x^2}{y^2} + \dfrac{1}{\sqrt{y}}\right)\mathbf{j}$

**12** $\mathbf{F}(x, y) = \dfrac{2x^{5/2} - 3y^{5/3}}{2x^{5/2}y^{2/3}}\mathbf{i} + \dfrac{3y^{5/3} - 2x^{5/2}}{3x^{3/2}y^{5/3}}\mathbf{j}$

In each of Problems 13–16, apply the method of Example 3 to find a potential function for the indicated vector field.

**13** The vector field of Problem 3

**14** The vector field of Problem 4

**15** The vector field of Problem 9

**16** The vector field of Problem 6

In each of Problems 17–22, show that the given line integral is independent of path in the entire $xy$-plane, and then calculate the value of the line integral.

**17** $\displaystyle\int_{(0,0)}^{(1,2)} (y^2 + 2xy)\,dx + (x^2 + 2xy)\,dy$

**18** $\displaystyle\int_{(0,0)}^{(1,1)} (2x - 3y)\,dx + (2y - 3x)\,dy$

**19** $\displaystyle\int_{(0,0)}^{(1,-1)} 2xe^y\,dx + x^2e^y\,dy$

**20** $\displaystyle\int_{(0,0)}^{(2,\pi)} \cos y\,dx - x \sin y\,dy$

**21** $\displaystyle\int_{(\pi/2,\pi/2)}^{(\pi,\pi)} (\sin y + y \cos x)\,dx + (\sin x + x \cos y)\,dy$

**22** $\displaystyle\int_{(0,0)}^{(1,-1)} (e^y + ye^x)\,dx + (e^x + xe^y)\,dy$

Find a potential function for each of the conservative vector fields in Problems 23–25.

**23** $\mathbf{F}(x, y, z) = yz\mathbf{i} + xz\mathbf{j} + xy\mathbf{k}$

**24** $\mathbf{F}(x, y, z) = (2x - y - z)\mathbf{i} + (2y - x)\mathbf{j} + (2z - x)\mathbf{k}$

**25** $\mathbf{F}(x, y, z) = (y \cos z - yze^x)\mathbf{i} + (x \cos z - ze^x)\mathbf{j} - (xy \sin z + ye^x)\mathbf{k}$

**26** Let $\mathbf{F} = (-y\mathbf{i} + x\mathbf{j})/(x^2 + y^2)$ for $x$ and $y$ not both zero. Calculate the values of $\int_C \mathbf{F} \cdot \mathbf{T}\,ds$ along both the upper and lower halves of the circle $x^2 + y^2 = 1$ from $(1, 0)$ to $(-1, 0)$. Is there a function $f = f(x, y)$ defined for $x$, $y$ not both zero such that $\nabla f = \mathbf{F}$? Why?

**27** Show that if the force field $\mathbf{F} = P\mathbf{i} + Q\mathbf{j}$ is conservative, then $\partial P/\partial y = \partial Q/\partial x$. Show that the force field of Problem 26 satisfies the condition $\partial P/\partial y = \partial Q/\partial x$, but nevertheless is *not* conservative.

**28** Suppose that the force field $\mathbf{F} = P\mathbf{i} + Q\mathbf{j} + R\mathbf{k}$ is conservative. Show that $\partial P/\partial y = \partial Q/\partial x$, $\partial P/\partial z = \partial R/\partial x$, and $\partial Q/\partial z = \partial R/\partial y$.

**29** Apply the result of Problem 28 together with Theorem 1 to show that $\int_C 2xy\,dx + x^2\,dy + y^2\,dz$ is not independent of path.

**30** Let

$$\mathbf{F} = \mathbf{F}(x, y, z) = yz\mathbf{i} + (xz + y)\mathbf{j} + (xy + 1)\mathbf{k}.$$

Define the function $f$ by $f(x, y, z) = \int_C \mathbf{F} \cdot \mathbf{T}\,ds$ where $C$ is the line segment from $(0, 0, 0)$ to $(x, y, z)$. Determine $f$ by evaluating this line integral, and then show that $\nabla f = \mathbf{F}$.

---

### 17-4

## Green's Theorem

Green's theorem relates a line integral around a simple closed plane curve $C$ to an ordinary double integral over the plane region $R$ bounded by $C$. Suppose that the curve $C$ is piecewise smooth—it consists of finitely many parametric arcs with continuous nonzero velocity vectors. Then $C$ has a unit tangent vector $\mathbf{T}$ except possibly at finitely many *corner points*. The **positive** or **counterclockwise** direction along $C$ is the one determined by a parametrization $\mathbf{r}(t)$ of $C$ such that the region $R$ remains on the *left* as the point $\mathbf{r}(t)$ traces the boundary curve $C$. That is, the vector obtained from the unit tangent vector $\mathbf{T}$ by a counterclockwise rotation through $90°$ always points *into* the region **R**, as shown in Fig. 17.9. The symbol

$$\oint_C P\,dx + Q\,dy$$

denotes a line integral along or around $C$ in this positive direction. A reversed arrow on the circle through the integral sign indicates a line integral around $C$ in the opposite direction, which we naturally call either the **negative** or the **clockwise** direction.

The following result first appeared (in an equivalent form) in a booklet on the applications of mathematics to electricity and magnetism, published privately in 1828 by the self-taught English mathematical physicist George Green (1793–1841).

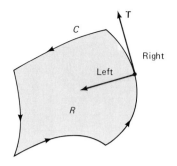

17.9   Positive orientation of the curve $C$: The region $R$ within $C$ is to the *left* of the unit tangent vector **T**.

> **Green's Theorem**
>
> Let $C$ be a piecewise-smooth simple closed curve that bounds the region $R$ in the plane. Suppose that the functions $P(x, y)$ and $Q(x, y)$ are continuous and have continuous first order partial derivatives in $R$. Then
>
> $$\oint_C P\,dx + Q\,dy = \iint_R \left(\frac{\partial Q}{\partial x} - \frac{\partial P}{\partial y}\right) dA. \qquad (1)$$

**Proof**   First we give a proof for the case in which the region $R$ is both horizontally simple and vertically simple. Then we shall indicate how to extend the result to more general regions.

Recall that if $R$ is vertically simple, then it has a description of the form $g_1(x) \le y \le g_2(x)$, $a \le x \le b$. The boundary curve $C$ is then the union of the four arcs $C_1, C_2, C_3$, and $C_4$ of Fig. 17.10, oriented as indicated there. Hence

$$\oint_C P\,dx = \int_{C_1} P\,dx + \int_{C_2} P\,dx + \int_{C_3} P\,dx + \int_{C_4} P\,dx.$$

The integrals along $C_2$ and $C_4$ are both zero, because on these two curves $x(t)$ is constant, so that $dx = x'(t)\,dt = 0$. It remains to compute the integrals along $C_1$ and $C_3$.

The point $(x, g_1(x))$ traces $C_1$ as $x$ increases from $a$ to $b$, while the point $(x, g_2(x))$ traces $C_3$ as $x$ *decreases* from $b$ to $a$. Hence

$$\oint_C P\,dx = \int_a^b P(x, g_1(x))\,dx + \int_b^a P(x, g_2(x))\,dx$$

$$= -\int_a^b [P(x, g_2(x)) - P(x, g_1(x))]\,dx$$

$$= -\int_a^b \int_{g_1(x)}^{g_2(x)} \frac{\partial P}{\partial y}\,dy\,dx$$

by the fundamental theorem of calculus. Thus

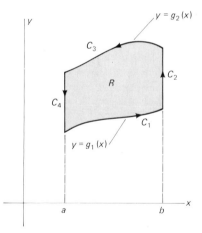

17.10   The boundary curve $C$ is the union of the four arcs $C_1, C_2, C_3$, and $C_4$.

$$\oint_C P\,dx = -\iint_R \frac{\partial P}{\partial y}\,dA. \qquad (2)$$

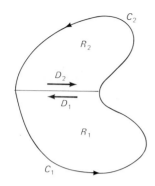

**17.11** Decomposing the region $R$ into two horizontally and vertically simple regions by means of a crosscut.

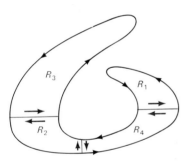

**17.12** Many important regions can be decomposed into simple regions by using two or more crosscuts.

In Problem 28 we ask you to show in a precisely similar way that

$$\oint_C Q \, dy = + \iint_R \frac{\partial Q}{\partial x} \, dA \tag{3}$$

if the region $R$ is horizontally simple. We then obtain Equation (1), the conclusion of Green's theorem, simply by adding Equations (2) and (3). ∎

The complete proof of Green's theorem for more general regions is beyond the scope of an introductory text. But the typical region $R$ that appears in practice can be subdivided into smaller regions $R_1, R_2, R_3, \ldots, R_k$ that are both vertically and horizontally simple. Green's theorem for the region $R$ then follows from the fact that it applies to each of the regions $R_1, R_2, R_3, \ldots, R_k$ (see Problem 29).

For example, the horseshoe-shaped region $R$ of Fig. 17.11 can be subdivided into the regions $R_1$ and $R_2$, each of which is both horizontally simple and vertically simple. We also subdivide the boundary $C$ of $R$ and write $C_1 \cup D_1$ for the boundary of $R_1$ and $C_2 \cup D_2$ for the boundary of $R_2$ (see Fig. 17.11). Applying Green's theorem separately to the regions $R_1$ and $R_2$, we get

$$\oint_{C_1 \cup D_1} P \, dx + Q \, dy = \iint_{R_1} \left( \frac{\partial Q}{\partial x} - \frac{\partial P}{\partial y} \right) dA$$

and

$$\oint_{C_2 \cup D_2} P \, dx + Q \, dy = \iint_{R_2} \left( \frac{\partial Q}{\partial x} - \frac{\partial P}{\partial y} \right) dA.$$

When we add these two equations, the result is Equation (1), Green's theorem for the region $R$, because the two line integrals along $D_1$ and $D_2$ cancel. This occurs because $D_1$ and $D_2$ represent the same curve with opposite orientations, so that

$$\int_{D_2} P \, dx + Q \, dy = - \int_{D_1} P \, dx + Q \, dy$$

by Equation (11) in Section 17-2. It therefore follows that

$$\int_{C_1 \cup D_1 \cup C_2 \cup D_2} P \, dx + Q \, dy = \int_{C_1 \cup C_2} P \, dx + Q \, dy = \oint_C P \, dx + Q \, dy.$$

Similarly, Green's theorem for the region shown in Fig. 17.12 could be established by subdividing it into the four simple regions as indicated.

**EXAMPLE 1**  Use Green's theorem to evaluate the line integral

$$\oint_C (2y + \sqrt{9 + x^3}) \, dx + (5x + e^{\arctan y}) \, dy,$$

where $C$ is the circle $x^2 + y^2 = 4$.

**Solution**  With $P(x, y) = 2y + (9 + x^3)^{1/2}$ and $Q(x, y) = 5x + \exp(\arctan y)$, we see that $\partial Q/\partial x - \partial P/\partial y = 5 - 2 = 3$. Since $R$ is a circular disk with area $4\pi$, Green's theorem therefore tells us that the given line integral is equal to

$$\iint_R 3 \, dA = 3(4\pi) = 12\pi.$$

**EXAMPLE 2**  Evaluate the line integral $\oint_C 3xy \, dx + 2x^2 \, dy$ where $C$ is the boundary of the region $R$ shown in Fig. 17.13; it is bounded above by the line $y = x$ and below by the parabola $y = x^2 - 2x$.

*Solution*  To evaluate the line integral directly, we would need to parametrize separately the line and the parabola. Instead, we apply Green's theorem with $P = 3xy$ and $Q = 2x^2$, so

$$\frac{\partial Q}{\partial x} - \frac{\partial P}{\partial y} = 4x - 3x = x.$$

Then

$$\oint_C 3xy \, dx + 2x^2 \, dy = \iint_R x \, dA$$

$$= \int_0^3 \int_{x^2 - 2x}^x x \, dy \, dx = \int_0^3 \left[ xy \right]_{x^2 - 2x}^x dx$$

$$= \int_0^3 (3x^2 - x^3) \, dx = \left[ x^3 - \tfrac{1}{4}x^4 \right]_0^3 = \tfrac{27}{4}.$$

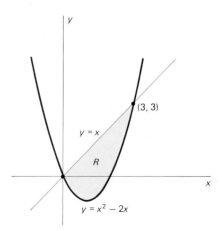

**17.13**  The region of Example 2

In Examples 1 and 2 we found the double integral easier to evaluate directly than the line integral. Sometimes the situation is reversed. The following consequence of Green's theorem illustrates the technique of evaluating a double integral $\iint_R f(x, y) \, dA$ by converting it into a line integral $\oint_C P \, dx + Q \, dy$. To do this, we must be able to find functions $P(x, y)$ and $Q(x, y)$ such that $\partial Q/\partial x - \partial P/\partial y = f(x, y)$. As in the proof of the following result, this is sometimes easy.

---

*Corollary to Green's Theorem*

The area $A$ of the region $R$ bounded by the piecewise-smooth simple closed curve $C$ is given by

$$A = \tfrac{1}{2} \oint_C - y \, dx + x \, dy = -\oint_C y \, dx = \oint_C x \, dy. \qquad (4)$$

---

*Proof*  With $P(x, y) = -y$ and $Q(x, y) \equiv 0$, Green's theorem gives

$$-\oint_C y \, dx = \iint_R 1 \, dA = A.$$

Similarly, with $P(x, y) \equiv 0$ and $Q(x, y) = x$, we obtain $A = \oint_C x \, dy$. With $P(x, y) = -y/2$ and $Q(x, y) = x/2$, Green's theorem gives

$$\tfrac{1}{2}\oint_C - y \, dx + x \, dy = \iint_R (\tfrac{1}{2} + \tfrac{1}{2}) \, dA = A. \qquad \blacksquare$$

**EXAMPLE 3**  Apply the corollary above to find the area bounded by the ellipse $x^2/a^2 + y^2/b^2 = 1$.

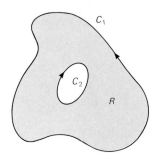

**17.14** An annular region—the boundary consists of two simple closed curves, one within the other.

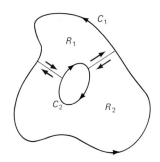

**17.15** Two crosscuts convert the annular region into the union of two ordinary regions.

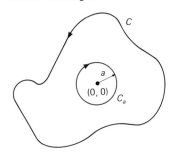

**17.16** Use the small circle $C_a$ if $C$ encloses the origin.

*Solution* With the parametrization $x = a \cos t$, $y = b \sin t$, $0 \leq t \leq 2\pi$, Equation (4) gives

$$A = \oint_{\text{ellipse}} x \, dy = \int_0^{2\pi} (a \cos t)(b \cos t) \, dt$$

$$= \tfrac{1}{2} ab \int_0^{2\pi} (1 + \cos 2t) \, dt = \pi ab.$$

By using the technique of subdividing a region into simpler ones, Green's theorem may be extended to regions which have boundaries that consist of two or more simple closed curves. For example, consider the annular region $R$ of Fig. 17.14, with boundary $C$ consisting of the two simple closed curves $C_1$ and $C_2$. The positive direction along $C$—the direction for which the region $R$ always lies on the left—is counterclockwise on the outer curve $C_1$ but clockwise on the inner curve $C_2$.

We subdivide $R$ into two regions $R_1$ and $R_2$ by using two crosscuts, as shown in Fig. 17.15. Applying Green's theorem to each of these subregions, we get

$$\iint_R \left( \frac{\partial Q}{\partial x} - \frac{\partial P}{\partial y} \right) dA = \iint_{R_1} \left( \frac{\partial Q}{\partial x} - \frac{\partial P}{\partial y} \right) dA + \iint_{R_2} \left( \frac{\partial Q}{\partial x} - \frac{\partial P}{\partial y} \right) dA$$

$$= \oint_{C_1} P \, dx + Q \, dy + \oint_{C_2} P \, dx + Q \, dy$$

$$= \oint_C P \, dx + Q \, dy.$$

Thus we obtain Green's theorem for the given region $R$. What makes this proof work is that the opposite line integrals along the two crosscuts cancel each other.

**EXAMPLE 4** Suppose that $C$ is a smooth simple closed curve that encloses the origin $(0, 0)$. Show that

$$\oint_C \frac{-y \, dx + x \, dy}{x^2 + y^2} = 2\pi,$$

but that this integral is zero if $C$ does *not* enclose the origin.

*Solution* With $P(x, y) = -y/(x^2 + y^2)$ and $Q(x, y) = x/(x^2 + y^2)$, a brief computation gives $\partial Q/\partial x - \partial P/\partial y = 0$ when $x$ and $y$ are not both zero. If the region $R$ bounded by $C$ does not contain the origin, then $P$ and $Q$ and their derivatives are continuous on $R$. Hence Green's theorem gives

$$\oint_C \frac{-y \, dx + x \, dy}{x^2 + y^2} = \iint_R 0 \, dA = 0.$$

If $C$ does enclose the origin, then we enclose the origin in a small circle $C_a$ interior to $C$ (as in Fig. 17.16) and parametrize this circle by $x = a \cos t$, $y = a \sin t$. Then Green's theorem, applied to the region $R$ between $C$ and $C_a$, gives

$$\oint_C \frac{-y \, dx + x \, dy}{x^2 + y^2} + \oint_{C_a} \frac{-y \, dx + x \, dy}{x^2 + y^2} = \iint_R 0 \, dA = 0.$$

Therefore,

$$\oint_C \frac{-y\,dx + x\,dy}{x^2 + y^2} = \oint_{C_a} \frac{-y\,dx + x\,dy}{x^2 + y^2}$$

$$= \int_0^{2\pi} \frac{(-a \sin t)(-a \sin t\,dt) + (a \cos t)(a \cos t\,dt)}{(a \cos t)^2 + (a \sin t)^2}$$

$$= \int_0^{2\pi} 1\,dt = 2\pi.$$

---

The result of Example 4 can be interpreted in terms of the polar coordinate angle $\theta = \tan^{-1}(y/x)$. Because

$$d\theta = \frac{-y\,dx + x\,dy}{x^2 + y^2},$$

the line integral of Example 4 measures the net change in $\theta$ as we go around the curve $C$ once in a counterclockwise direction. This net change is $2\pi$ if $C$ encloses the origin and is zero otherwise.

## THE DIVERGENCE AND FLUX OF A VECTOR FIELD

Now let us consider the steady flow of a thin layer of fluid in the plane (perhaps like a sheet of water spreading across the floor). Let $\mathbf{v}(x, y)$ be its velocity vector field and $\rho(x, y)$ the density of the fluid at the point $(x, y)$. The term *steady flow* means that $\mathbf{v}$ and $\rho$ depend only upon $x$ and $y$ and *not* upon time $t$. We want to compute the rate at which the fluid flows out of the region $R$ bounded by a simple closed curve $C$ (Fig. 17.17). We seek the net rate of outflow—the actual outflow minus the inflow.

Let $\Delta s_i$ be a short segment of the curve $C$, and let $(x_i^*, y_i^*)$ be an end point of $\Delta s_i$. Then the area of the portion of the fluid that flows out of $R$ across $\Delta s_i$ per unit time is approximately the area of the parallelogram of Fig. 17.17; this parallelogram is spanned by the segment $\Delta s_i$ and the vector $\mathbf{v}_i = \mathbf{v}(x_i^*, y_i^*)$. Suppose that $\mathbf{n}_i$ is the unit normal vector to $C$ at the point $(x_i^*, y_i^*)$, the normal pointing *out* of $R$. Then the area of this parallelogram is

$$(|\mathbf{v}_i| \cos \theta)\,\Delta s_i = \mathbf{v}_i \cdot \mathbf{n}_i\,\Delta s_i$$

where $\theta$ is the angle between $\mathbf{n}_i$ and $\mathbf{v}_i$.

We multiply by the density $\rho_i = \rho(x_i^*, y_i^*)$, then add these terms over those values of $i$ that correspond to a subdivision of the entire curve $C$. This gives the (net) total mass of fluid leaving $R$ per unit of time; it is approximately

$$\sum_{i=1}^n \rho_i \mathbf{v}_i \cdot \mathbf{n}_i\,\Delta s_i = \sum_{i=1}^n \mathbf{F}_i \cdot \mathbf{n}_i\,\Delta s_i$$

where $\mathbf{F} = \rho \mathbf{v}$. The line integral around $C$ that this sum approximates is called the *flux of the vector field* $\mathbf{F}$ *across the curve* $C$,

$$(\text{flux of } \mathbf{F} \text{ across } C) = \oint_C \mathbf{F} \cdot \mathbf{n}\,ds, \tag{5}$$

where $\mathbf{n}$ is the *outer* unit normal to $C$.

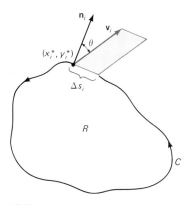

**17.17**   The area of the parallelogram approximates the fluid flow across $\Delta s_i$ in unit time.

In the present case of fluid flow with velocity vector **v**, the flux of **F** = ρ**v** is the rate at which the fluid is flowing out of $R$ across the boundary curve $C$, in units of mass per unit of time. But the same terminology is used for an arbitrary vector field **F** = $M$**i** + $N$**j**. For example, we may speak of the flux of an electric or gravitational field across a curve $C$.

From Fig. 17.18 we see that the outer unit normal **n** is equal to **T** × **k**. The unit tangent **T** to the curve $C$ is given by

$$\mathbf{T} = \frac{1}{v}\left(\mathbf{i}\frac{dx}{dt} + \mathbf{j}\frac{dy}{dt}\right) = \mathbf{i}\frac{dx}{ds} + \mathbf{j}\frac{dy}{ds}$$

since $v = ds/dt$. Hence

$$\mathbf{n} = \mathbf{T} \times \mathbf{k} = \left(\mathbf{i}\frac{dx}{ds} + \mathbf{j}\frac{dy}{ds}\right) \times \mathbf{k}.$$

Since **i** × **k** = −**j** and **j** × **k** = **i**, we find that

$$\mathbf{n} = \mathbf{i}\frac{dy}{ds} - \mathbf{j}\frac{dx}{ds}. \tag{6}$$

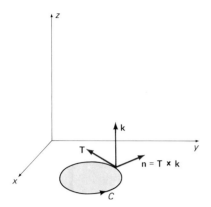

**17.18** Computing the outer unit normal **n** from the unit tangent vector **T**

Substitution of the expression in Equation (6) in the flux integral of Equation (5) gives

$$\oint_C \mathbf{F} \cdot \mathbf{n}\, ds = \oint_C (M\mathbf{i} + N\mathbf{j}) \cdot \left(\mathbf{i}\frac{dy}{ds} - \mathbf{j}\frac{dx}{ds}\right) ds$$

$$= \oint_C -N\, dx + M\, dy.$$

Applying Green's theorem to this last line integral with $P = -N$ and $Q = M$, we get

$$\oint_C \mathbf{F} \cdot \mathbf{n}\, ds = \iint_R \left(\frac{\partial M}{\partial x} + \frac{\partial N}{\partial y}\right) dA \tag{7}$$

for the flux of **F** = $M$**i** + $N$**j** across $C$.

The scalar function $\partial M/\partial x + \partial N/\partial y$ that appears in Equation (7) is the **divergence** of the two-dimensional vector field **F** = $M$**i** + $N$**j** as defined in Section 17-1 and denoted by

$$\operatorname{div}\mathbf{F} = \nabla \cdot \mathbf{F} = \frac{\partial M}{\partial x} + \frac{\partial N}{\partial y}. \tag{8}$$

When we substitute (8) in (7), we obtain a **vector form of Green's theorem:**

$$\oint_C \mathbf{F} \cdot \mathbf{n}\, ds = \iint_R \nabla \cdot \mathbf{F}\, dA, \tag{9}$$

with the understanding that **n** is the *outer* unit normal to $C$. Thus the flux of a vector field across a simple closed curve $C$ is equal to the double integral of its divergence over the region $R$ bounded by $C$.

If the region $R$ is bounded by a circle $C_r$ of radius $r$ centered at the point $(x_0, y_0)$, then

$$\oint_{C_r} \mathbf{F} \cdot \mathbf{n}\, ds = \iint_R \nabla \cdot \mathbf{F}\, dA = (\pi r^2)\nabla \cdot \mathbf{F}(\bar{x}, \bar{y})$$

**930**

for some point $(\bar{x}, \bar{y})$ in $R$; this is a consequence of the average value property of double integrals (see Problem 30 in Section 16.2). We divide by $\pi r^2$ and then let $r$ approach 0. Thus we find that

$$\mathbf{V} \cdot \mathbf{F}(x_0, y_0) = \lim_{r \to 0} \frac{1}{\pi r^2} \oint_{C_r} \mathbf{F} \cdot \mathbf{n} \, ds \tag{10}$$

because $(\bar{x}, \bar{y}) \to (x_0, y_0)$ as $r \to 0$.

In the case of our original fluid flow, with $\mathbf{F} = \rho \mathbf{v}$, Equation (10) tells us that the value of $\mathbf{V} \cdot \mathbf{F}$ at $(x_0, y_0)$ is a measure of the rate at which the fluid is "diverging away" from the point $(x_0, y_0)$.

**EXAMPLE 5** The vector field $\mathbf{F} = -y\mathbf{i} + x\mathbf{j}$ is the velocity field of a steady-state counterclockwise rotation about the origin. Show that the flux of $\mathbf{F}$ across any simple closed curve $C$ is zero.

*Solution* This follows immediately from Equation (9) because

$$\mathbf{V} \cdot \mathbf{F} = \frac{\partial}{\partial x}(-y) + \frac{\partial}{\partial y}(x) = 0.$$

## 17-4 PROBLEMS

In each of Problems 1–12, apply Green's theorem to evaluate the indicated integral.

**1** $\oint_C (x + y^2) \, dx + (y + x^2) \, dy$, where $C$ is the square with vertices $(\pm 1, \pm 1)$.

**2** $\oint_C (x^2 + y^2) \, dx - 2xy \, dy$, where $C$ is the boundary of the triangle bounded by the lines $x = 0$, $y = 0$, and $x + y = 1$.

**3** $\oint_C (y + e^x) \, dx + (2x^2 + \cos y) \, dy$, where $C$ is the boundary of the triangle with vertices $(0, 0)$, $(1, 1)$, and $(2, 0)$.

**4** $\oint_C (x^2 - y^2) \, dx + xy \, dy$, where $C$ is the boundary of the region that is bounded by the line $y = x$ and the parabola $y = x^2$.

**5** $\oint_C (-y^2 + e^x) \, dx + (\tan^{-1} y) \, dy$, where $C$ is the boundary of the region between the parabolas $y = x^2$ and $x = y^2$.

**6** $\oint_C y^2 \, dx + (2x - 3y) \, dy$, where $C$ is the circle $x^2 + y^2 = 9$.

**7** $\oint_C (x - y) \, dx + y \, dy$, where $C$ is the boundary of the region between the x-axis and the graph $y = \sin x$ for $0 \le x \le \pi$.

**8** $\oint_C e^x \sin y \, dx + e^x \cos y \, dy$, where $C$ is the right-hand loop of the lemniscate $r^2 = 4 \cos \theta$.

**9** $\oint_C y^2 \, dx + xy \, dy$, where $C$ is the ellipse $x^2/9 + y^2/4 = 1$.

**10** $\oint_C y(1 + x^2)^{-1} \, dx + \tan^{-1} x \, dy$, where $C$ is the oval $x^4 + y^4 = 1$.

**11** $\oint_C xy \, dx + x^2 \, dy$, where $C$ is the first-quadrant loop of $r = \sin 2\theta$.

**12** $\oint_C x^2 \, dy - y^2 \, dx$, where $C$ is the cardioid $r = 1 + \cos \theta$.

In each of Problems 13–16, use the corollary to Green's theorem to find the area of the indicated region.

**13** The circle bounded by $x = a \cos t$, $y = a \sin t$, $0 \le t \le 2\pi$.

**14** The region under one arch of the cycloid with parametric equations $x = a(t - \sin t)$, $y = a(1 - \cos t)$.

**15** The region bounded by the astroid $x = \cos^3 t$, $y = \sin^3 t$, $0 \le t \le 2\pi$.

**16** The region bounded by $y = x^2$ and $y = x^3$.

**17** Suppose that $f$ is a twice-differentiable scalar function of $x$ and $y$. Show that

$$\mathbf{V}^2 f = \text{div}(\mathbf{V} f) = \frac{\partial^2 f}{\partial x^2} + \frac{\partial^2 f}{\partial y^2}.$$

**18** Show that $f(x, y) = \ln(x^2 + y^2)$ satisfies **Laplace's equation** $\mathbf{V}^2 f = 0$ (except at the point $(0, 0)$).

**19** Suppose that $f$ and $g$ are twice-differentiable functions. Show that $\mathbf{V}^2(fg) = f\mathbf{V}^2 g + g\mathbf{V}^2 f + 2\mathbf{V} f \cdot \mathbf{V} g$.

**20** Suppose that the function $f(x, y)$ is twice continuously differentiable in the region $R$ bounded by the piecewise-smooth closed curve $C$. Show that

$$\oint_C \frac{\partial f}{\partial x} \, dy - \frac{\partial f}{\partial y} \, dx = \iint_R \mathbf{V}^2 f \, dx \, dy.$$

**21** Let $R$ be the plane region with area $A$ enclosed by the piecewise-smooth simple closed curve $C$. Use Green's theorem to show that the coordinates of the centroid of $R$ are

$$\bar{x} = \frac{1}{2A} \oint_C x^2 \, dy \quad \text{and} \quad \bar{y} = -\frac{1}{2A} \oint_C y^2 \, dx.$$

**22** Use the result of Problem 21 to find the centroid of:
(a) a semicircular region of radius $a$;
(b) a quarter-circular region of radius $a$.

**23** Suppose that a lamina shaped like the region of Problem 21 has constant density $\rho$. Show that its moments of inertia about the coordinate axes are

$$I_x = -\frac{\rho}{3} \oint_C y^3 \, dx \quad \text{and} \quad I_y = \frac{\rho}{3} \oint_C x^3 \, dy.$$

**24** Use the result of Problem 23 to show that the polar moment of inertia $I_0 = I_x + I_y$ of a circular lamina of radius $a$, centered at the origin, and of constant density $\rho$, is $\frac{1}{2}Ma^2$ where $M$ is the mass of the lamina.

**25** The loop of the folium of Descartes (with equation $x^3 + y^3 = 3xy$) appears in Fig. 17.19. Apply the corollary

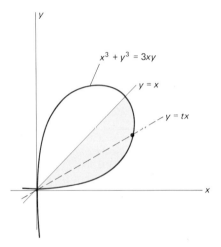

**17.19**  The loop of Problem 25

to Green's theorem to find the area of this loop. (*Suggestion:* Set $y = tx$ to discover the parametrization $x = 3t/(1 + t^3)$, $y = 3t^2/(1 + t^3)$. To obtain the area of the loop, use values of $t$ in the interval $[0, 1]$. This gives the half of the loop lying below the line $y = x$.)

**26** Find the area bounded by one loop of the curve $x = \sin 2t$, $y = \sin t$.

**27** Let $f$ and $g$ be functions with continuous second order partial derivatives in the region $R$ bounded by the piecewise-smooth simple closed curve $C$. Apply Green's theorem in vector form to show that

$$\oint_C f \nabla g \cdot \mathbf{n} \, ds = \iint_R [(f)(\nabla \cdot \nabla g) + \nabla f \cdot \nabla g] \, dA.$$

It was this formula rather than Green's theorem itself that appeared in Green's book.

**28** Complete the proof of the simple case of Green's theorem by showing directly that $\oint_C Q \, dy = \iint_R (\partial Q/\partial x) \, dA$ if the region $R$ is horizontally simple.

**29** Suppose that the bounded plane region $R$ is subdivided into the nonoverlapping subregions $R_1, R_2, R_3, \ldots, R_k$. If Green's theorem, Equation (1), holds for each of these subregions, explain why it follows that Green's theorem holds for $R$. State carefully any assumptions that you need to make.

In Problems 30–32, find the flux of the given vector field $\mathbf{F}$ outward across the ellipse $x = 4 \cos t$, $y = 3 \sin t$, $0 \leq t \leq 2\pi$.

**30** $\mathbf{F} = x\mathbf{i} + 2y\mathbf{j}$

**31** $\mathbf{F} = \dfrac{x\mathbf{i} + y\mathbf{j}}{x^2 + y^2}$

**32** $\mathbf{F} = -y\mathbf{i} + x\mathbf{j}$

---

## 17-5

### Surface Integrals

A surface integral is to surfaces in space what a line (or "curve") integral is to curves in the plane. Consider a curved thin metal sheet shaped like the surface $S$. Suppose that this sheet has a variable density, given at the point $(x, y, z)$ by the known continuous function $f(x, y, z)$, in units such as grams per square centimeter of surface. We want to define the surface integral

$$\iint_S f(x, y, z) \, dS$$

in such a way that—upon evaluation—it gives the total mass of the thin metal sheet. In case $f(x, y, z) \equiv 1$, the numerical value of the integral should also equal the surface area of $S$.

As in Section 16-8, we assume that $S$ is a parametric surface described by the function or transformation

$$\mathbf{r}(u, v) = \langle x(u, v), y(u, v), z(u, v) \rangle$$

for $(u, v)$ in a region $D$ in the $uv$-plane. We suppose throughout that the component functions of $\mathbf{r}$ have continuous partial derivatives, and also that the vectors $\mathbf{r}_u = \partial \mathbf{r}/\partial u$ and $\mathbf{r}_v = \partial \mathbf{r}/\partial v$ are nonzero and nonparallel at each interior point of $D$.

Recall now how we computed the surface area $A$ of $S$ in Section 16-8. We began with an inner partition of $D$ consisting of $n$ rectangles $R_1$, $R_2$, $R_3, \ldots, R_n$, each $\Delta u$ by $\Delta v$ in size. The images under $\mathbf{r}$ of these rectangles are curvilinear figures filling up most of the surface $S$, and these pieces of $S$ are themselves approximated by parallelograms $P_i$ of the sort shown in Fig. 17.20. This gave us the approximation

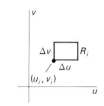

$$A \approx \sum_{i=1}^{n} \Delta P_i = \sum_{i=1}^{n} |\mathbf{N}(u_i, v_i)| \, \Delta u \, \Delta v. \tag{1}$$

Here, $\Delta P_i = |\mathbf{N}(u_i, v_i)| \, \Delta u \, \Delta v$ is the area of the parallelogram $P_i$ that is tangent to the surface $S$ at the point $\mathbf{r}(u_i, v_i)$. The vector

$$\mathbf{N} = \frac{\partial \mathbf{r}}{\partial u} \times \frac{\partial \mathbf{r}}{\partial v} \tag{2}$$

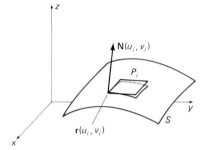

is normal to $S$ at $\mathbf{r}(u, v)$. If the surface $S$ now has a density function $f(x, y, z)$, we can approximate the total mass $M$ of the surface by first multiplying each parallelogram area $\Delta P_i$ in Equation (1) by the density $f(\mathbf{r}(u_i, v_i))$ at $\mathbf{r}(u_i, v_i)$, and then summing these estimates over all such parallelograms. Thus we obtain the approximation

$$M \approx \sum_{i=1}^{n} f(\mathbf{r}(u_i, v_i)) \, \Delta P_i = \sum_{i=1}^{n} f(\mathbf{r}(u_i, v_i)) |\mathbf{N}(u_i, v_i)| \, \Delta u \, \Delta v.$$

**17.20** Approximating surface area with parallelograms

This approximation is a Riemann sum for the **surface integral of the function $f$ over the surface $S$**, defined by

$$\iint_S f(x, y, z) \, dS = \iint_D f(\mathbf{r}(u, v)) |\mathbf{N}(u, v)| \, du \, dv \tag{3}$$

$$= \iint_D f(\mathbf{r}(u, v)) \left| \frac{\partial \mathbf{r}}{\partial u} \times \frac{\partial \mathbf{r}}{\partial v} \right| \, du \, dv.$$

To evaluate the surface integral $\iint_S f(x, y, z) \, dS$, we simply use the parametrization $\mathbf{r}$ to express the variables $x$, $y$, and $z$ in terms of $u$ and $v$ and formally replace the **surface area element** $dS$ by

$$dS = |\mathbf{N}(u, v)| \, du \, dv = \left| \frac{\partial \mathbf{r}}{\partial u} \times \frac{\partial \mathbf{r}}{\partial v} \right| \, du \, dv. \tag{4}$$

This converts the surface integral into an ordinary double integral over the region $D$ in the $uv$-plane.

In the important special case of a surface $S$ described by $z = h(x, y)$, $(x, y)$ in $D$, we may use $x$ and $y$ as the parameters (rather than $u$ and $v$). The surface area element takes the form

$$dS = \sqrt{1 + (\partial h/\partial x)^2 + (\partial h/\partial y)^2} \, dx \, dy \tag{5}$$

corresponding to Equation (9) in Section 16-8. The surface integral of $f$ over $S$ is then given by

$$\iint_S f(x, y, z) \, dS = \iint_D f(x, y, h(x, y)) \sqrt{1 + \left(\frac{\partial h}{\partial x}\right)^2 + \left(\frac{\partial h}{\partial y}\right)^2} \, dx \, dy. \quad (6)$$

Moments, centroids, and moments of inertia for surfaces are computed in the same way as for curves (Section 17-2), using surface integrals in place of line integrals. For example, suppose that the surface $S$ has density $\rho(x, y, z)$ at the point $(x, y, z)$. Then its moment $M_{xy}$ about the $xy$-plane and its moment of inertia $I_z$ about the $z$-axis are given by

$$M_{xy} = \iint_S z\rho(x, y, z) \, dS \quad \text{and} \quad I_z = \iint_S (x^2 + y^2)\rho(x, y, z) \, dS.$$

**EXAMPLE 1** Find the centroid of the unit-density hemispherical surface $z = \sqrt{a^2 - x^2 - y^2}$, $x^2 + y^2 \leq a^2$.

**Solution** By symmetry, $\bar{x} = 0 = \bar{y}$. A simple computation gives $\partial z/\partial x = -x/z$ and $\partial z/\partial y = -y/z$, so Equation (5) yields

$$dS = \sqrt{1 + \left(\frac{\partial z}{\partial x}\right)^2 + \left(\frac{\partial z}{\partial y}\right)^2} \, dx \, dy$$

$$= \sqrt{1 + \frac{x^2}{z^2} + \frac{y^2}{z^2}} \, dx \, dy$$

$$= \frac{1}{z}\sqrt{x^2 + y^2 + z^2} \, dx \, dy = \frac{a}{z} \, dx \, dy.$$

Hence

$$\bar{z} = \frac{M_{xy}}{A} = \frac{1}{2\pi a^2} \iint_D (z)\left(\frac{a}{z}\right) dx \, dy = \frac{1}{2\pi a} \iint_D 1 \, dx \, dy = \frac{a}{2}.$$

Note in the final step that $D$ is a circle of radius $a$ in the $xy$-plane; this simplifies the computation.

**EXAMPLE 2** Find the moment of inertia about the $z$-axis of the spherical surface $x^2 + y^2 + z^2 = a^2$, assuming that it has constant density $\rho = k$.

**Solution** The sphere of radius $a$ is parametrized in spherical coordinates by

$$x = a \sin \phi \cos \theta, \qquad y = a \sin \phi \sin \theta, \qquad z = a \cos \phi$$

for $0 \leq \phi \leq \pi$, $0 \leq \theta \leq 2\pi$. By Problem 18 in Section 16-8, the surface element is then $dS = a^2 \sin \phi \, d\phi \, d\theta$. We first note that

$$x^2 + y^2 = a^2 \sin^2\phi \cos^2\theta + a^2 \sin^2\phi \sin^2\theta = a^2 \sin^2\phi,$$

and it then follows that

$$I_z = \iint_S (x^2 + y^2)\rho \, dS = \int_0^{2\pi} \int_0^\pi k(a^2 \sin^2\phi) a^2 \sin\phi \, d\phi \, d\theta$$

$$= 2\pi k a^4 \int_0^\pi \sin^3\phi \, d\phi = \tfrac{8}{3}\pi k a^4 = \tfrac{2}{3}Ma^2.$$

In the last step we use the fact that the mass $M$ of the spherical surface with density $k$ is $4\pi ka^2$.

---

The surface integral $\iint_S f(x, y, z) \, dS$ is analogous to the line integral $\int_C f(x, y) \, ds$. There is a second type of surface integral which is analogous to the line integral of the form $\int_C P \, dx + Q \, dy$. To define the surface integral

$$\iint_S f(x, y, z) \, dx \, dy,$$

with $dx \, dy$ in place of $dS$, we replace the parallelogram area $\Delta P_i$ in Equation (1) by its projection into the $xy$-plane (see Fig. 17.21). To see how this works out, consider the *unit* normal to $S$,

$$\mathbf{n} = \frac{\mathbf{N}}{|\mathbf{N}|} = \mathbf{i} \cos \alpha + \mathbf{j} \cos \beta + \mathbf{k} \cos \gamma. \tag{7}$$

Since

$$\mathbf{N} = \begin{vmatrix} \mathbf{i} & \mathbf{j} & \mathbf{k} \\ \dfrac{\partial x}{\partial u} & \dfrac{\partial y}{\partial u} & \dfrac{\partial z}{\partial u} \\ \dfrac{\partial x}{\partial v} & \dfrac{\partial y}{\partial v} & \dfrac{\partial z}{\partial v} \end{vmatrix}$$

or—in the Jacobian notation of Section 16-9—

$$\mathbf{N} = \mathbf{i} \frac{\partial(y, z)}{\partial(u, v)} + \mathbf{j} \frac{\partial(z, x)}{\partial(u, v)} + \mathbf{k} \frac{\partial(x, y)}{\partial(u, v)},$$

we see that the components of the unit normal $\mathbf{n}$ are

$$\cos \alpha = \frac{1}{|\mathbf{N}|} \frac{\partial(y, z)}{\partial(u, v)}, \qquad \cos \beta = \frac{1}{|\mathbf{N}|} \frac{\partial(z, x)}{\partial(u, v)}, \qquad \cos \gamma = \frac{1}{|\mathbf{N}|} \frac{\partial(x, y)}{\partial(u, v)}. \tag{8}$$

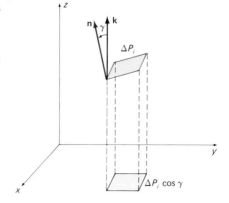

**17.21** Finding the area of the projected parallelogram

From Fig. 17.21 we see that the (signed) projection of the area $\Delta P_i$ into the $xy$-plane is $\Delta P_i \cos \gamma$. The corresponding Riemann sum motivates the *definition*

$$\iint_S f(x, y, z) \, dx \, dy = \iint_S f(x, y, z) \cos \gamma \, dS = \iint_D f(\mathbf{r}(u, v)) \frac{\partial(x, y)}{\partial(u, v)} \, du \, dv, \tag{9}$$

where the last integral is obtained by substituting for $\cos \gamma$ from (8) and for $dS$ from (4). Similarly, we *define*

$$\iint_S f(x, y, z) \, dy \, dz = \iint_S f(x, y, z) \cos \alpha \, dS = \iint_D f(\mathbf{r}(u, v)) \frac{\partial(y, z)}{\partial(u, v)} \, du \, dv \tag{10}$$

and

$$\iint_S f(x, y, z) \, dz \, dx = \iint_S f(x, y, z) \cos \beta \, dS = \iint_D f(\mathbf{r}(u, v)) \frac{\partial(z, x)}{\partial(u, v)} \, du \, dv. \tag{11}$$

Note that the symbols $z$ and $x$ in Formula (11) appear in the reverse of alphabetical order. It is important to write them in the correct order because the fact that $\partial(x, z)/\partial(u, v) = -\partial(z, x)/\partial(u, v)$ implies that

$$\iint_S f(x, y, z)\, dx\, dz = -\iint_S f(x, y, z)\, dz\, dx.$$

In an ordinary *double integral*, the order in which the differentials are written simply indicates the order of integration. But in a *surface integral*, it instead indicates the order of appearance of the corresponding variables in the Jacobians of Formulas (9) through (11).

The general surface integral of the second type is the sum

$$\iint_S P\, dy\, dz + Q\, dz\, dx + R\, dx\, dy = \iint_S (P \cos \alpha + Q \cos \beta + R \cos \gamma)\, dS \quad (12)$$

$$= \iint_D \left( P \frac{\partial(y, z)}{\partial(u, v)} + Q \frac{\partial(z, x)}{\partial(u, v)} + R \frac{\partial(x, y)}{\partial(u, v)} \right) du\, dv. \quad (13)$$

Here, $P$, $Q$, and $R$ are continuous functions of $x$, $y$, and $z$.

Suppose that $\mathbf{F} = P\mathbf{i} + Q\mathbf{j} + R\mathbf{k}$. Then the integrand in Equation (12) is simply $\mathbf{F} \cdot \mathbf{n}$, so we obtain the basic relation

$$\iint_S \mathbf{F} \cdot \mathbf{n}\, dS = \iint_S P\, dy\, dz + Q\, dz\, dx + R\, dx\, dy \quad (14)$$

between our two types of surface integrals. This formula is analogous to the earlier formula

$$\int_C \mathbf{F} \cdot \mathbf{T}\, ds = \int_C P\, dx + Q\, dy + R\, dz$$

for line integrals. On the other hand, Formula (13) gives the evaluation procedure for the surface integral of (12)—substitute for $x$, $y$, $z$ and their derivatives in terms of $u$ and $v$ and then integrate over the appropriate region $D$ in the $uv$-plane.

It turns out that only the *sign* of the right-hand surface integral in Equation (14) depends upon the parametrization of $S$. The unit normal on the left-hand side is the one that is provided by the parametrization of $S$ via Equations (8). In the case of a surface given by $z = h(x, y)$, with $x$ and $y$ used as the parameters $u$ and $v$, this will be the *upper* normal, as you will see in the computation of the following example.

Suppose that $S$ is the surface $z = h(x, y)$, $(x, y)$ in $D$. Then show that

$$\iint_S P\, dy\, dz + Q\, dz\, dx + R\, dx\, dy = \iint_D \left( -P \frac{\partial z}{\partial x} - Q \frac{\partial z}{\partial y} + R \right) dx\, dy, \quad (15)$$

where $P$, $Q$, and $R$ in the second integral are evaluated at $(x, y, h(x, y))$.

This is simply a matter of computing the three Jacobians in (13) with the parameters $x$ and $y$. We note first that $\partial x/\partial x = 1 = \partial y/\partial y$ and that $\partial x/\partial y = 0 = \partial y/\partial x$. Hence

$$\frac{\partial(y, z)}{\partial(x, y)} = \begin{vmatrix} \dfrac{\partial y}{\partial x} & \dfrac{\partial y}{\partial y} \\[2mm] \dfrac{\partial z}{\partial x} & \dfrac{\partial z}{\partial y} \end{vmatrix} = -\frac{\partial z}{\partial x},$$

$$\frac{\partial(z, x)}{\partial(x, y)} = \begin{vmatrix} \dfrac{\partial z}{\partial x} & \dfrac{\partial z}{\partial y} \\[2mm] \dfrac{\partial x}{\partial x} & \dfrac{\partial x}{\partial y} \end{vmatrix} = -\frac{\partial z}{\partial y},$$

and

$$\frac{\partial(x, y)}{\partial(x, y)} = \begin{vmatrix} \dfrac{\partial x}{\partial x} & \dfrac{\partial x}{\partial y} \\[2mm] \dfrac{\partial y}{\partial x} & \dfrac{\partial y}{\partial y} \end{vmatrix} = 1.$$

Equation (15) is an immediate consequence.

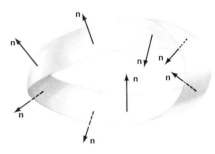

17.22  A Möbius strip—an example of a one-sided surface

One of the most important applications of surface integrals is to the computation of the flux of a vector field. In order to define the flux of the vector field **F** across the surface $S$, we assume that $S$ has a unit normal vector **n** that varies *continuously* from point to point of $S$. This condition excludes from our consideration one-sided (*nonorientable*) surfaces, such as the Möbius strip of Fig. 17.22. If $S$ is a two-sided (*orientable*) surface, then there are two possible choices for **n**. For example, if $S$ is a closed surface (such as a sphere or torus) that separates space, then we may choose for **n** either the outer normal or the inner normal. These two options are shown in Fig. 17.23. The unit normal defined by the formula in (7) may be either the outer normal or the inner normal; which of the two it is depends on how $S$ has been parametrized.

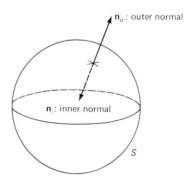

17.23  Inner and outer normals to a two-sided closed surface

Now we turn our attention to the flux of a vector field. Suppose that we are given the vector field **F**, the orientable surface $S$, and a continuous unit normal vector **n** on $S$. We define the **flux of F across $S$ in the direction of n** in analogy with Equation (5) of Section 17-4:

$$(\text{flux of } \mathbf{F}) = \iint\limits_{S} \mathbf{F} \cdot \mathbf{n} \, dS. \tag{16}$$

For example, if $\mathbf{F} = \rho\mathbf{v}$ where **v** is the velocity vector field corresponding to the steady flow in space of a fluid of density $\rho$ and **n** is the *outer* normal for a closed surface $S$ bounding a space region $T$, then the flux determined by the above equation is the rate of flow of the fluid *out of $T$* across its boundary surface $S$, in such units as grams per second.

A similar application is to the flow of heat, which is mathematically quite similar to the flow of a fluid. Suppose that a body has temperature $u = u(x, y, z)$ at the point $(x, y, z)$. Experiments indicate that the flow of heat in the body is described by the heat flow vector

$$\mathbf{q} = -K \, \nabla u. \tag{17}$$

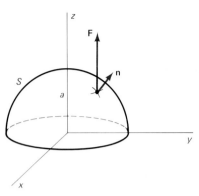

**17.24** The hemisphere $S$ of Example 4

The number $K$—normally, but not always, a constant—is the *heat conductivity* of the body. The vector $\mathbf{q}$ points in the direction of heat flow, and its length is the rate of flow of heat across a unit area normal to $\mathbf{q}$. This flow rate is measured in such units as calories per second per square centimeter. If $S$ is a closed surface within the body bounding a region $T$ and $\mathbf{n}$ denotes the outer unit normal to $S$, then

$$\iint_S \mathbf{q} \cdot \mathbf{n} \, dS = -\iint_S K \, \nabla u \cdot \mathbf{n} \, dS \qquad (18)$$

is the rate of flow of heat (in calories per second, for example) out of the region $T$ across its boundary surface $S$.

**EXAMPLE 4**  Calculate the flux $\iint_S \mathbf{F} \cdot \mathbf{n} \, dS$, where $\mathbf{F} = v_0 \mathbf{k}$ and $S$ is the hemispherical surface of radius $a$ shown in Fig. 17.24, with outward unit normal vector $\mathbf{n}$.

*Solution*  If we think of $\mathbf{F} = v_0 \mathbf{k}$ as the velocity vector field of a fluid flowing upward with constant speed $v_0$, then the flux in question may be interpreted as the rate of flow (in cubic centimeters per second, for example) of the fluid across $S$. To calculate this flux, we note that

$$\mathbf{n} = \frac{x\mathbf{i} + y\mathbf{j} + z\mathbf{k}}{\sqrt{x^2 + y^2 + z^2}} = \frac{1}{a}(x\mathbf{i} + y\mathbf{j} + z\mathbf{k}).$$

Hence

$$\mathbf{F} \cdot \mathbf{n} = v_0 \mathbf{k} \cdot \frac{1}{a}(x\mathbf{i} + y\mathbf{j} + z\mathbf{k}) = \frac{v_0 z}{a},$$

so

$$\iint_S \mathbf{F} \cdot \mathbf{n} \, dS = \iint_S \frac{v_0 z}{a} \, dS.$$

If we introduce spherical coordinates $z = a \cos \phi$, $dS = a^2 \sin \phi \, d\phi \, d\theta$ on the sphere, we get

$$\iint_S \mathbf{F} \cdot \mathbf{n} \, dS = \frac{v_0}{a} \int_0^{2\pi} \int_0^{\pi/2} (a \cos \phi)(a^2 \sin \phi) \, d\phi \, d\theta$$

$$= 2\pi a^2 v_0 \int_0^{\pi/2} \cos \phi \sin \phi \, d\phi = 2\pi a^2 v_0 \left[ \frac{1}{2} \sin^2 \phi \right]_0^{\pi/2};$$

thus

$$\iint_S \mathbf{F} \cdot \mathbf{n} \, dS = \pi a^2 v_0.$$

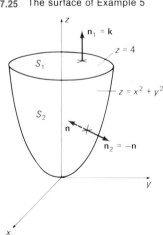

**17.25** The surface of Example 5

Note that this result is equal to the flux of $\mathbf{F} = v_0 \mathbf{k}$ across the disk $x^2 + y^2 \leqq a^2$ of area $\pi a^2$. If we think of the hemispherical region $T$ bounded by the hemisphere $S$ and the circular disk that forms its base, it should be no surprise that the rate of inflow of an incompressible fluid across the circular disk is equal to its rate of outflow across the hemisphere $S$.

**EXAMPLE 5**  Find the flux of the vector field $\mathbf{F} = x\mathbf{i} + y\mathbf{j} + 3\mathbf{k}$ out of the region $T$ bounded by the paraboloid $z = x^2 + y^2$ and the plane $z = 4$ (see Fig. 17.25).

*Solution* Let $S_1$ denote the circular top, which has outer unit normal $\mathbf{n}_1 = \mathbf{k}$; let $S_2$ be the parabolic part of the surface, with outer unit normal $\mathbf{n}_2$. The flux across $S_1$ is

$$\iint_{S_1} \mathbf{F} \cdot \mathbf{n}_1 \, dS = \iint_{S_1} 3 \, dS = 12\pi$$

because $S_1$ is a circle of radius 2.

The computation of Example 3 gives

$$\mathbf{N} = \left\langle -\frac{\partial z}{\partial x}, -\frac{\partial z}{\partial y}, 1 \right\rangle = \langle -2x, -2y, 1 \rangle$$

for a normal vector to the paraboloid $z = x^2 + y^2$. Then $\mathbf{n} = \mathbf{N}/|\mathbf{N}|$ is an upper, and thus an *inner*, unit normal to the surface $S_2$. The unit *outer* normal is therefore $\mathbf{n}_2 = -\mathbf{n}$, opposite to the direction of $\mathbf{N} = \langle -2x, -2y, 1 \rangle$. With parameters $(x, y)$ in the circular disk $x^2 + y^2 \leq 4$ in the $xy$-plane, the surface area element is $dS = |\mathbf{N}| \, dx \, dy$. Therefore, the outward flux across $S_2$ is

$$\iint_{S_2} \mathbf{F} \cdot \mathbf{n}_2 \, dS = -\iint_{S_2} \mathbf{F} \cdot \mathbf{n} \, dS = -\iint_{D} \mathbf{F} \cdot \frac{\mathbf{N}}{|\mathbf{N}|} \, |\mathbf{N}| \, dx \, dy$$

$$= -\iint_{D} \left[ (x)(-2x) + (y)(-2y) + (3)(1) \right] \, dx \, dy.$$

We change to polar coordinates and find that

$$\iint_{S_2} \mathbf{F} \cdot \mathbf{n}_2 \, dS = \int_0^{2\pi} \int_0^2 (2r^2 - 3) r \, dr \, d\theta = 2\pi \left[ \tfrac{1}{2} r^4 - \tfrac{3}{2} r^2 \right]_0^2 = 4\pi.$$

Hence the total flux of $\mathbf{F}$ out of $T$ is $16\pi$, or approximately 50.27.

## 17-5 PROBLEMS

In each of Problems 1–5, evaluate the surface integral $\iint_S f(x, y, z) \, dS$.

**1** $f(x, y, z) = xyz$; $S$ is the first-octant part of the plane $x + y + z = 1$.

**2** $f(x, y, z) = x + y$; $S$ is the part of the plane $z = 2x + 3y$ that lies within the cylinder $x^2 + y^2 = 9$.

**3** $f(x, y, z) = z$; $S$ is the part of the paraboloid $z = r^2$ that lies within the cylinder $r = 2$.

**4** $f(x, y, z) = (x^2 + y^2)z$; $S$ is the hemispherical surface $\rho = 1$, $z \geq 0$. (*Suggestion:* Use spherical coordinates.)

**5** $f(x, y, z) = z^2$; $S$ is the cylindrical surface parametrized by $x = \cos\theta$, $y = \sin\theta$, $z = z$, $0 \leq \theta \leq 2\pi$, $0 \leq z \leq 2$.

In Problems 6–10, use Formulas (13) and (14) to evaluate the surface integral $\iint_S \mathbf{F} \cdot \mathbf{n} \, dS$, where $\mathbf{n}$ is the upward-pointing unit normal to the given surface $S$.

**6** $\mathbf{F} = x\mathbf{i} + y\mathbf{j} + z\mathbf{k}$; $S$ is the first-octant part of the plane $2x + 2y + z = 3$.

**7** $\mathbf{F} = 2y\mathbf{i} + 3z\mathbf{k}$; $S$ is the part of the plane $z = 3x + 2$ that lies within the cylinder $x^2 + y^2 = 4$.

**8** $\mathbf{F} = z\mathbf{k}$; $S$ is the upper half of the spherical surface $\rho = 2$. (*Suggestion:* Use spherical coordinates.)

**9** $\mathbf{F} = y\mathbf{i} - x\mathbf{j}$; $S$ is the part of the cone $z = r$ that lies within the cylinder $r = 3$.

**10** $\mathbf{F} = 2x\mathbf{i} + 2y\mathbf{j} + 3\mathbf{k}$; $S$ is the paraboloid $z = 4 - x^2 - y^2$ above the $xy$-plane.

**11** The first-octant part of the spherical surface $\rho = a$ has unit density. Find its centroid.

**12** The conical surface $z = r$, $r \leq a$, has constant density $\delta = k$. Find its centroid and its moment of inertia about the $z$-axis.

**13** The paraboloid $z = r^2$, $r \leq a$, has constant density $\delta$. Find its centroid and its moment of inertia about the $z$-axis.

**14** Find the centroid of the part of the spherical surface $\rho = a$ that lies within the cone $r = z$.

**15** Find the centroid of the part of the spherical surface $x^2 + y^2 + z^2 = 4$ that lies inside the cylinder $x^2 + y^2 = 2x$ and above the $xy$-plane.

**16** Suppose that the toroidal surface of Example 4 in Section 16-8 has uniform density and total mass $M$. Show that its moment of inertia about the $z$-axis is $M(3a^2 + 2b^2)/2$.

**17** Let $S$ be the boundary of the region bounded by the coordinate planes and the plane $2x + 3y + z = 6$. Find the flux of $\mathbf{F} = x\mathbf{i} - y\mathbf{j}$ across $S$ in the direction of its outer normal.

**18** Let $S$ be the boundary of the solid that is bounded by the paraboloid $z = 4 - x^2 - y^2$ and the $xy$-plane. Find the flux across $S$ in the direction of the outer normal of the vector field $\mathbf{F} = 2x\mathbf{i} + 2y\mathbf{j} + 3\mathbf{k}$.

**19** Let $S$ be the boundary of the solid bounded by the paraboloids $z = x^2 + y^2$ and $z = 18 - x^2 - y^2$. Find the flux of $\mathbf{F} = z^2\mathbf{k}$ across $S$ in the direction of the outer normal.

**20** Let $S$ be the surface $z = h(x, y)$ for $(x, y)$ in the region $D$ in the $xy$-plane, and let $\gamma$ be the angle between $\mathbf{k}$ and

the upper normal vector $\mathbf{N}$ to $S$. Show that

$$\iint_S f(x, y, z)\, dS = \iint_D f(x, y, h(x, y)) \sec \gamma\, dx\, dy.$$

**21** Consider a homogeneous thin spherical shell $S$ of radius $a$ centered at the origin, with density $\delta$ and total mass $M = 4\pi a^2 \delta$. A particle of mass $m$ is located at the point $(0, 0, c)$ with $c > a$. Use the method and notation of Problem 34 in Section 16-7 to show that the gravitational force of attraction between the particle and the spherical shell is

$$F = \iint_S \frac{Gm\delta\, dS}{w^2} = \frac{GMm}{c^2}.$$

---

## 17-6

## The Divergence Theorem

The *divergence theorem* is for surface integrals what Green's theorem is for line integrals. It lets us convert a surface integral over a closed surface into a triple integral over the enclosed region, or vice versa. The divergence theorem is also known as *Gauss's theorem*, and in some eastern European countries it is called *Ostrogradski's theorem*. The German "prince of mathematics" Carl Friedrich Gauss (1777–1855) used it to study inverse-square force fields, while the Russian Michel Ostrogradski (1801–1861) used it to study heat flow. The work of each was done in the 1830s.

The surface $S$ is called **piecewise smooth** if it consists of finitely many smooth parametric surfaces. It is called **closed** if it is the boundary of a bounded space region. For example, the boundary of a cube is a closed piecewise-smooth surface, as are the boundary of a pyramid and the boundary of a solid cylinder.

---

> *The Divergence Theorem*
>
> Suppose that $S$ is a closed piecewise smooth surface bounding the space region $T$. Let $\mathbf{F} = P\mathbf{i} + Q\mathbf{j} + R\mathbf{k}$ be a vector field with component functions that have continuous first-order partial derivatives on $T$. Let $\mathbf{n}$ be the *outer* unit normal to $S$. Then
>
> $$\iint_S \mathbf{F} \cdot \mathbf{n}\, dS = \iiint_T \nabla \cdot \mathbf{F}\, dV. \qquad (1)$$

---

Equation (1) is a three-dimensional analogue of the vector form of Green's theorem (Equation (9) in Section 17-4). Note that the left-hand side in Equation (1) is the flux of $\mathbf{F}$ across $S$ in the direction of the outer unit normal $\mathbf{n}$. Recall that the *divergence* $\nabla \cdot \mathbf{F}$ of the vector field $\mathbf{F}$ is given in the three-dimensional case by

$$\operatorname{div} \mathbf{F} = \nabla \cdot \mathbf{F} = \frac{\partial P}{\partial x} + \frac{\partial Q}{\partial y} + \frac{\partial R}{\partial z}. \qquad (2)$$

If **n** is given in terms of its direction cosines, as $\mathbf{n} = \langle \cos\alpha, \cos\beta, \cos\gamma \rangle$, then we can write the divergence theorem in scalar form:

$$\iint_S (P\cos\alpha + Q\cos\beta + R\cos\gamma)\,dS = \iiint_T \left(\frac{\partial P}{\partial x} + \frac{\partial Q}{\partial y} + \frac{\partial R}{\partial z}\right) dV. \quad (3)$$

It is best to parametrize $S$ so that the normal vector given by the parametrization is the outer normal. For then Equation (3) can be written entirely in Cartesian form:

$$\iint_S P\,dy\,dz + Q\,dz\,dx + R\,dx\,dy = \iiint_T \left(\frac{\partial P}{\partial x} + \frac{\partial Q}{\partial y} + \frac{\partial R}{\partial z}\right) dV. \quad (4)$$

**Proof of the Divergence Theorem**  We shall prove the theorem only in the case in which the region $T$ is simultaneously $x$-simple, $y$-simple, and $z$-simple. This guarantees that every straight line parallel to a coordinate axis intersects $T$, if at all, in a single point or a line segment. It will suffice to derive separately the equations

$$\iint_S P\,dy\,dz = \iiint_T \frac{\partial P}{\partial x}\,dV,$$

$$\iint_S Q\,dz\,dx = \iiint_T \frac{\partial Q}{\partial y}\,dV, \qquad \text{and} \qquad (5)$$

$$\iint_S R\,dx\,dy = \iiint_T \frac{\partial R}{\partial z}\,dV.$$

For the sum of the Equations (5) is Equation (4).

Because $T$ is $z$-simple, it has the description $h_1(x, y) \leqq z \leqq h_2(x, y)$, $(x, y)$ in $D$. As in Fig. 17.26, we denote the lower surface $z = h_1(x, y)$ of $T$ by $S_1$, the upper surface $z = h_2(x, y)$ by $S_2$, and the lateral surface between $S_1$ and $S_2$ by $S_3$. In the case of some simple surfaces—such as a sphere—there may

**17.26**  A $z$-simple space region bounded by the surfaces $S_1$, $S_2$, and $S_3$

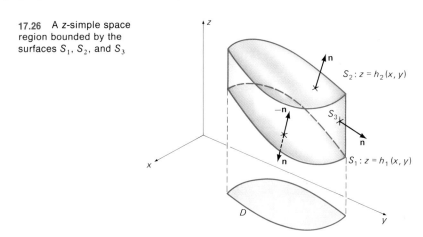

be *no* $S_3$ to consider. But if there is, we see that

$$\iint\limits_{S_3} R \, dx \, dy = \iint\limits_{S_3} R \cos \gamma \, dS = 0 \tag{6}$$

because $\gamma = 90°$ at each point of the vertical cylinder $S_3$.

On the upper surface $S_2$, the upper unit normal corresponding to the parametrization $z = h_2(x, y)$ is the given outer normal **n**, so the formula in Equation (15) of Section 17-5 yields

$$\iint\limits_{S_2} R \, dx \, dy = \iint\limits_{D} R(x, y, h_2(x, y)) \, dx \, dy. \tag{7}$$

But on the lower surface $S_1$, the unit upper normal corresponding to the parametrization $z = h_1(x, y)$ is the inner normal $-\mathbf{n}$, so we must reverse the sign. Thus

$$\iint\limits_{S_1} R \, dx \, dy = -\iint\limits_{D} R(x, y, h_1(x, y)) \, dx \, dy. \tag{8}$$

We add Equations (6), (7), and (8), and get

$$\iint\limits_{S} R \, dx \, dy = \iint\limits_{D} \left( R(x, y, h_2(x, y)) - R(x, y, h_1(x, y)) \right) dx \, dy$$

$$= \iint\limits_{D} \left( \int_{h_1(x,y)}^{h_2(x,y)} \frac{\partial R}{\partial z} \, dz \right) dx \, dy.$$

Therefore,

$$\iint\limits_{S} R \, dx \, dy = \iiint\limits_{T} \frac{\partial R}{\partial z} \, dV.$$

This is the third of the three equations in (5), and the other two may be derived in much the same way. ∎

The divergence theorem is established for more general regions by the device of subdivision of $T$ into simpler regions, ones for which the above proof holds. For example, suppose that $T$ is the shell between the concentric spheres $S_a$ and $S_b$ of radii $a$ and $b$, with $0 < a < b$. The coordinate planes separate $T$ into eight regions $T_1, T_2, T_3, \ldots, T_8$, each shaped like the one of Fig. 17.27. Let $\Sigma_i$ be the boundary of $T_i$, and let $\mathbf{n}_i$ be the outer unit normal to $\Sigma_i$. We apply the divergence theorem to each of these eight regions, and obtain

$$\iiint\limits_{T} \mathbf{\nabla} \cdot \mathbf{F} \, dV = \sum_{i=1}^{8} \iiint\limits_{T_i} \mathbf{\nabla} \cdot \mathbf{F} \, dV$$

$$= \sum_{i=1}^{8} \iint\limits_{\Sigma_i} \mathbf{F} \cdot \mathbf{n}_i \, dS \qquad \text{(divergence theorem)}$$

$$= \iint\limits_{S_a} \mathbf{F} \cdot \mathbf{n}_a \, dS + \iint\limits_{S_b} \mathbf{F} \cdot \mathbf{n}_b \, dS.$$

Here we write $\mathbf{n}_a$ for the inner normal on $S_a$ and $\mathbf{n}_b$ for the outer normal on $S_b$. The last equality holds because the surface integrals over the internal boundary surfaces (the ones in the coordinate planes) cancel in pairs—the

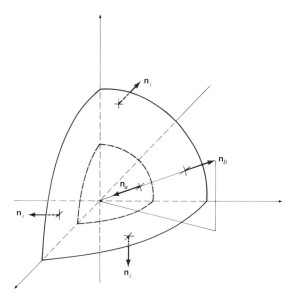

**17.27** One octant of the shell between $S_a$ and $S_b$

normals are oppositely oriented there. Since the boundary $S$ of $T$ is the union of the spherical surfaces $S_a$ and $S_b$, it follows that

$$\iiint_T \nabla \cdot \mathbf{F} \, dV = \iint_S \mathbf{F} \cdot \mathbf{n} \, dS.$$

This is the divergence theorem for the spherical shell $T$.

**EXAMPLE 1** Let $S$ be the surface (with outer unit normal $\mathbf{n}$) of the region $T$ that is bounded by the planes $z = 0$, $y = 0$, $y = 2$, and the paraboloid $z = 1 - x^2$ (see Fig. 17.28). Apply the divergence theorem to compute $\iint_S \mathbf{F} \cdot \mathbf{n} \, dS$ given

$$\mathbf{F} = (x + \cos y)\mathbf{i} + (y + \sin z)\mathbf{j} + (z + e^x)\mathbf{k}.$$

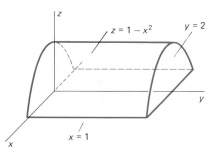

**17.28** The region of Example 1

**Solution** To evaluate the surface integral directly would be quite lengthy. But div $\mathbf{F} = 1 + 1 + 1 = 3$, so the divergence theorem may be applied easily and yields

$$\iint_S \mathbf{F} \cdot \mathbf{n} \, dS = \iiint_T \text{div } \mathbf{F} \, dV = \iiint_T 3 \, dV.$$

We examine Fig. 17.28 to find the limits for the volume integral and thus obtain

$$\iint_S \mathbf{F} \cdot \mathbf{n} \, dS = \int_{-1}^{1} \int_0^2 \int_0^{1-x^2} 3 \, dz \, dy \, dx$$

$$= 12 \int_0^1 (1 - x^2) \, dx = 8.$$

**EXAMPLE 2** Let $S$ be the surface of the solid cylinder $T$ that is bounded by the planes $z = 0$ and $z = 3$ and by the cylinder $x^2 + y^2 = 4$. Calculate the outward flux $\iint_S \mathbf{F} \cdot \mathbf{n} \, dS$ given

$$\mathbf{F} = (x^2 + y^2 + z^2)(x\mathbf{i} + y\mathbf{j} + z\mathbf{k}).$$

**Solution** If we denote by $P$, $Q$, and $R$ the component functions of the vector field $\mathbf{F}$, we find that

$$\frac{\partial P}{\partial x} = (2x)(x) + (x^2 + y^2 + z^2)(1) = 3x^2 + y^2 + z^2.$$

Similarly,

$$\frac{\partial Q}{\partial y} = 3y^2 + z^2 + x^2 \quad \text{and} \quad \frac{\partial R}{\partial z} = 3z^2 + x^2 + y^2,$$

so

$$\text{div } \mathbf{F} = 5(x^2 + y^2 + z^2).$$

Therefore, the divergence theorem yields

$$\iint_S \mathbf{F} \cdot \mathbf{n} \, dS = \iiint_T 5(x^2 + y^2 + z^2) \, dV.$$

Using cylindrical coordinates to evaluate the volume integral, we get

$$\iint_S \mathbf{F} \cdot \mathbf{n} \, dS = \int_0^{2\pi} \int_0^2 \int_0^3 5(r^2 + z^2) \, r \, dz \, dr \, d\theta$$

$$= 10\pi \int_0^2 \left[ r^3 z + \frac{1}{3} r z^3 \right]_{z=0}^3 \, dr$$

$$= 10\pi \int_0^2 (3r^3 + 9r) \, dr$$

$$= 10\pi \left[ \frac{3}{4} r^4 + \frac{9}{2} r^2 \right]_0^2 = 300\pi.$$

**EXAMPLE 3** Suppose that the region $T$ is bounded by the closed surface $S$ with a parametrization that gives the outer unit normal. Show that the volume $V$ of $T$ is given by

$$V = \tfrac{1}{3} \iint_S x \, dy \, dz + y \, dz \, dx + z \, dx \, dy. \tag{9}$$

**Solution** Equation (9) follows immediately from Equation (4) if we take $P(x, y, z) = x$, $Q(x, y, z) = y$, and $R(x, y, z) = z$. For example, if $S$ is the spherical surface $x^2 + y^2 + z^2 = a^2$ with outer unit normal

$$\mathbf{n} = \langle \cos \alpha, \cos \beta, \cos \gamma \rangle = \left\langle \frac{x}{a}, \frac{y}{a}, \frac{z}{a} \right\rangle,$$

then Equation (9) gives

$$V = \tfrac{1}{3} \iint_S x \, dy \, dz + y \, dz \, dx + z \, dx \, dy$$

$$= \tfrac{1}{3} \iint_S (x \cos \alpha + y \cos \beta + z \cos \gamma) \, dS$$

$$= \tfrac{1}{3} \iint_S \frac{x^2 + y^2 + z^2}{a} \, dS = \frac{a}{3} \iint_S 1 \, dS = \frac{1}{3} aA.$$

This result is consistent with the familiar formulas $V = \frac{4}{3}\pi a^3$ and $A = 4\pi a^2$.

## Stokes' Theorem

In Equation (9) of Section 17-4, we gave a vector form of Green's theorem,

$$\oint_C P\,dx + Q\,dy = \iint_R \left(\frac{\partial Q}{\partial x} - \frac{\partial P}{\partial y}\right) dA, \tag{1}$$

that amounted to a two-dimensional version of the divergence theorem. There is another vector form of Green's theorem, one that involves the **curl** of a vector field. Recall that if $\mathbf{F} = P\mathbf{i} + Q\mathbf{j} + R\mathbf{k}$ is a vector field, then curl $\mathbf{F}$ is the vector field given by

$$\text{curl } \mathbf{F} = \mathbf{V} \times \mathbf{F} = \begin{vmatrix} \mathbf{i} & \mathbf{j} & \mathbf{k} \\ \dfrac{\partial}{\partial x} & \dfrac{\partial}{\partial y} & \dfrac{\partial}{\partial z} \\ P & Q & R \end{vmatrix}$$

$$= \left(\frac{\partial R}{\partial y} - \frac{\partial Q}{\partial z}\right)\mathbf{i} + \left(\frac{\partial P}{\partial z} - \frac{\partial R}{\partial x}\right)\mathbf{j} + \left(\frac{\partial Q}{\partial x} - \frac{\partial P}{\partial y}\right)\mathbf{k}. \tag{2}$$

Note first that the **k**-component of $\mathbf{V} \times \mathbf{F}$ is the integrand of the double integral in Equation (1). We know from Section 17-2 that the line integral in (1) can also be written as $\oint_C \mathbf{F} \cdot \mathbf{T}\,ds$, where $\mathbf{T}$ is the positive-directed unit tangent. Consequently Green's theorem can be rewritten in the form

$$\oint_C \mathbf{F} \cdot \mathbf{T}\,ds = \iint_R (\text{curl } \mathbf{F}) \cdot \mathbf{k}\,dA. \tag{3}$$

Stokes' theorem is the generalization of Formula (3) that we get by replacing the plane region $R$ with a floppy two-dimensional version: an oriented bounded surface $S$ in three-dimensional space with boundary $C$ that consists of one or more simple closed space curves.

Recall that an *oriented* surface is one with a continuous unit normal vector $\mathbf{n}$. The positive orientation of the boundary $C$ of an oriented surface $S$ corresponds to the unit tangent vector $\mathbf{T}$ such that $\mathbf{n} \times \mathbf{T}$ always points *into* $S$ (see Fig. 17.31). You should check that for a plane region with unit normal $\mathbf{k}$, the positive orientation of its outer boundary curve is counterclockwise.

17.31 Vectors, surface, and boundary curve mentioned in the statement of Stokes' theorem

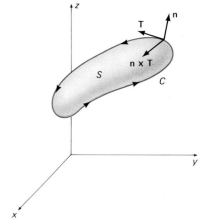

> ### Stokes' Theorem
>
> Let $S$ be an oriented, bounded piecewise smooth surface in space with positively oriented boundary $C$. Suppose that the components of the vector field $\mathbf{F}$ have continuous first order partial derivatives in a space region containing $S$. Then
>
> $$\oint_C \mathbf{F} \cdot \mathbf{T}\,ds = \iint_S (\text{curl } \mathbf{F}) \cdot \mathbf{n}\,dS. \tag{4}$$

Thus Stokes' theorem says that *the line integral around the boundary curve of the tangential component of* **F** *is equal to the surface integral of the normal component of curl* **F**. Compare Equations (3) and (4).

This result first appeared publicly as a problem posed by George Stokes (1819–1903) on a prize examination for Cambridge University students in 1854. It had been stated in an 1850 letter to Stokes from the physicist William Thomson (Lord Kelvin, 1824–1907).

In terms of the components of $\mathbf{F} = P\mathbf{i} + Q\mathbf{j} + R\mathbf{k}$ and those of curl **F**, Stokes' theorem can—with the aid of Equation (14) of Section 17-5—be recast into its scalar form:

$$\oint_C P\,dx + Q\,dy + R\,dz = \iint_S \left(\frac{\partial R}{\partial y} - \frac{\partial Q}{\partial z}\right) dy\,dz + \left(\frac{\partial P}{\partial z} - \frac{\partial R}{\partial x}\right) dz\,dx$$

$$+ \left(\frac{\partial Q}{\partial x} - \frac{\partial P}{\partial y}\right) dx\,dy. \tag{5}$$

Here we need the usual stipulation that the parametrization of $S$ corresponds to the given unit normal **n**.

To prove Stokes' theorem, it is therefore enough to establish the equation

$$\oint_C P\,dx = \iint_S \left(\frac{\partial P}{\partial z}\,dz\,dx - \frac{\partial P}{\partial y}\,dx\,dy\right) \tag{6}$$

and the corresponding two equations that are the $Q$ and $R$ "components" of Equation (5). Equation (5) itself then follows by adding the three results.

*Partial Proof*  Suppose first that $S$ is the graph of a function $z = f(x, y)$, $(x, y)$ in $D$, with an upper unit normal and with $D$ a region in the $xy$-plane bounded by the simple closed curve $J$. (See Fig. 17.32.) Then

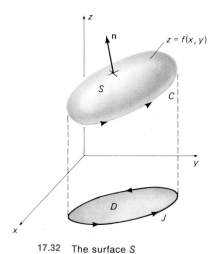

17.32  The surface $S$

$$\oint_C P\,dx = \oint_J P(x, y, f(x, y))\,dx$$

$$= \oint_J p(x, y)\,dx \qquad \text{(where } p(x, y) \equiv P(x, y, f(x, y)))$$

$$= -\iint_D \frac{\partial p}{\partial y}\,dx\,dy \qquad \text{(by Green's theorem)}.$$

We now use the chain rule to compute $\partial p/\partial y$ and find that

$$\oint_C P\,dx = -\iint_D \left(\frac{\partial P}{\partial y} + \frac{\partial P}{\partial z}\frac{\partial z}{\partial y}\right) dx\,dy. \tag{7}$$

Next we use Equation (15) of Section 17-5:

$$\iint_S P\,dy\,dz + Q\,dz\,dx + R\,dx\,dy = \iint_D \left(-P\frac{\partial z}{\partial x} - Q\frac{\partial z}{\partial y} + R\right) dx\,dy.$$

In this equation we replace $P$ by 0, $Q$ by $\partial P/\partial z$, and $R$ by $-\partial P/\partial y$. This gives

$$\iint\limits_{S} \left( \frac{\partial P}{\partial z} \, dz \, dx - \frac{\partial P}{\partial y} \, dx \, dy \right) = \iint\limits_{D} \left( -\frac{\partial P}{\partial z} \frac{\partial z}{\partial y} - \frac{\partial P}{\partial y} \right) dx \, dy. \qquad (8)$$

Finally we compare Equations (7) and (8) and see that we have established Equation (6). If the surface $S$ can also be written in the forms $y = g(x, z)$ and $x = h(y, z)$, then the $Q$ and $R$ "components" of Equation (5) can be derived in essentially the same way. This proves Stokes' theorem for the special case of a surface $S$ that can be represented as a graph in all three coordinate directions. Stokes' theorem may then be extended to a more general oriented surface in the now-familiar way—by subdividing it into simpler surfaces, to each of which the above proof is applicable. ∎

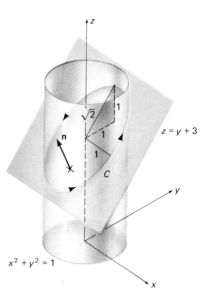

**EXAMPLE 1** Apply Stokes' theorem to evaluate $\oint_C \mathbf{F} \cdot \mathbf{T} \, ds$, where $C$ is the ellipse in which the plane $z = y + 3$ intersects the cylinder $x^2 + y^2 = 1$. Orient the ellipse counterclockwise as viewed from above. Take $\mathbf{F}(x, y, z) = 3z\mathbf{i} + 5x\mathbf{j} - 2y\mathbf{k}$.

**17.33** The ellipse of Example 1

**Solution** The plane, cylinder, and ellipse appear in Fig. 17.33. The given orientation of $C$ corresponds to the upward unit normal $\mathbf{n} = (-\mathbf{j} + \mathbf{k})/\sqrt{2}$ to the elliptical region $S$ in the plane $z = y + 3$ bounded by $C$. Now

$$\text{curl } \mathbf{F} = \begin{vmatrix} \mathbf{i} & \mathbf{j} & \mathbf{k} \\ \dfrac{\partial}{\partial x} & \dfrac{\partial}{\partial y} & \dfrac{\partial}{\partial z} \\ 3z & 5x & -2y \end{vmatrix} = -2\mathbf{i} + 3\mathbf{j} + 5\mathbf{k},$$

so

$$(\text{curl } \mathbf{F}) \cdot \mathbf{n} = (-2\mathbf{i} + 3\mathbf{j} + 5\mathbf{k}) \cdot \frac{1}{\sqrt{2}}(-\mathbf{j} + \mathbf{k}) = \frac{-3 + 5}{\sqrt{2}} = \frac{2}{\sqrt{2}}.$$

Hence, by Stokes' theorem,

$$\oint_C \mathbf{F} \cdot \mathbf{T} \, ds = \iint\limits_{S} (\text{curl } \mathbf{F}) \cdot \mathbf{n} \, dS$$

$$= \iint\limits_{S} \frac{2}{\sqrt{2}} \, dS = \frac{2}{\sqrt{2}} \text{ area } (S) = 2\pi,$$

because we can see from Fig. 17.33 that $S$ is an ellipse with semiaxes 1 and $\sqrt{2}$; thus its area is $\pi\sqrt{2}$.

**17.34** The parabolic surface of Example 2

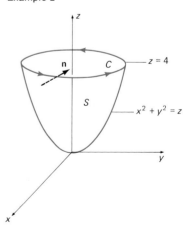

**EXAMPLE 2** Apply Stokes' theorem to evaluate $\iint_S (\nabla \times \mathbf{F}) \cdot \mathbf{n} \, dS$ where $\mathbf{F} = 3z\mathbf{i} + 5x\mathbf{j} - 2y\mathbf{k}$ and $S$ is the part of the parabolic surface $z = x^2 + y^2$ that lies below the plane $z = 4$ and has orientation given by the upper unit normal vector (see Fig. 17.34).

**Solution** We parametrize the boundary circle $C$ of $S$ by $x = 2 \cos t$, $y = 2 \sin t, z = 4$ for $0 \le t \le 2\pi$. Then $dx = -2 \sin t \, dt, dy = 2 \cos t \, dt$, and $dz = 0$. So Stokes' theorem yields

$$\iint_S (\nabla \times \mathbf{F}) \cdot \mathbf{n} \, dS = \oint_C \mathbf{F} \cdot \mathbf{T} \, ds$$

$$= \oint_C 3z \, dx + 5x \, dy - 2y \, dz$$

$$= \int_0^{2\pi} (3)(4)(-2 \sin t \, dt) + (5)(2 \cos t)(2 \cos t \, dt)$$

$$- (2)(2 \sin t)(0)$$

$$= \int_0^{2\pi} (-24 \sin t + 20 \cos^2 t) \, dt$$

$$= \int_0^{2\pi} (-24 \sin t + 10 + 10 \cos 2t) \, dt$$

$$= \left[ 24 \cos t + 10t + 5 \sin 2t \right]_0^{2\pi} = 20\pi.$$

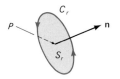

**17.35** A physical interpretation of the curl of a vector field

Just as the divergence theorem yields a physical interpretation of div $\mathbf{F}$ (the formula in Equation (10) in Section 17-6), Stokes' theorem yields a physical interpretation of curl $\mathbf{F}$. Let $S_r$ be a circular disk of radius $r$, centered at the point $P$ in space and perpendicular to the unit vector $\mathbf{n}$. Let $C_r$ be the boundary circle of $S_r$; see Fig. 17.35. Then Stokes' theorem and the average value property of double integrals together give

$$\oint_{C_r} \mathbf{F} \cdot \mathbf{T} \, ds = \iint_{S_r} (\text{curl } \mathbf{F}) \cdot \mathbf{n} \, dS = \pi r^2 [(\text{curl } \mathbf{F}) \cdot \mathbf{n}]_{P*}$$

for some point $P^*$ of $S_r$, where $[(\text{curl } \mathbf{F}) \cdot \mathbf{n}]_{P*}$ denotes the value of $[(\text{curl } \mathbf{F}) \cdot \mathbf{n}]$ at the point $P^*$. We divide this equality by $\pi r^2$ and then take the limit as $r \to 0$. This gives

$$[(\text{curl } \mathbf{F}) \cdot \mathbf{n}]_P = \lim_{r \to 0} \frac{1}{\pi r^2} \oint_{C_r} \mathbf{F} \cdot \mathbf{T} \, ds. \tag{9}$$

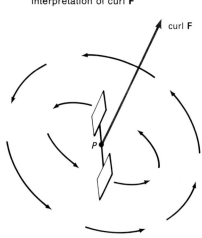

**17.36** The paddle-wheel interpretation of curl $\mathbf{F}$

This last formula has a natural physical meaning. Suppose that $\mathbf{F} = \rho \mathbf{v}$, where $\mathbf{v}$ is the velocity vector field of a steady-state fluid flow with constant density $\rho$. Then the value of the line integral

$$\Gamma(C) = \oint_C \mathbf{F} \cdot \mathbf{T} \, ds \tag{10}$$

measures the rate of flow of fluid mass *around* the curve $C$ and is therefore called the **circulation** of $\mathbf{F}$ around $C$. From (9) we see that

$$[(\text{curl } \mathbf{F}) \cdot \mathbf{n}]_P \approx \frac{\Gamma(C_r)}{\pi r^2}$$

if $C_r$ is a circle of very small radius $r$ centered at $P$ and perpendicular to $\mathbf{n}$. If $[\text{curl } \mathbf{F}]_P \ne \mathbf{0}$, it follows that $\Gamma(C_r)$ is greatest (for $r$ fixed and small) when the unit vector $\mathbf{n}$ points in the direction of $[\text{curl } \mathbf{F}]_P$. Hence the line through $P$ determined by $[\text{curl } \mathbf{F}]_P$ is the axis about which the fluid near $P$ is revolving the most rapidly. A tiny paddle wheel placed in the fluid at $P$ (see Fig. 17.36) would rotate fastest if its axis lay along this line. In Miscellaneous Problem 32 (at the end of this chapter), we ask you to show that—in the case of a fluid revolving steadily about a fixed axis with constant angular speed $\omega$ (in radians per second)—$|\text{curl } \mathbf{F}| = 2\rho\omega$. Thus $[\text{curl } \mathbf{F}]_P$ indicates both the direction *and* rate of rotation of the fluid near $P$. Because of this interpretation, the notation rot $\mathbf{F}$ for the curl is sometimes seen in older books, an abbreviation that we are happy has disappeared from general use.

If curl $\mathbf{F} = \mathbf{0}$ everywhere, the fluid flow and the vector field $\mathbf{F}$ itself are said to be **irrotational.** An infinitesimal straw placed in an irrotational fluid flow would be translated parallel to itself without rotating. It turns out that a vector field $\mathbf{F}$ defined on a simply connected region $D$ is irrotational if and only if it is conservative, which in turn is true if and only if the line integral $\int_C \mathbf{F} \cdot \mathbf{T} \, ds$ is independent of the path in $D$. (The region $D$ is said to be **simply connected** if every simple closed curve in $D$ can be continuously shrunk to a point while staying within $D$. The interior of a torus is an example of a space region that is *not* simply connected. It is true, though not obvious, that any piecewise-smooth simple closed curve in a simply connected region $D$ is the boundary of a piecewise smooth oriented surface in $D$.)

---

**Theorem** *Conservative and Irrotational Fields*

Let $\mathbf{F}$ be a vector field with continuous first order partial derivatives in a simply connected space region $D$. Then the vector field $\mathbf{F}$ is irrotational if and only if it is conservative; that is, $\nabla \times \mathbf{F} = \mathbf{0}$ if and only if $\mathbf{F} = \nabla\phi$ for some scalar function $\phi$ defined on $D$.

---

*Partial Proof* A complete proof of the *if* part of the theorem is easy; in Example 7 of Section 17-1 we showed that $\nabla \times (\nabla\phi) = \mathbf{0}$ for any twice-differentiable scalar function $\phi$.

Here is a sketch of how one might show the *only if* part of the proof of the theorem. Assume that $\mathbf{F}$ is irrotational. Let $P_0(x_0, y_0, z_0)$ be a fixed point of $D$. Given an arbitrary point $P(x, y, z)$ of $D$, we would like to define

$$\phi(x, y, z) = \int_{C_1} \mathbf{F} \cdot \mathbf{T} \, ds \tag{11}$$

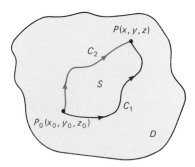

**17.37** Two paths from $P_0$ to $P$ in the simply connected space region $D$

where $C_1$ is a path in $D$ from $P_0$ to $P$. But it is essential that we show that any *other* path $C_2$ from $P_0$ to $P$ would give the *same* value for $\phi(x, y, z)$.

We may assume, as suggested by Fig. 17.37, that $C_1$ and $C_2$ intersect only at their end points. Then the simple closed curve $C = C_1 \cup (-C_2)$ bounds an oriented surface $S$, and Stokes' theorem gives

$$\int_{C_1} \mathbf{F} \cdot \mathbf{T} \, ds - \int_{C_2} \mathbf{F} \cdot \mathbf{T} \, ds = \oint_C \mathbf{F} \cdot \mathbf{T} \, ds = \iint_S (\nabla \times \mathbf{F}) \cdot \mathbf{n} \, dS = 0$$

because of our hypothesis that $\nabla \times \mathbf{F} \equiv \mathbf{0}$. This shows that the line integral $\int_C \mathbf{F} \cdot \mathbf{T} \, ds$ is *independent of path*, just as desired. In Problem 21 we ask you to complete this proof by showing that the function $\phi$ of Equation (11) is the one whose existence is claimed by the theorem. That is, if $\phi$ is defined as in Equation (11), then $\mathbf{F} = \nabla\phi$. ∎

**EXAMPLE 3** Show that the vector field $\mathbf{F} = 3x^2\mathbf{i} + 5z^2\mathbf{j} + 10yz\mathbf{k}$ is irrotational. Then find a potential function $\phi(x, y, z)$ such that $\nabla\phi = \mathbf{F}$.

---

*Solution* To show that **F** is irrotational, we calculate

$$\nabla \times \mathbf{F} = \begin{vmatrix} \mathbf{i} & \mathbf{j} & \mathbf{k} \\ \dfrac{\partial}{\partial x} & \dfrac{\partial}{\partial y} & \dfrac{\partial}{\partial z} \\ 3x^2 & 5z^2 & 10yz \end{vmatrix} = (10z - 10z)\mathbf{k} = \mathbf{0}.$$

Hence the theorem above implies that **F** has a potential function $\phi$. We can apply Equation (11) to actually find $\phi$. If $C_1$ is the straight line segment from $(0, 0, 0)$ to $(x_1, y_1, z_1)$ parametrized by $x = x_1 t$, $y = y_1 t$, $z = z_1 t$ for $0 \leq t \leq 1$, then (11) yields

$$\begin{aligned} \phi(x_1, y_1, z_1) &= \int_{C_1} \mathbf{F} \cdot \mathbf{T}\, ds \\ &= \int_{(0,0,0)}^{(x_1, y_1, z_1)} 3x^2\, dx + 5z^2\, dy + 10yz\, dz \\ &= \int_0^1 (3x_1^2 t^2)(x_1\, dt) + (5z_1^2 t^2)(y_1\, dt) + (10 y_1 t z_1 t)(z_1\, dt) \\ &= \int_0^1 (3x_1^3 t^2 + 15 y_1 z_1^2 t^2)\, dt \\ &= \left[ x_1^3 t^3 + 5 y_1 z_1^2 t^3 \right]_0^1, \end{aligned}$$

and thus

$$\phi(x_1, y_1, z_1) = x_1^3 + 5 y_1 z_1^2.$$

We may drop the subscripts because $(x_1, y_1, z_1)$ is actually an arbitrary point of space, and therefore we obtain the scalar potential function $\phi(x, y, z) = x^3 + 5yz^2$. As a check, we note that $\phi_x = 3x^2$, $\phi_y = 5z^2$, and $\phi_z = 10yz$, so $\nabla \phi = \mathbf{F}$ as desired.

---

APPLICATION  Suppose that **v** is the velocity vector field of a steady fluid flow that is both irrotational and incompressible—the density $\rho$ of the fluid is constant. Suppose that $S$ is any closed surface bounding a region $T$. Then, because of conservation of mass, the flux of **v** across $S$ must be zero; the mass of fluid within $S$ remains constant. Hence the divergence theorem gives

$$\iiint_T \operatorname{div} \mathbf{v}\, dV = \iint_S \mathbf{v} \cdot \mathbf{n}\, dS = 0.$$

Since this holds for *any* region $T$, it follows from the usual average value property argument that div **v** $= 0$ everywhere. The scalar function $\phi$ provided by the theorem above, for which $\mathbf{v} = \nabla \phi$, is called the **velocity potential** of the fluid flow. We substitute $\mathbf{v} = \nabla \phi$ into the equation div **v** $= 0$ and thereby obtain

$$\operatorname{div}(\nabla \phi) = \frac{\partial^2 \phi}{\partial x^2} + \frac{\partial^2 \phi}{\partial y^2} + \frac{\partial^2 \phi}{\partial z^2} = 0. \tag{12}$$

Thus the velocity potential $\phi$ of an irrotational and incompressible fluid flow satisfies *Laplace's equation*.

Now consider an *airfoil*, shaped like a (noncircular) cylinder perpendicular to the $xy$-plane and placed in a steady flow that is initially parallel to the $x$-axis, $\mathbf{v}_0 = u_0\mathbf{i}$, as shown in cross section in Fig. 17.38. The assumption of incompressibility is valid in the range of subsonic aerodynamics (Mach number less than 0.3). To determine the resulting flow around the airfoil, it suffices to solve Laplace's equation in (12) for the velocity potential $\phi$. The solution $\phi$ must satisfy the *boundary condition* that $\mathbf{v} = \nabla\phi \approx u_0\mathbf{i}$ at points far from the airfoil, and that $\mathbf{v}$ is tangential to the boundary curve $C$ of the airfoil. (Since the $z$-coordinate is not involved, we may project everything into the $xy$-plane.)

The solution of Laplace's equation ordinarily requires advanced techniques, but suppose that it has been carried out and that we now know the velocity vector field $\mathbf{v} = \nabla\phi$ of the fluid flow around the airfoil. Then a standard theorem of fluid dynamics says that the net force or *lift* $\mathbf{L}$ exerted on the airfoil by the fluid is perpendicular to the *free stream* velocity $\mathbf{v}_0 = u_0\mathbf{i}$, and its magnitude is $|\mathbf{L}| = \rho\mu_0\Gamma$, where $\Gamma = \oint_C \mathbf{v} \cdot \mathbf{T}\, ds$ is the circulation of $\mathbf{v}$ around $C$ (the boundary curve of the airfoil). Thus the question of "whether it'll fly" boils down to the evaluation of a line integral! For the "standard theorem" to which we refer above, see page 122 in W. F. Hughes and J. A. Brighton, *Theory and Problems of Fluid Dynamics*, Schaum's Outline (New York: McGraw-Hill, 1967).

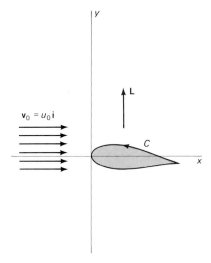

**17.38** An airfoil

This brief outline of a single application should indicate how the mathematics of this chapter forms the starting point for investigations in a number of areas, including acoustics, aerodynamics, electromagnetism, meteorology, and oceanography, among others. Indeed, the entire subject of vector analysis stems historically from its scientific applications rather than from abstract mathematical considerations. The modern form of the subject is due primarily to J. Willard Gibbs (1839–1903), the first great American physicist, and the English electrical engineer Oliver Heaviside (1850–1925).

## 17-7 PROBLEMS

In Problems 1–5, use Stokes' theorem for the evaluation of $\iint_S [\text{curl } \mathbf{F}] \cdot \mathbf{n}\, dS$.

**1** $\mathbf{F} = 3y\mathbf{i} - 2x\mathbf{j} + xyz\mathbf{k}$, and $S$ is the hemispherical surface $z = +(4 - x^2 - y^2)^{1/2}$ with upper unit normal.

**2** $\mathbf{F} = 2y\mathbf{i} + 3x\mathbf{j} + e^z\mathbf{k}$, and $S$ is the part of the paraboloid $z = x^2 + y^2$ below the plane $z = 4$ with upper unit normal.

**3** $\mathbf{F} = \langle xy, -2, \tan^{-1}x^2\rangle$, and $S$ is the part of the paraboloid $z = 9 - x^2 - y^2$ above the $xy$-plane with upper unit normal.

**4** $\mathbf{F} = yz\mathbf{i} + xz\mathbf{j} + xy\mathbf{k}$, and $S$ is the part of the cylinder $x^2 + y^2 = 1$ between the two planes $z = 1$ and $z = 3$ with outer unit normal.

**5** $\mathbf{F} = \langle yz, -xz, z^3\rangle$, and $S$ is the part of the cone $z = (x^2 + y^2)^{1/2}$ between the two planes $z = 1$ and $z = 3$ with upper unit normal.

In Problems 6–10, use Stokes' theorem to evaluate $\oint_C \mathbf{F} \cdot \mathbf{T}\, ds$.

**6** $\mathbf{F} = 3y\mathbf{i} - 2z\mathbf{j} + 4x\mathbf{k}$, and $C$ is the circle $x^2 + y^2 = 9$, $z = 4$ oriented counterclockwise as viewed from above.

**7** $\mathbf{F} = 2z\mathbf{i} + x\mathbf{j} + 3y\mathbf{k}$, and $C$ is the ellipse in which the plane $z = x$ meets the cylinder $x^2 + y^2 = 4$, oriented counterclockwise as viewed from above.

**8** $\mathbf{F} = y\mathbf{i} + z\mathbf{j} + x\mathbf{k}$, and $C$ is the boundary of the triangle with vertices $(0, 0, 0)$, $(2, 0, 0)$, and $(0, 2, 2)$, oriented counterclockwise as viewed from above.

**9** $\mathbf{F} = \langle y - x, x - z, x - y\rangle$, and $C$ is the boundary of the part of the plane $x + 2y + z = 2$ in the first octant, oriented counterclockwise as viewed from above.

**10** $\mathbf{F} = y^2\mathbf{i} + z^2\mathbf{j} + x^2\mathbf{k}$, and $C$ is the intersection of plane $z = y$ and the cylinder $x^2 + y^2 = 2y$, oriented counterclockwise as viewed from above.

In each of Problems 11–14, first show that the given vector field $\mathbf{F}$ is irrotational, and then apply the method of Example 3 to find a potential function $\phi(x, y, z)$ for $\mathbf{F}$.

**11** $\mathbf{F} = (3y - 2z)\mathbf{i} + (3x + z)\mathbf{j} + (y - 2x)\mathbf{k}$
**12** $\mathbf{F} = (3y^3 - 10xz^2)\mathbf{i} + 9xy^2\mathbf{j} - 10x^2z\mathbf{k}$
**13** $\mathbf{F} = (3e^z - 5y\sin x)\mathbf{i} + (5\cos x)\mathbf{j} + (17 + 3xe^z)\mathbf{k}$
**14** $\mathbf{F} = r^3\mathbf{r}$, where $\mathbf{r} = x\mathbf{i} + y\mathbf{j} + z\mathbf{k}$ and $r = |\mathbf{r}|$

**15** Suppose that $\mathbf{r} = x\mathbf{i} + y\mathbf{j} + z\mathbf{k}$ and that $\mathbf{a}$ is a constant vector. Show that
(a) $\nabla \cdot (\mathbf{a} \times \mathbf{r}) = 0$;
(b) $\nabla \times (\mathbf{a} \times \mathbf{r}) = 2\mathbf{a}$;
(c) $\nabla \cdot ([\mathbf{r} \cdot \mathbf{r}]\mathbf{a}) = 2\mathbf{r} \cdot \mathbf{a}$;
(d) $\nabla \times ([\mathbf{r} \cdot \mathbf{r}]\mathbf{a}) = 2(\mathbf{r} \times \mathbf{a})$.

**16** Show that $\iint_S (\text{curl } \mathbf{F}) \cdot \mathbf{n} \, dS$ has the same value for all oriented surfaces $S$ that have the same oriented boundary curve $C$.

**17** Suppose that $S$ is a closed surface. Show that

$$\iint_S (\text{curl } \mathbf{F}) \cdot \mathbf{n} \, dS = 0$$

in two different ways:
(a) by using the divergence theorem, with $T$ the region bounded outside by $S$;
(b) by using Stokes' theorem, with the aid of a nice simple closed curve $C$ on $S$.

Line integrals, surface integrals, and triple integrals of vector-valued functions are defined by componentwise integration. Such integrals appear in the following three problems.

**18** Suppose that $C$ and $S$ are as in the statement of Stokes' theorem and that $\phi$ is a scalar function. Show that

$$\oint_C \phi\mathbf{T} \, ds = \iint_S \mathbf{n} \times \nabla\phi \, dS.$$

(*Suggestion:* Apply Stokes' theorem with $\mathbf{F} = \phi\mathbf{a}$, where $\mathbf{a}$ is an arbitrary constant vector.)

**19** Suppose that $\mathbf{a}$ and $\mathbf{r}$ are as in Problem 15. Show that

$$\oint_C \mathbf{a} \times \mathbf{r} \cdot \mathbf{T} \, ds = 2\mathbf{a} \cdot \iint_S \mathbf{n} \, dS.$$

**20** Suppose that $S$ is a closed surface that bounds the region $T$. Show that

$$\iint_S \mathbf{n} \times \mathbf{F} \, dS = \iiint_T \nabla \times \mathbf{F} \, dV.$$

(*Suggestion:* Apply the divergence theorem to $\mathbf{F} \times \mathbf{a}$ where $\mathbf{a}$ is an arbitrary constant vector.) Note that the formulas of this problem, the divergence theorem, and Problem 20 in Section 17-6 all fit the pattern

$$\iint_S \mathbf{n} * (\ ) \, dS = \iiint_T \nabla * (\ ) \, dV,$$

where $*$ may denote either ordinary multiplication, the dot product, or the cross product, and either a scalar function or a vector-valued function is placed within the parentheses, as appropriate.

**21** Suppose that the line integral $\int_C \mathbf{F} \cdot \mathbf{T} \, ds$ is independent of path. If $\phi(x, y, z) = \int_{P_0}^{P} \mathbf{F} \cdot \mathbf{T} \, ds$ as in Equation (11), show that $\nabla\phi = \mathbf{F}$. (*Suggestion:* Note if $L$ is the line segment from $(x, y, z)$ to $(x + \Delta x, y, z)$, then

$$\phi(x + \Delta x, y, z) - \phi(x, y, z) = \int_L \mathbf{F} \cdot \mathbf{T} \, ds = \int_x^{x+\Delta x} P \, dx.)$$

**22** Let $T$ be the submerged body of Problem 21 in Section 17-6, with centroid

$$\mathbf{r}_0 = \frac{1}{V} \iiint_T \mathbf{r} \, dV.$$

The torque about $\mathbf{r}_0$ of Archimedes' buoyant force $\mathbf{B} = -W\mathbf{k}$ is given by

$$\mathbf{L} = \iint_S (\mathbf{r} - \mathbf{r}_0) \times (-\rho gz\mathbf{n}) \, dS.$$

(Why?) Apply the result of Problem 20 above to show that $\mathbf{L} = \mathbf{0}$. It follows that $\mathbf{B}$ acts along the vertical line through the centroid $\mathbf{r}_0$ of the submerged body. (Why?)

---

## CHAPTER 17 REVIEW:  Definitions, Concepts, Results

Use the list below as a guide to concepts that you may need to review.

**1** Definition and evaluation of the line integral $\int_C f(x, y, z) \, ds$

**2** Definition and evaluation of the line integral

$$\int_C P \, dx + Q \, dy + R \, dz$$

**3** Relationship between the two types of line integrals, and the line integral of the tangential component of a vector field

**4** Line integrals and independence of path

**5** Green's theorem

**6** Flux and the vector form of Green's theorem

**7** The divergence of a vector field

**8** Definition and evaluation of the surface integral $\iint_S f(x, y, z) \, dS$

**9** Definition and evaluation of the surface integral $\iint_S P \, dy \, dz + Q \, dz \, dx + R \, dx \, dy$

**10** Relationship between the two types of surface integrals, and the flux of a vector field across a surface

**11** The divergence theorem, in vector and in scalar notation

**12** Gauss's law and inverse-square force fields

**13** The curl of a vector field

**14** Stokes' theorem, in vector and in scalar notation

**15** The circulation of a vector field around a simple closed curve

**16** Physical interpretations of the divergence and the curl of a vector field

In each of Problems 1–5, evaluate the given line integral.

**1** $\int_C (x^2 + y^2)\, ds$, where $C$ is the straight-line segment from $(0, 0)$ to $(3, 4)$.

**2** $\int_C y^2\, dx + x^2\, dy$, where $C$ is the graph of $y = x^3$ from $(-1, -1)$ to $(1, 1)$.

**3** $\int_C \mathbf{F} \cdot \mathbf{T}\, ds$ where $\mathbf{F} = x\mathbf{i} + y\mathbf{j} + z\mathbf{k}$ and $C$ is the curve $x = e^{2t}$, $y = e^t$, $z = e^{-t}$, $0 \le t \le \ln 2$.

**4** $\int_C xyz\, ds$, where $C$ is the path from $(1, 1, 2)$ to $(2, 3, 6)$ consisting of three straight line segments, the first parallel to the $x$-axis, the second parallel to the $y$-axis, and the third parallel to the $z$-axis.

**5** $\int_C \sqrt{z}\, dx + \sqrt{x}\, dy + y^2\, dz$, where $C$ is the curve $x = t$, $y = t^{3/2}$, $z = t^2$, $0 \le t \le 4$.

**6** Apply Theorem 1 in Section 17-3 to show that the line integral $\int_C y^2\, dx + 2xy\, dy + z\, dz$ is independent of the path $C$ from $A$ to $B$.

**7** Apply Theorem 1 in Section 17-3 to show that the line integral $\int_C x^2 y\, dx + xy^2\, dy$ is not independent of the path $C$ from $(0, 0)$ to $(1, 1)$.

**8** A wire shaped like the circle $x^2 + y^2 = a^2$, $z = 0$, has constant density and total mass $M$. Find its moment of inertia about:
(a) the $z$-axis;
(b) the $x$-axis.

**9** A wire shaped like the parabola $y = \frac{1}{2}x^2$, $0 \le x \le 2$, has density function $\rho = x$. Find its mass and moment of inertia about the $y$-axis.

**10** Find the work done by the force field $\mathbf{F} = z\mathbf{i} - x\mathbf{j} + y\mathbf{k}$ in moving a particle from $(1, 1, 1)$ to $(2, 4, 8)$ along the curve $y = x^2$, $z = x^3$.

**11** Apply Green's theorem to evaluate the line integral $\oint_C x^2 y\, dx + xy^2\, dy$, where $C$ is the boundary of the region between the curves $y = x^2$ and $y = 8 - x^2$.

**12** Evaluate the line integral $\oint_C x^2\, dy$, where $C$ is the cardioid $r = 1 + \cos\theta$, by first applying Green's theorem and then changing to polar coordinates.

**13** Let $C_1$ be the circle $x^2 + y^2 = 1$ and $C_2$ the circle $(x - 1)^2 + y^2 = 9$. Show that, if $\mathbf{F} = x^2 y\mathbf{i} - xy^2\mathbf{j}$, then $\oint_{C_1} \mathbf{F} \cdot \mathbf{n}\, ds = \oint_{C_2} \mathbf{F} \cdot \mathbf{n}\, ds$.

**14** (a) Let $C$ be the straight line segment from $(x_1, y_1)$ to $(x_2, y_2)$. Show that

$$\frac{1}{2}\int_C -y\, dx + x\, dy = \frac{1}{2}(x_1 y_2 - x_2 y_1).$$

(b) Suppose that the vertices of a polygon are $(x_1, y_1)$, $(x_2, y_2), \ldots, (x_n, y_n)$, named in counterclockwise order around the polygon. Apply part (a) to show that the area of the polygon is

$$A = \frac{1}{2}\sum_{i=1}^{n}(x_i y_{i+1} - x_{i+1} y_i)$$

where $x_{n+1}$ means $x_1$ and $y_{n+1}$ means $y_1$.

**15** Suppose that the line integral $\oint P\, dx + Q\, dy$ is independent of path in the plane region $D$. Show that $\oint_C P\, dx + Q\, dy = 0$ for every piecewise smooth simple closed curve $C$ in $D$.

**16** Use Green's theorem to show that $\oint_C P\, dx + Q\, dy = 0$ for every piecewise smooth simple closed curve $C$ in the plane region $D$ if and only if $\partial P/\partial y = \partial Q/\partial x$ in $D$.

**17** Evaluate the surface integral $\iint_S (x^2 + y^2 + 2z)\, dS$ if $S$ is the part of the paraboloid $z = 2 - x^2 - y^2$ above the $xy$-plane.

**18** Suppose that $\mathbf{F} = (x^2 + y^2 + z^2)(x\mathbf{i} + y\mathbf{j} + z\mathbf{k})$ and that $S$ is the spherical surface $x^2 + y^2 + z^2 = a^2$. Evaluate $\iint_S \mathbf{F} \cdot \mathbf{n}\, dS$ without actually performing an antidifferentiation.

**19** Let $T$ be the solid bounded by the paraboloids $z = x^2 + 2y^2$ and $z = 12 - 2x^2 - y^2$ and suppose that $\mathbf{F} = x\mathbf{i} + y\mathbf{j} + z\mathbf{k}$. Find (by evaluation of surface integrals) the outward flux of $\mathbf{F}$ across the boundary of $T$.

**20** Give a reasonable definition—as a surface integral—of the average distance of the point $P$ from points of the surface $S$. Then show that the average distance of a fixed point of a spherical surface of radius $a$ from all points of the surface is $4a/3$.

**21** Suppose that the surface $S$ is the graph of the equation $x = g(y, z)$ for $(y, z)$ in the region $D$ of the $yz$-plane. Show that

$$\iint_S P\, dy\, dz + Q\, dz\, dx + R\, dx\, dy$$

$$= \iint_D \left( P - Q\frac{\partial x}{\partial y} - R\frac{\partial x}{\partial z} \right) dy\, dz.$$

**22** Suppose that the surface $S$ is the graph of the equation $y = g(x, z)$ for $(x, z)$ in the region $D$ of the $xz$-plane. Show that

$$\iint_S f(x, y, z)\, dS = \iint_D f(x, g(x, z), z)\sec\beta\, dx\, dz$$

where $\sec\beta = \sqrt{1 + (\partial y/\partial x)^2 + (\partial y/\partial z)^2}$.

**23** Let $T$ be a space region with volume $V$, boundary surface $S$, and centroid $(\bar{x}, \bar{y}, \bar{z})$. Use the divergence theorem to show that $\bar{z} = (1/2V)\iint_S z^2\, dx\, dy$.

**24** Apply the result of Problem 23 to find the centroid of the solid hemisphere $x^2 + y^2 + z^2 \le a^2$, $z \ge 0$.

Problems 25–27 outline the derivation of the heat equation for a body having temperature $u = u(x, y, z, t)$ at the point $(x, y, z)$ at time $t$. Denote by $K$ its heat conductivity and by $c$ its heat capacity, both assumed constant, and let $k = K/c$. Let $B$ be a small solid ball within the body, and let $S$ denote the boundary sphere of $B$.

**25** Deduce from the divergence theorem and Equation (18)

of Section 17-5 that the rate of flow of heat across $S$ *into* $B$ is $R = \iiint_B K\nabla^2 u \, dV$.

**26** The meaning of heat capacity is that, if $\Delta u$ is small, then $(c \, \Delta u) \, \Delta V$ calories of heat are required to raise the temperature of the volume $\Delta V$ by $\Delta u$ degrees. It follows that the rate at which the volume $\Delta V$ is absorbing heat is $c(\partial u/\partial t) \, \Delta V$ (why?). Conclude that the rate of flow of heat into $B$ is $R = \iiint_B c(\partial u/\partial t) \, dV$.

**27** Equate the results of Problems 25 and 26, apply the average value property of triple integrals, and then take the limit as the radius of the ball $B$ approaches zero. You should thereby obtain the heat equation $\partial u/\partial t = k\nabla^2 u$.

**28** For a *steady state* temperature function—one that is independent of time $t$—the heat equation reduces to Laplace's equation

$$\nabla^2 u = \frac{\partial^2 u}{\partial x^2} + \frac{\partial^2 u}{\partial y^2} + \frac{\partial^2 u}{\partial z^2} = 0.$$

(a) Suppose that $u_1$ and $u_2$ are two solutions of Laplace's equation in the region $T$, and that $u_1$ and $u_2$ agree on its boundary surface $S$. Apply Problem 17 in Section 17-6 to $f = u_1 - u_2$ to conclude that $\nabla f = \mathbf{0}$ at each point of $T$.
(b) From the fact that $\nabla f \equiv \mathbf{0}$ in $T$ and $f \equiv 0$ on $S$, conclude that $f \equiv 0$, so that $u_1 \equiv u_2$. Thus the steady-state temperatures within a region are *determined* by the boundary value temperatures.

**29** Suppose that $\mathbf{r} = x\mathbf{i} + y\mathbf{j} + z\mathbf{k}$ and that $\phi(r)$ is a scalar function of $r = |\mathbf{r}|$. Compute:
(a) $\nabla\phi(r)$;
(b) $\operatorname{div}[\phi(r)\mathbf{r}]$;
(c) $\operatorname{curl}[\phi(r)\mathbf{r}]$.

**30** Let $S$ be the upper half of the torus obtained by revolving the circle $(y - a)^2 + z^2 = b^2$ in the $yz$-plane around the $z$-axis, with upper unit normal. Describe how to subdivide $S$ to establish Stokes' theorem for it. How are the two boundary circles oriented?

**31** Explain why the method of subdivision does not suffice to establish Stokes' theorem for the Möbius strip of Fig. 17.22.

**32** (a) Suppose that a fluid or a rigid body is rotating with angular speed $\omega$ radians per second about the line through the origin determined by the unit vector $\mathbf{u}$. Show that the velocity of the point with position vector $\mathbf{r}$ is $\mathbf{v} = \boldsymbol{\omega} \times \mathbf{r}$, where $\boldsymbol{\omega} = \omega\mathbf{u}$ is the angular velocity vector. Note first that $|\mathbf{v}| = \omega|\mathbf{r}| \sin \theta$ where $\theta$ is the angle between $\mathbf{r}$ and $\boldsymbol{\omega}$.
(b) Use the fact that $\mathbf{v} = \boldsymbol{\omega} \times \mathbf{r}$ (established in Part (a)) to show that $\operatorname{curl} \mathbf{v} = 2\boldsymbol{\omega}$.

**33** Consider the steady flow past a circular cylinder as shown in Fig. 17.39. Let

$$\phi(x, y) = u_0 x\left(1 + \frac{a^2}{x^2 + y^2}\right) = u_0 x\left(1 + \frac{a^2}{r^2}\right).$$

(a) Show that $\phi(x, y)$ satisfies Laplace's equation.
(b) Show that, in polar coordinates,

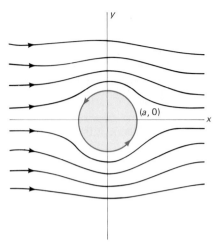

**17.39** Fluid flow around a cylindrical airfoil

$$\mathbf{v} = \nabla\phi = \frac{u_0 a^2}{r^2}\left[\left(\frac{r^2}{a^2} - \cos 2\theta\right)\mathbf{i} - (\sin 2\theta)\mathbf{j}\right].$$

(c) Show that $\lim\limits_{r \to \infty} (\mathbf{v} - u_0\mathbf{i}) = \mathbf{0}$ and that $\mathbf{v}$ is tangential to the circle $r = a$.
(d) Parts (a), (b), and (c) above establish that $\phi(x, y)$ as given is the velocity potential for the indicated flow. Now show that the lift

$$|\mathbf{L}| = \rho u_0 \oint_C \mathbf{v} \cdot \mathbf{T} \, ds$$

is zero, as symmetry indicates it should be.

**34** Consider an incompressible fluid flowing in space (no sources or sinks) with variable density $\rho(x, y, z, t)$ and velocity field $\mathbf{v}(x, y, z, t)$. Let $B$ be a small ball with radius $r$ and spherical surface $S$ centered at the point $(x_0, y_0, z_0)$. Then the amount of fluid within $S$ at time $t$ is

$$Q(t) = \iiint_B \rho \, dV,$$

and differentiation under the integral sign yields

$$Q'(t) = \iiint_B \frac{\partial \rho}{\partial t} \, dV.$$

(a) Consider fluid flow across $S$ to get

$$Q'(t) = -\iint_S \rho\mathbf{v} \cdot \mathbf{n} \, dS$$

where $\mathbf{n}$ is the outer unit normal to $S$. Now apply the divergence theorem to convert this to a volume integral.
(b) Equate your two volume integrals for $Q'(t)$, apply the mean value theorem for integrals, and finally take limits as $r \to 0$ to obtain the **continuity equation**

$$\frac{\partial \rho}{\partial t} + \nabla \cdot (\rho\mathbf{v}) = 0.$$

# 18

# Differential Equations

## Differential Equations and Mathematical Models

Because the derivative $dx/dt = f'(t)$ of the function $f$ may be regarded as the rate at which the quantity $x = f(t)$ changes with respect to the independent variable $t$, it is natural that equations involving derivatives are those that describe the changing universe. An equation involving an unknown function and one or more of its derivatives is called a *differential equation*, and the study of differential equations has two principal goals:

1 To discover the differential equation that describes a physical situation.
2 To find the appropriate solution of that equation.

Unlike algebra, in which we seek the unknown numbers that satisfy an equation such as $x^3 + 7x^2 - 11x + 41 = 0$, in solving a differential equation we are challenged to find the unknown *functions* $y = g(x)$ for which an identity such as

$$g'(x) - 2xg(x) = 0$$

—in Leibniz notation,

$$\frac{dy}{dx} - 2xy = 0$$

—holds on some interval of real numbers. Ordinarily we will want to find *all* solutions of the differential equation if possible.

The following three examples from earlier sections illustrate the process of translating scientific laws and principles into differential equations by interpreting rates of change as derivatives. In each of these examples the independent variable is time $t$, but there are numerous applications in which some quantity other than time is the independent variable.

EXAMPLE 1 (Section 7-6) Newton's law of cooling may be stated in the following form: The *time rate of change* (the rate of change with respect to time $t$) of the temperature $T(t)$ of a body is proportional to the difference between $T$ and the temperature $A$ of the surrounding medium. That is,

$$\frac{dT}{dt} = k(A - T) \tag{1}$$

where $k$ is a positive constant.

Thus the physical law is translated into a differential equation. If we are given the values of $k$ and $A$, we hope to find an explicit formula for $T(t)$, and then—with the aid of this formula—we can predict the future temperature of the body.

EXAMPLE 2 (Section 7-5) The *time rate of change* of a population $P(t)$ with constant birth and death rates is, in many simple cases, proportional to the size of the population. That is,

$$\frac{dP}{dt} = kP \tag{2}$$

where $k$ is the constant of proportionality.

**EXAMPLE 3** (Section 4-8)  Torricelli's law implies that the *time rate of change* of the volume $V$ of water in a draining tank is proportional to the square root of the depth $y$ of the water in the tank:

$$\frac{dV}{dt} = -k\sqrt{y} \tag{3}$$

where $k$ is constant. If the tank is a cylinder with cross-sectional area $A$, then $V = Ay$, and so $dV/dt = A(dy/dt)$. In this case Equation (3) takes the form

$$\frac{dy}{dt} = -h\sqrt{y} \tag{4}$$

where $h = k/A$.

---

Let us discuss Example 2 further. Note first that each function of the form

$$P(t) = Ce^{kt} \tag{5}$$

is a solution of the differential equation

$$\frac{dP}{dt} = kP \tag{2}$$

because

$$P'(t) = Cke^{kt} = k(Ce^{kt}) = kP(t)$$

for all real numbers $t$. Because substitution of each function of the form given in (5) into Equation (2) produces an identity, all these functions are solutions of Equation (2).

Thus even if the value of the constant $k$ is known, the differential equation $dP/dt = kP$ has *infinitely* many different solutions of the form $P(t) = Ce^{kt}$—one for each choice of the "arbitrary" constant $C$. This is typical of differential equations in general. It is also fortunate because it allows us to use additional information to select from all the solutions a particular one that fits the situation under study.

**EXAMPLE 4**  Suppose that $P(t)$ is the population of a bacterial colony at time $t$, that the population at time $t = 0$ (hours) was 1000, and that the population doubled after 1 h. This additional information about the function $P(t)$ yields the following equations:

$$1000 = P(0) = Ce^0 = C, \qquad 2000 = P(1) = Ce^k.$$

It follows that $C = 1000$ and that $k = \ln 2$. Thus the function $P(t)$ describing the population of this particular bacterial colony is known exactly:

$$P(t) = 1000e^{t \ln 2}.$$

Therefore, we can predict the population at any future time; for example, the population at time $t = 90$ min (1.5 h) will be $P(1.5) = 1000e^{(1.5)\ln 2}$, or about 2828 bacteria.

---

The condition $P(0) = 1000$ is called an *initial condition* because we normally write differential equations for which $t = 0$ is the "starting time." Fig. 18.1 shows a number of graphs of the form $P(t) = Ce^{kt}$ for which $k = \ln 2$. The graphs of all the solutions of

$$\frac{dP}{dt} = (\ln 2)P \approx (0.69315)P$$

in fact fill up the entire two-dimensional plane, and no two intersect. Moreover, the selection of any point on the $P$-axis amounts to a determination of the value $P(0)$. Because exactly one solution passes through each such point, we see in this case that an initial condition $P(0) = P_0$ may determine a unique solution agreeing with known data.

It is possible that none of these solutions fits the known information. In such a case we must suspect that the differential equation—a mathematical model of the physical phenomenon in question—may not adequately describe the real world. The solutions of Equation (2) are of the form $P(t) = Ce^{kt}$, where $C$ is a positive constant, but for *no* choice of the constants $k$ and $C$ does $P(t)$ accurately describe the actual growth of the human population of the world over the past hundred years. We must therefore write a more

**18.1** Graphs of $P(t) = C \exp(t \ln 2)$

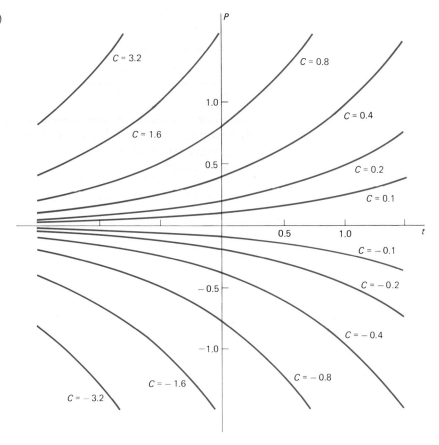

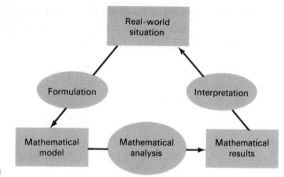

18.2 The process of mathematical modeling

complicated differential equation, one that takes into account the effects of population pressure on the birth rate, the declining food supply, and other factors. This should not be regarded as a failure of the model of Example 2, but as an insight into what additional factors must be considered in studying the growth of populations. Indeed, Equation (2) is quite accurate under certain circumstances—for example, the growth of a bacterial population under conditions of unlimited food and space.

This brief discussion of population growth illustrates the crucial process of *mathematical modeling* (see Fig. 18.2), which involves

1 The formulation of a real-world problem in mathematical terms; that is, the construction of a mathematical model,
2 The analysis or solution of the resulting mathematical problem, and
3 The interpretation of the mathematical results in the context of the original real-world situation—for example, answering the question originally posed.

In the population example, the real-world problem is that of determining the population at some future time. A *mathematical model* consists of a list of variables ($P$ and $t$) that describe the given situation together with one or more equations relating these variables ($dP/dt = kP$, $P(0) = P_0$) that are known or are assumed to hold. The mathematical analysis consists of solving these equations (here, for $P$ as a function of $t$). Finally, we apply these mathematical results to answer the original real-world question.

But in our population example we ignored the effects of such factors as varying birth and death rates. This made the mathematical analysis quite simple, perhaps unrealistically so. A satisfactory mathematical model is subject to two contradictory requirements: It must be sufficiently detailed to represent the real-world situation with relative accuracy, and yet it must be sufficiently simple to make the mathematical analysis practical. If the model is so detailed that it fully represents the physical situation, the mathematical analysis may be too difficult to carry out. If the model is too simple, the results may be so inaccurate as to be useless. Thus there is an inevitable trade-off between what is physically realistic and what is mathematically possible. The construction of a model that adequately bridges this gap between realism and feasibility is therefore the most crucial and delicate step in the process. Ways must be found to simplify the model mathematically without sacrificing essential features of the real-world situation.

Mathematical models have appeared throughout this book. The remainder of this section is devoted to simple examples and to standard terminology used in discussing differential equations and their solutions.

**EXAMPLE 5** Given the differential equation $y' = y^2$, it is easy to verify that the function $y = 1/(1 - x)$ is a solution on the interval $x < 1$. For

$$\frac{dy}{dx} = 1/(1 - x)^2 = y^2$$

if $x < 1$. The same computation is valid for the interval $x > 1$. Therefore, $y = 1/(1 - x)$ is a solution of the given equation on $(-\infty, 1)$ and on $(1, +\infty)$. There is no (continuous) solution of the equation $y' = y^2$ on the entire real line except for the constant function $y \equiv 0$ (though this is not obvious).

**EXAMPLE 6** Verify that the function

$$y = 2x^{1/2} - x^{1/2}\ln x$$

satisfies the differential equation

$$4x^2 y'' + y = 0 \tag{6}$$

for all $x > 0$.

*Solution* First we compute the derivatives

$$y' = -\tfrac{1}{2}x^{-1/2}\ln x$$

and

$$y'' = \tfrac{1}{4}x^{-3/2}\ln x - \tfrac{1}{2}x^{-3/2}.$$

Then substitution in Equation (6) yields

$$4x^2 y'' + y = 4x^2(\tfrac{1}{4}x^{-3/2}\ln x - \tfrac{1}{2}x^{-3/2})$$
$$+ 2x^{1/2} - x^{1/2}\ln x = 0$$

if $x$ is positive, and so the differential equation is satisfied for all $x > 0$.

---

The fact that we can write a differential equation is not enough to guarantee that it has a solution. For example, it is clear that the differential equation

$$(y')^2 + y^2 = -1 \tag{7}$$

has *no* (real-valued) solution because the sum of nonnegative numbers cannot be negative. For another variation on this theme, note that the equation

$$(y')^2 + y^2 = 0 \tag{8}$$

obviously has only the (real-valued) solution $y(x) \equiv 0$. In our previous examples, any differential equation having at least one solution indeed had infinitely many.

The *order* of a differential equation is the order of the highest derivative that appears in it. The differential equation of Example 6 is of second order, our previous examples are first order equations, and

$$y^{(4)} + x^2 y^{(3)} + x^5 y = \sin x$$

is a fourth order equation. The most general form of an *nth order* differential equation with independent variable $x$ and unknown function or dependent variable $y = y(x)$ is

$$F(x, y, y', y'', \ldots, y^{(n)}) = 0, \tag{9}$$

where $F$ is a specific real-valued function of $n + 2$ variables.

Our usage of the word "solution" has until now been somewhat informal. More precisely, we say that the function $y = u(x)$ is a **solution** of the differential equation in (9) **on the interval** $I$ provided that the derivatives $u', u'', \ldots, u^{(n)}$ exist and that

$$F(x, u, u', u'', \ldots, u^{(n)}) = 0$$

for all $x$ in $I$. When brevity is needed, we say that $y = u(x)$ **satisfies** the differential equation in (9) on $I$.

In both Equations (7) and (8), the appearance of $y'$ as an implicitly defined function causes complications. For this reason, we will ordinarily assume that any differential equation under study can be solved explicitly for the highest derivative that appears; that is, that the equation may be written in the form

$$y^{(n)} = G(x, y, y', y'', \ldots, y^{(n-1)}), \tag{10}$$

where $G$ is a real-valued function of $n + 1$ variables. In addition, we will always seek only real-valued solutions unless we warn the reader to the contrary.

All the differential equations we have mentioned so far are **ordinary** differential equations, meaning that the unknown function (dependent variable) depends upon only a *single* independent variable. For this reason only ordinary derivatives appear in the equation. If the dependent variable is a function of two or more independent variables, then partial derivatives are likely to be involved; if so, the equation is called a **partial** differential equation. For example, the temperature $u = u(x, t)$ of a long thin rod at the point $x$ at time $t$ satisfies (under appropriate simple conditions) the partial differential equation

$$\frac{\partial u}{\partial t} = k \frac{\partial^2 u}{\partial x^2},$$

where $k$ is a constant (called the *thermal diffusivity* of the rod). In this chapter we will be concerned only with *ordinary* differential equations and will refer to them simply as differential equations.

In the following two sections we concentrate our attention on first order differential equations of the general form

$$\frac{dy}{dx} = f(x, y). \tag{11}$$

We also will sample the wide range of applications of such equations. A typical mathematical model of an applied situation will be an **initial value problem,** consisting of a differential equation of the above form together with an *initial condition* $y(x_0) = y_0$. Note that we call $y(x_0) = y_0$ an initial condition whether or not $x_0 = 0$. To solve the initial value problem

$$\frac{dy}{dx} = f(x, y), \qquad y(x_0) = y_0 \tag{12}$$

means to find a differentiable function $y(x)$ that satisfies both conditions in Equation (12).

The central question of greatest interest to us is this: If we are given a differential equation known to have a solution satisfying a given initial condition, how do we actually *find* or *compute* that solution? And, once found, what can we *do* with it? We will see that a relatively few simple techniques are enough to enable us to solve a diversity of differential equations having impressive applications. In a single chapter we can only scratch the surface of a vast and venerable area of advanced mathematics whose applications permeate modern science and technology. This subject can be pursued further in Edwards and Penney, *Elementary Differential Equations with Applications* (Englewood Cliffs N.J.: Prentice-Hall, 1985).

## 18-1  PROBLEMS

In each of Problems 1–12, verify by substitution that each given function is a solution of the given differential equation.

1  $y' = 3x^2$;  $y = x^3 + 7$
2  $y' + 2y = 0$;  $y = 3e^{-2x}$
3  $y'' + 4y = 0$;  $y_1 = \cos 2x, y_2 = \sin 2x$
4  $y'' = 9y$;  $y_1 = e^{3x}, y_2 = e^{-3x}$
5  $y' = y + 2e^{-x}$;  $y = e^x - e^{-x}$
6  $y'' + 4y' + 4y = 0$;  $y_1 = e^{-2x}, y_2 = xe^{-2x}$
7  $y'' - 2y' + 2y = 0$;  $y_1 = e^x\cos x, y_2 = e^x\sin x$
8  $y'' + y = 3 \cos 2x$;  $y_1 = \cos x - \cos 2x,$
$y_2 = \sin x - \cos 2x$
9  $y' + 2xy^2 = 0$;  $y = 1/(1 + x^2)$
10  $x^2y'' + xy' - y = \ln x$;  $y_1 = x - \ln x,$
$y_2 = 1/x - \ln x$
11  $x^2y'' + 5xy' + 4y = 0$;  $y_1 = 1/x^2, y_2 = (\ln x)/x^2$
12  $x^2y'' - xy' + 2y = 0$;  $y_1 = x \cos(\ln x),$
$y_2 = x \sin(\ln x)$

In each of Problems 13–16, substitute $y = e^{rx}$ into the given differential equation to determine all values of $r$ for which $y = e^{rx}$ is a solution of the equation.

13  $3y' = 2y$       14  $4y'' = y$
15  $y'' + y' - 2y = 0$       16  $3y'' + 4y' - 4y = 0$

In each of Problems 17–20, show that $y(x)$ satisfies the given differential equation for all values of the constants $A$ and $B$. Then find values of $A$ and $B$ so that $y(0) = y'(0) = 1$.

17  $y'' + 3y' = 0$;  $y(x) = A + Be^{-3x}$
18  $y'' - 2y' + y = 0$;  $y(x) = Ae^x + Bxe^x$
19  $y'' = 4y$;  $y(x) = Ae^{2x} + Be^{-2x}$
20  $y'' - 4y' + 5y = 0$;  $y(x) = e^{2x}(A \cos x + B \sin x)$

In each of Problems 21–25, a function $y = g(x)$ is described by some geometric property of its graph. Write a differential equation of the form $y' = f(x, y)$ having the function $g(x)$ as its solution (or as one of its solutions).

21  The slope of the graph of $g$ at the point $(x, y)$ is the sum of $x$ and $y$.

22  The tangent line to the graph of $g$ at the point $(x, y)$ intersects the $x$-axis at the point $(x/2, 0)$.
23  Every straight line normal to the graph of $g$ passes through the point $(0, 1)$.
24  The graph of $g$ is normal to every curve of the form $y = k + x^2$ ($k$ is a constant) where they meet.
25  The line tangent to the graph of $g$ at $(x, y)$ passes through the point $(-y, x)$.

In each of Problems 26–30 write—in the manner of Equations (1) through (4) of this section—a differential equation that is a mathematical model of the situation described.

26  The time rate of change of a population $P$ is proportional to the square root of $P$.
27  The time rate of change of the velocity $v$ of a coasting motorboat is proportional to the square of $v$.
28  The acceleration $dv/dt$ of a certain sports car is proportional to the difference between 250 km/h and the velocity of the car.
29  In a city having a fixed population of $P$ persons, the time rate of change of the number $N$ of those persons who have heard a certain rumor is proportional to the number of those who have not yet heard the rumor.
30  In a city with a fixed population of $P$ persons, the time rate of change of the number $N$ of persons infected with a certain disease is proportional to the product of the number who have the disease and the number who do not.

In Problems 31–40, determine at least one solution of the given differential equation by inspection.

31  $y'' = 0$       32  $y' = y$
33  $xy' + y = 3x^2$       34  $(y')^2 + y^2 = 1$
35  $y' + y = e^x$       36  $y'' + y = 0$
37  $y' = 2xy$       38  $y^{(4)} + y' = 3x^2$
39  $y' - y = 1$       40  $y' - y = x$

**966**

The first order differential equation

$$\frac{dy}{dx} = H(x, y) \tag{1}$$

is called **separable** provided that $H(x, y)$ can be written as the product of a function of $x$ and a function of $y$ or, equivalently, as a quotient $H(x, y) = g(x)/f(y)$. In this case the variables $x$ and $y$ can be "separated"—isolated on opposite sides of an equation—by writing informally the equation

$$f(y)\, dy = g(x)\, dx,$$

which we understand to be compact notation for the differential equation

$$f(y)\frac{dy}{dx} = g(x). \tag{2}$$

It is easy to solve this special type of differential equation simply by integrating both sides with respect to $x$:

$$\int f(y(x))\frac{dy}{dx}\, dx = \int g(x)\, dx + C;$$

more concisely,

$$\int f(y)\, dy = \int g(x)\, dx + C. \tag{3}$$

All that's required is that the antiderivatives $F(y) = D_y^{-1}f(y)$ and $G(x) = D_x^{-1}g(x)$ can be found. To see that (2) and (3) are equivalent, note the following consequence of the chain rule:

$$D_x F(y(x)) = F'(y(x))y'(x) = f(y)\frac{dy}{dx} = g(x) = D_x G(x),$$

which in turn is equivalent to

$$F(y(x)) = G(x) + C \tag{4}$$

because two functions have the same derivative on an interval if and only if they differ by a constant on that interval.

**EXAMPLE 1** Solve the differential equation

$$x^2\frac{dy}{dx} = \frac{x^2 + 1}{3y^2 + 1}.$$

***Solution*** We multiply each side by the formal expression $(3y^2 + 1)\, dx/x^2$ to obtain

$$(3y^2 + 1)\, dy = \left(1 + \frac{1}{x^2}\right) dx.$$

Integration of both sides then gives

$$y^3 + y = x - \frac{1}{x} + C.$$

As Example 1 illustrates, it may not be possible or practical to solve Equation (4) explicitly for $y$ as a function of $x$. If so, we call (4) an *implicit solution* of the differential equation in (2). Because Equation (4) contains the arbitrary constant $C$, we also call it a **general solution** of (2). Given an initial condition $y(x_0) = y_0$, the choice $C_0 = F(y_0) - G(x_0)$ of $C$ yields the equation

$$F(y) = G(x) + C_0$$

that implicitly defines a *particular solution* (if any) of the initial value problem

$$f(y)\frac{dy}{dx} = g(x), \qquad y(x_0) = y_0.$$

For instance, we see from Example 1 that a solution of the initial value problem

$$x^2\frac{dy}{dx} = \frac{x^2 + 1}{3y^2 + 1}, \qquad y(1) = 2$$

is defined implicitly by the equation

$$y^3 + y = x - \frac{1}{x} + 10.$$

Thus the equation $K(x, y) = 0$ is called an **implicit solution** of a differential equation if it is satisfied (on some interval) by some solution $y = y(x)$ of the differential equation. But note that a particular solution $y = y(x)$ of $K(x, y) = 0$ may or may not satisfy a given initial condition. For example, differentiation of

$$x^2 + y^2 = 4$$

yields

$$x + y\frac{dy}{dx} = 0,$$

so $x^2 + y^2 = 4$ is an implicit solution of the differential equation $x + yy' = 0$. But only the first of the two explicit solutions $y = +(4 - x^2)^{1/2}$ and $y = -(4 - x^2)^{1/2}$ satisfies the initial condition $y(0) = 2$.

The argument preceding Example 1 shows that every particular solution of (2) satisfies (4) for some choice of $C$; *this* is why it is appropriate to call (4) a general solution of (2).

WARNING    Suppose, however, that we begin with the differential equation

$$\frac{dy}{dx} = g(x)h(y) \tag{5}$$

and divide by $h(y)$ to obtain the separated equation

$$\frac{1}{h(y)}\frac{dy}{dx} = g(x). \tag{6}$$

If $y_0$ is a root of the equation $h(y) = 0$—that is, if $h(y_0) = 0$—then the constant function $y(x) \equiv y_0$ is clearly a solution of (5) but may *not* be "contained" in the general solution of (6). Thus solutions of a differential equation may be "lost" upon division by a vanishing factor. (Indeed, false solutions may be "gained" upon multiplication by a vanishing factor. This phenomenon is similar to the introduction of extraneous roots in solving algebraic equations.)

In Section 18-3 we shall see that every particular solution of a *linear* first order differential equation is contained in its general solution. By contrast, it is common for a nonlinear first order differential equation to have both a general solution involving an arbitrary constant $C$ and one or several particular solutions that cannot be obtained by selecting a value for $C$. These exceptional solutions are frequently called *singular solutions*.

**EXAMPLE 2**  Find all solutions of the differential equation

$$\frac{dy}{dx} = 2x\sqrt{y - 1}.$$

*Solution*  We note first the constant solution $y(x) \equiv 1$. If $y \neq 1$, we can divide each side by $\sqrt{y - 1}$ to obtain

$$(y - 1)^{-1/2} \frac{dy}{dx} = 2x.$$

Integration gives

$$2\sqrt{y - 1} = x^2 + C;$$

upon solving for $y$ we get the general solution

$$y = 1 + \frac{1}{4}(x^2 + C)^2.$$

Note that no value of $C$ gives the particular solution $y \equiv 1$. It was lost when we divided by $\sqrt{y - 1}$.

**EXAMPLE 3**  Solve the initial value problem

$$\frac{dy}{dx} = xy + x - 2y - 2, \qquad y(0) = 2.$$

*Solution*  Sometimes a factorization is not obvious at first glance; here we have

$$\frac{dy}{dx} = (x - 2)(y + 1).$$

The constant function $y \equiv -1$ satisfies the differential equation but does not satisfy the initial condition, so we cannot lose the solution of the initial value problem by dividing by $y + 1$. We therefore perform that division and integrate:

$$\int \frac{dy}{y + 1} = \int (x - 2)\, dx;$$

$$\ln|y + 1| = \frac{1}{2}x^2 - 2x + C;$$

$$\ln(y + 1) = \frac{1}{2}x^2 - 2x + C.$$

In the final step we use the fact that $y + 1 > 0$ near the initial value $y = 2$. Now apply the exponential function to each side of the last equation. With the observation that $e^{\ln z} = z$, we get

$$y + 1 = \exp(\tfrac{1}{2}x^2 - 2x + C),$$

so that

$$y = A \exp(\tfrac{1}{2}x^2 - 2x) - 1$$

where $A = e^C$. Note in passing that we did not lose the constant solution $y \equiv -1$; it corresponds to $A = 0$. But to conclude the example, the initial condition $y(0) = 2$ implies that $A = 3$, so the desired solution is

$$y = 3 \exp(\tfrac{1}{2}x^2 - 2x) - 1.$$

## TORRICELLI'S LAW

Suppose that a water tank has a hole with area $a$ at its bottom from which water is leaking. Denote by $y(t)$ the depth of the water in the tank at time $t$ and by $V(t)$ the volume of water in the tank then. It is plausible—as well as true under ideal conditions—that the velocity of water exiting through the hole is

$$v = \sqrt{2gy}, \tag{7}$$

which is the velocity a drop of water would acquire in falling freely from the surface of the water to the hole. Under real conditions, taking into account the constriction of a water jet from an orifice, $v = c\sqrt{2gy}$, where $c$ is an empirical constant between 0 and 1 (usually about 0.6 for a small continuous stream of water). For simplicity we take $c = 1$ in the following discussion.

As a consequence of Equation (7), we have

$$\frac{dV}{dt} = -av = -a\sqrt{2gy}; \tag{8}$$

this is a statement of Torricelli's law for a draining tank. If $A(y)$ denotes the horizontal cross-sectional area of the tank at height $y$, then the method of volume by cross sections gives

$$V = \int_0^y A(y)\, dy,$$

so the fundamental theorem of calculus implies that $dV/dy = A(y)$ and therefore that

$$\frac{dV}{dt} = \frac{dV}{dy}\frac{dy}{dt} = A(y)\frac{dy}{dt}. \tag{9}$$

From Equations (8) and (9) we finally obtain

$$A(y)\frac{dy}{dt} = -a\sqrt{2gy}, \tag{10}$$

the most useful form of Torricelli's law.

**EXAMPLE 4** A hemispherical tank has top radius 4 ft and, at time $t = 0$, is full of water. At that moment a circular hole of diameter 1 in. is opened in the bottom of the tank. How long will it take for all the water to drain from the tank?

*Solution* From the right triangle in Figure 18.3, we see that

$$A(y) = \pi r^2 = \pi[16 - (4 - y)^2] = \pi(8y - y^2).$$

With $g = 32$ ft/s$^2$, Equation (10) becomes

$$\pi(8y - y^2)\frac{dy}{dt} = -\pi\left(\frac{1}{24}\right)^2 \sqrt{64y};$$

$$\int (8y^{1/2} - y^{3/2})\, dy = -\int \frac{dt}{72};$$

$$\frac{16}{3} y^{3/2} - \frac{2}{5} y^{5/2} = -\frac{1}{72} t + C.$$

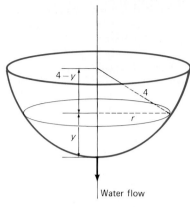

$4 - y$

$4$

$r$

$y$

Water flow

**18.3** Draining a hemispherical tank

Now $y(0) = 4$, so

$$C = \tfrac{16}{3}(4)^{3/2} - \tfrac{2}{5}(4)^{5/2} = \tfrac{448}{15}.$$

The tank is empty when $y = 0$, thus when

$$t = (72)\left(\frac{448}{15}\right) \approx 2150$$

s; that is, about 35 min 50 s. So it takes slightly less than 36 min for the tank to drain.

---

## 18-2 PROBLEMS

Find general solutions (implicit if necessary, explicit if convenient) of the differential equations in Problems 1–8.

**1** $\dfrac{dy}{dx} = xy^3$

**2** $y\dfrac{dy}{dx} = x(y^2 + 1)$

**3** $y^3\dfrac{dy}{dx} = (y^4 + 1)\cos x$

**4** $\dfrac{dy}{dx} = \dfrac{1 + \sqrt{x}}{1 + \sqrt{y}}$

**5** $\dfrac{dy}{dx} = \dfrac{(x - 1)y^5}{x^2(2y^3 - y)}$

**6** $(x^2 + 1)\dfrac{dy}{dx}\tan y = x$

**7** $\dfrac{dy}{dx} = 1 + x + y + xy$

**8** $x^2 y' = 1 - x^2 + y^2 - x^2 y^2$

Find explicit particular solutions of the initial value problems in Problems 9–16.

**9** $\dfrac{dy}{dx} = ye^x;\quad y(0) = 2e$

**10** $\dfrac{dy}{dx} = 3x^2(y^2 + 1);\quad y(0) = 1$

**11** $2y\dfrac{dy}{dx} = x(x^2 - 16)^{-1/2};\quad y(5) = 2$

**12** $\dfrac{dy}{dx} = 4x^3 y - y;\quad y(1) = -3$

**13** $\dfrac{dy}{dx} + 1 = 2y;\quad y(1) = 1$

**14** $y'\tan x = y;\quad y\left(\dfrac{\pi}{2}\right) = \dfrac{\pi}{2}$

**15** $x\dfrac{dy}{dx} - y = 2x^2 y;\quad y(1) = 1$

**16** $\dfrac{dy}{dx} = 2xy^2 + 3x^2 y^2;\quad y(1) = -1$

**17** At time $t = 0$, the bottom plug (at the vertex) of a full conical water tank 16 ft high is removed. After 1 h the water in the tank is 9 ft deep. When will the tank be empty?

**18** A tank is shaped like a vertical cylinder; it initially contains water to a depth of 9 ft, and a bottom plug is pulled at time $t = 0$ (h). After 1 h the depth has dropped to 4 ft. How long does it take for all the water to run out of this tank?

**19** Suppose that the tank of Problem 18 has a radius of 3 ft and that its bottom hole is circular with radius 1 in. How long will it take the water (initially 9 ft deep) to drain completely?

**20** A cylindrical tank with length 5 ft and radius 3 ft is situated with its axis horizontal. If a circular bottom hole with a radius of 1 in. is opened, and the tank is initially half full of benzene, how long will it take for the liquid to completely drain?

**21** A spherical tank with radius 4 ft is full of gasoline when a circular bottom hole with radius 1 in. is opened. How long will be required for all the gasoline to drain from the tank?

**22** (The Clepsydra, or water clock) A 12-h water clock is to be designed with the dimensions shown in Fig. 18.4, shaped like the surface obtained by revolving the curve $y = f(x)$ around the $y$-axis. What should be this curve *and*

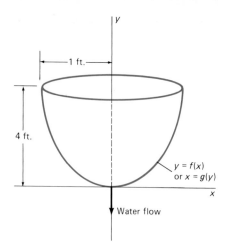

**18.4** The clepsydra

what should be the radius of the circular bottom hole in order that the water level will fall at the *constant* rate of 4 in./h?

**23** Suppose that a cylindrical tank initially containing $V_0$ gallons of water drains (through a bottom hole) in $T$ minutes. Use Torricelli's law to show that the volume of water in the tank after $t \leq T$ minutes is $V = V_0(1 - (t/T))^2$.

---

## 18-3

## Linear First Order Equations

In Section 18-2 we saw how to solve a separable differential equation by integrating *after* multiplying each side by an appropriate factor. For instance, to solve the equation

$$\frac{dy}{dx} = 2xy \qquad (y > 0), \tag{1}$$

we multiply each side by the factor $1/y$ to get

$$\frac{1}{y}\frac{dy}{dx} = 2x; \quad \text{that is,} \quad D_x(\ln y) = D_x(x^2). \tag{2}$$

Because each side of the equation in (2) is recognizable as a *derivative* (with respect to the independent variable $x$), all that remains is a simple integration, which yields $\ln y = x^2 + C$. For this reason, the function $\rho = 1/y$ is called an *integrating factor* for the original equation in (1). An **integrating factor** for a differential equation is a function $\rho(x, y)$ such that multiplication of each side of the differential equation by $\rho(x, y)$ yields an equation in which each side is recognizable as a derivative.

With the aid of the appropriate integrating factor, there is a standard technique for solving the **linear first order equation**

$$\frac{dy}{dx} + P(x)y = Q(x) \tag{3}$$

on an interval where the coefficient functions $P(x)$ and $Q(x)$ are continuous.

We multiply each side in Equation (3) by the integrating factor

$$\rho = \rho(x) = e^{\int P(x)\,dx}.\qquad(4)$$

The result is

$$e^{\int P(x)\,dx}\frac{dy}{dx} + P(x)e^{\int P(x)\,dx}y = Q(x)e^{\int P(x)\,dx}\qquad(5)$$

Because

$$D_x\left[\int P(x)\,dx\right] = P(x),$$

the left-hand side is the derivative of the *product* $y \cdot e^{\int P(x)\,dx}$, so (5) is equivalent to

$$D_x\left[y(x)e^{\int P(x)\,dx}\right] = Q(x)e^{\int P(x)\,dx}.$$

Integration of both sides of this equation gives

$$y(x)e^{\int P(x)\,dx} = \int \left(Q(x)e^{\int P(x)\,dx}\right) dx + C.$$

Finally solving for $y$, we obtain the general solution of the linear first order equation in (3):

$$y = y(x) = e^{-\int P(x)\,dx}\left[\int\left(Q(x)e^{\int P(x)\,dx}\right) dx + C\right].\qquad(6)$$

The formula in (6) need not be memorized. In a specific problem it generally is simpler to use the *method* by which we developed this formula. Begin by calculating the integrating factor given in Equation (4), $\rho = e^{\int P(x)\,dx}$. Then multiply each side of the differential equation by $\rho$, recognize the left-hand side of the resulting equation as the derivative of a product, integrate this equation, and finally solve for $y$. Moreover, given an initial condition $y(x_0) = y_0$, we can substitute $x = x_0$ and $y = y_0$ into (6) to solve for the value of $C$ yielding the particular solution of (3) that satisfies this initial condition.

The integrating factor $\rho(x)$ is determined only to within a multiplicative constant. If we replace $\int P(x)\,dx$ by $\int P(x)\,dx + c$ in (4), the result is

$$\rho(x) = e^{\int P(x)\,dx + c} = e^c e^{\int P(x)\,dx}.$$

But the constant factor $e^c$ does not affect the result of multiplying both sides of the differential equation in (3) by $\rho(x)$. Hence we may choose for $\int P(x)\,dx$ any convenient antiderivative of $P(x)$.

**EXAMPLE 1**  Solve the initial value problem

$$\frac{dy}{dx} - 3y = e^{2x}, \qquad y(0) = 3.$$

*Solution*  Here we have $P(x) = -3$ and $Q(x) = e^{2x}$, so the integrating factor is

$$\rho = e^{\int(-3)\,dx} = e^{-3x}.$$

Multiplication of each side of the given equation by $e^{-3x}$ yields

$$e^{-3x}\frac{dy}{dx} - 3e^{-3x}y = e^{-x},$$

which we recognize as

$$\frac{d}{dx}(e^{-3x}y) = e^{-x}.$$

Hence integration with respect to $x$ gives

$$e^{-3x}y = -e^{-x} + C,$$

so the general solution is

$$y = Ce^{3x} - e^{2x}.$$

Substitution of the initial condition $(x_0, y_0) = (0, 3)$ produces the result $C = 4$. Thus the desired particular solution is

$$y = 4e^{3x} - e^{2x}.$$

**EXAMPLE 2**  Find the general solution of

$$(x^2 + 1)\frac{dy}{dx} + 3xy = 6x.$$

*Solution*   After division of each side of the equation by $x^2 + 1$, we recognize the result

$$\frac{dy}{dx} + \frac{3x}{x^2 + 1}y = \frac{6x}{x^2 + 1}$$

as a first order linear equation in which we have $P(x) = 3x/(x^2 + 1)$ and $Q(x) = 6x/(x^2 + 1)$. Multiplication by

$$\rho = \exp\left(\int \frac{3x}{x^2 + 1}\,dx\right)$$

$$= \exp\left(\frac{3}{2}\ln(x^2 + 1)\right) = (x^2 + 1)^{3/2}$$

yields

$$(x^2 + 1)^{3/2}\frac{dy}{dx} + 3x(x^2 + 1)^{1/2}y = 6x(x^2 + 1)^{1/2},$$

and thus

$$D_x[(x^2 + 1)^{3/2}y] = 6x(x^2 + 1)^{1/2}.$$

Integration then yields

$$(x^2 + 1)^{3/2}y = \int 6x(x^2 + 1)^{1/2}\,dx = 2(x^2 + 1)^{3/2} + C.$$

Multiplication of both sides by $(x^2 + 1)^{-3/2}$ gives the general solution

$$y = 2 + C(x^2 + 1)^{-3/2}$$

---

The derivation above of the solution in (6) of the linear first order equation in (3) bears a closer examination. Suppose that the functions $P(x)$ and $Q(x)$ are continuous on the (possibly unbounded) open interval $I$. Then the antiderivatives

$$\int P(x)\,dx \quad \text{and} \quad \int\left(Q(x)e^{\int P(x)\,dx}\right)dx$$

exist on $I$. Our derivation of Equation (6) shows that *if $y = y(x)$ is a solution of Equation (3) on $I$, then $y(x)$ is given by the formula in (6) for some choice of the constant $C$.* Conversely, you may verify by direct substitution (Problem 16) that the function $y(x)$ given in Equation (6) satisfies Equation (3). Finally, given a point $x_0$ of $I$ and any number $y_0$, there is (as previously noted) a unique value of $C$ such that $y(x_0) = y_0$. Consequently, we have proved the following existence-uniqueness theorem.

---

**Theorem** *The Linear First Order Equation*

If the functions $P(x)$ and $Q(x)$ are continuous on the open interval $I$ containing the point $x_0$, then the initial value problem

$$\frac{dy}{dx} + P(x)y = Q(x), \qquad y(x_0) = y_0$$

has a unique solution $y(x)$ on $I$, given by the formula in Equation (6) with an appropriate value of the constant $C$.

---

## MIXTURE PROBLEMS

As a first application of linear first order equations, we consider a tank containing a solution—a mixture of solute and solvent—such as salt dissolved in water. There is both inflow and outflow, and we want to compute the *amount $x(t)$ of solute* in the tank at time $t$, given the amount $x(0) = x_0$ at time $t = 0$. Suppose that solution with a concentration of $c_i$ grams of solute per liter of solution flows into the tank at the constant rate of $r_i$ liters per second and that the solution in the tank—kept thoroughly mixed by stirring—flows out at the constant rate of $r_o$ liters per second.

To set up a differential equation for $x(t)$, we estimate the change $\Delta x$ in $x$ during the brief time interval $[t, t + \Delta t]$. The amount of solute that flows into the tank during $\Delta t$ seconds is $r_i c_i \Delta t$ grams. To check this, note how the cancellation of dimensions checks our computation:

$$(r_i \text{ liters/second})(c_i \text{ grams/liter})(\Delta t \text{ seconds})$$

yields a quantity measured in grams.

The amount of solute that flows out of the tank during the same time interval depends upon the concentration $c_o(t)$ in the tank at time $t$. But, as noted in Fig. 18.5, $c_o(t) = x(t)/V(t)$, where $V(t)$ denotes the volume (not constant unless $r_i = r_o$) of solution in the tank at time $t$. Then

$$\Delta x = \{\text{grams input}\} - \{\text{grams output}\}$$

$$\approx r_i c_i \, \Delta t - r_o c_o \, \Delta t$$

$$= r_i c_i \, \Delta t - r_o \frac{x}{V} \, \Delta t.$$

We now divide by $\Delta t$:

$$\frac{\Delta x}{\Delta t} \approx r_i c_i - r_o \frac{x}{V}.$$

Finally we take the limit as $\Delta t \to 0$; if all the functions involved are continuous and $x$ is differentiable, then the error in the approximations also approaches

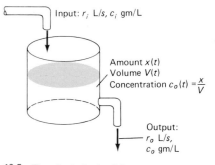

Input: $r_i$ L/s, $c_i$ gm/L

Amount $x(t)$
Volume $V(t)$
Concentration $c_o(t) = \dfrac{x}{V}$

Output:
$r_o$ L/s,
$c_o$ gm/L

**18.5** The single-tank mixture problem

zero, and we obtain the differential equation

$$\frac{dx}{dt} = r_i c_i - \frac{r_0}{V} x. \tag{7}$$

If $V_0 = V(0)$, then $V(t) = V_0 + (r_i - r_o)t$, so Equation (7) is a linear first order differential equation for the amount $x(t)$ of solute in the tank.

IMPORTANT    Equation (7) is *not* one you should commit to memory. It is the process we used to obtain that equation—examination of the behavior of the system over a short time interval $[t, t + \Delta t]$—that you should strive to understand because it is a very useful tool for obtaining all sorts of differential equations.

EXAMPLE 3    Lake Erie has a volume of 458 km$^3$, and its rate of inflow and outflow are both 175 km$^3$/year. Suppose that at the time $t = 0$ (years) its pollutant concentration is 0.05% and that thereafter the concentration of pollutants in the inflowing water is 0.01%. Assuming that the outflow is perfectly mixed lake water, how long will it take to reduce the pollution concentration in the lake to 0.02%?

*Solution*    Here we have $V = 458$ (km$^3$) and $r_i = r_o = 175$ (km$^3$/year). Let $x(t)$ denote the volume of pollutants in the lake at time $t$. We are given

$$x(0) = (0.0005)(458) = 0.2290 \quad \text{(km}^3\text{)},$$

and we want to find when

$$x(t) = (0.0002)(458) = 0.0916 \quad \text{(km}^3\text{)}.$$

The change $\Delta x$ in $\Delta t$ years is

$$\Delta x \approx (0.0001)(175)\,\Delta t - \left(\frac{x}{458}\right)(175)\,\Delta t$$

$$= (0.0175 - 0.3821x)\,\Delta t,$$

so our differential equation is

$$\frac{dx}{dt} + (0.3821)x = 0.0175.$$

The integrating factor is $e^{(0.3821)t}$; it yields

$$D_t(e^{(0.3821)t}x) = (0.0175)e^{(0.3821)t};$$

$$(e^{(0.3821)t})x = (0.0458)e^{(0.3821)t} + C.$$

Substitution of $x(0) = 0.2290$ into the last equation gives

$$C = 0.2290 - 0.0458 = 0.1832,$$

so the solution is

$$x(t) = 0.0458 + (0.1832)e^{-(0.3821)t}.$$

Finally, we solve the equation

$$0.0916 = 0.0458 + (0.1832)e^{-(0.3821)t}$$

for

$$t = \frac{-1}{0.3821} \ln \frac{0.0916 - 0.0458}{0.1832} \approx 3.63.$$

The answer to the problem, then, is this: After about 3.63 years.

**EXAMPLE 4**  A 120-gal. tank initially contains 90 lb of salt dissolved in 90 gal. of water. Brine containing 2 lb/gal. of salt flows into the tank at the rate of 4 gal./min, and the mixture flows out of the tank at the rate of 3 gal./min. How much salt does the tank contain when it is full?

*Solution*  The interesting feature of this example is that, due to the differing rates of inflow and outflow, the volume of brine in the tank increases steadily with $V(t) = 90 + t$ (gallons). The change $\Delta x$ in the amount $x$ of salt in the tank from time $t$ to time $t + \Delta t$ (minutes) is given by

$$\Delta x \approx (4)(2) \, \Delta t - 3 \left( \frac{x}{90 + t} \right) \Delta t,$$

so our differential equation is

$$\frac{dx}{dt} + \frac{3}{90 + t} x = 8.$$

The integrating factor is

$$\rho = \exp\left( \int \frac{3 \, dt}{90 + t} \right) = e^{3 \ln(90 + t)} = (90 + t)^3,$$

which gives

$$D_t[(90 + t)^3 x] = 8(90 + t)^3;$$

$$(90 + t)^3 x = 2(90 + t)^4 + C.$$

Substitution of $x(0) = 90$ gives $C = -(90)^4$, so the amount of salt in the tank at time $t$ is

$$x(t) = 2(90 + t) - \frac{(90)^4}{(90 + t)^3}.$$

The tank is full after 30 min, and when $t = 30$, we have

$$x(30) = 2(90 + 30) - \frac{(90)^4}{(120)^3} \approx 202$$

lb of salt.

---

18-3   **PROBLEMS**

Find general solutions of the differential equations in Problems 1–15. If an initial condition is given, find the corresponding particular solution. Throughout, primes denote derivatives with respect to $x$.

1  $xy' + y = 3xy;$  $y(1) = 0$
2  $xy' + 3y = 2x^5;$  $y(2) = 1$

3  $y' + y = e^x;$  $y(0) = 1$
4  $xy' - 3y = x^3;$  $y(1) = 10$
5  $y' + 2xy = x;$  $y(0) = -2$
6  $y' = (1 - y) \cos x;$  $y(\pi) = 2$
7  $(1 + x)y' + y = \cos x;$  $y(0) = 1$
8  $xy' = 2y + x^3 \cos x$

9 $y' + y \cot x = \cos x$

10 $y' = 1 + x + y + xy$; $y(0) = 0$

11 $xy' = 3y + x^4 \cos x$; $y(2\pi) = 0$

12 $y' = 2xy + 3x^2 \exp(x^2)$; $y(0) = 5$

13 $xy' + (2x - 3)y = 4x^4$

14 $(x^2 + 4)y' + 3xy = x$; $y(0) = 1$

15 $(x^2 + 1)y' + 3x^3 y = 6x \exp(-3x^2/2)$; $y(0) = 1$

16 (a) Show that $y_c(x) = Ce^{-\int P(x)\,dx}$ is a general solution of $y' + P(x)y = 0$.

(b) Show that

$$y_p(x) = e^{-\int P(x)\,dx}\left[\int (Q(x)e^{\int P(x)\,dx})\,dx\right]$$

is a particular solution of $y' + P(x)y = Q(x)$.

(c) If $y_c(x)$ is any general solution of $y' + P(x)y = 0$ and $y_p(x)$ is any particular solution of $y' + P(x)y = Q(x)$, then show that $y(x) = y_c(x) + y_p(x)$ is a general solution of $y' + P(x)y = Q(x)$.

17 A tank contains 1000 L of a solution consisting of 100 kg of salt dissolved in water. Pure water is pumped into the tank at the rate of 5 L/s, and the mixture—kept uniform by stirring—is pumped out at the same rate. How long will it be until only 10 kg of salt remain in the tank?

18 Consider a reservoir with a volume of 8 billion cubic feet and an initial pollutant concentration of 0.25%. There is a daily inflow of 500 million cubic feet of water with a pollutant concentration of 0.05% and an equal daily outflow of the well-mixed water of the reservoir. How long will it take to reduce the pollutant concentration in the reservoir to 0.10%?

19 Rework Example 3 for the case of Lake Ontario; the only differences are that this lake has a volume of 1636 km³ and an inflow-outflow rate of 209 km³/year.

20 A tank initially contains 60 gal. of pure water. Brine containing 1 lb of salt per gallon enters the tank at 2 gal./min, and the (perfectly mixed) solution leaves the tank at 3 gal./min; the tank is empty after exactly 1 h.

(a) Find the amount of salt in the tank after $t$ minutes.

(b) What is the maximum amount of salt ever in the tank?

21 A 400-gal. tank initially contains 100 gal. of brine containing 50 lb of salt. Brine containing 1 lb of salt per gallon enters the tank at the rate of 5 gal./s, and the mixed brine in the tank flows out at the rate of 3 gal./s. How much salt will the tank contain when it is full of brine?

22 Consider the *cascade* of two tanks shown in Fig. 18.6, with $V_1 = 100$ (gal.) and $V_2 = 200$ (gal.) the volumes of

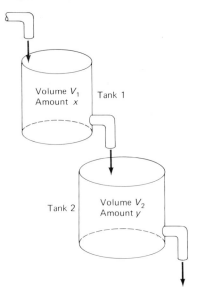

18.6   A cascade of two tanks

brine in the two tanks. Each tank initially contains 50 lb of salt. The three flow rates are each 5 gal./s, with pure water flowing into tank 1.

(a) Find the amount $x(t)$ of salt in tank 1 at time $t$.

(b) Suppose that $y(t)$ is the amount of salt in tank 2 at time $t$. Show first that

$$\frac{dy}{dt} = \frac{5x}{100} - \frac{5y}{200},$$

and then solve for $y(t)$, using the value of $x(t)$ found in part (a).

(c) Finally, find the maximum amount of salt ever in tank 2.

23 Suppose that in the cascade shown in Fig. 18.6, tank 1 initially contains 100 gal. of pure ethyl alcohol and tank 2 initially contains 100 gal. of pure water. Pure water flows into tank 1 at 10 gal./min, and the other two flow rates are also 10 gal./min.

(a) Find the amounts $x(t)$ and $y(t)$ of alcohol in the two tanks.

(b) Find the maximum amount of alcohol ever in tank 2.

---

## 18-4

## Complex Numbers and Functions

We include in this section the elementary algebra and calculus of complex numbers and functions—as much as is needed to solve higher order linear differential equations with constant coefficients (Section 18-5). Recall that a **complex number** is one of the form

$$z = a + bi \tag{1}$$

Euler's formula leads to important expressions for the sine and cosine functions in terms of the exponential function. The equations

$$e^{ix} = \cos x + i \sin x \quad \text{and} \quad e^{-ix} = \cos x - i \sin x,$$

which follow from (16) (put $\theta = x$, then $\theta = -x$), imply (add them, subtract them) the formulas

$$\cos x = \frac{e^{ix} + e^{-ix}}{2} \quad \text{and} \quad \sin x = \frac{e^{ix} - e^{-ix}}{2i}. \tag{20}$$

## COMPLEX-VALUED FUNCTIONS

A **complex-valued function** $F$ of the *real* variable $x$ associates with each real number $x$ (in its domain of definition) the complex number

$$z = F(x) = f(x) + ig(x). \tag{21}$$

The real-valued functions $f(x)$ and $g(x)$ are called the real and imaginary parts, respectively, of $F(x)$. If they are differentiable, we can define the **derivative** of $F$ to be

$$F'(x) = f'(x) + ig'(x). \tag{22}$$

Thus we merely differentiate the real and imaginary parts of $F$ separately. Similarly, we integrate a complex-valued function by separately integrating its real and imaginary parts, provided the integrals exist. That is,

$$\int_a^b F(x)\, dx = \int_a^b f(x)\, dx + i \int_a^b g(x)\, dx. \tag{23}$$

A particular *complex-valued* function that plays an important role in the solution of *real* differential equations is the exponential function $F(x) = e^{rx}$ where $r = a + bi$ is a complex number. By (19), we have

$$e^{rx} = e^{ax}(\cos bx + i \sin bx). \tag{24}$$

So the real and imaginary parts of $e^{rx}$ are $e^{ax} \cos bx$ and $e^{ax} \sin bx$, respectively.

**EXAMPLE 5**   Suppose that $r = a + bi$. Show that

$$De^{rx} = re^{rx}, \tag{25}$$

just as if $r$ were a real number.

*Solution*   From (24) we obtain

$$
\begin{aligned}
De^{rx} &= D(e^{ax}\cos bx) + iD(e^{ax}\sin bx) \\
&= [ae^{ax}\cos bx - be^{ax}\sin bx] + i[ae^{ax}\sin bx + be^{ax}\cos bx] \\
&= (a + bi)(e^{ax}\cos bx + ie^{ax}\sin bx) = re^{rx}.
\end{aligned}
$$

## THE FUNDAMENTAL THEOREM OF ALGEBRA

The fundamental theorem of algebra, easy to state but difficult to prove, is the underlying reason for the importance—indeed, the very unavoidability—of complex numbers in mathematics. The theorem tells us that every polynomial equation of degree $n$,

$$p(x) = a_n x^n + a_{n-1} x^{n-1} + \cdots + a_1 x + a_0 = 0, \tag{26}$$

has at least one (complex) root $r_1$. By the **factor theorem,** we can then write $p(x) = (x - r_1)q(x)$, where $q(x)$ is a polynomial of degree $n - 1$. By induction on $n$, it then follows that the $n$th degree polynomial $p(x)$ has the factorization

$$p(x) = a_n(x - r_1)(x - r_2) \cdots (x - r_n),$$

and the $n$ roots $r_1, r_2, r_3, \ldots, r_n$ (not necessarily distinct). Even if the coefficients $a_0, a_1, a_2, \ldots, a_n$ of the polynomial $p(x)$ are all real, the roots will be, in general, complex; this is one reason for the necessity of complex numbers. But if the coefficients of the polynomial $p(x)$ are all real, then it can be shown that the complex roots (if any) occur in **conjugate pairs** of the form $a \pm bi$.

## 18-4 PROBLEMS

Express the complex numbers given in Problems 1–6 both in the form $x + iy$ and in the form $r$ cis $\theta$.

1 $(3 - 4i)(3 + 4i)$

2 $(1 + 2i)(2 - 3i)$

3 $\dfrac{5 + 7i}{3 - 2i}$

4 $\dfrac{(1 + i)^2}{(1 - i\sqrt{3})^2}$

5 $(1 + i)^6$

6 $(-3 + 4i)^3$

7 Find the four fourth roots of $-1$.

8 Find the three cube roots of 27.

9 Find the six sixth roots of $-64$.

10 Find the five fifth roots of $-4 - 4i$.

11 Find the four roots of the equation $x^4 - 16 = 0$.

12 Find the four roots of the equation $x^4 - 4x^2 + 6 = 0$.

13 Find the six roots of the equation $x^6 + 4x^3 + 8 = 0$.

14 Expand $(\cos \theta + i \sin \theta)^3$. Then compare the result with that given by DeMoivre's formula, and thereby discover formulas for $\cos 3\theta$ and $\sin 3\theta$ in terms of $\sin \theta$ and $\cos \theta$.

15 Use DeMoivre's formula as in Problem 14 to discover formulas for $\cos 4\theta$ and $\sin 4\theta$ in terms of $\sin \theta$ and $\cos \theta$.

16 Suppose that $a$ is a positive real number. Find the 4 roots of the equation $x^4 + a^4 = 0$.

17 Prove that cis$(\alpha + \beta) = $ cis $\alpha$ cis $\beta$.

18 If $e^w = z$ then $z^i = e^{wi}$ by definition. Choose $w$ so that $e^w = i$, then conclude that

$$i^i = e^{-\pi/2} \approx 0.20788.$$

19 Obtain formulas for $\int e^{ax}\cos bx\, dx$ and $\int e^{ax}\sin bx\, dx$ by equating real and imaginary parts in the formula

$$\int e^{(a + bi)x}\, dx = \frac{e^{(a + bi)x}}{a + bi} + C.$$

20 Assume that $(e^z)^r = e^{rz}$ for every complex number $z = x + iy$ and real number $r$. Then deduce that De Moivre's formula in (14) holds for $n$ any *real number*.

21 Let $p, q, A,$ and $B$ be real numbers. Show that

$$Ae^{(p + iq)t} + Be^{(p - iq)t} = e^{pt}(C \cos qt + D \sin qt),$$

where $C$ and $D$ are complex numbers expressible in terms of $A$ and $B$.

22 Prove that $e^z \neq 0$ for all complex numbers $z$.

## 18-5

### Homogeneous Linear Equations with Constant Coefficients

Aside from second order equations that can be reduced by elementary substitutions to familiar first order equations, higher order differential equations ordinarily require specialized techniques for their solution. Higher order nonlinear equations can be very challenging; fortunately, many of the important equations arising from real-world applications are linear, and we now turn our attention to these. The general form of a **linear $n$th order differential equation** is

$$a_n(x)\frac{d^n y}{dx^n} + a_{n-1}(x)\frac{d^{n-1} y}{dx^{n-1}} + \cdots + a_1(x)\frac{dy}{dx} + a_0(x)y = F(x) \qquad (1)$$

where the *coefficients* $a_0(x), a_1(x), \ldots, a_n(x)$ and the function $F(x)$ are given. Equation (1) is called *linear* because only the *first* powers of the dependent variable $y(x)$ and of its derivatives are involved. It is linear in $y$ but not necessarily in the independent variable $x$—the coefficients and the function $F(x)$ need not be linear in $x$.

The linear differential equation (1) is called **homogeneous** provided that $F(x) \equiv 0$; otherwise it is said to be **inhomogeneous.** Thus the general form of a homogeneous linear equation is

$$a_n(x)\frac{d^n y}{dx^n} + a_{n-1}(x)\frac{d^{n-1} y}{dx^{n-1}} + \cdots + a_1(x)\frac{dy}{dx} + a_0(x)y = 0. \qquad (2)$$

We restrict our attention here to the important special case in which the coefficient functions in (2) are all *constants.* Thus we consider the **homogeneous linear differential equation**

$$a_n y^{(n)} + a_{n-1}y^{(n-1)} + \cdots + a_1 y' + a_0 y = 0 \qquad (3)$$

where the coefficients $a_n, a_{n-1}, \ldots, a_1, a_0$ are (real) constants.

The study of linear differential equations is greatly simplified by the following fact: If $y_1(x)$ and $y_2(x)$ are solutions of the homogeneous equation in (3), then so is the **linear combination**

$$y(x) = c_1 y_1(x) + c_2 y_2(x) \qquad (4)$$

for any choice of the *constants* $c_1$ and $c_2$. This is so because the linearity of the operation of differentiation implies that

$$a_k y^{(k)}(x) = c_1 a_k y_1^{(k)}(x) + c_2 a_k y_2^{(k)}(x)$$

for each $k = 0, 1, 2, \ldots, n$. So

$$\sum_{k=0}^{n} a_k y^{(k)}(x) = c_1 \sum_{k=0}^{n} a_k y_1^{(k)}(x) + c_2 \sum_{k=0}^{n} a_k y_2^{(k)}(x) = 0$$

provided that $y_1$ and $y_2$ are both solutions of (3).

**EXAMPLE 1**   According to the theorem in Section 8-6, the homogeneous second order linear equation

$$y'' + k^2 y = 0 \qquad (5)$$

has general solution

$$y(x) = c_1 \cos kx + c_2 \sin kx. \qquad (6)$$

That is, a function $y(x)$ satisfies Equation (5) if and only if it is of the form in (6). Thus the general solution $y(x)$ is a linear combination of the two particular solutions $y_1(x) = \cos kx$ and $y_2(x) = \sin kx$ of the differential equation $y'' + k^2 y = 0$.

---

Just as the general solution of the second order equation in (5) is a linear combination of two particular solutions, it is known that the general solution of the $n$th order homogeneous equation in (3) is a **linear combination**

$$y(x) = c_1 y_1(x) + c_2 y_2(x) + \cdots + c_n y_n(x) \qquad (7)$$

of $n$ particular solutions. That is, there exist $n$ particular solutions $y_1(x)$, $y_2(x), \ldots, y_n(x)$ of Equation (3) such that *every* solution of (3) is of the form in (7). See Section 2.2 of Edwards and Penney, *Elementary Differential Equations with Applications* (Englewood Cliffs, N.J.; Prentice-Hall, 1985). Our task is to find these particular solutions. Specifically, in order to construct the general solution in (7) we need to find $n$ particular solutions which are *linearly independent* in the sense that no one of them is a linear combination of the others.

## THE CHARACTERISTIC EQUATION

We first look for a *single* solution of Equation (3) and begin with the observation that

$$\frac{d^k}{dx^k}(e^{rx}) = r^k e^{rx}, \tag{8}$$

so any derivative of $e^{rx}$ is a constant multiple of $e^{rx}$. Hence, if we substituted $y = e^{rx}$ in Equation (3), each term would be a constant multiple of $e^{rx}$, with the constant coefficients dependent on $r$ and the coefficients $a_i$. This suggests that we try to find a value of $r$ so that all these multiples of $e^{rx}$ will have sum zero. If we succeed, then $y = e^{rx}$ will be a solution of Equation (3).

For instance, if we substitute $y = e^{rx}$ in the equation

$$y'' - 5y' + 6y = 0,$$

we obtain

$$r^2 e^{rx} - 5re^{rx} + 6e^{rx} = 0;$$

thus

$$(r - 2)(r - 3)e^{rx} = 0.$$

Hence $y = e^{rx}$ will be a solution if either $r = 2$ or $r = 3$. So in searching for a single solution, we actually have found *two* solutions: $y_1(x) = e^{2x}$ and $y_2(x) = e^{3x}$.

To carry out this technique in the general case, we substitute $y = e^{rx}$ in Equation (3). With the aid of Equation (8), we find the result to be

$$a_n r^n e^{rx} + a_{n-1} r^{n-1} e^{rx} + \cdots + a_2 r^2 e^{rx} + a_1 r e^{rx} + a_0 e^{rx} = 0;$$

that is,

$$e^{rx}(a_n r^n + a_{n-1} r^{n-1} + \cdots + a_2 r^2 + a_1 r + a_0) = 0.$$

Because $e^{rx}$ is never zero, we see that $y = e^{rx}$ will be a solution of Equation (3) precisely when $r$ is a root of the equation

$$a_n r^n + a_{n-1} r^{n-1} + \cdots + a_2 r^2 + a_1 r + a_0 = 0. \tag{9}$$

Equation (9) is called the **characteristic equation** of the differential equation in (3). If it has $n$ *distinct* (unequal) roots $r_1, r_2, \ldots r_n$, then the functions

$$e^{r_1 x}, e^{r_2 x}, \ldots, e^{r_n x}$$

turn out to be linearly independent. This gives the following result.

---

*Theorem 1*    *Distinct Real Roots*

If the $n$ roots $r_1, r_2, \ldots, r_n$ of the characteristic equation in (9) are real and distinct, then

$$y = c_1 e^{r_1 x} + c_2 e^{r_2 x} + \cdots + c_n e^{r_n x} \tag{10}$$

is a general solution of Equation (3).

---

A proof of this theorem can be found in Section 2.3 of the differential equations book by Edwards and Penney cited earlier.

**EXAMPLE 2**    Find a general solution of

$$y^{(3)} - y'' - 6y' = 0.$$

*Solution*    The characteristic equation of this differential equation is

$$r^3 - r^2 - 6r = 0,$$

which we solve by factoring:

$$r(r - 3)(r + 2) = 0.$$

The three roots are 0, 3, and $-2$. They are real and distinct, and therefore (because $e^0 = 1$) a general solution of the given differential equation is

$$y = c_1 + c_2 e^{3x} + c_3 e^{-2x}.$$

---

Note the two surprises: first, that the exponential function is so important in solving linear differential equations with constant coefficients; second, that *no calculus* is used in any explicit way. Theorem 1 changes a problem in differential equations into a problem involving the fundamental theorem of algebra.

If the roots of the characteristic equation in (9) are *not* distinct—that is, if there are repeated roots—then we do not yet have $n$ independent solutions of Equation (3). The problem in this case is to produce the "missing" independent solutions.

Suppose, for example, that $n = 2$ and that the characteristic equation has the double root $r = r_1$. This will happen precisely when the characteristic equation is a constant multiple of the equation

$$(r - r_1)^2 = r^2 - 2r_1 r + r_1^2 = 0.$$

Any differential equation with this characteristic equation is equivalent to

$$y'' - 2r_1 y' + r_1^2 = 0. \tag{11}$$

But it is easy to verify by direct substitution that $y = xe^{r_1 x}$ is a second solution (in addition to $y = e^{r_1 x}$) of Equation (11). It is clear that

$$y_1(x) = e^{r_1 x} \quad \text{and} \quad y_2(x) = xe^{r_1 x}$$

are independent functions, so the general solution of the differential equation in (11) is

$$y = c_1 e^{r_1 x} + c_2 x e^{r_1 x} = (c_1 + c_2 x)e^{r_1 x}.$$

Similarly, it is easy to verify by direct substitution that $e^{r_1 x}$, $xe^{r_1 x}$, and $x^2 e^{r_1 x}$ are all solutions of the third order equation

$$y''' - 3r_1 y'' + 3r_1^2 y' - r_1^3 y = 0, \tag{12}$$

in which $r = r_1$ is a triple root of the associated characteristic equation. Thus the general solution of (12) derived from this triple root is

$$y(x) = (c_1 + c_2 x + c_3 x^2)e^{r_1 x}.$$

By a general argument—see Section 2.3 of Edwards and Penney, *Elementary Differential Equations*—the following theorem can be established.

---

**Theorem 2  Repeated Roots**

If the characteristic equation in (9) has a repeated root $r_1$ of multiplicity $k$, then the part of the general solution of Equation (3) that corresponds to $r_1$ is

$$(c_1 + c_2 x + \cdots + c_k x^{k-1})e^{r_1 x}. \qquad (13)$$

---

**EXAMPLE 3**   Find the general solution of the differential equation

$$y^{(4)} + 3y^{(3)} + 3y'' + y' = 0.$$

*Solution*   The characteristic equation of this differential equation is

$$r^4 + 3r^3 + 3r^2 + r = r(r+1)^3 = 0.$$

It has the simple root $r_1 = 0$ and the triple root $r_2 = -1$. The particular solution corresponding to $r_1 = 0$ is $y_1 = c_1 e^0 = c_1$, while by Theorem 2 (with $k = 3$) the part of the general solution corresponding to $r_2 = -1$ is $(c_2 + c_3 x + c_4 x^2)e^{-x}$. Hence the desired general solution is

$$y = c_1 + (c_2 + c_3 x + c_4 x^2)e^{-x}.$$

---

The roots of the characteristic equation may be either real or complex. Because Equation (8) holds even when $r$ is complex, it follows that (whether $r$ is real, as before, or complex) $e^{rx}$ will be a solution of the differential equation in (3) if and only if $r$ is a root of its characteristic equation. If the complex conjugate pair of roots $r_1 = a + bi$ and $r_2 = a - bi$ are simple (non-repeated), then the corresponding part of the general solution of (3) is

$$\begin{aligned}
C_1 e^{(a+bi)x} + C_2 e^{(a-bi)x} &= C_1 e^{ax}(\cos bx + i \sin bx) \\
&\quad + C_2 e^{ax}(\cos bx - i \sin bx) \\
&= e^{ax}(c_1 \cos bx + c_2 \sin bx)
\end{aligned}$$

where $c_1 = C_1 + C_2$ and $c_2 = (C_1 - C_2)i$. Thus the conjugate pair of roots $a \pm bi$ leads to the linearly independent *real-valued* solutions $e^{ax}\cos bx$ and $e^{ax}\sin bx$. This yields the following result.

---

**Theorem 3  Complex Roots**

If the characteristic equation in (9) has an unrepeated pair of complex conjugate roots $a \pm bi$ (with $b \neq 0$), then the corresponding part of a general solution of the differential equation in (3) is of the form

$$e^{ax}(c_1 \cos bx + c_2 \sin bx). \qquad (14)$$

---

**EXAMPLE 4**   The characteristic equation of

$$y'' + b^2 y = 0 \qquad (b > 0)$$

is $r^2 + b^2 = 0$, with roots $\pm bi$. So Theorem 3 gives the general solution

$$y = c_1 \cos bx + c_2 \sin bx$$

with which we are already familiar.

**EXAMPLE 5**   Find the particular solution of

$$y'' - 4y' + 5y = 0$$

for which $y(0) = 1$ and $y'(0) = 5$.

*Solution*   The characteristic equation is $r^2 - 4r + 5 = 0$, with roots $2 + i$ and $2 - i$. Hence a general solution is

$$y = e^{2x}(c_1 \cos x + c_2 \sin x).$$

Then

$$y' = 2e^{2x}(c_1 \cos x + c_2 \sin x) + e^{2x}(-c_1 \sin x + c_2 \cos x),$$

so the initial conditions give

$$y(0) = c_1 = 1 \quad \text{and} \quad y'(0) = 2c_1 + c_2 = 5.$$

It follows that $c_2 = 3$, so the desired particular solution is

$$y = e^{2x}(\cos x + 3 \sin x).$$

**EXAMPLE 6**   Find a general solution of

$$y^{(4)} + 16y = 0.$$

*Solution*   The characteristic equation is

$$r^4 + 16 = (r^2 + 4i)(r^2 - 4i) = 0,$$

and its four roots are $\pm 2\sqrt{\pm i}$. Now $i = e^{i\pi/2}$ and $-i = e^{3i\pi/2}$, so

$$\sqrt{i} = (e^{i\pi/2})^{1/2} = e^{i\pi/4} = \frac{1 + i}{\sqrt{2}}$$

and

$$\sqrt{-i} = (e^{3i\pi/2})^{1/2} = e^{3i\pi/4} = \frac{-1 + i}{\sqrt{2}}.$$

Thus the four (distinct) roots of the characteristic equation are $r = \pm(1 \pm i)\sqrt{2}$. These two pairs of complex conjugate roots, $\sqrt{2} \pm i\sqrt{2}$ and $-\sqrt{2} \pm i\sqrt{2}$, give a general solution

$$y = e^{x\sqrt{2}}(c_1 \cos x\sqrt{2} + c_2 \sin x\sqrt{2}) + e^{-x\sqrt{2}}(c_3 \cos x\sqrt{2} + c_4 \sin x\sqrt{2})$$

of $y^{(4)} + 16y = 0$.

---

It turns out that, in the case of a complex conjugate pair of roots $a \pm bi$ of multiplicity $k$, the corresponding part of the general solution is a linear combination of the $2k$ functions

$$x^p e^{ax} \cos bx \quad \text{and} \quad x^p e^{ax} \sin bx$$

for $p = 0, 1, 2, \ldots, k - 1$.

## 18-5 PROBLEMS

Find a general solution of each of the differential equations given in Problems 1–20.

1 $y'' - 4y = 0$

2 $2y'' - 3y' = 0$

3 $y'' + 3y' - 10y = 0$

4 $2y'' - 7y' + 3y = 0$

5 $y'' + 6y' + 9y = 0$

6 $y'' + 5y' + 5y = 0$

7 $4y'' - 12y' + 9y = 0$

8 $y'' - 6y' + 13y = 0$

9 $y'' + 8y' + 25y = 0$

10 $5y^{(4)} + 3y^{(3)} = 0$

11 $y^{(4)} - 8y^{(3)} + 16y'' = 0$

12 $y^{(4)} - 3y^{(3)} + 3y'' - y' = 0$

13 $9y^{(3)} + 12y'' + 4y' = 0$

14 $y^{(4)} + 3y'' - 4y = 0$

15 $y^{(4)} - 8y'' + 16y = 0$

16 $y^{(4)} + 18y'' + 81y = 0$

17 $6y^{(4)} + 11y'' + 4y = 0$

18 $y^{(4)} = 16y$

19 $y^{(3)} + y'' - y' - y = 0$

20 $y^{(4)} + 2y^{(3)} + 3y'' + 2y' + y = 0$

Solve each of the initial value problems given in Problems 21–26.

21 $y'' - 4y' + 3y = 0;$ $y(0) = 7, y'(0) = 11$

22 $9y'' + 12y' + 4y = 0;$ $y(0) = 3, y'(0) = 4$

23 $y'' - 6y' + 25y = 0;$ $y(0) = 3, y'(0) = 1$

24 $2y^{(3)} - 3y'' - 2y' = 0;$ $y(0) = 1, y'(0) = -1,$ $y''(0) = 3$

25 $3y^{(3)} + 2y'' = 0;$ $y(0) = -1, y'(0) = 0, y''(0) = 1$

26 $y^{(3)} + 10y'' + 25y' = 0;$ $y(0) = 3, y'(0) = 4,$ $y''(0) = 5$

## Mechanical Vibrations

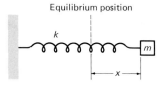

Equilibrium position

**18.10** The undamped mass-and-spring system

In Section 8-6, we discussed the motion of a mass $m$ on the end of a spring that exerts a restoring force $F_S = -kx$ on the mass when its displacement from the equilibrium position is $x$ (Fig. 18.10). Newton's law then gives $m(d^2x/dt^2) = F_S = -kx$; that is,

$$\frac{d^2x}{dt^2} + \omega^2 x = 0 \qquad \left(\omega^2 = \frac{k}{m}\right). \tag{1}$$

The general solution of (1) is

$$x(t) = A \cos \omega t + B \sin \omega t. \tag{2}$$

If the mass is released from rest at $x = x_0$, then the initial conditions $x(0) = x_0$ and $x'(0) = 0$ give

$$x(t) = x_0 \cos \omega t \tag{3}$$

as the motion that results. This equation describes a periodic oscillation or vibration with **amplitude** $x_0$, **period**

$$T = \frac{2\pi}{\omega} = 2\pi \sqrt{\frac{m}{k}}, \tag{4}$$

and **frequency** in hertz (cycles per second)

$$f = \frac{1}{T} = \frac{\omega}{2\pi} = \frac{1}{2\pi} \sqrt{\frac{k}{m}}. \tag{5}$$

For a specific particular solution of Equation (1), the values of the coefficients $A$ and $B$ in (2) are readily determined from given initial conditions, typically of the form $x(0) = x_0$ and $x'(0) = v_0$. Once this has been done, the identity

$$\cos(\alpha + \beta) = \cos \alpha \cos \beta - \sin \alpha \sin \beta \tag{6}$$

can be used to rewrite the solution in (2) in the form

$$x(t) = C \cos(\omega t - \alpha) \tag{7}$$

where $C = \sqrt{A^2 + B^2}$ is the **amplitude** of the motion, $\omega$ is its **circular frequency** in radians per second, and $\alpha$ is its **phase angle.** This process is illustrated in the following example.

**EXAMPLE 1**   A body that weighs $W = 16$ lb is attached to the end of a spring which is stretched 2 ft by a force of 100 lb. It is set in motion with initial position $x_0 = 0.5$ (ft) and initial velocity $v_0 = -10$ (ft/s). (Note that these data indicate that the body is displaced to the right and moving to the left at time $t = 0$.) Find the position function of the body, as well as the amplitude, frequency, period of oscillation, and phase angle of its motion.

*Solution*   We take $g = 32$ ft/s$^2$. The mass of the body is then $m = W/g = 0.5$ (slugs). The spring constant is $k = 100/2 = 50$ (lb/ft), so Equation (1) yields

$$\tfrac{1}{2}x'' + 50x = 0;$$

that is,

$$x'' + 100x = 0.$$

Consequently the circular frequency will be $\omega = 10$ (rad/s). So the body will oscillate with

$$\text{Frequency:}\quad \frac{10}{2\pi} \approx 1.59 \quad (\text{Hz})$$

and

$$\text{Period:}\quad \frac{2\pi}{10} \approx 0.63 \quad (\text{s}).$$

We now impose the initial conditions $x(0) = 0.5$ and $x'(0) = -10$ in the general solution

$$x = A \cos 10t + B \sin 10t,$$

and it follows that $A = 0.5$ and $B = -1$. So the position function of the body is

$$x(t) = \frac{1}{2} \cos 10t - \sin 10t.$$

Hence its amplitude of motion is

$$C = \sqrt{(1/2)^2 + (1)^2} = \frac{\sqrt{5}}{2} \approx 1.12 \quad (\text{ft}).$$

To find the phase angle we write

$$x = \frac{\sqrt{5}}{2}\left( \frac{1}{\sqrt{5}} \cos 10t - \frac{2}{\sqrt{5}} \sin 10t \right) = \frac{\sqrt{5}}{2} \cos(10t - \alpha),$$

using the identity in (6). Thus we require $\cos \alpha = 1/\sqrt{5} > 0$ and $\sin \alpha = -2/\sqrt{5} < 0$. Hence $\alpha$ is the fourth-quadrant angle

$$\alpha = 2\pi - \tan^{-1}\left( \frac{2/\sqrt{5}}{1/\sqrt{5}} \right) \approx 5.1760 \quad (\text{rad}).$$

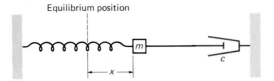

Equilibrium position

**18.11** A spring-and-mass system with damping force provided by a dashpot

In the form in which the amplitude and phase angle are made explicit, the position function is

$$x(t) \approx \frac{\sqrt{5}}{2} \cos(10t - 5.1760) \approx (1.1180) \cos(10t - 5.1760).$$

### DAMPED VIBRATIONS

Now suppose that the mass of the above discussion is attached, as indicated in Fig. 18.11, to a dashpot or shock absorber—as in the suspension system of an automobile—that exerts on it a force $F_R$ of resistance to motion that is proportional to the velocity $v$ of the mass. Then $F_R = -cv$, where the proportionality constant $c$ will depend on the viscosity of the fluid in the dashpot; the more viscous this fluid is, the larger $c > 0$ is. To take this resistive force into account, we replace Equation (1) by

$$m\frac{d^2x}{dt^2} = F_S + F_R = -kx - c\frac{dx}{dt}.$$

In order to analyze the motion of a mass-on-a-spring-with-dashpot, we therefore need to solve the homogeneous linear second order equation

$$mx'' + cx' + kx = 0; \tag{8}$$

alternatively,

$$x'' + 2px' + \omega_0^2 x = 0, \tag{9}$$

where $\omega_0 = \sqrt{k/m}$ is the corresponding *undamped* circular frequency and

$$p = \frac{c}{2m} > 0. \tag{10}$$

The characteristic equation $r^2 + 2pr + \omega_0^2 = 0$ of Equation (9) has roots

$$r_1 = -p + \sqrt{p^2 - \omega_0^2} \quad \text{and} \quad r_2 = -p - \sqrt{p^2 - \omega_0^2}. \tag{11}$$

Whether these roots are real or complex depends on the sign of

$$p^2 - \omega_0^2 = \frac{c^2}{4m^2} - \frac{k}{m} = \frac{c^2 - 4km}{4m^2}.$$

The **critical damping** $c_{CR}$ is given by

$$c_{CR} = \sqrt{4km},$$

and we distinguish three cases, according as $c > c_{CR}$, $c = c_{CR}$, or $c < c_{CR}$.

### OVERDAMPED CASE: $c > c_{CR}$ ($c^2 > 4km$)

Because $c$ is relatively large in this case, we are dealing with a strong resistance in comparison with a relatively weak force or a small mass. Then Equation (11) gives distinct real roots $r_1$ and $r_2$, both of which are negative.

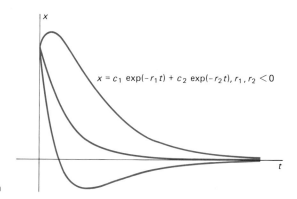

**18.12** Overdamped motion

$$x = c_1 \exp(-r_1 t) + c_2 \exp(-r_2 t), r_1, r_2 < 0$$

(Why?) The position function has the form

$$x(t) = c_1 \exp(r_1 t) + c_2 \exp(r_2 t). \tag{12}$$

It is easy to see that $x(t) \to 0$ as $t \to +\infty$ and that the body settles to its equilibrium position without any oscillations. Figure 18.12 shows some typical graphs of the position function for the overdamped case; we chose $x_0$ a fixed number and illustrated the effect of varying the initial velocity $v_0$. In every case the would-be oscillations are damped out.

**CRITICALLY DAMPED CASE: $c = c_{CR}$ ($c^2 = 4km$)**

In this case Equation (11) gives equal roots $r_1 = r_2 = -p$ of the characteristic equation, so the general solution is

$$x(t) = e^{-pt}(c_1 + c_2 t). \tag{13}$$

Because $e^{-pt} > 0$ for all $t$ and $c_1 + c_2 t$ has at most one positive zero, the body passes through its equilibrium position at most once, and it is clear that $x(t) \to 0$ as $t \to +\infty$. Some graphs of the motion in the critically damped case appear in Fig. 18.13, and they resemble those of the overdamped case in Fig. 18.12. In the critically damped case the resistance of the dashpot is just large enough to damp out any oscillations, but even a slight decrease in the resistance $c$ will bring us to the remaining case, the one that shows the most dramatic behavior.

**18.13** Critically damped motion

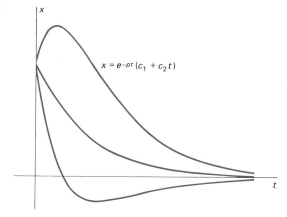

$$x = e^{-pt}(c_1 + c_2 t)$$

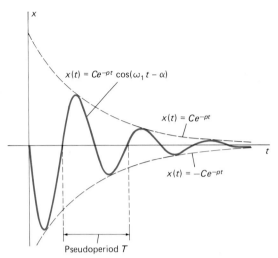

$x(t) = Ce^{-pt} \cos(\omega_1 t - \alpha)$

$x(t) = Ce^{-pt}$

$x(t) = -Ce^{-pt}$

Pseudoperiod $T$

**18.14** Underdamped motion

**UNDERDAMPED CASE: $c < c_{CR}$ ($c^2 < 4km$)**

The characteristic equation now has complex conjugate roots $-p \pm i\sqrt{\omega_0^2 - p^2}$, and the general solution is

$$x(t) = e^{-pt}(c_1 \cos \omega_1 t + c_2 \sin \omega_1 t), \tag{14}$$

where

$$\omega_1 = \sqrt{\omega_0^2 - p^2} = \frac{\sqrt{4km - c^2}}{2m}. \tag{15}$$

Using the cosine addition formula as in Example 1, we may rewrite (14) as

$$x(t) = Ce^{-pt}\cos(\omega_1 t - \alpha), \tag{16}$$

where $C = \sqrt{c_1^2 + c_2^2}$ and $\tan \alpha = c_2/c_1$.

The solution in (16) represents exponentially damped oscillations of the body about its equilibrium position. The graph of $x(t)$ lies between the curves $x = Ce^{-pt}$ and $x = -Ce^{-pt}$ and touches them when $\omega_1 t - \alpha$ is an integral multiple of $\pi$. The motion is not truly periodic, but it nevertheless is useful to call $\omega_1$ its **circular frequency,** $T_1 = 2\pi/\omega_1$ its **pseudoperiod** of oscillation, and $Ce^{-pt}$ its **time-varying amplitude.** Most of these quantities are shown on the typical graph of underdamped motion shown in Fig. 18.14. Note from Equation (15) that in this case $\omega_1$ is less than the undamped circular frequency $\omega_0$, so $T_1$ is larger than the period $T$ of oscillation of the same mass without damping on the same spring. Thus the damping of the dashpot has at least three effects: (1) It exponentially damps the oscillations, in accord with the time-varying amplitude; (2) It slows the motion; that is, the dashpot decreases the frequency of the motion; and (3) It delays the motion—this is the effect of the phase angle in Equation (16).

**EXAMPLE 2** The mass-and-spring of Example 1 is now also attached to a dashpot that provides 6 lb of resistive force for each foot per second of velocity. The mass is set in motion with the same initial position $x(0) = 0.5$ (ft) and the same initial velocity $x'(0) = v_0 = -10$ (ft/s). Find the position function of the mass, its new frequency and pseudoperiod, and its phase angle.

*Solution*   Rather than memorizing the various formulas given above, it is better practice in any particular case to set up the differential equation and then solve it directly. Recall that $m = \frac{1}{2}$, $k = 50$, and we are now given $c = 6$ in fps units. Hence Equation (8) is

$$\tfrac{1}{2}x'' + 6x' + 50x = 0;$$

that is,

$$x'' + 12x' + 100x = 0.$$

The roots of the characteristic equation $r^2 + 12r + 100 = 0$ are

$$\frac{-12 \pm \sqrt{144 - 400}}{2} = -6 \pm 8i,$$

so the general solution is

$$x(t) = e^{-6t}(A \cos 8t + B \sin 8t). \tag{17}$$

The new circular frequency is $\omega_1 = 8$ (rad/s) and the pseudoperiod and new frequency are

$$T_1 = \frac{2\pi}{8} \approx 0.79 \quad \text{(s)}$$

and

$$\frac{1}{T_1} = \frac{8}{2\pi} \approx 1.27 \quad \text{(Hz)}$$

(in contrast with 0.63 s and 1.59 Hz, respectively, in the undamped case).
From Equation (17) we compute

$$x'(t) = e^{-6t}(-8A \sin 8t + 8B \cos 8t) - 6e^{-6t}(A \cos 8t + B \sin 8t).$$

Our initial conditions therefore give the equations

$$x(0) = A = \frac{1}{2}$$

and

$$x'(0) = -6A + 8B = -10,$$

so $A = \frac{1}{2}$ and $B = -\frac{7}{8}$. Thus

$$x(t) = e^{-6t}(\tfrac{1}{2} \cos 8t - \tfrac{7}{8} \sin 8t),$$

so with $C = \sqrt{(\tfrac{1}{2})^2 + (\tfrac{7}{8})^2} = \sqrt{65}/8$, we have

$$x(t) = \frac{\sqrt{65}}{8} e^{-6t} \left( \frac{4}{\sqrt{65}} \cos 8t - \frac{7}{\sqrt{65}} \sin 8t \right).$$

We require $\cos \alpha = 4/\sqrt{65} > 0$ and $\sin \alpha = -7/\sqrt{65} < 0$, so $\alpha$ is the fourth-quadrant angle

$$\alpha = 2\pi - \tan^{-1}(7/4) \approx 5.2315 \quad \text{(rad)}.$$

Finally,

$$x(t) \approx \frac{\sqrt{65}}{8} e^{-6t} \cos(8t - 5.2315)$$

$$\approx (1.0078)e^{-6t} \cos(8t - 5.2315).$$

## 18-6 PROBLEMS

1 Determine the period and frequency of the simple harmonic motion of a 4-kg mass on the end of a spring with spring constant 16 N/m.

2 Determine the period and frequency of the simple harmonic motion of a body weighing 24 lb on the end of a spring with spring constant 48 lb/ft.

3 A mass of 3 kg is attached to the end of a spring that is stretched 20 cm by a force of 15 N. It is set in motion with initial position $x_0 = 0$ and initial velocity $v_0 = -10$ m/s. Find the amplitude, period, and frequency of the resulting motion.

4 A body weighing 8 lb is attached to the end of a spring that is stretched 1 in. by a force of 3 lb. At time $t = 0$ (s) the body is pulled 1 ft to the right (spring stretched) and set in motion with an initial velocity of 5 ft/s to the left.
(a) Find the displacement $x(t)$ of the body in the form $C \cos(\omega_0 + \alpha)$.
(b) Find the amplitude and period of motion of the body.

5 Consider a floating cylindrical buoy with radius $r$, height $h$, and density $\rho \leq 0.5$ (recall that the density of water is 1 gm/cm³). The buoy is initially suspended at rest with its bottom on the top surface of the water and is released at time $t = 0$. Thereafter, it is acted on by two forces: a downward gravitational force equal to its weight, $mg = \rho\pi r^2 hg$, and an upward force of buoyancy equal to the weight $\pi r^2 xg$ of water displaced, where $x = x(t)$ is the depth of the bottom of the buoy beneath the surface at time $t$. Conclude that the buoy undergoes simple harmonic motion about the equilibrium position $x_e = \rho h$ with period $p = 2\pi\sqrt{\rho h/g}$. Compute $p$ and the amplitude of the motion if $\rho = 0.5$ (gm/cm³), $h = 200$ (cm), and $g = 980$ (cm/s²).

6 A cylindrical buoy weighing 100 lb floats in water with its axis vertical (as in Problem 5). When depressed slightly and released, it oscillates up and down four times every 10 s. Assume that friction is negligible. Find the radius of the buoy.

The remaining problems in this section deal with free *damped* motion. In Problems 7–11, a mass $m$ is attached both to a spring (with given spring constant $k$) and a dashpot (with given damping constant $c$). The mass is set in motion with initial position $x_0$ and initial velocity $v_0$. Find the position function $x(t)$, and determine whether the motion is underdamped, critically damped, or overdamped. If it is underdamped, write $x(t)$ in the form $Ce^{-pt}\cos(\omega_1 t - \alpha)$.

7 $m = \frac{1}{2}, c = 3, k = 4$; $x_0 = 2, v_0 = 0$
8 $m = 3, c = 30, k = 63$; $x_0 = 2, v_0 = 2$
9 $m = 1, c = 8, k = 16$; $x_0 = 5, v_0 = -10$
10 $m = 2, c = 12, k = 50$; $x_0 = 0, v_0 = -8$
11 $m = 4, c = 20, k = 169$; $x_0 = -4, v_0 = 16$

12 A 12-lb weight is attached both to a vertically suspended spring that it stretches 6 in. and to a dashpot that provides 3 lb of resistive force for every foot per second of velocity.
(a) Suppose that the weight is pulled down 1 ft below its static equilibrium position and then released from rest at time $t = 0$. Find its position function $x(t)$.
(b) Find the frequency, time-varying amplitude, and phase angle of the motion of the weight.

13 This problem deals with a highly simplified model of a car weighing 3200 lb. Assume that the suspension system of the car acts like a single spring and its shock absorbers like a single dashpot, so that its vertical vibrations satisfy Equation (8) with appropriate values of the coefficients.
(a) Find the stiffness coefficient $k$ of the spring if the car undergoes free vibrations at 80 cycles/min when its shock absorbers are disconnected.
(b) With the shock absorbers connected, the car is set into vibration by driving it over a bump, and the resulting damped vibrations have a frequency of 78 cycles/min. After how long will the time-varying amplitude be 1% of its original value?

---

### 18-7

**Infinite Series Methods**

In Section 18-5 we saw that solving a homogeneous linear differential equation with constant coefficients can be reduced to the algebraic problem of finding the roots of its characteristic equation. There is no similar procedure for solving linear differential equations with *variable* coefficients, at least not routinely and in finitely many steps. In this section we discuss the use of infinite series for solving such equations.

The *power series method* for solving a differential equation consists of substituting the power series

$$y = \sum_{n=0}^{\infty} c_n x^n = c_0 + c_1 x + c_2 x^2 + c_3 x^3 + \cdots \tag{1}$$

in the differential equation and then attempting to determine what the coefficients $c_0, c_1, c_2, \ldots$ must be in order that the power series will satisfy the differential equation. This method is not always successful, but when it is we obtain an infinite series representation of a solution, in contrast to the "closed form" solutions that our previous methods have yielded.

Before we can substitute the power series in (1) in a differential equation, we must first know what to substitute for the derivatives $y', y'', \ldots$ . Recall that according to Theorem 3 in Section 12-8, the derivative of the series in (1) may be calculated by termwise differentiation. That is,

$$y' = \sum_{n=0}^{\infty} n c_n x^{n-1} = c_1 + 2c_2 x + 3c_3 x^2 + \cdots . \tag{2}$$

Similarly, the second derivative is given by

$$y'' = \sum_{n=0}^{\infty} n(n-1) c_n x^{n-2} = 2c_2 + 6c_3 x + 12c_4 x^2 + \cdots . \tag{3}$$

The series in (2) and (3) have the same radius of convergence as the original series in (1).

The process of determining the coefficients $c_0, c_1, c_2, \ldots$, in the series in (1) in order that it satisfy a given differential equation depends also on the following consequence of termwise differentiation (see Problem 40 in Section 12-8): If two power series represent the same function on an interval, then they are identical series. In particular, if $\sum a_n x^n = 0$ for all $x$ in some open interval, then it follows that $a_n = 0$ for all $n$. We will refer to this fact as the **identity principle** for power series.

**EXAMPLE 1**   Solve the equation $y' + 2y = 0$.

*Solution*   We substitute the series

$$y = \sum_{n=0}^{\infty} c_n x^n \quad \text{and} \quad y' = \sum_{n=1}^{\infty} n c_n x^{n-1},$$

and obtain

$$\sum_{n=1}^{\infty} n c_n x^{n-1} + 2 \sum_{n=0}^{\infty} c_n x^n = 0. \tag{4}$$

To compare coefficients here, we need the general term in each to be the term containing $x^n$. To accomplish this, we shift the index of summation in the first sum. To see how to do this, note that

$$\sum_{n=1}^{\infty} n c_n x^{n-1} = c_1 + 2c_2 x + 3c_3 x^2 + \cdots$$

$$= \sum_{n=0}^{\infty} (n+1) c_{n+1} x^n.$$

Thus we can replace $n$ by $n + 1$ if, at the same time, we start counting one step lower—that is, at $n = 0$ rather than at $n = 1$. This is a shift of $+1$ in the index of summation. The result of making this shift in (4) is the identity

$$\sum_{n=0}^{\infty} (n + 1)c_{n+1}x^n + 2 \sum_{n=0}^{\infty} c_n x^n = 0;$$

that is,

$$\sum_{n=0}^{\infty} [(n + 1)c_{n+1} + 2c_n]x^n = 0.$$

If this is true on some open interval, then it follows from the identity principle that

$$(n + 1)c_{n+1} + 2c_n = 0$$

for all $n \geq 0$; consequently

$$c_{n+1} = -\frac{2c_n}{n + 1} \tag{5}$$

for all $n \geq 0$. Equation (5) is a *recursion formula* from which we can successively compute $c_1, c_2, c_3, \ldots$ in terms of $c_0$; the latter will turn out to be the arbitrary constant we expect to find in a solution of a first order differential equation.

With $n = 0$, (5) gives

$$c_1 = -\frac{2c_0}{1}.$$

With $n = 1$, (5) gives

$$c_2 = -\frac{2c_1}{2} = +\frac{2^2 c_0}{1 \cdot 2} = +\frac{2^2 c_0}{2!}.$$

With $n = 2$, (5) gives

$$c_3 = -\frac{2c_2}{3} = -\frac{2^3 c_0}{1 \cdot 2 \cdot 3} = -\frac{2^3 c_0}{3!}.$$

By now it should be clear that after $n$ such steps, we will have

$$c_n = (-1)^n \frac{2^n c_0}{n!}, \qquad n \geq 1.$$

(This is easy to prove by induction on $n$.) Consequently our solution takes the form

$$y = \sum_{n=0}^{\infty} c_n x^n = \sum_{n=0}^{\infty} (-1)^n \frac{2^n c_0}{n!} x^n$$

$$= c_0 \sum_{n=0}^{\infty} \frac{(-2x)^n}{n!} = c_0 e^{-2x}.$$

In the final step we have used the familiar exponential series

$$e^t = \sum_{n=0}^{\infty} \frac{t^n}{n!} = 1 + t + \frac{t^2}{2!} + \frac{t^3}{3!} + \cdots$$

to identify our power series solution as the same solution $y = c_0 e^{-2x}$ we could have obtained immediately by the method of separation of variables.

---

We know that the power series obtained in Example 1 converges for all $x$ because it is an exponential series. More commonly, a power series solution is not recognizable in terms of the familiar elementary functions. When we get an unfamiliar power series solution, we need a way of ascertaining where it converges. After all, $y = \sum c_n x^n$ is merely an *assumed* form of the solution. The procedure illustrated in Example 1 for determining the coefficients $\{c_n\}$ is only a formal process and may or may not be valid. Its validity depends on the convergence of the initially *unknown* series $y = \sum c_n x^n$. Hence this formal process is justified only if in the end we can show that the power series obtained converges on some open interval. If so, it then represents a solution of the differential equation on that interval. Recall from Section 12-8 that the radius of convergence of the power series $\sum c_n x^n$ is given by

$$r = \lim_{n \to \infty} \left| \frac{c_n}{c_{n+1}} \right| \tag{6}$$

provided that this limit exists.

**EXAMPLE 2** Solve the equation $(x - 3)y' + 2y = 0$.

*Solution* As before, we substitute

$$y = \sum_{n=0}^{\infty} c_n x^n \quad \text{and} \quad y' = \sum_{n=1}^{\infty} n c_n x^{n-1}$$

to obtain

$$(x - 3) \sum_{n=1}^{\infty} n c_n x^{n-1} + 2 \sum_{n=0}^{\infty} c_n x^n = 0,$$

so that

$$\sum_{n=1}^{\infty} n c_n x^n - 3 \sum_{n=1}^{\infty} n c_n x^{n-1} + 2 \sum_{n=0}^{\infty} c_n x^n = 0.$$

In the first sum we can replace $n = 1$ by $n = 0$ with no effect on the sum. In the second sum we shift the index of summation by $+1$. This yields

$$\sum_{n=0}^{\infty} n c_n x^n - 3 \sum_{n=0}^{\infty} (n + 1) c_{n+1} x^n + 2 \sum_{n=0}^{\infty} c_n x^n = 0;$$

that is,

$$\sum_{n=0}^{\infty} [n c_n - 3(n + 1) c_{n+1} + 2 c_n] x^n = 0.$$

The identity principle then gives

$$n c_n - 3(n + 1) c_{n+1} + 2 c_n = 0,$$

from which we obtain the recursion formula

$$c_{n+1} = \frac{n + 2}{3(n + 1)} c_n.$$

We apply this formula with $n = 0$, $n = 1$, and $n = 2$ in turn, and find that

$$c_1 = \frac{2}{3} c_0, \qquad c_2 = \frac{3}{3 \cdot 2} c_1 = \frac{3}{3^2} c_0,$$

and

$$c_3 = \frac{4}{3 \cdot 3} c_2 = \frac{4}{3^3} c_0.$$

This is almost enough to make the pattern evident; it is not difficult to show by induction on $n$ that

$$c_n = \frac{n + 1}{3^n} c_0 \quad \text{if } n \geqq 1.$$

Hence our proposed power series solution is

$$y = c_0 \sum_{n=0}^{\infty} \frac{n + 1}{3^n} x^n. \tag{7}$$

Its radius of convergence is

$$r = \lim_{n \to \infty} \left| \frac{c_n}{c_{n+1}} \right| = \lim_{n \to \infty} \frac{3n + 3}{n + 2} = 3.$$

Thus the series in (7) converges if $-3 < x < 3$, but diverges if $|x| > 3$. In this particular example we can explain why. An elementary solution (obtained by separation of variables) of our differential equation is $y = 1/(3 - x)^2$. If we differentiate termwise the geometric series

$$\frac{1}{3 - x} = \frac{\frac{1}{3}}{1 - (x/3)} = \frac{1}{3} \sum_{n=0}^{\infty} \frac{x^n}{3^n},$$

we get a constant multiple of the series in (7). Thus this series (with the arbitrary constant $c_0$ appropriately chosen) represents the solution $y = 1/(3 - x)^2$ on the interval $-3 < x < 3$, and the "singularity" at $x = 3$ is the reason why the radius of convergence of our power series solution turned out to be $r = 3$.

**EXAMPLE 3**  Solve the equation $y'' + y = 0$.

*Solution*  If we assume a solution of the form

$$y = \sum_{n=0}^{\infty} c_n x^n,$$

we find that

$$y' = \sum_{n=1}^{\infty} n c_n x^{n-1} \quad \text{and} \quad y'' = \sum_{n=2}^{\infty} n(n - 1) c_n x^{n-2}.$$

Substitution for $y$ and $y''$ in the differential equation then yields

$$\sum_{n=2}^{\infty} n(n - 1) c_n x^{n-2} + \sum_{n=0}^{\infty} c_n x^n = 0.$$

We shift the index of summation in the left sum by $+2$ (by replacing $n = 2$ with $n = 0$ and $n$ with $n + 2$). This gives

$$\sum_{n=0}^{\infty} (n + 2)(n + 1)c_{n+2}x^n + \sum_{n=0}^{\infty} c_n x^n = 0.$$

The identity principle now leads to the identity

$$(n + 2)(n + 1)c_{n+2} + c_n = 0,$$

and thus we obtain the recursion formula

$$c_{n+2} = -\frac{c_n}{(n + 1)(n + 2)} \tag{8}$$

for $n \geq 0$. It is evident that this formula will determine the coefficients $c_n$ with even subscripts in terms of $c_0$, and those of odd subscript in terms of $c_1$; $c_0$ and $c_1$ are not predetermined and thus will be the two arbitrary constants we expect to find in a general solution of a second order equation.

When we apply the recursion formula in (8) with $n = 0, 2,$ and 4 in turn, we get

$$c_2 = -\frac{c_0}{2!}, \quad c_4 = \frac{c_0}{4!}, \quad \text{and} \quad c_6 = -\frac{c_0}{6!}.$$

Taking $n = 1, 3,$ and 5 in turn, we find that

$$c_3 = -\frac{c_1}{3!}, \quad c_5 = \frac{c_1}{5!}, \quad \text{and} \quad c_7 = -\frac{c_1}{7!}.$$

Again the pattern is clear; we leave it for you to show (by induction) that for $k \geq 1$,

$$c_{2k} = \frac{(-1)^k c_0}{(2k)!} \quad \text{and} \quad c_{2k+1} = \frac{(-1)^k c_1}{(2k + 1)!}.$$

Thus we get the power series solution

$$y = c_0\left(1 - \frac{x^2}{2!} + \frac{x^4}{4!} - \frac{x^6}{6!} + \cdots\right)$$

$$+ c_1\left(x - \frac{x^3}{3!} + \frac{x^5}{5!} - \frac{x^7}{7!} + \cdots\right)$$

—that is,

$$y = c_0 \cos x + c_1 \sin x.$$

Note that we have no problem with the radius of convergence here; the Taylor series for the sine and cosine functions converge for all $x$.

---

The solution of Example 3 can bear further comment. Suppose that we had never heard of the sine and cosine functions, let alone their Taylor series. We would then have discovered the two power series solutions

$$C(x) = \sum_{n=0}^{\infty} (-1)^n \frac{x^{2n}}{(2n)!} = 1 - \frac{x^2}{2!} + \frac{x^4}{4!} - \cdots \tag{9}$$

and

$$S(x) = \sum_{n=0}^{\infty} (-1)^n \frac{x^{2n+1}}{(2n+1)!} = x - \frac{x^3}{3!} + \frac{x^5}{5!} - \cdots \qquad (10)$$

of the differential equation $y'' + y = 0$. It is clear that $C(0) = 1$ and that $S(0) = 0$. After verifying that the two series in (9) and (10) converge for all $x$, we can differentiate them term by term to find that

$$C'(x) = -S(x) \quad \text{and} \quad S'(x) = C(x). \qquad (11)$$

Consequently, $C'(0) = 0$ and $S'(0) = 1$. Thus with the aid of the power series method (all the while knowing nothing about the sine and cosine functions), we have discovered that $y = C(x)$ is the unique solution of $y'' + y = 0$ that satisfies the initial conditions $y(0) = 1$ and $y'(0) = 0$, and that $y = S(x)$ is the solution that satisfies the initial conditions $y(0) = 0$ and $y'(0) = 1$. It follows that $C(x)$ and $S(x)$ are linearly independent, and—recognizing the importance of the differential equation $y'' + y = 0$—we can agree to call $C(x)$ the "cosine" function and $S(x)$ the "sine" function. Indeed, all the usual properties of these two functions can be established, using only their initial values (at $x = 0$) and the derivatives in (11); there is no need to refer to triangles or even to angles. (Can you use the series in (9) and (10) to show that

$$[C(x)]^2 + [S(x)]^2 = 1$$

for all $x$?) This demonstrates that the cosine and sine functions are fully determined by the differential equation $y'' + y = 0$ of which they are the natural linearly independent solutions.

This is by no means an uncommon situation. Many important special functions of mathematics occur in the first instance as power series solutions of differential equations and thus are in practice *defined* by means of these power series.

## 18-7   PROBLEMS

In each of Problems 1–10, find a power series solution of the given differential equation. Determine the radius of convergence of the resulting series, and use familiar power series to identify the series solution in terms of familiar elementary functions. (Of course, no one can prevent you from checking your work by also solving the equations by the methods of earlier sections!)

1  $y' = y$      2  $y' = 4y$
3  $2y' + 3y = 0$      4  $y' + 2xy = 0$
5  $y' = x^2y$      6  $(x - 2)y' + y = 0$
7  $(2x - 1)y' + 2y = 0$      8  $2(x + 1)y' = y$
9  $(x - 1)y' + 2y = 0$      10  $2(x - 1)y' = 3y$

In each of Problems 11–14, use the method of Example 3 to find two linearly independent power series solutions of the given differential equation. Determine the radius of convergence of each

series, and identify the general solution in terms of familiar elementary functions.

11  $y'' = y$      12  $y'' = 4y$
13  $y'' + 9y = 0$      14  $y'' + y = x$

In each of Problems 15–18, first derive a recursion formula giving $c_n$ for $n \geq 2$ in terms of $c_0$ or $c_1$ (or both). Then apply the given initial conditions to find the values of $c_0$ and $c_1$. Next determine $c_n$ (in terms of $n$, as in the text), and finally identify the particular solution in terms of familiar elementary functions.

15  $y'' + 4y = 0$;   $y(0) = 0$, $y'(0) = 3$
16  $y'' - 4y = 0$;   $y(0) = 2$, $y'(0) = 0$
17  $y'' - 2y' + y = 0$;   $y(0) = 0$, $y'(0) = 1$
18  $y'' + y' - 2y = 0$;   $y(0) = 1$, $y'(0) = -2$

**19** Establish the binomial series

$$(1 + x)^{\alpha} = 1 + \alpha x + \frac{\alpha(\alpha - 1)}{2!} x^2$$

$$+ \frac{\alpha(\alpha - 1)(\alpha - 2)}{3!} x^3 + \cdots \qquad (12)$$

by means of the following steps.

(a) Show that $y = (1 + x)^{\alpha}$ satisfies the initial value problem $(1 + x)y' = \alpha y$, $y(0) = 1$.

(b) Show that the power series method gives the binomial series in (12) as the solution of the initial value problem in part (a) and that this series converges if $|x| < 1$.

(c) Explain why the validity of the binomial series given in (12) follows from (a) and (b).

---

# Numerical Methods

It is the exception rather than the rule when a differential equation of the general form

$$\frac{dy}{dx} = f(x, y) \qquad (1)$$

can be solved exactly by simple methods such as those in Sections 18-2 and 18-3. Moreover, the infinite series methods of Section 18-7 tend to be unsuccessful when the function $f(x, y)$ is not linear in $y$—it is tedious even to square an infinite series.

Here we discuss *Euler's method* for computing numerical approximations to the solution of the initial value problem

$$\frac{dy}{dx} = f(x, y), \qquad y(a) = y_0 \qquad (2)$$

on an interval of the form $a \leq x \leq b$. We first choose a fixed *step size* $h > 0$ and consider the points

$$a = x_0, x_1, x_2, \ldots, x_n, \ldots$$

where $x_n = a + nh$, so that $x_n = x_{n-1} + h$ for each $n = 1, 2, 3, \ldots$. Our goal is to find suitable approximations $y_1, y_2, y_3, \ldots$ to the true values of the solution $y(x)$ of Equation (2) at the points $x_1, x_2, x_3, \ldots$. Thus we seek reasonably accurate approximations

$$y_n \approx y(x_n) \qquad (3)$$

for each $n$.

When $x = x_0$, the rate of change of $y$ with respect to $x$ is $y' = f(x_0, y_0)$. If $y$ continued to change at this same rate from $x = x_0$ to $x = x_1 = x_0 + h$, the change in $y$ would be exactly $hf(x_0, y_0)$. We therefore take

$$y_1 = y_0 + hf(x_0, y_0) \qquad (4)$$

as our approximation to the true value $y(x_1)$ of the solution at $x = x_1$. Similarly, we take

$$y_2 = y_1 + hf(x_1, y_1) \qquad (5)$$

as our approximation to $y(x_2)$. Having reached the $n$th approximate value $y_n \approx y(x_n)$, we take

$$y_{n+1} = y_n + hf(x_n, y_n) \qquad (6)$$

as our approximation to the true value $y(x_{n+1})$.

| N | Xn | Yn | Y actual |
|---|-----|---------|----------|
| 0 | 0.0 | 1.00000 | 1.00000 |
| 1 | 0.1 | 1.10000 | 1.11034 |
| 2 | 0.2 | 1.22000 | 1.24281 |
| 3 | 0.3 | 1.36200 | 1.39972 |
| 4 | 0.4 | 1.52820 | 1.58365 |
| 5 | 0.5 | 1.72102 | 1.79744 |
| 6 | 0.6 | 1.94312 | 2.04424 |
| 7 | 0.7 | 2.19743 | 2.32751 |
| 8 | 0.8 | 2.48718 | 2.65108 |
| 9 | 0.9 | 2.81590 | 3.01921 |
| 10 | 1.0 | 3.18748 | 3.43656 |

With $h = 0.01$:

| N | Xn | Yn | Y actual |
|-----|-----|---------|----------|
| 0 | 0.0 | 1.00000 | 1.00000 |
| 10 | 0.1 | 1.10924 | 1.11034 |
| 20 | 0.2 | 1.24038 | 1.24281 |
| 30 | 0.3 | 1.39570 | 1.39972 |
| 40 | 0.4 | 1.57773 | 1.58365 |
| 50 | 0.5 | 1.78926 | 1.79744 |
| 60 | 0.6 | 2.03339 | 2.04424 |
| 70 | 0.7 | 2.31353 | 2.32751 |
| 80 | 0.8 | 2.63343 | 2.65108 |
| 90 | 0.9 | 2.99727 | 3.01921 |
| 100 | 1.0 | 3.40963 | 3.43656 |

With $h = 0.001$:

| N | Xn | Yn | Y actual |
|------|-----|---------|----------|
| 0 | 0.0 | 1.00000 | 1.00000 |
| 100 | 0.1 | 1.11023 | 1.11034 |
| 200 | 0.2 | 1.24256 | 1.24281 |
| 300 | 0.3 | 1.39931 | 1.39972 |
| 400 | 0.4 | 1.58305 | 1.58365 |
| 500 | 0.5 | 1.79662 | 1.79744 |
| 600 | 0.6 | 2.04315 | 2.04424 |
| 700 | 0.7 | 2.32610 | 2.32751 |
| 800 | 0.8 | 2.64930 | 2.65108 |
| 900 | 0.9 | 3.01699 | 3.01921 |
| 1000 | 1.0 | 3.43385 | 3.43656 |

**18.15**  Using Euler's method to approximate the solution of (7)

---

*Algorithm*   *The Euler Method*

Given the initial value problem

$$y' = f(x, y), \qquad y(a) = y_0, \tag{2}$$

**Euler's method with step size $h$** consists in applying the iterative formula

$$y_{n+1} = y_n + hf(x_n, y_n) \qquad (n \geq 0) \tag{6}$$

to compute successively approximations $y_1, y_2, y_3, \ldots$ to the (true) values $y(x_1), y(x_2), y(x_3), \ldots$ of the (exact) solution $y = y(x)$ at the points $x_1$, $x_2, x_3, \ldots$, respectively.

Although the most important practical applications of Euler's method are to nonlinear differential equations, we will illustrate the method with the linear first order initial value problem

$$\frac{dy}{dx} = x + y, \qquad y(0) = 1, \tag{7}$$

in order that we may compare our approximate solution with the exact solution

$$y(x) = 2e^x - x - 1 \tag{8}$$

that is easy to find by the methods of Section 18-3.

**EXAMPLE 1**   Apply Euler's method to approximate the solution of the initial value problem in (7) on the interval $0 \leq x \leq 1$.

*Solution*   Here $f(x, y) = x + y$, so the iterative formula in (6) is

$$y_{n+1} = y_n + h(x_n + y_n). \tag{9}$$

The tables in Fig. 18.15 show the results we obtain, beginning with $y_0 = 1$ and using the step sizes $h = 0.1, h = 0.01$, and $h = 0.001$, respectively. With $h = 0.1$ the computation can be carried out easily with only a pocket calculator. With $h = 0.01$ and $h = 0.001$ a computer is needed, and we have shown the results only at intervals of $\Delta x = 0.1$. Note in each case that the error $y_{\text{actual}} - y_{\text{approx}}$ increases as $x$ increases—that is, as $x$ gets further and further from $x_0$. But each error decreases as $h$ decreases. The percentage errors at the final point $x = 1$ are $7.25\%$ with $h = 0.1, 0.78\%$ with $h = 0.01$, and only $0.08\%$ with $h = 0.001$.

**THE IMPROVED EULER METHOD**

To increase the accuracy of the basic Euler technique, we can regard the result

$$u_{n+1} = y_n + hf(x_n, y_n) \tag{10}$$

of Euler's formula as merely a first attempt at estimating the true value of $y(x_{n+1})$. We then take the average value

$$m_n = \tfrac{1}{2}[f(x_n, y_n) + f(x_{n+1}, u_{n+1})] \tag{11}$$

as an improved estimate of the rate of change of $y$ with respect to $x$ over the interval $[x_n, x_{n+1}]$. This yields

$$y_{n+1} = y_n + hm_n \qquad (12)$$

as our $(n + 1)$st approximate value.

The **improved Euler method** consists of applying the formulas in (10) through (12) at each step in the iteration to compute the successive approximations $y_1, y_2, y_3, \ldots$. It is an example of a *predictor-corrector method*. The formula in (10) is the *predictor* used as a first attempt at the next approximation. The formula in (12) is the *corrector* used to improve the value of the first attempt.

**EXAMPLE 2** Apply the improved Euler method to approximate the solution of the initial value problem in (7) on the interval $0 \leq x \leq 1$.

*Solution*  With $f(x, y) = x + y$, the formulas in (10) through (12) yield

$$u_{n+1} = y_n + h(x_n + y_n),$$

$$m_n = \tfrac{1}{2}(x_n + y_n + x_n + h + u_{n+1}),$$

$$y_{n+1} = y_n + hm_n.$$

The tables in Fig. 18.16 show the results we get using the step sizes $h = 0.1$, $h = 0.01$, and $h = 0.001$, respectively. The percentage errors at the final point $x = 1$ are 0.244% with $h = 0.1$ and 0.003% with $h = 0.01$, while with $h = 0.001$ the results shown are all accurate to five decimal places.

### A REAL-WORLD APPLICATION

Suppose that a skydiver steps out of an airplane at an altitude of 10,000 ft, and thereafter is subject both to gravitational acceleration ($g = 32$ ft/s$^2$) and to air resistance proportional to the square of his velocity $v$. We denote by $x(t)$ the (downward) distance the skydiver falls in $t$ seconds and assume that $v = dx/dt$ satisfies the differential equation

$$\frac{dv}{dt} = 32 - (0.005)v^2, \qquad (13)$$

giving the skydiver's acceleration $a = dv/dt$. In Problem 21 we ask you to separate the variables in (13) to derive the (exact) formulas

$$v(t) = 80 \tanh(0.4t) \qquad (14)$$

and

$$x(t) = 200 \ln(\cosh 0.4t). \qquad (15)$$

Here we want to solve Equation (13) numerically, with initial conditions $x(0) = v(0) = 0$.

Choosing a step size $h > 0$, we want to calculate approximations $v_n$ and $x_n$ to $v(t_n)$ and $x(t_n)$, respectively, where $t_n = nh$. We write

$$a_n = 32 - (0.005)v_n^2 \qquad (16)$$

With $h = 0.1$:

| N | Xn | Yn | Y actual |
|---|-----|---------|----------|
| 0 | 0.0 | 1.00000 | 1.00000 |
| 1 | 0.1 | 1.11000 | 1.11034 |
| 2 | 0.2 | 1.24205 | 1.24281 |
| 3 | 0.3 | 1.39847 | 1.39972 |
| 4 | 0.4 | 1.58180 | 1.58365 |
| 5 | 0.5 | 1.79489 | 1.79744 |
| 6 | 0.6 | 2.04086 | 2.04424 |
| 7 | 0.7 | 2.32315 | 2.32751 |
| 8 | 0.8 | 2.64558 | 2.65108 |
| 9 | 0.9 | 3.01236 | 3.01921 |
| 10 | 1.0 | 3.42816 | 3.43656 |

With $h = 0.01$:

| N | Xn | Yn | Y actual |
|---|-----|---------|----------|
| 0 | 0.0 | 1.00000 | 1.00000 |
| 10 | 0.1 | 1.11034 | 1.11034 |
| 20 | 0.2 | 1.24280 | 1.24281 |
| 30 | 0.3 | 1.39970 | 1.39972 |
| 40 | 0.4 | 1.58363 | 1.58365 |
| 50 | 0.5 | 1.79742 | 1.79744 |
| 60 | 0.6 | 2.04420 | 2.04424 |
| 70 | 0.7 | 2.32746 | 2.32751 |
| 80 | 0.8 | 2.65102 | 2.65108 |
| 90 | 0.9 | 3.01913 | 3.01921 |
| 100 | 1.0 | 3.43647 | 3.43656 |

With $h = 0.001$:

| N | Xn | Yn | Y actual |
|---|-----|---------|----------|
| 0 | 0.0 | 1.00000 | 1.00000 |
| 100 | 0.1 | 1.11034 | 1.11034 |
| 200 | 0.2 | 1.24281 | 1.24281 |
| 300 | 0.3 | 1.39972 | 1.39972 |
| 400 | 0.4 | 1.58365 | 1.58365 |
| 500 | 0.5 | 1.79744 | 1.79744 |
| 600 | 0.6 | 2.04424 | 2.04424 |
| 700 | 0.7 | 2.32750 | 2.32751 |
| 800 | 0.8 | 2.65108 | 2.65108 |
| 900 | 0.9 | 3.01921 | 3.01921 |
| 1000 | 1.0 | 3.43656 | 3.43656 |

**18.16**  Using the improved Euler method to approximate the solution of (7)

| T | APPROX V | ACTUAL V | APPROX X | ACTUAL X |
|---|---|---|---|---|
| 1 | 30.61 | 30.40 | 15.66 | 15.59 |
| 2 | 53.65 | 53.12 | 58.59 | 58.15 |
| 3 | 67.28 | 66.69 | 119.75 | 118.74 |
| 4 | 74.19 | 73.73 | 190.90 | 189.36 |
| 5 | 77.42 | 77.12 | 266.92 | 265.00 |
| 6 | 78.87 | 78.69 | 345.16 | 343.01 |
| 7 | 79.51 | 79.41 | 424.39 | 422.11 |
| 8 | 79.79 | 79.73 | 504.05 | 501.70 |
| 9 | 79.91 | 79.88 | 583.90 | 581.52 |
| 10 | 79.96 | 79.95 | 663.84 | 661.44 |
| 11 | 79.98 | 79.98 | 743.81 | 741.40 |
| 12 | 79.99 | 79.99 | 823.80 | 821.38 |
| 13 | 80.00 | 80.00 | 903.80 | 901.38 |
| 14 | 80.00 | 80.00 | 983.79 | 981.37 |
| 15 | 80.00 | 80.00 | 1063.79 | 1061.37 |

**18.17**   The free-falling skydiver

for the approximate acceleration at time $t_n$. Then Euler's method applied to Equation (13) yields

$$v_{n+1} = v_n + ha_n \qquad (17)$$

as a first approximation to the skydiver's velocity at time $t_{n+1}$. We implement the improved Euler method by using the average

$$\bar{v}_n = \tfrac{1}{2}(v_n + v_{n+1})$$

as a better estimate of the velocity over the interval $[t_n, t_{n+1}]$. This yields

$$x_{n+1} = x_n + \bar{v}_n h = x_n + hv_n + \tfrac{1}{2}h^2 a_n \qquad (18)$$

as the approximate position at time $t_{n+1}$.

The formulas in (16) through (18) were used with $h = 0.1$ to compute the approximate values shown in Fig. 18.17, which also lists for comparison (at intervals of 1 s) the exact values of $v$ and $x$ (obtained from Equations (14) and (15)). Note that after 15 s, the skydiver has fallen 1064 ft (taking the approximate value) and has reached a limiting velocity of 80 ft/s. If the skydiver's parachute fails to open, the skydiver will therefore hit the ground after

$$\frac{10{,}000 - 1064}{80} \approx 112$$

s. Thus the total time of descent is about 2 min 7s.

## 18-8   PROBLEMS

In each of Problems 1–10, use Euler's method with $h = 0.1$ to approximate the value of the given initial value problem on the interval $[0, 1]$. Also find the exact solution, and compare the approximate value of $y(1)$ with its exact value.

1   $y' = -y;\quad y(0) = 2$
3   $y' = y + 1;\quad y(0) = 1$
5   $y' = y - x - 1;\quad y(0) = 1$
6   $y' = -2xy;\quad y(0) = 2$
7   $y' = -3x^2 y;\quad y(0) = 3$
8   $y' = e^{-y};\quad y(0) = 0$
9   $y' = \tfrac{1}{4}(1 + y^2);\quad y(0) = 1$

10   $y' = \dfrac{2x}{1 + 3y^2};\quad y(0) = 1$

2   $y' = 2y;\quad y(0) = \tfrac{1}{2}$
4   $y' = x - y;\quad y(0) = 1$

11–20   Repeat Problems 1–10, except use the improved Euler method with step size $h = 0.1$.

21   Separate the variables in Equation (13) and then use the integral

$$\int \frac{dx}{1 - x^2} = \tanh^{-1}x + C$$

to derive the formulas in (14) and (15).

22   Suppose that a skydiver steps out of an airplane at an altitude of 8000 ft and falls freely for 30 s with acceleration $a = 32 - (0.005)v^2$. Then he opens his parachute, after which $a = 32 - (0.08)v^2$. Approximate his height and velocity at 1 s intervals after he opens the parachute. When will he hit the ground?

## *18-8   Optional Computer Application

The following IBM-PC program was used to compute the data that appear in Fig. 18.15.

```
100   REM   Program EULER
110   REM   This program uses Euler's method to
120   REM   approximate a solution of the equation
130   REM   y' = f(x, y)   on the interval   [0, 1].
```

```
140    REM      The function  f(x, y)  is defined in
150    REM      line 190.
160    REM
170    REM      Initialization:
180         DEFDBL  F, H, X, Y,  :  DEFINT  I, K, N
190         DEF  FNF(X, Y) = X + Y
200         INPUT  "  INITIAL VALUE  Y(0)  "; Y
210         INPUT  "  NUMBER OF SUBINTERVALS  ";  N
220         K = N/10  :  H = 1/N  :  X = 0.0
230         PRINT  "  Xn  ",  "  Yn  ",  "  Y actual  "  :  PRINT
240    REM
250    REM      Euler's iteration:
260         FOR  I = 0  TO  N
270            IF  I/K = I\K  THEN PRINT USING
               "#.#        #.#####        #.#####";
               X, Y, 2*EXP(X) - X - 1
280            Y = Y + H*FNF(X, Y)
290            X = X + H
300         NEXT  I
310    REM
320         END
```

The function $f(x, y) = x + y$ is defined in line 190, and the exact solution $y = 2e^x - x - 1$ appears in line 270. Line 280 is the iterative formula of Euler's method. The condition $I/K = I\backslash K$ in line 270 tests whether $I$ is a multiple of $K = N/10$ (because $I\backslash K$ denotes the integral part of $I/K$). Consequently, data are printed at intervals of $\Delta x = 0.1$, whatever number $N$ of subintervals we enter in line 210. We therefore can produce the three tables in Fig. 18.15 with three separate runs of the same program, using the values $N = 10$, $N = 100$, and $N = 1000$.

To compute the data of Fig. 18.16, it is necessary only to replace line 280 above with the two lines

```
280         U = Y + H*FNF(X, Y)
285         Y = Y + H*(0.5)*(FNF(X, Y) + FNF(X+H, U))
```

and run the modified program three times, just as before.

## CHAPTER 18 REVIEW:   Concepts and Methods

Use the list below as a guide to concepts and methods that you may need to review.

**1** General and particular solutions of differential equations

**2** Solution of separable first order equations

**3** Torricelli's law and draining tanks

**4** Solution of linear first order equations

**5** Mixture problems

**6** Complex numbers and complex-valued functions

**7** Euler's formula and DeMoivre's formula

**8** Solution of homogeneous linear equations with constant coefficients; the characteristic equation and the solutions corresponding to distinct roots, repeated roots, and complex roots

**9** Mechanical vibrations—amplitude, period, frequency, and phase angle

**10** The power series method of solution

**11** Euler's method and the improved Euler method of numerical approximation of solutions

## MISCELLANEOUS PROBLEMS

Find general solutions of the differential equations in Problems 1–18.

1  $xy' = 2x^2y^2 + 3y^2$

2  $yy' = xe^{x^2-y^2}$

3  $xy' + y = x^5$

4  $y' + y \tan x = \cos x$

5  $(x^2 + 1)y' + y = 1$

6  $(x + 1)y' = x + y + 2$

7  $2y'' + 3y' + y = 0$

8  $4y'' + 5y' - 6y = 0$

9  $y''' - 4y' = 0$

10  $4y''' + 4y'' + y' = 0$

11  $y''' + 3y'' + 3y' + y = 0$

12  $y''' + 9y' = 0$

13  $y'' + 5y' + 4y = 0$

14  $y'' + 13y' + 12y = 0$

15  $y^{(4)} = 16y$

16  $y^{(4)} + 16y = 0$

17  $y^{(4)} + 2y'' + y = 0$

18  $y^{(4)} - 3y'' - 4y = 0$

Solve the initial value problems in Exercises 19–23.

19  $y'' - 4y = 0$;  $y(0) = 1, y'(0) = 10$

20  $y'' + 2y' - 8y = 0$;  $y(0) = 8, y'(0) = -2$

21  $y'' + 25y = 0$;  $y(0) = 3, y'(0) = -10$

22  $y'' + 3y' = 0$;  $y(0) = 0, y'(0) = 2$

23  $y'' - 6y' + 13y = 0$;  $y(0) = 5, y'(0) = 9$

24  Show that the substitution $x = e^z, z = \ln x$ transforms the *Euler equation* $ax^2y'' + bxy' + cy = 0$ into the homogeneous linear equation

$$a\frac{d^2y}{dz^2} + (b - a)\frac{dy}{dz} + cy = 0$$

with constant coefficients.

Use the substitution $x = e^z$ to solve the Euler equations in Problems 25–27.

25  $2x^2y'' + 5xy' + y = 0$

26  $4x^2y'' + 8xy' + y = 0$

27  $x^2y'' + 6xy' + 4y = 0$

The differential equation $y' + P(x)y = Q(x)y^n$ is called a *Bernoulli equation* if $n \neq 0$ or 1. Use the substitution $v = y^{1-n}$ to solve the Bernoulli equations in Problems 28–30.

28  $xy' + y = x^3y^3$

29  $y' + y = y^2e^x$

30  $y' = ay(b - y)$ (here, $a$ and $b$ are positive constants; this is the logistic equation—compare your solution with Equation (6) in Section 7-8).

In each of Problems 31–36, find a power series solution of the given differential equation. Then identify the series solution in terms of familiar elementary functions.

31  $y' + y = 0$

32  $3y' - y = 0$

33  $(1 - x)y' = y$

34  $(3x - 1)y' = 6y$

35  $xy' = (1 + x)y$

36  $xy' = (2 - x)y$

# References for Further Study

References 2, 3, 6, and 10 may be consulted for historical topics pertinent to calculus. Reference 13 provides a more theoretical treatment of single-variable calculus topics than ours. References 4, 5, 7, 9, and 14 include advanced topics in multivariable calculus. Reference 11 is a standard work on infinite series. References 1, 8, and 12 are differential equations textbooks.

1   BOYCE, W. E., and R. C. DiPRIMA, *Elementary Differential Equations* (4th ed.) New York: John Wiley, 1986.

2   BOYER, C. B., *A History of Mathematics*. New York: John Wiley, 1968.

3   BOYER, C. B., *The History of the Calculus*. New York: Dover Publications, 1959.

4   BUCK, R. CREIGHTON, *Advanced Calculus* (3rd ed.) New York: McGraw-Hill, 1978.

5   COURANT, RICHARD, and FRITZ JOHN, *Introduction to Calculus and Analysis*. Vol. I, New York: Interscience, 1965. Vol. II, New York: Wiley-Interscience, 1974, with the assistance of Albert A. Blank and Alan Solomon.

6   EDWARDS, C. H., JR., *The Historical Development of the Calculus*. New York: Springer-Verlag, 1979.

7   EDWARDS, C. H., JR., *Advanced Calculus of Several Variables*. New York: Academic Press, 1973.

8   EDWARDS, C. H., JR., and DAVID E. PENNEY, *Elementary Differential Equations with Applications*. Englewood-Cliffs, N.J.: Prentice-Hall, 1985.

9   HILDEBRAND, FRANCIS B., *Advanced Calculus for Applications* (2nd ed.) Englewood Cliffs, N.J.: Prentice-Hall, 1976.

10  KLINE, MORRIS, *Mathematical Thought from Ancient to Modern Times*. New York: Oxford University Press, 1972.

11  KNOPP, KONRAD, *Theory and Application of Infinite Series*. New York: Hafner, 1981.

12  SIMMONS, GEORGE F., *Differential Equations with Applications and Historical Notes*. New York: McGraw-Hill, 1972.

13  SPIVAK, MICHAEL E., *Calculus* (2nd ed.) Berkeley: Publish or Perish, 1980.

14  TAYLOR, ANGUS E., and W. ROBERT MANN, *Advanced Calculus* (3rd ed.) New York: John Wiley, 1983.

# Appendices

## Existence of the integral

When the basic computational algorithms of the calculus were discovered by Newton and Leibniz in the latter half of the seventeenth century, the logical rigor that had been a feature of the Greek method of exhaustion was largely abandoned. When computing the area $A$ under the curve $y = f(x)$, for example, Newton took it as intuitively obvious that the area function existed, and proceeded to compute it as the antiderivative of the height function $f(x)$. Leibniz regarded $A$ as an infinite sum of infinitesimal area elements, each of the form $dA = f(x)\,dx$, but in practice computed the area

$$A = \int_a^b f(x)\,dx$$

by antidifferentiation just as Newton did—that is, by computing

$$A = \left[D^{-1}f(x)\right]_a^b.$$

The question of the *existence* of the area function—one of the conditions that a function $f$ must satisfy in order for its integral to exist—did not at first seem to be of much importance. Eighteenth-century mathematicians were mainly occupied (and satisfied) with the impressive applications of calculus to the solution of real-world problems and did not concentrate on the logical foundations of the subject.

The first attempt at a precise definition of the integral and a proof of its existence for continuous functions was that of the French mathematician Augustin-Louis Cauchy (1789–1857). Curiously enough, Cauchy was trained as an engineer, and much of his research in mathematics was in fields that we today regard as applications-oriented: hydrodynamics, waves in elastic media, vibrations of elastic membranes, polarization of light, and the like. But he was a prolific researcher, and his writings cover the entire spectrum of mathematics, with occasional essays into almost unrelated fields.

Around 1824, Cauchy defined the integral of a continuous function in a way that is familiar to us, as a limit of left-endpoint approximations:

$$\int_a^b f(x)\,dx = \lim_{\Delta x \to 0} \sum_{i=1}^{n} f(x_{i-1})\,\Delta x_i.$$

Of course, this is a much more complicated sort of limit than the ones we discussed in Chapter 2. Cauchy was not entirely clear about the nature of the limit process involved in the above equation, nor about the precise role that the hypothesis of continuity of $f$ played in proving that the above limit exists.

A complete definition of the integral, as we gave in Section 5-3, was finally produced in the 1850s by the German mathematician Georg Bernhard Riemann (1826–1866). Riemann was a student of Gauss; he met Gauss upon his arrival at Göttingen for the purpose of studying theology, when he was about 20 years old and Gauss was about 70. Riemann soon decided to study mathematics and became known as one of the truly great mathematicians of the century. Like Cauchy, he was particularly interested in applications of mathematics to the real world; his research particularly emphasized electricity, heat, light, acoustics, fluid dynamics, and—as you might infer from the fact that Wilhelm Weber was a major influence on Riemann's education—magnetism. Riemann also made significant contributions to mathematics itself, particularly in the field of complex analysis; a major conjecture of his, involving the zeta function

$$\zeta(s) = \sum_{n=1}^{\infty} \frac{1}{n^s}, \tag{1}$$

remains unsolved to this day, and has important consequences in the theory of the distribution of prime numbers because

$$\zeta(k) = \prod \left(1 - \frac{1}{p^k}\right)^{-1}$$

where the product is taken over all primes $p$. (The zeta function is defined in (1) for complex numbers $s$ to the right of the vertical line at $x = 1$ and is extended to other complex numbers by the requirement that it be differentiable.) Riemann died of tuberculosis shortly before his 40th birthday.

In this section we give a proof of the existence of the integral of a continuous function. We will follow Riemann's approach. Specifically, suppose that the function $f$ is continuous on the closed interval $[a, b]$. We will prove that the definite integral

$$\int_a^b f(x)\, dx$$

exists. That is, we will demonstrate the existence of a number $I$ satisfying the following condition: For every $\varepsilon > 0$, there exists $\delta > 0$ such that, for *every* Riemann sum $R$ associated with *any* partition $P$ with $|P| < \delta$,

$$|I - R| < \varepsilon.$$

(Recall that the mesh $|P|$ of the partition $P$ is the length of the longest subinterval in the partition.) In other words, every Riemann sum associated with every sufficiently "fine" partition is close to the number $I$. If this happens, then the definite integral

$$\int_a^b f(x)\, dx$$

is said to **exist,** and $I$ is its **value.**

Now we begin the proof. Suppose throughout that $f$ is a function continuous on the closed interval $[a, b]$. Given $\varepsilon > 0$, we need to show the existence of a number $\delta > 0$ and a number $I$ such that

$$\left| I - \sum_{i=1}^{n} f(x_i^*)\, \Delta x_i \right| < \varepsilon \tag{2}$$

for every Riemann sum associated with any partition $P$ of $[a, b]$ with $|P| < \delta$.

Given a partition $P$ of $[a, b]$ into $n$ *not necessarily equally long* subintervals, let $p_i$ be a point in the subinterval $[x_{i-1}, x_i]$ at which $f$ attains its minimum value $f(p_i)$; similarly, let $f(q_i)$ be its maximum value there. These numbers exist for $i = 1, 2, 3, \ldots, n$ because of the maximum value property of continuous functions (Theorem 4 in Appendix B).

In what follows we will denote the resulting lower and upper Riemann sums associated with $P$ by

$$L(P) = \sum_{i=1}^{n} f(p_i)\, \Delta x_i \tag{3a}$$

and

$$U(P) = \sum_{i=1}^{n} f(q_i)\, \Delta x_i, \tag{3b}$$

respectively. Then our first lemma is obvious.

---

*Lemma 1*

For any partition $P$ of $[a, b]$, $L(P) \leq U(P)$.

---

Now a definition. The partition $P'$ is called a *refinement* of the partition $P$ if each subinterval of $P'$ is contained in some subinterval of $P$. That is, $P'$ is obtained from $P$ by adding more points of subdivision.

---

*Lemma 2*

Suppose that $P'$ is a refinement of $P$. Then

$$L(P) \leq L(P') \leq U(P') \leq U(P). \tag{4}$$

---

***Proof*** The inequality $L(P') \leq U(P')$ is a consequence of Lemma 1. We will show that $L(P) \leq L(P')$; the proof that $U(P') \leq U(P)$ is similar.

The refinement $P'$ is obtained from $P$ by adding some more points of subdivision to $P$. So all we really need is to show that the Riemann sum cannot be decreased by adding a single point of subdivision. Thus we will suppose that the partition $P'$ is obtained from $P$ by dividing the $k$th subinterval $[x_{k-1}, x_k]$ of $P$ into two subintervals $[x_{k-1}, z]$ and $[z, x_k]$ by means of the new point $z$.

The only resulting effect on the corresponding Riemann sum is to replace the term

$$f(p_k) \cdot (x_k - x_{k-1})$$

in $L(P)$ by the two-term sum

$$f(u) \cdot (z - x_{k-1}) + f(v) \cdot (x_k - z),$$

where $f(u)$ is the minimum of $f$ on $[x_{k-1}, z]$ and $f(v)$ is the minimum of $f$ on $[z, x_k]$. So

$$\begin{aligned}
L(P') - L(P) &= f(u) \cdot (z - x_{k-1}) + f(v) \cdot (x_k - z) - f(p_k) \cdot (x_k - x_{k-1}) \\
&\geq f(p_k) \cdot (z - x_{k-1}) + f(p_k) \cdot (x_k - z) - f(p_k) \cdot (x_k - x_{k-1}) \\
&= f(p_k) \cdot (z - x_{k-1} + x_k - z - x_k + x_{k-1}),
\end{aligned}$$

and thus

$$L(P') - L(P) \geq 0.$$

Because this is all we needed to show, we have completed the proof of Lemma 2. ∎

To prove that all the Riemann sums for sufficiently "fine" partitions are close to some number $I$, we must give some construction of $I$. This is accomplished through the next lemma.

---

**Lemma 3**

Let $P_n$ denote the regular partition of $[a, b]$ into $2^n$ equal subintervals. Then the (sequential) limit

$$I = \lim_{n \to \infty} L(P_n) \qquad (5)$$

exists.

---

**Proof**  Note first that each $P_{n+1}$ is a partition of $P_n$, so that (by Lemma 2)

$$L(P_1) \leq L(P_2) \leq \cdots \leq L(P_n) \leq \cdots.$$

Therefore, $\{L(P_n)\}$ is a nondecreasing sequence of real numbers. Moreover,

$$L(P_n) = \sum_{i=1}^{2^n} f(p_i)\, \Delta x_i$$

$$\leq M \cdot \sum_{i=1}^{2^n} \Delta x_i = M \cdot (b - a),$$

where $M$ is the maximum value of $f$ on $[a, b]$.

Now Theorem 1 in Appendix B guarantees that a bounded monotone sequence of real numbers must converge. Thus the number

$$I = \lim_{n \to \infty} L(P_n)$$

exists. This establishes Equation (5), and thus the proof of Lemma 3 is complete. ∎

It is proved in advanced calculus that if $f$ is continuous on $[a, b]$, then—for every number $\varepsilon > 0$—there exists a number $\delta > 0$ such that

$$\left| f(u) - f(v) \right| < \varepsilon$$

for any two points $u$ and $v$ of $[a, b]$ such that

$$\left| u - v \right| < \delta.$$

This property of a function is called **uniform continuity** on the interval $[a, b]$. Thus the theorem from advanced calculus that we need to use states that every continuous function on a closed interval is uniformly continuous there.

NOTE  That $f$ is continuous on $[a, b]$ means that for each number $u$ in the interval and each $\varepsilon > 0$, there exists $\delta > 0$ such that if $v$ is a number in the interval with $\left| u - v \right| < \delta$, then $\left| f(u) - f(v) \right| < \varepsilon$. But *uniform* continuity is a more stringent condition; it means that given $\varepsilon > 0$, you can not only find a value $\delta_1$ that "works" for $u_1$, a value $\delta_2$ that works for $u_2$, and so on, but more: You can in fact find a universal value $\delta$ that works for *all* values of $u$ in the interval. This should not be obvious when you realize it is possible that $\delta_1 = 1$, $\delta_2 = \frac{1}{2}$, $\delta_3 = \frac{1}{3}$, and so on. In any case, it is clear that uniform continuity of $f$ on an interval implies its continuity there.

Remember that, throughout, we have a continuous function $f$ defined on the closed interval $[a, b]$.

---

**Lemma 4**

Suppose that $\varepsilon > 0$ is given. Then there exists a number $\delta > 0$ such that if $P$ is a partition with $|P| < \delta$ and $P'$ is a refinement of $P$, then

$$\left| R(P) - R(P') \right| < \frac{\varepsilon}{3} \qquad (6)$$

for any two Riemann sums $R(P)$ associated with $P$ and $R(P')$ associated with $P'$.

---

**Proof**  Because $f$ must be uniformly continuous on $[a, b]$, there exists a number $\delta > 0$ such that if

$$|u - v| < \delta,$$

then

$$\left| f(u) - f(v) \right| < \frac{\varepsilon}{3(b - a)}.$$

Suppose now that $P$ is a partition with $|P| < \delta$. Then

$$\left| U(P) - L(P) \right| = \sum_{i=1}^{n} \left| f(q_i) - f(p_i) \right| \Delta x_i$$

$$< \frac{\varepsilon}{3(b - a)} \sum_{i=1}^{n} \Delta x_i = \frac{\varepsilon}{3}.$$

This is valid because $|p_i - q_i| < \delta$, for $p_i$ and $q_i$ both belong to the same subinterval $[x_{i-1}, x_i]$ of $P$, and $|P| < \delta$.

Now, as shown in the figure below, we know that $L(P)$ and $U(P)$ differ by less than $\varepsilon/3$. We also know that

$$L(P) \leqq R(P) \leqq U(P)$$

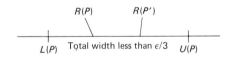

for every Riemann sum $R(P)$ associated with $P$. But

$$L(P) \leqq L(P') \leqq U(P') \leqq U(P)$$

by Lemma 2 because $P'$ is a refinement of $P$; moreover,

$$L(P') \leqq R(P') \leqq U(P')$$

for any Riemann sum $R(P')$ associated with $P'$.

The situation is illustrated in the figure: The two numbers $R(P)$ and $R(P')$ both belong to the interval $[L(P), U(P)]$ of length less than $\varepsilon/3$, and so Equation (6) follows, as desired. This concludes the proof of Lemma 4. ∎

> **Theorem** *Existence of the Definite Integral*
> If $f$ is continuous on the closed interval $[a, b]$, then the integral
> $$\int_a^b f(x)\,dx$$
> exists.

**Proof**  Suppose that $\varepsilon > 0$ is given. We must show the existence of a number $\delta > 0$ such that, for every partition $P$ with $|P| < \delta$, we have

$$|I - R(P)| < \varepsilon,$$

where $I$ is the number given in Lemma 3 and $R(P)$ is an arbitrary Riemann sum associated with $P$.

We choose the number $\delta$ provided by Lemma 4 such that

$$|R(P) - R(P')| < \frac{\varepsilon}{3}$$

if $|P| < \delta$ and $P'$ is a refinement of $P$.

By Lemma 3, we can choose an integer $N$ so large that

$$|P_N| < \delta \quad \text{and} \quad |L(P_N) - I| < \frac{\varepsilon}{3}. \tag{7}$$

Now, given an arbitrary partition $P$ such that $|P| < \delta$, let $P'$ be a common refinement of $P$ and $P_N$. You can obtain such a partition $P'$, for example, by using all the points of subdivision of both $P$ and $P_N$ to form the subintervals of $[a, b]$.

Because $P'$ is a refinement of both $P$ and $P_N$ and both of the latter have mesh less than $\delta$, Lemma 4 implies that

$$|R(P) - R(P')| < \frac{\varepsilon}{3} \quad \text{and} \quad |L(P_N) - R(P')| < \frac{\varepsilon}{3}. \tag{8}$$

Here, of course, $R(P)$ and $R(P')$ are (arbitrary) Riemann sums associated with $P$ and $P'$, respectively.

Now, given an arbitrary Riemann sum $R(P)$ associated with our partition $P$ having mesh less than $\delta$, we see that

$$\begin{aligned}|I - R(P)| &= |I - L(P_N) + L(P_N) - R(P') + R(P') - R(P)| \\ &\leq |I - L(P_N)| + |L(P_N) - R(P')| + |R(P') - R(P)|.\end{aligned}$$

In the last sum, each of the last two terms is less than $\varepsilon/3$ by virtue of the inequalities in (8). We also know, by (7), that the first term is also less than $\varepsilon/3$. Consequently,

$$|I - R(P)| < \varepsilon.$$

This establishes the theorem.  ∎

We close with an example that shows that some continuity hypothesis is required for integrability.

**EXAMPLE**  Suppose that $f$ is defined for $0 \leq x \leq 1$ as follows:

$$f(x) = \begin{cases} 1 & \text{if } x \text{ is irrational;} \\ 0 & \text{if } x \text{ is rational.} \end{cases}$$

Then $f$ is not continuous anywhere. (Why?) Given a partition $P$ of $[0, 1]$, let $p_i$ be a rational point and $q_i$ be an irrational point of the $i$th subinterval of $P$, for $i = 1, 2, 3, \ldots, n$. As before, $f$ attains its minimum value 0 at each $p_i$ and its maximum value 1 at each $q_i$. Also,

$$L(P) = \sum_{i=1}^{n} f(p_i)\, \Delta x_i = 0$$

while

$$U(P) = \sum_{i=1}^{n} f(q_i)\, \Delta x_i = 1.$$

Thus if we choose $\varepsilon = 1/2$, there is *no* number $I$ that can lie within $\varepsilon$ of both $L(P)$ and $U(P)$, no matter how small the mesh of $P$. It follows that $f$ is *not* Riemann integrable on $[0, 1]$.

---

# B

## The completeness of the real-number system

Here we present a self-contained treatment of those consequences of the completeness of the real-number system that are relevant to material of this text. Our principal objective is to prove the intermediate value theorem and the maximum value theorem. We begin with the least upper bound property of the real numbers, which we take as an axiom.

---

**Definition**  *Upper Bound and Lower Bound*

The set $S$ of real numbers is said to be **bounded above** if there is a number $b$ such that $x \leq b$ for every number $x$ in $S$. If so, the number $b$ is called an **upper bound** for $S$. Similarly, if there is a number $a$ such that $x \geq a$ for every number $x$ in $S$, then $S$ is said to be **bounded below** and $a$ is called a **lower bound** for $S$.

---

**Definition**  *Least Upper Bound and Greatest Lower Bound*

The number $\lambda$ is said to be a **least upper bound** for the set $S$ of real numbers provided that:

**1** $\lambda$ is an upper bound for $S$  and
**2** If $b$ is any upper bound for $S$, then $\lambda \leq b$.

Similarly, the number $\gamma$ is said to be a **greatest lower bound** for the set $S$ if $\gamma$ is not only a lower bound for $S$, but also $\gamma \geq a$ for every lower bound $a$ of $S$.

---

EXERCISE  Show that if a set $S$ has a least upper bound $\lambda$, then it is unique. That is, if $\lambda$ and $\mu$ are both least upper bounds for $S$, then $\lambda = \mu$.

It is easy to show that the greatest lower bound $\gamma$ of a set $S$, if any, is also unique. At this point you should construct examples to illustrate that a set with a least upper bound $\lambda$ may or may not contain $\lambda$, and similarly for its greatest lower bound. We now state the *completeness axiom* of the real number system.

> **Least Upper Bound Axiom**
>
> If the nonempty set $S$ of real numbers has an upper bound, then it has a least upper bound.

By working with the set $T$ consisting of the numbers $-x$ where $x$ is in $S$, it is not difficult to show the following consequence of the least upper bound axiom: If the nonempty set $S$ of real numbers is bounded below, then $S$ has a greatest lower bound. Because of this symmetry we need only one axiom, not two; results for least upper bounds also hold for greatest lower bounds provided some attention is paid to getting the inequalities the right way around.

The restriction that $S$ be nonempty is annoying but necessary. For if $S$ is the "empty" set of real numbers, then 73 is an upper bound for $S$, but $S$ has no least upper bound because $72, 71, 70, \ldots, 0, -1, -2, \ldots$ are all also upper bounds for $S$.

> **Definition**   *Increasing, Decreasing, and Monotone Sequences*
>
> The infinite sequence $x_1, x_2, x_3, \ldots, x_n, \ldots$ of real numbers is said to be **nondecreasing** if $x_n \leq x_{n+1}$ for every $n \geq 1$. This sequence is said to be **nonincreasing** if $x_n \geq x_{n+1}$ for every $n \geq 1$. If the sequence $\{x_n\}$ is either nonincreasing or nondecreasing, then it is said to be **monotone.**

The following theorem gives the **bounded monotone sequence property** of the set of real numbers. (Recall that a set $S$ of real numbers is said to be **bounded** if it is contained in an interval of the form $[a, b]$.)

> **Theorem 1**   *Bounded Monotone Sequences*
>
> Every bounded monotone sequence of real numbers converges.

***Proof***   Suppose that the sequence $S = \{x_n\} = \{x_1, x_2, x_3, \ldots, x_n, \ldots\}$ is bounded and nondecreasing. By the least upper bound axiom, $S$ has a least upper bound $\lambda$. We claim that $\lambda$ is the limit of the sequence $\{x_n\}$. For consider an open interval centered at $\lambda$; that is, an interval of the form $I = (\lambda - \varepsilon, \lambda + \varepsilon)$ where $\varepsilon > 0$. Some terms of the sequence must lie within $I$, or else $\lambda$ would not be an upper bound for $S$. But if $x_N$ is in $I$, then—because we are dealing with a nondecreasing sequence—$x_k$ must also lie in $I$ for all $k \geq N$. Because $\varepsilon$ is an arbitrary positive number, $\lambda$ is by definition (see Section 12-2) the limit of the sequence $\{x_n\}$. That is, a bounded nondecreasing sequence converges. A similar proof can be constructed for nonincreasing sequences by working with the greatest lower bound. ∎

So the least upper bound axiom implies the bounded monotone sequence property of the real numbers. With just a little trouble, you can prove that the two actually are logically equivalent—if you take the bounded monotone sequence property as an axiom, then the least upper bound property becomes a theorem. The **nested interval property** in the next theorem is also equivalent to the least upper bound property, but we shall prove only that it follows from the least upper bound property, because we have chosen the latter as the fundamental completeness axiom for the real number system.

> **Theorem 2**  *Nested Interval Property of the Real Numbers*
> Suppose that $I_1, I_2, I_3, \ldots, I_n, \ldots$ is a sequence of closed intervals (so that $I_n$ is of the form $[a_n, b_n]$ for each $n$) such that
>
> **1** $I_n$ contains $I_{n+1}$ for each $n \geq 1$      and
> **2** $\lim\limits_{n \to \infty} (b_n - a_n) = 0.$
>
> Then there exists exactly one real number $c$ such that $c$ is an element of $I_n$ for each $n$. Thus
>
> $$\{c\} = I_1 \cap I_2 \cap I_3 \cap \cdots.$$

***Proof***  It is clear from Property (2) that there is at most one such number $c$. The sequence $\{a_n\}$ of left-hand end points of the intervals is a bounded (by $b_1$) nondecreasing sequence and thus has a limit $a$ by the bounded monotone sequence property. Similarly, the sequence $\{b_n\}$ has a limit $b$. Because $a_n \leq b_n$ for all $n$, it follows easily that $a \leq b$. It is clear that $a_n \leq a \leq b_n$ for all $n \geq 1$, so $a$ belongs to every interval $I_n$; so does $b$, by the same argument. But then Property (2) above implies that $a = b$, and clearly this common value—call it $c$—is the number satisfying the conclusion of the theorem. ∎

We can now use these results to prove several important theorems of the text.

> **Theorem 3**  *Intermediate Value Property of Continuous Functions*
> If the function $f$ is continuous on $[a, b]$ and $f(a) < K < f(b)$, then $K = f(c)$ for some number $c$ in $(a, b)$.

***Proof***  Let $I_1 = [a, b]$. Suppose that $I_n$ has been defined. We describe (inductively) how to define $I_{n+1}$, and this shows in particular how to define $I_2, I_3$, and so forth. Let $a_n$ be the left-hand endpoint of $I_n$, $b_n$ its right-hand endpoint, and $m_n$ its midpoint. If $f(m_n) > K$, then $f(a_n) < K < f(m_n)$; in this case, let $a_{n+1} = a_n$, $b_{n+1} = m_n$, and $I_{n+1} = [a_{n+1}, b_{n+1}]$. If $f(m_n) < K$, then let $a_{n+1} = m_n$ and $b_{n+1} = b_n$. Thus at each stage we bisect $I_n$ and let $I_{n+1}$ be the half of $I_n$ on which $f$ takes on values both above and below $K$. Note that if $f(m_n)$ is ever actually equal to $K$, we simply let $c = m_n$ and stop.

It is easy to show that the sequence $\{I_n\}$ of intervals satisfies the hypotheses of Theorem 2. Let $c$ be the (unique) real number common to all the intervals $I_n$. We will show that $f(c) = K$, and this will conclude the proof. The sequence $\{b_n\}$ has limit $c$, so (by the continuity of $f$) the sequence $\{f(b_n)\}$ has limit $f(c)$. But $f(b_n) > K$ for all $n$, so the limit of $\{f(b_n)\}$ can be no less than $K$; that is, $f(c) \geq K$. By considering the sequence $\{a_n\}$, it follows that $f(c) \leq K$. Therefore $f(c) = K$. ∎

> **Lemma 1**
> If $f$ is continuous on the closed interval $[a, b]$, then $f$ is bounded there.

***Proof*** Suppose by way of contradiction that $f$ is not bounded on $I_1 = [a, b]$. Bisect $I_1$ and let $I_2$ be either half on which $f$ is unbounded—if $f$ is unbounded on both halves, let $I_2$ be the left half. In general, let $I_{n+1}$ be a half of $I_n$ on which $f$ is unbounded.

Again it is easy to show that the sequence $\{I_n\}$ of closed intervals satisfies the hypotheses of Theorem 2. Let $c$ be the number common to them all. Because $f$ is continuous, there is a number $\varepsilon > 0$ such that $f$ is bounded on the interval $(c - \varepsilon, c + \varepsilon)$. But for sufficiently large values of $n$, $I_n$ is a subset of $(c - \varepsilon, c + \varepsilon)$. This contradiction shows that $f$ must be bounded on $[a, b]$. ∎

---

> ***Theorem 4*** *Maximum Value Property of Continuous Functions*
>
> If the function $f$ is continuous on the closed interval $[a, b]$, then there exists a number $c$ in $[a, b]$ such that $f(x) \leq f(c)$ for all $x$ in $[a, b]$.

---

***Proof*** Consider the set $S = \{f(x) \mid a \leq x \leq b\}$. This set is bounded by the lemma above; let $\lambda$ be its least upper bound. Our strategy is to show that $\lambda$ is a value $f(c)$ of $f$.

With $I_1 = [a, b]$, bisect $I_1$ as before. Note that $\lambda$ is the least upper bound of the values of $f$ on at least one of the two halves of $I_1$; let $I_2$ be that half. Having defined $I_n$, let $I_{n+1}$ be the half of $I_n$ on which $\lambda$ is the least upper bound of the values of $f$. Let $c$ be the number common to all these intervals. It then follows from the continuity of $f$, much as in the proof of Theorem 3, that $f(c) = \lambda$. And it is clear that $f(x) \leq \lambda$ for all $x$ in $[a, b]$. ∎

The technique we have been using in these proofs is called the *method of bisection*. We now use it once again to establish the *Bolzano-Weierstrass property* of the real number system.

---

> ***Definition*** *Limit Point*
>
> Let $S$ be a set of real numbers. The number $p$ is said to be a **limit point** of $S$ if every open interval containing $p$ also contains points of $S$ other than $p$.

---

> ***Theorem 5*** *Bolzano-Weierstrass Theorem*
>
> Every bounded infinite set of real numbers has a limit point.

---

***Proof*** Let $I_0$ be a closed interval containing the bounded infinite set $S$ of real numbers. Let $I_1$ be one of the closed half-intervals of $I_0$ that contains infinitely many points of $S$. If $I_n$ has been chosen, let $I_{n+1}$ be one of the closed half-intervals of $I_n$ containing infinitely many points of $S$. An application of Theorem 2 yields a number $p$ common to all the intervals $I_n$. If $J$ is an open interval containing $p$, then $J$ contains $I_n$ for some sufficiently large value of $n$ and thus contains in fact infinitely many points of $S$. Therefore, $p$ is a limit point of $S$. ∎

Our final goal is in sight: We can now prove that a sequence of real numbers converges if and only if it is a Cauchy sequence.

> **Definition** *Cauchy Sequence*
>
> The sequence $\{a_n\}_1^\infty$ is said to be a **Cauchy sequence** if for every $\varepsilon > 0$, there exists an integer $N$ such that
>
> $$|a_m - a_n| < \varepsilon$$
>
> for all $m, n \geq N$.

> **Lemma 2** *Convergent Subsequences*
>
> Every bounded sequence of real numbers has a convergent subsequence.

**Proof**  If $\{a_n\}$ consists of only finitely many values, then the conclusion of the theorem follows easily. We therefore focus our attention on the case in which $\{a_n\}$ is an infinite set. It is easy to show that this set is also bounded, and thus we may apply the Bolzano-Weierstrass theorem to obtain a limit point $p$ of $\{a_n\}$. For each integer $k \geq 1$, let $a_{n(k)}$ be a term of the sequence $\{a_n\}$ such that

**1** $n(k + 1) > n(k)$ for all $k \geq 1$      and

**2** $|a_{n(k)} - p| < \dfrac{1}{k}$.

It is then easy to show that $\{a_{n(k)}\}$ is a convergent (to $p$) subsequence of $\{a_n\}$.

■

> **Theorem 6** *Convergence of Cauchy Sequences*
>
> A sequence of real numbers converges if and only if it is a Cauchy sequence.

**Proof**  It follows immediately from the triangle inequality that every convergent sequence is a Cauchy sequence. Thus suppose that the sequence $\{a_n\}$ is a Cauchy sequence.

Choose $N$ such that

$$|a_m - a_n| < 1$$

if $m, n \geq N$. It follows that if $n \geq N$, then $a_n$ lies in the closed interval $[a_N - 1, a_N + 1]$. This implies that the sequence $\{a_n\}$ is bounded, and thus by Lemma 2 it has a convergent subsequence $\{a_{n(k)}\}$. Let $p$ be the limit of this subsequence.

We claim that $\{a_n\}$ itself converges to $p$. For given $\varepsilon > 0$, choose $M$ such that

$$|a_m - a_n| < \frac{\varepsilon}{2}$$

if $m, n \geq M$. Next choose $K$ such that $n(K) \geq M$ and

$$|a_{n(K)} - p| < \frac{\varepsilon}{2}.$$

Then if $n \geqq M$,

$$|a_n - p| \leqq |a_n - a_{n(K)}| + |a_{n(K)} - p| < \varepsilon.$$

Therefore, $\{a_n\}$ converges to $p$ by definition. ∎

---

## Units of measurement and conversion factors

**MKS SCIENTIFIC UNITS**

*Length* in meters (m); *mass* in kilograms (kg); *time* in seconds (s).

*Force* in newtons (N); a force of 1 N provides an acceleration of 1 m/s² to a mass of 1 kg.

*Work* in joules (J); 1 J is the work done by a force of 1 N acting through a distance of 1 m.

*Power* in watts (W); a watt is 1 J/s.

**BRITISH ENGINEERING UNITS (fps)**

*Length* in feet (ft); *force* in pounds (lb); *time* in seconds (s).

*Mass* in slugs; 1 lb of force provides an acceleration of 1 ft/s² to a mass of 1 slug. A mass of $m$ slugs at the surface of the earth has a *weight* of $w = mg$ pounds, where $g \approx 32.17$ ft/s².

*Work* in ft-lb; *power* in ft-lb/sec.

**CONVERSION FACTORS**

1 in. = 2.54 cm = 0.0254 m, 1 m ≈ 3.2808 ft

1 mi = 5280 ft; 60 mi/h = 88 ft/sec

1 lb ≈ 4.4482 N; 1 slug ≈ 14.594 kg

1 hp = 550 ft-lb/s ≈ 745.7 W

Gravitational acceleration: $g \approx 32.17$ ft/s² ≈ 9.807 m/s²

Atmospheric pressure: 1 atm is the pressure exerted by a column of 76 cm of mercury; 1 atm ≈ 14.70 lb/in.² ≈ $1.013 \times 10^5$ N/m².

Heat energy: 1 BTU ≈ 778 ft-lb ≈ 252 cal, 1 cal ≈ 4.184 J

---

## Formulas from algebra and geometry

1  *Laws of Exponents*

$a^m a^n = a^{m+n}$,  $(a^m)^n = a^{mn}$,  $(ab)^n = a^n b^n$,  $a^{m/n} = \sqrt[n]{a^m}$;  in particular, $a^{1/2} = \sqrt{a}$.

If $a \neq 0$, then $a^{m-n} = \dfrac{a^m}{a^n}$, $a^{-n} = \dfrac{1}{a^n}$, and $a^0 = 1$.

2  *Quadratic Formula*

The quadratic equation

$$ax^2 + bx + c = 0 \qquad (a \neq 0)$$

has solutions

$$x = \frac{-b \pm \sqrt{b^2 - 4ac}}{2a}.$$

3  *Factoring*

$$a^2 - b^2 = (a - b)(a + b).$$
$$a^3 - b^3 = (a - b)(a^2 + ab + b^2).$$
$$a^4 - b^4 = (a - b)(a^3 + a^2b + ab^2 + b^3) = (a - b)(a + b)(a^2 + b^2).$$
$$a^5 - b^5 = (a - b)(a^4 + a^3b + a^2b^2 + ab^3 + b^4).$$

(The pattern continues.)

$$a^3 + b^3 = (a + b)(a^2 - ab + b^2).$$
$$a^5 + b^5 = (a + b)(a^4 - a^3b + a^2b^2 - ab^3 + b^4).$$

(The pattern continues for odd exponents.)

4  *Binomial Formula*

$$(a + b)^n = a^n + na^{n-1}b + \frac{n(n - 1)}{1 \cdot 2} a^{n-2}b^2$$

$$+ \frac{n(n - 1)(n - 2)}{1 \cdot 2 \cdot 3} a^{n-3}b^3 + \cdots + nab^{n-1} + b^n$$

if $n$ is a positive integer.

5  *Areas and Volumes* (See the figure and formulas below: $A$ denotes area; $b$, base; $B$, area of base; $C$, circumference; $h$, height; $\ell$, length; $r$, radius; $V$, volume; $w$, width.)

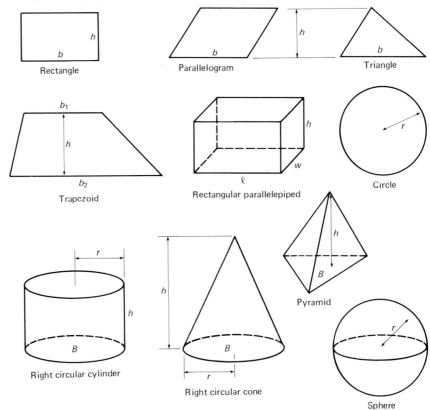

Rectangle:   $A = bh.$

Parallelogram:   $A = bh.$

Triangle: $A = \frac{1}{2}bh.$

Trapezoid: $A = \frac{1}{2}(b_1 + b_2)h.$

Circle: $C = 2\pi r$ and $A = \pi r^2.$

Rectangular parallelepiped: $V = \ell wh.$

Right circular cylinder: $V = \pi r^2 h = Bh.$

Right circular cone: $V = \frac{1}{3}\pi r^2 h = \frac{1}{3}Bh.$

Pyramid: $V = \frac{1}{3}Bh.$

Sphere: $V = \frac{4}{3}\pi r^3, A = 4\pi r^2.$

6  *Pythagorean Theorem*

In a right triangle with legs $a$ and $b$ and hypotenuse $c$, $a^2 + b^2 = c^2.$

$\sin(-\theta) = -\sin\theta, \cos(-\theta) = \cos\theta.$

$\sin^2\theta + \cos^2\theta = 1.$

$\sin 2\theta = 2\sin\theta\cos\theta$  and  $\cos 2\theta = \cos^2\theta - \sin^2\theta.$

$\sin(\alpha + \beta) = \sin\alpha\cos\beta + \cos\alpha\sin\beta.$

$\cos(\alpha + \beta) = \cos\alpha\cos\beta - \sin\alpha\sin\beta.$

$\tan(\alpha + \beta) = \dfrac{\tan\alpha + \tan\beta}{1 - \tan\alpha\tan\beta}.$

$\sin^2\dfrac{\theta}{2} = \dfrac{1}{2}(1 - \cos\theta).$

$\cos^2\dfrac{\theta}{2} = \dfrac{1}{2}(1 + \cos\theta).$

**Formulas from trigonometry**

For an arbitrary triangle:

$$\text{Law of sines} \qquad \frac{\sin A}{a} = \frac{\sin B}{b} = \frac{\sin C}{c}.$$

$$\text{Law of cosines} \qquad c^2 = a^2 + b^2 - 2ab\cos C.$$

**The Greek alphabet**

| A | $\alpha$ | alpha | I | $\iota$ | iota | P | $\rho$ | rho |
|---|---|---|---|---|---|---|---|---|
| B | $\beta$ | beta | K | $\kappa$ | kappa | $\Sigma$ | $\sigma$ | sigma |
| $\Gamma$ | $\gamma$ | gamma | $\Lambda$ | $\lambda$ | lambda | T | $\tau$ | tau |
| $\Delta$ | $\delta$ | delta | M | $\mu$ | mu | $\Upsilon$ | $\upsilon$ | upsilon |
| E | $\varepsilon$ | epsilon | N | $\nu$ | nu | $\Phi$ | $\phi$ | phi |
| Z | $\zeta$ | zeta | $\Xi$ | $\xi$ | xi | X | $\chi$ | chi |
| H | $\eta$ | eta | O | o | omicron | $\Psi$ | $\psi$ | psi |
| $\Theta$ | $\theta$ | theta | $\Pi$ | $\pi$ | pi | $\Omega$ | $\omega$ | omega |

## Tables

**TABLE 1**  *Trigonometric Functions in Radians*

| t | sin t | cos t | tan t | t | sin t | cos t | tan t |
|---|-------|-------|-------|---|-------|-------|-------|
| .0 | .00000 | 1.00000 | .00000 | 3.1 | .04158 | −.99914 | −.04162 |
| .1 | .09983 | .99500 | .10033 | π | .00000 | −1.00000 | .00000 |
| .2 | .19867 | .98007 | .20271 | 3.2 | −.05837 | −.99829 | .05847 |
| .3 | .29552 | .95534 | .30934 | 3.3 | −.15775 | −.98748 | .15975 |
| .4 | .38942 | .92106 | .42279 | 3.4 | −.25554 | −.96680 | .26432 |
| .5 | .47943 | .87758 | .54630 | 3.5 | −.35078 | −.93646 | .37459 |
| .6 | .56464 | .82534 | .68414 | 3.6 | −.44252 | −.89676 | .49347 |
| .7 | .64422 | .76484 | .84229 | 3.7 | −.52984 | −.84810 | .62473 |
| .8 | .71736 | .69671 | 1.02964 | 3.8 | −.61186 | −.79097 | .77356 |
| .9 | .78333 | .62161 | 1.26016 | 3.9 | −.68777 | −.72593 | .94742 |
| 1.0 | .84147 | .54030 | 1.55741 | 4.0 | −.75680 | −.65364 | 1.15782 |
| 1.1 | .89121 | .45360 | 1.96476 | 4.1 | −.81828 | −.57482 | 1.42353 |
| 1.2 | .93204 | .36236 | 2.57215 | 4.2 | −.87158 | −.49026 | 1.77778 |
| 1.3 | .96356 | .26750 | 3.60210 | 4.3 | −.91617 | −.40080 | 2.28585 |
| 1.4 | .98545 | .16997 | 5.79788 | 4.4 | −.95160 | −.30733 | 3.09632 |
| 1.5 | .99749 | 0.7074 | 14.10142 | 4.5 | −.97753 | −.21080 | 4.63733 |
| π/2 | 1.00000 | .00000 | ******** | 4.6 | −.99369 | −.11215 | 8.86017 |
| 1.6 | .99957 | −.02920 | −34.23253 | 4.7 | −.99992 | −.01239 | 80.71269 |
| 1.7 | .99166 | −.12884 | −7.69660 | 3π/2 | −1.00000 | .00000 | ******** |
| 1.8 | .97385 | −.22720 | −4.28626 | 4.8 | −.99616 | .08750 | −11.38487 |
| 1.9 | .94630 | −.32329 | −2.92710 | 4.9 | −.98245 | .18651 | −5.26749 |
| 2.0 | .90930 | −.41615 | −2.18504 | 5.0 | −.95892 | .28366 | −3.38052 |
| 2.1 | .86321 | −.50485 | −1.70985 | 5.1 | −.92581 | .37798 | −2.44939 |
| 2.2 | .80850 | −.58850 | −1.37382 | 5.2 | −.88345 | .46852 | −1.88564 |
| 2.3 | .74571 | −.66628 | −1.11921 | 5.3 | −.83227 | .55437 | −1.50127 |
| 2.4 | .67546 | −.73739 | −.91601 | 5.4 | −.77276 | .63469 | −1.21754 |
| 2.5 | .59847 | −.80114 | −.74702 | 5.5 | −.70554 | .70867 | −.99558 |
| 2.6 | .51550 | −.85689 | −.60160 | 5.6 | −.63127 | .77557 | −.81394 |
| 2.7 | .42738 | −.90407 | −.47273 | 5.7 | −.55069 | .83471 | −.65973 |
| 2.8 | .33499 | −.94222 | −.35553 | 5.8 | −.46460 | .88552 | −.52467 |
| 2.9 | .23925 | −.97096 | −.24641 | 5.9 | −.37388 | .92748 | −.40311 |
| 3.0 | .14112 | −.98999 | −.14255 | 6.0 | −.27942 | .96017 | −.29101 |
| — | — | — | — | 6.1 | −.18216 | .98327 | −.18526 |
| — | — | — | — | 6.2 | −.08309 | .99654 | −.08338 |
| — | — | — | — | 2π | .00000 | 1.00000 | .00000 |

# TABLE 2    *The Natural Exponential Function*

| $x$ | $e^x$ | $e^{-x}$ | $x$ | $e^x$ | $e^{-x}$ | $x$ | $e^x$ | $e^{-x}$ |
|---|---|---|---|---|---|---|---|---|
| **.00** | 1.00000 | 1.00000 | **.40** | 1.49182 | .67032 | **.80** | 2.22554 | .44933 |
| .01 | 1.01005 | .99005 | .41 | 1.50682 | .66365 | .81 | 2.24791 | .44486 |
| .02 | 1.02020 | .98020 | .42 | 1.52196 | .65705 | .82 | 2.27050 | .44043 |
| .03 | 1.03045 | .97045 | .43 | 1.53726 | .65051 | .83 | 2.29332 | .43605 |
| .04 | 1.04081 | .96079 | .44 | 1.55271 | .64404 | .84 | 2.31637 | .43171 |
| .05 | 1.05127 | .95123 | .45 | 1.56831 | .63763 | .85 | 2.33965 | .42741 |
| .06 | 1.06184 | .94176 | .46 | 1.58407 | .63128 | .86 | 2.36316 | .42316 |
| .07 | 1.07251 | .93239 | .47 | 1.59999 | .62500 | .87 | 2.38691 | .41895 |
| .08 | 1.08329 | .92312 | .48 | 1.61607 | .61878 | .88 | 2.41090 | .41478 |
| .09 | 1.09417 | .91393 | .49 | 1.63232 | .61263 | .89 | 2.43513 | .41066 |
| **.10** | 1.10517 | .90484 | **.50** | 1.64872 | .60653 | **.90** | 2.45960 | .40657 |
| .11 | 1.11628 | .89583 | .51 | 1.66529 | .60050 | .91 | 2.48432 | .40252 |
| .12 | 1.12750 | .88692 | .52 | 1.68203 | .59452 | .92 | 2.50929 | .39852 |
| .13 | 1.13883 | .87810 | .53 | 1.69893 | .58860 | .93 | 2.53451 | .39455 |
| .14 | 1.15027 | .86936 | .54 | 1.71601 | .58275 | .94 | 2.55998 | .39063 |
| .15 | 1.16183 | .86071 | .55 | 1.73325 | .57695 | .95 | 2.58571 | .38674 |
| .16 | 1.17351 | .85214 | .56 | 1.75067 | .57121 | .96 | 2.61170 | .38289 |
| .17 | 1.18530 | .84366 | .57 | 1.76827 | .56553 | .97 | 2.63794 | .37908 |
| .18 | 1.19722 | .83527 | .58 | 1.78604 | .55990 | .98 | 2.66446 | .37531 |
| .19 | 1.20925 | .82696 | .59 | 1.80399 | .55433 | .99 | 2.69123 | .37158 |
| **.20** | 1.22140 | .81873 | **.60** | 1.82212 | .54881 | **1.00** | 2.71828 | .36788 |
| .21 | 1.23368 | .81058 | .61 | 1.84043 | .54335 | 1.01 | 2.74560 | .36422 |
| .22 | 1.24608 | .80252 | .62 | 1.85893 | .53794 | 1.02 | 2.77319 | .36059 |
| .23 | 1.25860 | .79453 | .63 | 1.87761 | .53259 | 1.03 | 2.80107 | .35701 |
| .24 | 1.27125 | .78663 | .64 | 1.89648 | .52729 | 1.04 | 2.82922 | .35345 |
| .25 | 1.28403 | .77880 | .65 | 1.91554 | .52205 | 1.05 | 2.85765 | .34994 |
| .26 | 1.29693 | .77105 | .66 | 1.93479 | .51685 | 1.06 | 2.88637 | .34646 |
| .27 | 1.30996 | .76338 | .67 | 1.95424 | .51171 | 1.07 | 2.91538 | .34301 |
| .28 | 1.32313 | .75578 | .68 | 1.97388 | .50662 | 1.08 | 2.94468 | .33960 |
| .29 | 1.33643 | .74826 | .69 | 1.99372 | .50158 | 1.09 | 2.97427 | .33622 |
| **.30** | 1.34986 | .74082 | **.70** | 2.01375 | .49659 | **1.10** | 3.00417 | .33287 |
| .31 | 1.36343 | .73345 | .71 | 2.03399 | .49164 | 1.11 | 3.03436 | .32956 |
| .32 | 1.37713 | .72615 | .72 | 2.05443 | .48675 | 1.12 | 3.06485 | .32628 |
| .33 | 1.39097 | .71892 | .73 | 2.07508 | .48191 | 1.13 | 3.09566 | .32303 |
| .34 | 1.40495 | .71177 | .74 | 2.09594 | .47711 | 1.14 | 3.12677 | .31982 |
| .35 | 1.41907 | .70469 | .75 | 2.11700 | .47237 | 1.15 | 3.15819 | .31664 |
| .36 | 1.43333 | .69768 | .76 | 2.13828 | .46767 | 1.16 | 3.18993 | .31349 |
| .37 | 1.44773 | .69073 | .77 | 2.15977 | .46301 | 1.17 | 3.22199 | .31037 |
| .38 | 1.46228 | .68386 | .78 | 2.18147 | .45841 | 1.18 | 3.25437 | .30728 |
| .39 | 1.47698 | .67706 | .79 | 2.20340 | .45384 | 1.19 | 3.28708 | .30422 |

TABLE 2  *The Natural Exponential Function* (continued)

| $x$ | $e^x$ | $e^{-x}$ | $x$ | $e^x$ | $e^{-x}$ | $x$ | $e^x$ | $e^{-x}$ |
|---|---|---|---|---|---|---|---|---|
| **1.20** | 3.32012 | .30119 | **1.60** | 4.95303 | .20190 | **4.0** | 54.598 | .01832 |
| 1.21 | 3.35348 | .29820 | 1.61 | 5.00281 | .19989 | 4.1 | 60.340 | .01657 |
| 1.22 | 3.38719 | .29523 | 1.62 | 5.05309 | .19790 | 4.2 | 66.686 | .01500 |
| 1.23 | 3.42123 | .29229 | 1.63 | 5.10387 | .19593 | 4.3 | 73.700 | .01357 |
| 1.24 | 3.45561 | .28938 | 1.64 | 5.15517 | .19398 | 4.4 | 81.451 | .01228 |
| 1.25 | 3.49034 | .28650 | 1.65 | 5.20698 | .19205 | 4.5 | 90.017 | .01111 |
| 1.26 | 3.52542 | .28365 | 1.66 | 5.25931 | .19014 | 4.6 | 99.484 | .01005 |
| 1.27 | 3.56085 | .28083 | 1.67 | 5.31217 | .18825 | 4.7 | 109.947 | .00910 |
| 1.28 | 3.59664 | .27804 | 1.68 | 5.36556 | .18637 | 4.8 | 121.510 | .00823 |
| 1.29 | 3.63279 | .27527 | 1.69 | 5.41948 | .18452 | 4.9 | 134.290 | .00745 |
| **1.30** | 3.66930 | .27253 | **1.70** | 5.47395 | .18268 | **5.0** | 148.41 | .00674 |
| 1.31 | 3.70617 | .26982 | 1.71 | 5.52896 | .18087 | 5.1 | 164.02 | .00610 |
| 1.32 | 3.74342 | .26714 | 1.72 | 5.58453 | .17907 | 5.2 | 181.27 | .00552 |
| 1.33 | 3.78104 | .26448 | 1.73 | 5.64065 | .17728 | 5.3 | 200.34 | .00499 |
| 1.34 | 3.81904 | .26185 | 1.74 | 5.69734 | .17552 | 5.4 | 221.41 | .00452 |
| 1.35 | 3.85743 | .25924 | 1.75 | 5.75460 | .17377 | 5.5 | 244.69 | .00409 |
| 1.36 | 3.89619 | .25666 | 1.80 | 6.04965 | .16530 | 5.6 | 270.43 | .00370 |
| 1.37 | 3.93535 | .25411 | 1.85 | 6.35982 | .15724 | 5.7 | 298.87 | .00335 |
| 1.38 | 3.97490 | .25158 | 1.90 | 6.68589 | .14957 | 5.8 | 330.30 | .00303 |
| 1.39 | 4.01485 | .24908 | 1.95 | 7.02869 | .14227 | 5.9 | 365.04 | .00274 |
| **1.40** | 4.05520 | .24660 | **2.0** | 7.3891 | .13534 | **6.0** | 403.43 | .00248 |
| 1.41 | 4.09596 | .24414 | 2.1 | 8.1662 | .12246 | 6.1 | 445.86 | .00224 |
| 1.42 | 4.13712 | .24171 | 2.2 | 9.0250 | .11080 | 6.2 | 492.75 | .00203 |
| 1.43 | 4.17870 | .23931 | 2.3 | 9.9742 | .10026 | 6.3 | 544.57 | .00184 |
| 1.44 | 4.22070 | .23693 | 2.4 | 11.0232 | .09072 | 6.4 | 601.85 | .00166 |
| 1.45 | 4.26311 | .23457 | 2.5 | 12.1825 | .08208 | 6.5 | 665.14 | .00150 |
| 1.46 | 4.30596 | .23224 | 2.6 | 13.4637 | .07427 | 6.6 | 735.10 | .00136 |
| 1.47 | 4.34924 | .22993 | 2.7 | 14.8797 | .06721 | 6.7 | 812.41 | .00123 |
| 1.48 | 4.39295 | .22764 | 2.8 | 16.4446 | .06081 | 6.8 | 897.85 | .00111 |
| 1.49 | 4.43710 | .22537 | 2.9 | 18.1741 | .05502 | 6.9 | 992.27 | .00101 |
| **1.50** | 4.48169 | .22313 | **3.0** | 20.086 | .04979 | **7.0** | 1096.6 | .00091 |
| 1.51 | 4.52673 | .22091 | 3.1 | 22.198 | .04505 | 7.5 | 1808.0 | .00055 |
| 1.52 | 4.57223 | .21871 | 3.2 | 24.533 | .04076 | 8.0 | 2981.0 | .00034 |
| 1.53 | 4.61818 | .21654 | 3.3 | 27.113 | .03688 | 8.5 | 4914.8 | .00020 |
| 1.54 | 4.66459 | .21438 | 3.4 | 29.964 | .03337 | 9.0 | 8103.1 | .00012 |
| 1.55 | 4.71147 | .21225 | 3.5 | 33.115 | .03020 | 9.5 | 13360 | .00007 |
| 1.56 | 4.75882 | .21014 | 3.6 | 36.598 | .02732 | 10.0 | 22026 | .00005 |
| 1.57 | 4.80665 | .20805 | 3.7 | 40.447 | .02472 | 10.5 | 36316 | .00003 |
| 1.58 | 4.85496 | .20598 | 3.8 | 44.701 | .02237 | 11.0 | 59874 | .00002 |
| 1.59 | 4.90375 | .20393 | 3.9 | 49.402 | .02024 | 11.5 | 98716 | .00001 |

**TABLE 3** *Natural Logarithms*

| x | ln x | x | ln x | x | ln x | x | ln x | x | ln x |
|---|------|---|------|---|------|---|------|---|------|
|      |          | 0.50 | −0.69315 | 1.00 | 0.00000 | 1.50 | 0.40547 | 2.00 | 0.69315 |
| 0.01 | −4.60517 | 0.51 | −0.67334 | 1.01 | 0.00995 | 1.51 | 0.41211 | 2.10 | 0.74194 |
| 0.02 | −3.91202 | 0.52 | −0.65393 | 1.02 | 0.01980 | 1.52 | 0.41871 | 2.20 | 0.78846 |
| 0.03 | −3.50656 | 0.53 | −0.63488 | 1.03 | 0.02956 | 1.53 | 0.42527 | 2.30 | 0.83291 |
| 0.04 | −3.21888 | 0.54 | −0.61619 | 1.04 | 0.03922 | 1.54 | 0.43178 | 2.40 | 0.87547 |
| 0.05 | −2.99573 | 0.55 | −0.59784 | 1.05 | 0.04879 | 1.55 | 0.43825 | 2.50 | 0.91629 |
| 0.06 | −2.81341 | 0.56 | −0.57982 | 1.06 | 0.05827 | 1.56 | 0.44469 | 2.60 | 0.95551 |
| 0.07 | −2.65926 | 0.57 | −0.56212 | 1.07 | 0.06766 | 1.57 | 0.45108 | 2.70 | 0.99325 |
| 0.08 | −2.52573 | 0.58 | −0.54473 | 1.08 | 0.07696 | 1.58 | 0.45742 | 2.80 | 1.02962 |
| 0.09 | −2.40795 | 0.59 | −0.52763 | 1.09 | 0.08618 | 1.59 | 0.46373 | 2.90 | 1.06471 |
| 0.10 | −2.30259 | 0.60 | −0.51083 | 1.10 | 0.09531 | 1.60 | 0.47000 | 3.00 | 1.09861 |
| 0.11 | −2.20727 | 0.61 | −0.49430 | 1.11 | 0.10436 | 1.61 | 0.47623 | 3.10 | 1.13140 |
| 0.12 | −2.12026 | 0.62 | −0.47804 | 1.12 | 0.11333 | 1.62 | 0.48243 | 3.20 | 1.16315 |
| 0.13 | −2.04022 | 0.63 | −0.46204 | 1.13 | 0.12222 | 1.63 | 0.48858 | 3.30 | 1.19392 |
| 0.14 | −1.96611 | 0.64 | −0.44629 | 1.14 | 0.13103 | 1.64 | 0.49470 | 3.40 | 1.22378 |
| 0.15 | −1.89712 | 0.65 | −0.43078 | 1.15 | 0.13976 | 1.65 | 0.50078 | 3.50 | 1.25276 |
| 0.16 | −1.83258 | 0.66 | −0.41552 | 1.16 | 0.14842 | 1.66 | 0.50682 | 3.60 | 1.28093 |
| 0.17 | −1.77196 | 0.67 | −0.40048 | 1.17 | 0.15700 | 1.67 | 0.51282 | 3.70 | 1.30833 |
| 0.18 | −1.71480 | 0.68 | −0.38566 | 1.18 | 0.16551 | 1.68 | 0.51879 | 3.80 | 1.33500 |
| 0.19 | −1.66073 | 0.69 | −0.37106 | 1.19 | 0.17395 | 1.69 | 0.52473 | 3.90 | 1.36098 |
| 0.20 | −1.60944 | 0.70 | −0.35667 | 1.20 | 0.18232 | 1.70 | 0.53063 | 4.00 | 1.38629 |
| 0.21 | −1.56065 | 0.71 | −0.34249 | 1.21 | 0.19062 | 1.71 | 0.53649 | 4.10 | 1.41099 |
| 0.22 | −1.51413 | 0.72 | −0.32850 | 1.22 | 0.19885 | 1.72 | 0.54232 | 4.20 | 1.43508 |
| 0.23 | −1.46968 | 0.73 | −0.31471 | 1.23 | 0.20701 | 1.73 | 0.54812 | 4.30 | 1.45862 |
| 0.24 | −1.42712 | 0.74 | −0.30111 | 1.24 | 0.21511 | 1.74 | 0.55389 | 4.40 | 1.48160 |
| 0.25 | −1.38629 | 0.75 | −0.28768 | 1.25 | 0.22314 | 1.75 | 0.55962 | 4.50 | 1.50408 |
| 0.26 | −1.34707 | 0.76 | −0.27444 | 1.26 | 0.23111 | 1.76 | 0.56531 | 4.60 | 1.52606 |
| 0.27 | −1.30933 | 0.77 | −0.26136 | 1.27 | 0.23902 | 1.77 | 0.57098 | 4.70 | 1.54756 |
| 0.28 | −1.27297 | 0.78 | −0.24846 | 1.28 | 0.24686 | 1.78 | 0.57661 | 4.80 | 1.56862 |
| 0.29 | −1.23787 | 0.79 | −0.23572 | 1.29 | 0.25464 | 1.79 | 0.58222 | 4.90 | 1.58924 |
| 0.30 | −1.20397 | 0.80 | −0.22314 | 1.30 | 0.26236 | 1.80 | 0.58779 | 5.00 | 1.60944 |
| 0.31 | −1.17118 | 0.81 | −0.21072 | 1.31 | 0.27003 | 1.81 | 0.59333 | 5.10 | 1.62924 |
| 0.32 | −1.13943 | 0.82 | −0.19845 | 1.32 | 0.27763 | 1.82 | 0.59884 | 5.20 | 1.64866 |
| 0.33 | −1.10866 | 0.83 | −0.18633 | 1.33 | 0.28518 | 1.83 | 0.60432 | 5.30 | 1.66771 |
| 0.34 | −1.07881 | 0.84 | −0.17435 | 1.34 | 0.29267 | 1.84 | 0.60977 | 5.40 | 1.68640 |
| 0.35 | −1.04982 | 0.85 | −0.16252 | 1.35 | 0.30010 | 1.85 | 0.61519 | 5.50 | 1.70475 |
| 0.36 | −1.02165 | 0.86 | −0.15082 | 1.36 | 0.30748 | 1.86 | 0.62058 | 5.60 | 1.72277 |
| 0.37 | −0.99425 | 0.87 | −0.13926 | 1.37 | 0.31481 | 1.87 | 0.62594 | 5.70 | 1.74047 |
| 0.38 | −0.96758 | 0.88 | −0.12783 | 1.38 | 0.32208 | 1.88 | 0.63127 | 5.80 | 1.75786 |
| 0.39 | −0.94161 | 0.89 | −0.11653 | 1.39 | 0.32930 | 1.89 | 0.63658 | 5.90 | 1.77495 |
| 0.40 | −0.91629 | 0.90 | −0.10536 | 1.40 | 0.33647 | 1.90 | 0.64185 | 6.00 | 1.79176 |
| 0.41 | −0.89160 | 0.91 | −0.09431 | 1.41 | 0.34359 | 1.91 | 0.64710 | 6.10 | 1.80829 |
| 0.42 | −0.86750 | 0.92 | −0.08338 | 1.42 | 0.35066 | 1.92 | 0.65233 | 6.20 | 1.82455 |
| 0.43 | −0.84397 | 0.93 | −0.07257 | 1.43 | 0.35767 | 1.93 | 0.65752 | 6.30 | 1.84055 |
| 0.44 | −0.82098 | 0.94 | −0.06188 | 1.44 | 0.36464 | 1.94 | 0.66269 | 6.40 | 1.85630 |
| 0.45 | −0.79851 | 0.95 | −0.05129 | 1.45 | 0.37156 | 1.95 | 0.66783 | 6.50 | 1.87180 |
| 0.46 | −0.77653 | 0.96 | −0.04082 | 1.46 | 0.37844 | 1.96 | 0.67294 | 6.60 | 1.88707 |
| 0.47 | −0.75502 | 0.97 | −0.03046 | 1.47 | 0.38526 | 1.97 | 0.67803 | 6.70 | 1.90211 |
| 0.48 | −0.73397 | 0.98 | −0.02020 | 1.48 | 0.39204 | 1.98 | 0.68310 | 6.80 | 1.91692 |
| 0.49 | −0.71335 | 0.99 | −0.01005 | 1.49 | 0.39878 | 1.99 | 0.68813 | 6.90 | 1.93152 |

TABLE 3    *Natural Logarithms* (*continued*)

| x | ln x | x | ln x | x | ln x | x | ln x | x | ln x |
|---|------|---|------|---|------|---|------|---|------|
| 7.00 | 1.94591 | 30.0 | 3.40120 | 80.0 | 4.38203 | 400. | 5.99146 | 900. | 6.80239 |
| 7.10 | 1.96009 | 31.0 | 3.43399 | 81.0 | 4.39445 | 410. | 6.01616 | 910. | 6.81344 |
| 7.20 | 1.97408 | 32.0 | 3.46574 | 82.0 | 4.40672 | 420. | 6.04025 | 920. | 6.82437 |
| 7.30 | 1.98787 | 33.0 | 3.49651 | 83.0 | 4.41884 | 430. | 6.06379 | 930. | 6.83518 |
| 7.40 | 2.00148 | 34.0 | 3.52636 | 84.0 | 4.43082 | 440. | 6.08677 | 940. | 6.84588 |
| 7.50 | 2.01490 | 35.0 | 3.55535 | 85.0 | 4.44265 | 450. | 6.10925 | 950. | 6.85646 |
| 7.60 | 2.02815 | 36.0 | 3.58352 | 86.0 | 4.45435 | 460. | 6.13123 | 960. | 6.86693 |
| 7.70 | 2.04122 | 37.0 | 3.61092 | 87.0 | 4.46591 | 470. | 6.15273 | 970. | 6.87730 |
| 7.80 | 2.05412 | 38.0 | 3.63759 | 88.0 | 4.47734 | 480. | 6.17379 | 980. | 6.88755 |
| 7.90 | 2.06686 | 39.0 | 3.66356 | 89.0 | 4.48864 | 490. | 6.19441 | 990. | 6.89770 |
| 8.00 | 2.07944 | 40.0 | 3.68888 | 90.0 | 4.49981 | 500. | 6.21461 | 1000. | 6.90776 |
| 8.10 | 2.09186 | 41.0 | 3.71357 | 91.0 | 4.51086 | 510. | 6.23441 | — | — |
| 8.20 | 2.10413 | 42.0 | 3.73767 | 92.0 | 4.52179 | 520. | 6.25383 | — | — |
| 8.30 | 2.11626 | 43.0 | 3.76120 | 93.0 | 4.53260 | 530. | 6.27288 | — | — |
| 8.40 | 2.12823 | 44.0 | 3.78419 | 94.0 | 4.54329 | 540. | 6.29157 | — | — |
| 8.50 | 2.14007 | 45.0 | 3.80666 | 95.0 | 4.55388 | 550. | 6.30992 | — | — |
| 8.60 | 2.15176 | 46.0 | 3.82864 | 96.0 | 4.56435 | 560. | 6.32794 | — | — |
| 8.70 | 2.16332 | 47.0 | 3.85015 | 97.0 | 4.57471 | 570. | 6.34564 | — | — |
| 8.80 | 2.17475 | 48.0 | 3.87120 | 98.0 | 4.58497 | 580. | 6.36303 | — | — |
| 8.90 | 2.18605 | 49.0 | 3.89182 | 99.0 | 4.59512 | 590. | 6.38012 | — | — |
| 9.00 | 2.19722 | 50.0 | 3.91202 | 100. | 4.60517 | 600. | 6.39693 | — | — |
| 9.10 | 2.20827 | 51.0 | 3.93183 | 110. | 4.70048 | 610. | 6.41346 | — | — |
| 9.20 | 2.21920 | 52.0 | 3.95124 | 120. | 4.78749 | 620. | 6.42972 | — | — |
| 9.30 | 2.23001 | 53.0 | 3.97029 | 130. | 4.86753 | 630. | 6.44572 | — | — |
| 9.40 | 2.24071 | 54.0 | 3.98898 | 140. | 4.94164 | 640. | 6.46147 | — | — |
| 9.50 | 2.25129 | 55.0 | 4.00733 | 150. | 5.01064 | 650. | 6.47697 | — | — |
| 9.60 | 2.26176 | 56.0 | 4.02535 | 160. | 5.07517 | 660. | 6.49224 | — | — |
| 9.70 | 2.27213 | 57.0 | 4.04305 | 170. | 5.13580 | 670. | 6.50728 | — | — |
| 9.80 | 2.28238 | 58.0 | 4.06044 | 180. | 5.19296 | 680. | 6.52209 | — | — |
| 9.90 | 2.29253 | 59.0 | 4.07754 | ‡90. | 5.24702 | 690. | 6.53669 | — | — |
| 10.0 | 2.30259 | 60.0 | 4.09434 | 200. | 5.29832 | 700. | 6.55108 | — | — |
| 11.0 | 2.39790 | 61.0 | 4.11087 | 210. | 5.34711 | 710. | 6.56526 | — | — |
| 12.0 | 2.48491 | 62.0 | 4.12713 | 220. | 5.39363 | 720. | 6.57925 | — | — |
| 13.0 | 2.56495 | 63.0 | 4.14313 | 230. | 5.43808 | 730. | 6.59304 | — | — |
| 14.0 | 2.63906 | 64.0 | 4.15888 | 240. | 5.48064 | 740. | 6.60665 | — | — |
| 15.0 | 2.70805 | 65.0 | 4.17439 | 250. | 5.52146 | 750. | 6.62007 | — | — |
| 16.0 | 2.77259 | 66.0 | 4.18965 | 260. | 5.56068 | 760. | 6.63332 | — | — |
| 17.0 | 2.83321 | 67.0 | 4.20469 | 270. | 5.59842 | 770. | 6.64639 | — | — |
| 18.0 | 2.89037 | 68.0 | 4.21951 | 280. | 5.63479 | 780. | 6.65929 | — | — |
| 19.0 | 2.94444 | 69.0 | 4.23411 | 290. | 5.66988 | 790. | 6.67203 | — | — |
| 20.0 | 2.99573 | 70.0 | 4.24850 | 300. | 5.70378 | 800. | 6.68461 | — | — |
| 21.0 | 3.04452 | 71.0 | 4.26268 | 310. | 5.73657 | 810. | 6.69703 | — | — |
| 22.0 | 3.09104 | 72.0 | 4.27667 | 320. | 5.76832 | 820. | 6.70930 | — | — |
| 23.0 | 3.13549 | 73.0 | 4.29046 | 330. | 5.79909 | 830. | 6.72143 | — | — |
| 24.0 | 3.17805 | 74.0 | 4.30407 | 340. | 5.82895 | 840. | 6.73340 | — | — |
| 25.0 | 3.21888 | 75.0 | 4.31749 | 350. | 5.85793 | 850. | 6.74524 | — | — |
| 26.0 | 3.25810 | 76.0 | 4.33073 | 360. | 5.88610 | 860. | 6.75693 | — | — |
| 27.0 | 3.29584 | 77.0 | 4.34381 | 370. | 5.91350 | 870. | 6.76849 | — | — |
| 28.0 | 3.33220 | 78.0 | 4.35671 | 380. | 5.94017 | 880. | 6.77992 | — | — |
| 29.0 | 3.36730 | 79.0 | 4.36945 | 390. | 5.96615 | 890. | 6.79122 | — | — |

# Answers to Odd-Numbered Problems

# CHAPTER 1

**SECTION 1-2 (page 8)**

**1** 14   **3** $\frac{1}{2}$   **5** 25   **7** 27   **9** $\frac{22}{7} - \pi \approx 0.0013$   **11** $3 - x$   **13** $(-\infty, 2)$   **15** $[7, \infty)$   **17** $(-\frac{5}{3}, \infty)$   **19** $(-4, 1)$
**21** $(\frac{3}{2}, \frac{11}{2}]$   **23** $(-1, 4)$   **25** $(-\infty, -\frac{1}{3}) \cup (1, \infty)$   **27** $[0, \frac{2}{5}] \cup [\frac{6}{5}, \frac{8}{5}]$   **29** $(\frac{7}{3}, \frac{37}{15}]$   **31** $(-\infty, \frac{1}{5}) \cup (\frac{1}{5}, \infty)$
**33** $(-\infty, -2) \cup (4, \infty)$   **35** $(-\infty, \frac{1}{2}] \cup [\frac{3}{2}, \infty)$   **37** $4 \leqq p \leqq 8$   **39** $2 < I < 4$

**SECTION 1-3 (page 13)**

**1** (a) $-1/a$   (b) $a$   (c) $1/\sqrt{a}$   (d) $1/a^2$   **3** (a) $1/(a^2 + 5)$   (b) $a^2/(5a^2 + 1)$   (c) $1/(a + 5)$   (d) $1/(a^4 + 5)$   **5** $\frac{1}{3}$
**7** $3, -3$   **9** $100$   **11** $3h$   **13** $2ah + h^2$   **15** $-h/[a(a + h)]$   **17** $\{-1, 0, 1\}$   **19** $\{-1, 1\}$

| | Domain | Range |
|---|---|---|
| **21** | $\mathcal{R}$ | $(-\infty, 10]$ |
| **23** | $\mathcal{R}$ | $[0, +\infty)$ |
| **25** | $[\frac{5}{3}, +\infty)$ | $[0, +\infty)$ |
| **27** | $(-\infty, \frac{1}{2}]$ | $[0, +\infty)$ |
| **29** | $(-\infty, 3) \cup (3, +\infty)$ | $(-\infty, 0) \cup (0, +\infty)$ |
| **31** | $\mathcal{R}$ | $[3, +\infty)$ |
| **33** | $[0, 16]$ | $[0, 2]$ |
| **35** | $(-\infty, 0) \cup (0, +\infty)$ | $\{-1, 1\}$ |

**37** $C(A) = 2\sqrt{\pi A}, A \geq 0$   **39** $C(F) = \frac{5}{9}(F - 32), F \geq F_{\min}$   **41** $A(x) = x\sqrt{16 - x^2}, 0 \leq x \leq 4$
**43** $P(x) = 2x + (200/x)$   **45** $A(x) = 2\pi x^2 + (2000/x)$   **47** $V(x) = 4x(25 - x)^2$

**SECTION 1-4 (page 22)**

**1** On a line with slope 1   **3** Not on a line   **9** $m = \frac{2}{3}, b = 0$   **11** $m = 2, b = 3$   **13** $m = -\frac{2}{5}, b = \frac{3}{5}$   **15** $y = -5$
**17** $y = 2x - 7$   **19** $x + y = 6$   **21** $2x + y = 7$   **23** $x + 2y = 13$   **25** $\frac{4}{13}\sqrt{26}$
**33** $K = (5F + 2297)/9; F = -459.4$ when $K = 0$   **35** 1136 gal/week

**SECTION 1-5 (page 28)**

**1**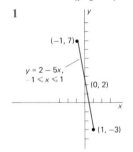

**3**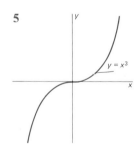

**5**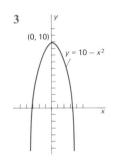

**7**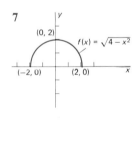

**9**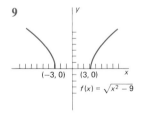

**11**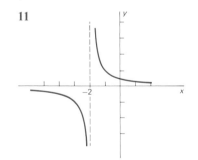

**13**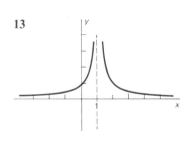

**15**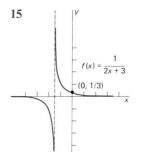

$f(x) = \dfrac{1}{2x + 3}$

(0, 1/3)

**17**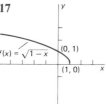

$f(x) = \sqrt{1-x}$

(0, 1)

(1, 0)

**19**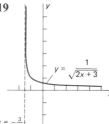

$y = \dfrac{1}{\sqrt{2x + 3}}$

$x = -\dfrac{3}{2}$

**21**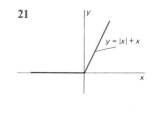

$y = |x| + x$

**23**

−5/2

**25**

2

**27**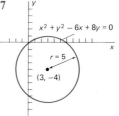

$x^2 + y^2 - 6x + 8y = 0$

$r = 5$

(3, −4)

**29** There are no points on the graph.

**31**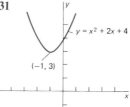

$y = x^2 + 2x + 4$

(−1, 3)

**33**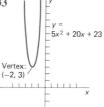

$y = -5x^2 + 20x + 23$

Vertex: (−2, 3)

**35**

**37**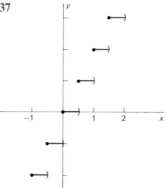

**39**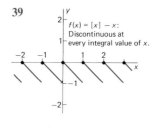

$f(x) = [x] - x$: Discontinuous at every integral value of $x$.

**41** 144 ft

**SECTION 1-6 (page 37)**

**1** 0   **3** $2x$   **5** 4   **7** $4x - 3$   **9** $4x + 6$   **11** $2 - (x/50)$   **13** $8x$   **15** $(0, 10)$   **17** $(1, 0)$   **19** $(50, 25)$
**21** $f'(x) \equiv 3$; $y = 3x - 1$   **23** $f'(x) = 4x - 3$; $y = 5x - 3$   **25** $f'(x) = 2x - 2$; $y = 2x - 3$
**27** $f'(x) = -1/x^2$; $x + 4y = 4$   **29** $f'(x) = -2/x^3$; $x + 4y = 3$   **31** $f'(x) = -2/(x - 1)^2$; $2x + y = 6$
**33** Tangent line: $4x + y + 4 = 0$; normal line: $4y - x = 18$
**35** Tangent line: $y = 11x - 13$; normal line: $x + 11y = 101$   **37** 4500 barrels per day, achieved with $x = 10$ wells
**39** $y = 12x - 36$

**CHAPTER 1 MISCELLANEOUS PROBLEMS (page 38)**

**1** $(-\infty, -5)$   **3** $(-\frac{15}{7}, \infty)$   **5** $(1, 2)$   **7** $(-\frac{1}{2}, 2]$   **9** $(-\infty, -\frac{1}{8}] \cup [\frac{7}{16}, \infty)$   **11** $(-\infty, -3) \cup (-3, 3) \cup (3, \infty)$
**13** $[0, \infty)$   **15** $(-\infty, \frac{2}{3}]$   **17** $[2, 4]$   **19** $A(P) = (P^2\sqrt{3})/36$   **21** $y = 2x + 11$   **23** $y = (x/2) - 5$   **25** $x + 2y = 11$

**27**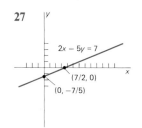

**29**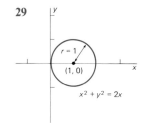

**31**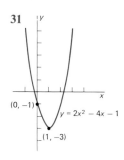

**33**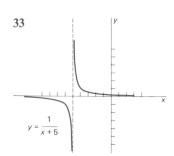

**35**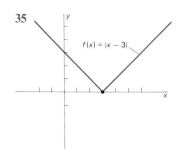

**37** $f'(x) = 4x;\ y = 4x + 1$
**39** $f'(x) = 6x + 4;\ y = 10x - 8$
**41** $f'(x) = 4x - 1;\ y = 3x - 3$
**43** $f'(x) = 4x + 3$
**45** $f'(x) = 1/(3 - x)^2$
**47** $f'(x) = 1 + (1/x^2)$
**49** $f'(x) = -2/(x - 1)^2$

**51** $y - 14 - 6\sqrt{5} = (6 + 2\sqrt{5})(x - 3 - \sqrt{5})$ and $y - 14 + 6\sqrt{5} = (6 - 2\sqrt{5})(x - 3 + \sqrt{5})$

# CHAPTER 2

### SECTION 2-1 (page 49)

**1** $\lim_{x\to 0} (3x^2 + 7x - 12) = \lim_{x\to 0} (3x^2) + \lim_{x\to 0} (7x) - \lim_{x\to 0} 12 = 3(\lim_{x\to 0} x)^2 + 7(\lim_{x\to 0} x) - 12 = 3(0)^2 + 7(0) - 12 = -12$   **3** $-1$   **5** $128$   **7** $(x + 1)/(x^2 - x - 2) = (x + 1)/[(x + 1)(x - 2)] = 1/(x - 2) \to -\frac{1}{3}$ as $x \to -1$
**9** $6$   **11** $16\sqrt{2}$   **13** $1$   **15** $(\sqrt{x + 4} - 2)/x = (x + 4 - 4)/x(\sqrt{x + 4} + 2) = 1/(\sqrt{x + 4} + 2) \to \frac{1}{4}$ as $x \to 0$   **17** $-\frac{1}{54}$
**19** $2$   **21** $4$   **23** $-\frac{3}{2}$   **25** $-32$   **27** $1$   **29** $0$

**31**

| $x$ | $(\sqrt{x + 9} - 3)/x$ |
|---|---|
| 0.1 | 0.16621 |
| 0.01 | 0.16662 |
| 0.001 | 0.16666 |
| −0.001 | 0.16667 |
| 0.0001 | 0.16667 |
| −0.0001 | 0.16667 |

The limit seems to be $\frac{1}{6}$.

**33**

| $x$ | $(1/x)[1/(x + 5) - \frac{1}{5}]$ |
|---|---|
| 0.1 | −0.03922 |
| 0.01 | −0.03912 |
| 0.001 | −0.03999 |
| 0.0001 | −0.04000 |
| −0.0001 | −0.04000 |

The limit seems to be $-\frac{1}{25}$.

**35**

| $h$ | $(1 + h)^{1/h}$ |
|---|---|
| 0.1 | 2.59374 |
| 0.01 | 2.70481 |
| 0.001 | 2.71692 |
| 0.0001 | 2.71815 |
| 0.00001 | 2.71827 |
| 0.000001 | 2.71828 |
| −0.000001 | 2.71828 |

The limit seems to be approximately 2.718.

**37** $f'(x) = -1/(2x^{3/2})$   **39** $f'(x) = 1/(2x + 1)^2$
**41** $f'(x) = (x^2 + 2x)/(x + 1)^2$

**A–26**

**1** 3   **3** Does not exist   **5** 0   **7** 0   **9** Does not exist   **11** −1   **13** 1   **15** −1   **17** 2   **19** −1
**21** $\lim_{x \to 1^-} f(x) = -\infty$, $\lim_{x \to 1^+} f(x) = +\infty$   **23** $\lim_{x \to -1^-} f(x) = +\infty$, $\lim_{x \to -1^+} f(x) = -\infty$
**25** $\lim_{x \to -2^-} f(x) = +\infty$, $\lim_{x \to -2^+} f(x) = -\infty$   **27** $\lim_{x \to 1^-} f(x) = \lim_{x \to 1^+} f(x) = +\infty$
**29** $\lim_{x \to -2^-} f(x) = +\infty$, $\lim_{x \to -2^+} f(x) = -\infty$
**31** $\lim_{x \to n^-} f(x) = \lim_{x \to n^+} f(x) = 2$

**33** $\lim_{x \to n^-} f(x) = (-1)^{n+1}$, $\lim_{x \to n^+} f(x) = (-1)^n$

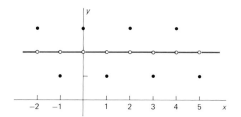

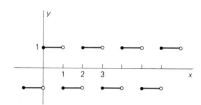

**35** $\lim_{x \to n^-} f(x) = 0$, $\lim_{x \to n^+} f(x) = 1$

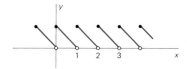

**1** $(f + g)(x) = x^2 + 3x - 2$, domain $\mathscr{R}$; $(fg)(x) = x^3 + 3x^2 - x - 3$, domain $\mathscr{R}$; $(f/g)(x) = (x + 1)/(x^2 + 2x - 3)$, domain $x \neq -3, x \neq 1$
**3** $(f + g)(x) = \sqrt{x} + \sqrt{x - 2}$, domain $[2, +\infty)$; $(fg)(x) = (x^2 - 2x)^{1/2}$, domain $[2, +\infty)$; $(f/g)(x) = \sqrt{x}/(x - 2)$, domain $(2, +\infty)$
**5** $(f + g)(x) = (x^2 - 1)^{1/2} + 1/(4 - x^2)^{1/2}$; $(fg)(x) = (x^2 - 1)^{1/2}/(4 - x^2)^{1/2}$; $(f/g)(x) = (x^2 - 1)^{1/2}(4 - x^2)^{1/2}$; Domain in each case: $(-2, -1] \cup [1, 2)$   **7** $f(g(x)) \equiv -17$, domain $\mathscr{R}$; $g(f(x)) \equiv 17$, domain $\mathscr{R}$
**9** $f(g(x)) = 1 + 1/(x^2 + 1)^2$, domain $\mathscr{R}$; $g(f(x)) = 1/[1 + (x^2 + 1)^2]$, domain $\mathscr{R}$   **11** $f(x) = x^2$, $g(x) = 2 + 3x$
**13** $f(x) = x^{1/2}$, $g(x) = 2x - x^3$   **15** $f(x) = x^{3/2}$, $g(x) = 5 - x^2$   **17** $f(x) = x^{-1}$, $g(x) = x + 1$
**19** $f(x) = x^{-1/2}$, $g(x) = x + 10$   **21** $g(x) = x + 1$   **23** $g(x) = (x^4 + 1)^{1/2}$ or $g(x) = -(x^4 + 1)^{1/2}$, etc.
**25** $g(x) = x + 4$   **27** No, because $f(g(0)) \neq 0$.
**29** Yes, if the domains are suitably restricted: Use $(-1, +\infty)$ for the domain of $f$, $(0, +\infty)$ for the domain of $g$. Not if the "natural" domains are used, because $f(g(-0.5)) = +0.5$   **31** $g(x) = (x + 5)/13$, domain $\mathscr{R}$
**33** $g(x) = ((x - 1)/3)^{1/3}$, domain $\mathscr{R}$   **35** $g(x) = (1 + x^{1/17})^{1/3}$, $-1 \leqq x \leqq 0$

**1** Continuous everywhere   **3** Continuous except at $x = -3$   **5** Continuous everywhere
**7** Continuous except at $x = 5$   **9** Continuous except at $x = 2$   **11** Continuous except at $x = 1$
**13** Continuous except at $x = 0$ and at $x = 1$   **15** Continuous on $(-2, 2)$   **17** Cannot be made continuous at $x = -3$
**19** Cannot be made continuous at $x = -2$; $f(2) = \frac{1}{4}$ makes $f$ continuous at $x = 2$
**21** Cannot be made continuous at $x = -1$ or at $x = 1$   **23** Cannot be made continuous at $x = 17$
**25** $f(0) = 0$ makes $f$ continuous at $x = 0$   **27** Maximum value: $f(-1) = 2$; no minimum value
**29** Minimum value: $f(0) = 0$; no maximum value   **31** Minimum value: $f(2) = 0$; maximum value: $f(4) = 2$
**33** Minimum value: $f(-1) = 0$; maximum value: $f(1) = 2$
**45** The function $f$ is continuous at points of the form $x = 3n$ where $n$ is an integer.

**1** 4   **3** 0   **5** $-\frac{5}{3}$   **7** $-2$   **9** 0   **11** 4   **13** 8   **15** $\frac{1}{6}$   **17** $-\frac{1}{54}$   **19** $-1$   **21** 1   **23** Does not exist   **25** $+\infty$   **27** $+\infty$
**29** $-\infty$   **31** 3   **33** $\frac{3}{2}$   **35** 0   **37** $\frac{9}{4}$   **39** 2   **41** $g(x) = (5 - x^2)/2$, $x \geqq 0$   **43** $g(x) = (1 + x^{-1/5})^{1/3}$, $x \neq 0$

**45** $g(x) = x^2 - x$   **47** $f(1) = \frac{1}{2}$ makes $f$ continuous at $x = 1$; $f$ cannot be made continuous at $x = -1$.
**49** $f(1) = \frac{3}{4}$ makes $f$ continuous at $x = 1$; $f$ cannot be made continuous at $x = -3$.

# CHAPTER 3

## SECTION 3-1 (page 89)

**1** $f'(x) = 4$   **3** $h'(z) = 25 - 2z$   **5** $dy/dx = 4x + 3$   **7** $dz/du = 10u + 3$   **9** $dx/dy = -10y + 17$   **11** $f'(x) = 2$
**13** $f'(x) = 2x$   **15** $f'(x) = -2/(2x + 1)^2$   **17** $f'(x) = 1/\sqrt{2x + 1}$   **19** $f'(x) = 1/(1 - 2x)^2$   **21** $s = 100$   **23** $s = 99$
**25** $s = 120$   **27** 64 ft   **29** 194 ft   **31** $dA/dC = C/2\pi$   **33** It skids for 10 s, a total distance of 500 ft.
**35** (a) 2.5 months   (b) 50 rodents per month   **37** $v(20) \approx 73$ ft/s, about 50 mi/h; $v(40) \approx 91$ ft/s, about 62 mi/h
**41** When $t = 30$, $dV/dt = -(\pi/3)(6.25)$, so the air is leaking out at approximately 6.545 in.$^3$/s.
**43** (a) $144\pi$ cm$^3$/h   (b) $156\pi$ cm$^3$/h   **45** $v = 0$ (soft touchdown) when $t = 2$ (s)

## SECTION 3-2 (page 99)

**1** $f'(x) = 6x - 1$   **3** $f'(x) = 2(3x - 2) + 3(2x + 3)$   **5** $h'(x) = 3x^2 + 6x + 3$
**7** $f'(y) = (2y - 1)(2y + 1) + 2y(2y + 1) + 2y(2y - 1)$   **9** $g'(x) = -1/(x + 1)^2 + 1/(x - 1)^2$
**11** $h'(x) = -3(2x + 1)/(x^2 + x + 1)^2$   **13** $g'(t) = 2t(t^3 + t^2 + 1) + (3t^2 + 2t)(t^2 + 1)$   **15** $g'(z) = -1/2z^2 + 2/3z^3$
**17** $g'(y) = 2(3y^2 - 1)(y^2 + 2y + 3) + 6y(2y)(y^2 + 2y + 3) + (2y + 2)(2y)(3y^2 - 1)$
**19** $g'(t) = ((t^2 + 2t + 1) - (t - 1)(2t + 2))/(t^2 + 2t + 1)^2$   **21** $v'(t) = -3/(t - 1)^4$
**23** $g'(x) = (3(x^3 + 7x - 5) - (3x)(3x^2 + 7))/(x^3 + 7x - 5)^2$
**25** $g'(x) = ((2x - 3)(3x^2 - 4x) - 2(x^3 - 2x^2))/(2x - 3)^2$   **27** $3x^2 - 30x^4 - 6x^{-5}$   **29** $2x + 4x^{-2} - 15x^{-4}$
**31** $3 + 1/(2x^3)$   **33** $-(x - 1)^{-2} - 1/(3x^2)$   **35** $(x^4 + 31x^2 - 10x - 36)/(x^2 + 9)^2$   **37** $(150x^{10} - 240x^5)/(15x^2 - 4)^2$
**39** $(x^2 + 2x)/(x + 1)^2$   **41** $12x - y = 16$   **43** $x + y = 3$   **45** $5x - \dot{y} - 10$   **47** $18x - y = -25$   **49** $3x + y = 0$
**51** (a) It contracts.   (b) $0.06427$ cm$^3$/°C   **53** $14,400\pi \approx 45,239$ cm$^3$/cm   **55** $y = 3x + 2$
**57** Suppose that such a line $L$ is tangent at the points $(a, a^2)$ and $(b, b^2)$. Show that $a = b$ is a consequence.
**59** $[(n - 1)/n]x_0$ for $x_0 \neq 0$   **65** $g'(x) = 17(x^3 - 17x + 35)^{16}(3x^2 - 17)$

## SECTION 3-3 (page 107)

**1** $15(3x + 4)^4$   **3** $-3(3x - 2)^{-2}$   **5** $3(x^2 + 3x + 4)^2(2x + 3)$   **7** $-4(2 - x)^3(3 + x)^7 + 7(2 - x)^4(3 + x)^6$
**9** $\{(3x - 4) - 9(x + 2)\}/(3x - 4)^4$   **11** $12\{1 + (1 + x)^3\}^3(1 + x)^2$   **13** $-6x^{-3}(x^{-2} + 1)^2$
**15** $48\{1 + (4x - 1)^4\}^2(4x - 1)^3$   **17** $-4x^{-5}(1 - x^{-4})^3 + 12x^{-9}(1 - x^{-4})^2$
**19** $-4x^{-5}(x^{-2} - x^{-8})^3 + 3x^{-4}(x^{-2} - x^{-8})^2(8x^{-9} - 2x^{-3})$   **21** $u = 2x - x^2, n = 3, f'(x) = 3(2x - x^2)^2(2 - 2x)$
**23** $u = 1 - x^2, n = -4, f'(x) = 8x(1 - x^2)^{-5}$   **25** $u = (x + 1)/(x - 1), n = 7, f'(x) = -14(x + 1)^6(x - 1)^{-8}$
**27** $g'(y) = 1 + 10(2y - 3)^4$   **29** $F'(s) = 3(s - s^{-2})^2(1 + 2s^{-3})$   **31** $f'(u) = 3(1 + u)^2(1 + u^2)^4 + 8u(1 + u)^3(1 + u^2)^3$
**33** $h'(v) = -2\{v - (1 - v^{-1})^{-1}\}^{-3}[1 + (v - 1)^{-2}]$   **35** $F'(z) = -10(3 - 4z + 5z^5)^{-11}(-4 + 25z^4)$
**37** $dy/dx = 4(x^3)^3(3x^2) = 12x^{11}$   **39** $dy/dx = 4x(x^2 - 1) = 4x^3 - 4x$
**41** $dy/dx = 4(x + 1)^3 = 4x^3 + 12x^2 + 12x + 4$   **43** $dy/dx = -(x^2 + 1)^{-2}(2x) = -2x/(x^2 + 1)^2$
**45** $f'(x) = 3x^2\cos(x^3)$   **47** $g'(z) = 3(\sin 2z)^2(2 \cos 2z)$   **49** $40\pi$ in.$^2$/s   **51** 40 in.$^2$/s   **53** 600 in.$^3$/h   **55** $-18$
**57** $400\pi \approx 1256.64$ cm$^3$/s   **59** 5 cm
**61** Total melting time: $2/(2 - \sqrt[3]{4}) \approx 4.85$ h; completely melted at about 2:50:50 P.M. on the same day.
**63** $du/dx = (du/dw)(dw/dx)$, but $du/dw = (du/dv)(dv/dx)$.

## SECTION 3-4 (page 113)

**1** $f'(x) = 3x^2/2\sqrt{x^3 + 1}$   **3** $f'(x) = 2x/\sqrt{2x^2 + 1}$   **5** $f'(t) = \frac{3}{2}\sqrt{2t}$   **7** $f'(x) = \frac{3}{2}(2x^2 - x + 7)^{1/2}(4x - 1)$
**9** $g'(x) = -4(1 - 6x^2)/3(x - 2x^3)^{7/3}$   **11** $f'(x) = (1 - 2x^2)/\sqrt{1 - x^2}$   **13** $f'(t) = -2t/(t^2 + 1)^{1/2}(t^2 - 1)^{3/2}$
**15** $f'(x) = 3(x^2 + 1)(x^2 - 1)^2/x^4$   **17** $f'(v) = -(v + 2)/2v^2\sqrt{v + 1}$   **19** $f'(x) = -2x/3(1 - x^2)^{2/3}$
**21** $-2x(3 - 4x)^{-1/2} + (3 - 4x)^{1/2}$   **23** $-2x(2x + 4)^{4/3} + \frac{8}{3}(1 - x^2)(2x + 4)^{1/3}$
**25** $-(2/t^2)(1 + 1/t)(3t^2 + 1)^{1/2} + 3t(1 + 1/t)^2(3t^2 + 1)^{-1/2}$   **27** $[2(3x + 4)^5 - 15(3x + 4)^4(2x - 1)]/(3x + 4)^{10}$
**29** $[(3x + 4)^{1/3}(2x - 1)^{-1/2} - (3x + 4)^{-2/3}(2x + 1)^{1/2}]/(3x + 4)^{2/3}$
**31** $\{3y^{2/3}[(1 + y)^{-1/2} - (1 - y)^{-1/2}] - 10[(1 + y)^{1/2} + (1 - y)^{1/2}]\}/(y^{8/3})$
**33** $[t + (t + t^{1/2})^{1/2}]^{-1/2}[1 + (t + t^{1/2})^{-1/2}(1 + t^{-1/2}/2)]/2$
**35** No horizontal tangent line; vertical tangent line at $(0, 0)$

**37** Horizontal tangent line at $(1/3, 2/(3\sqrt{3}))$; vertical tangent line at $(0, 0)$   **39** No horizontal or vertical tangent lines
**41** $\pi^2/32 \approx 0.3084$ s/ft   **43** $(2/\sqrt{5}, 1/\sqrt{5})$ and $(-2/\sqrt{5}, -1/\sqrt{5})$   **45** $x + 2y = 3$   **47** $3x + 2y = 5$ and $3x - 2y = -5$

## SECTION 3-5 (page 118)

**1** $-4$ and $11$   **3** $-5$ and $3$   **5** $0$ and $9$   **7** $-2$ and $52$   **9** $4$ and $5$   **11** Minimum: 1; maximum: 5
**13** Minimum: $-16$; maximum: 9   **15** Minimum: $-22$; maximum: 10   **17** Minimum: $-56$; maximum: 56
**19** Minimum: 5; maximum: 13   **21** Minimum: 0; maximum: 17   **23** Minimum: 0; maximum $\frac{3}{4}$
**25** Minimum: $-\frac{1}{6}$ at $x = 3$; maximum: $\frac{1}{2}$ at $x = -1$
**27** Maximum value: $f(1/\sqrt{2}) = \frac{1}{2}$; minimum value: $f(-1/\sqrt{2}) = -\frac{1}{2}$
**29** Maximum value: $f(3/2) = 3/(2^{4/3})$; minimum value: $f(3) = -3$

## SECTION 3-6 (page 126)

**1** Each is 25.   **3** 1250   **5** 500 in.$^3$   **7** 1152   **9** 250   **11** 11,250 yd$^2$   **13** 128   **15** Approximately 3.9665°C
**17** 1000 cm$^3$   **19** 0.25 m$^3$ (all cubes, no open-top boxes)   **21** 30,000 m$^2$   **23** Approximately 9259.26 in.$^3$
**25** Five presses   **27** The minimizing value of $x$ is $-2 + (10\sqrt{6})/3$ in. To the nearest integer, we use $x = 6$ in. of insulation for an annual savings of \$285.   **29** Charge either \$1.10 or \$1.15 for the largest revenue
**31** Radius $2R/3$, height $H/3$   **35** $(2000\pi\sqrt{3})/27$   **37** Maximum 4; minimum $(16)^{1/3}$   **39** $\sqrt{3}/2$
**41** Each plank is $(\sqrt{34} - 3\sqrt{2})/8 \approx 0.19854$ by $\frac{1}{2}\sqrt{7 - \sqrt{17}} \approx 0.84807$.
**43** Strike the shore $(2\sqrt{3})/3 \approx 1.1547$ km from the point nearest the island.   **45** $\sqrt{3}/3$
**47** The actual value of $x$ is approximately 3.45246.   **49** $8/\sqrt{5}$ ft and $10/\sqrt{5}$ ft

## SECTION 3-7 (page 134)

**1** $x/y$   **3** $-16x/25y$   **5** $-\sqrt{y/x}$   **7** $-(y/x)^{1/3}$   **9** $dy/dx = (3x^2 - 2xy - y^2)/(3y^2 + 2xy + x^2)$
**11** $dy/dx = -x/y$; $3x - 4y = 25$   **13** $dy/dx = (1 - 2xy)/x^2$; $3x + 4y = 10$
**15** $dy/dx = -(2xy + y^2)/(2xy + x^2)$; $y = -2$   **17** $dy/dx = (25y - 24x)/(24y - 25x)$; $3y = 4x$
**19** $dy/dx = -(y/x)^4$; $x + y = 2$   **21** $dy/dx = (5x^4y - y^2)/(3xy - 2x^5)$; the slope is $\frac{3}{2}$.
**23** None ($dy/dx$ is undefined at $(1, 1)$.)   **25** $(2, 2 + \sqrt{8})$ and $(2, 2 - \sqrt{8})$   **27** Tangent line equations: $y = \frac{1}{2}x \pm 3$
**29** Closest points: $(\sqrt{3}, -\sqrt{3})$ and $(-\sqrt{3}, \sqrt{3})$. Farthest points: $(3, 3)$ and $(-3, -3)$
**31** Farthest points: $(1, 1)$ and $(-1, -1)$   **33** 0 and 9   **35** $10\sqrt{2}$
**37** The closest points are the two points of intersection of the rectangular hyperbola $xy = 4$ with the folium; their approximate coordinates are $(2.76, 1.45)$ and $(1.45, 2.76)$. The two exact $x$-coordinates are $(12 \pm 4\sqrt{5})^{1/3}$.
**39** 500 in.$^3$   **41** 11,250 yd$^2$   **43** 30,000 m$^2$   **45** Approximately 9259.26 in.$^3$   **47** Radius $2R/3$, height $H/3$
**49** $(2000\pi\sqrt{3})/27$

## SECTION 3-8 (page 138)

**1** 2   **3** 5   **5** 4   **7** $4/(5\pi) \approx 0.25465$ ft/s   **9** $32\pi/125 \approx 0.80425$ m/h   **11** 20 cm$^2$/s   **13** 0.25 cm/s   **15** 6 ft/s
**17** 384 mi/h   **19** (a) About 0.047 ft/min   (b) about 0.083 ft/min   **21** 400/9 ft/s   **23** Increasing at $16\pi$ cm$^3$/s
**25** 6000 mi/h   **27** $(11\sqrt{21})/15 \approx 3.36$ ft/s downward
**29** They are closest at $t = 12$ min, and the distance between them then is $32\sqrt{13} \approx 115.38$ mi.
**31** $1/(162\pi) \approx 0.001965$ ft/s   **33** $10/(81\pi) \approx 0.0393$ in./min   **35** 600 mi/h   **37** $1/30$ ft/s

## SECTION 3-9 (page 148)

**1** 2.2361   **3** 2.5119   **5** 0.3028   **7** $-0.7402$   **9** 1.9786   **11** 4.7043   **13** 2.3393   **15** 1.9953   **17** 2.154435   **19** 1.802191
**21** (b) 1.25992   **23** 1.145048   **25** $x_{n+1} = \sqrt{\sqrt{\sqrt{10x_n}}}$; 1.389495   **27** 2.1147 and $-2.1147$
**29** $-1.5820$, 0.4021, and 1.3719   **31** $-0.4292$, 0.4698, and 4.9593   **33** 0.7549   **35** 1.62 and $-0.62$
**37** $-1.3578$, 0.7147, and 1.2570 are the three real solutions.   **39** 0.618033989   **41** 3.4525   **43** 0.2261

## CHAPTER 3 MISCELLANEOUS PROBLEMS (page 151)

**1** $2x - 6/x^3$   **3** $1/2\sqrt{x} - 1/3x^{4/3}$   **5** $7(x - 1)^6(3x + 2)^9 + 27(x - 1)^7(3x + 2)^8$   **7** $4(3x - 1/2x^2)^3(3 + 1/x^3)$
**9** $-y/x = -9/x^2$   **11** $-\frac{3}{2}(x^3 - x)^{-5/2}(3x^2 - 1)$   **13** $4x(1 + x^2)/(x^4 + 2x^2 + 2)^2$

**15** $\frac{7}{3}(\sqrt{x} + \sqrt[3]{2x})^{4/3}(1/2\sqrt{x} + \sqrt[3]{2}/3x^{2/3})$  **17** $-1/\sqrt{x+1}(\sqrt{x+1} - 1)^2$  **19** $dy/dx = (-2xy^2 + 1)/(2x^2y - 1)$

**21** $\frac{1}{2}[x + (2x + \sqrt{3x})^{1/2}]^{-1/2}[1 + \frac{1}{2}(2x + \sqrt{3x})^{-1/2}(2 + \sqrt{3}/2\sqrt{x})]$  **23** $-(y/x)^{2/3}$

**25** $-18(1 + 2/[1 + x]^3)^2/[1 + x]^4$  **27** $\{x(1 + x^{1/2})\}^{-1/2}(1 + 3x^{1/2}/2)/2$  **29** $(x + 2)(x + 1)^{-3/2}/2$

**31** $2x(1 - x^3)^{1/2} - \frac{3}{2}x^4(1 - x^3)^{-1/2}$  **33** $10(x + x^{-1})^9(1 - x^{-2})$  **35** $(\frac{4}{3})x^3(1 + x^4)^{-2/3}$  **37** $-\frac{3}{4}$  **39** $x = 0$

**41** $\frac{1}{2}$ ft/min  **43** $-x(x^2 + 25)^{-3/2}$  **45** $\frac{5}{3}(x - 1)^{2/3}$  **47** $g(x) = x^{-1}, f(x) = x^2 + 1, h'(x) = -2x(x^2 + 1)^{-2}$

**49** $dV/dS = \frac{1}{4}\sqrt{S/\pi}$  **51** 27  **53** The maximum area is $R^2$.

**55** Minimum area: $(36\pi V^2)^{1/3}$, obtained by making *one* sphere of radius $(3V/4\pi)^{1/3}$. Maximum area: $(8\pi)^{1/3}(3V)^{2/3}$, obtained by making two equal spheres, each of radius $\frac{1}{2}(3V/\pi)^{1/3}$.  **57** $(32\pi/81)R^3$  **59** $M/2$  **61** 36 ft$^3$  **63** $3\sqrt{3}$

**67** 2 mi from the point on the shore nearest the first town  **69** (a) The maximum height is $m^2v_0^2/[64(m^2 + 1)]$. (b) The maximum range occurs when $m = 1$, thus when $\alpha = \pi/4$.  **71** 2.6458  **73** 2.3714  **75** $-0.3473$  **77** 0.7402  **79** 1.2216

**81** $-0.8794$  **83** Approximately 1.54785 ft  **85** To five places, the three real solutions are $-2.72249, 0.80126$, and 2.30998.

**91** 4 in.$^2$/s  **93** $-50/(9\pi) \approx -1.7684$ ft/min  **95** 1 in./min

# CHAPTER 4

## SECTION 4-2 (page 161)

**1** $(6x + 8x^{-3}) dx$  **3** $[1 + (3x^2/2)(4 - x^3)^{-1/2}] dx$  **5** $[6x(x - 3)^{3/2} + (9/2)x^2(x - 3)^{1/2}] dx$

**7** $[(x^2 + 25)^{1/4} + (\frac{1}{2})x^2(x^2 + 25)^{-3/4}] dx$  **9** $dy = (\frac{3}{2})x^{-1/2}(1 + x^{1/2})^2 dx$

**11** $dy = \{2(1 + x^4)^{1/2} + 4x^4(1 + x^4)^{-1/2}\} dx$  **13** $dy = \{(1 + 2x)^{-1/2} - (1 + 2x)^{1/2}\}/(3x^2) dx$

**15** $dy = \{1 - (2x + 1)^{1/2}\}^{-2}(2x + 1)^{-1/2} dx$  **17** $\Delta y = 3, dy = 3$

**19** $\Delta y = \sqrt{11} - \sqrt{10} \approx 0.15435, dy = 1/(2\sqrt{10}) \approx 0.15811$

**21** $\Delta y = -1 + 1/(1 + 0.04) = -1/26 \approx -0.03846, dy = 0$  **23** $3 - (\frac{2}{27}) \approx 2.926$  **25** $2 - (\frac{1}{32}) \approx 1.969$

**27** $\frac{95}{1536} \approx 0.06185$  **29** 9.99  **31** 0.9  **33** $dy/dx = -x/y$  **35** $(y - x^2)/(y^2 - x)$  **37** $-2(1 + y)/x$  **39** $-4$ in.$^2$

**41** $-405\pi/2$ cm$^3$  **43** 5 ft  **45** 6 W  **47** $25\pi \approx 78.54$ in.$^3$  **49** $4\pi \approx 12.57$ m$^2$

## SECTION 4-3 (page 168)

**1** Increasing on the whole real line  **3** Increasing for $x < 0$, decreasing for $x > 0$

**5** Increasing for $x < \frac{3}{2}$, decreasing for $x > \frac{3}{2}$

**7** Increasing for $-1 < x < 0$ and for $x > 1$, decreasing for $x < -1$ and for $0 < x < 1$

**9** Increasing for $-2 < x < 0$ and for $x > 1$, decreasing for $x < -2$ and for $0 < x < 1$

**11** Increasing for $x < 2$, decreasing for $x > 2$

**13** Increasing for $x < -\sqrt{3}$, for $-\sqrt{3} < x < 1$, and for $x > 3$; decreasing for $1 < x < \sqrt{3}$ and for $\sqrt{3} < x < 3$

**15** $f(0) = 0 = f(2), f'(x) = 2x - 2, c = 1$  **17** $f(-1) = 0 = f(1), f'(x) = -4x/(1 + x^2)^2, c = 0$  **19** $f'(0)$ does not exist.

**21** $f(0) \neq f(1)$  **23** $c = -\frac{1}{2}$  **25** $c = \frac{35}{27}$  **27** The average slope is $\frac{1}{3}$, but $f'(x) = \pm 1$ where $f'$ exists.

**29** The average slope is 1, but $f'(x) = 0$ wherever it exists.

**31** If $g(x) = x^5 + 2x - 3$, then $g'(x) > 0$ for all $x$ in $[0, 1]$ and $g(1) = 0$. So $x = 1$ is the only root of the equation in the given interval.

**33** If $g(x) = x^4 - 3x - 20$, then $g(2) = -10$ and $g(3) = 52$. Because $g'(x) = 4x^3 - 3 \geqq 4(2^3) - 3 = 29 > 0$, $g$ is an increasing function and thus has exactly one zero in the interval $2 \leqq x \leqq 3$.

**35** Note that $f'(x) = 3(-1 + \sqrt{x + 1})/2$.

**37** Assume that $f'(x)$ has the form $a_0 + a_1x + \cdots + a_{n-1}x^{n-1}$. Construct a polynomial $p(x)$ such that $p'(x) = f'(x)$. Conclude that $f(x) = p(x) + C$ on the interval $a \leqq x \leqq b$.

## SECTION 4-4 (page 173)

**1** Local minimum at $x = 2$  **3** Local maximum at $x = 0$, local minimum at $x = 2$

**5** Neither a local maximum nor a local minimum at $x = 1$  **7** Local minimum at $x = -2$, local maximum at $x = 5$

**9** Local minima at $x = -1$ and $x = 1$, local maximum at $x = 0$

**11** Local maximum at $x = -1$, local minimum at $x = 1$  **13** Local minimum at $x = 1$  **15** Local maximum at $x = 0$

**17** $-10$ and 10  **19** $(1, 1)$  **21** 9 in. wide, 18 in. long, 6 in. high  **23** Radius $5/\sqrt[3]{\pi}$ in., height $10/\sqrt[3]{\pi}$ in.

**27** Base: 5 in. by 5 in.; height: 2.5 in.  **29** Radius: $(25/\pi)^{1/3} \approx 1.9965$ in.; height: 4 times that radius

**31** The nearest points are $(\sqrt{6}/2, 3/2)$ and $(-\sqrt{6}/2, 3/2)$; $(0, 0)$ is *not* the nearest point.   **33** 8 in.
**35** $L = [20 + 12\sqrt[3]{4} + 24\sqrt[3]{2}]^{1/2}$ (about 8.324 m)   **39** Height: $(6V)^{1/3}$; base edge: $(9V^2/2)^{1/6}$

**SECTION 4-5 (page 181)**

**1**

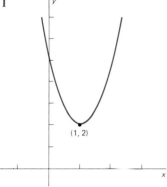

**3**

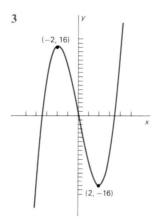

**5**

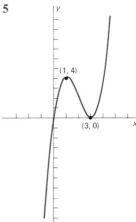

**7**

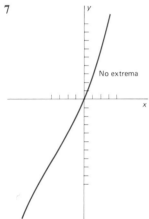

**9**

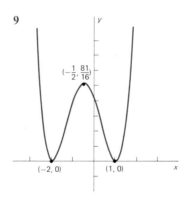

**11**

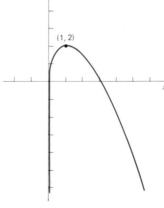

**13**

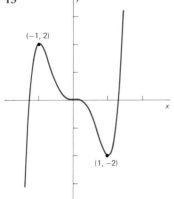

**15**

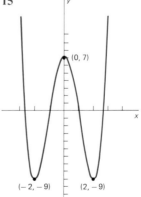

**17** Minimum: $(0.75, -10.125)$, increasing: $(0.75, +\infty)$, decreasing: $(-\infty, 0.75)$

**19** Minimum: $(1, -7)$, maximum: $(-2, 20)$, decreasing: $(-2, 1)$, increasing: $(-\infty, -2)$ and $(1, +\infty)$

**21** Maximum: $(0.6, 16.2)$, minimum: $(0.8, 16.0)$, increasing: $(-\infty, \frac{3}{5})$, and $(\frac{4}{5}, +\infty)$ decreasing: $(\frac{3}{5}, \frac{4}{5})$

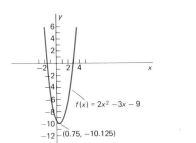

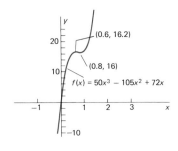

**23** Increasing: $(-1, 0)$ and $(2, +\infty)$, decreasing: $(-\infty, -1)$ and $(0, 2)$, maximum: $(0, 8)$, minima: $(-1, 3)$ and $(2, -24)$

**25** Maximum: $(-2, 64)$, minimum: $(2, -64)$, increasing: $(-\infty, -2)$ and $(2, +\infty)$, decreasing: $(-2, 2)$

**27** Increasing: $(-\infty, +\infty)$, decreasing: Nowhere, Extrema: None

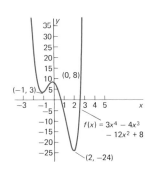

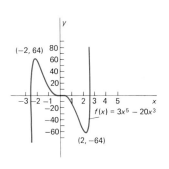

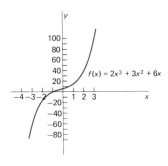

**29** Maxima: $(+\sqrt{2}, 16)$, minimum: $(0, 0)$, increasing: $(-\infty, -\sqrt{2})$ and $(0, \sqrt{2})$, decreasing: $(-\sqrt{2}, 0)$ and $(\sqrt{2}, +\infty)$

**31** Increasing: $(-\infty, 1)$, decreasing: $(1, +\infty)$, maximum: $(1, 3)$

**33** Decreasing: $(0.6, 1)$, increasing: $(-\infty, 0.6)$ and $(1, +\infty)$, minimum: $(1, 0)$, maximum: $(0.6, 0.3257)$ (ordinate approximate)

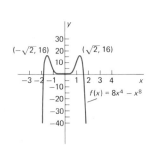

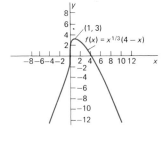

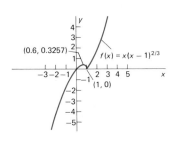

**1** $8x^3 - 9x^2 + 6$, $24x^2 - 18x$, and $48x - 18$   **3** $-8(2x - 1)^{-3}$, $48(2x - 1)^{-4}$, and $-384(2x - 1)^{-5}$

**5** $4(3t - 2)^{1/3}$, $4(3t - 2)^{-2/3}$, and $-8(3t - 2)^{-5/3}$   **7** $(y + 1)^{-2}$, $-2(y + 1)^{-3}$, and $6(y + 1)^{-4}$

**9** $-\frac{1}{4}t^{-3/2} - (1 - t)^{-4/3}$, $\frac{3}{8}t^{-5/2} - \frac{4}{3}(1 - t)^{-7/3}$, and $-\frac{15}{16}t^{-7/2} - \frac{28}{9}(1 - t)^{-10/3}$

**11** $7(2x + 3)^{5/2}$, $35(2x + 3)^{3/2}$, $105(2x + 3)^{1/2}$   **13** $2(t + 1)^{-2}$, $-4(t + 1)^{-3}$, $12(t + 1)^{-4}$

**15** $-2x(1 + x^2)^{-2}$, $8x^2(1 + x^2)^{-3} - 2(1 + x^2)^{-2}$, $24x(1 + x^2)^{-3} - 48x^3(1 + x^2)^{-4}$

**17** $-(2x + y)/(x + 2y)$ and $-18/(x + 2y)^3$

**19** $dy/dx = -(1 + 2x)/(3y^2)$ and $d^2y/dx^2 = -(\frac{2}{9})y^{-5}(3y^3 + 1 + 4x + 4x^2) = (\frac{2}{9})(y^{-2} - 21y^{-5})$

**21** $dy/dx = -(2x + y)/(x + 2y)$, $d^2y/dx^2 = 0$ [You should first obtain $-6(x^2 + xy + y^2)/(x + 2y)^3$].

**23** Local minimum value $-1$ at $x = 2$; no inflection points

**25** Local maximum value 3 at $x = -1$, local minimum value $-1$ at $x = 1$; inflection point $(0, 1)$

**27** No local extrema; inflection point at $(0, 0)$   **29** No local extrema; inflection point at $(0, 0)$

**31** Local maximum value $\frac{1}{16}$ at $x = \frac{1}{2}$, local minimum value $y = 0$ at $x = 0$ and at $x = 1$; abscissas of the inflection points at the roots of $6x^2 - 6x + 1 = 0$—the inflection points are approximately $(0.42, 0.06)$ and $(1.58, 0.83)$.

**33** Local maximum value 0 at $x = 1$, local minimum value (exactly) $-0.03456$ at $x = 1.4$; inflection points at $(2, 0)$, where $x = 1.4 + \sqrt{6}/10$, and where $x = 1.4 - \sqrt{6}/10$—the latter two points are about $(1.155, -0.015)$ and $(1.645, -0.019)$.

**35** Local maximum at $(-1, 10)$, local minimum at $(2, -17)$, inflection point at $(0.5, -3.5)$

**37** Local maximum at $(-2, 22)$ and at $(2, 22)$, local minimum at $(0, 6)$. Inflection points at $(\pm 2\sqrt{3}/3, \frac{134}{9})$

**39** Minima at $(-1, -6)$ and $(2, -32)$, maximum at $(0, -1)$. Inflection points at $(x, f(x))$ where $x = (1 \pm \sqrt{7})/3$ [approximately $(1.22, -19.36)$ and $(-0.55, -3.68)$]

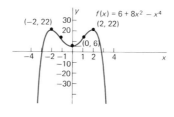

$f(x) = 2x^3 - 3x^2 - 12x + 3$

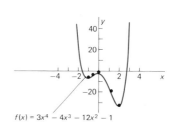

$f(x) = 6 + 8x^2 - x^4$

$f(x) = 3x^4 - 4x^3 - 12x^2 - 1$

**41** Local maximum at $(\frac{3}{7}, 0.008)$ (ordinate approximate), local minimum at $(1, 0)$. Inflection points at $(0, 0)$ and at $(x, f(x))$ where $x = (3 \pm \sqrt{2})/7$. The general shape of the graph is shown in the upper left; the lower right shows the behavior of $f$ on and near $[0, 1]$ with the vertical scale greatly magnified

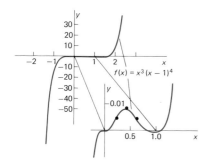

$f(x) = x^3(x - 1)^4$

**43** Inflection point at $(0, 1)$; no extrema

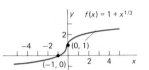

**45** Minimum at $(0, 0)$, inflection point at $(1, 4)$. Concave up for $x > 1$

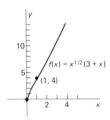

**47** Maximum at $(1, 3)$, inflection points at $(0, 0)$ and $(-2, -7.56)$ (ordinate approximate)

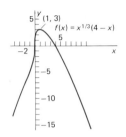

**49** $5(x + 1)^4$, $20(x + 1)^3$, $60(x + 1)^2$, $120(x + 1)$, $120$
**57** $a = 3PV^2 \approx 3583858.8$, $b = V/3 \approx 42.7$, $R = 8PV/(3T) \approx 81.80421$

## SECTION 4-7 (page 197)

**1** 1 **3** 3 **5** 2 **7** 1 **9** 4 **11** 0 **13** 2 **15** $+\infty$ (or does not exist)

**17** No critical points, no inflection points. Vertical asymptote: $x = 3$. Horizontal asymptote: $y = 0$ (the $x$-axis). No $x$-intercept; the $y$-intercept is $(0, -\frac{2}{3})$.

**19** No critical points, no inflection points. Vertical asymptote: $x = -2$. Horizontal asymptote: $y = 0$

**21** No critical points, no inflection points. Vertical asymptote: $x = \frac{3}{2}$. Horizontal asymptote: $y = 0$

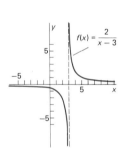

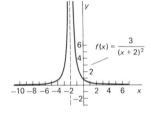

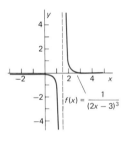

**23** Minimum at $(0, 0)$, inflection points at $(\pm\frac{1}{3}\sqrt{3}, \frac{1}{4})$. Horizontal asymptote: $y = 1$

**25** Local maximum at $(0, -\frac{1}{9})$, no inflection points. Vertical asymptotes. $x = \pm 3$. Horizontal asymptote: $y = 0$

**27** Local maximum: $(-\frac{1}{2}, -\frac{4}{25})$. Horizontal asymptote: $y = 0$. Vertical asymptotes: $x = -3$ and $x = 2$. No inflection points.

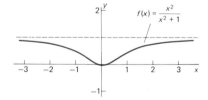

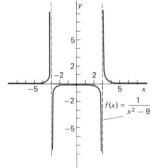

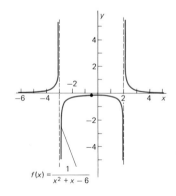

**29** Local minimum at (1, 2), local maximum at $(-1, -2)$. No inflection points. Vertical asymptote: $x = 0$. The line $y = x$ is also an asymptote

**31** Local minimum: (2, 4). Local maximum: (0, 0). Asymptotes: $y = x + 1$, $x = 1$. No inflection points

**33**

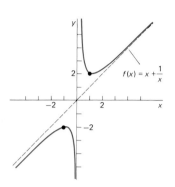

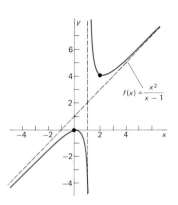

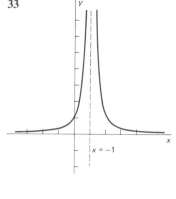

**35**

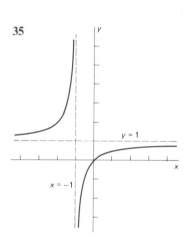

**37**

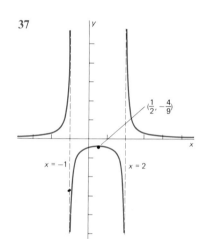

**39**

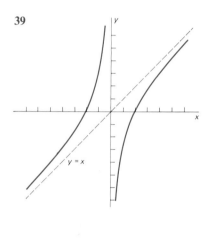

**41**

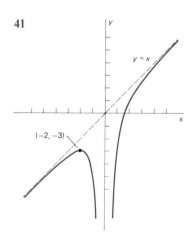

**43** Local minimum: (1, 3). Inflection point at $(\sqrt[3]{-2}, 0)$. Vertical asymptote: $x = 0$

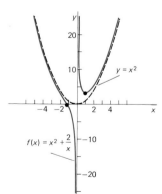

1 $x^3 + x^2 + x + C$  3 $x - \frac{2}{3}x^3 + \frac{3}{4}x^4 + C$  5 $-\frac{3}{2}x^{-2} + \frac{4}{5}x^{5/2} - x + C$  7 $t^{3/2} + 7t + C$  9 $\frac{3}{5}x^{5/3} - 16x^{-1/4} + C$
11 $x^4 - 2x^2 + 6x + C$  13 $7x + C$  15 $\frac{4}{5}x^{5/2} - 2x^{1/2} + C$  17 $-\frac{1}{6}(x - 10)^{-6} + C$  19 $\frac{1}{4}(x^3 + 2)^{4/3} + C$
21 $-(\frac{1}{11})(1 + x^{-1})^{11} + C$  23 $(\frac{1}{88})(x^8 + 9)^{11} + C$  25 $(\frac{4}{9})(1 + x^{3/2})^{3/2} + C$  27 $(\frac{1}{16})(x^4 + x^2)^8 + C$
29 $\frac{1}{7}x^7 + \frac{3}{5}x^5 + x^3 + x + C$  31 $-\frac{1}{3}(1 - x^2)^{3/2} + C$  33 $-3x^{-1} + 6x^{-2} + C$  35 $5x - x^2 + \frac{1}{4}x^4 + C$
37 $\frac{1}{3}x^3 - x^2 + x + C$  39 $(\frac{1}{5})(1 - x^{-1})^5 + C$  41 $\frac{1}{3}x^3 + 2x - x^{-1} + C$  43 $\frac{1}{10}x^{10} - \frac{1}{8}x^8 + C$
45 $-6/x - \frac{5}{3}x^3 + C$  47 $(2/3b)(a + bx)^{3/2} + C$  49 $(\frac{1}{18})(x + 1)^{18} + C$  51 $\frac{1}{12}(2t - 1)^6 + C$  53 $\frac{3}{8}(x^3 + 7)^{8/3} + C$
55 $\sqrt{x^2 + 1} + C$  57 $\frac{3}{5}x^{5/3} + 2x + 3x^{1/3} + C$  59 $(\frac{4}{5})(1 + x^{1/2})^{5/2} + C$  61 $(x^2 + 4x + 5)^{1/2} + C$
63 $-(1/4b)(a + bt^2)^{-2} + C$  65 $2(x^2 + 3x)^{1/2} + C$
67 (a) 250 tons  (b) $50\sqrt{29} \approx 269.258$ tons  (c) $\cos^{-1}(5/\sqrt{29}) \approx 0.38$, or about $21°48'5''$

1 $y = f(x) = x^2 + x + 3$  3 $\frac{2}{3}(x^{3/2} - 8)$  5 $2\sqrt{x + 2} - 5$  7 $y = \frac{3}{4}x^4 - 2/x + \frac{9}{4}$  9 $y = \frac{1}{4}(x - 1)^4 + \frac{7}{4}$
11 $y = 2(x - 13)^{1/2} - 2$  13 $y = -\frac{1}{3}(1 - x^2)^{3/2}$  15 $y(x) = 1/(1 - x)$  17 $y(x) = (x + 1)^{1/4}$  19 $y(x) = (2 + \frac{1}{3}x^{3/2})^2$
21 $x(t) = 25t^2 + 10t + 20$  23 $x(t) = \frac{1}{2}t^3 + 5t$  25 $x(t) = \frac{1}{3}(t + 3)^4 - 37t - 26$  27 $x(t) = \frac{1}{2}(t + 1)^{-1} + \frac{1}{2}t - \frac{1}{2}$
29 144 ft; 6 s  31 144 ft  33 5 s; 112 ft/s  35 $\sqrt{60}$ s (about 7.75 s); $32\sqrt{60}$ ft/s (about 247.87 ft/s)  37 120 ft/s
39 5 s; $-160$ ft/s  41 400 ft; 10 s  43 $(-5 + 2\sqrt{145})/4 \approx 4.77$ s; $16\sqrt{145} \approx 192.6655$ ft/s  45 544/3 ft/s
47 22 ft/s² ($s = -11t^2 + 88t$ is the distance traveled $t$ seconds after the brakes are applied)
49

| | m/s | km/h | mi/h |
|---|---|---|---|
| Moon | 2,376 | 8,554 | 5,315 |
| Mars | 4,983 | 17,940 | 11,147 |
| Jupiter | 58,996 | 212,387 | 131,971 |
| Sun | 614,267 | 2,211,361 | 1,374,076 |
| Ganymede | 2,562 | 9,222 | 5,730 |

51 Approximately 2.78 m  53 (b) 225 animals  (c) $20(\sqrt{2} - 1) \approx 8.284$

| | (a) $C(400)$ | (b) $C(400)/400$ | (c) $C'(400)$ | (d) $C(401) - C(400)$ |
|---|---|---|---|---|
| 1 | $2,200.00 | $5.50 | $3.00 | $3.00 |
| 3 | $24,640.00 | $61.60 | $43.20 | $43.20 |
| 5 | $2,705.00 | $6.76 | $4.99 | $4.99 |

7 Minimum value $53.31 for $x = 1414$
11 417 per week, to be sold at $35.83 each, for maximum profit of $3208.33
15 $C(x) = 50x + (0.001)x^2 + 990$; $C(200)/200 = $55.15
17 (b) 5 years

1 $dy = 3(4x - x^2)^{1/2}(2 - x) dx$  3 $dy = -[2/(x - 1)^2] dx$  5 $dy = 2x(x^2 + 1)^{-2} dx$  7 $10 + \frac{1}{60} \approx 10.0167$  9 132.5
11 $2 + \frac{1}{32} \approx 2.0313$  13 7.5  15 $10\pi \approx 31.416$ in³.  17 $\frac{1}{96}\pi \approx 0.0327$ s  19 $c = \sqrt{3}$  21 $c = 1$  23 $c = (\frac{11}{5})^{1/4}$
25 Minimum at $(3, -5)$, decreasing on $(-\infty, 3)$, increasing on $(3, +\infty)$  27 Increasing everywhere, no extrema  29 Increasing on $(-\infty, \frac{1}{4})$, decreasing on $(\frac{1}{4}, +\infty)$, maximum: $(\frac{1}{4}, 3/4 \sqrt[3]{4})$

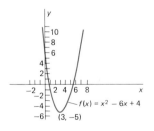

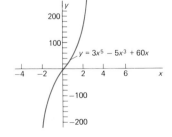

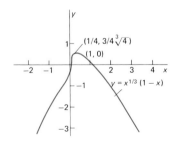

**31** $3x^2 - 2$, $6x$, $6$  **33** $-(t^{-2}) + 2(2t - 1)^{-2}$, $2t^{-3} - 8(2t + 1)^{-3}$, $-6t^{-4} + 48(2t + 1)^{-4}$

**35** $3t^{1/2} - 4t^{1/3}$, $\frac{3}{2}t^{-1/2} - \frac{4}{3}t^{-2/3}$, $-\frac{3}{4}t^{-3/2} + \frac{8}{9}t^{-5/3}$  **37** $-4(t - 2)^{-2}$, $8(t - 2)^{-3}$, $-24(t - 2)^{-4}$

**39** $-\frac{4}{3}(5 - 4x)^{-2/3}$, $-\frac{32}{9}(5 - 4x)^{-5/3}$, $-\frac{640}{27}(5 - 4x)^{-8/3}$  **41** $dy/dx = -(y/x)^{2/3}$, $d^2y/dx^2 = \frac{2}{3}y^{1/3}x^{-5/3}$

**43** $dy/dx = \frac{1}{2}(5y^4 - 4)^{-1}x^{-1/2}$, $d^2y/dx^2 = -[20x^{1/2}y^3 + (5y^4 - 4)^2]/[4x^{3/2}(5y^4 - 4)^3]$

**45** Minimum at $(2, -48)$, x-intercepts 0 and $2\sqrt[3]{4} \approx 3.1748$. Concave up everywhere, no inflection points, no asymptotes.

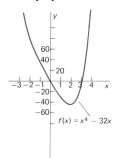

**47** Local maximum at $(0, 0)$, minima at $(\pm\frac{2}{3}\sqrt{3}, -\frac{32}{27})$ (about $(\pm1.15, -1.19)$). Inflection points at $(\pm\frac{2}{5}\sqrt{5}, -\frac{96}{125})$ (approximately $(\pm0.89, -0.77)$). No asymptotes.

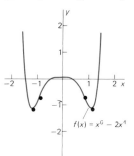

**49** Maximum at $(3, 3)$, inflection points at $(4, 0)$ and $(6, -6\sqrt[3]{2})$ (the ordinate is approximately $-7.56$). No asymptotes; vertical tangent at $(4, 0)$.

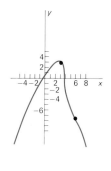

**51** Local maximum at $(0, -\frac{1}{4})$, horizontal asymptote $y = 1$, vertical asymptotes $x = \pm2$. No inflection points.

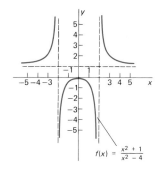

**53** The inflection point has abscissa the only real solution of $x^3 + 6x^2 + 4 = 0$, which is approximately $-6.10724$.

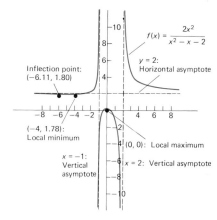

**55**

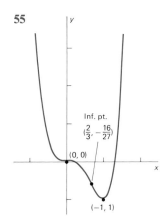

**57**

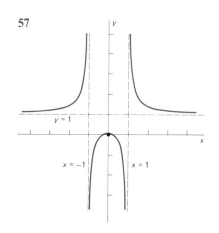

**59**

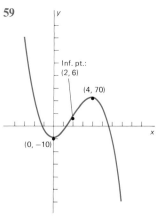

**61**

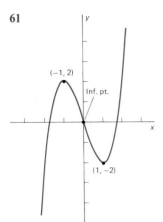

**63**

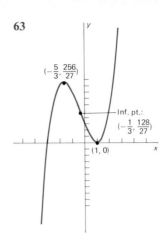

**65** $\frac{1}{3}x^3 + 2x^{-1} - \frac{5}{2}x^{-2} + C$  **67** $\frac{3}{4}(x^2 + 2x)^{2/3} + C$  **69** $\frac{1}{4}x^4 + \frac{1}{3}x^3 + \frac{1}{2}x^2 + x + C$  **71** $-\frac{1}{3}(1 + 2x^{3/2})^{-1} + C$
**73** $y(x) = \frac{1}{12}(2x + 1)^6 + \frac{23}{12}$  **75** $y(x) = \frac{3}{2}x^{2/3} - \frac{1}{2}$  **77** $y(x) = \frac{1}{16}(x + 1)^4$  **79** $y = 2/(6 - x^2)$  **81** $y = x$
**83** Maximum value 1, at $x = -1$
**85** The distance from the foot of the ladder to the base of the wall turns out to be $(1600)^{1/3} \approx 11.696$ ft; the length of
the shortest ladder is $4\{1 + (25)^{1/3}\}^{3/2} \approx 31.093$ ft.
**87** 6 s; 396 ft  **89** 120 ft/s  **91** Impact time: $t = 2\sqrt{10} \approx 6.32$ s; impact speed: $20\sqrt{10} \approx 63.25$ ft/s
**93** 176 ft  **95** $100(20/9)^{2/5} \approx 54.79$ mi/h
**97** Horizontal tangents when $x = 1 - \frac{1}{3}\sqrt{3} \approx 0.42$, $y = \pm\frac{1}{3}(12)^{1/4} \approx \pm0.62$. Intercepts: $(0, 0)$, $(1, 0)$, and $(2, 0)$. Vertical
tangent at each intercept. Inflection points corresponding to the (only) positive solution of $3x^2(x - 2)^2 = 4$—that is,
$x \approx 2.46789$, at which $y \approx \pm1.30191$. No asymptotes.

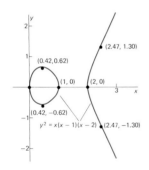

# CHAPTER 5

### SECTION 5-2 (page 234)

**1** 190  **3** 1165  **5** 224  **7** 1834  **9** 333,833,500  **11** $\frac{1}{2}$  **13** $\frac{1}{5}$  **15** $\frac{1}{3}n(2n - 1)(2n + 1)$  **17** 24  **19** $\frac{1}{3}$  **21** 132
**23** 90  **25** $\frac{1}{3}b^3$

### SECTION 5-3 (page 242)

**1** 0.44  **3** 1.45  **5** 19.5  **7** 58.8  **9** 0.24  **11** $\frac{137}{60} \approx 2.28333$  **13** 16.5  **15** 26.4  **17** 0.33  **19** $\frac{6086}{3465} \approx 1.75462$
**21** 18  **23** 40.575  **25** 1.83753  **27** 2  **29** 12  **31** 30

### SECTION 5-4 (page 249)

**1** 0  **3** 6  **5** $-\frac{1}{20}$  **7** $\frac{1}{4}$  **9** $\frac{16}{3}$  **11** 24  **13** 0  **15** $\frac{609}{4}$  **17** 0  **19** $\frac{93}{5}$  **21** $\frac{17}{6}$  **23** $\frac{28}{3}$  **25** $\frac{52}{5}$
**27** $(13\sqrt{13} - 8)/3 \approx 12.9574$  **29** 34  **31** 0  **33** $\frac{1}{2}$  **35** $\frac{2}{3}$  **37** $25\pi/4 \approx 19.635$  **41** $1000 + \int_0^{30} V'(t)\, dt = 160$ (gal.)

**43** Let $Q = (1/1.2 + 1/1.4 + 1/1.6 + 1/1.8)(0.2) \approx 0.5456$, and $I = \int_1^2 (1/x)\, dx$. Then $Q + 0.1 \leq I \leq Q + 0.2$. Hence $0.64 < I < 0.75$.

## SECTION 5-5 (page 256)

**1** 2 **3** $\frac{26}{3}$ **5** 0 **** $\frac{125}{4}$ **9** $\frac{14}{9}$ **11** $\frac{1}{2}$ **13** 4 **15** $\frac{1}{3}$ **17** $-\frac{22}{81} \approx -0.271605$ **19** $\frac{1}{3}$ **21** $\frac{35}{24} \approx 1.4583$ **23** $-\frac{3}{8}$
**25** 4 **27** $\frac{38}{3}$ **29** $\frac{61}{3}$ **31** $f'(x) = (x^2 + 1)^{17}$ **33** $h'(z) = (z - 1)^{1/3}$ **35** $f'(x) = -x - (1/x)$ **37** $G'(x) = (x + 4)^{1/2}$
**39** $G'(x) = (x^3 + 1)^{1/2}$ **41** $f'(x) = -1/x$ **43** $f'(x) = x^{-1/2}(1 + x)^{-1}/2$ **45** $f'(x) = 2x/(x^2 + 1)$
**47** The integral does not exist (examine the graph of $y = 1/x^2$).
**51** Average height: $800/3 \approx 266.67$ ft; average velocity: $-80$ ft/s **53** $5000/3 \approx 1666.67$ (gal.) **55** $a^2/6$

## SECTION 5-6 (page 261)

**1** $\frac{1}{6}(1 + x^4)^{3/2} + C$ **3** $(-2)(1 + x^{1/2})^{-1} + C$ **5** $\frac{1}{6}(1 + 4x^3)^{1/2} + C$ **7** $\frac{1}{30}(x^2 + 1)^{15} + C$
**9** $\frac{2}{5}(4 - x)^{5/2} - \frac{8}{3}(4 - x)^{3/2} + C$ **11** $(1 + x^4)^{1/2} + C$ **13** $\frac{1}{24}(4x - 3)^6 + C$ **15** $-\frac{1}{9}(2 - 3x^2)^{3/2} + C$
**17** $-\frac{1}{9}(x^3 + 5)^{-3} + C$ **19** $-\frac{1}{16}(2 - 4x^3)^{4/3} + C$ **21** $(-\frac{1}{3})(t^3 + 3t)^{-1} + C$ **23** $-\frac{3}{8}$ **25** 2 **27** $\frac{1192}{15} \approx 79.46667$
**29** $\frac{38}{3}$ **31** $a^3/3$ **33** $\frac{3}{4}(x^4 - 4x)^{1/3} + C$ **35** $\frac{4}{15}(6 - t^3)^{5/4} + C$

## SECTION 5-7 (page 268)

**1** $\frac{2}{3}$ **3** $\frac{16}{3}$ **5** $+\frac{1}{4}$ **7** 9 **9** $9 + \frac{16}{3}$ **11** $\frac{1}{4}$ **13** $\frac{1}{20}$ **15** $\frac{4}{9}$ **17** $\frac{32}{3}$ **19** $\frac{128}{3}\sqrt{2}$ **21** $\frac{500}{3}$ **23** $\frac{64}{3}$ **25** $\frac{500}{3}$ **27** $\frac{4}{3}$ **29** $\frac{1}{12}$
**31** $\frac{16}{3}$ **33** $\frac{16}{3}$ **35** $\frac{27}{2}$ **37** $\frac{320}{3}$ **39** $\frac{8}{15}$ **41** $\frac{37}{12}$ **45** 1

## SECTION 5-8 (page 280)

**1** $L_5 = 2.04$, $R_5 = 2.64$ **3** $L_4 \approx 1.07$, $R_4 \approx 1.17$
**5** $L_8 \approx 0.457419672$, which rounds to 0.46; $R_8 \approx 0.446696367$, which rounds to 0.45.
**7** 0.45194753 (which rounds to 0.45)
**9** (a) $T_8 \approx 1.0892$ (b) $S_8 \approx 1.0901$ **11** (a) 1.2846 (b) 1.2854 **13** (a) 3.0200 (b) 3.0717
**15** Both answers should be near 2435. **17** $L_{10} \approx 0.71877$, $R_{10} \approx 0.66877$ **19** $M_{10} \approx 0.692835$, $T_{10} \approx 0.693771$
**21** $n \geq 14$

## CHAPTER 5 MISCELLANEOUS PROBLEMS (page 283)

**1** 1700 **3** 2845 **5** $2(\sqrt{2} - 1) \approx 0.82843$ **7** $(2\pi/3)(2\sqrt{2} - 1) \approx 3.82945$ **13** $2x\sqrt{2x}/3 + 2/\sqrt{3x} + C$
**15** $-2/x - x^2/4 + C$ **17** $\frac{1}{3}(x^2 + 2x + 5)^{3/2} + C$ **19** $\frac{1}{4}(1 + x^{1/2})^8 + C$ **21** $-\frac{3}{8}(1 + u^{4/3})^{-2} + C$ **23** $\frac{38}{3}$
**25** $(4x^2 - 1)^{3/2}/3x^3 + C$ (use $u = 1/x$) **27** $\frac{1}{30}$ **29** $\frac{44}{15}$ **31** $\frac{125}{6}$ **33** Semicircle, center (1, 0), radius 1: area $\pi/2$
**35** $f(x) = \sqrt{4x^2 - 1}$ **37** $n \geq 9$. $L_{10} \approx 1.12767$, $R_{10} \approx 1.16909$. An estimate: their average is $A = 1.1483 \pm 0.05$
**39** $M_5 \approx 0.28667$, $T_5 \approx 0.28971$. They bound the true value of the integral because the second derivative of the integrand is positive for all $x > 0$.

# CHAPTER 6

## SECTION 6-1 (page 291)

**1** $-320$; 320 **3** $-50$; 106.25 **5** 65; 97 **7** 3; 3 **9** $\frac{64}{5}$; $\frac{64}{5}$ **11** 1 **13** $\frac{16}{3}$ **15** $\frac{98}{3}$ **17** 4 **19** $-1$
**21** $\int_{-1}^{1} [f(x)]^2\, dx$ **23** $\int_{-2}^{3} 2\pi x\sqrt{1 + [f(x)]^2}\, dx$ **25** 550 gal. **27** 385,000 **31** $\frac{3}{4}$

## SECTION 6-2 (page 298)

**1** $\pi/5$ **3** $8\pi$ **5** $8\pi/105$ **7** $3\pi/10$ **9** $512\pi/3$ **11** $16\pi/15$ **13** $\pi/2$ **15** $8\pi$ **17** $121\pi/210$ **19** $8\pi$ **21** $9\pi$ **23** $4\pi a^2 b/3$
**25** $16a^3/3$ **27** $\frac{4}{3}a^3\sqrt{3}$ **33** $16a^3/3$

## SECTION 6-3 (page 304)

**1** $8\pi$ **3** $625\pi/2$ **5** $16\pi$ **7** $\pi$ **9** $6\pi/5$ **11** $256\pi/15$ **13** $4\pi/15$ **15** $11\pi/15$ **17** $56\pi/5$ **19** $8\pi/3$ **21** $2\pi/15$ **23** $\pi/2$
**25** $16\pi/3$ **27** $64\pi$ **31** $4\pi a^2 b/3$ **33** $V = 2\pi^2 a^2 b$ **35** $V = 2\pi^2 a^3$ **37** (a) $V = \pi h^3/6$

## SECTION 6-4 (page 313)

In 1–19, the integrand is given, followed by the interval of integration.

**1** $\sqrt{1 + 4x^2}$; $[0, 1]$   **3** $\sqrt{1 + 36(x^4 - 2x^3 + x^2)}$; $[0, 2]$   **5** $\sqrt{1 + 4x^2}$; $[0, 100]$   **7** $\sqrt{1 + 16y^6}$; $[-1, 2]$
**9** $(1/x^2)\sqrt{x^4 + 1}$; $[1, 2]$   **11** $2\pi x^2\sqrt{1 + 4x^2}$; $[0, 4]$   **13** $2\pi(x - x^2)\sqrt{2 - 4x + 4x^2}$; $[0, 1]$
**15** $2\pi(2 - x)\sqrt{1 + 4x^2}$; $[0, 1]$   **17** $\pi\sqrt{4x + 1}$; $[1, 4]$   **19** $\pi(x + 1)\sqrt{4 + 9x}$; $[1, 4]$   **21** $\frac{22}{3}$   **23** $\frac{14}{3}$
**25** $\frac{123}{32} - 3.84375$   **27** $\frac{1}{27}(104\sqrt{13} - 125) \approx 9.25842$   **29** $\frac{1}{6}\pi(5\sqrt{5} - 1) \approx 5.3304$   **31** $339\pi/16 \approx 66.5625$
**33** $\frac{1}{9}\pi(82\sqrt{82} - 1) \approx 258.8468$   **35** $4\pi$   **37** $732.3929$
**41** Avoid the problem when $x = 0$ as follows: $L = 8\int_{1/2\sqrt{2}}^{1} x^{-1/3}\,dx = 6$.

## SECTION 6-5 (page 321)

**1** 30   **3** 9   **5** $\frac{6}{5}$   **7** 15 ft-lb   **9** $2.816 \times 10^9$ ft-lb (with $R = 4000$ mi, $g = 32$ ft/s²)   **11** $13{,}000\pi \approx 40{,}840.7$ ft-lb
**13** $125{,}000\pi/3 \approx 130{,}899.69$ ft-lb   **15** $156{,}000\pi \approx 490{,}088.454$ ft-lb   **17** $4{,}160{,}000\pi \approx 13{,}069{,}025.44$ ft-lb   **19** 8750 ft-lb
**21** 11250 ft-lb   **23** $25{,}000(1 - [0.1]^{0.4})$ in.-lb, or approximately 1253.9434 ft-lb   **25** $16\pi$ ft-lb   **27** $1{,}382{,}400\pi$ ft-lb
**29** Approximately 690.53 ft-lb

## SECTION 6-6 (page 330)

**1** $(2, 3)$   **3** $(1, 1)$   **5** $(\frac{4}{3}, \frac{2}{3})$   **7** $(\frac{3}{2}, \frac{6}{5})$   **9** $(0, \frac{8}{5})$   **11** $(\bar{x}, \bar{y}) = (0, \frac{8}{5})$   **13** $(\bar{x}, \bar{y}) = (\frac{3}{4}, \frac{9}{10})$   **15** $(\bar{x}, \bar{y}) = (-\frac{1}{2}, 2)$
**17** $(\bar{x}, \bar{y}) = (\frac{3}{5}, \frac{12}{35})$   **19** $(\bar{x}, \bar{y}) = (4r/3\pi, 4r/3\pi)$   **21** $(\bar{x}, \bar{y}) = (2r/\pi, 2r/\pi)$
**29** $\bar{y} = (4a^2 + 3\pi ab + 6b^2)/(12b + 3\pi a)$, $\bar{x} = 0$, $V = \frac{1}{3}\pi a(4a^2 + 3\pi ab + 6b^2)$

## SECTION 6-7 (page 335)

**1** $(\frac{4}{3}, 0, 0)$   **3** $(1, 0, 0)$   **5** $(\frac{15}{8}, 0, 0)$   **7** $(\frac{48}{11}, 0, 0)$   **11** $(0, 3b/8, 0)$   **15** 249.6 lb   **17** 748.8 lb   **19** 19,500 lb
**21** $700\rho/3 \approx 14{,}560$ lb   **23** About 32,574 tons
**25** (a) $(62.4)(12.5)(25) = 19{,}500$ lb   (b) $(62.4)(13)(9\pi) \approx 22{,}936.14$ lb   (c) $(62.4)(10 + 16/3\pi)(8\pi) \approx 18{,}345.231$ lb

## SECTION 6-8 (page 339)

**1** $\frac{1}{3}$   **3** $\frac{8}{3}$   **5** $\frac{21}{4}$   **7** $\frac{52}{9}$

## CHAPTER 6 MISCELLANEOUS PROBLEMS (page 340)

**1** $-\frac{3}{2}; \frac{31}{6}$   **3** $-1; 1$   **5** $\frac{124}{3}$   **7** 1   **9** 12 in.   **11** $41\pi/105$   **13** $10.625\pi \approx 33.379$ gm   **19** $f(x) = \sqrt{1 + 3x}$
**21** (a) $40\pi/3$;   (b) $104\pi/9$   **23** $\frac{10}{3}$   **25** $\frac{63}{8}$   **27** $52\pi/5$   **31** $\bar{x} = \frac{393}{352}$, $\bar{y} = \frac{268}{165}$   **33** $\bar{x} = \frac{111}{112}$, $\bar{y} = \frac{1136}{245}$
**35** $\bar{x} = \frac{16}{21}$, $\bar{y} = 0$   **37** $\bar{y} = 4b/3\pi$   **43** 1 ft   **45** $W = 4\pi R^4\rho$   **47** 10,454,400 ft-lb   **49** 36,400 tons
**51** Approximately 42,031 tons

# CHAPTER 7

## SECTION 7-1 (page 351)

**1** 128   **3** 64   **5** 1   **7** 16   **9** 16   **11** 4   **13** 3   **15** 3   **17** $3\ln 2$   **19** $\ln 2 + \ln 3$   **21** $3\ln 2 + 2\ln 3$
**23** $3\ln 2 - 3\ln 3$   **25** $3\ln 3 - 3\ln 2 - \ln 5$   **27** $2^{34} = 2^{81} \gg 2^{12} = (2^3)^4$
**29** 2 and 4 (There is a third value, approximately $-0.7666647$.)   **31** 6   **33** $-2$   **35** 1, 2   **37** 81   **39** 0   **41** $(x + 1)e^x$
**43** $(x^{1/2} + \frac{1}{2}x^{-1/2})e^x$   **45** $\dfrac{x - 2}{x^3}e^x$   **47** $1 + \ln x$   **49** $x^{-1/2}(1 + \frac{1}{2}\ln x)$   **51** $(1 - x)e^{-x}$   **53** $3/x$   **57** $\frac{1}{10}e^{x/10}$
**61** $P'(t) = 3^t\ln 3$   **63** $P'(t) = -2^{-t}\ln 2$

## SECTION 7-2 (page 359)

**1** $3/(3x - 1)$   **3** $1/(1 + 2x)$   **5** $(3x^2 - 1)/[3(x^3 - x)]$   **7** $\ln x$   **9** $-1/[x(\ln x)^2]$   **11** $6/(2x + 1) + 8x/(x^2 - 4)$
**13** $-x/(4 - x^2) - x/(9 + x^2)$   **15** $-15/4(1 - 5x) - 20x^3/3(1 + x^4)$   **17** $3/2(3x - 1)$   **19** $2(1 + \ln t)$
**21** $(1/x) + x/(x^2 + 1)$   **23** $(2x/3)(x^2 + 1)^{-1} + (3x^2/2)(x^3 - 1)^{-1}$   **25** $3t^{-1} - 2t(t^2 + 1)^{-1}$
**27** $(\frac{1}{2})t^{-1/2}(1 + \ln t)^2 + 2t^{-1/2}(1 + \ln t)$   **29** $\frac{1}{2}t^{-1} - \frac{1}{2}(t + 1)^{-1}$   **31** $(2/t) - 2t/(t^2 + 1)$   **33** $-(1 - x)^{-1} - x^{-1}$
**35** $\frac{1}{6}\ln(1 + 3x^2) + C$   **37** $\frac{1}{4}\ln|2x^2 + 4x + 1| + C$   **39** $\frac{1}{3}(\ln x)^3 + C$   **41** $\ln|x + 1| + C$   **43** $\ln(x^2 + x + 1) + C$
**45** $\frac{1}{2}(\ln x)^2 + C$   **47** $\frac{1}{2}\ln(x^4 + x^2) + C$   **49** $\frac{1}{3}\ln|x^3 - 3x^2 + 1| + C$   **51** 0   **53** 0   **55** 0
**59** $m \approx -0.2479$, $k \approx 291.7616$

**65** $y \to 0$ as $x \to 0^+$; also $y' \to 0$ as $x \to 0^+$. The point $(0, 0)$ is not on the graph. Intercept at $(1, 0)$, minimum at $(1/\sqrt{e}, -1/2e) \approx (0.61, -0.18)$, inflection point at $(1/e\sqrt{e}, -3/2e^3) \approx (0.22, -0.07)$. The figure is *not drawn to scale*.

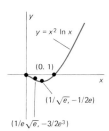

**67** $y \to -\infty$ as $x \to 0^+$; $y \to 0$ as $x \to +\infty$. Maximum at $(e^2, 2/e)$, inflection point at $(e^{8/3}, 8/3e^{4/3})$. The $x$-axis is a horizontal asymptote and the $y$-axis is a vertical asymptote. The only intercept is $(1, 0)$. The figure is *not drawn to scale*.

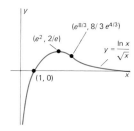

**71** Midpoint estimate: approximately 872.47. Trapezoidal estimate: approximately 872.60. The true value of the integral is approximately 872.5174.

**SECTION 7-3 (page 367)**

**1** $2e^{2x}$ **3** $2xe^{(x^2)} = 2xe^{x^2} = 2x \exp(x^2)$ **5** $-2x^{-3}\exp(x^{-2})$ **7** $(1 + \sqrt{t}/2)\exp(\sqrt{t})$ **9** $(1 + 2t - t^2)e^{-t}$
**11** $t^{-1}e^{1 + \ln t}$ **13** $e^{-x}/(1 - e^{-x}) = 1/(e^x - 1)$ **15** $(1 - e^{-x})/(x + e^{-x})$ **17** $-2e^{-2x}\ln\sqrt{x} + (1/2x)e^{-2x}$
**19** $15(e^t - \ln t)^4(e^t - t^{-1})$ **21** $-(5 + 12x)e^{-4x}$ **23** $(e^{-t} + te^{-t} - 1)/t^2$ **25** $11x^{10}$ **27** $(x - 2)/e^x$ **29** $e^x e^{e^x}$ **31** $1$
**33** $-[(x + 1)\ln(x + 1) + 1]/\{(x + 1)e^x[\ln(x + 1)]^2\}$ **37** $-\frac{1}{2}e^{1 - 2x} + C$ **39** $\frac{1}{9}\exp(3x^3 + 1) + C$
**41** $\frac{1}{2}\ln(1 + e^{2x}) + C$ **43** $e^{1 + \ln x} + C = ex + C$ **45** $\frac{1}{2}\ln(x^2 + e^{2x}) + C$ **47** $-\exp(-t^2/2) + C$ **49** $2\exp(\sqrt{x}) + C$
**51** $\ln(1 + e^x) + C$ **53** $-\frac{2}{3}\exp(-x^{3/2}) + C$ **55** $e^2$ **57** $e$ **59** $+\infty$ **61** $+\infty$
**63** Minimum and intercept at $(0, 0)$, local maximum at $(2, 4/e^2)$. Inflection points where $x = 2 \pm \sqrt{2}$. The $x$-axis is an asymptote. The figure is *not drawn to scale*.

**65** Maximum at $(0, 1)$, the only intercept. The $x$-axis is the only asymptote. Inflection points where $x = \pm\sqrt{2}/2$.

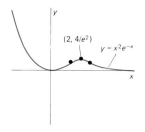

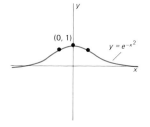

**67** $\frac{1}{2}\pi(e^2 - 1) \approx 10.0359$ **69** $(e^2 - 1)/(2e) \approx 1.1752$
**71** The solution is approximately 1.278464543. Note that if $f(x) = e^x - x + 1$, then $f'(x) < 0$ for all $x$.
**73** $f'(x) = 0$ only for $x = 0$ and $x = n$; $f$ is increasing on $(0, n)$ and decreasing on $(n, +\infty)$. Thus $x = n$ yields the absolute maximum value of $f(x)$ for $x \geq 0$. The $x$-axis is a horizontal asymptote, and there are inflection points with abscissas $x = n \pm \sqrt{n}$. The graph is *not drawn to scale*.

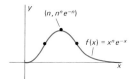

**77** $y = e^{-2x} + 4e^x$

**SECTION 7-4 (page 373)**

**1** $10^x \ln 10$ **3** $3^x 4^{-x} \ln 3 - 3^x 4^{-x} \ln 4 = (\frac{3}{4})^x \ln(\frac{3}{4})$ **5** $7^{\exp(x^2)}(2x \exp(x^2))(\ln 7)$ **7** $2^{x\sqrt{x}}(\ln 2)(\frac{3}{2})\sqrt{x}$ **9** $(1/x) 2^{\ln x} \ln 2$
**11** $17^x \ln 17$ **13** $-x^{-2} 10^{1/x} \ln 10$ **15** $(2^{2x} \ln 2)(2^x \ln 2)$ **17** $(1/\ln 3)[x/(x^2 + 4)]$ **19** $(\ln 2)/\ln 3 = \log_3 2$

**21** $1/[x(\ln 2)(\ln x)]$    **23** $\exp(\log_{10} x)/(x \ln 10)$    **25** $3^{2x}/(2 \ln 3)$    **27** $2(2^{\sqrt{x}}/\ln 2) + C$    **29** $(7^{x^3+1})/(3 \ln 7) + C$
**31** $(\ln x)^2/(2 \ln 2) + C$    **33** $dy/dx = \{x/(x^2 - 4) + 1/[2(2x + 1)]\}\sqrt{(x^2 - 4)\sqrt{2x + 1}}$    **35** $dy/dx = 2^x \ln 2$
**37** $dy/dx = (x^{\ln x})(2 \ln x)/x$    **39** $\frac{1}{3}[1/(x + 1) + 1/(x + 2) - 2x/(x^2 + 1) - 2x/(x^2 + 2)] \cdot y$
**41** $dy/dx = (\ln x)^{\sqrt{x}}[\frac{1}{2}x^{-1/2}\ln(\ln x) + (x^{1/2}\ln x)^{-1}]$    **43** $[3x/(1 + x^2) - 4x^2(1 + x^3)](1 + x^2)^{3/2}(1 + x^3)^{-4/3}$
**45** $[2x^3/(x^2 + 1) + 2x \ln(x^2 + 1)](x^2 + 1)^{(x^2)}$    **47** $\frac{1}{4}x^{-1/2}(2 + \ln x)(\sqrt{x})^{\sqrt{x}}$    **49** $e^x$    **51** $x^{\exp(x)}e^x(x^{-1} + \ln x)$
**57** Note that $\ln(x^x/e^x) = x \ln(x/e)$.

## SECTION 7-5 (page 381)

**1** $119.35; $396.24    **3** Approximately 3.8685 h    **5** The sample is about 686 years old.
**7** (a) 9.308%   (b) 9.381%   (c) 9.409%   (d) 9.416%   (e) 9.417%    **9** $44.52    **11** After an additional 32.26 days
**13** Approximately 35 years    **15** About $4.2521 \times 10^9$ years old    **17** 2.40942 min
**19** (a) 20.486 in.; 9.604 in.   (b) 3.4524 mi, or about 18,230 ft

## SECTION 7-6 (page 389)

**1** $y(x) = -1 + 2e^x$    **3** $y(x) = \frac{1}{2}(e^{2x} + 3)$    **5** $x(t) = 1 - e^{2t}$    **7** $x(t) = 27e^{5t} - 2$    **9** $v(t) = 10(1 - e^{-10t})$
**11** 4,870,238    **15** About 46 days after the rumor starts    **19** $400/(\ln 2) \approx 577$ ft    **23** $1,308,283
**25** The maximum height is $1600(1 - 2 \ln 1.5) \approx 302.5$ ft, attained after $10 \ln 1.5 \approx 4.055$ s aloft. The total time aloft is approximately 8.7422 s, and the impact speed is about 119.75 ft/s.
**27** After 5 min the boat has coasted $400 \ln 31 \approx 1373.6$ ft, and is then moving at $\frac{40}{31} \approx 1.29$ ft/s.

## SECTION 7-7 (page 395)

**1** $20 \ln 3 \approx 22$ years    **3** $50 \ln 1.8 \approx 29.4$ years    **5** (a) $43196.96   (b) $7634.94    **7** At $21672.83 per year
**9** (a) $1,334,109   (b) $1,265,014

## SECTION 7-8 (page 400)

**1** $x(t) = t^2 - t + 3$    **3** $x(t) = [5e^t - 1]^{1/2}$    **5** $x(t) = (18t - 3)/(6t + 1)$    **7** $x(t) = 1/(3 - 2\sqrt{t})$    **9** (b) 256
**11** $x(t) = (3 + e^{2t})/(3 - e^{2t})$    **13** $x(t) = 2(e^t - 1)/(2 - e^t)$    **15** After another 9.24 days

## CHAPTER 7 MISCELLANEOUS PROBLEMS (page 402)

**1** $1/2x$    **3** $(1 - e^x)/(x - e^x)$    **5** $\ln 2$    **7** $(2 + 3x^2)e^{-1/x^2}$    **9** $(1 + \ln(\ln x))/x$    **11** $(1/x) 2^{\ln x} \ln 2$
**13** $-(2/(x - 1)^2) \exp([x + 1]/[x - 1])$    **15** $\frac{3}{2}(1/(x - 1) + 8x/(3 - 4x^2))$    **17** $\dfrac{\exp\sqrt{1 + \ln x}}{2x\sqrt{1 + \ln x}}$    **19** $\dfrac{1}{2x} + \ln 3$
**21** $x^{-2}x^{1/x}(1 - \ln x)$    **23** $[(1 + \ln(\ln x))/x](\ln x)^{\ln x}$    **25** $-\frac{1}{2}\ln|1 - 2x| + C$    **27** $\frac{1}{2}\ln|1 + 6x - x^2| + C$
**29** $\ln(2 + e^x) + C$    **31** $(2/\ln 10)10^{\sqrt{x}} + C$    **33** $\frac{2}{3}(1 + e^x)^{3/2} + C$    **35** $6^x/\ln 6 + C$    **37** $x(t) = t^2 + 17$
**39** $x(t) = 1 + e^t$    **41** $x(t) = \frac{1}{3}(2 + 7e^{3t})$    **43** $x(t) = \exp(1 + e^t)$

**45** Horizontal asymptote: The $x$-axis. Maximum at $(\frac{1}{2}, 1/\sqrt{2e})$, inflection point above $x = (1 + \sqrt{2})/2$—approximately $(1.21, 0.33)$. Minimum and intercept at $(0, 0)$, with a vertical tangent there as well. The graph is *not drawn to scale.*

**47** Minimum at $(4, 2 - \ln 4)$, inflection point at $(16, 1.23)$ (ordinate approximate). The $y$-axis is a vertical asymptote. The graph continues to rise for large increasing $x$; there is no horizontal asymptote. The graph is *not drawn to scale.*

**49** Inflection point at $(\frac{1}{2}, 1/e^2)$. The horizontal line $y = 1$ and the $y$-axis are asymptotes. The point $(0, 0)$ is *not* on the graph. As $x \to 0^+$, $y \to 0$; as $x \to 0^-$, $y \to +\infty$. As $|x| \to \infty$, $y \to 1$. The graph is *not drawn to scale.*

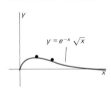

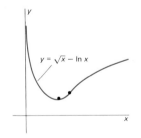

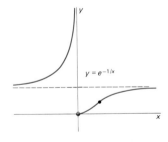

**51** Sell immediately!
**53** (b) The minimizing value is approximately 10.516. But because the batch size must be an integer, it turns out that 11 (rather than 10) minimizes $f(x)$. Thus the answer to part (c) is $977.85.

**57** 20 weeks   **59** (a) $925.20   (b) $1262.88
**61** About 22.567 h after the power failure; that is, at about 9:34 P.M. on the following evening.
**63** (a) 2000 gm   (b) $\frac{5}{4}\ln(1.56/0.76) \approx 0.8989$ s   **65** (b) $k = 1/600$; 240 s   (c) $120\ln 3 \approx 131.8$ s

# CHAPTER 8

## SECTION 8-2 (page 414)

**1** (a) $\frac{2}{9}\pi$   (b) $-\frac{3}{2}\pi$   (c) $\frac{7}{4}\pi$   (d) $\frac{7}{6}\pi$   (e) $-\frac{5}{6}\pi$   **3**

|  | (a), (d) | (b) | (c) |
|---|---|---|---|
| $x$ | $-\pi/3,\ 5\pi/3$ | $3\pi/4$ | $7\pi/6$ |
| $\sin x$ | $-\sqrt{3}/2$ | $\sqrt{2}/2$ | $-1/2$ |
| $\cos x$ | $\frac{1}{2}$ | $-\sqrt{2}/2$ | $-\sqrt{3}/2$ |
| $\tan x$ | $-\sqrt{3}$ | $-1$ | $\sqrt{3}/3$ |
| $\sec x$ | $2$ | $-\sqrt{2}$ | $-\frac{2}{3}\sqrt{3}$ |
| $\csc x$ | $-\frac{2}{3}\sqrt{3}$ | $\sqrt{2}$ | $-2$ |
| $\cot x$ | $-\frac{1}{3}\sqrt{3}$ | $-1$ | $\sqrt{3}$ |

**5** (a) The odd multiples of $\pi/2$: $x = (2k + 1)\pi/2$ where $k$ is an integer   (b) The even multiples of $\pi$: $x = 2k\pi$
(c) The odd multiples of $\pi$: $x = (2k + 1)\pi$   **7** $\sin x = -\frac{3}{5}$, $\cos x = -\frac{4}{5}$, etc.
**11** (a) $\frac{1}{2}$   (b) $-\sqrt{3}/2$   (c) $-\frac{1}{2}$   (d) $-\sqrt{3}/2$   **15** $x = \pi/3, 2\pi/3$   **17** $x = \pi/6, \pi/2, 5\pi/6$   **19** $x = \pi/8, 3\pi/8, 5\pi/8, 7\pi/8$
**21** $\theta^2/\sin\theta = \theta(\theta/\sin\theta) \to (0)(1) = 0$ as $\theta \to 0$   **23** Use the identity $(1 - \cos 2x)/2 = \sin^2 x$.
**25** $2x/[(\sin x) - x] = 2/[(1/x)(\sin x) - 1] \to -\infty$ as $x \to 0$ (because $(\sin x)/x \le 1$ if $x \ne 0$).   **27** 5   **29** 0   **31** $\frac{1}{3}$   **33** 0
**35** 1   **37** $\frac{1}{2}$   **39** 1   **41** $\frac{1}{3}$   **43** $\frac{1}{4}$   **45** $\frac{1}{4}$   **47** 0

## SECTION 8-3 (page 423)

**1** $6\sin x \cos x$   **3** $\cos x - x\sin x$   **5** $x^{-1}\cos x - x^{-2}\sin x$   **7** $\cos^3 x - 2\cos x \sin^2 x$   **9** $4(\cos t)(1 + \sin t)^3$
**11** $-(\sin t + \cos t)^{-2}(\cos t - \sin t)$   **13** $(3x^2 + 2)\sin x - 4x\cos x$   **15** $3\cos 2x \cos 3x - 2\sin 2x \sin 3x$
**17** $3t^2\sin^2 2t + 4t^3\sin 2t \cos 2t$   **19** $(\frac{5}{2})(\cos 3t + \cos 5t)^{3/2}(-3\sin 3t - 5\sin 5t)$   **21** $x^{-1/2}\sin(x^{1/2})\cos(x^{1/2})$
**23** $2x\cos(3x^2 - 1) - 6x^3\sin(3x^2 - 1)$   **25** $2\cos 2x \cos 3x - 3\sin 2x \sin 3x$   **27** $-\dfrac{3\sin 5x \sin 3x + 5\cos 5x \cos 3x}{\sin^2 5x}$
**29** $4x\sin(x^2)\cos(x^2)$   **31** $x^{-1/2}\cos 2x^{1/2}$   **33** $\sin(x^2) + 2x^2\cos(x^2)$   **35** $\dfrac{1}{2\sqrt{x}}\sin\sqrt{x} + \frac{1}{2}\cos\sqrt{x}$
**37** $(\frac{1}{2})x^{-1/2}(x - \cos x)^3 + 3x^{1/2}(x - \cos x)^2(1 + \sin x)$   **39** $-\{\sin(\sin x^2)\}(2x\cos x^2)$   **41** $f'(x) = -(1/x)\sin(\ln x)$
**43** $f'(x) = -(\sin x)/(\cos x)$   **45** $f'(t) = 2t\ln(\cos t) - (t^2\sin t)/(\cos t)$   **47** $2e^x\cos(2e^x)$   **49** $-e^{\cos t}\sin t$   **51** $\frac{1}{3}\sin 3x + C$
**53** $2/\pi$   **55** $-2\cos(t/2) + C$   **57** $\frac{15}{128}$   **59** $\frac{2}{5}(4\sqrt{2} - 1)$   **61** $\cos(1/t) + C$   **63** $\frac{1}{2}\ln(1 - \cos 2x) + C$
**65** $\frac{1}{2}e^{1 - \cos 2x} + C$   **69** $\pi/4$   **71** Approximately 0.4224 mi/s   **73** Approximately 158.67 mi/h   **75** $(8/\pi) - (2\pi/3)$
**77** $S_6 = 3.819403193$ (true value: about 3.8201978)   **79** $\pi/3$   **81** $8\pi R^2/3$   **83** $\frac{3}{4}\sqrt{3} \approx 1.299$
**85** *Suggestion:* $A(\theta) = s^2(\theta - \sin\theta)/(2\theta^2)$   **87** $-1.306440008, 1.977383029$, and $3.837467107$

## SECTION 8-4 (page 430)

**1** $2\cos(2x + 3)$   **3** $\frac{2}{3}\sec^2(2x/3)$   **5** $-(1/x^2)\sec(1/x)\tan(1/x)$   **7** $12\sec^2 4x \tan^2 4x$   **9** $\cot x$   **11** $(2/x)\tan(\ln x)\sec^2(\ln x)$
**13** $-\csc x$   **15** $-3[\sin^2(\csc x)][\cos(\csc x)][\csc x \cot x]$   **17** $\sec x \tan x$   **19** $[\exp(\sin x)](1 + \sec x \tan x)$
**21** $-4\sin x \cos x$   **23** $2^{\sec 5t}(5\ln 2)\sec 5t \tan 5t$   **25** $\frac{1}{3}(\sin x)^{-2/3}\cos x - \frac{1}{3}x^{-2/3}\cos(x^{1/3})$
**27** $(\sec^2 x - \sec x \tan x + \sec x)(1 + \sec x)^{-2}$   **29** $-12t^3\csc^2(1 + t^4)\cot^2(1 + t^4)$   **31** $2\tan(x/2) + C$
**33** $\frac{1}{2}x + \frac{1}{12}\sin 6x + C$   **35** $-\cos x + C$   **37** $\ln|1 + \sec x| + C$   **39** $\frac{1}{2}\sec x^2 + C$   **41** $\frac{1}{2}\exp(\sin 2x) + C$
**43** $\sec x + C$   **45** $\ln|\tan e^x| + C$   **47** $\frac{1}{6}\sin^6 x + C$   **49** $\frac{1}{8}\sec^4 2x + C$   **51** $\frac{1}{11}\sin^{11}x + C$   **53** $\frac{1}{4}(1 + \cos x)^{-4} + C$
**55** $\frac{1}{2}\sin e^{2x} + C$   **57** $\dfrac{2^{\sec x}}{\ln 2} + C$   **59** $\sin(\ln x) + C$

**61** $f'(x) = \sec^2 x$, which is always positive. $f''(x) = 2\sec^2 x \tan x$, which is zero at $(0, 0), (\pi, 0), (-\pi, 0)$, and so on; these are all $x$-intercepts and inflection points. There are vertical asymptotes at the odd multiples of $\pi/2$.

**63** Minima at $(-3\pi/4, -\sqrt{2})$, $(5\pi/4, -\sqrt{2})$, and so on; maxima at $(\pi/4, \sqrt{2})$, $(9\pi/4, \sqrt{2})$, and so on; $x$-intercepts at $-\pi/4$, $3\pi/4$, $7\pi/4$, and so on (these are also inflection points); the $y$-intercept is $(0, 1)$.

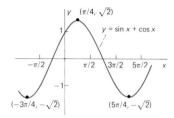

**65** $\pi/4$  **67** ln 4  **71** The two least solutions are approximately 0.86033359 and 3.42561846.

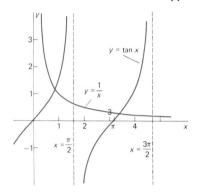

**73** $\alpha_1 \approx 2.0287578 \approx (1.29/2)\pi$, $\alpha_2 \approx 4.9131804 \approx (3.13/2)\pi$, $\alpha_3 \approx 7.9786657 \approx (5.08/2)\pi$, $\alpha_4 \approx 11.085538 \approx (7.06/2)\pi$

### SECTION 8-5 (page 438)

**1** (a) $\pi/6$  (b) $-\pi/6$  (c) $\pi/4$  (d) $-\pi/3$  **3** (a) 0  (b) $\pi/4$  (c) $-\pi/4$  (d) $\pi/3$  **5** $(100x^{99})(1 - x^{200})^{-1/2}$
**7** $1/[x|\ln x|\sqrt{(\ln x)^2 - 1}]$  **9** $(\sec^2 x)(1 - \tan^2 x)^{-1/2}$  **11** $e^x/\sqrt{1 - e^{2x}}$  **13** $-2/\sqrt{1 - x^2}$  **15** $-2/(x\sqrt{x^4 - 1})$
**17** $-1/[(1 + x^2)(\arctan x)^2]$  **19** $1/\{x[1 + (\ln x)^2]\}$  **21** $2e^x/(1 + e^{2x})$  **23** $[\cos(\arctan x)]/(1 + x^2)$
**25** $\{[(1 + 9x^2)(\arctan 3x) - 3]/[(1 + 9x^2)(\arctan 3x)^2]\}e^x$  **27** $(1 - 4x \arctan x)/(1 + x^2)^3$  **29** $1/(a^2 + x^2)$  **31** $\pi/4$
**33** $\pi/12$  **35** $\pi/12$  **37** $\frac{1}{2}\arcsin 2x + C$  **39** $\frac{1}{5}\text{arcsec}|x/5| + C$  **41** $\arctan(e^x) + C$  **43** $\frac{1}{15}\text{arcsec}|x^3/5| + C$
**45** Either $\sin^{-1}(2x - 1) + C$ or $2\sin^{-1}(\sqrt{x}) + C$  **47** $\frac{1}{50}\arctan(x^{50}) + C$  **49** $\arctan(\ln x) + C$  **51** $\pi/4$  **53** $\pi/2$
**55** $\pi/12$  **59** 8 ft  **63** $\pi$  **65** $A = 1 - \pi/3$, $B = 1 + 2\pi/3$

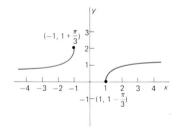

### SECTION 8-6 (page 446)

**1**

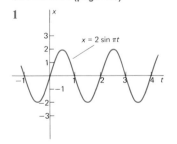

**3**

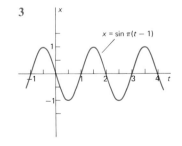

**5**

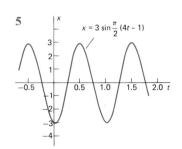

**A–44**

**7**

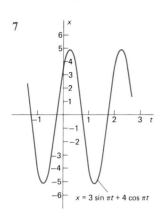

$x = 3\sin\pi t + 4\cos\pi t$

**9** 1  **11** $\frac{1}{2}\sqrt{5}$  **13** $5\pi$  **15** Period $2\pi$, amplitude 5
**19** Approximately 1.9795 mi; that is, about 10,452 ft

### SECTION 8-7 (page 452)

**1** $3\sinh(3x - 2)$  **3** $2x\tanh(1/x) - \text{sech}^2(1/x)$  **5** $-12\coth^2 4x\,\text{csch}^2 4x$  **7** $-e^{\text{csch}\,x}\text{csch}\,x\,\coth x$
**9** $(\cosh x)\cos(\sinh x)$  **11** $4x^3\cosh x^4$  **13** $-(1 + \text{sech}^2 x)/(x + \tanh x)^2$  **15** $\frac{1}{2}\cosh x^2 + C$  **17** $x - \frac{1}{3}\tanh 3x + C$
**19** $\frac{1}{6}\sinh^3 2x + C$  **21** $-\frac{1}{2}\text{sech}^2 x + C$  **23** $-\frac{1}{2}\text{csch}^2 x + C$  **25** $\ln(1 + \cosh x) + C$  **27** $\frac{1}{4}\tanh x + C$  **33** $\sinh a$

### SECTION 8-8 (page 459)

**1** $2/\sqrt{4x^2 + 1}$  **3** $1/2\sqrt{x}\,(1 - x)$  **5** $1/\sqrt{x^2 - 1}$  **7** $3\sqrt{\sinh^{-1}x}/2\sqrt{x^2 + 1}$  **9** $1/((1 - x^2)\tanh^{-1}x)$
**11** $\sinh^{-1}(x/3) + C$  **13** $\frac{1}{4}\ln\frac{9}{5} \approx 0.14695$  **15** $-\frac{1}{2}\text{sech}^{-1}|3x/2| + C$  **17** $\sinh^{-1}(e^x) + C$  **19** $-\text{sech}^{-1}(e^x) + C$
**29** Time: approximately 485 s; impact speed: about 20.656 ft/s
**35** (a) $T_0 \approx 259.16$ lb  (b) $T_{\max} \approx 278.70$ lb  (c) $H \approx 19.5$ ft

### CHAPTER 8 MISCELLANEOUS PROBLEMS (page 460)

**1** $\cos\sqrt{x}/2\sqrt{x}$  **3** $\sec^2(\ln x)/x$  **5** $-10(\csc 2x)(\csc 2x + \cot 2x)^5$  **7** $6\tan 3x$  **9** $e^{\arctan x}/(1 + x^2)$  **11** $1/(2\sqrt{x}\sqrt{1 - x})$
**13** $2x/(x^4 + 2x^2 + 2)$  **15** $e^x\sinh e^x + e^{2x}\cosh e^x$  **17** 0  **19** $x/|x|\sqrt{x^2 + 1}$  **21** $\frac{1}{2}\tan x^2 + C$  **23** $-\frac{1}{4}\csc^2 2x + C$
**25** $1/x - \tan(1/x) + C$  **27** $\frac{1}{2}\ln|\sec(x^2 + 1)| + C$  **29** $\sin^{-1}(e^x) + C$  **31** $\frac{1}{2}\arcsin(2x/3) + C$  **33** $\frac{1}{3}\arctan(x^3) + C$
**35** $\sec^{-1}|2x| + C$  **37** $\sec^{-1}(e^x) + C$  **39** $2\cosh\sqrt{x} + C$  **41** $\frac{1}{2}(\tan^{-1}x)^2 + C$  **43** $\frac{1}{2}\sinh^{-1}(2x/3) + C$  **45** $\pi^2/6$
**47** The zeros shown are at (approximately) $-0.197$, 1.373, and 2.944. The maximum shown is at (0.588, 13) and the
minima at $(-0.983, -13)$ and $(2.159, -13)$ (abscissas approximate)  **51** $x \approx 4.730041$  **55** Approximately 8 ft

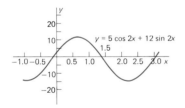

$y = 5\cos 2x + 12\sin 2x$

**57** (d) $\sqrt{Rg}$; about 17,691 mi/h

## CHAPTER 9

### SECTION 9-2 (page 466)

**1** $-\frac{1}{15}(2 - 3x)^5 + C$  **3** $\frac{1}{9}(2x^3 - 4)^{3/2} + C$  **5** $\frac{9}{8}(2x^2 + 3)^{2/3} + C$  **7** $-2\csc\sqrt{y} + C$  **9** $\frac{1}{6}(1 + \sin\theta)^6 + C$
**11** $e^{-\cot x} + C$  **13** $\frac{1}{11}(\ln t)^{11} + C$  **15** $\frac{1}{3}\sin^{-1}3t + C$  **17** $\frac{1}{2}\arctan(e^{2x}) + C$  **19** $\frac{3}{2}\arcsin(x^2) + C$  **21** $\frac{1}{15}\tan^5 3x + C$
**23** $\tan^{-1}(\sin\theta) + C$  **25** $\frac{2}{5}(1 + \sqrt{x})^5 + C$  **27** $\ln|\arctan t| + C$  **29** $\sec^{-1}e^x + C$
**31** $\frac{2}{7}(x - 2)^{7/2} + \frac{8}{5}(x - 2)^{5/2} + \frac{8}{3}(x - 2)^{3/2} + C$, which can be simplified to $\frac{2}{105}(x - 2)^{3/2}(15x^2 + 24x + 32) + C$
**33** $\frac{1}{3}(2x + 3)^{1/2}(x - 3) + C$  **35** $\frac{3}{10}(x + 1)^{2/3}(2x - 3) + C$

**37** The substitution $u = e^x$ leads to an integral in the form of (44) in the Table of Integrals (see the end papers); the answer is $\frac{1}{2}e^x\sqrt{9 + e^{2x}} + \frac{9}{2}\ln(e^x + \sqrt{9 + e^{2x}}) + C$.

**39** With $u = x^2$ and (47) in the Table of Integrals: $\frac{1}{2}(x^4 - 1)^{1/2} - \sec^{-1}x^2 + C$

**41** With $u = \ln x$ and (48) in the Table of Integrals: $\frac{1}{8}([\ln x](2[\ln x]^2 + 1)([\ln x]^2 + 1)^{1/2} - \ln|[\ln x] + ([\ln x]^2 + 1)^{1/2}|) + C$    **45** $\sin^{-1}(x - 1) + C$

### SECTION 9-3 (page 470)

**1** $\frac{1}{3}\cos^3 x - \cos x + C$    **3** $\frac{1}{3}\sin^3\theta - \frac{1}{5}\sin^5\theta + C$    **5** $\frac{1}{5}\sin^5 x - \frac{2}{3}\sin^3 x + \sin x + C$    **7** $\frac{2}{5}(\cos x)^{5/2} - 2(\cos x)^{1/2} + C$

**9** $-\frac{1}{14}\cos^7 2z + \frac{1}{5}\cos^5 2z - \frac{1}{6}\cos^3 2z + C$    **11** $\frac{1}{4}(\sec 4x + \cos 4x) + C$    **13** $\frac{1}{3}\tan^3 t + \tan t + C$

**15** $-\frac{1}{4}\csc^2 2x - \frac{1}{2}\ln|\sin 2x| + C$    **17** $\frac{1}{12}\tan^6 2x + C$    **19** $-\frac{1}{10}\cot^5 2t - \frac{1}{3}\cot^3 2t - \frac{1}{2}\cot 2t + C$

**21** $\frac{1}{4}\cos^4\theta - \frac{1}{2}\cos^2\theta + C$    **23** $\frac{2}{3}(\sec t)^{3/2} + 2(\sec t)^{-1/2} + C$    **25** $\frac{1}{3}\sin^3\theta + C$    **27** $\frac{1}{5}\sin 5t - \frac{1}{15}\sin^3 5t + C$

**29** $t + \frac{1}{3}\cot 3t - \frac{1}{9}\cot^3 3t + C$    **31** $-\frac{1}{5}\cos^{5/2}2t + \frac{2}{9}\cos^{9/2}2t - \frac{1}{13}\cos^{13/2}2t + C$    **33** $\frac{1}{2}\sin^2 x - \cos x + C$

**35** $-\cot x - \frac{1}{3}\cot^3 x - \frac{1}{2}\csc^2 x + C$    **39** $\frac{1}{4}\cos 2x - \frac{1}{16}\cos 8x + C$    **41** $\frac{1}{6}\sin 3x + \frac{1}{10}\sin 5x + C$

### SECTION 9-4 (page 476)

**1** $\frac{1}{2}xe^{2x} - \frac{1}{4}e^{2x} + C$    **3** $-t\cos t + \sin t + C$    **5** $(x/3)\sin 3x + \frac{1}{9}\cos 3x + C$    **7** $\frac{1}{4}x^4\ln x - \frac{1}{16}x^4 + C$

**9** $x\tan^{-1}x - \frac{1}{2}\ln(1 + x^2) + C$    **11** $\frac{2}{3}y^{3/2}\ln y - \frac{4}{9}y^{3/2} + C$    **13** $t(\ln t)^2 - 2t\ln t + 2t + C$

**15** $\frac{2}{3}x(x + 3)^{3/2} - \frac{4}{15}(x + 3)^{5/2} + C = \frac{2}{5}(x + 3)^{3/2}(x - 2) + C$

**17** $\frac{2}{9}x^3(x^3 + 1)^{3/2} - \frac{4}{45}(x^3 + 1)^{5/2} + C = \frac{2}{45}(x^3 + 1)^{3/2}(3x^3 - 2) + C$    **19** $-\frac{1}{2}(\csc\theta\cot\theta + \ln|\csc\theta + \cot\theta|) + C$

**21** $\frac{1}{3}x^3\tan^{-1}x - \frac{1}{6}x^2 + \frac{1}{6}\ln(1 + x^2) + C$    **23** $x\sec^{-1}\sqrt{x} - \sqrt{x - 1} + C$    **25** $(x + 1)\tan^{-1}\sqrt{x} - \sqrt{x} + C$

**27** $-x\cot x + \ln|\sin x| + C$    **29** $\frac{1}{2}x^2\sin(x^2) + \frac{1}{2}\cos(x^2) + C$    **31** $-2x^{-1/2}(2 + \ln x) + C$    **33** $x\sinh x - \cosh x + C$

**35** $x^2\cosh x - 2x\sinh x + 2\cosh x + C$    **37** $\pi(e - 2) \approx 2.25655$    **39** $\bar{x} = (e^2 - 1)/(4e - 8) \approx 2.22373$

**47** $6 - 2e \approx 0.563436$    **49** $6 - 2e \approx 0.563436$

### SECTION 9-5 (page 481)

**1** $-(1/x)\sqrt{1 - x^2} - \sin^{-1}x + C$    **3** $\ln|x + \sqrt{x^2 - 1}| - (1/x)\sqrt{x^2 - 1} + C$

**5** $\frac{1}{80}[(9 + 4x^2)^{5/2} - 15(9 + 4x^2)^{3/2}] + C = ((2x^2 - 3)/40)(9 + 4x^2)^{3/2} + C$

**7** $\sqrt{1 - 4x^2} - \ln|(1 + \sqrt{1 - 4x^2})/2x| + C$    **9** $\frac{1}{2}\ln|2x + \sqrt{9 + 4x^2}| + C$    **11** $\frac{25}{2}\sin^{-1}(x/5) - (x/2)\sqrt{25 - x^2} + C$

**13** $(x/2)\sqrt{x^2 + 1} - \frac{1}{2}\ln|x + \sqrt{1 + x^2}| + C$    **15** $\frac{1}{18}x\sqrt{4 + 9x^2} - \frac{2}{27}\ln|3x + \sqrt{4 + 9x^2}| + C$    **17** $x/\sqrt{1 + x^2} + C$

**19** $\frac{1}{256}(16x/(4 - x^2)^2 + 6x/(4 - x^2) + 3\ln|(2 + x)/\sqrt{4 - x^2}|) + C = (20x - 3x^3)/128(4 - x^2)^2 + \frac{3}{256}\ln|(2 + x)/(2 - x)| + C$

**21** $\frac{1}{2}x\sqrt{9 + 16x^2} + \frac{9}{8}\ln|4x + \sqrt{9 + 16x^2}| + C$    **23** $\sqrt{x^2 - 25} - 5\sec^{-1}(x/5) + C$

**25** $\frac{1}{8}x(2x^2 + 1)\sqrt{x^2 - 1} - \frac{1}{8}\ln|x + \sqrt{x^2 - 1}| + C$    **27** $-x/\sqrt{4x^2 - 1} + C$

**29** $-(1/x)\sqrt{x^2 - 5} + \ln|x + \sqrt{x^2 - 5}| + C$    **31** $\sinh^{-1}(x/5) + C$    **33** $\cosh^{-1}(x/2) - (1/x)\sqrt{x^2 - 4} + C$

**35** $\frac{1}{8}[x(1 + 2x^2)\sqrt{1 + x^2} - \sinh^{-1}x] + C$    **37** $(\pi/32)[18\sqrt{5} - \ln(2 + \sqrt{5})] \approx 3.8097$

**39** $\sqrt{5} - \sqrt{2} + \ln[(2 + 2\sqrt{2})/(1 + \sqrt{5})] \approx 1.222016$    **43** $2\pi[\sqrt{2} + \ln(1 + \sqrt{2})] \approx 14.4236$    **47** $\$6\frac{2}{3}$ million

### SECTION 9-6 (page 486)

**1** $\tan^{-1}(x + 2) + C$    **3** $11\tan^{-1}(x + 2) - \frac{3}{2}\ln(x^2 + 4x + 5) + C$    **5** $\sin^{-1}[(x + 1)/2] + C$

**7** $-2\sin^{-1}[(x + 1)/2] - \frac{1}{2}(x + 1)\sqrt{3 - 2x - x^2} - \frac{1}{3}(3 - 2x - x^2)^{3/2} + C$    **9** $\frac{5}{16}\ln|2x + 3| + \frac{7}{16}\ln|2x - 1| + C$

**11** $\frac{1}{3}\tan^{-1}[(x + 2)/3] + C$    **13** $\frac{1}{4}\ln|(1 + x)/(3 - x)| + C$    **15** $\ln(x^2 + 2x + 2) - 7\tan^{-1}(x + 1) + C$

**17** $\frac{2}{9}\sin^{-1}(x - \frac{2}{3}) - \frac{1}{9}\sqrt{5 + 12x - 9x^2} + C$

**19** $\frac{75}{4}\sin^{-1}(\frac{2}{5}|x - 2|) + \frac{3}{2}(x - 2)\sqrt{9 + 16x - 4x^2} + \frac{1}{6}(9 + 16x - 4x^2)^{3/2} + C$    **21** $(7x - 12)/(9\sqrt{6x - x^2}) + C$

**23** $-1/(16x^2 + 48x + 52) + C$    **25** $\frac{3}{2}\ln(x^2 + x + 1) - \frac{5}{3}\sqrt{3}\tan^{-1}(\frac{1}{3}\sqrt{3}[2x + 1]) + C$

**27** $\frac{1}{32}\ln[(x + 2)/(x - 2)] - x/[8(x^2 - 4)] + C$    **31** Approximately 3.69 mi    **33** (a) \$16.20 million    (b) \$16.66 million

### SECTION 9-7 (page 494)

**1** $\frac{1}{2}x^2 - x + \ln|x + 1| + C$    **3** $\frac{1}{3}\ln|(x - 3)/x| + C$    **5** $\frac{1}{5}\ln|(x - 2)/(x + 3)| + C$    **7** $\frac{1}{4}\ln|x| - \frac{1}{8}\ln(x^2 + 4) + C$

**9** $\frac{1}{3}x^3 - 4x + 8\tan^{-1}(x/2) + C$    **11** $x - 2\ln|x + 1| + C$    **13** $x + 1/(x + 1) + C$    **15** $\frac{1}{4}\ln|(x - 2)/(x + 2)| + C$

**17** $\frac{3}{2}\ln|2x - 1| - \ln|x + 3| + C$ **19** $\ln|x| + 2/(x + 1) + C$ **21** $\frac{3}{2}\ln|x^2 - 4| + \frac{1}{2}\ln|x^2 - 1| + C$

**23** $\ln|x + 2| + 4/(x + 2) - 2/(x + 2)^2 + C$ **25** $\frac{1}{2}\ln|x^2/(x^2 + 1)| + C$ **27** $\frac{1}{2}\ln|x^2/(x^2 + 4)| + \frac{1}{2}\arctan(x/2) + C$

**29** $-\frac{1}{2}\ln|x + 1| + \frac{1}{4}\ln(x^2 + 1) + \frac{1}{2}\tan^{-1}x + C$ **31** $\tan^{-1}(x/2) - \frac{3}{2}\sqrt{2}\tan^{-1}(x\sqrt{2}) + C$

**33** $-\frac{1}{8}\{\tan^{-1}|(x + 1)/2| + (2x + 14)/(x^2 + 2x + 5)\} + C$ or $\frac{1}{8}\{(3x^2 + 4x + 1)/(x^2 + 2x + 5) - \tan^{-1}[(x + 1)/2] + $
$C$, depending on the technique used to find the antiderivative

**35** $x + \frac{1}{2}\ln|x - 1| - 5/(2x - 2) + \frac{3}{4}\ln(x^2 + 1) + 2\tan^{-1}x + C$ **37** $\frac{1}{4}(1 - 2e^{2t})/(e^{2t} - 1)^2 + C$

**39** $\frac{1}{4}[\ln|3 + 2\ln t| + 1/(3 + 2\ln t)] + C$ **41** $\frac{1}{2}x^2 + \ln|x - 1| + \frac{1}{2}\ln(x^2 + x + 1) + \frac{1}{3}\sqrt{3}\tan^{-1}(\frac{1}{3}\sqrt{3}[2x + 1]) + C$

**43** $\frac{1}{4}\ln(x^4 + x^2 + 1) - \frac{1}{2}\sqrt{3}\arctan(\sqrt{3}/[2x^2 + 1]) + C$

## SECTION 9-8 (page 499)

**1** $\frac{1}{81}(\frac{2}{9}(3x - 2)^{9/2} + \frac{12}{7}(3x - 2)^{7/2} + \frac{24}{5}(3x - 2)^{5/2} + \frac{16}{3}(3x - 2)^{3/2}) + C$ **3** $2\sqrt{x} - 2\ln(1 + \sqrt{x}) + C$

**5** $[u = (x^2 - 1)^{1/3}]\ 3(x^2 - 3)/4(x^2 - 1)^{1/3} + C$ **7** $[x = u^4]\ x - \frac{4}{5}x^{5/4} + C$

**9** $[u = 1 + x^3]\ \frac{2}{9}(1 + x^3)^{3/2} - \frac{2}{3}(1 + x^3)^{1/2} + C$ **11** $[u = x^{1/3}]\ 3x^{1/3} - 3\tan^{-1}x^{1/3} + C$

**13** $[u = \sqrt{x + 4}]\ 2\sqrt{x + 4} - 2\ln(1 + \sqrt{x + 4}) + C$ **15** $(\sin\theta - 1)/\cos\theta + C$

**17** $[u = \tan(\theta/2)]\sqrt{2}\ln|(\sqrt{2} - 1 + \tan(\theta/2))/\sqrt{1 + 2\tan(\theta/2) - \tan^2(\theta/2)}| + C$ **19** $-\ln|2 + \cos\theta| + C$

**23** $\frac{8}{15}$ **25** $\sqrt{2} + \ln(1 + \sqrt{2}) \approx 2.295587$

## CHAPTER 9 MISCELLANEOUS PROBLEMS (page 500)

*Note:* Different techniques of integration may produce answers that appear to differ from those below; the difference should only be a constant.

**1** $2\arctan\sqrt{x} + C$ **3** $\ln|\sec x| + C$ **5** $\frac{1}{2}\sec^2\theta + C$ **7** $x\tan x - \frac{1}{2}x^2 + \ln|\cos x| + C$

**9** $\frac{2}{15}(2 - x^3)^{5/2} - \frac{4}{9}(2 - x^3)^{3/2} + C$ **11** $(x/2)\sqrt{25 + x^2} - \frac{25}{2}\ln|x + \sqrt{25 + x^2}| + C$

**13** $(2\sqrt{3}/3)\tan^{-1}((\sqrt{3}/3)[2x - 1]) + C$ **15** $(103\sqrt{29}/87)\tan^{-1}((\sqrt{29}/29)[3x - 2]) + \frac{5}{6}\ln(9x^2 - 12x + 33) + C$

**17** $\frac{2}{3}\arctan\frac{1}{3}\tan(\theta/2) + C$ **19** $\sin^{-1}(\frac{1}{2}\sin x) + C$ **21** $-\ln|\ln\cos x| + C$ **23** $(1 + x)\ln(1 + x) - x + C$

**25** $(x/2)\sqrt{x^2 + 9} + \frac{9}{2}\ln|x + \sqrt{x^2 + 9}| + C$ **27** $\frac{1}{2}(x - 1)\sqrt{2x - x^2} + \frac{1}{2}\sin^{-1}(x - 1) + C$

**29** $\frac{1}{3}x^3 + 2x - \sqrt{2}\ln|(x + \sqrt{2})/(x - \sqrt{2})| + C$ **31** $\frac{1}{2}((x^2 + x)/(x^2 + 2x + 2) - \tan^{-1}(x + 1)) + C$

**33** $\frac{1}{2}\tan\theta + C$ or $\sin 2\theta/2(1 + \cos 2\theta) + C$ **35** $\frac{1}{5}\sec^5 x - \frac{1}{3}\sec^3 x + C$

**37** $(x^2/8)[4(\ln x)^3 - 6(\ln x)^2 + 6(\ln x) - 3] + C$ **39** $\frac{1}{2}(e^x\sqrt{1 + e^{2x}} + \ln[e^x + \sqrt{1 + e^{2x}}]) + C$

**41** $\frac{1}{54}\sec^{-1}|x/3| + \sqrt{x^2 - 9}/18x^2 + C$ **43** $\ln|x| + \frac{1}{2}\arctan(2x) + C$ **45** $\frac{1}{2}(\sec x\tan x - \ln|\sec x + \tan x|) + C$

**47** $\ln|x + 1| - (2/3x^3) + C$

**49** $\frac{1}{9}(\ln|(x^2 + x + 1)/(x^2 - 2x + 1)| - 1/(x - 1) + 2\sqrt{3}\tan^{-1}((\sqrt{3}/3)[2x + 1]) + (x - 1)/(x^2 + x + 1)) + C$

**51** $\frac{1}{3}\ln|(3\sin\theta + 1 - \cos\theta)/(1 - \cos\theta - 3\sin\theta)| + C$ **53** $\frac{1}{3}(\sin^{-1}x)^3 + C$ **55** $\frac{1}{2}\sec^2 z + \ln|\cos z| + C$

**57** $\frac{1}{2}\arctan(e^{x^2}) + C$ **59** $-((x^2 + 1)/2)e^{-x^2} + C$ **61** $-(1/x)\arcsin x - \ln|(1 + \sqrt{1 - x^2})/x| + C$

**63** $\frac{1}{8}(\sin^{-1}x + x\sqrt{1 - x^2}(2x^2 - 1)) + C$ **65** $\frac{1}{4}(\ln|2x + 1| + 5/(2x + 1)) + C$ **67** $\frac{1}{2}\ln|e^{2x} - 1| + C$

**69** $2\ln|x + 1| + 3/(x + 1) - 5/(3(x + 1)^3) + C$ **71** $\frac{1}{2}\ln(x^2 + 1) + \tan^{-1}x - 1/2(x^2 + 1) + C$

**73** $\frac{1}{45}(x^3 - 1)^{3/2}(6x^3 + 4) + C$ **75** $\frac{2}{3}(1 + \sin x)^{3/2} + C$ **77** $\frac{1}{2}\ln|\sec x + \tan x| + C$ **79** $-2\sqrt{1 - \sin t} + C$

**81** $-2x + \sqrt{3}\tan^{-1}((\sqrt{3}/3)[2x + 1]) + ((2x + 1)/2)\ln|x^2 + x + 1| + C$ **83** $-(1/x)\tan^{-1}x + \ln|x/(1 + x^2)^{1/2}| + C$

**85** $\frac{1}{2}\ln(x^2 + 1) + 1/2(x^2 + 1) + C$ **87** $(x - 6)/2\sqrt{x^2 + 4} + C$ **89** $\frac{1}{3}(1 + \sin^2 x)^{3/2} + C$

**91** $(e^x/2)(x\sin x - x\cos x + \cos x) + C$ **93** $-\frac{1}{2}(x - 1)^{-2}\tan^{-1}x + \frac{1}{8}[(x^2 + 1)(x - 1)^{-2}] - \frac{1}{4}(x - 1)^{-1} + C$

**95** $\frac{1}{9}(11\sin^{-1}((3x - 1)/2) - 2\sqrt{3 + 6x - 9x^2}) + C$ **97** $\frac{1}{2}\cos^2\theta + \cos\theta + C$ **99** $x\sec^{-1}\sqrt{x} - \sqrt{x - 1} + C$

**101** $(\pi/4)(e^2 - e^{-2} + 4)$

**103** (a) $A_b = \pi(\sqrt{2} - e^{-b}\sqrt{1 + e^{-2b}} + \ln|(1 + \sqrt{2})/(e^{-b} + \sqrt{1 + e^{-2b}})|)$ (b) $\pi[\sqrt{2} + \ln(1 + \sqrt{2})] \approx 7.2118$

**105** $(\pi\sqrt{2}/2)[2\sqrt{14} - \sqrt{2} + \ln((1 + \sqrt{2})/(2\sqrt{2} + \sqrt{7}))] \approx 11.66353$ **109** $\frac{5}{4}\pi \approx 3.92699$

**111** The value of the integral is $\frac{1}{630}$. **113** $((5\sqrt{3} - 3)/2)\sqrt{2} + \frac{1}{2}\ln((1 + \sqrt{2})/(\sqrt{3} + \sqrt{2})) \approx 3.869983$

**115** The substitution is $u = e^x$ (a) $(2\sqrt{3}/3)\tan^{-1}((\sqrt{3}/3)[1 + 2e^x]) + C$

**119** $\frac{1}{4}\sqrt{2}\ln|(1 + \tan\theta - \sqrt{2}\tan\theta)/(1 + \tan\theta + \sqrt{2}\tan\theta)| - \frac{1}{2}\sqrt{2}\tan^{-1}(\sqrt{2}\cot\theta) + C$

# CHAPTER 10

## SECTION 10-1 (page 508)

**1** $x + 2y + 3 = 0$  **3** $4y + 25 = 3x$  **5** $x + y = 1$  **7** Center: $(-1, 0)$; radius: $\sqrt{5}$  **9** Center: $(2, -3)$; radius: 4
**11** Center: $(\frac{1}{2}, 0)$; radius 1  **13** Center: $(\frac{1}{2}, -\frac{3}{2})$; radius 3  **15** Center: $(-\frac{1}{3}, \frac{4}{3})$; radius: 2  **17** The point $(3, 2)$
**19** No points  **21** $(x + 1)^2 + (y + 2)^2 = 34$  **23** $(x - 6)^2 + (y - 6)^2 = \frac{4}{5}$
**25** Equation: $2x + y = 13$  **27** $(x - 6)^2 + (y - 11)^2 = 18$  **29** $(x/5)^2 + (y/3)^2 = 1$

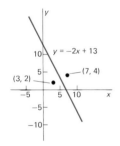
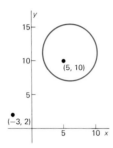
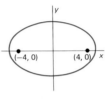

**31** $y - 7 + 4\sqrt{3} = (4 - 2\sqrt{3})(x - 2 + \sqrt{3})$, $y - 7 - 4\sqrt{3} = (4 + 2\sqrt{3})(x - 2 - \sqrt{3})$
**33** $y - 1 = 4(x - 4)$ and $y + 1 = 4(x + 4)$  **35** $h = p(e^2 + 1)/(1 - e^2)$, $a = \sqrt{|h^2 - p^2|}$, $b = a\sqrt{e^2 - 1}$

## SECTION 10-2 (page 514)

**1** (a) $(\sqrt{2}/2, \sqrt{2}/2)$  (b) $(1, -\sqrt{3})$  (c) $(1/2, -\sqrt{3}/2)$  (d) $(0, -3)$  (e) $(\sqrt{2}, \sqrt{2})$  (f) $(\sqrt{3}, -1)$  (g) $(-\sqrt{3}, 1)$

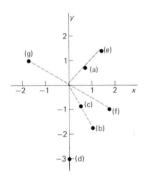

**3** $r\cos\theta = 4$  **5** $\theta = \tan^{-1}(\frac{1}{3})$  **7** $r^2\cos\theta\sin\theta = 1$  **9** $r = \tan\theta\sec\theta$  **11** $x^2 + y^2 = 9$  **13** $x^2 + 5x + y^2 = 0$
**15** $(x^2 + y^2)^3 = 4y^4$  **17** $x = 3$  **19** $x = 2; r = 2\sec\theta$  **21** $x + y = 1; r = 1/(\cos\theta + \sin\theta)$
**23** $y = x + 2; r = 2/(\sin\theta - \cos\theta)$  **25** $x^2 + y^2 + 8y = 0; r = -8\sin\theta$  **27** $x^2 + y^2 = 2x + 2y; r = 2(\cos\theta + \sin\theta)$
**29** Symmetric about the $x$-axis  **31** Symmetric about the $x$-axis  **33** Symmetric about the $x$-axis

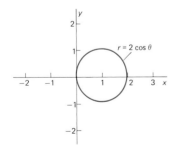
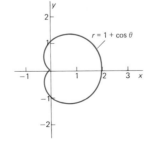
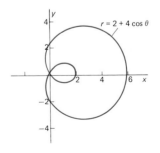

**35** Symmetric about the origin

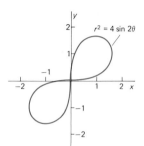

**37** Symmetric about both axes and about the origin

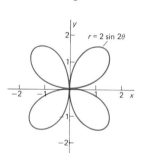

**39** Symmetric about the *x*-axis

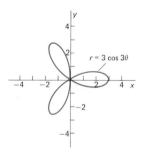

**41** Symmetric about the *y*-axis

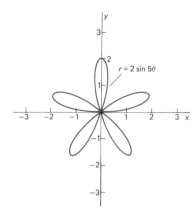

**43** No points of intersection    **45** $(0, 0), (1/2, \pi/6), (1/2, 5\pi/6), (1, \pi/2)$

**47** Four points: The pole, the point $r = 2, \theta = \pi$, and the two points $r = 2(\sqrt{2} - 1), \theta = \pm\cos^{-1}(3 - 2\sqrt{2})$.

**49** (a) $r\cos(\theta - \alpha) = p$

### SECTION 10-3 (page 518)

**1** $\pi$    **3** $3\pi/2$    **5** $9\pi/2$    **7** $4\pi$    **9** $19\pi/2$    **11** $\pi/2$ (one of *four* loops)    **13** $\pi/4$ (one of *eight* loops)    **15** 2 (one of *two* loops)

**17** 4 (one of *two* loops)    **19** $(2\pi + 3\sqrt{3})/6$    **21** $(5\pi - 6\sqrt{3})/24$    **23** $(39\sqrt{3} - 10\pi)/6$    **25** $(2 - \sqrt{2})/2$

**27** $(20\pi + 21\sqrt{3})/6$    **29** $\pi/2$

### SECTION 10-4 (page 523)

**1** $y^2 = 12x$

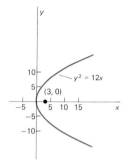

**3** $(x - 2)^2 = -8(y - 3)$

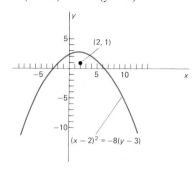

**5** $(y - 3)^2 = -8(x - 2)$

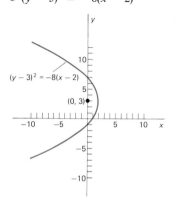

**7** $x^2 = -6(y + \frac{3}{2})$

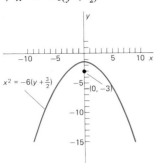

**9** $x^2 = 4(y + 1)$

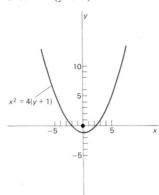

**11** $y^2 = 12x$, vertex: $(0, 0)$, axis: the x-axis

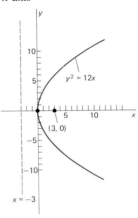

**13** $y^2 = -6x$, vertex: $(0, 0)$, axis: the x-axis

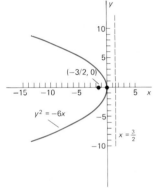

**15** $x^2 - 4x - 4y = 0$, vertex: $(2, -1)$, axis: $x = 2$

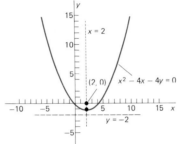

**17** $4x^2 + 4x + 4y + 13 = 0$, vertex: $(-\frac{1}{2}, -3)$, axis: $x = -\frac{1}{2}$

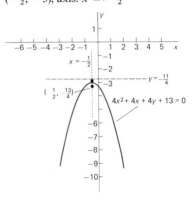

**23** About 0.693 days; that is, about 16 h 38 min  **27** $\alpha = \pi/12$, $\alpha = 5\pi/12$

**SECTION 10-5 (page 528)**

**1** $(x/4)^2 + (y/5)^2 = 1$  **3** $(x/15)^2 + (y/17)^2 = 1$  **5** $x^2/16 + y^2/7 = 1$  **7** $x^2/100 + y^2/75 = 1$  **9** $x^2/16 + y^2/12 = 1$
**11** $(x - 2)^2/16 + (y - 3)^2/4 = 1$  **13** $(x - 1)^2/25 + (y - 1)^2/16 = 1$  **15** $(x - 1)^2/81 + (y - 2)^2/72 = 1$

**17** Center: $(0, 0)$, foci: $(\pm 2\sqrt{5}, 0)$, major axis length: 12, minor axis length: 8

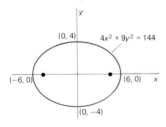

**19** Center: $(0, 4)$, foci: $(0, 4 \pm \sqrt{5})$, major axis length: 6, minor axis length: 4

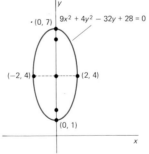

**21** About 3466.54 A.U., or $3.22 \times 10^{11}$ mi, or 20 light-days  **27** $(x - 1)^2/4 + y^2/(\frac{16}{3}) = 1$

**SECTION 10-6 (page 536)**

**1** $x^2/1 - y^2/15 = 1$  **3** $x^2/16 - y^2/9 = 1$  **5** $y^2/25 - x^2/25 = 1$  **7** $y^2/9 - x^2/27 = 1$  **9** $x^2/4 - y^2/12 = 1$
**11** $(x - 2)^2/9 - (y - 2)^2/27 = 1$  **13** $(y + 2)^2/9 - (x - 1)^2/4 = 1$

**A–50**

**15** Center: $(1, 2)$, foci: $(1 \pm \sqrt{2}, 2)$, asymptotes: $y - 2 = \pm(x - 1)$

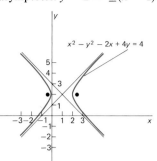

$x^2 - y^2 - 2x + 4y = 4$

**17** Center: $(0, 3)$, foci: $(0, 3 \pm 2\sqrt{3})$, asymptotes: $y = 3 \pm x\sqrt{3}$

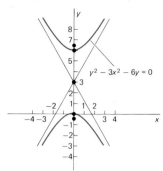

$y^2 - 3x^2 - 6y = 0$

**19** Center: $(-1, 1)$, foci: $(-1 \pm \sqrt{13}, 1)$, asymptotes: $y = \frac{3}{2}x + \frac{5}{2}$ and
$$y = -\frac{3}{2}x - \frac{1}{2}$$

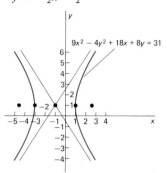

$9x^2 - 4y^2 + 18x + 8y = 31$

**21** There are no points on the graph if $c > 15$.   **23** $16x^2 + 50xy + 16y^2 = 369$

**27** The plane is about 16.42 mi north of $B$ and 8.66 mi west of $B$; that is, it is about 18.56 mi from $B$ at a bearing of $27°48'$ west of north.

**SECTION 10-7 (page 541)**

**1** $B^2 - 4AC = 0$: a parabola (or a degenerate case). Rotation angle: $\alpha = \pi/4$. Transformed equation: $(x')^2 = 1$.

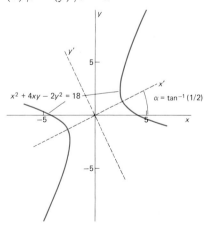

$x^2 + 2xy + y^2 = 2$

$\alpha = \pi/4$

$x' = 1$

$x' = -1$

**3** $B^2 - 4AC = 5 > 0$: hyperbola. Rotation angle: $\alpha = \pi/4$. Transformed equation: $(x')^2/2 - (y')^2/10 = 1$.

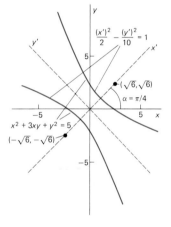

$\dfrac{(x')^2}{2} - \dfrac{(y')^2}{10} = 1$

$(\sqrt{6}, \sqrt{6})$

$\alpha = \pi/4$

$x^2 + 3xy + y^2 = 5$

$(-\sqrt{6}, -\sqrt{6})$

**5** $B^2 - 4AC = 24 > 0$: hyperbola. Rotation angle: $\alpha = \tan^{-1}(\frac{1}{2}) \approx 26°33'54''$. Transformed equation: $(x')^2/9 - (y')^2/6 = 1$.

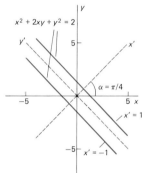

$x^2 + 4xy - 2y^2 = 18$

$\alpha = \tan^{-1}(1/2)$

**7** $B^2 - 4AC = -3 < 0$: ellipse. Rotation angle: $\alpha = \pi/4$. Transformed equation: $(x' - 1)^2/4 + (y' + 1)^2/(\frac{4}{3}) = 1$.

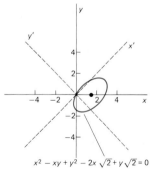

$x^2 - xy + y^2 - 2x\sqrt{2} + y\sqrt{2} = 0$

**9** $B^2 - 4AC = 5000 > 0$: hyperbola. Rotation angle: $\alpha = \tan^{-1}(\frac{4}{3}) \approx 0.9273$. Transformed equation: $(y' - 1)^2/(\frac{1}{2}) - (x' - 2)^2/1 = 1$.

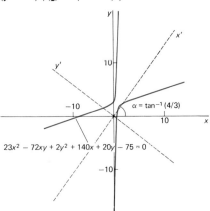

**11** $B^2 - 4AC = 334084 > 0$: hyperbola. Rotation angle: $\alpha = \tan^{-1}(\frac{8}{15}) \approx 0.49$ (a little over 28°). Transformed equation: $(x' - 1)^2/1 - (y')^2/1 = 1$.

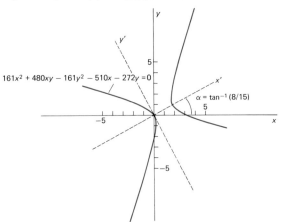

**SECTION 10-8 (page 546)**

**1** Parabola, directrix $x = 6$, vertex $(3, 0)$

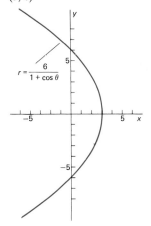

**3** Parabola, directrix $y = -3$, vertex $(0, -\frac{3}{2})$

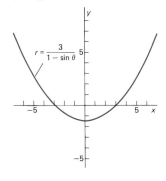

**5** Ellipse, eccentricity $\frac{1}{2}$, directrix $y = -6$

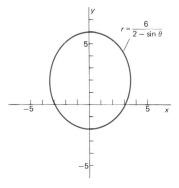

**7** Hyperbola, eccentricity

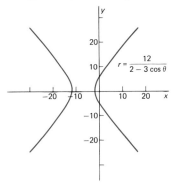

**9** Hyperbola, eccentricity $\sqrt{2}$, rotated $\pi/4$ from the "standard position"

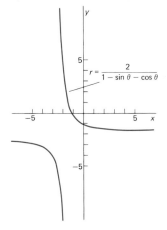

**11** $r = 3/(1 + \cos \theta)$   **13** $r = 2/(3 + \sin \theta)$   **15** $r = (3\sqrt{2}/2)/(1 + \cos(\theta - \pi/4))$   **17** $3x^2 + 32x + 64 = y^2$
**19** $9x^2 + 5y^2 = 24y + 36$
**21** $A = \sqrt{2}$, $\alpha = \pi/4$. Equation: $r = 2/(1 - \cos(\theta - \pi/4))$.   **25** 2000 mi
Parabola, focus at the pole, rotated 45° from such a
parabola with directrix $x' = -2$.

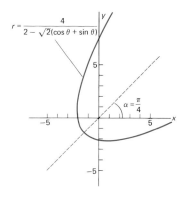

### CHAPTER 10 MISCELLANEOUS PROBLEMS (page 547)

**1** Circle, center $(1, -1)$, radius 2   **3** Circle, center $(3, -1)$, radius 1   **5** Parabola, vertex $(4, -2)$, opening downward
**7** Ellipse, center $(2, 0)$, major axis 6, minor axis 4   **9** Hyperbola, center $(-1, 1)$, vertical axis, foci at $(-1, 1 \pm \sqrt{3})$
**11** There are no points on the graph   **13** Hyperbola, axis inclined at 22.5° $(\pi/8)$ from horizontal
**15** Ellipse, major axis $2\sqrt{2}$, minor axis 1, rotated through angle $\alpha = \pi/4$, center at the origin
**17** Parabola, vertex $(0, 0)$, opening to the "northeast," axis at angle $\alpha = \tan^{-1}(\frac{3}{4})$ from the horizontal
**19** Circle, center $(1, 0)$, radius 1   **21** Straight line $y = x + 1$   **23** Horizontal line $y = 3$
**25** Two ovals tangent to each other and to the $y$-axis at       **27** Apple-shaped curve, symmetric about $y$-axis
$(0, 0)$

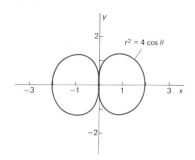

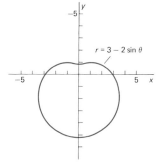

**29** Ellipse, one focus at $(0, 0)$, directrix $x = 4$, eccentricity $e = \frac{1}{2}$   **31** $(\pi - 2)/2$   **33** $(39\sqrt{3} - 10\pi)/6 \approx 6.02234$
**35** 2   **37** $5\pi/4$   **39** $r = 2p \cos(\theta - \alpha)$   **41** If $a > b$, the maximum is $2a$ and the minimum is $2b$.
**43** $b^2y = 4hx(b - x)$ or $r = b \sec \theta - (b^2/4h) \sec \theta \tan \theta$
**45** Suggestion: Let $\theta$ be the angle that $QR$ makes with the $x$-axis.
**49** Suggestion: $4x^2 + 12x + 9 = x^2 + y^2$, so $(2x + 3)^2 = r^2$. The curve is a hyperbola with one focus at the origin,
directrix $x = -\frac{3}{2}$, and eccentricity $e = 2$.
**51** $\frac{3}{2}$   **53** (a) $r = 31.11/(1 + 0.83 \cos \theta)$   (b) About 18.7 days

# CHAPTER 11

### SECTION 11-1 (page 555)

**1** $\frac{1}{2}$   **3** $\frac{2}{5}$   **5** 0   **7** 0   **9** $\frac{1}{2}$   **11** 2   **13** 0   **15** 1   **17** 1   **19** $\frac{3}{5}$   **21** $\frac{3}{2}$   **23** $\frac{1}{3}$   **25** $(\ln 2)/(\ln 3)$   **27** $\frac{1}{2}$   **29** 1   **31** $\frac{1}{3}$
**33** $-\frac{1}{2}$   **35** 1   **37** $\frac{1}{4}$   **39** $\frac{2}{3}$   **41** 6   **43** $\frac{4}{3}$   **45** $\frac{2}{3}$   **47** 0

## SECTION 11-2 (page 559)

**1** 1   **3** $\frac{3}{8}$   **5** $\frac{1}{4}$   **7** 1   **9** 0   **11** $-1$   **13** $-\infty$   **15** $+\infty$   **17** $-\frac{1}{2}$   **19** 0   **21** 1   **23** 1   **25** $e^{-1/6}$   **27** $1/\sqrt{e}$   **29** 1
**31** $e$   **33** $-\infty$

## SECTION 11-3 (page 569)

**1** $e^{-x} = 1 - x/1! + x^2/2! - x^3/3! + x^4/4! - x^5/5! + (e^{-z}/6!)x^6$ for some $z$ between 0 and $x$
**3** $\cos x = 1 - x^2/2! + x^4/4! - ((\sin z)/5!)x^5$ for some $z$ between 0 and $x$
**5** $\sqrt{1 + x} = 1 + (x/1!2) - (x^2/2!4) + (3x^3/3!8) - (5x^4/128)(1 + z)^{7/2}$ for some $z$ between 0 and $x$
**7** $\tan x = x/1! + 2x^3/3! + ((16 \sec^4 z \tan z + 8 \sec^2 z \tan^3 z)/4!)x^4$ for some $z$ between 0 and $x$
**9** $\sin^{-1}x = x/1! + (x^3/3!)(1 + 2z^2)/(1 - z^2)^{5/2}$ for some $z$ between 0 and $x$
**11** $e^x = e + (e/1!)(x - 1) + (e/2!)(x - 1)^2 + (e/3!)(x - 1)^3 + (e/4!)(x - 1)^4 + (e^z/5!)(x - 1)^5$ for some $z$ between 1 and $x$
**13** $\sin x = \frac{1}{2} + (\sqrt{3}/1!2)(x - \pi/6) - (1/2!2)(x - \pi/6)^2 - (\sqrt{3}/3!2)(x - \pi/6)^3 + ((\sin z)/4!)(x - \pi/6)^4$ for some $z$ between $\pi/6$ and $x$
**15** $1/(x - 4)^2 = 1 - 2(x - 5) + 3(x - 5)^2 - 4(x - 5)^3 + 5(x - 5)^4 - 6(x - 5)^5 + (7/(z - 4)^8)(x - 5)^6$ for some $z$ between 5 and $x$   **17** Six-place accuracy   **19** Five-place accuracy
**21** Problem 17 predicts $e^{1/3} \approx 1.3955826$; the true value is about 1.3956124 (the error is about 0.00003).
**25** The third-degree polynomial gives the estimate 0.81915.   **27** The third-degree polynomial gives the estimate 0.8829476.

## SECTION 11-4 (page 576)

**1** 1   **3** $+\infty$   **5** $+\infty$   **7** 1   **9** $+\infty$   **11** $-\frac{1}{2}$   **13** $\frac{9}{2}$   **15** $+\infty$   **17** Does not exist   **19** $2(e - 1)$   **21** $+\infty$   **23** $\frac{1}{4}$

## SECTION 11-5 (page 582)

**1** 50 ⌈exact⌋   **3** Bounds: 0.631813, 0.632341. True value: About 0.63212056
**5** Bounds: 1.974231, 2.014605. True value: Exactly 2   **7** Bounds: 1.8627403, 1.8627423. True value: About 1.8627417
**9** Bounds: 0.3862936, 0.3862968. True value: About 0.3862944   **11** $n \geq 104$   **13** $n \geq 5$   **15** $n \geq 26$   **17** $n \geq 4$
**19** $V = (h/6)(0 + 4\pi(r/2)^2 + \pi r^2) = \frac{1}{3}\pi r^2 h$   **21** $V = (a/6)(0 + 4\pi a/2 + \pi a) = \pi a^2/2$
**23** $n = 10$; $T_{10} \approx 0.6937714$, so $\ln 2 \approx 0.69$
**25** Use $n = 10$; $T_{10} \approx 1.112332$ (the true value of the integral is approximately 1.11144797)
**27** $S_2 - ES_2 = \frac{1}{6}(0 + \frac{4}{16} + 1) - 24/(180)(2^4) = \frac{1}{5}$

## CHAPTER 11 MISCELLANEOUS PROBLEMS (page 583)

**1** $\frac{1}{4}$   **3** $\frac{1}{2}$   **5** $\frac{7}{45}$   **7** 1   **9** $-\infty$   **11** $+\infty$   **13** $e^2$   **15** $-e/2$   **17** $+\infty$   **19** 1   **21** 2
**23** $e^{x+1} = e + ex/1 + ex^2/2! + ex^3/3! + ex^4/4! + ex^5/5! + (e^{z+1}/6!)x^6$ for some number $z$ between 0 and $x$
**25** $e^{-x^2} = 1 - (2/2!)x^2 + ((16z^4 - 48z^2 + 12)e^{-z^2}/4!)x^4$ for some number $z$ between 0 and $x$
**27** $x^{1/3} - 5 + (x - 125)/75 - (x - 125)^2/28125 + (x - 125)^3/6328125 - 10(x - 125)^4/243z^{11/3}$ for some $z$ between 125 and $x$
**29** $x^{-1/2} = 1 - (x - 1)/1!2 + 3(x - 1)^2/2!4 - 15(x - 1)^3/3!8 + 105(x - 1)^4/4!16 - 945(x - 1)^5/5!32z^{11/2}$ for some $z$ between 1 and $x$
**31** $x^4 = 1 + 4(x - 1) + 6(x - 1)^2 + 4(x - 1)^3 + (x - 1)^4 + 0$—the remainder ("0") is in fact zero in this case
**33** Five terms of the series for $f(x) = x^{1/10}$, center 1024, give the estimate 1.99526. The true value is approximately 1.995262315.
**35** Not even one-place accuracy because the error may be as large as 0.2.

# CHAPTER 12

## SECTION 12-2 (page 592)

**1** $\frac{2}{5}$   **3** $\frac{1}{2}$   **5** 1   **7** Does not converge   **9** 0   **11** 0   **13** 0   **15** 1   **17** 0   **19** 0   **21** 0   **23** 0   **25** $e$   **27** $1/e^2$   **29** 2
**31** 1   **33** Does not converge   **35** 0   **41** (b) 4

**1** $\frac{3}{2}$  **3** Diverges  **5** Diverges  **7** 6  **9** Diverges  **11** Diverges  **13** Diverges  **15** $\sqrt{2}/(\sqrt{2}-1)$  **17** Diverges  **19** $\frac{1}{12}$
**21** $e/(\pi - e)$  **23** Diverges  **25** $\frac{65}{12}$  **27** $\frac{167}{8}$  **29** $\frac{1}{4}$  **31** $\frac{47}{99}$  **33** $\frac{41}{333}$  **35** 314,156/99,999  **37** $S_n = \ln(n+1)$; diverges
**39** $S_n = \frac{3}{2} - 1/n - 1/(n+1)$; the sum is $\frac{3}{2}$.  **45** Computations with $S$ are meaningless because $S$ is not a number.
**47** 4.5 s  **49** (a) $M_n = (0.95)^n M_0$  (b) 0  **51** $\frac{1}{12}$  **53** Peter's probability is $\frac{36}{91}$; Paul's is $\frac{30}{91}$.

**1** Diverges  **3** Diverges  **5** Converges  **7** Diverges  **9** Converges  **11** Converges  **13** Converges  **15** Converges
**17** Converges  **19** Converges  **21** Diverges  **23** Diverges  **25** Converges  **27** Converges
**29** The terms are not nonnegative.  **31** The terms are not monotone decreasing.  **33** $n = 100$
**37** With $n = 6$, $1.03689 < S < 1.03699$
**39** About a million centuries  **41** Results with $n = 10$: $S_{10} \approx 1.08203658$ and $3.141566 < \pi < 3.141627$.

**1** Converges  **3** Diverges  **5** Converges  **7** Diverges  **9** Converges  **11** Converges  **13** Diverges
**15, 17, 19, 21, 23,** and **25** Converges  **27** Diverges  **29** Diverges  **31** Converges  **33** Converges  **35** Diverges

**1** Converges  **3** Converges  **5** Converges  **7** Converges  **9** Diverges  **11** Converges absolutely
**13** Converges conditionally  **15** Converges absolutely  **17** Converges absolutely  **19** Diverges
**21** Converges absolutely  **23** Converges conditionally  **25** Diverges  **27** Diverges  **29** Converges absolutely
**31** Converges absolutely  **33** $n = 1999$  **35** $n = 6$  **37** The first six terms give the estimate 0.6065.
**39** The first three terms give the estimate 0.0953.
**41** Results using the first ten terms: $S_{10} \approx 0.8179622$ and $3.1329 < \pi < 3.1488$.

**1** $x - \frac{1}{2}x^2 + \frac{1}{3}x^3 - \frac{1}{4}x^4 + \cdots$  **3** $1 - x + (1/2!)x^2 - (1/3!)x^3 + (1/4!)x^4 - \cdots$
**5** $(x-1) - \frac{1}{2}(x-1)^2 + \frac{1}{3}(x-1)^3 - \frac{1}{4}(x-1)^4 + \cdots$
**7** $\sqrt{2}/2 - (\sqrt{2}/1!2)(x - \pi/2) - (\sqrt{2}/2!2)(x - \pi/2)^2 + (\sqrt{2}/3!2)(x - \pi/2)^3 + (\sqrt{2}/4!2)(x - \pi/2)^4 - (\sqrt{2}/5!2)(x - \pi/2)^5 - (\sqrt{2}/6!2)(x - \pi/2)^6 + (\sqrt{2}/7!2)(x - \pi/2)^7 - \cdots$
**9** $\sinh x = x + (1/3!)x^3 + (1/5!)x^5 + (1/7!)x^7 + \cdots$
**11** $1/x = 1 - (x-1) + (x-1)^2 - (x-1)^3 + \cdots$
**13** $\sin x = (\sqrt{2}/2)[1 + (1/1!)(x - \pi/4) - (1/2!)(x - \pi/4)^2 - (1/3!)(x - \pi/4)^3 + (1/4!)(x - \pi/4)^4 + \cdots]$
**17** Five terms give the estimate 0.7966.  **19** Three terms give the estimate 0.2315266.
**21** Four terms give the estimate 0.487222.  **25** Seven terms give the estimate 0.2877.

**1** $[-1, 1)$  **3** $(-1, 1)$  **5** $[-1, 1]$  **7** $(\frac{2}{5}, \frac{4}{5})$  **9** $[\frac{5}{2}, \frac{7}{2}]$  **11** Converges only for $x = 0$  **13** $(-4, 2)$  **15** $[2, 4]$
**17** Converges only for $x = 5$  **19** $(-1, 1)$  **21** $f(x) = x^2 - (3/1!)x^3 + (3^2/2!)x^4 - (3^3/3!)x^5 + (3^4/4!)x^6 - \cdots$; the radius of convergence is $+\infty$.
**23** $f(x) = x^2 - (1/3!)x^6 + (1/5!)x^{10} - (1/7!)x^{14} + (1/9!)x^{18} - \cdots$; the radius of convergence is $+\infty$.
**25** $(1-x)^{1/3} = 1 - (1/3)x - (2/2!3^2)x^2 - (2 \cdot 5/3!3^3)x^3 - (2 \cdot 5 \cdot 8/4!3^4)x^4 - (2 \cdot 5 \cdot 8 \cdot 11/5!3^5)x^5 - \cdots$; $r = 1$
**27** $f(x) = 1 - 3x + (4 \cdot 3/2)x^2 - (5 \cdot 4/2)x^3 + (6 \cdot 5/2)x^4 - \cdots$; $r = 1$
**29** $f(x) = 1 - \frac{1}{2}x + \frac{1}{3}x^2 - \frac{1}{4}x^3 + \frac{1}{5}x^4 - \cdots$; $r = 1$
**31** $f(x) = (1/4)x^4 - (1/3!10)x^{10} + (1/5!16)x^{16} - (1/7!22)x^{22} + \cdots = \sum_{n=0}^{\infty} [(-1)^n/(2n+1)!(6n+4)]x^{6n+4}$.
**33** $f(x) = x - (1/1!4)x^4 + (1/2!7)x^7 - (1/3!10)x^{10} + (1/4!13)x^{13} - \cdots = \sum_{n=0}^{\infty} [(-1)^n/n!(3n+1)]x^{3n+1}$.
**35** $f(x) = (1/1!)x - (1/2!3)x^3 + (1/3!5)x^5 - (1/4!7)x^7 + (1/5!9)x^9 - \cdots = \sum_{n=1}^{\infty} [(-1)^{n+1}/n!(2n-1)]x^{2n-1}$.
**37** By using six terms, we find that $3.14130878 < \pi < 3.1416744$.

**1** $65^{1/3} = 4(1 + \frac{1}{64})^{1/3}$. The first four terms of the binomial series give $\sqrt[3]{65} \approx 4.020726$; answer: 4.021.
**3** Three terms of the usual sine series give 0.479427 with error less than 0.000002; answer: 0.479.
**5** Five terms of the usual arctangent series give 0.463684 with error less than 0.000045; answer: 0.464.   **7** 0.309
**9** 0.174   **11** 0.946   **13** 0.487   **15** 0.0976   **17** 0.470   **19** 0.747   **21** $-\frac{1}{2}$   **23** $\frac{1}{2}$   **25** 0
**31** The first five coefficients are 1, 0, $\frac{1}{2}$, 0, and $\frac{5}{24}$.

**5** 19 min 9 s   **7** 26,556.97 ft

## CHAPTER 12 MISCELLANEOUS PROBLEMS (page 652)

**1** 1   **3** 10   **5** 0   **7** 0   **9** No limit   **11** 0   **13** $+\infty$ or "no limit"   **15** 1   **17** Converges   **19** Converges   **21** Converges
**23** Diverges   **25** Converges   **27** Converges   **29** Diverges   **31** $(-\infty, +\infty)$   **33** $[-2, 4)$   **35** $[-1, 1]$
**37** It converges only for $x = 0$.   **39** $(-\infty, +\infty)$   **41** It converges for *no x*.   **43** It converges for all *x*
**45** Arithmetic computations involving *x* are meaningless because *x* is not a number.
**51** Seven terms of the binomial series give 1.084.   **53** Six terms of the "obvious" series give 0.747.
**55** Four terms give 0.444.

# CHAPTER 13

**1** $y = 2x - 3$   **3** $y^2 = x^3$   **5** $y = 2x^2 - 5x + 2$   **7** $y = 4x^2, x > 0$   **9** $(x/5)^2 + (y/3)^2 = 1$   **11** $x^2 - y^2 = 1$
**13** $y - 5 = \frac{9}{4}(x - 3)$   **15** $y = -(\pi/2)(x - \pi/2)$   **17** $x + y = 3$   **19** $\psi = \pi/6$ [constant]   **21** $\psi = \pi/2$
**25** $x = p/m^2, y = 2p/m$

**1** $\frac{22}{5}$   **3** $\frac{4}{3}$   **5** $(1 + e^\pi)/2 \approx 12.0703$   **7** $358\pi/35 \approx 32.13400$   **9** $16\pi/15 \approx 3.35103$   **11** $\frac{74}{3}$   **13** $\pi\sqrt{2}/4 \approx 1.11072$
**15** $(e^{2\pi} - 1)\sqrt{5} \approx 1195.1597$   **17** $(8\pi/3)(5\sqrt{5} - 2\sqrt{2}) \approx 69.96882$   **19** $(2\pi/27)(13\sqrt{13} - 8) \approx 9.04596$   **21** $16\pi^2$
**23** $5\pi^2 a^3$   **25** (a) $A = \pi ab$   (b) $V = \frac{4}{3}\pi ab^2$   **27** $\pi\sqrt{1 + 4\pi^2} + \frac{1}{2}\ln(2\pi + \sqrt{1 + 4\pi^2}) \approx 21.25629$   **29** $\frac{3}{8}\pi a^2$   **31** $\frac{12}{5}\pi a^2$
**33** $216\sqrt{3}/5$   **35** $243\pi\sqrt{3}/4$

**1** $\sqrt{5}; 2\sqrt{13}; 4\sqrt{2}; \langle -2, 0 \rangle; \langle 9, -10 \rangle$; no   **3** $2\sqrt{2}; 10; \sqrt{5}; \langle -5, -6 \rangle; \langle 0, 2 \rangle$; no
**5** $\sqrt{10}; 2\sqrt{29}; \sqrt{65}; 3\mathbf{i} - 2\mathbf{j}; -\mathbf{i} + 19\mathbf{j}$; no   **7** $4; 14; \sqrt{65}; 4\mathbf{i} - 7\mathbf{j}; 12\mathbf{i} + 14\mathbf{j}$; yes   **9** $\mathbf{u} = \langle -\frac{3}{5}, -\frac{4}{5} \rangle, \mathbf{v} = \langle \frac{3}{5}, \frac{4}{5} \rangle$
**11** $\mathbf{u} = \frac{8}{17}\mathbf{i} + \frac{15}{17}\mathbf{j}, \mathbf{v} = -\frac{8}{17}\mathbf{i} - \frac{15}{17}\mathbf{j}$   **13** $-4\mathbf{j}$   **15** $8\mathbf{i} - 14\mathbf{j}$   **17** Yes   **19** No   **21** (a) $15\mathbf{i} - 21\mathbf{j}$   (b) $\frac{5}{3}\mathbf{i} - \frac{7}{3}\mathbf{j}$
**23** (a) $(35\sqrt{58}/58)\mathbf{i} - (15\sqrt{58}/58)\mathbf{j}$   (b) $-(40\sqrt{89}/89)\mathbf{i} - (25\sqrt{89}/89)\mathbf{j}$   **25** $c = 0$   **33** $\mathbf{v}_a = (500 + 25\sqrt{2})\mathbf{i} + 25\sqrt{2}\mathbf{j}$
**35** $\mathbf{v}_a = -225\sqrt{2}\mathbf{i} + 275\sqrt{2}\mathbf{j}$

**1** 0; 0   **3** $2\mathbf{i} - \mathbf{j}; 4\mathbf{i} + \mathbf{j}$   **5** $6\pi\mathbf{i}; 12\pi^2\mathbf{j}$   **7** $\mathbf{j}; \mathbf{i}$   **9** $((2 - \sqrt{2})/2)\mathbf{i} + \sqrt{2}\mathbf{j}$   **11** $\frac{484}{15}\mathbf{i}$   **13** 11   **15** 0   **17** $\mathbf{i} + 2t\mathbf{j}; t\mathbf{i} + \cdot t^2\mathbf{j}$
**19** $\frac{1}{2}t^2\mathbf{i} + \frac{1}{3}t^3\mathbf{j}; (1 + \frac{1}{6}t^3)\mathbf{i} + \frac{1}{12}t^4\mathbf{j}$   **25** Begin with $(d/dt)(\mathbf{v} \cdot \mathbf{v}) = 0$.
**27** A replusive force acting directly away from the origin, with magnitude proportional to distance from the origin

**1** $v_0 \approx 411.047$ ft/s   **5** (a) $y_m = 100$ ft, $R = 400\sqrt{3}$ ft   (b) $y_m = 200, R = 800$   (c) $y_m = 300, R = 400\sqrt{3}$
**7** $v_0 = 1056$ ft/s   **9** $v_0 = 4\sqrt{4181} \approx 258.64$ ft/s; $\alpha = \tan^{-1}(0.82)$, approximately $39°21'6''$
**11** $M_S \approx 1.98885 \times 10^{30}$ kg; the ratio is about 333252 to 1   **13** Approximately 238,020 mi
**15** $\alpha = \tan^{-1}(55/108) \approx 26°59'16''$   **17** $20\sqrt{5}$ ft/s (about 30.5 mi/h)

**1** 0   **3** 1   **5** $40\sqrt{2}/41\sqrt{41} \approx 0.2155$   **7** At $(-\frac{1}{2}\ln 2, \frac{1}{2}\sqrt{2})$   **9** $\kappa_{max} = 5/9$ at $(\pm 5, 0)$; $\kappa_{min} = 3/25$ at $(0, \pm 3)$.
**11** $\mathbf{T} = (\sqrt{10}/10)\mathbf{i} + (3\sqrt{10}/10)\mathbf{j}$; $\mathbf{N} = (3\sqrt{10}/10)\mathbf{i} - (\sqrt{10}/10)\mathbf{j}$
**13** $\mathbf{T} = (\sqrt{57}/57)(3\mathbf{i} - 4\sqrt{3}\mathbf{j})$; $\mathbf{N} = -(\sqrt{57}/57)(4\sqrt{3}\mathbf{i} + 3\mathbf{j})$   **15** $\mathbf{T} = -(\sqrt{2}/2)\mathbf{i} - (\sqrt{2}/2)\mathbf{j}$; $\mathbf{N} = (\sqrt{2}/2)\mathbf{i} - (\sqrt{2}/2)\mathbf{j}$
**17** $a_T = 18t/\sqrt{9t^2 + 1}$, $a_N = 6/\sqrt{9t^2 + 1}$   **19** $a_T = t/(1 + t^2)^{1/2}$, $a_N = (2 + t^2)/(1 + t^2)^{1/2}$   **21** $\kappa = 1/a$
**23** $x^2 + (y - \frac{1}{2})^2 = \frac{1}{4}$   **25** $(x - \frac{3}{2})^2 + (y - \frac{3}{2})^2 = 2$   **27** Begin with $(d/dt)(\mathbf{v} \cdot \mathbf{v})$   **29** $\kappa = 1/|t|$
**31** $y = 3x^5 - 8x^4 + 6x^3$

**1** $\mathbf{v} = a\mathbf{u}_\theta$, $\mathbf{a} = -a\mathbf{u}_r$   **3** $\mathbf{v} = \mathbf{u}_r + t\mathbf{u}_\theta$, $\mathbf{a} = -t\mathbf{u}_r + 2\mathbf{u}_\theta$
**5** $\mathbf{v} = (12\cos 4t)\mathbf{u}_r + (6\sin 4t)\mathbf{u}_\theta$, $\mathbf{a} = (-60\sin 4t)\mathbf{u}_r + (48\cos 4t)\mathbf{u}_\theta$   **9** 36.65 mi/s; 24.13 mi/s   **11** 0.672 mi/s; 0.602 mi/s
**13** (a) 5 h 18 min 27 s;   (b) 7.35 sec

**CHAPTER 13 MISCELLANEOUS PROBLEMS (page 700)**

**1** The straight line $y = x + 2$   **3** The circle $(x - 2)^2 + (y - 1)^2 = 1$
**5** Equation: $y^2 = (x - 1)^3$

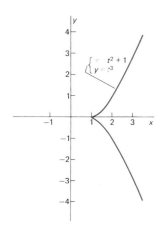

**7** $y - 2\sqrt{2} = -\frac{4}{3}(x - (3\sqrt{2}/2))$   **9** $y + (2/\pi)x = \pi/2$   **11** 24   **13** $3\pi$   **15** $(13\sqrt{13} - 8)/27 \approx 1.4397$   **17** $\frac{43}{6}$
**19** $(4\pi - 3\sqrt{3})/8 \approx 0.92128$   **21** $471295\pi/1024 \approx 1445.915$   **23** $(\pi\sqrt{5}/2)(e^\pi + 1) \approx 84.7919$
**25** $x = a\theta - b\sin\theta$, $y = a - b\cos\theta$. The graph shows the case $a = 1$, $b = 0.7$.

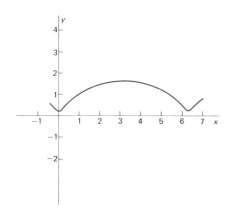

27 Suggestion: Compute $r^2 = x^2 + y^2$.   29 $6\pi^3 a^3$
35 There are two solutions: $\alpha \approx 0.033364$ radians (about $1°54'53''$) and $\alpha \approx 1.29116$ radians (about $73°58'40''$).
37 The curvature is zero when $x$ is a multiple of $\pi$, and reaches the maximum value 1 when $x$ is an odd multiple of $\pi/2$.
39 $\mathbf{T} = -(\pi/(\pi^2 + 4)^{1/2})\mathbf{i} + (2/(\pi^2 + 4)^{1/2})\mathbf{j}$, $\mathbf{N} = -(2/(\pi^2 + 4)^{1/2})\mathbf{i} - (\pi/(\pi^2 + 4)^{1/2})\mathbf{j}$   43 $y = \frac{15}{8}x - \frac{5}{4}x^3 + \frac{3}{8}x^5$
45 $f(x) = (\omega^2/2g)x^2$

# CHAPTER 14

## SECTION 14-1 (page 711)

1 $\langle 5, 8, -11 \rangle$; $\langle 2, 23, 0 \rangle$; 4; $\sqrt{51}$; $(\sqrt{5}/15)\langle 2, 5, -4 \rangle$   3 $2\mathbf{i} + 3\mathbf{j} + \mathbf{k}$; $3\mathbf{i} - \mathbf{j} + 7\mathbf{k}$; 0; $\sqrt{5}$; $(\sqrt{3}/3)(\mathbf{i} + \mathbf{j} + \mathbf{k})$
5 $4\mathbf{i} - \mathbf{j} - 3\mathbf{k}$; $6\mathbf{i} - 7\mathbf{j} + 12\mathbf{k}$; $-1$; $\sqrt{17}$; $(\sqrt{5}/5)(2\mathbf{i} - \mathbf{j})$   7 $\theta = \cos^{-1}(-13\sqrt{10}/50) \approx 2.536$
9 $\theta = \cos^{-1}(-34/\sqrt{3154}) \approx 2.221$   11 $\text{comp}_\mathbf{b}\mathbf{a} = \frac{2}{7}\sqrt{14}$; $\text{comp}_\mathbf{a}\mathbf{b} = \frac{4}{15}\sqrt{5}$   13 $\text{comp}_\mathbf{b}\mathbf{a} = 0 = \text{comp}_\mathbf{a}\mathbf{b}$
15 $\text{comp}_\mathbf{b}\mathbf{a} = -\frac{1}{10}\sqrt{10}$; $\text{comp}_\mathbf{a}\mathbf{b} = -\frac{1}{5}\sqrt{5}$   17 $(x + 2)^2 + (y - 1)^2 + (z + 5)^2 = 7$
19 $(x - 4)^2 + (y - 5)^2 + (z + 2)^2 = 38$   21 Center: $(-2, 3, 0)$; radius: $\sqrt{13}$   23 Center: $(0, 0, 3)$; radius: 5
25 A plane perpendicular to the $z$-axis at $z = 10$   27 All points in the coordinate planes
29 The point $(1, 0, 0)$   31 $\alpha = \cos^{-1}(\sqrt{6}/9) \approx 74.21°$, $\beta = \cos^{-1}(5\sqrt{6}/18) \approx 47.12°$, $\gamma = \beta$
33 $\alpha = \cos^{-1}(3\sqrt{2}/10) \approx 64.90°$, $\beta = \cos^{-1}(2\sqrt{2}/5) \approx 55.55°$, $\gamma = \cos^{-1}(\sqrt{2}/2) = 45°$   35 48   39 $A = \frac{3}{2}\sqrt{69} \approx 12.46$
41 $\cos^{-1}(\sqrt{3}/3) \approx 0.9553$, or about $54.7356°$
47 $2x + 9y - 5z = 23$; the plane through the midpoint of the segment AB perpendicular to AB   49 $60°$

## SECTION 14-2 (page 719)

1 $\langle 0, -14, 7 \rangle$   3 $\langle -10, -7, 1 \rangle$   7 $(\mathbf{a} \times \mathbf{b}) \times \mathbf{c} = \langle -1, 1, 0 \rangle$, $\mathbf{a} \times (\mathbf{b} \times \mathbf{c}) = \langle 0, 0, -1 \rangle$   11 $A = \frac{1}{2}\sqrt{2546} \approx 25.229$
13 (a) $V = 55$   (b) $V = 55/6$   17 $\frac{1}{38}\sqrt{9842} \approx 2.6107$

## SECTION 14-3 (page 724)

1 $x = t$, $y = 2t$, $z = 3t$   3 $x = 4 + 2t$, $y = 13$, $z = -3 - 3t$   5 $x = 3 + 3t$, $y = 5 - 13t$, $z = 7 + 3t$
7 $x + 2y + 3z = 0$   9 $x - z + 8 = 0$   11 $x = 2 + t$, $y = 3 - t$, $z = -4 - 2t$; $x - 2 = -y + 3 = (-z - 4)/2$
13 $x = 1$, $y = 1$, $z = 1 + t$; $x - 1 = 0 = y - 1$, $z$ arbitrary
15 $x = 2 + 2t$, $y = -3 - t$, $z = 4 + 3t$; $(x - 2)/2 = -y - 3 = (z - 4)/3$   17 $y = 7$   19 $7x + 11y = 114$
21 $3x + 4y - z = 0$   23 $2x - y - z = 0$   25 $\theta = \cos^{-1}(1/\sqrt{3}) \approx 54.736°$   27 The planes are parallel: $\theta = 0$.
29 $(x - 3)/2 = y - 3 = (-z + 1)/5$   31 $3x + 2y + z = 6$   33 $7x - 5y - 2z = 9$   35 $x - 2y + 4z = 3$   37 $\frac{10}{3}\sqrt{3}$
41 $D = 133/\sqrt{501} \approx 5.942$

## SECTION 14-4 (page 730)

1 $\mathbf{v} = 5\mathbf{k}$, $\mathbf{a} = 0$, $v = 5$   3 $\mathbf{v} = \mathbf{i} + 2t\mathbf{j} + 3t^2\mathbf{k}$, $\mathbf{a} = 2\mathbf{j} + 6t\mathbf{k}$, $v = (1 + 4t^2 + 9t^4)^{1/2}$
5 $\mathbf{v} = \mathbf{i} + 3e^t\mathbf{j} + 4e^t\mathbf{k}$, $\mathbf{a} = 3e^t\mathbf{j} + 4e^t\mathbf{k}$, $v = (1 + 25e^t)^{1/2}$
7 $\mathbf{v} = -(3 \sin t)\mathbf{i} + (3 \cos t)\mathbf{j} - 4\mathbf{k}$, $\mathbf{a} = -(3 \cos t)\mathbf{i} - (3 \sin t)\mathbf{j}$, $v = 5$   9 $\mathbf{r} = t^2\mathbf{i} + 10t\mathbf{j} - 2t^2\mathbf{k}$
11 $\mathbf{r} = 2\mathbf{i} + t^2\mathbf{j} + (5t - t^3)\mathbf{k}$   13 $\mathbf{r} = (10 + \frac{1}{6}t^3)\mathbf{i} + (10t + \frac{1}{12}t^4)\mathbf{j} + \frac{1}{20}t^5\mathbf{k}$
15 $\mathbf{r} = (1 - t - \cos t)\mathbf{i} + (1 + t - \sin t)\mathbf{j} + 5t\mathbf{k}$   17 $\mathbf{v} = 3\sqrt{2}(\mathbf{i} + \mathbf{j}) + 8\mathbf{k}$, $v = 10$, $\mathbf{a} = 6\sqrt{2}(-\mathbf{i} + \mathbf{j})$
21 Suggestion: Compute $(d/dt)(\mathbf{r} \cdot \mathbf{r})$.   23 Maximum height: 200 ft; speed: $\sqrt{6425} \approx 80.16$ ft/s

## SECTION 14-5 (page 736)

1 $10\pi$   3 $19(e - 1) \approx 32.647$   5 $2 + \frac{9}{10}\ln 3 \approx 3.5588$   7 $\kappa = \frac{1}{2}$   9 $\kappa = (\sqrt{2}/3)e^{-t}$   11 $a_T = 0$, $a_N = 0$
13 $a_T = (4t + 18t^3)/(1 + 4t^2 + 9t^4)^{1/2}$, $a_N = 2(1 + 9t^2 + 9t^4)^{1/2}/(1 + 4t^2 + 9t^4)^{1/2}$
15 $a_T = t/(t^2 + 2)^{1/2}$, $a_N = (t^4 + 5t^2 + 8)^{1/2}/(t^2 + 2)^{1/2}$
17 $\mathbf{T} = (\sqrt{2}/2)(\mathbf{i} + \mathbf{j} \cos t - \mathbf{k} \sin t)$, $\mathbf{N} = -\mathbf{j} \sin t - \mathbf{k} \cos t$; at $(0, 0, 1)$, $\mathbf{T} = (\sqrt{2}/2)(\mathbf{i} + \mathbf{j})$, $\mathbf{N} = -\mathbf{k}$.
19 $\mathbf{T} = (\sqrt{3}/3)(\mathbf{i} + \mathbf{j} + \mathbf{k})$, $\mathbf{N} = (\sqrt{2}/2)(-\mathbf{i} + \mathbf{j})$   21 $x = 2 + \frac{4}{13}s$, $y = 1 - \frac{12}{13}s$, $z = 3 + \frac{3}{13}s$
23 $x(s) = 3 \cos(s/5)$, $y(s) = 3 \sin(s/5)$, $z(s) = 4s/5$

## SECTION 14-6 (page 746)

1 A plane with intercepts $(\frac{20}{3}, 0, 0)$, $(0, 10, 0)$, and $(0, 0, 2)$   3 A vertical circular cylinder with radius 3
5 A vertical cylinder intersecting the $xy$-plane in the rectangular hyperbola $xy = 4$

**7** An elliptical paraboloid opening upward from its vertex at the origin

**9** A circular paraboloid that opens downward from its vertex at (0, 0, 4)

**11** Paraboloid, vertex at the origin, axis the $z$-axis, opening upward

**13** Cone, vertex the origin, axis the $z$-axis (both nappes)

**15** Parabolic cylinder perpendicular to the $xz$-plane, its trace there the parabola opening upward with axis the $z$-axis and vertex at $(0, -2)$

**17** Elliptical cylinder perpendicular to the $xy$-plane, its trace there the ellipse with center $(0, 0)$ and intercepts $(1, 0)$, $(0, 2)$, $(-1, 0)$, and $(0, -2)$   **19** Elliptical cone, vertex $(0, 0, 0)$, axis the $x$-axis

**21** Paraboloid, vertex at the origin, axis the $z$-axis, opening downward

**23** Hyperbolic paraboloid, saddle point at the origin, meets the $xz$-plane in a parabola with vertex the origin and opening downward, meets the $xy$-plane in a parabola with vertex the origin and opening upward, meets each plane parallel to the $yz$-plane in a hyperbola with directrices parallel to the $y$-axis

**25** Hyperboloid of one sheet, axis the $z$-axis, trace in the $xy$-plane the circle with center $(0, 0)$ and radius 3, traces in parallel planes larger circles and traces in planes parallel to the $z$-axis hyperbolas

**27** Elliptic paraboloid, axis the $y$-axis, vertex at the origin   **29** Hyperboloid of two sheets, axis the $y$-axis

**31** Paraboloid, axis the $x$-axis, vertex at the origin, equation $x = 2(y^2 + z^2)$

**33** Hyperboloid of one sheet (see Fig. 14.46) with equation $x^2 + y^2 - z^2 = 1$

**35** Paraboloid, vertex at the origin, axis the $x$-axis, equation $y^2 + z^2 = 4x$

**37** The surface resembles a rug with a soccer ball underneath. Its highest point is $(0, 0, 1)$; $z \to 0$ from above as $x$ or $y$ (or both) increase without bound. Its equation is $z = \exp(-x^2 - y^2)$.

**39** A circular cone with the $z$-axis as its axis of symmetry   **41** Ellipses with semiaxes 2 and 1   **43** Circles

**45** Parabolas that open downward   **47** Parabolas that open upward if $k > 0$, downward if $k < 0$

**51** The projection of the intersection has equation $x^2 + y^2 = 2y$; it is the circle with center $(0, 1)$ and radius 1.

**53** Equation: $5x^2 + 8xy + 8y^2 - 8x - 8y = 0$. Because $B^2 - 4AC < 0$, it is an ellipse. (In a $uv$-plane rotated approximately $55°16'41''$ from the $xy$-plane, the ellipse has center $(0.517, -0.453)$, minor axis 0.352 in the $u$-direction, major axis 0.774 in the $v$-direction.)

### SECTION 14-7 (page 751)

**1** Cylindrical coordinates $(0, 0, 5)$; spherical coordinates $(5, 0, 0)$

**3** Cylindrical coordinates $(\sqrt{2}, \pi/4, 0)$; spherical coordinates $(\sqrt{2}, \pi/2, \pi/4)$

**5** Cylindrical coordinates $(\sqrt{2}, \pi/4, 1)$; spherical coordinates $(\sqrt{3}, \tan^{-1}\sqrt{2}, \pi/4)$

**7** Cylindrical coordinates $(\sqrt{5}, \tan^{-1}(\frac{1}{2}), -2)$; spherical coordinates $(3, \pi/2 + \tan^{-1}(\sqrt{5}/2), \tan^{-1}(\frac{1}{2}))$

**9** Cylindrical coordinates $(5, \tan^{-1}(\frac{4}{3}), 12)$; spherical coordinates $(13, \tan^{-1}(\frac{5}{13}), \tan^{-1}(\frac{4}{3}))$

**11** Cylinder, radius 5, axis the $z$-axis   **13** The *plane* $y = x$   **15** The upper nappe of the cone $x^2 + y^2 = 3z^2$

**17** The $xy$-plane   **19** Cylinder, axis the vertical line $x = 0$, $y = 1$; its trace in the $xy$-plane is the circle $x^2 + (y - 1)^2 = 1$

**21** The vertical plane having trace the line $y = -x$ in the $xy$-plane   **23** The horizontal plane $z = 1$

**25** $r^2 + z^2 = 25$; $\rho = 5$   **27** $r(\cos \theta + \sin \theta) + z = 1$; $\rho(\sin \phi \cos \theta + \sin \phi \sin \theta + \cos \phi) = 1$

**29** $r^2 + z^2 = r(\cos \theta + \sin \theta) + z$; $\rho = \sin \phi \cos \theta + \sin \phi \sin \theta + \cos \phi$   **31** $z = r^2$

**33** (a) $1 \leq r^2 \leq 4 - z^2$   (b) $\csc \phi \leq \rho \leq 2$ (and, as a consequence, $\pi/6 \leq \phi \leq 5\pi/6$)   **35** About 3821 mi

**37** Just under 31 mi   **39** $0 \leq \rho \leq H \sec \phi$, $0 \leq \phi \leq \tan^{-1}(R/H)$, $\theta$ arbitrary

### CHAPTER 14 MISCELLANEOUS PROBLEMS (page 752)

**7** $x = 1 + 2t$, $y = -1 + 3t$, $z = 2 - 3t$; $(x - 1)/2 = (y + 1)/3 = (-z + 2)/3$   **9** $-13x + 22y + 6z = -23$

**11** $x - y + 2z = 3$   **15** 3   **17** $\kappa = 1/9$, $a_T = 2$, $a_N = 1$   **21** $3x - 3y + z = 1$   **27** $\rho = 2 \cos \phi$

**29** $\rho^2 = 2 \cos 2\phi$; shape: like an hour-glass with rounded ends

# CHAPTER 15

### SECTION 15-2 (page 765)

**1** All $(x, y)$   **3** Except on the line $x = y$   **5** Except on the coordinate axes $x = 0$ and $y = 0$

**7** Except on the lines $x = y$ and $x = -y$   **9** Except at the origin $(0, 0, 0)$   **11** The horizontal plane with equation $z = 10$

**13** A plane that makes a $45°$ angle with the $xy$-plane, intersecting it in the line $x + y = 0$

**15** A circular paraboloid opening upward from its vertex at the origin

**17** The upper hemispherical surface of radius 2 centered at the origin

**19** A circular cone opening downward from its vertex at $(0, 0, 10)$   **21** Straight lines of slope 1

**23** Ellipses centered at $(0, 0)$, each with major axis twice the minor axis and lying on the $x$-axis

**25** Vertical ($y$-direction) translates of the curve $y = x^3$   **27** Circles centered at the point $(2, 0)$

**29** Circles centered at the origin   **31** Circular paraboloids opening upward, each with its vertex on the $z$-axis

**33** Spheres centered at $(2, -1, 3)$

**35** Elliptical cylinders each of which has axis the vertical line through $(2, 1, 0)$ and with the length of the $x$-semiaxis twice that of the $y$-semiaxis

**37**

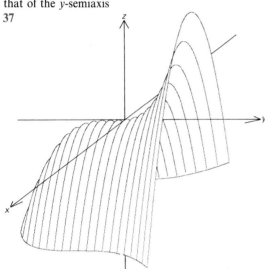

**39**

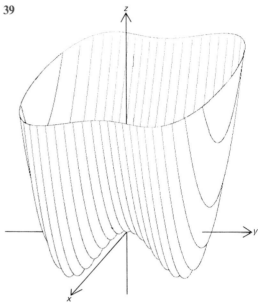

**41** Corresponding figures are: 15.22 and 15.30; 15.23 and 15.31; 15.24 and 15.28; 15.25 and 15.29; 15.26 and 15.33; 15.27 and 15.32

### SECTION 15-3 (page 774)

**1** 7   **3** $e$   **5** $\frac{5}{3}$   **7** 0   **9** 1   **11** $-\frac{3}{2}$   **13** 1   **15** $-4$   **17** $y$; $x$   **19** $y^2$; $2xy$

**29** Note that $f(x, y) = (x - \frac{1}{2})^2 + (y + 1)^2 - \frac{1}{4} \geq -\frac{1}{4}$ for all $(x, y)$.

### SECTION 15-4 (page 782)

**1** $\partial f/\partial x = 4x^3 - 3x^2y + 2xy^2 - y^3$, $\partial f/\partial y = -x^3 + 2x^2y - 3xy^2 + 4y^3$

**3** $\partial f/\partial x = e^x(\cos y - \sin y)$, $\partial f/\partial y = -e^x(\cos y + \sin y)$   **5** $\partial f/\partial x = -2y/(x - y)^2$, $\partial f/\partial y = 2x/(x - y)^2$

**7** $\partial f/\partial x = 2x/(x^2 + y^2)$, $\partial f/\partial y = 2y/(x^2 + y^2)$   **9** $\partial f/\partial x = yx^{y-1}$, $\partial f/\partial y = x^y\ln x$

**11** $\partial f/\partial x = 2xy^3z^4$, $\partial f/\partial y = 3x^2y^2z^4$, $\partial f/\partial z = 4x^2y^3z^3$   **13** $\partial f/\partial x = yze^{xyz}$, $\partial f/\partial y = xze^{xyz}$, $\partial f/\partial z = xye^{xyz}$

**15** $\partial f/\partial x = 2xe^y\ln z$, $\partial f/\partial y = x^2e^y\ln z$, $\partial f/\partial z = (x^2e^y)/z$   **17** $\partial f/\partial r = 4rs^2/(r^2 + s^2)^2$, $\partial f/\partial s = -4r^2s/(r^2 + s^2)^2$

**19** $\partial f/\partial u = e^v + we^u$, $\partial f/\partial v = e^w + ue^v$, $\partial f/\partial w = e^u + ve^w$   **21** $z_{xy} = z_{yx} = -4$   **23** $z_{xy} = z_{yx} = -4xye^{-y^2}$

**25** $z_{xy} = z_{yx} = -1/(x + y)^2$   **27** $z_{xy} = z_{yx} = 3e^{-3x}\sin y$   **29** $z_{xy} = z_{yx} = (-4x/y^3)\sinh(1/y^2)$   **31** $6x + 8y - z = 25$

**33** $z = -1$   **35** $27x - 12y - z = 38$   **37** $x - y + z = 1$   **39** $10x - 16y - z = 9$

**43** $f_{xyz} = (x^2y^2z^2 + 3xyz + 1)e^{xyz}$   **51** $(10, -7, -58)$

**53** (a) A decrease of about 2750 cm$^3$   (b) An increase of approximately 82.5 cm$^3$

### SECTION 15-5 (page 791)

**1** None   **3** $(0, 0, 5)$   **5** $(3, -1, -5)$   **7** $(-2, 0, -4)$   **9** $(-2, 0, -7)$ and $(-2, 1, -9)$

**11** $(0, 0, 0)$, $(1, 0, 2/e)$, $(-1, 0, 2/e)$, $(0, 1, 3/e)$, and $(0, -1, 3/e)$   **13** Minimum value: $f(1, 1) = 1$

**15** Maximum value: $f(1, 1) = f(1, -1) = 2$   **17** Minimum value: $f(4, -2) = f(-4, 2) = -16$
**19** Maximum value: $f(1, -2) = e^5$   **21** Maximum value: $f(1, 1) = 3$; minimum value: $f(-1, -1) = -3$
**23** Maximum value: $f(0, 2) = 4$; minimum value: $f(1, 0) = -1$
**25** Maximum value: $f(1/\sqrt{2}, 1/\sqrt{2}) = f(-1/\sqrt{2}, -1/\sqrt{2}) = 1$; minimum value: $f(1/\sqrt{2}, -1/\sqrt{2}) = f(-1/\sqrt{2}, 1/\sqrt{2}) = -1$
**27** Dimensions: 10 by 10 by 10   **29** 10 by 10 by 10 cm   **31** Maximum value: $\frac{4}{3}$; $x = 2$, $y = 1$, $z = \frac{2}{3}$
**33** Height: 10 ft; front, back 40 ft wide; sides 20 ft deep   **35** $V_{\max} = 5488$ in$^3$   **37** $V_{\max} = \frac{1}{2}$
**39** Maximum area: 900 (one square); minimum: 300 (three equal squares)


### SECTION 15-6 (page 798)

**1** $dw = (6x + 4y)\,dx + (4x - 6y^2)\,dy$   **3** $dw = (x\,dx + y\,dy)/\sqrt{1 + x^2 + y^2}$   **5** $dw = (-y\,dx + x\,dy)/(x^2 + y^2)$
**7** $dw = (2x\,dx + 2y\,dy + 2z\,dz)/(x^2 + y^2 + z^2)$   **9** $dw = (\tan yz)\,dx + xz(\sec^2 yz)\,dy + xy(\sec^2 yz)\,dz$
**11** $dw = -(yz\,dx + xz\,dy + xy\,dz)e^{-xyz}$   **13** $dw = (2u\,du - 2u^2v\,dv)e^{-v^2}$
**15** $dw = (x\,dx + y\,dy + z\,dz)/\sqrt{x^2 + y^2 + z^2}$   **17** $\Delta f \approx 0.014$ (true value: about $0.01422975$)   **19** $\Delta f \approx -0.0007$
**21** $\Delta f \approx 53/1300 \approx 0.04077$   **23** $\Delta f \approx 0.06$   **25** 191.1   **27** 1.4   **29** $x \approx 1.95$   **31** 8.18 in.$^3$   **33** 0.022 acres
**35** The period increases by about 0.0278 s.   **37** Approximately 303.8 ft


### SECTION 15-7 (page 807)

**1** $-(2t + 1)e^{-t^2 - t}$   **3** $6t^5 \cos t^6$   **5** $\partial w/\partial s = \partial w/\partial t = 2/(s + t)$   **7** $\partial w/\partial s = 0, \partial w/\partial t = 5e^t$
**9** $\partial r/\partial x = (y + z)\exp(yz + xz + xy), \partial r/\partial y = (x + z)\exp(yz + xz + xy), \partial r/\partial z = (x + y)\exp(yz + xz + xy)$
**11** $z_x = -(z/x)^{1/3}, z_y = -(z/y)^{1/3}$
**13** $z_x = -(yz(e^{xy} + e^{xz}) + (xy + 1)e^{xy})/(e^{xy} + xye^{yz}), z_y = -(x(x + z)e^{xy} + e^{xz})/(xye^{xz} + e^{xy})$
**15** $z_x = -c^2x/a^2z, z_y = -c^2y/b^2z$
**17** $\partial w/\partial x = \frac{1}{4}(x + y)^{1/4}[xy/(x - y)]^{1/2} + \frac{1}{4}(x - y)^{1/4}[xy/(x + y)]^{1/2} + \frac{1}{2}(x^2 - y^2)^{1/4}(y/x)^{1/2}$;
$\partial w/\partial y = -\frac{1}{4}(x + y)^{1/4}[xy/(x - y)]^{1/2} + \frac{1}{4}(x - y)^{1/4}[xy/(x + y)]^{1/2} + \frac{1}{2}(x^2 - y^2)^{1/4}(x/y)^{1/2}$
**19** $\partial w/\partial x = (y^3 - 3x^2y)(x^2 + y^2)^{-3} - y, \partial w/\partial y = (x^3 - 3xy^2)(x^2 + y^2)^{-3} - x$   **21** $x + y - 2z = 7$
**23** $5x + 5y + 11z = 31$   **29** $\partial w/\partial x = f'(u)[\partial u/\partial x] = f'(u)$, and so on
**31** Show that $w_u = w_x + w_y$. Then note that $w_{uv} = (\partial/\partial v)(w_u) = (\partial w_u/\partial x)(\partial x/\partial v) + (\partial w_u/\partial y)(\partial y/\partial v)$.


### SECTION 15-8 (page 816)

**1** $3\mathbf{i} - 7\mathbf{j}$   **3** 0   **5** $6\mathbf{j} - 4\mathbf{k}$   **7** $\mathbf{i} + \mathbf{j} + \mathbf{k}$   **9** $\frac{1}{2}(4\mathbf{i} - 3\mathbf{j} - 4\mathbf{k})$   **11** $8\sqrt{2}$   **13** $\frac{12}{13}\sqrt{13}$   **15** $-\frac{13}{20}$   **17** $-\frac{1}{6}$   **19** $-6\sqrt{2}$
**21** Maximum: $\sqrt{170}$; direction: $7\mathbf{i} + 11\mathbf{j}$   **23** Maximum: $14\sqrt{2}$; direction: $3\mathbf{i} + 5\mathbf{j} - 8\mathbf{k}$
**25** Maximum: $2\sqrt{14}$; direction: $\mathbf{i} + 2\mathbf{j} + 3\mathbf{k}$   **27** $29(x - 2) - 4(y + 3) = 0$   **29** $x + y + z = 1$   **35** $\frac{55}{6}\sqrt{2}$
**37** (a) Direction: $-\mathbf{i} - 2\mathbf{j}$; angle: about $43°44'$   (b) Angle: about $23°10'$   **39** $x - 2y + z + 10 = 0$


### SECTION 15-9 (page 825)

**1** Maximum: 4, at $(\pm 2, 0)$, minimum: $-4$, at $(0, \pm 2)$
**3** Maximum: 3, at $(\pm\frac{3}{2}\sqrt{2}, \pm\sqrt{2})$ [same signs]; minimum: $-3$, at $(\pm\frac{3}{2}\sqrt{2}, \mp\sqrt{2})$ [opposite signs]
**5** $\frac{126}{49}$ at $(\frac{9}{7}, \frac{6}{7}, \frac{3}{7})$ is the minimum; no maximum   **7** Maximum: 7, at $(\frac{36}{7}, \frac{9}{7}, \frac{4}{7})$, minimum: $-7$, at $(-\frac{36}{7}, -\frac{9}{7}, -\frac{4}{7})$
**9** Minimum: $\frac{25}{3}$, at $(-\frac{5}{3}, \frac{1}{3}, \frac{7}{3})$.   **21** Closest points: $(2, -2, 1)$ and $(-2, 2, 1)$   **25** $(2, 3)$ and $(-2, -3)$.
**29** Highest point: $(\frac{2}{5}\sqrt{5}, \frac{1}{5}\sqrt{5}, \sqrt{5} - 4)$; lowest point: $(-\frac{2}{5}\sqrt{5}, -\frac{1}{5}\sqrt{5}, -\sqrt{5} - 4)$
**31** Farthest: $((-15 - 9\sqrt{5})/20, (-15 - 9\sqrt{5})/10, (9 + 3\sqrt{5})/4)$,
nearest: $((-15 + 9\sqrt{5})/20, (-15 + 9\sqrt{5})/10, (9 - 3\sqrt{5})/10, (9 - 3\sqrt{5})/4)$


### SECTION 15-10 (page 832)

**1** Minimum: $(-1, 2, -1)$, no other extrema   **3** Saddle point: $(-\frac{1}{2}, -\frac{1}{2}, \frac{29}{4})$, no extrema
**5** Minimum: $(-3, 4, -9)$, no other extrema   **7** Saddle point at $(0, 0, 3)$; local maximum at $(-1, -1, 4)$
**9** No extrema   **11** Saddle point at $(0, 0, 0)$, local minima at $(-1, -1, -2)$ and $(1, 1, -2)$

**13** Saddle point at $(-1, 1, 5)$, local minimum at $(3, -3, -27)$
**15** Saddle point at $(0, -2, 32)$, local minimum at $(-5, 3, -93)$
**17** Saddle point at $(0, 0, 0)$, local minima at $(\pm 1, \pm 2, 2)$  **19** Saddle point at $(-1, 0, 17)$, local minimum at $(2, 0, -10)$
**21** Local minimum at $(0, 0, 0)$, saddle point at $(2, 0, 4e^{-2})$  **23** Local minimum  **25** Local maximum
**27** Minimum value 3 at $(1, 1)$ and at $(-1, -1)$  **29** See Problem 39 in Section 15-5 and its answer.
**31** The critical points are of the form $(m, n)$ where $m$ and $n$ are either both even integers or both odd integers. The critical point $(m, n)$ is a saddle point if $m$ and $n$ are both even. It is a local maximum if $m$ and $n$ are both of the form $4k + 1$ or both of the form $4k + 3$. It is a local minimum if one is of each form.
**33** Local minima at $(-1.879385, 0)$ and $(1.532089, 0)$, saddle point at $(0.347296, 0)$
**34** Saddle point at $(1.179509, 0)$; no local extrema
**35** Local minima at $(-1.879385, 1.879385)$ and $(1.532089, -1.532089)$; saddle point at $(0.347296, -0.347296)$
**36** Saddle point at $(1.179509, 1.179509)$; no extrema
**37** Saddle point at $(0, 0)$; local minima at $(3.624679, 3.984224)$ and $(3.624679, -3.984224)$

**SECTION 15-11 (page 839)**
**1** $m = \frac{113}{38} \approx 2.9737$, $b = \frac{472}{95} \approx 4.9684$  **3** $m = \frac{1497}{89} \approx 16.8202$, $b \approx -\frac{30711}{890} \approx -34.5067$  **5** 174; 41  **7** 107
**9** $A = \frac{29}{15}$, $B = -\frac{17}{6}$, $C = \frac{72}{5}$, Answer: 10.5  **11** $x = 122$, $y = 176$, Profit: \$4189.16
**13** Raise 40 head of cattle, 40 hogs. Profit: \$1120.
**15** (a) $A = 125$, $B = 375$, $C = 125$, $P = \$15625$, $Q = \$281,250$, $R = \$46875$. Note that $P + R = \$62500$
(b) $A = 0$, $B = 250$, $C = 250$, $P = 0$, $Q = \$125,000$, $R = \$187,500$, $P + R = \$187,500$.

**CHAPTER 15 MISCELLANEOUS PROBLEMS (page 841)**
**3** On the line $y = x$, $g(x, y) \equiv \frac{1}{2}$, except that $g(0, 0) = 0$  **5** $f(x, y) = x^2 y^3 + e^x \sin y + y + C$
**7** All points of the form $(a, b, \frac{1}{2})$ (so that $a^2 + b^2 = \frac{1}{2}$) together with $(0, 0, 0)$.
**9** The normal to the cone at $(a, b, [a^2 + b^2]^{1/2})$ meets the $z$-axis at $z = 2[a^2 + b^2]^{1/2}$.
**15** Base: $2\sqrt[3]{3}$ by $2\sqrt[3]{3}$, height $5\sqrt[3]{3}$  **17** $200 \pm 2$ ohms  **19** 3%  **21** $(\pm 4, 0, 0)$, $(0, \pm 2, 0)$, $(0, 0, \pm \frac{4}{3})$  **25** $3\mathbf{i} + 4\mathbf{j}$; 1
**27** The plane tangent to the graph at $(a, b, c)$ has $x$-intercept $a^{1/3}$  **31** Semiaxes: 1 and 2
**33** There is no such triangle of minimum perimeter, unless one is willing to consider as a triangle the figure with all sides of length zero—a single point on the circumference of the circle. The triangle of *maximum* perimeter is equilateral, with perimeter $3\sqrt{3}$.  **35** Closest point: $(\frac{1}{3}\sqrt{6}, \frac{1}{6}\sqrt{6})$, farthest point: $(-\frac{1}{3}\sqrt{6}, -\frac{1}{6}\sqrt{6})$  **39** Maximum: 1, minimum: $-\frac{1}{2}$
**41** Local minima: $-1$, at $(1, 1)$ and at $(-1, -1)$; horizontal tangent plane (not extrema): $(0, 0, 0)$, $(\pm\sqrt{3}, 0, 0)$
**43** Local minimum: $-8$ at $(2, 2)$, horizontal tangent plane at $(0, 0, 0)$
**45** Local maximum $\frac{1}{432}$ at $(\frac{1}{2}, \frac{1}{3})$. The points on the intervals $(-\infty, 0)$ and $(1, +\infty)$ on the $x$-axis are all local minima (value 0), and those on the interval $(0, 1)$ on the $x$-axis are local maxima (value 0). Saddle point at $(0, 1, 0)$.
**47** Saddle point at $(0, 0, 1)$; each point on the hyperbola $xy = \ln 2$ yields a global minimum.
**49** No extrema; saddle points at $(1, 1, 0)$ and $(-1, -1, 0)$.

# CHAPTER 16

**SECTION 16-1 (page 848)**
**1** 80  **3** $-78$  **5** $\frac{513}{4}$  **7** $-\frac{9}{2}$  **9** 1  **11** $(e - 1)/2$  **13** $2(e - 1)$  **15** $2\pi + \frac{1}{4}\pi^4$  **17** 1  **19** $2 \ln 2$  **21** $-32$
**23** $\frac{4}{15}(9\sqrt{3} - 4\sqrt{2} - 1)$

**SECTION 16-2 (page 854)**
**1** $\frac{5}{6}$  **3** $\frac{1}{2}$  **5** $\frac{1}{12}$  **7** $\frac{1}{20}$  **9** $-\frac{1}{18}$  **11** $(e - 2)/2$  **13** $\frac{61}{3}$  **15** $\int_0^4 \int_{-y^{1/2}}^{y^{1/2}} x^2 y \, dx \, dy = \frac{512}{21}$
**17** $\int_0^1 \int_{-y^{1/2}}^{y^{1/2}} x \, dx \, dy + \int_1^9 \int_{(1/2)(y-3)}^{y^{1/2}} x \, dx \, dy = \frac{32}{3}$  **19** $\int_0^4 \int_{2-(4-y)^{1/2}}^{y/2} 1 \, dx \, dy = \frac{4}{3}$  **21** $\int_0^\pi \int_0^y (\sin y/y) \, dx \, dy = 2$
**23** $\int_0^1 \int_0^x (1/(1 + x^4)) \, dy \, dx = \pi/8$

## SECTION 16-3 (page 858)

**1** $\frac{1}{6}$  **3** $\frac{32}{3}$  **5** $\frac{5}{6}$  **7** $\frac{32}{3}$  **9** $2\ln 2$  **11** $2$  **13** $2e$  **15** $\frac{1}{3}$  **17** $\frac{41}{60}$  **19** $\frac{4}{15}$  **21** $\frac{10}{3}$  **23** $19$  **25** $\frac{4}{3}$  **27** $abc/6$  **29** $\frac{2}{3}$
**33** $625\pi/2 \approx 981.748$  **35** $\frac{1}{6}(2\pi + 3\sqrt{3})R^3 \approx (1.913)R^3$

## SECTION 16-4 (page 866)

**3** $3\pi/2$  **5** $\frac{1}{6}(4\pi - 3\sqrt{3}) \approx 1.22837$  **7** $\frac{1}{2}(2\pi - 3\sqrt{3}) \approx 0.5435$  **9** $16\pi/3$  **11** $23\pi/8$  **13** $\frac{1}{4}\pi\ln 2$  **15** $16\pi/5$
**17** $\frac{1}{4}\pi(1 - \cos 1) \approx 0.36105$  **19** $2\pi$  **21** $4\pi$  **27** $2\pi$  **29** $\frac{1}{3}\pi a^3(2 - \sqrt{2}) \approx (0.6134)a^3$  **31** $\pi/4$  **35** $2\pi^2 a^2 b$

## SECTION 16-5 (page 873)

**1** $M = \frac{1}{24}, \bar{x} = \bar{y} = \frac{2}{5}$  **3** $M = \frac{256}{15}, \bar{x} = 0, \bar{y} = \frac{16}{7}$  **5** $M = \frac{1}{12}, \bar{x} = \bar{y} = \frac{9}{14}$  **7** $M = \frac{1}{3}, \bar{x} = 0, \bar{y} = \frac{22}{35}$
**9** $M = 2, \bar{x} = \pi/2, \bar{y} = \pi/8$  **11** $M = a^3, \bar{x} = \bar{y} = 7a/12$  **13** $M = \frac{128}{5}, \bar{x} = 0, \bar{y} = \frac{20}{7}$
**15** $M = \pi, \bar{x} = (\pi^2 - 4)/\pi \approx 1.87, \bar{y} = \pi/8 \approx 0.39$  **17** $M = \frac{1}{3}\pi a^3, \bar{x} = 0, \bar{y} = 3a/2\pi$
**19** $M = \frac{2}{3}\pi + \frac{1}{4}\sqrt{3}, \bar{x} = 0, \bar{y} = (36\pi + 33\sqrt{3})/(32\pi + 12\sqrt{3}) \approx 1.4034$  **21** $I_0 = [2\pi/(n + 4)]a^{n+4}$  **23** $I_0 = 3\pi k/2$
**25** $I_0 = \frac{1}{9}$  **27** $\hat{x} = \frac{2}{21}\sqrt{105}, \hat{y} = \frac{4}{3}\sqrt{5}$  **29** $\hat{x} = \hat{y} = \frac{1}{10}a\sqrt{30}$  **33** $484k/3$  **35** (a) $\frac{1}{2}\pi\rho(b^4 - a^4)$  (b) $4\sqrt{gh/13}$
**37** $(1, \frac{1}{4})$

## SECTION 16-6 (page 880)

**1** $18$  **3** $128$  **5** $\frac{31}{60}$  **7** $0$  **9** $12$  **11** $V = \int_0^3 \int_0^{2 - (2x/3)} \int_0^{6 - 2x - 3y} 1\, dz\, dy\, dx = 6$  **13** $\frac{128}{5}$  **15** $\frac{332}{105}$  **17** $\frac{256}{15}$  **19** $\frac{11}{30}$
**21** $(0, \frac{20}{7}, \frac{10}{7})$  **23** $(0, \frac{8}{7}, \frac{12}{7})$  **25** $\bar{x} = 0, \bar{y} = (44 - 9\pi)/(72 - 9\pi), \bar{z} = (9\pi - 16)/(72 - 9\pi)$  **27** $\frac{8}{7}$  **29** $\frac{1}{30}$  **31** $\frac{2}{3}a^5$
**33** $\frac{38}{45}ka^7$  **35** $k/3$  **37** $(9\pi/64, 9\pi/64, \frac{3}{8})$  **39** $24\pi$  **41** $\pi/6$

## SECTION 16-7 (page 886)

**1** $8\pi$  **5** $\frac{4}{3}\pi(8 - 3\sqrt{3})$  **7** $\frac{1}{2}\pi a^2 h^2$  **9** $\frac{1}{2}\pi a^4 h^2$  **11** $V = 81\pi/2$, centroid $(0, 0, 3)$  **13** $V = 24\pi$  **15** $V = \frac{1}{6}\pi(8\sqrt{2} - 7)$
**17** $I_x = \frac{1}{12}\pi\delta a^2 h(3a^2 + 4h^2)$  **19** $\pi/3$  **23** $\pi/3$
**25** $V = \frac{1}{3}\pi a^3(2 - \sqrt{2})$; centroid: $\bar{x} = 0 = \bar{y}, \bar{z} = \frac{3}{16}(2 + \sqrt{2})a \approx (0.6402)a$  **27** $I_x = \frac{7}{5}Ma^2$
**29** Description: The surface obtained by rotating the circle in the $xz$-plane with center $(a, 0)$ and radius $a$ about the $z$-axis—a doughnut with an infinitesimal hole. Volume: $2\pi^2 a^3$.
**31** $I_x = \frac{2}{15}(128 - 51\sqrt{3})\pi a^5$
**33** (a) $I_z = \frac{8}{15}\pi\delta(b^5 - a^5)$;  (b) $v = \sqrt{140gh/101}$. With $h = 50$ and $g = 32, v \approx 47.094$ ft/s: slower than a marble, faster than a quarter.

## SECTION 16-8 (page 892)

**1** $6\pi\sqrt{11}$  **3** $\frac{1}{6}\pi(17\sqrt{17} - 1)$  **5** $A = \frac{1}{2}[6\sqrt{2} + \ln(3 + 2\sqrt{2})] \approx 5.124$  **7** $A = 3\sqrt{14}$  **9** $A = \frac{2}{3}\pi(2\sqrt{2} - 1) \approx 3.829$
**11** $\frac{1}{6}\pi(65\sqrt{65} - 1)$  **15** $A = 8a^2$

## SECTION 16-9 (page 899)

**1** $x = (u + v)/2, y = (u - v)/2, J = -\frac{1}{2}$  **3** $x = (u/v)^{1/2}, y = (uv)^{1/2}, J = 1/2v$
**5** $x = (u + v)/2, y = \frac{1}{2}(u - v)^{1/2}, J = -1/4(u - v)^{1/2}$  **7** $A = \frac{3}{5}$  **9** $A = \ln 2$  **11** $A = (2 - \sqrt{2})/8$  **13** $V = 39\pi/2$
**15** $V = 8$  **17** $S$ is the region $3u^2 + v^2 \le 3$; the value of the integral is $2\pi\sqrt{3}(e^3 - 1)/3e^3$

## CHAPTER 16 MISCELLANEOUS PROBLEMS (page 900)

**1** $(2 - \sqrt{2})/3$  **3** $(e - 1)/2e$  **5** $(e^4 - 1)/4$  **7** $\frac{4}{3}$  **9** $V = 9\pi$; centroid: $(0, 0, \frac{9}{16})$  **11** $4\pi$  **13** $4\pi$  **15** $(\pi - 2)/16$
**17** Mass: $\frac{128}{15}$, centroid: $(\frac{32}{7}, 0)$  **19** Mass: $k\pi$, centroid: $(1, 0)$  **21** $(0, \frac{8}{5})$  **23** $(10\pi/3)(\sqrt{5} - 2) \approx 2.4721$  **25** $\frac{3}{10}Ma^2$
**27** $\frac{1}{5}M(b^2 + c^2)$  **29** $128\delta(15\pi - 26)/225 \approx (12.017)\delta$ ($\delta$ is [constant] density)  **31** $8\pi/3$  **39** $\frac{18}{7}$
**41** $(\pi/6)(37\sqrt{37} - 17\sqrt{17}) \approx 81.1418$  **45** $4\sqrt{2}$  **46** Approximately 3.49608  **49** $I_0 = 3\delta$  **51** $M = \frac{8}{15}\pi abc$

# CHAPTER 17

**SECTION 17-1 (page 910)**

1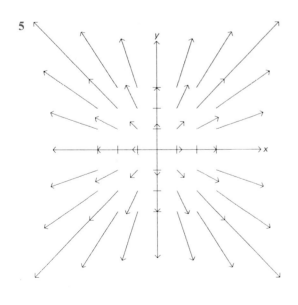

3

5

7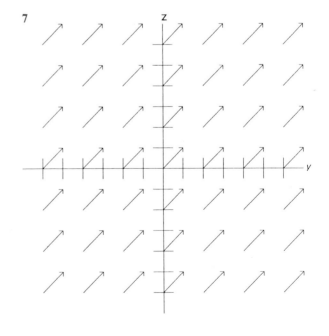

9

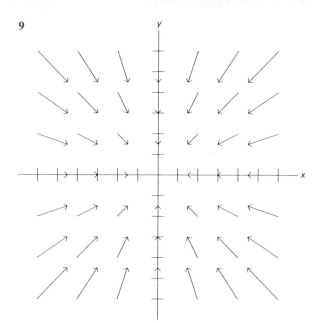

**11** $\mathbf{V} \cdot \mathbf{F} = 3, \mathbf{V} \times \mathbf{F} = \mathbf{0}$   **13** $\mathbf{V} \cdot \mathbf{F} = 0, \mathbf{V} \times \mathbf{F} = \mathbf{0}$   **15** $\mathbf{V} \cdot \mathbf{F} = x^2 + y^2 + z^2, \mathbf{V} \times \mathbf{F} = -2yz\mathbf{i} - 2xz\mathbf{j} - 2xy\mathbf{k}$
**17** $\mathbf{V} \cdot \mathbf{F} = 0, \mathbf{V} \times \mathbf{F} = (2y - 2z)\mathbf{i} + (2z - 2x)\mathbf{j} + (2x - 2y)\mathbf{k}$
**19** $\mathbf{V} \cdot \mathbf{F} = 3, \mathbf{V} \times \mathbf{F} = (x \cos xy - x \cos xz)\mathbf{i} + (y \cos yz - y \cos xy)\mathbf{j} + (z \cos xz - z \cos yz)\mathbf{k}$

**SECTION 17-2 (page 917)**

**1** $\frac{310}{3}, \frac{248}{3}, 62$   **3** $3\sqrt{2}, 3, 3$   **5** $\frac{49}{24}, \frac{3}{2}, \frac{4}{3}$   **7** $\frac{6}{5}$   **9** $315$   **11** $\frac{19}{60}$   **13** $\pi + 2\pi^2$   **15** $28$   **17** $\frac{1}{6}(14\sqrt{14} - 1) \approx 8.563867$
**19** $(0, 2a/\pi)$   **21** $10k\pi$; centroid: $(0, 0, 4\pi)$
**23** $M = \frac{1}{2}ka^3$; centroid $(2a/3, 2a/3)$; $I_x = I_y = \frac{1}{4}ka^5$
**25** (a) $\frac{1}{2}k \ln 2$;   (b) $-\frac{1}{2}k \ln 2$

**SECTION 17-3 (page 924)**

**1** $f(x, y) = x^2 + 3xy + y^2$   **3** $f(x, y) = x^3 + 2xy^2 + 2y^3$   **5** $f(x) = \frac{1}{4}x^4 + \frac{1}{3}y^3 + y \ln x$
**7** $f(x, y) = \sin x + x \ln y + e^y$   **9** $f(x, y) = x^3y^3 + xy^4 + \frac{1}{5}y^5$   **11** $f(x, y) = x^2/y + y^2/x^3 + 2\sqrt{y}$   **17** $6$   **19** $3/e$
**21** $-\pi$   **23** $f(x, y, z) = xyz$   **25** $f(x, y, z) = xy \cos z - yze^x$   **27** $\mathbf{F}$ is not conservative on any region containing $(0, 0)$.
**29** $Q_z = 0 \neq 2y = R_y$

**SECTION 17-4 (page 931)**

**1** $0$   **3** $3$   **5** $\frac{3}{10}$   **7** $2$   **9** $0$   **11** $\frac{16}{105}$   **13** $\pi a^2$   **15** $3\pi/8$   **25** $\frac{3}{2}$   **31** $2\pi$

**SECTION 17-5 (page 939)**

**1** $\sqrt{3}/120$   **3** $\frac{1}{60}\pi(1 + 391\sqrt{17}) \approx 84.4635$   **5** $16\pi/3$   **7** $24\pi$   **9** $0$   **11** $(a/2, a/2, a/2)$
**13** $\bar{x} = 0 = \bar{y}, \bar{z} = [1 + (24a^4 + 2a^2 - 1)\sqrt{1 + 4a^2}]/\{10[(1 + 4a^2)^{3/2} - 1]\}$,
$I_z = \frac{1}{60}\pi\delta[1 + (24a^4 + 2a^2 - 1)\sqrt{1 + 4a^2}]$   **15** $\bar{x} = 4/[3(\pi - 2)] \approx 1.16796, \bar{y} = 0, \bar{z} = \pi/[2(\pi - 2)] \approx 1.37597$
**17** Net flux: $0$   **19** Net flux: $1458\pi$

**SECTION 17-6 (page 947)**

**1** Both values: $4\pi$   **3** Both values: $24$   **5** Both values: $\frac{1}{2}$   **7** $2385\pi/2$   **9** $\frac{1}{4}$   **11** $730,125\pi/4$   **13** $16\pi$
**25** $\frac{1}{48}\pi(482,620 + 29,403 \ln 11)$

**SECTION 17-7 (page 955)**

**1** $-20\pi$   **3** $0$   **5** $-52\pi$   **7** $-8\pi$   **9** $-2$   **11** $\phi(x, y, z) = yz - 2xz + 3xy$   **13** $\phi(x, y, z) = 3xe^z + 5y \cos x + 17z$

1 $\frac{125}{3}$   3 $\frac{69}{8}$ (Use the fact that $\int \mathbf{F} \cdot \mathbf{T} \, ds$ is independent of the path.)   5 $\frac{2148}{5}$

9 $M = \frac{1}{3}(5\sqrt{5} - 1) \approx 3.3934$, $I_y = \frac{1}{15}(2 + 50\sqrt{5}) \approx 7.5869$   11 $\frac{2816}{7}$   17 $371\pi/30$   19 $72\pi$

29 (a) $\phi'(r)(\mathbf{r}/r)$   (b) $3\phi(r) + r\phi'(r)$   (c) $\mathbf{0}$

# CHAPTER 18

## SECTION 18-1 (page 966)

13 $r = \frac{3}{2}$   15 $r = -2, 1$   17 $A = \frac{4}{3}, B = -\frac{1}{3}$   19 $A = \frac{3}{4}, B = \frac{1}{4}$   21 $y' = x + y$   23 $y' = x/(1 - y)$

25 $y' = (y - x)/(x + y)$   27 $dv/dt = kv^2$   29 $dN/dt = k(P - N)$   31 $y = 1$ or $y = x$   33 $y = x^2$   35 $y = \frac{1}{2}e^x$

## SECTION 18-2 (page 971)

1 $y^2 = 1/(C - x^2)$   3 $\frac{1}{4}\ln(y^4 + 1) = \sin x + C$   5 $1/3y^3 - 2/y = 1/x + \ln|x| + C$   7 $\ln|1 + y| = x + \frac{1}{2}x^2 + C$

9 $y = 2\exp(e^x)$   11 $y^2 = 1 + \sqrt{x^2 - 16}$   13 $\ln(2y - 1) = 2(x - 1)$   15 $\ln y = x^2 - 1 + \ln x$

17 1 h 18 min 40 s after the plug is pulled   19 16 min 12 s   21 14 min 29 s

## SECTION 18-3 (page 977)

1 $xe^{-3x}y = C; y \equiv 0$   3 $e^x y = \frac{1}{2}e^{2x} + C; y = \frac{1}{2}(e^x + e^{-x})$   5 $y\exp(x^2) = \frac{1}{2}\exp(x^2) + C; y = \frac{1}{2}(1 - 5e^{-x^2})$

7 $(1 + x)y = \sin x + C; y = (1 + \sin x)/(1 + x)$   9 $y\sin x = \frac{1}{2}\sin^2 x + C$   11 $x^{-3}y = \sin x + C; y = x^3\sin x$

13 $x^{-3}e^{2x}y = 2e^{2x} + C$   15 $y = [\exp(-\frac{3}{2}x^2)][3(x^2 + 1)^{3/2} - 2]$   17 After about 7 min 41 s

19 After about 10.85 years   21 393.75 lb   23 (a) $x(t) = 50e^{-0.05t}$   (b) $y(t) = 150e^{-0.025t} - 100e^{-0.05t}$   (c) 56.25 lb

## SECTION 18-4 (page 984)

1 $25 + (0)i; 25 \text{ cis } 0$   3 $\frac{1}{13} + \frac{31}{13}i; (\sqrt{962}/13) \text{ cis}(\tan^{-1}31)$   5 $0 + (-8)i; 8 \text{ cis}(3\pi/2)$   7 $\pm\sqrt{2}/2 \pm (\sqrt{2}/2)i$

9 $\pm 2i, \pm\sqrt{3} \pm i$   11 $\pm 2, \pm 2i$   13 Let $P = \frac{1}{2}\sqrt{2 + \sqrt{3}}$ and $Q = \frac{1}{2}\sqrt{2 - \sqrt{3}}$; let $R = (2\sqrt{2})^{1/3}$. The solutions are then

given by $x = R(-Q + Pi), R(Q - Pi), R(Q + Pi)$, and $R(-P - Qi)$, and $R(\sqrt{2}/2 \pm (\sqrt{2}/2)i)$

15 $\cos 4\theta = \cos^4\theta - 6\cos^2\theta \sin^2\theta + \sin^4\theta, \sin 4\theta = 4\cos^3\theta \sin\theta - 4\cos\theta \sin^3\theta$

## SECTION 18-5 (page 990)

1 $y = c_1e^{2x} + c_2e^{-2x}$   3 $y = c_1e^{2x} + c_2e^{-5x}$   5 $y = c_1e^{-3x} + c_2xe^{-3x}$   7 $y = c_1e^{(3x/2)} + c_2xe^{(3x/2)}$

9 $y = e^{-4x}(c_1\cos 3x + c_2\sin 3x)$   11 $y = c_1 + c_2x + c_3e^{4x} + c_4xe^{4x}$   13 $y = c_1 + c_2e^{-(2x/3)} + c_3xe^{-(2x/3)}$

15 $y = c_1e^{2x} + c_2xe^{2x} + c_3e^{-2x} + c_4xe^{-2x}$   17 $y = c_1\cos(x/\sqrt{2}) + c_2\sin(x/\sqrt{2}) + c_3\cos(2x/\sqrt{3}) + c_4\sin(2x/\sqrt{3})$

19 $y = c_1e^x + c_2e^{-x} + c_3xe^{-x}$   21 $y = 5e^x + 2e^{3x}$   23 $y = e^{3x}(3\cos 4x - 2\sin 4x)$

25 $y = \frac{1}{4}(-13 + 6x + 9e^{-2x/3})$

## SECTION 18-6 (page 996)

1 Frequency: 2 rad/s; that is, $1/\pi$ Hz (cycles/s). Period: $\pi$ s   3 Amplitude: 2 m; frequency: 5 rad/s; period: $2\pi/5$ s

5 Amplitude: 100 cm; period: approximately 2 s   7 $x = 4e^{-2t} - 2e^{-4t}$; overdamped

9 $x = 5e^{-4t} + 10te^{-4t}$; critically damped   11 $x = \frac{1}{3}\sqrt{313}e^{-5t/2}\cos(6t - 0.8254)$

13 (a) $k \approx 7018.39$ lb/ft   (b) After about 2.49 s

## SECTION 18-7 (page 1002)

1 $y = c_0(1 + x + x^2/2! + x^3/3! + \cdots) = c_0e^x; \rho = \infty$

3 $y = c_0(1 - 3x/2 + 3^2x^2/2!2^2 - 3^3x^3/3!2^3 + 3^4x^4/4!2^4 - \cdots) = c_0e^{-3x/2}; \rho = \infty$

5 $y = c_0(1 + x^3/3 + x^6/2!3^2 + x^9/3!3^3 + \cdots) = c_0\exp(\frac{1}{3}x^3); \rho = \infty$

7 $y = c_0(1 + 2x + 4x^2 + 8x^3 + \cdots) = c_0/(1 - 2x); \rho = \frac{1}{2}$

9 $y = c_0(1 + 2x + 3x^2 + 4x^3 + \cdots) = c_0/(1 - x)^2; \rho = 1$

11 $y = c_0(1 + x^2/2! + x^4/4! + x^6/6! + \cdots) + c_1(x + x^3/3! + x^5/5! + x^7/7! + \cdots) = c_0\cosh x + c_1\sinh x; \rho = \infty$ for

each series.

13 $y = c_0(1 - 3^2x^2/2! + 3^4x^4/4! - 3^6x^6/6! + \cdots) + \frac{1}{3}c_1(3x - 3^3x^3/3! + 3^5x^5/5! - 3^7x^7/7! + \cdots) =$

$c_0\cos 3x + \frac{1}{3}c_1\sin 3x; \rho = \infty$ for each series

**A–66**

**15** $c_n = -[2^2/n(n-1)]c_{n-2}$ for $n \geq 2$; $c_0 = 0$ and $c_1 = 3$; $y = \frac{3}{2}\sin 2x$.
**17** $c_{n+1} = [1/n(n+1)](2nc_n - c_{n-1})$ for $n \geq 1$; $c_0 = 0$ and $c_1 = 1$; $y = xe^x$.

**SECTION 18-8 (page 1006)**

**1** Exact value: $y(1) \approx 0.73576$; approximate value: $y(1) \approx 0.69736$
**3** Exact value: $y(1) \approx 4.43656$; approximate value: $y(1) \approx 4.18748$
**5** Exact value: $y(1) \approx 0.28172$; approximate value: $y(1) \approx 0.40626$
**7** Exact value: $y(1) \approx 1.10364$; approximate value: $y(1) \approx 1.17838$
**9** Exact value: $y(1) \approx 1.16580$; approximate value: $y(1) \approx 1.65662$
**11** Exact value: $y(1) \approx 0.73576$; approximate value: $y(1) \approx 0.73708$
**13** Exact value: $y(1) \approx 4.43656$; approximate value: $y(1) \approx 4.42816$
**15** Exact value: $y(1) \approx 0.28172$; approximate value: $y(1) \approx 0.28592$
**17** Exact value: $y(1) \approx 1.10364$; approximate value: $y(1) \approx 1.10452$
**19** Exact value: $y(1) \approx 1.68580$; approximate value: $y(1) \approx 1.68538$

**CHAPTER 18 MISCELLANEOUS PROBLEMS (page 1008)**

**1** $y(x) = 1/(C - x^2 - 3\ln|x|)$ **3** $y(x) = \frac{1}{6}x^5 + C/x$ **5** $y(x) = 1 + C\exp(-\tan^{-1}x)$ **7** $y(x) = Ae^{-x} + Be^{-x/2}$
**9** $y(x) = C_1 + C_2e^{2x} + C_3e^{-2x}$ **11** $y(x) = (A + Bx + Cx^2)e^{-x}$ **13** $y(x) = C_1e^{-x} + C_2e^{-4x}$
**15** $y(x) = C_1e^{2x} + C_2e^{-2x} + C_3\cos 2x + C_4\sin 2x$ **17** $y(x) = (C_1 + C_2x)\cos x + (C_3 + C_4x)\sin x$
**19** $y(x) = 3e^{2x} - 2e^{-2x}$ **21** $y(x) = 3\cos 5x - 2\sin 5x$ **23** $y(x) = e^{3x}(5\cos 2x - 3\sin 2x)$ **25** $y(x) = A/x + B/\sqrt{x}$
**27** $y(x) = A/x + B/x^4$ **29** $y(x) = 1/[(C - x)e^x]$ **31** $y(x) = 1 - x + x^2/2! - x^3/3! + \cdots = e^{-x}$
**33** $y(x) = 1 + x + x^2 + x^3 + \cdots = 1/(1 - x)$ **35** $y(x) = x + x^2 + x^3/2! + x^4/3! + \cdots = xe^x$

# Index

Optical properties (*see* Fermat's principle, Reflection property, Snell's law)
Orbit, 692
  of Earth, 527, 698
  of Icarus, 548
  of Mercury, 696, 698, 699
  of Moon, 649, 650, 698
  of planet, 695
Orbital, 366
Order of differential equation, 382, 964
Order of integration (reversing), 853
Order of magnitude
  of exponential function, 364
  of $n^x/n!$, 626, 630
Ordinary differential equation, 965
Ordinate, 16
Oregon, 165
Orientable surface, 937
Orientation of a curve, 914
Origin, 16
Osculating circle, 688
Osculating plane, 753
Ostrogradski, Michel (1801–1861), 940
Ostrogradski's theorem, 940
Overdamping, 992
Ozone problem, 403

Palapa-B2, 697
paper crease problem, 425
Pappus (ca A.D. 300):
  first theorem, 327
  second theorem, 329
Parabola, 25, 205, 506, 519
  axis, 520
  compared with ellipse, 545
  as conic section, 506
  directrix, 519
  equation, 506
  focus, 519
  and hanging cable, 203–205
  polar form, 543
  reflection property, 521–522
  semicubical, 307
  tangent line property, 548
  vertex, 25, 520
Parabolic approximation, 270, 277–278
Parabolic bridge arch problem, 548
Parabolic cylinder, 731
Parabolic mirror, 522
Parabolic orbit, 545
Parabolic segment area, 270
Parabolic suspension cable, 651–652
Parabolic trajectories, 522–523
Paraboloid:
  elliptic, 740
  floating, 901
  hyperbolic, 743
  of revolution, 298, 305
  surface area, 312
  volume, 582
  volume of oblique segment, 879
Parachute, 388, 389
Parallelepiped volume, 718
Parallel lines, 20
Parallelogram area, 715, 754, A16
Parallelogram law (for vector addition), 669, 706
Parallel planes, 723
Parallel resistance problem, 841, 842
Parallel vectors, 708
Parameter(s), 656
  elimination of, 657
Parametric curve, 656
  arc length, 556
  graph, 545
  smooth, 659
  tangent line to, 659
Parametric equations, 656, 726

of cycloid, 496–497, 658–659
  and derivatives, 660–661
  of line, 720
  in polar coordinates, 666
Parametric surface, 888
Parametrization of a curve, 657
Paris, 446
Partial derivative(s), 775–782
  chain rule, 800
  equality of mixed, 781
  of $f(x, y)$, 775
  of $f(x, y, z)$, 791
  geometric interpretation, 777
  higher order, 781
  of implicitly defined function, 805
Partial differential equation, 947, 965
Partial fractions, 487–494
Partial integral, 846
*Partial Sums of Infinite Series, and How They Grow*, 601
Partial sums of series, 594
Particular antiderivative, 200
Particular solution, 396, 968
Partition, 230, 236
  inner, 850
  of interval, 230, 236
  mesh, 236, 845
  polar, 860
  of rectangle, 844
  refinement, A6
  regular, 240
  selection for, 236
  of solid region, 875
  spherical, 884
Parts, integration by, 471–472
Path, 919 (*see also* Curve in three dimensions)
Path-independent integral, 919
Paul (*see* Peter, Paul, and Mary)
Pendulum, 8
  simple, 443, 650–651
Pentobarbitol, 382
Period:
  of harmonic motion, 441, 990
  of pendulum, 8, 114, 152, 222, 445, 502–651, 799
  of satellite, 684
  of sine and cosine, 409
  of trigonometric functions, 409
Perpendicularity of vector product, 712
Perpendicular lines, 20
Perpendicular vectors, 672, 708
Perpetual annuity, 576
Perpetuity value, 576
Peter, Paul, and Mary, 598, 604
Petroleum reserve example, 392
Phase angle, 991
Phase delay (in harmonic motion), 441
Phenylethylamine, 390
*Philosophiae Naturalis Principia Mathematica*, 211
Physical interpretation of curl, 952
$\pi$, 228
  computation of, 228, 279–280, 281–282, 502, 608, 610, 624, 626, 641
  estimate by Simpson's approximation, 580
  series for, 608, 610, 624
  and $^{22}/_7$, 502
Piecewise smooth curve, 914, 924
Piecewise smooth surface, 940
Pinched torus, 893
Piston (reciprocating), 442
Piston engine, 897
Pizza, 57
Plane (coordinate), 16
Plane analytic geometry, 504
Plane(s) in space, 722–724
  normal vector to, 722

parallel, 723
  scalar equation of, 722
  tangent to surface, 778
  vector equation of, 722
Planetary motion example, 358
  Kepler's laws, 692
Point of discontinuity, 27
Point of inflection, 186
Point-slope equation, 19
Polar axis, 512
Polar coordinates, 509–514, 666
  arc length formula, 666–667
  area in 515–518
  of conic sections, 543–546
  and limits, 773
  of parametric curve, 666
Polar form of complex number, 979
Polar graph, 510
Polar moment of inertia, 869, 932
Polar partition, 860
Polar rectangle, 859
Polar unit vectors, 693
Pole, 509
Pollutant concentration problem, 367
Pollutant elimination, 386
Pollution potential forecast (*map*), 762
Polygon area, 957
Polynomial(s):
  approximations, 560–569
  continuity of, 64, 772
  derivative of, 93
  as derivative, 169
  graphing, 175–181
  *n*th derivative of, 190
  roots of, 983–984
  in several variables, 772
  Taylor, 563
Population growth, 376, 960–962
  example, 86
  with immigration, 383
  with limiting conditions, 396
  U.S., 86, 384, 398
Position function, 86, 206
Position vector, 669, 706, 726
Positive angle, 406
Positive number, 4
Positive orientation of boundary curve, 924
Positive-term series, 604
Postage problem, 13
Post office problem, 127, 792
Potential energy, 444, 576, 923
Potential function, 921, 922
Pound, A15
Power, 317, A15
Power function, 370
Power rule, 91, 92, 97, 370
  generalized, 103
Power series, 631–640 (*see also* Series)
  absolute convergence, 633
  addition of, 644
  algebra of, 644–645
  for arcsine, 640
  for arctangent, 627, 640
  binomial, 637, 1003
  convergence, 633
  for cosine, 626
  differentiation of, 639
  division of, 645, 647
  for hyperbolic cosine, 636
  for hyperbolic sine, 637
  identity principle, 997
  and indeterminate forms, 645–646
  integration of, 639
  interval of convergence, 633
  of inverse hyperbolic sine, 654
  of inverse hyperbolic tangent, 630
  Maclaurin, 636

## INVERSE TRIGONOMETRIC FORMS

69 $\int \sin^{-1} u \, du = u \sin^{-1} u + \sqrt{1 - u^2} + C$

70 $\int \tan^{-1} u \, du = u \tan^{-1} u - \frac{1}{2} \ln(1 + u^2) + C$

71 $\int \sec^{-1} u \, du = u \sec^{-1} u - \ln|u + \sqrt{u^2 - 1}| + C$

72 $\int u \sin^{-1} u \, du = \frac{1}{4}(2u^2 - 1) \sin^{-1} u + \frac{u}{4} \sqrt{1 - u^2} + C$

73 $\int u \tan^{-1} u \, du = \frac{1}{2}(u^2 + 1) \tan^{-1} u - \frac{u}{2} + C$

74 $\int u \sec^{-1} u \, du = \frac{u^2}{2} \sec^{-1} u - \frac{1}{2} \sqrt{u^2 - 1} + C$

75 $\int u^n \sin^{-1} u \, du = \frac{u^{n+1}}{n+1} \sin^{-1} u - \frac{1}{n+1} \int \frac{u^{n+1}}{\sqrt{1 - u^2}} \, du + C \quad \text{if } n \ne -1$

76 $\int u^n \tan^{-1} u \, du = \frac{u^{n+1}}{n+1} \tan^{-1} u - \frac{1}{n+1} \int \frac{u^{n+1}}{1 + u^2} \, du + C \quad \text{if } n \ne -1$

77 $\int u^n \sec^{-1} u \, du = \frac{u^{n+1}}{n+1} \sec^{-1} u - \frac{1}{n+1} \int \frac{u^n}{\sqrt{u^2 - 1}} \, du + C \quad \text{if } n \ne -1$

## HYPERBOLIC FORMS

78 $\int \sinh u \, du = \cosh u + C$

79 $\int \cosh u \, du = \sinh u + C$

80 $\int \tanh u \, du = \ln(\cosh u) + C$

81 $\int \coth u \, du = \ln|\sinh u| + C$

82 $\int \operatorname{sech} u \, du = \tan^{-1}|\sinh u| + C$

83 $\int \operatorname{csch} u \, du = \ln \left| \tanh \frac{u}{2} \right| + C$

84 $\int \sinh^2 u \, du = \frac{1}{4} \sinh 2u - \frac{u}{2} + C$

85 $\int \cosh^2 u \, du = \frac{1}{4} \sinh 2u + \frac{u}{2} + C$

86 $\int \tanh^2 u \, du = u - \tanh u + C$

87 $\int \coth^2 u \, du = u - \coth u + C$

88 $\int \operatorname{sech}^2 u \, du = \tanh u + C$

89 $\int \operatorname{csch}^2 u \, du = -\coth u + C$

90 $\int \operatorname{sech} u \tanh u \, du = -\operatorname{sech} u + C$

91 $\int \operatorname{csch} u \coth u \, du = -\operatorname{csch} u + C$

## MISCELLANEOUS ALGEBRAIC FORMS

92 $\int u(au + b)^{-1} \, du = \frac{u}{a} - \frac{b}{a^2} \ln|au + b| + C$

93 $\int u(au + b)^{-2} \, du = \frac{1}{a^2} \left[ \ln|au + b| + \frac{b}{au + b} \right] + C$

94 $\int u(au + b)^n \, du = \frac{(au + b)^{n+1}}{a^2} \left( \frac{au + b}{n + 2} - \frac{b}{n + 1} \right) + C \quad \text{if } n \ne -1, -2$

95 $\int \frac{du}{(a^2 \pm u^2)^n} = \frac{1}{2a^2(n-1)} \left( \frac{u}{(a^2 \pm u^2)^{n-1}} + (2n - 3) \int \frac{du}{(a^2 \pm u^2)^{n-1}} \right) \quad \text{if } n \ne 1$